Y0-CZF-619

Electronics: Circuits and Systems

The Howard W. Sams Engineering-Reference Book Series

PRACTICAL MICROWAVES
Thomas S. Laverghetta

SEMICONDUCTOR DEVICE TECHNOLOGY
Malcolm Goodge

AN INTRODUCTION TO THE ANALYSIS AND PROCESSING OF SIGNALS
Paul A. Lynn

RADIO HANDBOOK (22ND ED.)
William I. Orr

REFERENCE DATA FOR RADIO ENGINEERS (6TH ED.)

Other volumes in preparation

ELECTRONICS:
CIRCUITS AND SYSTEMS

Swaminathan Madhu
*Professor and Head, Department
of Electrical Engineering
Rochester Institute of Technology
Rochester, New York*

Howard W. Sams & Co., Inc.
A Publishing Subsidiary of ITT
4300 West 62nd Street, Indianapolis, Indiana 46268 U.S.A.

© 1985 by Howard W. Sams & Co., Inc.
A publishing subsidiary of ITT

FIRST EDITION
FIRST PRINTING—1985

All rights reserved. No part of this book shall be
reproduced, stored in a retrieval system, or transmitted
by any means, electronic, mechanical, photocopying,
recording, or otherwise, without written permission from
the publisher. No patent liability is assumed with respect
to the use of the information contained herein. While
every precaution has been taken in the preparation of this
book, the publisher assumes no responsibility for errors or
omissions. Neither is any liability assumed for damages
resulting from the use of the information contained
herein.

International Standard Book Number: 0-672-21984-0
Library of Congress Catalog Card Number: 84-50053

Edited by: *Pryor Associates*
Illustrated by: *Jill E. Martin*

Printed in the United States of America.

CONTENTS

Preface   20

Chapter 1 Basic Concepts of Signals and Systems   25

DC Signals   25
Resistors   26
Power Relationships for DC   26

AC Signals   27
Peak Value and RMS Value   28
Phase Angle   29
Power Relationships for AC   32

Periodic Nonsinusoidal Signals   35
Fourier Analysis   35
Amplitude Spectrum   37
Power in Periodic Signals   40
Power Spectrum of a Periodic Signal   43

Nonperiodic Signals   45
Energy Spectrum of a Pulse   46
Energy Signals and Power Signals   50

Two-Port Model of a System   52
Gain Quantities of a Two Port   52
Input Resistance of a Two Port   55
Useful Approximations   59

Decibels (dB)   61
Computational Advantages of the Decibel Unit   64
Voltage and Current Gains in Decibels   67

Summary Sheets   72

Answers to Drill Problems   74

Problems   77

Chapter 2 Review of Electric Circuits   83

Kirchhoff's Laws   83
Kirchhoff's Current Law   83
Kirchhoff's Voltage Law   84

Resistors   89
Resistors in Series   89
Voltage Dividers   91
Potentiometers   93
Resistors in Parallel   93
Current Dividers   94
Useful Approximations   96

Wheatstone Bridge   97

Capacitors   99
Energy Stored in a Capacitor   100
Voltage Continuity Condition   100
Charging of a Capacitor (RC Circuit)   100
Universal Charging Curve   104
Discharging of a Capacitor   105
An Application of the RC Circuit   106
Response of an RC Circuit to a Pulse   108
Series and Parallel Combinations of Capacitors   110

Inductors   111
Energy Stored in an Inductor   111
Current Continuity Condition   112
Current Buildup in an Inductor   112
Universal Current Buildup Curve   113
Decay of Current in an Inductor   115
Universal Current Decay Curve   116
An Application of an RL Circuit   117

AC Circuits   118
Phasor Representation of a Sinusoidal Function   118
Impedance   120
Resistors in AC Circuits   122
Capacitors in AC Circuits   122
Inductors in AC Circuits   123
Series RC Circuit   124

Filters   126
High-Pass Filter (HPF)   126
Cutoff Frequency of the HPF   128
Low-Pass Filter (LPF)   130
Frequency Response and Pulse Response   131
Applications of RC Filters   133

Thevenin Equivalent   134
   Maximum Power Transfer   140
   Use of Ideal Transformers in Matching   141

Summary Sheets   148

Answers to Drill Problems   152

Problems   154

Chapter 3 Semiconductor Devices and Integrated Circuits   166

Semiconductors   166
   Doped Semiconductors   167
   Thermal Generation and Recombination   170

PN Junction   170
   Depletion Region   171
   Potential Barrier   173
   Balance Between Diffusion and Drift Components of Current   173
   Metal Semiconductor Contacts   173
   PN Junction Under Forward Bias   175
   PN Junction Under Reverse Bias   175

PN Junction Diode   178

Bipolar Junction Transistor (BJT)   178
   NPN Transistor   180
   PNP Transistor   183
   Relationships Between the Currents in a Transistor   183
   Symbols of Transistors   185
   Output Characteristics of an NPN Transistor   185

Junction Field-Effect Transistor (JFET)   186
   N-Channel JFET   187
   Operation of the N-Channel JFET   187
   Pinch-Off Condition   190
   Currents in an N-Channel JFET   190
   P-Channel JFET   190
   Symbols and Output Characteristics   191

MOSFETs   192
   N-Channel Enhancement-Mode MOSFET (Enhancement NMOS)   193

Operation of the Enhancement NMOS   194
P-Channel Enhancement-Mode MOSFET (Enhancement PMOS)   195
N-Channel Depletion-Mode MOSFET (Depletion NMOS)   196
Symbols for MOSFET   197
Advantages of the MOSFET   199

Integrated Circuits (ICs)   **199**
Fabrication of an IC   200
Resistors in ICs   202
Capacitors in ICs   202
Ohmic and Nonohmic Contacts   202
Diodes and BJTs in ICs   204
An Example of IC Fabrication   204
Miscellaneous Transistors in ICs   207
Biasing Circuits for ICs Using the Current Mirror   212
Advantages and Disadvantages of ICs   213

Summary Sheets   **214**

Review Questions   **219**

Chapter 4 Diodes and Diode Circuits   223

Review of the PN Junction   **223**

Ideal Diode Approximation   **224**

Rectification of AC Power   **225**
Peak Inverse Voltage in a Diode   232
Filters for Power Supplies   233
Capacitor Filter   233

Miscellaneous Applications of Diodes   **236**
Clipping Circuits   236
Clamping Circuits   240
OR Gates   241

Zener Diode   **245**

Varactor Diode   **248**

Silicon-Controlled Rectifier (SCR)   **248**
SCR as an AC Switch   251
DC Static Switching Using SCRs   253
Triacs   255

Summary Sheets   259

Answers to Drill Problems   262

Problems   266

Chapter 5 BJT Amplifiers   274

Review of the BJT   274

Aspects of Analysis of an Amplifier   275

Two-Port Model of an Amplifier   276
Analysis of the Input Side   279
Analysis of the Output Side   279

Biasing of a BJT Amplifier   283
Analysis of the Biasing Circuit   284
Temperature Effects in a Transistor   290

BJT Amplifier   293
Common-Emitter Amplifier   293
Common-Base Amplifier   294
Common-Collector Amplifier   294

Common-Emitter Amplifier   295
Small-Signal Analysis of a Common-Emitter Amplifier   297
Small-Signal Parameters of a BJT   299
Small-Signal Model of a Transistor   300
Small-Signal Analysis   300

Common-Collector Amplifier   312
Input Resistance of an Emitter Follower   313
Output Resistance of an Emitter Follower   315
Application of the Emitter Follower to Matching   316

Common-Base Amplifier   317

Comparison of the Three BJT Amplifiers   319

Summary Sheets   320

Answers to Drill Problems   324

Problems   325

Chapter 6 FET Amplifiers   330

Principles and Properties of the FET   331

Transfer Characteristics of the FET   335

Determination of the Constant K 340

Biasing of the FET Amplifier 342
   Algebraic Determination of the Q Point 343
   Graphical Determination of the Q Point 345

Practical Considerations in Biasing a FET 353

Design of a Biasing Circuit 356

Common-Source FET Amplifier 361
   Input Side of the Circuit 364
   Output Side of the Circuit 364

Common-Drain Amplifier (Source Follower) 375

FET as a Voltage-Controlled Resistor 376

Handling Precautions for MOSFETs 377

Summary Sheets 379

Answers to Drill Problems 384

Problems 390

Chapter 7 Frequency Response and Pulse Response of Amplifiers 393

Variation of the Gain of an Amplifier With Frequency 394
   Frequency Response Curve of an Amplifier 394
   Factors that Affect Frequency Response 395

Determination of the Lower Cutoff Frequency f_1 397
   Lower Cutoff Frequency Due to the Input Coupling Capacitor 400
   Lower Cutoff Frequency Due to the Output Coupling Capacitor 400
   Lower Cutoff Frequency Due to the Bypass Capacitor 401

High-Frequency Model of an Amplifier 409
   Capacitance C_i 411
   Capacitance C_F 411
   Miller Effect 411
   Determination of the Upper Cutoff Frequency 413

Specifications of the Parameters of a Transistor 415
   BJT Parameters 416
   FET Parameters 417

Step-by-Step Procedure for the Complete Analysis of a BJT Amplifier 418

Pulse Response of an RC-Coupled Amplifier 434
 High-Pass RC filter 435
 Low-Pass RC Filter 440

Increasing the Bandwidth of an Amplifier 446

Summary Sheets 448

Answers to Drill Problems 452

Problems 453

Chapter 8 Operational Amplifiers and Feedback Amplifiers 456

Typical Stages of an Op Amp 457

Ideal Op Amps and Their Applications 457
 Parameters of Interest in an Amplifier 458
 Ideal Op Amp 458
 Circuits Using Op Amps 461
 Inverting Op Amp Circuits 461
 Noninverting Op Amp Circuits 467
 Summing Op Amp Circuits 473
 Voltage-Follower Circuits 476
 Voltage-to-Current Converter 478
 Sample and Hold Circuit 479
 Logarithmic Amplifiers 480
 Constant-Current Source Using an Op Amp 482
 Integrators Using an Op Amp 483

Differential Amplifiers 486
 Common Mode and Differential Mode 489
 Common-Mode Rejection Ratio (CMRR) 491
 Differential Amplifier Circuits 493
 Common-Mode Operation 495
 Differential-Mode Operation 497
 Realization of a Constant-Current Source 498

Comparators 501
 Window Detector Circuit 504
 Sine-Wave to Square-Wave Generator 506

Schmitt Trigger   506

Practical Operational Amplifiers   516

Feedback Amplifiers   518
  Some Typical Feedback Connections   *521*
  Effects of Negative Feedback   *523*
  Positive Feedback and Oscillators   *524*
  Stability of an Amplifier   *525*
  Gain Margin and Phase Margin   *526*

Summary Sheets   528

Answers to Drill Problems   533

Problems   536

Chapter 9 Optoelectronic Devices   543

Basic Principles and Relationships of Light   543
  Optical Radiation and Photons   *543*
  Flux and Illumination   *545*

Photodetectors   545
  Variation of Resistance With Illumination   *546*
  Speed of Response of a Photodetector   *550*
  Structure of a Photodetector   *554*

Photovoltaic Cells   554

Photodiodes   556
  Increasing the Efficiency of a Photodiode   *556*
  PIN Photodiode   *558*
  Speed of Response of Photodiodes   *559*
  Photodiode Amplifier Circuits   *559*

Phototransistors   560
  Speed of Response of Phototransistors   *561*
  Applications of Phototransistors   *561*
  Photo-FETs   *568*

Light-Emitting Diodes (LEDs)   569
  Bar Graph Display   *570*
  Drivers for LEDs   *570*
  XY Addressable LED Arrays   *572*

Optically Coupled Isolators   573

Summary Sheets   575

Answers to Drill Problems   579

Problems   579

Chapter 10 Logic Circuits   582

Principles of Logic Circuits   583
Basic Operations in Logic   584
Commercially Available OR and AND Gates   588
Positive and Negative Logic   591

NAND and NOR Operations   592
NAND Operations   593
Use of a NAND GATE as a NOT Gate   595
NOR Operations   596
Use of a NOR Gate as a NOT Gate   596

Exclusive OR and Exclusive NOR Functions   598
Exclusive OR Function   598
Exclusive NOR Function   601

Two-Level Positive Logic Networks   602
Analysis of Two-Level Positive Logic Networks   603
Design of Two-Level Positive Logic Networks   605
Two-Level Sum of Products Networks   610

Two-Level NAND Networks   611

Mixed Logic Networks   612
Design Procedure   616
Mixed Logic Design Versus Positive Logic Design   621
Summary of Mixed Logic Design Procedure   622
Analysis of Mixed Logic Networks   622

Karnaugh Maps   624
Two-Variable Map   624
Three-Variable Map   625
Four-Variable Map   627

Summary Sheets   630

Answers to Drill Problems   634

Problems   646

Chapter 11 Digital Logic Families   653

Transistor as a Switch   653
Cutoff Operation   *653*
Normal Active Operation   *654*
Saturation Operation   *654*
Reverse Active Operation   *656*

Propagation Delays   656

Noise Margins   658

Transistor-Transistor Logic   660
Case 1: At Least One Input Is Low   *660*
Case 2: All Inputs Are High   *662*
Open Collector TTL Gate   *663*
Loading Considerations   *663*
Totem Pole Output Stage   *665*
Interpretation and Use of the Data Sheet   *668*

Three-State Logic   673

Integrated Injection Logic   674

Emitter-Coupled Logic   678

Complementary MOSFET (CMOS) Logic   680
Case 1: Input Is Low   *681*
Case 2: Input Is High   *682*

Summary Sheets   686

Answers to Drill Problems   691

Problems   691

Chapter 12 Logic Packages and Memories   695

Multiplexers (MUX)   695
Expansion Capacity of a MUX   *698*
Use of a MUX in Logic Design   *700*
Summary of Design Procedure (Four-Input MUX)   *701*

Decoders and Encoders   704

Read-Only Memories   707
Capacity of a ROM   *710*
Programmable ROMs   *711*
Expansion of ROM Capacity   *711*

Random-Access Memory (RAM)   714
Static RAMs   716
Dynamic RAMs   719

Other Types of Memory Units   720

Summary Sheets   722

Answers to Drill Problems   724

Problems   727

Chapter 13 Arithmetic Logic, Counters, and Shift Registers   730

Binary Number System   730

Octal and Hexadecimal Number Systems   735

Adders   742
Half-Adders   743
Full-Adders   745

Representation of Negative Numbers   745
Ones Complement System   748
Twos Complement System   749

Addition and Subtraction in the Twos Complement System   750
Case 1: Addition of Two Positive Numbers   750
Case 2: Subtraction of a Smaller Number From a Larger Number   751
Case 3: Subtraction of a Larger Number From a Smaller Number   751
Case 4: Addition of Two Negative Numbers   751

BCD Code and BCD Adders   752

Arithmetic Logic Unit   752

Counters   754

Set-Reset Flip-Flops (SR Flip-Flop)   755
Clocked SR Flip-Flops   758
Edge-Triggered JK Flip-Flops   761
Master-Slave Flip-Flops   764
Special Cases of JK Flip-Flops   766
Summary of the Response Characteristics of Flip-Flops   768

Application of Flip-Flops 768
   Frequency Dividers 768
   Ripple Counters 769
   Synchronous Counters 771
   Switch Debouncing 778

Shift Register 780

Shift Register With Feedback 783

Summary Sheets 786

Answers to Drill Problems 791

Problems 795

Chapter 14 Analog Communication Techniques 800

Power and Energy in a Signal 801

Model of a Communication System 805

Modulation of a Signal 805
   Different Types of Modulation 807
   Time Domain and Frequency Domain 807

Frequency Translation 809

Double-Sideband Suppressed-Carrier AM (DSB-SC) 814

Generation of Product Signals 817

Single-Sideband AM (SSB) 818
   Generation of SSB-AM 820
   Demodulation of SSB 822

Vestigial-Sideband AM (VSB-AM) 824

Conventional AM (Double-Sideband—Large Carrier) 825
   Spectrum of an AM Signal 828
   Power Consideration in AM 829
   Demodulation of an AM Signal 834

Superheterodyne AM Receivers 835

Frequency Modulation 838
   Modulation Index in FM 839
   Spectrum of an FM Signal 840
   Bandwidth of an FM Signal 840

  Narrowband FM 843
  Direct Generation of FM 844
  Demodulation of an FM Signal 845
  FM Receivers 845

Noise in Communication Systems 845
  Noise Power Spectral Density 846
  White Noise 848
  Thermal Noise 848
  Equivalent Noise Temperature 849
  Shot Noise 849

Signal-to-Noise Ratio and Noise Figure 849
  Effect of Noise in Different Modulation Systems 850
  Deemphasis and Preemphasis in FM 850

Summary Sheets 852

Answers to Drill Problems 858

Problems 860

Chapter 15 Pulse Modulation and Digital Communication 864

Sampling Theorem 864

Informal Justification of the Sampling Theorem 865

Pulse-Amplitude Modulation 870

Time-Division Multiplexing 871

Pulse-Time Modulation 872

Pulse-Code Modulation 873
  Quantization Error in PCM 879
  Bandwidth of a PCM System 879
  Companding in PCM Systems 881
  Baseband PCM Transmission 882
  Radio-Frequency Transmission of PCM Signals 884
  Frequency-Shift Keying (FSK) 884
  Phase-Shift Keying 887

Delta Modulation 888
  Staircase Function 888
  Adaptive Delta Modulation 891

Information Theory   **891**
   Measure of Information   *891*
   Information Rate   *893*

Channel Capacity and Shannon's Theorem   **894**
   Noiseless Channel   *894*
   Noisy Channel   *895*
   Information Rate and Shannon's Theorem   *897*

Data Transmission   **900**

Digital-to-Analog and Analog-to-Digital Conversion   **901**
   Digital-to-Analog Conversion   *902*
   Analog-to-Digital Conversion   *906*
   Parallel Encoders   *906*

Summary Sheets   **913**

Answers to Drill Problems   **920**

Problems   **923**

Appendix A Answers to Problems   927

Appendix B Manufacturers' Data Sheets   937

Index   965

To Janice and my mother

PREFACE

The impact of electronics, especially integrated circuits and digital systems, is widely felt today even in areas that have been traditionally nonelectronic in nature. To cite but one example, electronic imaging is displacing the conventional form of photography in many image-processing and display applications. There are hardly any technical projects that do not utilize electronic devices and circuits to a significant extent. As a result of the widespread pervasiveness of electronics, a background in electronics has become an important (and even a critical) skill for technical persons whose major field of specialization is not in electronics. This book is written with the intention of fulfilling the need of engineers and scientists, who do not have a background in electronics, but who wish to acquire a well-grounded and reasonably thorough grasp of electronic devices, circuits, and systems (both analog and digital).

Intended Audience

Practicing engineers and scientists with a nonelectrical engineering degree can use this book as a self-study textbook for developing a sufficient background in electronics so as to interact effectively with electronics specialists. In order to facilitate self-study, a large number of drill problems (with solutions) are included in each chapter. The problems at the end of each chapter (for which answers are furnished at the end of the book) will help reinforce the material covered in that chapter.

Managers of engineering and scientific groups can use the book for learning an adequate amount of information and terminology so that they can understand and communicate with those working on electronic projects. The specially prepared summary sheets for each chapter will be particularly helpful for purposes of reference and review.

The book can be used as a textbook in electronics at colleges and universities for students majoring in nonelectrical engineering degree programs. The problems at the end of each chapter have been designed to serve as meaningful homework problems.

Level of Coverage

The material in this book has been used in the form of notes over the last several years for groups of engineers and scientists in industry, with a wide variety of backgrounds in education, age, and experience. My contact with these groups made it clear that a mere superficial survey of all the important topics of electronics would be neither adequate nor satisfying to such an audience. I also feel that there are topics, especially in digital systems, which could be learned in depth by any student with a scientific or engineering orientation and do not require a strong foundation in electrical engineering.

Consequently, a number of topics have been treated quite thoroughly and in sufficient detail so as to provide the reader with substantial, specific, and useful information.

I also found that the type of students I encountered in the previously mentioned groups was more interested in gaining a good grasp of the subject matter (e.g., how does the device work and what does the circuit do?) than getting snowed under by a flurry of mathematical equations and relationships. As a result, the amount of mathematical treatment has been pared to a minimum in this book. In terms of the mathematical background expected of the student, a good foundation in algebra (especially the solution of equations) and some knowledge of trigonometry are sufficient.

Question of Motivation

An extremely important consideration in teaching electronics to the non-specialist, especially to students who are *required* to take a course in electronics in order to fulfill their degree requirements, is providing the motivation that will keep the students interested in the subject. It is not enough to simply start with a discussion of the conventional techniques of circuit analysis followed by the discussion of the devices and their applications. I have tried to provide the necessary motivation by raising some important questions at the very beginning: what is it that electronic systems are trying to accomplish and what are they generally used for? By addressing such questions at the outset, students can develop a feeling for the importance of electronics in modern technology and the challenging problems that face an electronic engineer. Since the primary goal of electronic systems is the generation, shaping, transmission, and processing of electrical signals, the fundamental concepts and relationships involved in signals and a general introduction to amplifiers viewed as two-port networks are discussed in the first two chapters. The basic concepts and relationships of circuit components and circuit analysis are presented in conjunction with these topics rather than as a separate and isolated body of information. I have found that such an approach to the study of electronics motivates the students and builds up their interest in the material presented in the other chapters: they are able to see the relevance of the different topics to possible applications.

Coverage of Communication Systems

Computers and communication have become an integral part of our lives. A study of the principles of communication systems (analog and digital) has, consequently, become an essential requirement for anyone working in the area of electronics; it is no longer just an option. Two chapters have, therefore, been included covering the various aspects of communications in analog and digital systems as well as the conversion of analog signals to digital data and vice versa. Such topics are not normally found in textbooks written for the nonelectrical engineer and scientist.

Organization of the Material

The material in the book can be broadly divided into four parts. *Part I:* Introductory material on signals and circuits. *Part II:* Electronic devices and circuits (analog). *Part III:* Digital logic systems. *Part IV:* Communication systems.

Chapters 1 and 2 (which form Part I) discuss the basic concepts of circuit components, the different types of signals and their properties, some important circuit configurations (e.g., filters), and the two-port model of an amplifier. Concepts such as pulse response of circuits, bandwidth considerations, and frequency spectra are presented as well as gain considerations in amplifiers.

Part II starts with a discussion of the physical principles of operation of semiconductor devices in Chapter 3. The chapter covers pn junctions, bipolar junction transistors (BJTs) and field-effect transistors (FETs), and the basic principles of integrated circuits (ICs). Chapter 4 discusses the diode and its applications, the silicon-controlled rectifier (SCR), the triac, and their applications. The principles of amplifiers (in general) and the use of the BJT in different amplifier configurations are considered in Chapter 5. The small-signal model and calculations of gain and related quantities in the BJT amplifiers are presented in detail. Chapter 6 is a parallel of Chapter 5 on the topic of FETs. Chapter 7 is a study of the frequency response and pulse response of RC-coupled amplifiers (BJT as well as FET). An attempt has been made to unify the treatment so that the BJT and FET amplifiers can be analyzed simultaneously. Operational amplifiers (op amps) and differential amplifiers and their applications are discussed in Chapter 8. The discussion also includes the use of the op amp configuration in the positive (or regenerative) feedback mode in the Schmitt trigger circuit. Optoelectronic devices and their applications are treated in Chapter 9.

Chapter 10 marks the beginning of Part III, which is a discussion of digital (logic) systems. The basic concepts of logic components, circuits and their design are covered in Chapter 10. The use of mixed logic in design is treated in detail because of its importance and wide use in modern logic design. Chapter 11 examines the different digital logic families most commonly used in practice: transistor-transistor logic (TTL), emitter-coupled logic (ECL), integrated injection logic (I^2L), and complementary MOSFET (CMOS) logic gates. Questions such as speed of response and loading considerations are also discussed. The designer of modern digital systems relies heavily upon ready-made logic packages rather than discrete gates. Packages, such as multiplexers, read-only memories (ROMs), and random-access memories (RAMs) are discussed in Chapter 12. Chapter 13 considers adders, different notations for negative numbers in computer systems, counters, shift registers, and their applications.

Part IV consists of two chapters. Chapter 14 covers the area of analog (or continuous-wave) communication systems. The concept of frequency translation is used to explain the principles of modulation so as to minimize (or even avoid) the use of trigonometric identities. Chapter 15 concentrates on pulse modulation and digital data communications. Pulse-code modulation (PCM), basic principles of information theory, and the transmission of digital data are discussed in that chapter. The chapter also includes a discussion of analog-to-digital (A/D) conversion and digital-to-analog (D/A) conversion.

The entire material in the book can be covered in about 80 hours of classroom lectures. It is possible to develop courses of duration of one semester or one quarter by a careful selection of chapters to meet the needs of a specific group of students.

Acknowledgments

I am indebted to a number of people in the development and writing of this book. The material was initially developed for a course in electronics for a group of patent attorneys at Eastman Kodak Company, Rochester, NY. I wish to express my sincere appreciation to Ray Owens, who was a prime-mover in this project. The response of Ray and his colleagues made the many hours spent in writing the notes well worth the effort. Donald Dick of the Training Department at Eastman Kodak Company initially came up with the idea of the summary sheets in the form presented in this book. I am indebted to him for this unique approach in summarizing the information of each chapter.

During the preparation of the manuscript, I was constantly guided by the detailed and painstaking reviews of Professor Art Davis of the Department of Electrical Engineering at San Jose State University, San Jose, California. His comments and suggestions have been invaluable. It will not be an exaggeration to say that his constant encouragement, knowledge of the subject matter, and attention to detail have made the writing of this book a challenging and rewarding experience. I could not have wished or hoped for a more helpful and interested reviewer.

The idea of putting the volumes of notes I had developed into the form of a book was first suggested to me by Charlie Dresser, Manager of Engineering Books at Howard W. Sams & Co., Inc. He has been a most patient friend whose understanding and support have been a source of constant encouragement in the preparation of this book.

I also wish to thank Arlet Pryor for his careful checking of the manuscript and especially the solutions of the drill problems.

My wife, Janice, has shown an extraordinary amount of patience and understanding over the many months I have spent on the preparation of the manuscript. I sincerely appreciate her for having courageously borne the burden of a "book widow."

Swaminathan Madhu

CHAPTER 1

BASIC CONCEPTS OF SIGNALS AND SYSTEMS

The function of an electronic system is to modify a given signal to a form and amplitude suitable for use at the receiving end. For example, in a stereo amplifier system, a small electrical signal generated by the phonograph cartridge has to be amplified to a power level high enough to operate the coils in the loudspeakers. Since any electronic system, however complex it may be, operates on signals, we need to learn the important properties of signals. The first part of this chapter is a study of the basic concepts of signals especially from the point of view of analysis. Electronic systems can often be treated in terms of a model called the two port. The two-port model permits us to examine the important aspects of an electronic system without worrying about the details of the actual composition of the system itself. The two port will be discussed in the second part of this chapter.

DC SIGNALS

By a *signal*, we mean any voltage or current generated by a source that contains some information of importance and interest. Signals of practical importance occur in a wide variety. The simplest of these is the dc signal, and the most complex is one that varies with time in a random or unpredictable manner.

A *dc* (or *direct current*) *signal* has a constant value at all times. It is perhaps inappropriate to call a dc voltage or current a *signal* since most practical signals do not stay constant with time. But the importance of dc arises from the fact that dc power is necessary to drive electronic circuits. Also, in some cases such as instrumentation, measurement, and process-control systems, the signals remain constant over relatively long periods of time or vary at such a slow rate that they can be classified as dc.

Resistors

The most common component used in a dc circuit is the *resistor* whose symbol is shown in Fig. 1-1. R denotes the *resistance* of the component and is measured in *ohms* (Ω). If V is the voltage across the resistor, then we use + and − signs to indicate the polarities of the voltage: the terminal marked + is (or is assumed to be) at a higher potential than the other terminal. If a *current I flows from the + to the − terminal*, then the voltage V and current I are related by the equation:

(A) V = RI; natural current flow from positive (+) to negative (−).

(B) V = −(RI); a minus sign is necessary when the current is assumed to be flowing from the − to the + terminal in a resistor.

Fig. 1-1. The two versions of Ohm's Law.

$$V = RI \quad \text{(Ohm's Law)} \quad (1\text{-}1)$$

The natural flow of current in a resistor is from the terminal at the higher potential (the + terminal) to the terminal at the lower potential (the − terminal), as in Fig. 1-1A. But, in electronics, it is often customary to choose the current going the "wrong way," that is, from the − terminal to the + terminal as shown in Fig. 1-1B. In such situations, the Ohm's law equation has to be written in the form

$$V = -(RI) \quad (1\text{-}2)$$

The distinction between the two cases illustrated in Fig. 1-1, with the current flowing from + to − in one case and the current flowing from − to + in the other, must be carefully observed.

Power Relationships for DC

If the voltage across any component is V and the current through it is I, then

Electronics: Circuits and Systems

$$\text{power } P = VI \text{ watts} \quad (1\text{-}3)$$

When the voltage V and current I are constants, as in dc, the power P is also a constant. Using Ohm's law [Eq. (1-1)], the expression for power can be written also as,

$$P = RI^2$$

and

$$P = (V^2/R) \quad (1\text{-}4)$$

In any electronic circuit, dc power sources are needed for establishing the proper values of the operating voltages and currents in the different branches in the circuit. This is called *biasing* and will be discussed in detail in the chapters on amplifiers. The source of dc power can be a battery or can be obtained by converting the available ac power from a power outlet to dc by means of rectifier circuits.

AC SIGNALS

Signals that vary with time, called *time-varying signals*, are usually more interesting than dc. Time-varying signals fall into several categories: *sinusoidal (or ac) signals, periodic nonsinusoidal signals, nonperiodic signals*, and *random signals*.

A signal that varies with time in the form of the trigonometric sine function is referred to as a *sinusoid*, but more popularly as *alternating current* or *ac*. The waveform of an ac signal is shown in Fig. 1-2. The signal is repetitive, or *periodic*. If we start at time t = 0 and proceed to time t = T, we see that this portion of the waveform repeats itself.

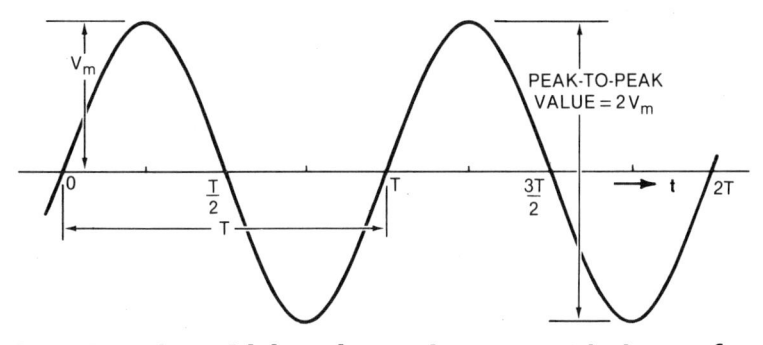

Fig. 1-2. A sinusoidal or alternating current (ac) waveform.

The duration T is called the *period* of the waveform and represents the time taken for one complete cycle.

The *number of cycles in one second* is called the *frequency*, *f*, of the waveform.

Since each cycle takes T seconds, the number of cycles in one second is (1/T). Thus, the frequency f and the period T of a sinusoidal waveform are related by the equation

$$f = \frac{1}{T} \tag{1-5}$$

The unit of frequency is the *hertz*, abbreviated *Hz*. Multiples of hertz are commonly used in electronics: *kilohertz (kHz)* and *megahertz (MHz)*. A signal of frequency 1 MHz has 10^6 cycles per second, and its period is $(1/10^6)$ or 1 microsecond.

Drill Problem 1.1: Calculate the period of each of the following sinusoidal waveforms whose frequencies are given: (a) 60 Hz; (b) 91.5 MHz.

Drill Problem 1.2: For each of the values of the period of an ac waveform given, calculate the frequency: (a) 50 ns; (b) 5 ms; (c) 5 s.

Peak Value and RMS Value

The maximum value attained by an ac signal, denoted by V_m in Fig. 1-2, is called its *peak value* or *amplitude*. Often, the *peak-to-peak value*, which is the *difference between the maximum and minimum* values of the signal, is used since it is more conveniently measured on the display of an oscilloscope. For the waveform in Fig. 1-2, the peak-to-peak value is $2V_m$.

In the measurement of an ac voltage by means of an ac voltmeter, the *rms*, or *root-mean-square*, value is more useful than the peak value. The rms value is, as the name implies, the (square) *root* of the *mean* (or the average) of the *square* of the signal. That is, the given signal is first squared (multiplied by itself), the average value of the squared signal is determined, and then the square root of the average value is found. Although this procedure may sound awfully complicated, it is the most natural thing for an ac meter to measure. The mechanism of an ac voltmeter responds to the square of the input voltage because the torque acting on the meter movement is proportional to the square of the input voltage. Since the input voltage varies with time, the torque also varies with time. The mechanical suspension attached to the movement cannot move back and forth as the torque varies because of its inertia, and it assumes a position of rest corresponding to the average value of the torque. Therefore, the deflection of the needle is determined by the average value of the square of the signal. The scale of the meter is calibrated in terms of the square root of the average (or mean) value of the square of the signal; that is, it reads the rms value directly.

Electronics: Circuits and Systems

For an ac signal, the rms and peak values are related by the equation,

$$\text{rms value} = \frac{1}{\sqrt{2}} \times \text{peak value}$$

$$= 0.707 V_m \qquad (1\text{-}6)$$

It is a common practice to specify the rms value of an ac voltage or current. For instance, the 110 V ac power supply that we normally use has an rms value of 110 V. Its peak value is, therefore, (110/0.707) = 156 V. The rms value is particularly useful in the calculation of power in ac signals.

Drill Problem 1.3: Calculate the rms value of each of the ac voltages or currents whose *peak* value is given: (a) 20 V; (b) 15 A.

Drill Problem 1.4: An ac voltage has an rms value of 220 V. Calculate its peak-to-peak value.

Drill Problem 1.5: A device is available in which the peak voltage cannot exceed 200 V. What is the rms value of the largest signal that can be applied to it? What is the corresponding peak-to-peak value?

Phase Angle

Consider two sinusoidal signals of the *same frequency* displaced in time relative to each other as in Fig. 1-3. There is seen to be a *time delay* of t_d between the two signals. Instead of using the time delay as a measure of the displacement between the signals, it is more convenient to convert it to an angle, called the *phase angle*, between the signals. Such a conversion of time to degrees is based on the fact that *one cycle of an ac signal corresponds to an angle of 360°*. That is,

$$T \text{ seconds} \longleftrightarrow 360°$$

The angle for any time interval is dependent upon the ratio of that time interval to the period T. Therefore, the time delay of t_d seconds between two signals will correspond to a phase angle given by,

$$\text{phase angle} = \frac{t_d}{T} \times 360° \qquad (1\text{-}7)$$

For example, if the time delay between two signals is one quarter of a period, or (T/4), the corresponding phase angle will be one quarter of 360°, or 90°.

When the time delay between two signals is one half of a period, or (T/2), the phase angle between them is 180°. In this case, one signal will be at its maximum while the other is at a minimum, and vice versa.

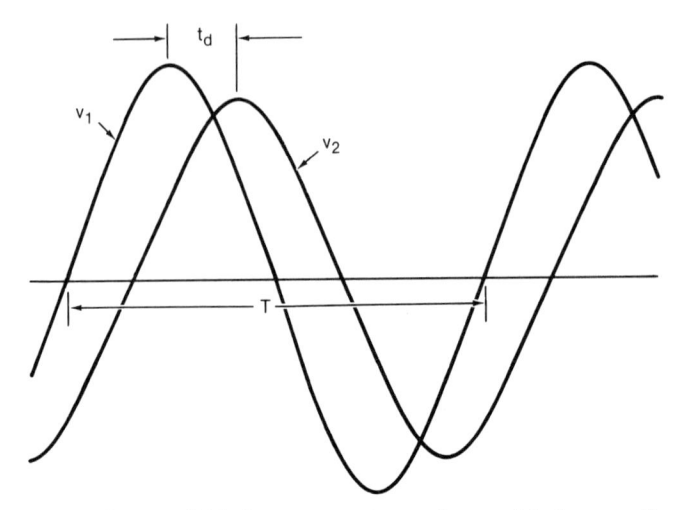

Fig. 1-3. Phase shift between two sinusoidal waveforms.

Example 1.1

Two waveforms of the same frequency are shown in Fig. 1-4, one representing a voltage and the other a current. Determine (a) the frequency, (b) the peak value of each, (c) the rms value of each, and (d) the phase angle between the two waveforms. Also indicate which one is leading the other.

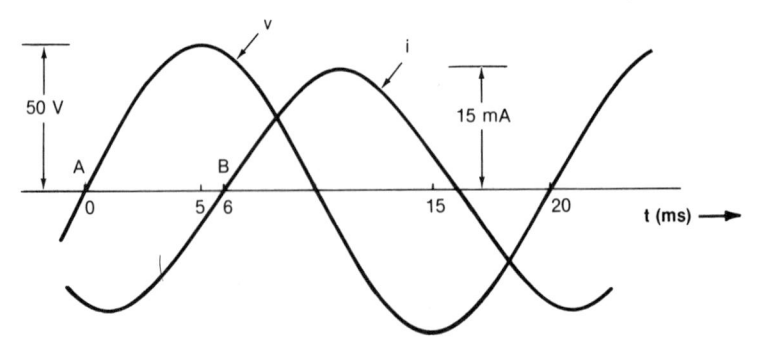

Fig. 1-4. Waveforms for Example 1.1.

Solution

(a) By inspection, the period is seen to be 20 ms. Therefore, the frequency is $f = (1/T) = 50$ Hz. (b) The peak value of the voltage signal is seen to be 50 V. The peak value of the current signal is seen to be 15 mA. (c) The rms value of each waveform is obtained by multiplying the peak value by 0.707. The rms value of the voltage is, therefore, 35.4 V, and the rms value of the current is 10.6 mA. (d) The time delay between the two waveforms is the interval between the two points A and B on the diagram, with A being the point at which the voltage waveform crosses the time axis with a positive slope and B being the point at which the current waveform crosses the time axis with a positive slope. (This method is the standard way of measuring the time delay between two signals since the points A and B can be easily identified on an oscilloscope display.) The time between the points A and B is given as 6 ms. Therefore,

$$\text{time delay} = 6 \text{ ms}$$

which leads to

$$\text{phase angle} = \frac{t_d}{T} \times 360°$$

$$= \frac{6}{20} \times 360°$$

$$= 108°$$

Since the point A occurs ahead of the point B, the *voltage is leading the current* by 108°.

Drill Problem 1.6: For the waveform shown in Fig. 1-5, determine the frequency, peak value, peak-to-peak value, and the rms value.

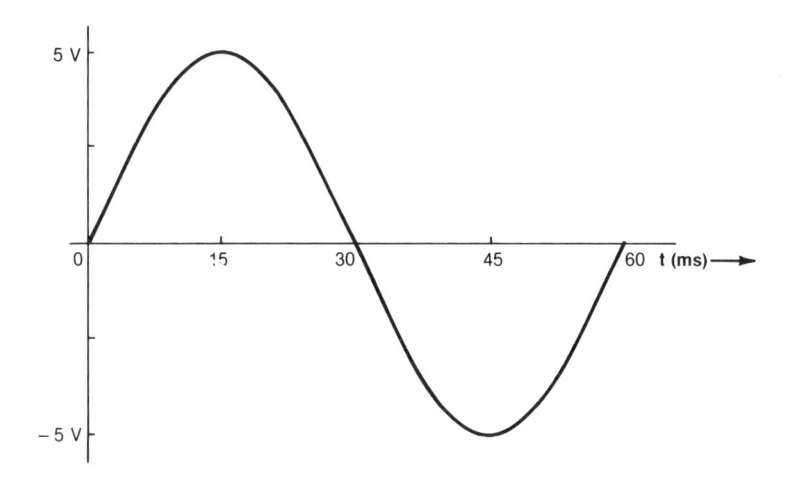

Fig. 1-5. Waveform for Drill Problem 1.6.

Drill Problem 1.7: Two ac waveforms of the same frequency are shown in Fig. 1-6. Determine the phase angle between them and state which waveform leads the other.

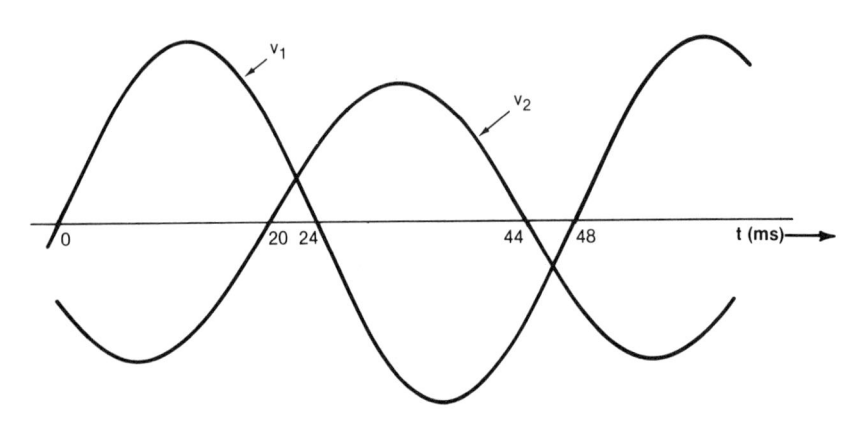

Fig. 1-6. Waveforms for Drill Problem 1.7.

Power Relationships for AC

The power in a signal given by the product VI varies with time in the case of an ac signal since both voltage and current vary with time. Since it is convenient to use a value that is a constant rather than time varying, the practical measure of power that is used is the *average value of the product VI* in the case of ac signals, which is called the *average power in an ac signal*, or quite simply *power*. Power in ac circuits is usually specified in terms of the average power as, for example, a 100-watt light bulb or a stereo amplifier delivering 75 watts per channel.

The power consumed by a resistor is the most important for our purposes since the load in an electronic circuit is usually a resistor. The (average) power consumed by a resistor subject to an ac signal is given by either of the following two expressions:

$$P = I_{rms}^2 R = \frac{V_{rms}^2}{R} \quad (1\text{-}8)$$

Note that *rms values must be used* in the calculation of the average power in a resistor. The expressions for power consumed by a resistor subject to an

Electronics: Circuits and Systems

ac signal are of the same form as those for a resistor subject to a dc signal, except that rms values are used in ac. But, proper care must be exercised in applying the expressions in Eq. 1-8. The power consumed by any resistor in *any ac circuit*, regardless of how complex that circuit may be, is always the *square of the rms value of the current through that resistor multiplied* by the value of that resistance; or it is the *square of the rms value of the voltage across that resistor divided* by the value of that resistance. That is, we need to know the current or the voltage for each *individual* resistor in an ac circuit in order to calculate the (average) power.

Example 1.2

The rms values of the currents I_1 and I_2 in the circuit of Fig. 1-7 are given as 5 A and 12 A, respectively. Calculate the power consumed by each of the two resistors R_1 and R_2.

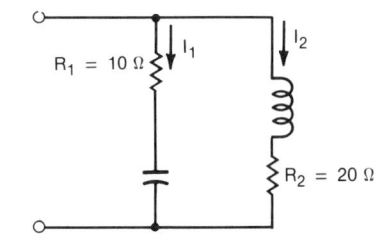

Fig. 1-7. Circuit for Example 1.2.

Solution

Since the current in each individual resistor is directly given, the power in each can be calculated by using $I_{rms}^2 R$. For R_1, the current is 5 A (rms), and the resistance is 10 Ω. The power consumed by R_1 is, therefore, $(5^2 \times 10) = 250$ W. Similarly, the power consumed by R_2 is 2880 W.

Example 1.3

The black box of Fig. 1-8 contains an amplifier whose output voltage V_o is 75 V (peak to peak). Calculate the power delivered to the load resistor R_L.

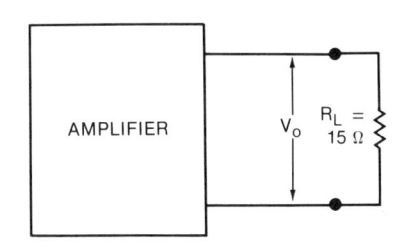

Fig. 1-8. Circuit for Example 1.3.

Solution

The peak-to-peak value of the voltage across the resistor is given and has to be converted to the corresponding rms value first. Since the peak-to-peak value is 75 V, the peak value is $(0.5 \times 75) = 37.5$ V. The rms value is, therefore, $(0.707 \times 37.5) = 26.5$ V.

Therefore, power consumed by the 15-Ω resistor is $(26.5^2/15) = 46.8$ W.

Now, suppose that the load on the amplifier of the previous example is not a single 15-Ω resistor but a series combination of a 15-Ω resistor and a capacitor as indicated in Fig. 1-9. Can the power consumed by the load be calculated by using the information given? No! The reason is that the voltage across the *15-Ω resistor* is no longer known, since the 75 V is across the resistor and capacitor together. Neither the voltage across the 15-Ω resistor nor the current through it can be determined from the data, and the power consumed by the load cannot be calculated.

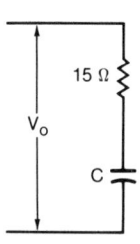

Fig. 1-9. Modified load for the circuit of Fig. 1-8.

Drill Problem 1.8: Calculate the power consumed by each of the resistors in the circuit of Fig. 1-10. If the power cannot be calculated in any of the resistors, state the reason.

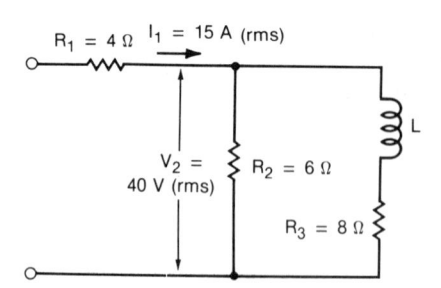

Fig. 1-10. Circuit for Drill Problem 1.8.

Drill Problem 1.9: If a loudspeaker is rated at 30 W and has a resistance of 8 Ω, what is the *peak-to-peak* voltage across it when operating at full power? If it was operating at one half the rated power, what will be the peak-to-peak voltage?

PERIODIC NONSINUSOIDAL SIGNALS

Periodic signals that do not have a sinusoidal waveform occur frequently in electronics and communications. Examples of such signals are *square (or rectangular) waves, sawtooth waves,* and *pulse trains.* Any periodic signal can be *decomposed* into a number of components, each of which is a *sinusoidal signal.* This property was first theoretically established by Fourier, and the analysis based on his work is called *Fourier analysis.*

Fourier Analysis

Consider the periodic pulse train shown in Fig. 1-11. The period of the pulse train is T = 60 ms. We define the reciprocal of T as the fundamental frequency f_0:

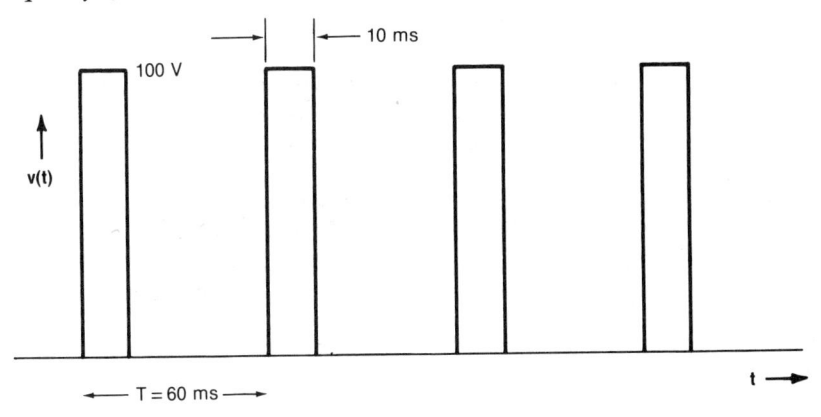

Fig. 1-11. A periodic pulse train.

$$\text{fundamental frequency } f_0 = \frac{1}{T} \text{ Hz} \qquad (1\text{-}9)$$

The fundamental frequency of the particular waveform shown in Fig. 1-11 is

$$f_0 = \frac{1}{60 \times 10^{-3}}$$

$$= 16.67 \text{ Hz}$$

Basic Concepts of Signals and Systems

According to Fourier, a periodic signal can be expressed in the form of a sum:

Given periodic signal = dc component + a sinusoidal component at f_0 + a sinusoidal component at $2f_0$ + a sinusoidal component at $3f_0$ + . . .

where,
the sum continues indefinitely.

The sinusoidal components are called *harmonics:* the component at the fundamental frequency f_0 is the first harmonic; that at $2f_0$ is the second harmonic, that at $3f_0$ is the third harmonic, and so on.

For the pulse train in Fig. 1-11, it can be written in the form of a sum:

given signal = dc component + a component at 16.67 Hz + a component at 33.34 Hz + a component at 50.01 Hz + . . .

Each of the harmonic components has an amplitude and phase angle that can be evaluated mathematically or by means of a computer once the signal waveform is known. The amplitudes of the harmonics of actual signals can also be determined in the laboratory by means of a *wave analyzer.* The amplitudes of the dc component and the various harmonics of the signal of Fig. 1-11 are shown listed in Table 1-1.

Table 1-1. Amplitudes of the DC Component and Various Harmonics of the Signal

Component Frequency (Hz)	Component Amplitude (V)
dc	16.67
16.67	31.83
33.34	27.57
50.01	21.22
66.68	13.78
83.35	6.37

The significance of the values in Table 1-1 is that if we were to draw a graph representing each of the components and add all these graphs point by point, the resulting waveform would begin to resemble the original signal of Fig. 1-11. The resemblance between the composite waveform and the actual signal will improve as more and more harmonics are added. For a more general case than the signal of Fig. 1-11, it is necessary to include not only the amplitudes of the various components but also their relative phase angles in order to add them and obtain the original signal.

Electronics: Circuits and Systems

Amplitude Spectrum

The amplitudes of the harmonics of a periodic signal (as well as its dc component), which can be obtained mathematically or by other procedures, are usually shown in a graphical form as in Fig. 1-12. The horizontal axis represents the values of frequency, and the heights of the lines represent the amplitudes of the different components present in the signal. The values used in Fig. 1-12 are pertinent to the pulse train of Fig. 1-11.

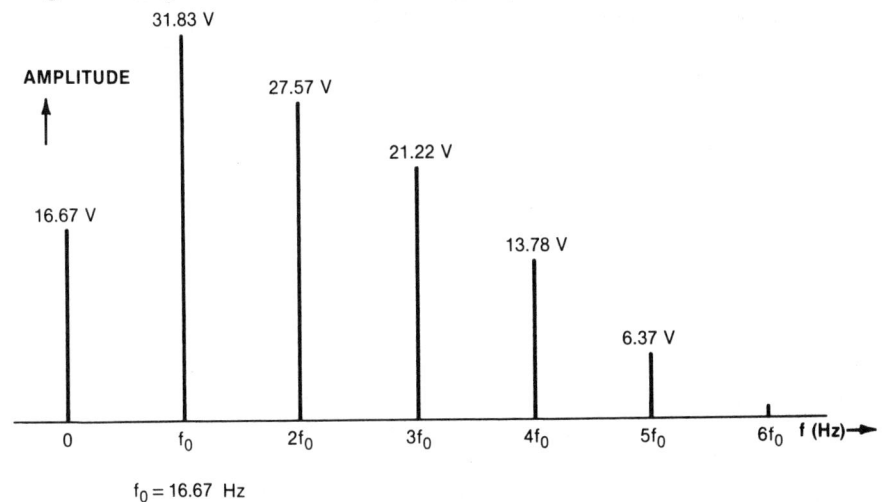

Fig. 1-12. Amplitude spectrum of a periodic signal.

$f_0 = 16.67$ Hz

Spectrum values: 16.67 V at 0; 31.83 V at f_0; 27.57 V at $2f_0$; 21.22 V at $3f_0$; 13.78 V at $4f_0$; 6.37 V at $5f_0$.

The graph showing the amplitudes of the different components against frequency as in Fig. 1-12 is called an *amplitude versus frequency spectrum*. It is usually called the *amplitude spectrum* and also, loosely, the *frequency spectrum*. The amplitude spectrum of any periodic signal contains important information about which frequencies are present in the signal and what their relative strengths are.

Even though our primary interest is not in the actual evaluation of the amplitude spectrum of a given signal, the following general properties of amplitude spectra should be kept in mind.

1. The amplitudes of the harmonics generally decrease as we go toward the higher harmonics. This relationship means that even though there is an infinite number of harmonics according to theory, only the first few harmonics are of sufficient importance from a practical point of view. Usually, the first three or four harmonics are the significant harmonics and the others can be ignored.

2. For a given waveform, an *increase* in the value of the period T does not affect the *relative amplitudes* of the harmonics, but the *lines get closer*

together on the spectrum, which is to be expected since an increase in T results in a decrease in the value of f_0 (= 1/T). Fig. 1-13 shows this crowding effect. In Fig. 1-13A, the period is 2 ms so that the value of f_0 = (1/T) = 500 Hz. The spacing between the spectral lines is, therefore, 500 Hz, but in Fig. 1-13B, the period is increased to 4 ms and the spacing between the spectral lines has decreased to 250 Hz. Any further increase in the period will cause a greater crowding of the spectral lines. In the limit where the period becomes infinite, the lines will get so close that they can no longer be distinguished as separate lines.

(A) Period of 2 ms.

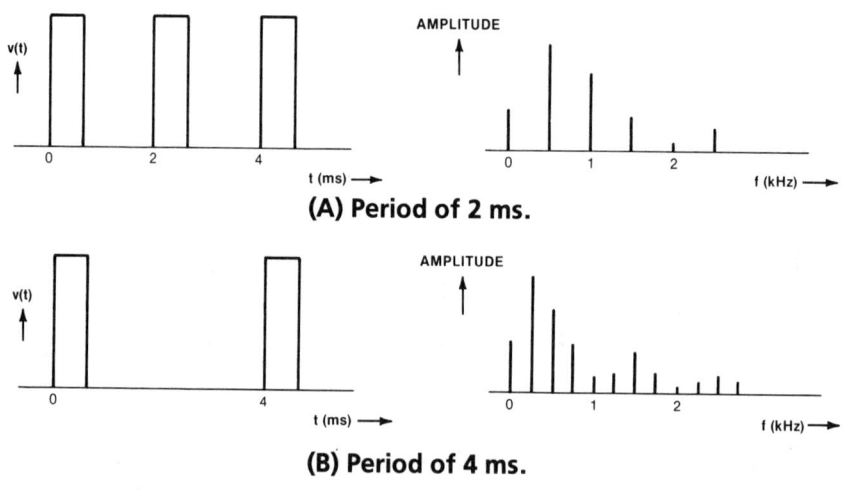

(B) Period of 4 ms.

Fig. 1-13. Effect of changing the period on the amplitude spectrum.

The amplitude spectrum of a signal is important because it can be used to determine the response of a circuit, such as an amplifier, to the given signal. For example, consider an input signal to an amplifier, which is a square wave as shown in Fig. 1-14A. The output of the amplifier will be found to have rounded-off edges as in Fig. 1-14B, instead of the sharp edges of the input square wave. The rounding off is due to the fact that the amplifier has suppressed the higher harmonics of the square wave, thus leading to a distorted output. The distortion can be reduced by making the amplifier pass higher frequencies (that is, by increasing its bandwidth). On the other hand, an amplifier with a smaller bandwidth will produce a more severe distortion. The decision on the bandwidth of an amplifier depends upon the amplitude spectrum of the input signal, the cost factor, and the type of application for which it is being used. These points will be discussed further at various points in this chapter.

Electronics: Circuits and Systems

(A) Input square wave.

(B) Output wave with rounded-off edges.

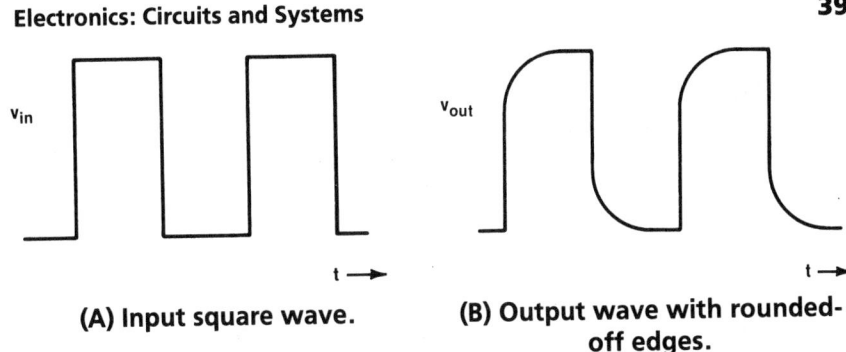

Fig. 1-14. Input and output waveforms of an amplifier with insuffient bandwidth.

Drill Problem 1.10: A periodic signal is found to have a period of 20 ns. What is the frequency of each of the following: (a) fundamental frequency; (b) its fifth harmonic; (c) its tenth harmonic?

Drill Problem 1.11: The amplitude spectrum of a signal is shown in Fig. 1-15. What is the fundamental frequency of the signal and its period?

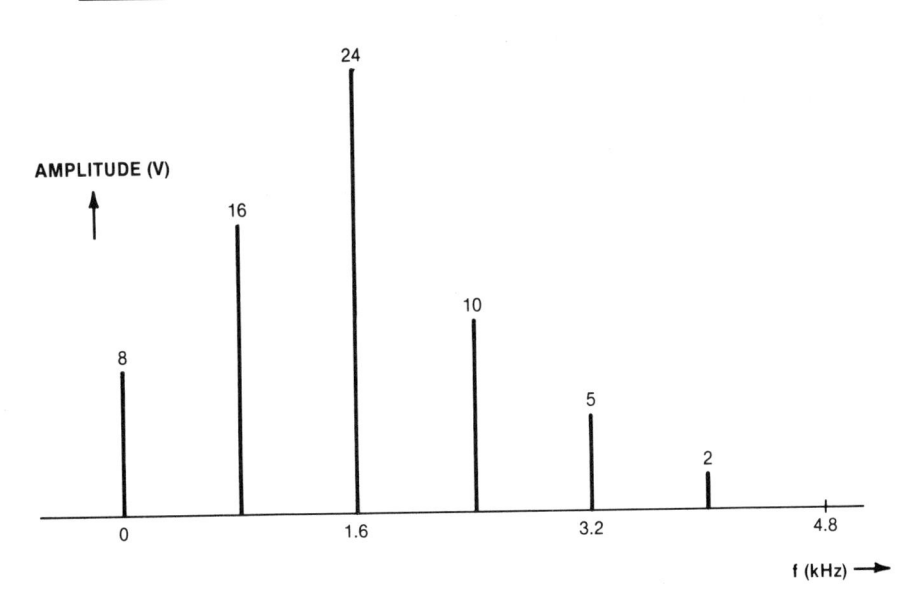

Fig. 1-15. Amplitude spectrum for Drill Problem 1.11.

Drill Problem 1.12: Suppose the period of the signal of the previous problem is *doubled*. Redraw the amplitude spectrum to fit the new period. Repeat for the case if the period is decreased to *one half* the original value.

Power in Periodic Signals

The power contained in a periodic signal can be determined by adding the power contributions from the individual components of the signal. That is, the average power in a periodic signal is given by the sum:

power in a periodic signal = power in the dc component + power in the fundamental frequency component + power in the second harmonic component + power in the third harmonic component + . . .

The contributions from each of the components are calculated by using Eqs. (1-3) and (1-8) along with Eq. (1-6). The procedure is illustrated by the following example.

Example 1.4

The components of a periodic signal are given in Table 1-2. Assume that the fifth and higher harmonics can be neglected in the calculation of power. Suppose the signal is applied to a 2-Ω resistor. Calculate the average power consumed by the resistor.

Table 1-2. Components of a Periodic Signal

Component Frequency (Hz)	Component Amplitude (V)
dc	2.0
100	1.5
200	0.8
300	0.35
400	0.09

Solution

In the calculation of the power contribution of each component, it is necessary to use the *rms* value of the voltage. For the dc component, the rms value and the amplitude are the same. For each of the other components, Eq. (1-6) is used to obtain the rms value: multiply the amplitude by 0.707.

Dc component:

$$\begin{aligned}\text{Power in the dc component} &= (V^2/R) \\ &= (2^2/2) \\ &= 2\text{ W}\end{aligned}$$

100-Hz component:

$$\begin{aligned}\text{rms value} &= 0.707 \times 1.5 \\ &= 1.06\text{ V}\end{aligned}$$

$$\begin{aligned}\text{power in the 100-Hz component} &= (V_{rms}^2/R) \\ &= (1.06^2/2) \\ &= 0.562\text{ W}\end{aligned}$$

200-Hz component:

$$\begin{aligned}\text{rms value} &= 0.566\text{ V} \\ \text{power} &= 0.160\text{ W}\end{aligned}$$

Similarly,

$$\text{power in the 300-Hz component} = 0.0306\text{ W}$$

$$\text{power in the 400-Hz component} = 2.024 \times 10^{-3}\text{ W}$$

The total power in the given signal is, therefore,

$$\begin{aligned}P &= (2 + 0.562 + 0.160 + 0.0306 + 2.024 \times 10^{-3}) \\ &= 2.755\text{ W}\end{aligned}$$

Drill Problem 1.13: Suppose the signal of Fig. 1-11 whose component amplitudes are given in Table 1-1 (or Fig. 1-12) is applied to a 9-Ω resistor. Assuming that the sixth and higher harmonics can be neglected, calculate the total average power consumed by the resistor.

Drill Problem 1.14: If the signal whose component amplitudes are given in Table 1-2 is applied to a 1-Ω resistor, calculate the power consumed by the resistor.

Drill Problem 1.15: If, in the previous problem, the resistance is changed to 4 Ω, what will be the new value of power? How does an increase (or decrease) in the load resistance affect the power consumed for a given signal input?

Returning to Example 1.4, the total power consumed by the load was 2.755

W. If we include only the contributions from the dc component, the fundamental frequency component, and the second harmonic component, the sum of these contributions is (2 + 0.562 + 0.160) = 2.722 W, which is 98.8% of the total power provided by the signal. That is, even if the highest two harmonics get suppressed (by using an amplifier of insufficient bandwidth), we can still recover 98.8% of the total available power. Therefore, even though theory dictates that *all* the harmonics of a signal must be included, practical considerations show that most of the power resides in the first few harmonics of a signal. In the design of amplifiers and communication systems, the ideal goal is to recover 100% of the power available in the input signal. But the effort and cost of building an amplifier to pass all frequencies may be too high, and it is necessary to strike a compromise between recovering the full signal power and the cost and complexity of the system. The usual decision is to settle for a certain (very large) percentage of the total power in the signal, say 95%. Then, the highest harmonic (starting with dc and working successively through the various harmonics) needed to reach the 95% value of the total power is used as the bandwidth needed for the system.

Example 1.5

A periodic signal has the component amplitudes listed in Table 1-3. It is fed to a 1-ohm load resistor. Determine the bandwidth needed to pass: (a) 90% of the total power; (b) 98% of the total power.

Table 1-3. A Periodic Signal and Its Component Amplitudes

Component Frequency (kHz)	Component Amplitude (V)
dc	0.0
1	2.0
2	1.0
3	0.75
4	0.50
5	0.4
6	0.32
7	0.18
8	0.08
9	0.04
10	0.025

Solution

The procedure is similar to that in Example 1.4. The values of the power contributions from the individual components are shown listed in Table 1-4.

Electronics: Circuits and Systems

Table 1-4. Values of the Power Contributions from Individual Components

Component Frequency (kHz)	Power in the Component (W)
dc	0
1	2.0
2	0.5
3	0.281
4	0.125
5	0.080
6	0.0512
7	0.0162
8	0.0032
9	0.0008
10	0.0003

The total power is seen to be 3.0577 W with 90% of the total at 2.7519 W, which is exceeded at 3 kHz. A bandwidth of 3 kHz is, therefore, sufficient to pass 90% of the total available power, and 98% of the total power is 2.9965 W. This value is exceeded at 6 kHz and, therefore, a 6-kHz bandwidth is needed to pass 98% of the total available power. Actually, if we use only 5 kHz as the bandwidth, then 97.7% of the total power is transmitted, and may be close enough to the required 98%, which means that a 5-kHz bandwidth is sufficient.

Drill Problem 1.16: Return to the calculations of Drill Problem 1-13 and calculate the bandwidth needed to transmit (a) 90% of the total power; (b) 98% of the total power.

Drill Problem 1.17: Repeat the calculations of the previous drill problem for the situation in Drill Problem 1-14.

The previous example and drill problems show the reason for the importance of the amplitude spectrum of a given periodic signal: the spectrum is the key to determining the bandwidth needed to transmit the signal.

Power Spectrum of a Periodic Signal

The amplitude spectrum of a periodic signal introduced earlier can be converted to a *power spectrum* in which each line represents the power contribution from a particular frequency component in the signal. The *standard load used in power spectrum, and in fact, for communication systems in general, is 1 Ω*.

In order to convert an amplitude spectrum to a power spectrum, we take each spectral line in the amplitude spectrum, calculate the power due to it by using *(V^2/R) for the dc component* and *(V_{rms}^2/R) for each of the harmonics*. Since $R = 1\ \Omega$, the previous calculations become, respectively, V^2 and V_{rms}^2. The power spectrum corresponding to the amplitude spectrum of Fig. 1-12 is shown in Fig. 1-16. The height of each line is in watts.

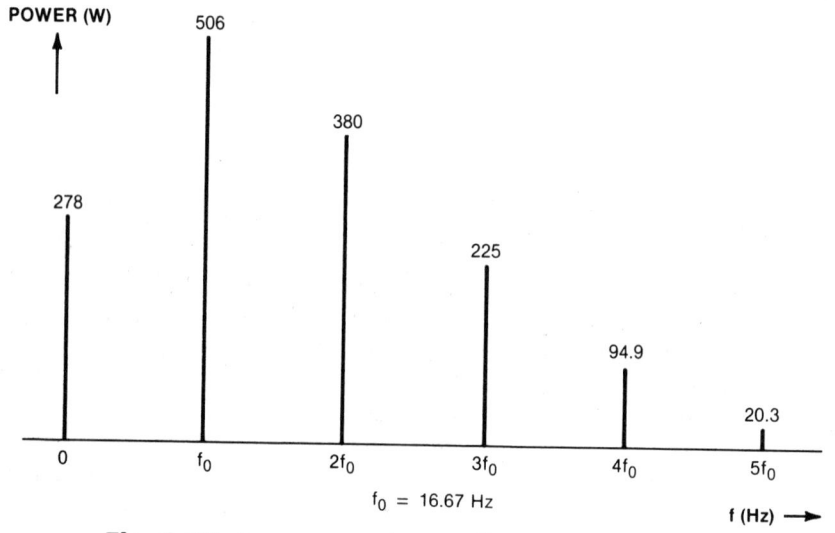

Fig. 1-16. Power spectrum of a periodic signal.

Power values: 278 (0), 506 (f_0), 380 ($2f_0$), 225 ($3f_0$), 94.9 ($4f_0$), 20.3 ($5f_0$); $f_0 = 16.67$ Hz.

Drill Problem 1.18: Obtain the power spectrum of the signal whose component amplitudes are given in Table 1-2.

Drill Problem 1.19: The amplitude spectrum of a periodic signal is shown in Fig. 1-17. Obtain the power spectrum and also calculate the total power transmitted if only dc and the first three harmonics were passed.

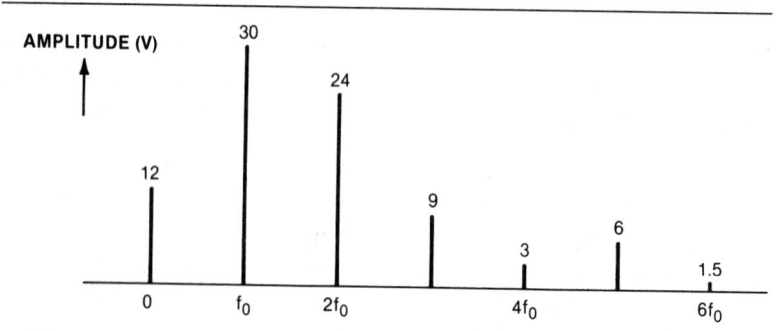

Fig. 1-17. Amplitude spectrum for Drill Problem 1.19.

Amplitude values (V): 12 (0), 30 (f_0), 24 ($2f_0$), 9 ($3f_0$), 3 ($4f_0$), 6 ($5f_0$), 1.5 ($6f_0$).

Electronics: Circuits and Systems

NONPERIODIC SIGNALS

A *nonperiodic* signal is one that does not occur in a predictably repetitive manner. Examples of nonperiodic signals are single pulses, groups of pulses not occurring at a specified rate, and so on. The discussion of periodic signals of the previous section can be extended to nonperiodic signals.

Consider a single rectangular pulse as shown in Fig. 1-18. This single pulse can be treated (conceptually) as if it were part of a *periodic pulse train*, except that the *period* of the pulse train is *infinity*. That is, we will have to wait forever before the next pulse shows up! The single pulse can, therefore, be thought of as a special case of a periodic signal with T = infinity.

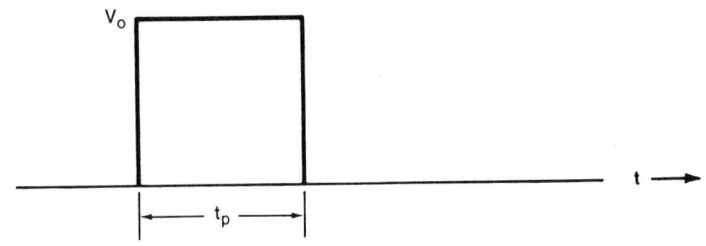

Fig. 1-18. A rectangular pulse.

In the discussion of periodic signals, we considered the effect of increasing the period T on the amplitude spectrum. It was mentioned that as the period is increased, the spectral lines kept getting closer together, and in the limit when T became infinite, the lines could no longer be distinguished as separate lines. That is, the spectrum becomes a *continuous curve* instead of a set of discrete lines. The amplitude spectrum of a pulse is, therefore, a continuous function whose expression is found to be of the form:

$$(V_o t_p) \frac{\sin(\pi f t_p)}{(\pi f t_p)}$$

where,

t_p is the duration of the pulse,
f is the frequency.

The amplitude spectrum is shown in Fig. 1-19.

The amplitude spectrum of a nonperiodic signal can be determined mathematically by using Fourier transforms or numerically by using computer programs. It can also be measured experimentally in the laboratory.

The amplitude spectrum of a nonperiodic signal serves a function similar to that of the periodic signal. It shows the relative strengths of the signal in different *ranges* of frequency (rather than at discrete frequencies as was the case

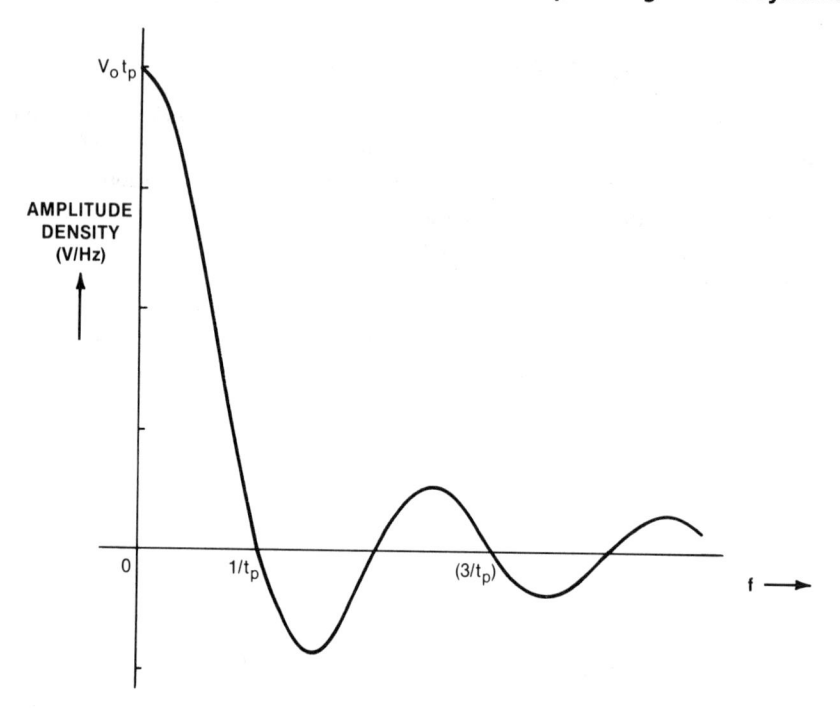

Fig. 1-19. Amplitude spectrum of a rectangular pulse.

in periodic signals). It is also important in determining the bandwidth needed for the transmission of a signal through a system. Fig. 1-20 shows two pulses of equal duration and equal maximum values and their amplitude spectra. It is seen that the spectrum of the triangular pulse falls off rapidly after about 500 Hz, whereas that of the rectangular pulse shows significant amplitudes at much higher frequencies. This implies that the bandwidth needed for the transmission of the triangular pulse will be much less than that for the rectangular pulse of the same duration and same maximum value. We will also see in the discussion of the energy spectrum that the uniformly smaller amplitude spectrum of the triangular pulse represents a smaller energy content of that signal relative to the rectangular pulse.

Energy Spectrum of a Pulse

The energy contained in a pulse can be calculated by taking the power (consumed by the standard load of 1 Ω) and integrating it over the duration of the pulse. For a rectangular pulse, shown in Fig. 1-18, we have a constant power value of V_o^2 for the duration of the pulse, and consequently, energy in a rectangular pulse is $V_o^2 \, t_p$ joules, which is seen to be a finite quantity.

The total energy contained in a pulse is not as important as the manner in

Electronics: Circuits and Systems

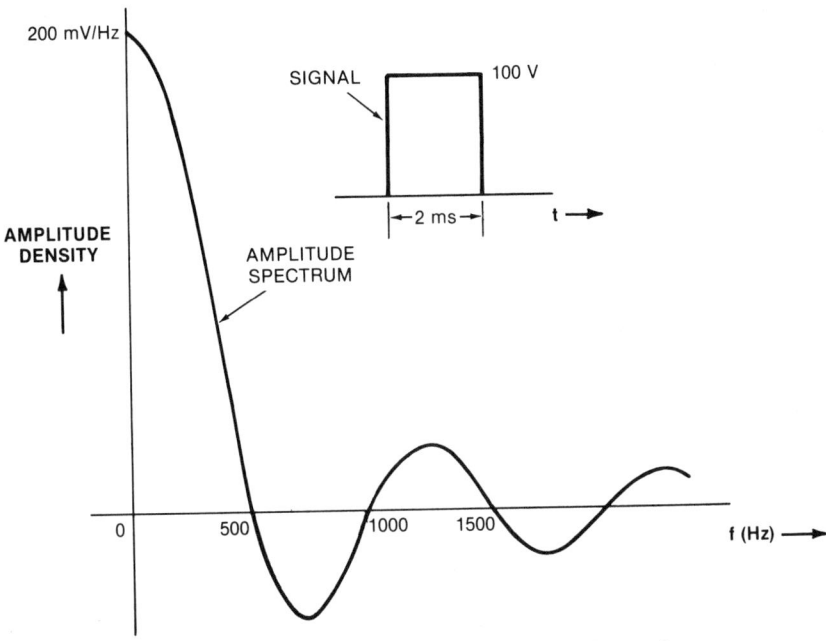

(A) Amplitude spectrum for a rectangular pulse.

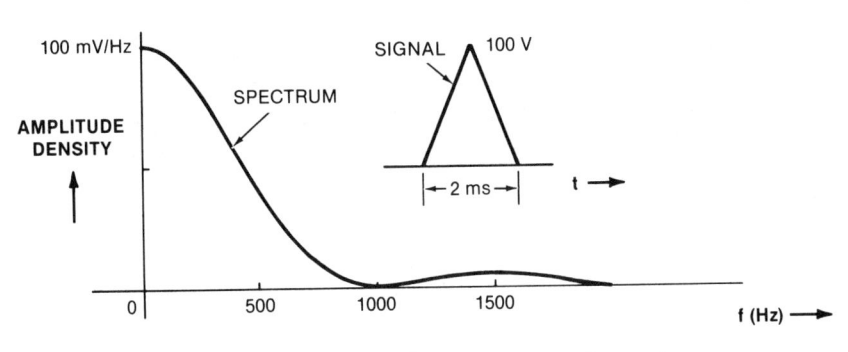

(B) Amplitude spectrum for a triangular pulse.

Fig. 1-20. Energy spectra of two different pulses.

which it is distributed with respect to frequency. The *energy spectrum* of a pulse is obtained by *squaring* its amplitude spectrum (point by point) and shows the distribution of energy as a function of frequency. The energy spectrum of a rectangular pulse is shown in Fig. 1-21, which was obtained by squaring the curve in Fig. 1-19. The units of the vertical axis in the diagram are *joules per hertz*.

48 Basic Concepts of Signals and Systems

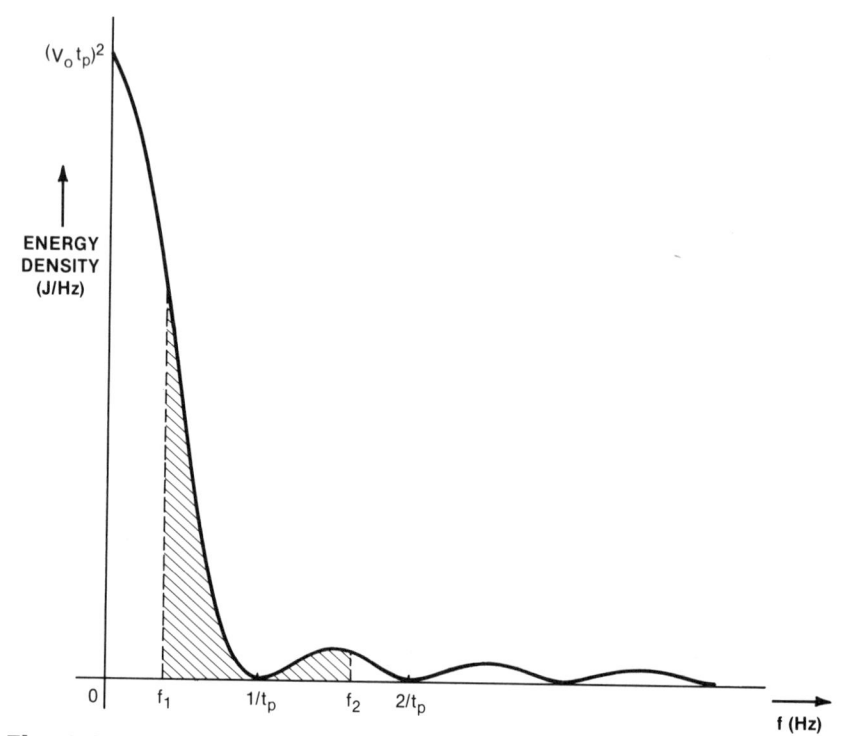

Fig. 1-21. Area under the energy density curve between f_1 and f_2 gives the energy contained in the signal in the band of frequencies from f_1 to f_2.

The energy spectrum has the property that the *energy contained in the signal in a given range of frequencies f_1 to f_2 is equal to the area under the energy spectrum over that frequency range*. For example, the shaded area in Fig. 1-21 gives the energy contained in the rectangular pulse in the band of frequencies between f_1 and f_2.

The bandwidth needed for the transmission of a nonperiodic signal can be estimated by using its energy spectrum. Suppose we decide that 90% of the total energy is to be passed by a system. Then, we can estimate the frequency range needed to produce 90% of the total area under the energy spectrum, and this frequency range will give us the desired bandwidth.

Example 1.6

The energy spectrum of an arbitrary signal is shown in Fig. 1-22. Calculate the energy contained in the following bands of frequency: (a) 0 to 4 kHz; (b) 4 kHz to 8 kHz. Also, calculate the total energy contained in the signal.

Electronics: Circuits and Systems

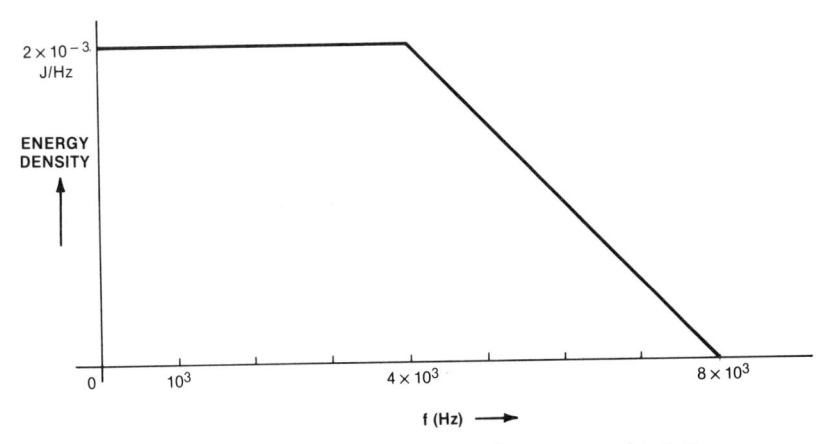

Fig. 1-22. Energy spectrum for Example 1-6.

Solution

The area shown shaded in Fig. 1-23A gives the energy in the range 0 to 4 kHz and is 8 J. The area shown shaded in Fig. 1-23B is the energy contained in the range 4 to 8 kHz and is 4 J.

The total energy in the signal is the total area under the energy spectrum and equals 12 J.

Drill Problem 1.20: Calculate the bandwidth needed to transmit 90% of the total energy contained in the signal of Example 1.6.

Consider once again the energy spectrum of the rectangular pulse as shown in Fig. 1-21. It is seen that a very large percentage of energy is in the band from dc ($f = 0$) to a frequency $f = (1/t_p)$. Therefore, if the bandwidth of a system is chosen as $(1/t_p)$, it can be considered adequate to pass a rectangular pulse of duration t_p seconds. This is in fact used as a rule of thumb in communication systems (such as television). We can see that the *bandwidth is inversely proportional to the duration of the pulse:* a wide pulse can be passed by a system with a narrow bandwidth, but a narrow pulse needs a system of a wide bandwidth. For example, a pulse of duration 1 ms will need a system (such as an amplifier) of bandwidth 1 kHz, whereas a pulse of duration 1 ns will need a system of bandwidth 1000 MHz! This type of inverse relationship between pulse duration and system bandwidth is valid even when the pulse is not rectangular in shape: *a narrower pulse needs a wider bandwidth.*

One particular energy spectrum, shown in Fig. 1-24, is of practical interest. It is a constant at all frequencies. The signal that possesses such an energy spectrum is called *white noise* (by analogy with white light whose optical

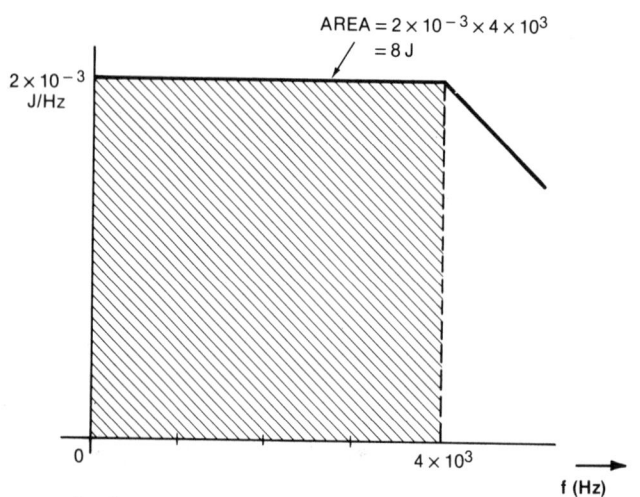

(A) Area equals the energy in the band of frequencies dc to 4 kHz.

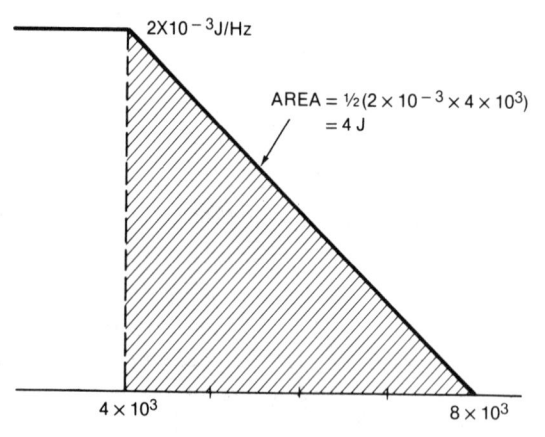

(B) Area equals the energy in the band 4 kHz to 8 kHz.

Fig. 1-23. The energy contained in two different bands of frequency ranges.

spectrum extends over all visible frequencies). White noise occurs in the analysis of communication systems.

Energy Signals and Power Signals

The similarity between the energy spectrum of a nonperiodic signal and the power spectrum of periodic signals is evident from the previous discussion. As

ENERGY DENSITY (J/Hz)

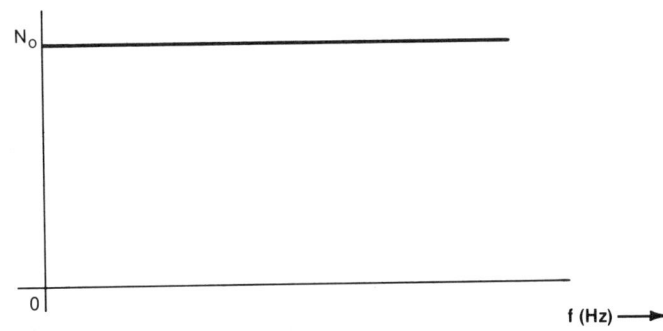

Fig. 1-24. Energy spectrum of white noise.

will be seen shortly, it makes more sense to use energy in the case of pulses and power in the case of periodic signals.

The power in *any* signal involves the use of the rms value of the voltage. As was discussed in connection with rms values, the determination of rms values requires an *averaging* process: the abbreviation *rms* stands for root mean square, and the term *mean* indicates that an average has to be obtained. The averaging is normally taken over a period T of a periodic signal. In the case of single pulses, the period is infinite, and the averaging will lead to a zero value since a division by infinity gives zero. Consequently, it would be meaningless to refer to the *average power* contained in single pulses, but pulses do have a finite amount of energy and their energy spectrum has special significance. Such signals with finite energy and a zero average power are referred to as *energy signals*.

For a periodic signal, the average power is not zero, since the rms value is nonzero. But, if we consider the energy contained in a periodic signal, it can be very large and even infinite, since energy equals the product of the average power and the time for which the signal is present. Consequently, it is more sensible to look at periodic signals in terms of the average power contained in them rather than their energy content. Therefore, the power spectrum of a periodic signal has a special significance. Such signals, which have finite average power, but potentially infinite amounts of energy, are referred to as *power signals*.

There is another category of signals that are not periodic but last indefinitely. The signals from a microphone placed in front of an orchestra, for example, are not periodic but can go on indefinitely. In such cases, the average power can be calculated by choosing an appropriately large but convenient length of time and finding the rms voltage over that period. The average power will be some finite nonzero value. But the energy content of such signals can be potentially infinite. Thus, signals of this type also belong to the class of power signals, and their power spectrum is of significance.

The energy spectrum of an energy signal and the power spectrum of a power signal can be determined experimentally through measurements in the laboratory, even when the signals cannot be described through mathematical functions. As was seen earlier, such spectra provide useful information about the distribution of energy or power over the different ranges of frequencies for a given signal. The spectra are also useful in the determination of the bandwidth needed for the transmission of the signals.

TWO-PORT MODEL OF A SYSTEM

Signals of different types and their important properties have been discussed in the preceding sections. The function of any electronic or communication system is to process a given signal and modify it to a form desired as the output of the system. A familiar example is the stereo amplifier whose input is the signal from the phonograph cartridge and whose output is the signal needed to activate the voice coil of the loudspeakers. In most cases, we are interested in the *overall effect* of the system rather than the detailed operation of the individual components in the system. In the case of an amplifier, for instance, the ratio of the power output to the input power may be the most important consideration and not the number of transistors and how they are interconnected. When the overall performance of a system is of primary interest, it is convenient to treat the system as a black box with a pair of input terminals (the *input port*) and a pair of output terminals (the *output port*). The word *port* is used here in the sense that we have an access to the system at these points. The black-box representation of a system is called its *two-port model*. Such a model is useful in the analysis and design of complex systems, because it permits the decomposition of complex subsystems, which can be analyzed or designed with less difficulty.

Gain Quantities of a Two Port

The diagram of a two port is shown in Fig. 1-25. V_{in} and V_{out} denote the input and the output voltages, and I_{in} and I_{out} denote the input and the output currents. The input and output powers are also of importance, and will be denoted by P_{in} and P_{out}.

The ratios of the output quantities to the input quantities are called *gains* and are important in describing the performance of a two port.

Voltage Gain is the ratio of the output voltage to the input voltage. That is,

$$\text{voltage gain} = \frac{V_{out}}{V_{in}} \tag{1-10}$$

Electronics: Circuits and Systems

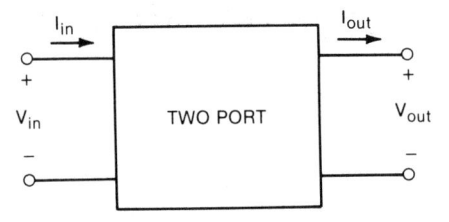

Fig. 1-25. Two port.

Current Gain is the ratio of the output current to the input current. That is,

$$\text{current gain} = \frac{I_{out}}{I_{in}} \quad (1\text{-}11)$$

Power Gain is the ratio of the output power to the input power. That is,

$$\text{power gain} = \frac{P_{out}}{P_{in}} \quad (1\text{-}12)$$

It should be mentioned that the term *gain* is used even in cases where the value of the gain may be actually less than 1.

The power gain can be shown to be equal to the product of the voltage gain and the current gain in the case of two ports containing no inductors or capacitors.

Since,

$$P_{out} = V_{out}I_{out}$$

and

$$P_{in} = V_{in}I_{in}$$

we get,

$$\frac{P_{out}}{P_{in}} = \frac{V_{out}I_{out}}{V_{in}I_{in}}$$

$$= \frac{V_{out}}{V_{in}} \times \frac{I_{out}}{I_{in}} \quad (1\text{-}13)$$

or *power gain = (voltage gain) × (current gain)*.

Example 1.7

A two port is shown in Fig. 1-26. Determine the various gains.

Basic Concepts of Signals and Systems

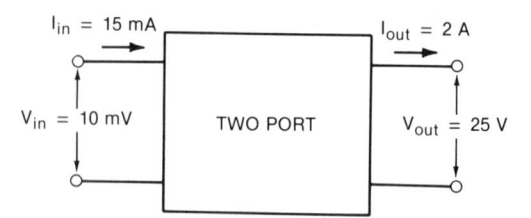

Fig. 1-26. Two port for Example 1.7.

Solution
The voltage gain is

$$\frac{25}{(10 \times 10^{-3})} = 2500$$

The current gain is

$$\frac{2}{(15 \times 10^{-3})} = 133$$

The power gain is the product of the voltage gain and the current gain and equals 3.32×10^5. The power gain could also be calculated by calculating the output power and the input power separately and then taking their ratio.

Drill Problem 1.21: If the input voltage of a two port is 5 mV, the input current is 15 mA, the output voltage is 2 V, and the output current is 5 A, calculate the three gains.

Example 1.8
The input voltage of a two port is 0.5 V, and the input current is 2 mA. The output voltage is 15 V. If the power gain of the two port is known to be 210, determine the output current and output power.

Solution
The voltage gain is $(15/0.5) = 30$. Since the power gain is the product of the voltage gain and the current gain, and the power gain is given as 210, the current gain is calculated to be $(210/30) = 7$.

Since the current gain = (output current/input current), and the input current is given as 2 mA, the output current is calculated to be $(7 \times 2) = 14$ mA.

The output power is the product of the output voltage and the output current and equals $(15 \times 14 \times 10^{-3}) = 210 \times 10^{-3}$ W.

Drill Problem 1.22: A two port with a power gain of 58.3 has an input

Electronics: Circuits and Systems

voltage of 17 V and an input current of 10 mA. The output current is 0.7 A. Calculate the output voltage and the output power.

Drill Problem 1.23: How much voltage gain is needed to produce an output power of 10 W, when the input voltage is 50 mV, the input current is 10 mA, and the current gain is 500?

Input Resistance of a Two Port

A two port is normally used in a system as shown in Fig. 1-27, where a signal source V_s with a resistance R_s is connected to the input side, and a load resistor R_L is connected to the output side.

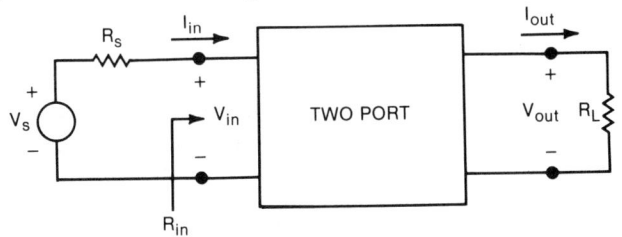

Fig. 1-27. Two-port system.

The two port and the load resistor together present an effective resistance to the input side, and this resistance is called the *input resistance* and denoted by R_{in}. This resistance relates the input voltage to the input current:

$$V_{in} = R_{in} I_{in} \qquad (1\text{-}14)$$

Even though an actual calculation of the input resistance can be made if the structure of the two port is known, it is usually possible to obtain a good practical estimate of the input resistance by knowing what the first active device is in the two port. For example, if the first stage of the amplifier in the two port is known to contain a bipolar junction transistor, then the input resistance is on the order of a few hundred ohms to a few kilohms. If the first stage of the two port is known to contain a field-effect transistor, then the input resistance is extremely high, a few hundred kilohms to a few megohms. Such "ball-park" figures of the input resistance are often sufficient to calculate the input current and input power approximately when the input voltage is known.

The concept of input resistance of a two port permits us to replace the two port and the load of Fig. 1-27 by a single resistance R_{in}, so that the two port system becomes the simple network of Fig. 1-28. This simple network is useful in the calculation of input quantities.

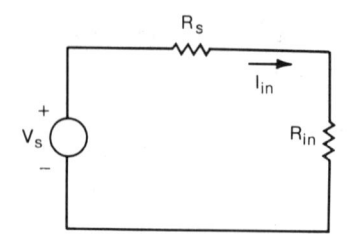

Fig. 1-28. Simplified model for the input side of a two-port system.

From the simple network of Fig. 1-28, we have

$$V_s = (R_s + R_{in})I_{in} \qquad (1\text{-}15)$$

which is useful in the calculations of the input quantities.

Example 1.9

For the two-port system of Fig. 1-27, assume that: $V_s = 100$ mV, $R_s = 1000\ \Omega$, $R_{in} = 4000\ \Omega$, $R_L = 500\ \Omega$, and voltage gain of the two port = 55.

Calculate (a) input voltage V_{in}; (b) output voltage; (c) output current; and (d) output power.

Solution

(a) For the input side, replace the system by a network similar to that in Fig. 1-28, as shown in Fig. 1-29. Then, using Eq. (1-15)

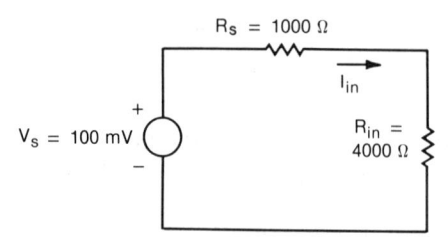

Fig. 1-29. Input side of the two-port system for Example 1.9.

$$I_{in} = \frac{V_s}{(R_s + R_{in})}$$

$$= 20 \times 10^{-6}\ \text{A}$$

The input voltage, V_{in}, is calculated by using Eq. (1-14)

$$V_{in} = R_{in} I_{in}$$

$$= 0.08\ \text{V}$$

Electronics: Circuits and Systems

(b) The output voltage is obtained by using the input voltage and the given voltage gain of 55.

$$V_{out} = 55 \times 0.08 \text{ V}$$
$$= 4.4 \text{ V}$$

(c) The output current is obtained by using Ohm's law on the load resistor:

$$V_{out} = R_L I_{out}$$

or

$$I_{out} = (4.4/500)$$
$$= 0.0088 \text{ A}$$

(d) The output power is obtained from the product of V_{out} and I_{out}, and equals 3.87×10^{-2} W.

Even though this might seem like an insignificant amount of power, note that the *power gain* of the two port is 24200.

Drill Problem 1.24: If the input resistance of the two port in the previous example is changed to 100 Ω, with all other values remaining unaffected, repeat the calculations.

Drill Problem 1.25: For a two-port system (as in Fig. 1-27), $V_s = 0.5$ V, $R_s = 1000$ Ω, $R_{in} = 500$ Ω, and $R_L = 2000$ Ω. The power gain of just the two port (that is, not including the signal source and R_s) is given as 10^4. Determine the output voltage, the voltage and current gains, and the power output of the system.

Example 1.10

In the two-port system of Fig. 1-30, the power output is to be 72 W, when the signal source produces $V_s = 110$ mV. Calculate the voltage gain, the current, and the power gain of the two port.

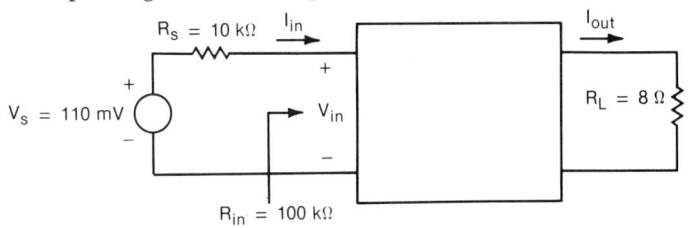

Fig. 1-30. Two-port system for Example 1.10.

Solution

The output voltage is calculated from the given power output and load resistance. Since $P_{out} = (V_{out}^2/R_L)$, we get

$$V_{out}^2 = 576$$

or

$$V_{out} = 24 \text{ V}$$

The output current is, by means of Ohm's law,

$$(V_{out}/R_L) = I_{out}$$
$$= 3 \text{ A}$$

Now, looking at the input side of the system, the total resistance seen by the source is

$$(R_s + R_{in}) = 10^4 + 10^5$$
$$= 11 \times 10^4 \text{ }\Omega$$

The input current is, therefore,

$$I_{in} = \frac{V_s}{(R_s + R_{in})}$$
$$= \frac{110 \times 10^{-3}}{11 \times 10^4}$$
$$= 10^{-6} \text{ A}$$

The input voltage V_{in} is given by

$$V_{in} = R_{in}I_{in}$$
$$= 10^5 \times 10^{-6}$$
$$= 0.1 \text{ V}$$

The various gains are, therefore,

$$\text{voltage gain} = (V_{out}/V_{in})$$
$$= 240$$
$$\text{current gain} = (I_{out}/I_{in})$$
$$= 3 \times 10^6$$
$$\text{power gain} = (\text{Voltage gain}) \times (\text{Current gain})$$
$$= 720 \times 10^6$$

Electronics: Circuits and Systems

Drill Problem 1.26: Assume that the two port of the previous example (with power gain being the same) is connected to a signal source of 150 mV with a resistance of $R_s = 10^6$ Ω. The load resistance is the same as before. Determine the value of the power output.

Useful Approximations

When the input resistance of a two port is very low or very high compared with the resistance R_s of the signal source, it is possible to make certain approximations that are useful in making quick calculations, which are reasonably accurate for all practical purposes.

High Input Resistance

When R_{in} is greater than *ten times* the value of R_s, we consider R_{in} to be *much greater* than R_s. For example, if R_s has a value of 1000 Ω, any value of R_{in} larger than 10^4 Ω will be considered as being much greater than R_s. In such cases, the total resistance $(R_s + R_{in})$ seen by the signal source is taken as approximately equal to R_{in}. That is, if

$$R_{in} > 10 R_s$$

then,

$$(R_{in} + R_s) \approx R_{in}$$

where,
the symbol $\approx$ means "approximately equal to." The input current is, therefore, approximately equal to (V_s/R_{in}). Also, since R_{in} is much larger than R_s, the voltage across the input of the two port is much greater than the voltage across R_s. That is, $V_{in} \approx V_s$.

In essence, *when the input resistance of a two port is high (much greater than R_s), we can ignore R_s for all practical purposes.* The error introduced by such an approximation will be less than 10% (which is usually considered a tolerable error).

Returning to Example 1.10, R_{in} is much greater than R_s for that system, and the approximate values of the input quantities (obtained by ignoring R_s) are

$$I_{in} = (V_s/R_{in})$$
$$= 1.1 \times 10^{-6} \text{ A}$$
$$V_{in} = V_s$$
$$= 110 \text{ mV}$$

The gain values, obtained by using this approximation, become

$$\text{voltage gain} = \frac{24}{110 \times 10^{-3}}$$

$$= 218$$

$$\text{current gain} = \frac{3}{1.1 \times 10^{-6}}$$

$$= 2.73 \times 10^6$$

$$\text{power gain} = (218)(2.73 \times 10^6)$$

$$= 595 \times 10^6$$

Comparing these values with 240, 3×10^6, and 720×10^6 obtained earlier, the error in each case will be found to be approximately 10%.

Low Input Resistance

When the input resistance of a two port is *smaller* than ($R_s/10$), it is considered to be much smaller than R_s. In such cases, we can essentially ignore the value of R_{in} and the input current is approximately (V_s/R_s). The amount of current drawn by the two port is dependent only upon R_s and is independent of the two port and the load. Therefore, the signal source is acting as an *ideal current source* when the input resistance of the two port is much smaller than R_s. (An ideal current source has the property that the current supplied by it is *independent of the load* connected to it.)

Example 1.11

The input resistance of the two port in Fig. 1-31 is 1 kΩ, and the source resistance R_s is 20 kΩ. Calculate the current supplied by the source (a) when approximation is used and (b) when no approximation is used. Calculate the percent of error.

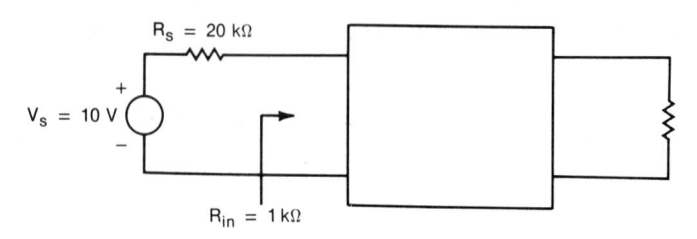

Fig. 1-31. Two-port system for Example 1.11.

Solution

Exact calculation:

$$I_{in} = \frac{10}{(10^3 + 20 \times 10^3)}$$

$$= 0.476 \times 10^{-3} \text{ A}$$

Approximate calculation:

$$I_{in} = \frac{V_s}{R_s}$$

$$= \frac{10}{20 \times 10^3}$$

$$= 0.5 \times 10^{-3} \text{ A}$$

The error is given by

$$\text{percent of error} = \frac{(\text{exact value}) - (\text{approximate value})}{\text{exact value}}$$

$$= -5.04\%$$

Drill Problem 1.27: A two port has an input resistance of 1000 Ω. A signal source of $V_s = 10$ V is connected to it. For each value of R_s listed, determine the input current, using an approximation where justifiable. Calculate the percent of error in the cases where approximation is used. (a) $R_s = 10$ Ω; (b) $R_s = 100$ Ω; (c) $R_s = 1$ kΩ; (d) $R_s = 15$ kΩ; (e) $R_s = 90$ kΩ.

DECIBELS (dB)

It has been found experimentally that the human ear responds to sound intensity on a *logarithmic* scale rather than a linear scale. That is, if the power content of a sound signal is increased by a factor of two, the ear does not perceive it as being twice as loud but as being louder by a factor of log 2 or 0.3. A new unit of measurement of power gain was introduced by Alexander Graham Bell to account for this peculiarity of the ear. The new unit has some advantages and is used generally in electronics and communications even when sound or hearing is not involved in such cases. The unit introduced is called the *bel* (a mutation of the inventor's name) and is defined by power gain in *bels* = log (power gain as a ratio)

For example, if the power gain of a system is 3.325×10^6 as a ratio, then

$$\text{power gain in } bels = \log(3.325 \times 10^6)$$
$$= 6.522$$

The bel is found to be too large a unit for most practical situations. For example, a power gain of 10^6 amounts only to 6 bels. It is more convenient to use a smaller unit, called the *decibel* (abbreviated *dB*), which equals one tenth a bel, that is,

1 bel = 10 dB or 1 dB = 0.1 bel

Therefore, in order to obtain a power gain in *decibels*, the power gain in bels should be multiplied by 10. That is,

power gain in *decibels* = 10 × log (power gain as a ratio)   (1-16)

For example, a power gain of 210 as a ratio will lead to

10 × log (210) = 10 × 2.322

= 23.22 dB

and a power gain of 10^6 (ratio) will give 60 dB.

It is important to distinguish between the two ways of expressing the power gain: (1) the power gain as a *ratio* of two power values as in (P_{out}/P_{in}), and (2) the power gain in decibels, which is obtained by taking the logarithm of that ratio and multiplying the logarithm by 10. In order to avoid any confusion, the symbol dB is always attached to the value of a gain when expressed in decibels. If the dB is missing, the gain value is taken as the ratio of two quantities.

To convert the value of a power gain in decibels to the corresponding ratio, divide the decibel value by 10, and then find the *antilogarithm* of the resulting quantity. That is,

power gain as a ratio = antilog (power gain in dB/10)   (1-17)

For example, given a power gain of 47.71 dB, divide it by 10, and find the antilogarithm of (47.71/10), which leads to a value of 5.902×10^4. Therefore, a power gain of 47.71 dB corresponds to a power gain of 5.902×10^4 (ratio).

Example 1.12

Convert each of the following power gains to decibels: (a) 500; (b) 2000; (c) 1.2×10^4.

Solution
(a) 10 × log (500) = 26.99 dB; (b) 10 × log (2000) = 33.01 dB; (c) 10 × log (1.2 × 10^4) = 40.79 dB.

Example 1.13
Convert each of the following decibel values of power gains to ratio value: (a) 84.51 dB; (b) 3.01 dB; (c) 100 dB.

Solution
(a) antilog (84.51/10) = 2.825 × 10^8; (b) antilog (3.01/10) = 2; (c) antilog (100/10) = 10^{10}.

When the value of a power gain is *less than 1*, the corresponding value in *decibels will be negative*. For example, a power gain of 0.25 gives [10 × log (0.25)] = −6.02 dB. Conversely, if the decibel value of a power gain is negative, the ratio value will be less than 1.

Table 1-5 shows some selected values of power gain and the corresponding decibel values. Note that the *multiplication of a power gain by a factor of 10 results in the decibel value going up by 10 dB*. Similarly, a *reduction in power gain by a factor of 10 results in the decibel value decreasing by 10 dB*. The table may be found useful in the solution of the following drill problems.

Table 1-5. Selected Values of Power Gain and the Corresponding Decibel Values

Power Gain (Ratio)	Power Gain (dB)
1	0
10	10
10^n	10n
2	3
20	13
3	4.77
5	6.99
7	8.45

Drill Problem 1.28: Convert each of the following power gains to decibels: (a) 2; (b) 2 × 10^4; (c) 0.2; (d) 2 × 10^{-6}.

Drill Problem 1.29: Convert each of the following decibel values of power gain to a corresponding ratio value: (a) 14.77 dB; (b) 114.77 dB; (c) 74.77 dB; (d) −5.23 dB; (e) −25.23 dB.

Drill Problem 1.30: An amplifier has a power gain of 46.99 dB. Convert it to a ratio value and then determine the power output when the input power is 0.5 mW.

Drill Problem 1.31: A loudspeaker requires 50-W power to operate and the phonograph cartridge produces a power of 10 μW. Calculate the power gain needed for the amplifier and express it in decibels.

Computational Advantages of the Decibel Unit

The decibel unit is advantageous in computations because of the following property of logarithms: the logarithm of a product of two numbers is equal to the sum of the logarithms of the two numbers. That is, given two numbers A and B,

$$\log (AB) = \log A + \log B \tag{1-18}$$

In electronics and communications it is frequently necessary to multiply the power gains of the various stages of a system, and the *multiplication reduces to addition if decibel units are used* for power gains rather than ratios.

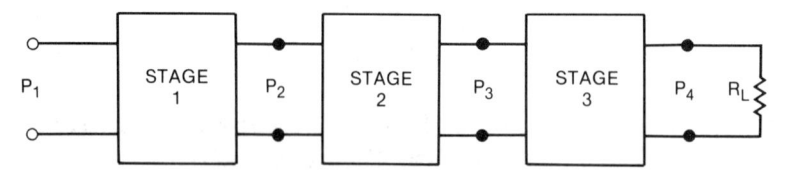

Fig. 1-32. Multistage amplifier.

Consider a multistage amplifier indicated in Fig. 1-32 in which the powers at the various ports are P_1, P_2, P_3, and P_4. Multistage amplifiers are necessary in electronic design when a single stage is not sufficient to produce the desired power gain. In the multistage system, the overall power gain is (P_4/P_1), and it can be seen that

$$\text{power gain } (P_4/P_1) = (P_4/P_3) \times (P_3/P_2) \times (P_2/P_1)$$

If, instead of using the ratio values of power gains as in the previous equation, the decibel values are used, we have

$$10 \log (P_4/P_1) = 10 \log [\, (P_4/P_3) \times (P_3/P_2) \times (P_2/P_1) \,]$$

which becomes, by using the property expressed in Eq. (1-18),

$$10 \log (P_4/P_1) = 10 \log (P_4/P_3) + 10 \log (P_3/P_2) + 10 \log (P_2/P_1)$$

or

(P_4/P_1) in decibels = (P_4/P_3) in decibels + (P_3/P_2) in decibels
+ (P_2/P_1) in decibels

The above result can be extended to any number of stages. *The overall power gain of a multistage system expressed in decibels is equal to the sum of the power gains of the individual stages all expressed in decibels.*

Since addition is much easier than multiplication, especially when a calculator is not handy, the use of the decibel makes calculations quicker than using the ratio values of power gains. The advantage of the decibel unit is particularly noticeable in design problems where a given overall gain has to be split into smaller values corresponding to the individual stages.

Example 1.14

A three-stage amplifier (like that in Fig. 1-32) has the following power gain values: 60 dB for stage 1, 36 dB for stage 2, and 3 dB for stage 3. (a) Calculate the overall power gain in decibels. (b) If the power output of the amplifier is given as 50 W, find the power input to each of the three stages.

Solution

(a) Since the power gain values of the individual stages are already in decibels, overall power gain = (60 + 36 + 3) = 99 dB. (b) To calculate the input power when the output power of a stage is given (or vice versa), it is *necessary* to have the power gain as a *ratio* and *not* as decibel (which is an important condition).

Stage 3:
Power gain (ratio) = antilog (3/10) = 2
Power output of third stage = 50 W (given)
Therefore,
Power input of third stage = (50/2) = 25 W

Stage 2:
Power gain (ratio) = antilog (36/10) = 3.98×10^3
Power output of second stage = Power input of third stage = 25 W
Therefore,

$$\text{power input of second stage} = (25/3.98 \times 10^3)$$
$$= 6.28 \times 10^{-3} \text{ W}$$

Stage 1:

Power gain (ratio) = antilog (60/10) = 10^6;

$$\text{power output of first stage} = \text{power input of second stage}$$
$$= 6.28 \times 10^{-3} \text{ W}$$
$$= 6.28 \text{ mW}$$

Therefore,

$$\text{power input of stage 1} = (6.28 \times 10^{-3})/10^6$$
$$= 6.28 \times 10^{-9} \text{ W}$$
$$= 6.28 \text{ nW}$$

Drill Problem 1.32: Change the power gain of the third stage of the amplifier in the previous example to -3 dB. Assume the power gains of the other stages to be the same as well as the given power output. Recalculate the results.

Drill Problem 1.33: A single-stage amplifier has a power gain (ratio) of 250. If three stages are cascaded, calculate the overall power gain in decibels.

Drill Problem 1.34: In a two-stage amplifier, the first stage has a power gain of 54.77 dB and the second stage has a power gain of -5.23 dB. The power input to the first stage is 50 μW. Calculate the power output of each amplifier stage.

Example 1.15

Single-stage amplifiers are available for a particular application, and the power gain (ratio) of each stage is 750. It is desired to design a multistage amplifier with an overall power gain (ratio) of 10^{10}. Calculate the number of amplifier stages to be cascaded to produce the desired power gain (as a minimum).

Solution

If n stages are needed to produce the desired gain, and if each stage had a gain of 750, as given, then it is necessary to find n such that $(750)^n =$ the total gain of 10^{10} desired. Even though it is possible to calculate n by solving the equation $(750)^n = 10^{10}$, it is much easier to proceed in terms of decibels. The gain of each stage is 10 log (750) = 28.8 dB. The overall gain desired is 10 log (10^{10}) = 100 dB. The total gain of n stages is equal to *n times the gain in decibels of each stage;* we have to make 28.8n = 100. Therefore,

Electronics: Circuits and Systems   67

$$n = (100/28.8)$$
$$= 3.48$$

or 3.48 stages are needed. Since fractional stages are unrealizable, the number n is rounded off to the next higher value. Therefore, four stages are needed.

Drill Problem 1.35: Using the amplifier stages of Drill Problem 1-33, calculate the number of stages needed to provide a gain of 9×10^{11} (ratio).

Voltage and Current Gains in Decibels

Even though the decibel was initially developed in connection with power gain, it is also used for expressing voltage and current gains of a two port. Since power is proportional to the square of a voltage, or the square of a current, the conversion of voltage or current gain to decibels has to take into account the squaring involved. If A is any number, then log (A^2) = 2 log A. Thus, a factor of 2 is introduced when a quantity is squared and then its logarithm calculated. Because of this, the following conversion procedures are used for voltage and current gains.

voltage gain *in decibels* = 20 log (voltage gain as a ratio)
current gain *in decibels* = 20 log (current gain as a ratio)

For example, a voltage gain of 120 gives 20 log (120) = 41.6 dB; and a current gain of 35 gives 20 log (35) = 30.9 dB.

The decibel values listed in Table 1-5 should each be *multiplied by a factor of 2* if the ratio values corresponded to *voltage or current gain*, instead of power gain. For example, the decibel value corresponding to a ratio of 7 becomes (2 × 8.45) = 16.9 dB if that ratio represented a voltage or current gain.

Drill Problem 1.36: Convert each of the following voltage gain (ratio) and current gain (ratio) values to decibels: (a) voltage gain of 30; (b) current gain of 70; (c) voltage gain of 0.5; (d) current gain of 0.3.

When decibel values of voltage or current gain are to be converted to ratio values, we have,

Voltage or current gain (ratio) = antilog (decibel value/20)

Drill Problem 1.37: Determine the ratio values of the gains whose decibel values are given: (a) voltage gain of 90 dB; (b) current gain of −3 dB; (c) voltage gain of −20 dB; (d) current gain of 60 dB.

Drill Problem 1.38: Convert each of the following gain values (ratios) to decibels paying careful attention to the type of gain in each case: (a) power gain of 562.5; (b) voltage gain of 562.5; (c) voltage gain of 32.8; (d) current gain of 32.8; (e) power gain of 0.5; (f) voltage gain of 0.707.

Drill Problem 1.39: Convert the following decibel values to ratio values of the gains, again paying careful attention to the type of gain in each case: (a) power gain of 30 dB; (b) voltage gain of 30 dB; (c) power gain of -20 dB; (d) current gain of -20 dB.

Example 1.16

A two port has an input voltage of 10 mV and an input current of 5 mA. The output voltage is found to be 5 V and the output current is 0.2 A.

Determine the voltage gain, current gain, and power gain—all in decibels.

Solution

The voltage and current gains can be calculated (as ratios) from the given information.

$$\text{voltage gain} = \frac{5}{10 \times 10^{-3}}$$

$$= 500 \text{ (ratio)}$$

$$\text{current gain} = \frac{0.2}{5 \times 10^{-3}}$$

$$= 40 \text{ (ratio)}$$

$$\text{power gain} = (500) \times (40)$$

$$= 2 \times 10^4$$

therefore,

$$\text{voltage gain in decibels} = 20 \log (500)$$

$$= 54 \text{ dB}$$

$$\text{current gain in decibels} = 20 \log (40)$$

$$= 32 \text{ dB}$$

$$\text{power gain in decibels} = 10 \log (2 \times 10^4)$$

$$= 43 \text{ dB}$$

It is important to note that the power gain in decibels is *not equal to* the *sum* of the decibel values of the voltage gain and the current gain, even though the power gain as a ratio is equal to the product of the ratio values of

Electronics: Circuits and Systems

the voltage gain and current gain. This is due to the different multiplication factors (20 in the case of voltage and current gains and 10 in the case of power gain) used in the conversion to decibels. It can be shown that the *decibel value of power gain is the average of the decibel values of the voltage and current gains*. In some cases, the decibel values of the three gains turn out to be the same, but this is not the general case. Consequently, care should be exercised in specifying which particular gain is being considered.

Example 1.17

A two-port system is shown in Fig. 1-33. Using the numerical values indicated in the diagram, calculate the values of the three gains of the system in decibels.

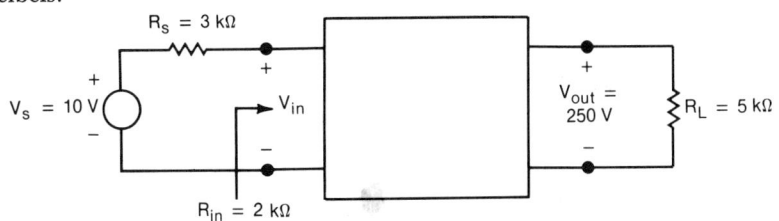

Fig. 1-33. Two-port system for Example 1.17.

Solution

The input current I_{in} is given by,

$$I_{in} = \frac{V_s}{(R_s + R_{in})}$$

$$= 2 \text{ mA}$$

The output current I_{out} is given by,

$$I_{out} = \frac{V_{out}}{R_L}$$

$$= 50 \text{ mA}$$

The current gain is, therefore, $(I_{out}/I_{in}) = 25$, which gives 20 log (25) = 28 dB.

The voltage gain can be defined as either (V_{out}/V_{in}) or (V_{out}/V_s). We will calculate both in this example.

$$\frac{V_{out}}{V_s} = \frac{250}{10}$$

$$= 25$$

which, converted to decibels, gives 28 dB.

The voltage V_{in} is obtained by subtracting from V_s the voltage drop across R_s, that is,

$$V_{in} = V_s - R_s I_{in}$$
$$= 10 - (2 \times 10^{-3}) \times (3 \times 10^3)$$
$$= 4 \text{ V}$$

Therefore, we get

$$\frac{V_{out}}{V_{in}} = \frac{250}{4}$$
$$= 62.5$$

which leads to 20 log (62.5) = 35.9 dB.

For power gain, the voltage gain (ratio value) is multiplied by the current gain (ratio value):

$$\text{power gain (ratio)} = 25 \times 25 = 625$$

which, converted to decibels, gives

$$10 \log (625) = 28 \text{ dB}$$

In this calculation, the voltage gain used is the ratio (V_{out}/V_s).

Drill Problem 1.40: An amplifier has a current gain of 25 dB and a voltage gain of 15 dB. Calculate its power gain in decibels.

Drill Problem 1.41: For the two-port system shown in Fig. 1-34, calculate the three gains in decibels.

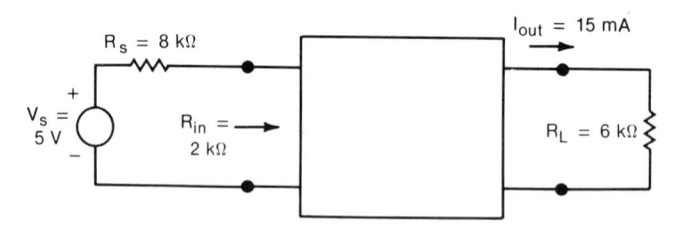

Fig. 1-34. Two-port system for Drill Problem 1.41.

As in the case of the power gain in decibels, the total voltage gain (or current gain) of a multistage system can be obtained by *adding* the decibel gains of the individual stages. For example, if an amplifier stage has a voltage gain

Electronics: Circuits and Systems

of 24 dB, and four such stages are cascaded together, then the overall voltage gain will be 96 dB.

Example 1.18

Amplifier stages are available with a voltage gain (ratio) of 48. How many stages should be cascaded together to obtain a total voltage gain of 144?

Solution

Converting the individual stage gain to decibels, we have

$$20 \times \log(48) = 33.62 \text{ dB}$$

Converting the overall gain to decibels, we have

$$20 \times \log(144) = 43.17 \text{ dB}$$

The number of stages needed is, therefore:

$$(43.17/33.62) = 1.28$$

That is, we need two stages.

Drill Problem 1.42: How many stages of the amplifier in the previous example will be needed to produce an overall voltage gain of 1440 (ratio value)?

Drill Problem 1.43: A three-stage amplifier has the following voltage gains: 20 dB for stage 1, 30 dB for stage 2, and −5 dB for stage 3. An input signal of 150 mV is applied to the first stage. Calculate the output voltage of each stage of the amplifier.

SUMMARY SHEETS

RESISTORS

Ohm's Law

$V = RI$

(current flow + to −)

$V = -RI$

(current flow − to +)

Power Equation

Power $P = (V^2/R) = I^2R$ watts

AC SIGNALS

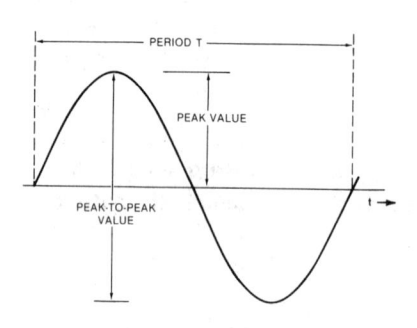

frequency $f = (1/T)$ Hz
rms value $= 0.707 \times$ peak value

Phase Angle

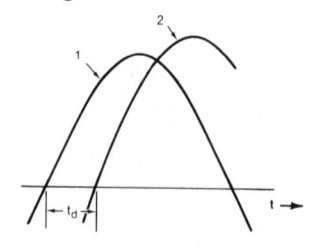

Signal 1 leads signal 2 by an angle $(t_d/T) \times 360°$

Power

Power $P = (V_{rms}^2/R)$
$I_{rms}^2 R$

PERIODIC NONSINUSOIDAL SIGNALS

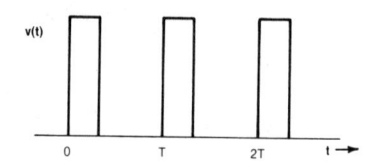

Fundamental frequency $f_0 = (1/T)$
nth harmonic $= nf_0$
$v(t) =$ component $+ f_0$ component $+$ all the harmonic components

Amplitude Spectrum

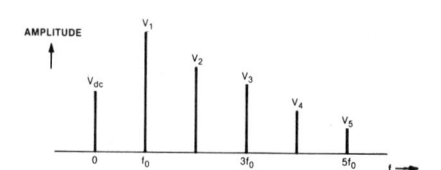

Power Spectrum

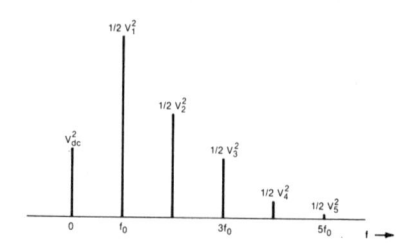

Power $P =$ sum of the powers due to the dc,

Electronics: Circuits and Systems

fundamental and all the harmonic components = the sum of the heights of the lines in the power spectrum.

PULSE SIGNALS

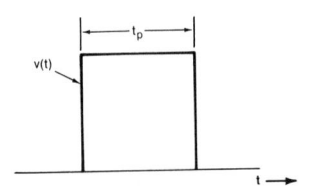

Amplitude Spectrum

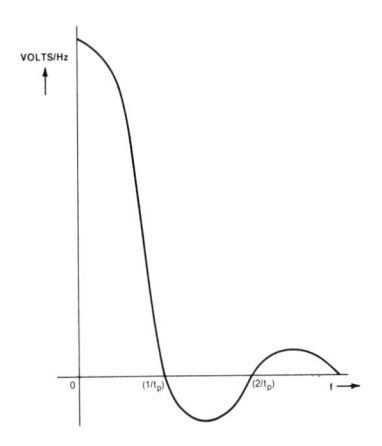

Energy Spectrum

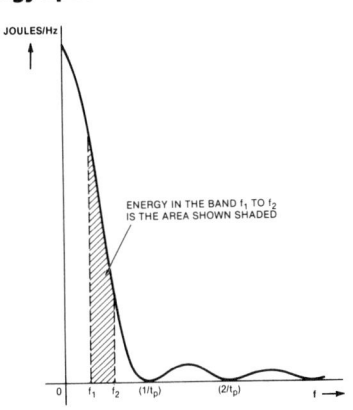

Energy in the band f_1 to f_2 is the area shown shaded.

TWO PORTS

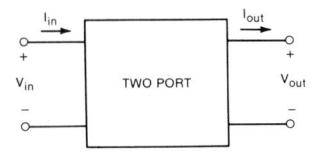

Voltage gain = (V_{out}/V_{in})
Current gain = (I_{out}/I_{in})
Power gain = (voltage gain) × (current gain)
Input resistance $R_{in} = (V_{in}/I_{in})$

Two-Port System

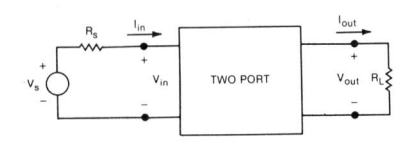

$$I_{in} = \frac{V_s}{R_s + R_{in}}$$

If $R_{in} \gg R_s$, $V_{in} \approx V_s$
If $R_{in} \ll R_s$, $I_{in} \approx (V_s/R_s)$

DECIBELS

Power gain in dB = 20 log (power gain)
Decibel value *negative* if power gain (ratio) value is less than 1.
Increase in power gain (ratio) by a *factor* of 10 → increase by 10 dB.
Multiple stages: Total power gain in decibels = sum of individual stage gains in decibels.
Voltage gain in decibels = 20 log (voltage gain)
Current gain in decibels = 20 log (current gain)
Power gain in decibels = *average* value of voltage gain and current gain in decibels.

ANSWERS TO DRILL PROBLEMS

1.1 (a) 16.7 ms; (b) 10.93 ns.
1.2 (a) 20 MHz; (b) 200 Hz; (c) 0.2 Hz.
1.3 (a) 14.14 V; (b) 10.6 A.
1.4 622.3 V
1.5 141.4 V (rms); 400 V (p-p).
1.6 f = 16.7 Hz; 5 V (peak); 10 V (p-p); 3.53 V_{rms}.
1.7 t_d = 20 ms; T = 48 ms; phase angle = 150°; v_1 leads.
1.8 Power consumed by R_1 = 900 W; by R_2 = 266.7 W. Power consumed by R_3 cannot be calculated from the data. Note that the current in it cannot be calculated from the data either.
1.9 V_{rms} = 15.49 V; V_{p-p} = 43.82 V; at half the rated power, V_{p-p} = 30.99 V.
1.10 (a) 50 MHz; (b) 250 MHz; (c) 500 MHz.
1.11 f_0 = 0.8 kHz or 800 Hz; T = 1.25 ms.
1.12 See Fig. 1-DP12.

(A)

[Spectrum plot: amplitudes 8, 16, 24, 10, 5, 2 at frequencies 0, 0.4, 0.8, 1.2, 1.6, 2.0 kHz]

(B)

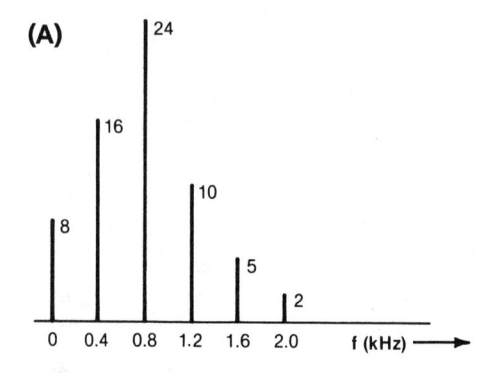

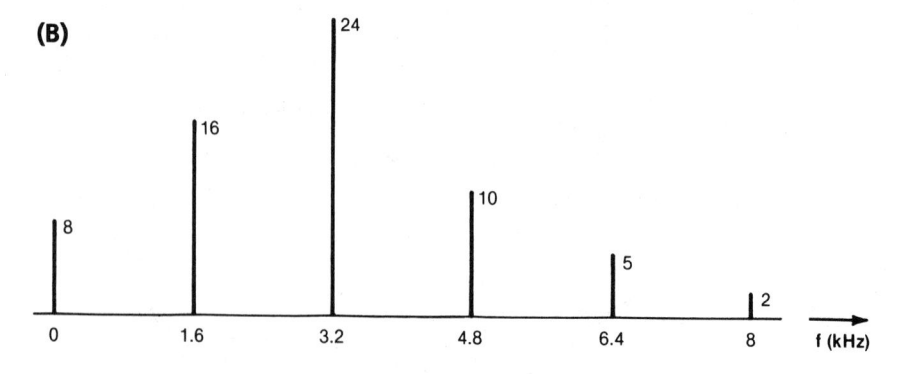

[Spectrum plot: amplitudes 8, 16, 24, 10, 5, 2 at frequencies 0, 1.6, 3.2, 4.8, 6.4, 8 kHz]

Fig. 1-DP-12.

Electronics: Circuits and Systems

1.13 $[30.88 + 56.27 + 42.22 + 25.01 + 10.55 + 2.254] = 167.2$ W
1.14 $[4 + 1.125 + 0.320 + 0.06125 + 0.00405] = 5.510$ W
1.15 Divide 5.510 by 4 to get 1.378 W. For a voltage signal, an increase in R causes a reduction in the power by a factor of R.
1.16 $3f_0$ or 50 Hz for 90% power; $4f_0$ or 66.7 Hz for 98% power.
1.17 f_0 or 100 Hz for 90% power; $2f_0$ or 200 Hz for 98% power.
1.18 See Fig. 1-DP18.

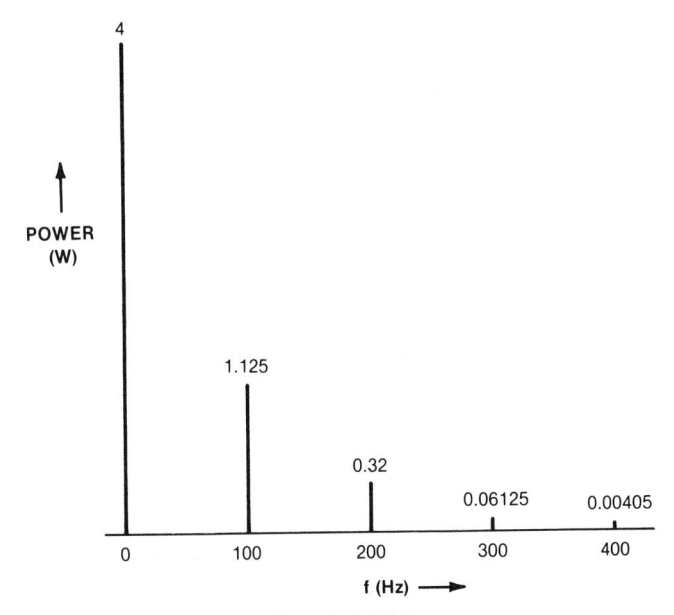

Fig. 1-DP18.

1.19 See Fig. 1-DP19 for the spectrum; power = 904.5 W.
1.20 90% of total energy = 10.8 J; referring to Fig. 1-DP20, the *unused* area (representing 10% of the total energy) is $(12 - 10.8) = 1.2$ J. The value of x can be calculated to be 2.19 kHz. Therefore, bandwidth needed for 90% of total energy = $(8 - 2.19) = 5.81$ kHz.
1.21 Voltage gain = 400; current gain = 333.3; power gain = 1.333×10^5.
1.22 $P_{in} = 170$ mW; $P_{out} = (170 \times 58.3)$ mW = 9.911 W; $I_{out} = 0.7$ A (given); $V_{out} = (9.911/0.7) = 14.16$ V.
1.23 Current gain = 500 (given) and $I_{in} = 10$ mA (given). $I_{out} = 5$ A; $P_{out} = 10$ W; $V_{out} = (10/5) = 2$ V; $V_{in} = 50$ mV (given); voltage gain = $(2/50 \times 10^{-3}) = 40$.
1.24 $I_{in} = (100/1100)$ mA = 9.091×10^{-2} mA; $V_{in} = 9.091$ mV; $V_{out} = 0.5$ V; $I_{out} = (0.5/500) = 10^{-3}$ A = 1 mA; $P_{out} = 0.5$ mW.
1.25 $I_{in} = 0.3333$ mA; $V_{in} = 0.167$ V; $P_{in} = 5.556 \times 10^{-5}$ W. $P_{out} = 0.5556$ W;

76   **Basic Concepts of Signals and Systems**

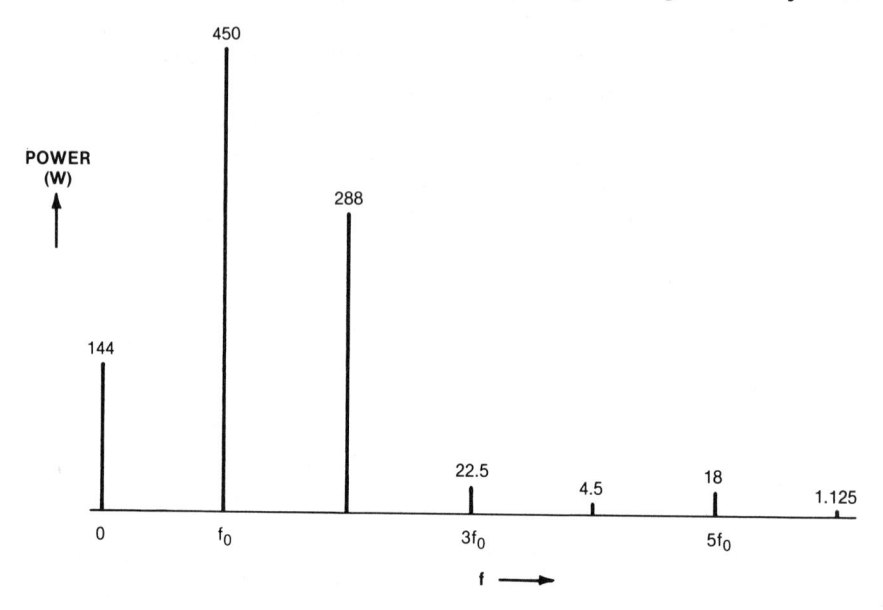

Fig. 1-DP19.

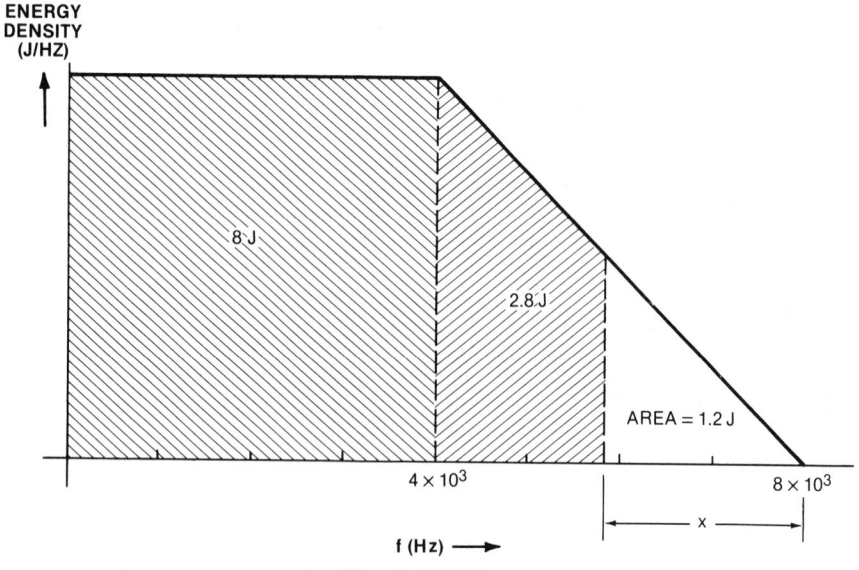

Fig. 1-DP20.

$V_{out} = 33.3$ V; $I_{out} = 16.67$ mA; voltage gain of the system = 66.6; voltage gain of a two port = 200; current gain = 50.

1.26 $I_{in} = 1.364 \times 10^{-4}$ mA; $V_{in} = 0.0136$ V; $P_{in} = 1.855$ nW; $P_{out} = 1.336$ W.

1.27 R_s may be neglected in (a) and (b); (a) $I_{in} = 10$ mA (approx); 9.90 mA (exact), error = -1%; (b) $I_{in} = 10$ mA (approx), 9.09 mA (exact), error =

Electronics: Circuits and Systems

-10%; (c) $I_{in} = 5$ mA (exact), R_{in} may be neglected in (d) and (e); (d) $I_{in} = 0.667$ mA (approx), 0.625 mA (exact), error $= -6.72\%$; (e) $I_{in} = 0.111$ mA (approx), 0.110 mA (exact); error $= -1.01\%$.

1.28 (a) 3 dB; (b) 43 dB; (c) -7 dB; (d) -57 dB.
1.29 (a) 30; (b) 3×10^{11}; (c) 3×10^7; (d) 0.3; (e) 3×10^{-3}.
1.30 Power gain $= 5 \times 10^4$ (ratio); $P_{out} = 25$ W.
1.31 67 dB.
1.32 Overall power gain $= 93$ dB. Stage 3: $P_{out} = 50$ W; power gain $= 0.5$, $P_{in} = 100$ W; Stage 2: $P_{out} = 100$ W, power gain $= 3.98 \times 10^3$, $P_{in} = 25.12 \times 10^{-3}$ W; Stage 1: $P_{out} = 25.12$ mW, power gain $= 10^6$, $P_{in} = 25.12 \times 10^{-9}$ W $= 25.12$ nW.
1.33 71.94 dB
1.34 Stage 1: power gain $= 3 \times 10^5$, $P_{out} = 15$ W; Stage 2: power gain $= 0.3$, $P_{out} = 4.5$ W.
1.35 Total gain $= 119.5$ dB; number of stages $= 5$.
1.36 (a) 29.5 dB; (b) 36.9 dB; (c) -6 dB; (d) -10.5 dB.
1.37 (a) 3.162×10^4; (b) 0.707; (c) 0.1; (d) 1000.
1.38 (a) 27.5 dB; (b) 55.0 dB; (c) 30.3 dB; (d) 30.3 dB; (e) -3 dB; (f) -3 dB.
1.39 (a) 1000; (b) 31.6; (c) 0.01; (d) 0.1.
1.40 20 dB.
1.41 $I_{in} = 0.5$ mA; $V_{out} = 90$ V; $(V_o/V_s) = 18$, which leads to 25.1 dB; $(I_{out}/I_s) = 30$, which leads to 29.5 dB; power gain $= 540$, which leads to 27.3 dB.
1.42 Overall gain $= 63.2$ dB. Number of stages is still 2.
1.43 Stage 1: Voltage gain $= 10$ (ratio), $V_{out} = 1.5$ V; Stage 2: Voltage gain $= 31.62$ (ratio), $V_{out} = 47.4$ V; Stage 3: Voltage gain $= 0.562$ (ratio), $V_{out} = 26.6$ V.

PROBLEMS

DC Signals

1.1 Calculate the value of the current in each of the resistors in Fig. 1-P1. The proper sign ($+$ or $-$) should be included in each answer.

Fig. 1-P1.

1.2 Calculate the value of the voltage in each of the resistors in Fig. 1-P2. The proper sign (+ or −) should be included in each answer.

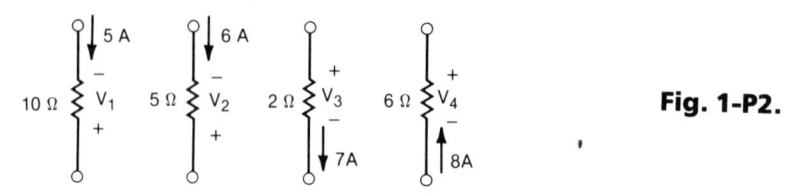

Fig. 1-P2.

1.3 Calculate the power consumed by each of the resistors in the last two problems.

1.4 Resistor R_1 has a current of 10 A and consumes a power of 200 W. What will be the power consumed if the voltage across the resistor is doubled? What should the voltage and current be for a power consumption of 150 W in the resistor?

AC Signals

1.5 Calculate the frequency of each ac voltage whose period is given: (a) 50 ms; (b) 10 ns; (c) 20 ps.

1.6 Calculate the period of each ac voltage whose frequency is given: (a) 12.5 Hz; (b) 1.25 MHz; (c) 200 kHz.

1.7 An ac voltage signal is found to take 10 ms to complete *one quarter* of a cycle. An ac signal is found to lag behind the given voltage by 2 ms, and the frequency of the current is the same as that of the voltage. Calculate the phase angle between the two signals.

1.8 Two sinusoidal voltages of frequency 500 kHz are found to have a phase angle of 48° between them. Calculate the time delay between the two signals.

1.9 The peak-to-peak value of the voltage across a device connected to an ac source is 100 V. Calculate the rms and peak values of the voltage.

1.10 Determine the rms and peak values of the current in each of the items in your home: (a) a 150-W light bulb; (b) a 1200-W toaster; (c) a 5-W night lamp. (Use 110 V.)

1.11 The diagrams of Fig. 1-P11 show several resistors (as components of circuit configurations). In each case, determine the power consumed by the resistor. If the power cannot be calculated from the data, state the reason.

Periodic Nonsinusoidal Signals

1.12 The third harmonic of a periodic signal is known to have a frequency of 60 kHz. Calculate the fundamental frequency and the frequency of the fifth harmonic. Also, calculate the period of the signal.

1.13 The amplitudes of the dc component and the first five harmonics of a

Electronics: Circuits and Systems

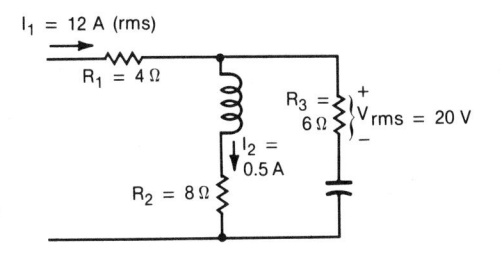

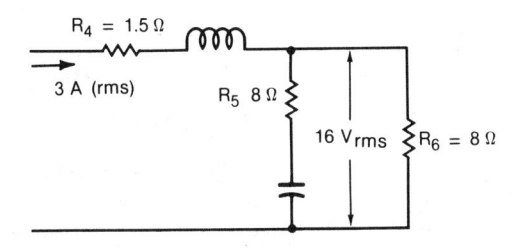

Fig. 1-P11.

periodic signal are, respectively, 3 V, 9 V, 4 V, 2.5 V, 1.2 V, and 0.5 V. Draw the amplitude spectrum of the signal when the period of the signal is (a) 0.5 ms; (b) 10 ns.

1.14 Suppose the signal in the previous problem is applied to the 8-Ω resistor. Assuming that the sixth and higher harmonics can be neglected, calculate the total average power consumed by the resistor. Repeat for a 3-Ω resistor fed by the signal.

1.15 Calculate the bandwidth needed to transmit the signal of the previous problem if (a) 70%; (b) 90%; and (c) 99% of the total power is to be transmitted. Does your answer depend upon the value of the load resistor?

1.16 The amplitude spectrum of a periodic signal has a dc component of 10 V and a fundamental frequency component of 8 V. Each successive harmonic has an amplitude that is one half of the preceding harmonic. Draw the power spectrum of the signal. Suppose the highest harmonic to be considered significant has a power content of at least 0.1 W (assuming a standard load of 1 Ω). Calculate the bandwidth needed to transmit 90% of the total average power available in the signal.

1.17 Draw the power spectrum of the signal in Problem 1-13.

Nonperiodic Signals

1.18 The energy spectrum of a certain pulse is as shown in Fig. 1-P18. Determine the energy contained in each of the following bands of frequencies: (a) dc to 1 kHz; (b) 3 kHz to 5 kHz; (c) dc to 6 kHz.

Fig. 1-P18.

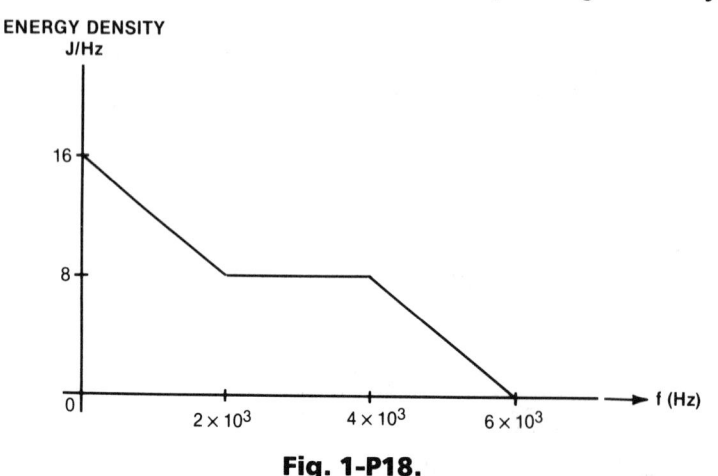

1.19 If the signal of the previous problem is passed through a filter, calculate the percentage of the total energy transmitted when the bandwidth of the filter is (a) 1 kHz; (b) 4 kHz. In each case, assume that the filter transmits all frequencies from dc to the specified bandwidth.

Two-Port Model

1.20 A two port has an input voltage of 24 V and an input power of 80 W. Its current gain is 4.8 and power gain is 84. Determine (a) the input current; (b) the output current; (c) the voltage gain; and (d) the output power.

1.21 An amplifier has a voltage gain of 0.9 and a current gain of 64. Calculate the input power needed to produce an output power of 10 W. If the load on the amplifier is an 8-Ω resistor, calculate the rms value of the voltage and current in the load.

1.22 A two port has a voltage gain of 40. Its input current is 2 mA. The output current (through a load resistor of 20 Ω) is 30 mA. Calculate the input voltage, input power, output voltage, output power, current gain, and power gain of the two-port.

1.23 A two-port system is shown in Fig. 1-P23. Calculate (a) the input voltage V_{in}; (b) the input current; (c) the output voltage; and (d) the output power.

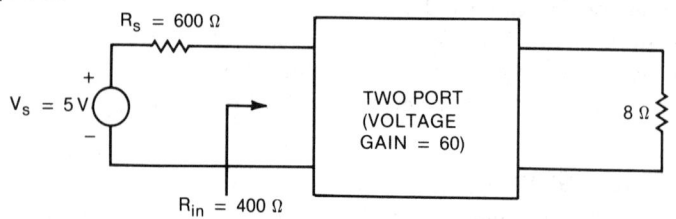

Fig. 1-P23.

Electronics: Circuits and Systems

1.24 A signal source V_s with a resistance of 1 kΩ is connected to the input of a two port with a voltage gain of 80. The load on the output side of the two port is a resistance of 16 Ω. If $V_s = 5$ V and the input resistance of the two port is 1 kΩ, calculate (a) the output voltage; (b) the output current; (c) the output power; (d) the power gain of the system; and (e) the current gain of the system.

1.25 The input resistance of a two port is 10 kΩ. A signal source $V_s = 10$ V is connected to the input. For each of the following values of the source resistance R_s, calculate the input voltage and input current, using approximations if appropriate: (a) 100 Ω; (b) 750 Ω; (c) 10 kΩ; (d) 500 kΩ; (e) 1 MΩ.

1.26 The first stage of an amplifier is a field-effect transistor (FET) and the input resistance of the amplifier is 10 MΩ. If R_s is the resistance of the signal source connected to the input of the amplifier, (a) calculate the range of values of R_s for which the current supplied by the signal source will be independent of R_{in} (that is, the source acts like a constant current source); (b) calculate the range of values of R_s for which the input voltage V_{in} of the amplifier will be independent of R_s (that is, the amplifier receives a constant voltage).

Decibels

1.27 Convert each of the following (ratio) values of power gain to decibels: (a) 13.2; (b) 1.32; (c) 0.132; (d) 1.32×10^{-5}.

1.28 Convert each of the following values of power gain to the ratio value: (a) 79.2 dB; (b) 7.92 dB; (c) −7.92 dB; (d) −79.2 dB.

1.29 An amplifier with a power gain of 36 dB produces an output power of 34 W. Calculate the input power.

1.30 Amplifier stages, each with a power gain of 28 dB, are available. Find the number of such stages needed to produce an output of 40 W when the input power to the amplifier is 2×10^{-6} W.

1.31 Convert each of the following gains to decibels: (a) power gain of 402; (b) voltage gain of 201; (c) current gain of 20.1; (d) power gain of 0.625; (e) current gain of 0.625.

1.32 Convert each of the following decibel values to ratios: (a) power gain of −4.96 dB; (b) voltage gain of −4.96 dB; (c) power gain of 4.96 dB; (d) current gain of 4.96 dB.

1.33 A three-stage amplifier has the following power gains: 60 dB (stage 1), 36 dB (stage 2), and 10 dB (stage 3). The output power is 50 W. Calculate the power input to the amplifier and each of its stages.

1.34 A three-stage amplifier has the following voltage gains: 48 dB (stage 1), 32 dB (stage 2), and −6 dB (stage 3). The input to the amplifier is 1.5 mV. Calculate the output voltage of each stage.

1.35 A two-port system is as shown in Fig. 1-P35. Calculate the decibel values of (a) the current gain; (b) the voltage gain of the two port; (c) the voltage gain of the system; (d) the power gain of the two port; and (e) the power gain of the system.

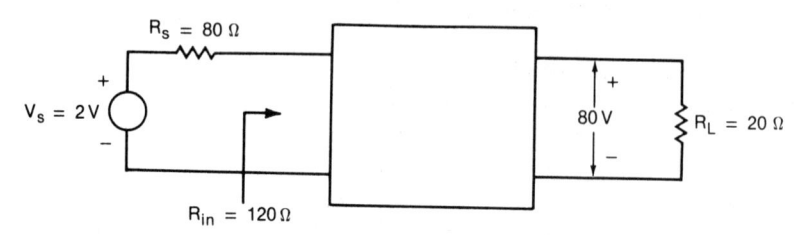

Fig. 1-P35.

1.36 The power gain of a two port system is 84 dB, and the current gain is 74 dB. The input voltage is 4 mV. The output power is measured across a load resistance of 8 Ω. Calculate (a) the input current; (b) the input power; (c) the output current; (d) the output power.

CHAPTER 2

REVIEW OF ELECTRIC CIRCUITS

The analysis of electronic circuits and communication systems requires a knowledge of basic circuit analysis. Even though you may not expect to analyze electronic circuits as thoroughly as a specialist, it is nevertheless important to have a clear understanding of the physical principles, properties, and basic relationships of electronic circuits. A review of the essential aspects of circuit analysis will be presented in this chapter, with the topics chosen on the basis of their relevance to electronic and communication circuits. The emphasis will be on the development of a "physical feel" for the response of circuits rather than on a completely rigorous mathematical development.

The three most commonly used *passive* elements in an electric circuit are the *resistor*, the *capacitor*, and the *inductor*. They are called passive since they cannot generate energy on their own even though capacitors and inductors can store energy that is available for later use. A discussion of the properties of these three elements and the behavior of circuits containing them will be discussed in this chapter.

KIRCHHOFF'S LAWS

The two fundamental laws that form the foundation of circuit analysis are *Kirchhoff's voltage law* (abbreviated *KVL*) and *Kirchhoff's current law* (*KCL*).

Kirchhoff's Current Law

KCL states that at a junction (of three or more branches) in an electric circuit, the *total current entering the junction must equal the total current leaving the junction*. For example, in the circuit of Fig. 2-1, the application of KCL leads to the equation:

$$I_1 + I_3 = I_2 + I_4 \tag{2-1}$$

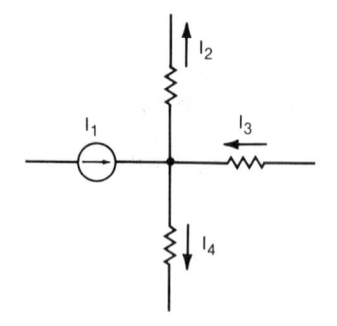

Fig. 2-1. KCL at a node.

since I_1 and I_3 are entering the junction while I_2 and I_4 are leaving it. KCL permits the calculation of the current in a branch when the currents in the other branches at a junction are known. If, in Fig. 2-1, it is given that $I_1 = 10$ A, $I_2 = 8$ A, and $I_3 = 7$ A, then $I_4 = 9$ A from Eq. (2-1).

Drill Problem 2.1: Calculate the value of the current labeled I_x in each of the circuits shown in Fig. 2-2.

(A)   (B)

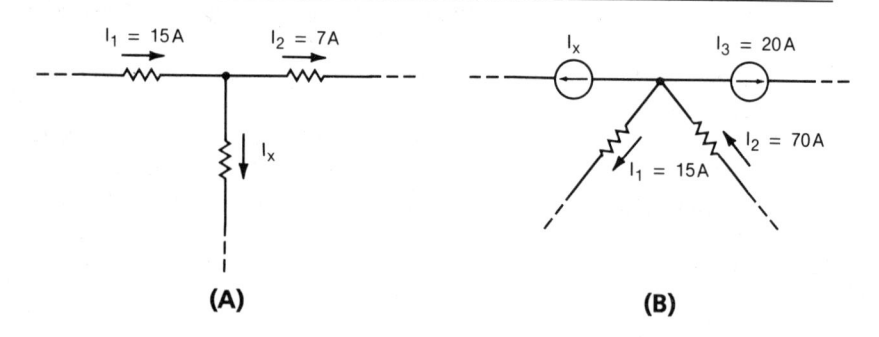

Fig. 2-2. Nodes for Drill Problem 2.1.

Kirchhoff's Voltage Law

KVL states that the *algebraic sum of the voltages in a closed path* in an electric circuit is equal to zero.

Consider the component shown as a box in Fig. 2-3. The actual nature of the component is of no interest to us at present, and we are only interested in the voltage across it and the polarity marks. The positive reference at termi-

Electronics: Circuits and Systems

nal A and the negative reference at terminal B mean that the voltage *increases* by V_1 as we go *from point B to point A*. Conversely, the voltage *drops* by V_1 if we go *from point A to point B*. That is, whenever we go through a component in a circuit *from the negative to the positive* terminal, there is an *increase* in voltage by the value of the voltage across that component, or we *add* that voltage. On the other hand, if we go through a component from the positive to the negative terminal, there is a *decrease* in voltage by the value of the voltage across the component, or we *subtract* that voltage.

Fig. 2-3. Arbitrary circuit component.

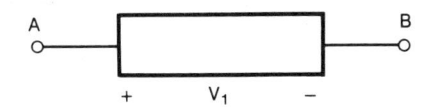

KVL states that if we go through a closed path in an electric circuit (that is, we start at some point, strike a path through the circuit, and return to the starting point) and add and subtract the component voltages as explained, the *net voltage change should be equated to zero*.

Example 2.1
In each circuit of Fig. 2-4, calculate the voltage marked V_x.

Solution

Circuit 1:
Starting from point P and going clockwise, the KVL equation becomes

$$V_1 + V_2 - V_3 + V_x = 0$$

and on substituting the given values, we get $V_x = 85$ V.

Circuit 2:
Starting from point Q and going clockwise, the KVL equation becomes

$$-V_1 - V_2 + V_x = 0$$

which gives $V_x = 36$ V.

In most of the circuits in this and the following chapters in the text, we will need to calculate the voltages at various points of a circuit with respect to some reference point (usually labeled *ground*).

Example 2.2
In each configuration shown in Fig. 2-5, calculate the voltage at point P with respect to ground.

86   Review of Electric Circuits

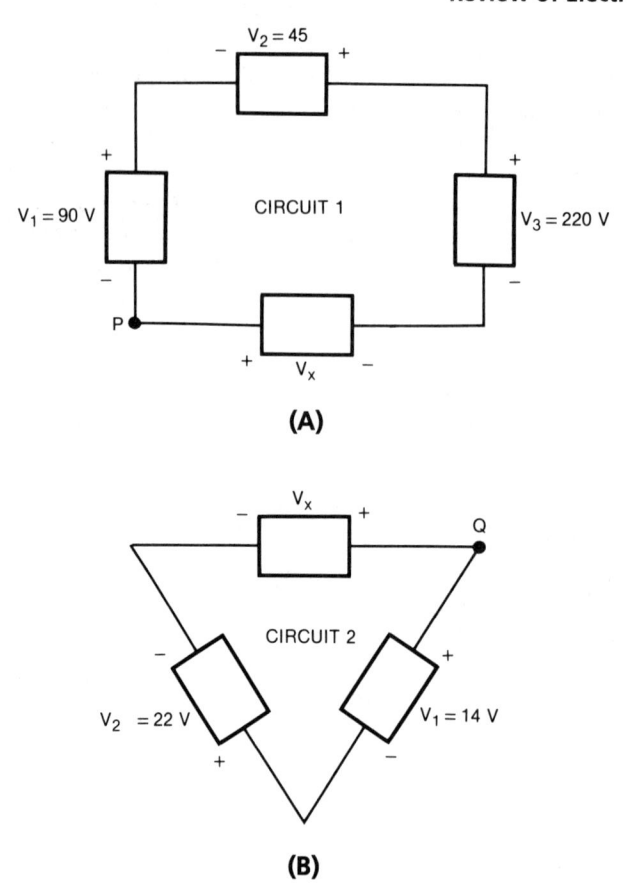

Fig. 2-4. Circuits for Example 2.1.

Solution

Circuit 1:
Starting from ground, the voltages at various points are $V_A = 17$ V; $V_B = V_A - 9 = 8$ V; $V_P = V_B + 13 = 21$ V.

Circuit 2:
Starting from ground, $V_A = -20$ V; $V_B = V_A - 15 = -35$ V; $V_P = V_B + 35 = 0$ V

Circuit 3:
It is more convenient to start from the point D in this case. (It is also possi-

Electronics: Circuits and Systems

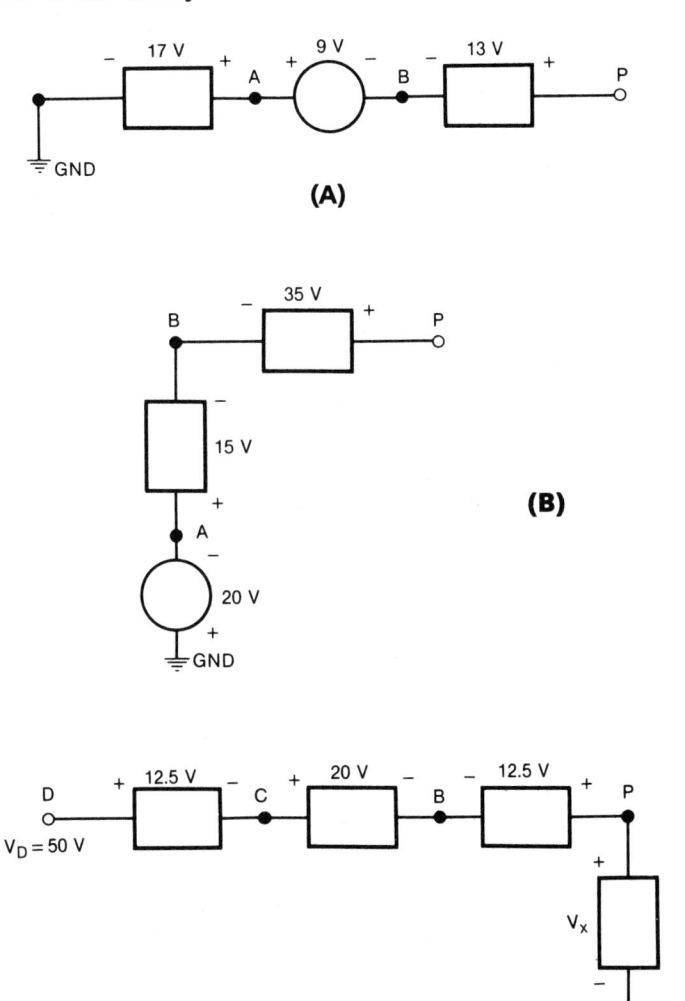

Fig. 2-5. Circuits for Example 2.2.

ble to start from ground and solve the problem.) $V_D = 50$ V (given); $V_C = V_D - 12.5 = 37.5$ V; $V_B = V_C - 20 = 17.5$ V; $V_P = V_B + 12.5 = 30$ V.

Drill Problem 2.2: Calculate the value of the voltage marked V_x in each of the circuits shown in Fig. 2-6.

---

88   **Review of Electric Circuits**

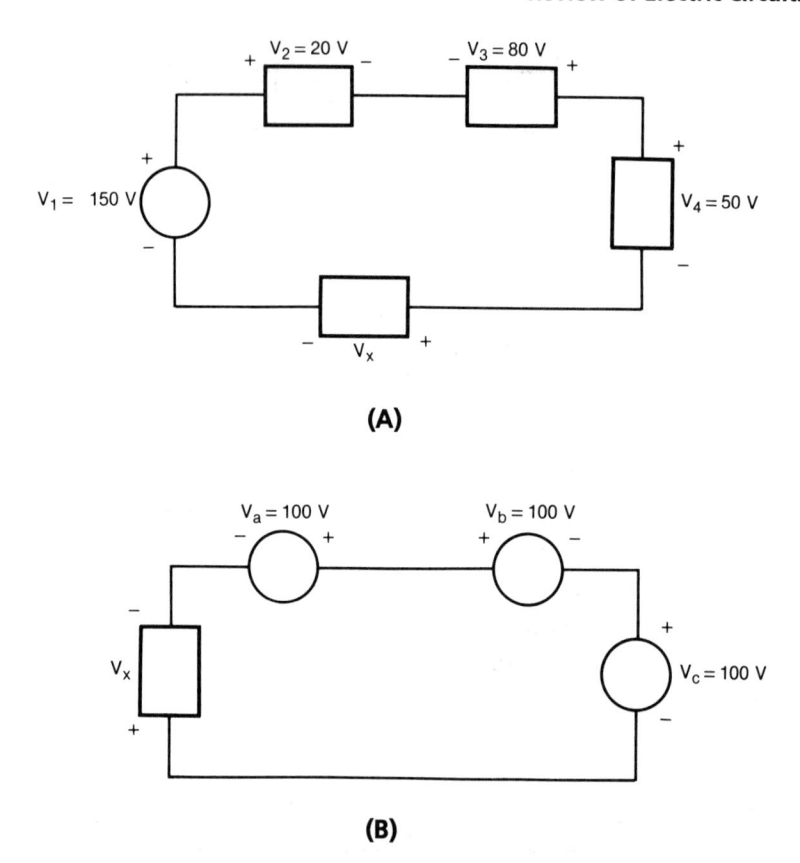

Fig. 2-6. Circuits for Drill Problem 2.2.

Drill Problem 2.3: Calculate the voltage at the point P in each of the circuits shown in Fig. 2-7.

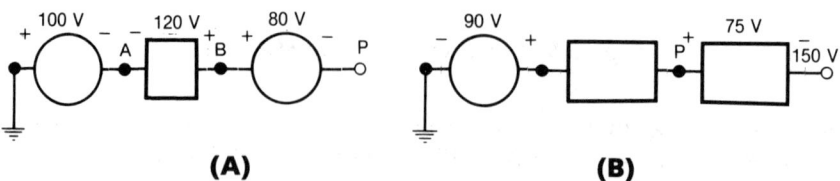

Fig. 2-7. Circuits for Drill Problem 2.3.

Electronics: Circuits and Systems

RESISTORS

Resistors in Series

Two or more resistors connected in such a way as to have the *same current* flowing through them are said to be *in series*. Such a connection of three resistors is shown in Fig. 2-8. The *total resistance* seen by the source V_s is given by

$$R_T = R_1 + R_2 + R_3 \qquad (2\text{-}2)$$

This relationship can be extended to any number of resistors. That is, the *total resistance of a number of resistors in series is equal to the sum of the individual resistances.*

Fig. 2-8. Resistors in series.

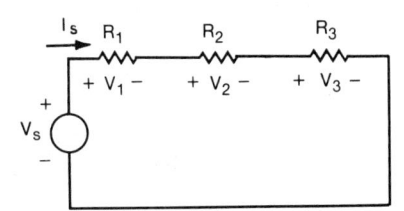

The current I_s supplied by the source is

$$I_s = (V_s/R_T) \qquad (2\text{-}3)$$

and the applied voltage is split among the resistors in proportion to the individual resistances. In Fig. 2-8, we have

$$V_1 = (R_1/R_T)V_s$$
$$V_2 = (R_2/R_T)V_s$$
$$V_3 = (R_3/R_T)V_s \qquad (2\text{-}4)$$

Example 2.3

Calculate the total resistance, the current supplied by the source, the voltage across each resistor, and the power consumed by each resistor in the circuit of Fig. 2-9.

Solution

Total resistance $R_T = (10 + 25 + 15) = 50\ \Omega$

Current supplied

$I_s = (V_s/R_T) = 3\ \text{A}$

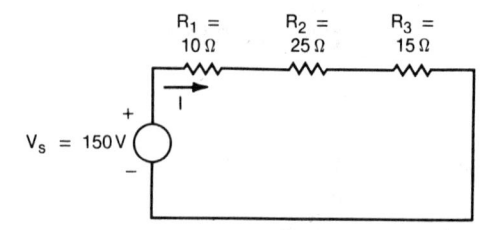

Fig. 2-9. Circuit for Example 2.3.

The voltages are

$V_1 = (R_1/R_T)V_s = 30$ V; $V_2 = 75$ V; $V_3 = 45$ V

The power consumed by the resistors is 90 W in R_1, 225 W in R_2, and 135 W in R_3.

Example 2.4

Suppose that the voltage across R_2 in the last example is to be 80 V. Calculate the new value of the source voltage V_s needed.

Solution

$$\text{Current in the circuit } I_s = (\text{Voltage across } R_2)/R_2$$
$$= (80/25)$$
$$= 3.2 \text{ A}$$
$$\text{Total resistance } R_T = 50 \ \Omega$$
$$\text{Source voltage } V_s = R_T I_s$$
$$= 160 \text{ V}$$

Drill Problem 2.4: Calculate the total resistance, the current supplied by the source, the voltage across each resistor, and the power consumed by each resistor in the circuit of Fig. 2-10.

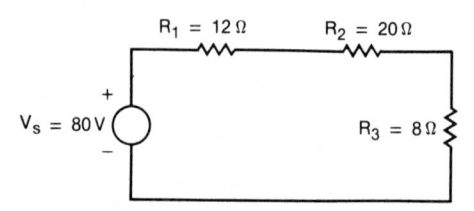

Fig. 2-10. Circuit for Drill Problem 2.4.

Electronics: Circuits and Systems 91

Drill Problem 2.5: Three resistors are in series: $R_1 = 10 \, \Omega$, $R_2 = 40 \, \Omega$, and $R_3 = 50 \, \Omega$. If the voltage across the 40-Ω resistor is 60 V, calculate the total voltage and the voltage across each of the other resistors. Suppose the power consumed by the 50-Ω resistor is to be changed to 200 W. Recalculate the previous values.

Voltage Dividers

The case of *two resistors in series* is commonly used for dividing an available voltage into two specified parts; it is, therefore, called a *voltage divider*. The underlying principle of a voltage divider is that the voltage across each resistor is proportional to its resistance. It is useful to remember that the *larger resistance* has the *higher voltage* across it.

Fig. 2-11. Voltage divider.

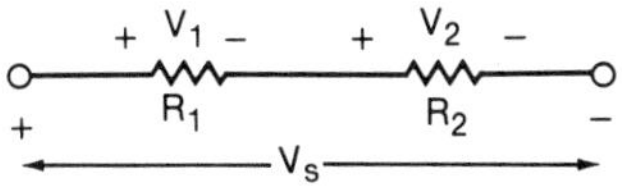

In the voltage divider of Fig. 2-11, the total resistance is $R_T = (R_1 + R_2)$ and

$$V_1 = \frac{R_1}{R_T} V_s$$

$$= \frac{R_1}{R_1 + R_2} V_s \quad (2\text{-}5)$$

$$V_2 = \frac{R_2}{R_T} V_s$$

$$= \frac{R_2}{R_1 + R_2} V_s \quad (2\text{-}6)$$

The last two equations (2-5 and 2-6) are known as the *voltage-divider formulas*.

Example 2.5

Two resistors are in series: $R_1 = 18 \, \Omega$ and $R_2 = 72 \, \Omega$. The voltage across the series combination is 100 V. Calculate the voltage across each resistor.

Solution

Using the voltage-divider formulas,

$$V_1 = \frac{18}{18 + 72} \times 100 = 20 \text{ V}$$

$$V_2 = \frac{72}{18 + 72} \times 100$$

$$= 80 \text{ V}$$

Example 2.6

In a voltage divider made up of two resistors R_1 and R_2, the total voltage is $V_s = 100$ V, and it is desired to make the voltage across R_2 equal 75 V. Determine the ratio (R_2/R_1).

Solution

Using Eq. (2-6),

$$V_2 = 75$$

$$= \frac{R_2}{R_1 + R_2} \times 100$$

Therefore,

$$\frac{R_2}{R_1 + R_2} = \frac{75}{100}$$

which gives,

$$R_2 = 3R_1$$

or,

$$(R_2/R_1) = 3$$

Note that the given information is not sufficient for the calculation of the *individual* resistances. Some additional information, such as the rated power dissipation of the available resistors, is needed for finding the values of the individual resistances.

Drill Problem 2.6: The total voltage across two resistors R_1 and R_2 in series is 28 V. If $R_1 = 10 \ \Omega$ and $R_2 = 4 \ \Omega$, find the voltage across each resistor.

Drill Problem 2.7: In a voltage divider made up of two resistors R_1 and R_2, the total voltage is 100 V. If $R_1 = 15 \ \Omega$, and the voltage across it is 35 V, find the value of R_2.

Drill Problem 2.8: The total resistance of a voltage divider is 90 Ω, and the total voltage across it is 360 V. If the voltage across one of the resistors is 78 V, calculate the values of the two resistances.

Potentiometers

A *potentiometer* (or "pot" as it is popularly called) is a voltage divider in which the two resistances can be *varied continuously* over a particular range of values, while the *total resistance remains fixed* at a constant value. Fig. 2-12 shows a potentiometer in which the contact C can be moved to any point between the fixed terminals A and B. The ratio of the two voltages V_{CB} to the total voltage V_{AB} is given by the voltage-divider relationship

Fig. 2-12. Potentiometer.

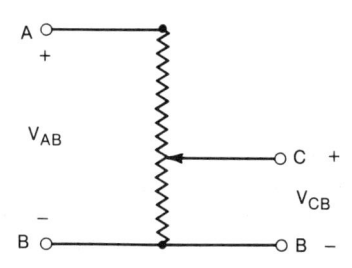

$$\frac{V_{CB}}{V_{AB}} = \frac{\text{resistance between contacts C and B}}{\text{total resistance between A and B}}$$

The value of V_{CB} can be varied from 0 to a maximum of V_{AB} continuously by moving the contact from the point B to point A. Potentiometers are available in a wide variety of values of the total resistance R_{AB}.

Resistors in Parallel

When two or more resistors are connected so that they have the *same voltage* across them, they are said to be in parallel. The special case of two resistors in parallel occurs quite commonly, and our discussion will be focused upon it. The parallel connection of three or more resistors can be dealt with by taking them two at a time.

The total resistance seen by a source in the parallel connection of two resistors R_1 and R_2, as shown in Fig. 2-13 is given by

Fig. 2-13. Resistors in parallel.

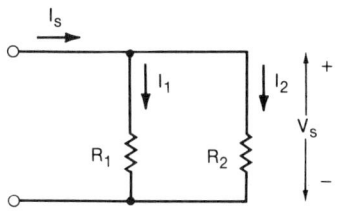

$$R_T = \frac{R_1 R_2}{R_1 + R_2} \tag{2-7}$$

It is important to remember that the *total resistance* of two parallel resistors is always *less than the smaller of the* two resistances.

Current Dividers

A parallel connection of two resistors divides the total current I_s into two parts I_1 and I_2, which are given by

$$I_1 = \frac{R_2}{R_1 + R_2} I_s$$

$$I_2 = \frac{R_2}{R_1 + R_2} I_s \qquad (2\text{-}8)$$

These two equations are called the *current-divider formulas*. It should be noted that the current in each resistor is proportional to the *other* resistance, and that the *larger current* flows through the *smaller resistance*.

The current-divider formulas Eq. (2-7) can be derived from the following two equations of the circuit of Fig. 2-13.

KCL equation:

$$I_s = I_1 + I_2 \qquad (2\text{-}9)$$

Ohm's law equations:

$$V_s = R_1 I_1$$
$$= R_2 I_2 \qquad (2\text{-}10)$$

Example 2.7

A current $I_s = 14$ A is fed to a parallel combination of two resistors $R_1 = 15\,\Omega$ and $R_2 = 75\,\Omega$. Find the current in each resistor, the voltage across each resistor, and the total resistance.

Solution

From Eq. (2-8),

$$I_1 = \frac{R_2}{R_1 + R_2} I_s$$

$$= 11.7 \text{ A}$$

and

$$I_2 = \frac{R_1}{R_1 + R_2} I_s$$

$$= 2.3 \text{ A}$$

I_2 could also have been calculated using Eq. (2-9).
The voltage across each resistor is obtained from Eq. (2-10).

$$V_s = R_1 I_1$$
$$= R_2 I_2$$
$$= 175 \text{ V}$$

The total resistance is, from Eq. (2-6):

$$R_T = \frac{R_1 R_2}{R_1 + R_2}$$
$$= 12.5 \text{ }\Omega$$

Example 2.8

If the total current in a current divider is $I_s = 20$ A, and the current through R_1 is to be 3 A, find the ratio (R_1/R_2).

Solution

Using the current divider formula,

$$I_1 = \frac{R_2}{R_1 + R_2} I_s$$

we get

$$\frac{R_2}{R_1 + R_2} = \frac{I_1}{I_s}$$

$$= 0.15$$

Therefore,

$$R_2 = 0.15 (R_1 + R_2)$$

or,

$$(R_1/R_2) = 5.67$$

Drill Problem 2.9: Two resistors, 12 Ω and 18 Ω, are parallel with a total current of 50 A entering the connection. Calculate the total resistance and the current in each resistor.

Drill Problem 2.10: A resistor of 24 Ω is in parallel with another resistor

R_2. If the total current is 18 A, and the current through R_2 is 12 A, find the value of R_2.

Drill Problem 2.11: Design a current divider to split a current of 32 A into 9 A and 23 A if the voltage across it is specified as 150 V.

Useful Approximations

In the case of two resistors *in series*, if one resistance is *much greater than* the other resistance, the total resistance can be taken as (approximately) equal to the *larger resistance*. The applied voltage can also be assumed to be fully across the larger resistance. The voltage across the smaller resistance is not zero but is small compared with the voltage across the larger resistance. The term "much greater than" is usually taken to mean "at least ten times as great as." The error introduced by such approximations is then less than 10%.

For example, if two resistors $R_1 = 10 \, \Omega$ and $R_2 = 200 \, \Omega$ are in series, we take the total resistance $R_T \approx 200 \, \Omega$, instead of the actual 210 Ω. If a voltage is applied to the series combination, then we take the voltage across R_2 as approximately equal to the applied voltage.

In the case of two resistors connected *in parallel*, if one resistance is much greater than the other, then the total resistance can be taken as approximately equal to the *smaller resistance*. If a current is fed to the parallel combination in such a case, then we can assume that *all the applied current goes through the smaller resistance*. The current in the larger resistance is not zero but is much smaller than the current in the other resistor.

The validity of these approximations for a parallel connection follows from the current-divider formulas, Eq. (2-8). Suppose R_1 is much greater than R_2. Then we can see, from Eq. (2-8), that I_1 is only a small fraction of the total current I_s, while I_2 is very nearly equal to I_s (within a 10% error). Thus, almost all the applied current flows through the smaller resistance, and it is as if the source does not even notice the presence of the larger resistance.

Approximations of this type are very convenient in making quick calculations in electronic circuits.

Drill Problem 2.12: Make suitable approximations when appropriate and find the total resistance and the current in each of the series connections shown in Fig. 2-14. Verify that the error introduced is less than 10% in the cases where an approximation was used.

Drill Problem 2.13: Make suitable approximations when appropriate and find the total resistance and the voltage in each of the parallel connections shown in Fig. 2-15. Verify that the error introduced is less than 10% in the cases where an approximation was used.

Electronics: Circuits and Systems

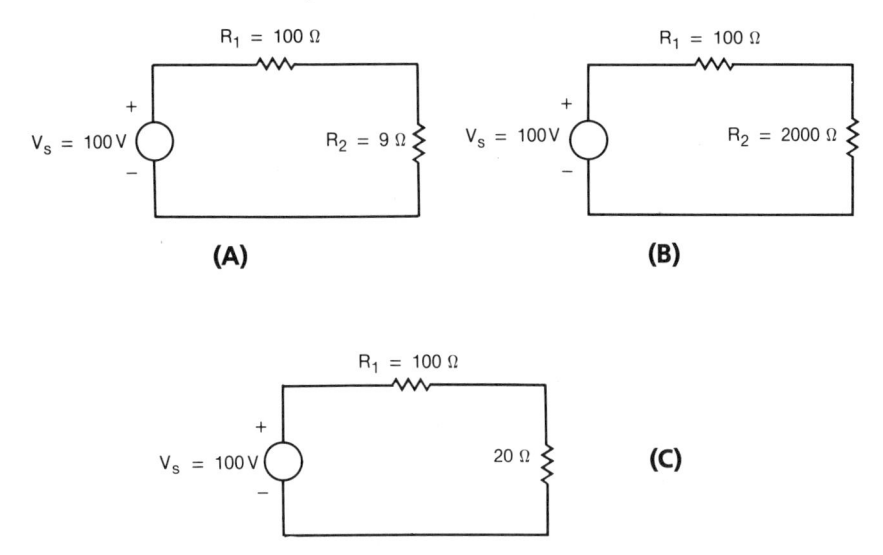

Fig. 2-14. Circuits for Drill Problem 2.12.

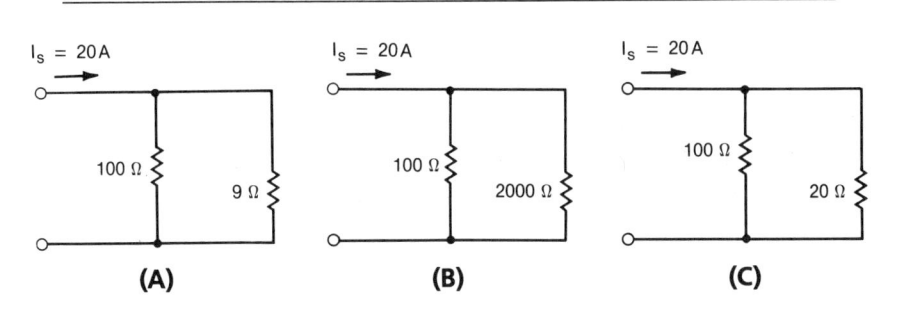

Fig. 2-15. Circuits for Drill Problem 2.13.

WHEATSTONE BRIDGE

An arrangement of four resistors, called a *Wheatstone bridge* (Fig. 2-16), is useful in the measurement of unknown resistances. The value of the resistance R_x is to be determined. The other three resistors are variable resistances whose values are known. In the measurement procedure, the three variable resistors are adjusted until there is a null deflection in the galvanometer G. The null deflection of the meter means that there is no current through that branch, and since the meter has a finite resistance, it also means that

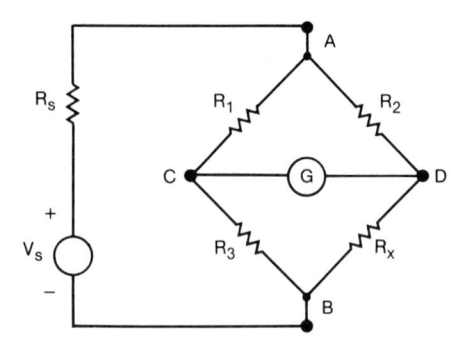

Fig. 2-16. Wheatstone bridge.

$$V_C = V_D \qquad (2\text{-}11)$$

where,

V_C and V_D are, respectively, the voltages at points C and D (with B as reference).

The resulting situation is then as shown in Fig. 2-17.

Fig. 2-17. Portion of a Wheatstone bridge used in analysis.

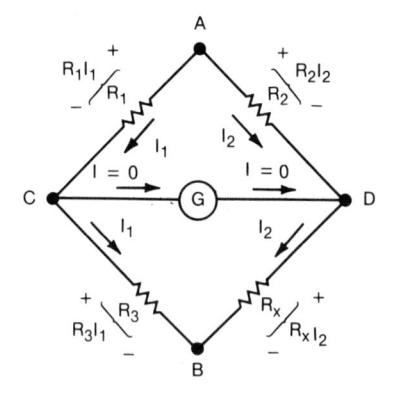

Starting from point B, and using KVL, V_C and V_D are given by

$$V_C = V_B + R_3 I_1 \qquad (2\text{-}12)$$

$$V_D = V_B + R_x I_2 \qquad (2\text{-}13)$$

Equating (2-12) and (2-13), as required by Eq. (2-11), we get

$$R_3 I_1 = R_x I_2 \qquad (2\text{-}14)$$

We can also start from point A and write the expressions for the voltages at points C and D:

Electronics: Circuits and Systems

$$V_C = V_A - R_1 I_1 \qquad (2\text{-}15)$$

$$V_D = V_A - R_2 I_2 \qquad (2\text{-}16)$$

Equating (2-15) and (2-16), as required by Eq. (2-11), we get

$$R_1 I_1 = R_2 I_2 \qquad (2\text{-}17)$$

Finally, dividing Eq. (2-14) by Eq. (2-17), we get the *condition for null deflection* (or "balancing the bridge"):

$$\frac{R_3}{R_1} = \frac{R_x}{R_2} \qquad (2\text{-}18)$$

Since all resistances except R_x are known, R_x can be determined. The degree of accuracy with which R_x can be measured can be improved by increasing the accuracy of the values of the other resistances. A Wheatstone bridge can also be used in the measurement of such quantities as temperature since the resistance of a wire depends upon its temperature.

CAPACITORS

A capacitor consists of two electrodes (metallic plates) separated by air or some insulator (called the *dielectric*). It stores electric charge on the plates, and the voltage, V, across the capacitor is proportional to the electric charge Q on the plates. The relationship between the charge and the voltage is

$$Q = CV \qquad (2\text{-}19)$$

where,

C denotes the capacitance of the capacitor.

Capacitance is measured in *farads*, but practical capacitors have capacitances in the range of *picofarads* (abbreviated pF and equal to 10^{-12} farad) or *microfarads* (abbreviated μF and equal to 10^{-6} farad).

The current through a capacitor is proportional to the *rate at which the voltage is varying with time*. That is, the current i through a capacitor C is given by

$$i = C \times \text{(rate of change of the voltage V)} \qquad (2\text{-}20)$$

or,

$$i = C \frac{dv}{dt}$$

It is important to note that *when the voltage across a capacitor is a constant, then the current through it is zero.*

Fig. 2-18 shows a capacitor with the voltage polarities and the current direction. When the current is flowing in the direction shown (+ to −), the capacitor is being *charged*. The current will flow in the opposite direction (− to +) when the capacitor is discharging.

Fig. 2-18. Capacitor.

Energy Stored in a Capacitor

The charging of a capacitor leads to a storage of energy in its electric field, and the energy is given by

$$w_C = \frac{1}{2} CV^2 \text{ joules} \tag{2-21}$$

The stored energy depends directly upon the voltage, V, on the capacitor. The stored energy is available for activating other components. For example, the energy can be discharged through a resistor. One familiar application of the charging of a capacitor and then discharging it is the electronic flash, where the discharge takes place through a gap between two electrical contacts.

Voltage Continuity Condition

In any physical system, there is some inertia that opposes change. The storing of energy or the removal of it in a physical system is affected by the inertia. Therefore, the variation of the energy of any physical system can only be gradual. That is, the energy in a physical system cannot change suddenly, and energy is a *continuous function of time*.

In a capacitor, the energy depends upon the voltage. Since the energy cannot change suddenly, it follows that the *voltage on a capacitor cannot change suddenly* either. *The voltage on a capacitor is a continuous function of time.* This property is known as the *voltage continuity condition* and plays an important part in the behavior of circuits containing capacitors.

Charging of a Capacitor (RC Circuit)

Fig. 2-19 shows an arrangement for charging a capacitor: a capacitor C is in series with a resistor, R, and a battery of voltage V_0. The switch S is open

Electronics: Circuits and Systems

before t = 0. Before t = 0, assume that there is no voltage across the capacitor. At t = 0, the switch is suddenly closed, and the capacitor starts charging. Since the capacitor voltage must be continuous, the voltage increases gradually from its initial value of zero to a final value of V_0 in the manner indicated in Fig. 2-20.

Fig. 2-19. Charging of a capacitor.

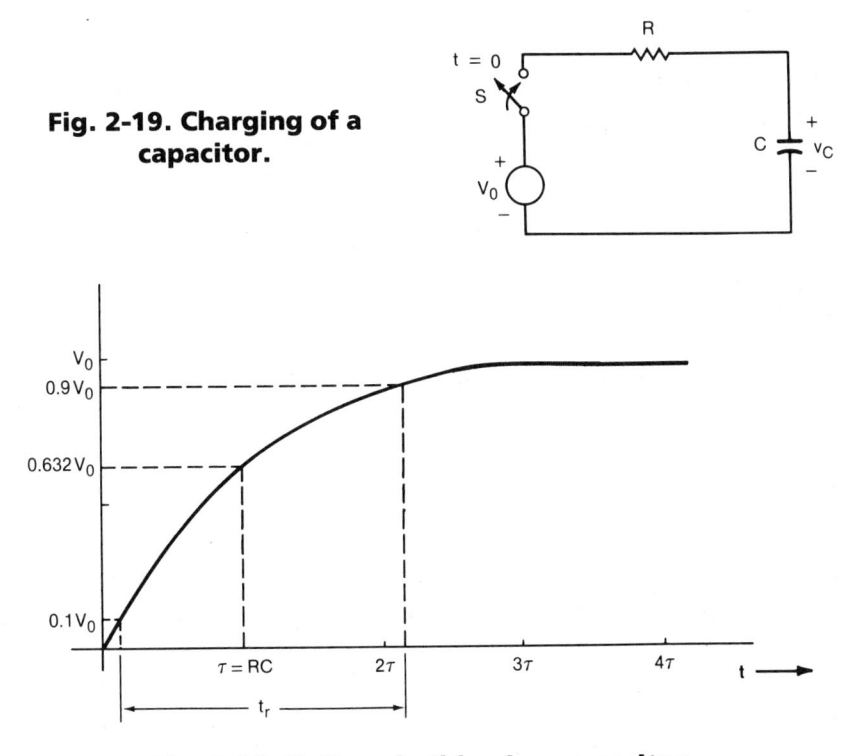

Fig. 2-20. Voltage buildup in a capacitor.

The shape of the charging curve in Fig. 2-20 is related to the exponential function. The voltage across the capacitor at any time is given by

$$v_C = V_0 - V_0 e^{-(t/RC)} \qquad (2\text{-}22)$$

When the capacitor voltage has reached a value of V_0, the capacitor voltage balances the voltage of the battery, and the voltage across the resistor becomes zero. No more current flows in the circuit, and the capacitor voltage is now a constant.

A measure of the speed with which the capacitor builds up in voltage is a parameter of the circuit called the *time constant*. The time constant, denoted by τ (the Greek letter *tau*) is given by

$$\tau = RC \text{ seconds} \tag{2-23}$$

and represents the time taken by an initially uncharged capacitor to reach 63.2% of the final value of V_0 as shown in Fig. 2-20.

An alternative measure of the speed of charging, which is more useful in laboratory measurements, is the *rise time* (also called the *10% to 90%* rise time). The rise time, t_r, is the time taken by the capacitor to go *from 10% of the final voltage to 90% of the final voltage*. It can be shown through Eq. (2-22) that the rise time and the time constant are related by the equation

$$\text{rise time} = 2.2 \times \text{(time constant)} \tag{2-24}$$

or,

$$t_r = 2.2RC \text{ seconds}$$

To show the effect of the time constant on the speed of charging of a capacitor, two curves are shown in Fig. 2-21 for two different time constants but the same final voltage V_0. It is seen that the *larger* the *time constant*, the *longer* it takes for the capacitor to reach its final voltage.

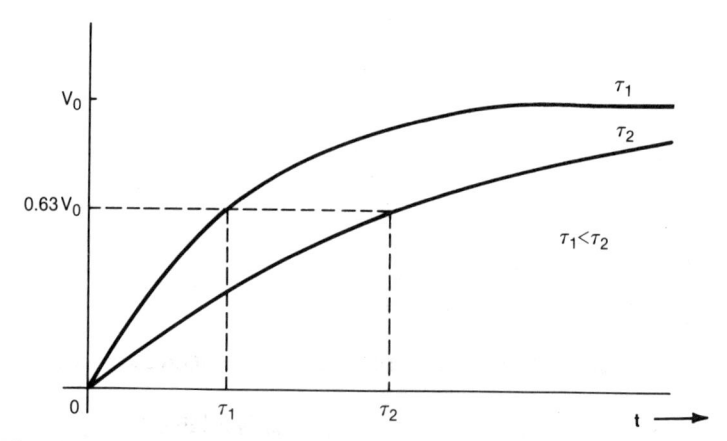

Fig. 2-21. Comparison of RC circuits with different time constants.

Theory demands that an infinite amount of time is needed for the capacitor to charge up to its final voltage, V_0, as can be seen from Eq. (2-22). Fortunately, however, from a practical standpoint, an interval of *five time constants* (5RC seconds) is sufficient to make the voltage reach about 99% of the final value. Therefore, the time for fully charging a capacitor is usually taken as five time constants. In some cases, an interval of three time constants is considered sufficient.

Electronics: Circuits and Systems

Example 2.9
A 50-pF capacitor is in series with a 6.5-kΩ resistor and a battery of voltage 150 V. Calculate the time constant, rise time, the final voltage on the capacitor, the approximate time taken to reach the final voltage, and the energy stored in the capacitor when it is fully charged.

Solution
Time constant $\tau = RC = 3.25 \times 10^{-7}$ s. Rise time $t_r = 2.2RC = 7.15 \times 10^{-7}$ s. Final voltage = 150 V. Approximate time to reach the final voltage = 16.25×10^{-7} s. Energy stored at full charge = $(1/2) CV_0^2 = 5.63 \times 10^{-7}$ J.

Example 2.10
A capacitor C is connected in series with a 22-MΩ resistor and a 25-V battery. The energy stored in it when fully charged is 1 mJ. Calculate the rise time.

Solution
Energy stored at full charge = $(1/2) CV_0^2 = 10^{-3}$ J and $V_0 = 25$ V. Therefore,

$$C = \frac{10^{-3}}{0.5 \times 25^2}$$

$$= 3.2 \times 10^{-6} \text{ F}$$

and R = 22 MΩ. Therefore, the rise time is

$$t_r = 2.2RC$$

$$= 154.9 \text{ s}$$

Drill Problem 2.14: A 250-μF capacitor is in series with a 150-kΩ resistor and a 500-V battery. Calculate the time constant, rise time, the final voltage on the capacitor, the time needed for full charge, and the energy stored at full charge.

Drill Problem 2.15: The rise time of a series RC combination is 10 ms. When connected to a 40-V battery, the energy at full charge is found to be 0.05 mJ. Calculate the values of R and C.

Drill Problem 2.16: If it is desired to fully charge a 25-pF capacitor in 20 ns, what series resistance should be used?

Universal Charging Curve

The charging curve of a capacitor of Fig. 2-20 can be redrawn by labeling the vertical axis in terms of V_0 (rather than absolute volts) and the horizontal axis in terms of the time constant τ (rather than absolute seconds); then it becomes a universal charging curve that can be used for any series RC circuit. The universal charging curve, Fig 2-22, is useful in making calculations about the charging of a capacitor. Eq. (2-22) can directly be used for such calculations, of course, but the universal curve is easier and quicker to use. In using the universal curve, it should be remembered that the capacitor must start from *zero initial voltage*.

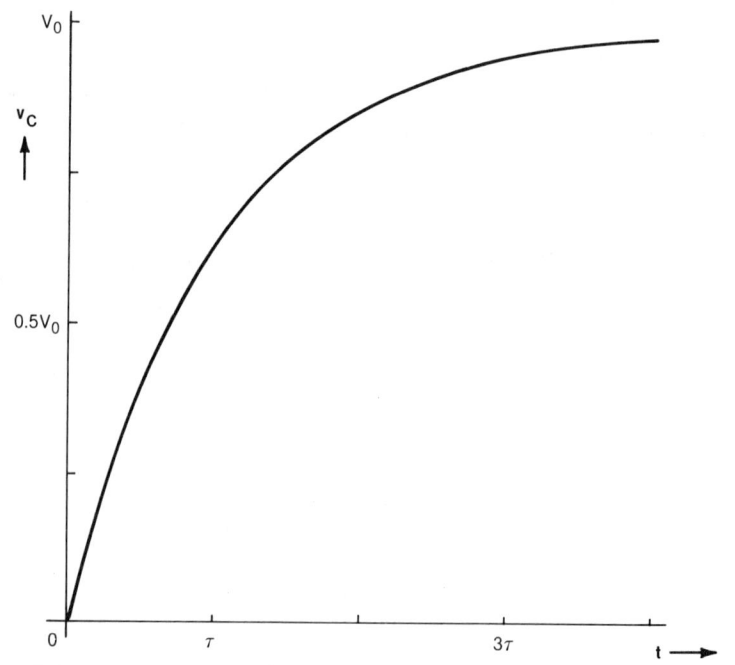

Fig. 2-22. Universal charging curve for a capacitor.

Example 2.11

A 10-pF capacitor is connected in series with a 150-MΩ resistor and a 250-V battery. The capacitor is initially uncharged. Determine (a) the voltage across the capacitor at the end of 3 ms; (b) the time taken for the capacitor voltage to reach 125 V.

Solution

From the data, $V_0 = 250$ V, and $\tau = 1.5 \times 10^{-3}$ s or 1.5 ms. (a) The given time value of 3 ms corresponds to two time constants. From the universal

Electronics: Circuits and Systems

charging curve, at two time constants, the capacitor voltage is $0.86V_0$, which is equal to 215 V. (b) A voltage of 125 V is equal to $0.5V_0$. From the universal charging curve, $0.5V_0$ occurs at $t = 0.7\tau$, or 1.05 ms.

Drill Problem 2.17: Find the time taken to charge a 350-pF capacitor connected in series with a 400-kΩ resistor to a 550-V battery. Find the time at which the capacitor voltage is 300 V. Also find the voltage at $t = 0.25$ ms.

Discharging of a Capacitor

A capacitor with an initial voltage of V_0 at $t = 0$ is connected to a resistor R and a switch as in Fig. 2-23. Until $t = 0$, the switch is open, and there is no current flow in the circuit. At $t = 0$, the switch is closed. This causes a flow of current that allows the dissipation of the energy initially stored in the capacitor by converting it to heat in the resistor. The stored energy and the voltage across the capacitor decrease as a function of time. Eventually, the energy in the capacitor is depleted and all action stops: the voltage across the capacitor, the voltage across the resistor, and the current all are reduced to zero.

Fig. 2-23. Discharging of a capacitor.

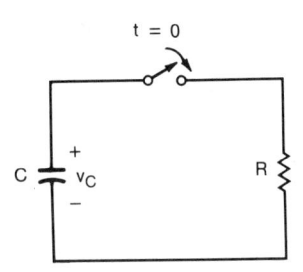

The discharging of the capacitor is controlled by the same time constant τ = RC as the charging of a capacitor. Fig. 2-24 shows a *universal discharging curve* where the vertical axis is labeled in terms of the *initial* voltage V_0 and the horizontal axis is labeled in terms of the time constant τ. As in the case of the charging of a capacitor, an interval of five time constants is considered sufficient to fully discharge the capacitor.

Example 2.12

A 10-pF capacitor with an initial voltage of 300 V is discharged through a 150-kΩ resistor. Calculate (a) the time taken for the voltage to reach 180 V; (b) the voltage at 4 μs.

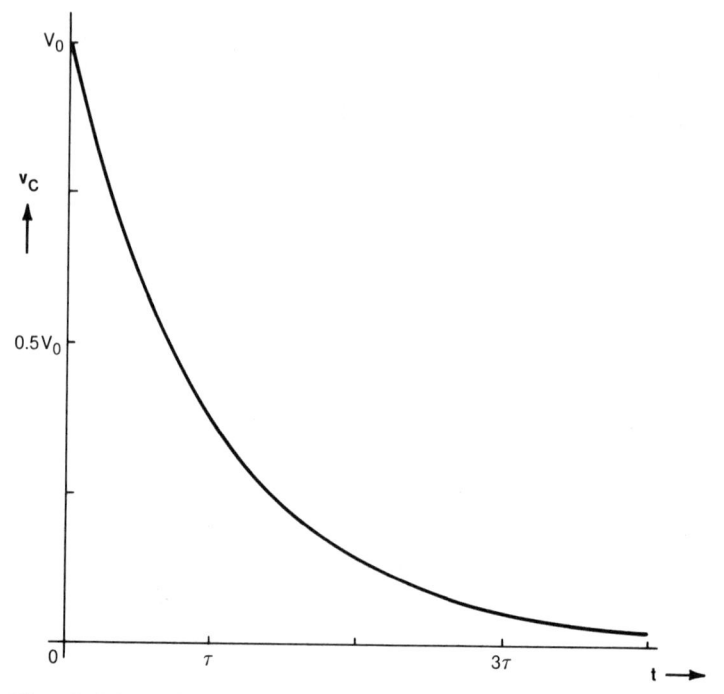

Fig. 2-24. Universal discharging curve for a capacitor.

Solution
From the data, $V_0 = 300$ V, and $\tau = (10 \times 10^{-12})(150 \times 10^{-3}) = 1.5$ μs.
(a) The given voltage of 180 V corresponds to $0.6V_0$, and this occurs at 0.5τ in the universal discharge curve. Therefore, the voltage is down to 180 V at 0.75 μs. (b) The given time t = 4 μs corresponds to $[4 \times (\tau/1.5)] = 2.67\tau$. From the universal discharge curve the voltage at t = 2.67τ is $0.07V_0 = 21$ V.

Drill Problem 2.18: A 16-μF capacitor with an initial voltage of 380 V is discharged through a 125-kΩ resistor. Calculate (a) the time taken for full discharge; (b) the time at which the capacitor voltage is down to 280 V; (c) the capacitor voltage at 1.6 s.

An Application of the RC Circuit
One of the many applications utilizing the charging and discharging in an RC circuit is the *automatic exposure control* mechanism of a camera. Even though the actual circuitry of such a mechanism is usually quite sophisticated

Electronics: Circuits and Systems

and complex, the principles of operation are relatively simple. The basic circuit consists of a capacitor; a photoconductor R_p, and a switch as shown in Fig. 2-25.

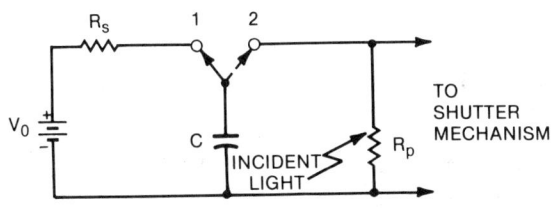

Fig. 2-25. Basic automatic exposure control circuit.

The photoconductor R_p has the property that its resistance is *inversely proportional to the intensity of light* incident on it. Thus, the time constant R_pC for the discharge of the capacitor when the switch is in position 2 is inversely proportional to the intensity of light incident on the camera. At low light levels, the time constant is *large*, and the capacitor discharges slowly; while at high light levels, the time constant is small, and the capacitor discharges rapidly.

The shutter mechanism is such that it will be open when the voltage V_2 across R_p is above a *threshold level* V_T and closed when V_2 is less than V_T.

Suppose the capacitor is fully charged up to V_0 (assumed to be larger than V_T) when the switch is in position 1, and then the switch is moved to position 2 by pressing a button on the camera. The voltage V_2 is now equal to the capacitor voltage. Since the capacitor voltage is initially equal to V_0, which is greater than the threshold level, the shutter opens. The capacitor starts discharging through R_p and when its voltage has decreased to V_T, the shutter will close. The time for which the shutter stays open equals the time taken for the capacitor to discharge from V_0 to V_T. At low light levels, the capacitor discharges slowly (since R_pC is large) and the shutter stays open a relatively long time. The opposite is true at bright light levels.

Example 2.13

In the circuit of Fig. 2-25, let $C = 5 \,\mu\text{F}$, $V_0 = 10$ V, and the threshold level $V_T = 2$ V. Assume that the value of R_p at a low light level is 10 kΩ, and at a bright light level is 1 kΩ.

Calculate the exposure times for the two cases.

Solution

The threshold level is 2 V, and $V_0 = 10$ V. Therefore, the shutter stays open as the capacitor discharges from V_0 to $V_T = 0.2V_0$. From the universal discharge curve, Fig. 2-24, it takes 1.6τ to effect this discharge.

Low light level:
$R_p = 10$ kΩ and $\tau = R_p C = 50$ ms. The shutter stays open for 1.6τ or 80 ms.

High light level:
$R_p = 1$ kΩ and $\tau = 5$ ms. The shutter stays open for 1.6τ or 8 ms.

For any light level between the two given in the previous example, the time constant may be assumed to vary linearly from 1 kΩ to 10 kΩ, with a corresponding variation of the exposure time continuously over a range of 8 ms to 80 ms.

Drill Problem 2.19: In Example 2.13 assume that the battery voltage has deteriorated down to 7.5 V with all the other data remaining the same. Recalculate the exposure times.

Response of an RC Circuit to a Pulse

Consider a rectangular pulse applied to an RC circuit as indicated in Fig. 2-26. Consider the following two cases. *Case 1:* The pulse duration is at least 5τ. *Case 2:* The pulse duration is less than 5τ. τ is the time constant RC of the circuit.

Case 1: ($t_p > 5RC$)

The capacitor voltage starts from zero at $t = 0$ and builds up toward V_0 (the voltage of the pulse). Since the capacitor can attain full charge in five time constants, the voltage reaches V_0 well before the input pulse is turned off. At $t = t_p$, the input pulse goes down to zero, and the capacitor starts discharging toward zero. The capacitor voltage will reach zero at five time constants after $t = t_p$. The resulting output pulse is shown in Fig. 2-26B. It is seen to be a rounded off version of the input. As the pulse duration is made very long compared with 5RC, the distortion in the output becomes less troublesome.

Case 2: ($t_p < 5RC$)

When the pulse duration is less than 5RC, the capacitor does not have sufficient time to reach the full voltage, V_0, since the pulse is turned off before five time constants have elapsed. At $t = t_p$, when the pulse is turned off, the capacitor discharges toward zero from whatever voltage it had reached at t_p. The output will be of the form shown in Fig. 2-26C. The output is significantly different in shape from the input.

This previous discussion shows that if the output pulse is to have a reasonably close similarity to the input pulse, the time constant of the RC circuit

Electronics: Circuits and Systems

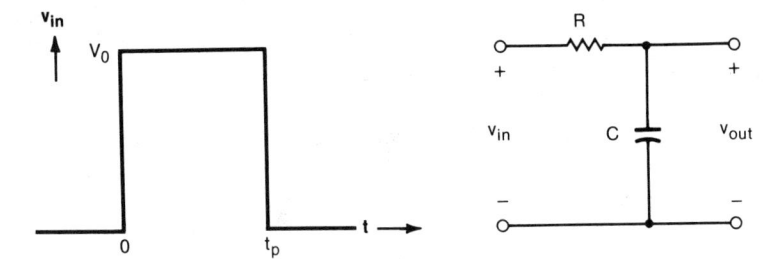

(A) Rectangular pulse applied to the RC circuit.

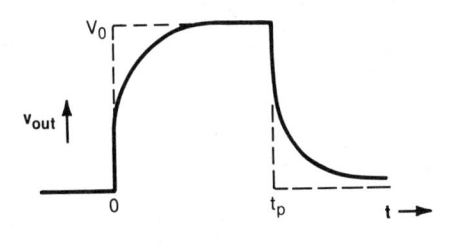

(B) Case 1, $t_p > 5\tau$.

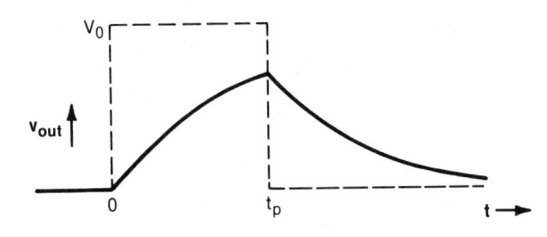

(C) Case 2, $t_p < 5\tau$.

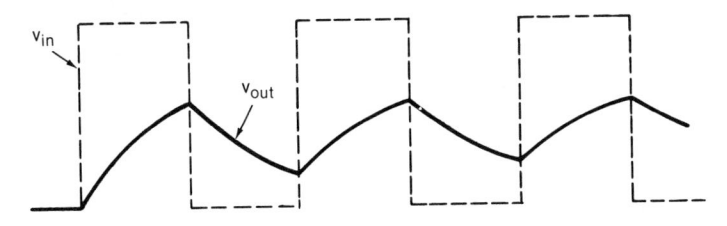

(D) Pulse train with each pulse duration $< 5\tau$.

Fig. 2-26. Pulse response of an RC circuit.

should be chosen so that 5RC *is considerably smaller than the pulse duration:* the smaller the value of RC, the more faithful is the output to the input. If, on the other hand, the time constant RC is made too large, then a significant distortion occurs. Such distortion becomes more severe as the time constant increases. A large time constant may be highly undesirable especially in the case of a series of pulses being applied to an RC circuit, as shown in Fig. 2-26D.

It was seen in Chapter 1 that the response of the RC circuit (referred to as a filter in that discussion) to an input pulse depended upon the *bandwidth* of the circuit: the wider the bandwidth, the more similar is the output to the input. In the present discussion such a dependence is related to the time constant instead of the bandwidth. We would suspect, therefore, that there must be a close correspondence between the time constant of an RC circuit and its bandwidth, which is indeed the case. It will be seen later that the bandwidth is inversely proportional to the time constant; and, consequently, the larger the time constant, the narrower is the bandwidth.

Series and Parallel Combinations of Capacitors

Series Connection

When two capacitors are connected in series, as shown in Fig. 2-27A, the *charge* on each capacitor will be the same in the steady-state dc condition. The total capacitance C_T is given by,

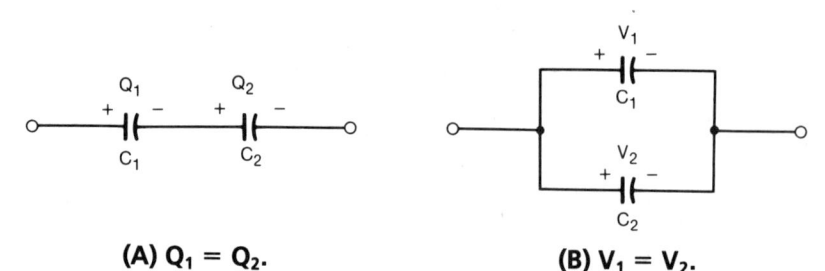

Fig. 2-27. Series and parallel combination of capacitors.
(A) $Q_1 = Q_2$.   (B) $V_1 = V_2$.

$$C_T = \frac{C_1 C_2}{C_1 + C_2}$$

Parallel Connection

When two capacitors are connected in parallel, as shown in Fig. 2-27B, the *voltage* on each capacitor is the same in the steady-state dc condition. The total capacitance, C_T, is given by,

Electronics: Circuits and Systems

$$C_T = C_1 + C_2$$

INDUCTORS

When a current flows through a conductor, it sets up a magnetic field around the conductor. In order to improve the linking of the magnetic flux (or "lines of force") with the conductor, it is usually wound in the form of a coil. The coupling between the magnetic flux and the wire is greatly improved when the coil is wound on a core made of special materials such as steel alloys and ferrites.

It was observed by Faraday that when the *magnetic flux* associated with a conductor *varies with time*, then a voltage is induced in the conductor and that the induced voltage is proportional to the *rate of change of the magnetic flux with time*. This phenomenon is called *electromagnetic induction* and plays an important part in electric and electronic systems. Now, if we consider a coil in which a current is flowing and the current varies with time, then the magnetic flux set up by the current will also vary with time. Therefore, a voltage will be induced in the coil due to a variation of the current flowing in it, which is known as a *self-induced* voltage. For an *ideal coil*, the magnetic flux is proportional to the current and, consequently, the *self-induced voltage is proportional to the rate of change of the current in the coil.* That is, if i is the current in the coil, then the voltage, v_L, induced in it is given by,

$$v_L = L \times \text{(rate of change of the current i)}$$

The coefficient L is called the *coefficient of self-induction*, or, more simply, the *inductance* of the coil. The unit of inductance is the *henry*, abbreviated H.

It is important to note that a self-induced voltage is present if and only if the current is varying with time. *If the current is a constant, then the induced voltage is zero.*

For an ideal inductor, the value of L should remain the same regardless of how much current flows in the coil. But, in practical coils, the value of L does not necessarily remain unaffected by the magnitude of the current. This effect is especially true for coils wound on cores made of special alloys. Also, a coil has a certain amount of resistance due to the resistance of the wire. In our discussion, we will assume the ideal situation: the value of L is a constant and the resistance is zero.

Energy Stored in an Inductor

An inductor stores energy in its magnetic field. The energy stored, w_L, when a current i is flowing through an inductance L is given by,

$$w_L = \frac{1}{2} L i^2 \; J \qquad (2\text{-}25)$$

The energy stored in the inductance is available to activate other components.

Current Continuity Condition

Since the energy stored in a coil depends upon the current, and since energy can change only continuously, it follows that the *current in an inductance can change only continuously*. This is analogous to the condition of voltage in a capacitor. The graph of the current through a coil will be a smooth and continuous curve and will not show any discontinuities. This condition is known as the *current continuity condition for an inductance* and plays an important part in the behavior of an inductance.

Current Buildup in an Inductor

Fig. 2-28 shows a circuit consisting of an inductance, a resistance, a battery, and a switch. Before $t = 0$, the switch is open, and the current is zero. At $t = 0$, the switch is suddenly closed, thus completing the circuit and making it possible for current to flow. Because of the condition of the continuity of current, the current increases gradually from zero to a final constant value. When the current reaches a constant value, the voltage across the inductance becomes zero, and all the applied voltage is across the resistor. Therefore, by Ohm's law, the final value of the current is equal to (V_0/R). The graph of Fig. 2-29 shows the variation of the current in the circuit as a function of time. Note the similarity to the graph of the charging of a capacitor discussed earlier.

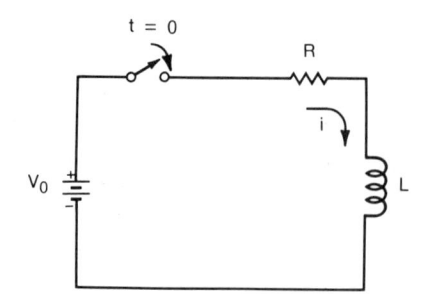

Fig. 2-28. Circuit for building up the current in an inductor.

There is a time constant associated with the buildup of the current in the RL circuit, similar to the time constant in the case of an RC circuit. The time constant is given by

Electronics: Circuits and Systems 113

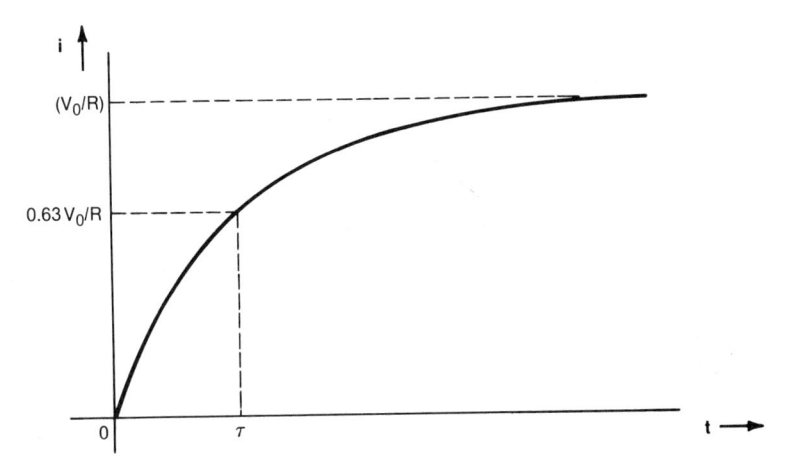

Fig. 2-29. Current buildup in an inductor.

$$\tau = \frac{L}{R} \quad (2\text{-}26)$$

If the inductance is large, then it stores more energy, which means that it will take longer for the current to build up. Thus, the time constant is proportional to L.

The time taken for the current to reach the final constant value (the dc steady state) is approximately five time constants.

Universal Current Buildup Curve

The universal charging curve for a capacitor, Fig. 2-22 can be used for the analysis of the buildup of the current in an inductor with a slight modification in the labeling of the vertical axis. The vertical axis is labeled in terms of the final steady-state current

$$I_0 = (V_0/R),$$

as shown in Fig. 2-30.

Example 2.14

In the circuit of Fig. 2-28, assume that $V_0 = 100$ V, $L = 0.5$ H, and $R = 40$ Ω. Calculate the time constant, the final steady-state current, the approximate time taken to reach the final value, and the energy stored in the inductor in the steady state.

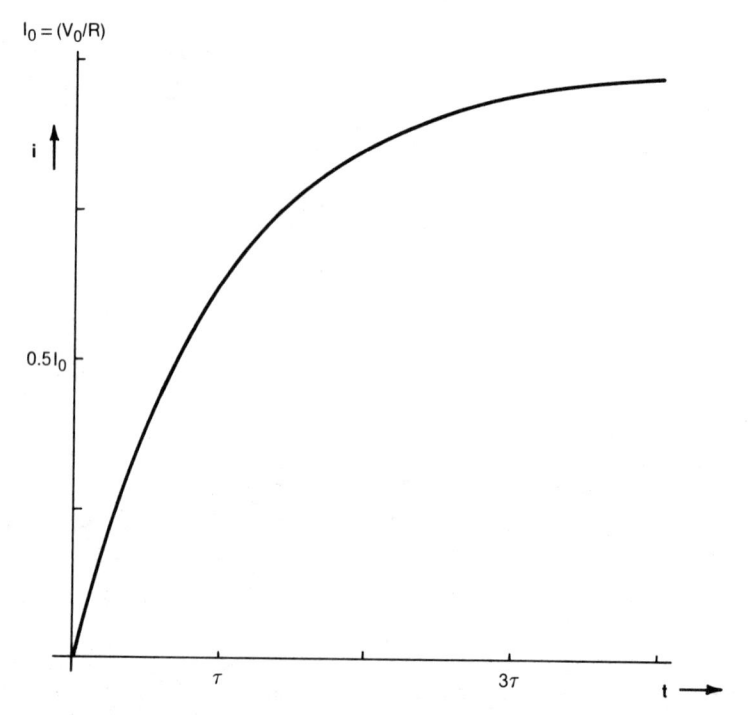

Fig. 2-30. Universal current build up curve for an inductor.

Solution

The time constant is $(L/R) = 0.0125$ s, or 12.5 ms. The final current in the inductor is $(V_0/R) = 2.5$ A. The approximate time taken for reaching the steady state is five time constants, or 62.5 ms. The energy stored in the inductor in the steady state is $(1/2)LI^2 = 1.56$ J.

Example 2.15

In the circuit of the previous example, what value should R be changed to in order to attain the steady state in 10 ms?

Solution

If the steady state is attained in 10 ms, then the time constant should be one fifth of 10 ms, or 2 ms. Therefore, $(L/R) = 2$ ms, and since $L = 0.5$, we get $R = 250 \, \Omega$.

Drill Problem 2.20: An inductance of 10 mH is in series with a resistance of 550 Ω and a battery of 300 V. Calculate the time constant, the

Electronics: Circuits and Systems

final value of the current, the time taken to reach steady state, and the energy stored in the inductance in the steady state.

Drill Problem 2.21: Given an inductor of 4 H, what value of resistance should be used in order to make the current build up to its final value in 25 ms?

Example 2.16

A 50-mH inductor in series with a 20-Ω resistor is connected to a 150-V battery in a circuit like that in Fig. 2-28. Determine the current in the inductor at the end of three time constants. Also, determine the time taken for the current to reach 30% of its final value.

Solution

The time constant of the circuit is (L/R) = 2.5 ms, The final steady-state current is I_0 = (V_0/R) = 7.5 A.

At three time constants, that is, at t = 7.5 ms, the universal current build-up curve (Fig. 2-30) shows that the current has reached 0.95 I_0, which is equal to 7.12 A in this circuit.

The time taken to reach 30% of the final value, or 0.3I_0, is 0.35τ from the universal curve, and this gives 0.875 ms.

Drill Problem 2.22: Find the time taken for the current in a circuit consisting of a 3-mH inductor, a 50-Ω resistor, and a 400-V battery to reach 3 A. Also find the current at t = 0.09 ms.

Decay of Current in an Inductor

Consider a circuit as shown in Fig. 2-31, where the switch is in position 1 until t = 0 and then moved suddenly to position 2. Before t = 0, the current in the inductor has been allowed to build up to a steady-state value of (V_0/R_1). Note that R_2 is not part of the circuit until t = 0. When the switch is moved to position 2, the battery and R_1 are no longer a part of the circuit, and a new closed path is established consisting of L and R_2 as indicated in Fig. 2-31B. The energy stored in the inductor due to the current in it at t = 0 is now available for dissipation through the resistor R_2. The current gradually decays from its initial value of (V_0/R_1) to a final value of zero. All the energy that had been stored in the inductor at t = 0 has now been dissipated by the resistor. The decay of the current is of the form shown in Fig. 2-32, which is seen to be similar to the discharge curve of a capacitor. The time constant controlling the current decay is (L/R_2), which is similar to the time constant of the current buildup in an inductor.

116   Review of Electric Circuits

(A) Switch in position 1.   **(B) Switch in position 2.**

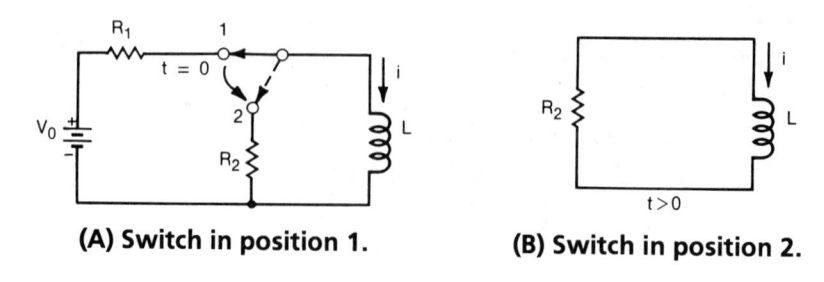

Fig. 2-31. Circuit arrangement for decay of the current in an inductor.

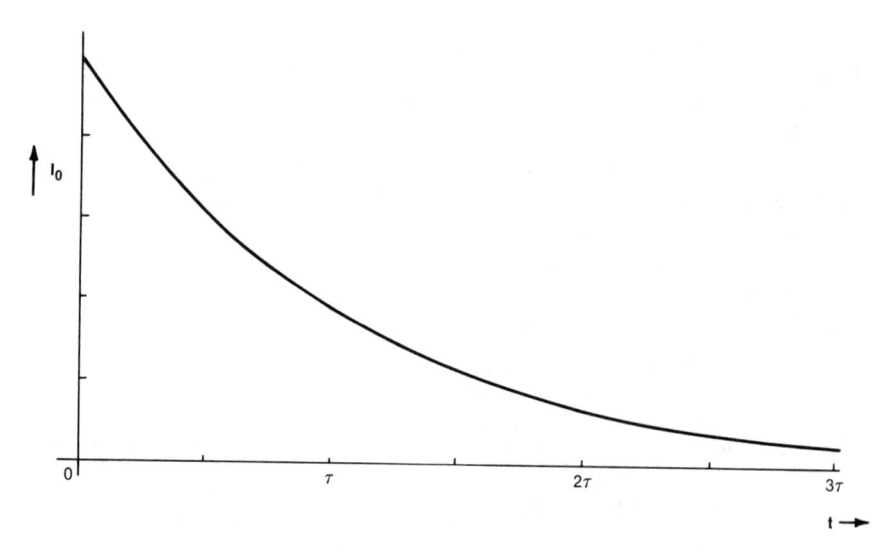

Fig. 2-32. Decay of current in an inductor.

Universal Current Decay Curve

The universal current decay curve for an inductor is similar to the discharge curve of a capacitor and is shown in Fig. 2-33.

Example 2.17

A 10-mH indicator has an initial current of 4 A. It is connected to a 40-Ω resistor that causes the current to decay to zero. Calculate the time taken for the current to decrease to 2.4 A. Also, calculate the current at the end of 0.2 ms.

Electronics: Circuits and Systems

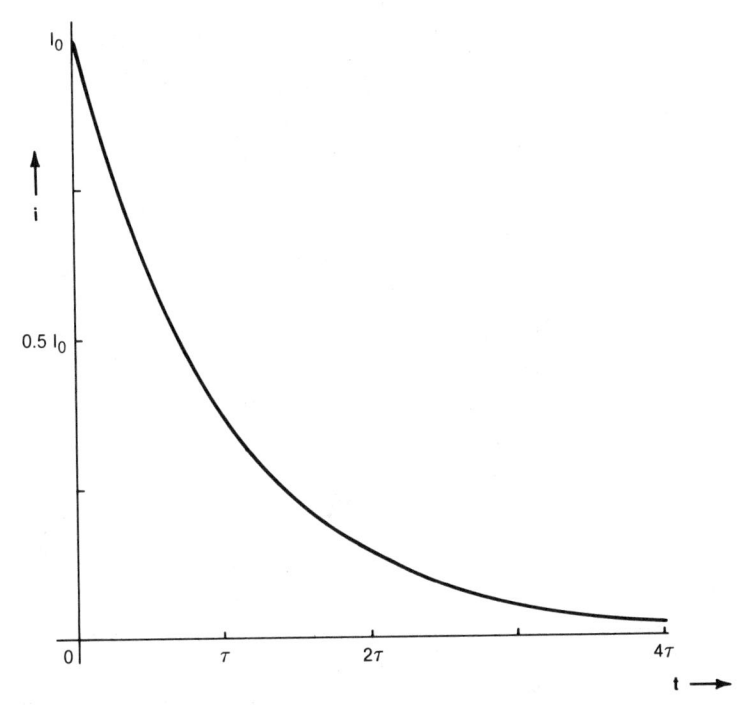

Fig. 2-33. Universal current decay curve for an inductor.

Solution

The current $I_0 = 4$ A. The time constant is $(L/R) = 0.25$ ms.

A current of 2.4 A corresponds to $0.6I_0$ in this case. From the universal current decay curve, Fig. 2-33, it is seen that this occurs at 0.5τ, or 0.125 ms.

A time interval of 0.2 ms corresponds to $(0.2/0.25)$ or 0.8τ. From the universal current decay curve, the current at 0.8τ is $0.45I_0$, or 1.8 A in this example.

Drill Problem 2.23: A 36-mH inductor has an initial current of 8 A, and the current is made to decay through a 2-Ω resistor. Calculate (a) the time at which the current has decreased to 4 A; (b) the value of the current at $t = 20$ ms; (c) the approximate time needed to make the current reach zero value.

An Application of an RL Circuit

Fig. 2-34 shows one of the applications of an RL circuit. The contacts of the relay close when the current in the coil reaches a specified value. Thus, there is a definite delay between the instant of closing the switch, S, and the instant at which the relay contacts close. Suppose in the given circuit, the current needed to activate the relay is 4 A. We wish to calculate the time delay between closing the switch and the activation of the relay.

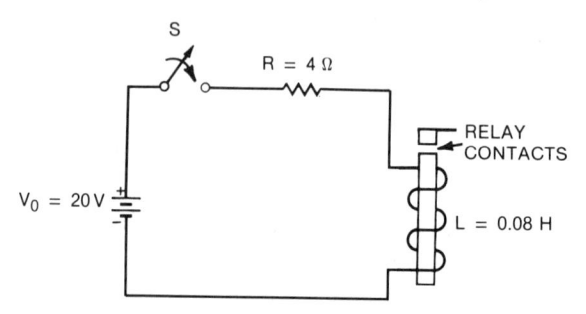

Fig. 2-34. Relay control circuit.

For the given circuit, $I_0 = (V_0/R) = 5$ A. The current of 4 A needed to close the relay contacts corresponds to $0.8I_0$. From the current buildup curve, Fig. 2-30, it is seen that the current builds up to $0.8I_0$ at $t = 1.6\tau$. For the circuit, $\tau = (L/R) = 20$ ms. Therefore, the time delay is $(1.6 \times 20) = 32$ ms.

Drill Problem 2.24: In the circuit of Fig. 2-34, calculate the time delay if the battery voltage was changed to 100 V but all the other data remain unchanged.

AC CIRCUITS

The basic parameters of a sinusoidal waveform and the concept of phase angle were discussed in the last chapter. A brief review of ac circuit analysis will be presented here.

Phasor Representation of a Sinusoidal Function

Instead of using a time function for a sinusoidal voltage or current, a notation called the *phasor notation* is commonly used and it offers many computational advantages. Given a sinusoidal voltage of *peak value*, V_m, and *phase angle*, θ, the phasor notation for the voltage is written as $V_m \underline{/\theta}$ (read as "V_m at an angle of θ"). The phasor can also be shown diagrammatically, Fig. 2-35, where the *length of the vector* represents the *magnitude*, V_m, and the angle the vector makes with the horizontal axis is the angle, θ, of the phasor. A *positive* angle is measured by moving *counterclockwise* from the horizontal axis, while a *negative* angle is measured by moving *clockwise* from the horizontal axis. The phasor with the larger angle leads the one with a smaller phase angle.

Electronics: Circuits and Systems

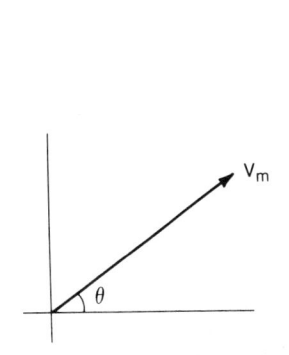

(A) V_m is the magnitude and θ is the angle.

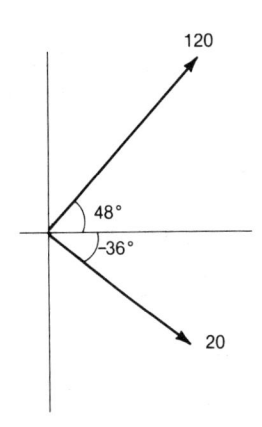

(B) Specific example.

Fig. 2-35. Phasor representations.

It is important to remember that when dealing with two or more phasors, they *must be of the same frequency*. Also, note that the information about the frequency has to be separately specified since the phasor itself does not include that information.

Example 2.18

A sinusoidal voltage has a peak value of 120 V and a phase angle of 48°. A sinusoidal current of the same frequency has a peak value of 20 A and a phase angle of −36°. Write them in phasor form, draw a phasor diagram showing both phasors, and calculate the phase angle between them.

Solution

Voltage phasor = 120 $\underline{/48°}$ V
Current phasor = 20 $\underline{/-36°}$ A

The phasor diagram is shown in Fig. 2-35B, and it is seen that the phase angle between the voltage and current is 84°. Also, since the phase angle of the voltage is positive and the phase angle of the current is negative, the *voltage is leading* the current.

Drill Problem 2.25: For each of the sets of quantities given, write the phasors and find the phase angle between them. Also, state which quantity leads. (Assume that the voltages and currents are at the same frequency in each set.) (a) voltage of amplitude 25 V and a phase angle of 45°, voltage of amplitude 95 V and a phase angle of 84°; (b) voltage of amplitude 160 V and a phase angle of 0°, current of amplitude 15 A and a phase angle of $-60°$; (c) two currents, both of the same amplitude of 2.5 A, one with a phase angle of $-53.1°$ and the other with a phase angle of $-36.9°$.

Drill Problem 2.26: An ac voltage has an amplitude of 84 V and a phase angle of $-24°$. Find a current phasor whose amplitude is one fourth that of the voltage and which leads the voltage by $60°$. Draw a phasor diagram.

Impedance

The impedance of a circuit, denoted by Z, is defined by

$$\text{impedance } Z = \frac{\text{voltage expressed as a phasor}}{\text{current expressed as a phasor}} \quad (2\text{-}27)$$

An impedance has both a *magnitude* and an *angle*.

$$\text{magnitude of } Z = \frac{\text{amplitude of the voltage phasor}}{\text{amplitude of the current phasor}} \quad (2\text{-}28)$$

and

$$\text{angle of } Z = (\text{angle of the voltage phasor}) - (\text{angle of the current phasor}) \quad (2\text{-}29)$$

Example 2-19

An ac voltage of amplitude 100 V and angle $0°$ is applied to a circuit. The resulting current has a peak value of 5 A and an angle of $30°$. Determine the impedance of the circuit.

Solution

Voltage phasor = 100 $\underline{/0°}$ V
Current phasor = 5 $\underline{/30°}$ A
Magnitude of Z = (100/5) = 20
Angle of Z = (0 − 30) = $-30°$
Therefore, the impedance of the circuit is Z = 20 $\underline{/-30°}$ Ω.

Drill Problem 2.27: For each of the voltage and current pairs given, determine the impedance: (a) voltage of peak value 30 V and phase angle of $-45°$; current of peak value 2.5 A and phase angle of $15°$; (b) voltage of amplitude 10 V and phase angle of $0°$; current of amplitude 40 A and phase angle of $-36°$.

Based on the definition of impedance, Eq. (2-27), we have the following

Electronics: Circuits and Systems

relationship for an ac circuit, which may be called the "Ohm's law" for ac circuits:

$$\text{voltage phasor} = (\text{impedance}) \times (\text{current phasor}) \qquad (2\text{-}30)$$

Example 2.20

In each of the following sets of values, one item, V, I, or Z, is missing. Calculate its value (both magnitude and angle).
(a) $V = 100 \underline{/36.9°}$ V, $I = 5 \underline{/0°}$ A;
(b) $V = 40 \underline{/0°}$ V, $Z = 10 \underline{/45°}$ kΩ;
(c) $I = 3 \underline{/0°}$ A, $Z = 5 \underline{/-53.1°}$ Ω.

Solution

Equations (2-28) and (2-29) are used to obtain the following values.

(a) magnitude of Z = (amplitude of V)/(amplitude of I)

$$= (100/5)$$

$$= 20$$

angle of Z = (angle of V) − (angle of I) = 36.9°

Therefore, $Z = 20 \underline{/36.9°}$ Ω

(b) amplitude of I = (amplitude of V)/(magnitude of Z)

$$= (40/10)$$

$$= 4 \text{ mA}$$

angle of I = (angle of V) − (angle of Z)

$$= -45°$$

Therefore, $I = 4 \underline{/-45°}$ mA

(c) amplitude of V = (amplitude of I) × (magnitude of Z)

$$= 3 \times 5$$

$$= 15 \text{ V}$$

angle of V = (angle of I) + (angle of Z) = −53.1°

Therefore, $V = 15 \underline{/-53.1°}$ V

Drill Problem 2.28: Calculate the missing item (V, I, or Z) in each of the following sets of values: (a) V = 50 $\underline{/10°}$ V, I = 4 $\underline{/-30°}$ A; (b) V = 20 $\underline{/30°}$ V, Z = 10 $\underline{/-53.1°}$ Ω; (c) I = 15 $\underline{/0°}$ A, Z = 20 $\underline{/-90°}$ Ω.

Resistors in AC Circuits

The impedance of a resistor R is,

$$Z = R \underline{/0°}$$

and does not depend upon the frequency of the voltage. The voltage and the current are *in phase* in a resistance. For example, if a voltage of 100 $\underline{/30°}$ V is applied to an 8-Ω resistor, the current will have an amplitude of (100/8) = 12.5 A and the same phase angle as the voltage. Therefore, I = 12.5 $\underline{/30°}$ A.

Capacitors in AC Circuits

The impedance of a capacitor *varies inversely with the frequency* of the voltage (or the current). If C is the capacitance and f the frequency of the applied voltage (or current), then the impedance of the capacitor is given by

$$Z_C = \frac{1}{2\pi fC} \underline{/-90°} \ \Omega \tag{2-31}$$

The magnitude of the impedance is (1/2πfC) and varies with the frequency. At low frequencies, the capacitor presents a high impedance; and at high frequencies, the capacitor presents a low impedance. The angle is always −90°. Therefore, the voltage and the current in a capacitor will always be 90° out of phase, with the *current leading the voltage*.

Example 2.21

A capacitor has a capacitance of 10 μF. The voltage across it is 150 $\underline{/0°}$ V. For each of the following frequencies (a) 100 Hz; (b) 50 × 10³ Hz; find the impedance and the current.

Solution

(a) f = 100 Hz

$$Z = \frac{1}{2\pi fC} \underline{/-90°} = \frac{1}{2\pi(100)(10 \times 10^{-6})} \underline{/-90°}$$

$$= 159 \underline{/-90°} \ \Omega$$

$$I = (150/159) \underline{/0 - (-90)} = 0.943 \underline{/90°} \ A$$

(b) f = 50 × 10³ Hz

Electronics: Circuits and Systems

$$Z = \frac{1}{2\pi(50 \times 10^3)(10 \times 10^{-6})} \underline{/-90°} = 0.318 \underline{/-90°}$$

$$I = (150/0.318) \underline{/0 - (-90)°} = 472 \underline{/90°} \text{ A}$$

Drill Problem 2.29: Given a capacitor of 500 pF and a voltage of 120 $\underline{/0°}$, find the impedance and the current at each of the following frequencies: (a) 10 kHz; (b) 10 MHz.

Inductors in AC Circuits

The impedance of an inductor *varies directly with the frequency* of the applied voltage or current. If L is the inductance and f is the frequency of the applied voltage or current, then the impedance of the inductor is given by

$$Z_L = (2\pi fL) \underline{/90°} \; \Omega \qquad (2\text{-}32)$$

The magnitude of the impedance is ($2\pi fL$) and varies with the frequency. At low frequencies, the inductor presents a low impedance, and at high frequencies it presents a high impedance. (This is exactly the opposite of a capacitor's behavior.) The angle is always 90°. Therefore, the voltage and the current in an inductor will always be 90° out of phase, with the *current lagging behind the voltage*.

Example 2.22

An inductor has an inductance of 15 mH. The voltage across it is 150 $\underline{/0°}$ V. For each of the following frequencies find the impedance and the current: (a) 100 Hz; (b) 50 kHz.

Solution

(a) f = 100 Hz

$$Z = (2\pi fL) \underline{/90°} = (2\pi)(100)(15 \times 10^{-3}) \underline{/90°}$$
$$= 9.42 \underline{/90°} \; \Omega$$

$$I = (150/9.42) \underline{/0-90} = 15.9 \underline{/-90°} \text{ A}$$

(b) f = 50 kHz

$$Z = (2\pi)(50 \times 10^3)(15 \times 10^{-3} \underline{/90°})$$
$$= 4712 \underline{/90°} \; \Omega$$

$$I = (150/4712) \underline{/0-90}$$
$$= 0.0318 \underline{/-90°} \text{ A}$$

Drill Problem 2.30: Given an inductor of 20 mH and a voltage of 120 $\underline{/0°}$ V, find the impedance and the current at each of the following frequencies: (a) 50 Hz; (b) 10 kHz.

Series RC Circuit

When two components are in series in an ac circuit, the total impedance is equal to the sum of the individual impedances. Impedances *cannot* be added like ordinary numbers since they have a magnitude as well as an angle. They have to be *added like vectors* as will be seen in the following discussion.

Consider the series RC circuit of Fig. 2-36A. The impedance of the resistor is R $\underline{/0°}$ Ω. A horizontal straight line of length R is drawn to represent this impedance. The impedance of the capacitor is $(1/2\pi fC)$ $\underline{/-90°}$ Ω. A vertical *downward* line is drawn (since the angle is $-90°$) to represent the impedance of the capacitor. The hypotenuse of the right-angled triangle obtained from the two straight lines gives the total impedance of the circuit: the *length* of the hypotenuse represents the *magnitude* of the total impedance and the *angle* the hypotenuse makes with the horizontal axis represents the *angle* of the total impedance. The resulting triangle shown in Fig. 2-36B is known as the *impedance triangle* and is a useful tool in solving simple series circuits. Using the well-known properties of a right triangle (Pythagoras's theorem), we get

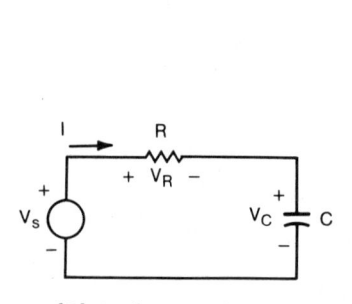

(A) Series RC circuits.　　**(B)** Impedance triangle.

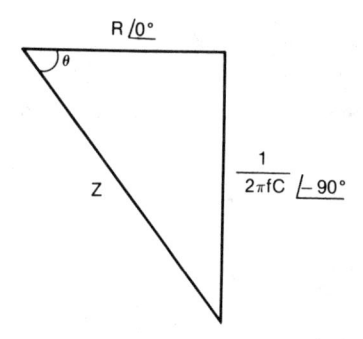

Fig. 2-36. Series RC circuit and impedance triangle.

$$\text{magnitude of } Z = \sqrt{R^2 + \frac{1}{(2\pi fC)^2}} \quad (2\text{-}33)$$

$$\tan \theta = \frac{(1/2\pi fC)}{R}$$

$$= \frac{1}{2\pi fRC} \quad (2\text{-}34)$$

The current in the circuit for a given voltage V_s can then be found by using

$$I = (V_s/Z) \quad (2\text{-}35)$$

The voltage across the components can be found from the equations

$$V_R = IZ_R$$

and

$$V_C = IZ_C \quad (2\text{-}36)$$

where,
Z_R and Z_C represent the impedances of the resistor and the capacitor, respectively.

Example 2.23

In the series circuit of Fig. 2-36A, let $R = 30\ \Omega$ and $C = 4\ \mu F$. A voltage of 120 $\underline{/0°}$ V at a frequency of 1 kHz is applied to the circuit. Find the total impedance, the current, and the voltage across each component.

Solution

Impedance of the resistor $Z_R = 30\ \underline{/0°}$

Impedance of the capacitor $Z_C = \dfrac{1}{2\pi fC}\ \underline{/-90°} = 39.8\ \underline{/-90°}$

The impedance triangle is shown in Fig. 2-37, and it is seen that:

Magnitude of $Z = \sqrt{30^2 + 39.8^2} = 49.8$
$\tan\theta = (39.8/30) = 1.33$
Angle of Z is $\theta = -53°$

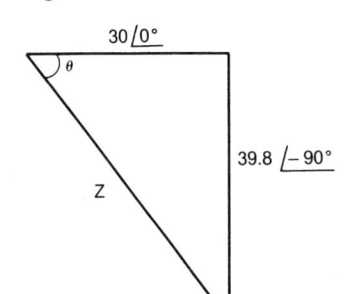

Fig. 2-37. Impedance triangle for Example 2.23.

Therefore,
$$Z = 49.8 \,/\!-53°\, \Omega$$

The current is obtained by using Eq. (2-35);
$$I = (120/49.8) \,/0 - (-53) = 2.41 \,/53°\, A$$

The voltages across the two components are given by Eq. (2-36):
$$V_R = IZ_R = (2.41) \times 30) \,/(53 + 0)° = 72.2 \,/53°\, V$$
$$V_C = IZ_C = (2.41) \times 39.8)/(53 - 90)° = 95.9 \,/\!-37°\, V$$

Drill Problem 2.31: Repeat the calculations of the previous example when the frequency is (a) 500 Hz; (b) 2000 Hz.

FILTERS

Signals occurring in practice contain components of more than one frequency and in some cases, occupy a wide bandwidth. For example, a television signal extends from dc to several megahertz. Signals are passed through *frequency selective networks* called *filters* that transmit freely a certain range of frequencies but attenuate or suppress frequencies outside that range. Filters are intentionally used in some cases to emphasize a particular band of frequencies or suppress unwanted frequencies. But in other cases, a two port has an inherent filter-like property. An amplifier, for example, has a finite bandwidth and cuts off frequencies at the low and the high ends. Two types of filters, the *low-pass filter* (or *LPF*) and the *high-pass filter* (or *HPF*) occur commonly in practice and can be designed in the form of simple series RC circuits of the type discussed in the last section.

High-Pass Filter (HPF)

An RC configuration that acts as a high-pass filter is shown in Fig. 2-38A. The output voltage of the filter will be seen to decrease as the frequency increases in this filter.

Consider the following numerical values: $R = 200\, \Omega$, $C = 1\, \mu F$, $V_s = 100 \,/0°$ V, and the following frequencies: (a) 10 Hz, (b) 200 Hz, (c) 5 kHz, and (d) 100 kHz. The output voltage V_o is measured across the resistor: $V_o = RI$. The calculations of the impedance, the current, and the voltage V_o are along the same lines as in Example 2.23, and the following results are obtained.

(a) f = 10 Hz
$$\text{Total impedance } Z = 1.59 \times 10^4 \,/\!-89.3°\, \Omega$$

Electronics: Circuits and Systems

(A) High-pass filter.

(B) High-pass filter response curve.

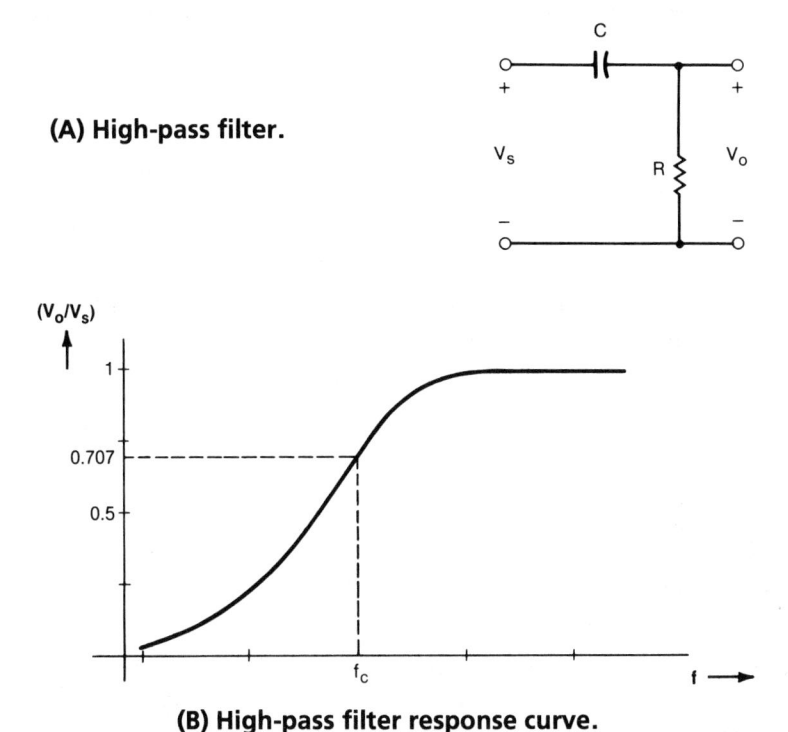

Fig. 2-38. High-pass filter and its response curve.

$$I = 6.29 \times 10^{-3} \,\underline{/89.3°} \text{ A}$$

$$V_o = 1.26 \,\underline{/89.3°} \text{ V}$$

The amplitude of the output voltage is only about 1% of the input voltage.

(b) f = 200 Hz

$$Z = 820 \,\underline{/-75.9°} \text{ Ω}$$

$$I = 0.122 \,\underline{/75.9°} \text{ A}$$

$$V_o = 24.4 \,\underline{/75.9°} \text{ V}$$

The amplitude of the output voltage has increased to nearly 25% of the input voltage.

(c) f = 5 kHz

$$Z = 202 \underline{/-9°}\ \Omega$$

$$I = 0.495 \underline{/9°}\ A$$

$$V_o = 99 \underline{/9°}\ V$$

The amplitude of the output voltage is nearly equal to the input voltage.

(d) f = 100 kHz

$$Z = 200 \underline{/-0.5°}\ \Omega$$

$$I = 0.5 \underline{/0.5°}\ A$$

$$V_o = 100 \underline{/0.5°}\ V$$

The amplitude of the output voltage is now equal to the input voltage.

In the high-pass filter, the capacitive impedance is very high (compared with the resistance) at low frequencies so that almost all the input voltage is across the capacitor. Only a fraction of the input is available at the output. As the frequency increases, the capacitive impedance decreases, while the resistance remains constant. A larger proportion of the input voltage appears at the output. At high frequencies, the capacitive impedance is negligibly small compared with the resistance, and almost all the input voltage appears at the output. If the frequency is sufficiently high, the capacitor is virtually a short circuit (zero impedance), and the output voltage is completely across the resistor. A graph of the magnitude of the *voltage gain*, that is, the ratio of the output voltage to the input voltage, as a function of frequency is shown in Fig. 2-38B. It is seen that the high-frequency voltage gain is close to one, while the gain is small at low frequencies. The filter passes the high-frequency components of the input signal with little or no loss, but severely attenuates the low-frequency components. It is, therefore, a *high-pass* filter.

Drill Problem 2.32: For the HPF of Fig. 2-37, assume R = 1 kΩ and C = 15 μF. Let $V_s = 100 \underline{/0°}$ V. For each of the following frequencies, calculate the output voltage: (a) 1 Hz; (b) 10 Hz; (c) 100 Hz.

Cutoff Frequency of the HPF

The frequency that can be considered as a threshold for the HPF is called the *cutoff frequency*. That is, the filter is said to pass all frequencies above the

Electronics: Circuits and Systems

cutoff frequency and suppress all frequencies below the cutoff frequency. The *cutoff frequency* is defined as the frequency at which the output voltage is 0.707 times the input voltage, or the *voltage gain is 0.707*. At this frequency, it can be shown that the *power output* of the circuit is *one half the power output at high frequencies*. Consequently, the cutoff frequency is also called the *half-power frequency*. In terms of decibels, a voltage gain of 0.707, or a power ratio of 0.5, becomes -3 dB. Therefore, a third name for the cutoff frequency is *the -3-dB frequency*.

The -3 dB frequency is the standard definition of the cutoff frequency in filters and other circuits whose voltage gain varies as a function of frequency.

For the HPF, the cutoff frequency f_c is given by the formula:

$$f_c = \frac{1}{2\pi RC} \text{ Hz} \tag{2-37}$$

Example 2.24

Calculate the cutoff frequency of the HPF of Example 2.23. Verify that the voltage gain is 0.707 at that frequency.

Solution

Using Eq. (2-37), with $R = 200\ \Omega$ and $C = 1\ \mu F$, $f_c = 795.8$ Hz.

At 795.8 Hz, the capacitive impedance is $200\ \underline{/-90°}$ and the resistive impedance is $200\ \underline{/0°}$. The impedances of the resistance and the capacitance have the *same magnitude* at the *cutoff frequency*. (This is a general property as will be shown shortly.)

The impedance of the circuit is $Z = 282.8\ \underline{/45°}\ \Omega$, $I = 0.3536\ \underline{/45°}$ A, and $V_o = 70.7\ \underline{/45°}$ V. Therefore, the voltage gain has a magnitude of 0.707 at 795.8 Hz.

At the cutoff frequency, the impedance of the capacitance has a magnitude given by,

$$Z_C = \frac{1}{2\pi f_c C}$$

which becomes, on using Eq. (2-37) for the cutoff frequency,

$$Z_C = R \tag{2-38}$$

That is, at the cutoff frequency, the impedance of the capacitance has a magnitude equal to the resistance in the filter. Also, since the impedance of the capacitance and the resistance are equal at the cutoff frequency, the *impedance triangle* will have *an angle of* $-45°$ for any HPF regardless of the values of R and C. The equality of Z_C and R, or the angle of the impedance triangle being 45° is frequently used in the determination of the cutoff frequency of the HPF.

Drill Problem 2.33: In a high-pass filter, R = 1600 Ω. Find the value of C such that the cutoff frequency occurs at 100 Hz.

Low-Pass Filter (LPF)

An interchange of the positions of R and C in the HPF leads to a low-pass filter (LPF) shown in Fig. 2-39A. Calculations similar to those used in the HPF will establish that the LPF passes low frequencies with little or no loss while it severely attenuates high frequencies. A graph of the voltage gain of the LPF as a function of frequency will be of the form shown in Fig. 2-39B.

(A) Low-pass filter.

(B) Low-pass response curve.

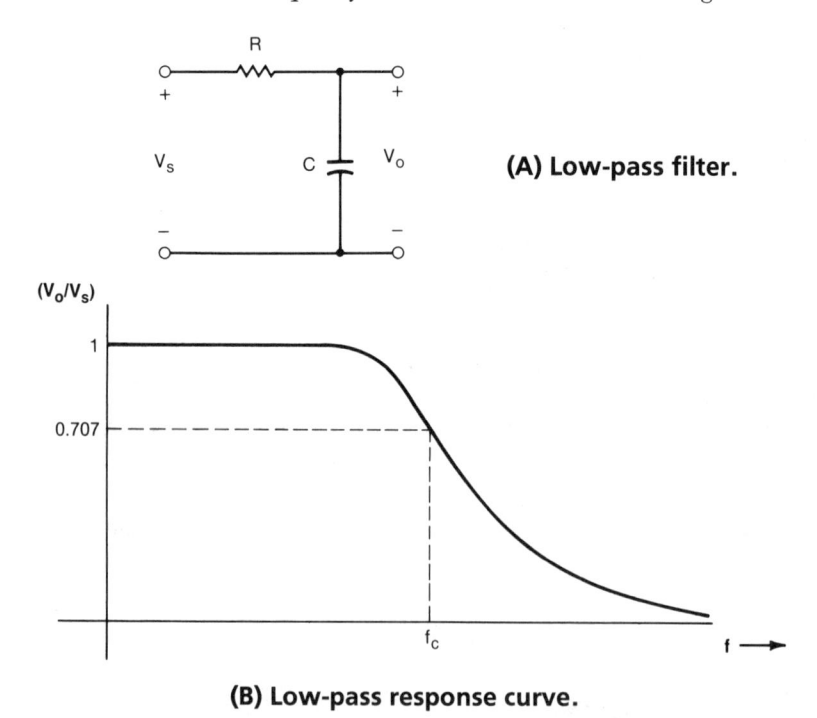

Fig. 2-39. Low-pass filter and its response curve.

The cutoff frequency of the LPF is defined in exactly the same manner as for the HPF: it is the *frequency at which the voltage gain becomes 0.707*. The formula for the cutoff frequency is the same as for the HPF, Eq. (2-37). In fact, *the discussion under the section on the cutoff frequency of the HPF*

Electronics: Circuits and Systems 131

applies completely to the LPF also, with the only exception being that the LPF passes all frequencies below the cutoff whereas the HPF passes all frequencies above the cutoff.

Example 2.25

An LPF has $R = 1\ M\Omega$ and $C = 1500\ pF$. Calculate the cutoff frequency. Show that the voltage gain is 0.707 and the impedance has an angle of $45°$ at the cutoff frequency.

Solution

Using Eq. (2-18), f_c is found to be 106 Hz. At 106 Hz, the impedance of the capacitance is $Z_C = 10^6\ \underline{/-90°}$. Thus, the impedance of the capacitance is equal to the resistance in magnitude. The angle of the impedance is therefore $-45°$ as shown in the impedance triangle in Fig. 2-40.

Fig. 2-40. Impedance triangle for Example 2.25.

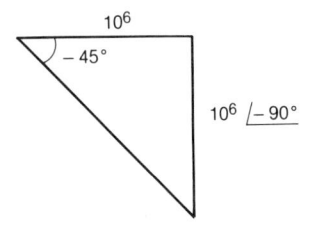

Drill Problem 2.34: For an LPF, $R = 500\ \Omega$ and $C = 10\ \mu F$. The input voltage is $V_s = 100\ \underline{/0°}$ V. Calculate the output voltage at (a) $f = 5$ Hz; (b) $f = 50$ Hz; (c) $f = 500$ Hz. Find the cutoff frequency and verify that the voltage gain is 0.707 at that frequency.

Frequency Response and Pulse Response

The response of an RC circuit to signals containing a number of frequencies, as was done in the study of filters, is called the *frequency response* and the graphs of Figs. 2-38 and 2-39 are known as the frequency response curves of RC filters. It was seen that the product RC appears in the formula for the cutoff frequency and affects the frequency response of a filter. In the earlier discussion on the charging and discharging of a capacitor and the *pulse response of an RC circuit*, the product RC appeared also and was found to affect the distortion in the output pulse. The *pulse response* is a study of the output as a *function of time*, whereas the *frequency response* is a study of the output as a *function of frequency*. Since the product RC appears in both cases, the two responses can be expected to be related to each other. That is, the pulse response of an RC circuit cannot be modified without modifying its frequency response also.

Consider a *low-pass filter* as shown in Fig. 2-41. Suppose a rectangular pulse is applied as the input. The output of the circuit will have the form shown in Fig. 2-42 and is a rounded-off version of the input pulse, the amount of rounding off depending upon the time constant RC of the circuit. A small time constant is necessary for a faithful reproduction of the input pulse, as was seen earlier. A small time constant RC will mean that the cutoff frequency f_c should be high. Therefore, the cutoff frequency of the LPF will have to be made high in order to improve the pulse response, and conversely, an LPF with a good pulse response will have a cutoff frequency that is quite high.

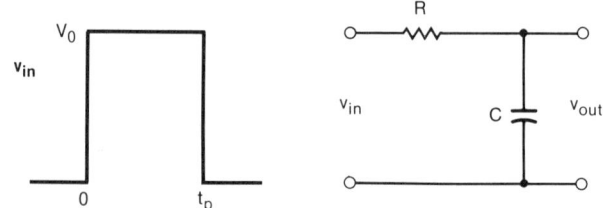

Fig. 2-41. Pulse input to an LPF.

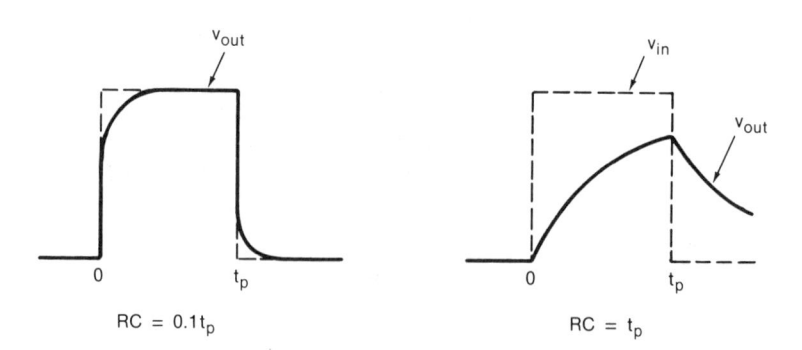

Fig. 2-42. Output pulses in an LPF.

Now, consider a high-pass filter as shown in Fig. 2-43 and let the input be a rectangular pulse. The output will have the form shown in Fig. 2-44. At $t = 0$, when the pulse jumps to V_0, the capacitor voltage does not change instantaneously (voltage-continuity condition) and all the voltage appears across the resistor. Then, the capacitor voltage builds up toward V_0, and the voltage across the resistor gradually decreases toward zero. When the pulse returns to zero at $t = t_p$, the output voltage takes a downward dip equal to the *change in the input voltage* because the capacitor voltage cannot change instantaneously. The output voltage at $t = t_p$ becomes negative. Then, the capacitor discharges through the resistance, and the output voltage eventually becomes zero. The output is again a distorted version of the input but the distortion is due to the *sag* (or *tilt*) in the output pulse. In order to make the sag as small as possible, the *time constant* RC of the circuit must be made *large*. In terms of frequency response, the HPF should have a very low cutoff frequency.

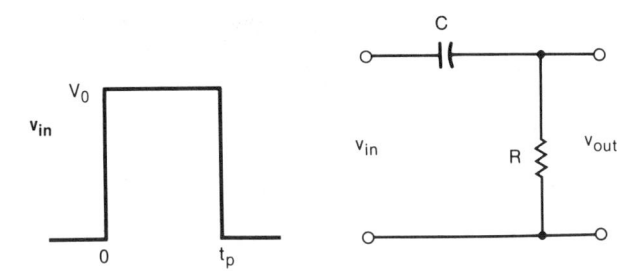

Fig. 2-43. Pulse input to an HPF.

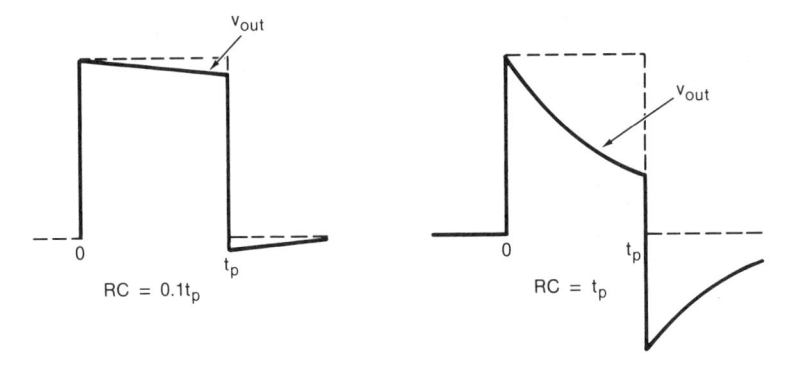

Fig. 2-44. Output pulses in an HPF.

Applications of RC Filters

Low-pass and high-pass filters are used frequently to filter out unwanted bands of frequencies or to emphasize some bands of frequencies. An example of their use is in loudspeaker systems where it is necessary to switch from the woofer (which responds well to low frequencies) to the tweeter (which responds well to high frequencies) and vice versa. An LPF is used to control the woofer, and an HPF is used to control the tweeter, and the arrangement is called the *crossover network*. We will see that communication systems use filters as an integral part of modulation and demodulation circuits.

Filters also appear when they are not exactly desirable. Electronic amplifiers exhibit *inherent* characteristics similar to low-pass and high-pass filters even though they are not purposely designed to do so. These characteristics affect the bandwidths of amplifiers (frequency response) and the shapes of output pulses of amplifiers (pulse response), and they will be studied in some detail in a later chapter.

THEVENIN EQUIVALENT

A complex circuit can be replaced (for the purposes of analysis) by a simple equivalent circuit consisting of a single voltage source in series with a single impedance. The procedure for obtaining such an equivalent of a given circuit is based on a theorem due to Thevenin; and the equivalent circuit is, therefore, called the *Thevenin equivalent*. Consider a black box as indicated in Fig. 2-45A with a pair of terminals A and B that provide access to the network in the black box. As seen from the terminals A and B, the network in the black box can (according to Thevenin's theorem) be replaced by a voltage source V_{Th} in series with an impedance Z_{Th}. V_{Th} is called the *Thevenin voltage* and Z_{Th} the *Thevenin impedance* with respect to the terminals A and B. The Thevenin voltage V_{Th} is the voltage that would appear across the terminals A and B if those terminals were *open circuited*. The Thevenin impedance Z_{Th} is the impedance that would be seen looking in from the terminals A and B if the sources in the network were made inactive (that is, each independent voltage source is replaced by a short circuit and each independent current source by an open circuit).

(A) Black box.   **(B) Equivalent circuit.**

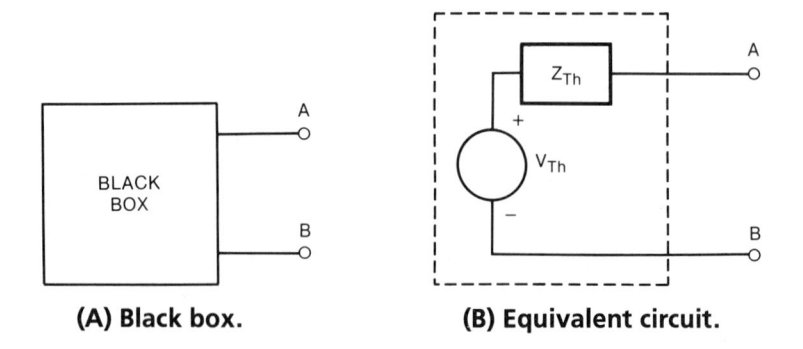

Fig. 2-45. Thevenin's equivalent circuit.

For our purposes, it is sufficient if we look at some simple circuit configurations containing voltage sources and resistors. In such cases, the procedure for finding the Thevenin equivalent consists of the following steps:

Electronics: Circuits and Systems   135

1. Open circuit the terminals A and B (unless they are already given open circuited).
2. Calculate the voltage across A and B. This may require only a voltage division in simple circuits. The voltage calculated is the Thevenin voltage V_{Th}.
3. Replace each voltage source in the given network by a short circuit.
4. Calculate (by using series and parallel combinations) the total resistance seen from the terminals A and B. The resistance so calculated gives the Thevenin resistance R_{Th}.
5. The Thevenin equivalent circuit is drawn from the voltage source V_{Th} in series with the Thevenin resistance R_{Th} as shown in Fig. 2-46.

Fig. 2-46. Thevenin equivalent.

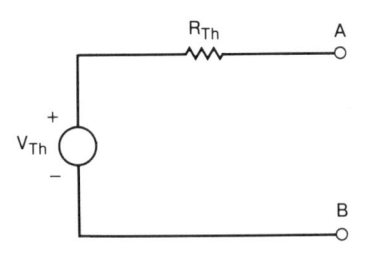

Example 2.26

Obtain the Thevenin equivalent of the circuit given in Fig. 2-47 as seen from the terminals A and B.

Fig. 2-47. Circuit for Example 2.26.

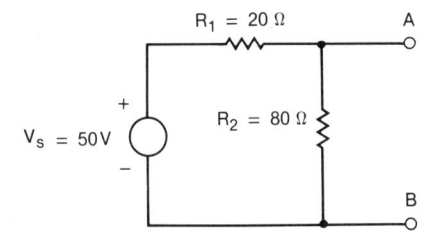

Solution

The terminals A and B are already open. So, we calculate the voltage across A and B, which is done by using a voltage division formula. Shown in Fig. 2-48A.

$$V_{Th} = \frac{R_2}{R_1 + R_2} V_s$$
$$= 40 \text{ V}$$

Replace the voltage source in the given network by a short circuit, leading

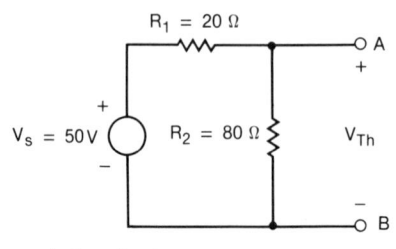

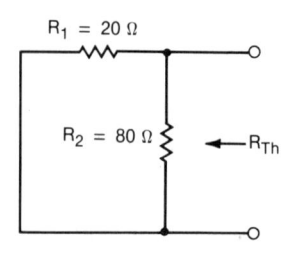

(A) Calculating V_{TH} using a voltage division formula.

(B) Replacing the voltage source by a short circuit to calculate R_{TH}.

Fig. 2-48. Calculation of V_{TH} and R_{TH} in Example 2.26.

to the circuit shown in Fig. 2-48B. Looking in from A and B in that circuit, we see that the two resistors R_1 and R_2 are in parallel. Therefore, the Thevenin resistance R_{Th} is given by

$$R_{Th} = \frac{R_1 R_2}{R_1 + R_2}$$

$$= 16 \, \Omega$$

Fig. 2-49 shows the Thevenin equivalent of the given circuit as seen from the terminals A and B.

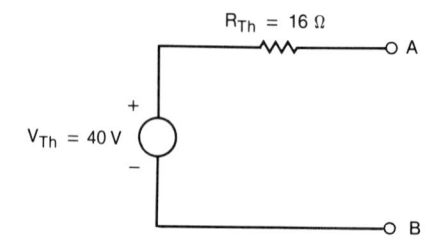

Fig. 2-49. Thevenin equivalent of the circuit in Example 2.26.

Drill Problem 2.35: For each of the circuits shown in Fig. 2-50 obtain the Thevenin equivalent as seen from the terminals A and B.

The Thevenin equivalent performs the useful function of simplifying the calculations in a circuit in certain situations. Consider the arrangement shown in Fig. 2-51, where the black box contains a fixed network, and a variable-load network is connected to the terminals A and B. If the variable load assumes a number of different configurations and values, then the solution of the given arrangement will have to be repeated a number of times, which could be unnecessarily repetitious. Suppose we replace the fixed net-

Electronics: Circuits and Systems

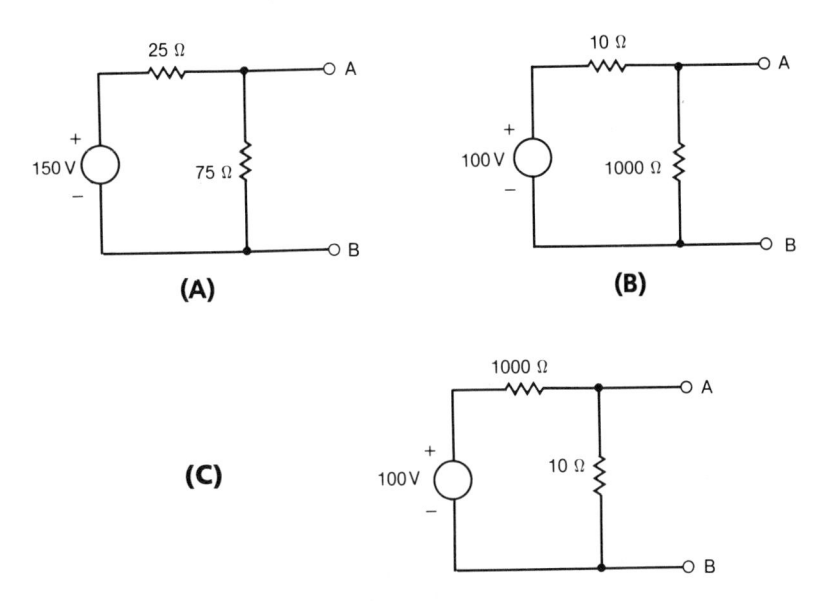

Fig. 2-50. Circuits for Drill Problem 2.35.

work in the black box by its Thevenin equivalent leading to the situation shown in Fig. 2-52. Then, as the load network is varied, the currents and voltages in it can be calculated by using the configuration in Fig. 2-52. Since the fixed network may be quite complex, the use of the simpler arrangement of Fig. 2-52 will be a lot less tedious than using the original arrangement.

Fig. 2-51. A fixed network with a variable load. (The slanted arrow is used to indicate a variable component.)

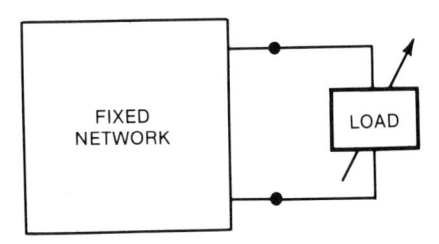

Fig. 2-52. Thevenin equivalent of a fixed network with a variable load.

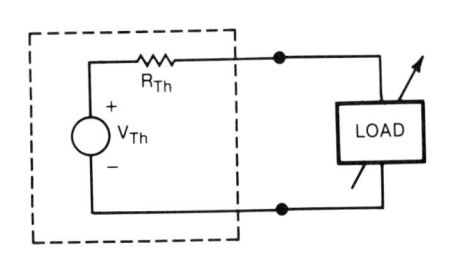

Example 2.27

The load resistor in the circuit of Fig. 2-53 assumes the following values: 10 Ω, 20 Ω, 40 Ω, and 100 Ω. Calculate the load current in each case.

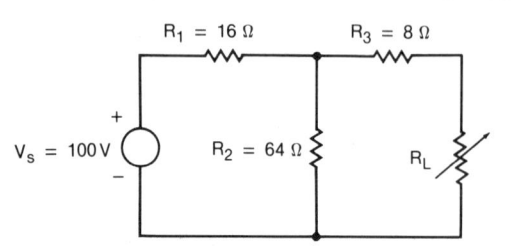

Fig. 2-53. Circuit for Example 2.27.

Solution

First, we replace the circuit with a Thevenin equivalent as seen from the terminals of R_L. When the load resistor is removed so as to open circuit the terminals A and B, the circuit is as shown in Fig. 2-54A. There is no current in R_3 due to the open circuit, and hence there is no voltage drop across it. The Thevenin voltage is, therefore, that across R_2 and is obtained by using voltage division.

$$V_{Th} = \frac{R_2}{R_2 + R_1} V_s$$

$$= 8 \text{ V}$$

To find R_{Th}, replace the voltage source V_s by a short circuit leading to the circuit of Fig. 2-54B. Then, the Thevenin resistance is given by

$$R_{Th} = (R_1 \| R_2) + R_3$$

$$= 20.8 \text{ Ω}$$

Now we replace the original network by its Thevenin equivalent and reconnect the load as in Fig. 2-54C. From this circuit, we have for the current in R_L

$$I_L = \frac{V_{Th}}{R_{Th} + R_L}$$

$$= \frac{8}{20.8 + R_L}$$

The values of I_L are 0.26 A, 0.196 A, 0.132 A, and 0.0662 A for R_L = 10 Ω, 20 Ω, 40 Ω, and 100 Ω, respectively.

Electronics: Circuits and Systems

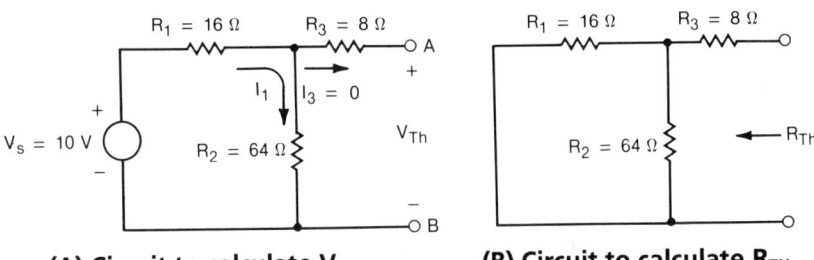

(A) Circuit to calculate V_{TH}.

(B) Circuit to calculate R_{TH}.

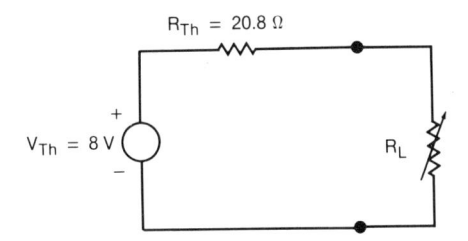

(C) Thevenin equivalent circuit with load reconnected.

Fig. 2-54. Determination of the Thevenin equivalent of the circuit in Example 2.27.

Drill Problem 2.36: The load resistor in the circuit of Fig. 2-55 is varied in steps of 2 Ω, starting at 2 Ω and stopping at 12 Ω. Calculate the voltage across R_L in each case.

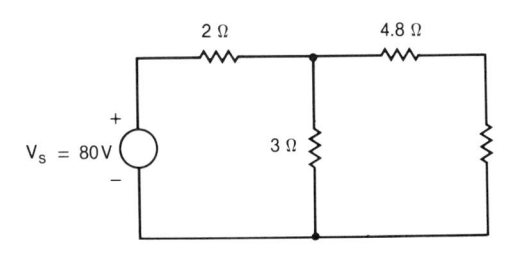

Fig. 2-55. Circuit for Drill Problem 2.36.

Thevenin equivalents are also useful in the analysis of multistage amplifiers. When we are analyzing a specific stage of the amplifier, it is convenient to replace the preceding stages by a Thevenin equivalent circuit and do the same thing for the following stages. By using such an approach, the analysis of a multistage amplifier is reduced to that of a single-stage amplifier

Maximum Power Transfer

One of the applications of Thevenin's theorem is in the consideration of when maximum power is transferred from a given network to a load. Consider a *fixed network* connected to a *variable-load resistance* as indicated in Fig. 2-56. We wish to adjust the load resistance such that the power output from it is a maximum. An example of such a situation is when the fixed network may represent a multistage amplifier (which is fixed in its structure and component values), and the load resistance has to be selected so as to maximize the power output. In such cases, the amplifier is first replaced by its Thevenin equivalent so that the situation reduces to that of Fig. 2-56. The load current in the circuit of Fig. 2-56 is

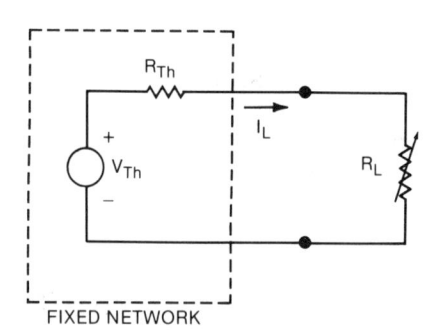

Fig. 2-56. Fixed network with a variable load resistor.

$$I_L = \frac{V_{Th}}{R_L + R_{Th}} \qquad (2\text{-}39)$$

and the power output is

$$P_L = I_L^2 R_L \qquad (2\text{-}40)$$

Combining the last two equations, we get

$$P_L = \frac{V_{Th}^2 R_L}{(R_{Th} + R_L)^2} \qquad (2\text{-}41)$$

A plot of the power P_L, as the load resistance is varied, is of the form shown in Fig. 2-57, and it is seen that the maximum occurs when

$$R_L = R_{Th} \qquad (2\text{-}42)$$

Electronics: Circuits and Systems

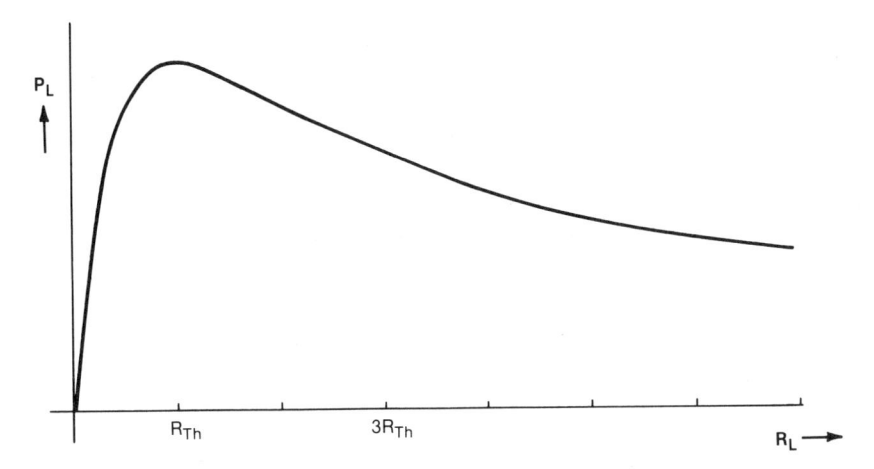

Fig. 2-57. Power in the load as a function of the load resistance in the circuit of Fig. 2-56.

(This result can be derived mathematically by taking the derivative (dP_L/dR_L), equating it to zero, and solving for R_L.)

This result is known as the *maximum power transfer theorem: given a network with a fixed value of Thevenin resistance, the load resistance must be made equal to the Thevenin resistance in order to transfer maximum power from the network to the load.* In Example 2.27, for instance, the load resistance should be made equal to 20.8 Ω if we wish to obtain as much power out of it as possible.

When the load resistance is made equal to the Thevenin resistance so as to maximize the power output, we say that the load is *matched* to the network. In electronic and communication systems, the network of the system cannot be readily altered by the user. Also, the devices and configurations used in their design place limits on the range of values that their Thevenin resistance can have. Consequently, the user of an electronic or communication system has very little control over the Thevenin resistance of the system itself. The only recourse available to the user is, therefore, to choose the load resistance so as to match it to the system.

Use of Ideal Transformers in Matching

It sometimes happens that the available range of load resistances is such as to make matching impossible when the load is directly connected to the given network. For example, an amplifier may have an output resistance (that is, the Thevenin resistance seen from the output terminals) in the order of a few thousand ohms while the load resistance may be in the order of ten ohms. In such cases, matching can be accomplished by interposing an *ideal transformer* between the network and the load. (Other methods of accomplishing matching in such situations will be discussed in Chapter 5.)

An ideal transformer has two coils magnetically coupled to each other such that when a voltage is applied to one coil, a voltage is induced in the other due to electromagnetic induction.

An ideal transformer (Fig 2-58) with N_1 turns in coil 1 and N_2 turns in coil 2 has the following relationships for its currents and voltages:

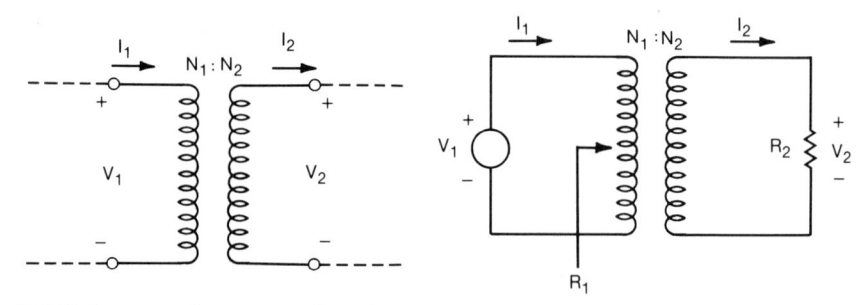

(A) Voltage ratio proportional to the number of turns.

(B) Reflected resistance.

Fig. 2-58. Ideal transformers.

$$\text{voltage ratio:} \quad (V_1/V_2) = (N_1/N_2) \qquad (2\text{-}43)$$

$$\text{current ratio:} \quad (I_1/I_2) = (N_2/N_1) \qquad (2\text{-}44)$$

That is, the voltage ratio is directly proportional to the ratio of the number of turns. *The larger voltage appears across the coil with the larger number of turns.* The situation with the current is just the opposite to that of the voltage.

Now consider a load resistance R_2 connected to coil 2 and a voltage source V_1 applied to coil 1, as in Fig. 2-58B. The resistance *seen by the source* is called the *reflected resistance*, denoted by R_1 in the diagram.

The resistance R_1 is related to V_1 and I_1 by the equation:

$$R_1 = (V_1/I_1)$$

The resistance R_2 is related to V_2 and I_2 by the equation:

$$R_2 = (V_2/I_2)$$

From the last two equations, we get,

Electronics: Circuits and Systems

$$\frac{R_1}{R_2} = \frac{V_1}{I_1}\frac{I_2}{V_2}$$

$$= \frac{V_1}{V_2}\frac{I_2}{I_1}$$

$$= \frac{(V_1/V_2)}{(I_1/I_2)} \quad (2\text{-}45)$$

The ratios (V_1/V_2) and (I_1/I_2) are related to the turns ratio of the transformer, Eqs. (2-43) and (2-44):

$$(V_1/V_2) = (N_1/N_2)$$

and

$$(I_1/I_2) = (N_2/N_1)$$

so that Eq. (2-45) becomes

$$\frac{R_1}{R_2} = (N_1/N_2)^2 \quad (2\text{-}46)$$

That is, the reflected resistance seen from the side of the transformer with N_1 turns is equal to the product of $(N_1/N_2)^2$ and the actual resistance connected to the side with N_2 turns.

By a similar argument, the reflected resistance seen from the side of the transformer with N_2 turns is equal to the product of $(N_2/N_1)^2$ and the actual resistance connected to the side with N_1 turns.

The two situations are shown in Fig. 2-59. It is convenient to remember that *the reflected resistance is larger when seen from the side with the larger number of turns*, and that it is the *square* of the turns ratio that multiplies the actual resistance.

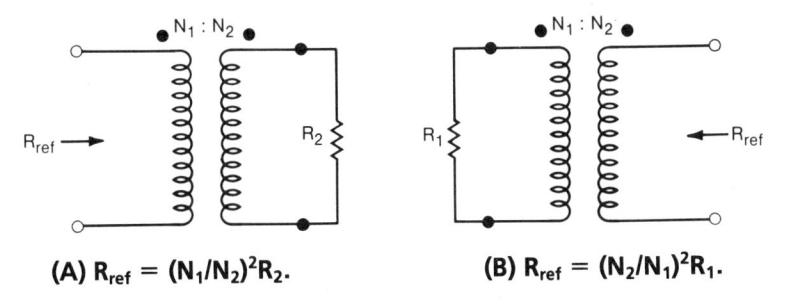

(A) $R_{ref} = (N_1/N_2)^2 R_2$. (B) $R_{ref} = (N_2/N_1)^2 R_1$.

Fig. 2-59. Reflected resistance in an ideal transformer.

Example 2.28

For each of the cases shown in Fig. 2-60, calculate the reflected resistance R_{ref} indicated.

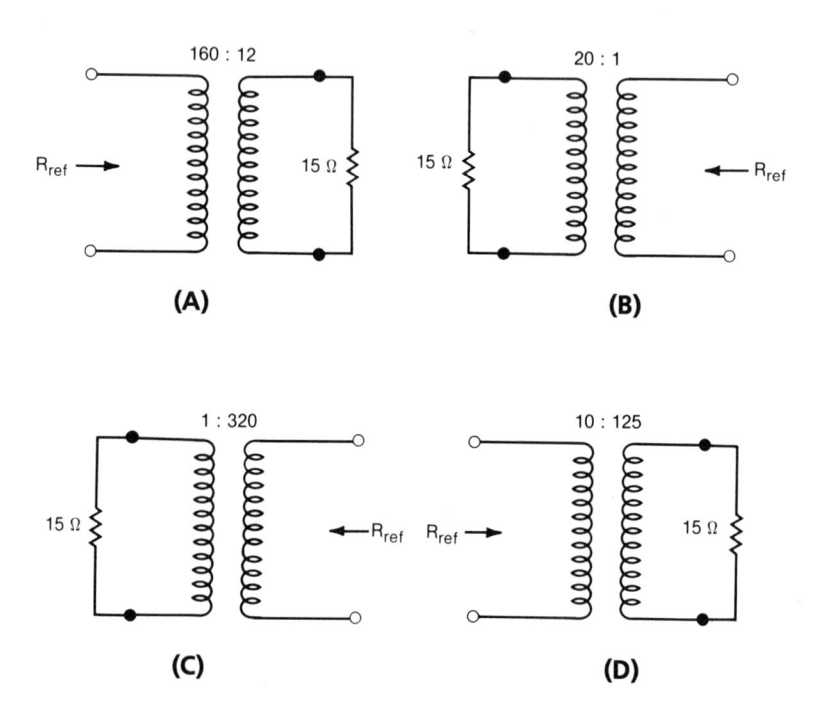

Fig. 2-60. Ideal transformer connections for Example 2.28.

Solution
(a) $R_{ref} = (160/12)^2 \times 15$
$= 2667 \; \Omega$
(b) $R_{ref} = (1/20)^2 \times 15$
$= 3.75 \times 10^{-2} \; \Omega$
(c) $R_{ref} = (320/1)^2 \times 15$
$= 1.54 \times 10^6 \; \Omega$
(d) $R_{ref} = (10/125)^2 \times 15$
$= 9.6 \times 10^{-2} \; \Omega$

Drill Problem 2.37: Calculate the reflected resistance R_{ref} in each of the cases shown in Fig. 2-61.

Electronics: Circuits and Systems

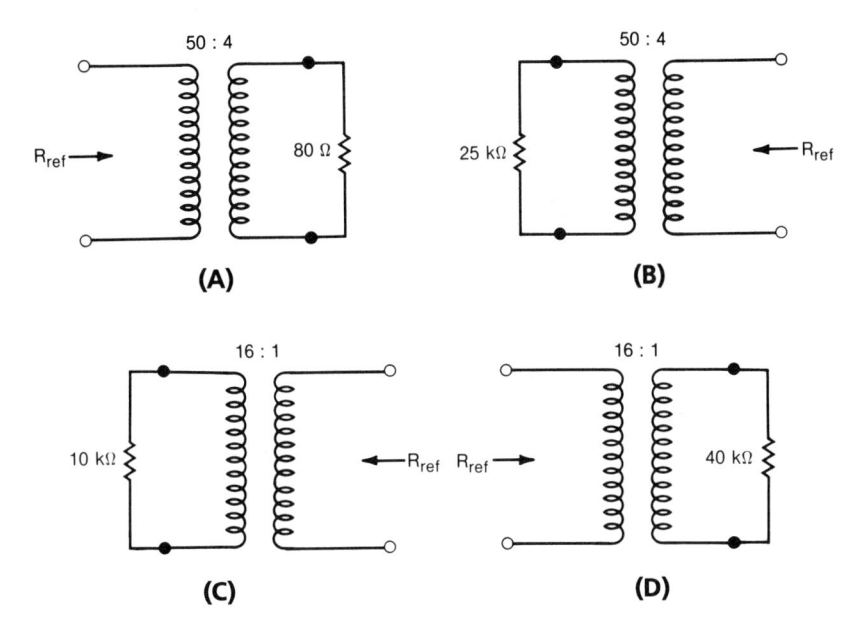

Fig. 2-61. Ideal transformer connections for Drill Problem 2.37.

Example 2.29

The output resistance of an amplifier is 40 kΩ. A loudspeaker of resistance 16 Ω is available. Find the turns ratio of the transformer needed to match the speaker to the amplifier.

Solution

The arrangement should be as shown in Fig. 2-62. The reflected resistance seen from the side with N_1 turns should be 40 kΩ. Therefore, we have

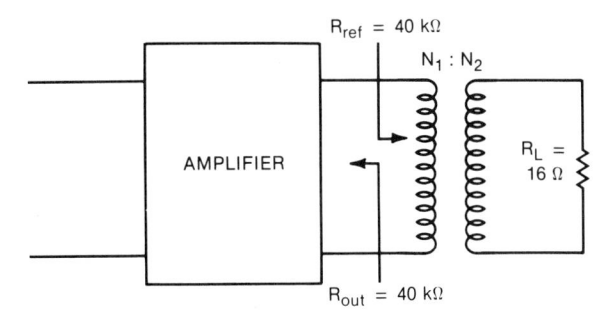

Fig. 2-62. Matching of an amplifier to a load.

$$40 \times 10^3 = (N_1/N_2)^2 (16)$$

which leads to,

$$(N_1/N_2) = 50$$

Drill Problem 2.38: Repeat the calculations of the last example for a loudspeaker of resistance (a) 4 Ω; (b) 8 Ω.

Example 2.30

A two port has an input resistance of 4 kΩ and an output resistance of 20 kΩ. The input is to be fed by a signal source with a resistance of $R_s = 100$ kΩ, and the output is to be connected to a load of 300 Ω. Find the turns ratios of the two transformers needed to have a completely matched system.

Solution

The arrangement is of the form shown in Fig. 2-63. For the input side, the reflected resistance seen from the side with N_a turns is to be equal to R_s or 100 kΩ. Therefore, we have,

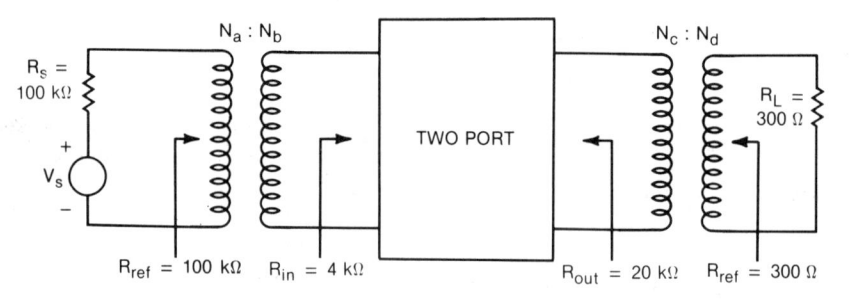

Fig. 2-63. Matching of a two-port system.

$$R_s = 100 \times 10^3$$
$$= (N_a/N_b)^2 R_{in}$$
$$= (N_a/N_b)^2 (4 \times 10^3)$$

which gives,

$$(N_a/N_b) = 5$$

For the output side, the reflected resistance seen from the side with N_d turns should be equal to R_L or 300 Ω. Therefore,

$$R_L = 300$$
$$= (N_d/N_c)^2 R_{out}$$
$$= (N_d/N_c)^2 (20 \times 10^3)$$

which gives,

$$(N_d/N_c) = 0.122$$

(The results obtained give only the turns *ratio* of each transformer. The actual number of turns is determined on the basis of various other factors such as cost, frequency range over which the system is to operate, and the material of the core, among others.)

Drill Problem 2.39: Repeat the calculations of the last example if the source resistance is changed to 10 kΩ and the load resistance is changed to 16 Ω.

Among the disadvantages of a transformer (used in impedance matching) are its weight and cost. A high-quality transformer is bulky and expensive. But the more important problems associated with a transformer are its nonlinearity and frequency response characteristics. A practical transformer departs from the ideal characteristics assumed earlier and tends to saturate at high currents. Consequently, instead of responding linearly to all input signal levels, it introduces distortions due to nonlinearity. The range of frequencies over which a transformer (that is, its bandwidth) is usually limited also. The transformer is often eliminated by using other techniques of impedance matching whenever possible. But, the transformer has certain distinct advantages: its ruggedness and ability to handle high power.

SUMMARY SHEETS

KIRCHHOFF'S LAWS

Kirchhoff's Current Law
At any junction in an electric circuit
Σ (currents entering) = Σ (currents leaving)

Kirchhoff's Voltage Law
In any closed path in an electric circuit
Σ (voltage drops) = Σ (voltage rises)

RESISTORS

Resistors in series

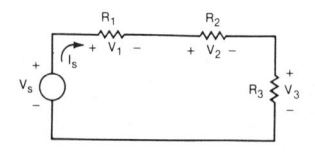

$R_T = R_1 + R_2 + R_3$.
$I_s = (V_s/R_T)$
$V_1 = (R_1/R_T)V_s \quad V_2 = (R_2/R_T)V_s$
$V_3 = (R_3/R_T)V_s$

Voltage Divider

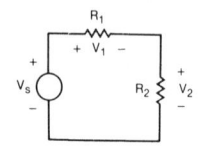

$$V_1 = \frac{R_1}{R_1 + R_2}V_s$$

$$V_2 = \frac{R_2}{R_1 + R_2}V_s$$

The *larger* resistance has the *larger* voltage across it.

Potentiometers

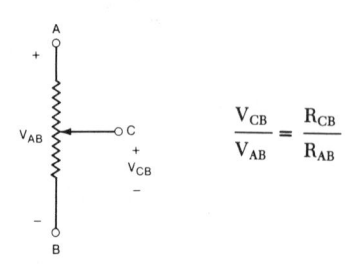

$$\frac{V_{CB}}{V_{AB}} = \frac{R_{CB}}{R_{AB}}$$

Resistors in Parallel

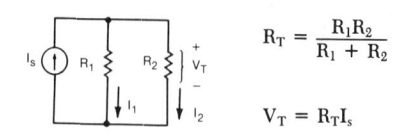

$$R_T = \frac{R_1 R_2}{R_1 + R_2}$$

$$V_T = R_T I_s$$

Current Division

$$I_1 = \frac{R_2}{R_1 + R_2}I_s \qquad I_2 = \frac{R_1}{R_1 + R_2}I_s$$

The *smaller* resistance has the *larger* current through it.

Approximations
Series resistors: When one resistance is much larger than the other, ignore the *smaller* resistance.
Parallel resistors: When one resistance is much larger than the other, ignore the *larger* resistance. ("much larger than" usually means a factor of 10 or more.)

CAPACITORS

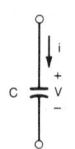

$i = C \times$ (rate of change of V)
Energy stored in a capacitor = $\frac{1}{2} CV^2$ joules.
Voltage continuity condition: The voltage in a capacitor cannot change suddenly; it can only change continuously.

Charging of a Capacitor

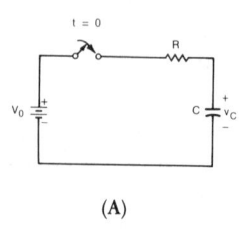

(A)

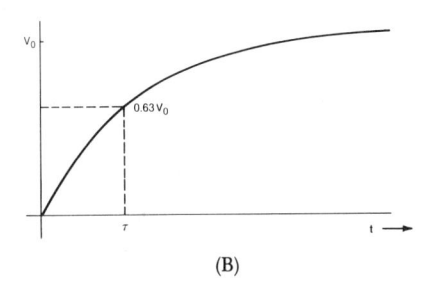

Electronics: Circuits and Systems

Pulse Response of an RC Circuit

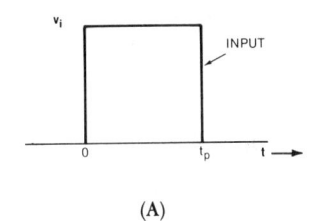

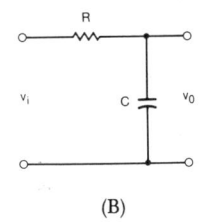

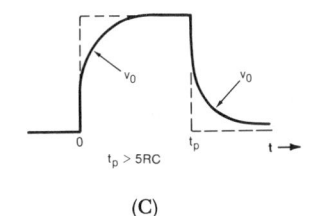

(B) [charging curve, 0.63 V₀ at τ]

(A) [input pulse]

$v_C = V_0 - V_0 e^{-(t/RC)}$
Steady state value of $v_c = V_0$
Time constant $\tau = RC$ seconds.
Rise time = Time taken for v_c to increase from 10% of V_0 to 90% of $V_0 = 2.2\,RC$ seconds.
Larger the time constant → It takes longer to charge the capacitor to V_0.
Time taken to attain steady state = 5 time constants.

(B) [RC circuit diagram]

Discharging of a Capacitor

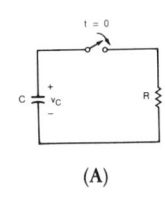

(A) [discharge circuit]

(C) $t_p > 5RC$

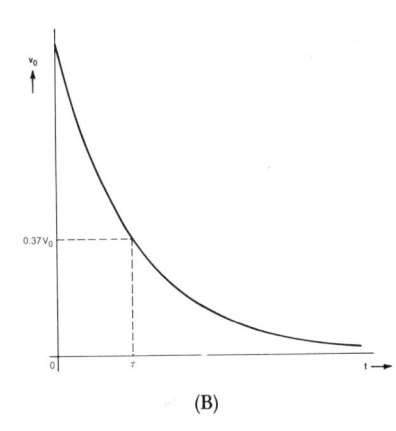

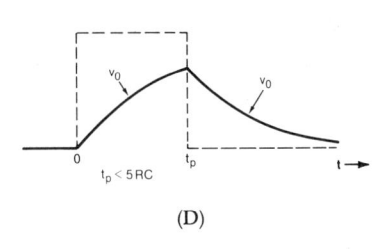

(B) [discharge curve, 0.37 V₀ at τ]

(D) $t_p < 5RC$

Bandwidth of an RC circuit is inversely proportional to RC.

INDUCTORS

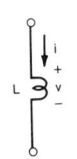

[inductor symbol]

$v_C = V_0 e^{-(t/RC)}$
Steady state value of $v_C = 0$
Time constant same as for charging. Time needed to discharge completely = 5 time constants.

v = L × (rate of change of i)
Energy stored in an inductor = $\frac{1}{2} Li^2$ joules

Current Continuity Condition

The current in an inductor cannot change suddenly; it can only change continuously.

Current Buildup in an Inductor

Steady state value of $i = (V_0/R)$
Time constant $= (L/R)$ seconds
Larger time constant → It takes longer to build up the current to (V_0/R)
Time taken to attain steady state = 5 time constants

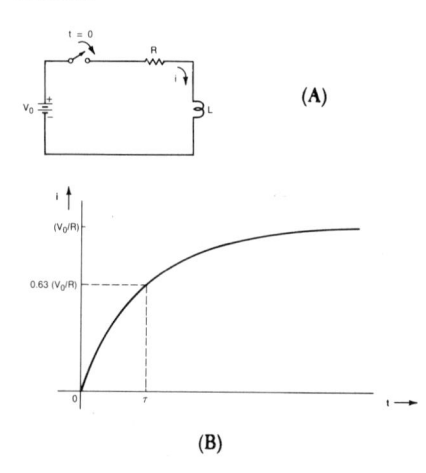

(A)

(B)

Current Decay in an Inductor

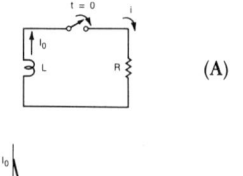

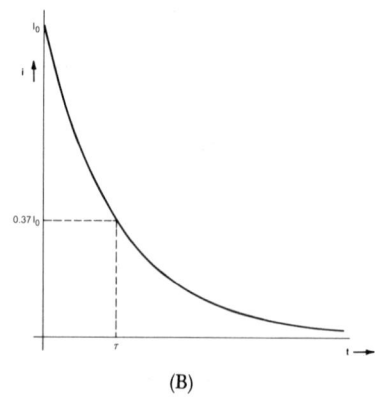

(A)

(B)

Steady-state value of $i = 0$
Time constant same as for the buildup
Time needed for the current to decay to zero is 5 time constants

AC CIRCUITS

Phasor Representation

$V_m \underline{/\theta}$ represents a sinusoidal voltage of peak value V_m and a phase angle of θ.

Impedance Z

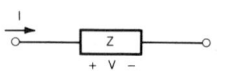

$$Z = \frac{V \text{ (in phasor form)}}{I \text{ (in phasor form)}}$$

Magnitude of Z:
$$|Z| = \frac{\text{amplitude of V}}{\text{amplitude of I}}$$

angle of Z = (angle of V) − (angle of I).

Ohm's Law for AC

V (phasor) = Z × I (phasor)

Resistors in AC Circuits

$$Z = R \underline{/0^0}$$

V and I are in phase with each other.

Capacitors in AC Circuits

$$Z = \frac{1}{2\pi fC} \underline{/-90^\circ}$$

The current *leads* the voltage by 90° in a capacitance.

Series RC Circuits

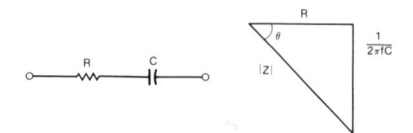

Impedance triangle.

Electronics: Circuits and Systems

FILTERS

High-Pass Filter (HPF)

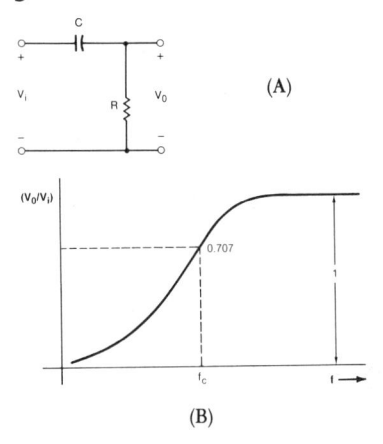

(A)

(B)

Cutoff (or -3 dB) frequency $f_c = (1/2\pi RC)$

Low-Pass Filter (LPF)

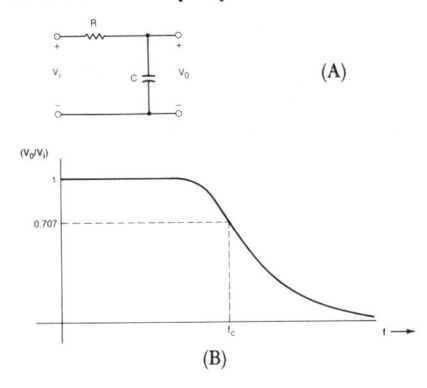

(A)

(B)

Cutoff (or -3 dB) frequency f_c given by the same formula as for the HPF.

Frequency Response and Pulse Response of Filters

Pulse response: variation of the filter output as a function of *time*.
Frequency response: variation of the filter output as a function of frequency.

LPF: To reduce the rounding off of the pulse output, the cutoff frequency f_c has to be increased.

HPF: To reduce the sag in the pulse output, the cutoff frequency of the HPF has to be decreased.

THEVENIN EQUIVALENT CIRCUIT

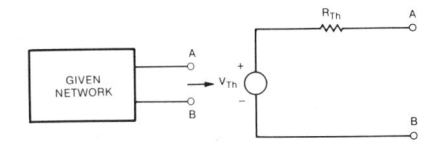

V_{Th} = open circuit voltage across the terminals A-B of the given network

R_{Th} = resistance measured at terminals A-B when the independent sources in the network have been made inactive

Maximum Power Transfer

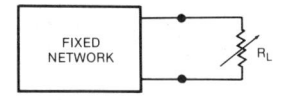

For a network with a fixed R_{Th} and a variable load resistance R_L, we must make $R_L = R_{Th}$ in order to obtain maximum power output.

IDEAL TRANSFORMERS

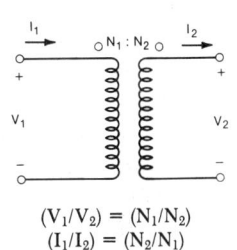

$(V_1/V_2) = (N_1/N_2)$
$(I_1/I_2) = (N_2/N_1)$

The higher voltage appears on the side with the larger number of turns.

Reflected Resistance

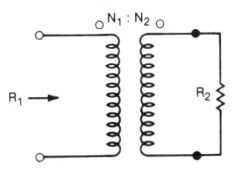

$R_1 = (N_1/N_2)^2 R_2$

The reflected resistance is larger when seen from the side with with the larger number of turns.

ANSWERS TO DRILL PROBLEMS

2.1 (a) 8 A; (b) 35 A.

2.2 (a) $V_1 - V_2 + V_3 - V_4 - V_x = 0$; $V_x = 160$ V; (b) $-V_x + V_a - V_b - V_c = 0$; $V_x = -100$ V.

2.3 (a) $V_P = -100 + 120 - 80 = -60$ V; (b) $V_P = 150 + 75 = 225$ V.

2.4 $R_T = 40$ Ω; $I = 2$ A; $V_1 = 24$ V; $V_2 = 40$ V; $V_3 = 16$ V; $P_1 = 48$ W; $P_2 = 80$ W; $P_3 = 32$ W.

2.5 $I = (60/40) = 1.5$ A; $R_T = 100$ Ω; Voltages: 150 V, 15 V, 60 V, 75 V; 50 $I^2 = 200$; $I = 2$ A; Voltages: 200 V, 20 V, 80 V, 100 V.

2.6 $V_1 = (10/14) \times 28 = 20$ V; $V_2 = (4/14) \times 28 = 8$ V.

2.7 $[(15/R_2 + 15) \times 100] = 35$; $15 = 0.35 R_2 + 5.25$; $R_2 = 27.9$ Ω.

2.8 $(R_1/90) \times 360 = 78$; $R_1 = 19.5$ Ω; $R_2 = (90 - 19.5) = 70.5$ Ω.

2.9 $R_T = 7.2$ Ω; $I_{12} = (18/30) \times 50 = 30$ A; $I_{18} = 20$ A.

2.10 $[24/(R_2 + 24)] \times 18 = 12$; $R_2 = 12$ Ω.

2.11 $9 R_1 = 23 R_2 = 150$; $R_1 = 16.7$ Ω. $R_2 = 6.52$ Ω.

2.12 (a) $R_T \approx 100$ Ω, $I \approx 1$ A, exact $I = 0.917$ A, error = 9%; (b) $R_T \approx 2000$ Ω, $I \approx 0.05$ A, exact $I = 0.04762$ A, error = 5%; (c) No approximations, $R_T = 120$ Ω, $I = 0.833$ A.

2.13 (a) $R_T \approx 9$ Ω, $V \approx 180$ V, exact $R_T = 8.26$ Ω, exact $V = 165$ V, error = 9%; (b) $R_T \approx 100$ Ω, $V \approx 2000$ V, exact $R_T = 95.2$ Ω, exact $V = 1905$ V, error = 5%; (c) No approximations, $R_T = 16.7$ Ω, $V = 333$ V.

2.14 $\tau = 37.5$ s; $t_r = 82.5$ s; $V_0 = 500$ V; time for full charge = 187.5 s, energy at full charge = 31.25 J.

2.15 $\tau = (t_r/2.2) = 4.55$ ms; energy = $(CV_0^2/2) = 0.05 \times 10^{-3}$ C = 6.25 $\times 10^{-8}$ = 62.5 nF, R = 72.7 kΩ.

2.16 $\tau = (20/5) = 4$ ns; $R = (4 \times 10^{-9}/25 \times 10^{-12}) = 160$ Ω.

2.17 $\tau = 0.14$ ms, time for full charge = 0.56 ms, 300 V = $0.545 V_0$ occurs at $0.8\tau = 0.112$ ms, t = 0.25 ms = 1.79τ gives $0.84 V_0 = 462$ V.

2.18 $\tau = 2$ s; (a) time for full discharge = 10 s; (b) 280 V = $0.737 V_0$ occurs at $0.3\tau = 0.6$ s; (c) 1.6 s = 0.8τ gives $0.45 V_0 = 171$ V.

2.19 Time to discharge from $V_0 = 7.5$ V to 2 V (= $0.267 V_0$) is 1.3τ. When $R_p = 10$ kΩ, $\tau = 50$ ms and shutter is open for 65 ms. When $R_p = 1$ kΩ, $\tau = 5$ ms and shutter is open for 6.5 ms.

2.20 $\tau = 1.82 \times 10^{-5}$ s, $I_0 = 0.545$ A. Time for steady state = 9.10×10^{-5} s. Energy stored = 1.49 mJ.

2.21 $\tau = 5$ ms = (L/R), R = 0.8 kΩ.

2.22 $\tau = 6 \times 10^{-5}$ s, $I_0 = 8$ A, 3 A = $0.375 I_0$ occurs at $0.47\tau = 2.82 \times 10^{-5}$ s. At 0.09 ms = 1.5τ, current = $0.775 I_0 = 6.2$ A.

2.23 $\tau = 18$ ms. 4 A = $0.5 I_0$ occurs at $0.7\tau = 12.6$ ms. At 20 ms = 1.1τ, current = $0.325 I_0 = 2.6$ A.

2.24 $I_0 = 25$ A. $\tau = 20$ ms. 4 A = $0.16 I_0$ occurs at $0.2\tau = 4$ ms.

Electronics: Circuits and Systems

2.25 (a) $V_1 = 25 \underline{/45°}$ V. $V_2 = 95 \underline{/84°}$ V. V_2 leads by $39°$; (b) $V_1 = 160 \underline{/0°}$ V. $I_1 = 15 \underline{/-60°}$ A. V_1 leads by $60°$; (c) $I_1 = 2.5 \underline{/-53.1°}$ A. $I_2 = 2.5 \underline{/-36.9°}$ A. I_2 leads by $16.2°$.

2.26 $I = 21 \underline{/36°}$ A.

2.27 (a) $Z = (30/2.5) \underline{/-45° - 15°} = 12 \underline{/-60°}$ Ω; (b) $Z = (10/40) \underline{/0° - (-36°)} = 0.25 \underline{/36°}$ Ω.

2.28 (a) $Z = 12.5 \underline{/40°}$ Ω; (b) $I = 2 \underline{/83.1°}$ A; (c) $V = 300 \underline{/-90°}$ V.

2.29 (a) $Z_C = 3.18 \times 10^4 \underline{/-90°}$ Ω. $I = 3.77 \times 10^{-3} \underline{/90°}$ A; (b) $Z_C = 31.8 \underline{/-90°}$ Ω; $I = 3.77 \underline{/90°}$ A.

2.30 (a) $Z_L = 6.28 \underline{/90°}$ Ω $I = 19.1 \underline{/-90°}$ A; (b) $Z_L = 1257 \underline{/90°}$ Ω, $I = 0.0955 \underline{/-90°}$ A.

2.31 (a) 500 Hz: $Z_C = 79.6 \underline{/-90°}$ Ω, $Z = 85 \underline{/-69.3°}$ Ω, $I = 1.41 \underline{/69.3°}$ A, $V_R = 42.4 \underline{/69.3°}$ V, $V_C = 112 \underline{/-20.7°}$ V; (b) 2000 Hz: $Z_C = 19.9 \underline{/-90°}$ Ω, $Z = 36 \underline{/-33.5°}$ Ω, $I = 3.33 \underline{/33.5°}$ A, $V_R = 100 \underline{/33.5°}$ V, $V_c = 66.2 \underline{/-56.5°}$ V.

2.32 (a) 1 Hz: $Z_C = 10.6 \times 10^3 \underline{/-90°}$ Ω, $Z = 10.7 \times 10^3 \underline{/-84.6°}$ Ω, $I = 9.38 \times 10^{-3} \underline{/84.6°}$ A, $V_0 = 9.38 \underline{/84.6°}$ V; (b) 10 Hz: $Z_C = 1.06 \times 10^3 \underline{/-90°}$ Ω, $Z = 1.46 \times 10^3 \underline{/-46.7°}$ Ω, $I = 0.0686 \underline{/46.7°}$ A, $V_0 = 68.6 \underline{/46.7°}$ V; (c) 100 Hz: $Z_C = 0.106 \times 10^3 \underline{/-90°}$ Ω, $Z = 1.00 \times 10^3 \underline{/-6.06°}$ Ω, $I = 9.95 \times 10^{-2} \underline{/6.06°}$ A. $V_0 = 99.5 \underline{/6.06°}$ V,

2.33 $C = (1/2\pi f_c R) = 0.995$ μF.

2.34 (a) 5 Hz: $Z_C = 3183 \underline{/-90°}$ Ω, $Z = 3222 \underline{/-81.1°}$ Ω, $I = 3.1 \times 10^{-2} \underline{/81.1°}$ A. $V_0 = 98.8 \underline{/-8.9°}$ V; (b) 50 Hz: $Z_C = 318.3 \underline{/-90°}$ Ω. $Z = 593 \underline{/-32.5°}$ Ω. $I = 0.169 \underline{/32.5°}$ A. $V_0 = 53.7 \underline{/-57.5°}$ V; (c) 500 Hz: $Z_C = 31.83 \underline{/-90°}$ Ω, $Z = 501 \underline{/-3.6°}$ Ω, $I = 0.2 \underline{/3.6°}$ A, $V_0 = 6.35 \underline{/-86.4°}$ V, cutoff frequency $f_c = 31.83$ Hz; $Z_C = 500 \underline{/-90°}$ Ω, $Z = 707 \underline{/-45°}$ Ω, $I = 0.1414 \underline{/45°}$ A, $V_0 = 70.7 \underline{/-45°}$ V.

2.35 (a) $V_{Th} = 112.5$ V, $R_{Th} = 18.75$ Ω; (b) $V_{Th} \approx 100$ V, $R_{Th} \approx 10$ Ω; (c) $V_{Th} = 0.99$ V, $R_{Th} \approx 10$ Ω.

2.36 $V_{Th} = 48$ V, $R_{Th} = 6$ Ω, $V_L = 12$ V, 19.2 V, 24 V, 27.4 V, 30 V, 32 V.

2.37 (a) $(50/4)^2 \times 80 = 12.5$ kΩ; (b) $(4/50)^2 \times 25 \times 10^3 = 160$ Ω; (c) 39.1 Ω; (d) 10.2 MΩ.

2.38 (a) $(N_1/N_2)^2 = 10^4$, $(N_1/N_2) = 100$; (b) $(N_1/N_2)^2 = 5000$. $(N_1/N_2) = 70.7$.

2.39 $(N_a/N_b)^2 \times 4000 = 10 \times 10^3$, $(N_a/N_b) = 1.58$, $(N_c/N_d)^2 \times 16 = 20 \times 10^3$, $(N_c/N_d) = 35.4$.

PROBLEMS

Kirchhoff's Laws

2.1 Calculate the current marked I_x in each of the circuits in Fig. 2-P1.

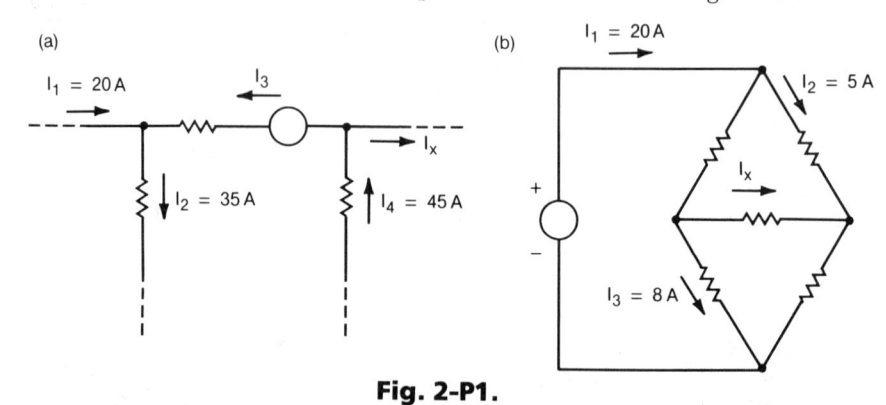

Fig. 2-P1.

2.2 Calculate the voltage marked V_x in each of the circuits in Fig. 2-P2.

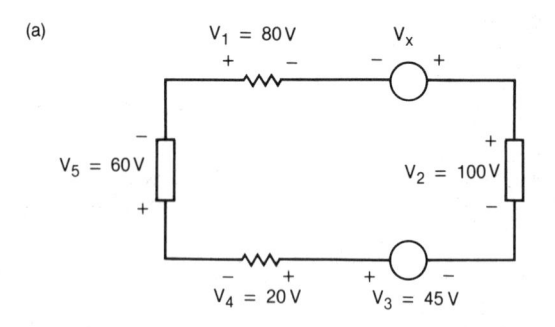

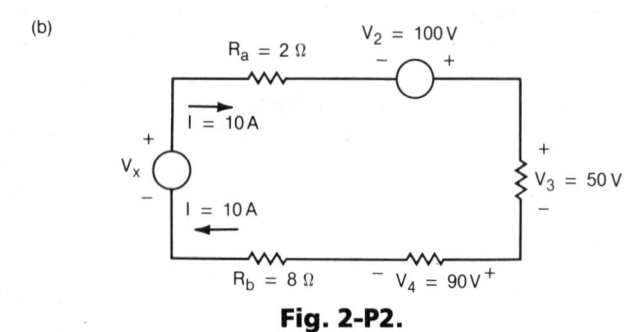

Fig. 2-P2.

2.3 Calculate the voltages at points A, B, C, D in each of the circuits in Fig. 2-P3, measured with respect to ground.

Electronics: Circuits and Systems

(a)

$V_1 = 20\,V$, $V_2 = 40\,V$, $V_3 = 80\,V$, $V_4 = 50\,V$

(b)

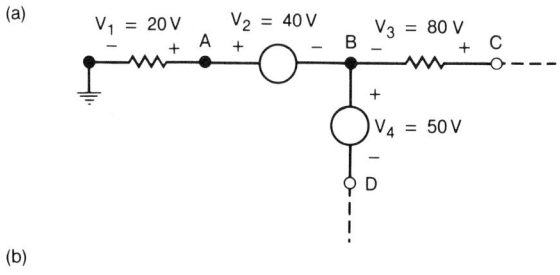

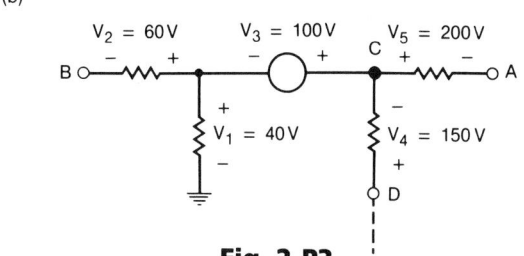

$V_2 = 60\,V$, $V_3 = 100\,V$, $V_5 = 200\,V$, $V_1 = 40\,V$, $V_4 = 150\,V$

Fig. 2-P3.

2.4 Calculate the voltage at point P in each of the circuits in Fig. 2-P4, measured with respect to ground.

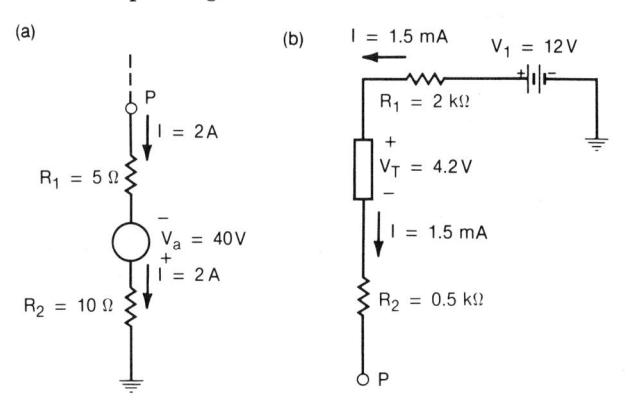

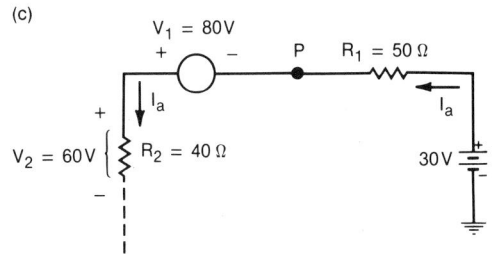

(a) $R_1 = 5\,\Omega$, $I = 2\,A$, $R_2 = 10\,\Omega$, $V_a = 40\,V$, $I = 2\,A$

(b) $I = 1.5\,mA$, $V_1 = 12\,V$, $R_1 = 2\,k\Omega$, $V_T = 4.2\,V$, $I = 1.5\,mA$, $R_2 = 0.5\,k\Omega$

(c) $V_1 = 80\,V$, $V_2 = 60\,V$, $R_2 = 40\,\Omega$, $R_1 = 50\,\Omega$, $30\,V$

Fig. 2-P4.

2.5 Calculate the current marked I_x in each of the circuits in Fig. 2-P5.

156   Review of Electric Circuits

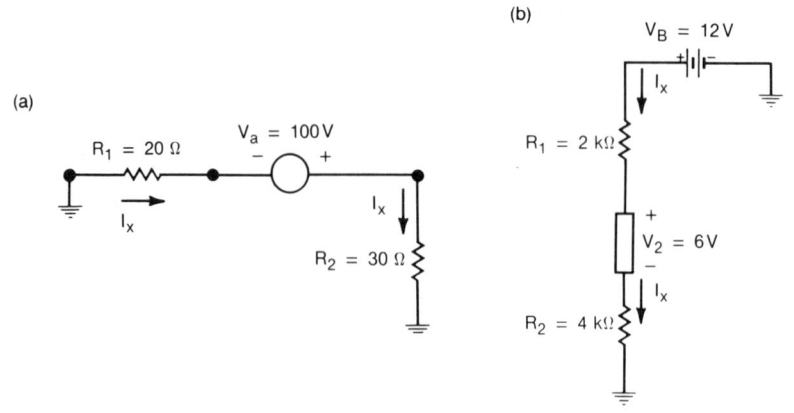

Fig. 2-P5.

Resistors

2.6   Calculate the voltages marked V_1 and V_2 in each of the connections shown in Fig. 2-P6.

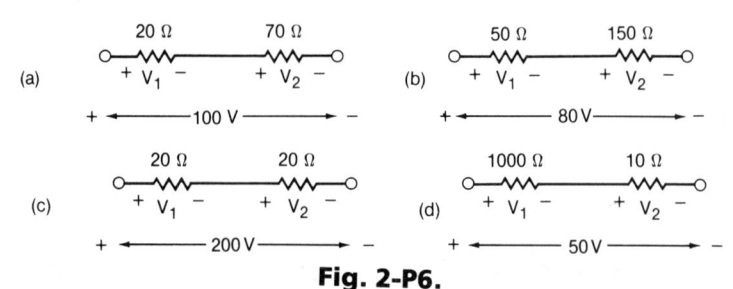

Fig. 2-P6.

2.7   Calculate the voltages marked V_1 and V_2 and the current I in each of the circuits of Fig. 2-P7.

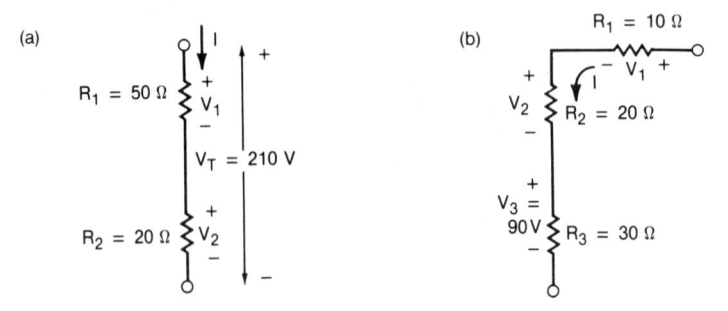

Fig. 2-P7.

2.8   Calculate the voltages marked V_2 and V_T and the current I in each of the circuits of Fig. 2-P8.

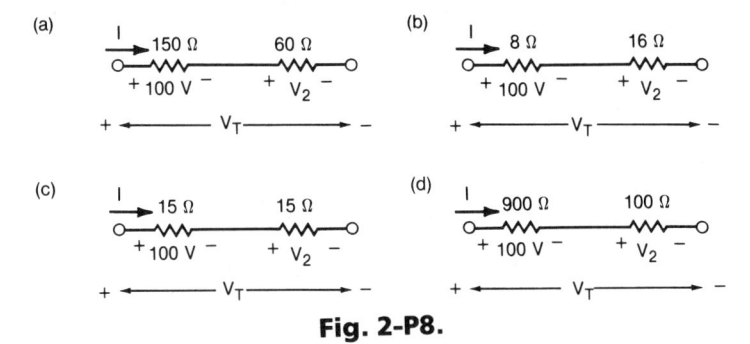

Electronics: Circuits and Systems 157

Fig. 2-P8.

2.9 Design a voltage divider to meet the following specifications: available voltage source = 150 V, voltage across one of the resistors = 43 V, current in the circuit not to exceed 2 A.

2.10 For each of the parallel combinations in Fig. 2-P10, calculate the total resistance.

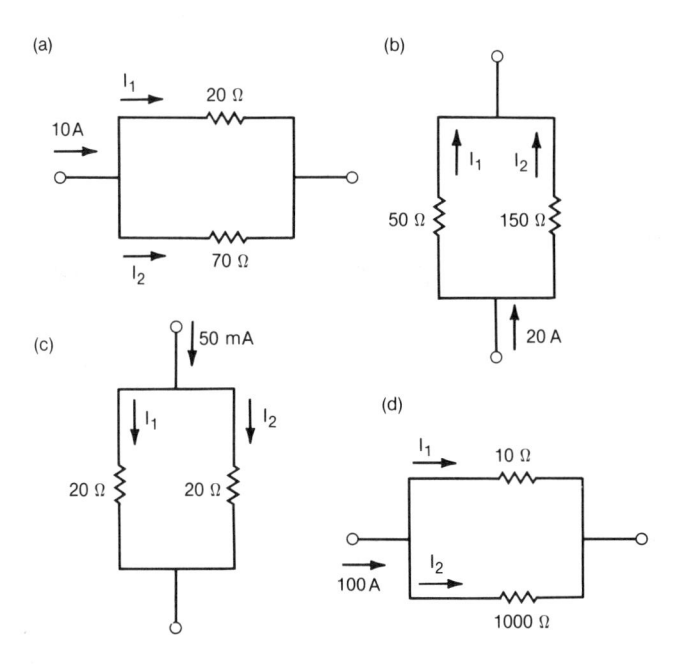

Fig. 2-P10.

2.11 Calculate the currents I_1 and I_2 in each of the connections of Fig. 2-P10.

2.12 Calculate the currents marked I_2 and I_T and the voltage V_T in each of the circuits of Fig. 2-P12.

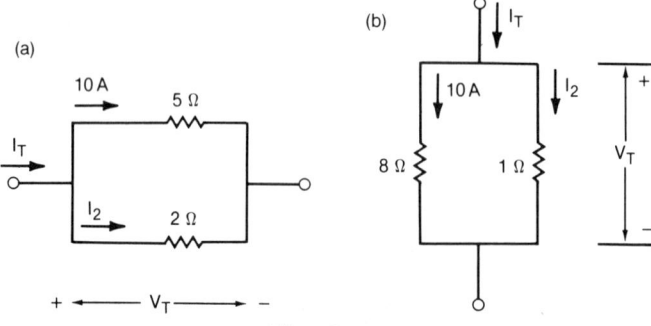

Fig. 2-P12.

2.13 Design a current divider to split a current of 18 A into 7 A and 11 A. The voltage is not to exceed 250 V.

2.14 Use valid approximations and calculate the total resistance, the current, and the voltage across the smaller resistance in each of the circuits shown in Fig. 2-P14.

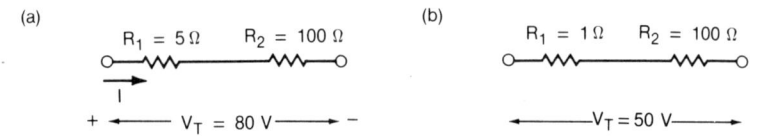

Fig. 2-P14.

2.15 Use valid approximations and calculate the total resistance, the voltage, and the current in the larger resistance in each of the circuits shown in Fig. 2-P15.

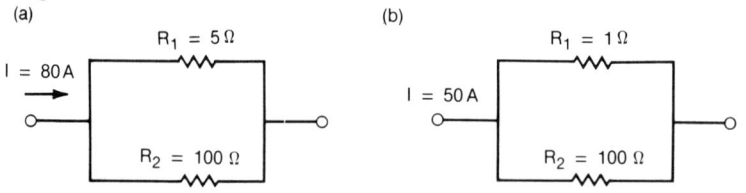

Fig. 2-P15.

Capacitors

2.16 A 50-μF capacitor has a charge of 15×10^{-6} coulomb. Calculate the voltage on the capacitor and the energy stored.

2.17 The energy stored in a capacitor is 6 J, and the voltage is 550 V. Calculate the value of the capacitance and the charge.

2.18 A capacitor of 40 μF and a 15-kΩ resistor are connected in series with a 160-V battery and a switch. The switch closes at t = 0, and the capacitor starts charging. Calculate (a) the time constant; (b) the final voltage on

Electronics: Circuits and Systems

the capacitor; (c) the time taken to reach full charge; (d) the energy stored at full charge.

2.19 Design a series RC circuit that takes 15 ms to reach the final voltage of 200 V if the energy stored in the capacitor is 5 J at full charge.

2.20 In the circuit of Fig. 2-P20, calculate the time taken for the capacitor voltage to reach (a) 10 V; (b) 20 V; (c) 40 V.

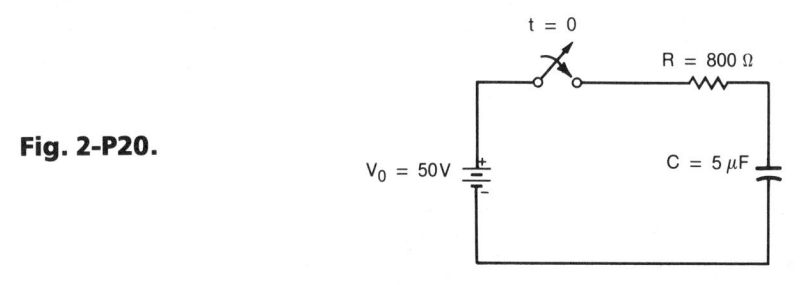

Fig. 2-P20.

2.21 In the circuit of the previous problem, calculate the voltage at each of the following instants of time: (a) 2 ms; (b) 4 ms; (c) 24 ms.

2.22 The voltage across the capacitor in Fig. 2-P22 is made to activate an external circuit when its voltage reaches a value of 25 V. Calculate the delay between the closing of the switch and the activation of the external circuit for each of the following values of V_s: (a) 50 V; (b) 75 V; (c) 200 V.

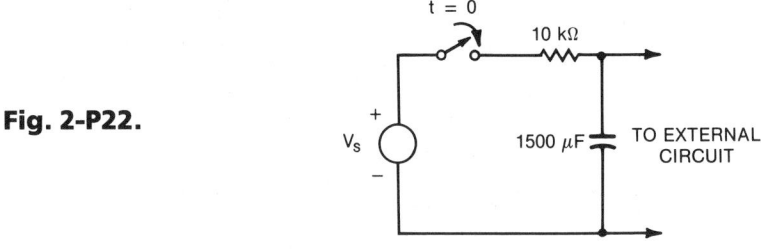

Fig. 2-P22.

2.23 A capacitor with an initial voltage of 50 V is connected to a resistance of 100 kΩ through which it discharges. If it takes 50 ms for a complete discharge, calculate the capacitance.

2.24 In the circuit of the previous problem, find the instant of time at which each of the following voltages is reached: (a) 20 V; (b) 40 V.

2.25 A capacitor of 250 pF has an initial energy stored equal to 10 mJ. It is connected to a resistance of 5 MΩ, through which it discharges starting at t = 0. Calculate the voltage across it at each of the following instants of time: (a) 0; (b) 0.5 ms; (c) 2.5 ms.

2.26 In the arrangement of the previous problem, calculate the time taken for the energy stored to decrease to (a) 50% of its initial value; (b) 10% of its initial value.

2.27 In the arrangement of Fig. 2-P27, the voltage on the capacitor is used to activate and deactivate an external mechanism. When the capacitor voltage reaches a value of 10 V, it activates the external mechanism. At the same instant of time, it also pulls the switch to position 2 (through some coupling not shown) so that the capacitor starts discharging (from an initial value of 10 V). When the capacitor has discharged to a voltage of 2 V, the external mechanism is deactivated, while the capacitor continues to discharge to zero. Calculate the interval for which the external mechanism remains activated.

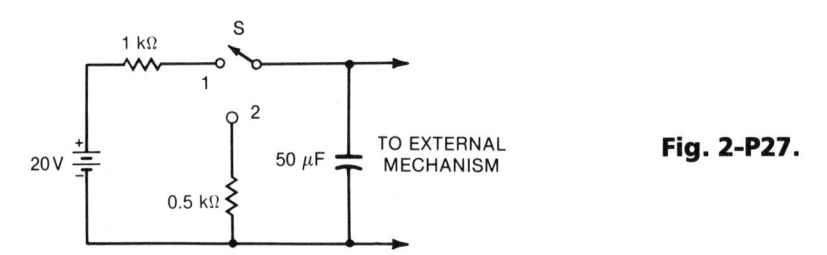

Fig. 2-P27.

2.28 The pulse shown in Fig. 2-P28 is applied to the given RC circuit. Draw a neat and carefully plotted graph of the output voltage as a function of time for each of the following cases: (a) $\tau = 5$ ms; (b) $\tau = 20$ ms.

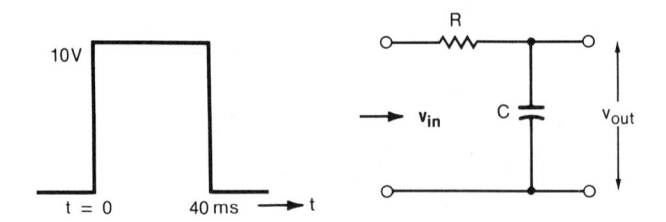

Fig. 2-P28.

Inductors

2.29 A 50-mH inductor has a current of 10 A flowing in it. Calculate the energy stored. Calculate the value of the current that will increase the energy stored by a factor of 10.

2.30 A 160-mH inductor in series with a 4-Ω resistor is connected to an 80-V battery through a switch that closes at t = 0. Calculate (a) the time constant; (b) the steady-state current; (c) the time taken to reach the steady state.

2.31 A series RL circuit connected to a 12-V battery has a steady-state current

Electronics: Circuits and Systems

of 50 mA. If the time taken to reach the steady state is 30 ms, calculate R and L.

2.32 In the circuit of Fig. 2-P32, calculate the time at which the current reaches a value of (a) 2 A; (b) 4 A; (c) 8 A.

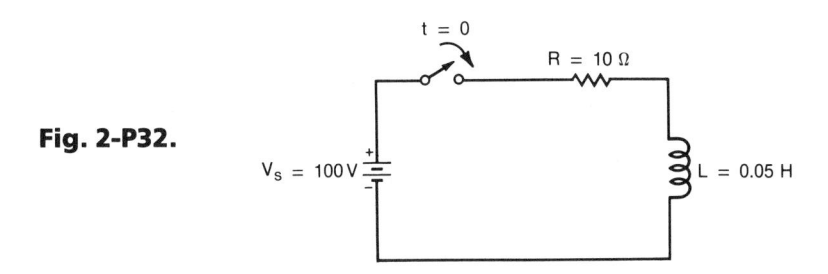

Fig. 2-P32.

$t = 0$, $V_s = 100$ V, $R = 10\ \Omega$, $L = 0.05$ H

2.33 An 8-μH inductor has an initial current of 25 A and the current is made to decay through a 55-Ω resistor starting at t = 0. Determine (a) the initial energy stored in the inductor; (b) the time taken for the stored energy to decrease to 40% of its initial value.

2.34 In the circuit of the previous problem, find the value of the stored energy at (a) t = 0.1 μs; (b) t = 0.32 μs.

AC Circuits

2.35 Write the phasor for each of the ac voltages or currents: (a) voltage of amplitude 120 V with a phase angle of 0°; (b) current of amplitude 16 A lagging the voltage in (a) by 43°; (c) current with twice the amplitude of that in (b) and leading that current by 80°.

2.36 Given the voltages $V_1 = 220\ \underline{/-45°}$ V, and $V_2 = 140\ \underline{/30°}$ V, and the currents $I_a = 4\ \underline{/-36°}$ A and $I_b = 9\ \underline{/74°}$ A (all being at the same frequency), calculate the phase angle between every possible pair of the previous functions. In each case, state which function leads.

2.37 In each set of values given, calculate the missing item (V, I, or Z). Assume that the voltage and current are at the same frequency in each set. (a) $V_1 = 160\ \underline{/75°}$ V, $I_1 = 15\ \underline{/0°}$ A; (b) $V_2 = 480\ \underline{/-48°}$ V, $I_2 = 25\ \underline{/-48°}$ A; (c) $V_3 = 60\ \underline{/30°}$ V, $Z_3 = 6\ \underline{/-30°}\ \Omega$; (d) $I_4 = 14\ \underline{/70°}$ A, $Z_4 = 6\ \underline{/45°}\ \Omega$; (e) $I_5 = 14\ \underline{/-70°}$ A, $Z_5 = 6\ \underline{/45°}\ \Omega$.

2.38 Calculate the impedance of a 50-pF capacitor at each of the following frequencies: (a) 100 Hz; (b) 10^4 Hz; (c) 10^6 Hz; (d) 10^{12} Hz.

2.39 Calculate the impedance of a 40-mH inductor at each of the following frequencies: (a) 100 Hz; (b) 10^3 Hz; (c) 10^8 Hz.

2.40 A 50-μF capacitor and a 60-Ω resistor are in series across a voltage source of 100 $\underline{/0°}$ V. Calculate the total impedance, the current, and the voltage across each component for each of the following frequencies: (a) 256 Hz; (b) 50 Hz; (c) 5 Hz.

2.41 A 16-mH inductor and a 100-Ω resistor are in series across a voltage source of 100 /0° V. Calculate the total impedance, the current, and the voltage across each component for each of the following frequencies: (a) 100 Hz; (b) 1 kHz; (c) 10 kHz.

2.42 A 50-Ω resistor is in series with a 500-pF capacitor. Calculate the frequency at which the angle of the total impedance of the circuit is (a) 22.5°; (b) 45°; (c) 67.5°. In each case, calculate the magnitude of the total impedance also.

2.43 A 50-Ω resistor is in series with a 10-mH inductor. Repeat the calculations of the previous problem for this circuit.

Filters

2.44 A high-pass filter is made up of R = 150 kΩ and C = 1500 pF. Calculate the ratio of the output to input voltages at (a) 100 Hz; (b) 1 kHz; (c) 10 kHz.

2.45 Calculate the cutoff frequency of the filter in the previous problem. Calculate the output voltage (in terms of the input voltage) at that frequency. Verify that the angle between the input and output voltages is 45°.

2.46 A low-pass filter is made up of R = 500 Ω and C = 10 μF. Calculate the ratio of the output to input voltages at (a) 5 Hz; (b) 50 Hz; (c) 500 Hz.

2.47 Calculate the cutoff frequency of the LPF of the previous problem. Calculate the ratio of the output voltage to the input voltage at that frequency and verify that the phase angle between them is 45°.

2.48 Calculate the cutoff frequency of each of the following series RC circuits: (a) time constant of the RC circuit = 10 ns; (b) rise time of the RC circuit is 100 ns.

2.49 A low-pass filter has a cutoff frequency of 1 kHz. Suppose an input pulse of duration 2 ms is applied to the input. Sketch the output voltage as a function of time. Repeat when the input pulse has a duration of 0.2 ms. Assume the pulse amplitude to be 100 V.

Thevenin Equivalent and Maximum Power Transfer

2.50 Obtain the Thevenin equivalent of the circuit in Fig. 2-P50 as seen from (a) the terminals marked A and B; (b) the terminals marked C and D.

Electronics: Circuits and Systems

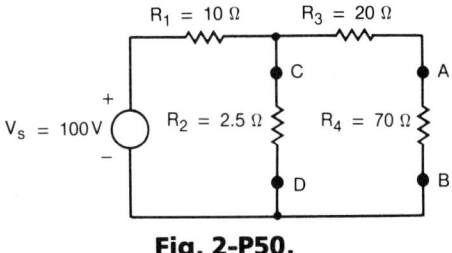

Fig. 2-P50.

$R_1 = 10\ \Omega$, $R_3 = 20\ \Omega$, $V_s = 100\ \text{V}$, $R_2 = 2.5\ \Omega$, $R_4 = 70\ \Omega$

2.51 Obtain the Thevenin equivalent of each of the circuits in Fig. 2-P51 as seen from the terminals A and B.

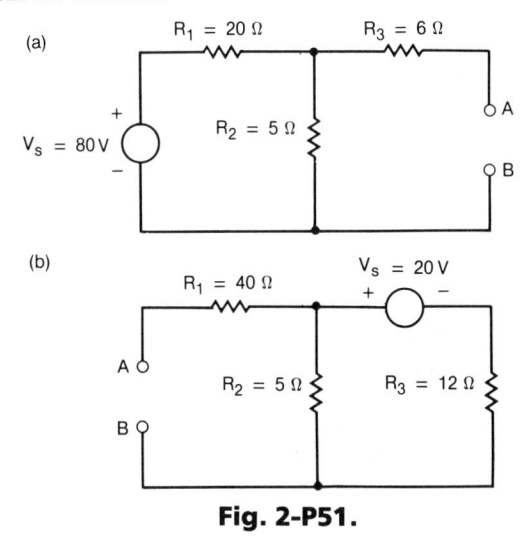

(a) $R_1 = 20\ \Omega$, $R_3 = 6\ \Omega$, $V_s = 80\ \text{V}$, $R_2 = 5\ \Omega$

(b) $R_1 = 40\ \Omega$, $V_s = 20\ \text{V}$, $R_2 = 5\ \Omega$, $R_3 = 12\ \Omega$

Fig. 2-P51.

2.52 The load resistance in the circuit of Fig. 2-P52 is varied in steps of 100 Ω starting from 50 Ω and stopping at 550 Ω. Calculate the voltage across the load for each case.

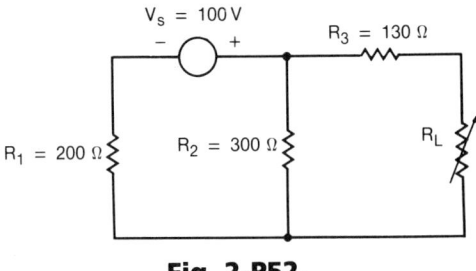

$V_s = 100\ \text{V}$, $R_3 = 130\ \Omega$, $R_1 = 200\ \Omega$, $R_2 = 300\ \Omega$, R_L

Fig. 2-P52.

2.53 Calculate the value of R_L in each of the circuits of Fig. 2-P53 so as to maximize the power output.

164   **Review of Electric Circuits**

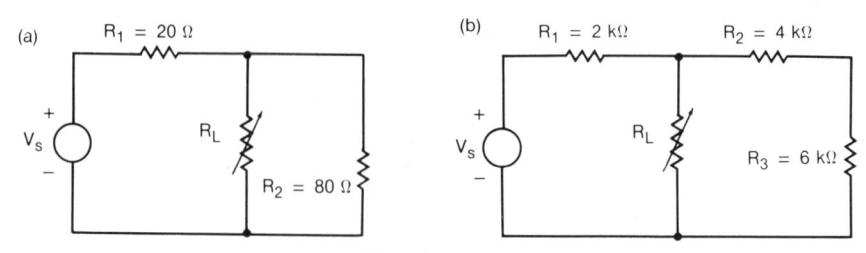

Fig. 2-P53.

2.54 In each of the cases shown in Fig. 2-P54, calculate the reflected resistance R_{ref}.

Fig. 2-P54.

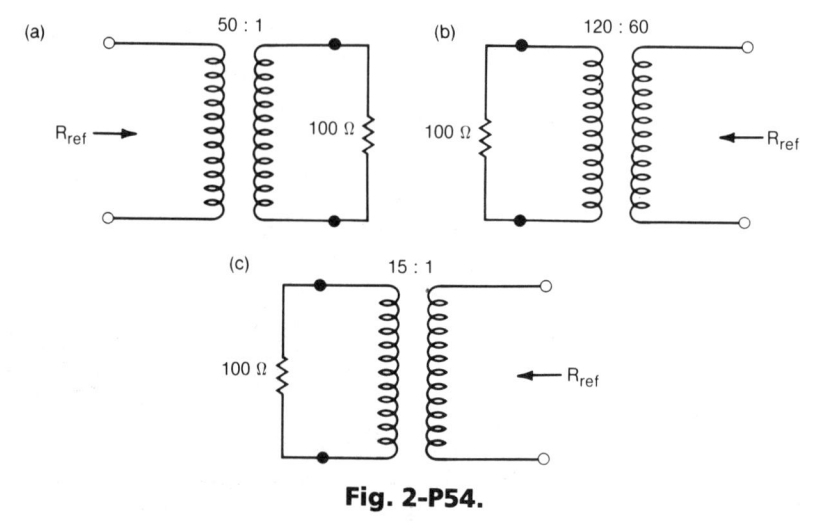

2.55 In each of the cases shown in Fig. 2-P55, calculate the turns ratio (N_2/N_1) or (N_1/N_2) as the case may be so as to make $R_{ref} = 1$ kΩ.

2.56 A two port has an input resistance of $R_{in} = 1$ MΩ and an output resistance of $R_{out} = 60$ kΩ. A signal source with a resistance of 100 kΩ is to be connected to the input port and a load resistance of 8 Ω is to be connected to the output port. Determine the turns ratios of the transformers needed for complete matching.

Fig. 2-P55.

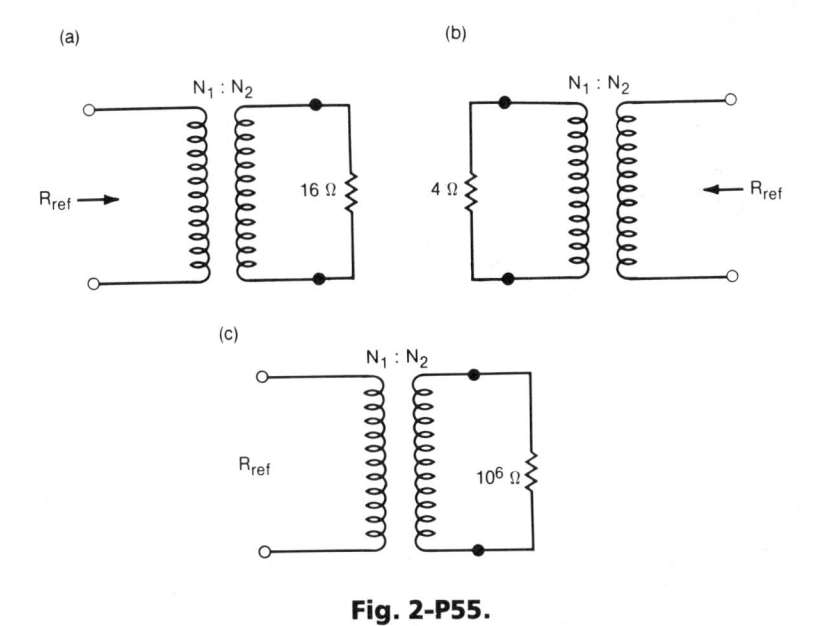

CHAPTER 3

SEMICONDUCTOR DEVICES AND INTEGRATED CIRCUITS

Modern electronic devices depend on the properties of certain types of materials called semiconductors, in which the concentrations and the polarity of the current carriers can be controlled. The flow of current in a semiconductor device is confined to extremely small areas, and it is, therefore, possible to fabricate such devices and circuits in microscopic dimensions. Integrated circuits contain highly complex electronic circuits in an area of only a few square millimeters.

This chapter starts with a discussion of the physical principles of operation of the semiconductor devices, which is followed by a study of the basic amplifier configurations. The chapter concludes with a discussion of the principles of operation and special features of integrated circuits.

SEMICONDUCTORS

Solids can be classified into three categories based on their ability to conduct electricity: *conductors, insulators,* and *semiconductors.* The differences in the electrical characteristics arise from the manner in which bonding forces are set up in the structure of the solid state of various materials.

The atom of an element can be modeled by a nucleus surrounded by concentric spherical shells with electronic orbits. The distribution of the electrons in the outermost shell of the atom of an element dictates the properties of the element. When the outermost shell is incomplete; that is, it does not have the full quota of electrons that can be accommodated, the number of electrons in that shell determines the *valence* of the element. When a solid is formed, there is a merger of the shells and orbits, and bonding forces are developed so as to maintain the solid-state structure.

In the case of conductors, such as copper, the electrons in the outermost shells of the individual atoms are only loosely bound to the parent atoms and form a supply of free electrons that are free to move even under the influence of a small external voltage. The conductivity is very high because of the abundant supply of free electrons.

In the case of insulators, the crystalline structure keeps the electrons in the outermost shells tightly bound to the parent atoms. The energy required to release the electrons from such a tight bond is quite high, and it needs a high voltage to cause any current flow. The insulators have extremely low conductivities.

A semiconductor lies between the conductors and the insulators in its ability to conduct electricity. In silicon (Si) and germanium (Ge), which are the two commonly used semiconductor materials, the outermost shell is incomplete and contains four electrons. The valence of these two elements is four. The crystalline structure of Si and Ge is due to the *covalent bond* developed between neighboring atoms. That is, each of the four valence electrons of an atom pairs off with one of the four valence electrons of a neighboring atom, as indicated in Fig. 3-1. The large circles with the plus signs represent the ions of Si or Ge. The small circles with the minus signs denote the valence electrons. The covalent bond is indicated by the curved lines between the valence electrons. The covalent bond is sufficiently strong, and it requires a significant amount of energy to release the valence electrons from the bond. A voltage of about 0.7 V in Ge and 1.1 V in Si is required to release the electrons. The energy corresponding to these voltages is not available from the thermal agitation of the atoms under normal room temperature. Consequently, the number of free electrons in Si or Ge is negligibly small, and the material is essentially an insulator. The conductivity of Si and Ge can be increased by the addition of other material in a carefully controlled manner, and such a process is called *doping*.

Doped Semiconductors

Doping of a semiconductor consists of making the atoms of a material (other than Si or Ge) take the place of some of the atoms of the original material in the crystalline structure. The new material is called a *substitutional impurity*. The concentration of the impurity atoms is extremely small compared with the original (Si or Ge) atoms in a crystal so that the crystal retains its original characteristics. The doping process involves the diffusion of a gas containing the impurity through a chamber in which the pure Si or Ge crystals are placed. Diffusion occurs at a high temperature, and the impurity atoms displace some of the original atoms.

N-Type Impurity

Phosphorus is a *pentavalent element;* that is, it has a valence of five, and its outermost shell has five valence electrons. When a pentavalent element is

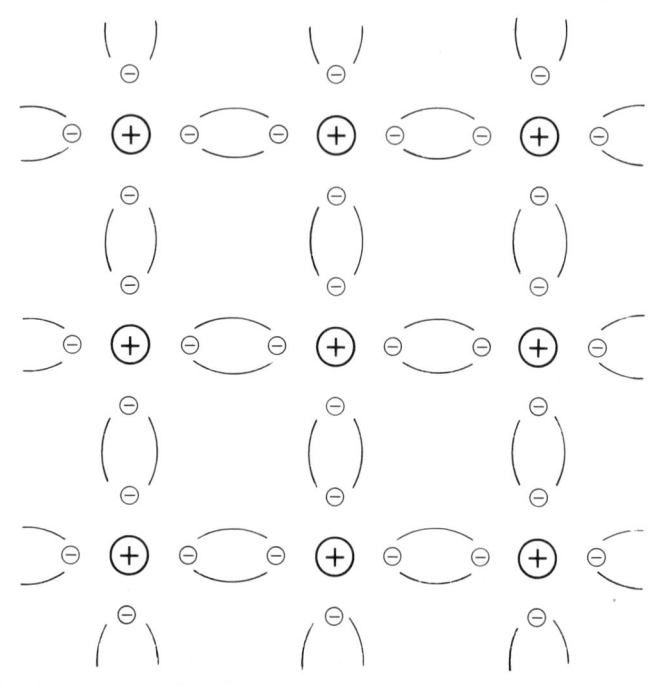

Fig. 3-1. Convalent bonds in a silicon crystal. Valence electrons are shared by neighboring atoms.

used as the impurity, and one of its atoms takes the place of a silicon atom in a crystal, four of the valence electrons are needed to maintain the covalent bonding of the crystalline structure. The fifth valence electron is not needed for the bond and hence is free, which is indicated in Fig. 3-2A. Each pentavalent atom gives rise to, or *donates*, a free electron to the material. The impurity is, therefore, called a *donor impurity*. Since the semiconductor has a supply of *free electrons*, which are negatively charged carriers of electricity, it is called an *n-type semiconductor*. The conductivity of the material can be controlled by controlling the concentration of the impurity diffused into the silicon crystal.

P-Type Impurity

Boron is a *trivalent element;* that is, it has a valence of three and has three electrons in the outermost shell of each atom. When a trivalent atom takes the place of a silicon atom in a silicon crystal, the three valence electrons are not sufficient to provide the bonding, and there is, consequently, a *vacancy*, as indicated in Fig. 3-2B. The vacancy is called a *hole*. It is found that the hole is free to move freely just like an electron, except that it moves like a *positively* charged carrier. The trivalent element is called an *acceptor impurity*

Electronics: Circuits and Systems

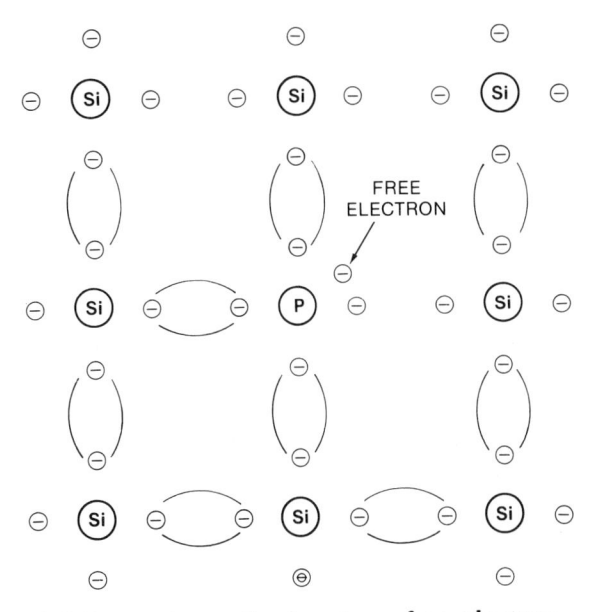

(A) N-type impurity donates a free electron.

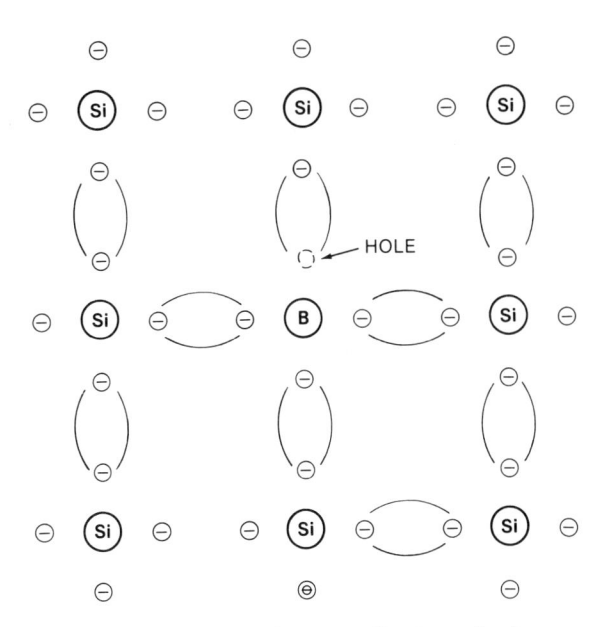

(B) P-type impurity contributes a hole.

Fig. 3-2. Effect of doping with an impurity.

since it has to borrow (or "accept") an electron from some other atom to complete the covalent bond. The semiconductor has a supply of positively charged carriers of current, or holes, and is called a *p-type semiconductor*. The conductivity of the material can be controlled by controlling the concentration of the impurity diffused into the silicon crystal.

This discussion has used silicon as the material, but the same applies to germanium also. Integrated circuits, however, use silicon crystals because they have more desirable properties (high purity, for example) than germanium.

The terms *intrinsic* and *extrinsic* are used in referring to semiconductors. An *extrinsic* semiconductor owes its electrical characteristics to the impurities injected into it, whereas an *intrinsic* semiconductor owes its electrical characteristics to electrons and holes generated due to thermal agitation from the silicon atoms of the crystal. An undoped silicon crystal is an intrinsic semiconductor.

Thermal Generation and Recombination

The electrons in the covalent bonds in a crystal gain energy due to thermal agitation caused by the ambient temperature. The distribution of the thermal energy among the electrons is of a statistical nature with some electrons having significantly higher energy than the average energy. Such electrons are able to overcome the bonding force and become free. There is a generation of a number of free electrons even at normal room temperature. When an electron leaves its bond, it leaves a vacancy there causing the formation of a hole. In any semiconductor (undoped or doped) electron-hole pairs are being continually generated due to thermal agitation.

The electron-hole pairs created by thermal energy do not, however, last indefinitely. They recombine either with their original counterparts or with other holes and electrons available in the material. When a recombination occurs, the original thermal energy that caused the generation of the pair is released. In many cases, the energy released is noticed in the form of heat. But in some cases, the energy released is in the form of a radiation either in the visible range or just outside the visible range. *Light-emitting devices* are based upon such materials.

There is a continual generation and recombination of electron-hole pairs in any semiconductor, which will be seen to have a significant effect on the behavior of semiconductor devices.

PN JUNCTION

The importance of doped semiconductors is due to the fact that the *polarity* as well as the *concentration* of the current carriers in a sample can be controlled by a suitable choice of the impurity and the level of doping. The basic device that exploits these two important aspects of semiconductors is the *pn junction*.

Even though an actual pn junction is not fabricated in the manner described in the following thought-experiment, this description is a convenient way of visualizing the electrical processes that occur in a pn junction.

Imagine a p-type sample and an n-type sample placed next to each other with a partition between them. The majority of carriers in the p-type are holes that are positive charges. When a hole leaves the parent atom, it leaves a negatively charged ion behind. The negative ions are too massive compared with the holes and can be considered immobile. There is also a certain number of free electrons on the p-side due to thermal generation of electron-hole pairs. Therefore, the p-type sample has a large number of holes, a small number of electrons, and a large number of negative ions. The total positive charge due to the holes is balanced by the total negative charge due to the electrons and the negative ions in the p-type sample, and it is electrically neutral. A similar situation exists in the n-type sample. There are a large number of electrons, a small number of holes, and a large number of positive ions (created by electrons leaving their parent atoms). The n-type sample is electrically neutral also: the total negative charge due to the free electrons is balanced by the total positive charge of the holes and the positive ions.

Depletion Region

Suppose that the partition is now removed bringing the two samples together. There is a *concentration gradient* of the holes (more numerous on the p side than on the n side), and the holes tend to diffuse from the p side to the n side. Similarly, there is a concentration gradient of electrons (more numerous on the n side than on the p side), and the electrons tend to diffuse from the n side to the p side. Such a diffusion is indicated in Fig. 3-3A. As the diffusion progresses, the electrical neutrality of the two sides changes. The p side becomes more and more negative electrically since it is losing holes and gaining electrons. The n side becomes more and more positive since it is losing electrons and gaining holes. In the neighborhood of the junction, a layer of positive ions that have lost electrons is formed on the n side, and a layer of negative ions that have lost holes is formed on the p side as indicated in Fig. 3-3B. The positive ion layer exerts a repulsive force on any holes that tend to migrate into the n region, and the negative ion layer exerts a repulsive force on any electrons that tend to migrate into the p region. Eventually, the force of repulsion is strong enough to stop the diffusion. The region of ions around the area of the junction is called the *depletion region* since it has been *depleted of free electrons and holes*. It is also called the *transition region* since it represents the transition from a p-type to an n-type semiconductor.

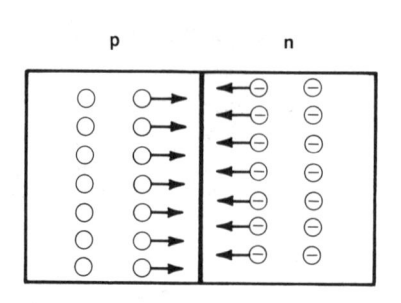

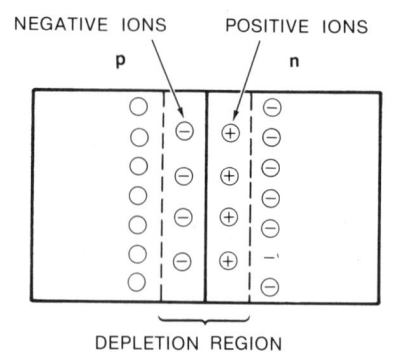

172   **Semiconductor Devices and Integrated Circuits**

(A) P side has a supply of holes, and n side has a supply of free electrons before the junction is formed.

(B) A space charge region (depletion region) is formed at the junction.

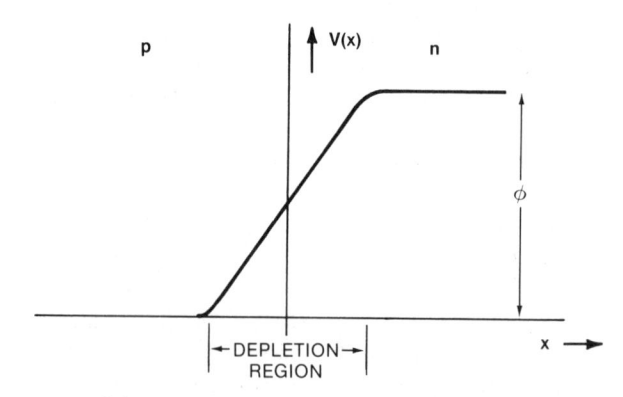

(C) Potential barrier at the junction.

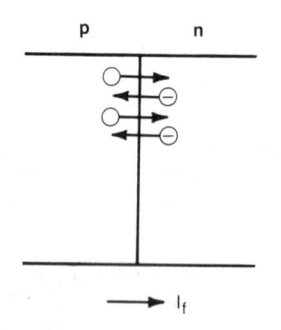

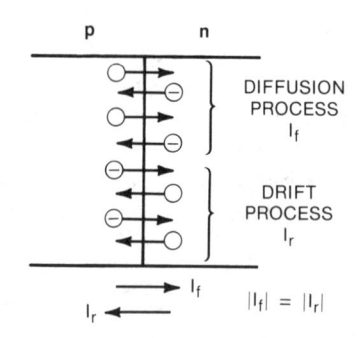

(D) Diffusion of carriers across the junction.

(E) Balance of drift and diffusion components of current: $I_r + I_f = 0$.

Fig. 3-3. Formation of a pn junction.

Potential Barrier

The charge due to the ions in the depletion region acts just like the charge stored on a capacitor and causes a voltage to appear across the region. If x is the distance measured from the junction, then the voltage due to the space charge in the depletion region varies with the distance in the form shown in Fig. 3-3C. The total voltage across the junction due to the space charge is denoted by ϕ. The potential distribution at the junction acts as a *potential barrier* that inhibits further diffusion of electrons from n to p and holes from p to n. Any electron or hole has to have enough energy to overcome the height of the potential barrier, ϕ, in order to diffuse to the other side. Statistically, a small number of electrons and holes do possess enough energy to overcome the potential barrier, and there is, consequently, a small amount of diffusion that persists in spite of the potential barrier, which gives rise to a small diffusion current component I_f from p to n as indicated in Fig. 3-3D. (Note that electrons going from n to p represent a *current* going from p to n.)

Balance Between Diffusion and Drift Components of Current

There is also a constant generation of electron-hole pairs in both regions due to thermal agitation as was discussed earlier. There is a certain amount of recombination of such pairs, but some of the electrons and holes created in this manner can drift to the other side. Consider an *electron* thermally generated in the *p-region*. It may recombine with a hole there, but it also faces a *favorable potential barrier* in the depletion region since the ion layer on the n side is positive and will attract electrons. Therefore, some of the electrons thermally generated in the p-region tend to drift into the n-region because of the favorable potential barrier. Similarly, a hole thermally generated in the n-region can drift into the p-region because of the favorable potential barrier. These two components due to the drifting of thermally generated electrons (from p to n) and holes (from n to p) cause a current I_r as indicated in Fig. 3-3E.

When a pn junction has been formed and it has no external connections, there can be no net current into or out of the material. But this is a *dynamic* equilibrium condition: there are two components of current inside the material I_f and I_r which *cancel each other*, as indicated in Fig. 3-3E. *I_f is due to the diffusion of electrons and holes that have a sufficient energy to overcome the potential barrier, and I_r is due to the drifting of electrons and holes thermally generated and taking advantage of the favorable potential barrier.*

Metal Semiconductor Contacts

It is necessary to place metal contacts on the two ends of a pn junction so that external circuit components can be connected to it. When a metal contact is formed on a semiconductor, the contact may be *ohmic* or *nonohmic* (which is also called *rectifying*). A nonohmic contact has the property that it conducts current more easily in one direction than in the other, and an ohmic contact conducts currents equally freely in either direction. Nonohmic contacts are used in certain special situations (such as the formation of *Schottky diodes* to be discussed later) but normally the metal-semiconductor contacts should be ohmic. When a metal-semiconductor contact is made, there is a *contact potential* created at the junction between the semiconductor and the metal. A pn junction with two external contacts is shown in Fig. 3-4. There are three voltages present: a voltage V_n due to the contact potential between the metal contact and the n-type materal, a voltage ϕ due to the potential barrier at the pn junction, and a voltage V_p due to the contact potential between the p-type material and the metal contact on it. The values and polarities of these three voltages are found to satisfy the equation:

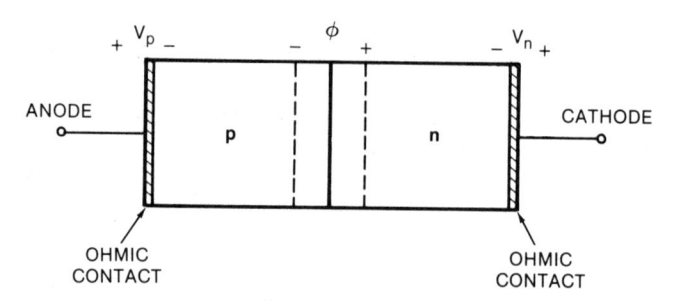

(A) Pn junction with two external contacts.

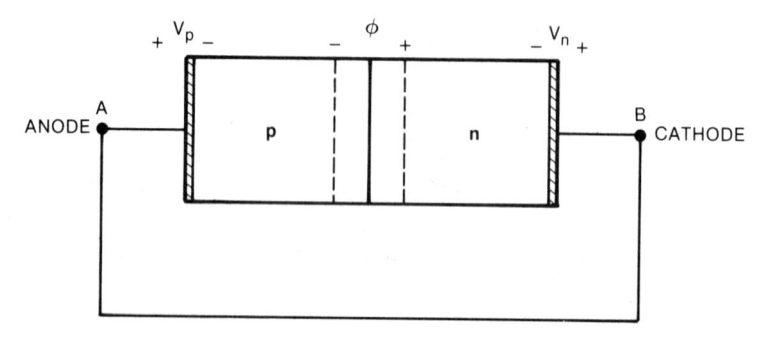

(B) Two external contacts connected with wire but no current flow.

Fig. 3-4. Contact potentials and junction potential at a pn junction.

Electronics: Circuits and Systems

$$V_p - \phi - V_n = 0 \qquad (3\text{-}1)$$

That is, the *net potential difference between the two terminals A and B is zero*. Therefore, if we now connect the two terminals A and B externally by a wire as indicated in Fig. 3-4B, no current will flow through it, again verifying that there is a balance between the diffusion component I_f and the drift component I_r in the pn junction.

PN Junction Under Forward Bias

When a battery is connected externally to the leads of a pn junction, the junction is said to be *biased*. If the *positive* terminal of the battery is connected to the *p-side* terminal of the device and the *negative* terminal of the battery is connected to the *n-side* terminal of the device, the junction is then *forward biased*, as indicated in Fig. 3-5A. Consider the voltages in the device of Fig. 3-5B: the voltage V_n between the n-type material and its metal contact and the voltage V_p between the p-type material and its metal contact have the same value as under no bias. Therefore, the applied voltage V_B must affect the voltage across the junction V_j. Using KVL, we have

$$V_B - V_p + V_j + V_n = 0 \qquad (3\text{-}2)$$

Combining Eqs. (3-1) and (3-2), we obtain,

$$V_j = \phi - V_B \qquad (3\text{-}3)$$

That is, the *potential barrier* at the junction has been *lowered* (from its height under no bias) *by the voltage V_B of the external forward bias*, as indicated in Fig. 3-5C. The lower potential barrier at the junction permits the diffusion of a larger number of holes from p to n and a larger number of electrons from n to p. Thus, the *diffusion component* of the current I_f *increases due to the forward bias*. The current I_f is found to *increase exponentially with the applied voltage V_B*. The drift component of the current I_r remains the same as under no bias since it depends only upon the temperature. The drift component becomes negligibly small compared with the diffusion component when the pn junction is forward biased. The total current in the pn junction is, therefore, essentially equal to I_f.

PN Junction Under Reverse Bias

If the *positive* terminal of a battery is connected to the *n-side* lead of the pn junction and the *negative* terminal of the battery is connected to the *p-side* lead of the pn function, the device is then *reverse biased*, as indicated in Fig. 3-6A. The voltages in the device of Fig. 3-6B then satisfy the KVL equation

$$-V_B - V_p + V_j + V_n = 0 \qquad (3\text{-}4)$$

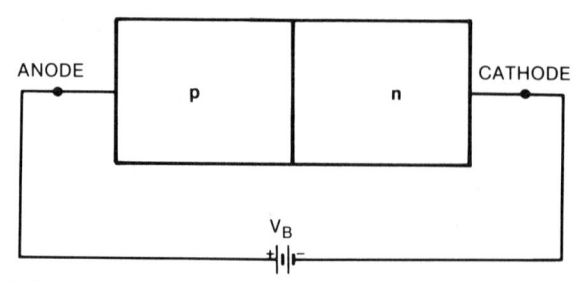

Semiconductor Devices and Integrated Circuits

(A) External bias connections for forward bias.

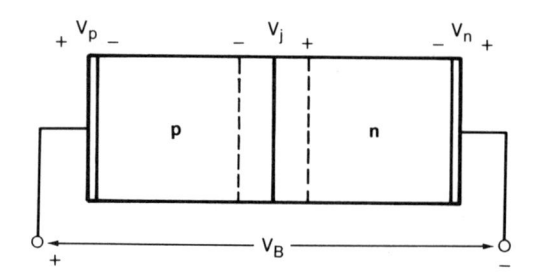

(B) Voltages in the device.

(C) Lowered potential barrier.

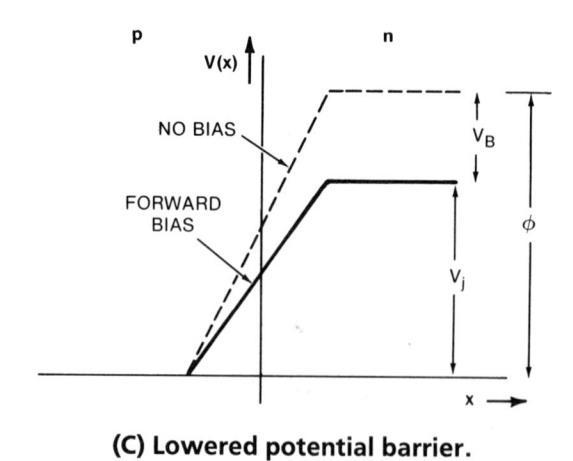

Fig. 3-5. Forward-biased pn junction.

which, in conjunction with Eq. (3-1), leads to

$$V_j = \phi + V_B \qquad (3\text{-}5)$$

That is, the *potential barrier* at the junction has been *raised* (relative to its

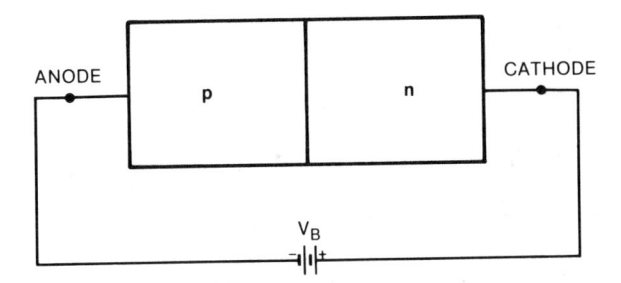

(A) External bias connections for reverse bias.

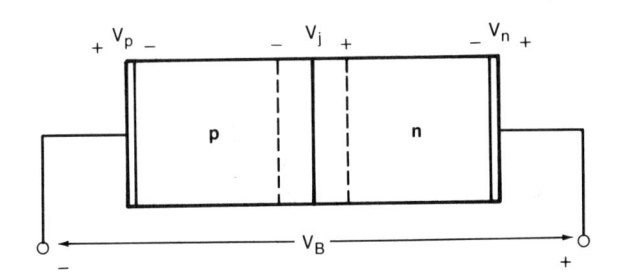

(B) Voltages in the device.

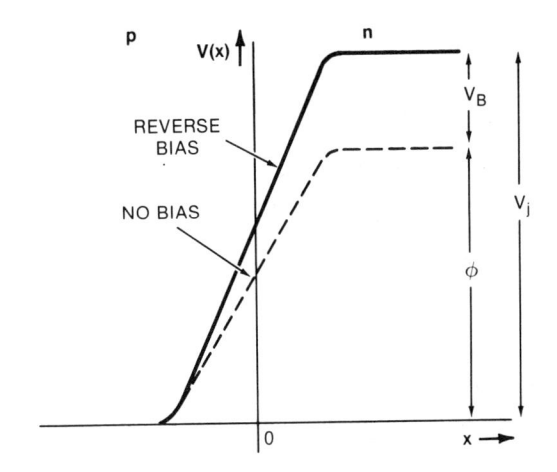

(C) Heightened potential barrier.

Fig. 3-6. Reverse-biased pn junction.

height under no bias) *by the voltage V_B of the external reverse bias*, as indicated in Fig. 3-6C. The higher potential barrier at the junction causes a decrease in the diffusion of holes from p to n and electrons from n to p to the point where the diffusion component of the current I_f becomes essentially zero. The drift component I_r again remains the same as under no bias. The total current in the device is, therefore, essentially equal to I_r and is extremely small. The pn junction under reverse bias acts as an extremely high resistance.

PN JUNCTION DIODE

The preceding discussion shows that a pn junction possesses *unilateral* characteristics: it presents an extremely low resistance to the flow of current when biased in the forward direction (p-side positive) but an extremely high resistance when biased in the reverse direction (p-side negative). The pn junction acts as a *diode*. The symbol of a diode is shown in Fig. 3-7A. The terminal on the *p-side* is called the *anode*, and the terminal on the *n-side* is called the *cathode*. The anode in the symbol of a diode is identified by the large arrow. The *forward current* direction in a diode is in the *direction of the arrow*, that is *from p to n*.

The current-voltage characteristics of a diode are shown in Fig. 3-7B. When the voltage v is positive, the diode is forward biased, and at room temperature (traditionally taken as 300°K and represents a rather warm room), the current i is related to the applied voltage by

$$i \approx I_0 e^{(v/0.0259)} \qquad (3\text{-}6)$$

where,

I_0 is the *reverse saturation current* and is in the order of microamperes or smaller.

Because of the exponential relationship between the current and the voltage when the diode is forward biased, the curve is seen to become essentially a vertical line after about v = 0.7 V.

When the voltage v is negative, the diode is reverse biased and there is a small current in the reverse direction. This reverse current soon becomes a constant and equal to $-I_0$. The condition of a negligibly small reverse current of $-I_0$ amperes will persist until the reverse voltage becomes large enough to cause a breakdown of the junction. The breakdown of a pn junction under reverse bias is discussed in the next chapter.

BIPOLAR JUNCTION TRANSISTOR (BJT)

A bipolar junction transistor (BJT) is formed by the diffusion of impurities in a semiconductor so as to create two pn junctions side by side. The device is

Electronics: Circuits and Systems

(A) Symbol of a diode.

(B) Current-voltage characteristics.

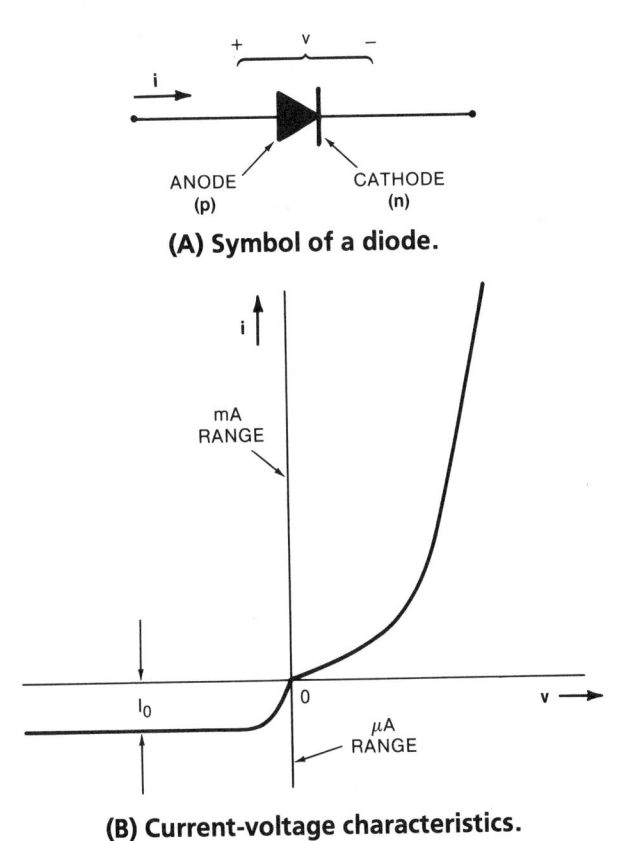

Fig. 3-7. Diode symbol and current-voltage characteristics of a diode.

called a *bipolar junction* transistor because the current flow involves carriers of *both polarities* (holes and electrons), and *pn junctions* are the key to the operation of the device. A BJT can be fabricated either as a *pnp transistor* or an *npn transistor* as shown in Fig. 3-8. The central region is called the *base*. One of the junctions is *forward biased* and is called the *emitter-base junction*. The other junction is *reverse biased* and is called the *collector-base* junction. The three terminals are referred to as the *emitter, base,* and *collector terminals*.

In practice, npn transistors are more frequently used than pnp transistors since the former can be fabricated with a larger current gain than the latter. This preference is particularly true in the case of integrated circuits. In discussing the principles of operation of the npn transistor, we will focus on the flow of the electrons (which are the majority carriers in the emitter region).

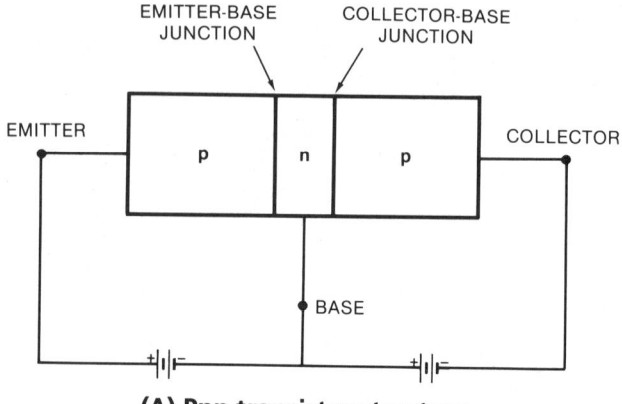

(A) Pnp transistor structure.

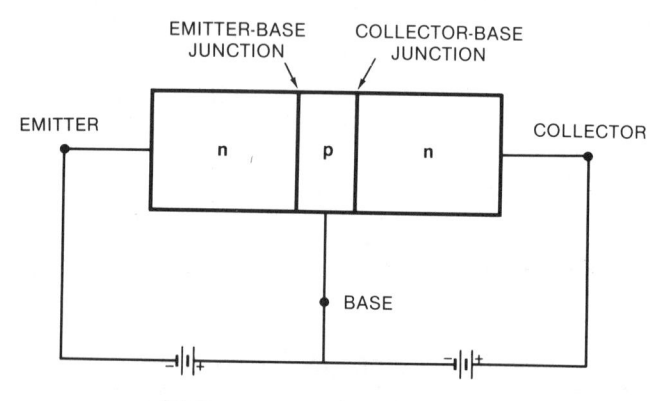

(B) Npn transistor structure.

Fig. 3-8. Pnp and npn transistor structures.

NPN Transistor

The following points made earlier about the potential barrier at a pn junction should be kept in mind in order to understand the principles of operation of the npn transistor: (1) the potential hill is *higher on the n side* than on the p side; (2) the height of the potential barrier is reduced under forward bias and increased under reverse bias; (3) an electron can easily *diffuse up the hill* since it is attracted by the more positive potential.

In the discussion of the principles of operation of the npn transistor, let us start by considering the emitter-base junction and the collector-base junction as if they were separate from each other. The emitter-base junction is forward biased and the potential hill is lower in the base region (which is p) than in

Electronics: Circuits and Systems

the emitter region (which is n) as indicated in Fig. 3-9A. The height of the hill is low due to the forward bias and permits electrons to diffuse easily from the emitter to the base. The collector-base junction is reverse biased and the potential hill is lower in the base region than in the collector region (which is n) as indicated in Fig. 3-9B. The hill is high due to the reverse bias. An *electron in the base region can easily move into the collector region* since an electron can easily move *up* a potential hill.

Now, consider the npn transistor as a whole. The potential distribution will now be as shown in Fig. 3-9C. A large number of *electrons* is *injected from the emitter into the base* due to the low potential barrier at that junction, which is the reason for the name *emitter:* it emits electrons that enter the base. The electrons entering the base can either leave through the base terminal or they can take advantage of the *favorable* potential barrier at the collector-base junction and enter the collector. By making the base region extremely thin and having a sufficiently strong reverse bias at the collector-base junction, a very large portion (90% or more) of the electrons injected into the base can be made to go through to the collector. The name *collector* indicates that it *collects* the electrons injected from the emitter to the base (just like a vacuum cleaner that collects dust particles). By making the base as thin as possible and increasing the area of cross-section of the collector region, it is possible to make well over 99% of the injected electrons enter the collector and less than 1% of them leave through the base terminal.

In terms of currents (rather than electrons), there are three currents in the transistor: an emitter current I_E, a base current I_B and a collector current I_C. *The emitter current is the primary current and gives rise to the other two currents.* That is,

$$I_E = I_B + I_C \tag{3-7}$$

Since these currents are due to the flow of *electrons*, the direction of each is opposite to that of the direction in which the electrons flow, as shown in Fig. 3-9D.

There is another component of current in the npn transistor that is analogous to the reverse current in a diode. Recall that there was a current in the diode caused by thermal agitation and that the direction of the current was from n to p. That current was also seen to be significant only when the pn junction was reverse biased. A similar current exists in the npn transistor due to the thermal generation of electron-hole pairs. Even though such a current exists at both the junctions, it is much more significant in the collector-base junction, which is reverse biased than in the forward biased emitter-base junction. Thermally generated holes in the collector drift into the base, and thermally generated electrons in the base drift into the collector. Such a drift of holes and electrons gives rise to a current component from the *collector to*

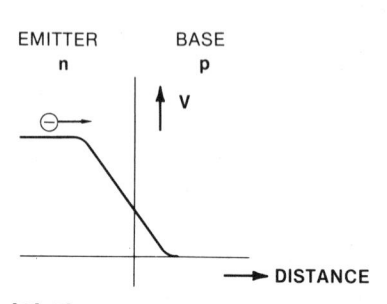

(A) Electrons are injected from emitter to base due to the forward bias.

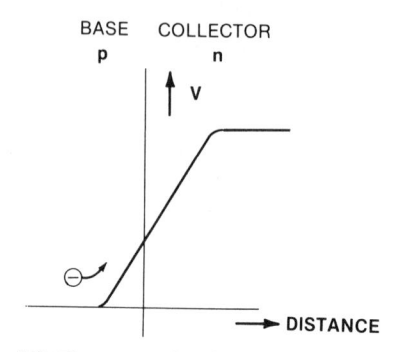

(B) Electrons in the base face a favorable potential hill to enter the collector.

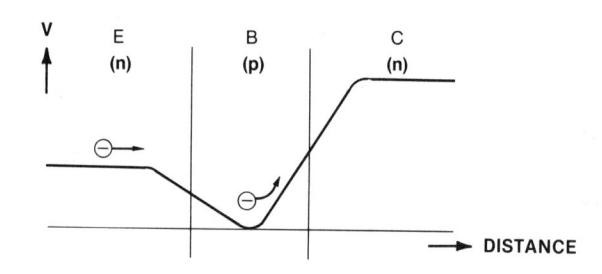

(C) Combination of the potential hills of the two junctions in the complete transistor.

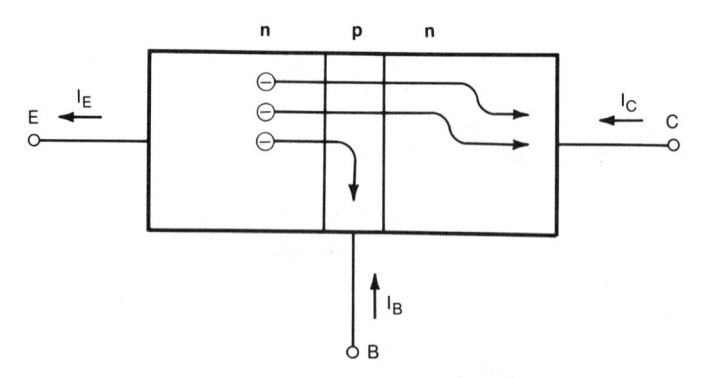

(D) Currents in the transistor.

Fig. 3-9. Principles of operation of an npn transistor.

Electronics: Circuits and Systems

the base as indicated in Fig. 3-10. This current is called the *collector cutoff current* and denoted by I_{CO}. It is of the order of less than a microampere at room temperature. But I_{CO} depends upon the temperature (since it depends upon thermally generated electron-hole pairs) and increases as the temperature increases: *it is found to double in value for every 10°C rise in temperature.* Therefore, I_{CO} cannot be neglected at high temperatures, and its effect on the operation of the transistor will be considered in Chapter 5.

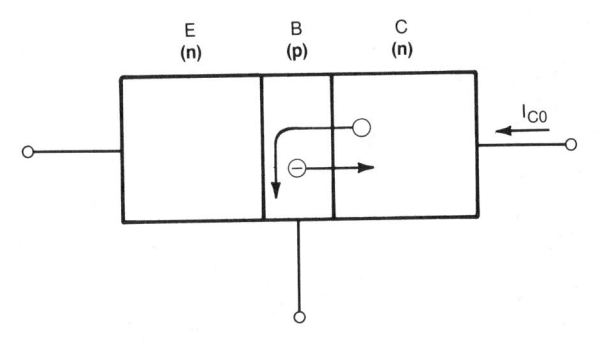

Fig. 3-10. Collector cutoff current.

Summarizing the operation of the npn transistor, a large number of electrons is injected from the emitter due to the forward bias at the emitter-base junction. A small fraction of the injected electrons leaves the transistor through the base terminal. The remainder of the injected electrons leaves through the collector terminal. There is an additional component of collector current due to thermal agitation, which is negligible except at high temperatures (relative to room temperature).

PNP Transistor

The discussion of the npn transistor can be readily modified to that of the pnp transistor by interchanging the words *electrons* and *holes*. Refer to Fig. 3-11. A large number of holes is injected from the emitter, a small percentage of which leave through the base terminal, while a large percentage leave through the collector terminal. The current directions in the pnp transistor are seen to be opposite to those in the npn since the polarities of the carriers are opposite in the two transistors. Just as in the npn transistor, there is a collector cutoff current I_{CO} in the pnp transistor due to thermal generation of electron-hole pairs.

Relationships Between the Currents in a Transistor

Regardless of whether an npn or a pnp transistor is being considered, the following relationships are equally valid. The only difference between the

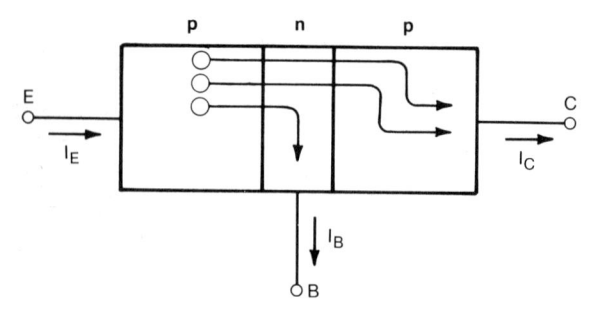

Fig. 3-11. Pnp transistor.

pnp and npn is that all the current *directions* are reversed when going from one type to the other.

The basic relationship between the emitter current, base current, and the collector current is

$$I_E = I_B + I_C \tag{3-7}$$

The collector current I_C has two components: a component due to the emitter current I_E and the collector cutoff current I_{CO}. Using the parameter α to denote the fraction of I_E that enters the collector, we can write

$$I_C = I_{CO} + \alpha I_E \tag{3-8}$$

The value of α is around 0.99 (and higher) for npn transistors fabricated using integrated-circuit (IC) technology, and it is around 0.8 for pnp transistors fabricated in the same technology. Except at high temperatures, the component I_{CO} in Eq. (3-8) can be neglected in comparison with αI_E, and we will use the approximate equation

$$I_C = \alpha I_E \tag{3-9}$$

in our discussion.

Combining Eqs. (3-7) and (3-9), we can obtain the following relationship between the collector current and the base current.

$$I_C = I_B \left(\frac{\alpha}{1 - \alpha} \right) \tag{3-10}$$

A parameter β is defined as

$$\beta = \frac{\alpha}{1 - \alpha} \tag{3-11}$$

so that Eq. (3-10) becomes

$$I_C = \beta I_B \tag{3-12}$$

Electronics: Circuits and Systems

The quantity β may be thought of as the current gain from the base to the collector. (A more precise name for it is *short-circuit forward current gain*.) From Eq. (3-11), it can be seen that, since α is close to 1, the value of β must be very large. For instance when α varies from 0.99 to 0.995, β varies from 100 to 200.

Eqs. (3-7), (3-9), and (3-12) are important and are used frequently in the study of transistor circuits.

Symbols of Transistors

The symbols of an npn and a pnp transistor are shown in Fig. 3-12, in which the directions of the three currents are also shown. In the transistor symbol, the *lead with the arrow* denotes the *emitter*, and the *arrow is directed from p to n*. The direction of the emitter current corresponds to the direction of the arrow on the emitter lead, and the other two currents in the transistor flow in the directions dictated by the emitter current: the emitter current splits into the base current and collector current.

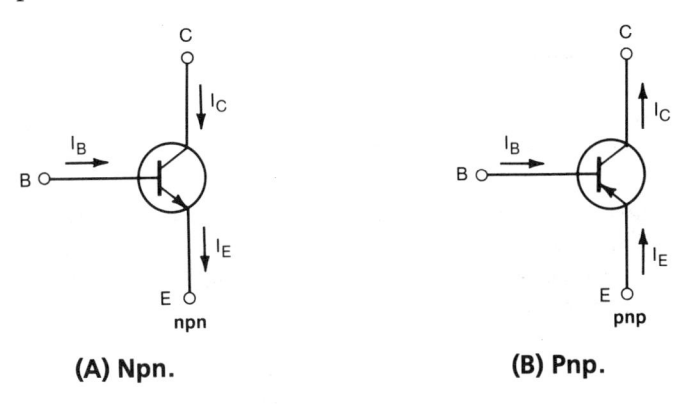

(A) Npn.   **(B) Pnp.**

Fig. 3-12. Symbols and current directions of BJT.

Output Characteristics of an NPN Transistor

The variation of the collector current I_C as a function of the voltage from collector to emitter V_{CE} of an npn transistor is shown by the set of characteristics, called the *output characteristics*, in Fig. 3-13. The output characteristics can be obtained from a manufacturer's data sheet or measured in the laboratory by means of a special-purpose oscilloscope. The following points should be noted in the output characteristics:

1. Except for a slight slope, the collector current is essentially a constant and independent of the voltage V_{CE}. It depends only upon the base current I_B Eq. (3-8).

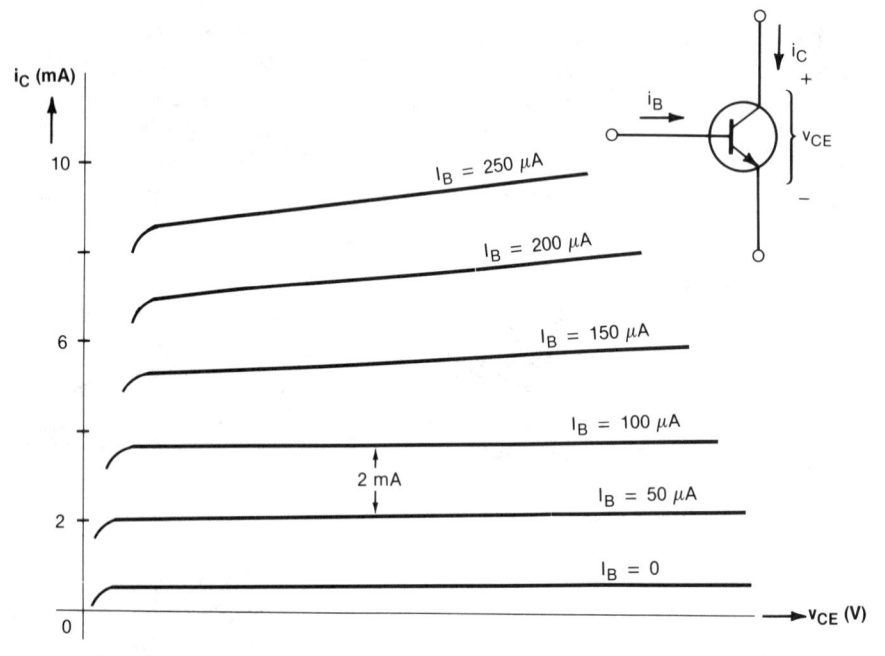

Fig. 3-13. Output characteristics of an npn transistor.

2. The curves can be used to estimate the value of β of a transistor. For example, in the characteristics of Fig. 3-13, it is seen that an increase of the base current from 50 μA to 100 μA produces a change of 2 mA in the collector current. Therefore,

$$\beta = \frac{\text{change in } I_C}{\text{change in } I_B}$$

$$= \frac{2 \times 10^{-3}}{50 \times 10^{-6}}$$

$$= 40$$

JUNCTION FIELD-EFFECT TRANSISTOR (JFET)

The chief difference between the bipolar junction transistor and the field-effect transistor is that in the FET, the current flow is due to carriers of *one polarity* (that is, holes *or* electrons). The name *field-effect transistor* arises from the fact that the current flow in a FET is due to an *electric field* that is present in the device. Field-effect transistors can be divided into two broad categories: *junction* field-effect transistors (JFET) and *metal-oxide semiconductor* field-effect transistors (MOSFET). There are further subdivisions: *n-channel* and *p-channel* within these categories. The abbreviations NMOS and PMOS denote, respectively, n-channel MOS and p-channel MOS devices.

N-Channel JFET

The general structure of an *n-channel JFET* and its terminals are shown in Fig. 3-14. The channel for the flow of current is an n-type semiconductor. One end of the channel acts as the *source* of electrons (analogous to the emitter of an npn transistor), and the other end acts as the *drain* for the electrons (analogous to the collector of an npn transistor). An external battery is connected between the source and drain terminals so as to make the drain positive relative to the source and facilitate the flow of electrons from the source to the drain. Two heavily doped p-type regions are diffused into the channel. These two p-type regions are called *gates* and provide control over the flow of the current in the channel, as will be seen shortly. The two gate terminals are tied together and connected to a battery such that the *gates are always reverse biased with respect to the channel.* The bias connections for an n-channel JFET are shown in Fig. 3-15.

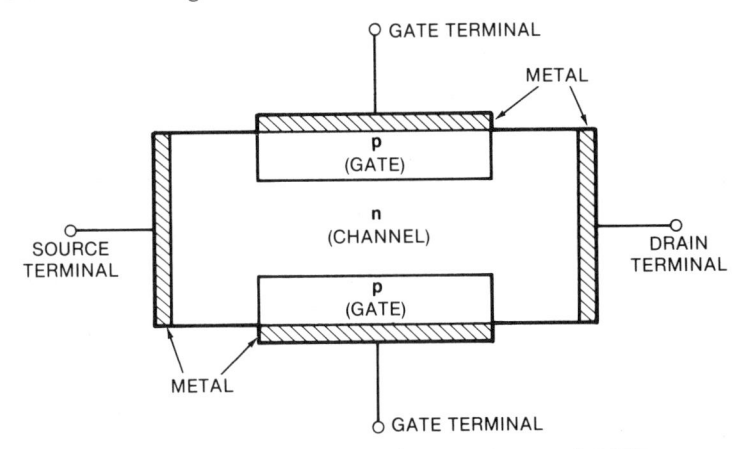

Fig. 3-14. Structure of an n-channel JFET.

Operation of the N-Channel JFET

If there were no gates in the FET, and a battery connected so as to make the drain positive relative to the source, then there would be a flow of electrons from the source to the drain, with the quantity of flow being proportional to the voltage from the source to the drain. That is, the device would act just like an ordinary resistance. But the presence of the gates and the

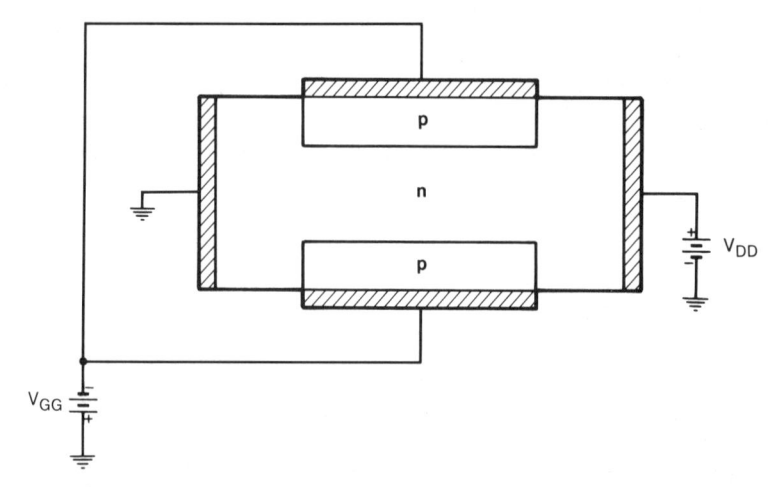

Fig. 3-15. Bias connections of an n-channel JFET.

reverse bias at the channel-gate junction provide a means of *controlling* the flow of electrons from the source to the drain. To understand this controlling action, it is first necessary to examine the behavior of a reverse-biased pn junction.

As was discussed earlier in this chapter, a depletion region exists at a pn junction and has the following properties.

1. There are no free carriers (of either polarity) in a depletion region since they have been swept away by a diffusion of holes from p to n and electrons from n to p. The depletion region offers an extremely high resistance to the flow of current owing to the absence of free carriers.

2. The width of the depletion region is larger under reverse bias than under forward bias. Moreover, the stronger the reverse bias, the wider the depletion region.

3. When the doping levels of the two regions are different, the depletion region penetrates *more deeply into the lightly doped* region.

Now, consider the reverse-biased channel-gate junction of a JFET in conjunction with the above properties. A depletion region, whose width can be controlled by the applied reverse bias, exists at the channel-gate junction. Since a JFET is fabricated with a *much heavier doping in the gate regions* than the channel, most of the depletion region lies in the *channel* itself. Since the depletion region offers an extremely high resistance to the flow of current (or free electrons), the electrons can only use the portion of the channel not blocked off by the depletion region. The depletion region in the JFET is of the form shown in Fig. 3-16. Note that the depletion region is *wider* near the drain than at the source because the reverse bias between the channel and the gate is not uniform throughout the channel. Since the voltage at the drain is

Electronics: Circuits and Systems　　　　　　　　　　　　　　　　　　**189**

$+V_{DD}$ while the source terminal is grounded, the voltage between the gate and channel is close to $-V_{GG}$ near the source end of the channel but increases to $(-V_{GG} - V_{DD})$ near the drain end of the channel.

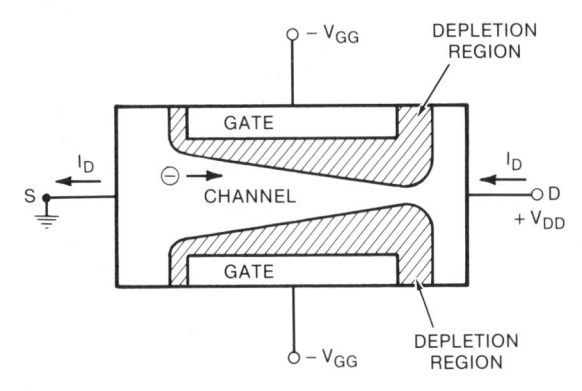

(A) Electron flow restricted to the gap in the channel.

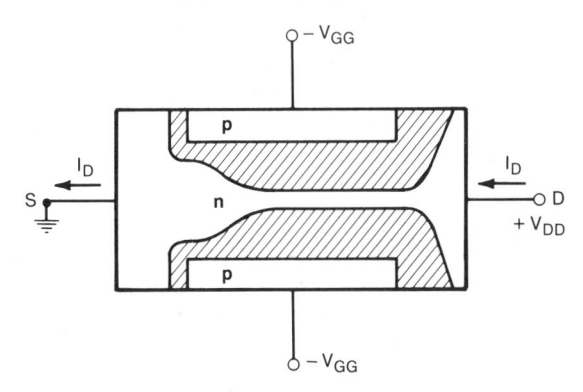

(B) Pinch-off condition.

Fig. 3-16. Depletion region in an n-channel JFET.

Since the flow of electrons is restricted to the portion of the channel not blocked off by the depletion region, it is possible to control the quantity of electron flow from the source to the drain by varying the reverse bias applied to the gate. The more negative the gate voltage, the stronger is the reverse bias, the wider the depletion region, and the narrower the opening in the channel for the flow of electrons. *For a given voltage from the source to the drain, the current decreases as the gate is made more negative.*

Pinch-Off Condition

As the gate voltage is made more and more negative, the depletion region eventually widens to an extent that it almost completely chokes off the channel. Actually, the depletion layer does not completely block the flow of free electrons but leaves an extremely narrow gap in the channel, and the gap stays at that width for any further increase in reverse bias. This condition is known as *pinch-off* and is indicated in Fig. 3-16B. The current in the JFET reaches a constant value and becomes independent of the voltage between the source and the drain when the pinch-off condition occurs.

Pinch-off condition can also be made to occur by keeping the gate voltage constant and increasing the drain-to-source voltage V_{DD} to a sufficiently high value. Increasing V_{DD} has the effect of increasing the reverse bias between the channel and the gate, since the reverse bias near the drain is $(-V_{GG} - V_{DD})$, as mentioned earlier.

Currents in an N-Channel JFET

Since the current in an n-channel JFET is due to the flow of *electrons from the source to the drain*, the *direction of the current* is taken as from the *drain to the source* as indicated in Fig. 3-16A. Also, there is no flow of electrons out of the gate terminal, and the gate current is zero. Therefore, the current entering the drain terminal is the same as that leaving the source terminal. As a matter of fact, there is a negligibly small amount of gate current if the gate voltage is made positive.

P-Channel JFET

The previous discussion of the principles of operation of the n-channel JFET can be readily modified to cover the p-channel JFET by taking into account the following differences. The channel of the p-channel JFET is made of a lightly doped p-type semiconductor, and its gates use heavily doped n-type material. The drain is kept at a positive potential with respect to the source by means of an external battery so that the holes in the channel flow from the source to the drain. The channel-gate junction is reverse biased by applying a positive potential to the gate. The operation of the p-channel JFET is similar to that of the n-channel JFET. With no bias applied to the gate, the device operates like an ordinary resistance. As a reverse bias is applied to the channel-gate junction (by making the gate positive), a depletion region appears in the channel, and the channel becomes constricted. As the reverse bias increases, the current flow decreases. Pinch-off occurs when a sufficiently high positive voltage is applied to the gate (or when the drain-to-source voltage is made sufficiently negative). The current in the device remains constant when pinch-off occurs and becomes independent of the voltage between the source and the drain. Fig. 3-17 shows the conditions pertinent to the p-channel JFET. The *current direction* in the p-channel device is seen to be *from the source to the drain.*

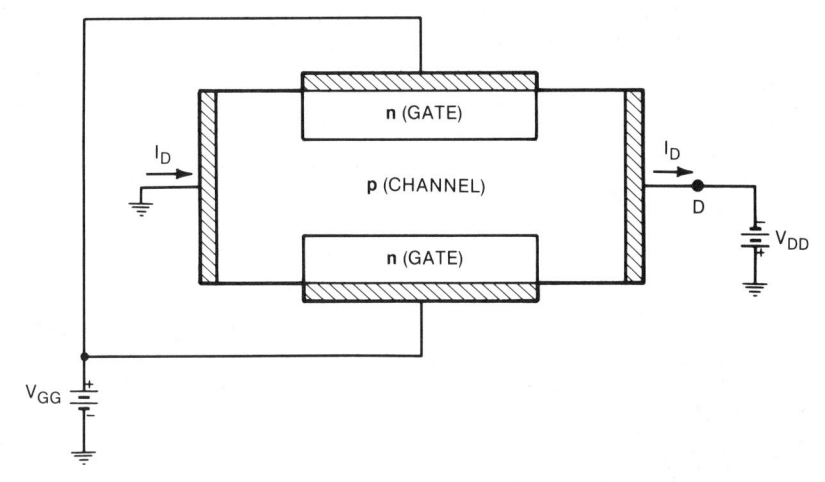

Fig. 3-17. Structure of a p-channel JFET.

Symbols and Output Characteristics

The symbols used for the n-channel and p-channel JFETs are shown in Fig. 3-18. The direction of the arrowhead on the gate lead provides the clue to whether the device is n-channel or p-channel: the arrow points *into an n-channel* and *away from* a *p-channel.*

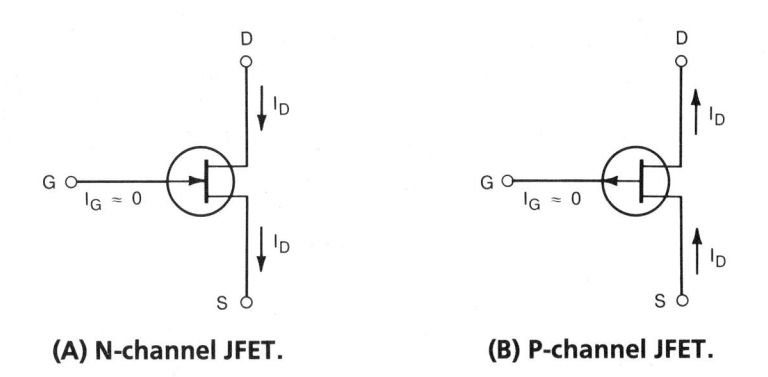

(A) N-channel JFET.   (B) P-channel JFET.

Fig. 3-18. Symbols of JFET.

The variation of the drain current I_D as a function of the drain-to-source voltage V_{DS} for the different values of the gate-to-source voltage V_{GS} is shown in Fig. 3-19, for an n-channel JFET. The following features should be noted in the output characteristics.

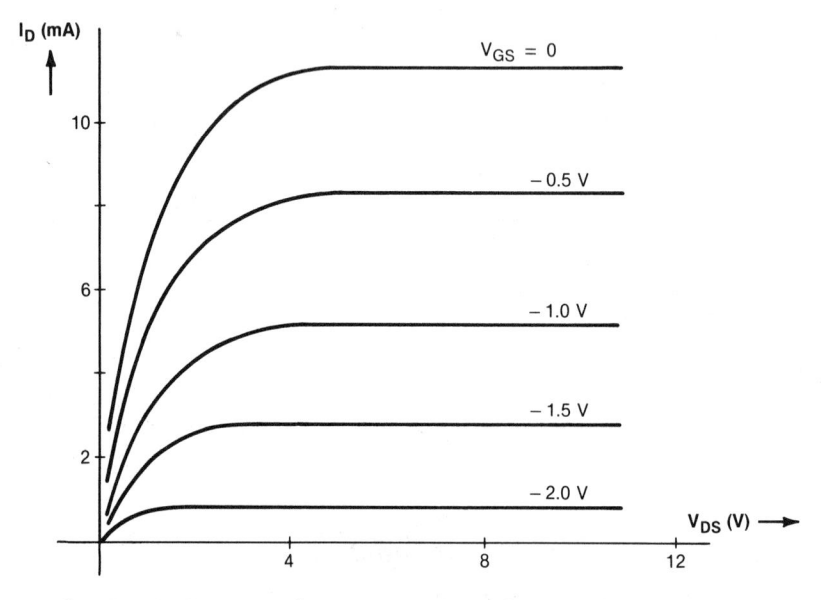

Fig. 3-19. Output characteristics of an n-channel JFET.

1. For any given value of V_{GS}, the current at first increases as V_{DS} increases but soon levels off corresponding to the onset of pinch-off.

2. Except for the portion where the current increases with the drain-to-source voltage, the slope of each characteristic is almost zero, implying that the resistance of the device (the ratio of the drain-to-source voltage to the drain current) is extremely high for a JFET—much higher than the resistance of a BJT.

The output characteristics of a JFET and a BJT (Fig. 3-13) are seen to have strong similarities.

MOSFETs

The *metal-oxide semiconductor field-effect transistor*, or the MOSFET, is a very attractive device from a practical point of view because it lends itself most readily to fabrication in integrated-circuit technology. It is possible to attain very high densities (number of transistors in a given area of a chip) using MOSFET devices. MOSFETs can be broadly classified into two categories: *enhancement-mode* MOSFETs and *depletion-mode* MOSFETs. Each of these categories is further subdivided into two types: p-channel MOSFETs (PMOS) and n-channel MOSFETs (NMOS).

N-Channel Enhancement-Mode MOSFET (Enhancement NMOS)

The foundation on which an n-channel enhancement-mode MOSFET is fabricated is a lightly doped p-type material, called the *p-substrate*. There is only a small number of holes in the p-substrate because of the light doping. Two heavily doped (n+) regions, one of which will act as the *source* and the other as the *drain*, are diffused into the p-substrate as indicated in Fig. 3-20. Metallic contacts are provided for the source and the drain terminals. A silicon dioxide layer, with openings for the source and drain terminals, is put on top of the substrate slightly overlapping the heavily doped n-regions. A layer of metal is formed on top of the oxide layer, and this metallic layer serves as the gate electrode of the device. The manner in which the device is fabricated (metal on top of the oxide on top of the semiconductor) gives rise to its name. Since the oxide serves as an insulator between the gate and the semiconductor, the device is also called the *insulated-gate field-effect transistor* (or *IGFET*).

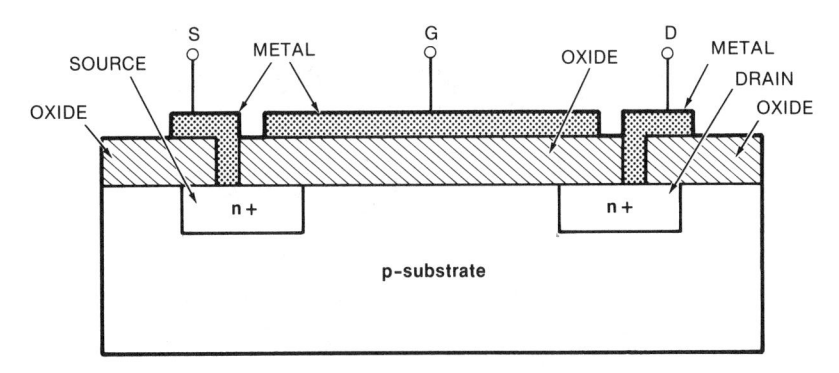

Fig. 3-20. Structure of an enhancement-mode NMOS. Note the absence of a channel between the two n+ regions.

The MOSFET is quite similar to a capacitor: the gate acts as one plate of a capacitor, the oxide serves as the dielectric, and the semiconductor serves as the second plate of the capacitor. The use of a semiconductor instead of a metal electrode for the second plate of a capacitor results in the controlling action that differentiates the MOSFET from an ordinary capacitor.

The current flow in an n-channel enhancement-mode MOSFET will be seen to be due to electrons moving from the source to the drain, and these electrons will be provided by means of a channel induced in the device by a voltage applied to the gate.

Operation of the Enhancement NMOS

Consider first the solution with the source grounded, the gate not connected to any potential, and the drain connected to a positive potential V_{DD} as shown in Fig. 3-21A. Since the p-substrate is lightly doped, there are not enough holes to cause any significant current flow under these conditions. The current is essentially zero.

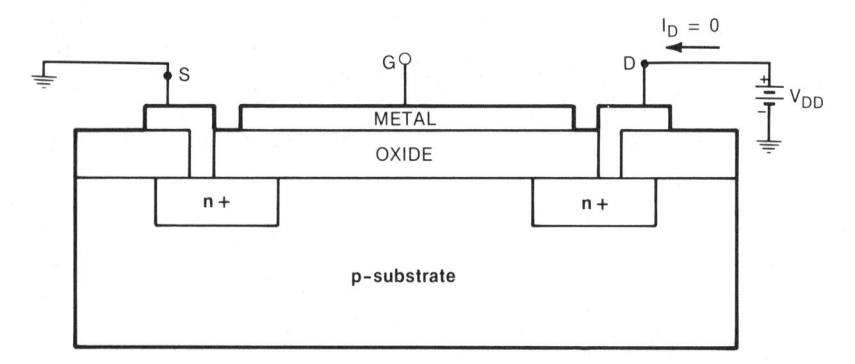

(A) With no bias at the gate, no electron flow occurs.

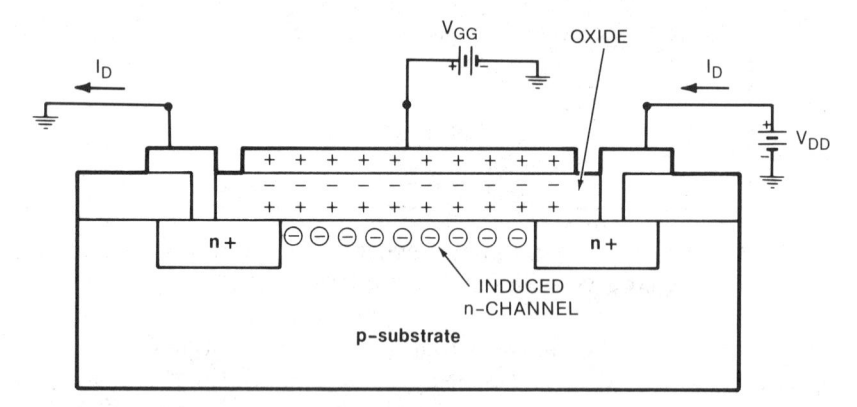

(B) With a positive bias at the gate, a channel is induced that facilitates the flow of electrons from the source to the drain.

Fig. 3-21. Flow of electrons in an enhancement NMOS.

Now, suppose that the gate is connected to a *positive* potential as shown in Fig. 3-21B, leaving the drain and the source connected as before. The positive voltage on the gate induces a negative charge on the upper layer of the oxide and a positive charge on the lower layer of the oxide, just as a positive electrode of a capacitor induces charges in the dielectric. The positive induced charge on the lower layer of the oxide exerts an attractive force on the electrons in the substrate and makes them escape their bonds from their parent atoms. These electrons are confined to a thin layer just below the oxide and form an *induced channel*. The electrons in the induced channel are now subject to the voltage between the source and the drain and move from the former to the latter, causing the flow of current. Since the electrons are *minority* carriers in the p-substrate, the action of the device depends upon the flow of minority carriers.

A current flow is possible if and only if the gate has a positive potential (with respect to the source), which is the reason for the name *enhancement type:* the gate voltage enhances the availability of free electrons for current flow. In fact, the current flow becomes significant only when the gate voltage is sufficiently positive with respect to the source. That is, V_{GS} must exceed a minimum value, called the *threshold voltage*, for the current to flow. The threshold voltage is usually in the range of 1 to 4 V.

For a given gate voltage (above the threshold) the current increases as the drain-to-source voltage increases. But soon the current levels off since the induced channel has only a finite number of electrons available for current flow. This condition is analogous to the pinch-off condition of the JFET.

The output characteristics of an enhancement-mode NMOS device are shown in Fig. 3-22. For a given drain-to-source voltage, V_{DS}, the drain current increases as the voltage V_{GS} increases, since more electrons are attracted into the induced channel by the higher potential of the gate. For a given V_{GS}, the drain current increases at first but soon levels off.

P-Channel Enhancement-Mode MOSFET (Enhancement PMOS)

By interchanging the p-regions and n-regions of the enhancement-mode NMOS, the structure can be made into an enhancement-mode PMOS, as indicated in Fig. 3-23A. The external bias connections to the enhancement PMOS are exactly the opposite of those in the enhancement NMOS. The operation of the enhancement PMOS is illustrated in Fig. 3-23B. Enhancement PMOS structures are directly used only infrequently in practice since they cannot be fabricated with as high packing densities in integrated circuits as in the enhancement NMOS. But the most important use of the enhancement PMOS is in the composite MOS structure called the *complementary MOSFET* (or CMOS). An NMOS and a PMOS are interconnected in the CMOS device so as

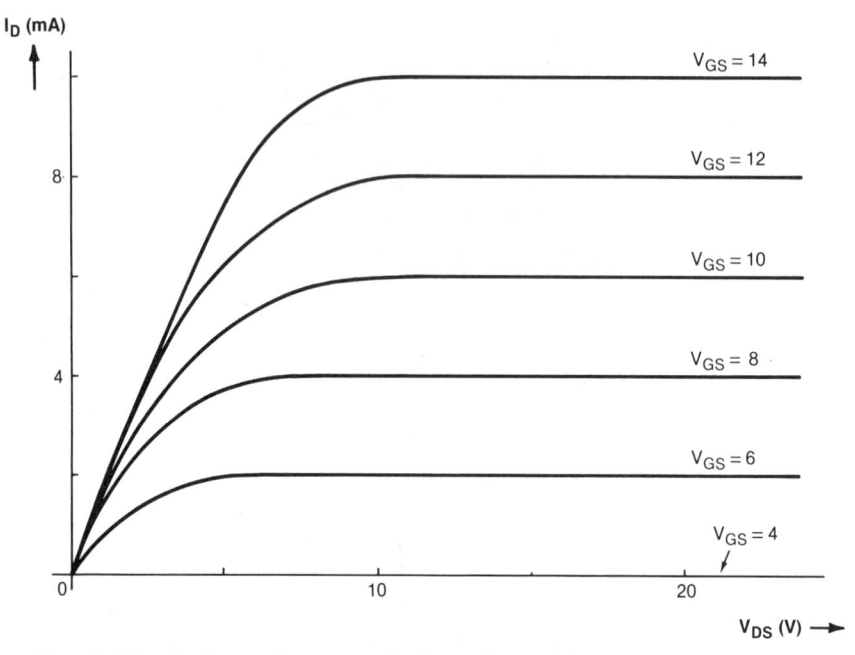

Fig. 3-22. Output characteristics of an enhancement NMOS with a threshold voltage of 4 V.

to exploit the property that the NMOS and PMOS require different polarities of gate voltage in order to conduct. The CMOS will be discussed later in this chapter and also the chapter on digital logic circuit families.

N-Channel Depletion-Mode MOSFET (Depletion NMOS)

The structure of an n-channel *depletion* MOSFET is almost exactly the same as the n-channel enhancement MOSFET with the only difference being that *an actual channel of n-type material is diffused* between the source and the drain during the fabrication. Therefore, an n channel is always available (with its free electrons) in the depletion-mode device. The structure of an n-channel depletion-mode MOSFET is shown in Fig. 3-24A.

The gate of the depletion-mode NMOS is kept at a *negative* potential (unlike that of the enhancement NMOS). As indicated in Fig. 3-24B the negative gate voltage induces a layer of positive charge in the upper edge of the oxide and a layer of negative charge in the lower edge. The induced negative charge exerts a *repulsive* force on the electrons in the channel, thus, creating a *depletion region* in the channel. The current flow is, therefore, restricted to the portion of the channel not blocked off by the depletion region, exactly similar to the JFET. As the gate voltage is made more negative, the depletion

Electronics: Circuits and Systems

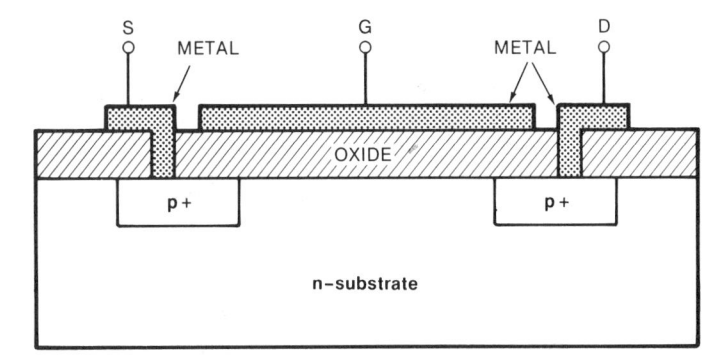

(A) Structure of an enhancement PMOS.

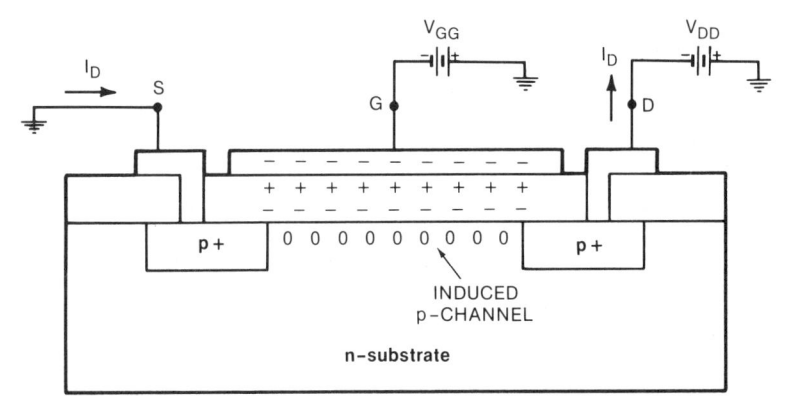

(B) Operation of an enhancement PMOS.

Fig. 3-23. Structure and operation of an enhancement PMOS.

layer extends farther into the channel and eventually a pinch-off condition sets in. The operation of the depletion-mode MOSFET is thus exactly the same as the JFET. The output characteristics of an NMOS depletion type will be exactly the same form as the n-channel JFET.

Symbols for MOSFET

The present practice is to use the same symbol for both enhancement-type and depletion-type devices. These are shown in Fig. 3-25. The arrow on the source terminal indicates the direction of the current flow: from source to drain in the case of p-channel devices (since holes flow from the source to the drain), and from drain to source in the case of n-channel devices (since electrons flow from the source to the drain).

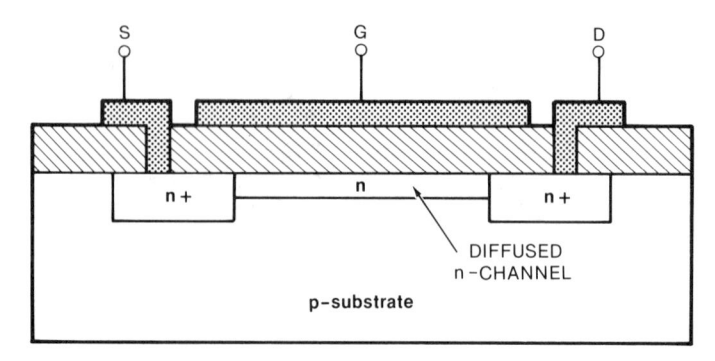

(A) Structure of a depletion NMOS.

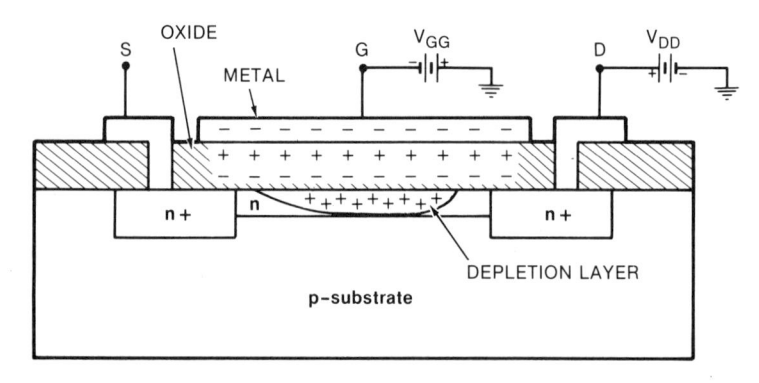

(B) Operation of a depletion NMOS.

Fig. 3-24. Structure and operation of a depletion NMOS.

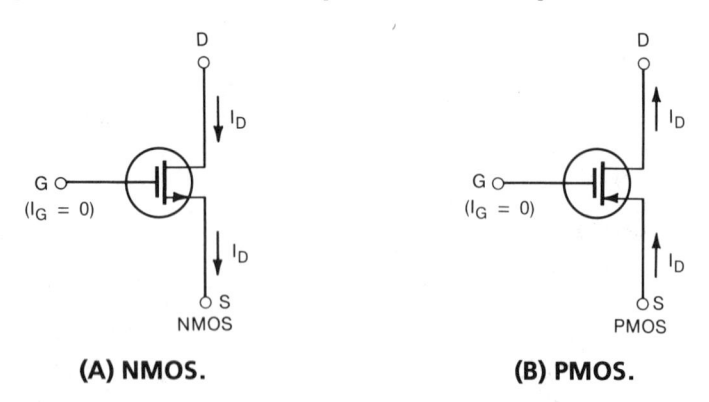

(A) NMOS.

(B) PMOS.

Fig. 3-25. Symbols for NMOS and PMOS. (No distinction is made between enhancement and depletion devices.)

Electronics: Circuits and Systems

The symbols of Fig. 3-26 show those that were prevalent until recently and still used fairly commonly in other literature. A distinction is made in these symbols between the depletion-type and the enhancement-type devices by using a broken line for the enhancement devices.

(A) Depletion.   **(B) Enhancement.**

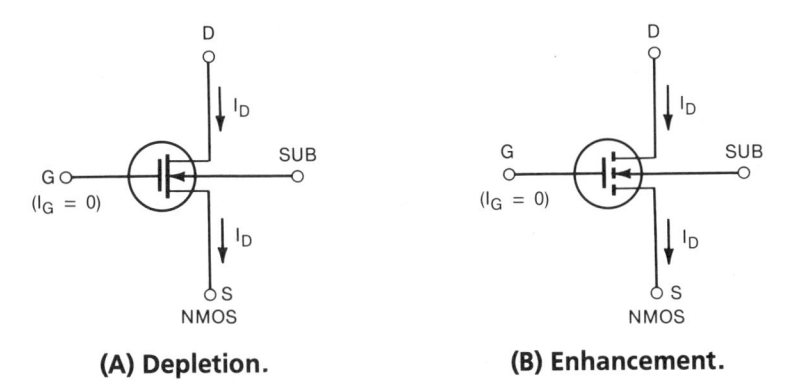

Fig. 3-26. Alternative symbols of NMOS. Distinction is made between enhancement and depletion devices.

Advantages of the MOSFET

The gate draws virtually no current in a MOSFET due to the presence of the oxide that serves as an excellent insulator. The input resistance (as seen from the gate) of a MOSFET is as high as 10^{12} Ω to 10^{15} Ω compared with 10^5 Ω for a JFET and a few kilohms for a BJT. A signal source connected to the gate of a MOSFET supplies almost no current, and hence, the input power is close to zero, and since the output power is a finite quantity, the power gain of a MOSFET is almost infinite! Also, MOSFETs need a much smaller area of a chip in integrated circuits than other types of transistors and lead to high packing densities in integrated circuits. Consequently, the MOS device is the one that is used in large-scale integration.

Some other MOS-type devices will be discussed in the following section on ICs.

INTEGRATED CIRCUITS (ICs)

When an electrical network is fabricated upon or within a single supporting layer, the *substrate*, it is called an *integrated circuit*. The electrical network usually includes active devices (such as transistors).

Thin-film and *thick-film integrated circuits* use an insulator as a substrate. Resistors and capacitors can be fabricated in thick-film and thin-film ICs but not active devices. Active devices (such as transistors) are externally attached to such ICs. Even though the inability to fabricate active devices on the substrate is a serious drawback, thick- and thin-film ICs are widely used since their manufacture does not require ultra-clean rooms and expensive equipment like the monolithic IC.

Monolithic integrated circuits, which are usually simply referred to as ICs, use a p-type semiconductor as the substrate, and both active and passive elements are directly fabricated on the substrate. IC technology has revolutionized the world of electronic systems and computers to a point never even dreamed of before their advent. For example, it is now possible to put a chip no larger than a fingernail in a computer that would have occupied a very large room in the 1950s. The manufacture of ICs requires expensive and elaborate equipment and facilities: ultra-clean rooms, furnaces with very precisely controlled temperatures, and photolithographic facilities of extremely high precision, to mention a few.

The following terms are used in connection with ICs: *small-scale integration* (or *SSI*), *medium-scale integration* (or *MSI*), *large-scale integration* (or *LSI*), and *very large-scale integration* (or *VLSI*). The distinction between the different scales of integration is based on the number of components on a single IC chip and also on the complexity of the function (or functions) that the circuit or system in the chip can perform. SSI chips contain tens of components and each circuit in an SSI chip may be, for example, a single logic gate. MSI chips contain hundreds of components, and a chip may contain a moderately complex logic system (such as a ROM). LSI chips contain several thousands of components, and each chip will contain a highly complex logic system such as memories with large capacities. VLSI chips contain tens of thousands of components, and each chip will be a digital system of extremely high complexity in itself. The technology of ICs is a constantly changing field with new developments and higher densities on a single chip being announced regularly. It is one of the most dynamic and competitive areas of technological innovations at present. Consequently, any detailed discussion of IC technology will almost definitely be outdated before it is published. No attempt will, therefore, be made here to present a comprehensive treatment of ICs. We will only examine the basic underlying principles and steps of processing involved in the fabrication of ICs.

Fabrication of an IC

Before considering the steps involved in the fabrication of a specific electronic circuit, we will describe one operation that is repeated many times during the fabrication of a single circuit. This operation is the diffusion of impurities into selected openings, or *windows*, in the semiconductor. The windows will be of different shapes and sizes and occupy different areas of a chip. The

Electronics: Circuits and Systems

creation of the windows involves four steps: *mask generation, oxidation, photolithography* and *etching*, and *diffusion*.

1. Mask Generation

The fabrication of each circuit requires the generation of a number of masks. A mask contains a pattern of clear and opaque areas. Once the pattern of a mask has been determined, it is first drawn on a large sheet of *Mylar*™ coated with a red plastic, called *rubylith*. The pattern is then photographed and reduced by a ratio of 500 to 1 by a series of reductions by a factor of 5 or 10 to 1. Then, the reduced version of the mask (which is only about 50 mils square) is repeated hundreds of times on a single glass plate of 3 or 4 inches in diameter.

2. Oxidation

The silicon wafer, on which the integrated circuit is to be fabricated, is coated with a very thin layer of silicon dioxide (called simply the *oxide*). The oxide layer can be selectively etched away so as to create the windows needed for diffusion. In a number of circuits, the oxide serves as the insulating region needed for the circuit itself.

3. Photolithography and Etching

A thin layer of *photoresist* is first spread on the wafer. Then the mask plate is placed on the wafer, and the wafer is exposed to ultraviolet light. The light polymerizes the photoresist emulsion under the clear areas of the mask while not affecting the photoresist emulsion under the opaque areas. The wafer is then developed and washed using photochemicals. This process removes the *unpolymerized* emulsion. The polymerized emulsion in the remainder of the wafer is then fixed by baking. The wafer now contains certain areas covered by polymerized emulsion and other areas not so protected.

The wafer is then immersed in an acid that etches away the oxide areas that are *not* covered by the polymerized emulsion. These areas are the windows through which diffusion can take place. It can be seen that the *windows* will correspond to those areas that were *opaque in the mask*.

4. Diffusion

P-type and n-type impurities are diffused into the openings in the wafer created by photolithography and etching. The diffusion is carried out in a furnace through which the wafer moves at a slow speed, and gases containing the desired impurities pass through the furnace. Precise temperature control is extremely critical in the diffusion process.

Resistors in ICs

The fabrication of a resistor in an IC is based on the principle of a *sheet resistance*, that is, a resistive material in the form of a sheet of uniform thickness. For such a sheet, the resistance R is given by (Fig. 3-27A)

$$R = (\text{constant}) \times (L/w) \qquad (3\text{-}13)$$

where,

L is the length in the direction of the current flow,
w is the width perpendicular to that direction.

For a *square* cut out of the sheet, L = w, and the resistance of the square is the constant that appears in Eq. (3-13). This constant is, therefore, called the *resistance per square*. It should be noted that for any *square*, regardless of how small or large it is, the resistance is the same, as indicated in Fig. 3-27B.

In order to fabricate a specified value of resistance, it is only necessary to determine how many squares are needed. For example, if the resistance per square of a sheet resistor is 200 Ω/sq, then five squares are needed to obtain 1 kΩ as indicated in Fig. 3-27C.

In integrated circuits, a p-type layer is used as the sheet resistor. The resistance per square can be controlled through the amount of p-type impurity diffused into the semiconductor. A specified resistance is then fabricated by making the p-type layer have the right number of squares. A 200 Ω/sq layer, for example, will need 25 squares to produce a 5-kΩ resistor. But, if the layer is fabricated as a long thin strip (of length 25 mils and width 1 mil for instance), it will take up too much space (or "real estate") on the chip. It is, therefore, more advantageous to fabricate it in the form of a folded strip as shown in Fig. 3-27D.

Capacitors in ICs

A capacitor needs two electrodes separated by an insulating layer (which is the dielectric). In integrated circuits, capacitors are fabricated using the structure of the MOS device: a metal electrode, an oxide layer that serves as the dielectric, and a semiconductor region that acts as the second electrode. One version of IC capacitors is the structure shown in Fig. 3-28. The n+ layer is used so as to provide an ohmic contact. An alternative (and more commonly used) structure of MOS capacitors uses *polycrystalline silicon* (referred to as *polysilicon* or *poly*) as the top and bottom electrodes.

Ohmic and Nonohmic Contacts

It might appear that if a terminal is to be attached to a semiconductor material (as for example the base terminal of an npn transistor) it should be a simple matter to routinely bond a metallic lead (of aluminum) to the semiconductor. Unfortunately, a metal-semiconductor may exhibit *rectifying* charac-

Electronics: Circuits and Systems

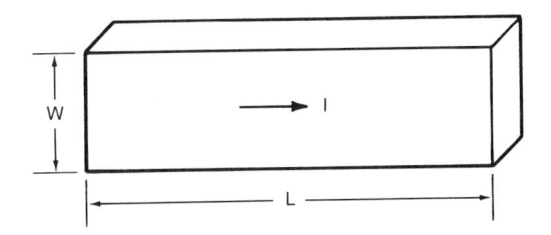

(A) Length and width used to determine resistance.

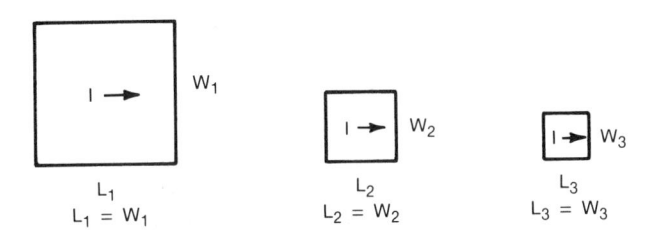

(B) The resistances of the different squares are equal to one another.

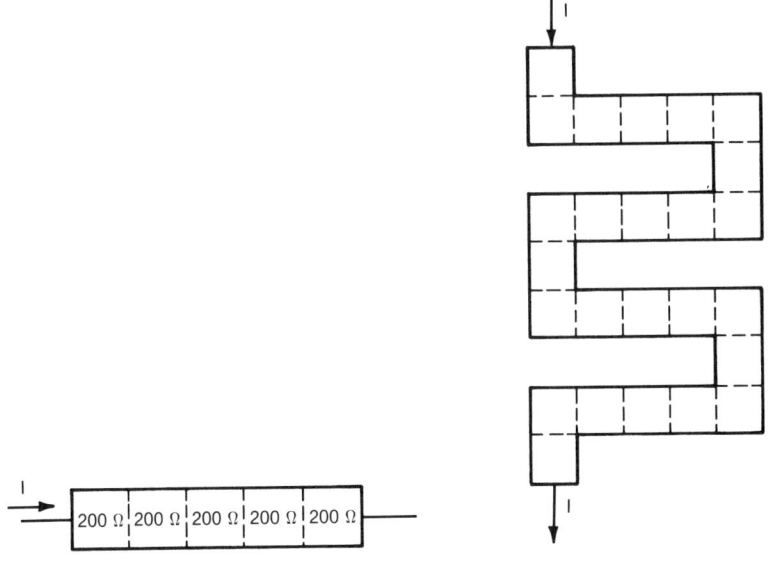

$L_1 = W_1$

$L_2 = W_2$

$L_3 = W_3$

(C) Five squares needed to obtain 1 kΩ.

(D) Folding conserves space on the chip.

Fig. 3-27. Sheet resistance.

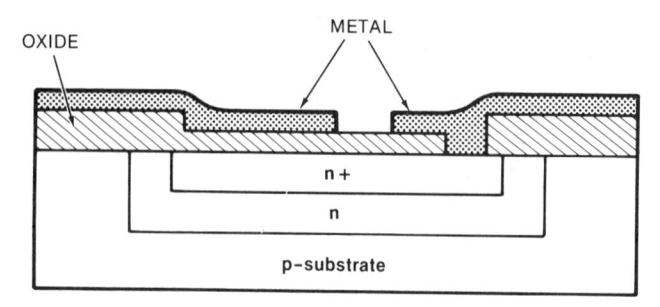

Fig. 3-28. IC capacitor.

teristics, that is, act like a diode. Aluminum acts as a p-type impurity when in contact with silicon. If aluminum is directly deposited on an n-type semiconductor, a *pn junction* is formed with the rectifying properties of a diode. Such a device is called the *Schottky diode*, which is an extremely useful device. But, if we wish to have an *ohmic contact* without the rectifying characteristics, then it is necessary to first form a heavily doped n-type (denoted by n+) layer on the n-type region and then deposit the aluminum on it. The n+ region is able to cancel the Schottky barrier and make the contact ohmic. It is important to note that such an n+ region is needed only when attaching aluminum to an n-type semiconductor. No such step is necessary in depositing aluminum on p-type materials.

Diodes and BJTs in ICs

A diode requires a p-region and an n-region, while a transistor requires two n-regions and one p-region. (Npn transistors are predominant in ICs. Pnp transistors are not common since it is not possible to attain high values of β for pnp devices in ICs.) The required n- and p-regions are formed by diffusion of the proper type of impurities. When terminals are to be attached to diodes and transistors, the precaution about ohmic contacts, mentioned previously, should be observed.

An Example of IC Fabrication

Consider the simple electronic circuit of Fig. 3-29. Looking at the various components of the circuit, the transistor needs three regions: p-type base, n-type emitter, and n-type collector. The transistor will, therefore, need three diffusions. The diode needs two diffusions: a p-type for the anode and an n-type for the cathode. Both of the diode diffusions can be combined with two of those needed for the transistor. The resistor needs one p-type diffusion that is combined with the p-type diffusion needed for the transistor. The capacitor needs one n-type diffusion that is combined with one of the n-type diffusions of the transistor. Finally, all the components have to be interconnected

Electronics: Circuits and Systems

through metallic leads so as to form the circuit, which requires the *metallization* process.

Fig. 3-29. Simple electronic circuit used in the discussion of the fabrication of ICs.

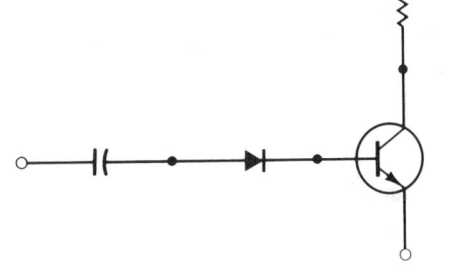

The first step in the fabrication of the circuit is the generation of the masks: five are needed for this example, and they are shown in Fig. 3-30. The reasoning behind the patterns should become clear as we look at the fabrication in detail. After the masks have been generated, most of the process involves the creation of windows and diffusion using oxidation, photolithography and etching, and diffusion described earlier.

The following steps are also shown in Fig. 3-31, and the sketches should be referred to while reading the following description.

1. Wafer

A silicon wafer, 3 or 4 inches in diameter, doped lightly with a p-type impurity forms the *substrate*, called the *p-substrate*, in which the circuit will be fabricated. The wafer will contain several thousands of identical circuits, each a replica of the given circuit, when the fabrication is complete. Each circuit will occupy an area about 50 mils square.

2. Epitaxial Growth

A thin film of a single crystal silicon is grown on the wafer using an epitaxial reactor. This layer has an n-type impurity and is called the *n-epi*. A portion of the n-epi will form the collector region of the transistor.

3. Isolation Diffusion

Electrical isolation between the n-type regions of the different components of the circuit is necessary, and it is achieved through the creation of pn junctions (which will be kept under reverse bias when the circuit is used eventually). P-type regions, called the *isolation regions*, are created in the n-epi through the use of Mask 1 to form the windows and a p-type diffusion.

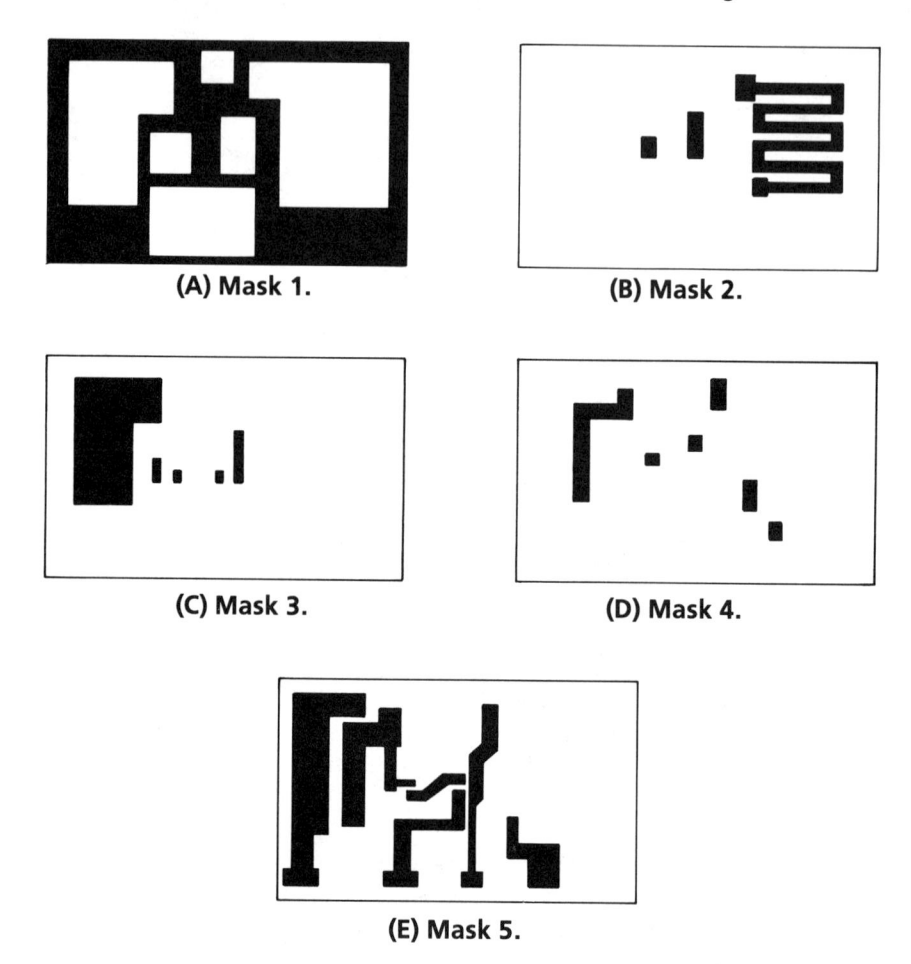

206   **Semiconductor Devices and Integrated Circuits**

(A) Mask 1.

(B) Mask 2.

(C) Mask 3.

(D) Mask 4.

(E) Mask 5.

Fig. 3-30. Masks used in the fabrication of the circuit of Fig. 3-29. (To be studied in conjunction with the steps indicated in Fig. 3-31.)

4. Base Diffusion

The base region of the transistor and the anode of the diode are formed through one p-type diffusion using Mask 2. This diffusion also forms the p-type region needed for the resistor. The pattern of the resistor is seen to be a long strip folded so as to occupy a small area in the mask.

5. Emitter Diffusion

The emitter of the transistor, the cathode of the diode, and one plate of the capacitor are formed by an n-type diffusion using Mask 3. This n-type diffusion is also made to penetrate into the collector region so as to create the heavily doped, n+ regions needed later for making ohmic contacts (as was discussed earlier).

At this point, all the semiconductor (and oxide) regions needed for the circuit are completed.

6. Pre-Ohmic Etching

In order to provide access to the various parts of the circuit, Mask 4 is used to create a set of windows. At the end of this step, the entire wafer is covered with an oxide layer except where the metallic contacts have to be attached to the semiconductor areas.

7. Metallization

A thin layer of aluminum is evaporated over the entire surface of the wafer. Then the metallic layer is selectively etched by using Mask 5 and photolithography.

It might be instructive to return to the diagrams of the masks and examine them in the *reverse* order to understand the reasoning behind the pattern in each case. Remember that the *dark areas* of the mask correspond to the *windows* through which diffusion takes place.

After the circuit has been fabricated on the wafer, the individual chips are separated and tested through automatic testing equipment. The good chips are then mounted on a *lead frame*, thin wires (of gold) are attached as leads from the metallic contacts on the chip to the contact areas on the lead frame, and the chip is then encapsulated in epoxy molds. In some circuits (such as reprogrammable digital circuits), a transparent window is provided in the mold.

In complex circuits, the generation of the masks is through the use of highly sophisticated computer-aided design programs.

Miscellaneous Transistors in ICs

Apart from the BJT and the FET discussed earlier, certain types of transistors have become practical and widely used due to IC technology. Some of these are discussed in the following.

Complementary MOSFET (CMOS)

As was discussed before, the enhancement MOS devices can be either p-channel or n-channel devices. A chip in which both these types are fabricated together leads to the complementary MOSFET, referred to as the CMOS device. A cross-section of the integrated circuit fabrication of the CMOS is shown in Fig. 3-32A, and the schematic form of the circuit is shown in Fig. 3-32B. This configuration acts as a logic inverter and will be discussed further in the chapter on digital logic circuit families.

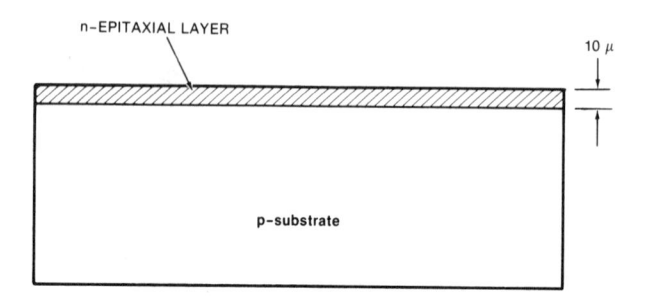

208   Semiconductor Devices and Integrated Circuits

(A) Epitaxial growth.

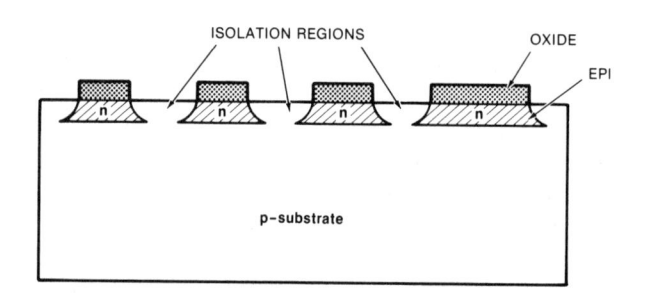

(B) Isolation diffusion.

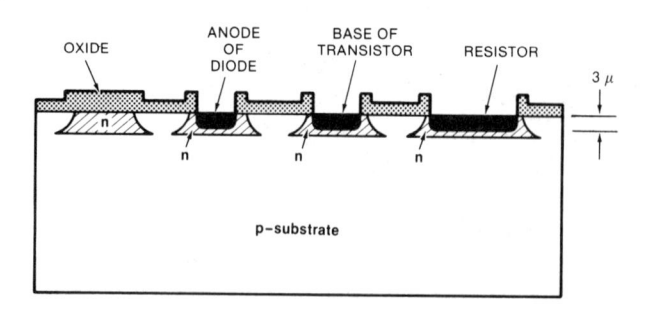

(C) Base diffusion.

Fig. 3-31. Steps in the fabrication of the

Electronics: Circuits and Systems

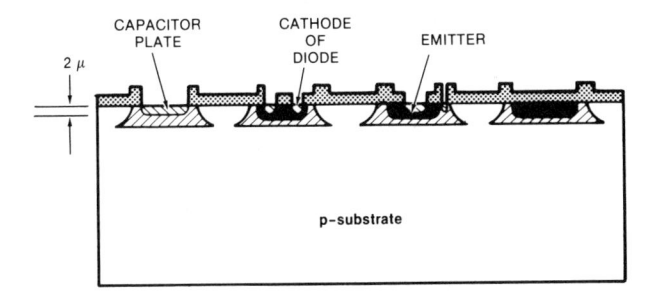

(D) Emitter diffusion.

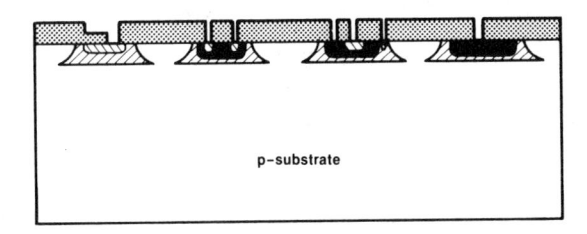

(E) Pre-ohmic etching.

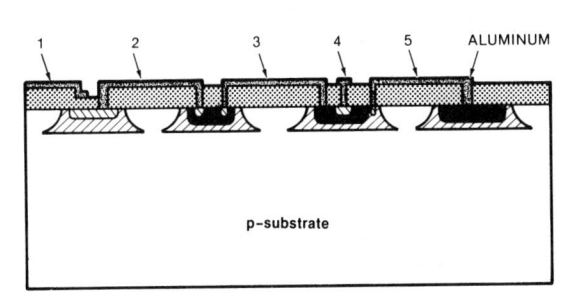

(F) Metallization.

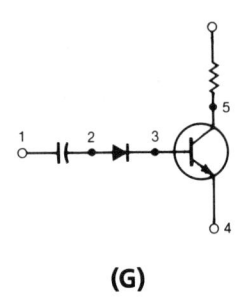

(G)

IC form of the circuit of Fig. 3-29.

(A) Structure.

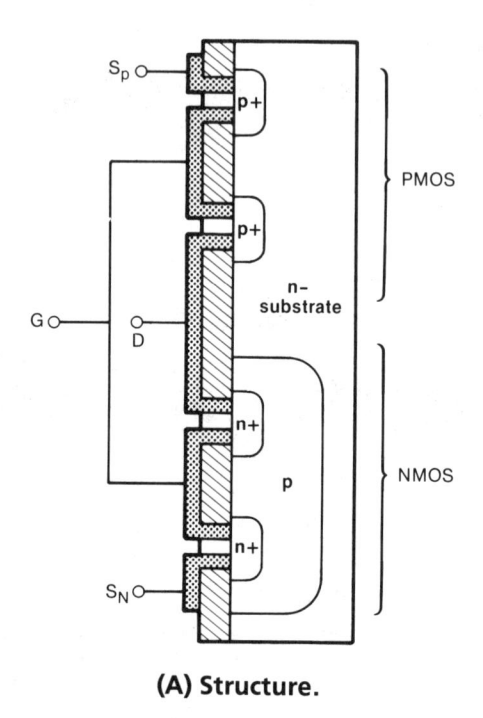

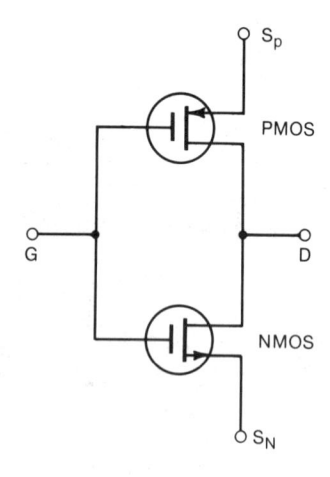

(B) Schematic diagram.

Fig. 3-32. Structure and schematic diagram of a CMOS.

Electronics: Circuits and Systems 211

Schottky Diodes and Transistors

As was mentioned earlier, aluminum acts as a p-type impurity in a semiconductor, and if it is deposited directly on an n-type semiconductor, a pn junction results. Such a pn junction is known as the *Schottky diode*, and it has found wide use in logic circuits because of its high speed of response. The symbol of the Schottky diode and the IC fabrication are shown in Fig. 3-33.

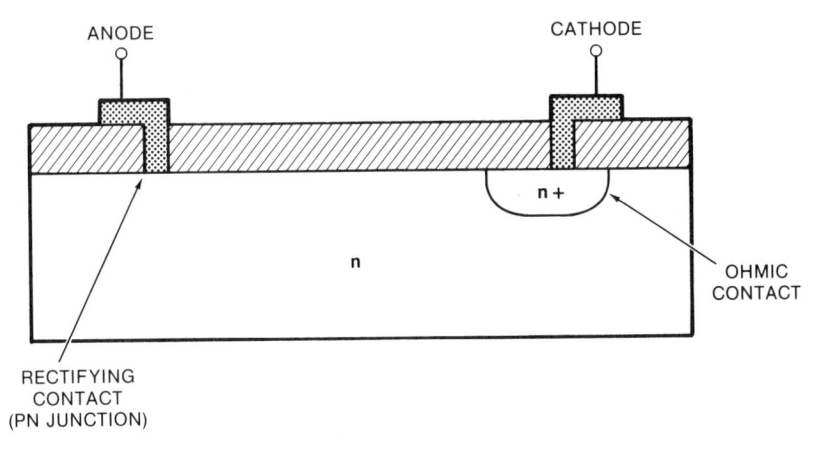

(A) Structure.

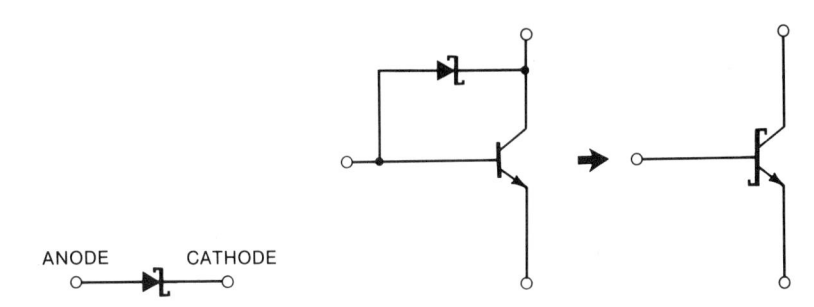

(B) Symbol of a Schottky diode.

(C) Schematic arrangement and symbol of a Schottky diode.

Fig. 3-33. Schottky diode and transistor.

The combination of a Schottky diode and an npn transistor, indicated schematically in Fig. 3-33C, is known as a *Schottky transistor*. The Schottky transistor is used widely in logic circuits because of its high speed of response.

Biasing Circuits for ICs Using the Current Mirror

As will be seen in the next chapter, the normal biasing for transistor circuits involves the use of resistors. Also, bypass capacitors are necessary due to considerations based upon frequency response. In integrated circuits, resistors and capacitors are wasteful since they occupy large areas of the chip. Also, the accuracy of fabricating a specified value of resistance is extremely poor with very high variations. It is easier to fabricate a pair of resistances whose *ratio* remains quite accurate even though the individual values may vary beyond tolerable limits. Because of these considerations, the biasing of transistors in ICs uses a basic circuit called the *current mirror*.

The basic principle of the current mirror, shown in Fig. 3-34, is that if two transistors are *identical* with *equal emitter-based voltages*, then their *collector currents will also be equal* to each other independent of other connections. In the circuit of Fig. 3-34, the transistors T_1 and T_2 are fabricated to be identical in their characteristics. Also, the base-to-emitter voltages are made equal by connecting the bases together and the emitters together. Therefore,

$$I_{C1} = I_{C2}$$

Since the transistor T_1 has its base and collector tied together, it acts *as a diode* with both the collector and the base being at the same voltage. If V_{B1} is taken as 0.7 V, as we usually do for a forward biased diode, then the current I_{C1} is given by

$$I_{C1} = (V_{CC} - 0.7)/R_1 \approx V_{CC}/R_1$$

if V_{CC} is sufficiently large. Neglecting the base currents (which is justified for high β transistors), we have

$$I_{C2} \approx (V_{CC}/R_1)$$

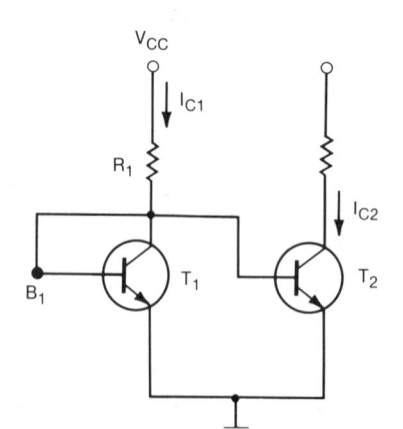

Fig. 3-34. Current mirror.

Electronics: Circuits and Systems

The collector current of T_2 is the same as the current in the resistor R_1 and hence the name "current mirror."

Advantages and Disadvantages of ICs

The most obvious advantage and the reason for the phenomenal success of the ICs is that highly complex systems can be packaged in a small space, which not only makes the systems physically small but reduces the usual problems that occur in the interconnection of circuits in discrete systems. Among the many developments that are taking place in IC technology is the work currently in progress on *three-dimensional IC* fabrication. ICs are fabricated in a thin silicon film on top of an insulator and by interweaving insulating and silicon layers, it is possible to have a stacking of circuits on a single chip. Advances like this open up completely new possibilities in the processing of signals and information.

The disadvantages (or constraints that deserve careful attention in the design of IC systems) are rather low limits on the maximum values of resistances and capacitances that can be fabricated, poor pnp and PMOS transistors, limitations on the frequency response due to parasitic capacitances, low power dissipation capabilities, and the inability to fabricate inductors and transformers. These disadvantages, however, are so outweighed by the advantages and new territories opened up by IC technology that microelectronics is one of the most exciting fields of research and development at present.

SUMMARY SHEETS

SEMICONDUCTORS

Semiconductors have low conductivity owing to the covalent bond between neighboring atoms in a crystal.

Doping

N-type: Pentavalent element (arsenic for example) used as a substitutional impurity. Each impurity atom donates one free electron. Material becomes an n-type semiconductor.

P-type: Trivalent element (e.g., boron) used as a substitutional impurity. The vacancy created in the bonding structure is called a *hole* and acts like a free positively charged carrier. Each impurity atom donates a hole. Material becomes a p-type semiconductor.

Thermal Generation and Recombination

Electron-hole pairs are created due to thermal agitation (in undoped as well as doped semiconductor materials). These electrons and holes eventually recombine causing a release of energy in the form of heat or radiation.

PN JUNCTION

Depletion Region

Space charge layer built up at the junction due to the diffusion of holes from p to n and electrons from n to p, when the junction is first formed. The space charge is positive on the n side and negative on the p side.

Potential Barrier

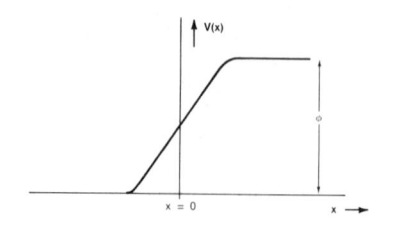

An electrostatic potential hill is set up by the space charge in the depletion region. The potential barrier inhibits the diffusion of electrons from n to p and holes from p to n. The height of the potential barrier depends upon the doping levels.

Drift Component of Current

Some of the electrons and holes generated due to thermal agitation *drift* by taking advantage of the favorable potential hill: electrons drift from p to n and holes from n to p. The drift component of current (which is directed from n to p) depends only upon the temperature.

Balance Between Diffusion and Drift

When there is no external bias applied to the junction, the diffusion component of the current (which is from p to n) is exactly cancelled by the drift component of current (which is from n to p) so that there is no net current flow across the junction.

Metal-Semiconductor Contacts

Contact potential appears at contacts between metals and semiconductors. A metal-semiconductor contact can be *ohmic* or *nonohmic* (also called rectifying). An ohmic contact permits the free flow of current in either direction, while a nonohmic contact shows a low resistance in one direction and a high resistance in the other. Nonohmic contacts form the basis of *Schottky* diodes.

PN JUNCTION UNDER EXTERNAL BIAS

Forward Bias

The positive terminal of a battery is connected to the p side of the junction and the negative terminal to the n side of the junction. The potential hill at the junction is lowered by an amount equal to the external (forward) bias. The diffusion component of current (due to

Electronics: Circuits and Systems

holes diffusing from p to n and electrons from n to p) increases to a large value, although the drift component is not affected. There is a significant flow of current in the direction p to n. The forward current increases *exponentially* with the forward bias.

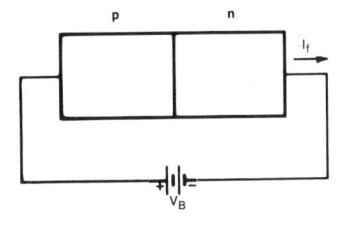

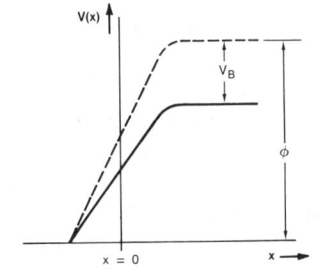

Reverse Bias

The negative terminal of a battery is connected to the p side of the junction and the positive terminal to the n side. The potential hill is raised by an amount equal to the applied reverse bias. The diffusion component of current is reduced to zero, but the drift component is not affected. There is an extremely small (order of microamperes or less) flow of current in the reverse direction (from n to p).

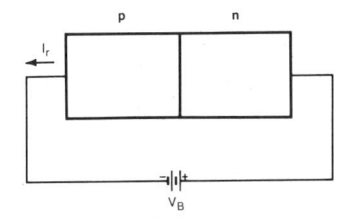

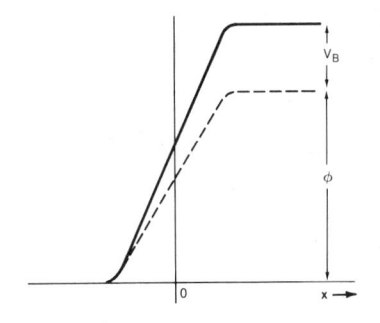

PN Diode

When $v > 0$, the diode is forward biased and

$$i = I_0 e^{v/0.0259}$$

where,

I_0 is called the *reverse saturation* current.

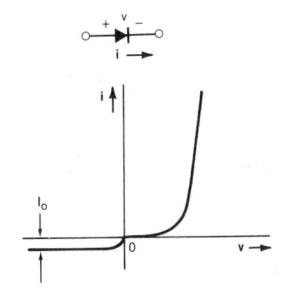

When $v < 0$, the diode is reverse biased and $i = -I_0$.

The diode acts as a unilateral device permitting an easy flow of current in the direction p to n, but opposing the flow of current in the direction n to p.

BIPOLAR JUNCTION TRANSISTORS

NPN Transistor

Electrons are injected from the emitter into the base due to the forward bias at the emitter-base junction. An extremely high percentage of these electrons is collected by the collector since the base region is quite narrow and the reverse bias at the collector-base junction is favorable to the flow of electrons from the base to the collector.

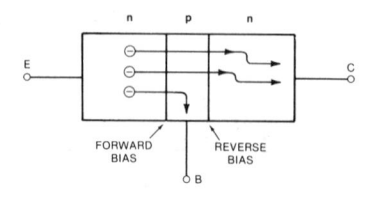

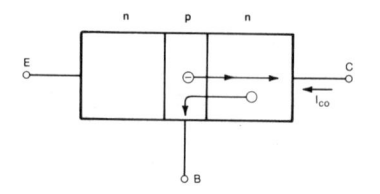

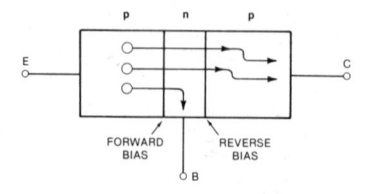

Symbols and Current Relationships

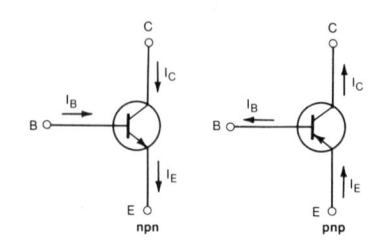

$I_E = I_B + I_C$

$I_C = I_{CO} + \alpha I_E$

I_{CO} can be neglected except at high temperatures and

$I_C = \alpha I_E$

For most transistors, α is so close to 1 that we can use $I_C = I_E$.

I_C and I_B are related by

$I_C = \beta I_B$

β is a measure of the current gain of the transistor. The symbol h_{FE} is used for β also.

These relationships are valid for both npn and pnp transistors.

Collector Cutoff Current

Electrons and holes are generated due to thermal agitation in the collector and base regions. Some of the electrons drift from the base to the collector and holes drift from the collector to the base. This current I_{CO}, the collector cutoff current, is negligibly small at normal temperatures but increases by a factor of two for every 10°C increase in temperature.

PNP Transistor

Holes are injected from the emitter into the base due to the forward bias at the emitter-base junction. A high percentage of these holes is collected by the collector, since the base region is quite narrow and the reverse bias at the collector-base junction is favorable to the flow of holes from the base to the collector.

A collector cutoff current I_{CO} is present for the same sort of reasons as in the npn transistor and behaves in the same manner.

JUNCTION FIELD-EFFECT TRANSISTORS (JFET)

N-Channel JFET

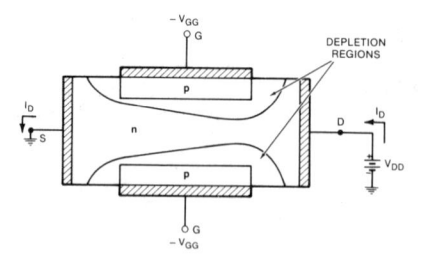

Drain is made positive with respect to the source; gate is made negative with respect to the source (reverse bias). Depletion region extends deeply into the channel and is wider near the drain than near the source.

The current decreases as the gate is made more negative for a given drain-to-source voltage. Current flow in the device can be controlled by varying the gate-to-source voltage.

Electronics: Circuits and Systems

Pinch-off Condition

As the gate becomes sufficiently negative (with respect to the source), the channel is pinched off and the current becomes independent of the drain-to-source voltage.

P-Channel JFET

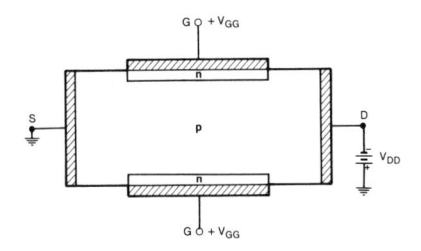

Drain is made negative with respect to the source; gate is made positive with respect to the source (reverse bias). Depletion region extends deeply into the channel and is wider near the drain than near the source.

The current decreases as the gate is made more positive for a given drain-to-source voltage. Current flow in the device can be controlled by varying the gate-to-source voltage.

A pinch-off condition occurs in a manner similar to that in the n-channel JFET.

Symbols and Currents in JFETS

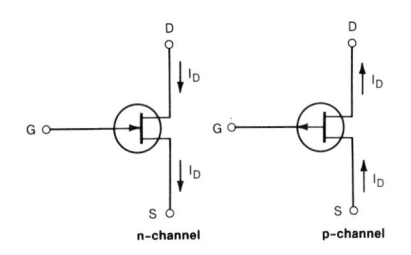

Gate current is essentially zero. Current flow is from *drain to source* in the *n-channel* JFET (since electrons flow from source to drain). Current flow is from *source to drain* in the p-channel JFET.

The drain current increases at first as the (magnitude of the) drain-to-source voltage is increased but then levels off with the onset of the pinch-off condition.

THE MOSFET

Enhancement NMOS

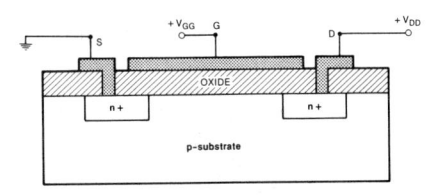

No actual type channel is fabricated between the source and the drain. There is no current flow when the gate is at zero bias. When a positive potential is applied to the gate, a channel containing free electrons is induced between the source and the drain, and current can flow. The *threshold voltage* is the minimum gate voltage needed to cause significant current flow. As the gate voltage increases above the threshold voltage, the current increases.

Enhancement PMOS

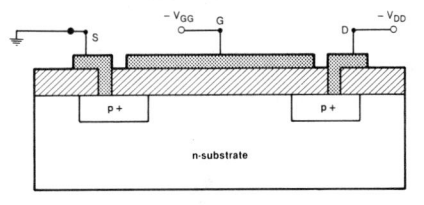

As in the enhancement NMOS, no actual channel is fabricated between the source and the drain, and there is no current flow when the gate is at zero bias.

When a negative potential is applied to the gate, a channel containing holes is induced between the source and the drain, and current can flow. The threshold voltage is negative, and the current flow is significant only when the gate voltage is at a more negative value than the threshold voltage. As the gate voltage becomes more negative than the threshold level, the current increases.

Depletion NMOS

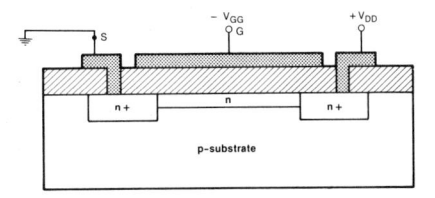

An actual n-type channel is fabricated between the source and the drain, and current can flow even when the gate is at zero bias. The gate is operated at a negative potential with respect to the source, and the operation of the device is similar to the n-channel JFET.

Depletion PMOS

As in the depletion NMOS, an actual channel is fabricated but with a p-type material between the source and the drain. Current can flow even when the gate is at zero bias. The gate is operated at a positive potential with respect to the source, and the operation of the device is exactly similar to the p-channel JFET.

Symbols of a MOSFET

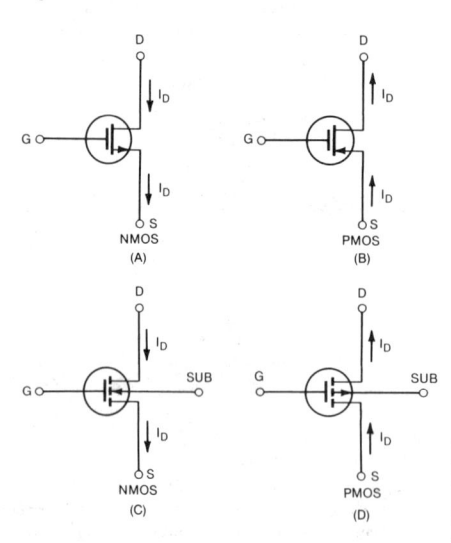

Symbols C and D are used only for the enhancement mode MOS, while A and B are used with either the enhancement or the depletion mode MOS devices. The present practice is to use A and B in preference to the other two.

INTEGRATED CIRCUITS

Resistors in ICs

Sheet resistance: For a resistive material of uniform thickness, the resistance of a square shaped sheet is the same regardless of the size of the square. The sheet resistance is expressed in ohms/square.

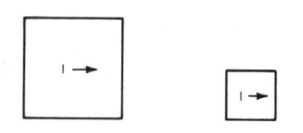

To obtain a desired resistance, the number of squares needed to produce that resistance is laid out.

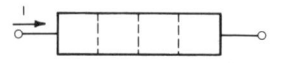

Schottky Diodes

A metal to n-type semiconductor contact can exhibit rectifying diode-like properties. Schottky diodes are fabricated by using aluminum on n-type semiconductor materials. They have a very high speed of response, and they are widely used in digital circuits.

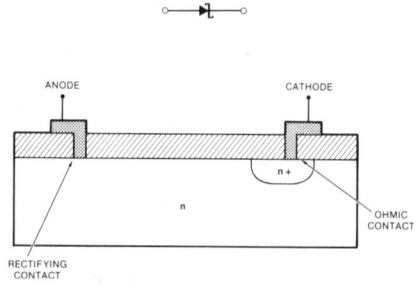

Current Mirror

In the current mirror, the collector current of T_2 is made to be the same as that in T_1 by making V_{BE} the same for both. The current mirror concept is extensively used as a biasing circuit in ICs.

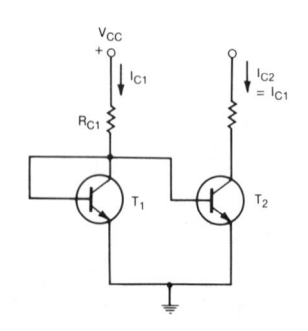

Electronics: Circuits and Systems

REVIEW QUESTIONS

Semiconductors
3.1 The conductivity of a *metal* is due to the presence of a large number of _____ .

3.2 What are the two commonly used semiconductor materials? How many valence electrons/atom do they have?

3.3 What are the approximate voltage levels needed to break the covalent bonds in a semiconductor material?

3.4 What type of carrier is made available by an n-type impurity? Name one or more elements that can be used as an n-type impurity.

3.5 What type of carrier is made available by a p-type impurity? Name one or more elements that can be used as a p-type impurity?

3.6 State the difference between an *extrinsic* and an *intrinsic semiconductor.*

3.7 What is meant by *thermal generation* and *recombination* in a semiconductor? What is the one controlling factor in this process?

PN Junction
3.8 Which kinds of charges are present in the depletion region of a pn junction: free electrons, holes, positive ions, negative ions?

3.9 What polarity of space charge (positive or negative) is present on the p-side of the pn junction?

3.10 What gives rise to the *forward current* in a pn junction? What gives rise to the *reverse current* in a pn junction?

3.11 Which terminal of a battery should be connected to the p-side terminal of a pn junction to provide forward bias?

3.12 How does forward bias affect the height of the potential barrier? How does it affect the magnitude of the forward current? How does it affect the magnitude of the reverse current? How do the magnitudes of the forward current and the reverse current compare?

3.13 Answer the questions of 3-12 for reverse bias of a pn junction.

3.14 What is the range of values of the current in a forward-biased pn junction? What is the range in a reverse-biased junction?

3.15 What are the approximating assumptions made for an ideal diode? What are the two resulting models?

Bipolar Junction Transistor
3.16 Which junction in a BJT is forward biased and which is reverse biased?

3.17 What are the three currents in a BJT? How are these related?

3.18 What gives rise to the collector cutoff current I_{CO}? How does I_{CO} vary with temperature?

3.19 What is the parameter that relates the *collector current* to the *emitter current*? What parameter relates the *collector current* to the *base current*? How are these two parameters related to each other?

3.20 Which of two BJTs will be the better one to use: one with a value of α as close to 1 as possible or one with a value of α not very close to 1?

3.21 Draw a symbol of an npn transistor. Show how you would connect two batteries to provide the proper bias.

3.22 Each of the diagrams in Fig. 3-35 contains a BJT. In each case, the values of two of the currents are given. Identify the base, emitter, and the collector terminals. State whether each is a pnp or an npn transistor. Calculate the value of β for each.

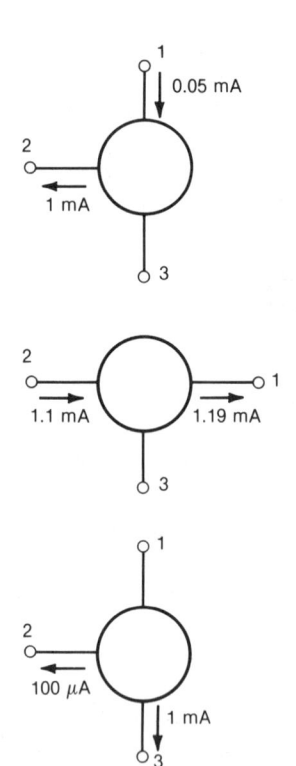

Fig. 3-35. Three diagrams for Review Question 3-22.

Field-Effect Transistor

3.23 What is the main difference between a *BJT* and an *FET* in terms of the flow of current?

3.24 Which type of carrier is active in the operation of a p-channel JFET? Which is active in an n-channel JFET?

Electronics: Circuits and Systems

3.25 Name the three terminals of a FET. What are the analogous terminals in the BJT?
3.26 What is the restriction on the gate voltage (in relation to the source voltage) in a JFET?
3.27 How is the width of the channel available for carrier flow controlled in a JFET?
3.28 In a p-channel JFET, does the current increase or decrease as the gate voltage is increased? Why?
3.29 What happens to the current flow in the pinch-off condition?
3.30 What is the direction of the current flow in a *p-channel JFET* (drain to source or source to drain)? What is the direction in an *n-channel JFET*?
3.31 To increase the current in an n-channel JFET, should the gate voltage be made more negative or less negative?
3.32 Draw the symbol of an n-channel JFET. Show how you would connect two batteries so as to provide the proper bias.

MOSFET

3.33 What are the two types of MOSFETS (in terms of the *mode* of fabrication)?
3.34 What is the important distinction in the structure between an enhancement-mode MOSFET and a depletion-mode MOSFET?
3.35 What type of carrier is active in the current flow in an n-channel MOSFET?
3.36 An induced channel is created in an n-channel enhancement-mode MOSFET. What polarity should the gate voltage be for this?
3.37 What polarity should the gate voltage be in an n-channel depletion-mode MOSFET?
3.38 Is there an induced channel in the depletion-mode MOSFET?
3.39 Draw the symbol or symbols for an NMOS device.

Integrated Circuits

3.40 Name the three types of integrated circuits.
3.41 What is a *mask* in IC fabrication? What is the significance of the pattern on each mask?
3.41 Does the clear area in a mask create windows in the silicon wafer (for diffusion of impurities), or is it the opaque area that does it?
3.42 What are the uses of an oxide layer in IC fabrication?
3.43 Which area of a wafer is polymerized: the area exposed to the ultraviolet or the area not so exposed because of the opaque portion of a mask?
3.44 For a sheet of resistive material of uniform thickness, does the resistance of a square area vary due to the dimensions of the square?

3.45 If it is desired to make a 10-kΩ resistor with a sheet of resistance 500 Ω square, and if the width of the resistor is to be 1 mil, what will be its length?

3.46 Why is the resistor in an IC fabricated in a folded-up fashion instead of as a single long strip?

3.47 What forms the dielectric of a capacitor in ICs? How are the two plates (electrodes) formed?

3.48 What is a *Schottky diode*?

3.49 How is a rectifying metal semiconductor diode avoided in IC fabrication? Is this precaution necessary in both the cases: attaching a metal contact to a p-type semiconductor and attaching a metal contact to an n-type semiconductor?

3.50 Name the different typical stages in the fabrication of an IC. In each of these stages what processes are repeatedly used?

3.51 What is a *current mirror*?

3.52 State the advantages and disadvantages of ICs.

CHAPTER 4

DIODES AND DIODE CIRCUITS

A *diode* is a two-terminal device that conducts freely in one direction but offers a very high resistance to current flow in the opposite direction; it is a unilateral device. Diodes are used in a variety of circuits in electronics and communications, such as for example, in the conversion of ac voltages to dc, logic circuits, and demodulators. It was seen in the last chapter that the pn junction has the unilateral property and acts as a diode. Even though we will use pn junction diodes in the following discussion, not all diodes are made of semiconductor junctions. But the principles and properties of the diode circuits presented in this chapter are applicable to all diodes.

REVIEW OF THE PN JUNCTION

A brief summary of the essential properties of the pn junction is presented diagrammatically in Fig. 4-1. Under *no bias*, Fig. 4-1A, the diffusion of holes from p to n and electrons from n to p is balanced by the drift of electrons from p to n and holes from n to p; the net current is zero. The reverse current (due to the drift of holes from n to p and electrons from p to n) depends upon the temperature. Under *forward bias*, Fig. 4-1B, the diffusion component of the current increases considerably while the reverse current stays at the same value as under no bias. The current is essentially equal to the forward current set up by the flow of holes from p to n and electrons from n to p. The forward current increases exponentially as a function of the voltage applied to the diode. Under *reverse bias*, the diffusion component of the current reduces to zero, and the only current flow in the device is the small reverse current due to the drift of holes from n to p and electrons from p to n, as indicated in Fig. 4-1C. The voltage-current characteristics and the symbol of a diode are shown in Figs. 4-1D and E.

(A) With no bias, the diffusion component I_f and the drift component I_r cancel each other and $I = 0$.

(B) With a forward bias (p-side made positive), the diffusion component I_f is much greater than the drift component and $I \approx I_f$.

(C) With a reverse bias (p-side made negative), the diffusion component becomes negligibly small and $I \approx -I_r$.

(D) I-v characteristics.

(E) Symbol of a diode.

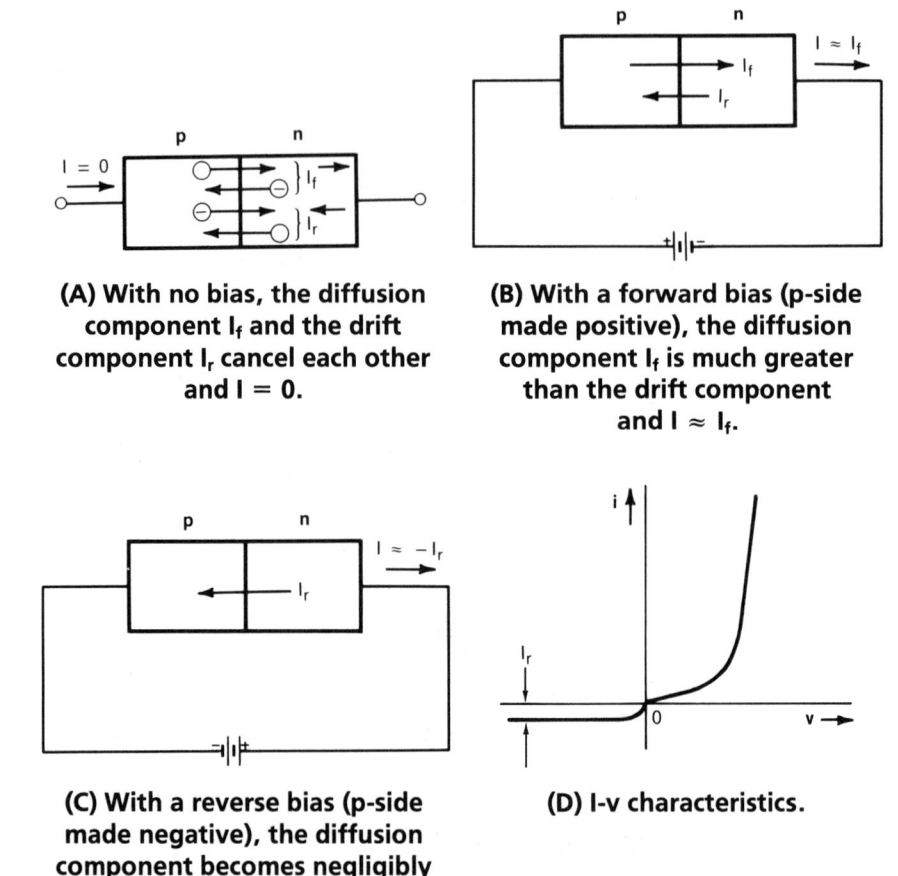

Fig. 4-1. Summary of the properties of a pn junction.

Ideal Diode Approximation

For purposes of analysis, it is convenient to replace the v-i characteristic of the diode, Fig. 4-1E, by the two straight line segments as shown in Fig. 4-2A. This approximation is called the *idealized-diode characteristic*. According to the ideal diode characteristic, we have the following two situations. (1) When

Electronics: Circuits and Systems

the voltage v is *less than* V_D, the diode acts like an *open circuit* and *does not conduct* any current. (2) When the voltage v is *greater than* V_D, then the diode conducts freely with a *zero resistance* but has a *constant voltage drop* of V_D across it. Based on these two operating conditions, an ideal diode can be represented by two simple models, one valid for $v < V_D$ and the other for $v > V_D$, as shown in Fig. 4-2B.

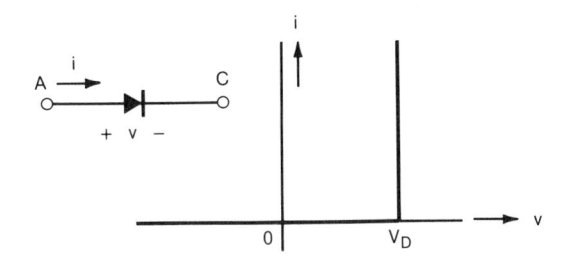

(A) Idealization of the i-v characteristics of a diode.

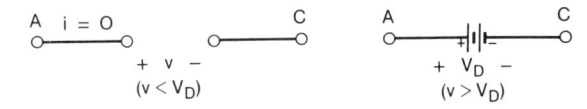

(B) Models of an ideal diode under reverse and forward bias.

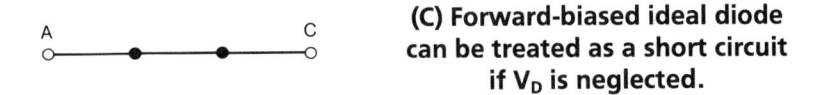

(C) Forward-biased ideal diode can be treated as a short circuit if V_D is neglected.

Fig. 4-2. Properties of an *ideal* diode.

The *voltage V_D* that acts as the breakpoint between conduction and no conduction in the ideal diode is usually taken as *0.7 V*. In a number of practical circuits, where the applied voltage is much greater than V_D, the voltage V_D can be neglected and the diode treated as a simple *short circuit under forward bias*, as indicated in Fig. 4-2C.

RECTIFICATION OF AC POWER

The dc power needed to operate electronic circuits is often obtained by converting available ac power to dc power by means of diode circuits. Such a conversion is called *rectification*. The basic rectifier is the *half-wave rectifier* shown in Fig. 4-3. When the input voltage v_s is less than V_D (needed to make the diode conduct), there is no current through the diode and the output voltage is zero. When the input voltage v_s exceeds V_D, the diode conducts and there is a voltage drop of V_D across it. The output voltage now follows the input voltage (except for the diode drop V_D). These two conditions are shown in Figs. 4-3B and C. If the voltage drop across the diode (when it is conducting) is neglected, the output voltage is seen to be simply the *positive half-cycle* portions of the input voltage, as shown in Fig. 4-3D.

When a dc voltmeter is used to measure a given voltage, it responds to the *average value* of the waveform of that voltage. The average value of a periodic waveform (also called its *dc component*) is defined by

$$V_{dc} = \frac{\text{area under the waveform in one period}}{T} \qquad (4\text{-}1)$$

where,

T represents the period of the waveform.

In a normal sinusoidal voltage, such as the input v_s in the circuit of Fig. 4-3, the area of each positive half-cycle is cancelled by the area of each negative half-cycle so that the net area in a period becomes zero. A dc voltmeter will read zero when measuring v_s. But in the case of a rectified half wave like the output v_0 of the circuit of Fig. 4-3, there are no negative half cycles, which means that the net area in a period is not zero. The output v_0, therefore, has a dc component that can be shown to be related to the peak value V_m by the equation

$$V_{dc} = (V_m/\pi)$$
$$= 0.318 V_m \quad \text{(half-wave rectifier)}$$

For example, if the input to a half-wave rectifier is the household ac power supply with a peak value of $V_m = 160$ V (approximately), the dc component in the output will be $(0.318 \times 160) = 50.9$ V.

Even though the half-wave rectifier converts an ac waveform with zero dc component into a voltage with a significant dc component, it is clear that the output is nowhere near an ideal dc voltage. On the contrary, it is a time-varying *periodic* signal. As was discussed in Chapter 1, such a waveform has a dc component, a fundamental frequency component, and second and higher harmonics. When the input is a 60-Hz ac supply, the output will contain a dc component, a 60-Hz component, a 120-Hz component, and so on. If we used a half-wave rectifier of the form in Fig. 4-3 to provide dc power to drive an electronic circuit, there would be a terrible hum due to the harmonic components.

Electronics: Circuits and Systems 227

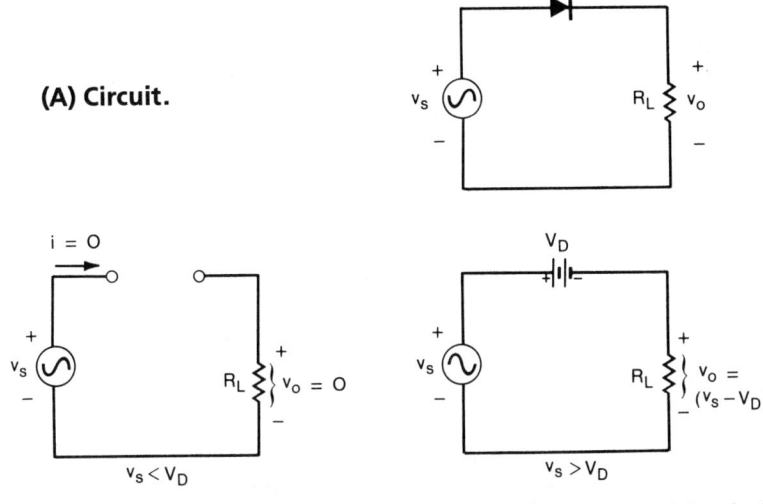

(A) Circuit.

(B) Circuit during negative half cycle.

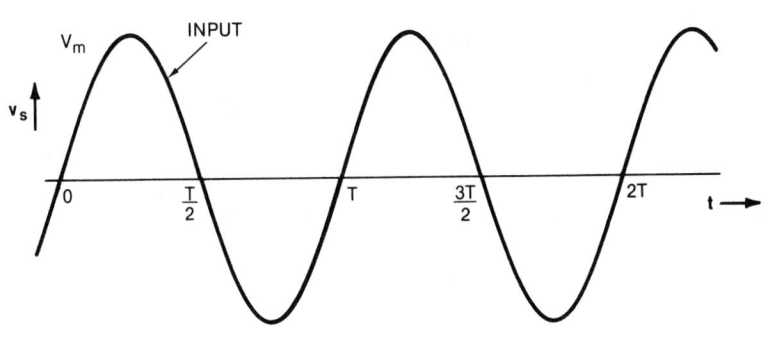

(C) Circuit during positive half cycle.

(D) Input waveform.

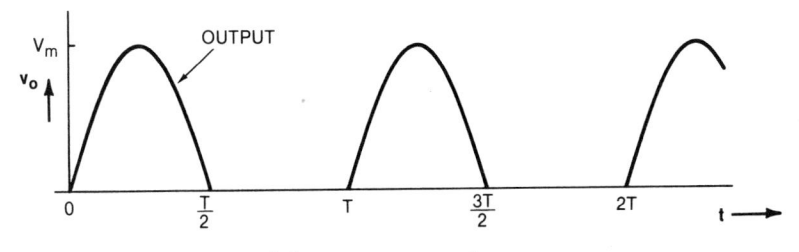

(E) Output waveform.

Fig. 4-3. Half-wave rectifier.

Example 4.1

(The half-wave rectifier circuit can be used with any time-varying input signal.) Determine the output waveform of the half-wave rectifier when the input is the triangular waveform shown in Fig. 4-4.

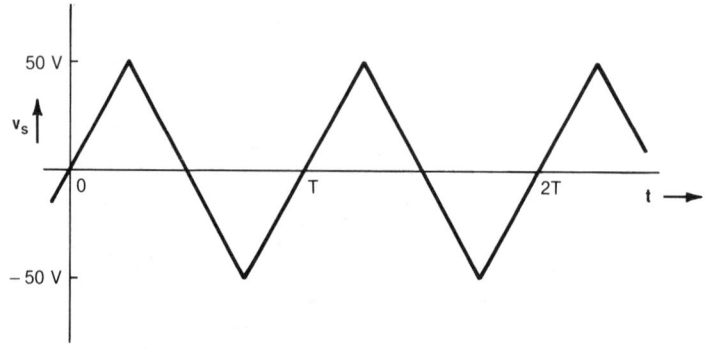

Fig. 4-4. Input signal for Example 4.1.

Solution

The output waveform consists of the positive half cycles as shown in Fig. 4-5.

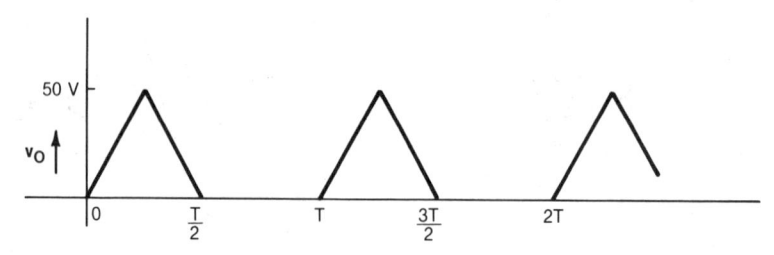

Fig. 4-5. Output waveform in Example 4.1.

$$V_{dc} = \frac{\text{area in the interval from } t = 0 \text{ to } t = T}{T}$$

$$= \frac{(1/2) \times 50 \times (T/2)}{T}$$

$$= 12.5 \text{ V}$$

Drill Problem 4.1: A voltage signal v_a has the waveform shown in Fig. 4-6. If it is applied to the input of a half-wave rectifier, sketch the output voltage waveform and calculate the dc component.

Electronics: Circuits and Systems

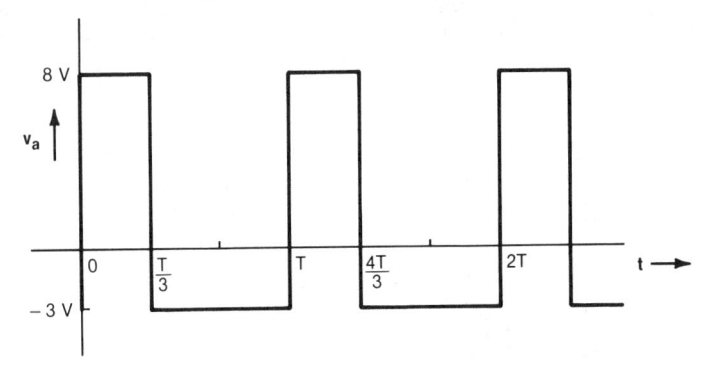

Fig. 4-6. Input signal for Drill Problem 4.1.

Drill Problem 4.2: Suppose the signal v_a of Fig. 4-6 is connected to a half-wave rectifier with the polarities reversed as indicated in Fig. 4-7. Sketch the output voltage waveform and calculate the dc component.

Fig. 4-7. Circuit connections for Drill Problem 4.2.

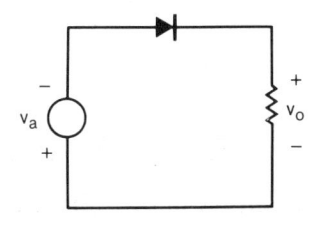

A *full-wave rectifier* presents an improvement over the half-wave rectifier and produces a dc component twice that in the half-wave circuit. One form of the full-wave rectifier is shown in Fig. 4-8. Neglecting the voltage drop V_D across the diode while conducting, the operation of the circuit is as follows.

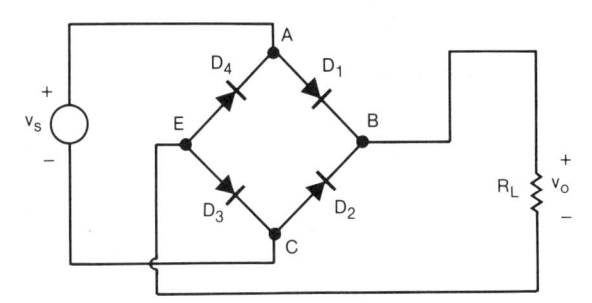

Fig. 4-8. Full-wave rectifier circuit.

When v_s is positive, point A in the circuit is at a higher potential than B or E, and C is at a lower potential than B or E. The diodes D_1 and D_3 are for-

230   **Diodes and Diode Circuits**

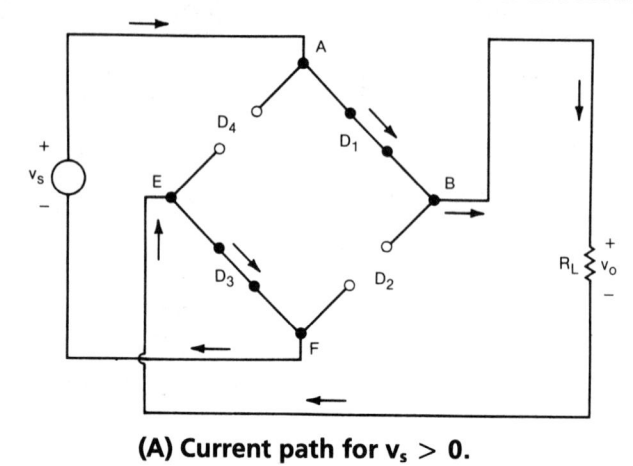

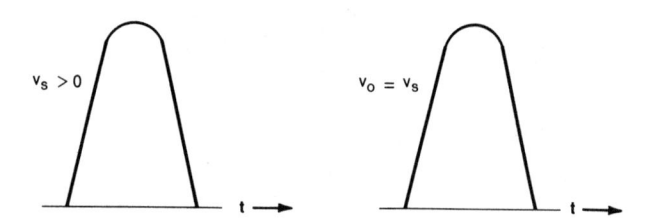

(A) Current path for $v_s > 0$.

(B) Conditions in the full-wave rectifier when input v_s is positive.

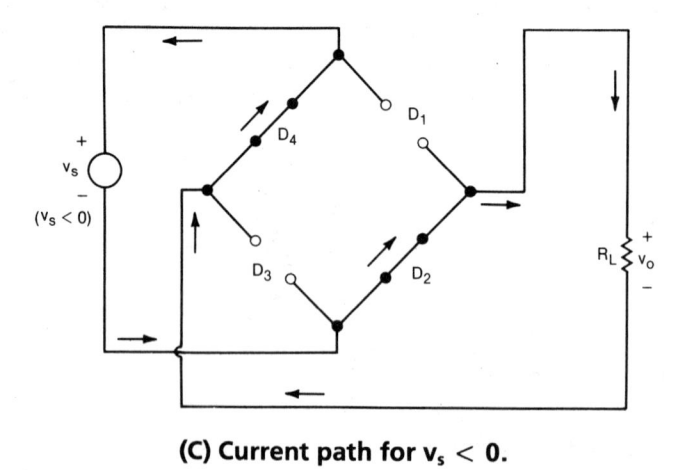

(C) Current path for $v_s < 0$.

Fig. 4-9. Full-wave

Electronics: Circuits and Systems

ward biased, and D_2 and D_4 are reverse biased. The flow of current is as shown in Fig. 4-9A, and the output voltage will be as shown in Fig. 4-9B. It is a replica of the positive half cycle of the input.

When v_s is negative, point A is at a lower potential than B or E, and C is at a higher potential than B or E. The diodes D_1 and D_3 are reverse biased, and D_2 and D_4 are forward biased. The flow of current is as shown in Fig. 4-9C. It is important to note that as far as the load resistor is concerned, the direction of the current flow in the present situation is the *same* as when v_s was positive (that is, the same as in Fig. 4-9A). Therefore, the voltage across the load resistor is again *positive*. The circuit has taken the negative half cycle of the input wave and flipped it over to be positive as indicated in Fig. 4-9D.

The output voltage of the full-wave rectifier will therefore consist of a series of successive half cycles as shown in Fig. 4-9E. The positive half cycles of the input voltage are reproduced exactly while the negative half-cycles of the input have been flipped over and made positive. The dc component of the full-wave rectified sinusoid is twice that of the half-wave rectified sinusoid since it has twice as much area in T seconds as the latter, that is,

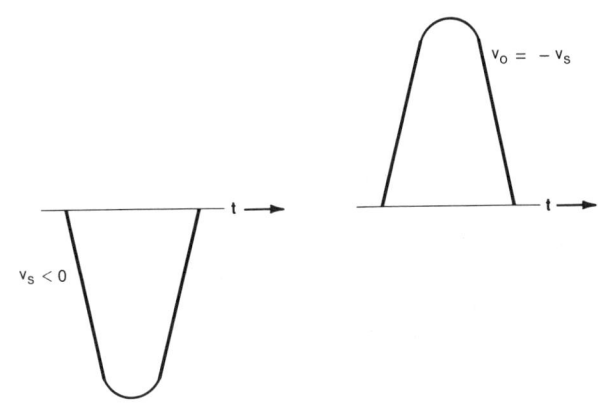

(D) Conditions when input v_s is negative.

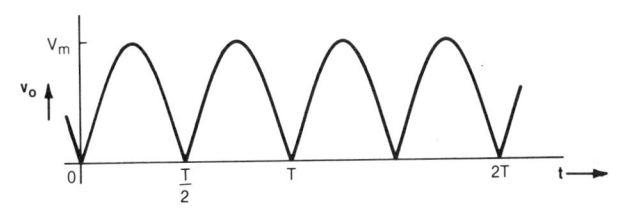

(E) Output waveform.

rectification.

$$V_{dc} = (2V_m/\pi)$$
$$= 0.636V_m$$

If the input to the full-wave rectifier is the 60-Hz household power supply, for example, the dc component of the output will be approximately (0.636 × 160) = 102 V. The output of a full-wave rectifier is periodic and will contain a number of harmonic components in addition to the dc component. But it can be shown that (in the case of the 60-Hz input signal) the output has a *fundamental frequency of 120 Hz* and not 60 Hz. Therefore, it contains 120 Hz and multiples of 120 Hz, with the 120-Hz component being dominant. If the full-wave rectifier circuit of Fig. 4-8 is used to drive an electronic circuit, then a noticeable 120-Hz hum will be present—a familiar experience to anyone who has a cheap or malfunctioning power supply in an amplifier.

Example 4.2

The triangular waveform of Fig. 4-4 is applied to a full-wave rectifier. Draw the output waveform and calculate the dc component.

Solution

The output waveform is shown in Fig. 4-10. The dc component is twice that calculated in Example 4.1. $V_{dc} = 25$ V.

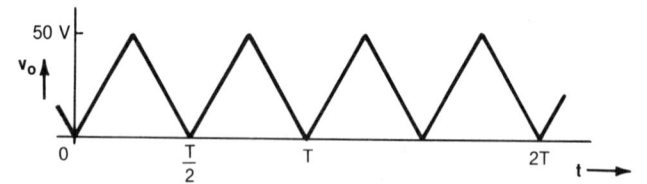

Fig. 4-10. Input signal for Example 4.2.

Drill Problem 4.3: The signal of Fig. 4-6 is applied to a full-wave rectifier. Sketch the output waveform and calculate the dc component in the output.

Drill Problem 4.4: Repeat Drill Problem 4.2 for a full-wave rectifier.

Peak Inverse Voltage in a Diode

When a diode is in the nonconducting state in a rectifier circuit, there is a voltage across it even though the current in it is zero. Consider the diode in the half-wave rectifier circuit when it is not conducting, as shown in Fig. 4-11. Since there is no voltage drop in the resistor (due to the absence of current), we can see that $v_d = -v_s$. The diode voltage increases as the input vol-

Electronics: Circuits and Systems

tage becomes more negative, and the maximum of the diode voltage occurs when the input is at its negative peak. *The highest voltage that appears across the diode when it is not conducting is known as the peak inverse voltage* (often abbreviated *piv*). For the half-wave rectifier, and also for the full-wave rectifier of Fig. 4-8, the peak inverse voltage on a diode is V_m. The peak inverse voltage is an important parameter since the diode must be able to withstand that value of voltage without breaking down.

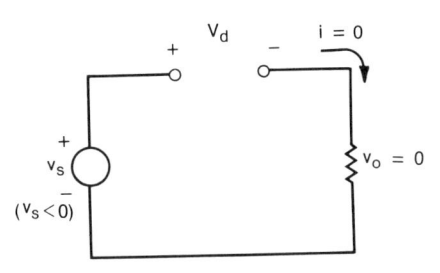

Fig. 4-11. Peak inverse voltage in a half-wave rectifier circuit.

Drill Problem 4.5: Calculate the peak inverse voltage across a diode in the cases of Examples 4.1 and 4.2.

Filters for Power Supplies

Neither the half-wave nor the full-wave rectifier is adequate for providing the dc power needed for electronic circuits because their harmonic components produce an audible hum. Filtering is essential in order to suppress the harmonics. The frequency response characteristics of capacitors and inductors are utilized to attenuate the harmonics and produce a voltage or current that is essentially a constant with negligibly small variations. One of the filters will be briefly discussed here.

Capacitor Filter

The simplest filter uses a capacitor to smooth out the variations in the output voltage of a rectifier. The circuit is shown in Fig. 4-12A.

Looking at the operation of the circuit from the viewpoint of *impedance*, a capacitor presents an open circuit for dc and the current in the circuit due to the dc component of the voltage passes only through the resistor. At 60 Hz and 120 Hz, the impedance of the capacitor can be made very low compared with the load resistance by choosing a sufficiently large value of C. The capacitor then serves as a bypass for the ac components of current in the cir-

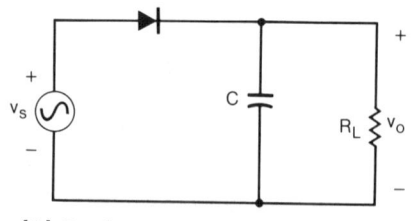

(A) Basic capacitor filter circuit.

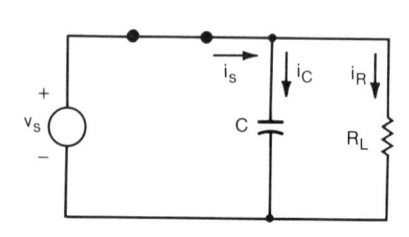

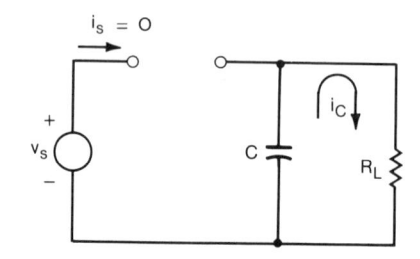

(B) Conditions when the diode is conducting; the capacitor charges up to V_m.

(C) Conditions when the diode is open: the capacitor discharges through the load resistor until its voltage becomes less than the input voltage v_s.

Fig. 4-12. Filtering a rectified sinusoid.

cuit. The result is that the output voltage will be predominantly dc with only a small proportion of ac components.

Looking at the circuit from the viewpoint of the charging and discharging of an RC circuit, when the input voltage v_s is greater than the capacitor voltage v_0, the diode conducts and a current flows that charges the capacitor. When the input voltage v_s is less than the capacitor voltage v_0, the diode stops conducting. Now the capacitor and the resistor are isolated from the input side, and the capacitor discharges through the resistor R_L. These two cases are shown in Figs. 4-12B and C.

When the diode is *conducting*, the capacitor voltage follows the input voltage and charges up to the peak value V_m. When the input falls below V_m, the diode stops conducting. The capacitor starts discharging through R_L. When the capacitor voltage falls below the value of v_s, the diode again starts conducting. This cycle repeats itself leading to the output voltage waveform of Fig. 4-13A. The output can be approximately represented by the triangular waveform of Fig. 4-13B. By choosing the time constant $R_L C$ sufficiently large, it is possible to make the variation ΔV_0 to be within a specified limit.

The capacitive filter has some disadvantages, such as the presence of cur-

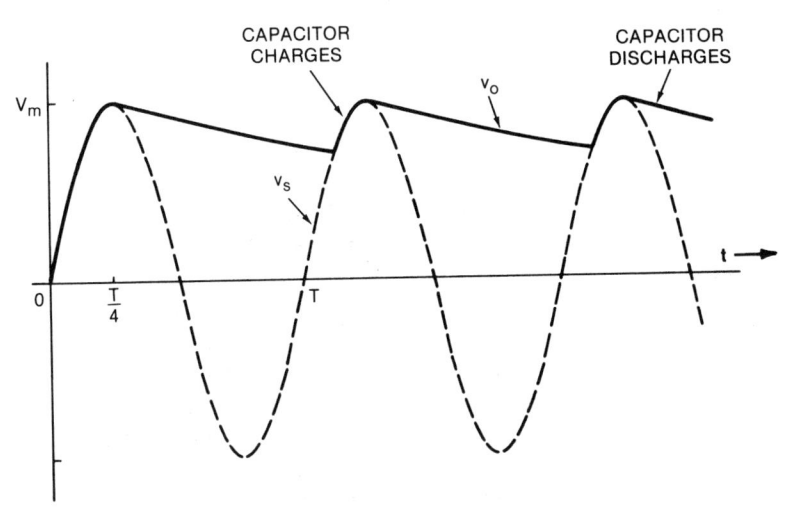

(A) Actual output voltage tracks the sinusoid while charging and is exponential while discharging.

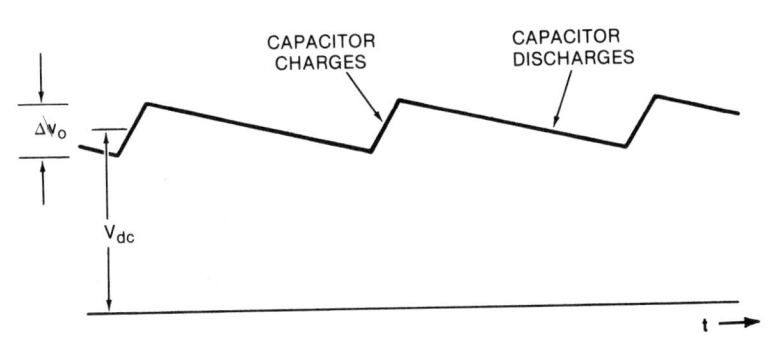

(B) Approximation to the output waveform, where both the charging and discharging portions are treated as straight line segments.

Fig. 4-13. Output waveforms of a capacitor filter.

rent spikes through the diode. Improvements are made by using a combination of an inductor (which smooths out the current flow) and a capacitor (which smooths out the voltage output).

Drill Problem 4.6: Calculate the peak inverse voltage across the diode in the capacitor filter.

MISCELLANEOUS APPLICATIONS OF DIODES

Clipping Circuits

It is sometimes necessary to clip the positive or the negative peaks or both of an input signal. For example, in radio communication, spurious peaks are created by the presence of noise, and these need to be removed for proper reception of the signal. A circuit to clip the positive peaks of a signal is shown in Fig. 4-14A. When the voltage at point A is greater than the battery voltage V_B by V_D (the diode drop when conducting), the diode conducts and the circuit is as shown in Fig. 4-14B. In this case, the output voltage becomes equal to ($V_B + V_D$) and remains a constant regardless of the shape of the input signal in that range. When the voltage at point A is less than the battery voltage V_B by V_D, the diode becomes an open circuit as shown in Fig. 4-14C, and the output voltage is equal to the input voltage since no current flows in the circuit. The output voltage for a specific set of values and input signal is shown in Fig. 4-15.

(A) Circuit.

(B) Conditions when the diode is conducting.

(C) Conditions when the diode is open.

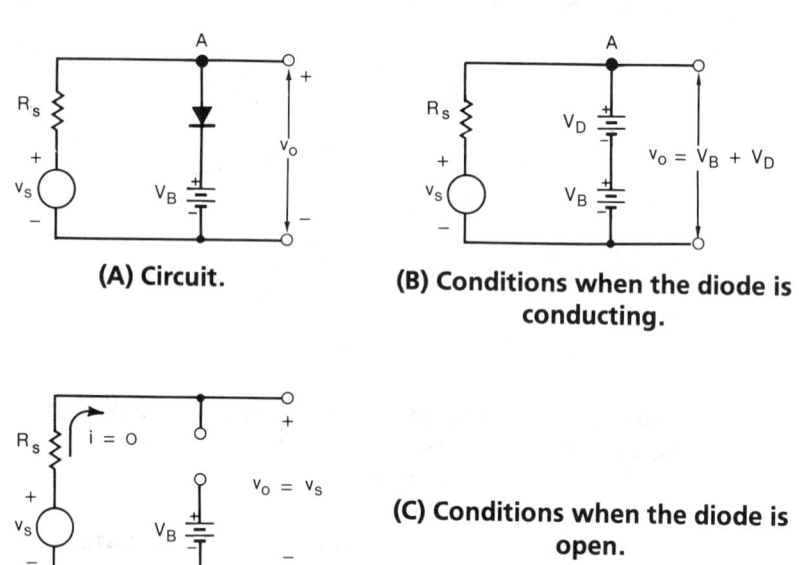

Fig. 4-14. Positive peak clipping circuit.

The level of clipping can be controlled by the battery voltage V_B.

Drill Problem 4.7: The circuit of Fig. 4-16 clips negative peaks of an input signal. Determine and sketch the output voltage.

Electronics: Circuits and Systems   237

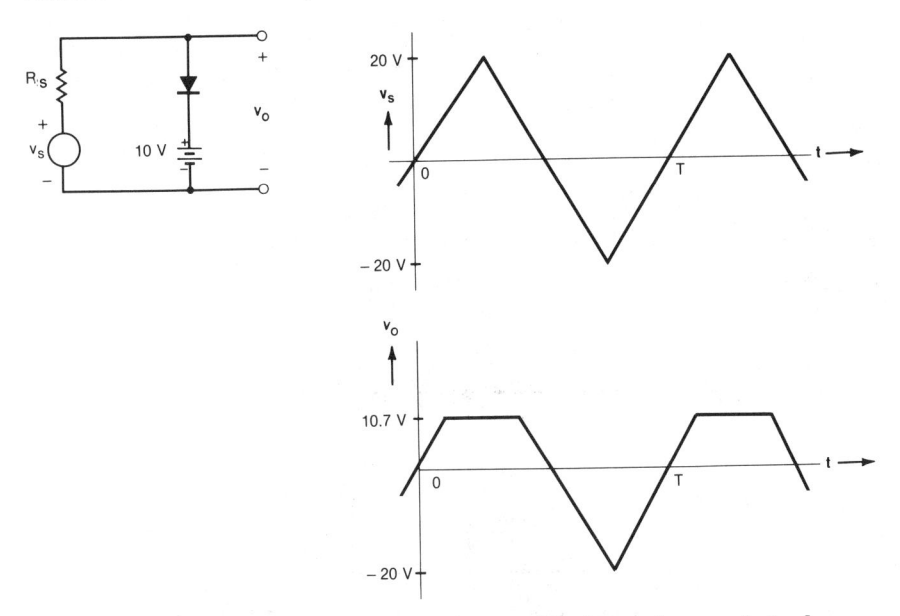

Fig. 4-15. Positive peak clipping circuit and associated waveforms.

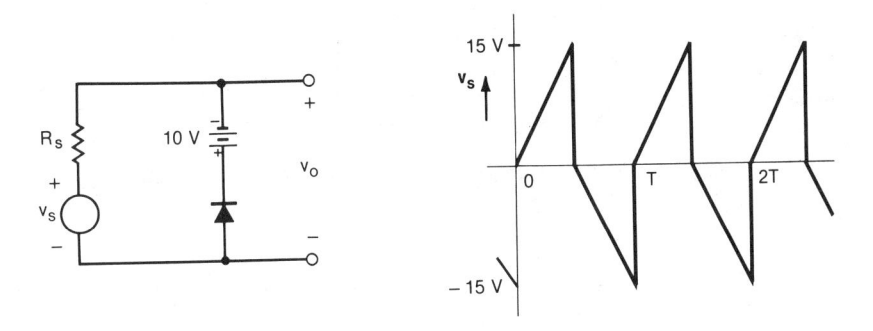

Fig. 4-16. Clipping circuit and input signal for Drill Problem 4.7.

A combination of the circuits of Figs. 4-14 and 4-16 leads to the one shown in Fig. 4.17, which clips both positive and negative peaks. There are three conditions (or *states*) of the circuit as shown in Figs. 4-17B, C, D. The output of the circuit for a specific input signal is shown in Fig. 4-17E.

238 Diodes and Diode Circuits

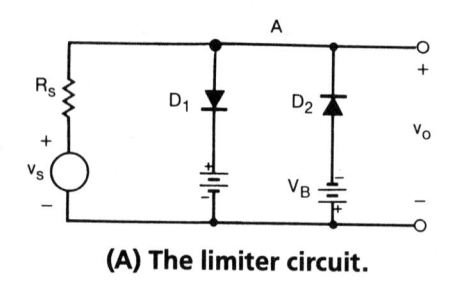

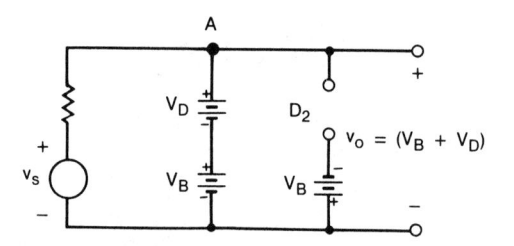

(A) The limiter circuit.

(B) Conditions when diode D_1 is on and D_2 is off.

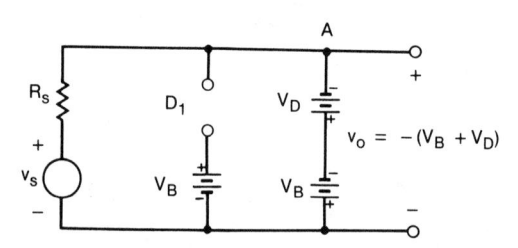

$v_o = (V_B + V_D)$

(C) Conditions when diode D_1 is off and D_2 is on.

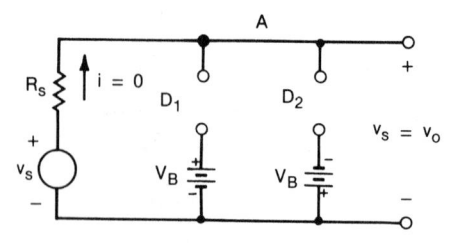

$v_o = -(V_B + V_D)$

(D) Conditions when both diodes are off.

$i = 0$, $v_s = v_o$

Fig. 4-17. Circuit to clip both

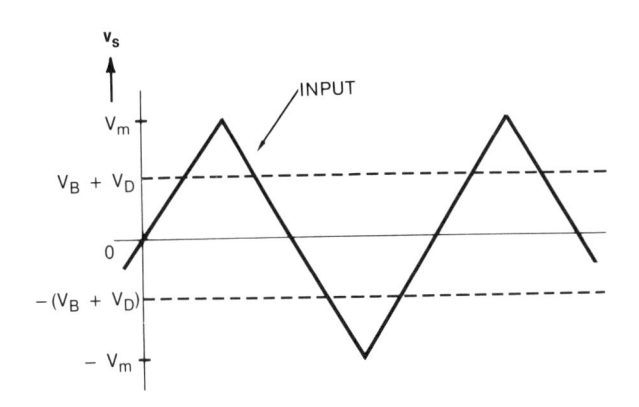

(E) Input and output voltage waveforms.

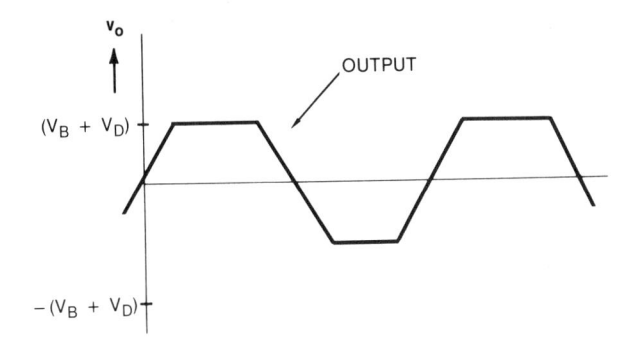

positive and negative peaks.

The circuit of Fig. 4-17 is also known as a *limiter* and is used in the demodulation of FM (frequency-modulated) signals to get rid of spurious spikes caused by noise introduced during the transmission through the atmosphere.

Drill Problem 4.8: Fig. 4-18 shows the input to the limiter of Fig. 4-17A. Determine and sketch the output waveform.

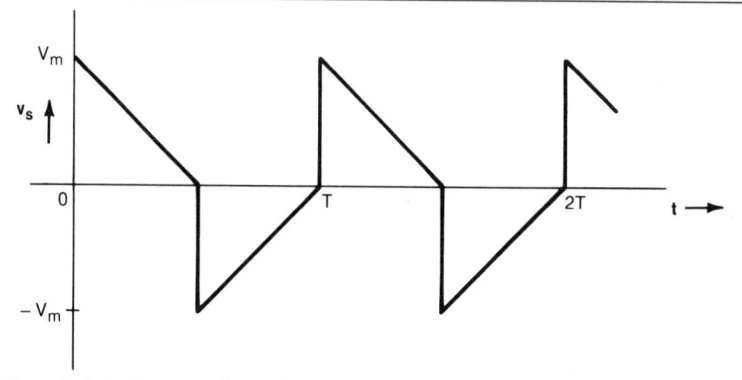

Fig. 4-18. Input signal to a limiter for Drill Problem 4.8.

Clamping Circuits

In the circuit of Fig. 4-19, the voltage across the capacitor starts at zero at t = 0, increases to V_m, and then it stays there. The diode does not conduct after the first quarter of the cycle of the input sine wave. The voltage across the diode is, therefore, given by

$$V_o = V_s - V_C$$
$$= V_s - V_m$$

(The voltage across an *open-circuit* diode is *not* zero, but is determined by the external connections made to it.)

The output is found to follow the input waveform (after the first quarter cycle) but the maximum value of the output is *clamped at zero*. A clamping circuit of the type shown here is used when it is necessary to ensure that the peak value of an output voltage does not exceed a specified value.

Drill Problem 4.9: A modified version of the clamping circuit is shown in Fig. 4-20. Using the signal in Fig. 4-19 as the input, determine and sketch the output waveform.

Electronics: Circuits and Systems 241

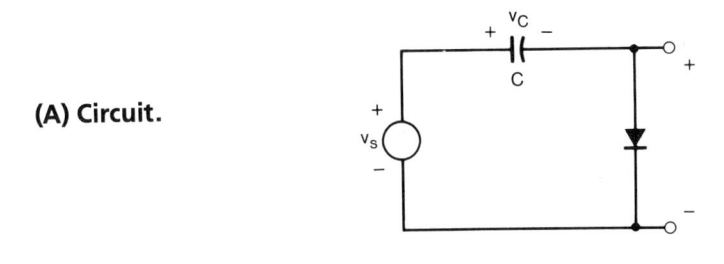

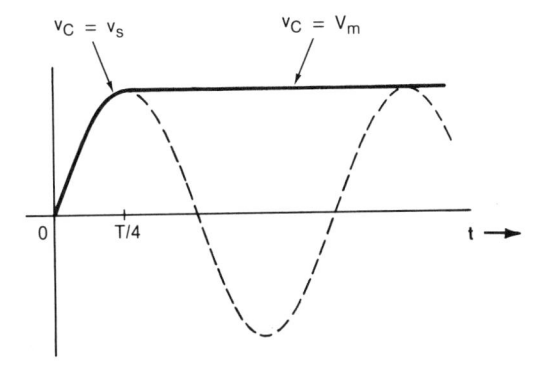

(A) Circuit.

(B) Voltage across the capacitor.

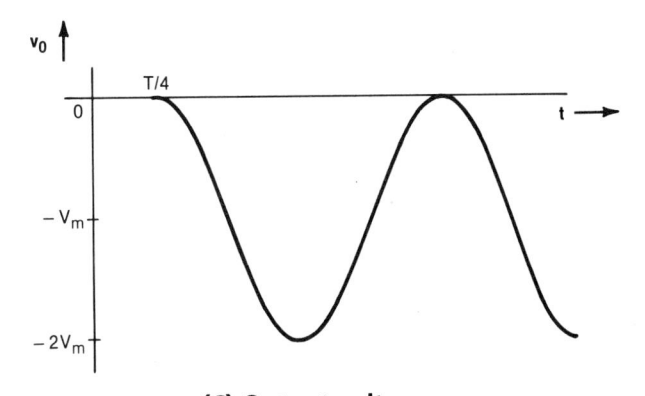

(C) Output voltage.

Fig. 4-19. Clamping circuit.

OR Gates

Even though the definition of the logic function OR will not be presented until a later chapter, a circuit that is used for performing the operation can be

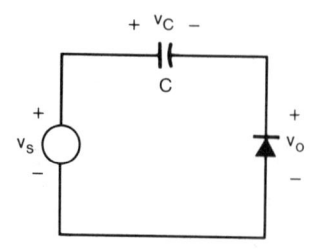

Fig. 4-20. Clamping circuit for Drill Problem 4.9.

discussed here. An OR circuit (Fig. 4-21) has the property that *its output is equal to the higher of the two inputs.*

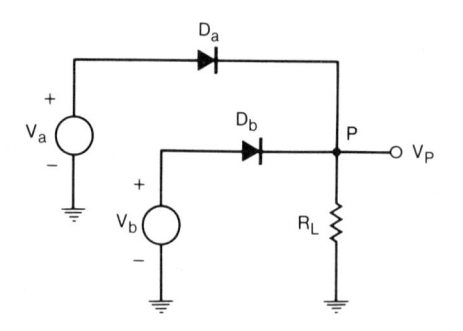

Fig. 4-21. OR circuit.

Suppose $V_a = 10$ V and $V_b = 0$. Then the diode D_a is forward biased, and point P is at 9.3 V (after subtracting the 0.7 V drop across the diode, D_a). If we make $V_a = 0$ and $V_b = 10$ V, then the diode D_b is forward biased, and the voltage at point P is again 9.3 V. In both these cases, only one diode conducts, and the other is reverse biased. Now, if we make both V_a and V_b equal to 10 V, both the diodes are forward biased, and $V_p = 9.3$ V. But if both V_a and V_b are made zero, $V_p = 0$. Thus, the output voltage assumes the same value as the higher of the two input voltages (except for the drop due to the diode when conducting).

Fig. 4-22 shows the inputs v_a and v_b and the corresponding output v_P of an OR gate.

Drill Problem 4.10: If the input V_a to an OR circuit is changed to that shown in Fig. 4-23, while V_b remains the same as in Fig. 4-22, determine and sketch the output voltage.

Electronics: Circuits and Systems

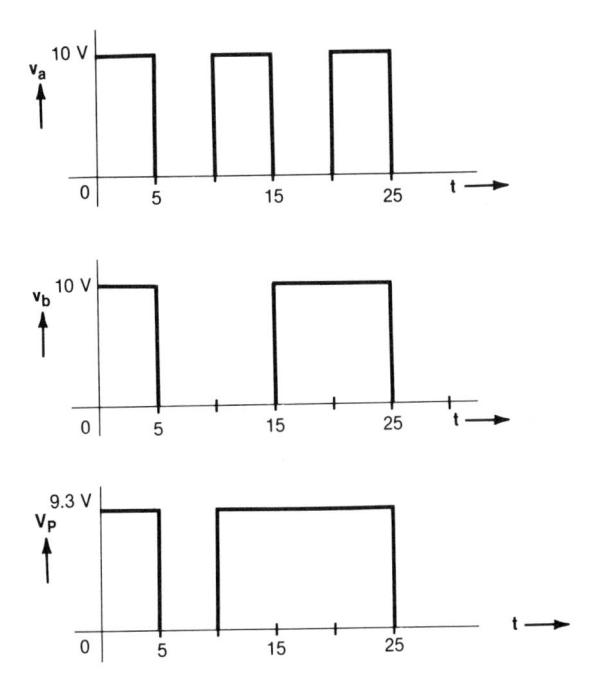

Fig. 4-22. Input signals and output of the OR circuit.

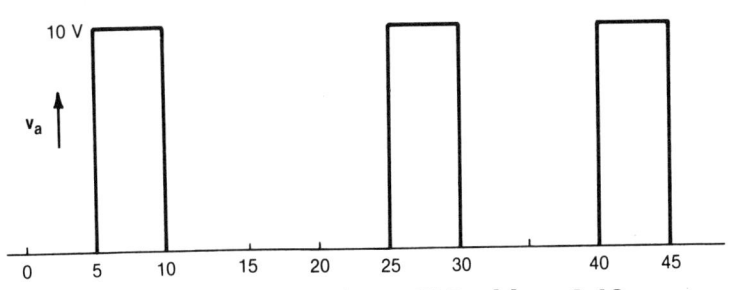

Fig. 4-23. Input v_a for Drill Problem 4.10.

A modification of the OR circuit leads to an AND circuit, which has the property that its output is equal to the *lower* of the two inputs. This circuit is considered in the following drill problems.

Drill Problem 4.11: If the input signals of Fig. 4-22 are applied to the AND gate of Fig. 4-24, determine and sketch the output waveform.

Diodes and Diode Circuits

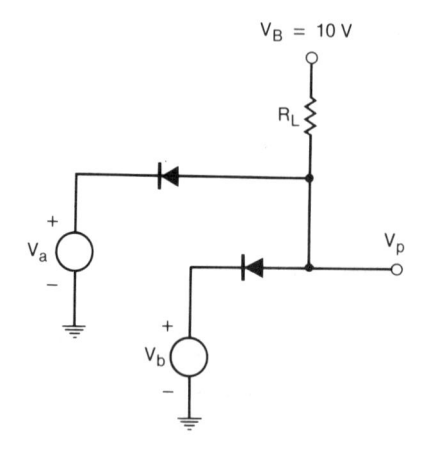

Fig. 4-24. The AND circuit for Drill Problem 4.11.

Drill Problem 4.12: Two input signals (shown in Fig. 4-25) are applied to (a) an OR gate and (b) an AND gate. Determine and sketch the output waveform in each case.

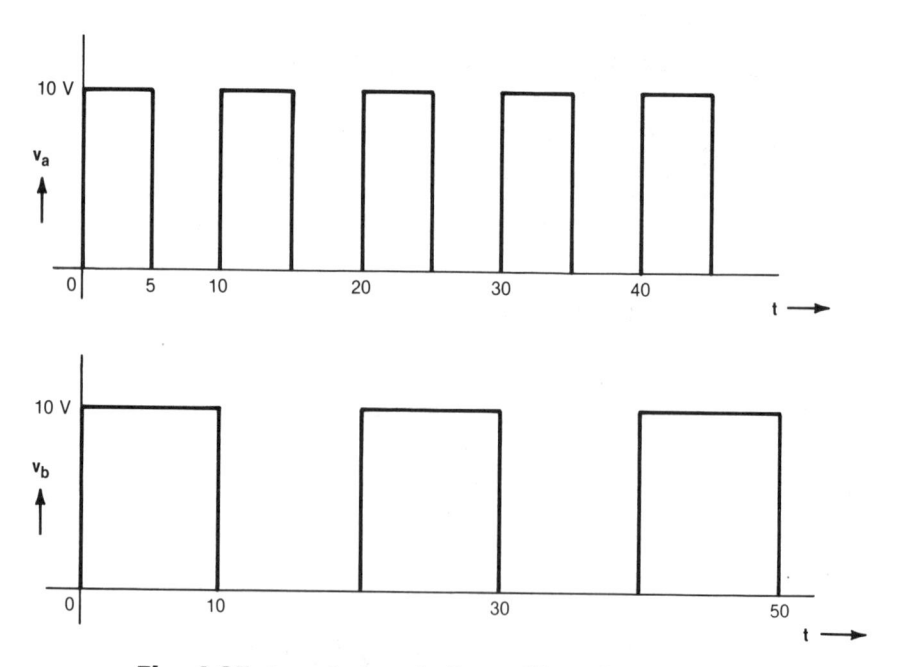

Fig. 4-25. Input signals for Drill Problem 4.12.

Electronics: Circuits and Systems

ZENER DIODE

It was mentioned in the last chapter that when a pn junction is reverse biased, a small reverse current exists due to the drift of thermally generated electrons and holes. Since the number of electrons and holes generated by thermal agitation at normal operating temperatures is small, the current is usually insignificant.

But, electrons and holes can be released from the covalent bonds of the atoms by applying energy through an electrical voltage also. If the reverse voltage is increased to a sufficiently high value, then electrons and holes are generated due to the high energy imparted by the voltage. In fact, electrons and holes released from covalent bonds have enough energy to cause other electrons and holes to break their bonds. A chain reaction is set in motion leading to the so called *avalanche breakdown* of the diode. The reverse current increases rapidly and can easily destroy the diode unless the current is limited to a safe value by an external circuit.

Fig. 4-26 shows the current-voltage characteristics of a diode in the reverse breakdown mode. It is seen that there is only a slight variation of the voltage across the diode in this region. Because of the essentially constant voltage, the diode can be used as a voltage regulator by operating it in the reverse breakdown mode. A diode fabricated to operate in the reverse breakdown mode is called the *zener diode*, whose symbol is shown in Fig. 4-26B. Note the *polarity* of the voltage V_Z across the diode and the *direction* of the current i_Z. For practical purposes, V_Z can be assumed to be a constant. Zener diodes are available with V_Z ranging from a few volts to 200 V. The *maximum power dissipation* of a zener diode is specified by the manufacturer and is from 0.2 W to 1 W. An external resistance must be used to ensure that the power dissipation does not exceed the rated maximum.

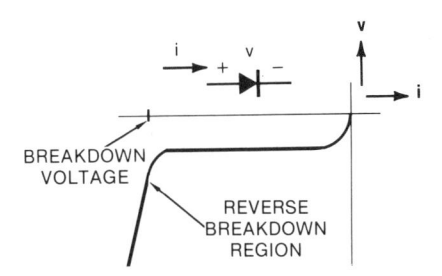

(A) I-v characteristics of a diode in the reverse breakdown region.

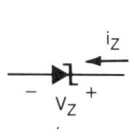

(B) Symbol of a zener diode. Note the voltage polarity and current direction used.

Fig. 4-26. Zener diode.

Fig. 4-27 shows a voltage regulator circuit using a zener diode. The input voltage is an unregulated dc power supply. The zener diode has an operating voltage V_Z, which is less than the (minimum value of the) input voltage. Note the manner in which it is connected: *cathode* at the positive end of V_{in} and *anode* at ground. The voltage across the load is maintained at a constant value of V_Z while the input V_{in} may vary over wide limits.

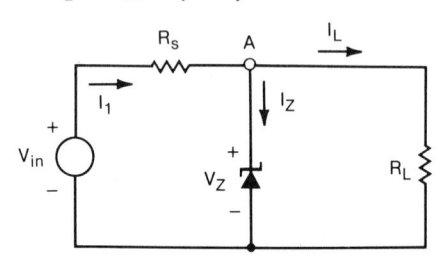

Fig. 4-27. Voltage-regulator circuit.

The value of V_Z and the maximum power dissipation of the zener diode are usually known in the regulator circuit. The problem is to choose the proper value of the resistor R_s so as to ensure that the current in the diode does not exceed the value corresponding to the maximum power dissipation.

The voltage at point A of the given circuit is always V_Z because of the zener diode. Therefore,

$$\text{voltage across } R_s = (V_{in} - V_Z)$$

$$\text{current in } R_s = (V_{in} - V_Z)/R_s$$

In the worst case, the load resistor may not be present, which means that all the current supplied by V_{in} will go through the zener diode. Therefore,

$$\text{Maximum current in the diode} = \text{current in } R_s$$

$$= (V_{in} - V_Z)/R_s$$

Now, if the power dissipation is to be less than a specified P_{max}, then

$$\text{maximum rated current in the diode} = (P_{max}/V_Z)$$

so that we have the equation:

$$\frac{P_{max}}{V_Z} = \frac{V_{in} - V_Z}{R_s}$$

which gives the *minimum value* of R_s needed.

Having determined the value of R_s, it is possible to calculate the range of

Electronics: Circuits and Systems

values of the input voltage V_{in} over which the regulator will function properly. This point will be discussed in the following example.

Example 4.3

In the regulator circuit of Fig. 4-27, let the zener diode have $V_Z = 100$ V and a maximum power dissipation of 1 W. Assume that $V_{in} = 180$ V and calculate the value of R_s. For the value of R_s calculated, find the range of values of V_{in} for which the circuit will function properly if the *minimum* current in the zener diode is specified as 1.5 mA.

Solution

Since $P_{max} = 1$ W and $V_Z = 100$ V, the maximum rated current in the diode $= (P_{max}/V_Z) = 0.01$ A, or 10 mA. The largest current in R_s flows when there is no load resistance, and the diode current is at its maximum value. Therefore, since

$$I_1 = (V_{in} - V_Z)/R_s$$

we get

$$R_s = (80 \text{ V}/10 \text{ mA})$$
$$= 8 \text{ k}\Omega$$

Consider the minimum specified current of 1.5 mA through the diode. Again, when there is no load, I_1 equals the diode current and

$$V_{in} = V_Z + R_s I_1$$
$$= 100 + (8 \times 1.5)$$
$$= 112 \text{ V}$$

When the diode current is at its maximum, $V_{in} = 180$ V, as seen earlier. Therefore, the circuit will function properly for a range of values of V_{in} from 112 V to 180 V.

Drill Problem 4.13: Determine the current I_1 in the previous circuit for each of the following values of V_{in} and discuss the results: (a) $V_{in} = 105$ V; (b) $V_{in} = 85$ V.

Drill Problem 4.14: Given a zener diode with $V_Z = 6.2$ V and $P_{max} = 0.4$ W, find the value of R_s needed in the regulator circuit when $V_{in} = 12$ V. If the minimum current in the diode is to be 7.5 mA, find the range of input voltage.

VARACTOR DIODE

The presence of the depletion region at a pn junction and the dependence of its width on the voltage across the junction gives rise to a useful device called the *varactor diode*.

When a pn junction is *reverse biased*, the depletion region at the junction has a width that depends upon the junction voltage. The depletion region is formed by an accumulation of space charge. Because of the space charge and the voltage across the depletion region, the depletion region has properties similar to a *capacitance*. The junction capacitance is *inversely* proportional to the *width of the depletion region*.

It is found that the width of the depletion region (under reverse bias) is proportional to V_j^k, where V_j denotes the junction voltage and k is a constant, which is 0.333 or 0.5 depending upon the manner in which the pn junction is fabricated.

Since the junction capacitance is inversely proportional to the width of the depletion region, which in turn is proportional to V_j^k, it is seen that

junction capacitance is proportional to ($1/V_j^k$)

Thus, the pn junction under reverse bias acts as a *voltage-controlled capacitor*. When a pn diode is used in this mode, it is called a *varactor* (a word coined from "variable reactor"). See Fig. 4-28.

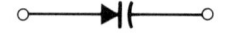

Fig. 4-28. Symbol of a varactor diode.

One important application of the varactor is in controlling the resonant frequency of a tuned circuit (parallel RLC circuit) as a function of the voltage of the signal applied to it. The use of a varactor instead of, or in addition to, a capacitor in a tuned circuit gives rise to a *voltage-tunable* resonant circuit. Such a circuit can be used in the generation of frequency-modulated (FM) signals.

SILICON-CONTROLLED RECTIFIER (SCR)

The *silicon-controlled rectifier (SCR)* acts like a diode but has an external control terminal that controls the turning on (that is, starting of the conduction) of the device. The SCR is made up of four semiconductor regions forming a *pnpn* device as shown in Fig. 4-29A. Three terminals are available: the anode and cathode as in a regular diode and a *gate* terminal, which is attached to the *lower* p-region. The voltage applied to the gate controls the start of current flow in the SCR as will be seen in the following discussion.

Electronics: Circuits and Systems 249

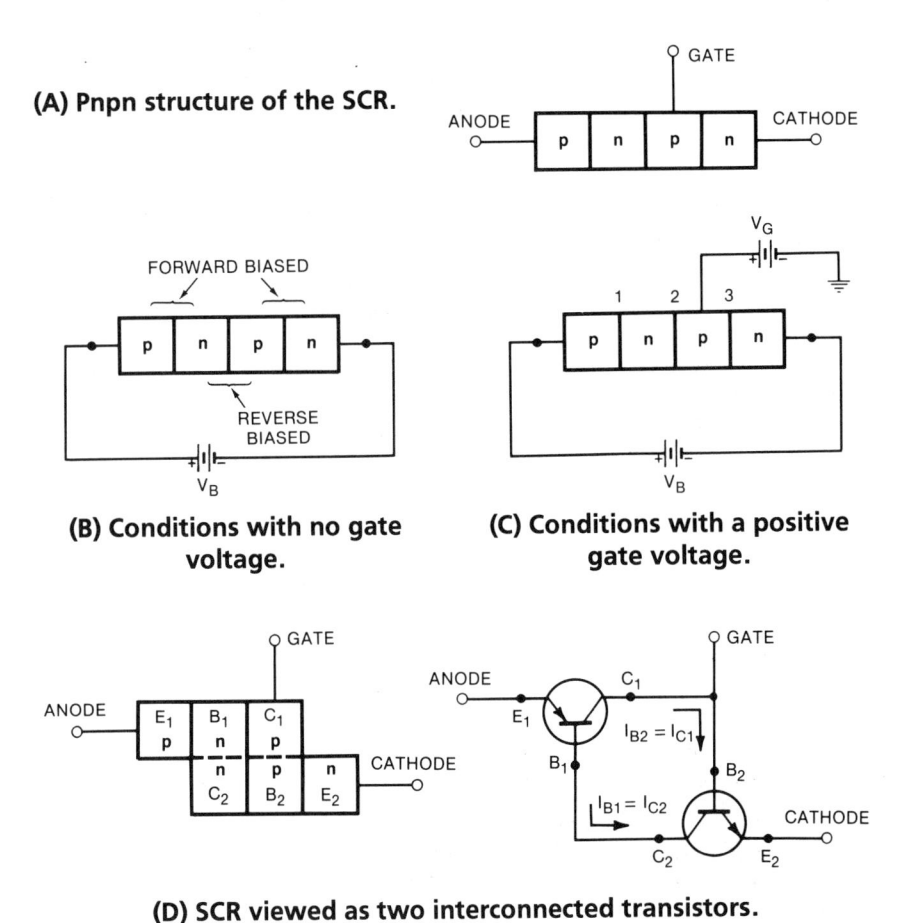

(A) Pnpn structure of the SCR.

(B) Conditions with no gate voltage.

(C) Conditions with a positive gate voltage.

(D) SCR viewed as two interconnected transistors.

Fig. 4-29. Silicon-controlled rectifier (SCR).

Consider first the case when there is no voltage applied to the gate, as indicated in Fig. 4-29B. A voltage V_B is applied to the SCR. Then, two of the three pn junctions in the structure are forward biased while the third is reverse biased. Almost all the applied voltage is across the reverse-biased junction due to its high resistance, and the forward bias at the other two junctions is insufficient to cause any significant current flow. The current in the SCR will remain at the level of the reverse current in a diode unless the external voltage V_B is increased to the point where an avalanche breakdown of the reverse-biased junction occurs.

Now consider a *positive* voltage applied to the gate as in Fig. 4-29C, so as to apply a sufficiently high forward bias to junction 3, which not only causes that junction to start conducting but also sets up a *regenerative action* in the whole device so that a large current flows from the anode to the cathode. In order to explain how this regenerative action occurs, think of the SCR as a composite of two bipolar transistors: a pnp and an npn structure with some common semiconductor regions as indicated in Fig. 4-29D. The forward biased junction in each transistor serves as its emitter-base junction while the reverse biased junction in each serves as its collector-base junction. This viewpoint leads to certain relationships between the currents in the SCR: the base current I_{B1} of the pnp portion is the same as the collector current I_{C2} in the npn portion; the collector current I_{C1} of the pnp portion is the same as the base current I_{B2} in the npn portion. Also, since the collector current and the base current in a transistor are related through the current gain β, we have $I_{C1} = \beta_1 I_{B1}$ and $I_{C2} = \beta_2 I_{B2}$. That is,

$$I_{B2} = I_{C1}$$
$$= \beta_1 I_{B1}$$

and

$$I_{B1} = I_{C2}$$
$$= \beta_2 I_{B2}$$

The last two sets of relationships show how a regenerative action takes place in the SCR. The base current I_{B2} set up by the positive voltage applied to the gate gets multiplied by β_2 to produce the collector current I_{C2}, which becomes the base current I_{B1}, which gets multiplied by β_1 to produce the collector current I_{C1}, which becomes the new base current I_{B2}. This cycle repeats itself. At each point where a collector current is influenced by the corresponding base current, there is a multiplication by the current gain β. The currents keep on increasing even if β is only slightly greater than 1. Eventually, the current in the SCR becomes large and is limited only by the external connections to it. Thus, a positive voltage applied to the gate triggers the SCR into a free conducting mode. It can be seen that, once the regenerative chain of events gets going, the gate voltage loses control of the flow of current in the SCR. Therefore, even if we remove the positive voltage at the gate (by grounding the gate, for example), the current flow in the SCR will continue. The only way the current flow can be stopped is by making the *anode-to-cathode voltage* across the SCR zero or negative.

The SCR acts like a diode that can be turned on by applying a (sufficiently large) positive voltage to its gate terminal. It is a diode with a trigger. The symbol of an SCR is shown in Fig. 4-30. The speed with which an SCR turns on or off is of the order of a fraction of a millisecond, which is slow compared with other switching circuits using bipolar transistors (to be discussed in a

Electronics: Circuits and Systems

later chapter). But the SCR can handle large voltages (up to several thousand volts) and large currents (hundreds of amperes), and it finds numerous applications in control circuits involving ac and dc power supplies.

Fig. 4-30. Symbol of an SCR.

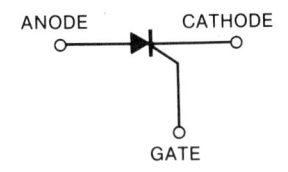

SCR as an AC Switch

A circuit that can be used to control the instant at which an output voltage appears in a circuit is shown in Fig. 4-31.

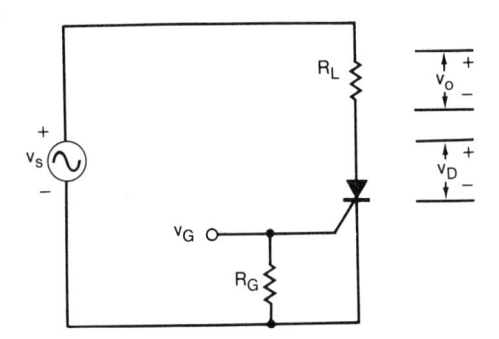

Fig. 4-31. Ac switch using the SCR.

The operation of the circuit is shown in the waveforms of Fig. 4-32. The SCR is off until $t = t_1$ since no voltage is applied to the gate. At $t = t_1$, the gate voltage turns the SCR on, and almost all the input voltage (except for the roughly 1-V drop in the SCR) is now across the load. At $t = t_2$, the input voltage becomes zero, and there is no more forward bias from the anode to the cathode of the SCR. The SCR stops conducting, and the output voltage becomes zero. When the input voltage is negative, the reverse bias across the SCR prevents it from conducting even if a positive voltage is applied to the gate (as at $t = t_3$). The SCR turns on again at $t = t_4$ when there is a forward bias across it and a positive voltage is applied to the gate. The circuit permits the precise control of the phase angle (of the input sinusoidal waveform) at which the output voltage appears in the circuit and is used in dc power control circuits.

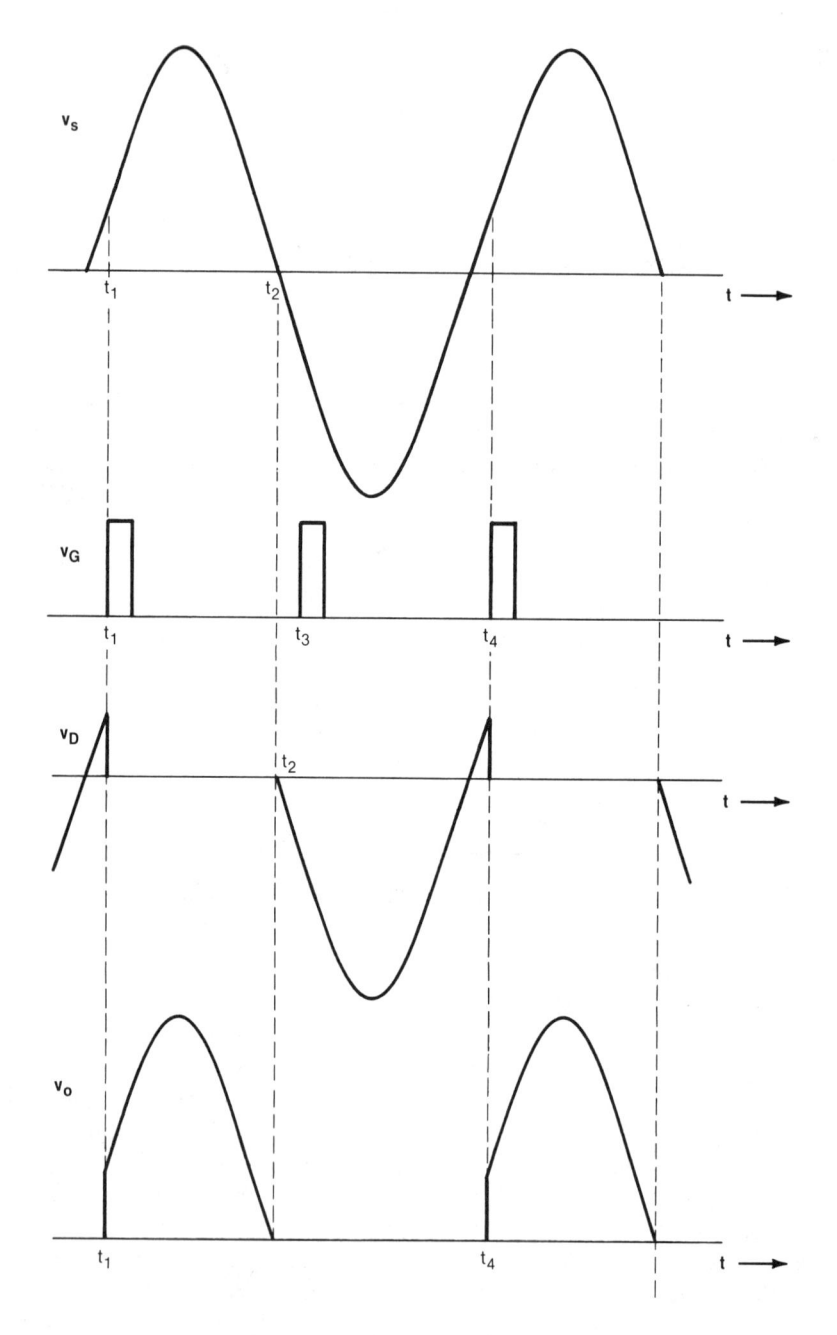

Fig. 4-32. Waveforms pertinent to the circuit of Fig. 4-31.

Electronics: Circuits and Systems 253

DC Static Switching Using SCRs

It is sometimes necessary to automatically turn the current in a load on or off by means of an external control signal that may be in the form of "on" or "off" pulses. When the current handling capacity of the load is several amperes or more, SCRs are more practical than the logic switches of the type discussed in a later chapter. An example of such a situation will be the turning on and off of a motor in a furnace as determined by the signals from a thermostat. The circuit of Fig. 4-33A acts as a dc switch. When a positive pulse is applied to the gate of SCR 1, it starts conducting. Neglecting the small voltage across a conducting SCR, the voltage V_{dc} is now across the load, and a load current flows, as indicated in Fig. 4-33B. The voltage V_{dc} is also across the R_1C_1 series circuit, and the capacitor charges up to V_{dc} with a time constant = R_1C_1. When the capacitor is fully charged, as in Fig. 4-33B, it is seen that the voltage across the (nonconducting) SCR 2 is V_{dc} as indicated. Now, suppose a positive pulse is applied to the gate of SCR 2. Since that SCR has a forward bias of V_{dc} (due to the voltage across C_1), it will start conducting, making SCR 1 stop conducting as can be seen from the following argument: When SCR 2 conducts, the voltage across it essentially becomes zero. Referring to Fig. 4-33C, the voltage across C_1 is now across SCR 1. Therefore, the voltage across SCR 1 is V_{dc} but the *anode is negative* while the *cathode is positive*. (Kirchhoff's voltage law can be used to verify the correctness of the polarity.) Since SCR 1 is reverse biased, it stops conducting, and the load current is now diverted through C_1 and SCR 2 as indicated in Fig. 4-33C. The current I_L charges the capacitor to a voltage V_{dc} with a polarity shown in Fig. 4-33D: the left side of the capacitor now becomes positive. That is, the voltage across the capacitor is opposite to what it was in the previous diagram. When the capacitor is fully charged to V_{dc}, it reaches a dc steady state and stops conducting. The capacitor becomes an open circuit, and there is no path for current flow in the circuit. The load current stops flowing. It can be seen in Fig. 4-33D that the voltage across the load is zero (by using KVL), which again shows that there is no load current.

The speed with which the load current is turned off in the previous circuit depends upon the speed with which the capacitor reaches the steady-state condition of Fig. 4-33D. The time constant governing the previous charging will depend upon the load resistance and C_1.

The term *thyristor* is used to denote any semiconductor switch that has two stable states (*on* and *off* states) and the switching depends upon a regenerative feedback action in a pnpn structure. The SCR is the most well known in the family of thyristors but there are several other members: light-activated SCR (LASCR), silicon-controlled switch (SCS), and the TRIAC to mention a few.

254 Diodes and Diode Circuits

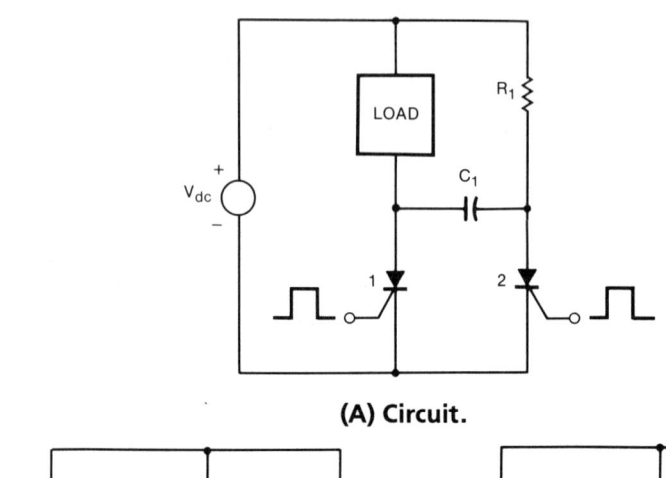

(A) Circuit.

(B) SCR 1 is conducting, load current flows, and the capacitor has charged to V_{dc}.

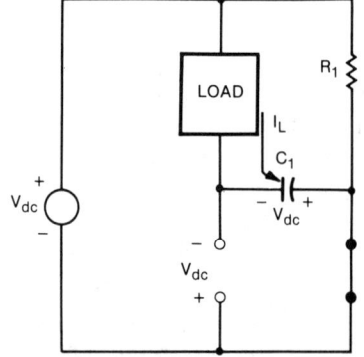

(C) SCR 2 is conducting, SCR 1 has turned off, and the capacitor now charges toward V_{dc} in the opposite direction to its present voltage.

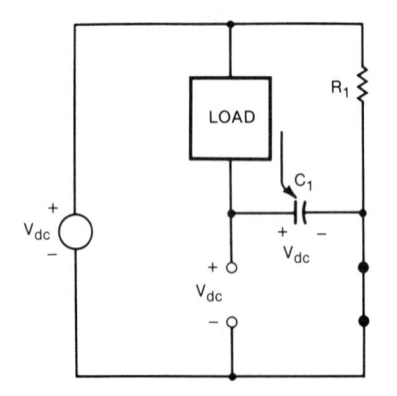

(D) Steady-state conditions with SCR 2 on. No current flows since SCR 1 is off and the fully charged capacitor is like an open circuit.

Fig. 4-33. Dc static switch using two SCRs.

Triacs

The *triac* is a *bidirectional* switch since it can conduct in either of two directions unlike the SCR, which is a unidirectional switch. Because of its bidirectional nature, the triac offers certain advantages over the SCR in some applications.

The name "triac" is an abbreviation that stands for *triode ac* switch. The triac is a three-terminal device with two main terminals T_1 and T_2 and a gate terminal as indicated in the symbol of Fig. 4-34A. The current flow in the device can be either from T_2 to T_1 or from T_1 to T_2 depending upon the external bias connections to the triac. The terminal T_1 is normally used as the *reference* for voltage specifications. The two preferred modes of operation of the triac are (1) terminal T_2 is made positive with reference to T_1, and a positive voltage pulse is applied to the gate, in which case the device conducts and current flows from T_2 to T_1 as indicated in Fig. 4-34B; (2) terminal T_2 is made negative with reference to T_1 and a negative voltage pulse is applied to the gate, in which case the device conducts and current flows from T_1 to T_2 as indicated in Fig. 4-34C. (The other two modes are obtained by reversing the gate pulse voltage in the previous two modes, but they are not normally recommended.)

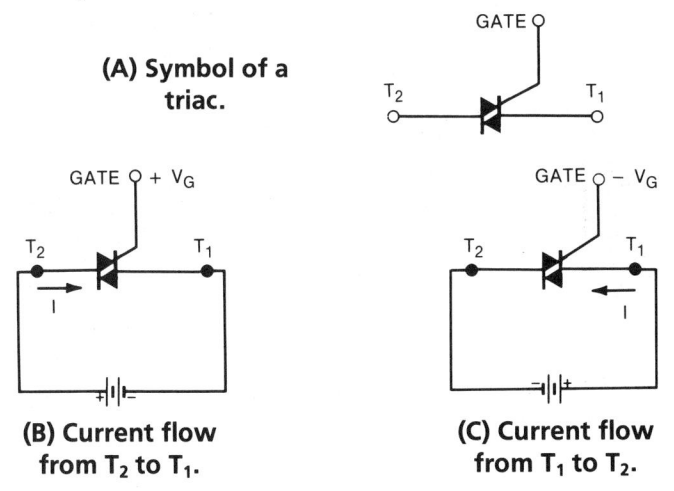

(A) Symbol of a triac.

(B) Current flow from T_2 to T_1.

(C) Current flow from T_1 to T_2.

Fig. 4-34. Triac.

A triac circuit, which is analogous to the SCR circuit of Fig. 4-31, is shown in Fig. 4-35A. For the gate pulses and input ac waveform shown in Fig. 4-35B, the output voltage waveform will be as shown in Fig. 4-35C.

(A) Circuit.

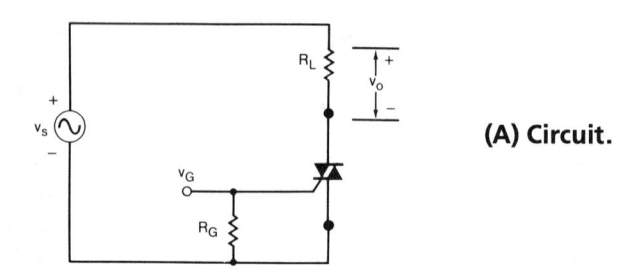

(B) Input signal and gate voltage.

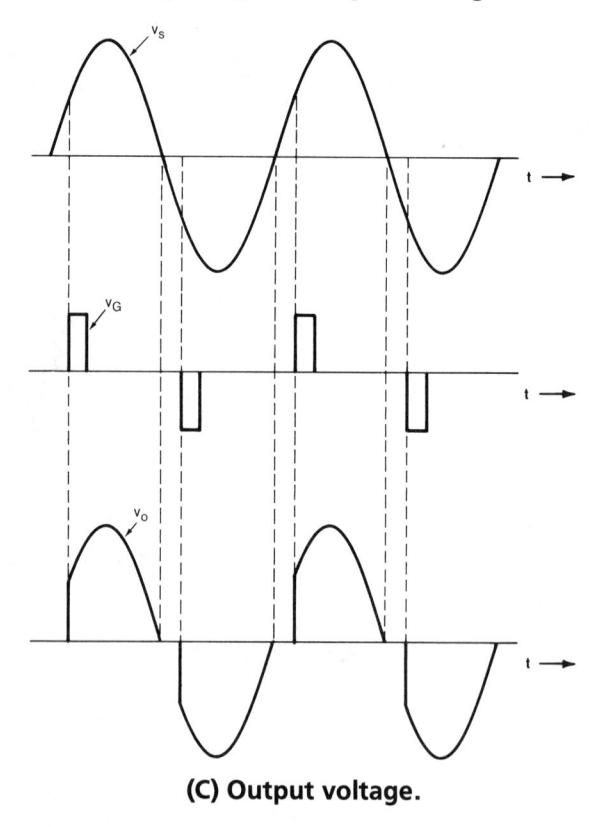

(C) Output voltage.

Fig. 4-35. Ac switch using a triac.

Electronics: Circuits and Systems 257

Even though the triac offers an advantage over the SCR due to its ability to conduct in either direction, it has some disadvantages. For example, the triac is designed to operate at 50 to 60 Hz or at 400 Hz, while the SCR can operate at frequencies in the 25 kHz to 30 kHz range.

Thyristors are widely used because they are immune to vibration and shock, and they need only a pulse (at the gate) to switch them into the conducting state. Among their many applications are motor speed control circuits, flasher circuits, temperature and air-conditioning control circuits, and phase control circuits.

Drill Problem 4.15: The triangular waveform of Fig. 4-36A is applied to the input of the SCR circuit of Fig. 4-31. The gate voltage waveform is shown in Fig. 4-36B. Determine and sketch the output voltage v_o as a function of time.

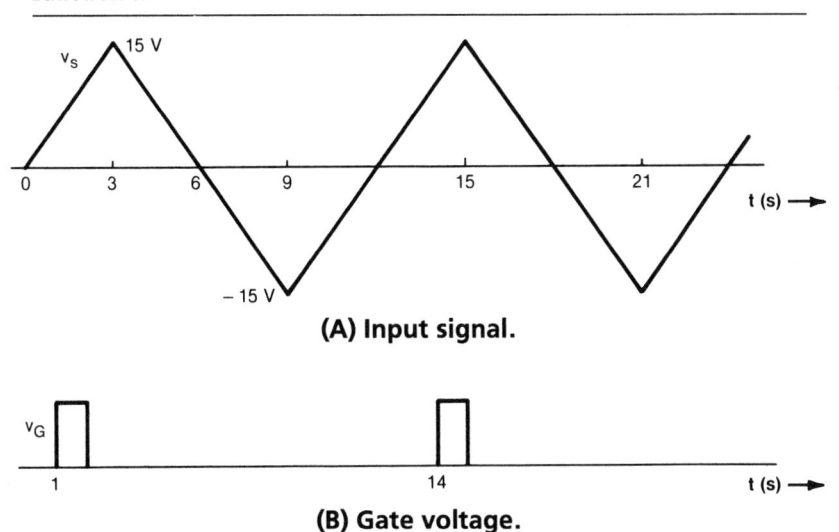

(A) Input signal.

(B) Gate voltage.

Fig. 4-36. Input signal and gate voltage for Drill Problem 4.15.

Drill Problem 4.16: Keeping the input to the circuit of Fig. 4-31 the same as in the previous problem, suppose the voltage applied to the gate is as shown in Fig. 4-37. Determine and sketch the output voltage v_o.

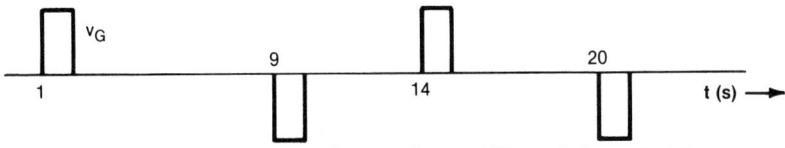

Fig. 4-37. Gate voltage for Drill Problem 4.16.

Drill Problem 4.17: The input voltage and the voltage applied to the gate of the circuit of Fig. 4-35 (using the triac) are as in Drill Problem 4.15. Determine and sketch the output voltage v_o.

Drill Problem 4.18: Repeat Drill Problem 4.17 for the situation described in Drill Problem 4.16.

SUMMARY SHEETS

IDEAL DIODE

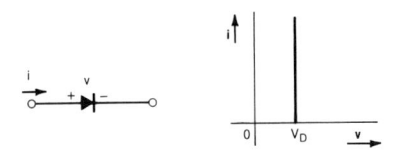

When $v < V_D$, the diode is an open circuit and $i = 0$. When v exceeds V_D, the diode acts like a battery of 0.7 V (silicon diodes).

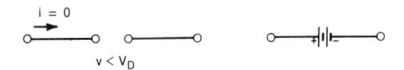

In some cases, V_D can be neglected and the diode treated as a short circuit.

RECTIFICATION OF AC

Half-Wave Rectifier

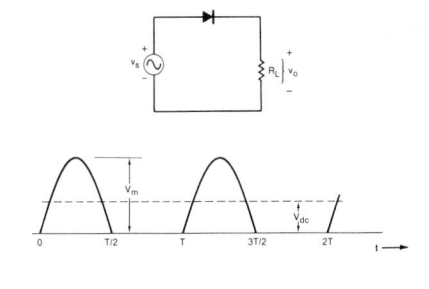

$$V_{dc} = (V_m/\pi) = 0.318 V_m$$

Full-Wave Rectifier

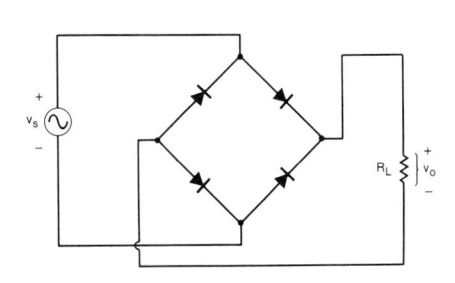

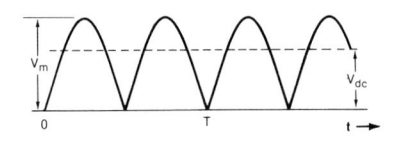

$$V_{dc} = (2V_m/\pi) = 0.636 V_m$$

Peak Inverse Voltage (PIV)

PIV = Maximum voltage that appears across a diode when it is not conducting. PIV = V_m in the two rectifier circuits shown.

Capacitor Filter

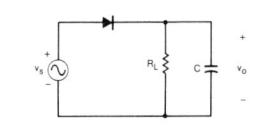

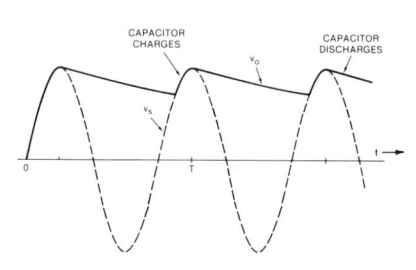

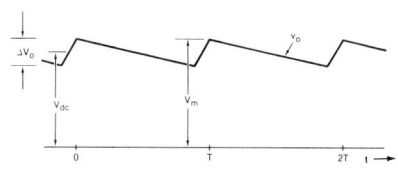

$$V_{dc} = V_m - (\Delta V_o/2)$$
When,
$$R_L C \gg T$$

260   **Diodes and Diode Circuits**

MISCELLANEOUS DIODE CIRCUITS

Positive Clipping

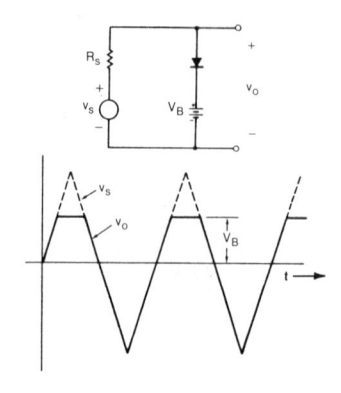

Negative Clipping

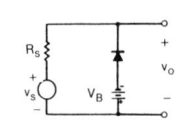

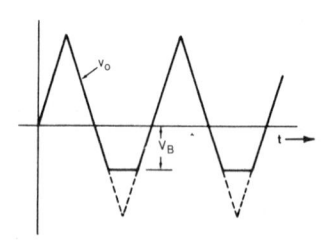

Limiter (Positive and Negative Clipping)

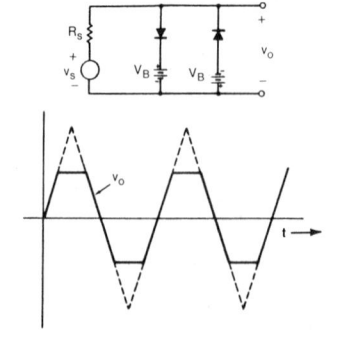

Clamping

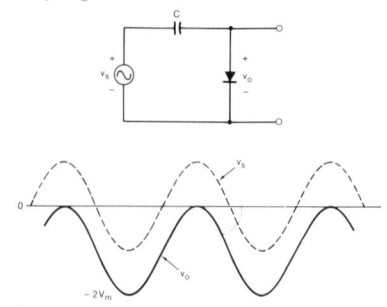

OR Gate

The value of V_o at any instant equals the *larger* of the two inputs at that instant.

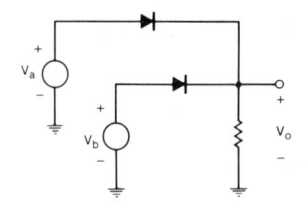

AND Gate

The value of V_o at any instant equals the *smaller* of the two inputs at that instant.

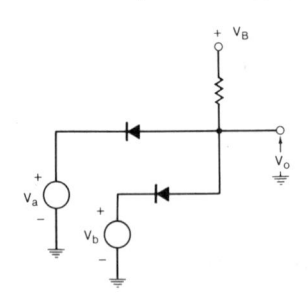

ZENER DIODE

When the diode current I_Z is between a specified maximum $I_{Z(max)}$ and minimum $I_{Z(min)}$, the zener diode acts as a constant voltage source at V_Z volts. If I_Z goes below the specified mini-

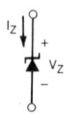

Electronics: Circuits and Systems

mum, the zener diode acts like an ordinary reverse biased diode, and its voltage is no longer a constant. If I_Z goes above the specified maximum, the power dissipation will be excessive.

Voltage Regulation

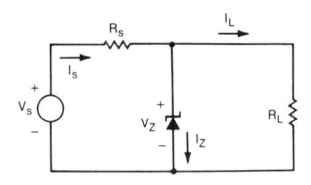

$I_L = (V_Z/R_L)$
$I_s = I_Z + I_L$
$I_s = (V_s - V_Z)/R_s$

VARACTOR DIODES

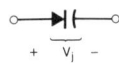

Junction capacitance is proportional to $(1/V_j^k)$, where k is a constant whose value is in the range 0.333 to 0.5. The varactor diode can be used as a voltage-controlled capacitance.

SCR (Silicon-Controlled Rectifier)

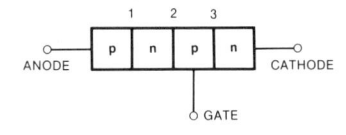

SCR Operation

When there is no voltage applied to the gate, almost all the applied voltage is across the reverse-based junction 2, and no current flows. When a positive voltage is applied sufficiently forward biased and sets up a current. A regenerative action ensues, which makes the forward current build up to a large value (limited only by the external connections to the device). Once the SCR has started conducting, it cannot be turned off by removing the gate voltage. It can be turned off only by making the anode-to-cathode voltage zero or negative.

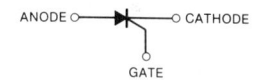

SCR as an AC Switch

The SCR conducts when there is a forward bias from anode to cathode and a positive pulse is applied to the gate. The SCR stops conducting when the anode-to-cathode voltage becomes zero.

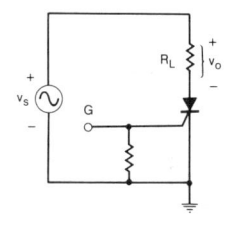

THYRISTOR

The term *thyristor* is used to denote any semiconductor switch with two stable states (ON and OFF) and the switching depends upon a regenerative feedback action in a pnpn structure.

Triac

The triac can conduct in either direction unlike the SCR, which can conduct only in one direction.

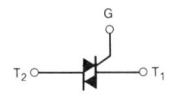

When the gate voltage is positive and the terminal T_2 is at a higher potential than T_1, the triac will conduct with current flowing in the direction T_2 to T_1. When the gate voltage is negative, and the terminal T_1 is at a higher potential than T_2, then the triac will conduct in the direction T_1 to T_2.

Triacs are confined to low frequencies (less than 1 kHz) while SCRs can operate up to 25 or 30 kHz.

ANSWERS TO DRILL PROBLEMS

4.1 See Fig. 4-DP1.

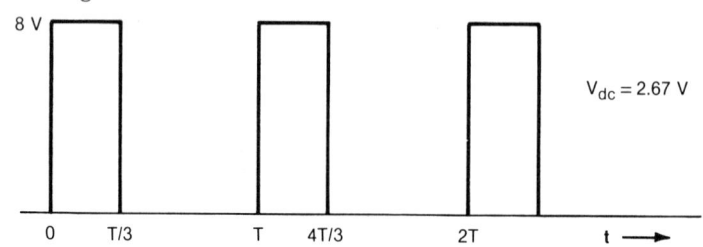

Fig. 4-DP1. $V_{dc} = 2.67$ V

4.2 See Fig. 4-DP2.

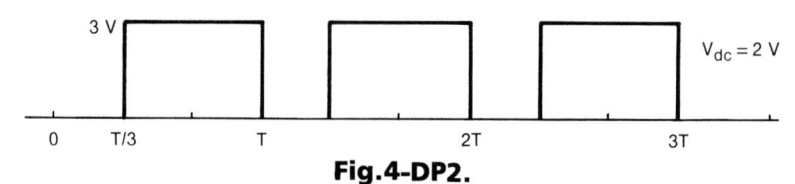

Fig. 4-DP2. $V_{dc} = 2$ V

4.3 See Fig. 4-DP3.

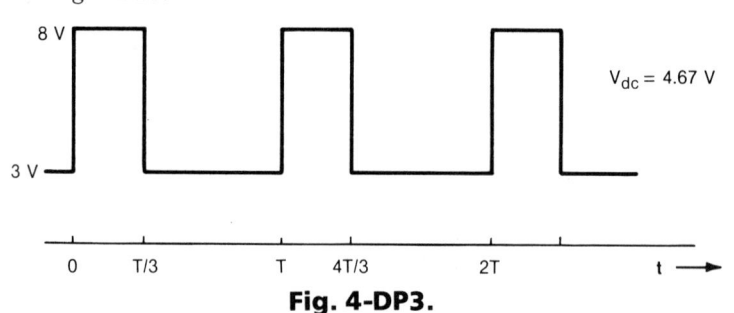

Fig. 4-DP3. $V_{dc} = 4.67$ V

4.4 Same as in Fig. 4-DP3.
4.5 50 V in both cases.
4.6 $2V_m$.
4.7 See Fig. 4-DP7.

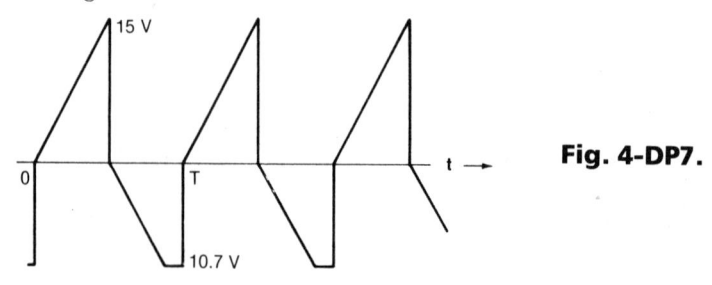

Fig. 4-DP7.

Electronics: Circuits and Systems   **263**

4.8 See Fig. 4-DP8.

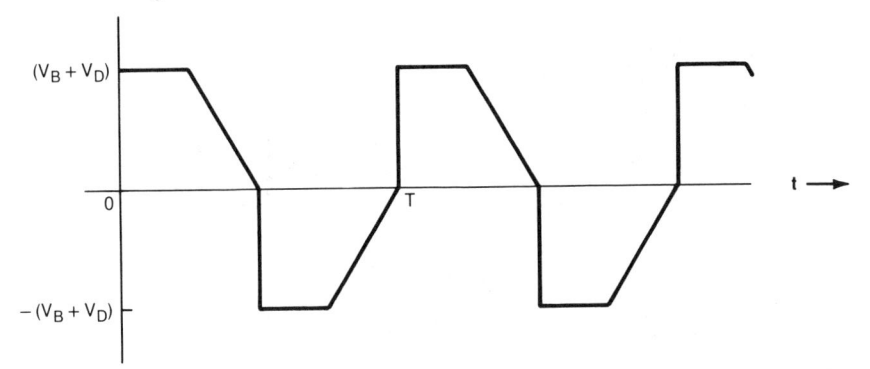

Fig. 4-DP8.

4.9 See Fig. 4-DP9.

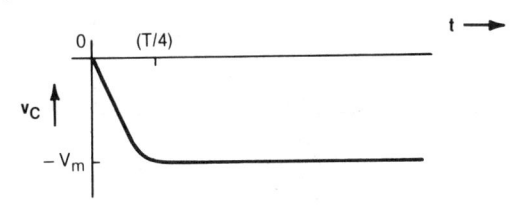

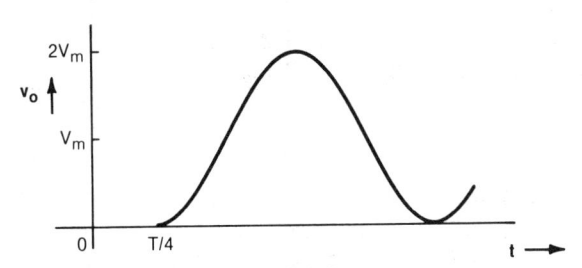

Fig. 4-DP9.

4.10 See Fig. 4-DP10.

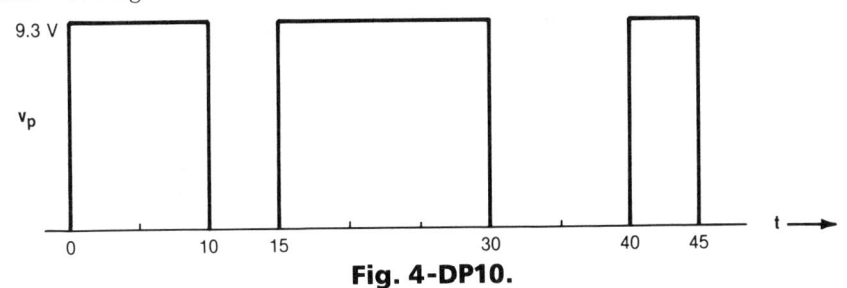

Fig. 4-DP10.

4.11 See Fig. 4-DP11.

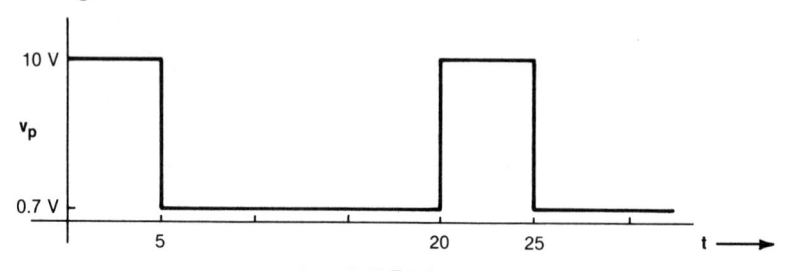

Fig. 4-DP11.

4.12 See Fig. 4-DP12.

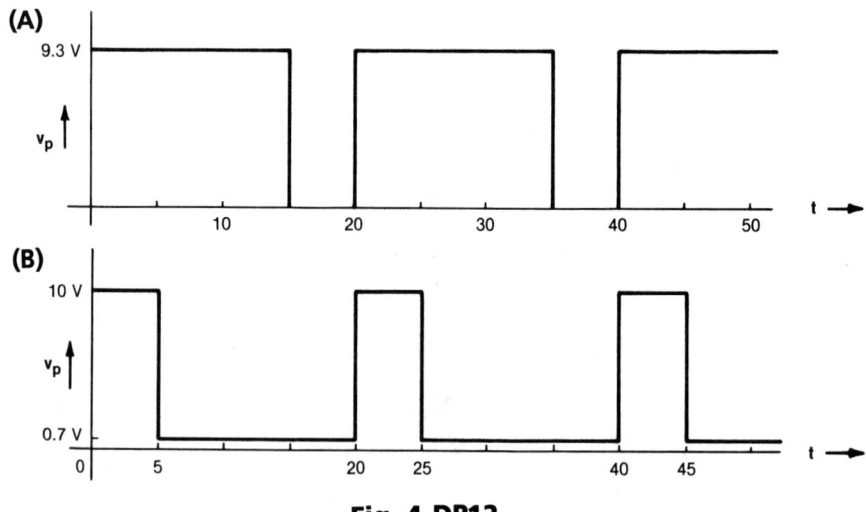

Fig. 4-DP12.

4.13 (a) 0.625 mA (less than the specified minimum current for the zener diode). (b) I_1 will have to be negative if V_Z is to be at 100 V, which is physically impossible. The voltage across the zener diode will be less than V_{in} = 85 V, and it will not function as a zener diode but as a reverse-biased pn junction.

4.14 I_{max} = 64.5 mA. R_s = 90 Ω. When I_1 = 7.5 mA, V_{in} = 6.88 V. The range of V_{in} is from 6.88 V to 12 V.

4.15 See Fig. 4-DP15.

Electronics: Circuits and Systems   265

4.16 Same as in the previous drill problem. The gate pulses present when the SCR is reverse biased cannot trigger it into conduction.
4.17 Same as in Fig. 4-DP15.
4.18 See Fig. 4-DP18.

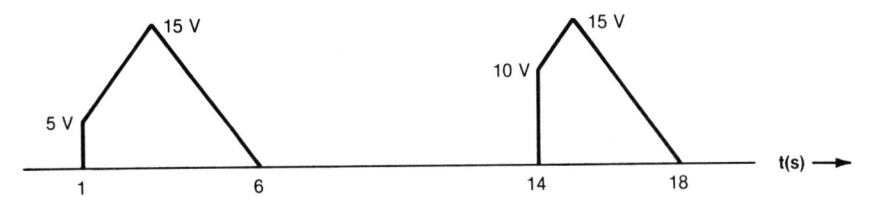

Fig. 4-DP15.

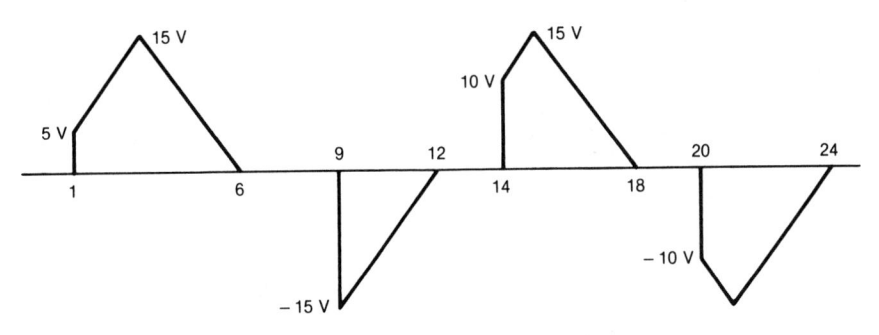

Fig. 4-DP18.

Diodes and Diode Circuits

PROBLEMS

Rectification of AC Power

4.1 The waveforms of Fig. 4-P1 are the inputs to a half-wave rectifier circuit (Fig. 4-3). Determine and sketch the output voltage waveform and calculate the dc component of the output in each case.

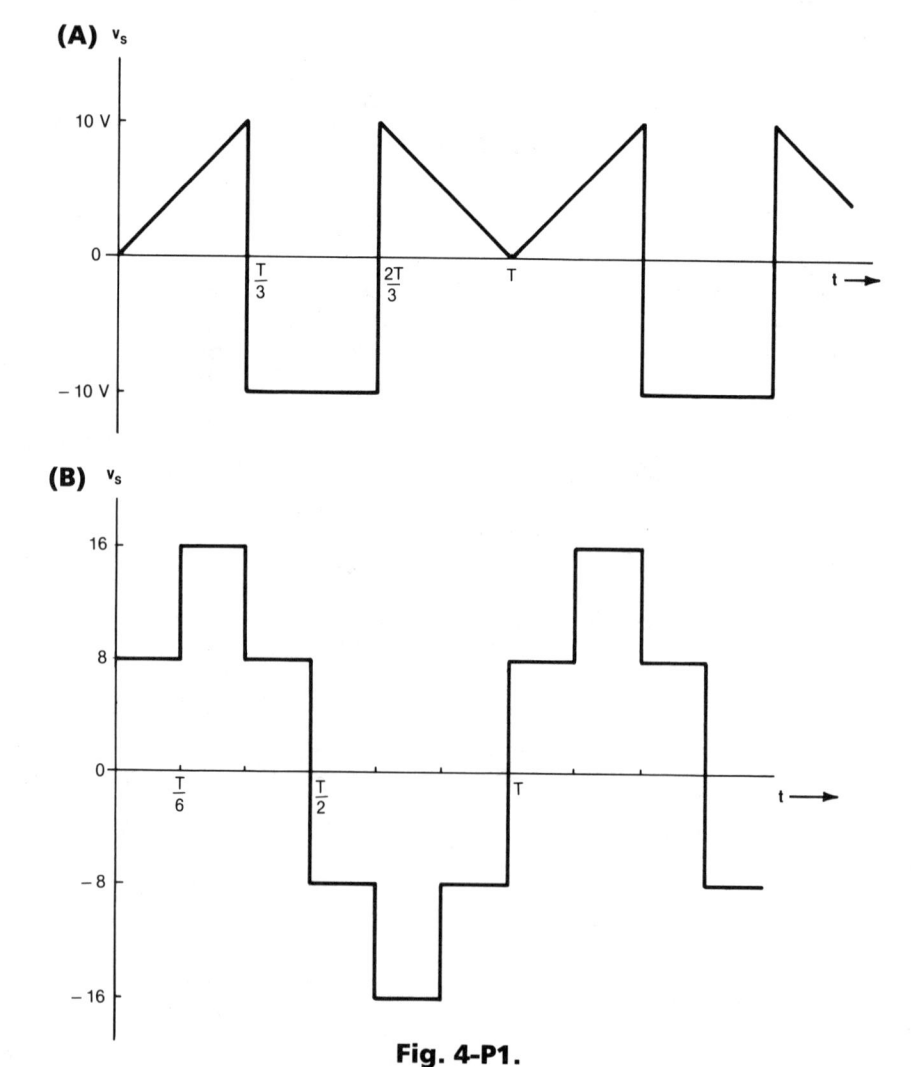

Fig. 4-P1.

4.2 Suppose each of the signals of the previous problem are fed to the half-

Electronics: Circuits and Systems

wave rectifier of Fig. 4-7. Repeat the sketches and the calculations of the dc components.

4.3 Suppose each of the signals of the previous problem are fed to a full-wave rectifier (Fig. 4-8). Determine and sketch the output voltage waveform and calculate the dc component of the output in each case.

4.4 Calculate the peak inverse voltage across the diode (or each diode in the case of a full-wave rectifier) in each situation of the last three problems.

Miscellaneous Applications of Diodes

4.5 A clipping circuit is shown in Fig. 4-P5A and the input to it is shown in Fig. 4-P5B. Determine and sketch the output waveform. Neglect diode drops in this and the following problems.

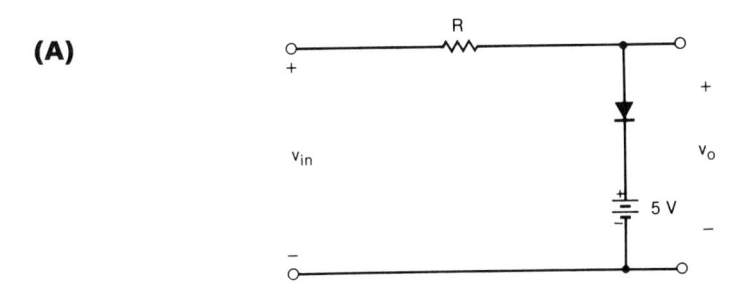

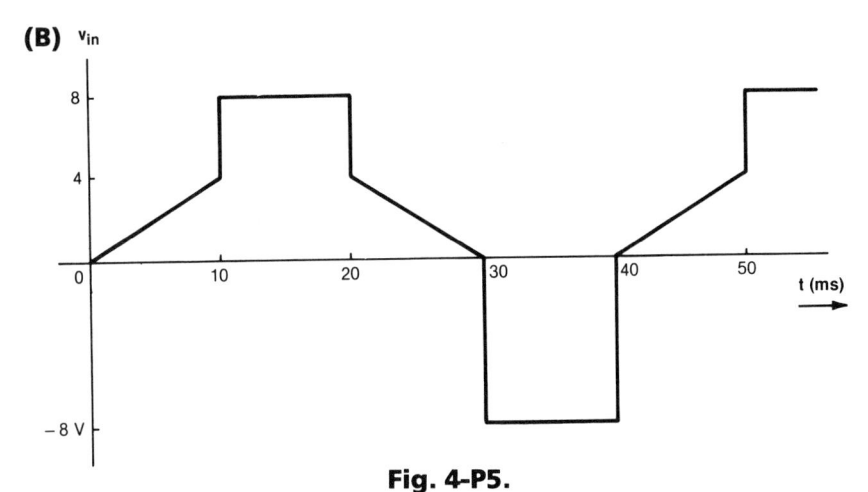

(A)

(B)

Fig. 4-P5.

4.6 Suppose the 5-V battery of the circuit in Fig. 4-P5 is reversed. Determine and sketch the output waveform.

4.7 A clipping circuit (with unequal clipping levels) is shown in Fig. 4-P7, and its input is the same as in Problem 4-5. Determine and sketch the output voltage.

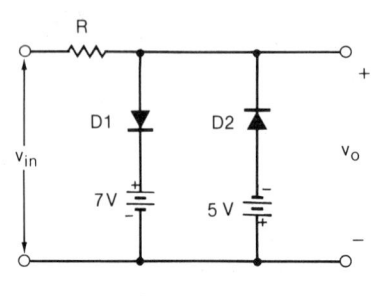

Fig. 4-P7.

4.8 In the circuit of the previous problem, determine and sketch the voltage across *each diode* as a function of time.

4.9 A clamping circuit is modified by the addition of a 5-V battery as shown in Fig. 4-P9. The input to the circuit is a sinusoidal voltage of peak value 10 V. Determine and sketch (a) the voltage across the capacitor; (b) the output voltage.

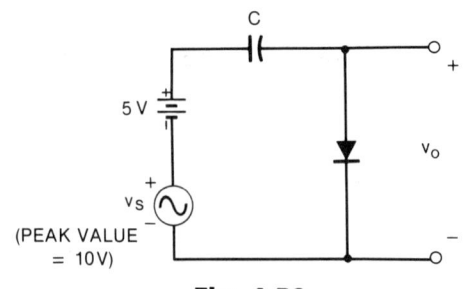

Fig. 4-P9.

4.10 The two signals shown in Fig. 4-P10 are fed to an OR gate (Fig. 4-21). Determine and sketch the output voltage.

4.11 Repeat Problem 4-10 for the two input signals shown in Fig. 4-P11.

4.12 The two signals of Fig. 4-P10 are fed to an AND gate (Fig. 4-24). Determine and sketch the output voltage.

4.13 Repeat Problem 4.12 for the two input signals of Fig. 4-P11.

4.14 The circuit of Fig. 4-P14 is a cascaded connection of an OR gate and an AND gate. The input signals are as shown in Fig. 4-P14B. Determine and sketch the waveforms at points A and B and the output voltage.

4.15 The circuit of Fig. 4-P15 is a cascaded connection of an AND gate and an OR gate. The inputs are the signals of Fig. 4-P14. Determine and sketch the waveforms at points C and D and the output voltage.

Electronics: Circuits and Systems

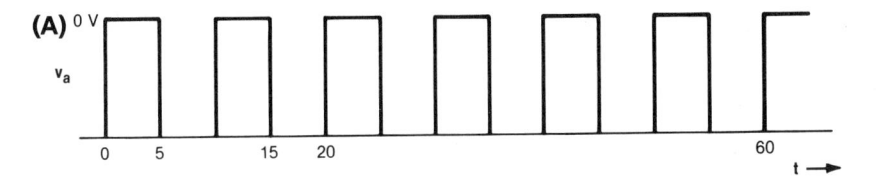

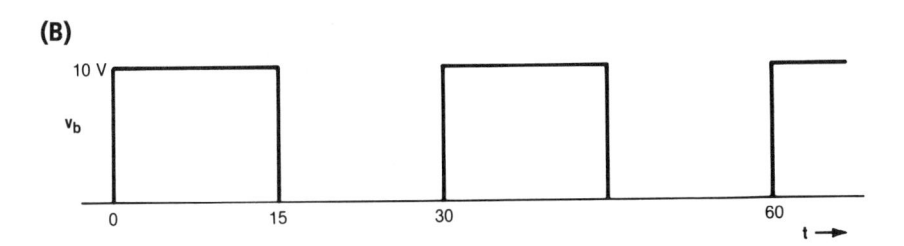

Fig. 4-P10.

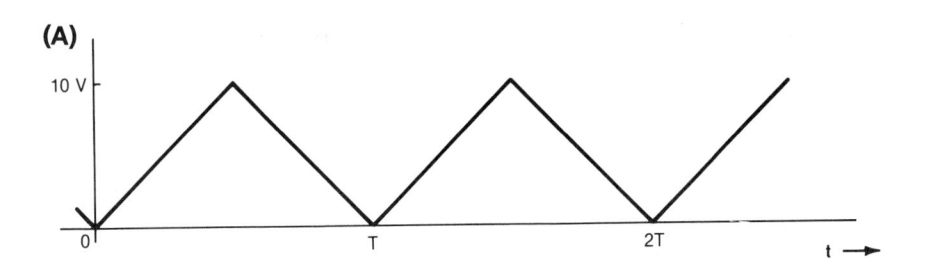

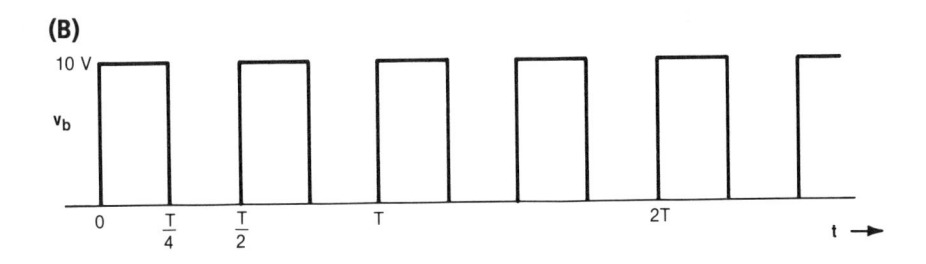

Fig. 4-P11.

270 Diodes and Diode Circuits

(A)

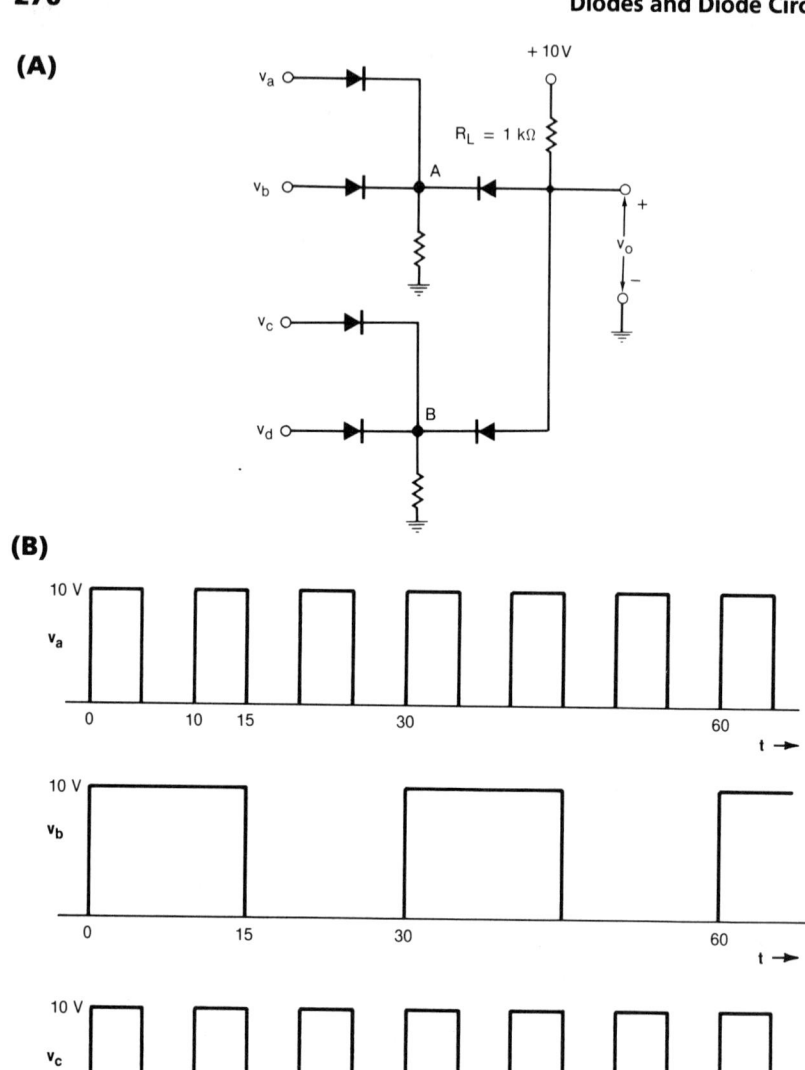

(B)

Fig. 4-P14.

Electronics: Circuits and Systems

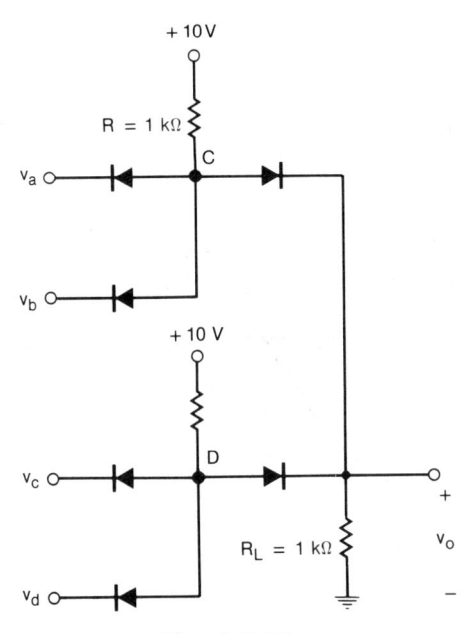

Fig. 4-P15.

Zener Diode

4.16 A zener diode is rated at 6.9 V and 0.2 W and is used in the regulator circuit of Fig. 4-27. Determine the value of R_s if it is given that $V_{in} = 10$ V. If the minimum current in the diode is specified as 1 mA, find the range of values of V_{in} for which the regulator will function properly (using the value of R_s calculated previously).

4.17 Replace the zener diode of the previous problem by one with $V_Z = 3.3$ V and $P_{max} = 1$ W. The circuit is the same as before otherwise. Discuss the effect produced by the change in the diode.

4.18 The circuit of Fig. 4-P18 uses two zener diodes (each with $V_x = 10$ V) connected back to back. Discuss the operation of the circuit and calculate the output voltage when the input is (a) below -10 V; (b) between -10 V and $+10$ V; (c) greater than $+10$ V. If the input to the circuit is the triangular waveform of Fig. 4-4, sketch the output.

Fig. 4-P18.

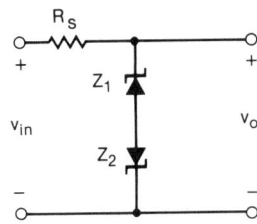

SCRs

4.19 The input voltage and the voltage applied to the gate of the SCR in the circuit of Fig. 4-31 are as shown in Fig. 4-P19. Determine and sketch the output voltage.

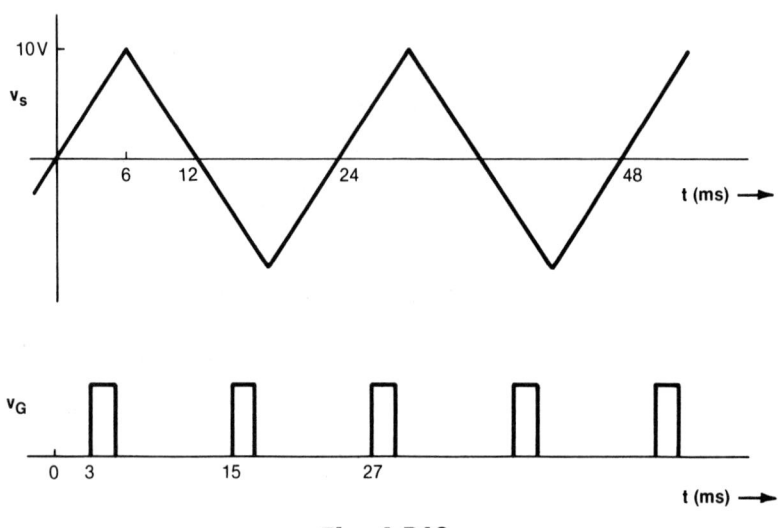

Fig. 4-P19.

4.20 The circuit of Fig. 4-P20 can be used to introduce a specified delay between the closing of the switch S_1 and the onset of current in the load resistor R_L. If a minimum gate voltage of 2 V is needed to trigger the SCR in the given circuit, calculate the delay introduced.

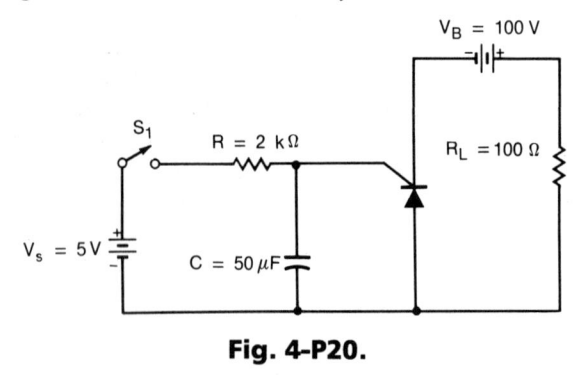

Fig. 4-P20.

4.21 Discuss the operation of the circuit shown in Fig. 4-P21 by considering (a) no pulses applied to the gate; (b) a positive pulse applied to the gate when the input voltage v_s is *positive*; (c) a positive pulse applied to the gate when the input voltage v_s is *negative*. In each case, indicate the direction of the current in the load resistor (if and when it flows). For the particular input signal and the gate pulse train shown in Fig. 4-P21B, sketch the voltage waveform across the load resistor. (Neglect the diode drops when they are conducting.)

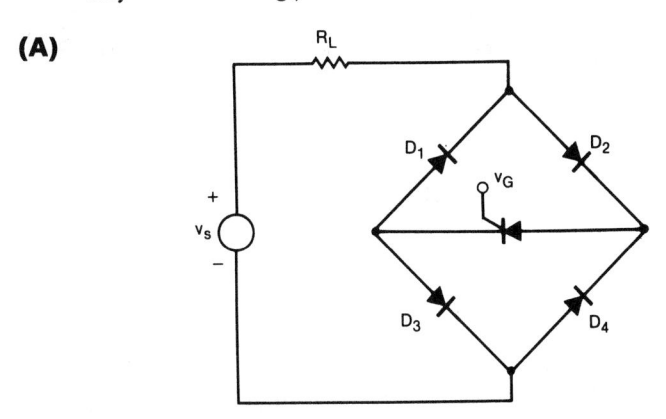

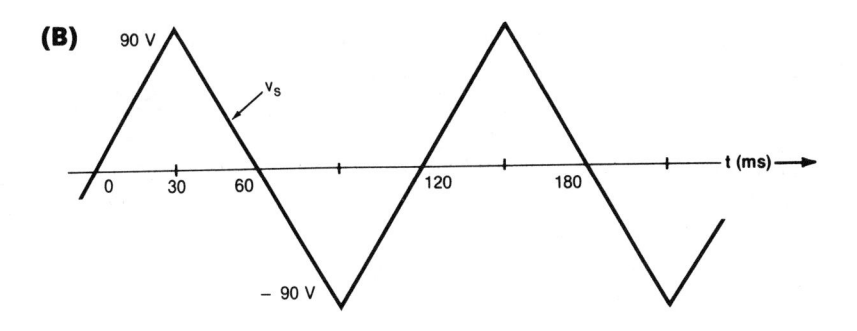

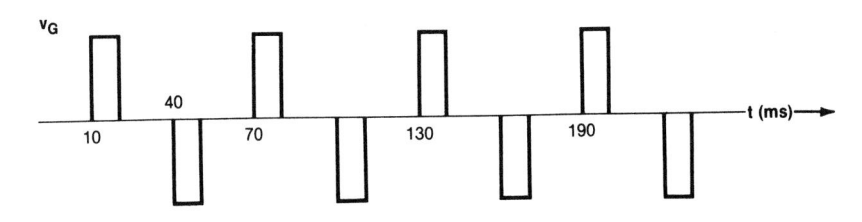

Fig. 4-P21.

CHAPTER 5

BJT AMPLIFIERS

Most of the signals of interest are found to occur at very low voltage levels. For example, the signal produced by a phonograph cartridge is a few millivolts, the signal hitting the antenna of a radar system is a few microvolts, and the signal coming from a star is even smaller. Amplification is absolutely essential for raising the level of the available signal to a level where it can be heard through a loudspeaker, seen on the screen of an oscilloscope, or recorded and measured by various measuring instruments.

It is important to realize, however, that the function of an amplifier is not merely to produce a sufficiently high voltage gain. Even a transformer can be used to step up voltages. But, the *unique feature of an amplifier is that it produces a power gain that is greater than one*, whereas a transformer has a power gain that is less than one. In fact, even though we customarily speak of the voltage gain of an amplifier, it is really the power gain that is being exploited. Power gains of several tens of decibels can be realized through a properly designed amplifier.

Bipolar junction transistors (BJTs) as well as field-effect transistors (FETs) are used in amplifiers. Each device has its own special characteristics as was discussed in Chapter 3, and consequently, the amplifiers display different characteristics in their performance. We will concentrate on the BJT amplifier in this chapter and discuss the FET amplifier in the next chapter.

REVIEW OF THE BJT

The symbol and the currents in an npn transistor are shown in Fig. 5-1A. The arrow on the emitter shows the direction of the current flow in the device. It should be remembered that the base current and the collector current add up to account for the emitter current. That is,

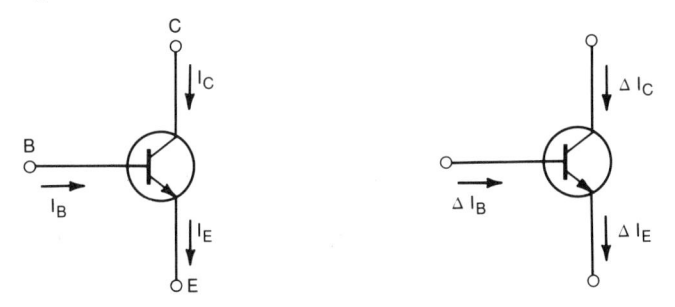

(A) Symbol and currents.   **(B) Incremental currents in a transistor.**

Fig. 5-1. The npn transistor.

$$I_E = I_B + I_C \qquad (5\text{-}1)$$

We recall that due to the forward bias at the emitter-base junction, a very large number of electrons are injected from the emitter into the base. But only a very small percentage of these flow out of the device through the base terminal, and a very large percentage (more than 90%) flows out through the collector. That is, the ratio of the collector current to the emitter current is close to one in a good transistor, and the ratio of the collector current to the base current is quite high (20 or more).

Now suppose we wish to increase the base current by an amount ΔI_B as indicated in Fig. 5-1B. This increase can be brought about by increasing the emitter current by an amount ΔI_E. But, since only a small portion of the emitter current forms the base current, the ratio of ΔI_E to ΔI_B must be high. In fact, ΔI_E is approximately β times ΔI_B. Therefore, a small change in the base current forces a much larger change in the emitter current and a corresponding increase in the collector current. As an example, if a transistor has β of 50, then a *change* of 0.01 mA in the base current will result in a *change* of 0.5 mA in the emitter and the collector currents.

If the *base current* in a transistor is controlled by *an input signal*, and the *output* is taken at the *collector terminal*, then a *small change in the input signal causes a large change in the output signal*. This is the basis of the amplification available from a transistor.

ASPECTS OF ANALYSIS OF AN AMPLIFIER

There are two aspects to the analysis and design of an amplifier (and these apply to amplifiers using any type of active device and not just BJTs).

The first consideration is that some dc biasing is necessary to make the transistor work. In a BJT amplifier, the emitter-base junction has to be forward biased, and the collector-base junction has to be reverse biased. Moreover, the bias voltages should be at appropriate levels so that the voltages and currents stay within an allowed range even when signals are applied to the amplifier. These considerations come under *transistor biasing*, which will be discussed in detail in a following section.

The second consideration is the response of the amplifier to the signal that is to be amplified. The input signal is usually a time-varying signal, and the output signal is also a time-varying signal. In most applications, it is necessary to make the output signal a faithful reproduction of the input signal; that is, the distortion should be negligible. Such an amplifier is said to be *linear*. Since the characteristics of an active device show some amount of nonlinearity, it is necessary to keep the variation of the input signal restricted to a small range. The analysis of an amplifier when confined to operate in a linear fashion is, therefore, called *small-signal analysis*. The small-signal analysis of an amplifier is performed through the use of an *equivalent circuit*, or *model*, of the transistor as will be discussed shortly.

TWO-PORT MODEL OF AN AMPLIFIER

For *small-signal analysis*, an amplifier can be represented as a two-port system as shown in Fig. 5-2. The box denoting the two-port contains one or more transistors, and we will look at the contents of the box later in this section. For the time being, we will concentrate on the terminal quantities. The following quantities are of interest in the small-signal analysis of an amplifier.

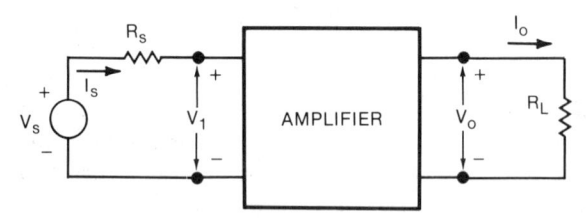

Fig. 5-2. Two-port system.

1. *Voltage gain:*

$$\text{Voltage gain} = (V_o/V_s) \qquad \text{Ratio}$$

$$\text{Voltage gain } in\ decibels = 20 \log (V_o/V_s) \qquad \text{Decibels}$$

2. *Current gain:*

$$\text{Current gain} = (I_o/I_s) \quad \text{Ratio}$$

$$\text{Current gain } \textit{in decibels} = 20 \log (I_o/I_s) \quad \text{Decibels}$$

3. *Power gain:*

$$\text{Power gain} = (\text{voltage gain}) \times (\text{current gain}) \quad \text{Ratio}$$

$$\text{Power gain } \textit{in decibels} = 10 \log (\text{power gain}) \quad \text{Decibels}$$

It is important to recall that (1) the decibel value of the power gain is *not* equal to the sum of the decibel values of the voltage gain and current gain; (2) the multiplication factor used to obtain the decibel value of the *power gain* is *10* while that used for the voltage and current gains is 20; (3) the decibel values of the voltage gain, current gain, and power gain are usually different from one another for a given two-port system.

Drill Problem 5.1: Convert the following voltage gains to decibels: (a) 76.4; (b) 0.985; (c) 2.5×10^4.

Drill Problem 5.2: Convert the following current gains to decibels: (a) 64.1; (b) 1.6×10^3; (c) 0.925.

Drill Problem 5.3: Take each of the voltage gain values in Drill Problem 5.1 and the corresponding value of the current gain in Drill Problem 5.2. Find the power gain in each case and convert it to decibels.

4. *Input resistance:*
The input resistance is given by

$$R_{in} = (V_1/I_s)$$

Note that the resistance R_s of the source is not included as part of the input resistance.

5. *Output resistance:*
If the signal source V_s is deactivated (by replacing it with a short circuit) and the resistance is measured from the output terminals by using a test source (or some suitable measuring instrument) as indicated in Fig. 5-3, the resistance so measured is called the *output resistance*. It is important to realize that the output resistance is not always easy to obtain by an inspection of the circuit.

The small-signal analysis (which we will simply refer to as "ac analysis" from now on) of an amplifier involves the calculation of several or all of the

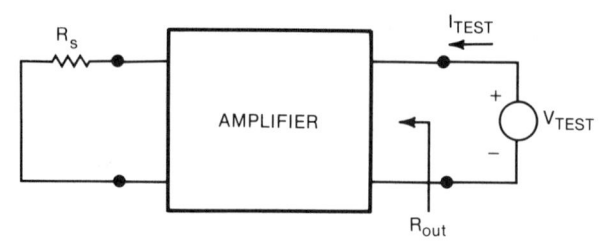

Fig. 5-3. Output resistance of an amplifier.

previous quantities and also some others that will be introduced in Chapter 7. In many cases, the calculation of the voltage gain is the most important.

Now we consider what might be placed inside the two-port box of Fig. 5-2 in the case of an amplifier.

A transistor is an *active device.* That is, it can generate an output signal that has a higher power content than the input signal. In the ac analysis of an amplifier containing active devices, the devices are replaced by *equivalent circuits*, or *models*. In most practical situations, a simple model is sufficient to give results of adequate accuracy.

The simple model of any active device consists of a resistor R_1 at the input port and a *dependent current source* (also called a *controlled current source*) at the output port as shown in Fig. 5-4. A resistance R_2 in parallel with the dependent current source completes the model. The dependent source supplies a current of $(g_m V_1)$ A. (In the case of the BJT, the current supplied by the dependent source can also be expressed in terms of β times the base current of the transistor.) Since the current delivered by the source *depends* upon the voltage across R_1, it is called a dependent source. The constant g_m is called the *transconductance* and is a measure of the control exerted by the input side of the two port on the current supplied by the dependent source. The unit of transconductance is the *reciprocal of the ohm* and is called *siemens*, symbolized with S. As an example of the behavior of the dependent source, suppose $g_m = 40$ millisiemens (mS). Then, any input voltage V_1 applied to the two port will be multiplied by (40×10^{-3}), and a current of that value will be supplied by the dependent source: if $V_1 = 5$ V, for example, then the current source will deliver 200 mA.

Assuming for the sake of simplicity that the two port contains only one active device, we can replace the two port by the model of Fig. 5-4, leading to the circuit shown in Fig. 5-5. The input side and the output side are seen to be linked only by the presence of the dependent source, and so they can be analyzed separately.

Electronics: Circuits and Systems

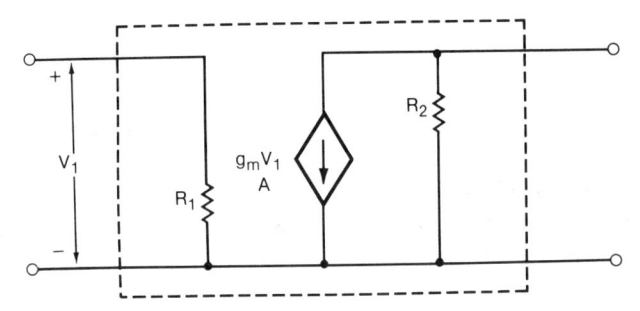

Fig. 5-4. Model of an active device.

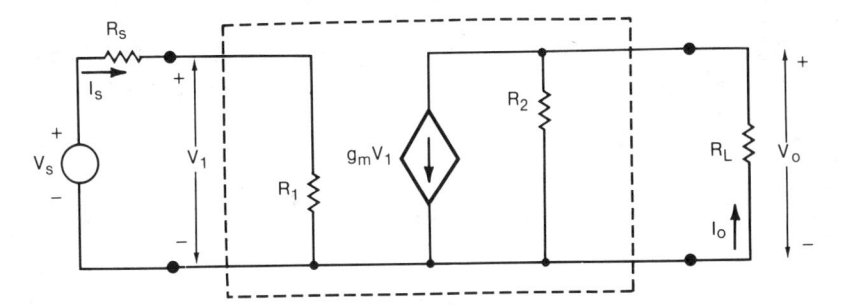

Fig. 5-5. Two-port system for analysis.

Analysis of the Input Side

The aim of the analysis of the input side is to determine the voltage V_1 and the current I_s in terms of V_s. The current I_s is given by

$$I_s = \frac{V_s}{R_1 + R_s} \quad (5\text{-}2)$$

and the voltage V_1 is given by using the voltage divider rule:

$$V_1 = \frac{R_1}{R_1 + R_s} V_s \quad (5\text{-}3)$$

Analysis of the Output Side

The aim of the analysis of the output side is to determine the voltage V_o and the current I_o in terms of V_1. These relationships can then be combined with Eqs. (5-2) and (5-3) to obtain the current and voltage gains.

On the output side, the resistances R_2 and R_L are in parallel. They can be combined into a single resistance R_{eq}:

$$R_{eq} = \frac{R_2 R_L}{R_2 + R_L} \tag{5-4}$$

which leads to the simplified circuit of Fig. 5-6. The output voltage V_o is then that across R_{eq} due to the current $(g_m V_1)$ through it. But, note that the current flows from the *reference negative* terminal of V_o to the *reference positive* terminal, and it is, therefore, necessary to use Ohm's law with a *minus* sign (as discussed in Chapter 1). That is,

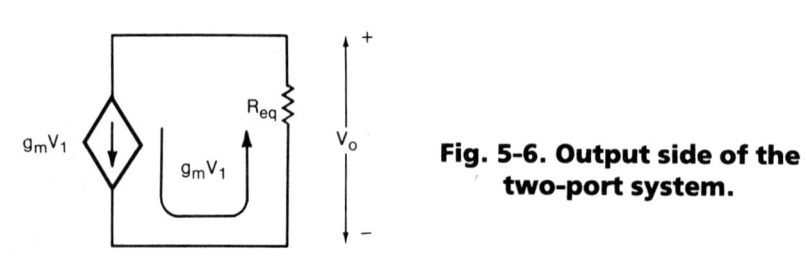

Fig. 5-6. Output side of the two-port system.

$$V_o = -(g_m V_1 R_{eq})$$

or,

$$V_o = (-g_m R_{eq}) V_1 \tag{5-5}$$

The current I_o is the current through R_L (Fig. 5-5) and since the voltage across R_L is V_o, we have

$$I_o = -\frac{V_o}{R_L}$$

which leads to

$$I_o = \frac{g_m R_{eq}}{R_L} V_1 \tag{5-6}$$

Equations (5-2), (5-3), (5-5), and (5-6) form the basis of the calculations of the various quantities of interest in an amplifier.

Example 5.1

Calculate the voltage gain, current gain, and power gain of the two-port system shown in Fig. 5-7.

Electronics: Circuits and Systems

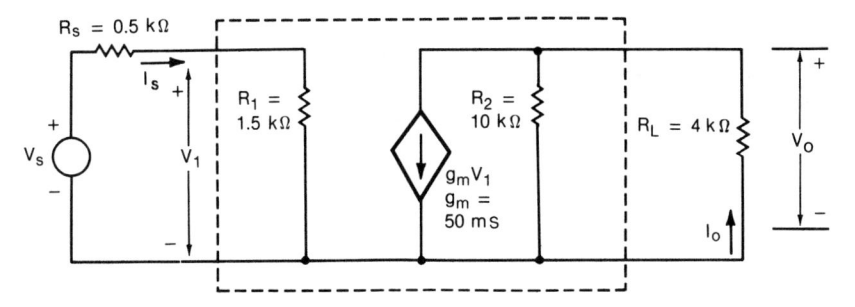

Fig. 5-7. Two-port system for Example 5.1.

Solution

Input side calculations:

$$I_s = \frac{V_s}{R_1 + R_s}$$

$$= 0.5 \times 10^{-3} V_s \tag{5-7}$$

$$V_1 = \frac{R_1}{R_1 + R_s} V_s$$

$$= 0.75 V_s \tag{5-8}$$

Output side calculations:

The parallel combination of R_2 and R_L gives

$$R_{eq} = \frac{R_2 R_L}{R_2 + R_L}$$

$$= 2.86 \times 10^3 \, \Omega$$

From Eq. (5-5),

$$V_o = g_m R_{eq} V_1 = -143 V_1 \tag{5-9}$$

and, using Ohm's law or Eq. (5-6),

$$I_o = -(V_o/R_L) = 3.58 \times 10^{-2} V_1 \tag{5-10}$$

Voltage gain:

From Eqs. (5-8) and (5-9),

$$\text{Voltage gain} = (V_o/V_s) = -107 \quad (40.6 \text{ dB})$$

Current gain:
From Eqs. (5-10) and (5-7),

$$\frac{I_o}{I_s} = \frac{3.58 \times 10^{-2} V_1}{0.5 \times 10^{-3} V_s}$$

$$= 71.6 \frac{V_1}{V_s}$$

which, in conjunction with Eq. (5-8) gives

$$\text{Current gain} = (I_o/I_s)$$
$$= (71.6)(0.75)$$
$$= 53.7 \qquad (34.6 \text{ dB})$$

Power gain:

$$\text{Power gain} = \text{voltage gain} \times \text{current gain}$$
$$= 5.76 \times 10^3 \qquad (37.6 \text{ dB})$$

The *currents* in an electronic circuit are almost always expressed *in milliamperes* and the *resistances in kilohms*. Since milliamperes multiplied by kilohms gives volts, it is convenient to use the following convention in all calculations: *all current values are consistently expressed in milliamperes and all resistance values in kilohms*. When this convention is used, it is unnecessary to place the 10^{-3} factor (for currents) and the 10^3 factor (for resistances) in the calculations in a problem. As a demonstration of this convention, the calculations of the previous example are repeated:

$$\text{input current } I_s = \frac{V_s}{2}$$

$$= 0.5 V_s \text{ mA}$$

$$\text{input voltage } V_1 = \frac{1.5}{2} V_s$$

$$= 0.75 V_s$$

Total resistance on the output side

$$R_{eq} = \frac{10 \times 4}{10 + 4}$$

$$= 2.86 \text{ k}\Omega$$

$$\text{output voltage } V_o = -(50 \times 2.86)V_1$$
$$= -143V_1$$
$$\text{output current } I_o = -(V_o/R_L)$$
$$= (143V_1/4)$$
$$= 35.8V_1 \text{ mA}$$

The remainder of the calculations will be the same as before. We will be using this convention in all the following examples and problems in the text.

Drill Problem 5.4: In the circuit of the previous example, suppose the signal source $V_s = 15$ mV and R_s is changed to 1 kΩ. Also, let R_L be changed to 8 kΩ. Calculate the new values of the three gains, the output voltage value and the output current value.

Drill Problem 5.5: The two-port model of an amplifier is made up of $R_1 = 500$ kΩ, $g_m = 12$ mS, and $R_2 = 10$ kΩ. A signal source of $V_s = 150$ mV and $R_s = 10$ kΩ is connected to the input side and a load resistance $R_L = 0.5$ kΩ is connected to the output side. Calculate the three gains of the amplifier system in decibels.

Drill Problem 5.6: Repeat the calculations of the previous drill problem if the resistance of the signal source is changed to $R_s = 500$ kΩ, and the load resistance to $R_L = 10$ kΩ.

BIASING OF A BJT AMPLIFIER

Consider an amplifier when there is no input signal. Then the currents and voltages in it are due only to the dc battery (or batteries) providing the proper biasing conditions. Such a state of affairs is called the *quiescent condition* (or the *no-signal condition*). The value of the collector current I_C and the collector-emitter voltage V_{CE} under quiescent conditions define the *quiescent operating point* (referred to simply as the *Q point*). (A knowledge of I_C and V_{CE} and some parameters of the transistor are sufficient to determine any other quantity in the amplifier under quiescent conditions.)

The function of the biasing circuit of a BJT amplifier is to place the Q point at a desirable level. For the time being, we will skip the considerations on how a desirable Q point is selected and return to it later. As was seen earlier, it is necessary to make the emitter-base junction forward biased and the collector-base junction reverse biased. But, instead of using two separate batteries at the two junctions, it is more economical to have a *single battery* to provide both the biasing conditions.

There are essentially two standard biasing circuits for a BJT, one of which is shown in Fig. 5-8. In order to simplify the schematic diagrams of electronic circuits, the battery V_{CC} (indicated by the dashed lines) is not fully shown. The positive terminal of V_{CC} is indicated by the label $+V_{CC}$ at the top of the resistor R_C and the negative terminal of the battery is assumed to be connected to ground. A similar convention is used for any dc power supply used in an electronic circuit. As will be seen shortly, the two resistors R_{B1} and R_{B2} essentially serve as a voltage divider that splits the voltage V_{CC} into two parts determined by the values of R_{B1} and R_{B2}. Starting from the emitter terminal, we see that the base can be at a higher potential than the emitter (so as to forward bias the emitter-base junction) and the collector at a higher potential than the base (so as to reverse bias the collector-base junction) by a proper choice of the values of the resistors in the biasing circuit.

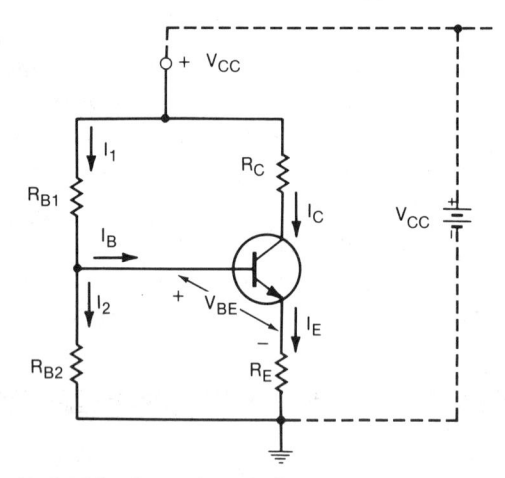

Fig. 5-8. Biasing circuit for an npn transistor.

Analysis of the Biasing Circuit

Since the emitter-base junction is forward biased, we can assume that the pn junction there has a constant voltage drop of 0.7 V (just as we did in diode circuits). Therefore,

$$V_{BE} = 0.7 \text{ V} \tag{5-11}$$

with the polarities shown in Fig. 5-8.

The three currents I_E, I_B, and I_C are related by the equation:

$$I_E = I_B + I_C \tag{5-12}$$

Electronics: Circuits and Systems 285

The collector current I_C and the base current are related by the equation

$$I_C = \beta I_B \tag{5-13}$$

These three equations form the basic tools of the analysis of the biasing circuit.

In the case of circuits where the β of the transistor is reasonably high, the following approximation can be made, the base current I_B is negligibly small compared with the current I_1 through R_{B1}. With this approximation, we have $I_1 \approx I_2$, and the two resistors R_{B1} and R_{B2} are effectively *in series* with each other as indicated in Fig. 5-9A. The voltage at the point B is then obtained by using the voltage-divider rule.

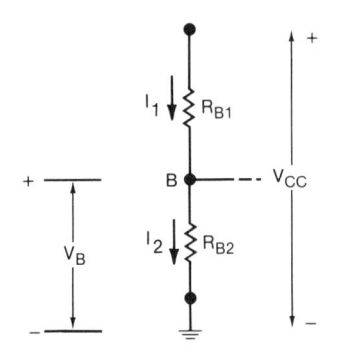

(A) R_{B1} and R_{B2} act as a voltage divider to produce voltage V_B.

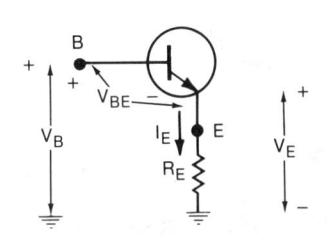

(B) Voltage at the emitter is 0.7 V less than V_B.

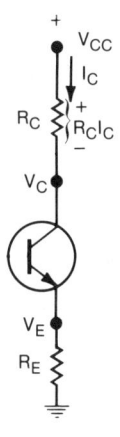

(C) Calculation of V_{CE}.

Fig. 5-9. Q point calculations.

$$V_B = \frac{R_{B2}}{R_{B1} + R_{B2}} V_{CC} \qquad (5\text{-}14)$$

Since the voltage drop from base to emitter is $V_{BE} = 0.7$ V, we have for the voltage at the emitter terminal

$$V_E = V_B - V_{BE}$$
$$= V_B - 0.7 \qquad (5\text{-}15)$$

The voltage V_E is across the resistor R_E (Fig. 5-9B) and so the emitter current I_E is given by

$$I_E = (V_E/R_E) \qquad (5\text{-}16)$$

If the β of the transistor is high (say 10 or more), then

$$I_C \approx I_E \qquad (5\text{-}17)$$

The collector current has now been determined. The other piece of information needed for the Q point is the collector to emitter voltage V_{CE}. Referring to Fig. 5-9C, the voltage at the collector is less than V_{CC} due to the voltage drop in R_C. Therefore,

$$V_C = V_{CC} - R_C I_C \qquad (5\text{-}18)$$

and the collector-to-emitter voltage is given by

$$V_{CE} = V_C - V_E \qquad (5\text{-}19)$$

In solving a numerical problem, it is easier to calculate each bit of information as outlined in the previous discussion rather than obtain and use one big formula.

Example 5.2

Determine the Q point of the transistor in the circuit shown in Fig. 5-10.

Solution

Base voltage: (treating R_{B1} and R_{B2} as a voltage divider)

$$V_B = \frac{R_{B2}}{R_{B1} + R_{B2}} V_{CC}$$
$$= \frac{20}{220} \times 25$$
$$= 2.27 \text{ V}$$

Electronics: Circuits and Systems

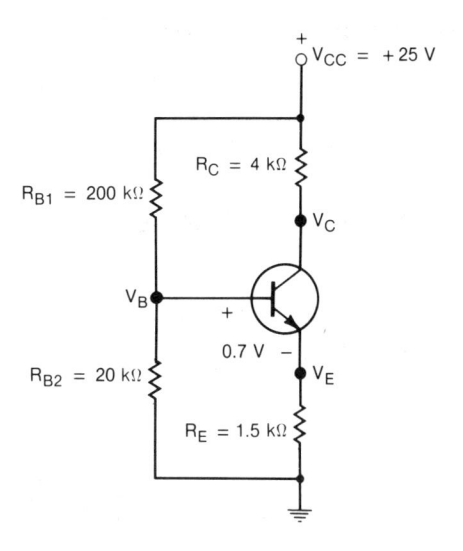

Fig. 5-10. Circuit for Example 5.2.

Emitter voltage:

$$V_E = V_B - 0.7$$
$$= (2.27 - 0.7)$$
$$= 1.57 \text{ V}$$

Emitter current:

$$I_E = (V_E/R_E)$$
$$= (1.57/1.5)$$
$$= 1.05 \text{ mA}$$

Collector current:

$$I_C \approx I_E = 1.05 \text{ mA}$$

Collector voltage:

$$V_C = V_{CC} - R_C I_C$$
$$= (25 - 4.2)$$
$$- 20.8 \text{ V}$$

Collector-to-emitter voltage:

$$V_{CE} = V_C - V_E$$
$$= (20.8 - 1.57)$$
$$= 19.23 \text{ V}$$

Therefore, the Q point of the transistor in the given circuit is $I_C = 1.05$ mA, $V_{CE} = 19.2$ V.

Drill Problem 5.7: Determine the Q point of the transistor in each circuit of Fig. 5-11.

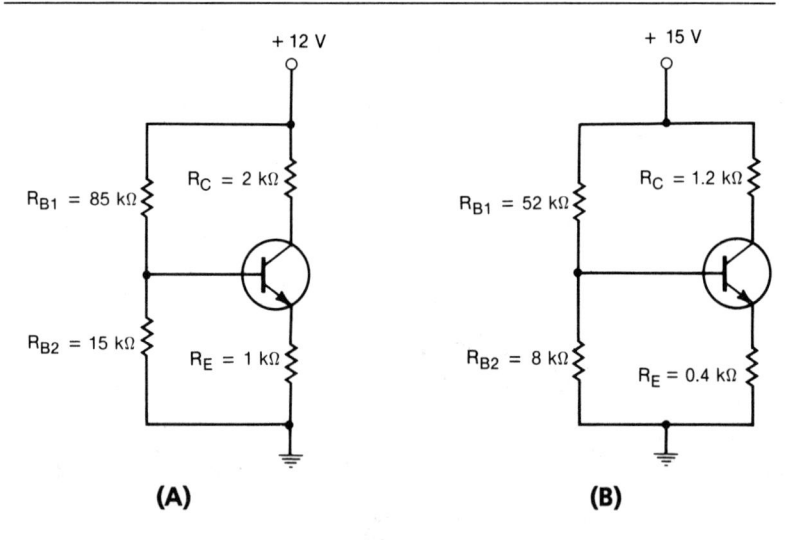

Fig. 5-11. Circuits for Drill Problem 5.7.

Example 5.3

In the circuit of Fig. 5-12, the Q point of the transistor is to be $I_C = 2$ mA and $V_{CE} = 6$ V. Assuming high β, find the values of R_C and R_{B2}.

Solution

Assuming high β, we have $I_E \approx I_C = 2$ mA. Therefore, the emitter voltage is

$$V_E = R_E I_E$$
$$= 1.2 \times 2$$
$$= 2.4 \text{ V}$$

Electronics: Circuits and Systems

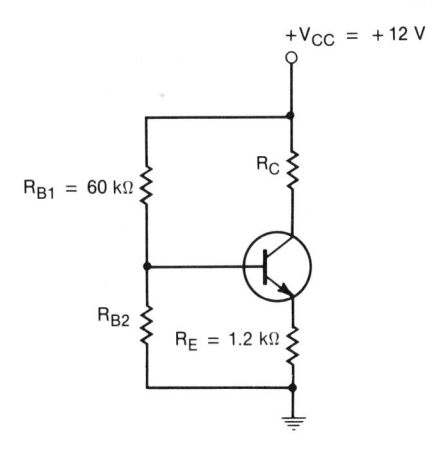

Fig. 5-12. Circuit for Example 5.3.

Referring to Fig. 5-13, the voltage at the collector is

$$V_C = V_E + V_{CE}$$
$$= 8.4 \text{ V}$$

(A)   **(B)**

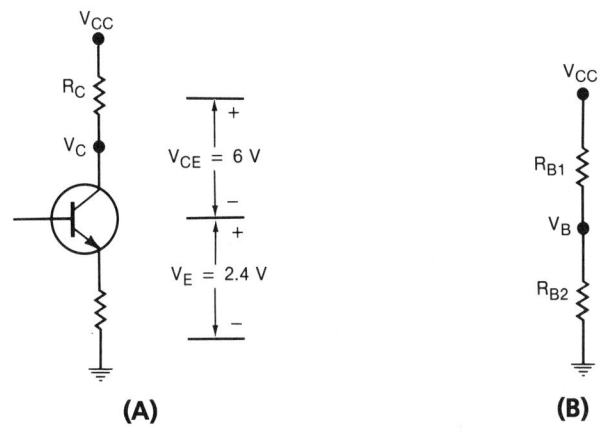

Fig. 5-13. Calculations in Example 5.3.

and

$$\text{Voltage across } R_C = V_{CC} - V_C = (12 - 8.4) = 3.6 \text{ V}$$

Since the current through R_C is given as 2 mA, we have

$$R_C = (\text{voltage across } R_C)/I_C$$
$$= (3.6/2) = 1.8 \text{ k}\Omega$$

In order to determine the value of R_{B2}, consider the voltage at the base, which is 0.7 V higher than the emitter voltage. Therefore,

$$V_B = (2.4 + 0.7)$$
$$= 3.1 \text{ V}$$

Using the voltage-divider equation (5-14),

$$V_B = \frac{R_{B2}}{R_{B1} + R_{B2}} V_{CC}$$

which leads to

$$\frac{R_{B2}}{R_{B1} + R_{B2}} = \frac{3.1}{12}$$
$$= 0.258$$

and since $R_{B1} = 60k$,

$$R_{B2} = 20.9 \text{ k}\Omega$$

Drill Problem 5.8: If in the circuit of Fig. 5-11A, it is necessary to change the Q point to $V_{CE} = 7.5$ V and $I_C = 2$ mA, find the new values of R_C and R_{B2} needed assuming that all the other components remain the same.

Drill Problem 5.9: If in the circuit of Fig. 5-11B, the Q point is to be changed to $V_{CE} = 5$ V and $I_C = 3$ mA, find the new values of R_C and R_{B2} needed, assuming that all the other components remain the same.

Temperature Effects in a Transistor

The effect of an increase in the temperature of a transistor is to increase the thermal activity inside the device. The increase in the thermal agitation results in an increase in the value of the collector cutoff current I_{CO}. The value of I_{CO} increases by a factor of two (that is, it doubles) for every 10°C increase in the temperature. The increase in I_{CO} results in an increase in the collector current for a given base current. As indicated in Fig. 5-14, there is an upward shift of the output current-voltage characteristics of the transistor.

Suppose a particular biasing circuit makes $I_B = 50$ μA when no signal is present. Let the Q point at the temperature T_1 be at Q_1. When the temperature increases to T_2, the value of I_B is still at 50 μA (since it is determined by the biasing circuit) and the Q point shifts upward to a new position Q_2 on the

Electronics: Circuits and Systems

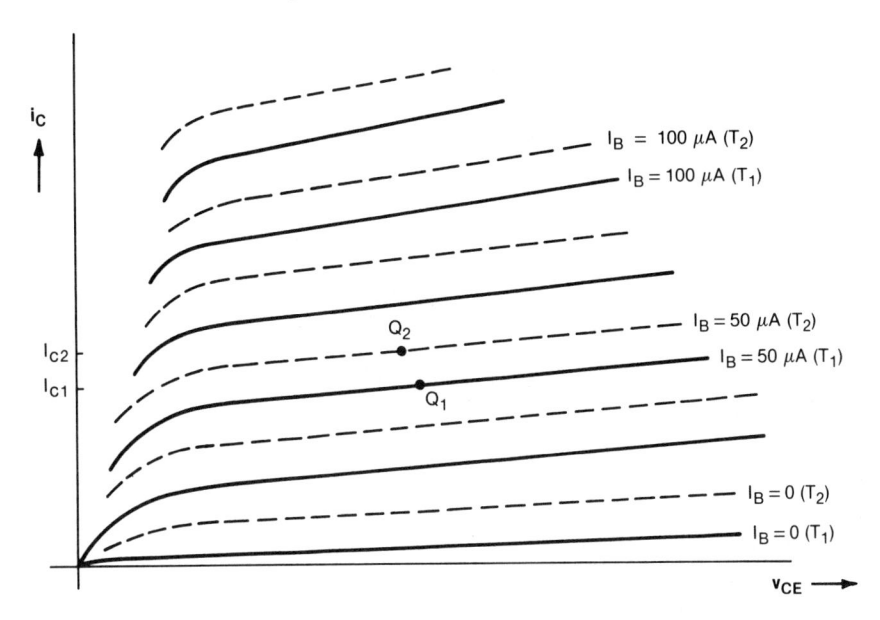

Fig. 5-14. Shifting of transistor characteristics due to an increase in temperature.

new $I_B = 50$ μA curve. The collector current with no signal is significantly larger at T_2 than at T_1. The higher value of I_C at the quiescent operating point will limit the upward swing: if the collector current were to increase due to an input signal then the amount by which it can increase without producing distortion will be decreased by the higher position of Q_2.

Distortion is, however, not the only undesirable effect of an increase in the temperature of a transistor. When the temperature increases (which will happen due to the current flowing in a transistor), the value of I_{CO} increases. This increase leads to an upward shift of the Q point, which means that the collector current is now higher than before. This, in turn, causes a further increase in temperature. A vicious circle is set up and unless some external control is placed on the temperature of the circuit, a *thermal runaway* can result and ruin the circuit. In power amplifiers, where the power level is high, it is necessary to provide cooling of the transistor by proper radiation of the heat from the device. In small signal amplifiers, the biasing circuit itself provides some control over the increase in the collector current.

Consider the output characteristics shown in Fig. 5-15. Suppose the value of I_B is 50 μA at the temperature T_1 and the Q point is at Q_1. The corresponding value of I_C is I_{C1}. If the temperature now increases to T_2, the Q point will shift to Q_2 and the collector current to I_{C2} *if the base current is maintained at the old value* of 50 μA. But suppose the biasing circuit is designed so that when the temperature went up to T_2, the value of I_B is *reduced* to, say, 30 µA. Then, the new Q point will be at Q_3 (on the $I_B = 30$ µA curve), and the collector current will increase only to I_{C3} at the new temperature instead of I_{C2}.

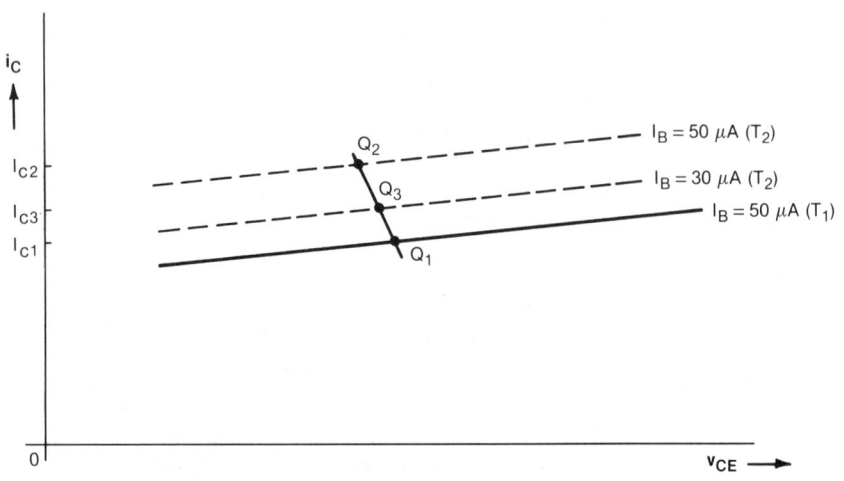

Fig. 5-15. Reducing the shift in the Q point due to high temperature.

The trick is, therefore, to cause a reduction in the value of the base current as the temperature increases so that the shift in the Q point is not as great as it would be. This is accomplished by the presence of the resistor R_E in the biasing circuit of Fig. 5-8 (the relevant portion of which is repeated in Fig. 5-16) in the following manner. Consider the current I_E through R_E in Fig. 5-16. If the collector current increases then I_E also increases. The voltage at point E increases since $V_E = R_E I_E$, resulting in a decrease in the voltage V_{BE} from the base to the emitter; that is, the forward bias decreases. There is, therefore, a reduction in the currents in the transistor. Any attempt on the part of the collector current to increase is fed back to the input side (base emitter) through the resistor R_E in such a way as to decrease the forward bias and hence the emitter and the collector currents. The resistor R_E is said to provide *negative feedback*. The biasing circuit discussed earlier (Fig. 5-8) exercises some control over the increase in the collector current caused by an increase in the temperature. In applications where a high stability of the Q point is necessary, more elaborate biasing circuits than the one considered previously are used. In power amplifiers with high power dissipation, the biasing circuit itself cannot provide a sufficient control over the Q point when the temperature increases, and heat sinks of sufficiently large areas are indispensable in such amplifiers.

Electronics: Circuits and Systems 293

Fig. 5-16. Stabilization of the Q point through negative feedback due to R_E.

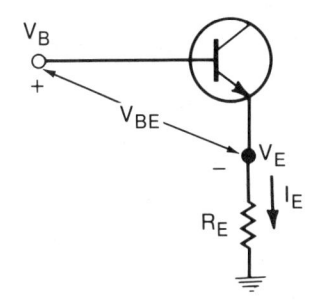

BJT AMPLIFIER

In an amplifier, a signal source V_s with a resistance R_s is connected to its input port, and the output port has a load resistor connected to it as was shown in Fig. 5-2. When a BJT is used as the amplifying device, there is a choice of several input ports and several output ports, leading to three possible configurations of a BJT amplifier. It should be noted that the biasing components are not included in the following configurations since we are only considering the *ac behavior* of the amplifiers.

Common-Emitter Amplifier

When the input port is formed by the base and emitter terminals and the output port is formed by the collector and emitter terminals, as indicated in Fig. 5-17, the emitter becomes common to the input and output ports. The resulting configuration is, therefore, called a *common-emitter amplifier*.

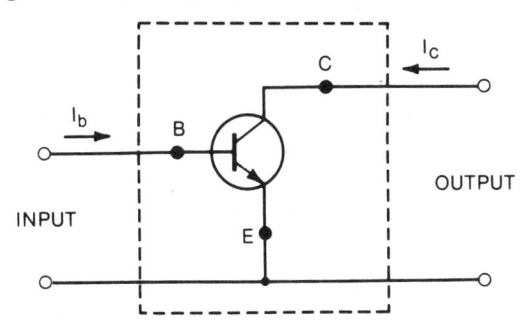

Fig. 5-17. Basic common-emitter amplifier.

The common-emitter amplifier has a high current gain and a high voltage gain. The high current gain is due to the fact that the input current is the base current and the output current is the collector current, which is many times as large as the base current. The input voltage is the small forward bias across the emitter-base junction and the output voltage is due to the large collector current flowing through the load resistor and the voltage gain is, consequently, high.

Because of its high voltage gain and high current gain, the common-emitter amplifier is the most widely used configuration of a BJT amplifier.

Common-Base Amplifier

When the input port of the amplifier is formed by the emitter and base terminals and the output port is formed by the collector and base terminals, as indicated in Fig. 5-18, the base terminal becomes common to the input and output ports. The resulting configuration is, therefore, called a *common-base amplifier*.

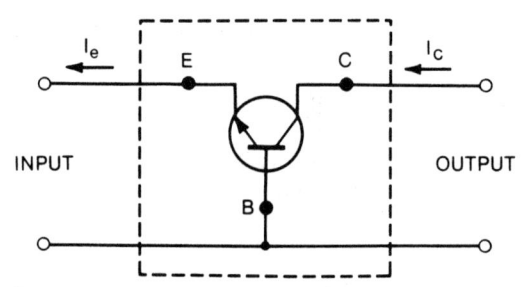

Fig. 5-18. Basic common-base amplifier.

The common-base amplifier has a current gain of less than one since the input current is the emitter current and the output current is the collector current (which is less than the emitter current). But it has a high voltage gain since the input voltage is the small forward bias across the emitter-base junction and the output is due to the collector current flowing through the load resistor. The common-base amplifier is useful, however, because of certain special characteristics. It has a much smaller input impedance (by a factor of β) than the common-emitter amplifier. The bandwidth of the common-base configuration is much wider than that of the common-emitter configuration. But the most important feature of the common-base connection is that it is less sensitive to an increase in temperature than the common-emitter connection. The reason is that the effect of the collector cutoff current in a common-base connection is only $(1/\beta)$ times the effect of the collector cutoff current in a common-emitter connection. Therefore, the shift is the Q point due to an increase in temperature is much smaller in the common-base connection than in the common-emitter connection.

Common-Collector Amplifier

When the input port is formed by the base and collector terminals and the output port is formed by the emitter and collector terminals, as indicated in

Electronics: Circuits and Systems

Fig. 5-19, the collector becomes common to the input and output ports. The resulting configuration is, therefore, called a *common-collector amplifier*. It is also called the *emitter follower* since the output voltage follows the emitter voltage (both being essentially the same).

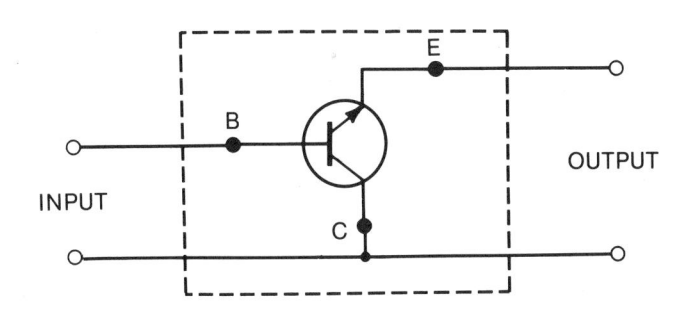

Fig. 5-19. Basic common-collector amplifier.

The common-collector amplifier has a high current gain since the input current is the base current and the output current is the emitter current. But its voltage gain is less than one since the input voltage is across the (reverse-biased) base-to-collector junction and the output voltage is from the collector to emitter, which is only slightly different from the base-collector voltage. In spite of its low voltage gain, the common-collector amplifier is frequently used as an interstage between an amplifier and a load resistor in order to provide the impedance matching needed for maximum power transfer.

COMMON-EMITTER AMPLIFIER

A common-emitter amplifier is shown in Fig. 5-20. The portion inside the dashed lines in the diagram is the *biasing circuit* discussed earlier in the section "Analysis of the Biasing Circuit." The remaining components are important for the *small-signal* (or *ac*) *operation* of the amplifier.

The source V_s produces an ac signal that is to be amplified and R_s represents its internal resistance. The $V_s - R_s$ combination may be an actual source (like the cartridge of a phonograph) or it may be a model of a preceding amplifier stage in a multistage amplifier. The ac components of current and voltage produced by the signal source are superimposed on the dc currents and voltages set up by the biasing circuit in the amplifier.

The two capacitors C_1 and C_2 are called *coupling capacitors* since they couple the transistor to the input signal and the load resistor, respectively. They

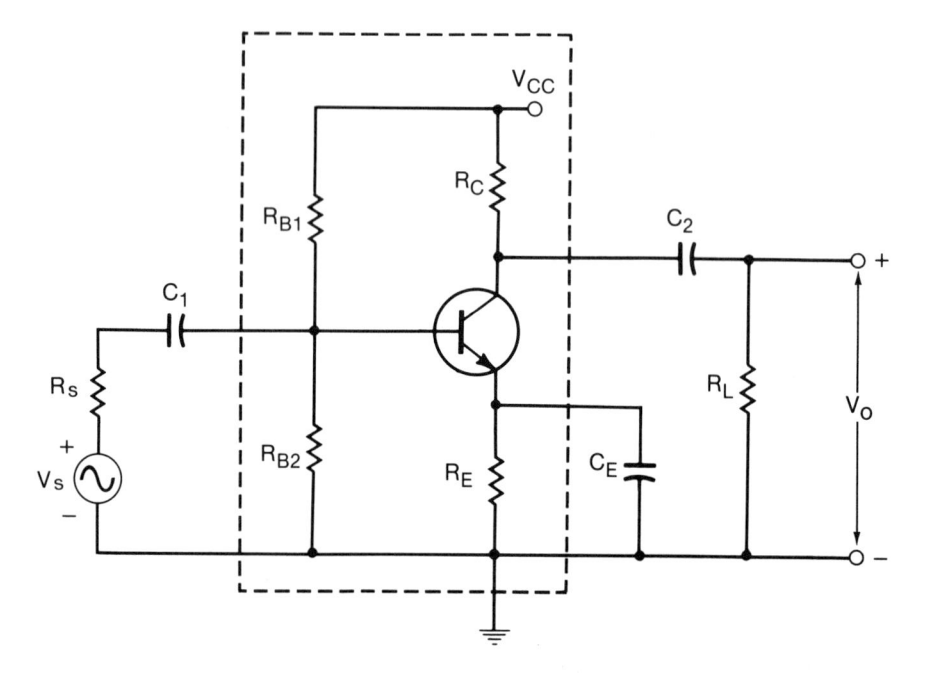

Fig. 5-20. Common-emitter amplifier circuit.

are necessary in order to provide *dc isolation* between the transistor biasing circuit and the resistor R_s on the one hand and the resistor R_L on the other. If the coupling capacitors were not used, then the resistors R_s and R_L would affect the dc operating conditions (that is, the Q point) of the transistor. Since it is desirable to have a control of the Q point of the transistor through the biasing circuit and prevent it from shifting due to the addition of R_s or R_L, dc isolation has to be provided between the transistor biasing circuit and the two resistors R_s and R_L externally connected to the transistor. The capacitors act as open circuits for dc and provide the needed isolation. They pass ac components of current freely, and the amplifier responds to the ac signal as if the capacitors were short circuits.

The capacitor C_E is called the *emitter bypass capacitor*. It was pointed out earlier that the presence of the resistor R_E provided a negative feedback and helped stabilize the Q point against changes in the temperature. If the resistor R_E were not bypassed, then it would provide *a negative feedback for the ac components of the current and voltage also*. Such a negative feedback will result in a reduction of the output voltage due to the ac signal. It is, therefore, necessary to bypass the resistor R_E during small-signal (ac) operation. The capacitor is chosen to have a very low impedance for ac so that it provides a short-circuit both for the ac component of the emitter current.

Small-Signal Analysis of a Common-Emitter Amplifier

The signal source causes ac components of currents and voltages to be present in an amplifier. These are superimposed on the dc voltages and currents that are present in the amplifier due to the biasing circuit. Therefore, the voltages and currents in an amplifier are a combination of a dc component and an ac (small-signal) component. The dc components are already known from the Q point calculations discussed in connection with biasing of a transistor amplifier. They will not change by the addition of the signal source, R_s, and R_L due to the dc isolation provided by the coupling capacitors. So, only the ac components set up by the signal source need to be determined. In fact, since we are really amplifying the *ac signal*, we are primarily interested in the ac components in the amplifier as far as voltage gain, current gain, and power gain are concerned. The small-signal analysis of an amplifier concentrates, therefore, on the calculation of the *ac components* of currents and voltages in the amplifier and the resulting gains. The dc components are ignored in such an analysis.

The circuit that is needed for small-signal analysis is obtained from the circuit of the common-emitter transistor amplifier of Fig. 5-20 by a series of steps as outlined in Fig. 5-21.

Fig. 5-21A is obtained by replacing the battery V_{CC}, the coupling capacitors, and the bypass capacitor of the original circuit by short circuits. Since the dc components are of no interest in small-signal analysis, the dc battery does not enter the calculations in small-signal analysis and is replaced by a short-circuit. The three capacitors have negligibly small impedances for ac components and are, consequently, replaced by short circuits also.

Fig. 5-21B is obtained from Fig. 5-21A by consolidating the two ground terminals into a single bottom line.

The resistors R_{B1} and R_{B2} are seen to be in parallel in Fig. 5-21B, and so they can be combined into a single resistance.

$$R_B = \frac{R_{B1}R_{B2}}{R_{B1} + R_{B2}}$$

Similarly, the resistors R_C and R_L are also seen to be in parallel in Fig. 5-21B and they can be combined into a single resistance

$$R_{eq} = \frac{R_C R_L}{R_C + R_L}$$

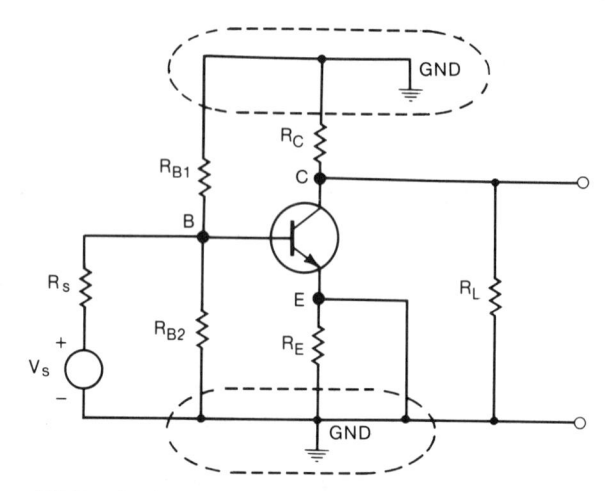

(A) Replacing components with short circuits.

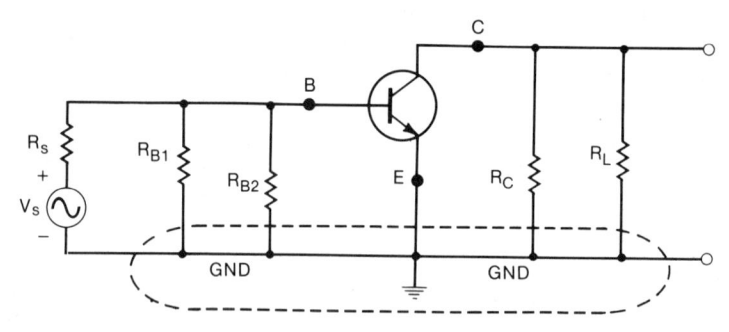

(B) Consolidating ground terminals.

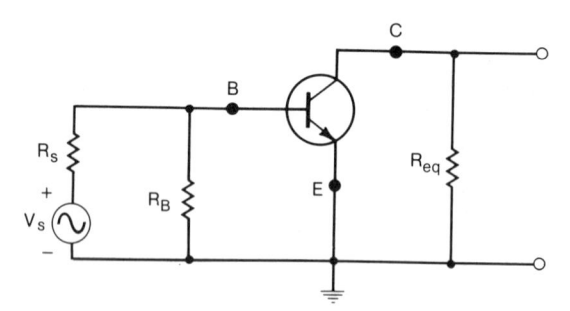

(C) Combining resistors in parallel.

Fig. 5-21. Modifications of the common-emitter amplifier for small-signal conditions.

Electronics: Circuits and Systems

for the calculation of the gains and the input resistance.

When the two previous equivalent resistances are used, the amplifier circuit is reduced to the form shown in Fig. 5-21C. In this diagram, the transistor acts like a two-port network that can be replaced by an appropriate model.

Small-Signal Parameters of a BJT

A transistor has an input resistance that is specified by the manufacturer as h_{ie}. The input resistance can also be expressed in terms of two component resistances $r_{b'e}$ and $R_{bb'}$. That is,

$$\text{input resistance of a BJT} = h_{ie} = r_{bb'} + r_{b'e} \qquad (5\text{-}20)$$

The component $r_{bb'}$, called the base spreading resistance, is usually of the order of 10 Ω to 50 Ω, while the input resistance of a BJT is in the range of 250 Ω to 8 kΩ. Therefore, we can assume that $h_{ie} \approx r_{b'e}$ for all practical purposes.

The current gain from base to collector under *ac conditions* is denoted by h_{fe} and the manufacturer provides a range of values of h_{fe} (measured at 1 kHz) for each transistor; h_{fe} has the same significance as β used in our earlier discussion except that β is measured at dc.

A third parameter of the BJT is the *transconductance* g_m, which is a measure of the effect of the voltage from base to emitter on the collector current of the transistor. Transconductance g_m, being a ratio of current to voltage, has the unit of *siemens* (S), but its value is usually expressed in *millisiemens* (mS). The transconductance g_m of a transistor can be calculated if the collector current at the Q point is known.

$$g_m \text{ (in millisiemens)} = \frac{I_C \text{ at the Q point in mA}}{0.0259} \qquad (5\text{-}21)$$

or,

$$g_m \approx 40 I_C \text{ mS} \quad (I_C \text{ in mA}) \qquad (5\text{-}22)$$

is a convenient approximation.

The transconductance g_m is related to the input resistance $r_{b'e}$ of the transistor through the equation

$$g_m \, r_{b'e} = h_{fe} \qquad (5\text{-}23)$$

Thus, once h_{fe} and the Q point of a BJT are known, the two parameters g_m and $r_{b'e}$ can be calculated by using Eqs. (5-21) and (5-23).

Drill Problem 5.10: The manufacturer's data sheet for a particular BJT provides the following information. The minimum value of the small-signal current gain h_{fe} is 50, and the maximum value is 300, and the measurements being made at $I_C = 1.0$ mA. Calculate the values of the transconductance g_m and the input resistance $r_{b'e}$ of the transistor corresponding to the maximum and minimum given.

Drill Problem 5.11: For the same BJT as in the previous drill problem, another set of values of h_{fe} is given with measurements being made at $I_C = 10$ mA: minimum $h_{fe} = 75$ and maximum $h_{fe} = 375$. Recalculate the values of g_m and $r_{b'e}$.

The symbol r_π is also frequently used in place of $r_{b'e}$ in the literature.

Small-Signal Model of a Transistor

Two models, which are equivalent to each other, are available for replacing a BJT in the small-signal analysis of an amplifier. In the model of Fig. 5-22A, the dependent current source is controlled by the base current I_b by means of the current gain parameter h_{fe}, whereas in the model of Fig. 5-22B, it is controlled by the base-to-emitter voltage V_{be} through the transconductance g_m. The equivalence between the two models can be verified by using $V_{be} = r_{b'e} I_b$ and Eq. (5-23).

The second model has the advantage that it is of the same form as that of a FET (to be introduced in the next chapter) and is also useful in high-frequency analysis of a transistor. Our analysis will, therefore, use the model of Fig. 5-22B for a BJT.

Now, we return to Fig. 5-21C and replace the transistor therein by its small-signal model. This leads to the circuit shown in Fig. 5-23, which is used for the calculation of the gains and other quantities of interest in a BJT common-emitter amplifier.

Small-Signal Analysis

As was done in the analysis following Fig. 5-5 earlier in this chapter, the input side and output side of Fig. 5-23 can be analyzed separately.

Input side

The circuit of the input side is shown in Fig. 5-24A. The voltage V_{be} is across the parallel combination of the resistors R_B and $r_{b'e}$. Using the voltage-divider formula, we get

$$V_{be} = \frac{(R_B \| r_{b'e})}{R_s + (R_B \| r_{b'e})} V_s \qquad (5\text{-}24)$$

The input current I_s supplied by the signal source V_s is given by

$$I_s = \frac{V_s}{R_s + (R_B \| r_{b'e})} \qquad (5\text{-}25)$$

Electronics: Circuits and Systems

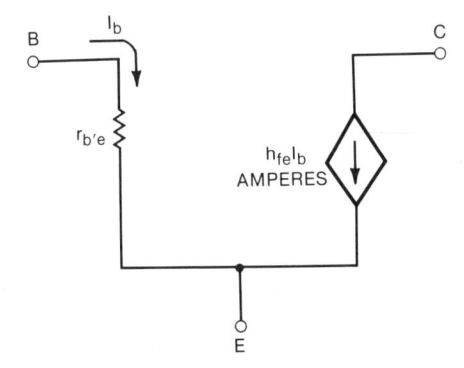

(A) Controlled by the current gain parameter.

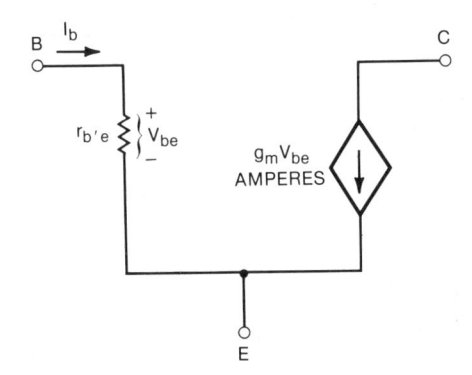

(B) Controlled by the base-to-emitter voltage.

Fig. 5-22. Models of a transistor.

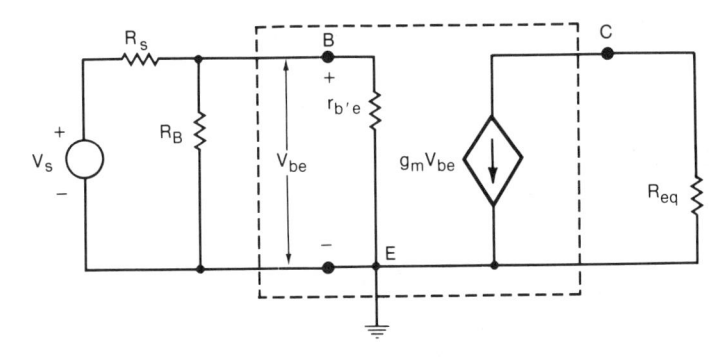

Fig. 5-23. Small-signal model of a common-emitter amplifier.

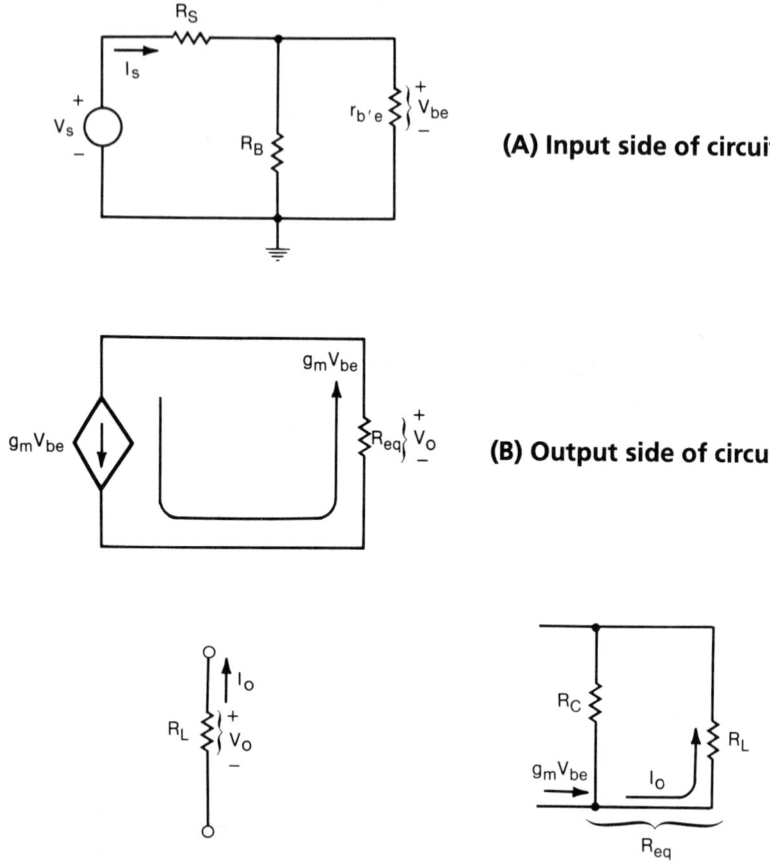

302 BJT Amplifiers

(A) Input side of circuit.

(B) Output side of circuit.

(C) Applying Ohm's law to the load resistor.

(D) Using current division.

Fig. 5-24. Calculation of gains in a common-emitter amplifier.

Output side

The circuit of the output side is shown in Fig. 5-24B. The output voltage V_o is across the resistance R_{eq} (which is the parallel combination of R_C and R_L), and is given by

$$V_o = -(g_m V_{be})R_{eq}$$

or,

$$V_o = -g_m R_{eq} V_{be} \qquad (5\text{-}26)$$

Electronics: Circuits and Systems

The output current can be found by applying Ohm's law to the load resistor R_L (refer to Fig. 5-24C).

$$I_o = -\frac{V_o}{R_L} \qquad (5\text{-}27)$$

An alternative method of finding I_o is to use current division (Fig. 5-24D); the current $g_m V_{be}$ splits between the two resistors R_C and R_L and the component through T_L is

$$I_o = \frac{R_C}{R_C + R_L}(g_m V_{be}) \qquad (5\text{-}28)$$

Eqs. (5-24) through (5-27), or (5-28), form the basic equations needed for the small-signal analysis of the common-emitter amplifier. These can be readily set up by looking at the input and output sides of a given amplifier circuit and need not be committed to memory. Even though it is possible to obtain expressions for the various gains of the amplifier, it is preferable to start with the basic equations each time and avoid having to look up some cumbersome formulas.

Example 5.4

A common-emitter amplifier circuit is shown in Fig. 5-25. If the transistor parameters are $r_{b'e} = 1.5$ kΩ and $g_m = 50 \times 10^{-3}$ S, find the voltage gain, current gain, and power gain of the amplifier.

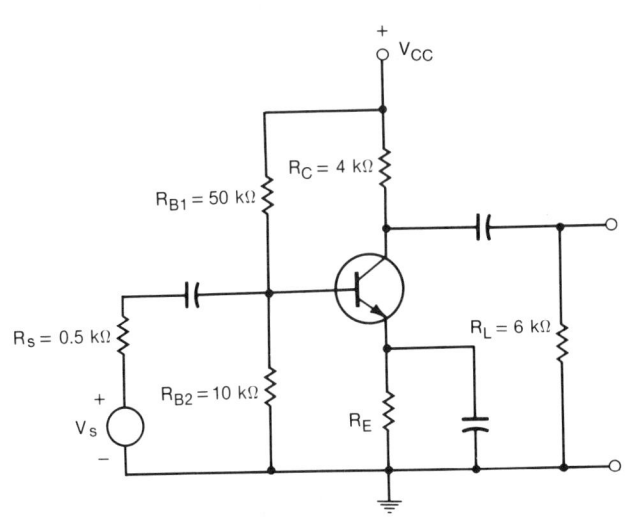

Fig. 5-25. Amplifier circuit for Example 5.4.

Solution

The given circuit is modified to a form suitable for small-signal analysis as shown in Fig. 5-26. These diagrams are exactly analogous to those in Figs. 5-21 and 5-23.

Input side calculations:

The two resistors R_B and $r_{b'e}$ are combined in parallel to give

$$R_B \parallel r_{b'e} = 1.27 \text{ k}\Omega$$

The voltage V_{be} is across the 1.27 kΩ and is obtained by using the voltage-divider formula (Fig. 5-26E):

$$V_{be} = \frac{1.27}{R_s + 1.27} V_s$$

$$= 0.718 V_s \tag{5-29}$$

The input current is given by

$$I_s = \frac{V_s}{0.5 + 1.27}$$

$$= 0.565 V_s \text{ mA} \tag{5-30}$$

Output side calculations:

The output voltage is given by

$$V_o = -(g_m V_{be}) R_{eq}$$
$$= -(50 \times 10^{-3}) \times (2.4 \times 10^3) V_{be}$$
$$= -120 V_{be} \tag{5-31}$$

The output current is obtained by using Ohm's law on the load resistor R_L:

$$I_o = -\frac{V_o}{R_L} = \frac{120 V_{be}}{6} \text{ mA}$$

$$= 20 V_{be} \text{ mA} \tag{5-32}$$

We now have all the data needed for the gain calculations.

Electronics: Circuits and Systems

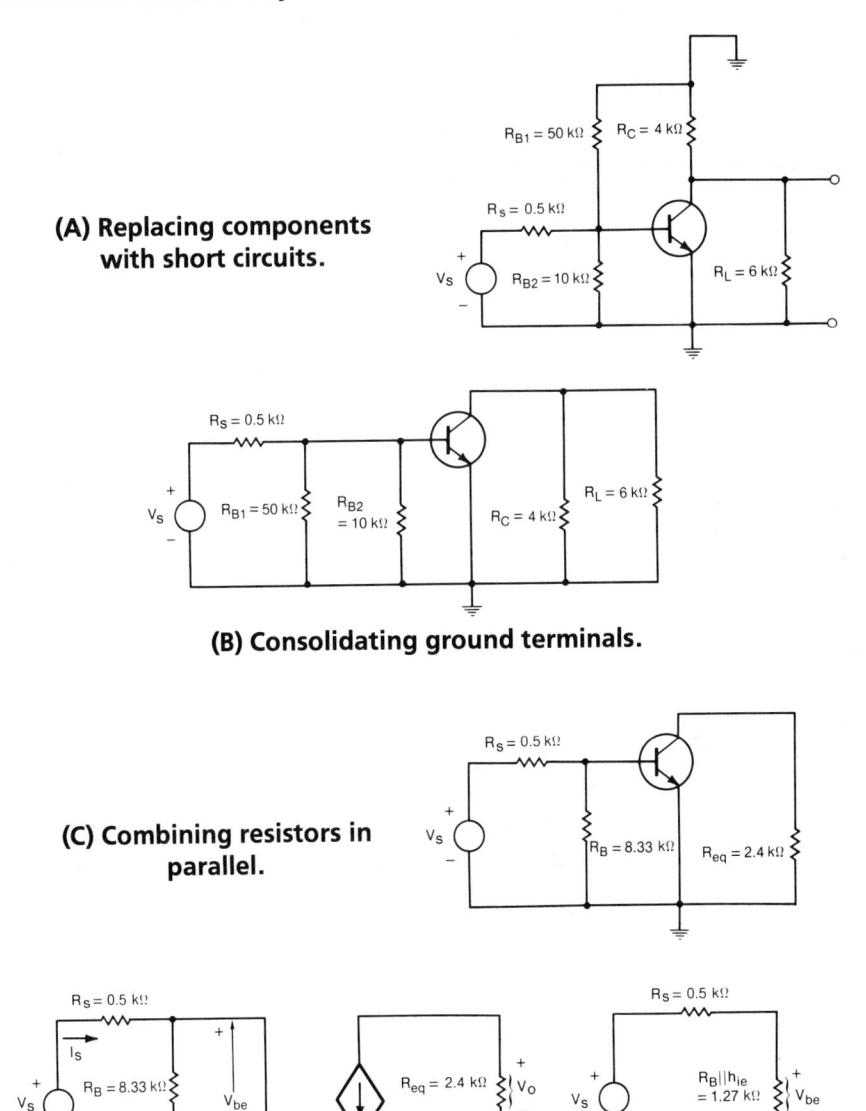

(A) Replacing components with short circuits.

(B) Consolidating ground terminals.

(C) Combining resistors in parallel.

(D) Dividing the circuit into the input side and the output side.

(E) Simplified input circuit.

Fig. 5-26. Development of the small-signal model for the circuit of Example 5.4.

Voltage gain:
Combining Eqs. (5-31) and (5-29),

$$V_o = -120V_{be}$$
$$= -120 \times (0.718V_s)$$
$$= -86.2V_s$$

Therefore,

$$\text{voltage gain} = \frac{V_o}{V_s}$$
$$= -86.2$$

The minus sign simply indicates that there is a phase reversal between the input signal and the output voltage: when the input increases, the output decreases and when the input decreases, the output increases. It can be overlooked in most calculations, and *must* be ignored when finding the decibel value of the voltage gain.

$$\text{voltage gain in decibels} = 20 \log (86.2)$$
$$= 38.7 \text{ dB}$$

Current Gain:
Combining Eqs. (5-32) and (5-29),

$$I_o = 20V_{be}$$
$$= 20 \times (0.718V_s)$$
$$= 14.4V_s \qquad (5\text{-}33)$$

From Eq. (5-30), we have

$$I_s = 0.565V_s$$

Therefore,

$$\frac{I_o}{I_s} = \frac{14.4V_s}{0.565V_s}$$
$$= 25.5$$

current gain = 25.4 (or 28.1 dB)

power gain = (voltage gain ratio) × (current gain ratio)
$$= 2189 \text{ (or 33.4 dB)}$$

Electronics: Circuits and Systems

Drill Problem 5.12: Suppose the transistor parameters in the last example are changed to $r_{b'e} = 0.8$ kΩ and $g_m = 80$ mS. Recalculate the three gains.

Drill Problem 5.13: The circuit of an amplifier (already modified for small-signal analysis) is shown in Fig. 5-27. Calculate the voltage gain of the amplifier in decibels if $r_{b'e} = 2$ kΩ and $g_m = 100$ mS.

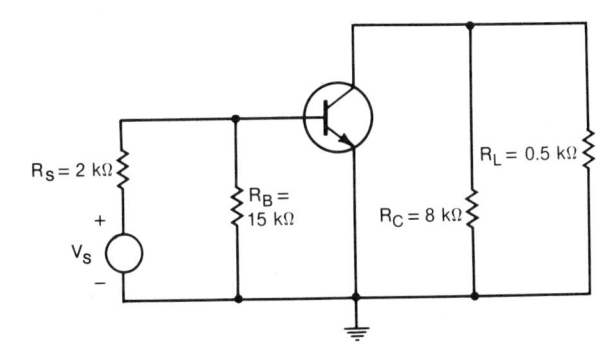

Fig. 5-27. Amplifier circuit for Example 5.13.

The following example illustrates the complete analysis (both dc and small-signal) of a common-emitter amplifier.

Example 5.5

Analyze the amplifier shown in Fig. 5-28. Assume that $h_{fe} = 100$ for the

Solution

DC analysis:

The relevant portion of the circuit for dc analysis is shown in Fig. 5-29. Use the voltage-divider formula to find the voltage at the base V_B.

$$V_B = \frac{R_{B2}}{R_{B1} + R_{B2}} V_{CC}$$

$$= \frac{10}{10 + 90} \times 12$$

$$= 1.2 \text{ V}$$

308   **BJT Amplifiers**

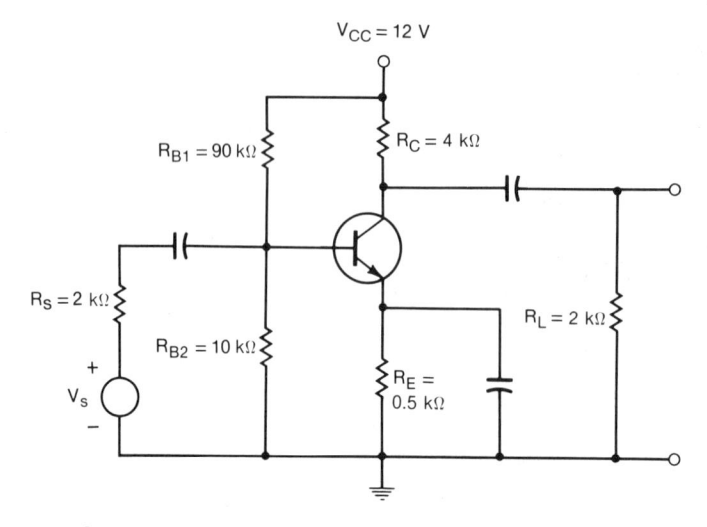

Fig. 5-28. Amplifier circuit for Example 5.5.

Subtract 0.7 V from V_B to get the emitter voltage V_E.

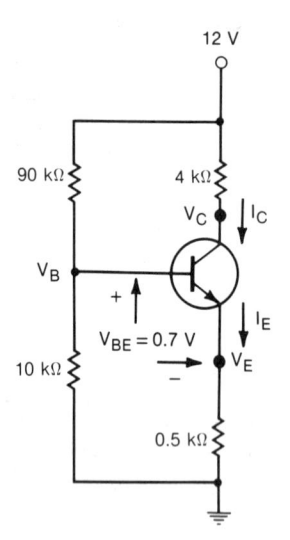

$$V_E = V_B - 0.7$$
$$= 0.5 \text{ V}$$

Fig. 5-29. Q point calculations in the circuit of Example 5.5.

Calculate I_E, the emitter current, using Ohm's law.

$$I_E = (V_E/R_E)$$
$$= 1 \text{ mA}$$

Electronics: Circuits and Systems

The collector current I_C is taken to be the same as I_E.

Calculate the voltage from collector to emitter V_{CE} by subtracting the drops in R_C and R_E from V_{CC}.

$$V_{CE} = V_{CC} - R_C I_C - R_E I_E$$

$$= 7.5 \text{ V}$$

The Q point is, therefore, $I_C = 1 \text{ mA}$, $V_{CE} = 7.5 \text{ V}$.

AC analysis:

Replace the coupling and bypass capacitors by short circuits. Replace the battery by a short circuit. Consolidate the two grounds (one at the top and one at the bottom) into a single bottom line. Combine R_{B1} and R_{B2} in parallel into a single resistance R_B. Combine R_C and R_L in parallel into a single resistance R_{eq}. These are shown in the diagrams of Fig. 5-30.

Use Eqs. (5-21) and (5-23) to calculate the parameters g_m and $r_{b'e}$:

$$g_m = \frac{I_C \text{ in mA (at Q point)}}{0.0259} \text{ mS}$$

$$= 38.6 \text{ mS}$$

$$r_{b'e} = (h_{fe}/g_m)$$

$$= (100/38.6)$$

$$= 2.59 \text{ k}\Omega$$

Input side calculations:

Combine R_B and $r_{b'e}$ in parallel:

$$R_{in} = (R_B \parallel r_{b'e})$$

$$= 2.01 \text{ k}\Omega$$

Calculate the voltage, V_{be}, using the voltage divider formula:

$$V_{be} = \frac{R_{in}}{R_{in} + R_s} V_s$$

$$= (2.01/4.01)V_s$$

$$= 0.501 V_s \quad (5\text{-}34)$$

The input current is given by

$$I_s = \frac{V_s}{R_{in} + R_s}$$

$$= 0.249 V_s \quad (5\text{-}35)$$

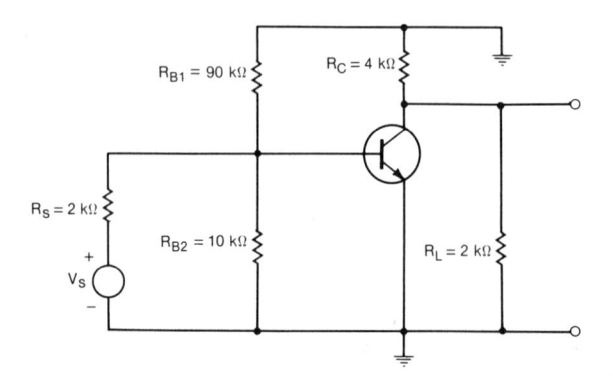

(A) Replacing components with short circuits.

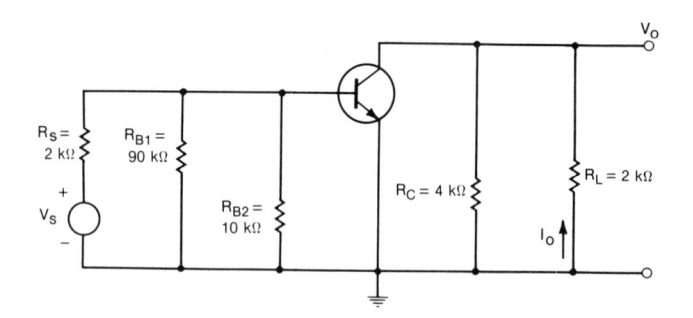

(B) Combining ground terminals.

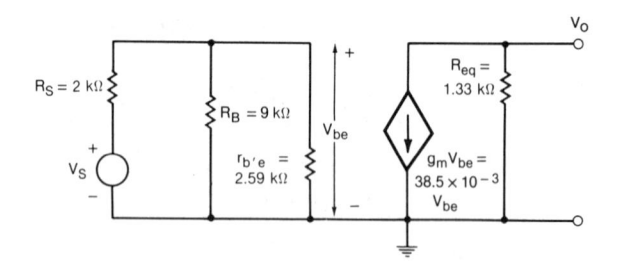

(C) Combining resistors and dividing the circuit.

Fig. 5-30. Small-signal model of the circuit of Example 5.5.

Output side calculations:

$$\text{output voltage } V_o = -(g_m R_{eq})V_{be}$$
$$= -51.3 V_{be} \quad (5\text{-}36)$$
$$\text{output current } I_o = -(V_o/R_L)$$
$$= (51.3/2)V_{be}$$
$$= 25.7 V_{be} \quad (5\text{-}37)$$

Gain calculations:

From Eqs. (5-36) and (5-34),

$$\text{voltage gain} = (-51.3)(0.501)$$
$$= -25.7 \text{ (or 28.2 dB)}$$

From Eqs. (5-37) and (5-34),

$$I_o = (25.7)(0.501)V_s$$
$$= 12.9 V_s$$

and combining this with Eq. (5-35), we get

$$\text{current gain} = 12.9/0.249$$
$$= 51.8 \text{ (or 34.3 dB)}$$
$$\text{power gain} = (25.7)(51.8)$$
$$= 1331 \text{ (or 31.2 dB)}$$

Drill Problem 5.14: If a BJT with $h_{fe} = 200$ is used in the amplifier of the previous example, recalculate the gains.

Drill Problem 5.15: Assume that a signal source $V_s = 150$ mV with $R_s = 50$ kΩ is used in the amplifier of Fig. 5-28 with the h_{fe} of the transistor at 300. Calculate the gains.

Apart from the gains of the common-emitter amplifier, the input resistance and output resistance are also of interest. The input resistance of the common-emitter amplifier is given by

$$R_{in} = (R_B \| r_{b'e}) \quad (5\text{-}38)$$

and this was seen to be 2.01 kΩ in the amplifier of the previous example.

The output resistance of the common-emitter amplifier is simply the resistance R_C (the resistor between the collector terminal and the power supply V_{CC}).

COMMON-COLLECTOR AMPLIFIER

When the input port is formed by the base and the collector and the output port by the emitter and the collector (under small-signal or ac conditions), the collector becomes common to the two ports and the resulting amplifier is a *common-collector amplifier*, which is also called an *emitter follower* (as mentioned earlier).

The circuit of a common-collector amplifier is shown in Fig. 5-31. Note that the *collector is tied directly to the battery* without an intervening resistor like R_C as in the common-emitter amplifier. The direct connection of the collector to the power supply in the schematic circuit diagram is the clue to the presence of an emitter follower.

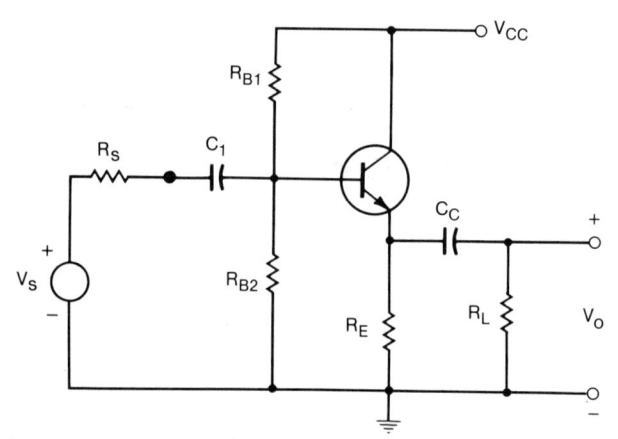

Fig. 5-31. Common-collector (emitter follower) circuit.

The circuit is modified for ac analysis by replacing the battery and coupling capacitors by short circuits and then consolidating the two grounds by a single bottom line as shown in Fig. 5-32. The resulting circuit shows that the collector is at *ground potential for ac conditions* and is thus common to both the input and output ports.

There is no bypass capacitor (like C_E in the common-emitter amplifier) in an emitter follower. If a bypass capacitor were placed across R_E, then it would act as a short circuit for ac components of current and the ac output voltage will be zero, which will defeat the purpose of the whole circuit.

The analysis of a common-collector amplifier is along lines similar to those of a common-emitter amplifier. We will, however, skip the detailed analysis

Electronics: Circuits and Systems 313

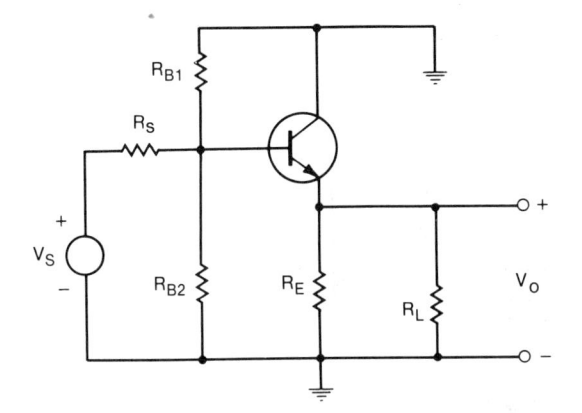

(A) Replacing components with short circuits.

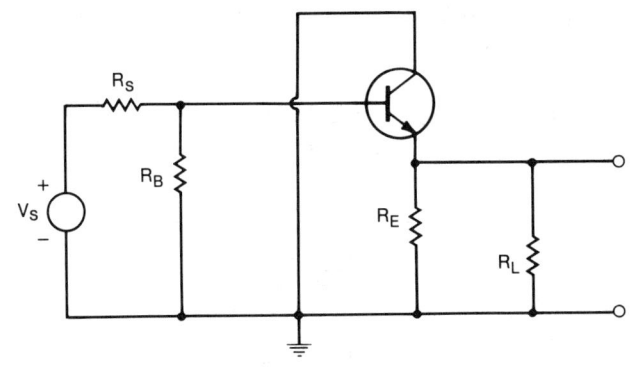

(B) Combining ground terminals.

Fig. 5-32. Modification of the emitter follower for small-signal conditions.

since the primary interest in an emitter follower is not its gain but its input and output resistances. The voltage gain of an emitter follower is slightly less than one and can be taken as equal to one for all practical purposes.

Input Resistance of an Emitter Follower

The input resistance of an emitter follower is much higher than that of a common-emitter amplifier. Consider the equivalent circuit of the emitter follower in Fig. 5-33A. The input resistance is

$$R_{in} = (V_{in}/I_b) \tag{5-39}$$

BJT Amplifiers

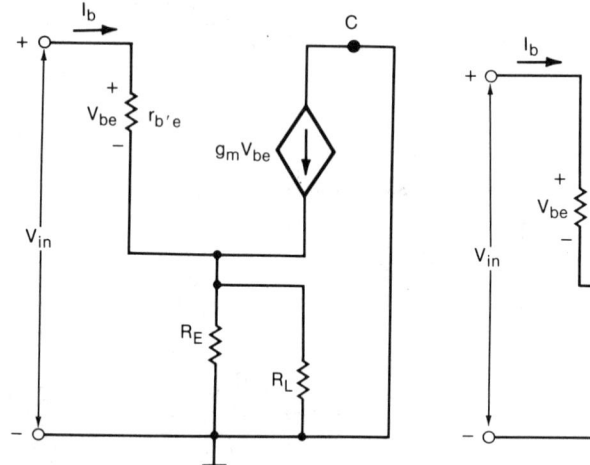

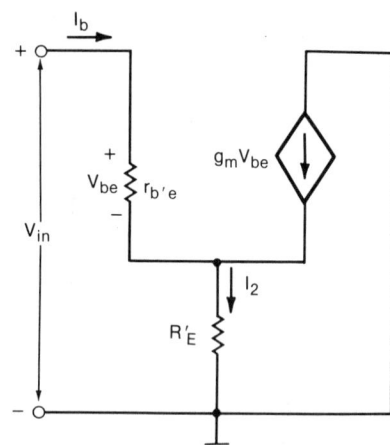

(A) Equivalent circuit of the emitter follower.

(B) Parallel resistors combined.

Fig. 5-33. Determination of the input resistance of the emitter follower.

It is seen that the current through the parallel combination ($R_E \parallel R_L$) (Fig. 5-33B) is $I_b + g_m V_{be}$.

Denote the parallel combination of R_E and R_L by R'_E (Fig. 5-33B). The current through R'_E is, by using Kirchhoff's current law,

$$I_2 = I_b + g_m V_{be}$$

which can also be written as, since $V_{be} = r_{b'e} I_b$ and $g_m r_{b'e} = h_{fe}$,

$$I_2 = I_b + h_{fe} I_b \qquad (5\text{-}40)$$

The voltage V_{in} is the sum of the voltages across R'_E and $r_{b'e}$. That is,

$$V_{in} = R'_E I_2 + r_{b'e} I_b$$

which, on using Eq. (5-40), becomes

$$V_{in} = [(1 + h_{fe})R'_E + r_{b'e}] I_b$$

Therefore, the input resistance is

$$R_{in} = (V_{in}/I_b)$$
$$= [(1 + h_{fe})R'_E + r_{b'e}] \qquad (5\text{-}41)$$

Since $r_{b'e}$ is in the order of a few thousand ohms while $(1 + h_{fe})R'_E$ is in the order of 100 kΩ, the previous expression can be approximated by

Electronics: Circuits and Systems

$$R_{in} \approx (1 + h_{fe})R'_E \qquad (5\text{-}42)$$

It can be seen that the input resistance of the emitter follower is quite high due to the multiplying factor $(1 + h_{fe})$. For example, with $R_E = 4$ kΩ, $R_L = 2$ kΩ, the $h_{fe} = 100$, the input resistance $R_{in} = 135$ kΩ, which should be compared with the input resistance of a common-emitter amplifier in the range of a few thousand ohms.

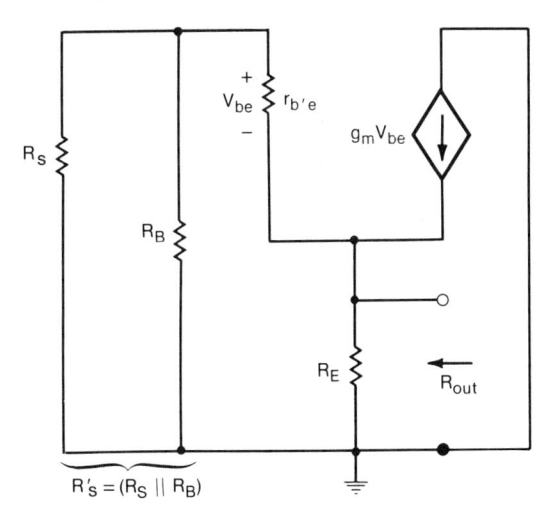

Output Resistance of an Emitter Follower

The output resistance of an amplifier is simply the Thevenin resistance seen looking in from the output terminals. That is, the signal source V_s is replaced by a short circuit (but the *dependent* source $g_m V_{be}$ is left in place), and the resistance seen from the terminal's emitter to ground is determined. Because of the presence of the dependent source, the Thevenin resistance is not simply $[(R'_s + r_{b'e}) \parallel R_E]$ in Fig. 5-34. Using the necessary analysis procedure, it can be shown that the output resistance is given by the *approximate* expression

$$R_{out} \approx \frac{R'_s + r_{b'e}}{(1 + h_{fe})} \qquad (5\text{-}43)$$

which is quite low due to the factor $(1 + h_{fe})$ in the denominator. For example, with $R'_s = 10$ kΩ, $r_{b'e} = 2.5$ kΩ, and $h_{fe} = 100$, the output resistance has a value of 124 Ω. Compare this with the output resistance of the common-emitter amplifier, which equals R_C and is in the range of a few thousand ohms.

Fig. 5-34. Output resistance of an emitter follower.

Application of the Emitter Follower to Matching

The primary application of an emitter follower is as a matching interstage between a source of high resistance and a load of low resistance. It was mentioned in Chapter 2 that, in order to transfer maximum power from a *fixed network* to a variable-load resistance, it is necessary to make the load resistance equal to the Thevenin resistance of the fixed network.

Suppose a signal source V_s has a resistance R_s, which is about 100 kΩ. If the signal source is connected to the input of an amplifier, and the input resistance of the amplifier is much smaller than the value of R_s, then the voltage input to the amplifier, V_{in} of Fig. 5-35A, will be much smaller than V_s. Therefore, the overall voltage gain of the amplifier (from the signal source to the output) will be quite small. It will be desirable to make the input resistance of the amplifier to be in the same neighborhood as R_s. For R_s around 100 kΩ, a common-emitter amplifier with an input resistance of around 2 kΩ will not be a suitable choice, but an emitter follower with its high input resistance will be a good first stage of the amplifier as indicated in Fig. 5-35B.

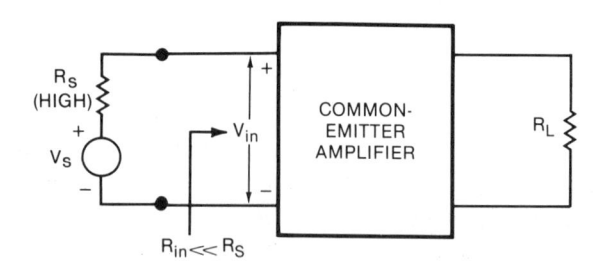

(A) One-stage amplifier.

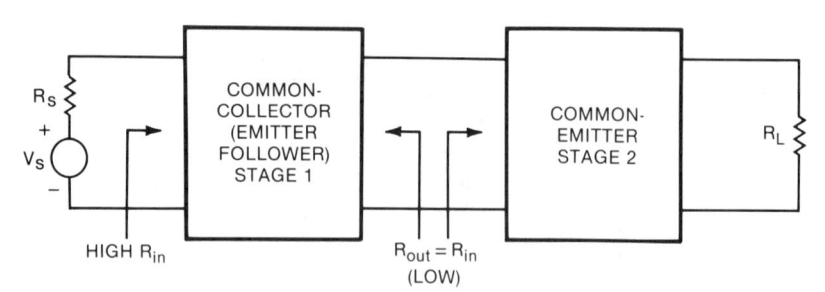

(B) Two-stage amplifier.

Fig. 5-35. Use of an emitter follower stage for matching a high resistance signal source to a common-emitter amplifier.

Now consider the output side of an amplifier where a load resistance is to be connected. The amplifier can be represented by its Thevenin equivalent (as seen from the load) as indicated in Fig. 5-36A. The Thevenin resistance is the output resistance of the amplifier. Since the amplifier is a fixed network, the load resistance must be made equal to R_{out} in order to have maximum power transfer to the load. When the load resistance is small (as for example 8 Ω to 16 Ω in a loudspeaker), we will need an amplifier with a low output resistance. We see that an emitter follower is capable of providing the desired low output resistance as indicated in Fig. 5-36B.

(A) Thevenin equivalent circuit.

(B) Use of an emitter follower stage as the last stage.

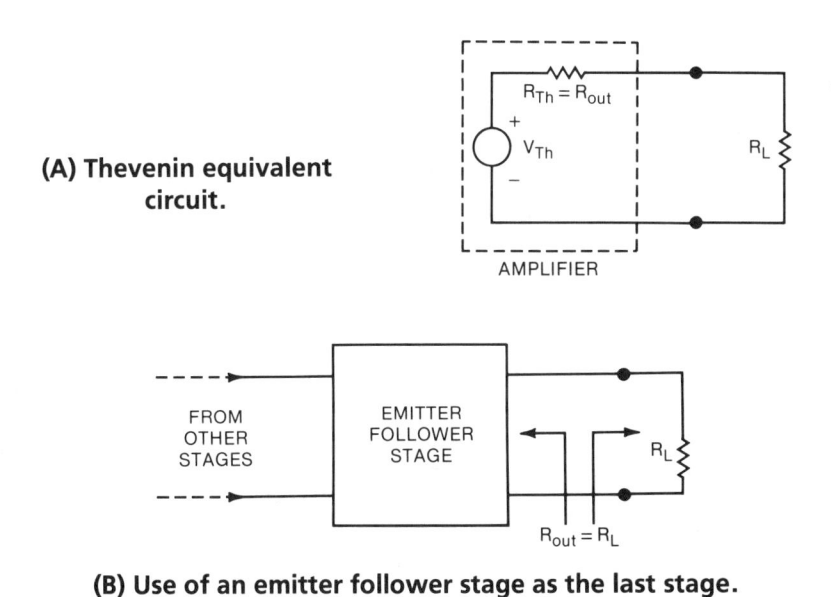

Fig. 5-36. Use of an emitter follower stage for matching a low load resistance to an amplifier.

The use of an emitter follower as the last stage of a multistage amplifier eliminates the need for matching transformers to drive low resistance loads.

COMMON-BASE AMPLIFIER

In the common-base amplifier, the input port is formed by the emitter and base terminals, and the output port is formed by the collector and base terminals (under small signal or ac conditions). The circuit of a common-base amplifier is shown in Fig. 5-37A, and the modified version for small signals is shown in Fig. 5-37B. The common-base amplifier has a current gain of less than one, since the input current is the emitter current and the output current is the collector current, which is always smaller than the emitter current. But its voltage gain is high being the same as that for a common-emitter amplifier. The power gain of the common-base amplifier is not as high as that of the common-emitter amplifier.

(A) Circuit.

(B) Modified version for small signals.

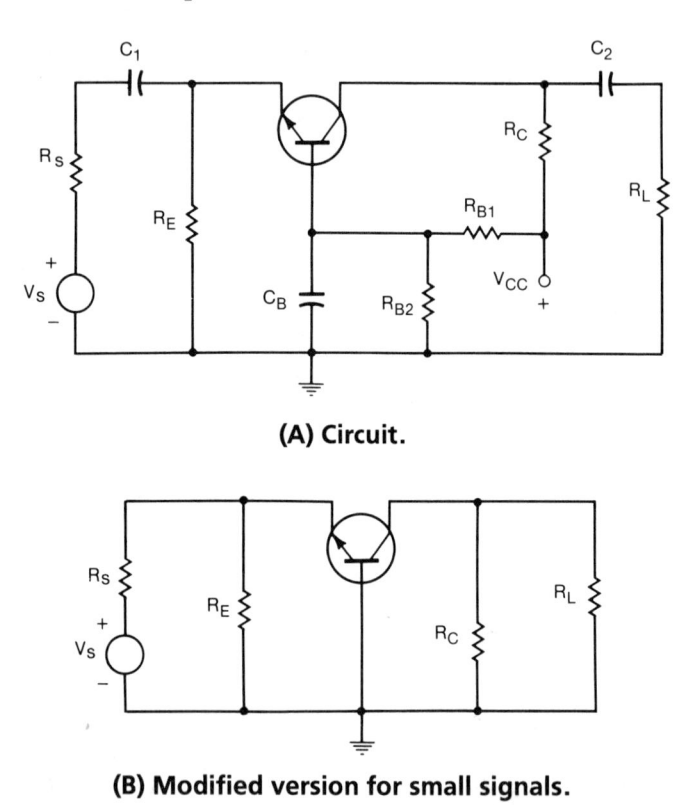

Fig. 5-37. Common-base amplifier circuit.

The common-base configuration has some properties that make it a useful stage in a number of applications in spite of its small power gain. It has an extremely low input resistance: $R_{in} = r_{b'e}/(h_{fe} + 1)$. It is less sensitive to temperature variations than the common-emitter stage. The collector cutoff current that affects the shifting of the Q point with temperature in the common-base amplifier is *smaller* than the cutoff current that affects the common-emitter amplifier by a factory of $(\beta + 1)$. Consequently, for a given change in temperature, the shift in the Q point of a common-base stage will be much smaller than in a common-emitter stage. Another feature of the common-base stage involves its high-frequency response. As will be seen in Chapter 7, the high-frequency response of an amplifier is limited by a Miller effect capacitance. In the case of the common-base amplifier, the Miller effect is absent, and it has a much wider bandwidth than the common-emitter amplifier.

Comparison of the Three BJT Amplifiers

Voltage Gain
The common-emitter and the common-base configurations have a high voltage gain (both being about the same), while the common-collector configuration has a voltage gain of less than one.

Current Gain
The common-emitter and common-collector configurations have a high current gain (both being about the same), while the common-base configuration has a current gain of less than one.

Power Gain
All three configurations have a large power gain but the common emitter has a much higher power gain than the other two types.

Input Resistance
The common-base configuration has a much smaller input resistance than the common-emitter connection, while the common-collector configuration has a much higher input resistance than the common-emitter connection.

Output Resistance
The output resistance of the common-emitter or common-base stage is equal to R_C (the resistance connected between the collector and the power supply) and is of the order of a few kilohms. The output resistance of the common collector is quite small and can be made to be in the range of a few ohms.

The common-emitter connection is the most commonly used amplifier stage because of its large voltage, current and power gains. The common-base amplifier is used when its advantages (less sensitivity to temperature and wider bandwidth) are to be exploited. The common-collector amplifier is used essentially for coupling low-resistance loads to an amplifier stage with a large output resistance.

SUMMARY SHEETS

BJT REVIEW

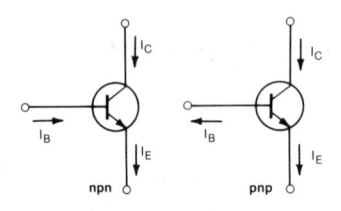

$I_E = I_B + I_C$

$I_E \approx I_C = \beta I_B$

Basis of amplification: A small change in I_B results in a large change in I_E and I_C. If the input signal is applied to the base and the output is taken at the collector, then a small change in the input signal causes a large change in the output signal.

TWO-PORT MODEL OF AN AMPLIFIER

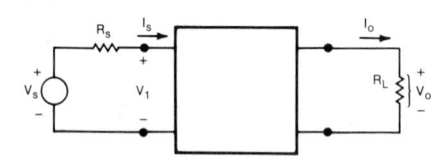

1. Voltage gain = (V_o/V_s)
2. Current gain = (I_o/I_s)
3. Power gain = (voltage gain) × (current gain)

To obtain decibel values, use 20 times the logarithm for voltage and current gains, and 10 times the logarithm for power gain. The decibel values of the three gains will usually be different. The decibel value of the power gain will be the *average* value of the voltage and current gains in decibels.

Model of an Active Device

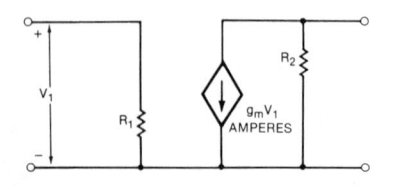

"$g_m V_1$" is a dependent current source whose current depends upon the voltage V_1 across R_1. The constant g_m is called the *transconductance* of the device.

Analysis

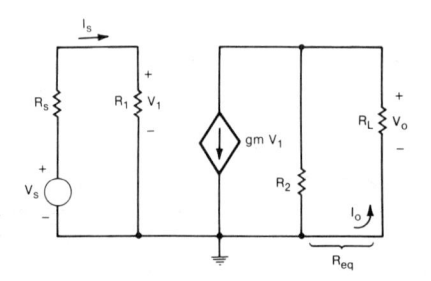

Input Side Calculations

$$V_1 = \frac{R_1}{R_1 + R_s} V_s \qquad I_s = \frac{V_s}{R_1 + R_s}$$

Output Side Calculations

$V_o = -g_m R_{eq} V_1$ where $R_{eq} = (R_2 \parallel R_L)$

$I_o = -(V_o/R_L) = g_m(R_{eq}/R_L)V_1$

The voltage and current gains can be calculated from these equations.

Units

Currents are consistently expressed in *milliamps* (mA) and resistances in *kilohms* (kΩ). This will give voltage in volts.

BIASING OF A BJT

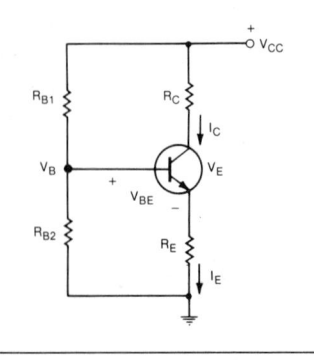

Electronics: Circuits and Systems

Temperature Effects

I_{co} doubles for every increase in the temperature by 10°C. As the temperature increases, the output characteristics of a BJT shift upward, with a corresponding upward shift of the Q point, which could cause a distortion in the output signal.

Thermal Runaway

An increase in the temperature causes an increase in the collector current, which in turn causes a further increase in the temperature. This cyclic chain can result in a thermal runaway in the transistor leading to very large currents.

Stabilization of the Q Point

The increase in the Q point value of the I_C can be made smaller by decreasing the base current when the temperature increases. The emitter resistor R_E provides a negative feedback that makes the base current decrease when the collector current tries to increase.

It is necessary in the case of large signal (power) amplifiers to provide a means of dissipating the heat by heat sinks and proper ventilation.

BJT Amplifier Configurations

1. *Common-Emitter Amplifier:* The input signal is applied to the base and the output is taken at the collector, and the emitter is common to the input and output sides of the amplifier.

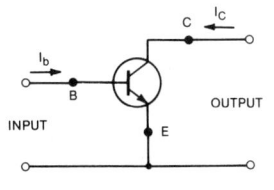

2. *Common-Base Amplifier:* The input signal is applied to the emitter and the output is taken at the collector, and the base is common to the input and output sides of the amplifier.

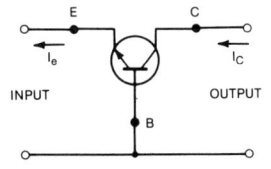

3. *Common-Collector Amplifier: (Emitter Follower)* The input is applied to the base, and the output is taken at the emitter, and the collector is common to the input and output sides of the amplifier. Since the output voltage (being the same as the emitter voltage) *follows* the emitter voltage, the circuit is known as the emitter follower.

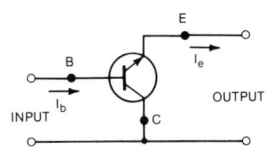

Common-Emitter Amplifier

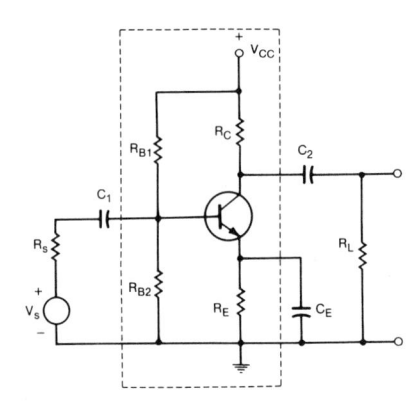

The portion in the dashed box is the biasing circuit.

C_1 and C_2 are coupling capacitors to provide *dc isolation* of the stage from the other components or circuits connected to it. C_E is a bypass capacitor that provides a very low impedance path for the ac components of the emitter current so that there is no negative feedback for ac.

Small-Signal Analysis

To obtain the small signal equivalent circuit, replace all the capacitors by short circuits and also the battery V_{CC} by a short circuit. Combine R_C and R_L in parallel to get R_{eq}. Combine R_{B1} and R_{B2} in parallel to get R_B.

BJT Amplifiers

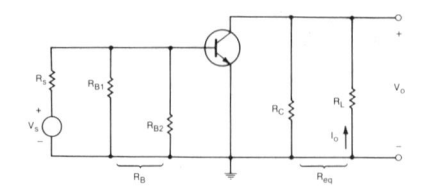

Small-Signal Parameters

The h_{fe} is specified by the manufacturer. (This has the same meaning as β except that it is measured for ac.)
Calculate,

$$g_m = \frac{I_C \text{ at Q point in mA}}{0.0259} \text{ mS}$$

$$\approx 40 I_C \text{ mS}$$

and

$$r_{b'e} = (h_{fe}/g_m) \text{ k}\Omega$$

Small Signal Models of a BJT

Either of the previous models can be used, and we use the latter.

Small-Signal Analysis

Input Side:

$$V_{be} = \frac{(r_{b'e} \| R_B)}{R_s + (r_{b'e} \| R_B)} V_s$$

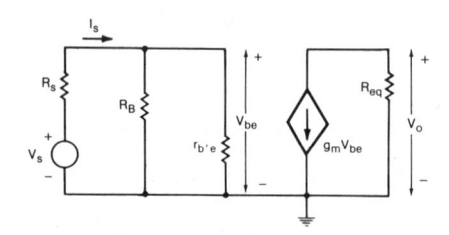

$$I_s = \frac{V_s}{R_s + (r_{b'e} \| R_B)}$$

Output Side:

$$V_o = -g_m R_{eq} V_{be}$$

$$I_o = -(V_o/R_L) \text{ or } \frac{R_C}{R_C + R_L}(g_m V_{be})$$

These equations give results that are used for the calculation of the different gains.

The voltage gain will have a *negative* sign, which indicates that there is 180° phase reversal between the input and the output voltages: when the input voltage goes up, the output voltage goes down. The minus sign must be ignored when calculating decibel values.
Input resistance: $R_i = (r_{b'e} \| R_B)$
Output resistance: $R_o = R_C$

Common-Collector Amplifier (Emitter Follower)

The collector terminal is directly connected to the battery. There is no emitter bypass capacitor.

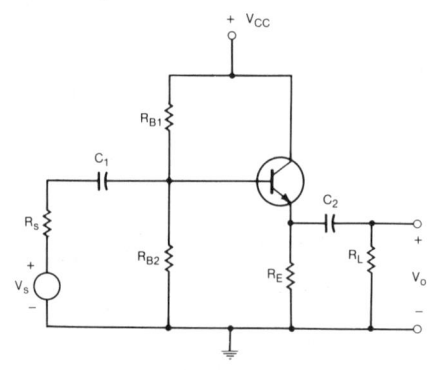

Voltage gain is taken as 1 (for convenience).
Input resistance: $R_i = (h_{fe} + 1) R'_E$ where,

$$R'_E = (R_E \| R_L)$$

Output resistance:

$$R_o = \frac{R'_s + r_{b'e}}{(h_{fe} + 1)}$$

where,

$$R'_s = (R_s \| R_B) \text{ and } R_B = (R_{B1} \| R_{B2})$$

The emitter follower has a high input resistance and a low output resistance. It is used as a matching interstage between, for example, a common-emitter amplifier with a high output resistance and a load of low resistance.

Common-Base Amplifier

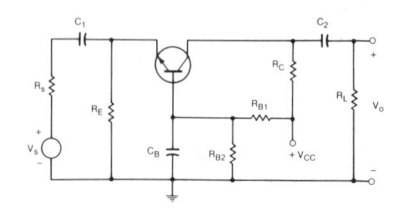

Input resistance:

$$R_i = \frac{r_{b'e}}{(h_{fe} + 1)}$$

Comparison of the Three BJT Amplifiers

The common-emitter amplifier has a high voltage gain, a high current gain, and a high power gain. Its input resistance is in the range of a few kilohms and so is its output resistance ($=R_C$). It is the most commonly used configuration.

The common-collector amplifier (emitter follower) has a voltage gain of less than one, a good current gain, and a power gain that is much smaller than that of a common-emitter stage. It has a high input resistance (tens to hundreds of kilohms) and a low output resistance (as low as a few ohms).

It is most commonly used for matching purposes in order to obtain maximum power transfer.

The common-base amplifier has a good voltage gain, a current gain of less than one, and not as high a power gain as the common-emitter stage. It is the only configuration with a low input resistance, however. It is used when a large bandwidth is desired. It is also much less sensitive to temperature increases than the common-emitter stage.

ANSWERS TO DRILL PROBLEMS

5.1 (a) 37.7 dB; (b) −0.131 dB; (c) 88.0 dB.

5.2 (a) 36.1 dB; (b) 64.1 dB; (c) −0.677 dB.

5.3 (a) 36.9 dB; (b) 32.0 dB; (c) 43.6 dB.

5.4 $I_s = 6\ \mu A$. $V_1 = 9$ mV. $R_{eq} = 4.44$ kΩ. $V_o = -2$ V. $I_o = 0.25$ mA. Voltage gain = −133 (42.5 dB). Current gain = 41.7 (32.4 dB). Power gain = 5546 (37.4 dB).

5.5 $I_s = 1.961 \times 10^{-3}\ V_s$ mA. $V_1 = 0.98\ V_s$. $R_{eq} = 0.476$ kΩ. $V_o = 5.7\ V_1$. $I_o = 11.4\ V_1$ mA. Voltage gain = 5.82 (or 15.3 dB). Current gain = 5600 (or 75 dB). Power gain = 3.26×10^4 (or 44.9 dB).

5.6 $I_s = 10^{-3}\ V_s$ mA. $V_1 = 0.5V_s$. $R_{eq} = 5$ kΩ. $V_o = 60V_1$. $I_o = 6\ V_1$ mA. Voltage gain = 30 (or 29.5 dB). Current gain = 3000 (or 69.5 dB). Power gain = 9×10^4 (or 49.5 dB). Note that the power gain has increased (by 5 dB) because of the matching of the input side to the signal source and the load to the output side.

5.7 (a) $V_B = 1.8$ V. $V_E = 1.1$ V. $I_E = I_C = 1.1$ mA. $V_{CE} = 8.7$ V; (b) $V_B = 2$ V. $V_E = 1.3$ V. $I_E = 3.25$ mA = I_C. $V_{CE} = 9.8$ V.

5.8 $R_{B2} = 24.7$ kΩ. $R_C = 1.25$ kΩ.

5.9 $R_{B2} = 7.55$ kΩ. $R_C = 2.93$ kΩ.

5.10 $g_m = 38.6$ mS. $r_{b'e} = 1.29$ kΩ, 7.77 kΩ.

5.11 $g_m = 386$ mS. $r_{b'e} = 0.194$ kΩ, 0.971 kΩ.

5.12 $V_{be} = 0.593V_s$. $I_s = 0.813V_s$ mA. $V_o = -192\ V_{be}$. $I_o = 32\ V_{be}$. Voltage gain = −114 (or 41.1 dB). Current gain = 23.3 (or 27.4 dB). Power gain = 2656 (or 34.2 dB).

5.13 $V_{be} = 0.469V_s$. $I_s = 0.266\ V_s$ mA. $R_{eq} = 0.471$ kΩ. $V_o = 47.1\ V_{be}$. $I_o = 94.1\ V_{be}$ mA. Voltage gain = 22.1 (or 26.9 dB). Current gain = 166 (or 44.4 dB). Power gain = 3669 (or 35.6 dB).

5.14 Q point same as before. $I_C = 1$ mA. $g_m = 38.6$ mS. $r_{b'e} = 5.18$ kΩ. $R_{in} = 3.29$ kΩ. $V_{be} = 0.621V_s$. $I_s = 0.189V_s$ mA. $V_o = -51.3\ V_{be}$. $I_o = 25.7\ V_{be}$. Voltage gain = 31.8 (or 30.0 dB). Current gain = 84.6 (or 39.5 dB). Power gain = 2690 (or 34.3 dB). The higher value of h_{fe} has resulted in larger gains.

5.15 Q point same as before. $I_C = 1$ mA. $g_m = 38.6$ mS. $r_{b'e} = 7.77$ kΩ. $R_{in} = 4.17$ kΩ. $V_{be} = 11.5$ mV. $I_s = 2.77\ \mu A$. $V_o = -51.3$. $V_{be} = -590$ mV. $I_o = 0.295$ mA. Voltage gain = 3.93 (or 11.9 dB). Current gain = 106.5 (or 40.5 dB). Power gain = 418.5 (or 26.2 dB). Even though the value of h_{fe} has increased, the high resistance of the signal source as compared with the input resistance of the amplifier has resulted in smaller gains (except the current gain).

PROBLEMS

Two-Port Model of an Amplifier

5.1 Convert each of the following gain values to decibels. (a) Voltage gain = 342.5 (b) Power gain = 342.5 (c) Current gain = 0.673. (d) Voltage gain = 0.982. (e) Power gain = 4×10^5.

5.2 Convert each of the following decibel values to its corresponding ratio: (a) Voltage gain = 56 dB. (b) Power gain = 56 dB. (c) Voltage gain = 0 dB. (d) Power gain = 37.8 dB. (e) Current gain = -6.92 dB.

5.3 The voltage gain of an amplifier is 6.94 dB. When its input current is 0.15 mA, the output current is found to be 3.25 mA. Calculate the power gain in decibels.

5.4 Calculate the three gains of the two-port system model of an amplifier shown in Fig. 5-P4. Express the gains in decibels.

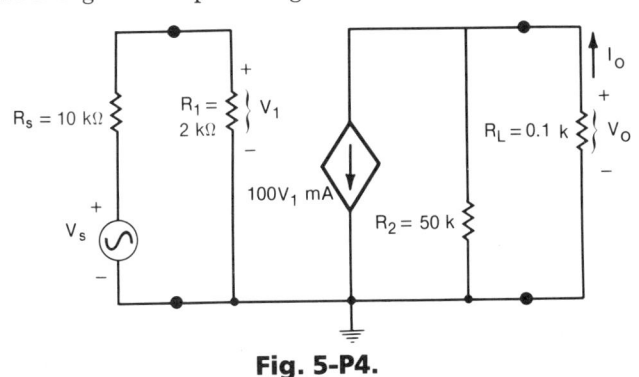

Fig. 5-P4.

5.5 In the previous problem, change R_s and R_L so that $R_s = R_1$ and $R_L = R_2$ (so as to have a matched system). Recalculate the three gains and express them in decibels. Compare these values with those in the previous problem.

Biasing of a BJT Amplifier

5.6 Determine the Q point of the BJT in the circuit shown in Fig. 5-P6 for each of the following sets of values:
(a) $R_{B1} = 95$ kΩ; $R_{B2} = 15$ kΩ; $R_E = 1.5$ kΩ; $R_C = 4.7$ kΩ; $V_{CC} = 15$ V.
(b) $R_{B1} = 9.5$ kΩ; $R_{B2} = 1.5$ kΩ; $R_E = 0.5$ kΩ; $R_C = 3.5$ kΩ; $V_{CC} = 12.5$ V.

5.7 Determine the values of R_{B1} and R_E in the circuit of Fig. 5-P6 so as to place the Q point at $I_C = 3.6$ mA and $V_{CE} = 12$ V. Assume $R_C = 0.75$ kΩ, $R_{B2} = 10$ kΩ, and $V_{CC} = 17.5$ V.

5.8 Suppose in the circuit of Fig. 5-10A (Example 5.2 in the text) a cut is made with a pair of pliers in the R_{B1} branch; (b) a cut is made with a pair of

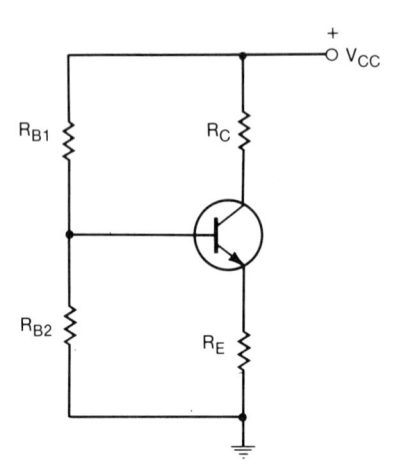

Fig. 5-P6.

pliers in the R_{B2} branch; (c) a short circuit develops across R_{B2}; (d) a short circuit develops across R_{B1}. (Assume that each of these situations occurs independently of the others.) In each case, discuss what happens to the Q point of the transistor and how its operation is affected.

5.9 Determine the Q point of the transistor in the circuit of Fig. 5-P9. Note that R_C is missing in this circuit. This is the circuit of an emitter follower.

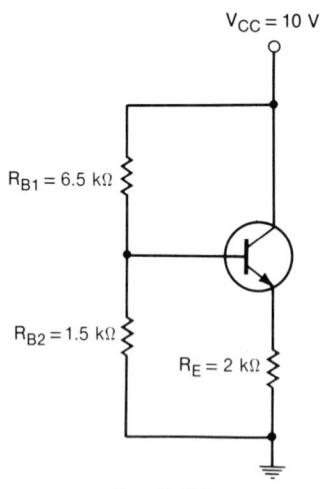

Fig. 5-P9.

Common-Emitter Amplifier

5.10 Given that the transistor parameters in the circuit of Fig. 5-P10 are $r_{b'e} = 0.75$ kΩ and $g_m = 240$ mS, determine the three gains of the amplifier and express them in decibels.

Electronics: Circuits and Systems　　　　　　　　　　　　　　　　**327**

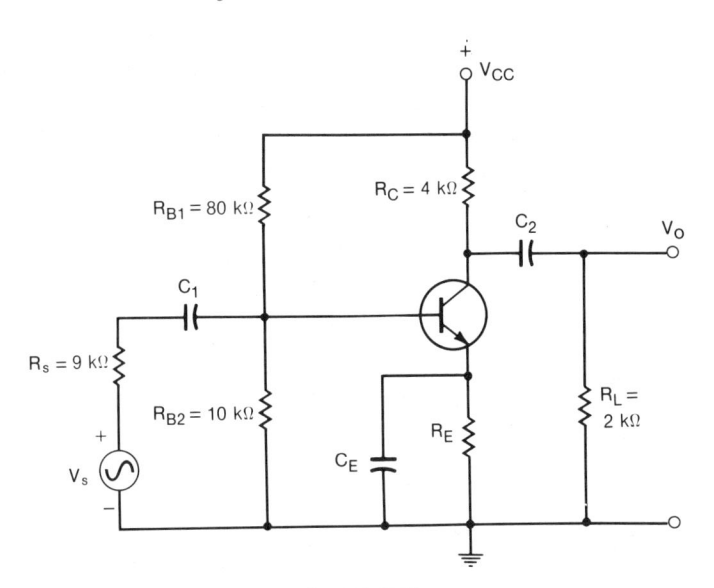

Fig. 5-P10.

5.11　Recalculate the gains of the amplifier in the previous problem if the resistance R_s is changed to 0.75 kΩ and $R_L = 100$ kΩ.

5.12　The transistor in the circuit of Fig. 5-P12 has $h_{fe} = 250$ and $r_{b'e} = 2.5$ kΩ. Calculate the voltage gain.

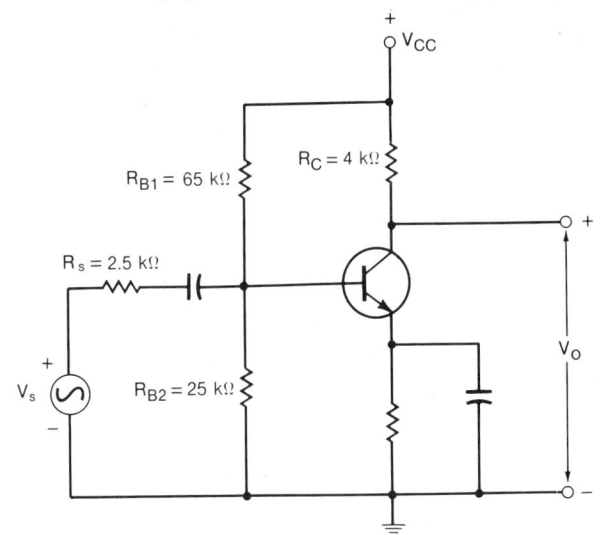

Fig. 5-P12.

5.13 Use the following values in the circuit of Fig. 5-P13 and calculate the three gains of the amplifier: $R_s = 0.25$ kΩ, $R_{B1} = 85$ kΩ, $R_{B2} = 15$ kΩ, $R_C = 5$ kΩ, $R_E = 0.8$ kΩ, and $R_L = 30$ kΩ. $h_{fe} = 100$, and $V_{CC} = 15$ V.

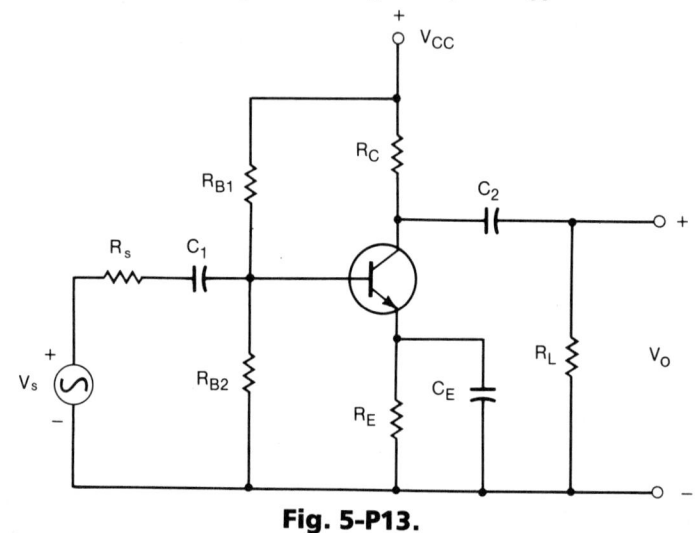

Fig. 5-P13.

5.14 Make the value of R_s of the previous problem equal to R_{in} of the amplifier and $R_L = 5$ kΩ. Other values remain the same as before. Repeat the calculation of the gains.

5.15 Change the value of R_E to 2.4 kΩ in the previous problem. Other values remain the same. R_s = the new R_{in}. Repeat the calculations.

5.16 Use the following component values in the circuit of Fig. 5-P13 and calculate the three gains. $R_s = 10$ kΩ, $R_{B1} = R_{B2} = 50$ kΩ, $R_C = 2$ kΩ, $R_E = 24$ kΩ, $V_{CC} = 20$ V, $R_L = 24$ kΩ, $h_{fe} = 200$.

Emitter Follower

5.17 Calculate the input and output resistances of the emitter follower circuit shown in Fig. 5-P17.

5.18 In the system of Fig. 5-P18, the emitter follower is used as a matching interstage. The input resistance of the emitter follower should be equal to R_s and its output resistance equal to R_L. Determine the transistor parameters: h_{fe} and $r_{b'e}$.

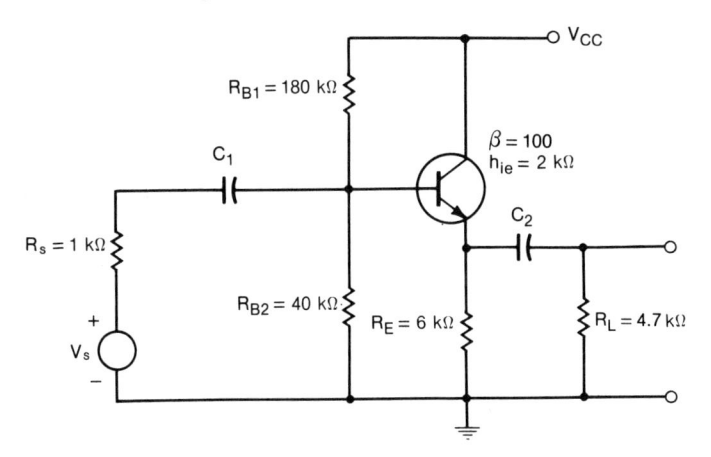

Fig. 5-P17.

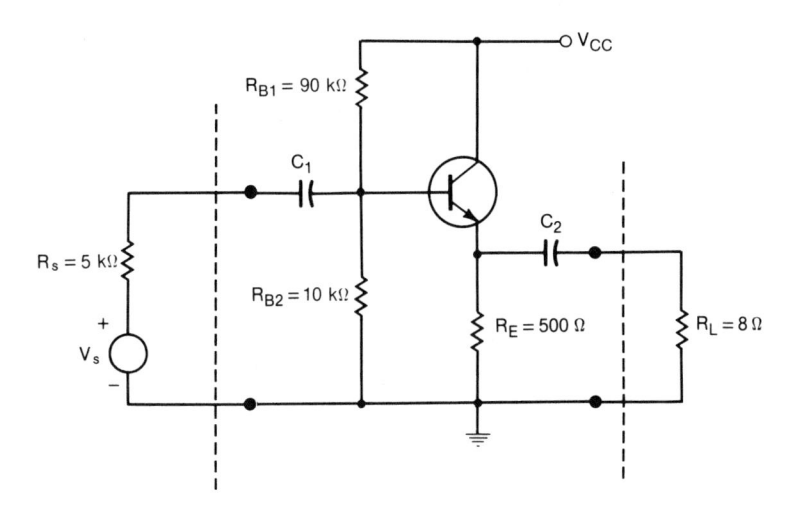

Fig. 5-P18.

CHAPTER 6

FET AMPLIFIERS

The field-effect transistor (FET) has an extremely high input impedance and a better frequency response than the BJT. But the small-signal gain of a FET amplifier is not as high as that of a BJT amplifier. The FET thus offers an attractive alternative to the BJT in those analog applications that can benefit from its high input impedance and better frequency response. FETs also can be used as constant current sources and voltage-controlled resistors in certain special applications. The more important set of uses of a FET is in the area of digital circuits and systems. MOSFET devices can be fabricated at extremely high densities in IC technology so that complex networks and systems can be built on a single chip. In the CMOS configuration, FETs lead to logic circuits that consume only small amounts of power. The most common use of the FET is, therefore, in LSI digital circuits and systems such as semiconductor memories, microprocessors, shift registers, and other such components. In fact, it is not too much to claim that the MOSFET in conjunction with IC technology has been responsible for the microminiaturzation of digital systems.

In analog applications, the FET is limited by its small gain. A more serious problem exists when using discrete FETs in amplifiers. There is a large spread (by as much as 5 to 1) in the values of its parameters even when using two FETs with the same part number. The design of FET amplifiers must, therefore, allow for a substantial degree of uncertainty in the characteristics of a given device. The most reliable design of amplifiers using FETs involves the use of *matched pairs* fabricated in the form of integrated circuits rather than individual FETs. Such a design is used in differential amplifiers discussed in Chapter 8.

PRINCIPLES AND PROPERTIES OF THE FET

The physical principles of operation of the FET were discussed in Chapter 3. It is not necessary to have a detailed understanding of all the physical principles and processes for studying the use of FETs in electronic circuits. It is more important to be fully aware of the terminal characteristics and special features of the FET.

It was seen in Chapter 3 that there are two major classifications of FETs: the *junction FET* (or *JFET*), and the *metal-oxide-semiconductor FET* (or *MOSFET*). There are two kinds of MOSFETs: the *depletion-mode* MOSFET and the *enhancement-mode* MOSFET. On top of all this division and subdivision, we also have the possibility that any FET can be an *n-channel* device or a *p-channel* device. Such a wide assortment can be quite bewildering to anyone who is not an electronics specialist. But, by concentrating on certain basic similarities between all the FETs and the analogy between the BJT and FET, we can avoid much of the confusion and learn to love and appreciate circuits using FETs.

Any field-effect transistor depends upon the flow of carriers through a channel. If the channel is n-type, then the current is due to the flow of electrons, and if the channel is p-type, then the current is due to the flow of holes. There are two terminal electrodes attached to the channel: the source terminal and the drain terminal. The flow of *carriers* is always from the source to the drain. A third terminal electrode, called the gate, exerts control over the flow of carriers from the source to the drain.

In order to avoid jumping back and forth between n-channel and p-channel devices (and making a lot of repetitious statements) we will concentrate on the n-channel device. Suitable modifications of the discussion can be readily made to convert it for the p-channel device. In the n-channel device, there is a flow of *electrons* from the source to the drain, which means that the *conventional current flow is from the drain to the source.*

There is a definite analogy between the BJT and the FET as presented:

FET		BJT
N-channel FET	⟷	Npn transistor
Source terminal	⟷	Emitter terminal
Drain terminal	⟷	Collector terminal
Gate terminal	⟷	Base terminal

To a large extent, we can ignore whether a given FET is a JFET, a depletion-mode MOSFET, or an enhancement-mode MOSFET even though there are differences that will be discussed shortly. (An n-channel MOS is usually referred to as NMOS.) The symbols used for n-channel FETs are shown in Fig. 6-1 along with the symbol of an npn transistor for comparison.

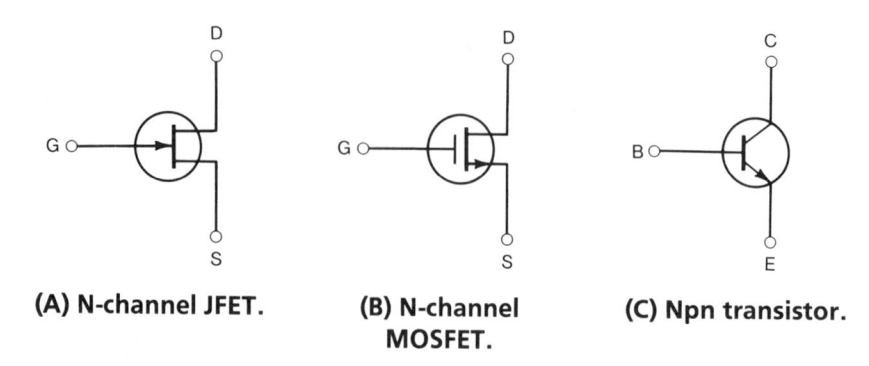

(A) N-channel JFET. **(B) N-channel MOSFET.** **(C) Npn transistor.**

Fig. 6-1. Symbols of n-channel FETS and npn BJT.

The analogy between the n-channel FET and the npn transistor provides the clue to the bias connection for the drain of the FET in an amplifier: the drain of the FET, like the collector of an npn transistor, is connected to the positive end of a power supply. Differences occur in the bias connections for the gate terminal in comparison with the bias connections of the base terminal of the BJT, as will be seen in the following discussion.

The *output characteristics*, that is, the drain current I_D as a function of the drain-to-source voltage V_{DS}, of the three types of n-channel FETs are shown in Fig. 6-2.

The following common features among the three should be noted.

1. For a given gate-to-source voltage V_{GS}, the drain current increases *linearly* with the drain-to-source voltage V_{DS} for very small values of V_{DS}. Such a range of operation is called the *linear region*, and the FET acts as a *voltage-controlled resistor*.

2. As V_{DS} increases (for a given value of V_{GS}), there is at first a region where the drain current increases *nonlinearly* with the drain-to-source voltage, which is usually referred to as the *triode region*.

3. As V_{DS} increases further (again for a given value of V_{GS}), a value of V_{DS} is reached at which the characteristic curve levels off. The *drain current becomes a constant* and *independent of the drain-to-source voltage*. The FET now acts as a *voltage-controlled constant current source* (where the controlling voltage is V_{GS}). This is the region used in amplifiers.

Linear region: I_D is proportional to V_{DS}

Triode region: I_D increases nonlinearly with V_{DS}

Saturation region: I_D is a constant independent of V_{DS}

4. When the gate-to-source voltage V_{GS} goes *below* a certain value, the device stops conducting and $I_D = 0$. We will refer to this *minimum value of V_{GS} needed for conduction* as the threshold voltage V_T. That is,

$$V_{GS} > V_T \text{ for the FET to conduct}$$

(What we refer to as the *threshold voltage* V_T in our discussion is also referred to in the case of the JFET and the depletion-mode MOSFET as the *pinch-off voltage with $V_{GS} = 0$* and denoted by the symbol V_P or V_{po} in other literature. We prefer the use of a single name and symbol for all FETs for the sake of simplicity.)

5. The beginning of the *saturation region* (for a given V_{GS}) occurs at $V_{DS} = (V_{GS} - V_T)$. That is,

$$\text{A FET is in the saturation region when } V_{DS} \geq (V_{GS} - V_T)$$

For example, in Fig. 6-2A, the threshold voltage of the FET is $V_T = -6$ V. Therefore, for any given value of V_{GS}, the saturation region begins when V_{DS} equals $[V_{GS} - (-6)] = (V_{GS} + 6)$ V. The characteristic curve for $V_{GS} = -1$ V, for instance, becomes flat at $V_{DS} = (-1 + 6) = 5$ V.

6. For a given value of V_{DS} (imagine a vertical straight line on the graphs of Fig. 6-2), the *drain current increases* as the *gate-to-source voltage becomes more positive* (or less negative).

Now we consider the points in which the three devices *differ*.

1. The threshold voltage V_T is *positive* for the *enhancement NMOS* but *negative for the other two* types.

Enhancement NMOS: V_T positive
Depletion NMOS and n-channel JFET: V_T negative

2. V_{GS} is always held positive for the enhancement NMOS (since it must be greater than V_T for the device to conduct). It is usually negative for the depletion NMOS but can be made slightly positive if desired. V_{GS} is always held less than or equal to zero for the JFET (and never made positive).

V_{GS} always positive for enhancement NMOS
V_{GS} never positive for n-channel JFET
V_{GS} usually negative for depletion NMOS

It is seen that the JFET and the depletion NMOS are very similar in their terminal characteristics, especially if we ignore the occasional positive values of V_{GS} permitted for the latter device. These two types can be treated as a single type for all practical purposes.

We notice that there *is* a difference between the n-channel JFET and the npn transistor (as also between the depletion NMOS and the npn transistor). The gate-to-source voltage of either FET is usually negative whereas the analogous base-to-emitter voltage of the npn transistor has to be positive when it is conducting.

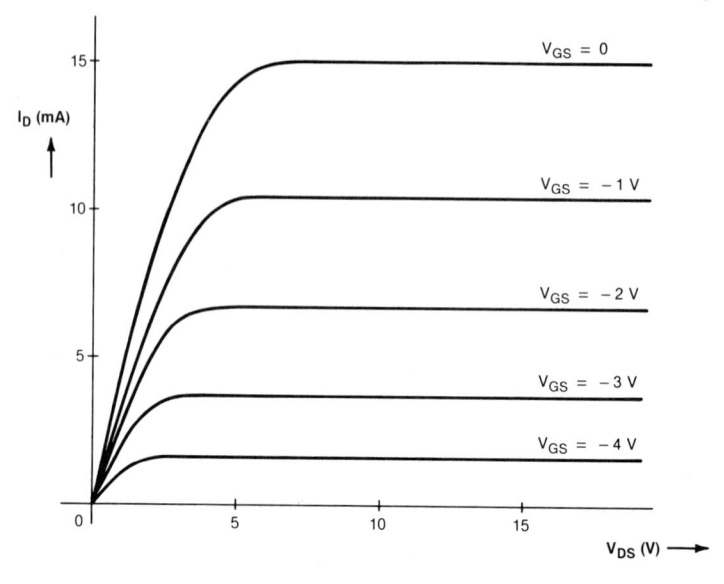

(A) N-channel JFET.

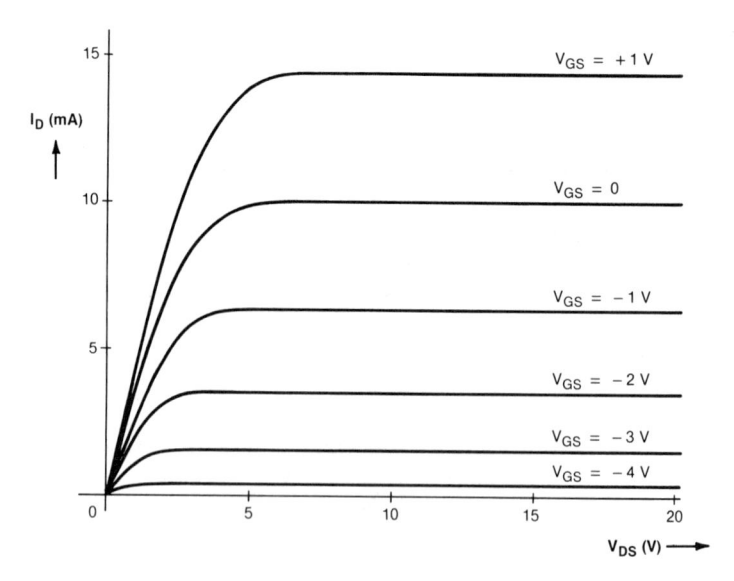

(B) N-channel depletion MOSFET.

Fig. 6-2. Output characteristics

TRANSFER CHARACTERISTICS OF THE FET

In the analysis and design of biasing circuits of FET amplifiers, it is convenient to use the *transfer characteristics* of the FET. These are curves that display the relationship between the drain current I_D and the gate-to-source voltage V_{GS} when the *FET is operating in the saturation region.* Typical transfer characteristics of the three types of FETs are shown in Fig. 6-3.

The transfer characteristics reemphasize the points already made about V_{GS} and the threshold voltage V_T: V_T is always positive for the enhancement NMOS but negative for the other two types; V_{GS} is held positive for the enhancement NMOS but negative for the other two types.

The transfer characteristics also exhibit the essential similarity between the behavior of the three types. The JFET and the depletion NMOS are identical in their behavior (when their threshold voltages are chosen as being equal as well as another parameter K to be discussed later). The difference between the enhancement NMOS and the other two types is merely a matter of *shifting the transfer characteristics horizontally in the proper direction.*

All three transfer characteristics follow the equation:

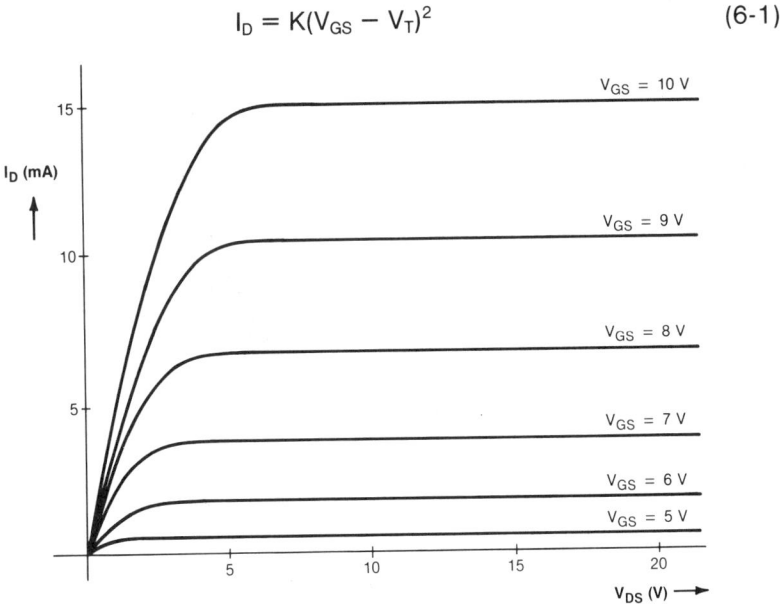

$$I_D = K(V_{GS} - V_T)^2 \qquad (6\text{-}1)$$

(C) N-channel enhancement NOSFET.

of n-channel FETs.

(A) N-channel JFET.

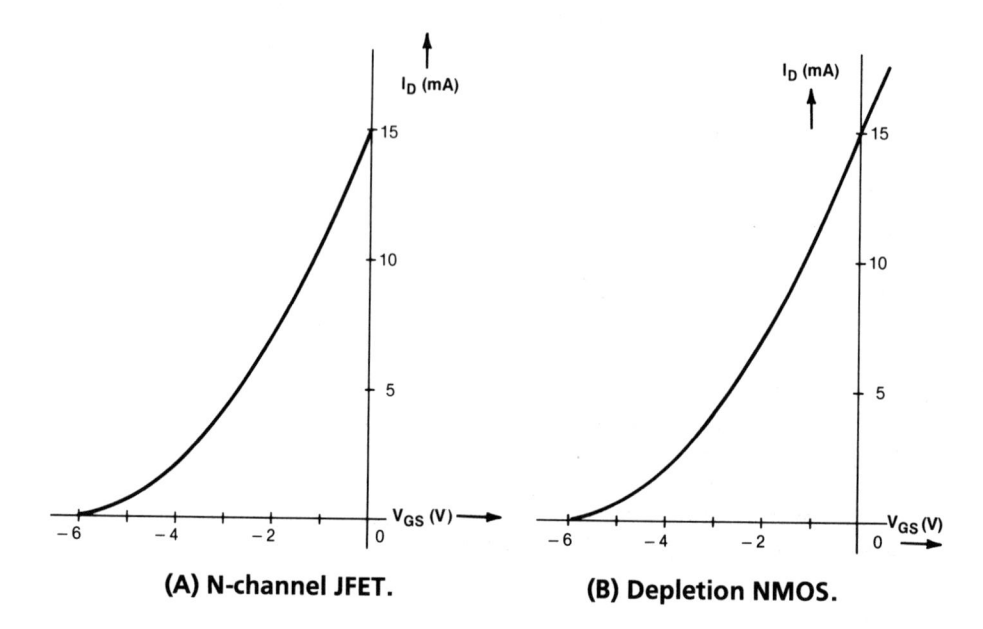

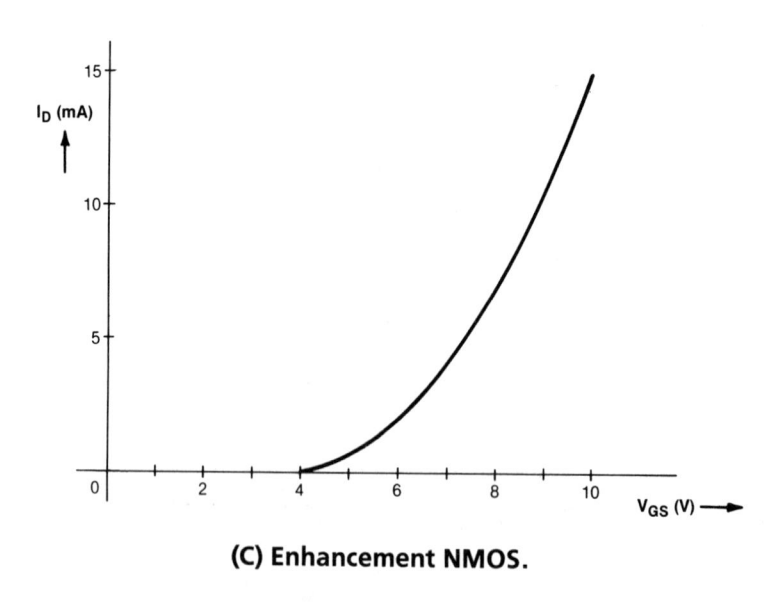

(B) Depletion NMOS.

(C) Enhancement NMOS.

Fig. 6-3. Transfer characteristics of n-channel FETs.

where,

K is a constant (usually expressed in mA/volt²).

This equation is the *general form of the equation for the behavior of all three types of FETs in the saturation region of operation.*

If the value of the constant K and the threshold voltage V_T are known for a FET, then its transfer characteristic can be drawn by using Eq. (6-1).

Example 6.1:

A given FET is known to have a threshold voltage $V_T = 2$ V, and the constant $K = 0.5$ mA/V². Plot its transfer characteristic.

Solution

Using the given values of V_T and K, Eq. (6-1) becomes

$$I_D = 0.5(V_{GS} - 2)^2 \text{ mA}$$

Table 6-1. Sample Data Point Calculations

V_{GS} (V)	I_D (mA)
2	0
4	2
5	4.5
6	8

Table 6-1 shows a few sample data point calculations and the transfer characteristic is shown in Fig. 6-4.

Example 6.2

A FET with $V_T = -5$ V is found to have a drain current of 10 mA when $V_{GS} = 0$. Obtain the equation of the transfer characteristic and plot it.

Solution

Using $V_T = -5$ V, and substituting $I_D = 10$ at $V_{GS} = 0$, Eq. (6-1) leads to

$$10 = K(0 + 5)^2 \text{ mA}$$

or,

$$K = 0.4 \text{ mA/V}^2$$

Therefore, the equation for the transfer characteristic becomes [from Eq. (6-1) with $K = 0.4$ and $V_T = -5$]

$$I_D = 0.4(V_{GS} + 5)^2$$

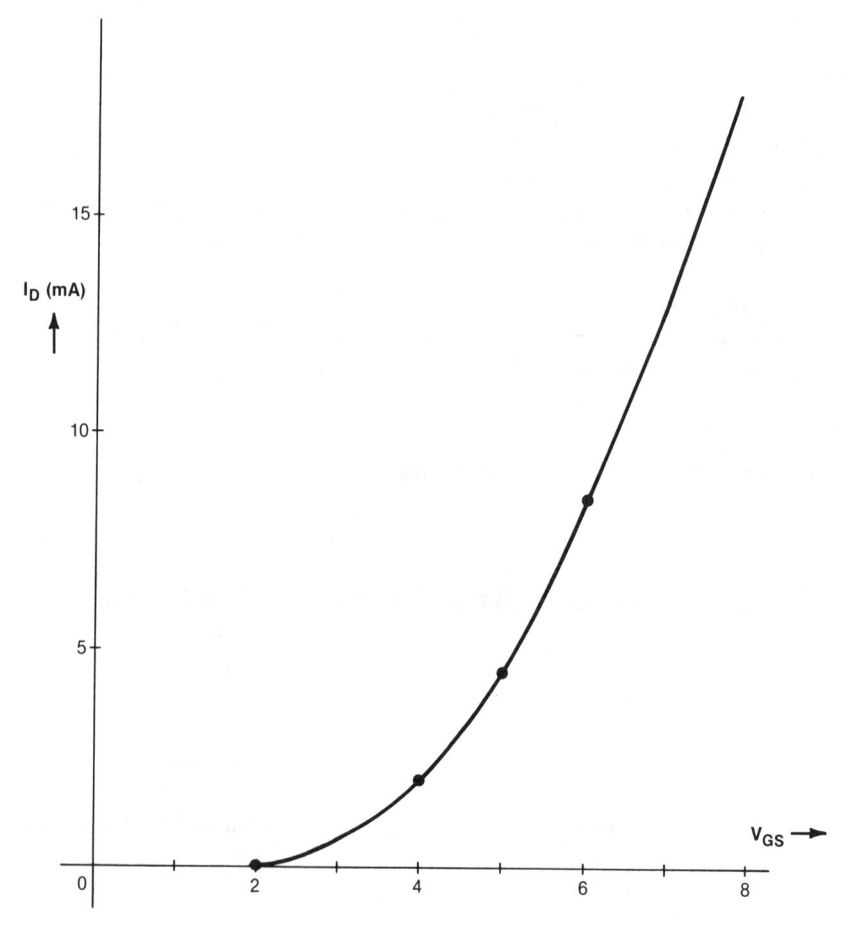

Fig. 6-4. Transfer characteristic for Example 6.1.

The transfer characteristic is shown in Fig. 6-5. Since V_T is negative, the given device must be either a JFET or a depletion NMOS. Therefore, the transfer characteristic is confined to negative values of V_{GS}.

Example 6.3

An enhancement NMOS has $V_T = 3$ V. At $V_{GS} = 4.5$ V, the current I_D is 8 mA. Calculate (a) the drain current when $V_{GS} = 6$ V; (b) the gate-to-source voltage needed to obtain a drain current of 16 mA.

Solution

First, we calculate the constant K using the data: $I_D = 8$ mA at $V_{GS} = 4.5$ V. Using these values in Eq. (6-1), we get

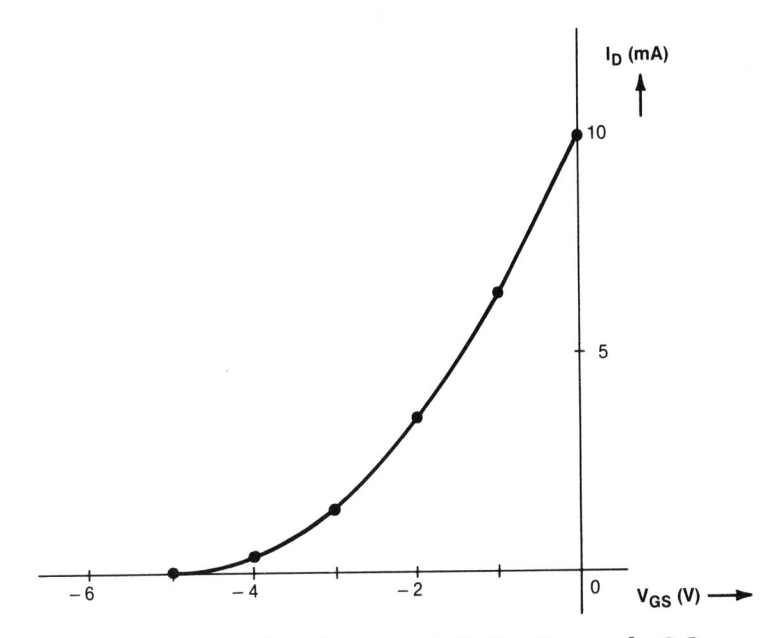

Fig. 6-5. Transfer characteristic for Example 6.2.

$$K = 3.56 \text{ mA/V}^2$$

Therefore, Eq. (6-1) becomes

$$I_D = 3.56(V_{GS} - 3)^2 \text{ mA} \qquad (6\text{-}2)$$

(a) When $V_{GS} = 6$ is used in Eq. (6-2), we get $I_D = 32$ mA.
(b) Using $I_D = 16$ mA, Eq. (6-2) leads to

$$(V_{GS} - 3)^2 = (16/3.56)$$

or,

$$V_{GS} = 5.12 \text{ V}$$

Note that the device is not linear. That is, when the drain current is reduced by a factor of 2 (from 32 to 16 mA) the voltage V_{GS} is not reduced by the same factor.

Drill Problem 6.1: Plot the transfer characteristic of a FET whose $V_T = -8$ V and $K = 0.6$ mA/V^2.

Drill Problem 6.2: A FET with $V_T = 0.5$ V has a drain current of 10 mA at $V_{GS} = 5$ V. Obtain the equation of the transfer characteristic of the device and plot it.

Drill Problem 6.3: A depletion NMOS has a threshold voltage of -10 V. At $V_{GS} = 0$, $I_D = 15$ mA. Calculate (a) I_D when $V_{GS} = 1$ V; (b) V_{GS} needed to make I_D one half the value obtained in (a).

DETERMINATION OF THE CONSTANT K

The constant K that occurs in the equation for the transfer characteristics of a FET depends upon the dimensions of the device and the manufacturing process itself. It is not specified directly by the manufacturer (unlike the threshold voltage, which is). It is determined from other data provided by the manufacturer.

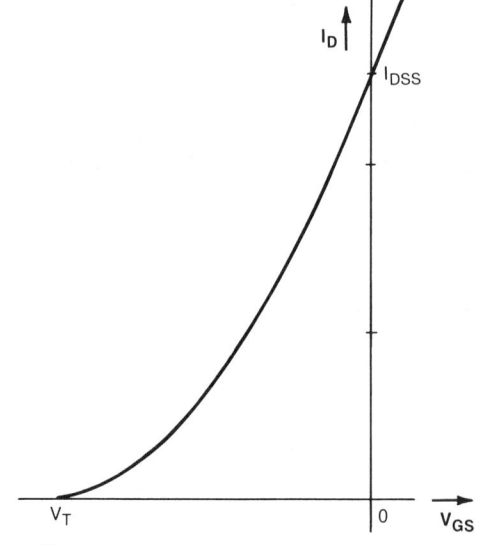

Fig. 6-6. Definition of the drain saturation current I_{DDS}.

The manufacturer specifies a particular value of the drain current I_{DSS} for the JFET and the *depletion* NMOS:

I_{DSS} is the drain current at $V_{GS} = 0$

Using I_{DSS} in Eq. (6-1) for the transfer characteristics, we get

$$K = \frac{I_{DSS}}{V_T^2}$$
(6-3)

Electronics: Circuits and Systems

Therefore, the transfer characteristics of a *JFET* and a *depletion NMOS* are fully defined from a knowledge of I_{DSS} and V_T. (Refer back to Example 6.2 and Drill Problem 6.3, where the data actually pertained to I_{DSS} even though we did not call it I_{DSS} then.)

For an *enhancement NMOS*, I_{DSS} is zero since no drain current flows when $V_{GS} = 0$ for such a device. (V_{GS} has to be positive for the device to conduct.) Therefore, K cannot be calculated by means of Eq. (6-3) for the enhancement NMOS. The manufacturer usually provides the value of the drain current $I_{D(ON)}$ at some particular value of the gate-to-source voltage $V_{GS(ON)}$. This information is used in Eq. (6-1) to determine the value of K for a given enhancement NMOS in a manner similar to that used in Example 6.3 and Drill Problem 6.2. (Note also that when $V_{GS} = (V_T + 1)$, Eq. (6-1) gives K = I_D. Thus, a measurement of I_D when the gate-to-source voltage is kept at 1 V above the threshold level will give the value of the constant K.)

Example 6.4

If a JFET has $V_T = -8$ V and $I_{DSS} = 12$ mA, obtain the equation of its transfer characteristics.

Solution

Using Eq. (6-3), we get

$$K = \frac{12}{(-8)^2} = 0.1875 \text{ mA/V}^2$$

and the equation for the transfer characteristic becomes

$$I_D = 0.1875(V_{GS} + 8)^2 \text{ mA}$$

Drill Problem 6.4: Obtain the transfer characteristic equation of a FET with $V_T = -4$ V and $I_{DSS} = 12$ mA.

Drill Problem 6.5: The manufacturer's specifications for an enhancement NMOS are $I_{D(on)} = 10$ mA at $V_{GS(on)} = 5$ V and $V_T = 1.8$ V. Obtain the equation of the transfer characteristic and plot it.

You will find that in the literature, Eqs. (6-1) and (6-3) for the JFET and depletion NMOS are combined to lead to a *special equation* for those two types:

$$I_D = I_{DSS}(1 - \frac{V_{GS}}{V_T^2})$$

But in our discussion, we will use Eq. (6-1) so that there is a single equation for all three types of FETs.

BIASING OF THE FET AMPLIFIER

In analogy with the BJT, we can see that there are three configurations for a FET amplifier as shown in Table 6-2.

Table 6-2. Three Configurations for a FET Amplifier

FET Amplifier	*BJT Amplifier*
Common source	Common emitter
Common drain (or Source follower)	Common collector (or Emitter follower)
Common gate	Common base

It will be found that there is a strong similarity in the response of each configuration of a FET amplifier with the corresponding BJT amplifier, with some changes brought about by the special features of the FET: extremely high input impedance and nearly zero input current.

A biasing arrangment for an n-channel FET amplifier (useful for *any of the three types:* JFET, depletion NMOS, enhancement NMOS) is shown in Fig. 6-7A. The npn transistor amplifier circuit is also shown for comparison.

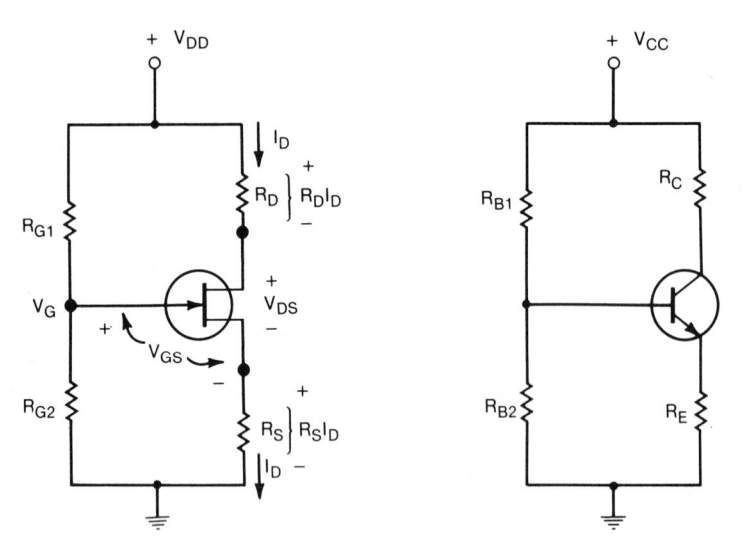

(A) Biasing circuit of an n- channel FET.

(B) Biasing circuit of an npn transistor.

Fig. 6-7. Comparison of the similarities between two biasing circuits.

Electronics: Circuits and Systems

Algebraic Determination of the Q Point

The quiescent operating point (or Q point) of the FET is defined by the drain current I_D and the drain-to-source voltage V_{DS}.

The voltage at the gate terminal with respect to *ground* is given by

$$V_G = \frac{R_{G2}}{R_{G1} + R_{G2}} V_{DD} \qquad (6\text{-}4)$$

since the two resistors R_{G1} and R_{G2} act as a voltage divider across V_{DD}. (The same current flows in the two resistors.)

The gate current for a FET is zero (or nearly so and we can ignore it). Therefore, the current through the source terminal is I_D also as indicated in Fig. 6-7. The voltage at the source terminal with respect to ground is equal to the voltage drop across R_S, $R_S I_D$.

The gate-to-source voltage V_{GS} is, therefore, given by

$$V_{GS} = V_G - R_S I_D \qquad (6\text{-}5)$$

where,

V_G is given by Eq. (6-4).

Eq. (6-4) is a statement of the constraint imposed on the values of V_{GS} and I_D by the biasing circuit.

It is assumed that the *FET is operating in the saturation region in an amplifier*, and consequently, the transfer characteristic equation is

$$I_D = K(V_{GS} - V_T)^2 \qquad (6\text{-}6)$$

A solution of Eqs. (6-5) and (6-6) will lead to the determination of the current I_D. Once I_D is known, the drain-to-source voltage V_{DS} in the given circuit can be calculated by using

$$V_{DS} = V_{DD} - I_D R_D - I_D R_S \qquad (6\text{-}7)$$

The solution of Eqs. (6-5) and (6-6) by using algebraic procedures involves the solution of a quadratic equation as will be seen in the following example. There is also a graphical approach to solving the two equations, which will be discussed later.

Example 6.5

Calculate the Q point of the FET amplifier shown in Fig. 6-8 if the parameters of the FET are $V_T = -3$ V and $I_{DSS} = 6$ mA.

Solution

First, let us obtain the transfer characteristic equation. Using $I_{DSS} = 6$ mA and $V_T = -3$ V in Eq. (6-3), we have

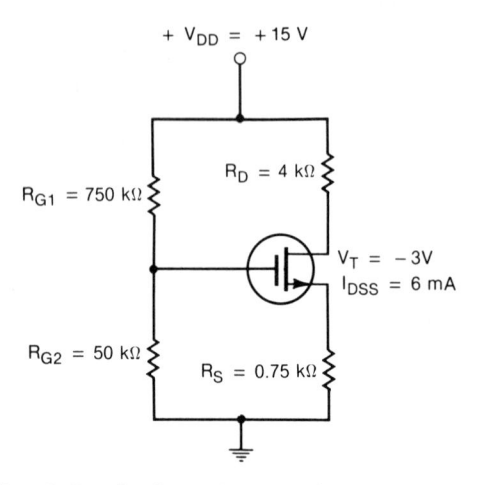

Fig. 6-8. Biasing circuit of Example 6.5.

$$K = (6/3^2) = 0.667 \text{ mA/V}^2$$

so that the equation for the transfer characteristic becomes

$$I_D = 0.667[V_{GS} - (-3)]^2 \text{ mA}$$
$$= 0.667(V_{GS} + 3)^2 \text{ mA} \tag{6-8}$$

Using the voltage-divider relationship, Eq. (6-4), we have for the voltage at the gate (with respect to ground):

$$V_G = \frac{R_{G2}}{R_{G1} + R_{G2}} V_{DD} = 0.9375 \text{ V}$$

Therefore, the gate-to-source voltage is [Eq. (6-5)]

$$V_{GS} = V_G - R_S I_D = 0.9375 - 0.75 I_D \tag{6-9}$$

where,

I_D is in mA.

Substitute for I_D in Eq. (6-9) from Eq. (6-8).

$$V_{GS} = 0.9375 - 0.75 \times [0.667 (V_{GS} + 3)^2]$$

which leads to, after some manipulation,

$$0.5 V_{GS}^2 + 4 V_{GS} + 3.5625 = 0 \tag{6-10}$$

The quadratic equation (6-10) is solved using the usual formula.

Electronics: Circuits and Systems

$$V_{GS} = \frac{-4 \pm \sqrt{4^2 - (4 \times 0.5 \times 3.5625)}}{2 \times 0.5}$$

$$= -4 \pm 2.98$$

There are two possible values: $(-4 - 2.98) = -6.98$ V and $(-4 + 2.98) = -1.02$ V. The former value is clearly not possible since V_{GS} *cannot be more negative than the threshold value.* Therefore, we choose

$$V_{GS} = -1.02 \text{ V}$$

Substituting the above value of V_{GS} in either Eqs. (6-8) or (6-9) leads to

$$I_D = 2.61 \text{ mA}$$

The drain-to-source voltage is obtained by subtracting the voltage drops across R_D and R_S from V_{DD}.

$$V_{DS} = V_{DD} - R_D I_D - R_S I_D = 2.60 \text{ V}$$

Therefore, the Q point of the given circuit is: $I_D = 2.61$ mA, $V_{DS} = 2.60$ V. Since we assumed that the FET is operating in the saturation region, when we used Eq. (6-8), it is necessary to verify that the assumption has not been contradicted. Therefore, we check to make sure that $V_{DS} > (V_{GS} - V_T)$ at the Q point, which is indeed the case in the present problem.

Drill Problem 6.6: Obtain the Q point of a FET amplifier (Fig. 6-7) when $V_{DD} = 16$ V, $R_D = 2.5$ kΩ, $R_S = 1.5$ kΩ, $R_{G1} = 4.5$ MΩ, $R_{G2} = 1$ MΩ. Assume that for the FET, $K = 4$ mA/V^2 and $V_T = +1.5$ V.

Drill Problem 6.7: Recalculate the Q point of the amplifier of the previous drill problem if V_{DD} is changed to 24 V.

Graphical Determination of the Q Point

Instead of solving Eqs. (6-5) and (6-6) of the FET amplifier circuit algebraically as was done earlier, they can be solved graphically. The graphical procedure is often preferable (unless you have a calculator with a built-in program for solving quadratic equations). In the graphical procedure, two graphs are drawn: one representing Eq. (6-5) and the other representing Eq. (6-6). We already know that the graph of Eq. (6-6) is the transfer characteristic of the device and how to draw it when the constant K and the threshold voltage are given. Now, let us consider Eq. (6-5), which is repeated here:

$$V_{GS} = V_G - R_S I_D \tag{6-5}$$

Eq. (6-5) can be rewritten as

$$I_D = \frac{V_G}{R_S} - \frac{V_{GS}}{R_S}$$

For a given circuit, R_S is a known constant and so is V_G since it can be calculated by using the voltage divider formula, Eq. (6-4). Therefore, the previous equation is of the form

$$I_D = \text{(some constant)} - \frac{V_{GS}}{\text{(some other constant)}}$$

When the last equation is plotted on a graph paper with I_D chosen along the vertical axis and V_{GS} along the horizontal axis, it leads to a straight line (called the *bias line* or *load line*) as shown in Fig. 6-9. The bias line has the following properties:

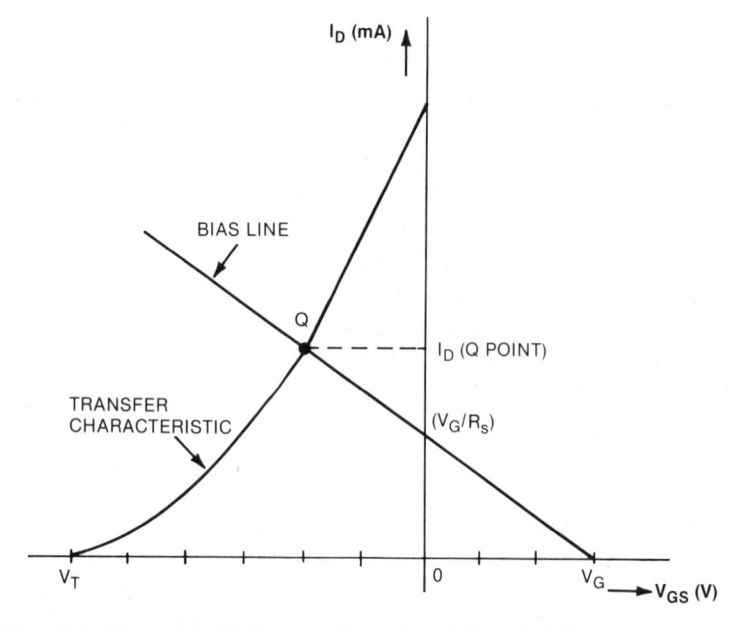

Fig. 6-9. Graphical determination of the Q point of a FET.

Slope is negative and $= (-1/R_S)$
Cuts the vertical axis at $I_D = (V_G/R_S)$
Cuts the horizontal axis at $V_{GS} = V_G$

where,

V_G is given by Eq. (6-4).

Electronics: Circuits and Systems

Now we have two graphs for a given FET amplifier circuit: one is the transfer characteristic, Eq. (6-6), and the other is the bias line due to Eq. (6-5) as shown in Fig. 6-9. The Q point of the circuit must satisfy both Eqs. (6-5) and (6-6), and must lie on both the graphs. Therefore, *the Q point is the point of intersection of the two graphs.*

The graphical determination of the Q point of a FET amplifier circuit involves the following steps.

1. Plot the transfer characteristic of the given FET by using Eq. (6-6).
2. Calculate V_G by using the voltage divider formula, Eq. (6-4).
3. Mark a point on the I_D axis of the graph at (V_G/R_S) and a point on the V_{GS} axis at V_G. Join the two points by a straight line and extend it as needed.
4. The Q point is the point at which the bias line intersects the transfer characteristic.

Example 6.6

Obtain the Q point of the FET amplifier circuit shown in Fig. 6-10. Assume $K = 0.8 \text{ mA/V}^2$ and $V_T = 0.5 \text{ V}$.

Fig. 6-10. Biasing circuit of Example 6.6.

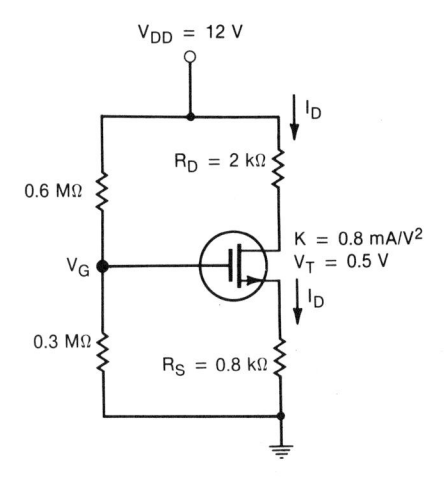

Solution

First we draw the transfer characteristic (as was done in Example 6.1) from the equation

$$I_D = 0.8(V_{GS} - 0.5)^2 \text{ mA}$$

Next, we calculate V_G (voltage-divider formula):

$$V_G = \frac{0.3}{0.3 + 0.6} \times V_{DD} = 4 \text{ V}$$

Eq. (6-5) becomes

$$V_{GS} = 4 - 0.8 I_D$$

and the straight line for this equation is obtained by taking $V_{GS} = V_G = 4$ on the horizontal axis and $I_D = (V_G/R_S) = 5$ mA on the vertical axis as shown in Fig. 6-11.

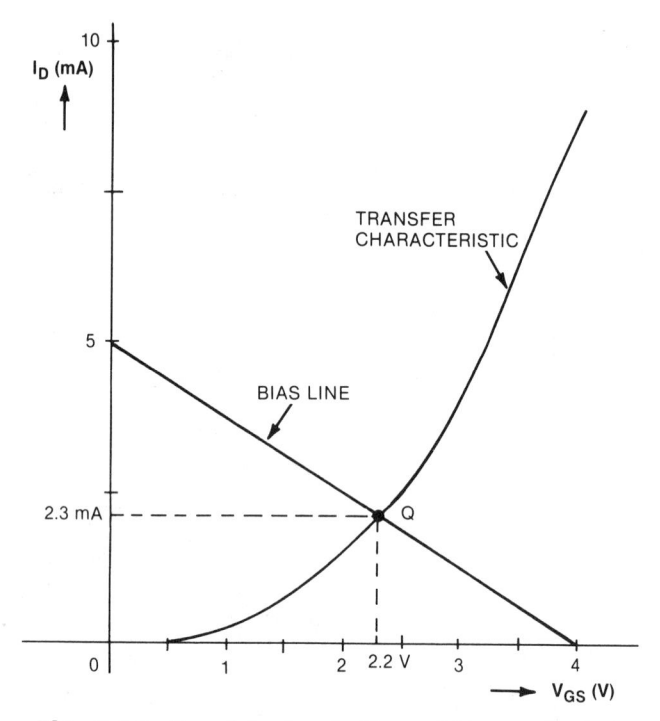

Fig. 6-11. Graphical solution of Example 6.6.

The intersection of the transfer characteristic with the bias line drawn gives us the Q point. It is seen that the value of I_D at the Q point is 2.3 mA.

To find the value of V_{DS} (the drain-to-source voltage) at the Q point, we subtract the drops in R_D and R_S from V_{DD}.

$$V_{DS} = 12 - (2 \times 2.3) - (0.8 \times 2.3)$$
$$= 5.56 \text{ V}$$

Electronics: Circuits and Systems   349

Therefore, the Q point of the circuit is $I_D = 2.3$ mA, $V_{DS} = 5.56$ V, $V_{GS} = 2.2$ V. [V_{DS} is seen to be $> (V_{GS} - V_T)$.]

Example 6.7

Determine the Q point of the circuit in Fig. 6-12 where the JFET has the parameters $I_{DSS} = 10$ mA and $V_T = -4$ V.

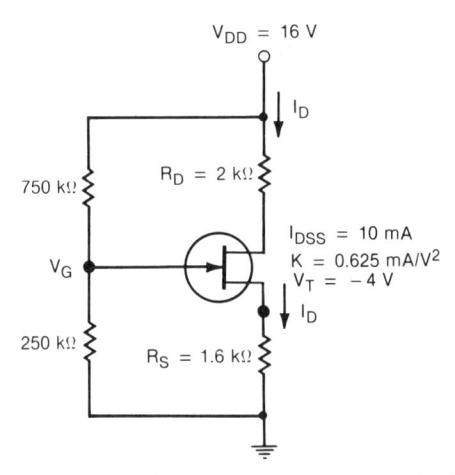

Fig. 6-12. Biasing circuit of Example 6.7.

Solution

The value of K for the transfer characteristic of the FET is given by Eq. (6-3):

$$K = (I_{DSS}/V_T^2) = 0.625$$

Therefore, the equation of the transfer characteristic is

$$I_D = 0.625[V_{GS} - (-4)]^2$$
$$= 0.625(V_{GS} + 4)^2 \text{ mA}$$

The transfer characteristic is shown in Fig. 6-13.
The voltage V_G at the gate is

$$V_G = \frac{250}{250 + 750} \times 16 = 4 \text{ V}$$

A straight line is drawn from $V_{GS} = V_G = 4$ V on the horizontal axis to $I_D = (V_G/R_S) = 2.5$ mA on the vertical axis.

350   **FET Amplifiers**

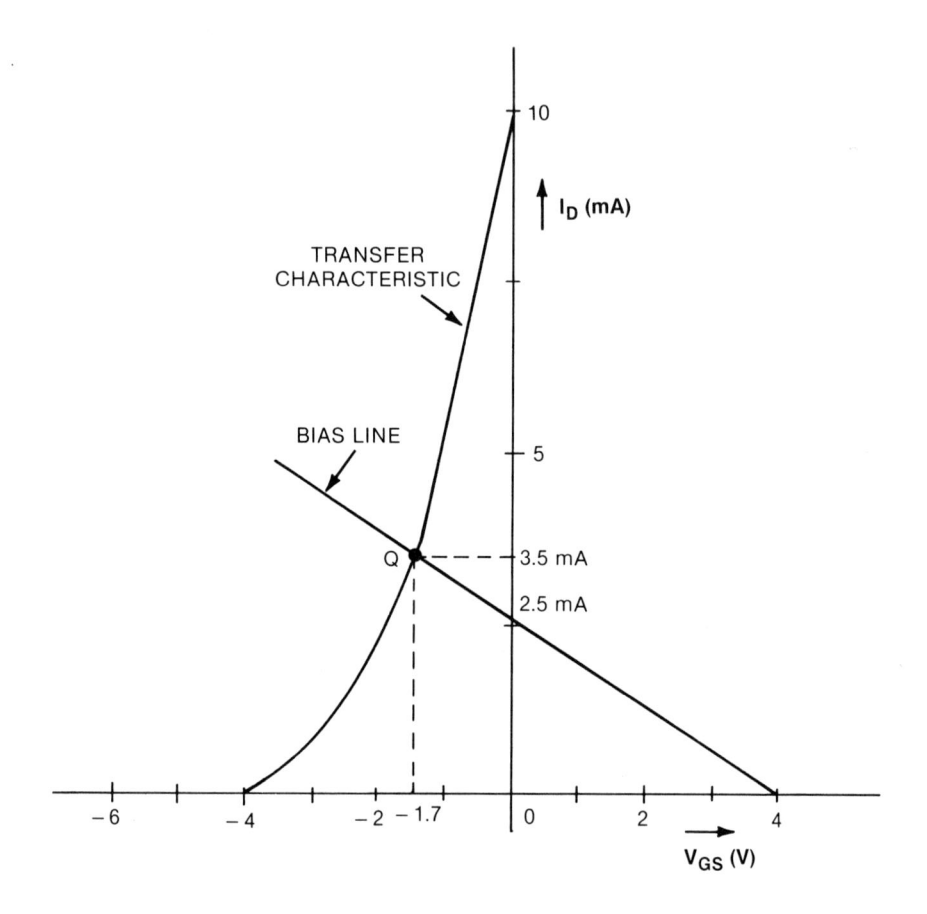

Fig. 6-13. Graphical solution of Example 6.7.

The bias line intersects the transfer characteristic at the Q point. The value of I_D is measured from the graph and seen to be 3.5 mA

The value of V_{DS} is obtained by subtracting the drops across R_D and R_S from V_{DD}, and is found to be 3.4 V.

Therefore, the Q point is at $I_D = 3.5$ mA, $V_{DS} = 3.4$ V, and $V_{GS} = -1.7$ V. V_{DS} is seen to be $> (V_{GS} - V_T)$.

Drill Problem 6.8: In the biasing circuit shown in Fig. 6-14, assume K $= 0.75$ mA/V^2 and $V_T = -5$ for the FET. Use $R_{G1} = 1$ MΩ, $R_{G2} = 0.25$ MΩ, $R_D = 1$ kΩ, $R_S = 0.8$ kΩ, $V_{DD} = 15$ V. Determine the Q point using the graphical procedure.

Electronics: Circuits and Systems

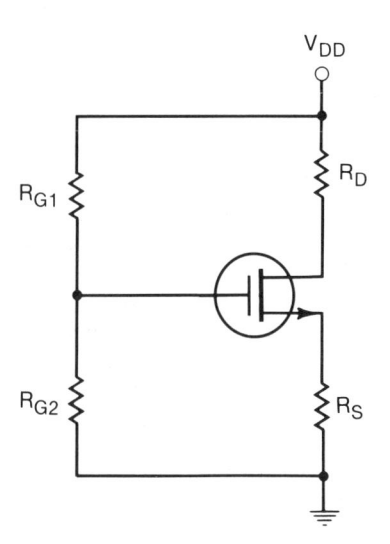

Fig. 6-14. Biasing circuit of Drill Problem 6.8.

Drill Problem 6.9: In the biasing circuit shown in Fig. 6-14, assume that $I_{DSS} = 12$ mA and $V_T = -6$ V. Determine the Q point if $R_{G1} = 750$ kΩ, $R_{G2} = 250$ kΩ, $R_D = 1$ kΩ, $R_S = 2$ kΩ, $V_{DD} = 16$ V.

Drill Problem 6.10: Suppose that in the circuit of the last drill problem, a short circuit is accidentally placed across the resistor R_{G1}. Determine the effect on the Q point.

Drill Problem 6.11: Suppose that in the circuit of Drill Problem 6.9, a short circuit accidentally develops across the resistor R_{G2}. Determine the effect on the Q point.

Drill Problem 6.12: Suppose that in the circuit of Drill Problem 6.9, an open circuit develops in the R_{G1} branch. Determine the effect on the Q point.

Drill Problem 6.13: Suppose that in the circuit of Drill Problem 6.9, an open circuit develops in the R_{G2} branch. Determine the effect on the Q point.

It should be kept in mind that either the algebraic approach (which involves the solution of a quadratic equation but does not require the availability of graph paper) or the graphical approach (which needs a graph sheet but saves the bother of solving a quadratic equation) can be used for any of the FET biasing circuits. The graphical procedure does not give as accurate a result since there are bound to be errors in plotting and reading a graph. But errors of that kind are quite permissible since electronic design is not an exact science. The graphical procedure is preferred in the *design* of a biasing circuit for a given FET, since it allows us to weigh the possible values of circuit components quickly. This point will be discussed in detail in the next section.

A self-biasing circuit that can be used with JFET and depletion NMOS (but not with enhancement NMOS) is considered in the following example.

Example 6.8

Determine the Q point of the FET in the biasing circuit of Fig. 6-15. Assume $K = 0.4$ mA/V^2 and $V_T = -5$ V for the FET.

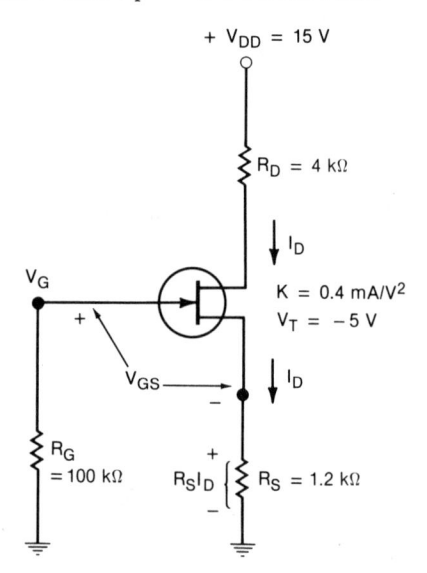

Fig. 6-15. Biasing circuit of Example 6.8. (Such a circuit cannot be used with enhancement MNOS.)

Solution

The *transfer characteristic equation* is

$$I_D = 0.4[V_{GS} - (-5)]^2$$
$$= 0.4(V_{GS} + 5)^2 \text{ mA}$$

Note that the only difference between the circuit configuration of Fig. 6-15 and the biasing circuits considered earlier is that the gate is not connected to the dc power supply (either directly or indirectly through a voltage divider). Since the gate current in the FET is zero, the voltage at the gate terminal is zero. That is,

$$V_G = 0$$

and the equation relating V_{GS} and I_D due to the biasing circuit becomes

$$V_{GS} = 0 - R_s I_D$$
$$= -R_s I_D \qquad (6\text{-}11)$$

The bias line intersects the horizontal axis at $V_{GS} = V_G = 0$, and the vertical axis at $I_D = (V_G/R_S) = 0$ also. That is, instead of getting two distinct points for drawing the bias line as we did in the earlier biasing circuit, we get only one: the origin in the graph. An additional point is needed for drawing the bias line, and this is obtained by using Eq. (6-11) itself in conjunction with some arbitrary value of I_D. For example, suppose we choose an arbitrary value of 4 mA for I_D. Then, Eq. (6-11) leads to

$$V_{GS} = -R_s I_D$$
$$= -4.8 \text{ V}$$

Therefore, the point whose coordinates are $V_{GS} = -4.8$ V and $I_D = 4$ mA must also lie on the bias line. Joining this point with the point at the origin gives us the bias line. An alternative approach is to use the *slope* information from Eq. (6-11). The slope of the bias line is $(-1/R_S)$ mA/V.

The bias line is seen (Fig. 6-16) to intersect the transfer characteristic at $I_D = 2.2$ mA. Then

$$V_{DS} = V_{DD} - R_D I_D - R_s I_D$$
$$= 3.56 \text{ V}$$

Therefore, the Q point is given by $I_D = 2.2$ mA, $V_{DS} = 3.56$ V, and $V_{GS} = -2.64$ V, obtained by using the value of I_D in Eq. (6-11). $V_{DS} > (V_{GS} - V_T)$ and so the FET is in the saturation region.

The problem could also be solved algebraically. The transfer characteristic equation is combined with Eq. (6-11) to obtain a quadratic equation, which is then solved for V_{GS}. The quadratic equation in this case is found to be

$$0.48 V_{GS}^2 + 5.8 V_{GS} + 12 = 0.$$

Drill Problem 6.14: Recalculate the Q point of the circuit in the previous example if R_S is changed to 2.5 kΩ.

PRACTICAL CONSIDERATIONS IN BIASING A FET

We have assumed in the previous discussion that the values of the parameters V_T and K (or I_{DSS}) of a FET are precisely known so that the transfer char-

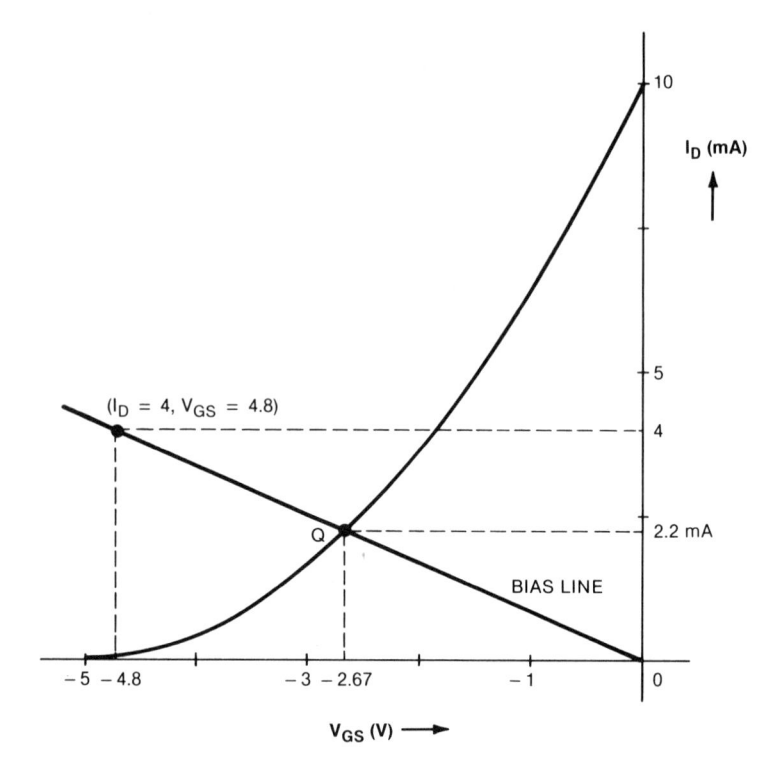

Fig. 6-16. Graphical solution of Example 6.8.

acteristic can be drawn for a given FET. Unfortunately, this is not the case in reality. The manufacturer specifies a *maximum* and a *minimum* for each parameter so that even for two units with the same type-number, there is a wide variation in the parameter values. As an extreme example, the manufacturer's specifications for the 2N5484 n-channel JFET are I_{DSS} has a minimum value of 1 mA and a maximum value of 5 mA (a spread by a factor of 5); the *magnitude* of V_T has a minimum value of 0.3 V and a maximum value of 3 V (a spread by a factor of 10!). This means that if we replace the FET in an amplifier by another having the same type-number, the Q point can be expected to shift to a totally different value. A shift in the Q point affects the response of the amplifier to an input signal and can cause a distorted output.

The effect of the spread of parameters on the Q point is shown in Fig. 6-17. Two transfer characteristics are shown one using the maximum value of I_{DSS} (denoted by I'_{DSS}) and the maximum magnitude of V_T (denoted by V'_T), and the other using the minimum value of I_{DSS} (denoted by I''_{DSS}) and the minimum magnitude of V_T (denoted by V''_T). The transfer characteristic of any

Electronics: Circuits and Systems

given FET (with that type-number) will necessarily fall between these two extreme curves.

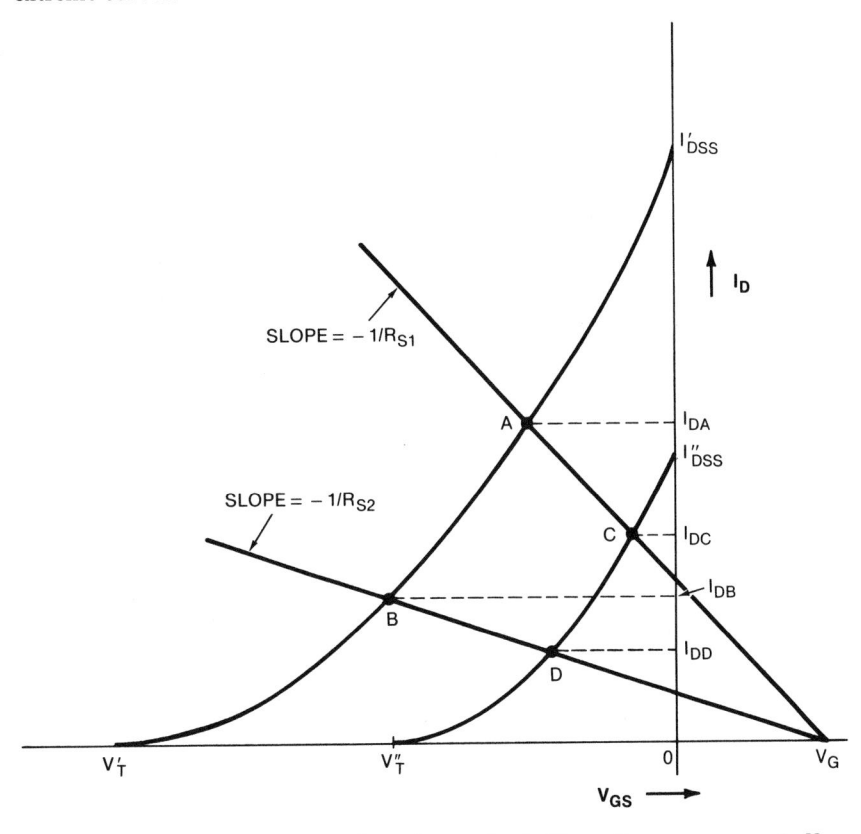

Fig. 6-17. Extreme transfer characteristics of a commercially available FET.

Two bias lines are shown in order to compare the effect of slope on the shift in the Q point. For the steeper bias line shown [with slope = $-(1/R_{S1})$], the Q points are seen to be at points A and C corresponding to drain currents I_{DA} and I_{DC}. For the bias line with the smaller slope [slope = $-(1/R_{S2})$], the Q points are at B and D corresponding to drain currents I_{DB} and I_{DD}. It is seen that the bias line with the *smaller slope* results in a *smaller change* in the drain current produced by a change in the parameter values of the FET. The slope of the bias line can be reduced either by increasing R_S (the resistor between the source terminal of the FET and ground) or by increasing V_G, the voltage at the gate terminal above ground. But R_S cannot be increased indefi-

356   **FET Amplifiers**

nitely since it will also affect the value of V_{DS} of the FET. An increase in V_G will usually require a larger power supply V_{DD}, since V_G is proportional to V_{DD}. Some compromising considerations are necessary in designing the biasing circuit, which stabilizes the Q point of the FET against variations of its parameter values.

DESIGN OF A BIASING CIRCUIT

Let us now examine the problem of designing the biasing circuit for a given FET. Suppose the manufacturer's specifications for a particular n-channel JFET are

$$I_{DSS\,(min)} = 4\text{ mA} \quad I_{DSS\,(max)} = 16\text{ mA}$$
$$|V_T|_{(min)} = 2\text{ V} \quad |V_T|_{(max)} = 8\text{ V}$$

(Note that the manufacturer always specifies the *magnitude* of the threshold voltage. When dealing with *n-channel* JFET and depletion NMOS, we read such specifications as the *negative* of the given values. Thus, V_T has the extreme values of -2 V and -8 V for the present example.)

Suppose it is desirable to keep the Q point value of the drain current between the limits 2 mA and 3.5 mA in the circuit under consideration. We wish to determine the values of the resistors in the biasing circuit to meet the given conditions.

Fig. 6-18 shows the extreme characteristics of the described FET. The two extreme transfer characteristics are drawn by pairing $I_{DSS} = 4$ mA with $V_T = -2$ V, and $I_{DSS} = 16$ mA with $V_T = -8$ V. (We pair the minimum I_{DSS} with the minimum *magnitude* V_T and the maximum I_{DSS} with the maximum *magnitude* V_T.)

Locate a Q point on each transfer characteristic: Q' on the *upper* one determined by the *upper limit* 3.5 mA of the drain current, and Q'' on the *lower* one determined by the *lower limit* 2 mA of the drain current. This will ensure that the Q point value will lie between the two specified limits even if the actual FET transfer characteristics lie anywhere between the extremes given.

Join the two Q points, Q'' and Q', by a straight line (which is now the bias line) and extend it to the right to cut the V_{GS} axis at V_G. The value of V_G is read from the graph and is found to be 4.6 V in the present example.

Calculate the slope of the bias line from the graph. Here the line goes from 0 to 2 mA with a corresponding voltage *change* of approximately 5.1 V. Therefore,

$$\text{slope of bias line} = -(2/5.1) = -0.392 \text{ mA/V}$$

Electronics: Circuits and Systems

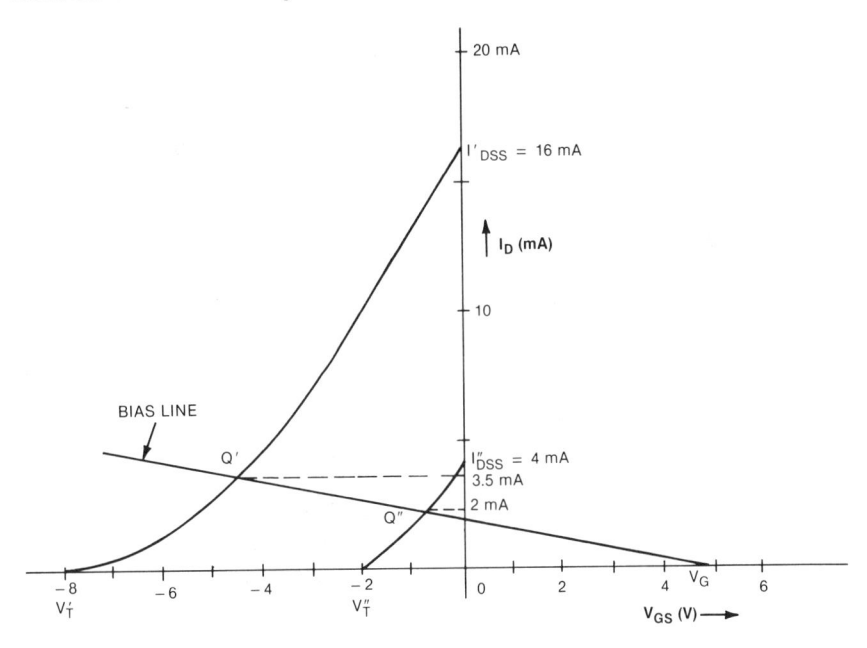

Fig. 6-18. Extreme transfer characteristics of a FET with the threshold voltage between −2 V and −8 V, and the drain saturation current between 4 mA and 16 mA.

and, since the slope $= -(1/R_S)$, we get

$$R_S = (1/0.392) = 2.55 \text{ k}\Omega$$

In order to proceed further with the design, we need to know the value of the power supply battery V_{DD} and the Q point value of the drain-to-source voltage V_{DS} of the FET. (In a real design situation, the Q point value of V_{DS} is chosen from a consideration of the *ac load* on the amplifier.) Here, let us assume $V_{DD} = 24$ V and V_{DS} at the Q point $= 12$ V.

The voltage V_G is obtained from the voltage-divider circuit (R_{G1} and R_{G2}) across the voltage V_{DD}. That is, (refer to Fig. 6-14)

$$V_G = \frac{R_{G2}}{R_{G1} + R_{G2}} V_{DD}$$

Therefore,

$$\frac{R_{G2}}{R_{G1} + R_{G2}} = \frac{V_G}{V_{DD}} = \frac{4.6}{24} = 0.192$$

which leads to the relationship

$$R_{G1} = 4.21 R_{G2}$$

Any two resistances R_{G1} and R_{G2} satisfying the previous relationship will suffice. R_{G2} is usually chosen to be quite large, typically around 100 kΩ. If R_{G2} is taken as 100 kΩ in the present example, then R_{G1} becomes 421 kΩ.

In order to determine the value of R_D, we need to know the voltage across it, and the relevant equation is (Fig. 6-19)

$$\text{voltage across } R_D = V_{DD} - V_{DS} - R_S I_D$$

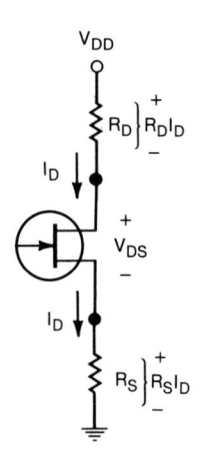

Fig. 6-19. Calculation of the voltage across R_D in the design example.

Using the average value of the two extreme values of I_D in the previous equation, we have

$$\text{voltage across } R_D = 24 - 12 - (2.55 \times 2.75)$$
$$= 4.99 \text{ V}$$

which gives

$$R_D = (4.99/I_D)$$
$$= 1.81 \text{ kΩ}$$

This completes the design of the biasing circuit and the final circuit is shown in Fig. 6-20.

Example 6.9

Given an n-channel JFET with the following specifications: $|V_T|$ is between 3 V and 6 V. I_{DSS} is between 5 mA and 10 mA. The power supply is $V_{DD} = 18$ V. Take V_{DS} at the Q point as one half of V_{DD}. Design a biasing circuit so that the drain current at the Q point lies in the range 3.5 mA to 4.5 mA.

Electronics: Circuits and Systems　　　　　　　　　　　　　　　　　**359**

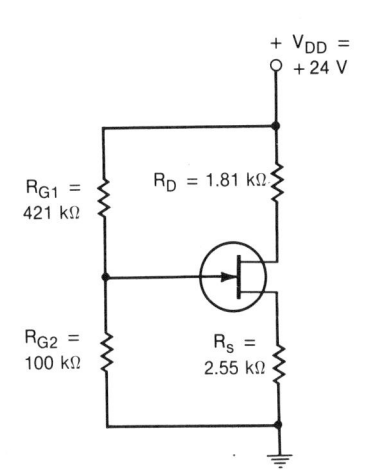

Fig. 6-20. Biasing circuit of the design example.

Solution

The values of K are 0.556 when $I_{DSS} = 5$ mA and $V_T = -3$ V; and 0.278 when $I_{DSS} = 10$ mA and $V_T = -6$ V. The two transfer characteristics are drawn (Fig. 6-21). The two Q points are obtained by taking $I_D = 3.5$ mA on the lower curve and $I_D = 4.5$ mA on the upper curve. The bias line obtained by joining the two Q points cuts the V_{GS} axis at $V_G = 4.75$ V. The slope of the bias line is found to be

$$-(3.5/5.25) = -0.667 \text{ mA/V}$$

Therefore,

$$R_S = (1/0.667)$$
$$= 1.5 \text{ k}\Omega$$

Using

$$\frac{R_{G2}}{R_{G1} + R_{G2}} = \frac{V_G}{V_{DD}} = 0.264$$

we get

$$R_{G1} = 2.79 R_{G2}$$

If R_{G2} is taken as 100 kΩ, then $R_{G1} = 279$ kΩ.
If we take V_{DS} at the Q point to be one half of V_{DD},

$$V_{DS} = 9 \text{ V}$$

The voltage drop across $R_S = R_S I_D$, and using the mean value of $I_D = (3.5$

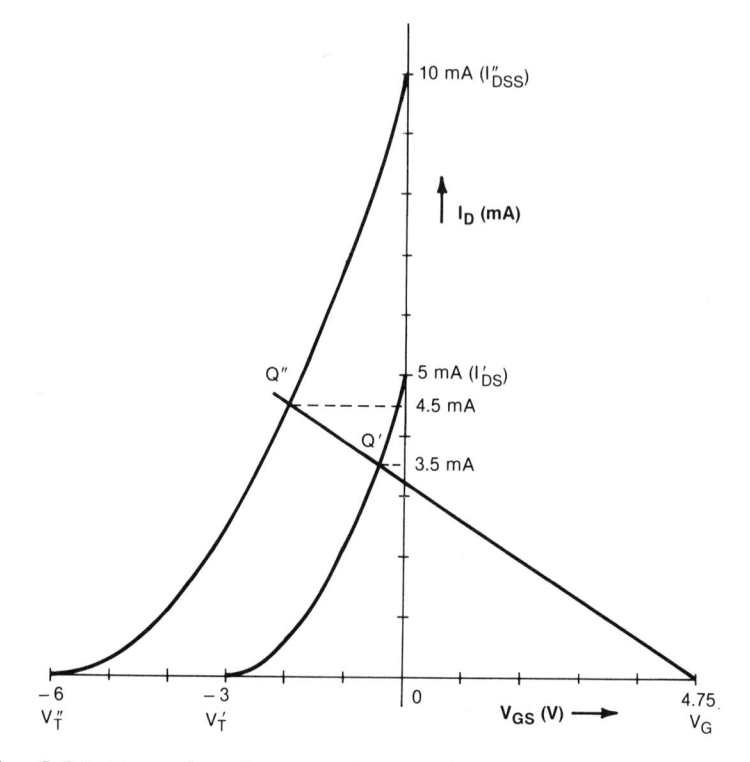

Fig. 6-21. Transfer characteristics of the FET of Example 6.9.

+ 4.5)/2 = 4 mA, we have $R_S I_D = 6$ V.
Therefore,

$$\text{voltage across } R_D = V_{DD} - V_{DS} - R_S I_D$$
$$= 3 \text{ V}$$

and

$$R_D = (3/I_D)$$
$$= 0.75 \text{ k}\Omega$$

The complete biasing circuit is shown in Fig. 6-22.

Drill Problem 6.15: Repeat Example 6.9 if the range of the drain current is to be between 2.5 and 4.5 mA.

Drill Problem 6.16: Suppose that the n-channel JFET of Example 6.9 is used with a biasing circuit with $V_G = 0$ and $R_S = 2.5$ kΩ. Calculate the

Electronics: Circuits and Systems

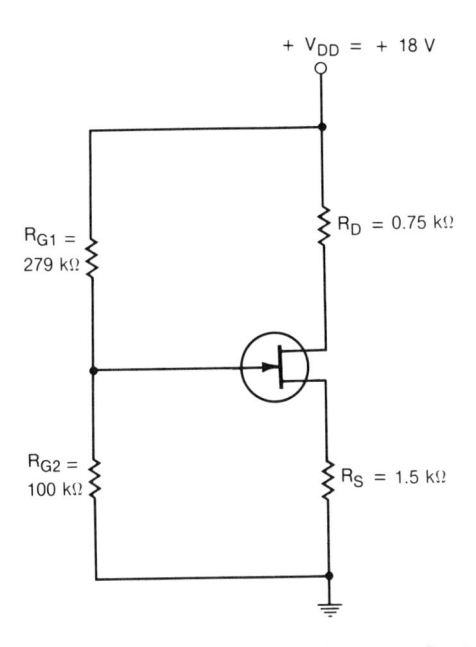

Fig. 6-22. Biasing circuit of Example 6.9.

upper and lower limits of the drain current at the Q point under these conditions.

Drill Problem 6.17: In the biasing circuit of Fig. 6-22 (last example), change R_S to 2 kΩ and V_{DD} to 24 V. Calculate the new upper and lower limits of I_D and V_{DS} at the Q point. How do these compare with the values obtained in Example 6.9?

COMMON-SOURCE FET AMPLIFIER

The common-source FET amplifier is analogous to the common-emitter BJT amplifier and is perhaps the most frequently used configuration of the FET amplifier. The common-source amplifier is shown in Fig. 6-23, where bypass and coupling capacitors have been added along with a signal source and a load resistor. The capacitors perform the same function as their counterparts did in the common-emitter amplifier: the coupling capacitors C_1 and C_2 provide dc isolation of the amplifier from the signal source and the load resistor, while the bypass capacitor C_S provides a low impedance path for ac currents so that the gain of the amplifier is not reduced by the presence of the resistor R_S.

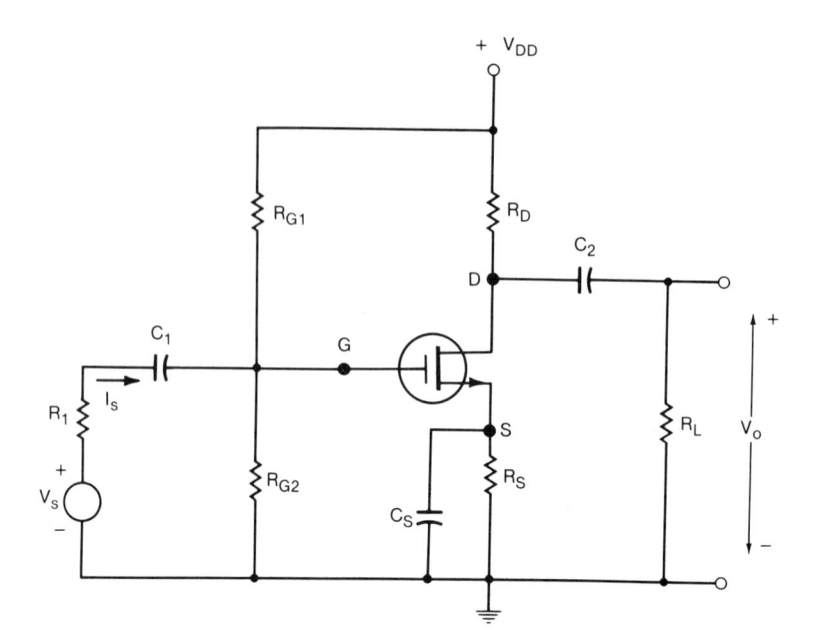

Fig. 6-23. A common-source FET amplifier circuit.

The small-signal model of a FET is similar to the small-signal model of the BJT used in the last chapter, except that the input resistance of a FET is *infinite*. That is, instead of having a resistance $r_{b'e}$ as in the BJT model, we simply have an open circuit between the gate terminal and the source terminal of the FET, as shown in Fig. 6-24A.

The transconductance g_m of a FET is a measure of the change in the drain current produced by a given change in the gate-to-source voltage. If, in Fig. 6-24B, ΔI_D is the change in the drain current produced by a change ΔV_{GS} in the gate-to-source voltage about the Q point, then

$$g_m = \frac{\Delta I_D}{\Delta V_{GS}} \text{ siemens}$$

which is, for small signals, the slope of the transfer characteristic curve at the Q point. Evaluating the slope of the transfer characteristic curve from its equation, Eq. (6-1), we get the following formula for the transconductance.

$$g_m = 2K(V_{GS} - V_T) \text{ siemens} \quad (6\text{-}12)$$

where,

V_{GS} is the gate-to-source voltage *at the Q point,*

V_T is the threshold voltage of the FET.

Electronics: Circuits and Systems

(A) The model.

(Figure: FET small-signal model with V_{gs} between G and S, and current source $g_m V_{gs}$ AMPERES between D and S.)

(B) Definition of the transconductance g_m.

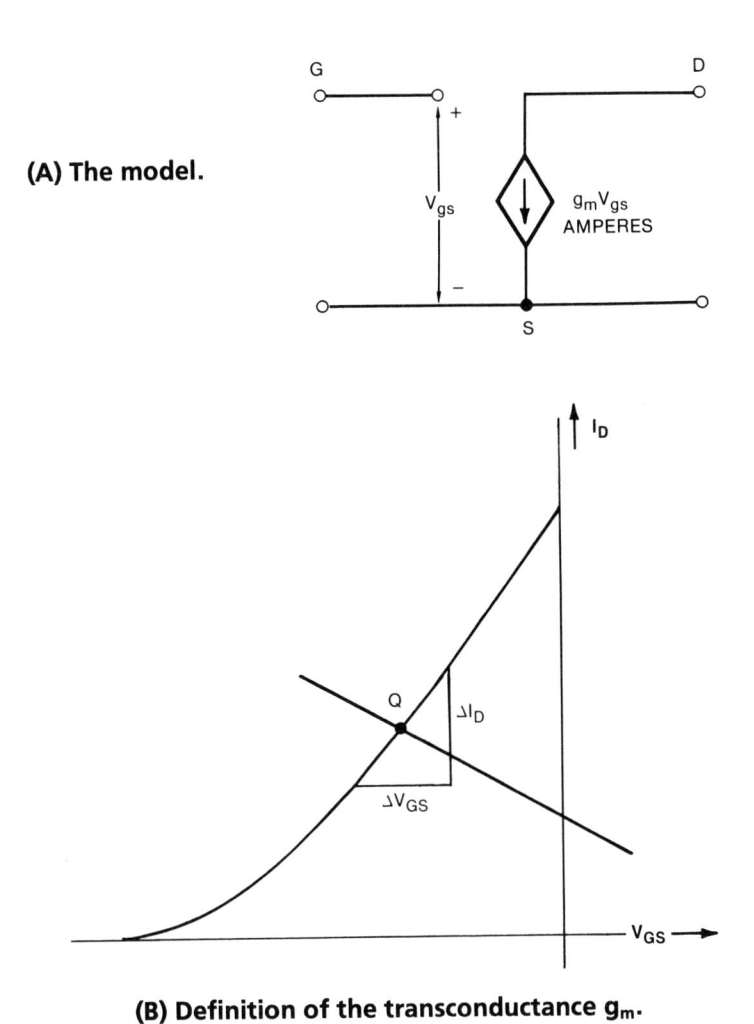

Fig. 6-24. Model of a FET for small-signal analysis.

If K is expressed in mA/V^2, which is usually the case, then the value of g_m obtained from Eq. (6-12) will be in *millisiemens (mS)*.

Example 6.10
Calculate the transconductance of the FET in each of the examples: 6.5, 6.6, and 6.7.

Solution

In Example 6.5, we had $K = 0.667$ mA/V^2, $V_T = -3$ V, and the value of V_{GS} at the Q point was found to be -1.02 V. Therefore, Eq. (6-12) gives

$$g_m = 2 \times 0.667 \times (-1.02 + 3)^2$$

$$= 5.23 \text{ mS}$$

In Example 6.6, $K = 0.8$ mA/V^2, $V_T = 0.5$ V. From the graph of Fig. 6-11, $V_{GS} = 2.2$ V at the Q point. Therefore, $g_m = 2.72$ mS.

In Example 6.7, $K = 0.625$ mA/V^2, $V_T = -4$ V. From the graph of Fig. 6-13, $V_{GS} = -1.6$ V. Therefore, $g_m = 3$ mS.

Drill Problem 6.18: Calculate the transconductance values of the FETs in Drill Problems 6.6, 6.7, 6.8, and 6.9.

It should be noted that the values of g_m for a FET are much smaller than the values of g_m for a BJT, and the gain of a FET amplifier will consequently be much smaller than that of a BJT amplifier.

When we are considering the small-signal analysis of a FET amplifier, we treat the circuit in the same manner as we did for the BJT amplifier. The capacitors are replaced by short circuits and the power supply is replaced by a short circuit so that the circuit assumes the form shown in Fig. 6-25. Replacing the FET by its model (from Fig. 6-24) and redrawing the circuit of Fig. 6-25, the circuit for small-signal analysis is as shown in Fig. 6-26. The resistances R_{G1} and R_{G2} are in parallel and combined into a single resistance R'_G. The resistances R_D and R_L are in parallel and can be combined into a single resistance R_{eq}, which leads to the circuit of Fig. 6-27.

Input Side of the Circuit

Using the voltage-divider relationship, the voltage V_{gs} is given by

$$V_{gs} = \frac{R'_G}{R_1 + R'_G} V_s \tag{6-13}$$

The input current supplied by the signal source is

$$I_s = \frac{V_s}{R_1 + R'_G} \tag{6-14}$$

Output Side of the Circuit

The current through R_{eq} is $g_m V_{gs}$ (going from the reference negative to reference positive, however). Therefore, the output voltage is

Electronics: Circuits and Systems 365

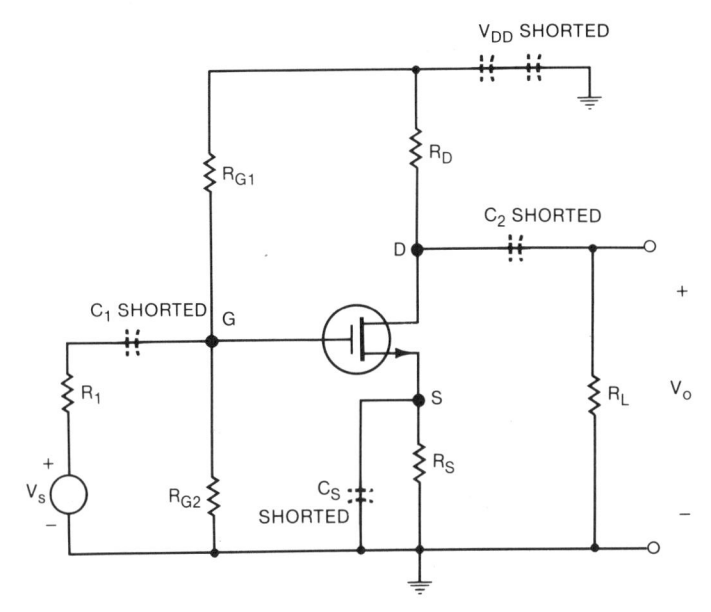

Fig. 6-25. Modification of the common-source amplifier circuit for ac (small-signal) analysis.

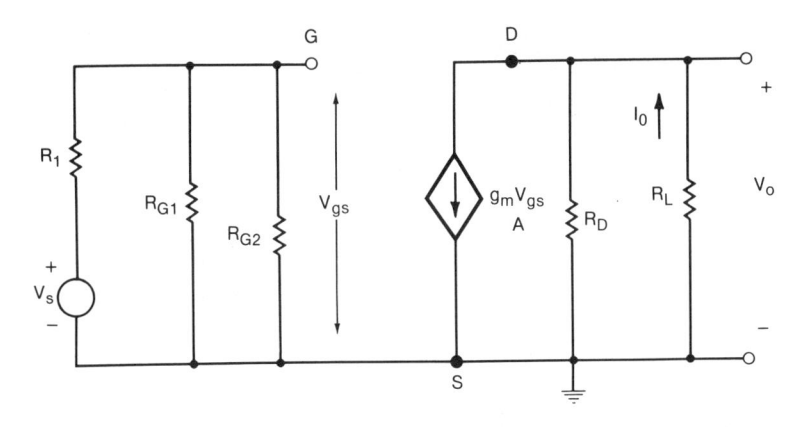

Fig. 6-26. Circuit for small-signal analysis of the common-source amplifier.

$$V_o = -g_m V_{gs} R_{eq}$$
$$= -(g_m R_{eq}) V_{gs} \tag{6-15}$$

The output current I_o can be found from V_o by using

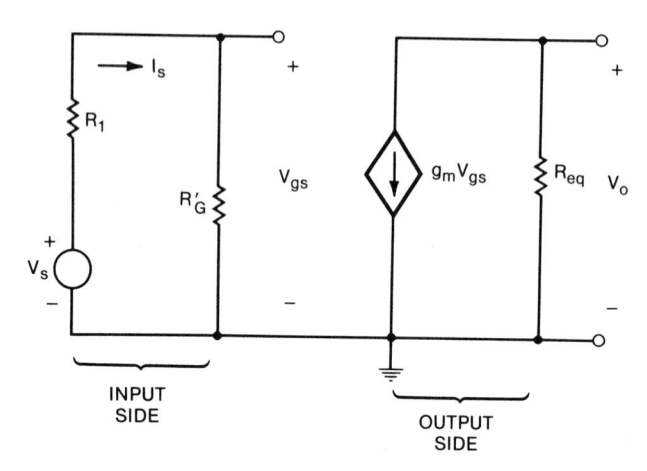

Fig. 6-27. Simplification of the small-signal equivalent circuit.

$$I_o = -\frac{V_o}{R_L}$$

It can also be found by using current division: the current $g_m V_{gs}$ splits between R_L and R_D (Fig. 6-28):

$$I_o = -g_m V_{gs} \frac{R_D}{R_L + R_D} \qquad (6\text{-}16)$$

Equations (6-13) through (6-16) form the basic equations needed for the small-signal analysis of the common-source FET amplifier. (The procedure and equations are exact analogs of the analysis of the common-emitter amplifier discussed in Chapter 5.)

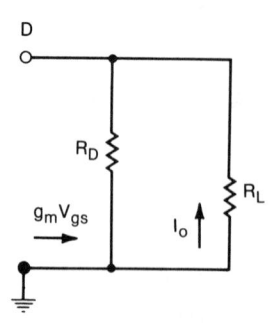

Fig. 6-28. Current division to obtain the output current I_o.

Electronics: Circuits and Systems

Example 6.11

A common-source FET amplifier is shown in Fig. 6-29. The transconductance of the FET is 8 mS. Find the voltage gain, current gain, and power gain of the amplifier.

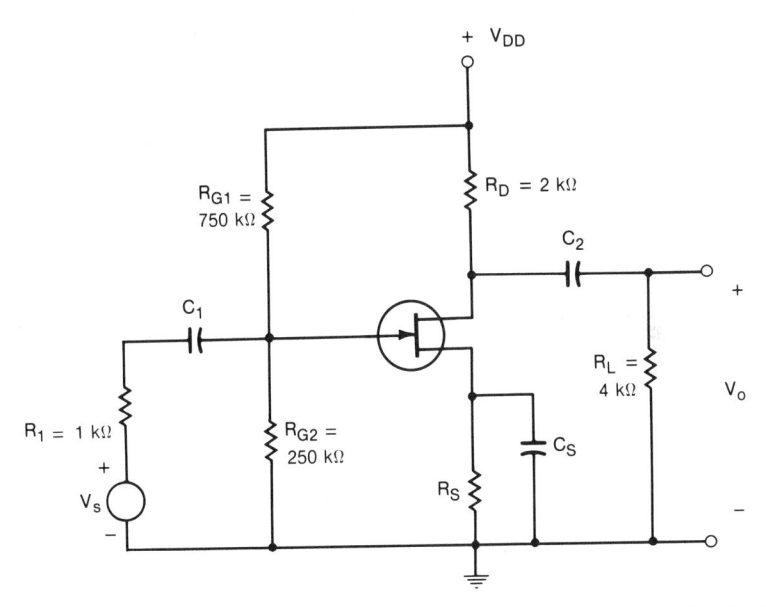

Fig. 6-29. Common-source amplifier circuit of Example 6.11.

Solution

The given amplifier circuit is modified to a form suitable for small-signal analysis as shown in Fig. 6-30.

Input side calculations:

$$V_{gs} = \frac{R'_G}{R_1 + R'_G} V_s \approx V_s$$

$$I_s = \frac{V_s}{R_1 + R'_G}$$

$$= 5.33 \times 10^{-3} V_s \text{ mA}$$

368   **FET Amplifiers**

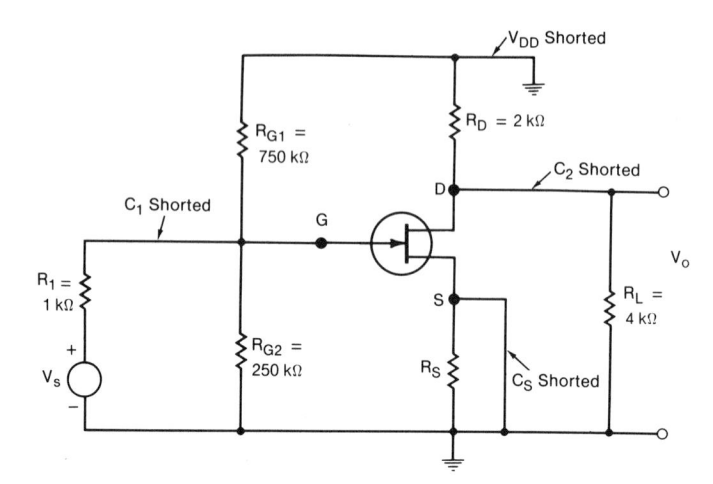

(A) Capacitors and power supply battery have been replaced by short circuits.

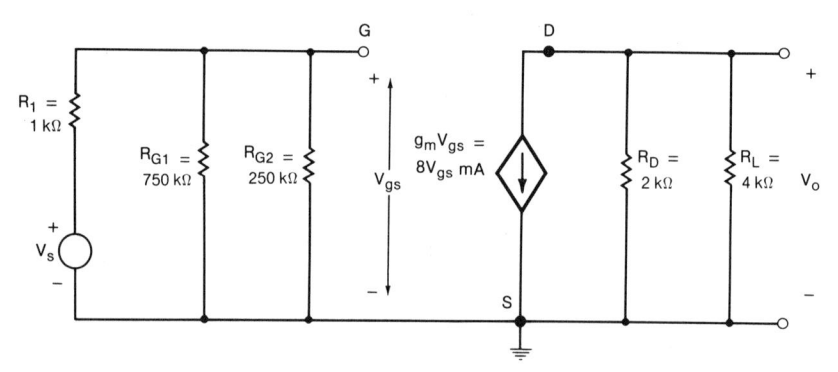

(B) Two ground points have been consolidated, and FET replaced by its model.

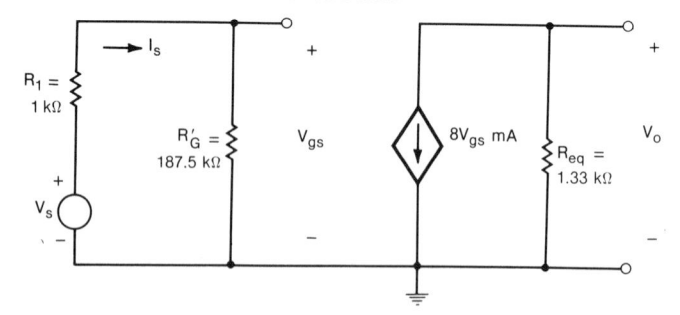

(C) Circuit has been reduced by combining parallel resistances.

Fig. 6-30. Conversion of the amplifier circuit to its small-signal equivalent circuit.

Electronics: Circuits and Systems

Output side calculations:

$$V_o = -g_m R_{eq} V_{gs}$$
$$= 8 \times 1.333 V_{gs}$$
$$= -10.7 \, V_{gs} \, V$$

$$I_o = -\frac{V_o}{R_L}$$
$$= -\frac{10.7}{4} V_{gs}$$
$$= 2.67 V_{gs} \, mA$$

Gains:

$$\text{voltage gain} = (V_o/V_s)$$
$$= \frac{-10.7 V_{gs}}{V_{gs}}$$
$$= -10.7 \text{ (or 20.6 dB)}$$

$$\text{current gain} = (I_o/I_s)$$
$$= \frac{2.67 V_{gs}}{5.33 \times 10^{-3} V_s}$$
$$= 500 \text{ (or 54 dB)}$$

$$\text{power gain} = (\text{voltage gain}) \times (\text{current gain})$$
$$= 5350 \text{ (or 37.3 dB)}$$

Drill Problem 6.19: The circuit of a FET amplifier is shown in Fig. 6-31. If the transconductance of the FET is 2 mS, calculate the three gains.

The complete analysis (dc and small signal) of a common-source FET amplifier is illustrated through the following example.

Example 6.12

Analyze the FET amplifier shown in Fig. 6-32 if $K = 0.3 \, mA/V^2$ and $V_T = -5$ V for the transistor.

370   **FET Amplifiers**

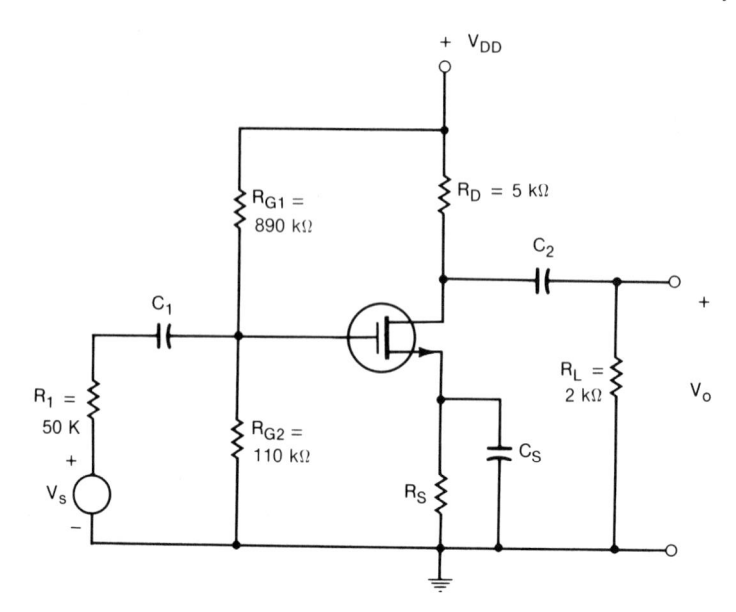

Fig. 6-31. Amplifier circuit of Drill Problem 6.19.

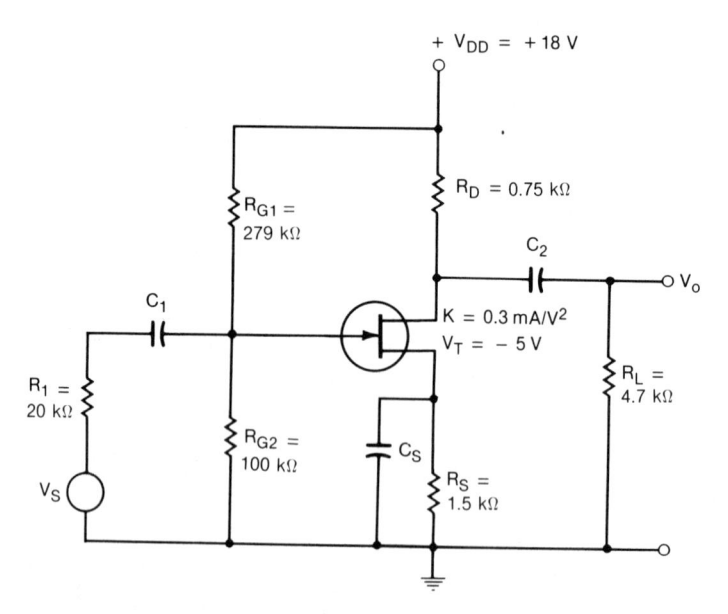

Fig. 6-32. Amplifier circuit of Example 6.12.

Electronics: Circuits and Systems

Solution

DC analysis:
The relevant portion of the circuit for dc analysis is shown in Fig. 6-33.

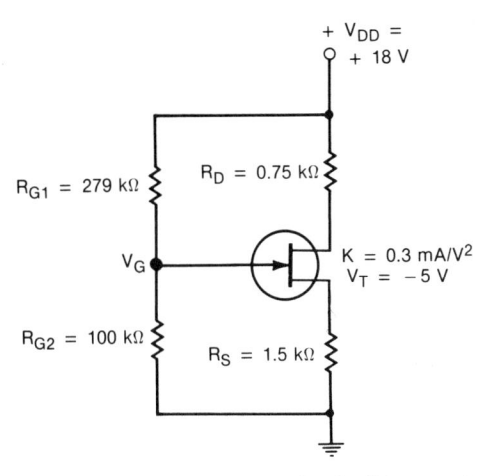

Fig. 6-33. Circuit for dc analysis (Example 6.12).

Use the voltage-divider formula to find the voltage at the gate V_G.

$$V_G = \frac{R_{G2}}{R_{G1} + R_{G2}} V_{DD}$$

$$= \frac{100}{100 + 279} \times 18$$

$$= 4.75 \text{ V}$$

Use the equation

$$I_D = K(V_{GS} - V_T)^2 \text{ mA}$$

$$= 0.3(V_{GS} + 5)^2 \text{ mA}$$

and plot the transfer characteristic (Fig. 6-34).
Draw the bias line by taking a point $V_{GS} = V_G = 4.75$ V on the horizontal axis and $I_D = (V_G/R_S) = 3.2$ mA on the vertical axis.
Read off I_D and V_{GS} at the Q point (intersection of the bias line and the transfer characteristic):

$$I_D = 4.1 \text{ mA}$$

$$V_{GS} = -1.25 \text{ V}$$

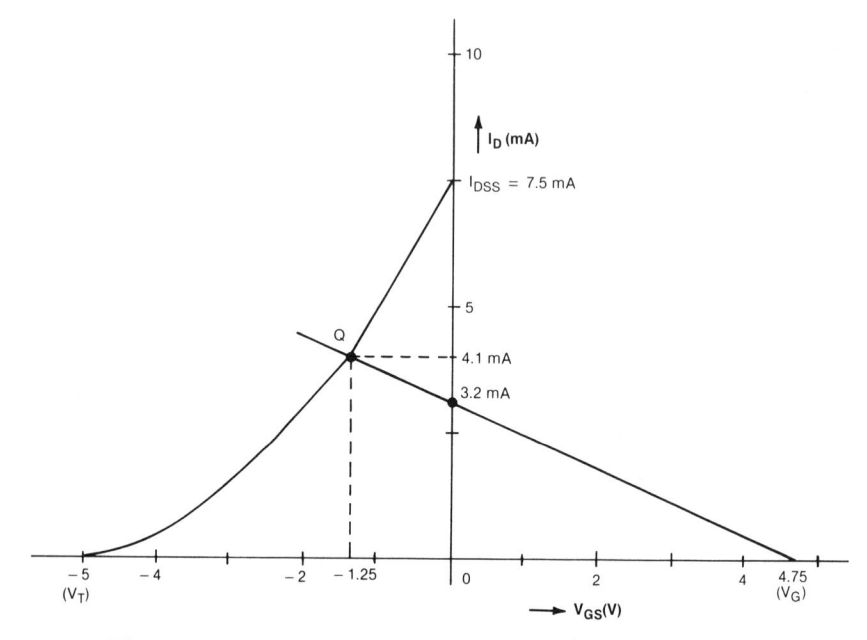

Fig. 6-34. Graphical determination of the Q point (Example 6.12).

Calculate the drain-to-source voltage V_{DS} by subtracting the drops in R_D and R_S from V_{DD}.

$$V_{DS} = V_{DD} - R_D I_D - R_S I_D$$
$$= 18 - 0.75 \times 4.1 - 1.5 \times 4.1)$$
$$= 8.78 \text{ V}$$

Therefore, Q point is at

$$I_D = 4.1 \text{ mA}$$
$$V_{GS} = -1.25 \text{ V}$$
$$V_{DS} = 8.78 \text{ V}$$

AC analysis:

Replace the coupling capacitors and the bypass capacitor by short circuits. Replace the battery by a short circuit. Consolidate the two grounds into one bottom line. Combine R_{G1} and R_{G2} in parallel into a single resistance R'_G.

Electronics: Circuits and Systems

$$R'_G = (R_{G1} \| R_{G2})$$
$$= 73.6 \text{ k}\Omega$$

Combine R_D and R_L into a single resistance, R_{eq}.

$$R_{eq} = (R_D \| R_L)$$
$$= 0.647 \text{ k}\Omega$$

Calculate the transconductance g_m using V_{GS} at the Q point.

$$g_m = 2K(V_{GS} - V_T)$$
$$= 2 \times 0.3(-1.25 + 5)$$
$$= 2.25 \text{ mS}$$

The small-signal equivalent circuit of the given amplifier is shown in Fig. 6-35.

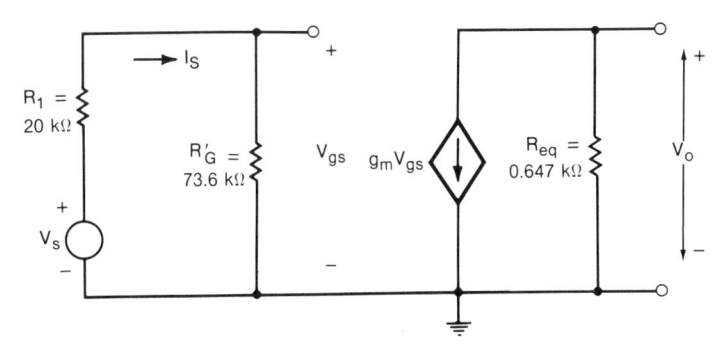

Fig. 6-35. Small-signal equivalent circuit of the amplifier of Example 6.12.

Input side calculations:

$$V_{gs} = \frac{R'_G}{R_1 + R'_G} V_s \text{ (voltage-divider formula)}$$
$$= (73.6/93.6)V_s$$
$$= 0.786 V_s$$

$$I_s = \frac{V_s}{R_1 + R'_G}$$

$$= 0.0107 V_s \text{ mA}$$

Output side calculations:

$$V_o = -g_m R_{eq} V_{gs}$$
$$= -1.14 V_{gs}$$
$$I_o = -\frac{V_o}{R_L}$$
$$= 0.244 V_{gs} \text{ mA}$$

Gain calculations:

$$V_o = -1.46 V_{gs}$$
$$= -1.46 \times 0.786 V_s$$
$$\text{voltage gain} = 1.15 \text{ (or } 1.21 \text{ dB)}$$
$$I_o = 0.311 V_{gs}$$
$$= 0.311 \times 0.786 V_s$$
$$= 0.244 V_s \text{ mA}$$

and (from input side calculations)

$$I_s = 0.0107 V_s \text{ mA}$$

$$\text{current gain} = (I_o/I_s)$$
$$= \frac{0.244 V_s}{0.0107 V_s}$$
$$= 22.8 \text{ (or } 27.2 \text{ dB)}$$

$$\text{power gain} = \text{voltage gain} \times \text{current gain}$$
$$= 26.2 \text{ (or } 14.2 \text{ dB)}$$

Drill Problem 6.20: Analyze the amplifier of Fig. 6-32 but change the parameters and component values to $K = 0.48$ mA/V^2, $V_T = +1.5$ V, $V_{DD} = 24$ V, $R_D = 4.7$ kΩ. Other components stay the same as before.

The input resistance of a FET itself is essentially infinite. Therefore, the input resistance of a common-source FET amplifier is R_G'. The input resistance of a FET amplifier can be made high by choosing R_G' high, whereas that of a BJT amplifier is limited by the value of $r_{b'e}$ of the transistor.

COMMON-DRAIN AMPLIFIER (SOURCE FOLLOWER)

When the drain of a FET is made common to the input side and the output side, the resulting amplifier configuration is a *common-drain amplifier*. Since the output is taken off the source terminal, the output *follows* the source voltage and the amplifier is also called the *source follower*. The *source follower is analogous to the BJT emitter follower*.

The circuit of a source follower is shown in Fig. 6-36. Note that the drain terminal is directly connected to the power supply V_{DD}. For *small signals*, therefore, the drain is at *ground* potential. Also, there is no bypass capacitor across the resistance R_s for reasons already explained in connection with the common-emitter amplifier in the last chapter.

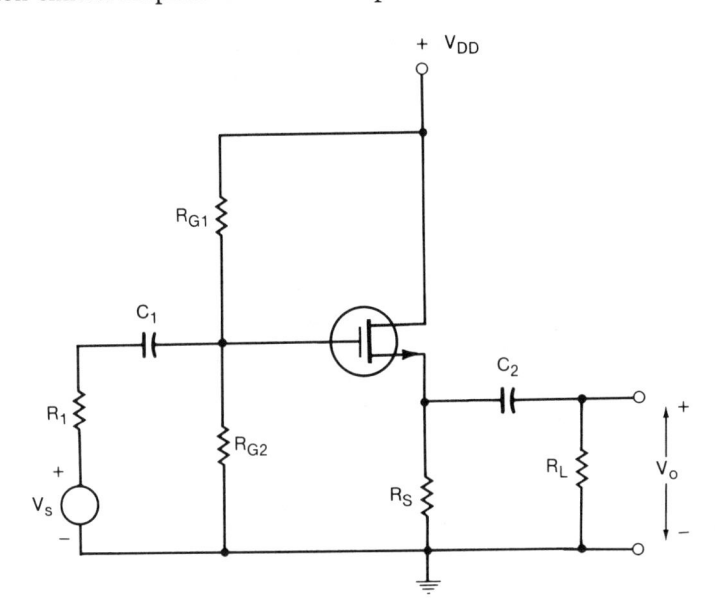

Fig. 6-36. The common-drain FET amplifier. (Source follower circuit.)

The analysis of a common-drain amplifier is quite similar to that of the common-source amplifier. But the primary point of interest in using a source follower is not its gain values but its input and output resistances. *The voltage gain of a source follower is always less than one*, but it has some current gain and power gain. When the value of the parallel combination of R_S and R_L is much greater than $(1/g_m)$, the voltage gain can be taken as approximately one.

The output resistance of the common-drain amplifier can be shown to be

$$R_{out} = R_S \mathbin\| (1/g_m) \qquad (6\text{-}17)$$

For example, if $R_S = 0.75$ kΩ and $g_m = 12$ mS, the output resistance is

$$R_{out} = [0.75 \mathbin\| (1/12)]$$

$$= \frac{0.75 \times 0.0833}{(0.75 + 0.0833)}$$

$$= 0.075 \text{ kΩ or } 75 \text{ Ω}$$

The output resistance of a source follower can be made quite low. The *input resistance* of a common-drain amplifier is the same as for the common-source amplifier: R'_G.

The source follower can be used as an interstage between other amplifier stages so as to provide impedance matching in the same manner as the emitter follower.

FET AS A VOLTAGE-CONTROLLED RESISTOR

It was seen at the beginning of this chapter that there is a *linear region* in the output characteristics of the FET, where the drain current I_D varies linearly with the drain-to-source voltage V_{DS}. The linear relationship between the two quantities is valid for values of V_{DS} *much less than* $(V_{GS} - V_T)$. Depending upon the particular FET, the linear region extends up to $[0.1\,(V_{GS} - V_T)]$ or somewhat higher. In the linear region, the FET acts as a resistance whose value can be controlled by the voltage V_{GS}; that is, it can be used as a *voltage-controlled variable resistor*. Also, the drain can be kept either positive or negative with respect to the source when using the FET in this mode, which gives it the *bilateral property* of an ordinary resistor.

The resistance R_{DS} of the FET in the linear region is given by

$$R_{DS} = \frac{1}{2K(V_{GS} - V_T)}$$

$$= \frac{1}{g_m}$$

This relationship shows explicitly the dependence of the resistance R_{DS} on V_{GS}. FETs for use as voltage-controlled resistors are commercially available with resistances at $V_{GS} = 0$ ranging from 20 Ω to 8 kΩ. They have a *dynamic range* of 10 to 1 (that is, their resistance can be varied by a factor of 1 to 10 by varying the voltage V_{GS}).

Of the many applications of a voltage-controlled resistor, one is their use in an *automatic gain control* circuit of a multistage amplifier. Consider the portion of an amplifier system shown in Fig. 6-37A. R_E is the resistance in the emitter leg of a BJT amplifier. It was pointed out in the last chapter that the presence of R_E decreases the gain of an amplifier: the larger the value of R_E the smaller the gain of the amplifier. If a variable resistance R_v is placed across a fixed resistance $R_{E'}$ then the gain depends upon the effective resistance in the emitter branch: $R'_E = (R_E \parallel R_v)$. As R_v increases, R'_E increases, and the gain decreases. The variable resistance R_v in an automatic gain control circuit (Fig. 6-37B) is provided by the drain and source terminals of a FET. The output voltage of the amplifier system controls the gate voltage of the FET. As the output voltage increases, the gate voltage increases, and the drain-to-source resistance (R_v) increases causing the gain to decrease. The capacitor C isolates the dc bias conditions of the BJT.

HANDLING PRECAUTIONS FOR MOSFETS

An integrated-circuit MOSFET package can be easily damaged by the static electricity generated by the person handling it when building an electronic circuit. The extremely thin layer of oxide used in the MOSFET can be punctured by the few kilovolts picked up by a person walking across a carpet or wearing static-prone clothing. It is, therefore, necessary to carefully observe a number of precautions in shipping and handling MOSFETs. They are shipped in conductive foams. When building a circuit, it is important to ground soldering irons, the workbench, and the wrists of the builder.

(A) Variable resistance in parallel with RE.

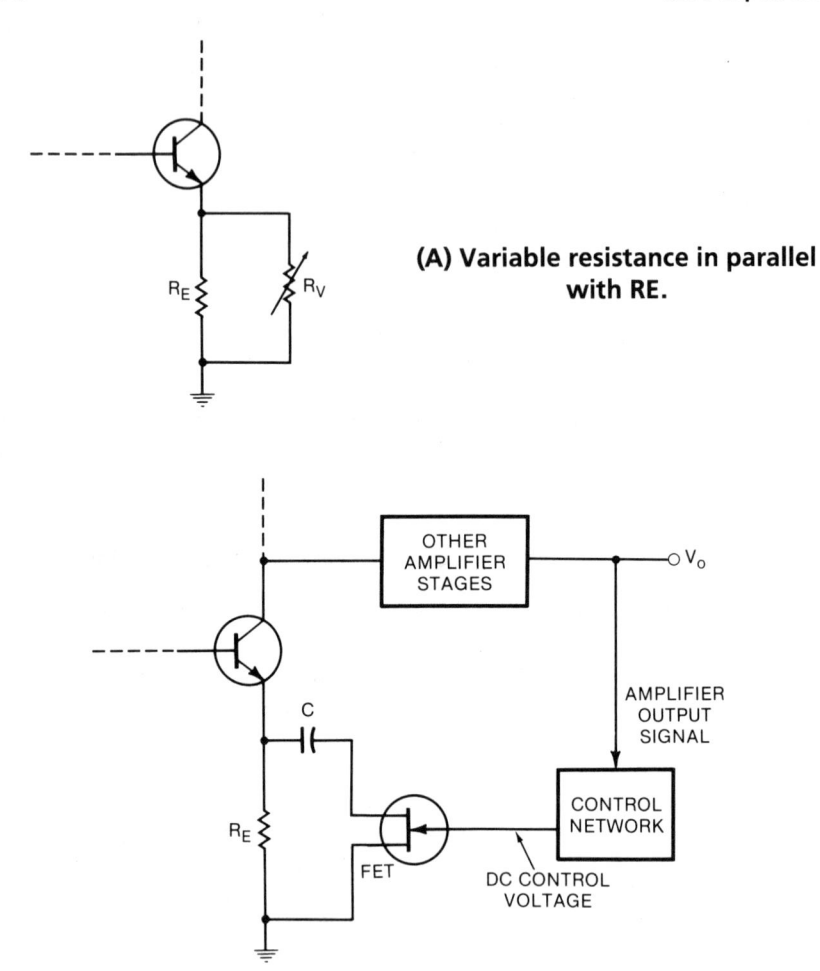

(B) FET provides R_V in the automatic gain control circuit.

Fig. 6-37. Use of a voltage-controlled resistor in an automatic gain control circuit.

SUMMARY SHEETS

PROPERTIES OF THE N-CHANNEL FET

Analogy Between FET and BJT

FET	BJT
n-channel FET	npn transistor
source terminal	emitter terminal
gate terminal	base terminal
drain terminal	collector terminal

Symbols of N-Channel FETs

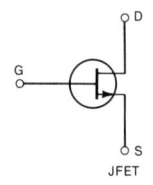

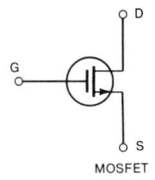

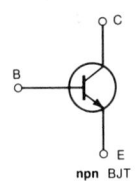

JFET

MOSFET

npn BJT

Output Characteristics

JFET and DEPLETION MOS

ENHANCEMENT MOS

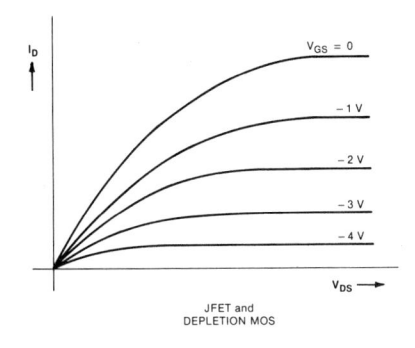

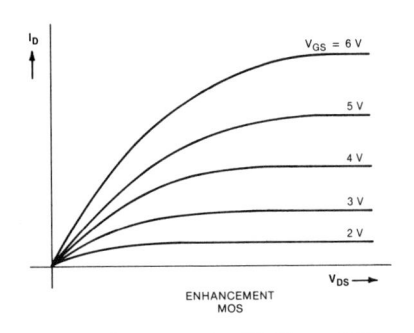

Features Common to All Three FETs

1. There are three regions in the output characteristics:

(a) *Linear region:* When V_{DS} is small, the drain current is proportional to the voltage V_{DS}. The FET can be used as a voltage-controlled resistor in this region.

(b) *Triode region:* The drain current increases with the voltage V_{DS} but not linearly.

(c) *Saturation region:* I_D remains constant and independent of V_{DS}. The FET acts as a voltage-controlled constant current source, with the value of the current being determined by V_{GS}. This is the region in which amplifiers are designed to operate.

2. The FET conducts only when V_{GS} is greater than a threshold voltage V_T. (The threshold voltage is also known as the pinch-off voltage V_P for JFET and depletion MOS.)

3. The FET operates in the saturation region when $V_{DS} \geq (V_{GS} - V_T)$.

4. The drain current increases as V_{GS} increases for a given value of V_{DS}. The relationship between I_D and V_{GS} is quadratic in the saturation region.

Differences Between the Three FETs

1. The threshold voltage $V_T > 0$ for the enhancement NMOS but < 0 for the JFET and depletion NMOS.

2. V_{GS} is always positive for the enhancement NMOS (for it to operate). V_{GS} is negative for n-channel JFETs. V_{GS} is usually negative for the depletion NMOS (and it may be made slightly positive).

Note the difference between n-channel JFET and npn transistors: the gate-to-source voltage is *negative* for the n-channel JFET amplifiers while the base-to-emitter voltage is positive for the npn amplifier.

TRANSFER CHARACTERISTICS

When a FET is operating in the *saturation* region, I_D and V_{GS} are related by the equation:

$$I_D = K(V_{GS} - V_T)^2$$

K is a constant, usually expressed in mA/V^2 so that I_D is in mA. (Note that in the saturation region, I_D is independent of V_{DS}.)

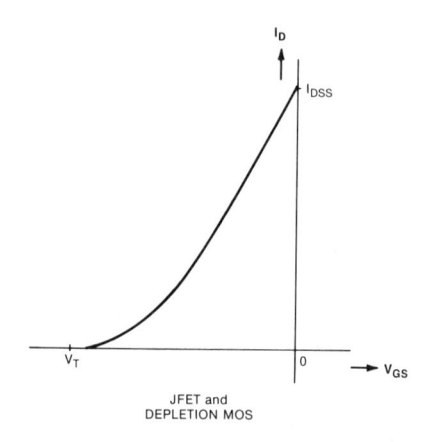

Evaluation of the Constant K

For the *JFET and depletion MOS*, the value of I_D when $V_{GS} = 0$, which is denoted by I_{DSS}, is available. Then K is calculated by using

$$K = I_{DSS}/V_T^2$$

The equation of the transfer characteristics is often written as

$$I_D = I_{DSS}\left(1 - \frac{V_{GS}}{V_T^2}\right)$$

for these two types.

For the *enhancement NMOS*, I_{DSS} has no meaning. The manufacturer specifies some value of V_{GS} (denoted $V_{GS(on)}$) and the corresponding value of I_D. This information is used to evaluate K. (Also, K can be measured by using the fact that $K = I_D$ when $V_{GS} = V_T + 1$.)

JFET and DEPLETION MOS

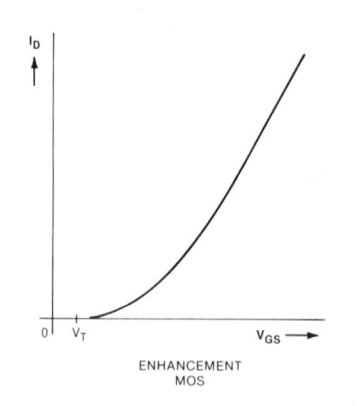

ENHANCEMENT MOS

The curves show the essential similarity between the three FET types. The curve for the enhancement type is obtained by shifting the curve to the right of the origin.

FET AMPLIFIER CONFIGURATIONS

1. Common-Source Amplifier: The source terminal is common to the input and output sides. The signal is fed to the gate, and the output is taken at the drain. This configuration is similar to the common-emitter BJT amplifier.

2. Common-Drain Amplifier: The drain terminal is common to the input and output sides. The signal is fed to the gate, and the output is taken at the source terminal. This configuration is similar to the common-collector BJT amplifier. It is referred to as the *source follower* also.

3. Common-Gate Amplifier: The gate terminal is common to the input and output sides. The signal is fed to the source terminal, and the output is taken at the drain. This configuration is similar to the common-base BJT amplifier.

Electronics: Circuits and Systems

BIASING OF A FET AMPLIFIER

Algebraic Determination of the Q Point

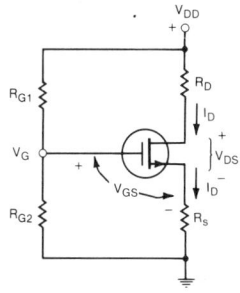

Use the voltage-divider formula to get V_G.

$$V_G = \frac{R_{G2}}{R_{G1} + R_{G2}} V_{DD}$$

Combine the two equations:

$$V_{GS} = V_G - R_S I_D$$

and

$$I_D = K(V_{GS} - V_T)^2$$

into a single quadratic equation for V_{GS} and solve for V_{GS}. Use the value of V_{GS} to calculate I_D and

$$V_{DS} = V_{DD} - (R_S + R_D)I_D$$

The Q point is defined by I_D and V_{DS}. Check to see if $V_{DS} \geq (V_{GS} - V_T)$; that is, the FET is in the saturation region, since saturation is assumed when using the previous equation for I_D in terms of V_{GS}.

Graphical Determination of the Q Point

Draw the transfer characteristic of the FET from the equation

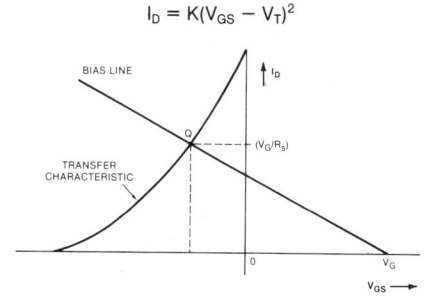

$$I_D = K(V_{GS} - V_T)^2$$

Draw the *bias line* by taking a point at V_G (obtained from the voltage-divider formula) on the V_{GS} axis and a point at (V_G/R_S) on the vertical axis. The point of intersection of the bias line and the transfer characteristic is the Q point.

PRACTICAL CONSIDERATIONS IN BIASING

The parameters of a FET vary widely even for the same type number. Two extreme transfer characteristics can be drawn for a given FET by using the extreme values of the parameters specified by the manufacturer.

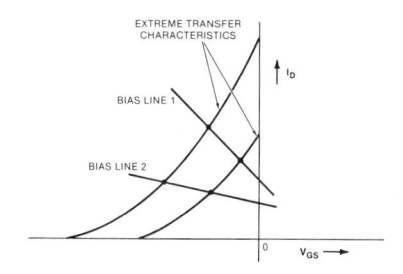

The slope of the bias line must be small to keep the variation of the value of I_D at the Q point small.

In the design of biasing circuits, assume a permissible range of I_D for the quiescent condition. Join these two points and extend to form the bias line.

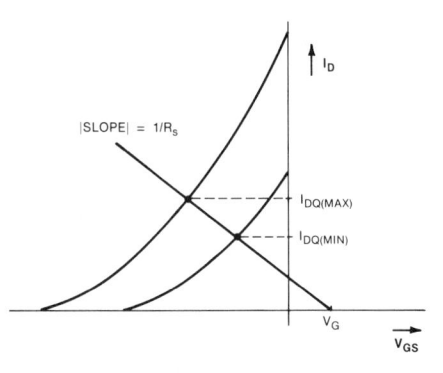

$R_S = |1/\text{slope of the bias line}|$

$V_G = $ point of intersection of the bias line with the V_{GS} axis.

Use the value of V_{DD} and V_G to calculate the

ratio (R_{G1}/R_{G2}). A value of V_{DS} at the Q point is needed for the determination of R_D.

COMMON-SOURCE FET AMPLIFIER

Small-Signal Model of a FET

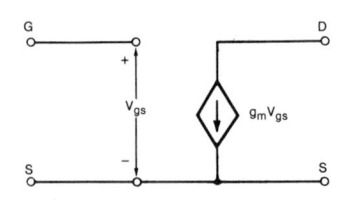

g_m = transconductance.

$$g_m = 2K(V_{GS} - V_T) \quad \text{mS}$$

Amplifier Circuit

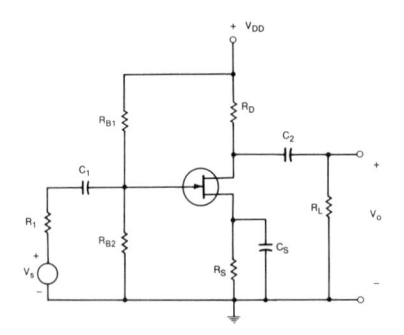

Small-Signal Equivalent Circuit

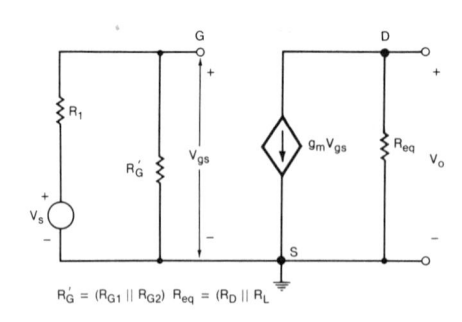

$R'_G = (R_{G1} \| R_{G2}) \quad R_{eq} = (R_D \| R_L)$

$R'_G = (R_{G1} \| R_{G2}) \qquad R_{eq} = (R_D \| R_L)$

Analysis

Input side:

$$V_{gs} = \frac{R'_G}{R'_G + R_1}$$

and

$$I_s = \frac{V_s}{R'_G + R_1}$$

Output side:

$$V_o = -g_m R_{eq} V_{gs}$$

and

$$I_o = -(V_o/R_L) \text{ or } -g_m V_{gs} \frac{R_L}{R_L + R_D}$$

Calculate the voltage gain and current gain from the four previous equations.

Input resistance: $R_{in} = R'_G$
Output resistance: $R_o = R_D$

The common-source FET amplifier does not have as high a voltage gain as the common-emitter BJT amplifier since g_m is much smaller for an FET (a few mS) than for a BJT (tens of mS).

COMMON-DRAIN AMPLIFIER (SOURCE FOLLOWER)

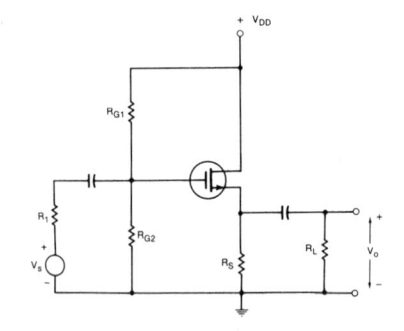

Note that the drain terminal is directly connected to the power supply.

The output resistance is the most important feature of this connection.

$$R_o = R_S \| (1/g_m)$$

FET AS A VOLTAGE-CONTROLLED RESISTOR

When $V_{DS} \ll (V_{GS} - V_T)$, I_D is proportional to V_{DS}. The FET can be used as a linear

Electronics: Circuits and Systems

resistor in the range $V_{DS} = 0$ to about $[0.1 \, (V_{GS} - V_T)]$. The value of the resistance depends upon V_{GS}—hence it is a voltage-controlled resistor. In this range,

$$R_{DS} = (1/g_m) = \frac{1}{2K(V_{GS} - V_T)}$$

Commercially available devices have R_{DS} between 20 Ω and 8 kΩ (at $V_{GS} = 0$) and have a dynamic range of 10 to 1.

By using the FET in parallel with the resistor R_E in a common-emitter BJT amplifier, the amount of negative feedback can be varied. Such a circuit leads to automatic gain control.

PRECAUTIONS IN HANDLING MOSFETS

The thin oxide layer can easily puncture due to static electricity developed in the person assembling the circuit. Proper grounding precautions must be observed in order not to damage the MOSFET package.

ANSWERS TO DRILL PROBLEMS

6.1 $I_D = 0.6(V_{GS} + 8)^2$ mA. See Fig. 6-DP1.

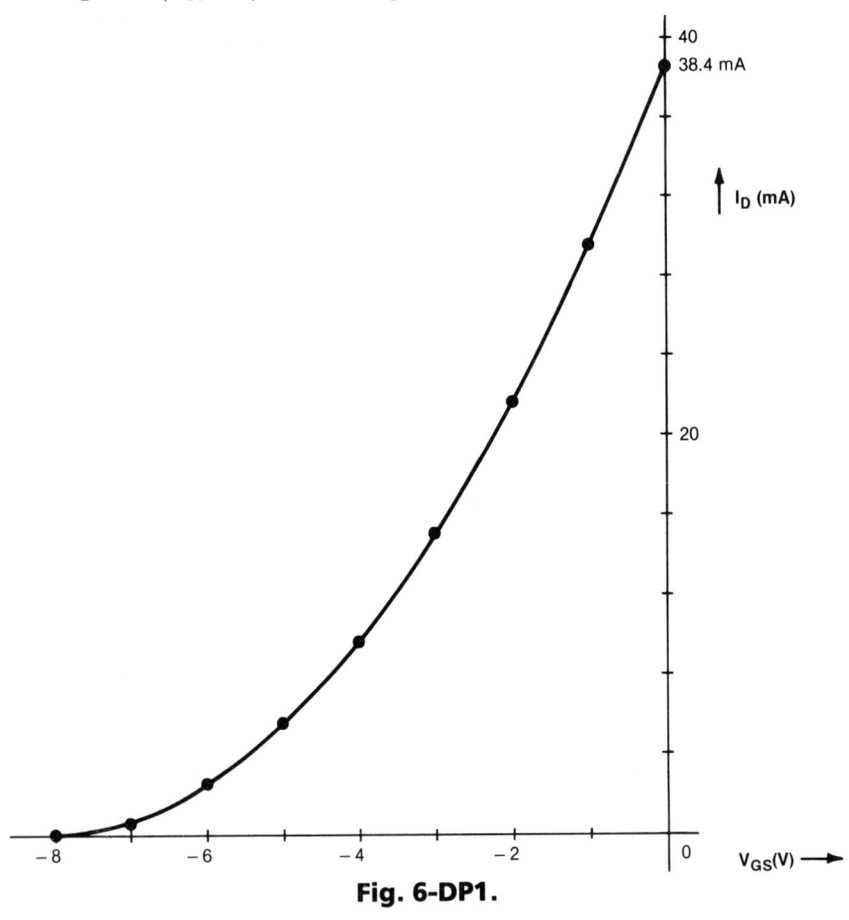

Fig. 6-DP1.

6.2 $I_D = 0.494(V_{GS} - 0.5)^2$ mA. See Fig. 6-DP2.
6.3 $I_D = 0.15(V_{GS} + 10)^2$ mA. (a) 18.2 mA; (b) -2.22 V.
6.4 $I_D = 0.75(V_{GS} + 4)^2$ mA.
6.5 $I_D = 0.977(V_{GS} - 1.8)^2$ mA. See Fig. 6-DP5.
6.6 $V_G = 2.91$ V; $6V_{GS}^2 - 17V_{GS} + 10.59 = 0$. $V_{GS} = 1.91$ V. Q point: $I_D = 0.672$ mA, $V_{DS} = 13.3$ V.
6.7 $V_G = 4.36$ V. $6V_{GS}^2 - 17V_{GS} + 9.14 = 0$. $V_{GS} = 2.11$ V. Q point: $I_D = 1.49$ mA. $V_{DS} = 18.1$ V.
6.8 $V_G = 3$ V. $(V_G/R_S) = 3.75$ mA. See Fig. 6-DP8. Q point: $I_D = 6.5$ mA, $V_{DS} = 3.3$ V, $V_{GS} = -2.1$ V.

Electronics: Circuits and Systems

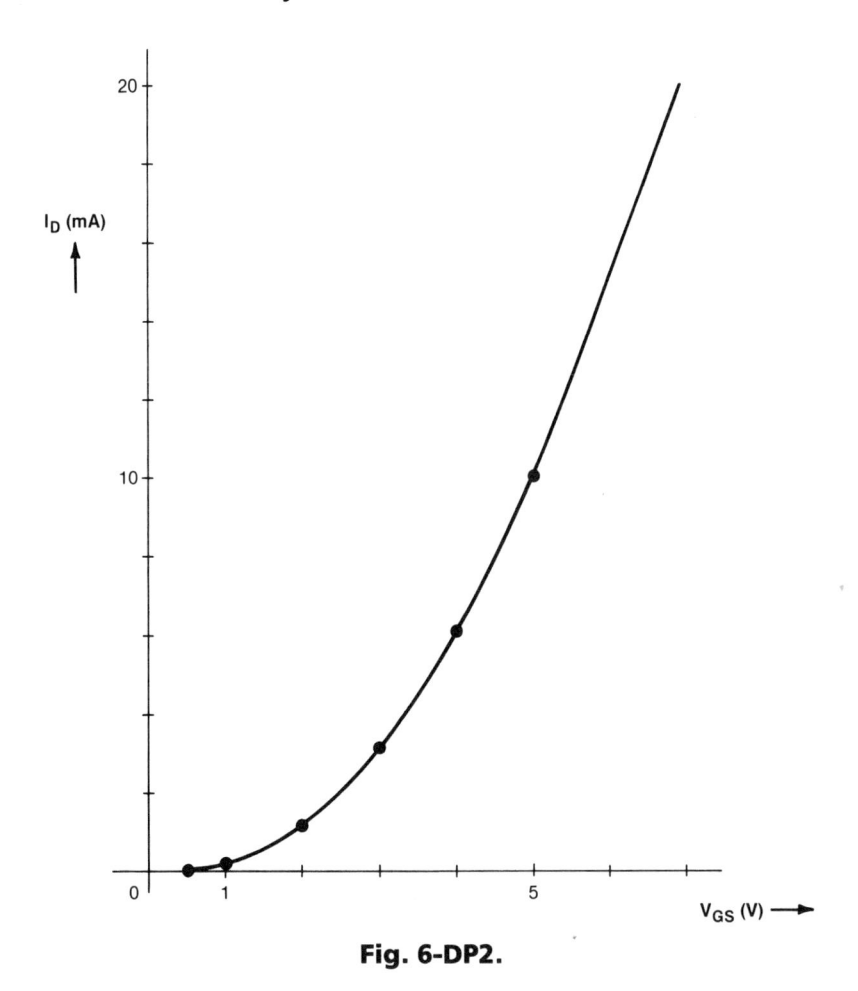

Fig. 6-DP2.

6.9 $V_G = 4$ V. $(V_G/R_S) = 2$ mA. See Fig. 6-DP9. Q point: $I_D = 3.4$ mA, $V_{DS} = 5.8$ V, $V_{GS} = -2.8$ V.

6.10 V_G becomes equal to $V_{DD} = 16$ V. $(V_G/R_S) = 8$ mA. See Fig. 6-DP9 (upper bias line). I_D is 8.5 mA at the Q point but this is impossible since it will make V_{DS} negative. What actually happens is that the FET is not in the saturation region and that it is operating with a *large positive* value of V_{GS}, which makes its behavior unpredictable.

6.11 V_G becomes zero. A graphical procedure leads to $V_{GS} = -3.7$ V at the Q point, $I_D = 1.8$ mA, and $V_{DS} = 10.6$ V. V_{DS} is found to be greater than $(V_{GS} - V_T)$, which puts the FET in the saturation region. At least from the Q point consideration, the circuit will perform as an amplifier.

6.12 The situation here is the same as in Drill Problem 6.11.

Fig. 6-DP5.

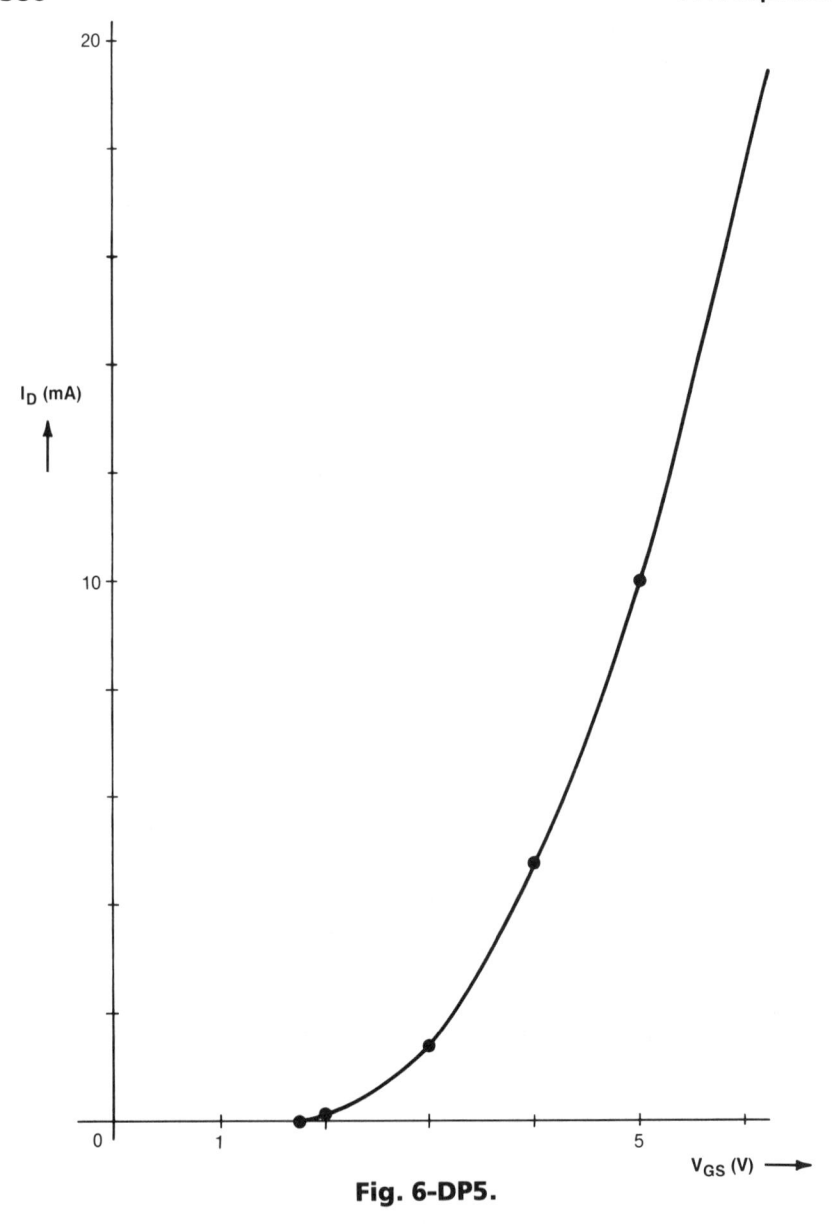

6.13 Since there is no gate current, $V_G = V_{DD} = 16$ V and we have the same situation as in Drill Problem 6.10.

6.14 $V_{GS} = -2.5 I_D$. $V_{GS}^2 + 11 V_{GS} + 25 = 0$. $V_{GS} = -3.2$ V. Q point: $I_{DS} = 1.28$ mA, $V_{DS} = 6.68$ V.

6.15 $V_G = 0.5$ V from Fig. 6-DP15 (and the appropriate bias line therein) and

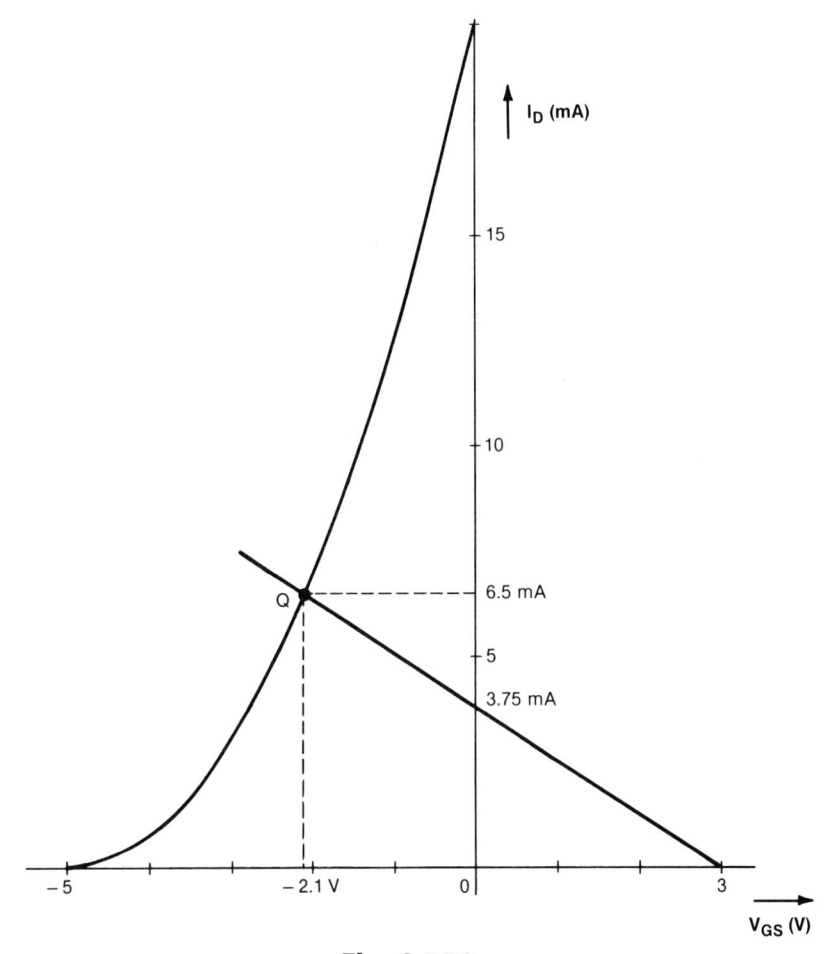

Fig. 6-DP8.

$(V_G/R_S) = 0.9$ mA. $R_S = 0.556$ kΩ. $R_{G1} = 35$ R_{G2}. If V_{DS} is taken as 9 V, $R_D = 2.02$ k.

6.16 Refer to Fig. 6-DP15 and the appropriate bias line in it. I_D at the Q point lies between 0.75 mA and 1.5 mA.

6.17 $V_G = 6.33$ V. $(V_G/R_S) = 3.16$ mA. Refer to Fig. 6-DP15 and the appropriate bias line therein. The two extreme Q points are $I_D = 3.5$ mA, $V_{DS} = 14.4$ V; and $I_D = 4.2$ mA, $V_{DS} = 12.4$ V. The variation is significantly less than in Example 6.9.

6.18 3.28 mS, 4.88 mS, 4.35 mS, 2.13 mS.

6.19 $R'_G = 97.9$ kΩ. $R_{eq} = 1.43$ kΩ. $V_{gs} = 0.662 V_s$. Voltage gain = −1.89 (or 5.54 dB). $I_o = (1.89/2)V_s$. Current gain = 140 (or 42.9 dB). Power gain = 265 (or 24.2 dB).

388   **FET Amplifiers**

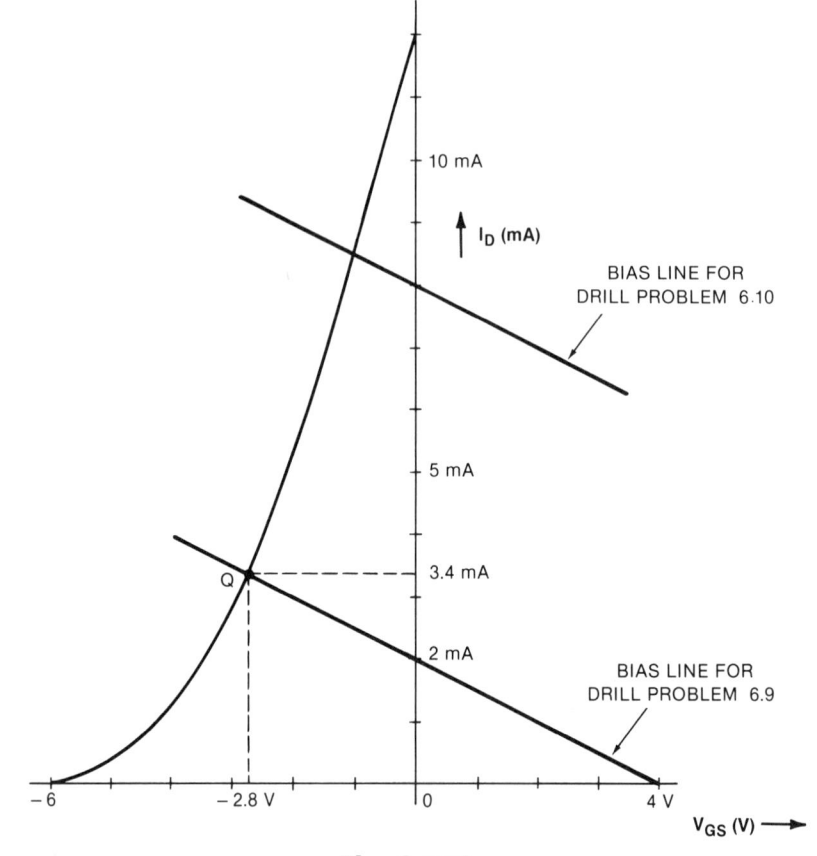

Fig. 6-DP9.

6.20 $V_G = 6.33$ V. The Q point is found to be at $V_{GS} = 3.49$ V. $g_m = 1.91$ mS. $R'_G = 73.6$ kΩ. $R_{eq} = 2.35$ kΩ. $V_{gs} = 0.786 V_s$. Voltage gain = -3.53 (or 11 dB). $I_o = (3.53/4.7)V_s$. Current gain = 70.3 (or 36.9 dB). Power gain = 248 (or 24 dB).

Electronics: Circuits and Systems **389**

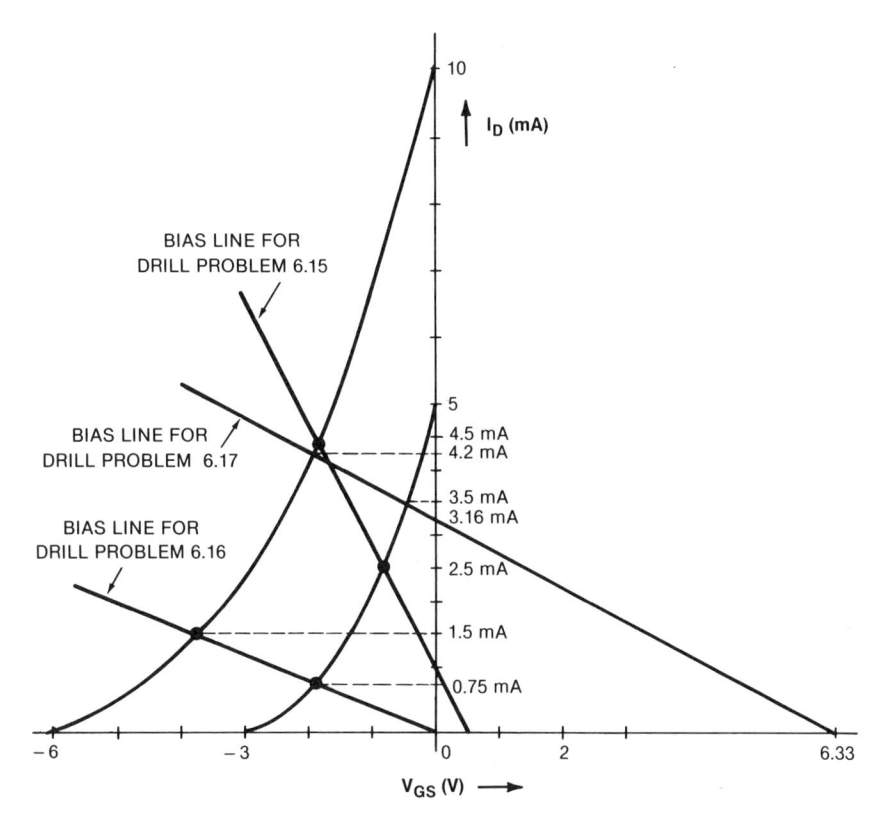

Fig. 6-DP15.

PROBLEMS

Principles and Properties of the FET

6.1 In each of the following cases, determine the range of values of V_{DS} for which the FET operates in the saturation region. (a) $V_T = -6$ V, $V_{GS} = -2.5$ V. (b) $V_T = 1.5$ V, $V_{GS} = 4.5$ V. (c) $V_T = 1.5$ V, $V_{GS} = 6$ V. (d) $V_T = -4$ V, $V_{GS} = 0$.

6.2 In each of the following cases, determine the range of values of V_{GS} (assuming V_{DS} to be fixed) in which the FET operates in the saturation region. (a) $V_T = -6$ V, $V_{DS} = 6$ V. (b) $V_T = 1.5$ V, $V_{DS} = 4$ V. (c) $V_T = -6$ V, $V_{DS} = 10$ V.

Transfer Characteristics of the FET

6.3 Plot the transfer characteristics of a FET with a threshold voltage of $+1.5$ V and $K = 0.75$ mA/V^2. Calculate the value of V_{GS} required to produce a drain current of 5 mA. Calculate the drain current at one half and twice the value of V_{GS} calculated previously.

6.4 A FET with a threshold voltage of -7.5 V is known to require $V_{GS} = -5$ V to produce $I_D = 8$ mA. Obtain the equation of the transfer characteristic. Calculate the value of V_{GS} needed to produce a drain current of 4 mA. Calculate the drain current when $V_{GS} = 0$.

6.5 A depletion NMOS has a threshold voltage of -7.5 V. The drain current is 10 mA when $V_{GS} = (V_T/2)$. Calculate the value of V_{GS} needed to produce a drain current of 8 mA. Calculate the drain current when $V_{GS} = -2.5$ V.

6.6 An enhancement NMOS has the following specifications. When $V_{GS} = 6$ V, $I_D = 12$ mA; when $V_{GS} = 3$ V, $I_D = 0.75$ mA. Obtain the equation of the transfer characteristic of the FET.

6.7 The value of the drain saturation current of a FET is 14 mA. It is also known that when $V_{GS} = -4$ V, $I_D = 8$ mA. Obtain the equation of the transfer characteristic.

6.8 The symbol of a *p-channel* JFET is shown in Fig. 6-P8 with the direction of the drain current as indicated. The *equation* of the transfer characteristic for the p-channel device is the same as for the n-channel device with the following constraints: the threshold voltage is positive and the device is usually operated with V_{GS} *less than* V_T, and V_{GS} is usually confined to positive values. Plot the transfer characteristic of a p-channel transistor with a threshold voltage of 8 V and $I_{DSS} = 16$ mA.

6.9 An enhancement *PMOS* has a *negative* threshold voltage V_T and is usually operated with V_{GS} less than the threshold voltage. (V_{GS} is confined to negative values.) Plot the transfer characteristic of an enhancement PMOS with the following specifications. $V_T = -0.5$ V. At $V_{GS} = -5$ V, I_D is found to be 8 mA.

Electronics: Circuits and Systems 391

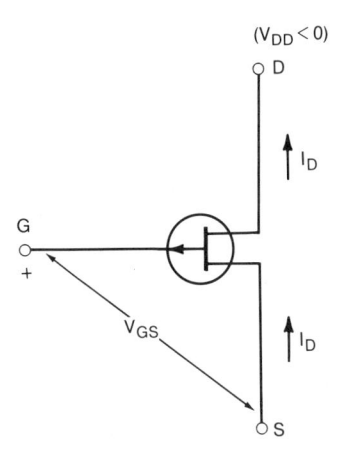

```
                    (V_DD < 0)
                       ○ D
                       
                       ↑ I_D
                       
    G                  
    ○────────┤├        
    +       ╱          
         V_GS          ↑ I_D
                       
                       ○ S
```

Fig. 6-P8.

Biasing of the FET Amplifier

6.10 The components and parameters of a biasing circuit (such as the one in Fig. 6-7) are $V_{DD} = 17.5$ V, $R_{G1} = 890$ kΩ, $R_{G2} = 100$ kΩ, $R_D = 4$ kΩ, $R_S = 3$ kΩ, $I_{DSS} = 15$ mA, and $V_T = -6$ V. Determine the Q point by using the *algebraic* procedure.

6.11 Repeat Problem 6.10 if the FET parameters are changed to $K = 0.333$ mA/V^2 and $V_T = +1$ V, with all the components remaining at the same value as before.

6.12 A FET with $K = 0.75$ mA/V^2 and $V_T = -6$ V is required to operate with the Q point at $I_D = 12$ mA and $V_{DS} = 5$ V. If, in the biasing circuit, $V_{DD} = 15$ V, $R_{G1} = 4 R_{G2}$, and $R_{G2} = 85$ kΩ, determine the values of the resistors R_D and R_S by using the *algebraic* procedure.

6.13 Repeat Problem 6.10 by using the graphical procedure.

6.14 Repeat Problem 6.11 by using the graphical procedure.

6.15 Repeat Problem 6.12 by using the graphical procedure.

6.16 Draw the biasing circuit of a *p-channel* FET that corresponds to the one in Fig. 6-7 for an n-channel FET. Calculate the Q point for the following component and parameter values. $V_T = -1$ V. $K = 4$ mA/V^2. $R_{G1} = 4$ MΩ. $R_{G2} = 1$ MΩ. $R_D = 0.5$ kΩ. $R_S = 0.5$ kΩ. $|V_{DD}| = 20$ V. Use the algebraic procedure.

6.17 Repeat Problem 6.16 by using the graphical procedure.

6.18 Repeat Problem 6.17 if the FET is replaced by one with $V_T = 8$ V and $K = 0.25$ mA/V^2.

Practical Considerations in Biasing a FET

6.19 A *depletion* NMOS has the following specifications. $|V_T| = 4$ V (min) and 8 V (max). $I_{DSS} = 9$ mA (min) and 16 mA (max). With a power supply

$V_{DD} = 20$ V, the drain current at the Q point is to be between 8 and 11 mA. Design a suitable biasing circuit. Assume $V_{DS} = 10$ V at the Q point.

6.20 An *enhancement* NMOS has the following specifications. $V_T = 0.5$ V (min) and 2 V (max). K = 0.5 mA/V^2 (min) and 1.5 mA/V^2 (max). (Match the *minimum* V_T with the *maximum* K, and the *maximum* V_T with the *minimum* K to get the two extreme transfer characteristics.) Using a power supply of 24 V, design a biasing circuit so that the drain current at the Q point is between 4 and 7 mA. Assume $V_{DS} = 10$ V at the Q point.

6.21 If, in the biasing circuit shown in Fig. 6-33, the device specifications are given as $I_{DSS} = 6$ mA (min) and 10 mA (max), and $|V_T| = 4$ V (min) and 8 V (max). Determine the values of I_D, V_{GS} and V_{DS} at the two extreme Q points.

6.22 Suppose a break develops in the resistor R_{G1} of the previous problem (so as to cause an open circuit in that branch). Recalculate the Q points.

Common-Source FET Amplifier

6.23 In the FET amplifier circuit of Fig. 6-23, assume the following values. $R_1 = 10$ kΩ. $R_{G1} = 800$ kΩ. $R_{G2} = 150$ kΩ. $g_m = 4$ mS. $R_D = 1.2$ kΩ. $R_L = 4.8$ kΩ. Determine the values of the three gains and express them in decibels.

6.24 A signal source V_s and a resistor $R_1 = 15$ kΩ are connected (through a coupling capacitor) to the input of the circuit of Fig. 6-8. A load resistor of $R_L = 1.2$ kΩ is connected (by means of a coupling capacitor) to the drain terminal. Assume the source resistor R_S is bypassed by means of a suitable capacitor. Determine the values of the three gains and express them in decibels.

6.25 Suppose the signal source, R_1, and R_L of the previous problem are used in conjunction with the circuit of Fig. 6-10 (with capacitors as needed). Calculate the three gains and express them in decibels.

6.26 The actual FET used in the circuit of Fig. 6-12 has the specifications: $|V_T| = 3$ V (min) and 5 V (max), $I_{DSS} = 6$ mA (min) and 12 mA (max). The signal source, R_1, and R_L of the previous two problems are connected to this circuit (along with capacitors as needed). Calculate the maximum and minimum values of each of the three gains of the amplifier.

6.27 Suppose the manufacturer's specifications of the FET in the amplifier of Fig. 6-32 are $|V_T| = 4$ V (min) and 9 V (max), $I_{DSS} = 6.8$ mA (min) and 11.5 mA (max). Calculate the minimum and maximum values of each of the three gains of the amplifier.

CHAPTER 7

FREQUENCY RESPONSE AND PULSE RESPONSE OF AMPLIFIERS

Any signal that is of sufficient interest to be passed through an amplifier is seldom made up of a single frequency but consists of a range of frequencies. The range of frequencies may extend from a low frequency (which may be just a few hertz) to a high frequency (which may be hundreds of megahertz) depending upon the source of the signal. If the output of an amplifier is to be a faithful reproduction of the input signal, it is necessary for the voltage gain of the amplifier to remain constant throughout the range of frequencies present in the input signal. Any variation in the gain will result in either an undesirable emphasis of certain frequencies or an undesirable attenuation of certain frequencies.

The gain of an amplifier, however, does not remain constant at all frequencies. There is a range of frequencies, called the *midband*, over which the gain of the amplifier is reasonably constant; but at low as well as at high frequencies, the gain decreases. The range of frequencies over which the gain remains essentially constant is called the *bandwidth* of the amplifier, and the two boundaries of the midband are called the *lower and upper cutoff frequencies*. The response of an amplifier to signals of different frequencies is referred to as *frequency response*.

In a number of applications, the signal applied to an amplifier is not a continuous signal, but a pulse or a series of pulses. When a pulse is applied to an amplifier, the output should be an acceptable replica of the input. The response of an amplifier to an input pulse is referred to as its *pulse response*. The pulse response of an amplifier may result in the rounding off of sharp corners of an input pulse or a sag (that is reduction in the voltage level) even though the input pulse has a constant amplitude.

The frequency response and the pulse response of an amplifier are closely related to each other since they both depend upon the behavior of

RC circuits, as will be seen in the following sections. We will first concentrate on the frequency response of an amplifier.

VARIATION OF THE GAIN OF AN AMPLIFIER WITH FREQUENCY

The amplifier configuration (the common-emitter BJT and the common-source FET amplifier) studied in the last two chapters is called an *RC-coupled amplifier*, since the coupling at the input and the output sides is through resistors and capacitors. Since the RC-coupled amplifier is one of the most commonly used configurations, its frequency response will be studied in detail.

Frequency Response Curve of an Amplifier

A graph showing the variation of the magnitude of the voltage gain (on the vertical axis) as a function of the frequency (on the horizontal axis) is referred to as the *frequency response curve* of an amplifier. The range of frequencies of interest in an amplifier may vary from a few hertz to several megahertz. A logarithmic scale of frequency is used in the graph in order to accommodate such a wide range of frequencies. A more important reason for using the logarithmic scale is that if the *magnitude of the voltage gain in decibels is plotted as a function of the logarithm of the frequency*, then the frequency response curve possesses certain standard properties that are extremely useful in the study of the frequency response of amplifiers. (This result was first established by Bode and his name is usually associated with such plots.) The frequency response curve of an RC-coupled amplifier is, therefore, drawn by using the *decibel value of the voltage gain* on the vertical axis and using the *logarithmic scale for the frequency*, as indicated in Fig. 7-1.

The gain of the amplifier in the midband is called its *midband gain*. The voltage gains calculated for the various amplifiers in the last two chapters *were* midband gains. The gain decreases on either side of the midband. Two frequencies, the *lower cutoff frequency* f_1 and the *upper cutoff frequency* f_2, are selected so that at *either frequency, the gain has dropped by 3 dB from the midband value.* The two cutoff frequencies are taken as the boundaries of the midband, and the useful bandwidth of the amplifier is defined as

$$\text{bandwidth} = (f_2 - f_1)$$

(The cutoff frequencies are also called *half-power frequencies* and -3 dB *frequencies.*)

For a properly designed single-stage RC-coupled amplifier, the frequency response curve becomes a straight line of slope -20 dB/decade on either side

Electronics: Circuits and Systems

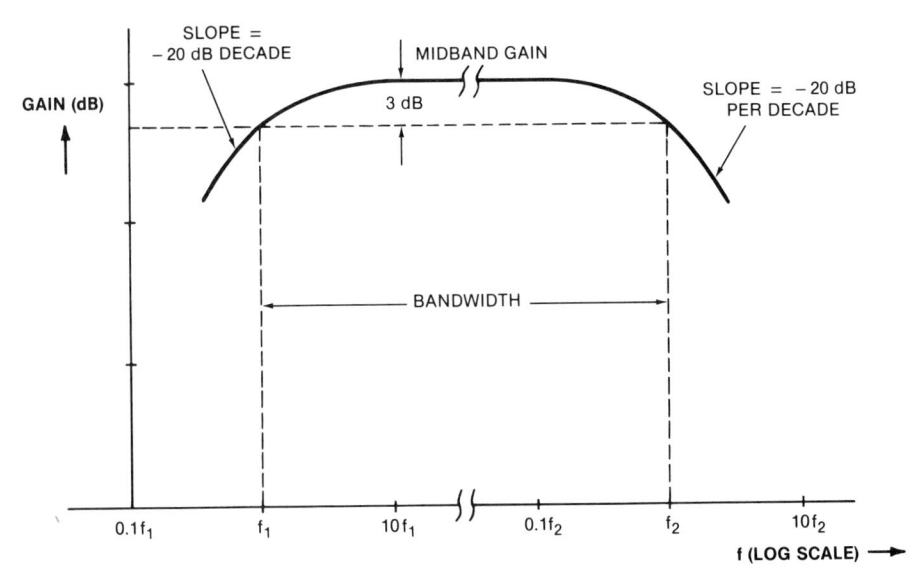

Fig. 7-1. Typical frequency response curve of an RC-coupled amplifier.

of the midband, and this slope value is valid down to $0.1f_1$ (or even lower) at the low-frequency end and up to $10f_2$ (or even higher) at the high-frequency end. For example, if the gain of an amplifier is 25 dB at f_2, then it decreases to 5 dB at $10f_2$.

The frequency response curve of an amplifier (in the range of interest) can be drawn if the midband gain and the two cutoff frequencies are known. Therefore, the study of the frequency response of an amplifier reduces to the determination of those three quantities. In fact, once those three quantities are known, it is unnecessary to actually draw the frequency response curve. The calculation of the midband gain was dealt with in the last two chapters, and we will now consider the determination of the cutoff frequencies.

Factors that Affect Frequency Response

The presence of capacitances in an amplifier circuit is responsible for the variation of its gain with frequency. The decrease of gain at low frequencies is caused by the two coupling capacitors C_1 and C_2 and the bypass capacitor C_3 (Fig. 7-2). The decrease of the gain at high frequencies is due to certain capacitive effects that occur at high frequencies in the devices (BJTs and FETs) themselves.

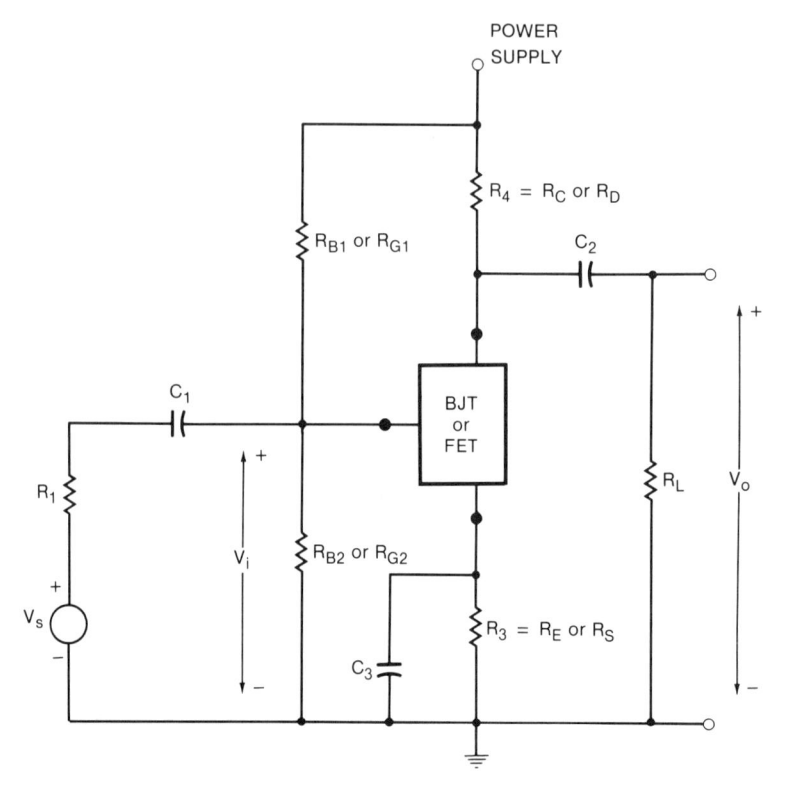

Fig. 7-2. RC-coupled amplifier. Either a BJT or an FET can be the active device. Symbols applicable to both cases are shown.

Effect of the Coupling Capacitors

At sufficiently high frequencies, the impedance of each coupling capacitor is low enough that we can neglect the voltage drop in it. They act like short circuits for all practical purposes, and in our midband calculations, we did in fact replace them by short circuits. As the frequency decreases, the impedance of each coupling capacitor increases. Soon a point is reached where the voltage drop across a coupling capacitor can no longer be ignored. The significant voltage drop across the input coupling capacitor C_1 reduces the input voltage available to the device, V_i. Since the input at the device is reduced, the output voltage is correspondingly reduced also. The significant voltage drop across the output coupling capacitor C_2 at low frequencies also reduces the output voltage developed across the load resistor R_L. Thus, the output voltage (for a given input signal voltage V_s) is made smaller than at midband by the presence of each of the coupling capacitors.

Effect of the Bypass Capacitor

The function of the bypass capacitor C_3 (which is the emitter bypass capacitor in BJT amplifiers and the source bypass capacitor in FET amplifiers) is to serve as a very low impedance path for ac currents in the device. When the bypass is absent or ineffective, all or a portion of the ac current in the device flows through the resistor R_3, which causes a reduction in the output voltage due to negative feedback. In the midband and higher frequencies, the impedance of the bypass capacitor is low enough to be ignored, and the capacitor can be treated as a zero impedance (or short circuit) as was done in the last two chapters. But at low frequencies, the impedance of the bypass capacitor becomes comparable to (or even higher than) the resistance in parallel with it. A portion (which may be quite substantial) of the device current now starts flowing through the resistor R_3, and there is a reduction in the output voltage.

Effect of Interelectrode Capacitances

A capacitive effect is present between each pair of terminals of a BJT or a FET. This effect can be modeled by showing a capacitance between the terminals of the device as shown in Fig. 7-3. One of the capacitive effects, the junction capacitance of the depletion region of a pn junction, was discussed in Chapter 4. The other capacitive effects are due to the transit time of electrons and holes, that is, the time taken for the holes and electrons to flow through the device. At low frequencies and in the midband, the impedances of the interelectrode capacitances are high, and they can be treated as open circuits. Therefore, they have no effect on the gain in the midband or lower frequencies. At high frequencies, they have low impedances and divert part or all of the currents in the device. For example, the low impedance of $C_{b'e}$ or C_{gs} causes a reduction in the input to the device by diverting part of the current available from the signal source as indicated in Fig. 7-4A. The capacitance, $C_{b'c}$ or C_{dg}, forms a low-impedance link between the input terminal (base or gate) and the output terminal (collector or drain), which tends to reduce the voltage gain of the amplifier to a value of 1 (Fig. 7-4B).

The cutoff frequencies of an amplifier will be found to depend upon the various capacitances and the resistances in the amplifier circuit. We will see a very close similarity between the following discussion and the response of RC filters studied in Chapter 2.

DETERMINATION OF THE LOWER CUTOFF FREQUENCY f_1

At low frequencies, the coupling capacitors and the bypass capacitor have to be retained in the circuit. The small-signal low-frequency equivalent cir-

398   **Frequency Response and Pulse Response of Amplifiers**

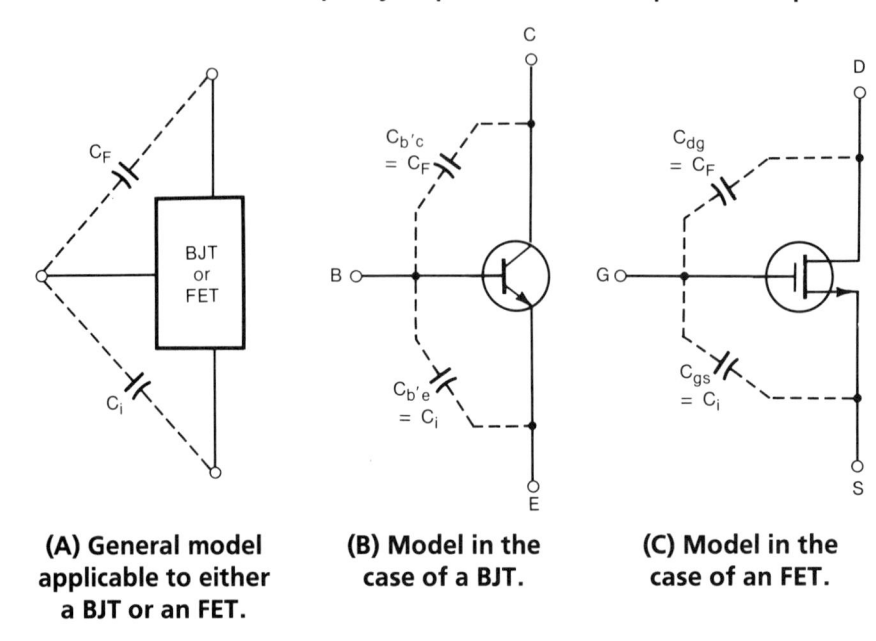

(A) General model applicable to either a BJT or an FET.

(B) Model in the case of a BJT.

(C) Model in the case of an FET.

Fig. 7-3. Interelectrode capacitances in a transistor.

cuit of an RC-coupled amplifier assumes the (rather formidable) form shown in Fig. 7-5. If a precise formula for the lower cutoff frequency were needed, it would be necessary to analyze the circuit of Fig. 7-5 as it is. Such an analysis will be time consuming, tedious, and probably more precise than needed in practice.

The simpler approach (which leads to results that are sufficiently accurate for practical purposes) is to take only one capacitor at a time and treat it as *the one* capacitor responsible for the lower cutoff frequency. The other two capacitors are treated as short circuits. In such a method, three values of the lower cutoff frequency are obtained: one due to the input coupling capacitor acting alone, one due to the output coupling capacitor acting alone, and one due to the bypass coupling capacitor acting alone.

In a number of amplifiers, the *highest* of the three values will be higher by a factor of ten (or more) than the other two values. For example, the three values obtained might be 100 Hz, 8.75 Hz, 2 Hz, and the highest value (100 Hz) is at least ten times as large as the next lower value of 8.75 Hz. In such cases, we choose the *highest of the three values* calculated as *the* lower cutoff frequency since the gain is already down by 3 dB at that frequency. Since the other two frequencies are sufficiently farther down, they do not cause any

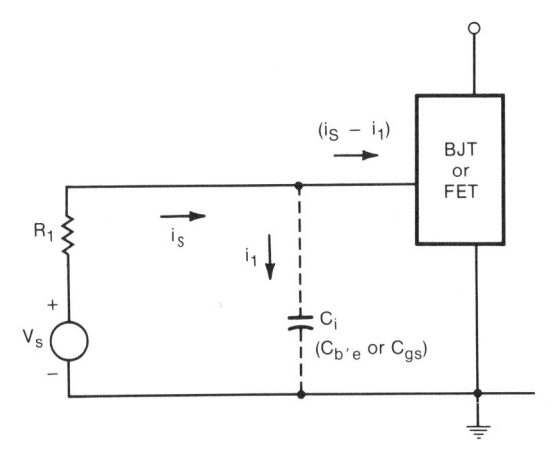

(A) C_i acts as a low impedance path for currents and reduces the input voltage of the device at high frequencies.

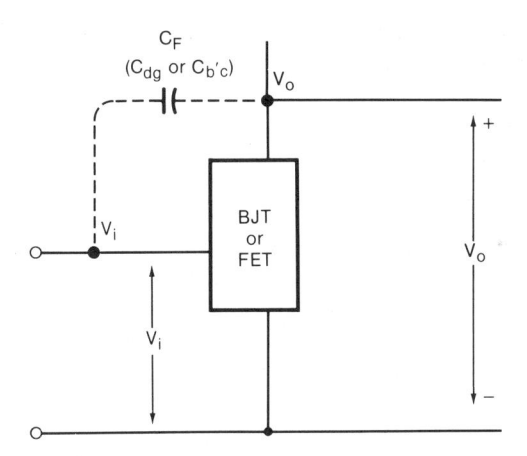

(B) C_F acts as a low impedance link between the output and input terminals of the transistor and tends to make V_o closer to V_i at high frequencies.

Fig. 7-4. Effect of the interelectrode capacitances at high frequencies.

additional attenuation of the gain. In some amplifiers, the highest frequency is not much higher than the other two cutoff frequencies calculated. In such cases, some estimates need to be made in order to arrive at a value of the lower cutoff frequency. We will consider this case later.

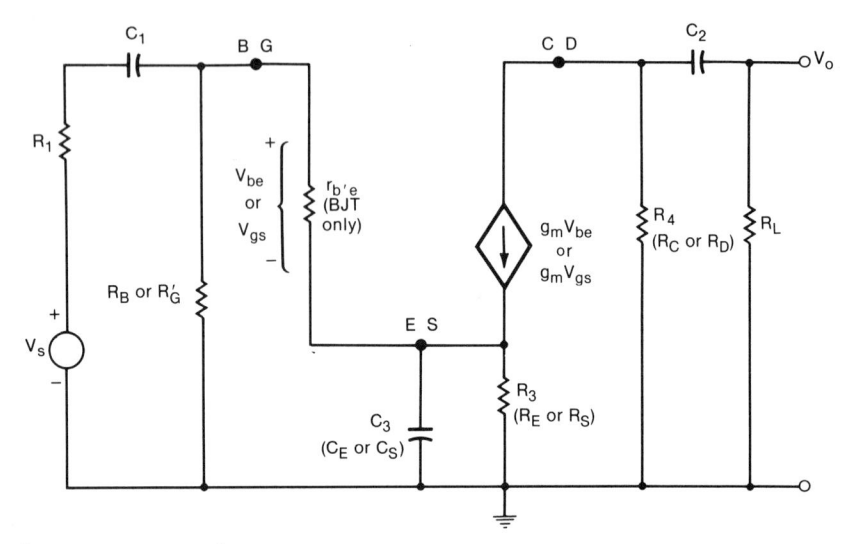

Fig. 7-5. Equivalent circuit of an RC-coupled amplifier for low frequencies. Symbols are shown for both BJT and FET.

Lower Cutoff Frequency Due to the Input Coupling Capacitor

Treat the other two capacitors (C_2 and C_3) as short circuits. The equivalent circuit of Fig. 7-5 reduces to the form shown in Fig. 7-6A.

The lower cutoff frequency due to C_1 is given by

$$f_1 = \frac{1}{2\pi C_1 (R_1 + R_{in})} \quad \text{due to the input coupling capacitor} \tag{7-1}$$

where,

$R_{in} = R'_G$ for the FET amplifier,
$R_{in} = R_B \| h_{ie}$ for the BJT amplifier.

Lower Cutoff Frequency Due to the Output Coupling Capacitor

Treat the capacitors C_1 and C_3 as short circuits. The equivalent circuit of Fig. 7-5 reduces to the form shown in Fig. 7-6B.

The lower cutoff frequency due to C_2 is given by

$$f_1 = \frac{1}{2\pi C_2 (R_4 + R_L)} \quad \text{due to the output coupling capacitor} \tag{7-2}$$

where,
$R_4 = R_C$ for the BJT amplifier,
$R_4 = R_D$ for the FET amplifier.

Lower Cutoff Frequency Due to the Bypass Capacitor

Treat the two coupling capacitors as short circuits. The equivalent circuit of Fig. 7-5 reduces to the form shown in Fig. 7-6C.

For this calculation, it is necessary to treat the FET and BJT cases separately because the presence of $r_{b'e}$ in the BJT amplifier affects the result.

For the *FET amplifier* the lower cutoff frequency due to the bypass capacitor is given by

$$f_1 = \frac{1}{2\pi C_3 [R_3 || (1/g_m)]} \quad \text{due to the bypass capacitor} \quad (7\text{-}3)$$

In the case of the BJT amplifier, the presence of the resistor $r_{b'e}$ makes the resistance of the portion to the left of the dashed line in Fig. 7-6C appear as a "reflected resistance" R_{ref} given by

$$R_{ref} = \frac{r_{b'e} + (R_B || R_1)}{g_m r_{b'e}} \quad (7\text{-}4)$$

and this resistance is in parallel with R_3. The lower cutoff frequency due to the bypass capacitor is then given by

$$f_1 = \frac{1}{2\pi C_3 (R_3 || R_{ref})} \quad (7\text{-}5)$$

If, in a given amplifier circuit, the component values satisfy the condition

$$(R_B || R_1) \text{ much less than } r_{b'e}$$

then it can be seen that Eq. (7-5) for the BJT becomes the same as Eq. (7-3) for the FET, and the lower cutoff frequency formula is the same for the two amplifiers.

Example 7.1

A common-emitter BJT amplifier is shown in Fig. 7-7. Assume $r_{b'e} = 6 \text{ k}\Omega$ and $g_m = 50$ mS for the transistor. Determine the lower cutoff frequency of the amplifier.

Solution
Lower cutoff due to the input coupling capacitor:
Using the circuit of Fig. 7-6A, with the values:
$R_s = 1 \text{ k}\Omega$ $R_{in} = (R_B || r_{b'e}) = (16.67 || 6) = 4.41 \text{ k}\Omega$
$C_1 = 4 \ \mu\text{F}$

402   Frequency Response and Pulse Response of Amplifiers

Note that if we use *resistances in kilohms* and *capacitances in microfarads*, then the *frequencies* calculated by using any of the previous formulas are obtained in *kilohertz*. We will generally use this convention in order to avoid carrying factors like 10^3 and 10^{-6} in the various calculations. Be careful to keep track of units, however!

From Eq. (7-1), the cutoff frequency due to the input coupling capacitor is

$$f_1 = \frac{1}{2\pi C_1 (R_1 + R_{in})}$$

$$= 0.00736 \text{ kHz}$$

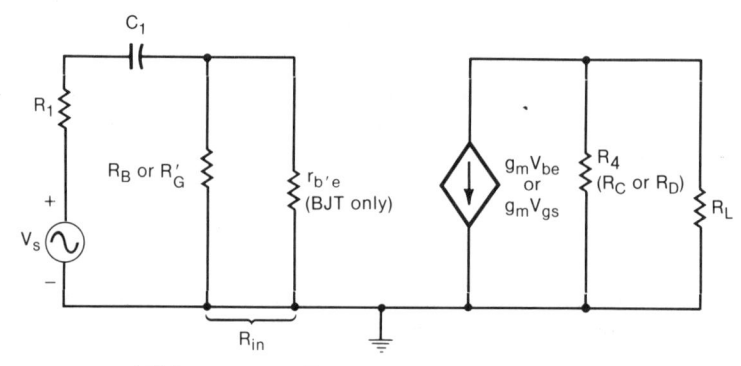

$$= 7.36 \text{ Hz}.$$

(A) Input coupling capacitor acting alone.

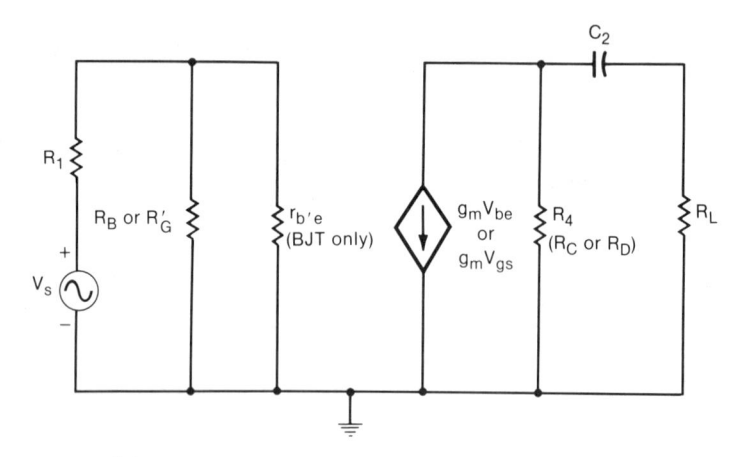

(B) Output coupling capacitor acting alone.

Fig. 7-6. Calculation of the lower cutoff

Lower cutoff frequency due to the output coupling capacitor:
Using Fig. 7-6B with $R_c = 5$ kΩ, $R_L = 3$ kΩ, $C_2 = 10$ μF, Eq. (7-2) leads to

$$f_1 = \frac{1}{2\pi C_2 (R_4 + R_L)}$$

$$= 1.99 \times 10^{-3} \text{ kHz}$$

$$= 1.99 \text{ Hz}.$$

Lower cutoff frequency due to the bypass capacitor:
First, calculate the reflected resistance R_{ref} from Eq. (7-4). Since $r_{b'e} = 6$ kΩ, $R_B = 16.67$ kΩ, $R_1 = 1$ kΩ, and $g_m = 50$ mS, we get

$$R_{ref} = \frac{r_{b'e} + (R_B || R_1)}{g_m r_{b'e}} = 0.0231 \text{ k}\Omega$$

and the lower cutoff frequency due to the bypass capacitor becomes (with $C_3 = 70$ μF and $R_3 = R_E = 5$ kΩ)

$$f_1 = \frac{1}{2\pi C_3 (R_3 || R_{ref})}$$

$$= 0.0989 \text{ kHz}$$

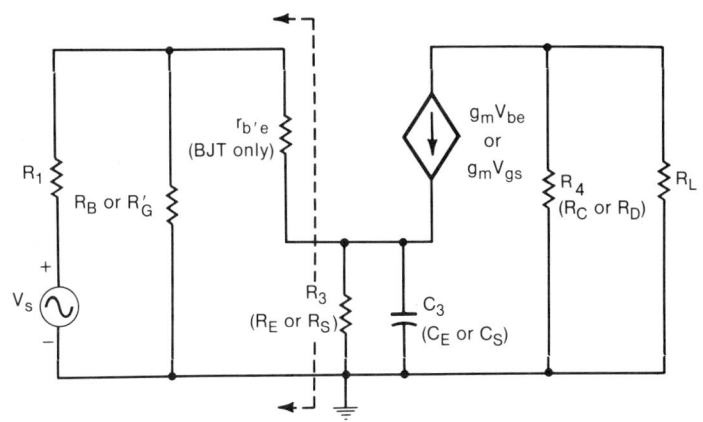

$$= 98.9 \text{ Hz}.$$

(C) Bypass capacitor acting alone. The resistance seen to the left of the dashed line is R_{ref}, which applies to the BJT amplifiers only.

frequency of an amplifier.

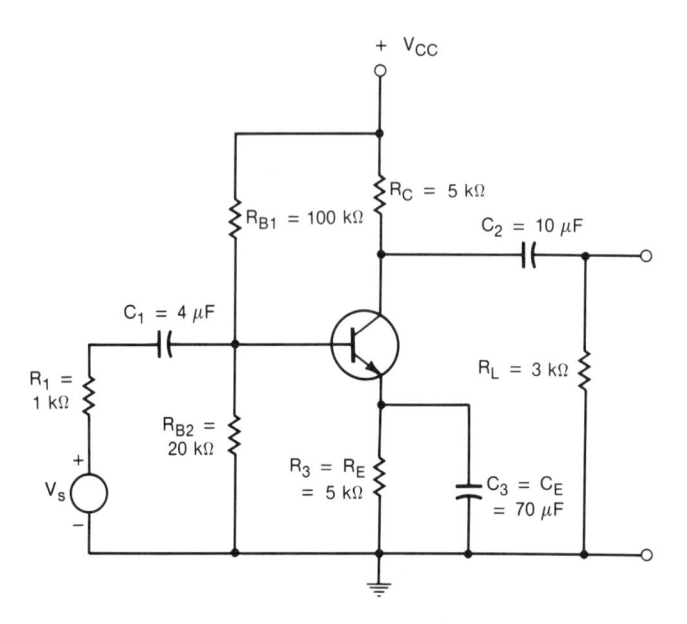

Fig. 7-7. Amplifier circuit for Example 7.1.

The three values of the lower cutoff frequency are, therefore, 7.36 Hz, 1.99 Hz, and 98.9 Hz.

The highest value obtained is 98.9 Hz, which is at least ten times greater than the next lower value of 7.36 Hz. Therefore, we take the lower cutoff frequency of the amplifier as 98.9 Hz. The bypass capacitor is seen to dominate the frequency response at the low end in this amplifier in this example.

Example 7.2

A common-source FET amplifier is shown in Fig. 7-8. Determine the value of the bypass capacitor so that the cutoff frequency due to it is one tenth of the cutoff frequency due to the coupling capacitors.

Solution

Cutoff due to the input coupling capacitor:

Using the circuit of Fig. 7-6A with the values $R_1 = 1$ kΩ, $R_{in} = R'_G = (R_{G1} \| R_{G2}) = (400 \| 180) = 124$ kΩ and $C_1 = 2$ μF, Eq. (7-1) leads to

$$f_1 = 0.637 \text{ Hz}$$

Electronics: Circuits and Systems

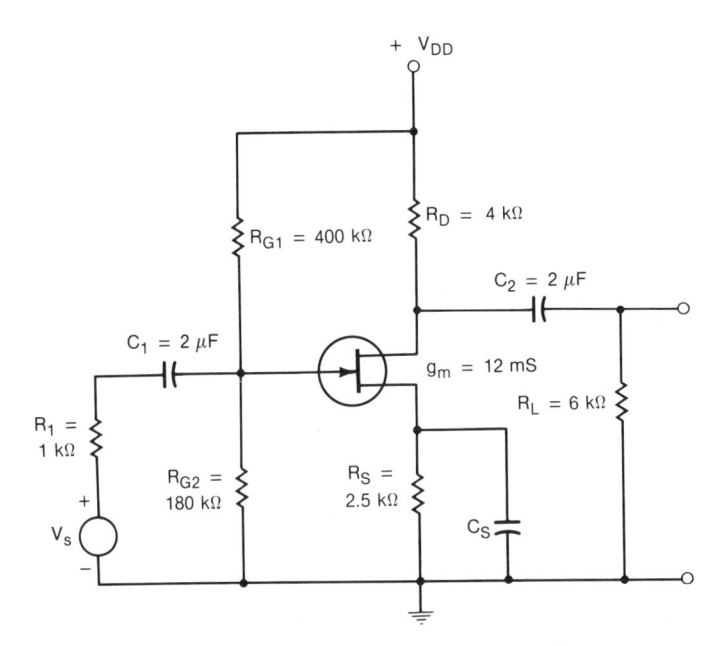

Fig. 7-8. Amplifier circuit of Example 7.2.

Cutoff due to the output coupling capacitor:

Using the circuit of Fig. 7-6B with the values $R_4 = R_D = 4$ kΩ, $R_L = 6$ kΩ, and $C_2 = 2$ μF, Eq. (7-2) leads to

$$f_1 = 7.96 \text{ Hz}$$

The higher value of the two frequencies obtained previously is seen to be at least ten times greater than the other. Therefore, we can choose the higher value of 7.96 Hz as *the* lower cutoff frequency due to the coupling capacitors. Then, according to the statement of the problem, we have to make the cutoff frequency due to the bypass capacitor *no more than one tenth* of the lower cutoff frequency already selected. That is,

$$f_1 \text{ due to the bypass capacitor } = (1/10) \times 7.96 \text{ Hz}$$
$$= 0.796 \text{ Hz or } 7.96 \times 10^{-4} \text{ kHz}.$$

(Remember that we used frequencies in kilohertz in our calculations.)

Using the circuit of Fig. 7-6D with $R_3 = R_S = 2.5$ k, $(1/g_m) = (1/12) = 8.33 \times 10^{-2}$ k, and $f_1 = 7.96 \times 10^{-4}$ kHz, we get from Eq. (7-3)

$$7.96 \times 10^{-4} = \frac{1}{2\pi C_S \times (2.5 \,||\, 8.33 \times 10^{-2})}$$

where,

C_S is in μF. This gives

$$C_S = 2400 \ \mu F$$

which is a large capacitor indeed. (This point is discussed further in the following comments.)

The following general features should be noted from the previous two examples.

1. The bypass capacitor is the worst culprit of the three capacitors in affecting the lower cutoff frequency. It dominates the low-frequency response. *If we wish* to make either coupling capacitor control the lower cutoff frequency, the bypass capacitor needed becomes extremely (if not impractically) large. As a practical matter, large capacitors (in thousands of microfarads) are available as discrete components that are physically small. Therefore, in the case of amplifiers built up of discrete components, large bypass capacitors do not pose too much of a problem, and the lower cutoff frequency can be extended down to low values. But in the case of ICs large capacitors are not economically feasible, and capacitive coupling must be avoided for IC amplifier stages if good low-frequency response is desired. We will see in Chapter 8 that direct coupling is used in operational amplifiers that are the most commonly available IC amplifier packages.

2. Even though the capacitors affect the lower cutoff frequency, the resistances in the circuit also affect it to an equal extent. In the case of BJT amplifiers, where the input resistance of the device is small (a few kilohms), the lower cutoff frequency due to the coupling capacitors tends to be much higher than in FET amplifiers with much larger input resistances (hundreds of kilohms). Therefore, it is possible to achieve better low-frequency response with a FET amplifier than a BJT amplifier.

Drill Problem 7.1: In Example 7.1, suppose the lower cutoff frequency is to be changed to 200 Hz. Assuming all components except the emitter bypass capacitor remain the same, find the new value of C_E.

Drill Problem 7.2: The lower cutoff frequency due to the emitter bypass capacitor in a BJT amplifier is 100 Hz. It is desired to make the lower cutoff frequencies equal to 8 Hz due to each coupling capacitor. The relevant component values are $R_1 = 2$ kΩ, $R_{B1} = 95$ kΩ, $R_{B2} = 15$ kΩ, $r_{b'e} = 4.5$ kΩ, $R_C = R_L = 4.7$ kΩ. Determine the values of the coupling capacitors.

Drill Problem 7.3: In Example 7.2, suppose the transconductance g_m of

Electronics: Circuits and Systems 407

the FET is reduced to 4 mS with all other values remaining unchanged. Recalculate the cutoff frequency due to the bypass capacitor.

Drill Problem 7.4: Calculate the lower cutoff frequency of the FET amplifier shown in Fig. 7-9.

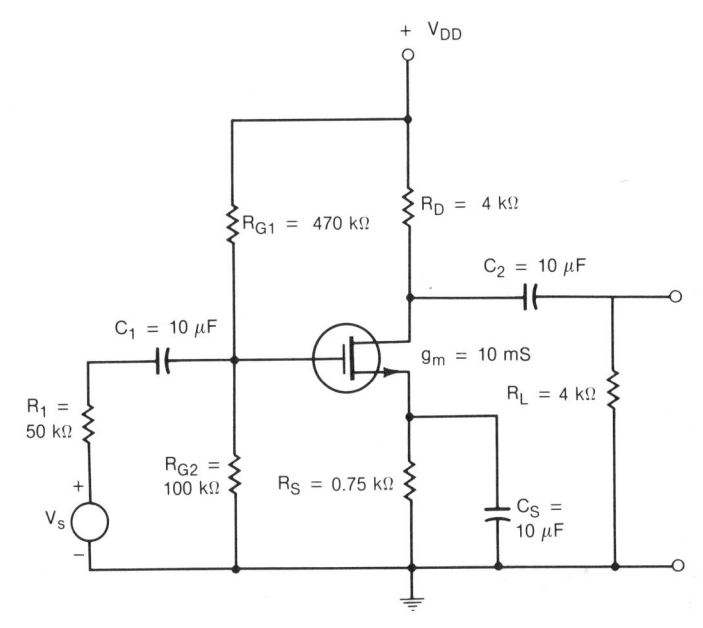

Fig. 7-9. Amplifier circuit of Drill Problem 7.4.

In the previous examples and drill problems, it was found that the highest value of the lower cutoff frequencies (calculated by considering each capacitor separately) was found to be at least a factor of ten higher than either of the other two values. In such cases, it is correct to choose the highest of the three values as the lower cutoff frequency of the amplifier since the other two are sufficiently low enough not to have any effect on the gain. The question arises about what should be done when the highest of the three calculated frequencies is not at least a factor of ten higher than either of the other two values. It is necessary in such cases to use a graphical procedure to estimate the actual lower cutoff frequency. For each value of cutoff frequency obtained (by considering one capacitor at a time), we draw a frequency response curve as follows. A horizontal line is drawn from some high frequency down to (10 f_1). Three points are plotted: one at 2 f_1, which is 1 dB below the level of the horizontal line, one at f_1, which is 3 dB below the level of the horizontal line, and one at ($f_1/2$), which is 7 dB below the level of the *horizontal line*. Join these three points and the horizontal line by a smooth curve as indicated in Fig. 7-10. The three response curves (obtained from the three calculations) are then added point by point in the neighborhood of the cutoff frequency to determine where the gain is down by 3 dB when the three effects are combined.

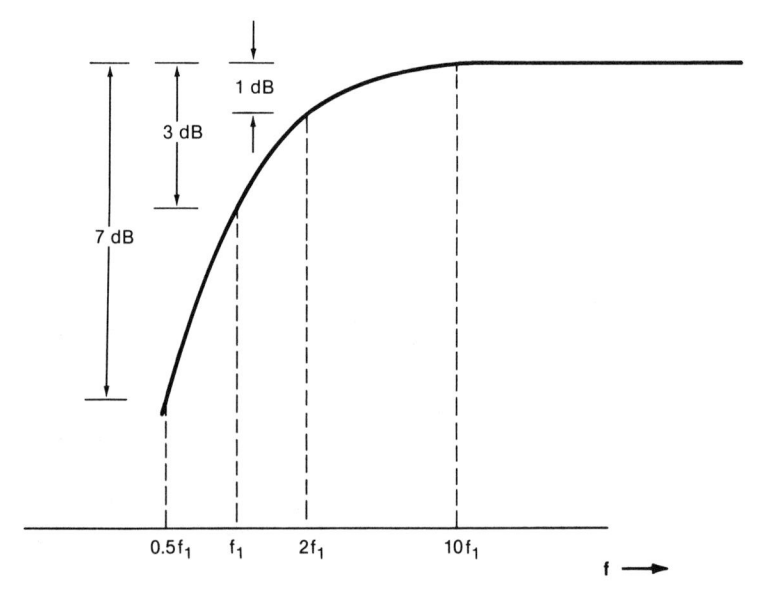

Fig. 7-10. Sketching the frequency response curve (low frequencies).

Example 7.3

Suppose the lower cutoff frequencies of an amplifier are found to be (when each capacitor is considered alone): 25 Hz, 50 Hz, and 80 Hz. Estimate the lower cutoff frequency of the amplifier.

Solution

The three frequency response curves are shown in Fig. 7-11. It is seen that the sum of the three curves is down by 3 dB at approximately 110 Hz. Therefore, $f_1 = 110$ Hz for the amplifier.

It is important to remember that we are only seeking a reasonable estimate of the cutoff frequency of the amplifier by the previous procedure. Therefore, there is not much to be gained by refining the curves to too fine a level. In any electronic circuit, the results obtained by analysis on paper can rarely be

Electronics: Circuits and Systems

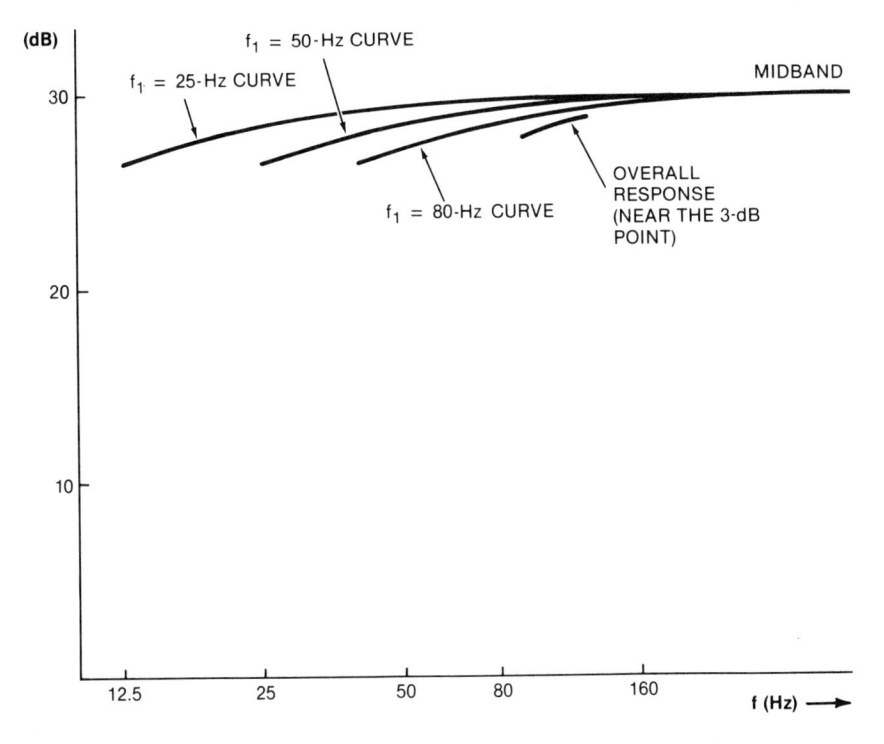

Fig. 7-11. Frequency response curves of Example 7.3.

expected to yield 100% accuracy because of the variation of component values and various approximating assumptions used.

Drill Problem 7.5: Estimate the lower cutoff frequency of an amplifier in which the cutoff frequency due to each coupling capacitor is 60 Hz and that due to the bypass capacitor is 120 Hz.

HIGH-FREQUENCY MODEL OF AN AMPLIFIER

An attempt is made in the following discussion of high-frequency response of amplifiers to unify the analysis of BJT and FET amplifiers, since the basic principles of analysis and the factors that affect the high-frequency response are the same for both amplifiers. In order to have a unified treatment, the same symbol is used for each interelectrode capacitance whether it applies to a BJT or to a FET. But, in order to conform to the standard notation used in the literature, the standard symbols are also presented at the appropriate places.

Frequency Response and Pulse Response of Amplifiers

The interelectrode capacitances of a transistor affect the high-frequency response of an amplifier. The model of a transistor (both BJT and FET) is of the form shown in Fig 7-12A. When the high-frequency model of the device is used in the circuit model of RC-coupled amplifier, the resulting equivalent circuit is of the form shown in Fig. 7-12B. There are two capacitances C_i and C_F in the high-frequency model.

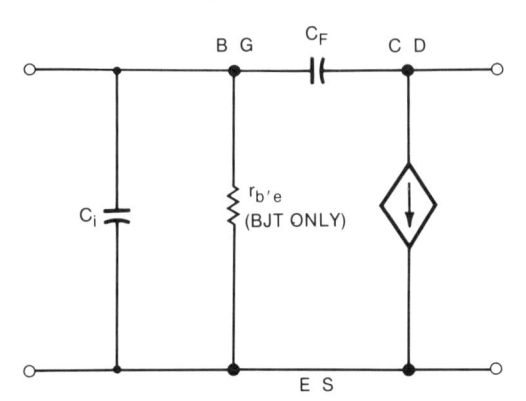

(A) High-frequency model of a transistor (either BJT or FET).

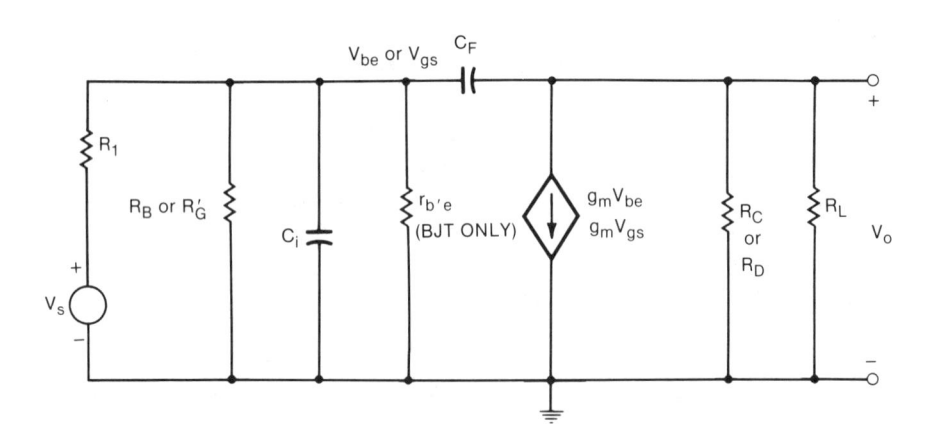

(B) High-frequency equivalent circuit of an amplifier. Symbols are shown for both BJTs and FETs. $C_F = C_{b'c}$ for BJTs and C_{dg} for FETs. $C_i = C_{b'e}$ for BJTs and C_{gs} for FETs. (A drain-to-source capacitance C_{ds}, which is present in the FETs, is disregarded in our calculations, since its effect is negligible in a common-source amplifier.)

Fig. 7-12. High-frequency response.

Capacitance C_i

The capacitance C_i corresponds to the base-to-emitter capacitance in a BJT, and is usually denoted by any of the following symbols: $C_{b'e}$, C_e, or C_π. The capacitance C_i corresponds to the gate-to-source capacitance in a FET, and is usually denoted by C_{gs}. At high frequencies, C_i acts as a low-impedance path for the current supplied by the signal v_s, which reduces the voltage available at the input of the device with a corresponding loss in the output voltage.

Capacitance C_F

The capacitance C_F corresponds to the collector-to-base capacitance in a BJT, and is usually denoted by any of the following symbols: $C_{b'c}$, C_μ, or C_c. In a FET, the capacitance C_F corresponds to the drain-to-gate capacitance, and is usually denoted by C_{dg}. The capacitance forms a link between the input and output terminals of the device. Therefore, it provides *feedback* from the output side to the input side: if the output voltage changes, the current in the capacitance C_F changes, and, as a result, the current in the resistor R_{in} (R_G') changes. Thus, any change in the output voltage is fed back as a change in the input voltage of the device. Because of the feedback effect, the capacitance C_F turns out to be equivalent to a much larger capacitance than its actual value when viewed from the input side of the amplifier.

Miller Effect

The fact that the feedback capacitance C_F has a larger effect on the frequency response of the amplifier than can be expected from its actual value is referred to as the *Miller effect*. In order to facilitate the analysis of the circuit of Fig. 7-12B (which is a rather complicated circuit), it is desirable to replace the actual feedback capacitance by an *equivalent* capacitance C_F' connected between the input terminals as indicated in Fig. 7-13B. This equivalent capacitance is known as the Miller equivalent capacitance, or simply the *Miller capacitance*. The relationship between the Miller equivalent capacitance C_F' and the actual feedback capacitance C_F is given by

$$C_F' = C_F(1 + g_m R_{eq}) \qquad (7\text{-}6)$$

where,

g_m is the transconductance of the device and R_{eq} is the equivalent resistance on the output side of the small-signal model of the amplifier.

Eq. (7-6) can be derived in the following manner. Consider the current I_F through the capacitance C_F in Fig. 7-13A and I_F' in C_F' in Fig. 7-13B. We want these two currents to be equal in order to make the equivalence valid. Now,

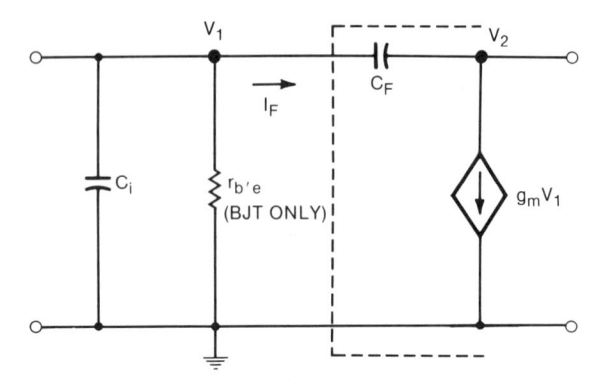

412 Frequency Response and Pulse Response of Amplifiers

(A) The actual feedback capacitance is C_F.

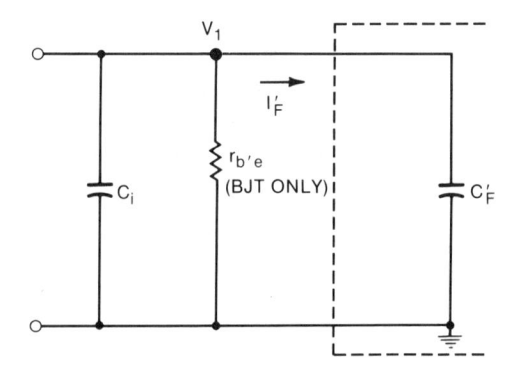

(B) C_F can be replaced by an equivalent Miller capacitance C_F' for purposes of analysis.

Fig. 7-13. Miller effect.

$$|I_F| = \frac{1}{(1/2\pi f C_F)} (V_1 - V_2) \tag{7-7}$$

since $(1/2\pi f C_F)$ is the impedance and $(V_1 - V_2)$ is the voltage across C_F. Eq. (7-7) can be rewritten in the form:

$$|I_F| = \frac{1}{2\pi f C_f}(1 - \frac{V_2}{V_1})V_1 \tag{7-8}$$

The current I_F' in C_F' of Fig. 7-13B is given by

$$|I_F'| = \frac{1}{(1/2\pi f C_F')} V_1 = 2\pi f C_F' V_1 \tag{7-9}$$

Electronics: Circuits and Systems

Since $I_F = I'_F$, we get

$$C'_F = C_F\left(1 - \frac{V_2}{V_1}\right) \tag{7-10}$$

but the quantity (V_2/V_1) is the voltage gain, and from our discussion of the last two chapters, we have

$$V_2 = -g_m R_{eq} V_1 \tag{7-11}$$

where,

$V_1 = V_{b'e}$ for the BJT and V_{gs} for the FET.

Eq. (7-6) for the Miller capacitance is seen to follow from Eqs. (7-10) and (7-11).

The circuit of Fig. 7-13B will be used for the determination of the higher cutoff frequency of an amplifier. In order to particularize our analysis to the BJT and the FET amplifiers, the correspondence between symbols is summarized in the following tabular form, and this table should be consulted while solving specific amplifier circuit problems:

Symbol Used	BJT Amplifier	FET Amplifier
C_i	$C_{b'e}$	C_{gs}
C_F	$C_{b'c}$	C_{dg}
R_{in}	$(R_B \| r_{b'e})$	R'_G
R_{eq}	$(R_C \| R_L)$	$(R_D \| R_L)$

It is customary to include a capacitance C_{ds} between the drain and source terminals of the high-frequency model of a FET, but its value is small compared with the Miller capacitance and can be disregarded in the determination of the upper cutoff frequency of a common-source amplifier.

Determination of the Upper Cutoff Frequency

Since the capacitances C_i and C'_F are in parallel in Fig. 7-13B, they can be combined into a single capacitance, the total input capacitance C_{in}.

$$C_{in} = C_i + C'_F = C_i + C_F(1 + g_m R_{eq}) \tag{7-12}$$

Then, the *upper cutoff frequency* of the amplifier is given by the expression

$$f_2 = \frac{1}{2\pi C_{in}(R_1 \| R_{in})} \tag{7-13}$$

Refer to Fig. 7-14.

If the capacitance C_{in} is expressed in *microfarads* and the resistances in

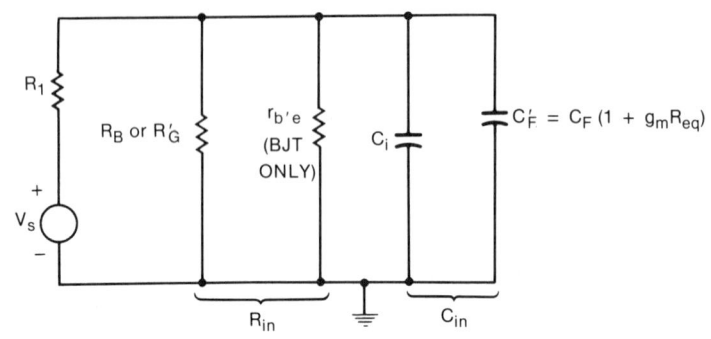

Fig. 7-14. Model useful for the calculation of the higher cutoff frequency of an amplifier.

kilohms, then f_2 will be in *kilohertz* in the previous formula. Be careful, however, since the capacitances in the high-frequency model are usually in *picofarads!*

Example 7.4

Calculate the upper cutoff frequency of the BJT amplifier whose high frequency equivalent circuit is shown in Fig. 7-15(A).

Solution

The circuit is redrawn using the following calculations.

$$\begin{aligned} R_{eq} &= (R_c \parallel R_L) \\ &= (5 \parallel 2) \\ &= 1.43 \text{ k}\Omega \\ R_{in} &= (R_B \parallel r_{b'e}) \\ &= (16.67 \parallel 2) \\ &= 1.79 \text{ k}\Omega \\ C'_F &= C_F(1 + g_m R_{eq}) \\ &= 218 \text{ pF} \end{aligned}$$

The input side of the amplifier assumes the form shown in Fig. 7-15B. The upper cutoff frequency is

$$f_2 = \frac{1}{2\pi C_{in}(R_1 \parallel R_{in})}$$

Electronics: Circuits and Systems   415

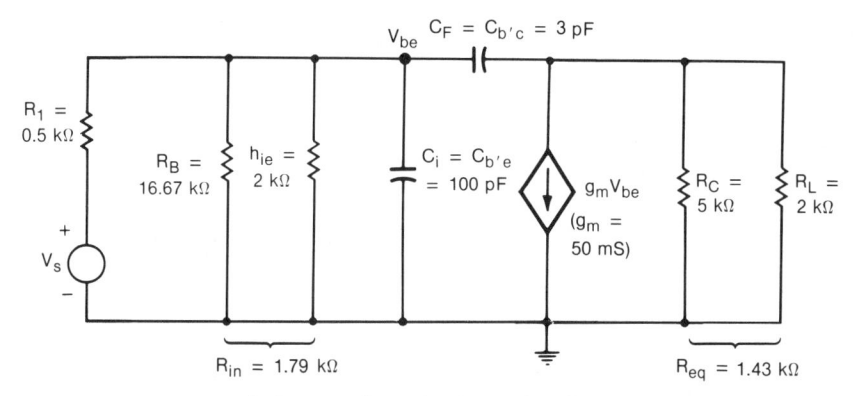

(A) Actual equivalent circuit.

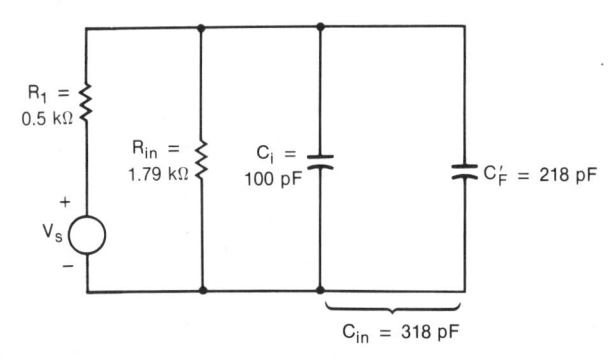

(B) Simpler model obtained by using the Miller effect.

Fig. 7-15. High-frequency equivalent circuit for Example 7.4.

Using $C_{in} = (218 + 100) = 318$ pF, and $(R_1 \| R_{in}) = 0.391$ k, we get

$$f_2 = 1.28 \text{ MHz}$$

as the upper cutoff frequency.

Drill Problem 7.6: The high-frequency model of a FET amplifier is shown in Fig. 7-16; $g_m = 8$ mS. Calculate the upper cutoff frequency.

SPECIFICATIONS OF THE PARAMETERS OF A TRANSISTOR

Now we have all the tools needed for a complete analysis of an amplifier: the dc analysis to determine the Q point, the small-signal analysis to deter-

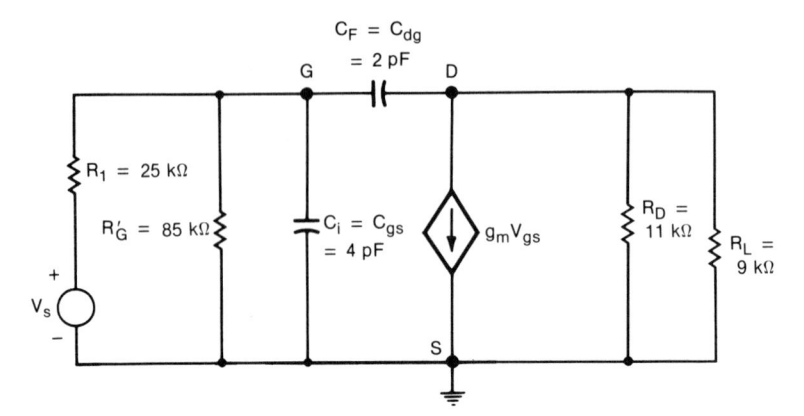

Fig. 7-16. High-frequency equivalent circuit for Drill Problem 7.6.

mine the midband gains, the low-frequency analysis to determine the lower cutoff frequency and the high-frequency analysis to determine the upper cutoff frequency. In order to perform the complete analysis, we need to know the *parameters* of the transistor itself (besides the values of the resistances and other elements connected to the device in the circuit). The parameters needed for the analysis are $r_{b'e}$, β, and the interelectrode capacitances $C_{b'c}$ and $C_{b'e}$ of a BJT; g_m, the constant K for the transfer characteristic, and the interelectrode capacitances C_{dg} and C_{gs} of a FET. The data sheets from a manufacturer of a transistor do not always provide all the parameters directly and some initial calculations are usually needed.

BJT Parameters

The data sheet from a manufacturer usually provides the following three pieces of information for a BJT: (1) the β of the transistor, usually specified as h_{FE}; (2) the collector-to-base capacitance $C_{b'c}$; and (3) a frequency f_T in megahertz called the *current-gain bandwidth product*.

The transconductance of a BJT is calculated from the value of the collector current at the Q point by using the equation: [Eq. (5-18)]

$$g_m = \frac{I_C \text{ at the Q point in milliamperes}}{0.026} \text{ mS} \qquad (7\text{-}14)$$

The resistance h_{ie} of a BJT is calculated from the values of g_m and β (or h_{FE}) through the equation [Eq. (5-17)]

$$r_{b'e} = (\beta/g_m)$$

$$= (h_{FE}/g_m) \qquad (7\text{-}15)$$

Electronics: Circuits and Systems

The interelectrode capacitance $C_{b'c}$ the collector-to-base capacitance, given by the data sheet is the feedback capacitance C_F we have used in the model of Fig. 7-12.

The interelectrode capacitance $C_{b'e}$ the base-to-emitter capacitance, or C_i in the model of Fig. 7-12, is calculated from the value of g_m and the frequency f_T given by the manufacturer:

$$C_i = C_{b'e}$$
$$= \frac{g_m}{2\pi f_T} \qquad (7\text{-}16)$$

The data provided by the manufacturer in conjunction with Eqs. (7-14), (7-15) and (7-16) furnish the set of parameters needed for the analysis of a BJT amplifier.

FET Parameters

The data sheet from a manufacturer of a FET usually provides the following pieces of information: (1) the threshold voltage V_T (also given as V_p, the pinch-off voltage; (2) I_{DSS} or some similar information that leads to the calculation of the constant K for the transfer characteristic equation; (3) a capacitance C_{iss}; and (4) a capacitance C_{rss}.

The transconductance of a FET is calculated by using the equation [Eq. (6-12)]:

$$g_m = 2K(V_{GS} - V_T) \qquad (7\text{-}17)$$

where,

V_{GS} is the gate-to-source voltage at the Q point.

The capacitance given as C_{rss} (the *smaller* of the two capacitances on the data sheet) is the drain-to-gate capacitance C_{dg}, which is the same as the feedback capacitance C_F used in the model of Fig. 7-12.

$$C_F = C_{dg}$$
$$= C_{rss} \qquad (7\text{-}18)$$

The capacitance between gate and source, C_{gs} (or C_i in the model of Fig. 7-12), is obtained from C_{iss} and C_{rss} on the data sheet:

$$C_i = C_{gs}$$
$$= C_{iss} - C_{rss} \qquad (7\text{-}19)$$

Eqs. (7-17), (7-18) and (7-19) provide the set of parameters needed for the analysis of a FET amplifier.

STEP-BY-STEP PROCEDURE FOR THE COMPLETE ANALYSIS OF A BJT AMPLIFIER

The step-by-step procedure for the complete analysis of a common-emitter BJT amplifier will be illustrated through the following example.

Example 7.5

The transistor used in the amplifier of Fig. 7-17 is a 2N5089 npn silicon transistor (Motorola, Inc.). The data sheet from the manufacturer provides the following information: h_{FE} (measured at $I_C = 1$ mA and $V_{CE} = 5$ V) = 450 (minimum); $f_T = 300$ MHz (at $I_C = 1$ mA); $C_{b'c} = 1.8$ pF (typical) and 4 pF (max). Analyze the amplifier.

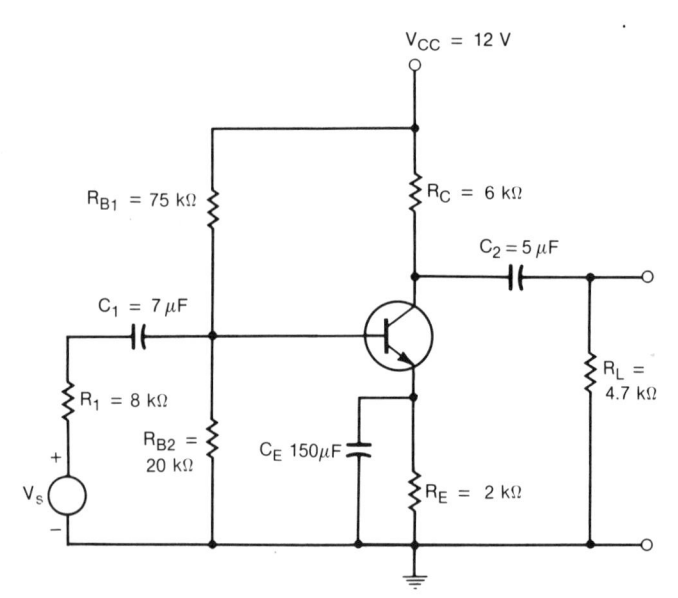

Fig 7-17. Amplifier circuit of Example 7.5.

Solution

Note: We have selected the *minimum* h_{FE} specified by the manufacturer since is leads to a *worst-case* analysis. Thus, any discrepancy between calculated results and actually measured results in the amplifier will be an error on the safe side.

DC analysis:

Step 1: Draw the relevant portion of the circuit valid for dc analysis. For this all capacitors are treated as open circuits.

Electronics: Circuits and Systems

The dc circuit for the present amplifier is shown in Fig. 7-18.

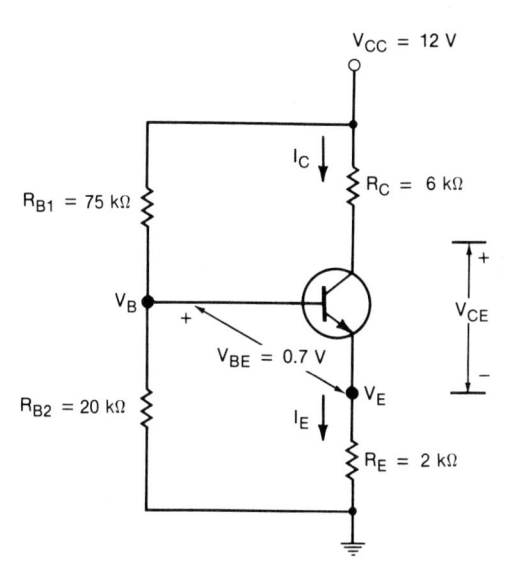

Fig. 7-18. Dc analysis of the amplifier of Example 7.5.

Step 2: Calculate the voltage at the base using the voltage-divider formula.

$$V_B = \frac{R_{B2}}{R_{B1} + R_{B2}} V_{CC}$$

$$= 2.53 \text{ V}$$

Step 3: Subtract 0.7 V from V_B to get the emitter voltage V_E.

$$V_E = V_B - 0.7$$

$$= 1.83 \text{ V}$$

Step 4: Calculate the emitter current I_E using Ohm's law.

$$I_E = \frac{V_E}{R_E}$$

$$= 0.915 \text{ mA}$$

The collector current is assumed to be the same as I_E.

$$I_E = I_C$$

$$= 0.915 \text{ mA}$$

420 Frequency Response and Pulse Response of Amplifiers

Step 5: Calculate the collector-to-emitter voltage V_{CE} by subtracting the voltage drops in R_C and R_E from V_{CC}.

$$V_{CE} = V_{CC} - I_C R_C - I_E R_E$$
$$= 4.68 \text{ V}$$

This completes the dc analysis.

Small-signal analysis (midband):

Step 1: Calculate the transconductance g_m from the value of I_C obtained previously at the Q point.

$$g_m = \frac{I_C \text{ in mA}}{0.026}$$
$$= 35.2 \text{ mS}$$

Step 2: Calculate the input resistance $r_{b'e}$ of the BJT.

$$r_{b'e} = \frac{h_{FE}}{g_m}$$
$$= 12.8 \text{ k}\Omega$$

Step 3: Draw the small-signal equivalent circuit (midband). For this, the two coupling capacitors and the bypass capacitor are replaced by short circuits. Note that the two resistors R_{B1} and R_{B2} are in parallel for ac currents. Similarly, the two resistors R_C and R_L are in parallel for ac currents also. The small-signal equivalent circuit in the midband is as shown in Fig. 7-19, where, $R_B = (R_{B1} \| R_{B2})$ and $R_{eq} = (R_C \| R_L)$.

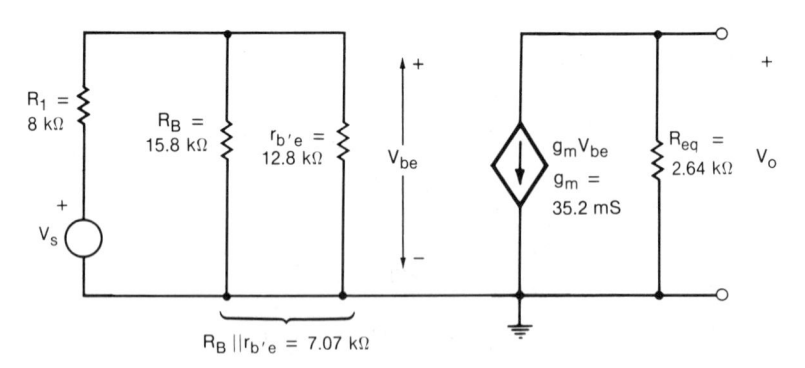

Fig. 7-19. Small-signal model of the amplifier of Example 7.5 for midband gain calculation.

Electronics: Circuits and Systems

Step 4: Calculate the voltage V_{be} using the voltage-divider formula.

$$V_{be} = \frac{(R_B \parallel r_{b'e})}{R_s + (R_B \parallel r_{b'e})} V_s$$

For the present circuit,

$$R_B \parallel r_{b'e} = (15.8 \parallel 12.8)$$
$$= 7.07 \text{ k}\Omega$$

and

$$V_{be} = \frac{7.07}{15.07} V_s$$
$$= 0.469 V_s$$

Step 5: Calculate the output voltage V_o.

$$V_o = -(g_m V_{be}) R_{eq}$$
$$= -(35.2)(2.64) V_{be}$$
$$= -92.9 V_{be}$$

Step 6: Calculate the midband voltage gain (V_o/V_s) using the results of Steps 5 and 6. Convert it to decibels.

$$V_o = -92.9 V_{be}$$
$$= -92.9 \times 0.469 V_s$$

$$\text{voltage gain} = -92.9 \times 0.469$$
$$= -43.6$$
$$= 32.8 \text{ dB (midband gain)}$$

This completes the small-signal analysis (midband voltage gain). Other quantities such as current gain and power gain can also be found using the results obtained in the previous steps as needed.

Frequency response—lower cutoff frequency:

Step 1: Calculate the lower cutoff frequency due to the input coupling capacitor C_1 (Fig. 7-20A).

$$f_1 = \frac{1}{2\pi C_1 (R_1 + R_{in})}$$

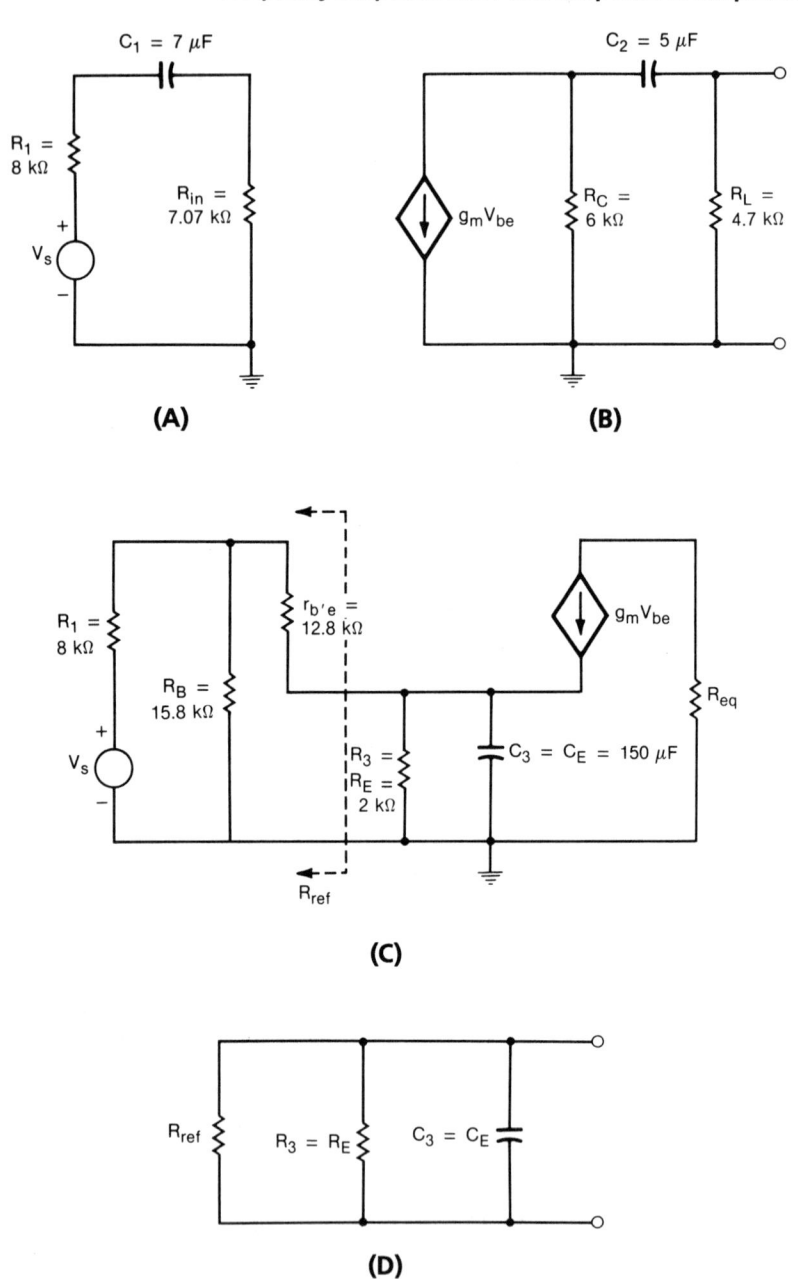

Fig. 7-20. Calculation of the lower cutoff frequency in Example 7.5.

where,
$R_{in} = (R_B \| r_{b'e})$.

$$R_{in} = (R_B \| r_{b'e})$$
$$= (15.8 \| 12.8)$$
$$= 7.07 \text{ k}\Omega$$
$$f_1 = 1.51 \text{ Hz}$$

Step 2: Calculate the lower cutoff frequency due to the output coupling capacitor C_2 (Fig. 7-20B).

$$f_1 = \frac{1}{2\pi C_2 (R_c + R_L)}$$
$$= 2.97 \text{ Hz}$$

Step 3: Calculate the lower cutoff frequency due to the emitter bypass capacitor. (Figs. 7-20C and D). The reflected resistance R_{ref} for this calculation is from Eq. (7-4):

$$R_{ref} = \frac{r_{b'e} + (R_B \| R_1)}{g_m r_{b'e}}$$
$$= 0.0402 \text{ }\Omega$$

The lower cutoff frequency due to the bypass capacitor is given by

$$f_1 = \frac{1}{2\pi C_E (R_3 \| R_{ref})}$$

where,
$R_3 = R_E = 2 \text{ k}\Omega$.
Therefore, since

$$C_E = 150 \text{ }\mu\text{F}$$

and

$$(R_3 \| R_{ref}) = 0.0394 \text{ k}\Omega,$$
$$f_1 = 26.9 \text{ Hz}$$

Since the value of 26.9 Hz is larger by a factor of more than ten than the other two lower cutoff frequencies calculated here, we conclude that

$$\text{lower cutoff frequency} = 26.9 \text{ Hz}$$

Frequency response—higher cutoff frequency:

Step 1: Calculation of the Miller Capacitance C'_F). The feedback capacitance C_F equals the collector-to-base capacitance $C_{b'c}$ specified by the manufacturer. Calculate the Miller Capacitance C'_F:

$$C'_F = C_{b'c}(1 + g_m R_{eq})$$

where,

R_{eq} is $(R_C \| R_L)$.

For the present example, $C_{b'c}$ is given as 1.8 pF (typical) and 4 pF (max). We use the maximum value to be on the safe side in our calculations.

$$C_{b'c} = 4 \text{ pF}$$

and

$$R_{eq} = (6 \| 4.7)$$
$$= 2.64 \text{ k}\Omega$$
$$C'_F = 376 \text{ pF}$$

Step 2: Calculate the capacitance C_i, which equals the base-to-emitter capacitance $C_{b'e}$.

$$C_i = C_{b'e}$$
$$= \frac{g_m}{2\pi f_T}$$

(In some cases, $C_{b'e}$ is directly specified by the manufacturer.)

$$C_i = C_{b'e}$$
$$= \frac{35.2 \times 10^{-3}}{2\pi(300 \times 10^6)}$$
$$= 1.867 \times 10^{-11} \text{ F}$$
$$= 18.67 \text{ pF}$$

Step 3: Draw the high-frequency equivalent circuit of the input side (Fig. 7-21).

Calculate the upper cutoff frequency f_2. Using

$$C_{in} = C_{b'e} + C_F$$
$$= 394.4 \text{ pF}$$

Electronics: Circuits and Systems

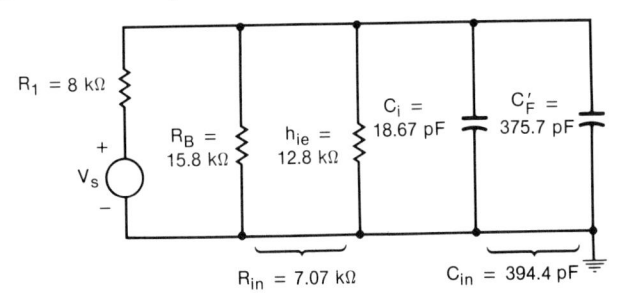

Fig. 7-21. Calculation of the upper cutoff frequency in Example 7.5.

and

$$R_{in} = (R_B \| r_{b'e})$$
$$= 7.07 \text{ k}\Omega$$

we get,

$$f_2 = 107.5 \text{ kHz}$$

which is the upper cutoff frequency of the amplifier.

This completes the analysis of the amplifier.

The frequency response of the amplifier is shown in Fig. 7-22. The mid-band gain is 32.8 dB, and the gain is down to $(32.8 - 3) = 29.8$ dB at the two cutoff frequencies.

$$\text{bandwidth of the amplifier} = (107.5 \times 10^3 - 37.9)$$
$$= 107.5 \text{ kHz}$$

The bandwidth is essentially equal to the upper cutoff frequency.

The step-by-step procedure for the complete analysis of the FET amplifier is illustrated through the following example.

Example 7.6

The FET used in the amplifier of Fig. 7-23 is a 2N5459 n-channel FET with the following specifications from the manufacturer:

$I_{DSS} = 4$ mA (min), 16 mA (max). V_T (specified as V_P by the manufacturer) = -2 V (min), -8 V (max). $C_{iss} = 7$ pF (max) $C_{rss} = 3$ pF (max). Analyze the amplifier.

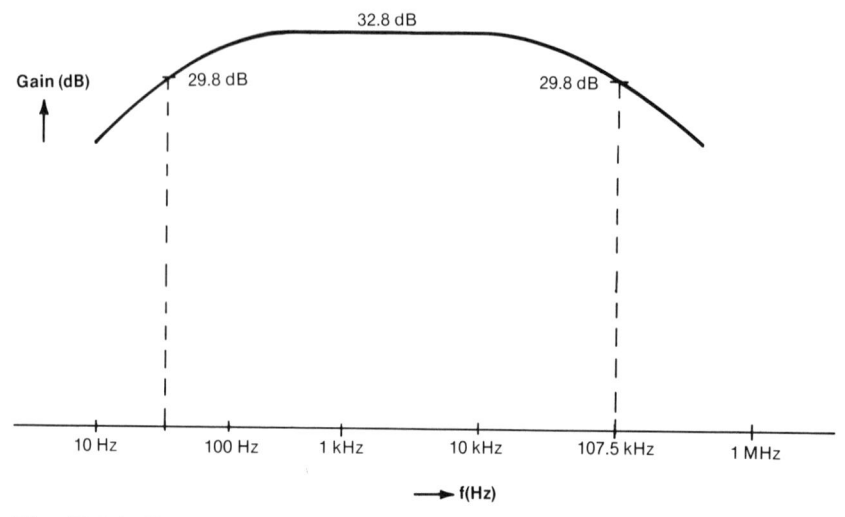

426   Frequency Response and Pulse Response of Amplifiers

Fig. 7-22. Frequency response curve in the neighborhood of the useful band of the amplifier of Example 7.5.

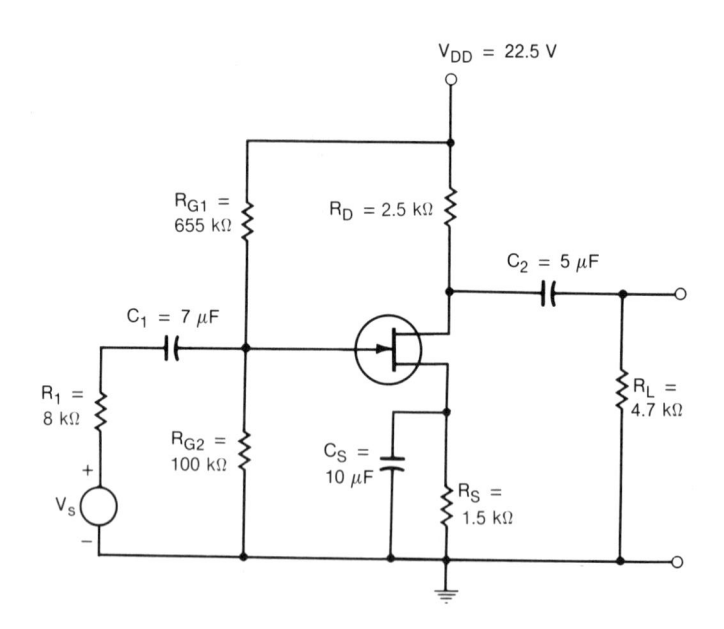

Fig. 7-23. Amplifier circuit of Example 7.6.

Electronics: Circuits and Systems

Solution

DC analysis:
Step 1: Draw the relevant portion of the circuit valid for dc analysis. All capacitors are open circuit for dc. The dc circuit is shown in Fig. 7-24.

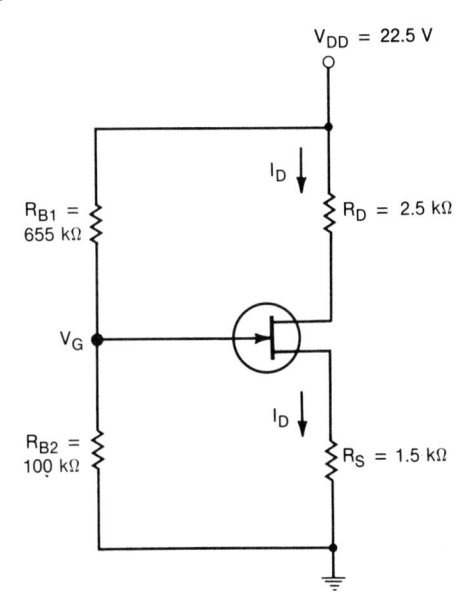

Fig. 7-24. Dc analysis of the amplifier of Example 7.6.

Step 2: Calculate the voltage at the gate terminal using the voltage-divider formula.

$$V_G = \frac{R_{G2}}{R_{G1} + R_{G2}} V_{DD}$$

$$= 2.98 \text{ V}$$

Step 3: Evaluate the two values of the constant K of the transfer characteristic using the two pairs of values of I_{DSS} and V_T. Plot the two (extreme) transfer characteristics.

$$K = (I_{DSS}/V_T^2) \text{ mA/V}^2.$$

When $I_{DSS} = 4$ mA, $V_T = -2$ V, $K = (I_{DSS}/V_T^2) = 1$ mA/V².
When $I_{DSS} = 16$ mA, $V_T = -8$ V, $K = 0.25$ mA/V². The two transfer characteristics are shown in Fig. 7-25.

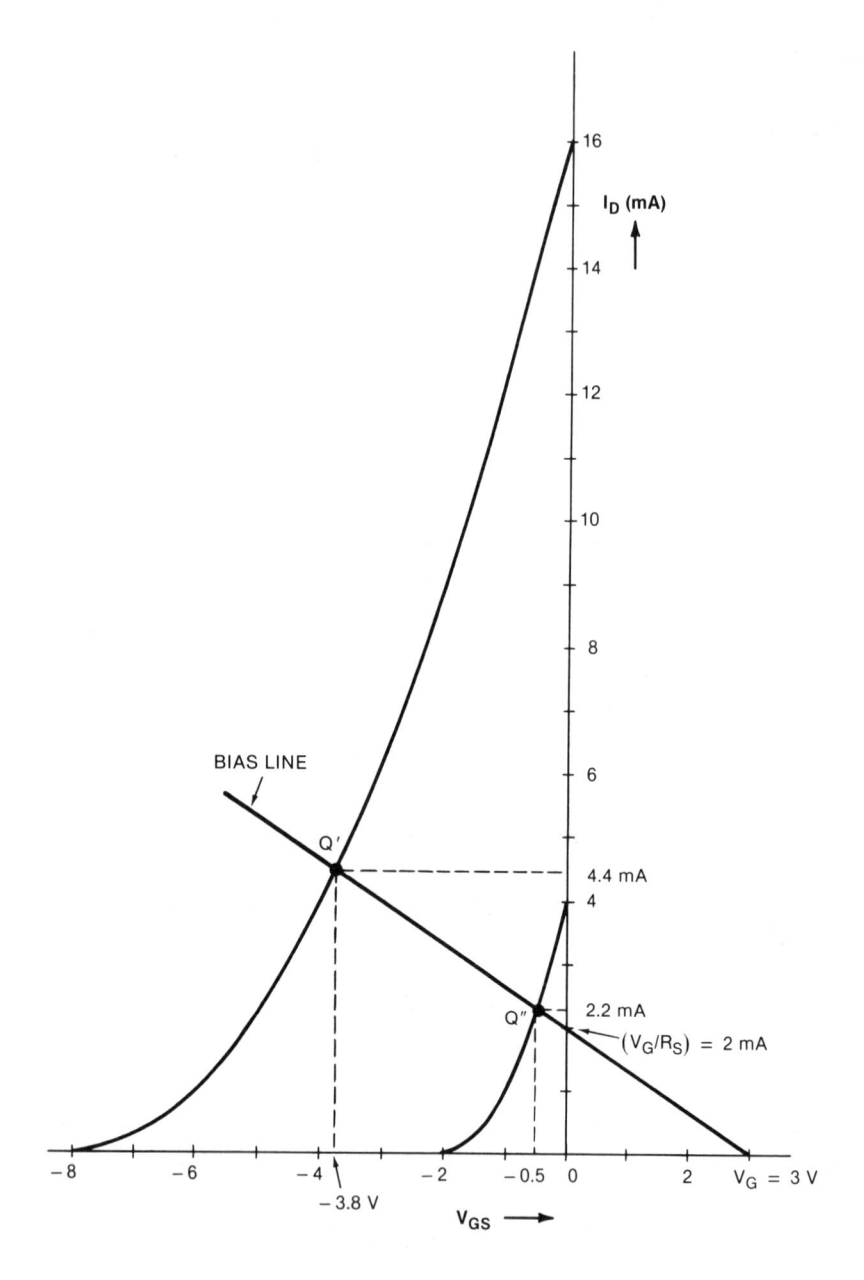

Fig. 7-25. Calculation of the Q points of the amplifier in Example 7.6.

Electronics: Circuits and Systems

Step 4: Draw the bias line by taking a point on the V_{GS} axis at V_G (from Step 2) and a point on the I_D axis at (V_G/R_S) and joining the two points.

Read off the values of I_D and V_{GS} at the two Q points (points of intersection of the bias line with the transfer characteristics).

$$I_D = 2.2 \text{ mA}$$
$$V_{GS} = -0.5 \text{ V}$$

(lower characteristic)

$$I_D = 4.4 \text{ mA}$$
$$V_{GS} = -3.8 \text{ V}$$

(upper characteristic)

Step 5: Calculate V_{DS} corresponding to each of the Q points by subtracting the drops in R_D and R_S from V_{DD}.

$$V_{DS} = 22.5 - 2.2 \times 2.5 - 2.2 \times 1.5$$
$$= 13.7 \text{ V}$$
$$V_{DS} = 22.5 - 4.4 \times 2.5 - 4.4 \times 1.5$$
$$= 4.9 \text{ V}$$

This completes the dc analysis of the amplifier. Two Q points (the two extremes) are obtained in the dc analysis.

Small-signal analysis (midband):

Step 1: Calculate the two values of transconductance g_m using the formula

$$g_m = 2K(V_{GS} - V_T) \text{ mS}$$

where,
 K = 1,
 $V_{GS} = -0.5$ V,
 $V_T = -2$ V.
Then,

$$g_m = 3 \text{ mS}$$

where,
 K = 0.25,
 $V_{GS} = -3.8$ V,
 $V_T = -8$ V.

430 Frequency Response and Pulse Response of Amplifiers

Then,

$$g_m = 2.1 \text{ mS}$$

Step 2: Draw the small-signal equivalent circuit (midband). The coupling and bypass capacitors are replaced by short circuits. The two resistors R_{G1} and R_{G2} are in parallel for ac currents and combined as R'_G. Similarly, the two resistors R_D and R_L are in parallel for ac currents and combined into a single resistance R_{eq}. The small-signal equivalent is shown in Fig. 7-26.

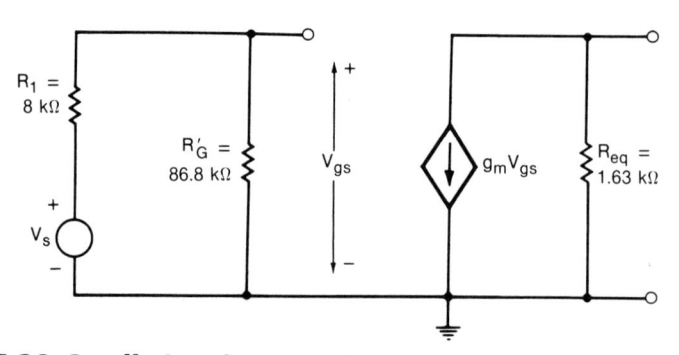

Fig. 7-26. Small-signal model of the amplifier of Example 7.6 for midband gain calculation.

Step 3: Calculate the voltage V_{gs} using the voltage-divider rule.

$$V_{gs} = \frac{R'_G}{R_1 + R'_G} V_s$$

$$= 0.916 V_s$$

Step 4: Calculate the output voltage V_o.

$$V_o = -g_m V_{gs} R_{eq}$$

When,
$g_m = 3$ mS:

$$V_o = -4.89 V_{gs}$$

When,
$g_m = 2.1$ mS:

$$V_o = -3.42 V_{gs}$$

Step 5: Calculate the midband gain (V_o/V_s) from the results of the last two steps. Convert it to decibels.

Electronics: Circuits and Systems

When,
  $g_m = 3$ mS:
$$V_o = -4.89 \times 0.916 V_s$$
therefore,
$$\text{Voltage gain} = -4.48 \text{ or } 13 \text{ dB}$$

When,
  $g_m - 2.1$ mS:
$$V_o = -3.42 \times 0.916 V_s$$
therefore,
$$\text{voltage gain} = -3.13 \text{ or } 9.91 \text{ dB}$$

This completes the small-signal (midband) analysis.

Frequency response—lower cutoff frequency:

Step 1: Calculate the lower cutoff frequency due to the input coupling capacitor C_1 (Fig. 7-27A).

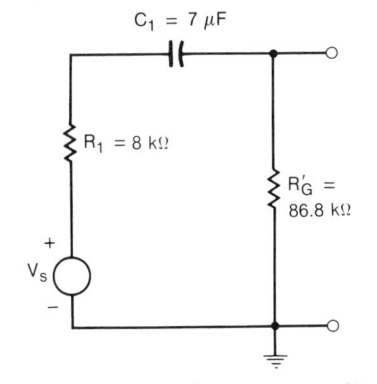

(A) Circuit with input coupling capacitor.

(B) Circuit with output coupling capacitor.

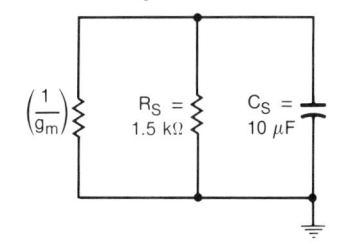

(C) Circuit with source bypass capacitor.

Fig. 7-27. Calculation of the lower cutoff frequency of the amplifier of Example 7.6.

$$f_1 = \frac{1}{2\pi C_1 (R_1 + R'_G)}$$

where,
R'_G is $(R_{G1} \| R_{G2})$.
Using $C_1 = 7\ \mu F$, $R'_G = 86.8\ k\Omega$, and $R_1 = 8\ k\Omega$, we get

$$f_1 = 0.24\ Hz$$

Step 2: Calculate the lower cutoff frequency due to the output coupling capacitor C_1 (Fig. 7-27B).

$$f_1 = \frac{1}{2\pi C_1 (R_D + R_L)}$$

$$= 4.42\ Hz$$

Step 3: Calculate the lower cutoff frequency due to the source bypass capacitor. The resistance for this calculation is R_S in parallel with $(1/g_m)$, which accounts for the effect of the FET (Fig. 7-27C).

$$f_1 = \frac{1}{2\pi C_S [R_S \| (1/g_m)]}$$

When,
$g_m = 3\ mS$:

$$[R_S \| (1/g_m)] = 0.272\ k$$

$$f_1 = 58.4\ Hz$$

When,
$g_m = 2.1\ mS$:

$$[R_s \| (1/g_m)] = 0.361\ k\Omega$$

$$f_1 = 44.1\ Hz$$

Since the cutoff frequency due to the source bypass capacitor is larger than that due to either of the other two capacitors by a factor of ten or more, we choose the lower cutoff frequency of the amplifier as either 44.1 Hz or 58.4 Hz. It is safer to use the higher of the two values:

$$\text{lower cutoff frequency} = 58.4\ Hz$$

Higher cutoff frequency:

Step 1: (Calculation of the Miller capacitance C'_F) The feedback capacitance C_F equals the drain-to-gate capacitance C_{dg} which equals the C_{rss} specified by the manufacturer. Calculate the Miller capacitance C'_F:

Electronics: Circuits and Systems

$$C'_F = C_{rss}(1 + g_m R_{eq})$$

where,
$R_{eq} = (R_D \| R_L)$

When,
$g_m = 3$ mS:

$$C_{rss} = 3 \text{ pF}$$
$$R_{eq} = 1.63 \text{ k}\Omega$$
$$C'_F = 17.7 \text{ pF}$$

When,
$g_m = 2.1$ mS:

$$C_{rss} = 3 \text{ pF}$$
$$R_{eq} = 1.63 \text{ k}\Omega$$
$$C'_F = 13.3 \text{ pF}$$

Step 2: Calculate the capacitance C_i, which equals the gate-to-source capacitance C_{gs}:

$$C_{gs} = C_{iss} - C_{rss}$$

using the specified values of C_{iss} and C_{rss}. (If no value of C_{iss} is given, assume $C_{gs} = 0$.)

$$C_i = C_{gs}$$
$$= 7 - 3$$
$$= 4 \text{ pF}$$

Step 3: Draw the high-frequency equivalent circuit of the input side (Fig. 7-28).

Calculate the upper cutoff frequency f_2.

$$f_2 = \frac{1}{2\pi C_{in}(R_1 \| R'_G)}$$

When,
$g_m = 3$ mS:

$$C_{in} = 17.7 + 4 = 21.7 \text{ pF}$$

and

$$f_2 = 1.00 \text{ MHz}$$

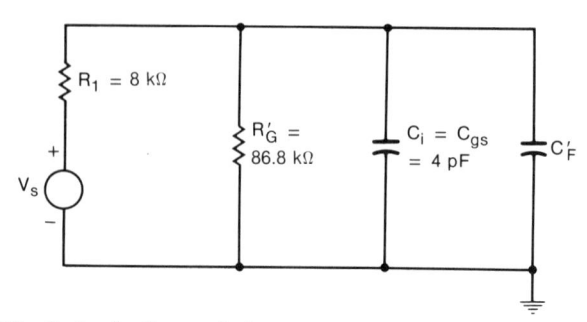

Fig. 7-28. Calculation of the upper cutoff frequency of the amplifier of Example 7.6.

When,
$g_m = 2.1$ mS:

$$C_{in} = 13.3 + 4$$
$$= 17.3 \text{ pF}$$

and

$$f_2 = 1.26 \text{ MHz}$$

Taking the worst case, the upper cutoff frequency is 1.00 MHz for the given amplifier.

$$\text{bandwidth} = 1 \text{ MHz}$$

The frequency response curve will be of the same *form* as the one shown for the BJT example.

PULSE RESPONSE OF AN RC-COUPLED AMPLIFIER

When a rectangular pulse is applied to the input of an RC-coupled amplifier, the shape of the output pulse is affected by the same capacitances in the amplifier that affected the frequency response. The coupling capacitors and the bypass capacitor (which caused the rolloff of the gain at low frequencies) produce a *sag* (or *tilt*) in the output as shown in Fig. 7-29B. The interelectrode capacitances of the device (which led to the decrease of the gain at high frequencies) introduce a rounding off of the output pulse as shown in Fig. 7-29C. The frequency response and the pulse response of an amplifier are, therefore, very closely related to each other. When studying frequency response, we were examining the variation of the output of an amplifier as a function of the *frequency* of the input signal. In pulse response our interest is in the variation of the output as a function of *time*. The pulse response is obtained from the *transient response* of the RC circuits present in an amplifier, and the ideas and principles discussed in Chapter 2 will be found to be useful in the study of the pulse response of an RC-coupled amplifier.

It is convenient to use the concepts and relationships of RC filters to relate the frequency response and the pulse response of an amplifier.

(A) Rectangular input pulse.

(B) Sag introduced by the coupling capacitors.

(C) Rounding off produced by interelectrode capacitances.

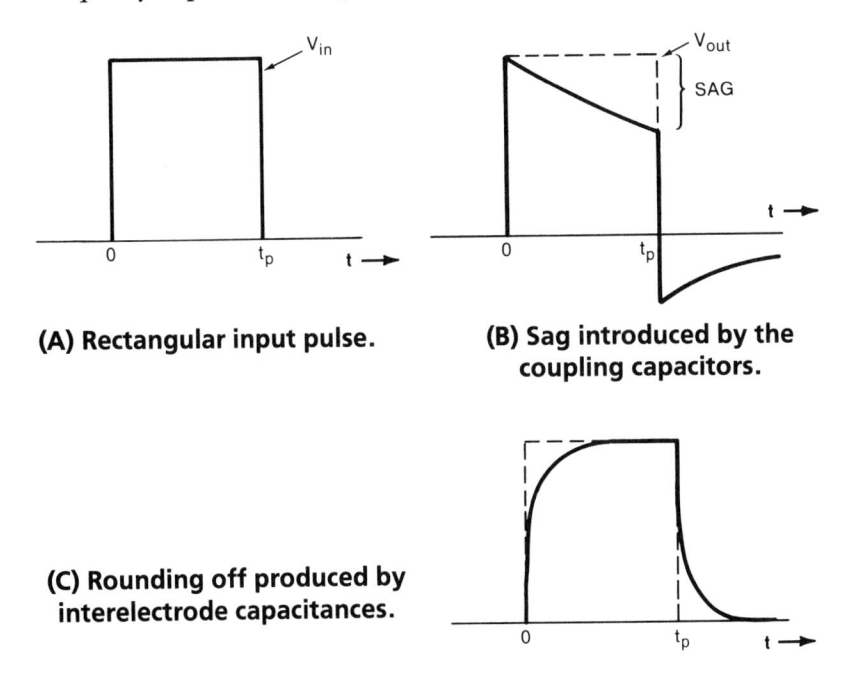

Fig. 7-29. Pulse response of an RC-coupled amplifier.

High-Pass RC Filter

An RC circuit connected as shown in Fig. 7-30A transmits the high-frequency components of the input signal freely, but attenuates the low-frequency components. Such a filter is called a *high-pass filter* (abbreviate HPF). The impedance of the capacitor C is small at high frequencies, the voltage drop across it is, therefore, negligible compared with that across the resistor R, and the output voltage (measured across R) is equal to the input signal at high frequencies. At low frequencies, the impedance of the capacitor is large, the voltage drop across it is also large compared with that across the resistor, and the output voltage is much less than the input at low frequencies. The *frequency response* of an HPF is shown in Fig. 7-30B and is seen to be exactly the same form as the lower half of the frequency response of an amplifier (Fig. 7-1). The cutoff frequency f_1 (at which the gain of the filter is 3 dB below the constant gain value) is given by

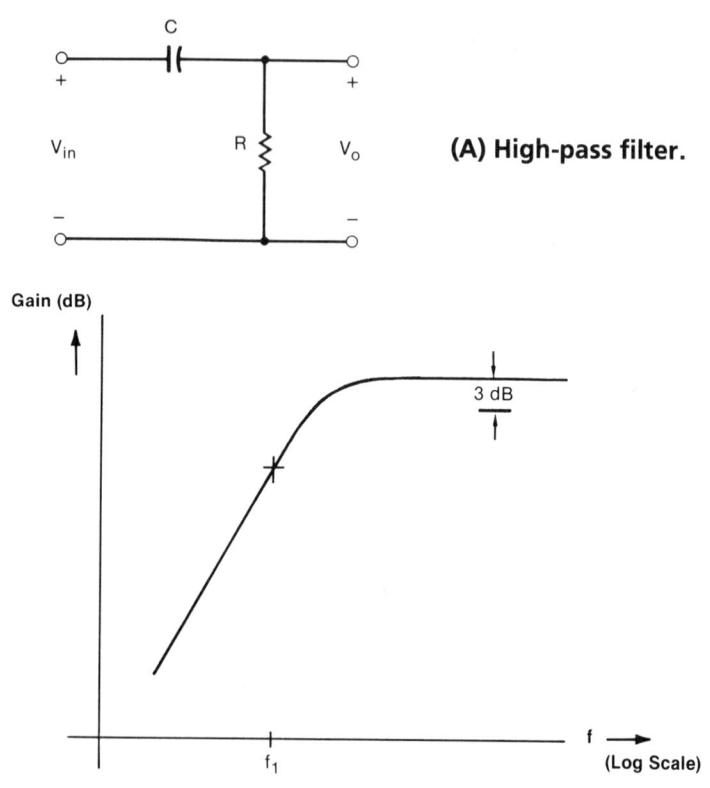

(A) High-pass filter.

(B) Frequency response curve of the high-pass filter.

Fig. 7-30. High-pass filter and its frequency response curve.

$$f_1 = \frac{1}{2\pi RC} \quad (7\text{-}20)$$

Consider the high-pass filter when the input is a rectangular pulse as shown in Fig. 7-31A. Suppose the voltage on the capacitor is zero before the pulse is applied. Using the voltage continuity condition (discussed in Chapter 2), it can be seen that the voltage across the capacitor will gradually build up toward V_1 at a rate controlled by the time constant RC. The output voltage (which is across the resistor) will, therefore, jump to V_1 at t = 0 and then

Electronics: Circuits and Systems

decay exponentially toward zero. The response of the filter in the interval from $t = 0$ to $t = t_p$ is, therefore, of the form shown in Fig. 7-31B.

If the duration of the pulse, t_p, is much smaller than the time constant RC, then the output will exhibit a *sag* (also called a *tilt*) rather than a complete decay. We will assume that the pulse duration is much less than RC so that the output shows only a sag rather than a substantial decay. For such cases, the *percent sag* is defined as (Fig. 7-31C):

$$\text{percent sag} = \frac{V_1 - V_1'}{V_1} \times 100$$

and the sag can be calculated by using the equation:

$$\text{percent sag} = \frac{t_p}{RC} \times 100$$

Combining the last expression with that for the cutoff frequency, Eq. (7-20), we have

$$\text{percent sag} = (2\pi f_1 t_p) \times 100$$

where,
f_1 is the cutoff frequency of the HPF,
t_p is the duration of the input pulse.

The complete pulse output (including the response after the pulse disappears) will be of the form shown in Fig. 7-31D. Note that there is a small excursion of V_0 below the time axis at $t = t_p$, which is caused by the requirement that the voltage across the capacitor cannot change but the input voltage changes instantaneously by V_1 volts to zero.

Let us now modify the results of the HPF to the pulse response of an RC-coupled amplifier. In the region of the lower half of the midband and below, the amplifier behaves just like a high-pass filter. The *lower cutoff frequency* f_1 of the amplifier (determined by the procedures described in the preceding sections) corresponds to the cutoff frequency of an HPF. The coupling and bypass capacitors that make the amplifier act like an HPF will cause a sag in the output of the amplifier when a rectangular pulse is applied as the input. If the pulse duration is small compared with the time constant RC, then the percent sag is given by

$$\text{percent sag} = (2\pi f_1 t_p) \times 100 \qquad (7\text{-}21)$$

where,
f_1 is the lower cutoff frequency of the amplifier,
t_p is the duration of the pulse.

The requirement that the pulse duration be small compared with the time constant RC can be written in the form:

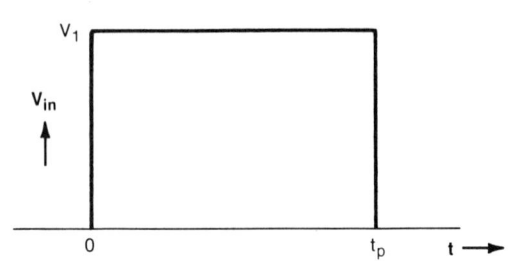

(A) Rectangular pulse input.

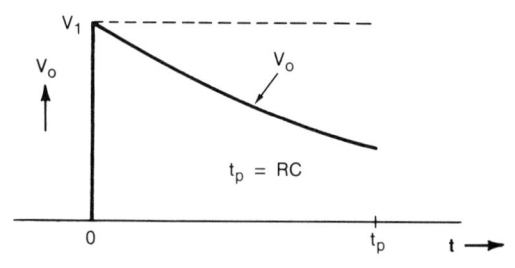

(B) Actual exponential decay from t = 0 to t = t_p.

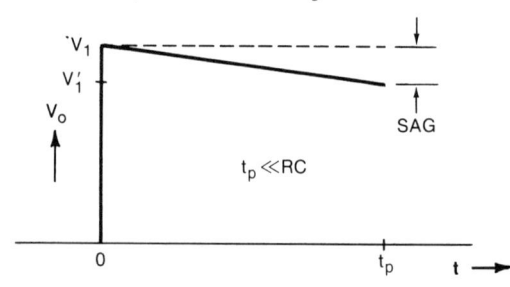

(C) Sag can be approximated by a straight line if the pulse duration is much less than the time constant.

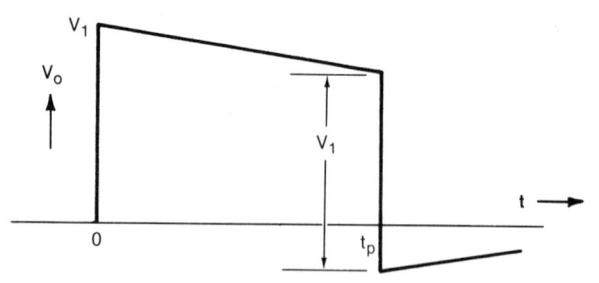

(D) Output goes negative at the end of the pulse because of the jump of the input signal by V_1 and the voltage continuity condition of the capacitor.

Fig. 7-31. Sag introduced by an HPF.

Electronics: Circuits and Systems

$$t_p \ll (1/2\pi f_1) \quad (7\text{-}22)$$

In order to keep the sag small (say less than 10%) it is necessary to *decrease the lower cutoff frequency* of the amplifier to a sufficiently low value as determined by Eq. (7-21). Conversely, for a given amplifier (that is, f_1 already fixed), Eq. (7-21) can be used to calculate the longest duration of a pulse that will be transmitted with a specified sag.

Example 7.7

The lower cutoff frequency of an amplifier is 45 Hz. Calculate the percent sag for a rectangular pulse input of duration: (a) $t_p = 1$ ms; (b) $t_p = 0.5$ ms; (c) $t_p = 0.1$ ms.

Solution:

A straightforward application of Eq. (7-21) leads to (a) 28.3%; (b) 14.4%; (c) 2.83%.

[Note: Any sag of greater than 20% implies that the relationship of Eq. (7-21) is not really valid since the condition (7-22) is probably not satisfied.]

Example 7.8

A pulse train is applied as the input to an amplifier. The pulse train has a repetition rate of 50 kHz. Find the value of the lower cutoff frequency the amplifier should have in order to have a sag of no more than 5% in the output.

Solution

A pulse repetition rate of 50 kHz means that the duration of each pulse should be

$$t_p = (T/2)$$

where,

T is the period of the pulse train.
Refer to Fig. 7-32.

$$\text{period } T = (1/50 \times 10^3)$$

$$= 0.02 \times 10^{-3} \text{ s}$$

$$\text{pulse duration } t_p = 0.01 \times 10^{-3}$$

Using Eq. (7-21), for a 5% sag,

$$5 = (2\pi f_1 \times 0.01 \times 10^{-3}) \times 100$$

Frequency Response and Pulse Response of Amplifiers

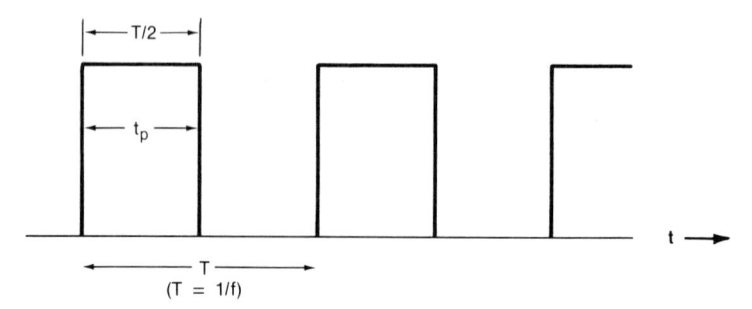

Fig. 7-32. Illustrating the relationship for pulse duration in pulse train, $t_p = (T/2)$.

or

$$f_1 = 796 \text{ Hz}$$

Drill Problem 7.7: If a pulse train of 10 kHz were applied as an input to the amplifier of the last example, calculate the percent sag that will result.

Drill Problem 7.8: Calculate the sag that will be caused if the pulse duration is $t_p = 5$ ms when the pulse is applied to an amplifier with a lower cutoff frequency given by (a) 10 Hz; (b) 5 Hz; (c) 0.5 Hz.

It should be noted that the sag is of concern when the input pulse train to an amplifier has a *low-repetition rate*. (Compare the results of Example 7.8 and Drill Problem 7.7). In most practical applications, the pulse repetition rate is high (tens of kilohertz to many megahertz), and the sag is, therefore, not necessarily a serious problem.

Low-Pass RC Filter

An RC circuit connected as shown in Fig. 7-33A transmits the low-frequency components of the input signal freely, but attenuates the high-frequency components. Such a filter is called a *low-pass filter* (abbreviated LPF). The impedance of the capacitor C is large at low frequencies, the voltage drop across it is large compared with that across the resistor, and the output voltage is equal to the input signal. At high frequencies, the impedance of the capacitor is small, the voltage across it is negligible compared with that across the resistor, and the output voltage is much less than the input at high frequencies. The *frequency response* of an LPF is shown in Fig. 7-33B and is seen to be exactly the same form as the upper half of the frequency response of an amplifier (Fig. 7-1). The cutoff frequency f_2 (at which the gain of the filter is 3 dB below the constant gain value) is given by

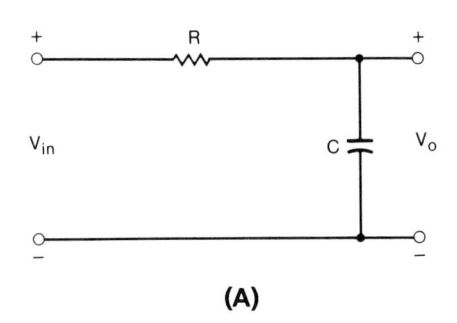

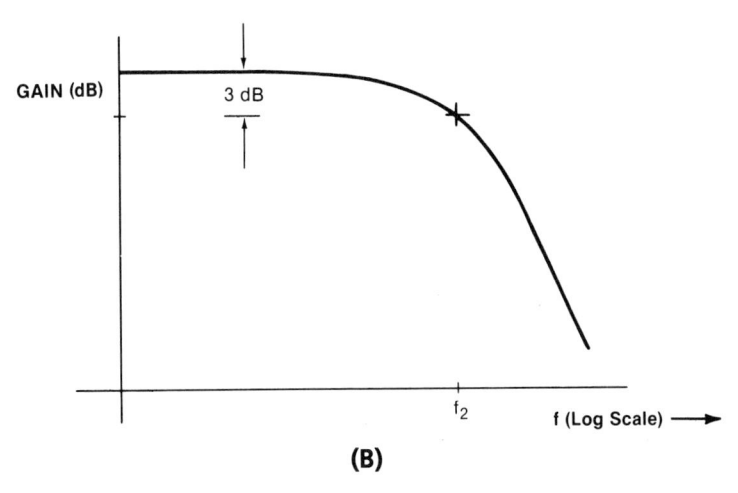

Fig. 7-33. Low-pass filter and its frequency response curve.

$$f_2 = \frac{1}{2\pi RC} \qquad (7\text{-}23)$$

Consider the LPF when the input is a rectangular pulse as shown in Fig. 7-34A. Suppose the voltage on the capacitor is zero before the pulse is applied. The voltage continuity condition of a capacitor implies that the output voltage will gradually build up toward V_1 at a rate controlled by the time constant RC. The response of the LPF in the interval from $t = 0$ to $t = t_p$ is of the form shown in Fig. 7-34B. Instead of jumping instantaneously to V_1 like the input pulse, the output takes a certain length of time to reach V_1. A measure of the *speed of response* of the LPF to an input pulse is the *10% to 90% rise time*, t_R, which is defined as the time taken for the output voltage to rise from 10% of its final value (that is, $0.1V_1$) to 90% of its final value (that is, $0.9V_1$). Refer to Fig. 7-34C.

It can be shown that if RC is the time constant of the LPF, then the rise time is given by

$$t_R = 2.2RC \text{ s} \tag{7-24}$$

Let us now modify the results of the LPF to the pulse response of an amplifier. In the region of the upper half of the midband and above, the amplifier behaves like a low-pass filter. The upper cutoff frequency f_2 of the amplifier (determined by the procedures described in the preceding sections of this chapter) corresponds to the cutoff frequency of an LPF. The interelectrode capacitances that make the amplifier act like an LPF will cause the output of the amplifier to rise gradually even if the input is a pulse that jumps instantaneously from one value to another. The 10% to 90% rise time t_R is a measure of the speed of response of the amplifier to an input pulse. Combining Eq. (7-24) for the rise time and Eq. (7-23) for the upper cutoff frequency, we get,

$$\text{rise time } t_R = \frac{2.2}{2\pi f_2}$$

$$= \frac{0.35}{f_2} \text{ s} \tag{7-25}$$

If the speed of response of an amplifier is to be increased, the rise time has to be made smaller, which means that the upper cutoff frequency must be made larger.

That is, the higher the upper cutoff frequency of an amplifier, the faster is its response to an input pulse. Since the *bandwidth* of an amplifier is approximately equal to its upper cutoff frequency (refer to Examples 7.6 and 7.7), we can make the following statement: *The wider the bandwidth the faster the response.*

If the response of the amplifier is not sufficiently fast, then it is possible that the output voltage never reaches the final value V_1 before the pulse returns to zero. Two cases are shown in Fig. 7-35 to compare the outputs of two amplifiers for a given input pulse. The distortion in the case of Fig. 7-35B is seen to be quite severe.

Example 7.9

Calculate the rise time of an amplifier if its upper cutoff frequency is (a) 100 MHz; (b) 10 MHz; (c) 1 kHz.

(A) Rectangular input pulse.

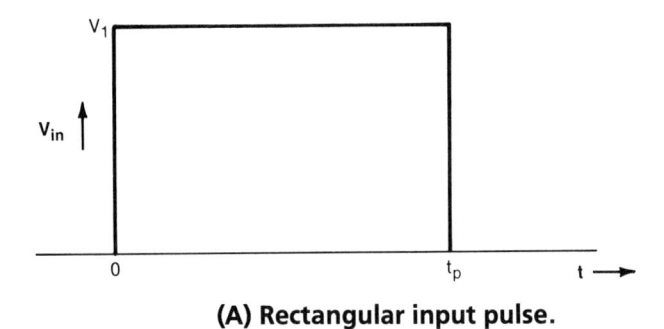

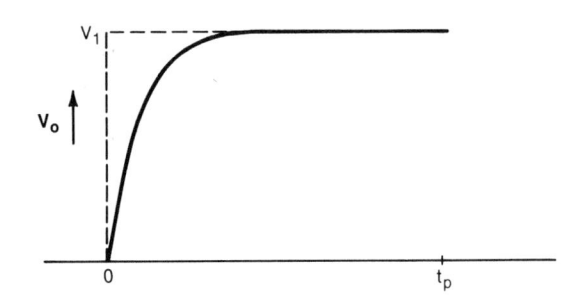

(B) Response of LPF from $t = 0$ to $t = t_p$.

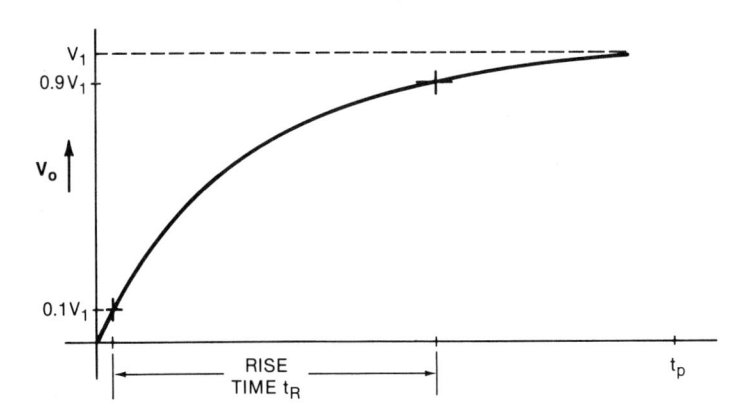

(C) Rise time is used as a measure of the speed of response of the LPF.

Fig. 7-34. Rounding off introduced by an LPF.

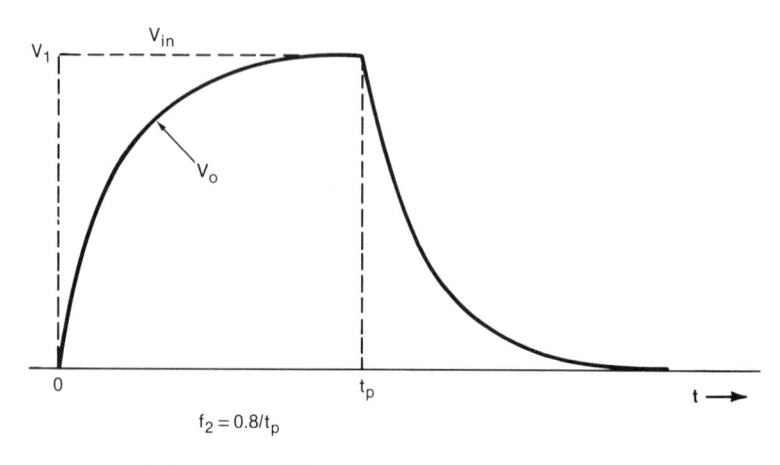

(A) Bandwidth is ten times larger than B.

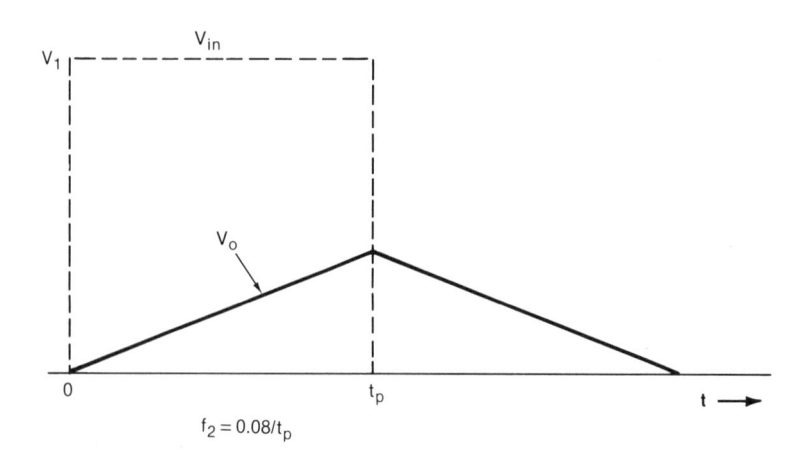

(B) Bandwidth is ten times smaller than A.

Fig. 7-35. Effect of an amplifier's bandwidth on the shape of its output pulse.

Solution
A straightforward application of Eq. (7-25) gives (a) 3.5 ns; (b) 35 ns; (c) 0.35 ms.

Electronics: Circuits and Systems 445

Example 7.10

A pulse train of repetition rate 40 MHz is applied to an amplifier. If the rise time is to be 50% of the pulse duration, calculate the upper cutoff frequency needed.

Solution

The duration of the pulse is $t_p = (T/2)$ (as in Example 7.7).

$$T = \frac{1}{40 \times 10^6}$$

$$= 2.5 \times 10^{-8} \text{ s}$$

$$t_p = 1.25 \times 10^{-8} \text{ s}$$

The rise time is to be 50% of t_p. Therefore, $t_R = 0.625 \times 10^{-8}$ s. Using Eq. (7-25), we have

$$0.625 \times 10^{-8} = (2.2/2\pi f_2)$$

Therefore,

$$f_2 = 56 \text{ MHz}$$

Drill Problem 7.9: Calculate the rise time of an amplifier if its upper cutoff frequency is (a) 50 kHz; (b) 895 MHz.

Drill Problem 7.10: If an input pulse train of repetition rate of 10 MHz is applied to the amplifier of Example 7.10, calculate the rise time as a percentage of pulse duration.

Drill Problem 7.11: Calculate the maximum repetition rate of an input pulse train that can be handled by an amplifier with a 625-kHz bandwidth if the rise time is to be no more than 10% of the pulse duration.

It can be seen that the smaller the rise time, the wider the bandwidth of the amplifier. A zero rise time requires an infinite bandwidth! Some criterion is necessary to determine what is an allowable distortion in the output of an amplifier. A commonly used *rule of thumb* is to make the *bandwidth* f_2 equal to the *reciprocal of the pulse width* t_p. That is,

$$t_p = (1/f_2)$$

in which case the resulting rise time is

$$t_R = (2.2/2\pi f_2)$$

$$= 0.35 t_p$$

The rise time is 35% of the pulse width in such cases.

The level of distortion of the output pulse of an amplifier is not always an overriding consideration. In a number of cases (especially in digital communication systems), it is only necessary to ascertain whether a pulse is present or not. (Once the presence of a pulse is sensed, the distorted pulse can be reshaped to a more desirable form by special circuits situated at various points in the system.) In such cases, the bandwidth can be much smaller than when the output pulse must closely resemble the input pulse.

Drill Problem 7.12: Recalculate the maximum repetition rate that can be handled by the amplifier of Drill Problem 7.11 using the rule of thumb in Eq. (7-26).

INCREASING THE BANDWIDTH OF AN AMPLIFIER

Amplifiers of wide bandwidths (called *wideband amplifiers*) are needed in systems where the input signals have high frequency contents (such as TV signals, for example) and also where the inputs are pulse trains of high repetition rates. Several methods are available for increasing the bandwidth of an amplifier. A careful selection of the transistor itself is helpful: a transistor with low values of interelectrode capacitances will give rise to large bandwidths. In the case of a BJT, a high value of f_T is a requirement. Transistors designated RF transistors are available for such purposes. The Miller capacitance can be reduced by reducing the product $(g_m R_{eq})$. In such cases, we are trading a wide bandwidth for a low gain.

An alternative method of reducing the Miller capacitance is to use a *common-base amplifier.* In the common-base amplifier, the feedback capacitance C_F is absent since there is no interelectrode capacitance between the input (the emitter) and the output (the collector) terminals. The Miller effect is, therefore, not a problem in common-base amplifiers, and their total input capacitance is quite small (a few pF). The higher cutoff frequency of a common-base stage is much greater than that of a common-emitter stage. A popular configuration for large bandwidth is the *cascode amplifier* (Fig. 7-36) that uses a common-emitter stage as the first stage and a common-base amplifier as the second stage. The low input impedance of the common-base stage decreases the value of R_{eq} seen by the common-emitter stage, which reduces the Miller capacitance of the first stage and increases its higher cutoff frequency. Since there is no Miller effect in the second stage (being common base), the bandwidth of the cascode amplifier is much wider than can be obtained from the single-stage common-emitter amplifier. The cascode amplifier has a high gain, a large bandwidth, a high input impedance, and provides good isolation between the input and output. It is widely used in high-frequency applications.

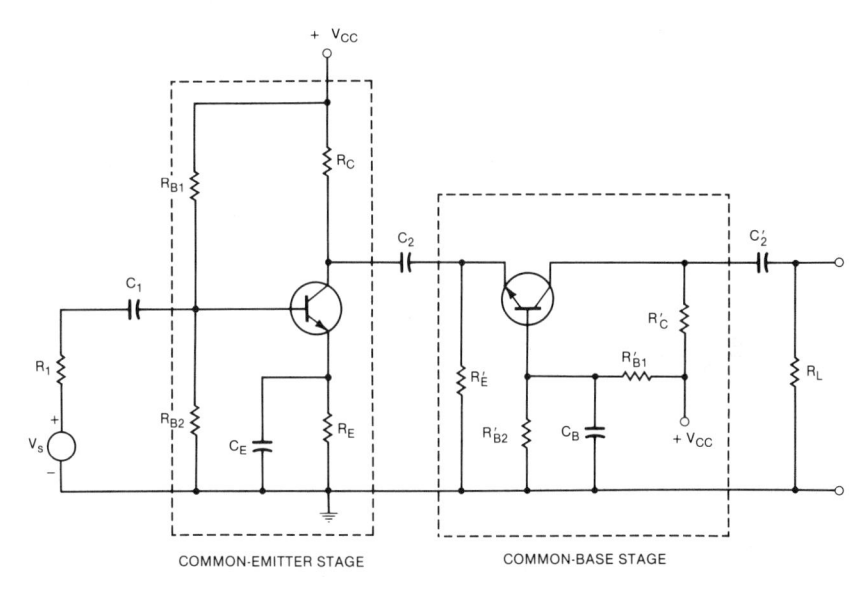

Fig. 7-36. Cascode amplifier (a common-emitter stage followed by a common-base amplifier).

An inductor can be used in series with the resistor R_C of a BJT so as to compensate for the effect of the interelectrode capacitances, which is called *shunt peaking*. The gain of such an amplifier is made to "peak," that is, go *higher than the midband value* at and near the upper cutoff frequency so that the frequency at which the gain drops by 3 dB is extended past the upper cutoff frequency of an ordinary RC-coupled amplifier.

SUMMARY SHEETS

FREQUENCY RESPONSE
Variation of voltage gain (in decibels) as a function of frequency (log scale) is frequency response.

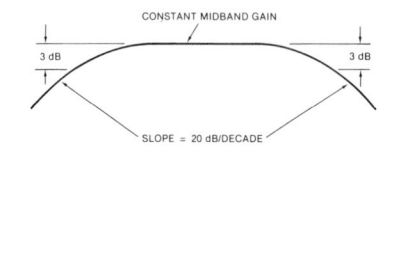

f_1 = lower cutoff frequency

f_2 = upper cutoff frequency

FACTORS AFFECTING FREQUENCY RESPONSE

Coupling Capacitors
Input coupling capacitor C_1 has a high impedance at low frequencies and causes a reduction in the input voltage V_{in} of the amplifier.

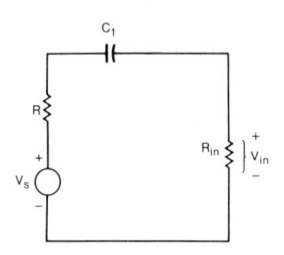

Output coupling capacitor C_2 has a high impedance at low frequencies and causes a reduction in the output voltage.

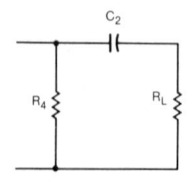

Bypass Capacitors
The *bypass capacitor* acts as a high impedance at low frequencies. The resistor R_3 is no longer perfectly bypassed and produces a negative feedback, which causes a reduction in the gain of the amplifier.

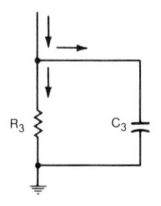

Interelectrode Capacitances
Interelectrode capacitances affect the high frequency response. C_i (which is $C_{b'e}$ for a BJT, C_{gs} for a FET) diverts part of the input current available from the signal source, thus reducing the input current to the device. C_F (which is $C_{b'c}$ for a BJT and C_{dg} for a FET) provides a low-impedance link between the input and output thus forcing the voltage gain toward a value of 1.

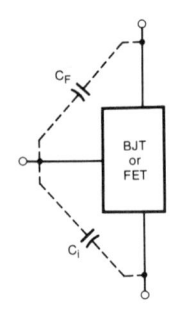

LOWER CUTOFF FREQUENCY f_1
Due to each of the capacitors acting alone f_1 is first calculated.

f_1 Due to C_1

$$f_1 = \frac{1}{2\pi C_1(R_1 + R_{in})}$$

$R_{in} = (R_B \| h_{ie})$ [BJT]

$R_{in} = R_G$ [FET]

Electronics: Circuits and Systems

f₁ Due to C₂

$$f_1 = \frac{1}{2\pi C_2(R_4 + R_L)}$$

$R_4 = R_C$ [BJT] and R_D [FET]

f₁ Due to Bypass Capacitor

FET:

$$f_1 = \frac{1}{2\pi C_S[R_S \| (1/g_m)]}$$

BJT: Reflected resistance

$$R_{ref} = \frac{r_{b'e} + (R_B \| R_1)}{g_m r_{b'e}}$$

$$f_1 = \frac{1}{2\pi C_E(R_E \| R_{ref})}$$

Overall Value of f₁

If the highest of the three calculated values of f_1 is at least ten times larger than the other two, then the highest of the three calculated values of f_1 is taken as the lower cutoff frequency of the amplifier. If this condition is not satisfied, then a graphical procedure is needed to estimate the lower cutoff frequency of the amplifier.

UPPER CUTOFF FREQUENCY f₂—MILLER EFFECT

The capacitance C_F produces a negative feedback from the output to the input side. Consequently, the effect of C_F as seen from the input side is much larger than that due to its actual value.

C_F can be replaced by an equivalent Miller capacitance C_F' between the input terminal and ground:

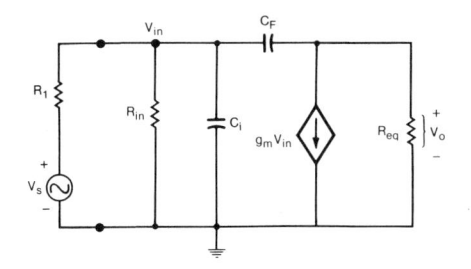

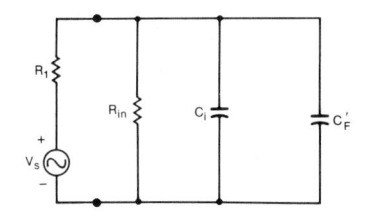

$$C_F' = (1 + g_m R_{eq})C_F$$

$$R_{eq} = (R_C \| R_L) \text{ [BJT]}$$

$$R_{eq} = (R_D \| R_L) \text{ [FET]}$$

Total input capacitance

$$C_{in} = C_i + C_F'$$

$$C_i = C_{b'e} \text{ [BJT] and } C_{gs} \text{ [FET]}$$

Upper cutoff frequency is

$$f_2 = \frac{1}{2\pi C_{in}(R_1 \| R_{in})}$$

SPECIFICATIONS OF PARAMETERS

BJT

Manufacturer gives values of h_{FE}, $C_{b'c}$, and f_T.
Calculate

$$g_m = \frac{I_C \text{ at Q point in mA}}{0.026} \approx 40\, I_C \text{ mS}$$

$$r_{b'e} = (h_{FE}/g_m)\, k\Omega$$

$$C_F = C_{b'c} \text{ (given)}$$

$$C_i = C_{b'e} = (g_m/2\pi f_T)$$

FET

Manufacturer gives the values of the threshold voltage (V_T or V_P), I_{DSS} or some information to evaluate the constant K of the transfer characteristic equation, a capacitance C_{rss} and a capacitance C_{iss}.
Calculate

$$g_m = 2K(V_{GS} - V_T)$$

where,
V_{GS} is the gate-to-source voltage at the Q point.

$$C_F = C_{dg} = C_{rss}$$

(the smaller of the two capacitances given).

$$C_i = C_{gs} = (C_{iss} - C_{rss})$$

PULSE RESPONSE
Variation of the output as a function of time.

High-Pass Filter (HPF)
An amplifier acts as an HPF at low frequencies. Cutoff frequency of the filter f_1 = lower cutoff frequency of the amplifier.

The output pulse (for a pulse input) exhibits a sag. If the pulse duration $t_p \ll RC$, then

$$\text{percent sag} = 100 \times (t_p/RC)$$
$$= 100 \times (2\pi f_1 t_p)$$

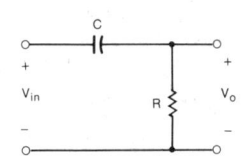

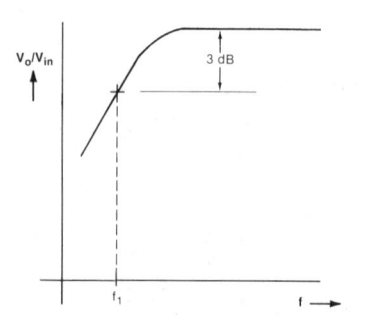

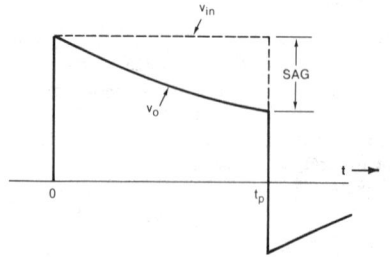

Smaller the value of $f_1 \longleftrightarrow$ the smaller the sag in the output pulse.

Low-Pass Filter (LPF)
An amplifier acts as an LPF at high frequencies. Cutoff frequency of the filter f_2 = upper cutoff frequency of the amplifier.

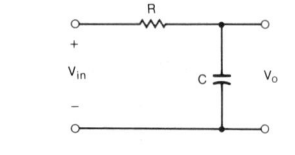

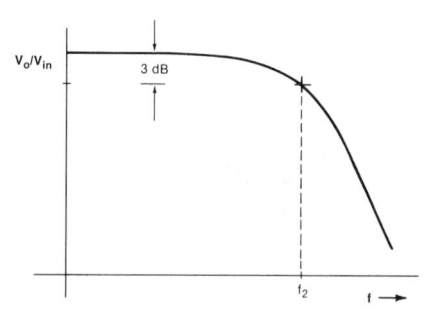

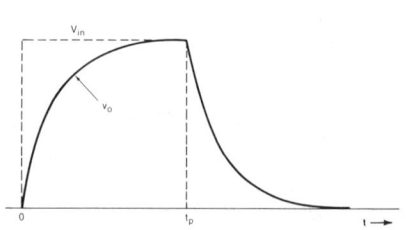

The output pulse (for a pulse input) shows a rounding-off effect. 10% to 90% rise time

$$t_R = 2.2RC = 0.35f_2$$

Higher the upper cutoff frequency the shorter $\longleftrightarrow$ the rise time.

Speed of response increases as the bandwidth of the amplifier increases.

Rule of thumb for bandwidth for a given pulse duration:

$$\text{bandwidth} = (1/t_p)$$

which makes the rise time

$$t_R = 0.35 t_p$$

INCREASING THE BANDWIDTH

1. Select a transistor with a higher f_T.
2. Reduce the Miller effect by reducing $g_m R_{eq}$; that is, trade off some gain to obtain a wider bandwidth.
3. Use a common-base amplifier stage. The input resistance of a common-base amplifier can be made quite small, and there is no Miller effect in it. Therefore, the upper cutoff frequency can be made much higher than in a common-emitter stage.

A *cascode amplifier* is a common-emitter stage followed by a common-base stage. Since R_{in} of the common-base stage is very low, R_{eq} of the first stage is low, which increases the bandwidth. The cascode amplifier has a high gain, a wide bandwidth, a high R_{in} and provides a good isolation between the input and the output.

ANSWERS TO DRILL PROBLEMS

7.1 $R_{ref} = 0.0231$ kΩ. $R_3 = R_E = 5$ kΩ. If $f_1 = 200$ Hz, then $C_E = 34.4$ μF.

7.2 $R_B = 13$ kΩ. $R_{in} = 3.34$ kΩ. If $f_1 = 8$, then $C_1 = 3.73$ μF and $C_2 = 2.12$ μF.

7.3 $f_1 = 0.292$ Hz.

7.4 $R'_G = 82.5$ kΩ = R_{in}. $f_1 = 0.120$ Hz due to C_1. $f_1 = 1.99$ Hz due to C_2. $f_1 = 180$ Hz due to C_S. Therefore, the lower cutoff frequency = 180 Hz.

7.5 160 Hz. (approx).

7.6 $C'_F = 81.2$ pF. $C_{in} = 85.2$ pF. $f_2 = 96.7$ kHz.

7.7 $f_1 = 796$ Hz. With $t_p = (T/2) = 0.05$ ms, percent sag = 25%. (Sag too high for using the approximate formula.)

7.8 (a) 31.4% (b) 15.7% (c) 1.57%.

7.9 (a) 7 μs (b) 0.391 ns.

7.10 $t_p = 50$ ns and $t_R = 6.25$ ns. $(t_R/t_p) = 12.5\%$.

7.11 $t_p = 10 \times t_R = 5.6$ μs. $T = 2t_p = 11.2$ μs. Repetition rate = 89.3 kHz.

7.12 $t_p = (1/f_2) = 1.6$ μs. $T = 3.2$ μs. Repetition rate = 312 kHz.

PROBLEMS

Lower Cutoff Frequency

7.1 (a) Use a BJT with $g_m = 50$ mS and $r_{b'e} = 2 k\Omega$ in the circuit of Fig. 7-2, and calculate the lower cutoff frequency when $R_1 = 100$ kΩ, $R_{B1} = 90$ kΩ, $R_{B2} = 10$ kΩ, $R_E = 2$ kΩ, $R_c = 5$ kΩ, $R_L = 5$ kΩ, $C_1 = C_2 = 25$ μF, and $C_E = 500$ μF. (b) Recalculate the lower cutoff frequency if R_1 is reduced to 0.5 kΩ while the other components remain as in part (a). (c) Recalculate the lower cutoff frequency if R_L is increased to 100 kΩ while the other components remain as in part (a). Comment on the effect of R_1 and R_L on the lower cutoff frequency.

7.2 (a) Use an FET with $g_m = 5$ mS in the circuit of Fig. 7-2, and calculate the lower cutoff frequency when $R_1 = 100$ kΩ, $R_{G1} = 9$ MΩ, $R_{G2} = 1$ MΩ, $R_S = 2$ kΩ, $R_D = 5$ kΩ, $R_L = 5$ kΩ, $C_1 = C_2 = 25$ μF, and $C_S = 500$ μF. (b) Recalculate the lower cutoff frequency if R_1 is reduced to 0.5 kΩ while the other components remain as in part (a). (c) Recalculate the lower cutoff frequency if R_L is increased to 100 kΩ while the other components remain as in part (a). Comment on the effect of R_1 and R_L on the lower cutoff frequency. Compare the effect here with that in the BJT amplifier of the previous problem.

7.3 In the amplifier circuit of Fig. 7-7, keep the same resistances as given. Choose values of C_1, C_2, and C_E so that the lower cutoff frequency due to $C_1 = 20$ Hz, due to $C_2 = 50$ Hz and due to $C_E = 100$ Hz. Estimate the lower cutoff frequency of the amplifier.

7.4 Repeat Problem 7.3 for the FET amplifier of Fig. 7-8.

Upper Cutoff Frequency

7.5 The feedback capacitance in an amplifier model is $C_F = 20$ pF. Calculate the Miller equivalent capacitance C_F' in each of the following cases. (a) $g_m = 50$ mS, $R_c = 5$ kΩ, $R_L = 0.25$ kΩ (b) $g_m = 5$ mS, $R_D = 5$ kΩ, $R_L = 0.25$ kΩ. (c) $g_m = 50$ mS, $R_c = 5$ kΩ, $R_L = 100$ kΩ. (d) $g_m = 5$ mS, $R_D = 5$ kΩ, $R_L = 100$ kΩ.

7.6 In the high-frequency equivalent circuit of the BJT amplifier of Fig. 7-15, (a) increase the value of R_1 to 100 kΩ and recalculate the upper cutoff frequency; (b) increase the value of R_L to 100 kΩ (with R_1 changed back to 0.5 kΩ as in the original circuit) and recalculate the upper cutoff frequency. Comment on the effect of R_1 and R_L on the upper cutoff frequency.

7.7 In the high-frequency equivalent circuit of the FET amplifier of Fig. 7-16, (a) reduce the value of R_1 to 0.25 kΩ and recalculate the upper cutoff frequency; (b) increase the value of R_L to 100 kΩ and recalculate the upper cutoff frequency (with R_1 returned to its original value of 25 kΩ). Comment on the effect of R_1 and R_L on the upper cutoff frequency. Compare these effects with the case of the BJT amplifier.

Complete Response

7.8 In the common-emitter amplifier circuit of Fig. 7-17, change the components to have the following values: $R_1 = 0.6$ kΩ, $R_{B1} = 90$ kΩ, $R_{B2} = 10$ kΩ, $R_E = 1$ kΩ, $R_C = 4$ kΩ, $R_L = 100$ kΩ, $C_1 = 5$ μF, $C_2 = 20$ μF, $C_E = 250$ μF, and $V_{CC} = 15$ V. The transistor has the following specifications: $h_{FE} = 200$, $f_T = 300$ MHz, $C_{b'c} = 10$ pF. Perform a complete analysis of the amplifier.

7.9 In the common-source amplifier of Fig. 7-23, change the components to the following values: $R_1 = 0.6$ kΩ, $R_{G1} = 840$ kΩ, $R_{G2} = 160$ kΩ, $R_D = 5$ kΩ, $R_S = 2$ kΩ, $R_L = 100$ kΩ, $C_1 = C_2 = 10$ μF, and $C_S = 50$ μF. $V_{DD} = 22.5$ V. The FET has the following specifications: $k = 1.1$ mA/V^2, $V_T = +0.5$ V. $C_{iss} = 10$ pF and $C_{rss} = 2$ pF. Perform a complete analysis of the amplifier.

7.10 In a common-emitter amplifier (with the same configuration as the one in Fig. 7-17), it is desired to have the Q point at $I_C = 1.5$ mA and $V_{CE} = 6$ V. The same transistor as in Example 7-5 is to be used. If it is given that $V_{CC} = 12$ V, $R_C = 3$ kΩ, $R_1 = 2$ kΩ, $R_{B2} = 10$ kΩ, and $R_L = 10$ kΩ, calculate the values of the other resistors. Calculate the midband gain. Choose the bypass capacitor so as to have a lower cutoff frequency of 500 Hz, and the coupling capacitors sufficiently large so that the lower cutoff frequency is determined by the bypass capacitor. Calculate the upper cutoff frequency of the amplifier.

7.11 A figure of merit, called the gain-bandwidth product, is useful in amplifiers. Taking the upper cutoff frequency as the bandwidth of the amplifier and the voltage gain as $g_m R_{eq}$, the gain-bandwidth product = $g_m R_{eq} f_2$. Make the following approximations: $C_F' \gg C_{b'e}$ and $g_m R_{eq} \gg 1$, in order to arrive at a simple expression for the gain-bandwidth product. This will show that an attempt to increase the midband gain of an amplifier will tend to decrease its bandwidth, and conversely, an attempt to increase the bandwidth will tend to decrease the midband gain.

Pulse Response

7.12 An amplifier has a lower cutoff frequency of 20 Hz. What is the minimum repetition rate of a pulse train so that the sag is no more than 10%? Assume that each pulse occupies one half the period of the pulse train.

7.13 A pulse is of duration 100 ns. When transmitted through a single-stage RC-coupled amplifier, the rise time of the output pulse is to be 70% of its duration. Calculate the bandwidth of the amplifier.

7.14 The bandwidth of an amplifier is 6 MHz. Calculate the maximum repetition rate of a pulse train (with evenly spaced pulses as in Fig. 7-32) that can be transmitted through the amplifier so that each pulse reaches its *maximum* value at ($t_p/2$), where t_p is the duration of each pulse.

7.15 In a television picture tube, the time taken for each horizontal scan by the electron beam is 63.5 μs. The total number of picture elements on the screen is 150×10^3. The screen has an aspect ratio of 4:3 (that is, there are four picture elements in the horizontal direction to every three picture elements in the vertical direction). Use the rule of thumb for the relationship between pulse duration and bandwidth given in this chapter and calculate the bandwidth needed for the transmission of a TV signal. Recalculate the bandwidth for the number of picture elements changed to 250×10^3 and to 50×10^3.

7.16 Discuss a figure of merit analogous to the gain-bandwidth product (Problem 7.11) that is applicable to the pulse response of an amplifier.

CHAPTER 8

OPERATIONAL AMPLIFIERS AND FEEDBACK AMPLIFIERS

The *operational amplifier* (usually referred to as the *op amp*) is one of the most widely used electronic components. It is a convenient building block in the design of analog electronic systems offering both economy and versatility.

The name *operational amplifier* arises from the fact that such amplifiers were originally developed for performing such specific mathematical operations as addition, subtraction, differentiation, and integration of electrical signals. The advent of integrated-circuit (IC) technology has, however, led to the widespread use of the op amp by making it versatile, readily available, and inexpensive.

One of the most important features of an op amp circuit is the use of *negative feedback*. When a component, such as a resistor, is placed as a link between the output side and input side of an amplifier, a portion of the output signal is fed back to the input side through the link. Such an amplifier is called a *feedback amplifier*. If the connection between the output and input sides is such that the feedback causes a reduction in the gain of the amplifier, then we have *negative feedback*. Even though a reduction in the gain of an amplifier results from negative feedback, the amplifier gain is more stable. Negative feedback has other advantages and is frequently employed in amplifier circuits.

Another important feature of an op amp is that it is *dc coupled:* the signal source is directly connected to the input terminals of the op amp instead of through a coupling capacitor like the amplifiers of the preceding chapters. The frequency response of an op amp extends down to dc as a consequence of the dc coupling.

The gain of an op amp is extremely high: values of hundreds of thousands for the gain are common in commercially available op amps. The high values of gain are realized by using several stages of amplification in fabricating an op amp.

Electronics: Circuits and Systems

Another advantage of the op amp is that it is possible to learn how to use it and analyze circuits containing it by relatively simple calculations even though the actual op amp circuit may be quite complex. Such ease is made possible by using an idealized model of the op amp, which will be discussed in a later section.

TYPICAL STAGES OF AN OP AMP

Even though the number and types of stages used in a commercially available op amp vary widely from one unit to the next, the general configuration is typically of the form indicated in the block diagram of Fig. 8-1.

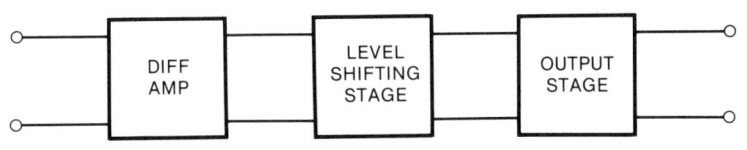

Fig. 8-1. Typical stages of an op amp.

The initial stage of an op amp is a specially configured amplifier called the *differential amplifier* (or the *diff amp* as it is usually called). The diff amp has two input terminals and can have one or two output terminals. The output of the diff amp is dependent upon the *difference* between the two inputs fed to it, which explains its name.

The function of the level shifting stage is to ensure that the output of the op amp is zero when there is no signal applied to it. The level shifting stage adjusts the output voltage of the diff-amp stage to realize this condition.

The output stage of an op amp can be quite complex and is designed to perform a variety of functions. Among the functions performed by the output stages are impedance transformation to make the output impedance of the op amp very low and the provision of the necessary capacity to drive the load connected to the op amp.

These stages will be discussed in some detail in later sections of this chapter. First, we will consider the properties and applications of the op amp.

IDEAL OP AMPS AND THEIR APPLICATIONS

As mentioned earlier, circuits containing op amps can be understood (and op amp circuits designed) without a great deal of complex algebraic manipulations and calculations if we use a model of an op amp based on certain *idealizing assumptions* about its characteristics. The idealized model leads to results that are quite adequate for most practical applications.

Parameters of Interest in an Amplifier

Before discussing the ideal op amp, it will be worthwhile to examine the quantities of interest in amplifiers in general. In Fig. 8-2, the model of an amplifier is shown within the dashed lines and consists of an input resistance R_i, a controlled voltage source AV_i, where A represents the voltage gain and an output resistance R_o.

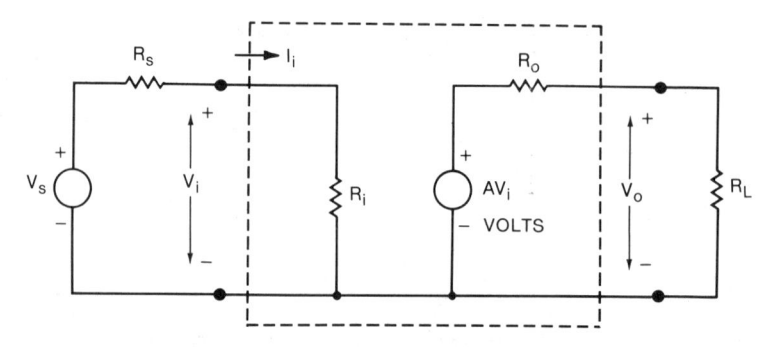

Fig. 8-2. Model of a general amplifier circuit.

Consider now the effects of the three parameters, the input resistance R_i, the gain A, and the output resistance R_o, on the performance of the amplifier.

The input resistance R_i affects the current drawn by the amplifier. The larger the input resistance, the smaller the current into the amplifier.

The output resistance R_o affects the output voltage V_o across the load since

$$V_o = (AV_i) \frac{R_L}{R_L + R_o} \tag{8-1}$$

by using the voltage-divider formula. As R_o is made much smaller than R_L, the output voltage becomes nearly equal to AV_i and independent of the actual value of the load resistance.

The gain of the amplifier A also affects the output voltage as can be seen from Eq. (8-1). The larger the voltage gain, the larger the output voltage for a given V_i. Conversely, for a desired output voltage, a large gain will permit us to manage with a small input voltage. These three parameters are pushed to certain limits in the ideal op amp as will be seen in the following discussion.

Ideal Op Amp

Fig. 8-3A shows the symbol of an op amp. It is seen that there are two input terminals, one marked + and called the *noninverting* input terminal, and the other marked − and called the *inverting* input terminal. The model

Electronics: Circuits and Systems

of an amplifier used in Fig. 8-2 can be placed inside the triangle representing the op amp, as indicated in Fig. 8-3B. The gain A in the case of the op amp is called the *open-loop* voltage gain since it is the gain available when there is no feedback connection between the input and the output (which would create a closed loop).

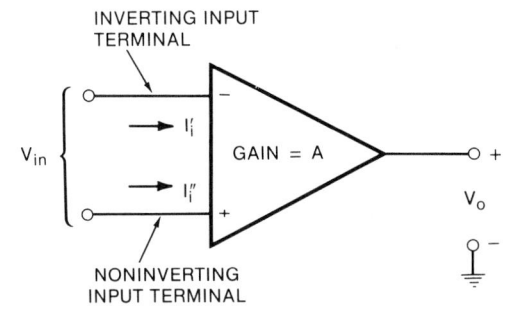

(A) Symbol of an op amp.

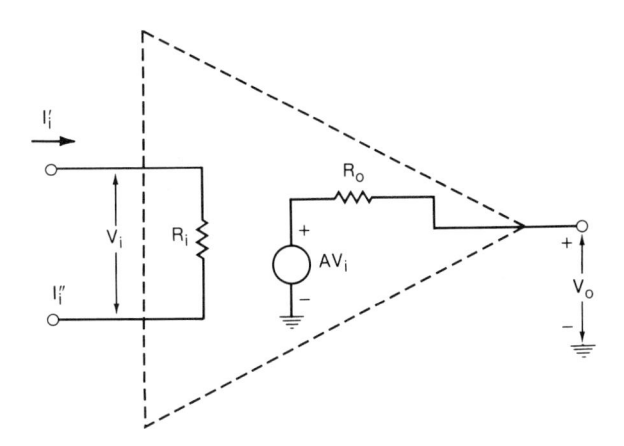

(B) Op amp model (nonideal).

Fig. 8-3. An op amp.

The following three idealizing assumptions are made for an *ideal op amp*.
1. The input resistance of an op amp is infinite. That is, $R_i = \infty$.
2. The output resistance of an op amp is *zero*. That is, $R_o = 0$.
3. The gain of an op amp is infinite, that is, $A = \infty$.

Let us examine the implications of these three assumptions.

1. $R_i = \infty$: As was seen earlier, the larger the input resistance, the smaller the input current to the amplifier. In the limit when $R_i = \infty$, the input cur-

rent is zero. That is, in an ideal op amp, there is no input current in either terminal.

$$I'_i = I''_i = 0 \text{ in an ideal op amp}$$

2. $R_o = 0$: Equation (8-1) showed that as R_0 is reduced, the output voltage approaches the value of (AV_i) and becomes essentially independent of the load itself. In the limit when $R_o = 0$, Eq. (8-1) reduces to $V_o = AV_i$ and the output voltage becomes completely independent of the load resistance. That is, we can load the ideal op amp as much as we wish but the output voltage will not be affected!

3. $A = \infty$: For a given finite output voltage, a large value of the gain implies that we need only a small input voltage. But what happens when the gain is assumed to be infinite? Since $V_o = AV_i$ when $R_o = 0$ in Eq. (8-1), an infinite gain will lead to an *infinite output voltage* if the input voltage is even negligibly small. If we want to have a *finite output* voltage, then the only way the equation $V_o = AV_i$ can be satisfied is by making $V_i = 0$! (Note that the product of zero and infinity is an indeterminate number but can be finite.) Therefore, the assumption of infinite gain is that

$$V_i = 0 \text{ in an ideal op amp}$$

When these three idealizing assumptions are used, the model of the op amp in Fig. 8-3B leads to the representation shown in Fig. 8-4. Note that $V_i = 0$, $I'_i = I''_i = 0$, and the output voltage V_o does not depend upon the load (which is therefore conveniently omitted). [The (AV_i) source has been replaced by a source of voltage V_o.] The actual value of the output voltage V_o can be calculated by considering connections made to the op amp as will be discussed in the next section.

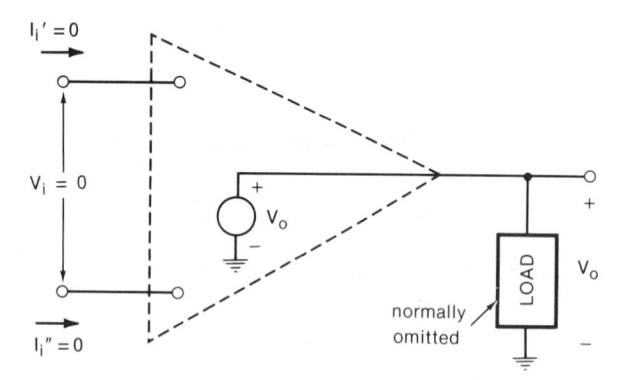

Fig. 8-4. Model of an ideal op amp. The load is not usually included since the output voltage is independent of the load in an ideal op amp.

Electronics: Circuits and Systems

When the consequences of the three idealizing assumptions are combined, we have the op amp as a component that draws no input current, and that needs no input voltage to produce a finite output voltage, and the output voltage is unaffected by the size of the load connected to the amplifier. What a selfless and magnanimous creature it is! How close does a real op amp resemble the ideal op amp? Even though the gain of an op amp is not infinite, it is about 2×10^5 or more, and for all practical purposes, the gain may be treated as infinite. The input impedance can be made essentially infinite by using a MOSFET as the device in the first stage. Even if a MOSFET is not used, the input impedance can be made 1 MΩ or higher by proper circuit design. The input impedance of commercially available op amps is large enough that the assumption of infinite input impedance is fairly reasonable. The output impedance of available op amps is not zero, but it can be made as low as a few ohms through negative feedback. Such values are small compared with the usual range of load values used. Consequently, the assumption of the zero output impedance is also sufficiently close to the real values in a practical op amp.

The ideal model of an op amp is, therefore, quite adequate for making calculations on a given op amp circuit or designing op amp circuits to perform specific functions.

Circuits Using Op Amps

When an op amp is used as a component in a circuit, it is necessary to provide a feedback link between the output and the input. The only exception to this rule is when we use it as a *comparator* to be discussed later in this chapter. The feedback link is often a resistor R_F connected in one of two ways as shown in Fig. 8-5.

When the feedback resistor is connected to the *inverting* input terminal, as in Fig. 8-5A, *negative feedback* results (as will be explained in the discussion of the inverting op amp). When the feedback resistor is connected to the *noninverting* input terminal, as in Fig. 8-5B, positive feedback results. We will confine ourselves to *negative* feedback connections for the time being and take up positive feedback when we study the *Schmitt trigger*.

Remember that for *negative feedback*, the resistor R_F must be connected between the output and the *inverting* input terminal.

Inverting Op Amp Circuits

The basic op amp circuit, and probably the most widely used connection of the op amp, is the *inverting op amp circuit* shown in Fig. 8-6. In this connection, the source V_s is connected to the *inverting* input terminal through a resistor R_1. R_F is the feedback resistor providing negative feedback.

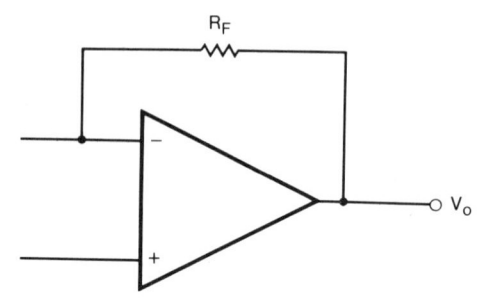

Operational Amplifiers and Feedback Amplifiers

(A) Negative feedback.

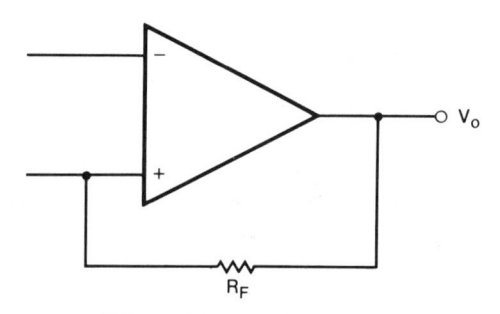

(B) Positive feedback.

Fig. 8-5. Negative and positive feedback connections in an op amp.

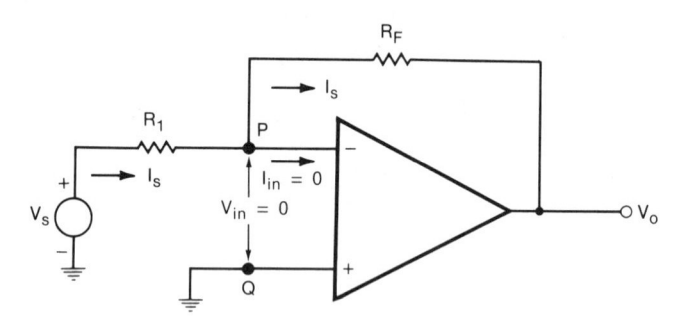

Fig. 8-6. An inverting op amp circuit.

Using the idealizing assumptions, we have $V_{in} = 0$ and $I_{in} = 0$. We can ignore the load since V_o is not affected by the load.

$V_{in} = 0$ implies that the voltage at point P (the inverting input terminal) must be equal to the voltage at point Q (the noninverting input terminal).

Since the voltage at Q is zero (by direct grounding), it follows that

$$V_P = 0$$

The point P is called a *virtual ground*.

$$I_{in} = 0$$

implies that none of the input current I_s goes into the op amp. All the current flowing through R_1 is diverted to the resistor R_F: the current in R_F is equal to the current in R_1.

Consider Fig. 8-7A. The voltage across R_1 equals V_s since V_P is zero (due to the virtual ground at P). Therefore,

$$I_s = (V_s/R_1) \qquad (8\text{-}2)$$

Consider Fig. 8-7B. The voltage across R_F is V_0 since V_P is zero due to the virtual ground. Therefore,

$$I_s = -(V_o/R_F) \qquad (8\text{-}3)$$

where,

the minus sign is necessitated by the current flowing from the lower potential (P) to the higher potential V_o in Fig. 8-7B.

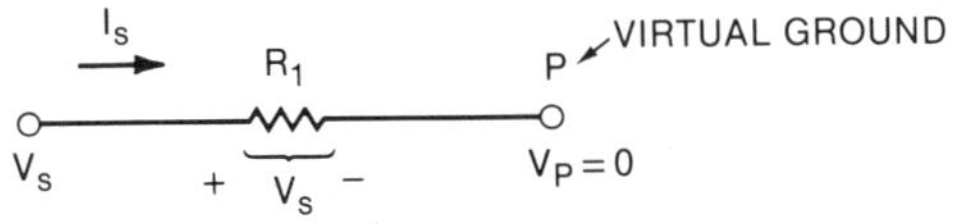

(A) Feedback resistor connected to inverting input terminal.

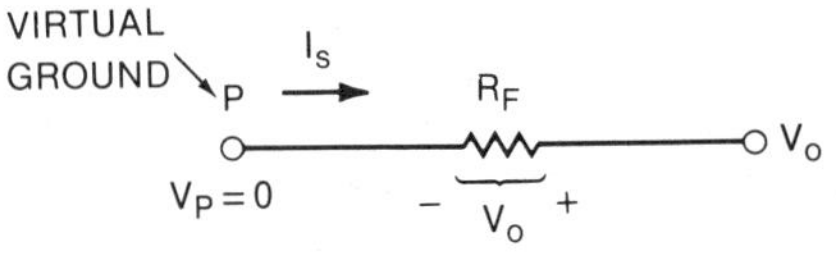

(B) Feedback resistor connected to noninverting input terminal.

Fig. 8-7. Current-voltage relationships in the inverting op amp circuit.

From Eqs. (8-2) and (8-3), we have

$$-(V_o/R_F) = (V_s/R_1)$$

or

$$\frac{V_o}{V_s} = -\frac{R_F}{R_1} \qquad (8\text{-}4)$$

The gain of the inverting op amp circuit is, therefore, the ratio of the feedback resistance to the resistance between the source and the inverting input terminal. The minus sign in Eq. (8-4) indicates that there is a phase reversal and that the output waveform will be an inverted version of the input signal.

The gain expression shows that the gain depends only upon the two resistances R_1 and R_F and not on the actual structure of the op amp or the values of its components, which is a result of the idealizing assumptions used.

The fact that the connection of the resistor R_F between the inverting input and the output terminals leads to *negative* feedback can be shown as follows.

Assume a nonideal op amp with a nonzero voltage V_i as in Fig. 8-8. If V_i were to increase, the output voltage V_o would increase. But the larger V_o would cause an increase in the current I_F through R_F and R_1. The resulting larger voltage drop in R_1 would cause a reduction in the input voltage V_i with a corresponding reduction in V_o. Thus, the presence of R_F tends to stabilize the output voltage by shifting the input voltage V_i in a *direction opposite to the change in the output voltage* V_o. This is the principle of negative feedback.

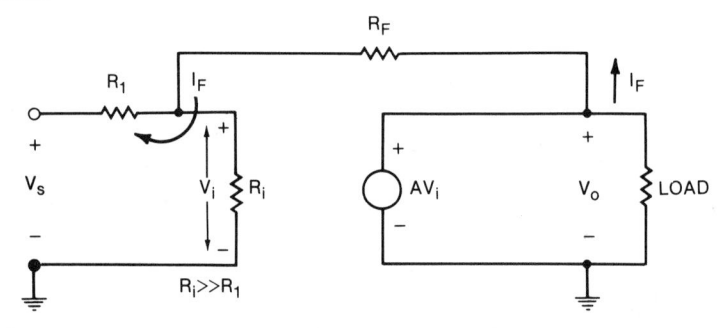

Fig. 8-8. Negative feedback when R_F is connected to the inverting input of the op amp.

Example 8.1

Calculate the gain of the op amp circuit shown in Fig. 8-9.

Solution

A direct use of the gain expression (8-4) gives

$$V_o/V_s = -(R_F/R_1)$$
$$= -10$$

Electronics: Circuits and Systems

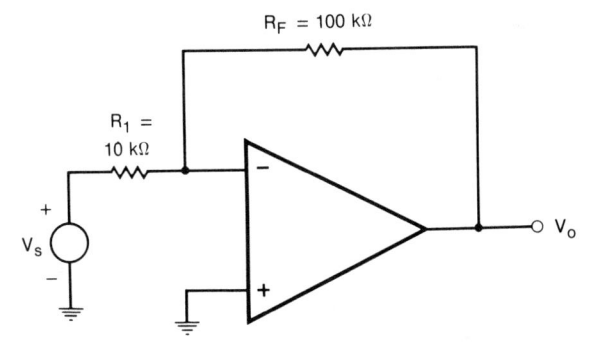

Fig. 8-9. Op amp circuit for Example 8.1.

Example 8.2

If the input signal applied to the op amp circuit of the previous example is as shown in Fig. 8-10A, sketch the output waveform.

Solution

The gain is -10 (as calculated earlier), and so we have

$$V_o = -10V_s$$

The output will be an inverted version of the input multiplied by a factor of 10. The output is shown in Fig. 8-10B.

Example 8.3

A signal source produces 50 mV and has an external resistance of 20 kΩ. Calculate the value of R_F needed so that the op amp output is 0.6 V.

Solution

The source resistance (20 kΩ) can be taken as the value of R_1.
The gain required is

$$|(V_o/V_s)| = (0.6/50 \times 10^{-3})$$

$$= 12$$

Therefore,

$$(R_F/R_1) = 12$$

and

$$R_F = 240 \text{ k}\Omega$$

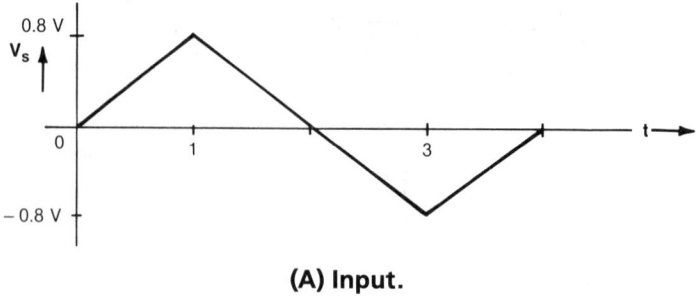

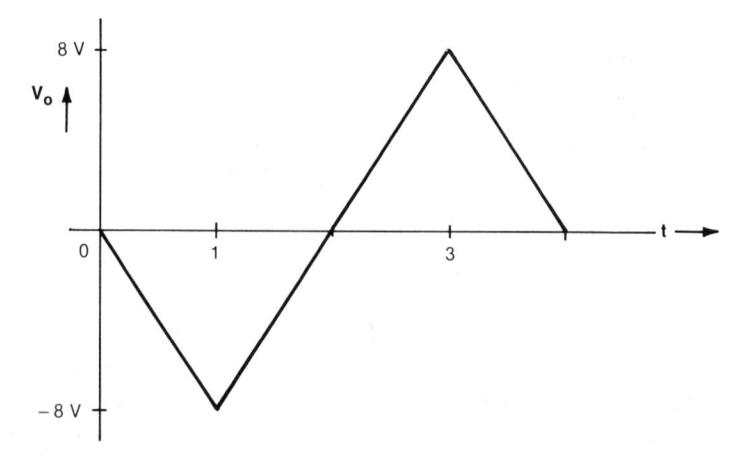

466   **Operational Amplifiers and Feedback Amplifiers**

(A) Input.

(B) Output.

Fig. 8-10. Waveforms in the circuit of Example 8.2.

Drill Problem 8.1: Calculate the gain of each of the following inverting op amp circuits: (a) $R_F = 1$ MΩ, $R_1 = 50$ kΩ; (b) $R_F = 100$ kΩ, $R_1 = 500$ kΩ; (c) $R_F = 10$ kΩ, $R_1 = 10$ kΩ.

Drill Problem 8.2: The signal shown in Fig. 8-11 is fed to an op amp with $R_1 = 25$ kΩ and $R_F = 100$ kΩ. Sketch the output voltage.

Drill Problem 8.3: A source of 10 mV is connected to the inverting input of an op amp through a 10-kΩ resistance. Calculate the feedback resistance needed to provide an output voltage of 3 V.

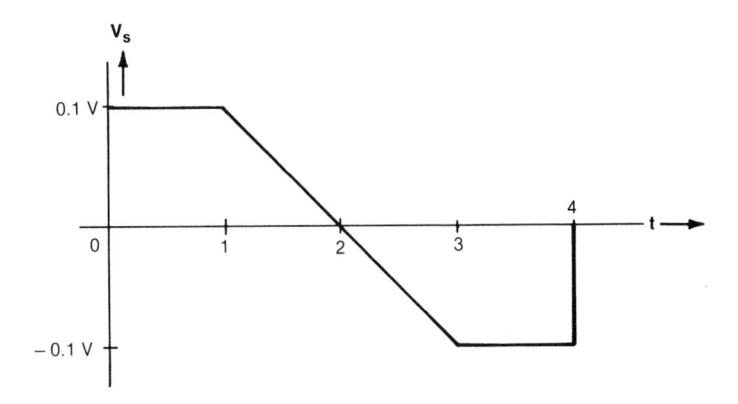

Fig. 8-11. Input signal for Drill Problem 8.2.

Noninverting Op Amp Circuits

The phase reversal between the input and the output that exists in the inverting op amp circuit can be eliminated by changing the connection of the source as shown in Fig. 8-12.

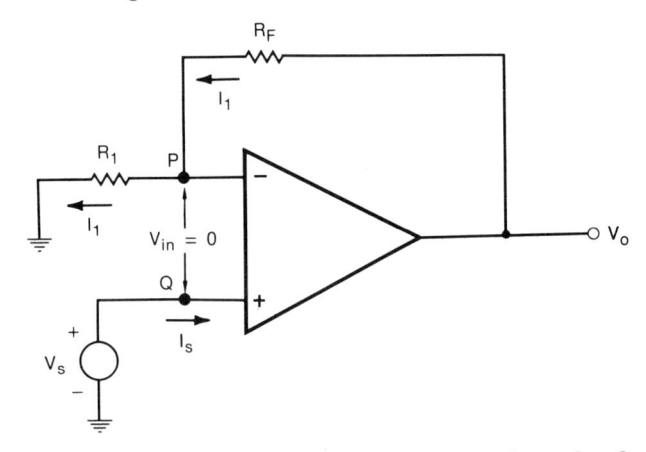

Fig. 8-12. Noninverting op amp circuit. Note that the feedback resistor is still connected to the *inverting* input.

It is important to note that in order to have *negative feedback*, the feedback resistor must be still connected to the *inverting* input terminal. The function of R_1 is to provide the correct voltage division and lead to the correct proportion of feedback. The ratio (R_F/R_1) will be found to appear in the gain expression for this circuit also.

Using the idealizing assumptions, we have

468   Operational Amplifiers and Feedback Amplifiers

1. $V_{in} = 0$: Voltage at point P is equal to the voltage at Q.

$$V_P = V_Q = V_S$$

Note that the point P is *not* a virtual ground in this circuit.

2. $I_i = 0$: The op amp draws no current. Therefore, there is *no current supplied by the source* V_s!

$$I_s = 0$$

The third idealizing assumption (zero output impedance) lets us ignore the load connected to the op amp.

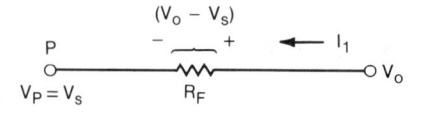

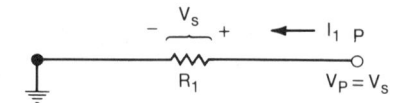

(A) Voltage conditions and current flow for R_F.

(B) Voltage conditions and current flow for R_1.

Fig. 8-13. Current-voltage relationships in the noninverting op amp circuit.

There is one current I_1 flowing through R_F, and this current also flows through R_1 since the op amp draws no current. From Fig. 8-13A, we see

$$\text{voltage across } R_F = (V_o - V_s)$$

in the direction of I_1 shown. Therefore,

$$I_1 = \frac{V_o - V_s}{R_F} \tag{8-5}$$

From Fig. 8-13B, it is seen that

$$I_1 = \frac{V_s}{R_1} \tag{8-6}$$

From Eqs. (8-5) and (8-6), we get

$$\frac{V_o - V_s}{R_F} = \frac{V_s}{R_1}$$

which leads to

$$\frac{V_o}{V_s} = 1 + \frac{R_F}{R_1} \tag{8-7}$$

Electronics: Circuits and Systems

The *gain is always positive*, which implies that there is no reversal of phase between the input and the output: the output will be an amplified replica of the input with no inversion. The gain is also seen to be greater than 1 for all nonzero values of R_F. This should be contrasted with the case of the inverting amplifier where the gain could be made less than one. (Recall the answer to Drill Problem 8.1B.)

Frequently, the noninverting op amp circuit is shown drawn in the form shown in Fig. 8-14. The op amp is reoriented in this diagram so that the noninverting input terminal is on top. But this circuit is exactly the same as the one in Fig. 8-12 discussed previously. This brings up the following important point: The "+" and "−" markings on the op amp must be clearly shown in a diagram, and we should pay careful attention to the manner in which components are connected to the two terminals in analysis. For negative feedback, the feedback resistor R_F and a resistor R_1 are connected together at the terminals marked "−".

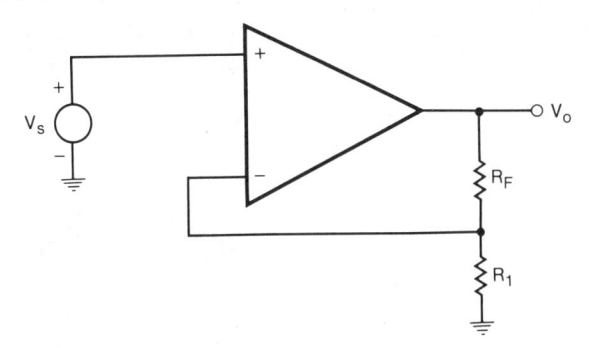

Fig. 8-14. Alternative orientation of the noninverting op amp circuit.

Example 8.4
Calculate the gain of the op amp circuit shown in Fig. 8-15.

Solution
This is a noninverting op amp circuit with $R_F = 9\ R_1$. Using the gain expression, Eq. (8-7), we get

$$\frac{V_o}{V_s} = 1 + \frac{R_F}{R_1}$$

$$= 10$$

This is a very useful circuit and known as the *gain-of-10 amplifier*.

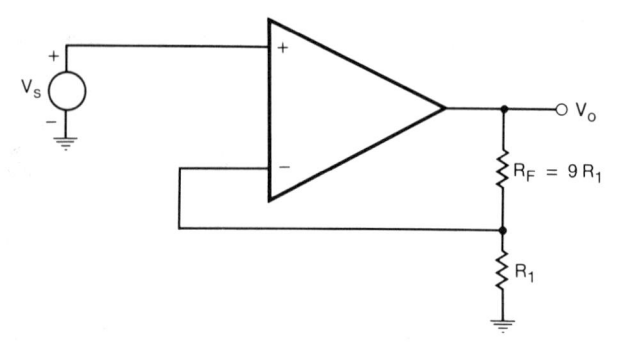

Fig. 8-15. "Gain-of-10" amplifier.

Example 8.5

A signal source of 10 mV is connected to the noninverting input terminal of an op amp. If it is desired to have an output of 0.5 V and R_1 is given as 80 kΩ, calculate the value of the feedback resistance R_F.

Solution

The gain needed is

$$(V_o/V_s) = (0.5/10 \times 10^{-3})$$
$$= 50$$

Therefore, we have

$$50 = 1 + (R_F/R_1)$$

or $(R_F/R_1) = 49$. Therefore,

$$R_F = (49 \times 80)$$
$$= 3920 \text{ k}\Omega$$

It is possible to connect more than one signal source to an op amp circuit in such a way as to have both inverting and noninverting action simultaneously. The following example illustrates one such case.

Example 8.6

For the op amp circuit shown in Fig. 8-16, determine the output voltage in terms of V_a and V_b.

Electronics: Circuits and Systems

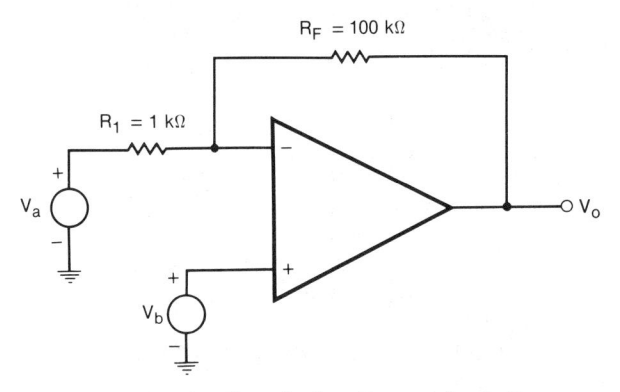

Fig. 8-16. Circuit for Example 8.6.

Solution

We solve the problem by considering each of the signal sources acting one at a time. (This is known as the principle of superposition.) Then the total output is obtained by adding the individual outputs.

V$_a$ acting alone:

The circuit is now as shown in Fig. 8-17A and is seen to be an inverting op amp. Therefore, the output V$_{o1}$ is related to V$_a$ by the equation:

$$\frac{V_{o1}}{V_a} = -\frac{R_F}{R_1}$$

$$= -100$$

so that

$$V_{o1} = -100V_a$$

V$_b$ acting alone:

The circuit now appears as shown in Fig. 8-17B and is a noninverting op amp. Therefore,

$$\frac{V_{o2}}{V_b} = 1 + \frac{R_F}{R_1}$$

$$= 101$$

so that

$$V_{o2} = 101V_b$$

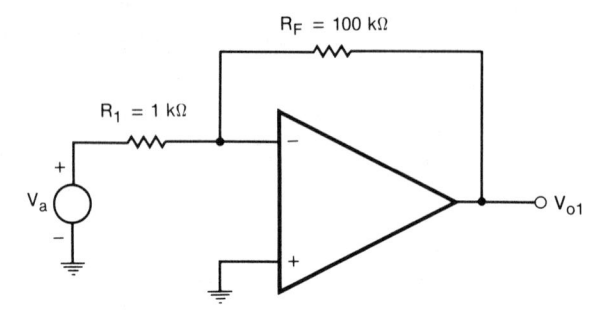

Operational Amplifiers and Feedback Amplifiers

(A) Only the source V_a is active.

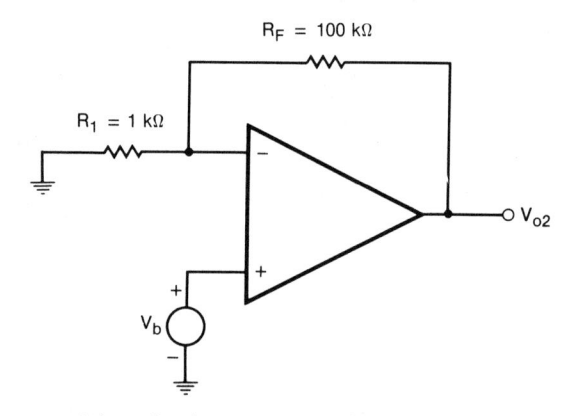

(B) Only the source V_b is active.

Fig. 8-17. Use of superposition to solve Example 8.6.

When both the sources are acting simultaneously, the output is the sum of V_{o1} and V_{o2}:

$$V_o = -100 V_a + 101 V_b$$

For example, if $V_a = 10$ mV and $V_b = 20$ mV, the output will be 1.02 V.

Drill Problem 8.4: Calculate the gain of a noninverting op amp circuit in which (a) $R_1 = 15$ kΩ and $R_F = 5$ kΩ; (b) $R_1 = 5$ kΩ and $R_F = 15$ kΩ.

Drill Problem 8.5: Suppose the resistor values given in the previous drill problem were used in an inverting op amp circuit. Calculate the gains and compare them with the results obtained for the noninverting case.

Drill Problem 8.6: Calculate the output of the op amp in each circuit shown in Fig. 8-18.

(A)

(B)

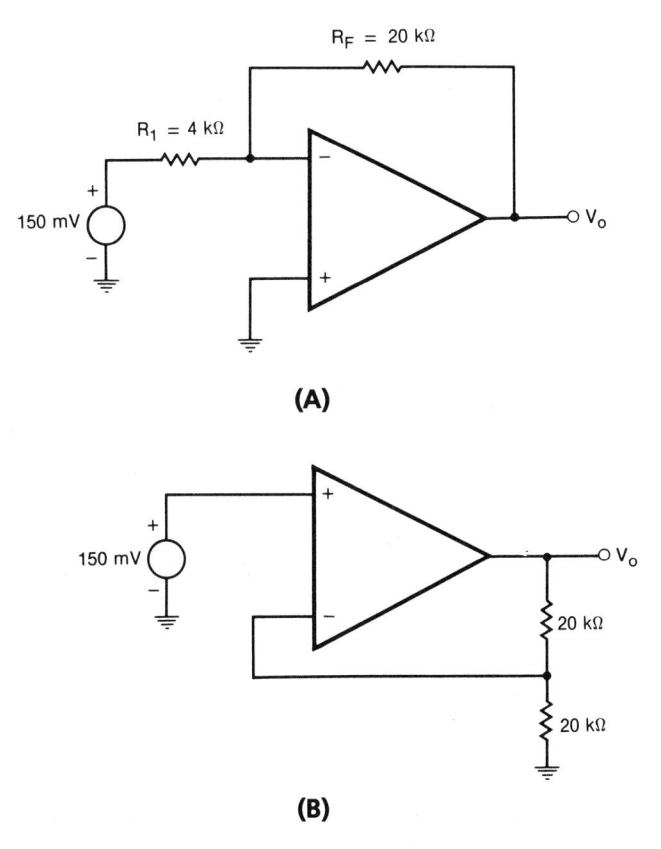

Fig. 8-18. Circuits for Drill Problem 8.6.

Drill Problem 8.7: In the circuit shown in Fig. 8-19, calculate the value of V_a needed to make the output voltage $V_o = 0$.

Summing Op Amp Circuits

The *weighted sum* of two or more signals can be performed by means of an op amp circuit. That is, given three signals V_1, V_2, V_3, for example, it is possible to obtain an output

$$V_o = aV_1 + bV_2 + cV_3$$

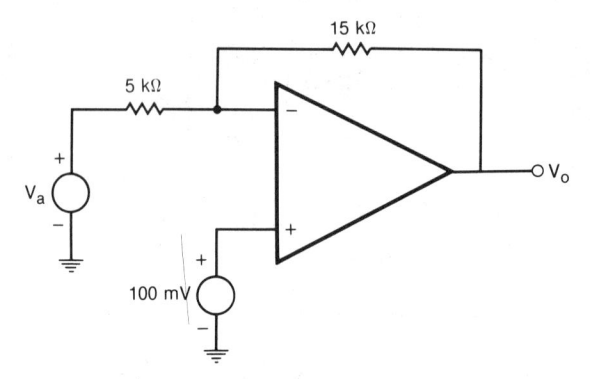

474   Operational Amplifiers and Feedback Amplifiers

Fig. 8-19. Circuits for Drill Problem 8.7.

where,
a, b, and c are some constants.
The circuit is known as the summing amplifier and is shown in Fig. 8-20.

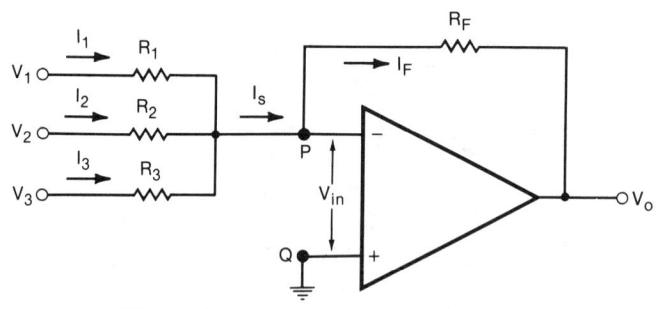

Fig. 8-20. Summing amplifier.

The point P is a virtual ground since the point Q is at ground and $V_{in} = 0$ for an op amp. Therefore, $V_P = 0$ and the voltages across the three resistors are V_1 across R_1, V_2 across R_2, and V_3 across R_3. The three currents I_1, I_2, I_3 are given by

$$I_1 = (V_1/R_1)$$
$$I_2 = (V_2/R_2)$$
$$I_3 = (V_3/R_3)$$

Since the op amp draws no current,

$$I_F = I_s$$
$$= I_1 + I_2 + I_3$$

Electronics: Circuits and Systems

$$= \frac{V_1}{R_1} + \frac{V_2}{R_2} + \frac{V_3}{R_3} \qquad (8\text{-}8)$$

The voltage across the feedback resistor R_F is V_o since the right side is at V_o and the left side is the virtual ground. The current I_F goes from the lower to the higher potential in R_F, and we have

$$V_o = -R_F I_F$$

or, using Eq. (8-8), we get

$$V_o = -R_F \left(\frac{V_1}{R_1} + \frac{V_2}{R_2} + \frac{V_3}{R_3} \right)$$

$$= -\left(\frac{R_F}{R_1} V_1 + \frac{R_F}{R_2} V_2 + \frac{R_F}{R_3} V_3 \right) \qquad (8\text{-}9)$$

Therefore, the output of the circuit in Fig. 8-20 is the *weighted sum* of the three input voltages, with the weights being (R_F/R_1), (R_F/R_2), and (R_F/R_3). There is a phase reversal since the op amp is being used in the inverting mode. As a special case, let $R_1 = R_2 = R_3 = R_F$. Then,

$$V_o = -(V_1 + V_2 + V_3)$$

or we have an inverting *adder* circuit.

Example 8.7

Calculate the output voltage of the summing op amp circuit shown in Fig. 8-21.

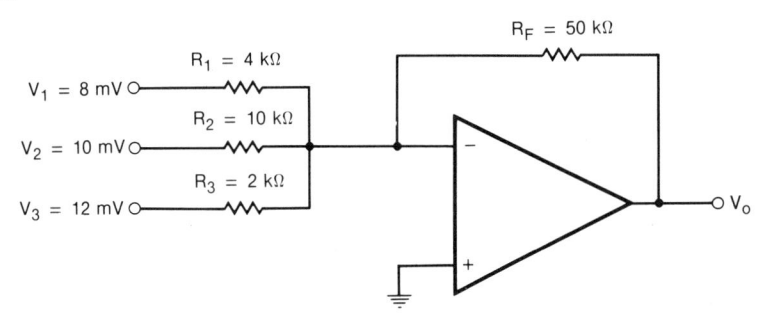

Fig. 8-21. Summing amplifier for Example 8.7.

Solution

Using Eq. (8-9), we get

$$V_o = -(\frac{50}{4}V_1 + \frac{50}{10}V_2 + \frac{50}{2}V_3)$$

$$= -450 \text{ mV}$$

Example 8.8

Determine the values of the resistances R_1, R_2, R_3 in the circuit of Fig. 8-22 so as to make the output

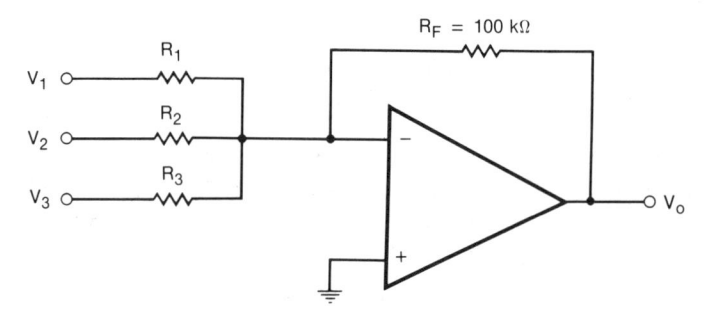

$$V_o = -(2V_1 + 8V_2 + 6V_3)$$

Fig. 8-22. Summing amplifier for Example 8.8.

Solution

Equation (8-9) becomes since $R_F = 100$ kΩ,

$$V_o = -(\frac{100}{R_1}V_1 + \frac{100}{R_2}V_2 + \frac{100}{R_3}V_3)$$

where,
the resistance values are in kΩ. Therefore, $(100/R_1) = 2$, or $R_1 = 50$ kΩ. $(100/R_2) = 8$, or $R_2 = 12.5$ kΩ. $(100/R_3) = 6$, or $R_3 = 16.7$ kΩ.

Drill Problem 8.8: Calculate the output voltage in a summing op amp circuit if $V_1 = 0.25$ V, $R_1 = 10$ kΩ; $V_2 = 0.75$ V, $R_2 = 20$ kΩ; $V_3 = 1$ V, $R_3 = 10$ kΩ; and $R_F = 5$ kΩ.

Drill Problem 8.9: Determine the values of R_1, R_2, R_3 in a summing op amp circuit so as to make $V_o = -(V_1 + 0.5 V_2 + 1.5 V_3)$ if $R_F = 50$ kΩ.

Voltage-Follower Circuits

A circuit that is quite useful even though it may not at first appear to be significant is the *voltage follower*, shown in Fig. 8-23. It is seen that $R_F = 0$ in

Electronics: Circuits and Systems

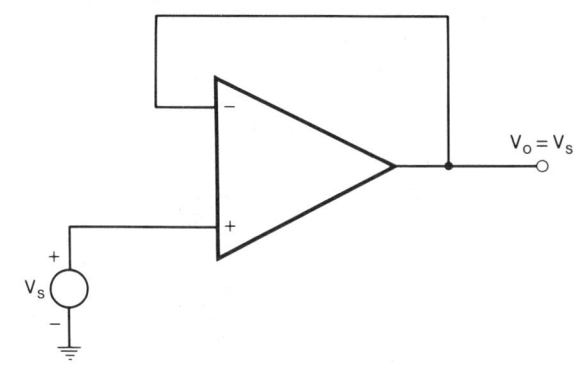

Fig. 8-23. Voltage follower.

this circuit and that the input is connected to the noninverting input terminal.

Using the gain expression for the noninverting op amp circuit, Eq. (8-7), with $R_F = 0$, we get,

$$(V_o/V_s) = 1$$

or the output voltage is exactly equal to the input signal. Since the output voltage "follows" the input signal, the circuit is called a voltage follower.

Why do we need a voltage follower? Why should we not simply use the input signal V_s directly? The voltage follower performs a useful function as a *buffer stage* between two amplifier stages when it is necessary to isolate the two amplifier stages from each other. Consider two amplifier stages connected directly in cascade as indicated in Fig. 8-24. The direct connection makes the voltages and currents in one stage directly affect those in the other, which is particularly the situation in *dc-coupled* amplifiers where capacitive coupling (which was used to isolate amplifier stages in the earlier chapters for dc conditions) cannot be used.

Now, consider the two amplifier stages connected through a voltage follower that acts as the buffer stage. The input impedance of the voltage follower is infinity (since it is an op amp), and the load on amplifier stage 1 is an open circuit. Therefore, amplifier stage 1 behaves as if nothing is connected to it. The output impedance of the buffer stage is zero (since it is an op amp). Therefore, it acts like an ideal voltage source feeding amplifier stage 2. Thus, the amplifier stage 2 simply sees a voltage V_{i2} with no internal impedance. It is not affected by the internal voltages and currents of the amplifier stage 1. Also, since $V_{i2} = V_{o1}$ due to the voltage follower, the input to amplifier stage 2 is the output of amplifier stage 1 just as it was in the direct connection of the two amplifier stages.

Thus, the voltage follower serves as a buffer between two amplifier stages

478　　　　　Operational Amplifiers and Feedback Amplifiers

(A) Without buffer stage.

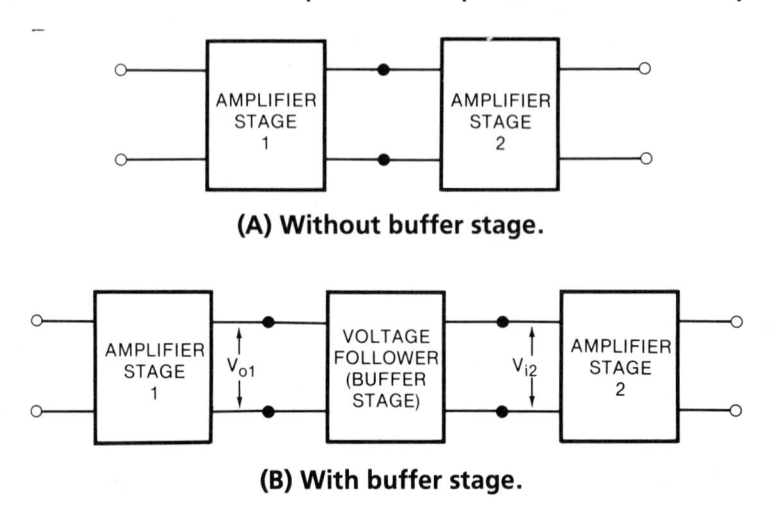

(B) With buffer stage.

Fig. 8-24. Use of the voltage follower as a buffer stage.

and isolates the two stages without altering the output voltage of amplifier stage 1 or the input voltage of amplifier stage 2. Such a buffer is indispensable in the direct connection of dc-coupled amplifier stages.

Voltage-to-Current Converter

It is sometimes desirable to produce an output current that is directly proportional to an input signal, which would be the case if we wish to send a current through a coil that is proportional to a given voltage. A circuit to convert an input voltage signal to a proportionate load current is shown in Fig. 8-25. The load impedance Z_L itself serves as the feedback link in this circuit. The op amp is being used in the noninverting mode.

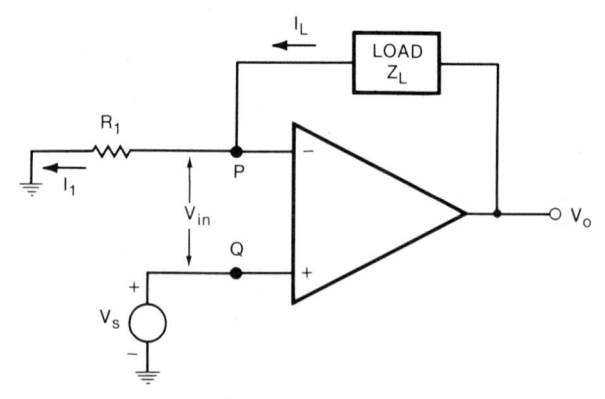

Fig. 8-25. Voltage-to-current converter.

Electronics: Circuits and Systems

$$\text{voltage across } R_1 = \text{voltage at point P}$$
$$= \text{voltage at point Q}$$
$$= V_s$$

since $V_{in} = 0$ for the op amp. Therefore,

$$I_1 = (V_s/R_1)$$

But, since the op amp draws no current, the load current $I_L = I_1$. Therefore, the output current of the circuit is

$$I_L = (V_s/R_1)$$

The current in the load depends only upon the input voltage V_s (since R_1 is a constant) and is independent of the nature or value of the load impedance Z_L itself. It is also to be noted that the signal source supplies zero current.

Sample and Hold Circuit

In a number of systems, it is necessary to record the signal voltage at a particular instant of time and store it for a certain period. For example, a signal in a communication system is sampled at various instants of time, and the sampled value is held by storing it as the voltage in a capacitor until the next sampling instant. Such a circuit is called a *sample and hold circuit*, one version of which is shown in Fig. 8-26. The MOSFET is a p-channel device that can conduct only when there is a *negative voltage* on the *control gate*. Whenever a negative pulse is applied to the control gate, the MOSFET conducts and the capacitor charges to the input voltage value at the instant of the occurrence of the pulse. When the pulse disappears, the capacitor voltage remains constant (since the op amp presents an open circuit as does the MOSFET when it is off) until the occurrence of the next control pulse. An input signal v_s, the periodic pulse train applied to the control gate, and the output voltage of the sample and hold circuit are shown in Fig. 8-27.

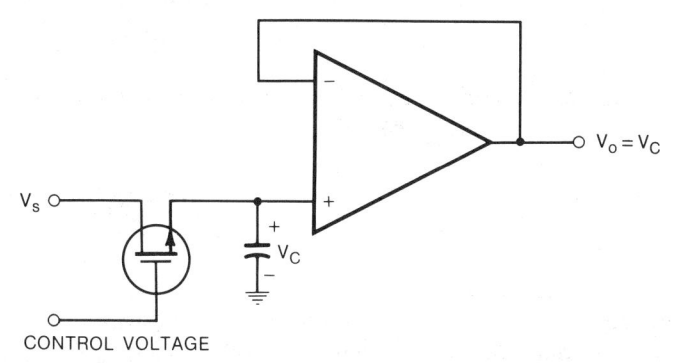

Fig. 8-26. Sample and hold circuit.

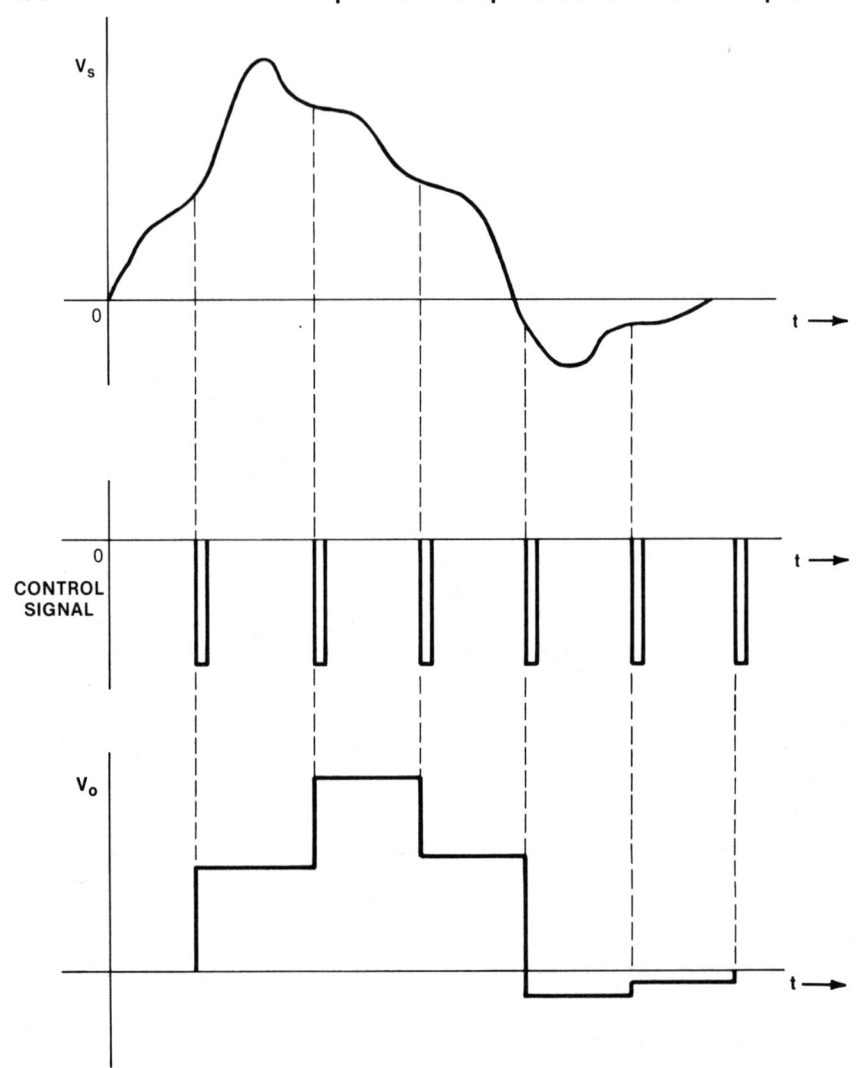

Fig. 8-27. Input, control, and output waveforms in a sample and hold circuit.

The op amp in the sample and hold circuit should have a FET as the input device to prevent the decaying of the capacitor voltage.

Logarithmic Amplifiers

The voltage across a pn junction is proportional to the *logarithm* of the current. That is, in the diode of Fig. 8-28, the voltage V_d is related to the current I_d by the equation

Fig. 8-28. Diode.

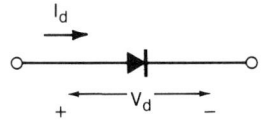

$$V_d = K_1 \log I_d + K_2 \quad (8\text{-}10)$$

where,

K_1 and K_2 are constants (whose values can be determined from a knowledge of the diode characteristics).

For example, for a typical pn diode, $K_1 = 0.0596$, and $K_2 = 0.358$, and

$$V_d = 0.0596 \log I_d + 0.358.$$

An op amp circuit in which a diode is used as the feedback link (or part of a feedback network) has an output voltage that is proportional to the logarithm of the input voltage. Such an amplifier is called a *logarithmic amplifier*, and an elementary version of a logarithmic amplifier is shown in Fig. 8-29.

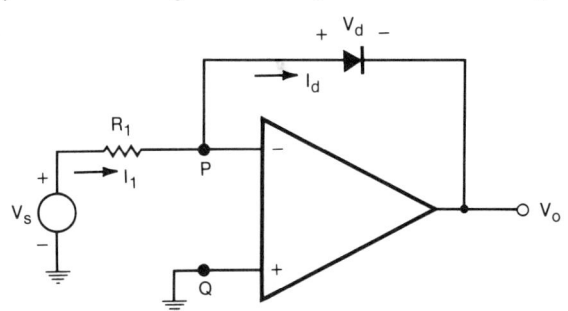

Fig. 8-29. Logarithmic amplifier.

The point P is a virtual ground (since the point Q is at ground). Therefore, the voltage across the diode V_d is given by

$$V_d = -V_o \quad (8\text{-}11)$$

The voltage across the resistor R_1 is V_s (again due to the virtual ground at P) and the current $I_1 = (V_s/R_1)$. Since the op amp draws no current, the diode current $I_d = I_1$. Therefore,

$$I_d = (V_s/R_1) \quad (8\text{-}12)$$

Combining Eqs. (8-11), (8-10), and (8-12), we have

$$V_o = -V_d$$
$$= -(K_1 \log I_d + K_2)$$

$$= -[K_1 \log(V_s/R_1) + K_2]$$

The output voltage of the amplifier of Fig. 8-29 is seen to depend upon the *logarithm* of the input voltage, rather than the input voltage itself. For example, if the input voltage were changed by a *factor of ten*, there is a *linear change* of K_1 volts in the output voltage, as can be verified from Eq. (8-19). Since K_1 is typically 5.96 mV, the amplifier output changes by 5.96 mV for every tenfold change in the input signal. A sophisticated version of the logarithmic amplifier is available, in which a matched pair of transistors is used instead of a single diode, and two op amps are employed. For such an amplifier, the change in the output voltage is nearly 3.6 V for every tenfold change in the input signal.

Constant-Current Source Using an Op Amp

The combination of a zener diode and an op amp leads to a constant current source; that is, the current in a load is constant independent of the fluctuations in the input signal. Such a circuit is shown in Fig. 8-30.

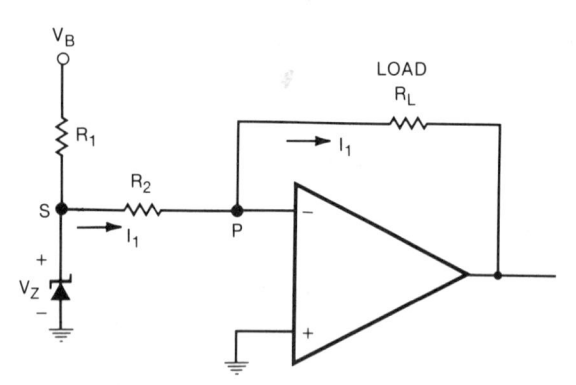

Fig. 8-30. Constant-current source circuit.

The voltage at point S in the circuit remains constant at the zener voltage V_Z of the diode regardless of the fluctuations in the supply voltage V_B, provided $V_B > V_Z$. The point P is a virtual ground so that the voltage across R_2 is equal to the voltage at point S, or V_Z. Therefore,

$$I_1 = (V_Z/R_2)$$

Since the op amp draws no current, the current in the load resistor is I_1 also. Therefore, the load current in the circuit is (V_Z/R_2), which is a constant since V_Z and R_2 are constant. As an example, if $V_Z = 6$ V and $R_2 = 0.3$ kΩ, then the load current remains constant at 20 mA so long as the supply voltage V_B is greater than 6 V.

Integrators Using an Op Amp

One of the most important uses of an op amp is to perform integration of an input signal; that is, the output of the op amp is to be the integral of the input. Using a capacitor as the feedback link, the op amp becomes an *integrator*.

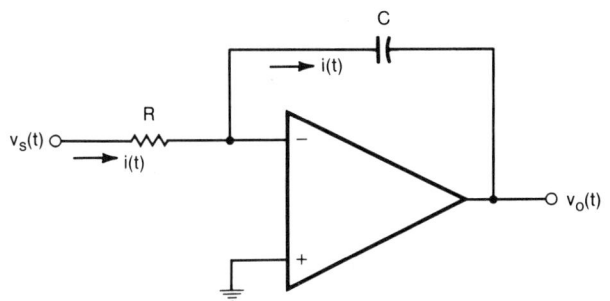

Fig. 8-31. Integrator circuit.

In the circuit of Fig. 8-31, the point P is a virtual ground. Therefore, the voltage across the resistor R is $v_s(t)$. The current in it is, therefore,

$$i(t) = \frac{v_s(t)}{R}$$

The current in the capacitor is $i(t)$ as given by the previous equation since the op amp draws no current. The voltage across the capacitor is equal to the output voltage $v_o(t)$, again due to the virtual ground at point P. Using the voltage-current relationship for a capacitor:

$$\text{voltage on a capacitor } v_C(t) = \frac{1}{C} \times [\text{time integral of } i(t)]$$

we get,

$$v_o(t) = -\frac{1}{C} \times [\text{time integral of } i(t)]$$

which becomes, on using Eq. (8-20),

$$v_o = -\frac{1}{RC} \times [\text{time integral of } v_s(t)]$$

The output of the circuit is, therefore, proportional to the integral of the input signal, and the circuit acts as an integrator.

To understand the significance of a time integral, consider the signal shown in Fig. 8-32 as the input to the integrator. When we integrate a waveform, it involves finding the area under the given waveform up to the given instant of time.

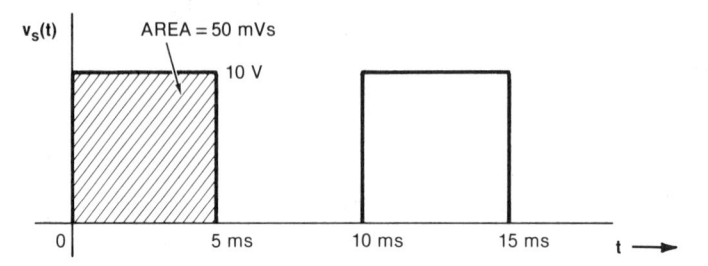

Fig. 8-32. Input signal to an integrator.

In the case of the signal of Fig. 8-32, the area under a pulse increases linearly as we go from the beginning of the pulse to its end. When the pulse ends, the area remains constant since there is no further contribution to the area from the pulse. When the next pulse appears, the area again increases linearly (starting from the already existing value) until the second pulse also disappears. After that, the area remains constant. The time integral of the input signal of Fig. 8-32 is, therefore, as shown in Fig. 8-33.

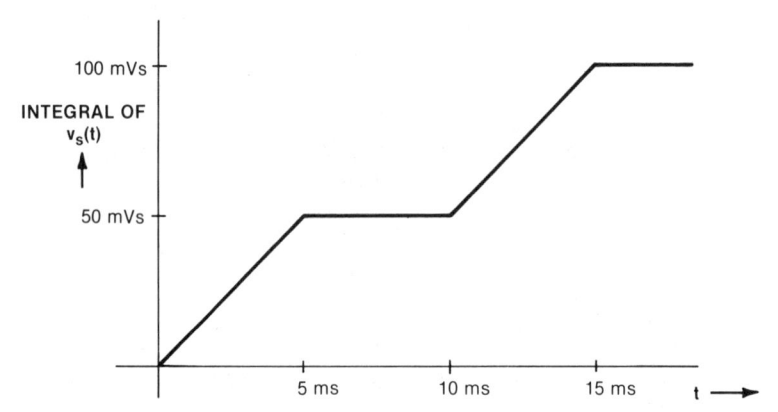

Fig. 8-33. Integral of the signal in Fig. 8-32.

The *output* of the op amp is obtained from the time integral shown in Fig. 8-33 by multiplying it by the factor $[-(1/RC)]$ and is shown in Fig. 8-34.

A special application of the integrator is when the input is simply a constant voltage as indicated in Fig. 8-35. The area under a constant voltage signal is a continuous, linearly increasing function as shown in Fig. 8-36. Such a function is called a *ramp function*. The ramp function is useful in oscilloscopes as the sweep function that moves the electron beam from the left side

Electronics: Circuits and Systems

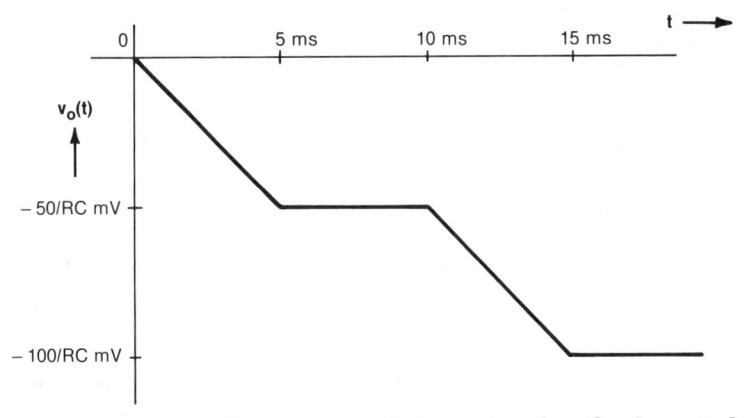

Fig. 8-34. Output of an op amp integrator for the input signal of Fig. 8-32.

of the screen to the right side at a uniform rate. The op amp integrator with a constant voltage input can be used as the sweep generator of an oscilloscope.

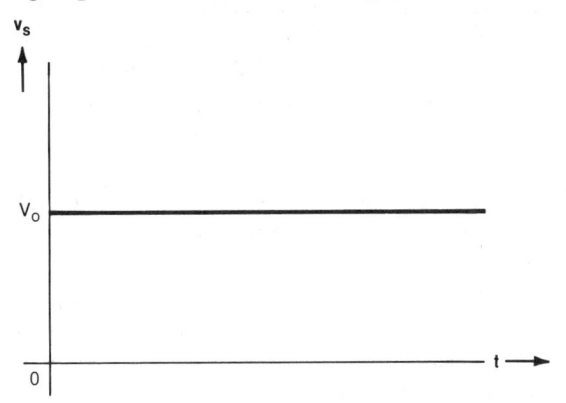

Fig. 8-35. Constant input signal for the integrator.

The factor (1/RC) that appears in the output of an integrator is a measure of the *speed* at which the integration of the input is taking place. Remember that the product RC has the dimensions of time. If RC = 1, for example, then the integration is taking place in real time; that is, the output of the integrator is exactly equal to the integral of the input (except for the inversion indicated by the minus sign).

A *differentiator*, or a circuit that differentiates an input signal, can be built by interchanging the positions of R and C in the integrator circuit. But, as a practical matter, the differentiator is not a desirable circuit. It is very sensitive to random disturbances (or noise) in the input signal and is not used in practice.

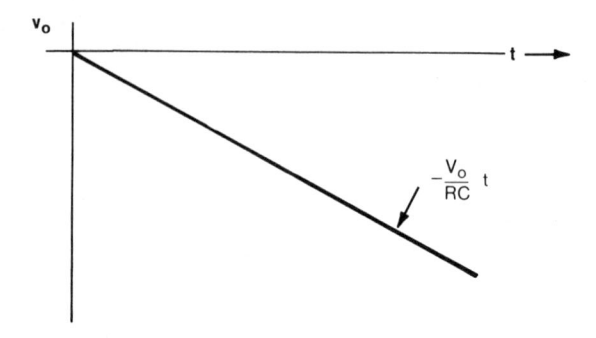

Fig. 8-36. Ramp output of an integrator with a constant input.

Integrators and summing amplifiers can be interconnected in the form of an appropriate network to solve the differential equations of linear electrical and mechanical systems. Such a network is called an *analog computer* and is useful in the study of the real-time response of electrical and mechanical systems. For example, the study of complex electrical power systems under fault conditions, and the study of the vibrations of an automobile under different conditions of suspension and external forces can be studied by means of an analog computer. The analog computer was the primary, if not the only, application of op amps until the successful fabrication of op amps using IC technology in the early 1960s.

The versatility of the op amp as a significant component of many electronic circuits and systems is evident from the previous discussion of what is only a small sample of its many uses. A few more applications will be discussed later in this chapter and also in the problems at the end of this chapter.

Let us now examine the stages of the op amp in some detail to appreciate how the important characteristics of an op amp are attained and how the real op amp differs from an ideal op amp.

DIFFERENTIAL AMPLIFIERS

The differential amplifier forms the first stage of an op amp. It has an output that depends upon the difference between the two signals fed to its (two) input terminals. Consider the diff amp in the form of a black box as indicated in Fig. 8-37, where v_{i1} and v_{i2} denote the two input signals applied, respectively, to the noninverting and inverting input terminals.

For an *ideal diff amp*, the output voltage v_0 should depend only upon the difference between the two inputs. That is, the output can be related to the two inputs by the equation

$$v_0 = A_d (v_{i1} - v_{i2}) \tag{8-13}$$

Electronics: Circuits and Systems

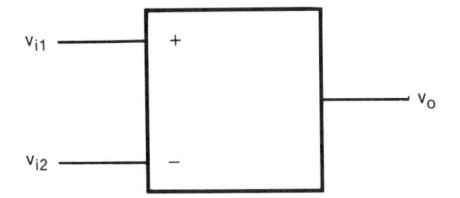

Fig. 8-37. Black box to represent a diff amp.

where,

A_d represents the gain of the amplifier.

A_d is called the *differential-mode gain* of the diff amp since it refers to the amplification of the *difference signal* $(v_{i1} - v_{i2})$.

Example 8.9

An ideal diff amp has a gain $A_d = 100$. For each of the sets of values of the inputs given, calculate the output voltage: (a) $v_{i1} = v_{i2} = 10$ mV; (b) $v_{i1} = 10$ mV, $v_{i2} = -10$ mV; (c) $v_{i1} = 990$ mV, $v_{i2} = 1010$ mV.

Solution

Using $A_d = 100$, Eq. (8-13) becomes: $v_o = 100 \, (v_{i1} - v_{i2})$. (a) The two inputs are equal. Therefore, $v_o = 0$. (b) $(v_{i1} - v_{i2}) = 20$ mV, $v_o = 2$ V. (c) $(v_{i1} - v_{i2}) = -20$ mV. $v_o = -2$ V.

(It should be noted that the *plus* and *minus* marks on the diff amp do not mean that only a positive signal may be connected to the plus terminal or only a negative signal can be connected to the minus terminal.)

Example 8.10

An ideal diff amp has a gain of $A_d = 100$. The two input signals have waveforms as shown in Fig. 8-38. Determine and sketch the output voltage waveform.

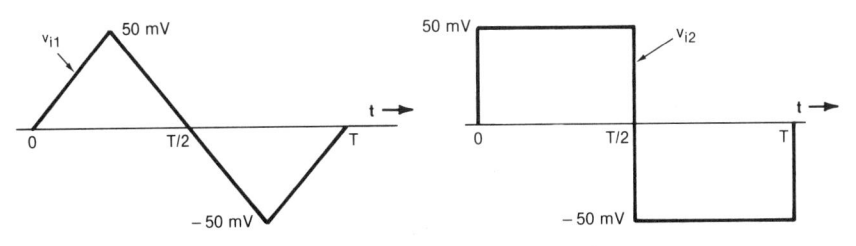

Fig. 8-38. Input signals to a diff amp.

Solution

A graphical procedure is best suited to this problem. Since $(v_{i1} - v_{i2})$ is needed in the solution, the v_{i2} waveform is first inverted and then added graphically to the v_{i1} waveform. The diagrams of Fig. 8-39 show the procedure and the final output waveform.

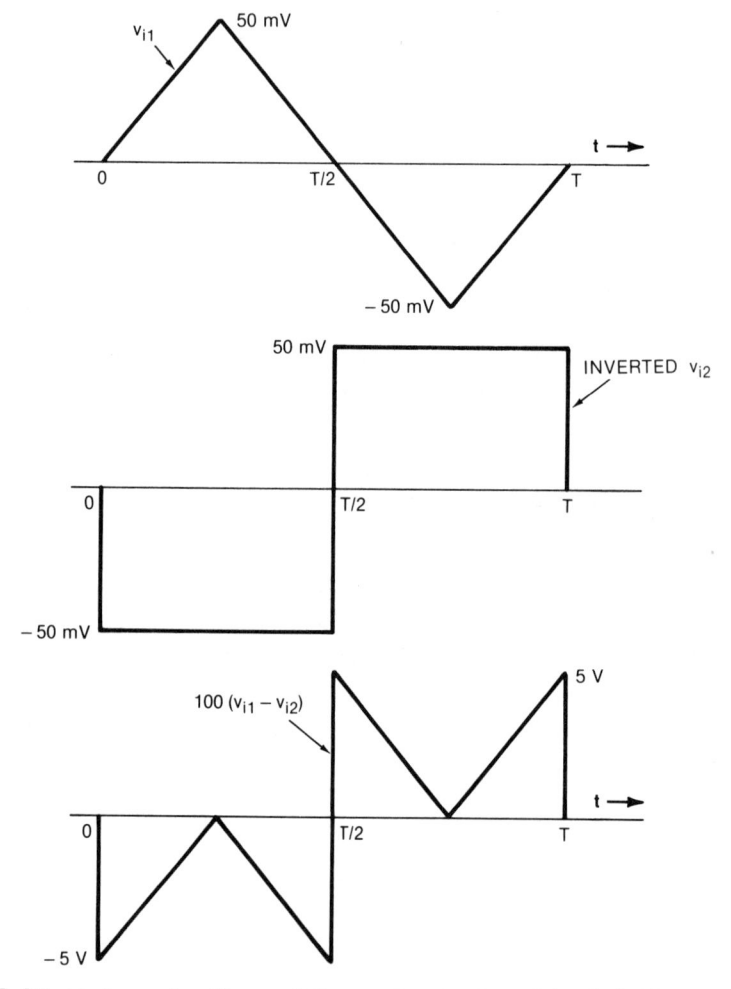

Fig. 8-39. Determination of the output of an ideal diff amp for the input signal of Fig. 8-38.

Drill Problem 8.10: An ideal diff amp has a gain of $A_d = 400$. For each of the following sets of input signals, calculate the output voltage. (a) v_{i1}

= 850 mV, v_{i2} = 950 mV; (b) v_{i1} = −950 mV, v_{i2} = −850 mV; (c) v_{i1} = −50 mV, v_{i2} = 50 mV.

Drill Problem 8.11: The input signals to a diff amp with a gain of 10 have the waveforms shown in Fig. 8-40. Determine and sketch the output voltage waveform.

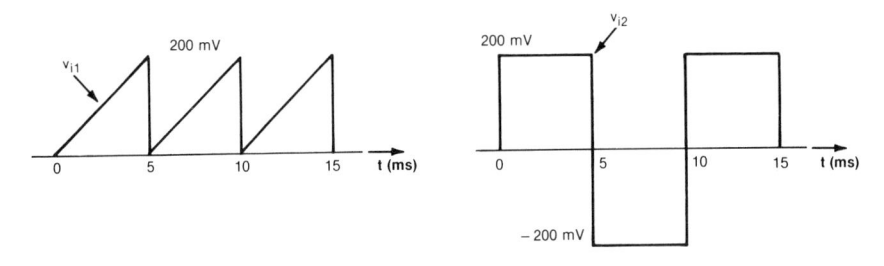

Fig. 8-40. Input signals for Drill Problem 8.11.

The ideal requirement that the output voltage of a diff amp depends only upon the difference between the two input signals can be met fairly closely in a physically realizable, or practical, diff amp. But, in a practical diff amp, the output voltage is found to have *two* components: one that is proportional to the difference between the input signals (as in the ideal requirement) and the other proportional to the *average value of the two input signals*. For example, if the two input signals are v_{i1} = 850 mV and v_{i2} = 950 mV, then the output voltage will have one component caused by the difference signal (950 − 850) mV and another component caused by the average value of the two signals given by

$$\frac{850 + 950}{2} = 900 \text{ mV}$$

Common Mode and Differential Mode

The output voltage of a practical amplifier can be split into two components: the *differential-mode* component and the *common-mode* component. The differential-mode component is due to the difference between the two input signals and the common-mode component is due to the average value of the two input signals. In the previous example, the component caused by the (850 to 950) or −100 mV is the differential-mode component and the component caused by the [(850 + 950)/2] or 900 mV is the common-mode component.

The output of a practical diff amp can be expressed as the sum of two components: v_{od} the differential-mode component and v_{oc} the common-mode component. Therefore,

$$v_o = v_{od} + v_{oc}$$

The differential-mode component v_{od} depends upon the difference between the two input signals and is given by

$$v_{od} = A_d(v_{i1} - v_{i2}) \tag{8-14}$$

where,

A_d is the *differential-mode gain* of the amplifier, which is, of course, the same as the output of an *ideal* diff amp.

The common-mode component v_{oc} depends upon the average value of the two input signals and is given by

$$v_{oc} = A_c[(v_{i2} + v_{i1})/2] \tag{8-15}$$

where,

A_c is the *common-mode gain* of the amplifier. The *common-mode gain of an ideal diff amp is zero*, and there is no common-mode output in an ideal diff amp.

The total output voltage of the diff amp is, therefore, given by

$$v_o = A_d(v_{i1} - v_{i2}) + \frac{1}{2} A_c(v_{i2} + v_{i1}) \tag{8-16}$$

The common-mode gain A_c of a well-designed diff amp can be made extremely small compared with the differential-mode gain.

Example 8.11

A diff amp has a differential-mode gain of 100 and a common-mode gain of 5. For each of the following sets of input voltages, calculate the output voltage. In each case, compare the result with the output of an ideal diff amp. (a) $v_{i1} = 25$ mV, $v_{i2} = -25$ mV. (b) $v_{i1} = 1025$ mV, $v_{i2} = 975$ mV.

Solution

(a) $v_{i1} - v_{i2} = 50$ mV, and $[(v_{i2} + v_{i1})/2] = 0$. $v_o = A_d (50) + A_c (0)$ mV = 5 V. The output of an *ideal* diff amp will also be 5 V for the input signals given.

(b) $v_{i1} - v_{i2} = 50$ mV. $[(v_{i2} + i_{i1})/2] = 1000$ mV. $v_o = A_d (50) + A_c (1000)$ mV = 10 V. The output of an *ideal* diff amp will, in this case, be only 5 V. The output of the real diff amp is seen to be substantially different from that of the ideal because of the common-mode component.

Electronics: Circuits and Systems

Drill Problem 8.12: The differential-mode gain of a diff amp is $A_d = 500$ and its common-mode gain is 0.5. For each set of inputs given, determine the output voltage and compare with that of an ideal diff amp whose $A_d = 500$. (a) $v_{i1} = 950$ mV, $v_{i2} = 850$ mV. (b) $v_{i1} = -850$ mV, $v_{i2} = -950$ mV.

Drill Problem 8.13: The differential-mode gain of a diff amp is $A_d = 10^4$. The input voltages are 250 μV and 350 μV. If the common-mode component in the output is to be 10% of the differential-mode component in the output, find the common-mode gain of the diff amp.

Common-Mode Rejection Ratio (CMRR)

It can be seen from the previous discussion that if a diff amp is to behave like an ideal diff amp, then its common-mode gain A_c should be zero. But, in a real diff amp, the common-mode gain is not exactly zero even though it will be very small compared with the differential-mode gain. A figure of merit, called the *common-mode rejection ratio* (abbreviated *cmrr*) is used to indicate how close a given diff amp is to the ideal. The cmrr is defined by

$$\text{cmrr} = (A_d/A_c) \tag{8-17}$$

The larger the cmrr, the closer is the given diff amp to the ideal. For an ideal diff amp, the cmrr is infinity. It is customary to express the cmrr in decibels.

$$\text{cmrr in decibels} = 20 \log (A_d/A_c) \tag{8-18}$$

The typical value of cmrr for commercially available diff amps (and op amps) lies in the range 60 dB to 120 dB. For a cmrr of 120 dB, the ratio (A_d/A_c) is 10^6. That is, the diff amp may have a differential-mode gain of 3×10^6 and a common-mode gain of 3.

Example 8.12

The diff-mode gain of an amplifier is 10^6 and its common-mode gain is 8. Determine the cmrr and express it in decibels.

Solution

$$\text{cmrr} = \frac{10^6}{8} = 1.25 \times 10^5$$

$$\text{cmrr in decibels} = 20 \log (1.25 \times 10^5)$$

$$= 102 \text{ dB}$$

Example 8.13

The diff-mode gain of an amplifier is 10^5, and its cmrr is 75 dB. Determine the common-mode gain.

Solution

$$\text{cmrr} = \text{antilog}\,[(\text{cmrr in decibels})/20]$$
$$= \text{antilog}\,(3.75)$$
$$= 5.62 \times 10^3$$

Therefore, the common-mode gain is

$$A_c = (A_d/\text{cmrr})$$
$$= \frac{10^5}{5.62 \times 10^3}$$
$$= 17.8$$

Example 8.14

The diff-mode gain of a diff amp is 10^3 and the cmrr is 2000. The input signals are $v_{i1} = 105$ mV and $v_{i2} = 95$ mV. Calculate the output voltage.

Solution

Since $A_d = 10^3$ and cmrr = 2000, we get

$$A_c = \frac{10^3}{2000}$$
$$= 0.5$$

The differential-mode input is $(v_{i1} - v_{i2}) = 10$ mV, and the common-mode input is $[(v_{i2} + v_{i1})/2] = 100$ mV. The output voltage is, therefore,

$$v_o = A_d(10) + A_c(100)\ \text{mV}$$
$$= 10.05\ \text{V}$$

Drill Problem 8.14: The diff-mode gain of a diff amp is 50×10^4, and its common-mode gain is 10. Determine the cmrr and express it in decibels.

Drill Problem 8.15: The diff-mode gain of a diff amp is 8000, and its cmrr is 40 dB. Determine the common-mode gain.

Drill Problem 8.16: The diff-mode gain of a diff amp is 15×10^5, and its cmrr is 500. The input voltages are $v_{i1} = 105\ \mu V$ and $v_{i2} = 95\ \mu V$. Calculate the output voltage.

Differential Amplifier Circuits

The basic diff amp circuit (Fig. 8-41) consists of an identical pair of transistors coupled through their emitters. The resistors R_{c1} and R_{c2} are usually made equal to each other (even though this is not a necessary condition). The input signals are applied to the two base terminals, and the output can be taken at either collector terminal (and ground) or between the two collectors. We will take the output as the voltage from the collector of T_2 to ground. The circuit (with a single-ended output as we are considering) can be made to have an *infinitely high cmrr* by replacing the resistor R_E of Fig. 8-41 by an *ideal constant current source* I_0 as shown in Fig. 8-42. Deferring the question of how a constant current source can be realized in practice, let us see how it produces an infinite cmrr.

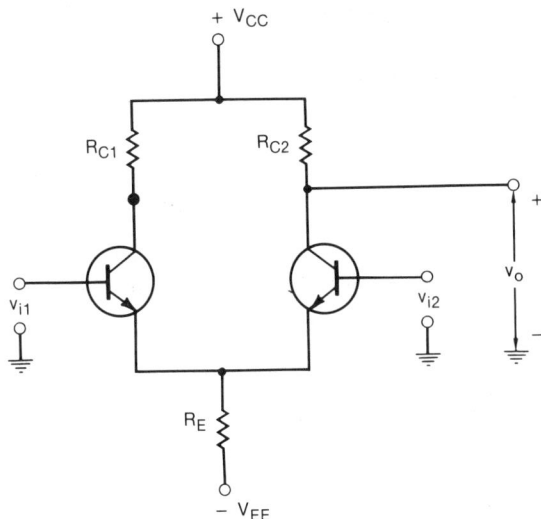

Fig. 8-41. Basic diff amp circuit.

The presence of the constant current source imposes the following constraint: the sum of the two emitter currents is always a constant, that is,

$$I_{E1} + I_{E2} = I_0$$
$$= \text{a constant} \qquad (8\text{-}19)$$

If the β's of the two transistors are sufficiently high, then the collector cur-

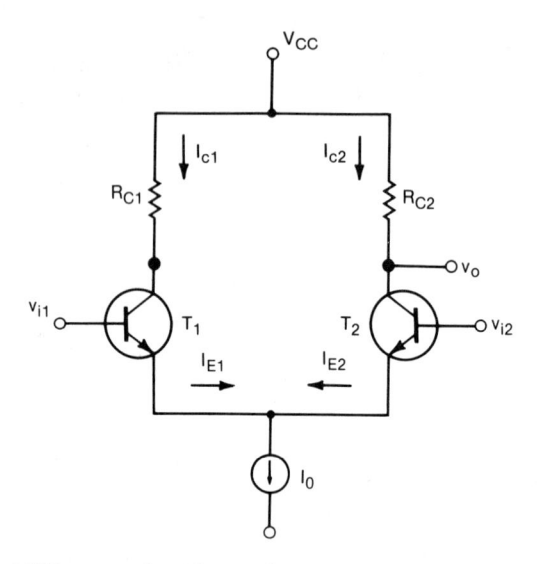

Fig. 8-42. Diff amp circuit with a constant-current source in the emitter branch.

rent will be approximately equal to the emitter current. As a result, Eq. (8-19) will become

$$I_{C1} + I_{C2} = I_0$$
$$= \text{a constant} \qquad (8\text{-}20)$$

Consider first the quiescent (or no signal) condition of the diff amp circuit. Assuming identical transistors and $R_{c1} = R_{c2}$, we can see that the Q point values must be the same for both the transistors, therefore,

$$I_{C1Q} = I_{C2Q}$$

Also, since the sum of the two collector currents must be I_0, we have

$$I_{C1Q} = I_{C2Q}$$
$$= \frac{I_0}{2} \qquad (8\text{-}21)$$

The two collector voltages V_{c1Q} and V_{c2Q} will be equal to each other, and the output voltage is

$$V_o = V_{C2Q}$$
$$= V_{CC} - (R_{C2}I_0/2) \qquad (8\text{-}22)$$

Now suppose two signals are applied to the two bases. The presence of the

Electronics: Circuits and Systems

signals will try to produce a change in the collector currents. If we denote the *changes* in the collector currents by ΔI_{C1} and ΔI_{C2}, then the new values of the collector currents are given by

$$I_{C1} = I_{C1Q} + \Delta I_{C1}$$
$$= \frac{I_0}{2} + \Delta I_{C1}$$

and

$$I_{C2} = I_{C2Q} + \Delta I_{C2}$$
$$= \frac{I_0}{2} + \Delta I_{C2}$$

Adding these two expressions, we have for the new sum of the two collector currents:

$$I_{C1} + I_{C2} = I_0 + (\Delta I_{C1} + \Delta I_{C2}) \tag{8-23}$$

But the sum of the collector currents is constrained to be equal to I_0 under all conditions. Therefore, $I_{C1} + I_{C2} = I_0$ in the last equation. Therefore, Eq. (8-23) reduces to

$$\Delta I_{C1} + \Delta I_{C2} = 0 \tag{8-24}$$

which is the condition that must be satisfied by any variations in the collector currents of the diff amp.

The *change* in the output voltage v_0 is due to the change in the voltage drop across R_{c1} by the new value of the collector current I_{c2}, that is,

$$\Delta v_o = R_2 \Delta I_{C2} \tag{8-25}$$

The two Eqs. (8-24) and (8-25) form the basis of the behavior of the diff amp under the application of the external signals. We will consider the two modes of operation of a diff amp: the common-mode operation and the differential-mode operation.

Common-Mode Operation

When both the bases are connected to the *same* signal source so that

$$v_{i1} = v_{i2}$$
$$= \frac{V_s}{2}$$

as indicated in Fig. 8-43A, then the diff amp is operating *strictly in the common mode*, since the differential-mode component $(v_{i1} - v_{i2}) = 0$.

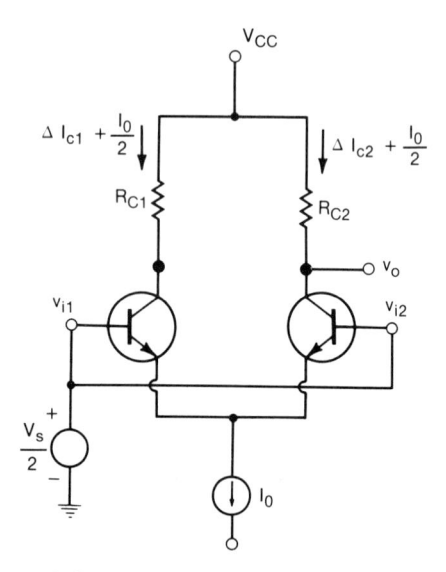

(A) Common mode: $v_{i1} = v_{i2}$.

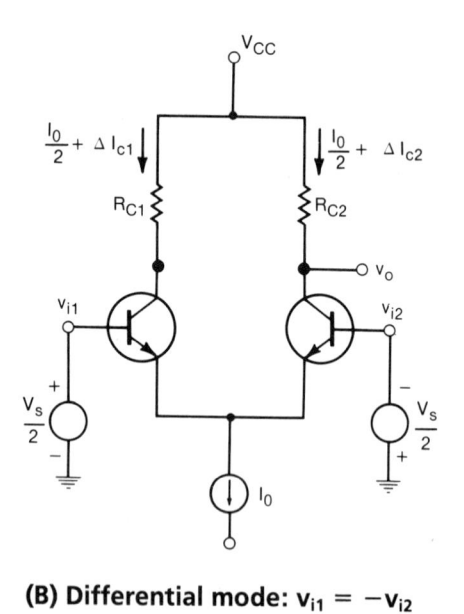

(B) Differential mode: $v_{i1} = -v_{i2}$

Fig. 8-43. Common-mode and differential-mode inputs to the ideal diff amp.

Electronics: Circuits and Systems

Since both the bases are activated by identical signals, the *changes* in the two collector currents must both be in the same direction. That is, ΔI_{C1} and ΔI_{C2} must both be positive (when V_s is positive), or they must both be negative (when V_s is negative). Then, the condition stated in Eq. (8-24):

$$\Delta I_{C1} + \Delta I_{C2} = 0$$

requires that the sum of two positive quantities or two negative quantities must be identically zero! The only way that such a condition can be met is to make

$$\Delta I_{C1} = \Delta I_{C2}$$
$$= 0 \qquad (8\text{-}26)$$

That is, in the *common-mode* operation, the *collector currents do not change* but stay put at their Q point values. The change in the output voltage, Δv_o, Eq. (8-25), becomes

$$\Delta v_o = 0 \qquad (8\text{-}27)$$

There is *no change* in the output voltage v_O from its Q point value when a common-mode signal is applied to the diff amp. It is important to remember that the two results, Eqs. (8-26) and (8-27), are brought about by the presence of the *constant current source* that forces the sum of the two collector currents to be a constant I_0.

Differential-Mode Operation

When one of the inputs is made exactly the opposite of the other so that

$$v_{i2} = -v_{i1}$$

as indicated in Fig. 8-43B, the diff amp is operating *strictly in the differential mode* since the common-mode component $[(v_{i1} + v_{i2})/2] = 0$.

Since the two input signals are in in opposite directions, the changes in the collector currents will also be in opposite directions. That is ΔI_{c1} and ΔI_{c2} will be of opposite signs. Eq. (8-24) then becomes

$$\Delta I_{C1} = -\Delta I_{C2} \qquad (8\text{-}28)$$

and the change in the output voltage is given by Eq. (8-25):

$$\Delta v_o = R_{C2}(\Delta I_{c2})$$

In the differential mode, therefore, the two collector currents change by an equal amount but in opposite directions. There is a change in the output voltage that is determined by the change in the collector current I_{C2}.

It is seen that when a constant current condition is imposed on the sum of the two collector currents, there is a finite nonzero gain in the *differential* mode but zero gain in the common mode, and consequently, the *cmrr* becomes infinite.

When a differential amplifier is operating in a mixed mode (that is, the inputs v_{i1} and v_{i2} are such that neither the common-mode component nor the differential-mode component is zero), the output v_o will fluctuate about its Q point value, and these variations will depend only upon the differential-mode component of the inputs. The variations in the output will be *in phase* with the signal v_{i1} but 180° out of phase with the signal v_{i2}.

If the single output of a diff amp is taken off the other collector, v_{o1} as indicated in Fig. 8-44, then a discussion similar to the previous discussion applies. The voltage v_{o1} will stay at its Q point and show no change when an input signal is applied under a *strictly common-mode* operation. When it is operating in a *strictly differential-mode* operation, the output v_{o1} will change in proportion to the voltage $(v_{i1} - v_{i2})$. The change in v_{o1} will be 180° out of phase with v_{i1} but in phase with v_{i2}.

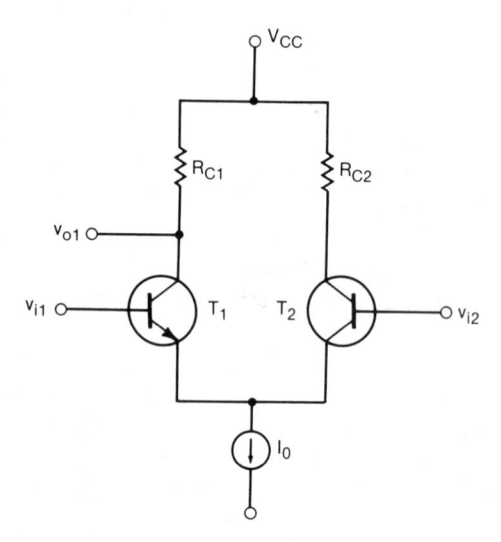

Fig. 8-44. Output taken from T₁ in the diff amp.

Realization of a Constant-Current Source

The preceding discussion shows that in order to have an ideal diff amp with an infinite cmrr, it is necessary to have a constant-current source that fixes the sum of the two collector currents at a constant value of I_0. Such a constant-current source can be approximated by the circuit of Fig. 8-45. The collector current I_C in the circuit can be made to be essentially a constant under all

Electronics: Circuits and Systems

operating conditions. The circuit of Fig. 8-45 replaces the current source symbol used in the diff amp circuits of Figs. 8-42 through 8-44.

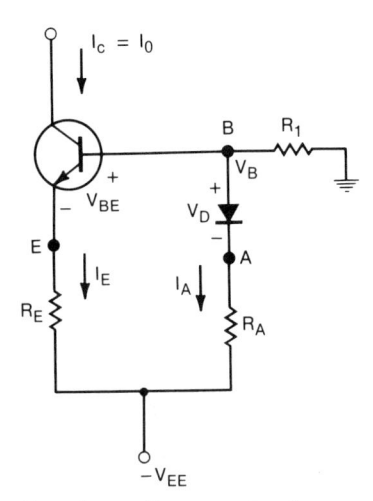

Fig. 8-45. Realization of a constant-current source.

The emitter current (and consequently the collector current) in a BJT depends essentially upon the emitter-base voltage V_{BE}. But the voltage V_{BE} varies with temperature at the rate of -2.5 mV/°C, and this causes the collector current to vary also. In the circuit of Fig. 8-45, a diode is used to compensate for the variation of V_{BE}. The voltage at point E is V_{BE} below the voltage at point B, and the voltage at point A is V_D below the point B. If the diode voltage V_D and the emitter-base voltage V_{BE} are equal (which is usually the case) then the voltages at points E and A are equal. Therefore,

$$R_E I_E = R_A I_A$$

since the other two ends of R_E and R_A are tied together. Therefore,

$$I_E = (R_A/R_E)I_A$$

and the current I_E does not depend upon V_{BE}. But this analysis is not completely accurate, and a methodical solution of the circuit of Fig. 8-45 shows that we need a series connection of N diodes instead of a single diode, as shown in Fig. 8-46.

It can be shown that in the circuit of Fig. 8-46, if the number of diodes is chosen to satisfy the equation:

$$N = 1 + \frac{R_A}{R_1} \tag{8-29}$$

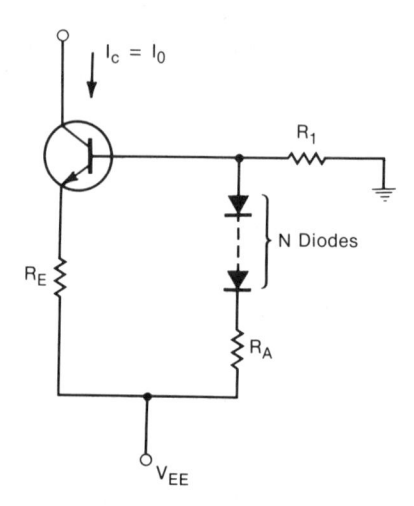

500   Operational Amplifiers and Feedback Amplifiers

Fig. 8-46. Use of several diodes in the constant-current source circuit.

then the current I_E becomes independent of V_{BE} or V_D. The collector current I_C in such a circuit has a constant value of I_0 given by

$$I_0 = \frac{R_A}{R_E(R_1 + R_A)} V_{EE} \tag{8-30}$$

A typical diff amp will contain a circuit of the type shown in Fig. 8-46 connected to the junction of the emitters of the two transistors. Fig. 8-47 shows a diff amp circuit in which the constant current source circuit is delineated by the dashed lines. Since $R_A = 1.5$ K and $R_1 = 3.2$ K, Eq. (8-42) gives a value of $N = 1.47$, or we need two diodes. The constant current I_0 in the circuit is, from Eq. (8-43), 0.145 V_{EE} mA. A battery of 15 V for V_{EE} will provide a current of $I_0 = 2.18$ mA.

The diff amp has numerous applications based upon its two chief characteristics: the dependence of the output on the difference between two input signals and high gain. High values of gain can be obtained by cascading two stages: the outputs from the two collectors of the first stage are used as the inputs to the bases of the second-stage transistors. One of the most frequent uses of the diff amp is, as mentioned earlier, as the first stage of an op amp. Other uses include phase splitters, comparators, modulators, Schmitt triggers. The use of a diff amp as a comparator is extremely important since it acts as the bridge between analog signals and digital circuits, and we will examine this application in some detail.

Electronics: Circuits and Systems

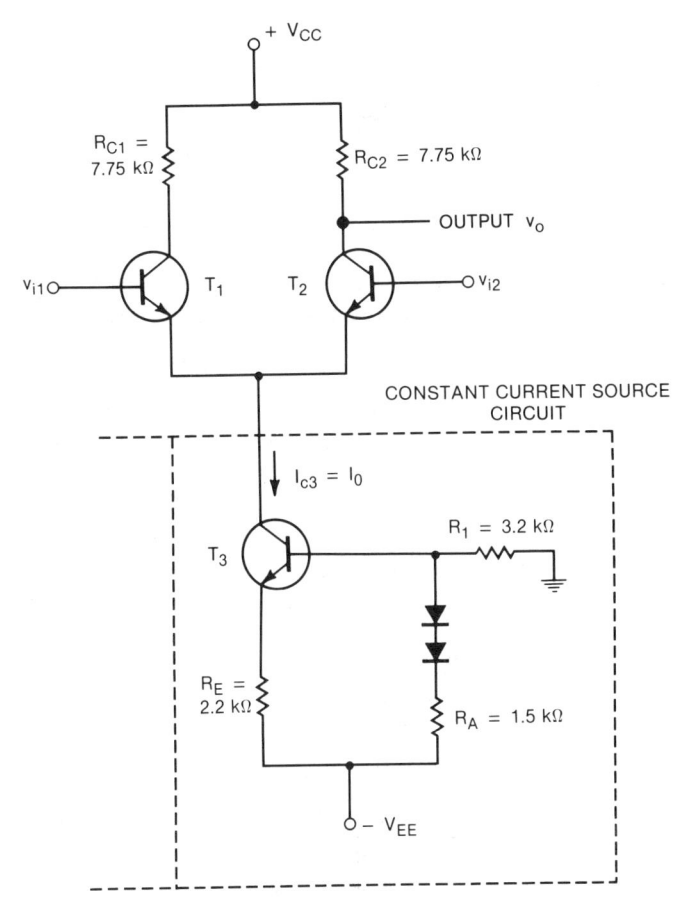

Fig. 8-47. Diff amp circuit.

COMPARATORS

A *comparator* is a circuit that has the property that its output voltage switches polarity when the input signal exceeds a certain threshold level. The threshold level is called the *reference voltage* V_R. As indicated in Fig. 8-48, when the input V_i is less than V_R, the output voltage V_o is negative. When the input V_i is greater than V_R, the output voltage V_o is positive. The comparator can be used to compare any two input signals, and its output indicates which of the two inputs is higher than the other at any instant.

Let us now consider how a diff amp can be designed to serve as a comparator. Neglecting the common-mode component, the output v_o (Fig. 8-42 or 8-47) fluctuates about its Q point value in the following manner: when v_{i1} is

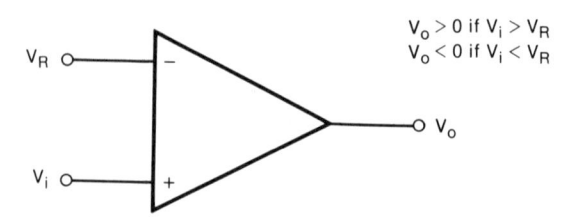

Fig. 8-48. Comparator. The output switches depending upon whether the input is greater or less than V_R.

greater than v_{i2}, the output v_o is above the Q point value, and when v_{i1} is less than v_{i2}, the output is below the Q point value. Also, the output voltage v_o in Fig. 8-47 cannot go below zero (which happens when the current through R_{c2} is high enough to make the voltage across R_{C2} equal to V_{CC}), and it cannot go above V_{CC} (which happens when the transistor T_2 does not conduct). Therefore, if v_o is plotted as a function of $(v_{i1} - v_{i2})$, the resulting *transfer characteristic* will be of the form shown in Fig. 8-49. There is a *linear region* where the output varies linearly with $(v_{i1} - v_{i2})$, and there are two saturation regions, one below and one above the linear region. The transfer characteristic is of the general shape required of the output of a comparator, except for the linear region and the fact that the output v_o is always positive.

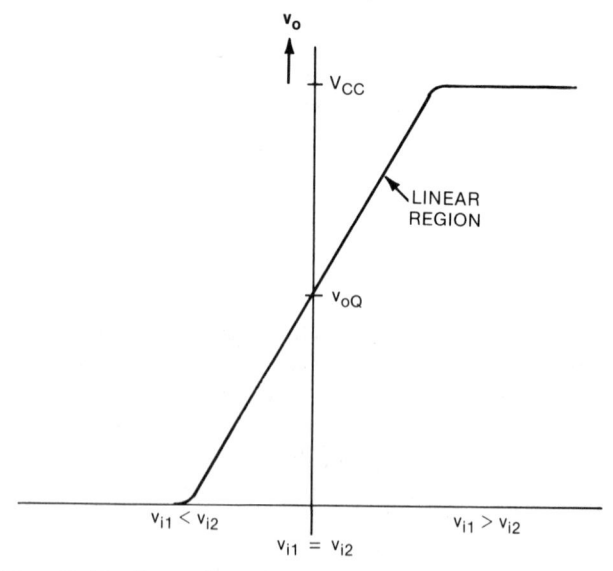

Fig. 8-49. Transfer characteristic of a diff amp.

Electronics: Circuits and Systems

By properly designing the diff amp, the linear region of the transfer characteristic can be limited to a few millivolts and can, therefore, be neglected.

Differential amplifier configurations are designed for use as comparators, in which the output voltage swings from a negative value $-V_o$ to a positive value $+V_o$. Such units need two power supplies: $-V_{CC}$ and $+V_{CC}$ and the two output levels are quite close to the battery voltages. Many comparators are available commercially, with typical output swings from -5 V to $+5$ V and the linear region confined to 1 or 2 mV on either side of the crossover point in the transfer characteristics. Fig. 8-50 shows the notation used for a comparator along with the transfer characteristics.

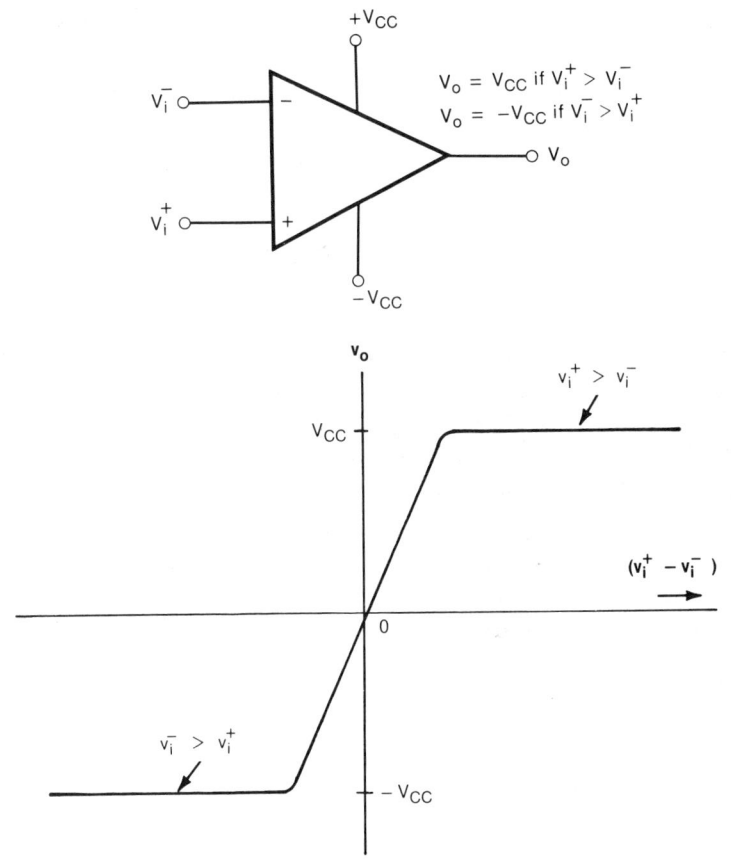

Fig. 8-50. Comparator symbol and its transfer characteristic.

In our discussion, we will ignore the linear region and assume ideal operation:

(Input to the + terminal) > (input to the − terminal) makes the output = +V$_{CC}$

(Input to the + terminal) < (input to the − terminal) makes the output = −V$_{CC}$

Window Detector Circuit

An application of the comparator is a circuit that detects whether an input signal is *between* two prescribed values. Such a circuit is called a *window detector*, Fig. 8-51. As a practical example, the input signal V$_A$ may be from a photosensitive device on a camera. The camera's operation may need to be modified depending on whether the incident light is too low, too bright, or in between those extremes.

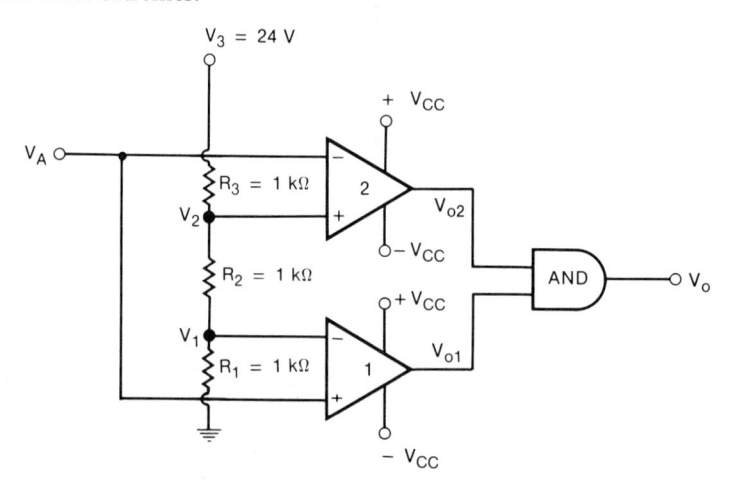

Fig. 8-51. Window detector circuit.

The reference voltages of the two comparators in the circuit are derived through a voltage-divider network and are given by

$$V_1 = \frac{R_1}{R_1 + R_2 + R_3} V_B$$

$$= 8 \text{ V}$$

$$V_2 = \frac{R_1 + R_2}{R_1 + R_2 + R_3} V_B$$

$$= 16 \text{ V}$$

The reference voltage V$_1$ is connected to the inverting *(minus)* input of

Electronics: Circuits and Systems

comparator 1. The output of comparator 1 will be positive ($+V_{CC}$) when the signal $V_A > V_1$ or 8 V, and negative ($-V_{CC}$) otherwise. The reference voltage V_2 is connected to the noninverting (plus) input terminal of comparator 2. The output of comparator 2 will be positive when $V_A < V_2$ or 16 V and negative otherwise, that is,

$$V_A < 8 : V_{o1} \text{ negative}, V_{o2} \text{ positive}$$

$$8 < V_A < 16 : V_{o1} \text{ and } V_{o2} \text{ are both positive}$$

$$V_A > 16 : V_{o1} \text{ positive}, V_{o2} \text{ negative}.$$

The outputs of the two comparators are passed through an AND gate (which is a logic gate that will be discussed in detail later in the text). The AND gate has the property that *the output of the AND gate is high when both inputs are high,* and *the output of the AND gate is low when either or both inputs are low.*

That is, the output of the AND gate will be high (say 5 V) when both V_{o1} and V_{o2} are positive and low (say 0 V) when either V_{o1} or V_{o2} or both are negative. We can see that the output of the AND gate will be high when the input signal V_A is between 8 V and 16 V, and low for all other signal levels.

Drill Problem 8.17: Suppose in the window detector circuit of Fig. 8-51, we accidentally connect V_1 to the noninverting (plus) terminal of comparator 1 and V_2 to the inverting (minus) terminal of comparator 2. Discuss the operation of the circuit. The signal is fed into the other terminal of each comparator.

Another example of the use of a comparator is a *low light-level detector* shown in Fig. 8-52. The resistor R_p is a photoresistor whose resistance is high when the light level is low.

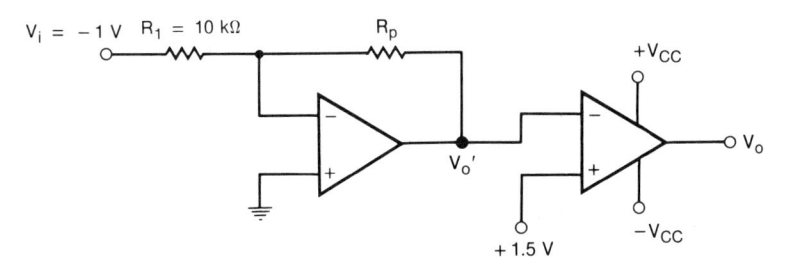

Fig. 8-52. Low light-level detector.

The output of the op amp in Eq. (8-4) is

$$V'_o = -(R_p/R_1)V_i$$

$$= (R_p/10) \text{ V}$$

where,

R_p is in kilohms. The output of the comparator will be positive when V'_o is less than the reference voltage of 1.5 V.
Therefore, we have

$(R_p/10) < 1.5$ V or $R_p < 15$ kΩ: V_o is positive

$(R_p/10) > 1.5$ V or $R_p > 15$ kΩ: V_o is negative

The circuit will have a negative output (which can be used to turn on a warning light) when the light level is low enough to make $R_p > 15$ kΩ.

Drill Problem 8.18: Modify the circuit of Fig. 8-52 so that the output of the circuit is *positive* when the light level is low enough to make $R_p > 50$ kΩ.

Sine-Wave to Square-Wave Generator

A comparator can be used to convert an input sinusoidal waveform to a square waveform. Such a circuit is shown in Fig. 8-53. Since the reference voltage is ground, the output is positive and a constant when the signal is positive. The output is negative and a constant when the signal is negative.

SCHMITT TRIGGER

The combination of a comparator and *positive feedback* leads to a useful circuit called the *Schmitt trigger*. The output of the Schmitt trigger is at one of two levels similar to that of a comparator. But the switching from high output to low output takes place at a *different* threshold level than that at which the transition from low to high output occurs. There are two different threshold levels in a Schmitt trigger circuit whereas there is only one in a comparator.

The arrangement of a Schmitt trigger is shown in Fig. 8-54. The connections are similar to the op amp circuits studied earlier in this chapter with the important difference that the *feedback* from the output to the input (through a voltage-divider network) is to the *noninverting* (or plus) input terminal of the comparator. This means that if the output voltage were to increase, then the voltage V_R fed back also increases, which in turn makes the output voltage increase even further, and the cycle repeats. Therefore, we have a *positive* or *regenerative feedback* in the present circuit. The increase in the output voltage does not continue indefinitely, however. The output quickly reaches a saturation value, which is denoted by V_{hi}. Similarly, if the output of the comparator were to decrease, a regenerative cycle is set up that drives the output

(A)

(B)

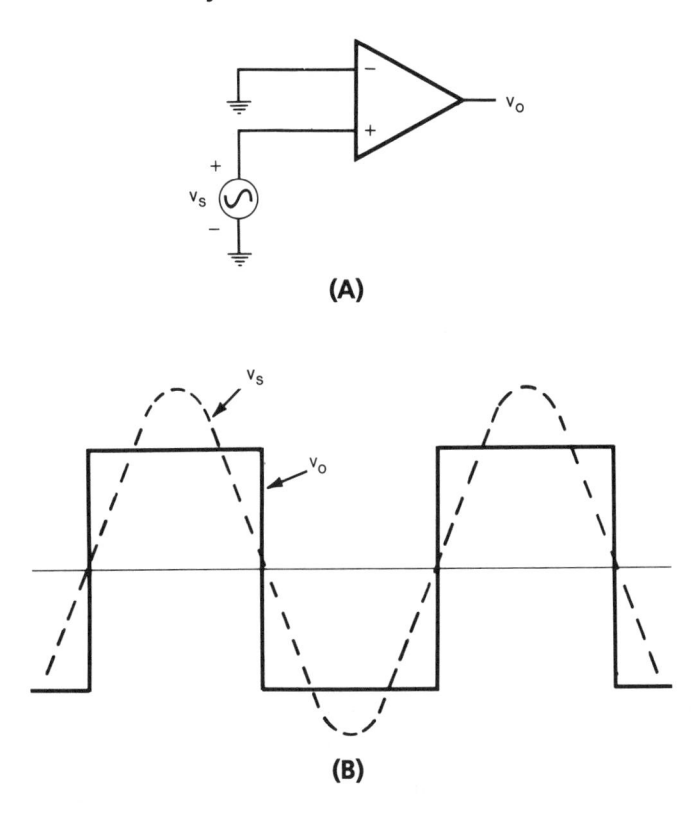

Fig. 8-53. Sine-wave to square-wave generator.

voltage quickly down to a saturation value of V_{low}. Thus, the output has two possible values V_{hi} and V_{low}, and the transition between these two levels is essentially instantaneous.

Since the reference voltage V_R is connected to the plus terminal of the comparator, we have

$$V_o = V_{hi} \text{ implies that } V_i < V_R$$
$$V_o = V_{low} \text{ implies that } V_i > V_R$$

Now, consider the reference voltage itself. It is not a fixed voltage but derived from the output voltage through a voltage-divider network. Since the output voltage has two possible values, so does the reference voltage. The availability of two different threshold levels results in a unique response characteristic of the Schmitt trigger as the following discussion will show.

Consider the voltage-divider network, Fig. 8-55. Since the voltage across the combination of R_1 and R_2 is $(V_o - V_B)$ and the comparator (which is the same as an op amp) draws no input current, we have

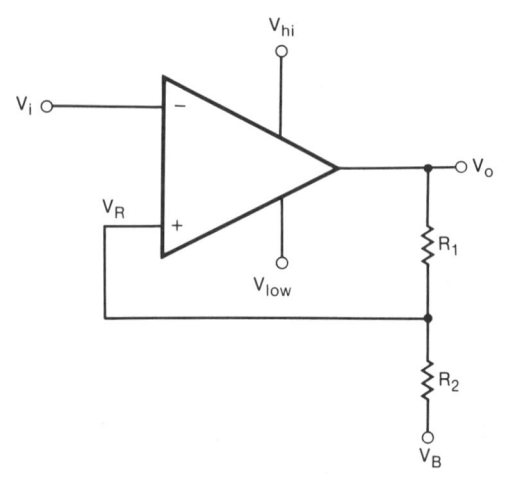

Fig. 8-54. Schmitt trigger.

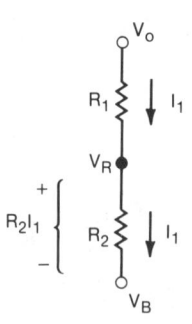

Fig. 8-55. Calculation of the reference voltage in a Schmitt trigger.

$$I_1 = \frac{V_o - V_B}{R_1 + R_2}$$

and

$$V_R = V_B + R_2 I_1$$

$$= \frac{R_1}{R_1 + R_2} V_B + \frac{R_2}{R_1 + R_2} V_o \qquad (8\text{-}31)$$

Of the two terms in the previous expression, the first one has a constant value but the second one is affected by the value of V_o. Let $V_{R(hi)}$ denote the value of V_R when $V_o = V_{hi}$ and $V_{R(low)}$ is the value of V_R when $V_o = V_{low}$.

Electronics: Circuits and Systems 509

$V_{R(hi)}$ is called the *upper threshold level* and $V_{(low)}$, the *lower threshold level* of the Schmitt trigger.

The circuit has two stable states of operation.

High-output state: $V_o = V_{hi}$ which makes $V_R = V_{R(hi)}$. The input signal V_i must be *less* than $V_{R(hi)}$ for this state to persist.

Low-output state: $V_o = V_{low}$ which makes $V_R = V_{R(low)}$. The input signal V_i must be *greater* than $V_{R(low)}$ for this state to persist.

Now consider the *switching* of the circuit between the two states. Suppose the circuit is initially in the high-output state, as indicated in Fig. 8-56A. The signal V_i must be less than $V_{R(hi)}$ for the state to be maintained. When the input signal exceeds $V_{R(hi)}$ the high-output state can no longer be maintained, and the circuit switches instantaneously to the low-output state as indicated in the diagram. The voltage V_R also switches to $V_{R(low)}$ at that instant.

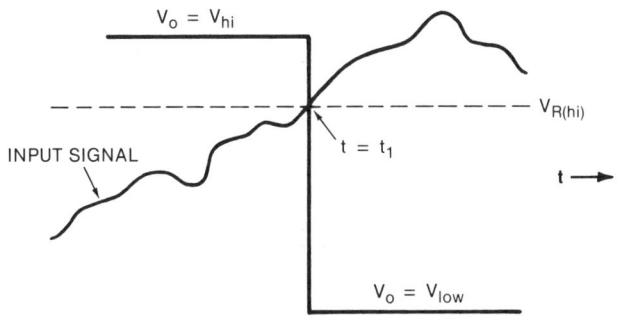

(A) Output switches from high to low only when the input crosses the high threshold with a positive slope.

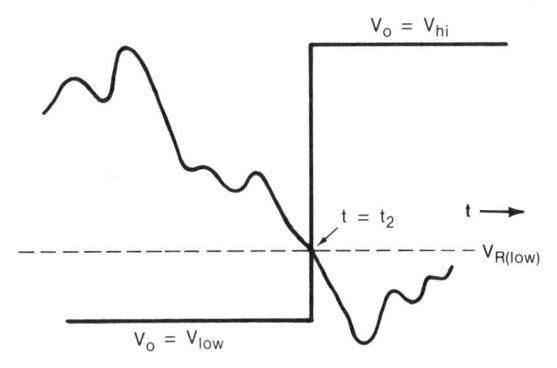

(B) Output switches from low to high only when the input crosses the low threshold with a negative slope.

Fig. 8-56. Switching of the output in a Schmitt trigger.

Switching from high input to low output occurs when the input is initially below $V_{R(hi)}$ and then goes above $V_{R(hi)}$.

Now suppose the circuit is initially in the low-output state as indicated in Fig. 8-56B. The signal V_i must be greater than $V_{R(low)}$ for this state to be maintained. When the input signal falls below $V_{R(low)}$ the low-output state can no longer be maintained and the circuit instantaneously switches to the high-output state as indicated. The voltage V_R also switches to $V_{R(hi)}$ at that instant.

Switching from low output to high output occurs when the input is initially above $V_{R(low)}$ and then falls below $V_{R(low)}$.

The response of the Schmitt trigger can be summarized as follows: (Refer to Fig. 8-57.)

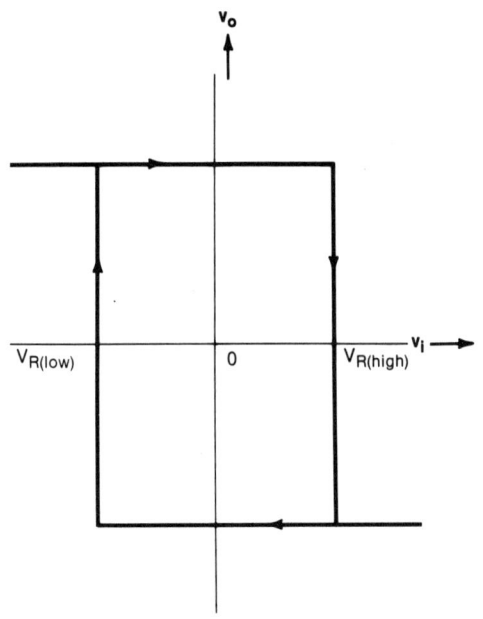

Fig. 8-57. Hysteretic property of the output-input characteristic of a Schmitt trigger.

There are two threshold levels: $V_{R(hi)}$ associated with the high-output state and $V_{R(low)}$ associated with the low-output state. Whenever the input *crosses the $V_{R(hi)}$ level with a positive slope*, the circuit switches from a high output to a low output. Whenever the input *crosses the $V_{R(low)}$ level with a negative slope*, the circuit switches from a low output to a high output. Switching between the two states can occur only under the precise fulfillment of the two conditions stated previously: the slope as well as the crossing are critical.

Fig. 8-58 shows an input signal and the corresponding output waveform.

Electronics: Circuits and Systems

Pay special attention to the instants at which switching between states occurs. The fact that in a Schmitt trigger the response depends upon the *past history of the input signal* is usually referred to as *hysteresis*.

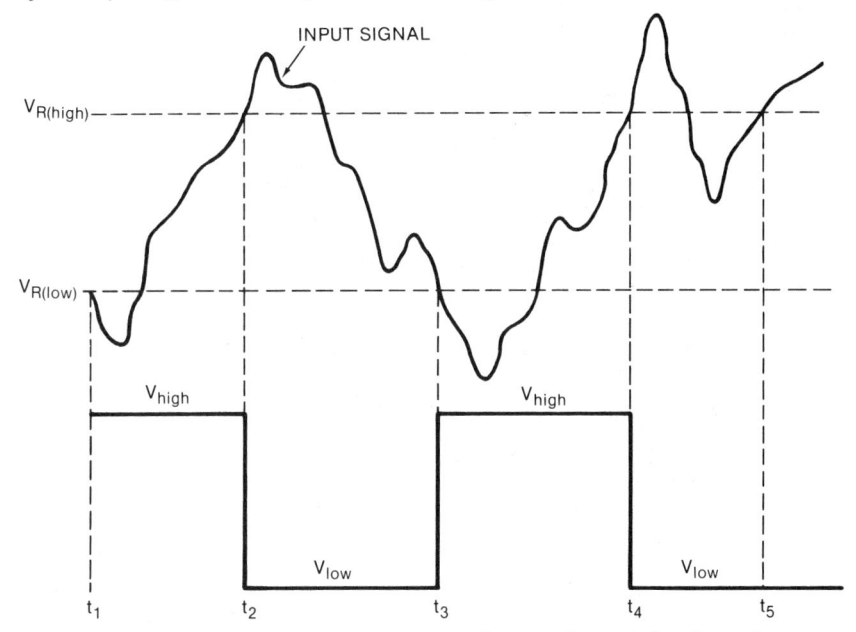

Fig. 8-58. Input and output waveforms in a Schmitt trigger.

In Fig. 8-58 at $t = t_1$, the input crosses the lower threshold with a negative slope, which causes the output to switch to V_{hi}. The same situation occurs at t_3. At $t = t_2$ and t_4, the input crosses the upper threshold with a positive slope, which causes the output to switch to V_{low}. The other instants of time at which the signal crosses either threshold level do not cause the output to switch since the slope is not of the proper sign.

Example 8.15

Determine the two threshold levels of the Schmitt trigger circuit shown in Fig. 8-59A. Draw the output waveform if the input signal is as shown in Fig. 8-59B.

Solution

Using Eq. (8-45), we have for the threshold level

$$V_R = \frac{R_1}{R_1 + R_2} V_B + \frac{R_2}{R_1 + R_2} V_o$$
$$= -2 + 0.2 V_o$$

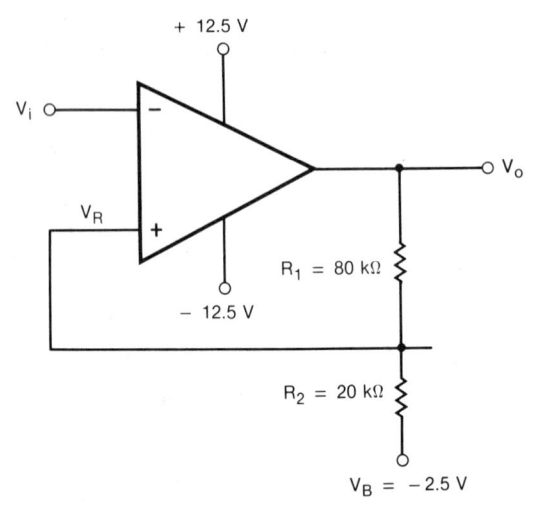

Operational Amplifiers and Feedback Amplifiers

(A) Schmitt trigger circuit.

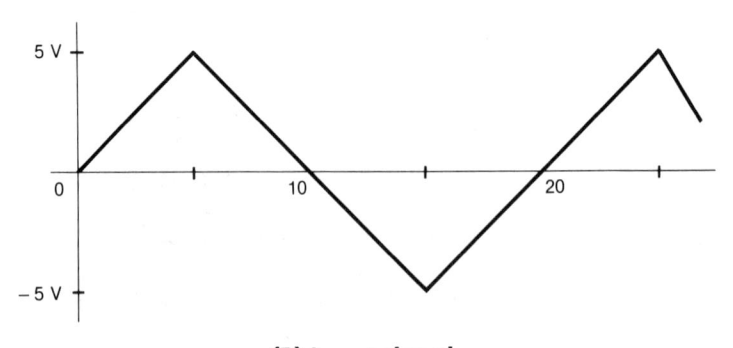

(B) Input signal.

Fig. 8-59. Circuit for Example 8.15.

Therefore,

$$V_{R(hi)} = -2 + 0.2 \times 12.5$$
$$= 0.5 \text{ V (upper threshold)}$$
$$V_{R(low)} = -2 - 0.2 \times 12.5$$
$$= -4.5 \text{ V (lower threshold)}$$

The output waveform for the given input is shown in Fig. 8-60.

Electronics: Circuits and Systems

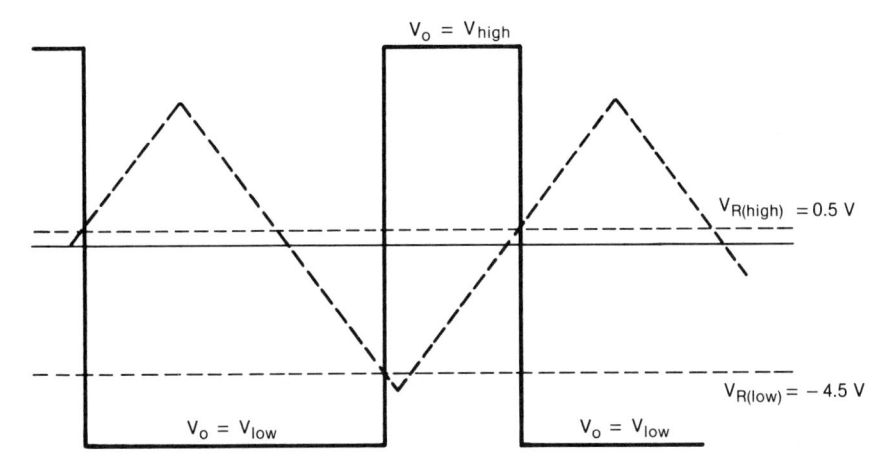

Fig. 8-60. Output waveform of Example 8.15.

Drill Problem 8.19: Change the battery V_B of the circuit of Fig. 8-59 so that $V_{R(low)} = 0$. Calculate the new value of $V_{R(hi)}$ and redraw the output waveform for the same input.

Drill Problem 8.20: Repeat Drill Problem 8.19 by changing the battery V_B to make $V_{R(low)} = -5$ V.

Example 8.16

The input to the RC circuit in the circuit of Fig. 8-61 is a rectangular pulse as shown. Determine and sketch the output waveform of the circuit. Assume that the capacitor has zero voltage before $t = 0$.

Solution

First calculate the two threshold levels of the Schmitt trigger.

$$V_R = 3.75 + 0.25 V_o$$

which gives $V_{R(hi)} = 8.75$ V and $V_{R(low)} = 3.75$ V.

The capacitor starts with zero voltage at $t = 0$ and charges up gradually to a value of 20 V due to the input pulse. At the end of the pulse, the capacitor starts discharging from 20 V toward a final value of zero. The capacitor voltage, which is the input signal to the Schmitt trigger, is shown in Fig. 8-62A. The charging and discharging curves of an RC circuit that were presented in Chapter 2 will be used in the calculations of the present circuit.

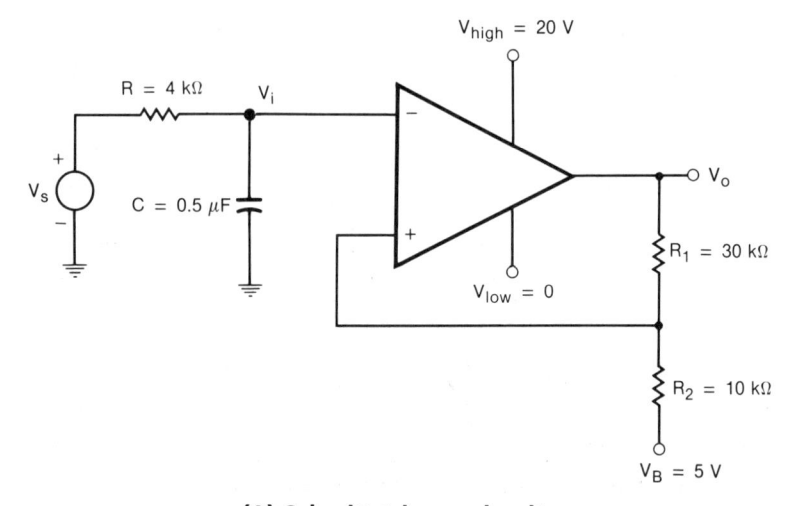

514   Operational Amplifiers and Feedback Amplifiers

(A) Schmitt trigger circuit.

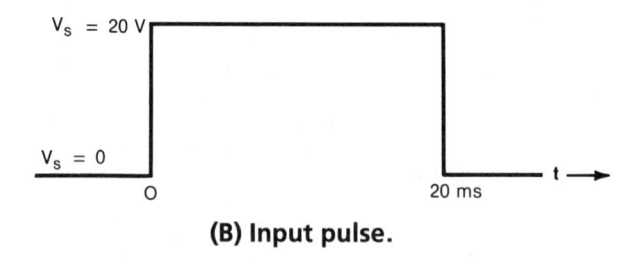

(B) Input pulse.

Fig. 8-61. Circuit and input pulse for Example 8.16.

We have to determine what state the Schmitt trigger is in before $t = 0$. Suppose we *assume* that it is in the low-output state before $t = 0$. Then, the pertinent threshold level will be $V_{R(low)}$, which is 3.75 V. But in order to have a *low-output state*, V_i must be *greater* than $V_{R(low)}$ which is not true in the present case since $V_i = 0 < V_{R(low)}$. Therefore, the circuit cannot be in the low-output state and must be in the high-output state. Let us see if this is consistent with the facts. For the circuit to be in the high-output state, V_i must be *less than* $V_{R(hi)}$, which is indeed the case.

The circuit is, therefore, in the *high-output state* before $t = 0$.

As the capacitor voltage builds up, V_i increases, and at some time t_1, it exceeds $V_{R(hi)}$ with a positive slope. The circuit switches to the low-output state at that instant. The value of t_1 is the instant at which the capacitor voltage reaches $V_{R(hi)} = 8.75$ V. By using the charging curve of Chapter 2, we get $t_1 = 1.2$ ms.

Electronics: Circuits and Systems

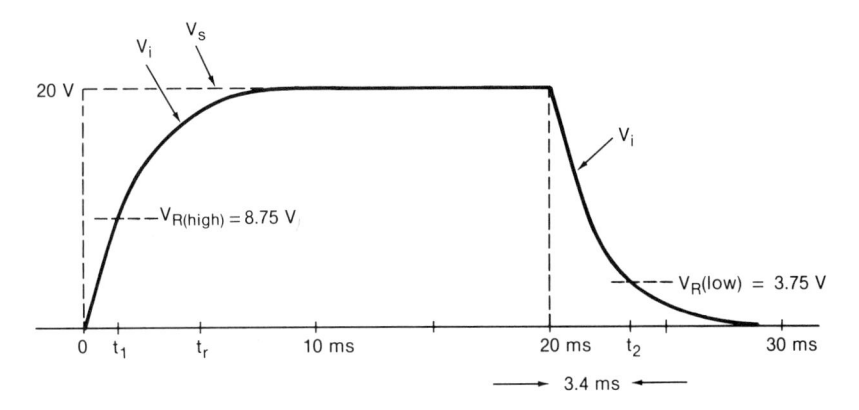

(A) Input signal to Schmitt trigger.

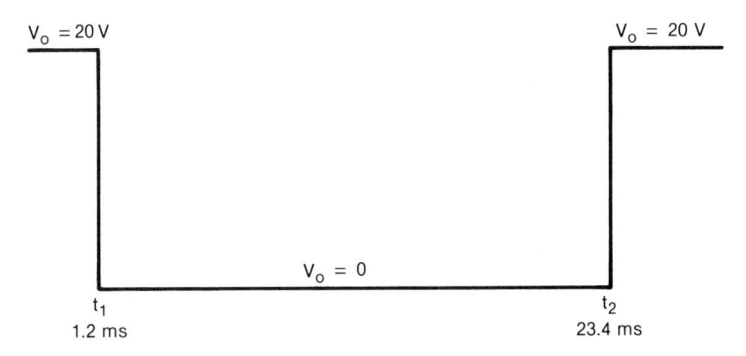

(B) Output is a rectangular pulse.

Fig. 8-62. Capacitor voltage (input to the trigger) and the trigger output of Example 8.16.

The low-output state persists through the presence of the input pulse. When the pulse disappears, the capacitor discharges and V_i begins to decrease toward zero. At some time t_2, V_i falls below $V_{R(low)}$ with a negative slope. The circuit switches to the high-output state at that instant. The value of t_2 is the instant at which the capacitor has discharged to $V_{R(low)} = 3.75$ V starting with an initial value of 20 V. By using the discharge curve of Chapter 2, we find $t_2 = 3.4$ ms *after* the pulse has become zero.

The output of the circuit is, therefore, a rectangular pulse (actually a trough) as shown in Fig. 8-62B.

Drill Problem 8.21: The voltage shown in Fig. 8-63 is applied to the RC circuit in the last example. Determine and sketch the output of the Schmitt trigger circuit.

516   Operational Amplifiers and Feedback Amplifiers

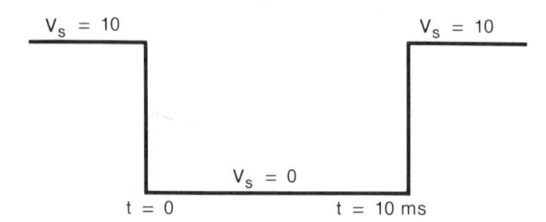

Fig. 8-63. Input pulse for Drill Problem 8.21.

The Schmitt trigger circuit derives its many uses from its property that it can convert a gradual variation of the input signal to a sharp instantaneous step in the output voltage. The hysteretic property of the trigger prevents the accidental change in the output due to a spike in the input that may be caused by noise.

PRACTICAL OPERATIONAL AMPLIFIERS

There are literally hundreds of op amps available commercially since an op amp is the workhorse of the analog electronic circuit designer. A very popular op amp is the μA 741 (manufactured by Fairchild Camera and Instrument Corp.). The circuit of the μA 741 op amp is shown in Fig. 8-64 in order to show how complex the circuit can be. Fortunately for us, it is not necessary to be able to analyze such a circuit to use the op amp itself. We can think of the whole complex circuit being hidden inside the triangle symbol we have been using for an op amp.

A practical op amp is fairly close to the ideal op amp but there are some limitations and departures from the ideal that should be kept in mind. The following is a list of such limitations and departures.

1. Input Offset Voltage: The commercial op amps have an input offset voltage that is typically from a fraction of a millivolt to 5 mV, which is the amount of input voltage that must be applied between the input terminals to get a zero-output voltage.

2. Input Offset Voltage Drift: The input offset voltage varies with the temperature at which the op amp operates, and this variation is specified as the drift. The drift is typically a few microvolts/°C.

3. Input Offset Current: To obtain a zero-output voltage, it is necessary to have a slightly different current entering the input terminals of a commercial op amp, which is specified as the input offset current. It is typically in the tens of nanoamps range but it may be as low as a few picoamps.

4. Input Offset Current Drift: The input offset current drift is analogous to the drift of the input offset voltage.

5. Slew Rate: The slew rate is the rate at which the output voltage can

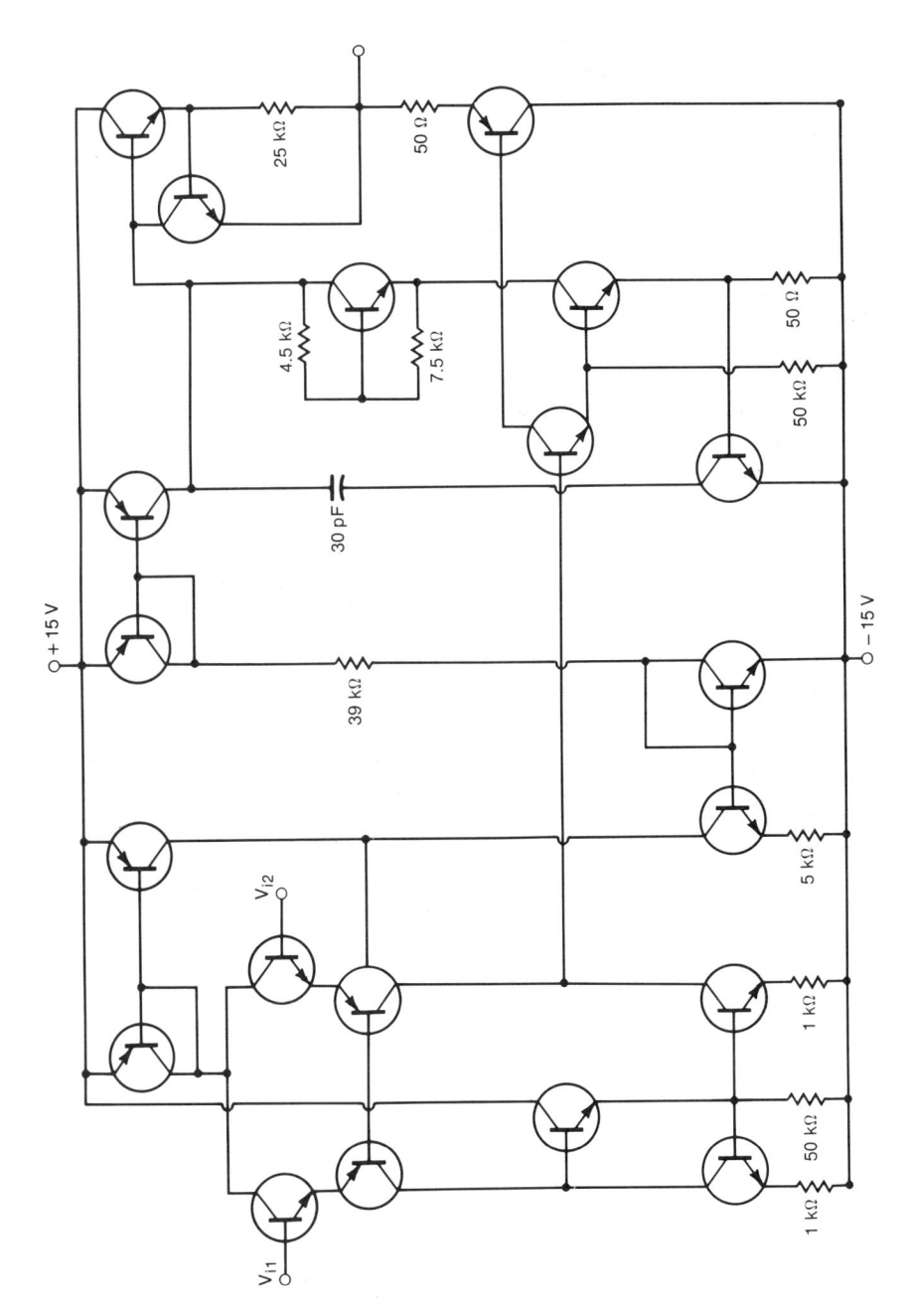

Fig. 8-64. The A 741 op amp circuit.

change in an op amp. That is, even though we can expect an *instantaneous* jump in the output voltage in an ideal op amp, there is a limit to the speed with which the output voltage of an op amp can rise in reality. The fastest rise is measured by the slew rate. The slew rate is specified in volts per microsecond. It is usually in the range of less than 1 V/μs, but there are op amps in which the slew rate is in the hundreds of volts/per microsecond. A very high speed op amp is available with a slew rate of 6000 V/μs.

6. *Frequency Response:* The frequency response of an op amp extends down to dc at the low-frequency end since there are no coupling capacitors used. The high-frequency response is affected by the interelectrode capacitances as discussed in the last chapter. The manufacturer specifies the gain-bandwidth product in the form of a frequency f_T, which is around 1 MHz for the 741 op amp.

7. *CMRR:* The common-mode rejection ratio, which was defined earlier in this chapter, is typically 100 dB.

FEEDBACK AMPLIFIERS

The concept of feedback has already been introduced in the discussion of op amps and the Schmitt trigger. We will present a more general discussion of feedback and the general properties of an amplifier with feedback.

A basic feedback arrangement is shown in Fig. 8-65. The output voltage V_o of the amplifier is multiplied by a quantity β_f, and the resulting voltage $\beta_f V_o$ is connected in series with the signal. The quantity β_f may be a constant or a function of frequency depending upon the feedback network.

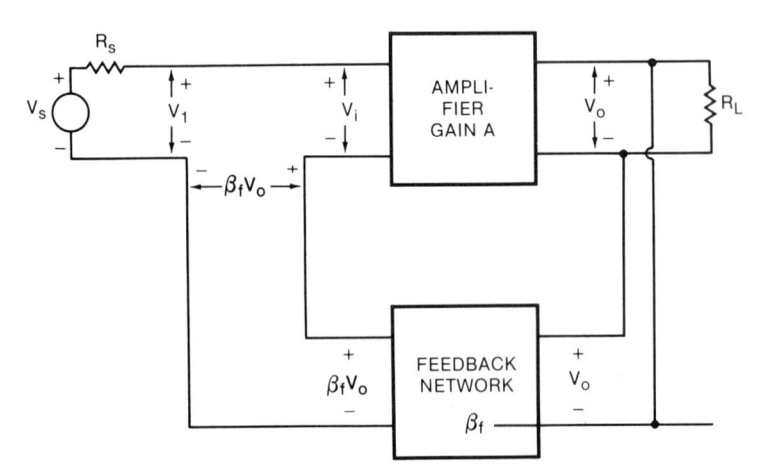

Fig. 8-65. Basic feedback amplifier arrangement.

The input voltage to the amplifier is

$$V_i = V_1 - (\beta_f V_o) \qquad (8\text{-}32)$$

If $\beta_f V_o$ is positive, then the effect of feedback is to *reduce* the input to the amplifier and the feedback is said to be *negative*. If $\beta_f V_o$ is negative, then the effect of the feedback is to *increase* the input to the amplifier and the feedback is said to be *positive*.

The overall gain with feedback is defined by

$$A_f = \frac{V_o}{V_1} \qquad (8\text{-}33)$$

and using Eq. (8-33), it is possible to show that

$$A_f = \frac{A}{1 + A\beta_f} \qquad (8\text{-}34)$$

where,

A is the *gain* (V_o/V_i) *without feedback* (that is, when $\beta_f = 0$).

If $A = 10^4$ and $\beta_f = 0.1$, for example, the gain with feedback is, from Eq. (8-34), $A_f = 9.99$, or approximately 10. The gain is reduced by the presence of negative feedback (in which case $A\beta_f$ is positive).

The reduction of the gain with negative feedback is advantageous, however, since it makes the gain less dependent on the devices in the amplifier. Suppose the gain A without feedback changes from 10^4 to 10^3 due to variation in the characteristics of the devices. Then, the gain with feedback (using $\beta_f = 0.1$) changes from 9.99 to 9.9, which is hardly noticeable. In fact, if β_f is made sufficiently large, Eq. (8-34) reduces to

$$A_f = -(1/\beta_f)$$

when,

$$A\beta_f \gg 1 \qquad (8\text{-}35)$$

Drill Problem 8.22: An amplifier has a gain without feedback of 1800. Calculate the gain with feedback when (a) $\beta_f = 0.1$; (b) $\beta_f = 0.9$.

Drill problem 8.23: Suppose the amplifier of the previous drill problem has a gain of 600 without feedback. Recalculate the gains with feedback and calculate the percentage change in the gain in each case.

A block diagram is often more useful than a detailed schematic diagram to analyze a feedback amplifier. The amplifier and the feedback network are shown as boxes and the input and output signal paths are shown by single

520   Operational Amplifiers and Feedback Amplifiers

lines as in Fig. 8-66. The voltages are assumed to be measured with reference to the ground even though the ground itself may not be shown explicitly. A small circle is used to indicate the addition (or subtraction) of two or more voltage signals.

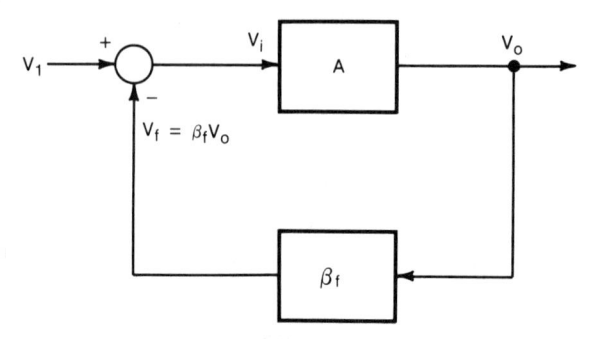

Fig. 8-66. Block diagram representation of a feedback amplifier.

In Fig. 8-66, $V_i = V_1 - V_f$ and $V_f = \beta_f V_o$. These two equations in conjunction with $V_o = AV_i$ lead to Eq. (8-34) for the gain with feedback.

Consider now the somewhat more elaborate arrangement of Fig. 8-67. V_1 is the signal to be amplified, and V_N is a noise entering the system due to some noise sources. The equations of the system are

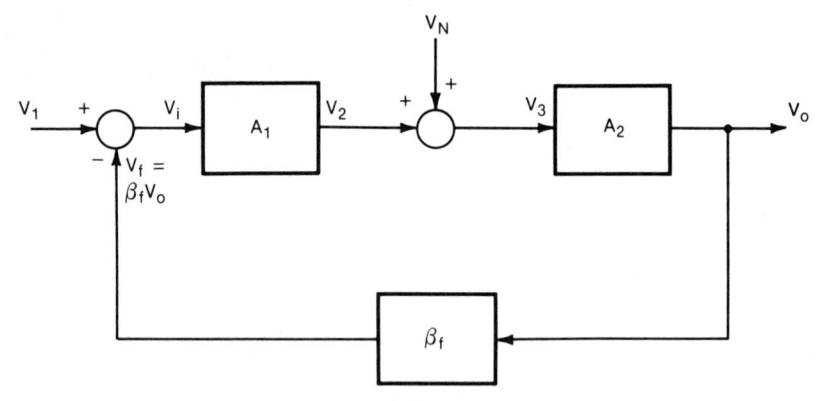

Fig. 8-67. Two-stage amplifier system with feedback.

$$V_o = A_1 A_2 V_i + A_2 V_N$$
$$V_i = V_1 - V_f$$
$$= V_1 - \beta_f V_o$$

Electronics: Circuits and Systems

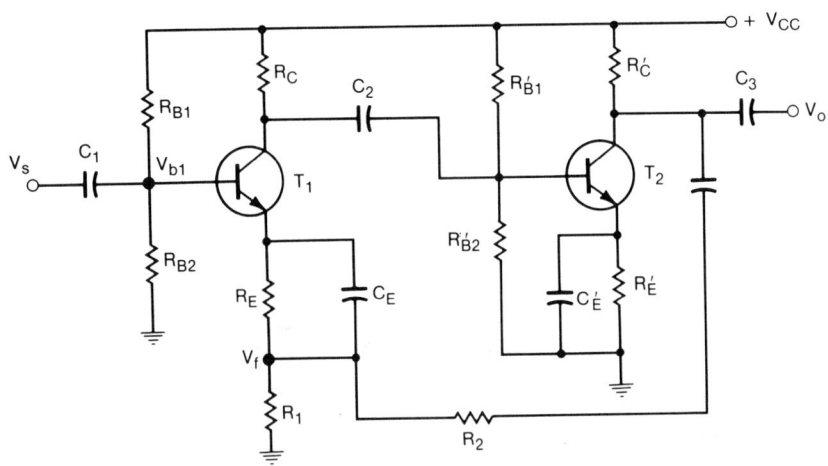

Fig. 8-68. Typical feedback amplifier circuit.

Combining the previous two equations (and after some algebraic manipulation), we get

$$V_o = \frac{A_1 A_2}{1 + A_1 A_2 \beta_f} V_i + \frac{A_2}{1 + A_1 A_2 \beta_f} V_N$$

The output has two components: one due to the input signal V_i and the other due to the noise input V_N. By making A_1 very large, it is possible to reduce the percentage of the noise component present in the output signal, which can be done by making the first stage amplifier (with gain A_1) as noise free as possible. The presence of negative feedback can, therefore, be used to reduce the distortion due to the noise sources in an amplifier system.

Some Typical Feedback Connections

In the circuit of Fig. 8-68, the output voltage of the second stage is fed back into the first stage through a voltage-divider network made up of R_1 and R_2. If we examine the circuit by replacing all capacitors by short circuits (for small-signal analysis), we see from Fig. 8-69A that the base-to-emitter voltage of T_1 is $V_i = (V_{b1} - V_f)$, which indicates the presence of a negative feedback. Consider the portion of the circuit shown in Fig. 8-69B. If the emitter current of T_1 is small compared with the current in R_2, then we have the two resistors R_1 and R_2 in series and

$$V_f = \frac{R_1}{R_1 + R_2} V_o$$

or

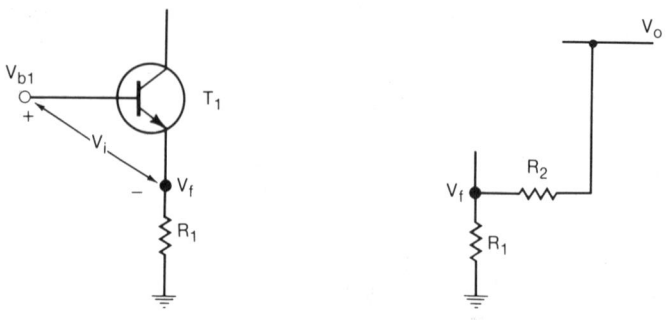

(A) V_i is the base-to-emitter voltage.

(B) R_1 and R_2 shown in series for small emitter current.

Fig. 8-69. Calculation of the feedback ratio in the circuit of Fig. 8-68.

$$\beta_f = [R_1/(R_1 + R_2)].$$

The circuit of Fig. 8-68 can be put in the block diagram form as shown in Fig. 8-70.

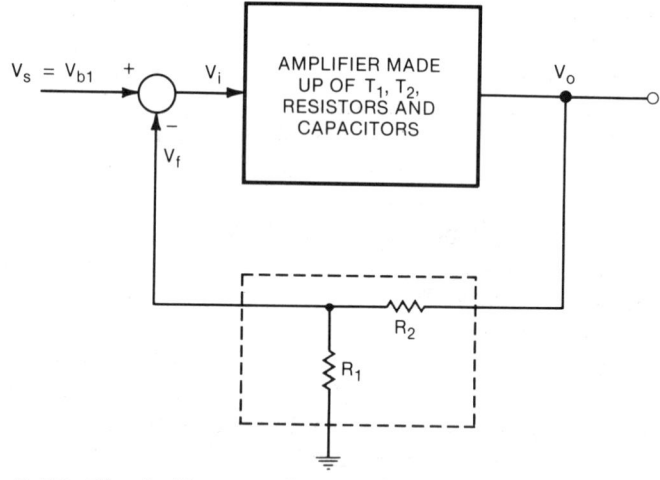

Fig. 8-70. Block diagram form of the circuit of Fig. 8-68.

Drill Problem 8.24: If $R_1 = 100\ \Omega$ and $R_2 = 4.7\ k\Omega$ in the circuit of Fig. 8-68, calculate the gain with feedback when the gain without feedback is 60.

Electronics: Circuits and Systems

The circuit of Fig. 8-68, in which the feedback was a voltage V_f at the input, is known as having a *voltage feedback*. A circuit in which there is a *current feedback* is shown in Fig. 8-71. It is seen that a portion I_2 of the output current I_{C2} is fed back to the input side. The input current to the transistor is $(I_s - I_2)$.

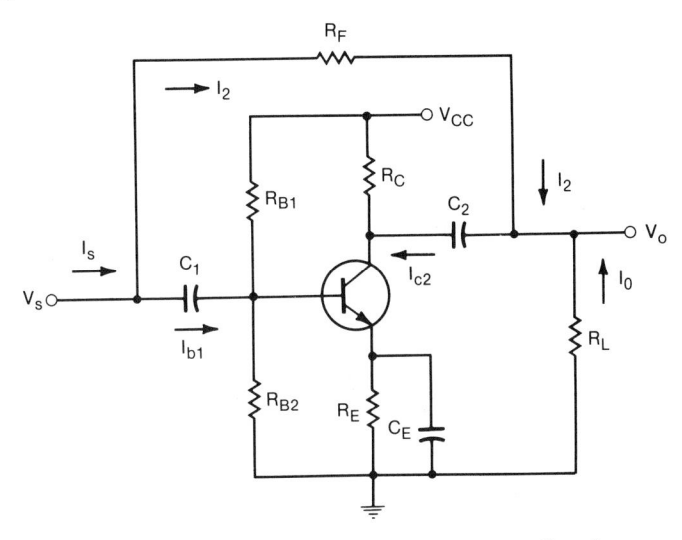

Fig. 8-71. Amplifier with current feedback.

Effects of Negative Feedback

It was seen previously that the gain of an amplifier is reduced by negative feedback. There is, however, a corresponding increase in the bandwidth of the amplifier. The lower cutoff frequency of the amplifier moves down by a factor $(1 + A\beta_f)$, and the upper cutoff frequency moves up by the same factor. The bandwidth increases by the factor $(1 + A\beta_f)$ also.

For example, if an amplifier without feedback has a gain = 10^3, lower cutoff frequency = 500 Hz, and upper cutoff frequency = 600 kHz, and a negative feedback is added with $\beta_f = 0.05$, then we get a gain with feedback = 20, the lower cutoff frequency moves down to 10 Hz, and the upper cutoff frequency moves up to 30 MHz.

Negative feedback affects the input and output resistances of an amplifier. But the actual effect depends upon whether voltage or current feedback is used and also upon whether such a feedback is in series or in shunt with the input quantities. As an example, in the case of the emitter follower (BJT) and source follower (FET) circuits, the input resistance increases and output resistance decreases due to the use of negative voltage-series feedback.

Positive Feedback and Oscillators

If in the expression for the gain with feedback, Eq. (8-34), repeated here,

$$A_f = \frac{A}{1 + A\beta_f}$$

we choose β_f such that $A\beta_f = -1$, then the gain with feedback becomes *infinite*. That is, the gain with feedback can be made infinitely large by choosing

$$\beta_f = -(1/A) \qquad (8\text{-}36)$$

This corresponds to positive feedback since $A\beta_f$ is negative.

An infinite gain for an amplifier means that the amplifier can function with no input signal, which is similar to the situation of an ideal op amp except that here we can actually make the gain infinite instead of simply approximating it. When the gain of an amplifier is made equal to infinity at a particular frequency (by making the feedback network sensitive to frequency), the amplifier becomes an *oscillator:* it produces an output that is a sinusoid of the single frequency.

To have oscillations, positive feedback is necessary, which means that the component of voltage (or current) fed back to the input side of the circuit must be *in phase* with the input signal to the first transistor. Therefore, some sort of a phase-shift network is required. Once the phase relationship is satisfied, then it is necessary to make $(A\beta_f) = -1$, which means that the *magnitude* of $(A\beta_f)$ must be 1. As a practical matter, the product $(A\beta_f)$ is chosen to have a magnitude of *greater* than 1 to start with, providing a larger positive feedback than needed for an oscillator. Therefore, the oscillations begin building up in amplitude but the *nonlinear* characteristic of the transistor makes the amplitude reach a constant value due to saturation. Therefore, the oscillations start with a small amplitude, increase at first and eventually stabilize at some particular value.

Oscillators can be built by using the resonance properties of LC circuits or by introducing phase shift through the use of RC circuits. The principle of a *phase-shift oscillator* is shown in block diagram form in Fig. 8-72 where each box marked 60° represents an RC network that produces a phase shift of 60°. The single-stage amplifier has a phase shift of 180° (recall the phase reversal of output with respect to the input in a common-emitter amplifier) and the additional 180° phase shift produced by the RC networks leads to a positive feedback.

The circuit of a phase-shift oscillator is shown in Fig. 8-73, where each R_1C_1 combination is designed to produce a 60° phase shift. The frequency of oscillations of the phase-shift oscillator can be shown to be

$$f = \frac{1}{\sqrt{6}\, R_1 C_1} \text{ Hz}$$

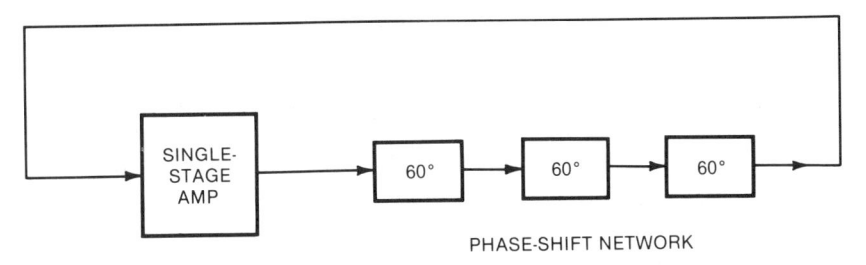

Fig. 8-72. Block diagram form of a phase-shift oscillator.

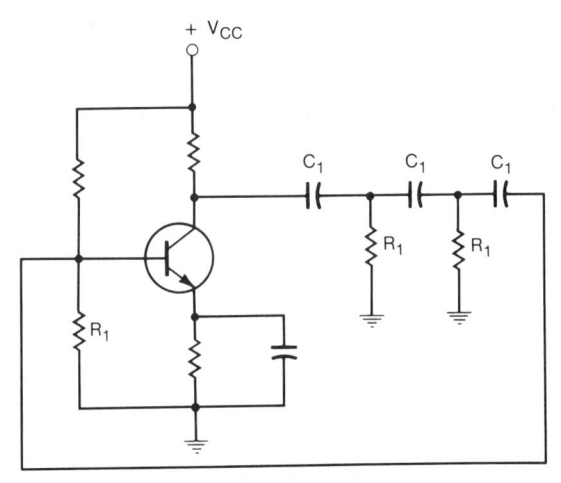

Fig. 8-73. Phase-shift oscillator circuit.

Stability of an Amplifier

An amplifier is said to be *stable* if it continues to operate as an amplifier at all frequencies instead of suddenly acting as an oscillator. In some amplifiers, positive feedback may inadvertently appear at some frequency, and the amplifier may behave like an oscillator. Such an amplifier is said to be unstable, and the spurious oscillations are called *ringing*.

There are two conditions for oscillations to occur as was discussed in the last section: the component (voltage or current) fed back from the output to the input must be *in phase* with the input signal of the first stage; the magnitude of $(A\beta_f)$ must be at least equal to one. In a single-stage amplifier the phase condition cannot be satisfied since the phase angle between the input and the output is 180° in the midband and becomes 270° at high frequencies.

Consider now a three-stage amplifier. In the midband, each stage produces a phase reversal so that effectively the output is 180° out of phase with the input in the midband. At high frequencies, each stage produces an additional phase shift of 90° so that the overall phase shift increases by 270° over the midband value. That is, the phase shift in a three-stage amplifier is 180° in the midband and increases to 450° in the high-frequency end. Therefore, it is possible to find a frequency at which the total phase shift is 360° and the output is in phase with the input. If feedback is provided at such a frequency, oscillations can occur if the magnitude of $(A\beta_f)$ at that frequency becomes greater than or equal to one. A high-gain, three-stage amplifier runs the risk of becoming unstable at some high frequency.

In order to prevent ringing in a three-stage amplifier, the gain of the amplifier is chosen so that the magnitude of $(A\beta_f)$ is significantly less than one at the frequency where each stage produces an additional shift of 60°.

Gain Margin and Phase Margin

It can be seen from the previous discussion that the trouble spot (where an amplifier may start ringing) is when the magnitude of $(A\beta_f)$ is one and its phase angle is 180°. When the magnitude of $(A\beta_f)$ is one, its phase angle must be made less than 180°, and when the phase angle of $(A\beta_f)$ is 180°, its magnitude must be kept less than one. The designer uses two safety margins to prevent oscillations occurring in an amplifier: the *gain margin* and the *phase margin*.

If $|A\beta_f|$ is expressed in decibels, then we want it to be below 0 dB at the frequency where its phase angle becomes 180°. The value by which $|A\beta_f|$ in decibels is below 0 dB at that frequency is called the *gain margin*, as indicated in Fig. 8-74A. The larger the gain margin, the more stable the amplifier. A value of 10 dB is usually used for the gain margin.

At the frequency where $A\beta_f$ is 0 dB, we want its phase angle to be less than 180°. The value by which the angle of $(A\beta_f)$ differs from 180° at that frequency is called the *phase margin*, as indicated in Fig. 8-74B. The larger the phase margin, the more stable the amplifier. A value of 30° to 60° is usually used for phase margin.

In practice, a variety of compensation networks are used to introduce the desired gain and phase margins in an amplifier to ensure its stable (nonoscillatory) operation.

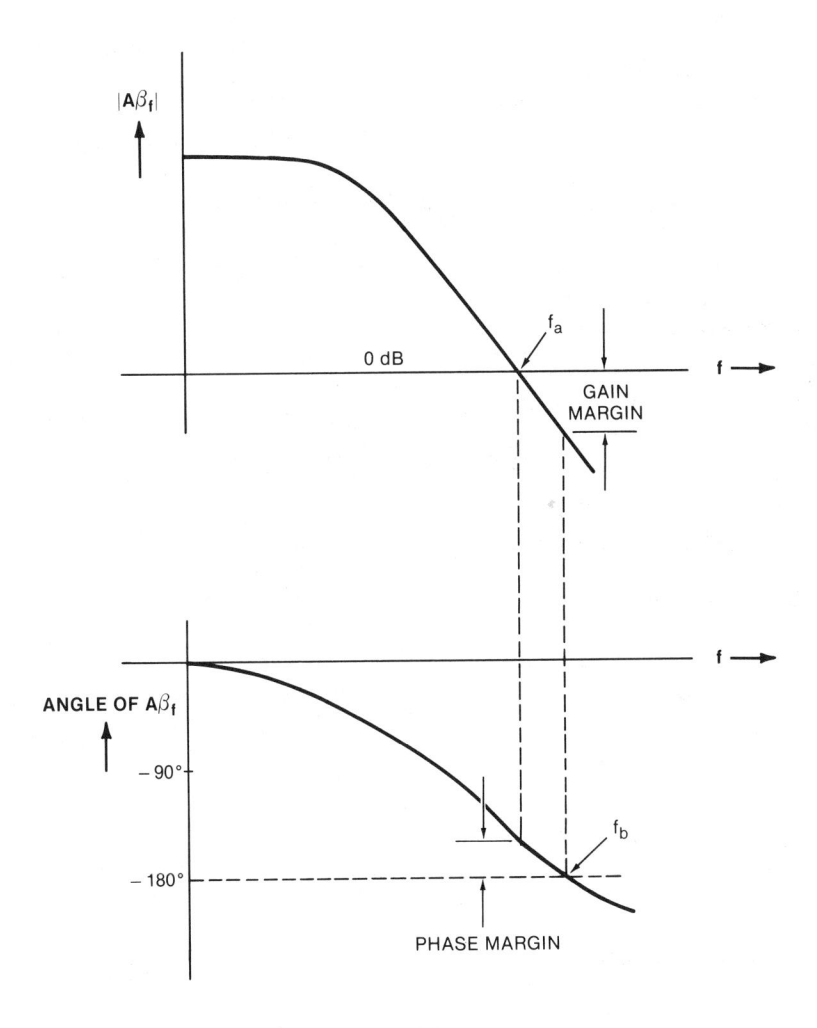

Fig. 8-74. Gain and phase margins.

SUMMARY SHEETS

OP AMPS

Symbol of an Op Amp

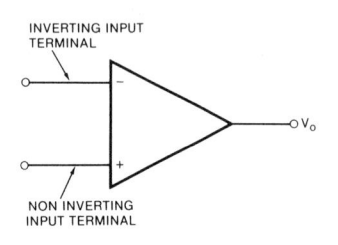

Properties of an Ideal Op Amp
1. $R_i = \infty$: No input current to the op amp at either input terminal.
2. $R_o = 0$: No variation of the output voltage V_o due to loading.
3. $A = \infty$: Voltage between the inverting and noninverting input terminals is zero.

OP AMP CIRCUITS

Inverting Op Amp Circuit

$(V_o/V_i) = -(R_F/R_1)$

Noninverting Op Amp Circuit

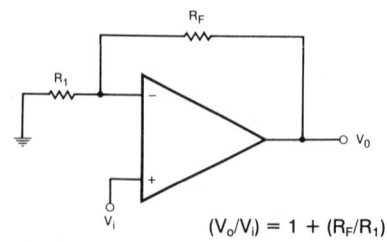

$(V_o/V_i) = 1 + (R_F/R_1)$

Gain is always greater than one.

Summing Op Amp

$$V_o = -\left(\frac{R_F}{R_1}V_1 + \frac{R_F}{R_2}V_2 + \frac{R_F}{R_3}V_3\right)$$

If $R_F = R_1 = R_2 = R_3$, then $V_o = -(V_1 + V_2 + V_3)$

Voltage Follower

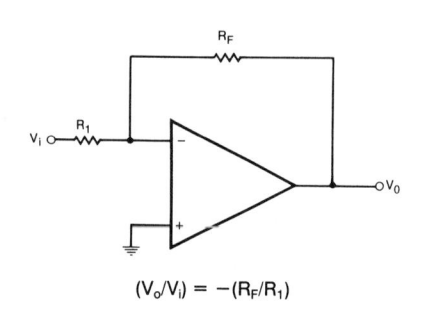

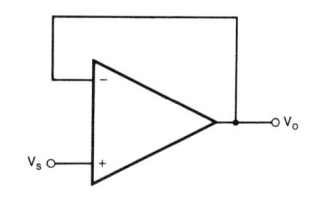

$V_o = V_s$

This is used for isolating two stages.

Voltage-to-Current Converter

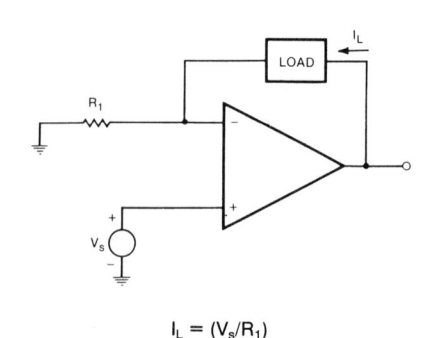

$I_L = (V_s/R_1)$

Electronics: Circuits and Systems

Sample and Hold Circuit

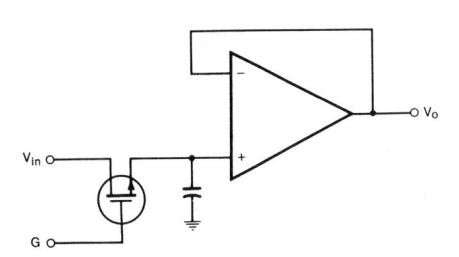

Whenever a negative pulse is applied to the gate of the MOSFET, it conducts and the capacitor charges to the input voltage. The capacitor voltage remains constant between input pulses.

Logarithmic Amplifier

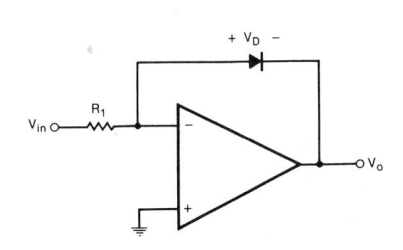

The change in the output voltage is proportional to the logarithm of the change in the input voltage.

Constant-Current Source

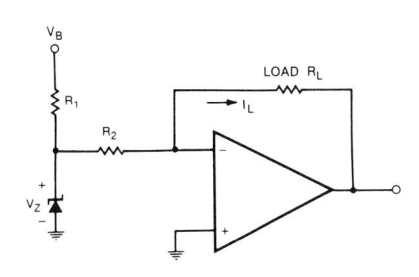

$I_L = (V_Z/R_2)$, a constant since V_Z is a constant.

Integrator

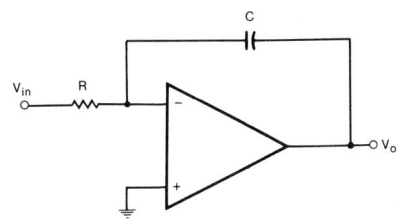

The output v_o is proportional to the (negative of the) time integral of the input voltage.

DIFFERENTIAL AMPLIFIER (Diff Amp)

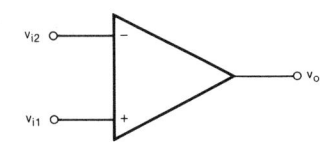

In an ideal diff amp, the output voltage is proportional to the difference between the two inputs.

$$v_o = A_d = (v_{i1} - v_{i2})$$

where,

A_d is the diff-mode gain.

In a nonideal diff amp, the output has two components: one due to the differential-mode input and the other due to the common-mode input.

$$v_o = A_d(v_{i1} - v_{i2}) + A_c[\frac{1}{2}(v_{i1} + v_{i2})]$$

Common-Mode Rejection Ratio (CMRR)

$$cmrr = (A_d/A_c)$$

$$cmrr = 20 \log (A_d/A_c) \text{ dB}$$

Cmrr is typically 60 to 120 dB.

Emitter-Coupled Diff Amp

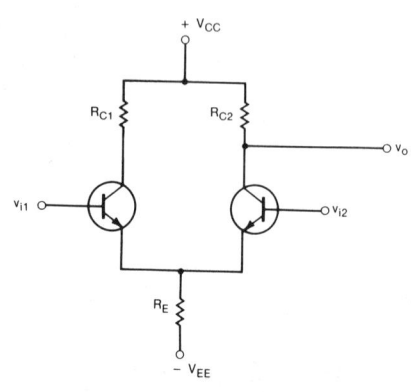

A constant current source used in place of R_E makes the common-mode gain zero and cmrr infinite.

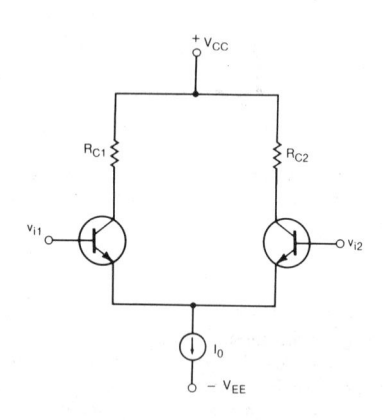

Realization of the Constant Current Source

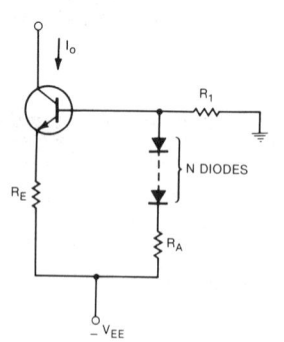

The number of diodes can be chosen to make I_0 a constant:

If

$$N = 1 + (R_A/R_1)$$

$$I_0 = \frac{R_A}{(R_1 + R_A)R_E} V_{EE}$$

COMPARATORS

When $V_i^- > V_i^+$,

V_0 is low $(= -V_{cc})$

When $V_i^+ < V_i^-$,

V_0 is high $(= +V_{cc})$

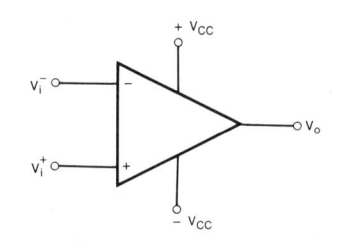

SCHMITT TRIGGER

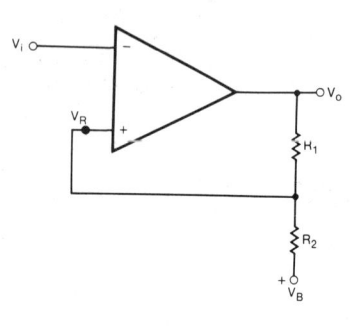

Upper Threshold Level

$$V_{R(hi)} = \frac{R_1}{R_1 + R_2} V_B + \frac{R_2}{R_1 + R_1} V_{o(hi)}$$

Lower Threshold Level

$$V_{R(low)} = \frac{R_1}{R_1 + R_2} V_B + \frac{R_2}{R_1 + R_2} V_{o(low)}$$

Electronics: Circuits and Systems

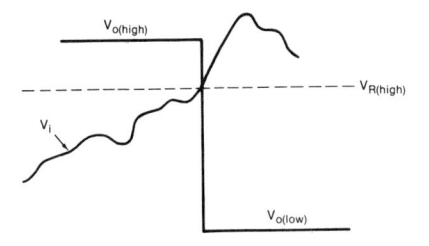

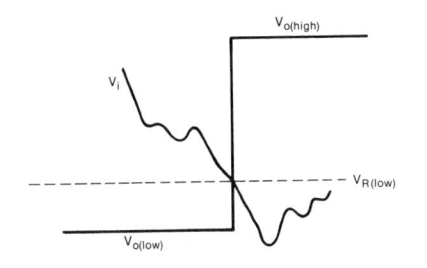

When the input signal crosses the *upper threshold level* with a *positive slope*, the output switches from *high* to *low*.

When the input signal crosses the *lower threshold level* with a *negative slope*, the output switches from *low* to *high*.

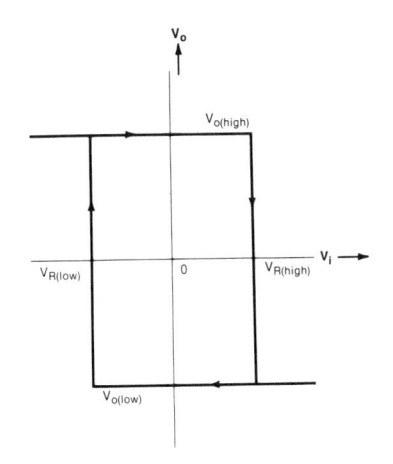

Hysteretic Behavior of the Schmitt Trigger

FEEDBACK AMPLIFIERS

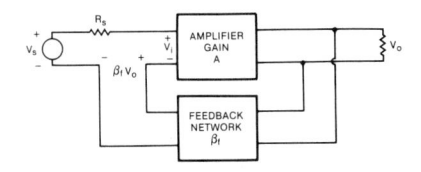

A = gain without feedback

A_f = gain with feedback

$$A_f = \frac{A}{1 + A\beta_f}$$

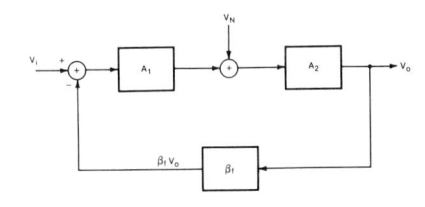

$$V_o = \frac{A_1 A_2}{1 + A_1 A_2 \beta_f} V_i + \frac{A_2}{1 + A_1 A_2 \beta_f} V_N$$

The effect of the noise can be reduced by using negative feedback, and a stage with a very high gain A_1 to precede the noise source.

Effects of Negative Feedback

(1) Reduction of the gain. (2) Increase of the bandwidth. (3) Changes in the input and output resistances. The actual nature of the change depends upon the type of feedback connection used.

Positive Feedback and Oscillations

When $A\beta_f = -1$, the gain of the amplifier with feedback becomes infinite and it can be made to function as an oscillator. For oscillations, the component of the voltage or current fed back to the input side must be *in phase* with the input signal to the first stage and the magnitude of $(A\beta_f)$ must be one. In actual oscillators, the initial value of $(A\beta_f)$ is made to be greater than one so that the amplitude of the oscillations increases at first. But the nonlinearity of the transistor causes the oscillations to stabilize at a constant amplitude.

Phase-Shift Oscillator

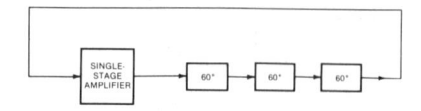

The three-stage phase-shift network produces a 180° phase shift so that the component fed back to the input is in phase with the input. The frequency of oscillations is

$$f = 1/(\sqrt{6}R_1C_1)$$

Stability of Amplifiers

An amplifier that tends to oscillate at some frequency is said to be *unstable*. Such a behavior is called *ringing*. Ringing is probable in a three-stage amplifier since the gain and phase shift could assume values that lead to the positive feedback condition.

Gain Margin and Phase Margin

These are safety margins used to ensure that ringing does not occur.

The value by which $|A\beta_f|$ in decibels is below 0 at the frequency where the phase shift of the amplifier becomes 180°.

The value by which $|A\beta_f|$ in decibels is below 0 dB at the frequency where the phase angle of $(A\beta_f)$ is 180° is known as the *gain margin*.

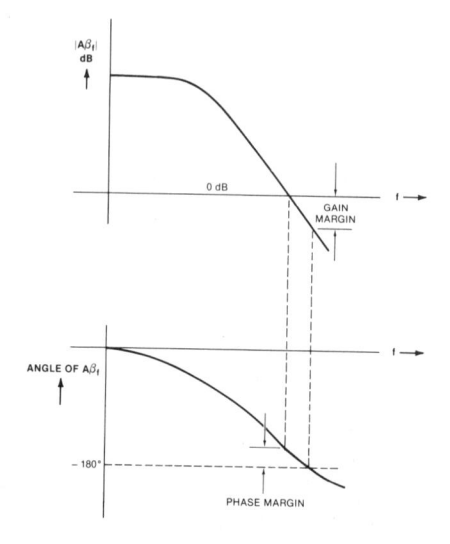

The value by which the angle of $(A\beta_f)$ is less than 180° at the frequency where $|A\beta_f|$ is 0 dB is called the *phase margin*.

A gain margin of 10 dB and a phase margin of 30° to 60° are usually used to ensure stability in an amplifier.

ANSWERS TO DRILL PROBLEMS

8.1 (a) −20; (b) −0.2; (c) −1.
8.2 Gain = −4. See Fig. 8-DP2.

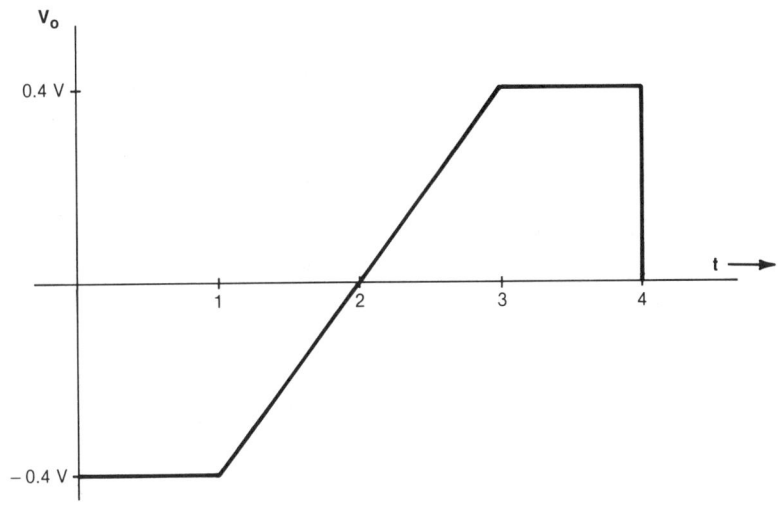

Fig. 8-DP2.

8.3 $R_F = (300R_1) = 3$ MΩ.
8.4 (a) 1.33; (b) 4.
8.5 (a) −0.33; (b) −3.
8.6 (a) −750 mV; (b) 300 mV.
8.7 $(-3V_a + 400) = 0$; $V_a = 133$ mV.
8.8 $V_0 = -0.5V_1 - 0.25V_2 - 0.5V_3 = -0.8125$ V.
8.9 $R_1 = 50$ kΩ; $R_2 = 100$ kΩ; $R_3 = 33.3$ kΩ.
8.10 All are −40 V.
8.11 See Fig. 8-DP11.
8.12 (a) 50.45 V (Ideal would be 50 V); (b) 49.55 V (Ideal would be 50 V).
8.13 $A_c = 333$.
8.14 94 dB.
8.15 $A_c = 80$.
8.16 $A_c = 3000$; $V_o = 15.3$ V.
8.17 $V_A < 8$ V: V_{o1} positive and V_{o2} negative; $8 < V_A < 16$: V_{o1} and V_{o2} are both negative; $V_A > 16$: V_{o1} negative and V_{o2} positive. The AND gate is useless in this case since its output will always be low. You need a NAND gate.
8.18 See Fig. 8-DP18. In this circuit, $V'_o = -(R_p/10)$ V and when $R_p > 50$ k, $V'_o < -5$ V and V_o becomes positive.
8.19 $V_B = 3.125$ V; $V_{R(low)} = 0$ and $V_{R\,(hi)} = 5$ V. The output for the given

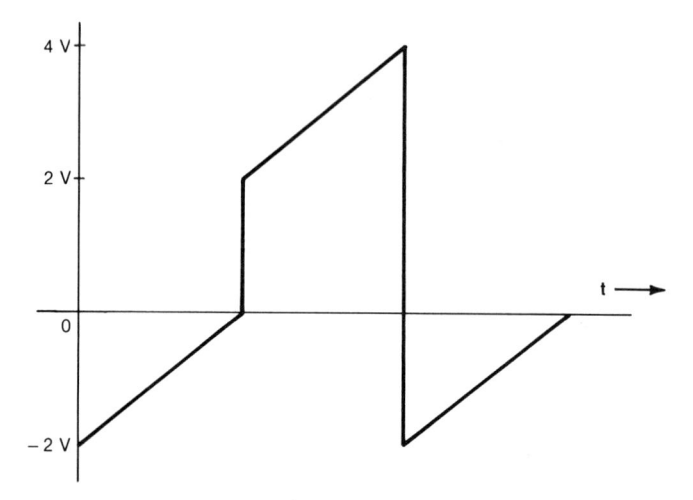

Fig. 8-DP11.

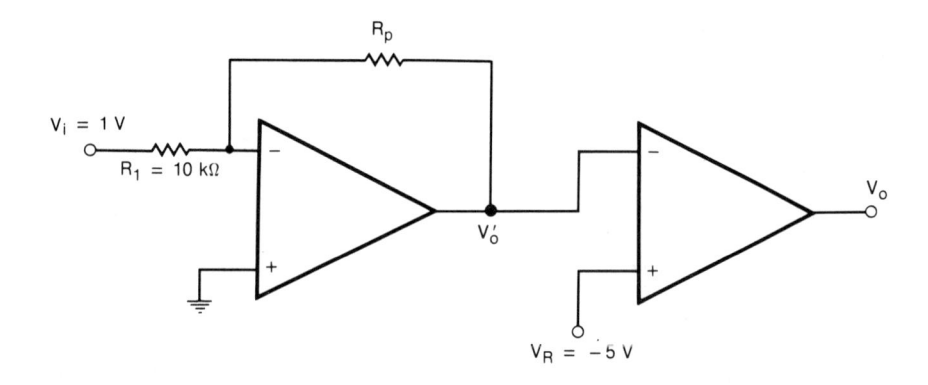

Fig. 8-DP18.

signal will always stay at high and never switch to low since the signal never crosses $V_{R(hi)}$ with a positive slope.

8.20 $V_B = -3.125$ V: $V_{R(low)} = -5$ V, $V_{R(hi)} = 0$. In this case, the output stays at low and never switches to high.

8.21 See Fig. 8-DP21. The trigger starts with a low-output state before $t = 0$.

8.22 (a) 9.94; (b) 1.110.

8.23 (a) 9.84; (change is 1% of previous value); (b) 1.109 (change is less than 0.1% from the previous value).

8.24 Gain with feedback = 26.7.

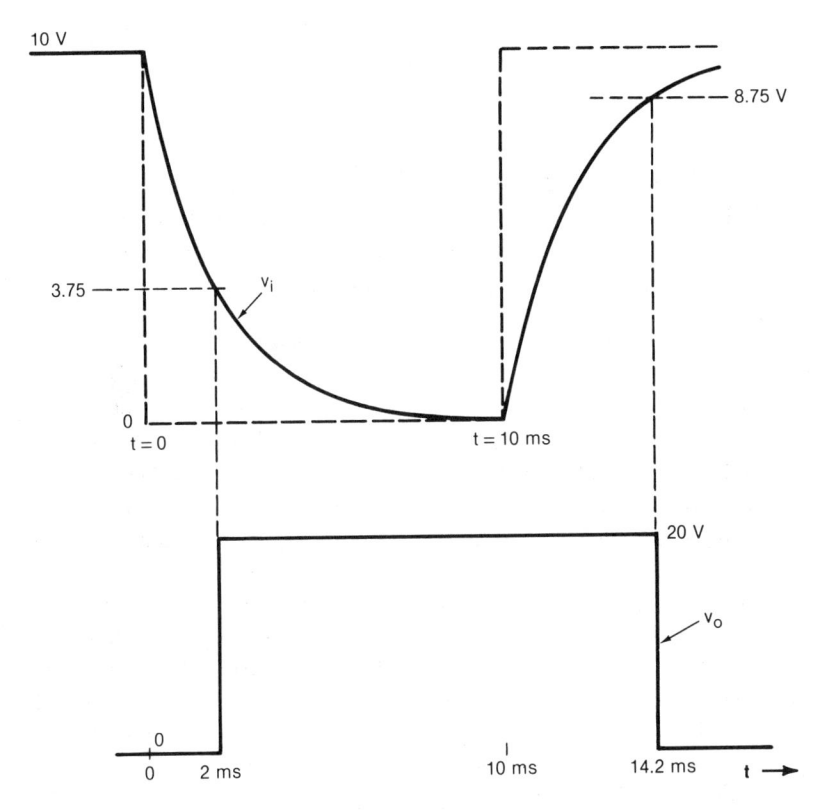

Fig. 8-DP21.

PROBLEMS

Ideal Op Amp and its Applications

8.1 Calculate the gains of the inverting op amp circuits with values of R_1 and R_F given by (a) $R_1 = 100\ \Omega$, $R_F = 5\ k\Omega$; (b) $R_1 = 5\ k\Omega$, $R_F = 100\ \Omega$; (c) $R_1 = 2\ M\Omega$, $R_F = 0.5\ M\Omega$; (d) $R_1 = 0.5\ M\Omega$, $R_F = 2\ M\Omega$.

8.2 Suppose the values of R_1 and R_F of the previous problem are used in a noninverting op amp circuit, calculate the gains.

8.3 A signal source that generates 20 mV and has an internal resistance of 500 kΩ is connected to the inverting input of an op amp. A feedback resistor R_F is connected between it and the output terminal of the op amp. If the desired output voltage is 5 V, calculate the value of R_F needed.

8.4 If the signal source of the previous problem is now moved to the noninverting input terminal of the op amp, and the same output voltage is desired, redesign the circuit to meet this new situation.

8.5 Two signal sources v_a and v_b are connected, respectively, to the inverting and noninverting inputs of an op amp, respectively. Each has an internal resistance of 100 kΩ. Determine the value of the feedback resistance R_F to make the output voltage zero when $v_a = 0.5$ V and $v_b = 0.25$ V. Suppose now the two voltage sources are interchanged. What will be the output voltage? With the sources in the new positions, can you find a value of R_F so that the output is again zero?

8.6 The arrangement of Fig. 8-P6 shows an adder-subtractor using an op amp. Obtain an expression for the output voltage.

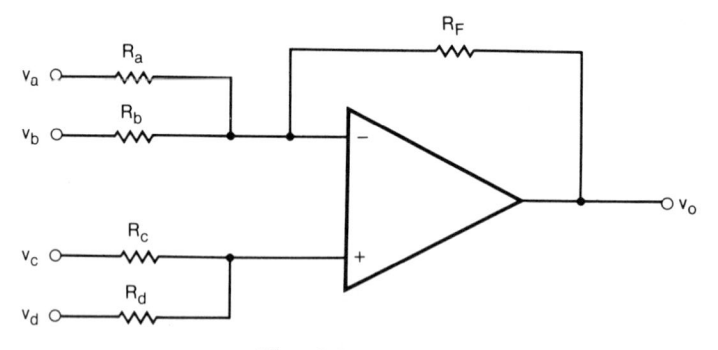

Fig. 8-P6.

8.7 The circuit of Fig. 8-P7 is known as a *Howland current source*. Calculate the value of the current I_L in the load for the following values of the resistors: $R_1 = 10\ k\Omega$, $R_2 = 20\ k\Omega$, $R_3 = 40\ k\Omega$, and $R_4 = 20\ k\Omega$.

Electronics: Circuits and Systems

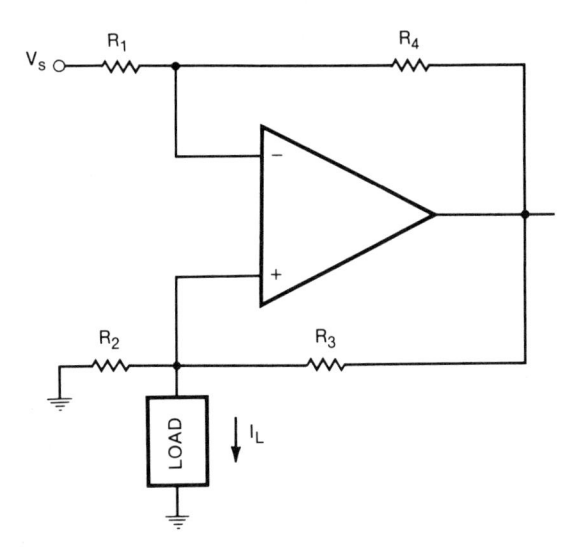

Fig. 8-P7.

8.8 Consider the Howland current source of the previous problem, but without any values of resistors specified. Obtain the relationship between the resistance that must be satisfied in order to make the load current $I_L = -(V_s/R_2)$.

8.9 Consider the circuit shown in Fig. 8-P9, which can be used as a constant current source. Analyze it and obtain the relationship between the current I_0 and the voltage V_1. Assume that the emitter and collector currents in the transistor are approximately equal.

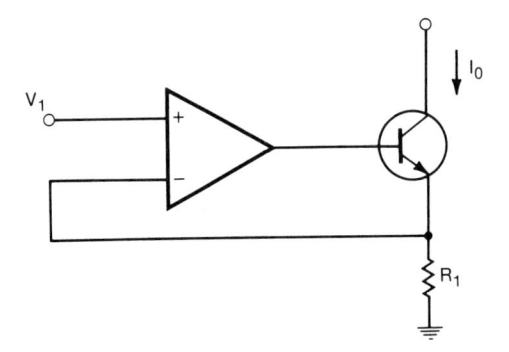

Fig. 8-P9.

8.10 An alternative constant current source arrangement is shown in Fig. 8-P10. Again obtain the relationship between the current I_0 and the voltage V_1.

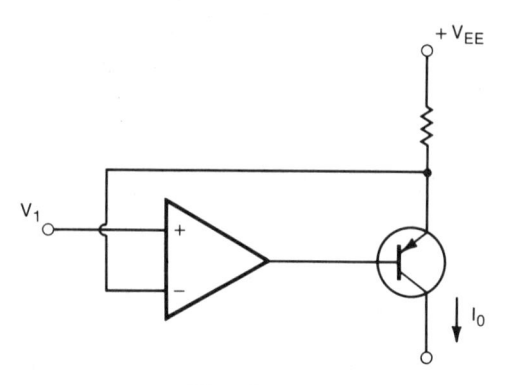

538   Operational Amplifiers and Feedback Amplifiers

Fig. 8-P10.

8.11 Discuss the operation of the circuit of Fig. 8-P11. Determine and sketch its output waveform if the input is a sinusoidal signal. Assume the diode to be ideal.

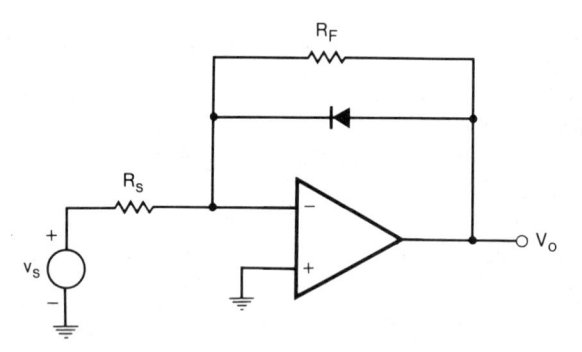

Fig. 8-P11.

8.12 Discuss the operation of the circuit shown in Fig. 8-P12. For what range of values of the input voltage will the circuit act as a constant current source?

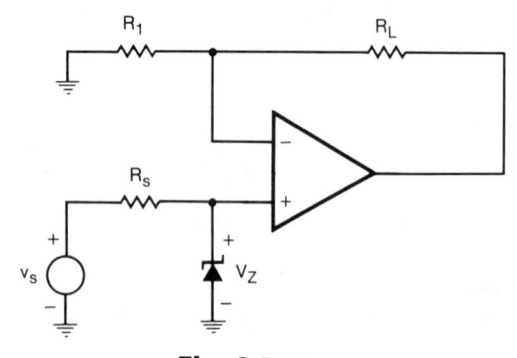

Fig. 8-P12.

Diff Amp

8.13 An *ideal* diff amp has a gain of $A_d = 500$. Calculate its output voltage in each of the following cases: (a) $v_{i1} = -10$ mV, $v_{i2} = 10$ mV; (b) $v_{i1} = 10$ mV, $v_{i2} = -10$ mV; (c) $v_{i1} = 10$ mV $= v_{i2}$; (d) $v_{i1} = -10$ mV $= v_{i2}$.

8.14 An ideal diff amp has a gain of $A_d = 100$. For the input signals shown in Fig. 8-P14, determine and sketch the output voltage waveform.

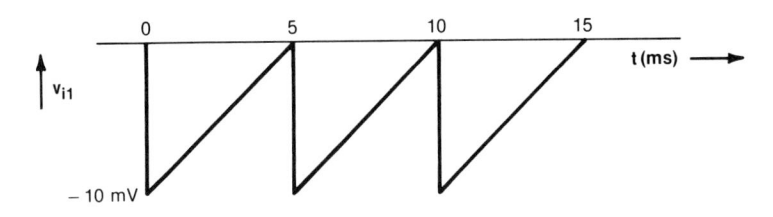

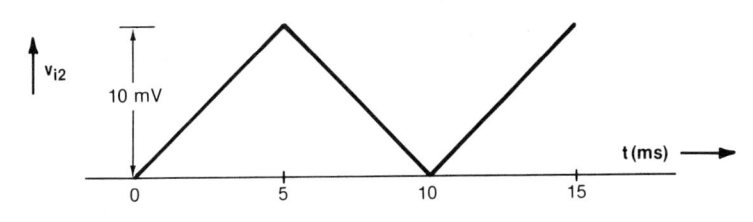

Fig. 8-P14.

8.15 For the sets of input voltages given in Problem 8.13, recalculate the output voltages if the diff amp is nonideal and has a common-mode gain of 50 and a differential-mode gain of 500.

8.16 Suppose the amplifier in Problem 8.14 had a common-mode gain of 50 and a differential-mode gain of 100. Sketch the new output voltage waveforms for the same set of input signals.

8.17 Repeat the previous problem if the two sources are interchanged.

8.18 The differential-mode gain of an amplifier is 10^4. For each of the cmrr values given, calculate its common-mode gain. (a) 120 dB; (b) 5×10^4.

8.19 The common-mode gain of an amplifier is 100. For each of the following cmrr values, calculate its differential-mode gain: (a) 110 dB; (b) 6×10^3.

Comparators

8.20 Discuss the operation of the comparator circuit shown in Fig. 8-P20. Consider the two cases V_i greater than V_R and less than V_R and compute the output voltage. For the particular input waveform given, sketch the output.

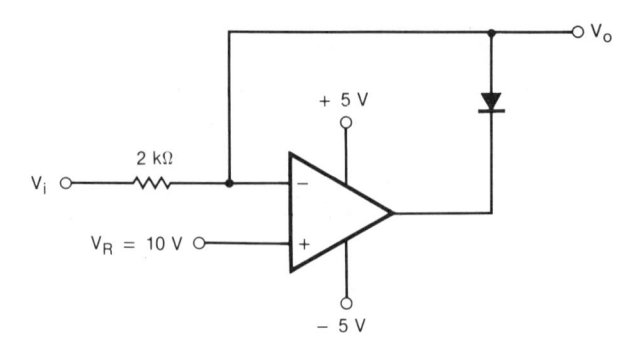

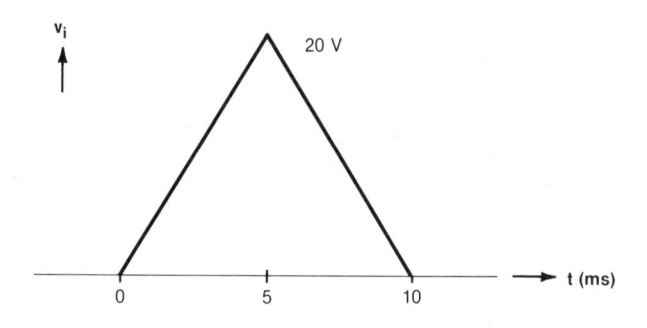

540   **Operational Amplifiers and Feedback Amplifiers**

Fig. 8-P20.

8.21 Design a window detector that will give a high output when the input signal is below 6 V or above 15 V. Assume a battery of 24 V is available. Also assume that only an AND gate can be used (as was done in Fig. 8-51). The smallest resistor to be used should be 1 kΩ.

8.22 Discuss the operation of the low light-level detector of Fig. 8-52 in the text if the resistors R_1 and R_p are interchanged.

8.23 Design a system with four comparators to meet the following specifications. The reference voltages of the comparators are derived through a voltage-divider network fed by a 10-V battery and their values are to be 2 V, 4 V, 6 V, and 8 V. A signal is fed to all the comparators. When the signal is below 2 V, all the comparators are to have a low output. When it is between 2 V and 4 V, one comparator is to have a high output, when it is between 4 V and 6 V, two comparators are to have high outputs, when it is between 6 V and 8 V, three comparators are to have high outputs, when it is above 8 V, all four comparators are to have high outputs. (This is the basis of bar-graph displays that will be discussed in the next chapter.)

Schmitt Trigger

8.24 Calculate the upper and lower threshold levels of a Schmitt trigger circuit with $R_1 = 5$ kΩ, $R_2 = 15$ kΩ, and $V_B = 20$ V. Suppose the input to the circuit is a sinusoidal voltage of peak value 20 V. Determine and sketch output waveform if $V_{hi} = 5$ V and $V_{low} = -5$ V.

8.25 Repeat the previous problem if the values of R_1 and R_2 are interchanged.

8.26 Design a Schmitt trigger circuit with its upper threshold voltage equal to that of a 5-V battery provided and its lower threshold voltage to be 2 V below the upper threshold. A symmetric triangular waveform of period 1 ms and maximum and minimum values of, respectively, $+6$ and -6 V is applied as input to the circuit. Sketch the output voltage.

8.27 In the Schmitt trigger circuit with the RC circuit of Fig. 8-61 in the text, assume that $R = 100$ kΩ, $C = 0.1$ μF. The source V_s is replaced by the arrangement indicated in Fig. 8-P27. Determine and sketch the output waveform if the switch is closed at $t = 0$ and released after 50 ms. (Assume that the switch was open for a long time before $t = 0$.)

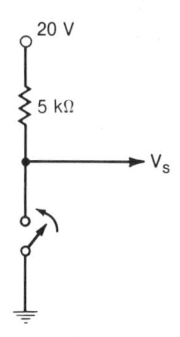

Fig. 8-P27.

Feedback

8.28 An amplifier has a gain of 3×10^3 without feedback. Calculate its gain with feedback if β_f is made equal to (a) 0.01; (b) 0.1; (c) 0.9.

8.29 An amplifier has a nominal gain without feedback of 400. When one of the devices is replaced, the gain without feedback could change by as much as 50%. Use of the necessary amount of feedback so that the change in gain is only 1% of the nominal value. Calculate the value of β_f.

8.30 The term "number of decibels of feedback" is used to refer to the ratio (gain without feedback/gain with feedback) expressed in decibels. Calculate this value in the three cases of Problem 8.28.

8.31 Consider the block diagram shown in Fig. 8-P31 which is a three-stage amplifier with feedback. The sources marked V_{N1} and V_{N2} may be considered as noise sources. (a) Obtain an expression for the output voltage

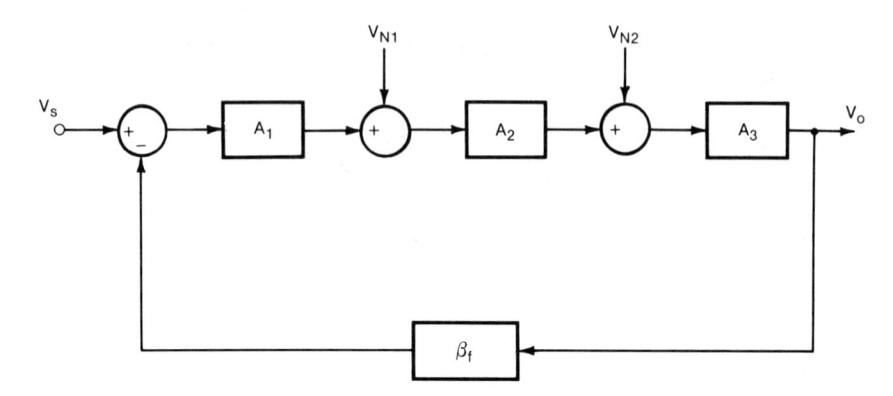

Fig. 8-P31.

V_o. (b) Calculate the ratio of the component in the output voltage due to the signal V_s to the component due to the noise source V_{N1}. (b) Calculate the ratio of the component in the output voltage due to the signal V_s to the component due to the noise source V_{N2}. (c) Suppose the feedback link was cut (by a pair of pliers). Recalculate the ratios in (b) and (c) and compare. How does feedback affect them?

CHAPTER 9

OPTOELECTRONIC DEVICES

There are some electronic devices whose response is governed by the incident light energy; that is, they are *photosensitive*. There are other devices that respond to an electrical signal by emitting light; that is, they are *photoemissive*. Photosensitive and photoemissive devices are called *optoelectronic* or *electro-optic* devices and used in a wide variety of applications, such as automatic exposure control circuits in cameras, industrial photoelectric controls, character recognition systems, and digital displays. Some optoelectronic devices and their applications will be discussed in this chapter.

BASIC PRINCIPLES AND RELATIONSHIPS OF LIGHT

Since light is an important ingredient in optoelectronic circuits, we will briefly consider the properties of light and some important basic relationships in the theory of light.

Optical Radiation and Photons

Optical radiation takes place in the form of an *electromagnetic wave* similar to radio waves. If f denotes the frequency of the electromagnetic radiation, then the wavelength λ is given by

$$f = (c/\lambda) \tag{9-1}$$

where,
c is the velocity of light and equals 2.998×10^8 meters per second.
The unit of wavelength in optical engineering is the *nanometer*, abbreviated *nm*, which equals 10^{-9} m. Another unit for wavelength is the *angstrom*,

abbreviated Å, which equals 10^{-10} m. That is, 10 angstroms make up 1 nm. The range of wavelengths of electromagnetic radiation that is considered as *optical* is from 10 nm to 10^5 nm. The range of wavelengths considered as *light* is the portion that is visible to the eye, which is from 380 nm to 750 nm. The wavelengths below 380 nm are in the untraviolet region, and those above 750 nm are in the infrared region.

From the viewpoint of quantum theory, light should be considered as comprising packets of radiant energy, or *photons*. Light has a dual nature: on the one hand it acts like a wave with a certain frequency and wavelength, and on the other it behaves like discrete packets each of which has a certain amount of energy. The energy associated with a photon is given by

$$W = hf \text{ J} \tag{9-2}$$

where,

h is called *Planck's constant* and has a value of 6.63×10^{-34} J/s or W/s². As an example, light of wavelength 600 nm has a frequency

$$f = (c/\lambda)$$
$$= 5 \times 10^{14} \text{ Hz}$$

and the energy association with photons of that frequency is

$$W = hf = 6.63 \times 10^{-34} \times 5 \times 10^{14}$$
$$= 3.32 \times 10^{-19} \text{ J}$$

Even though energy is usually expressed in joules (as in the previous example), it is often more convenient to use the unit of *electronvolt*, abbreviated *eV*. The electronvolt is defined as the *energy that would be acquired by an electron when accelerated through 1 volt*. Since the magnitude of the charge on an electron is 1.6×10^{-19} coulomb, the relationship between an electronvolt and joule is given by

$$1 \text{ eV} = 1.6 \times 10^{-19} \text{ J} \tag{9-3}$$

The light of wavelength 600 nm (considered earlier) has an energy of 3.32×10^{-19} J, which equals $(3.32 \times 10^{-19}/1.6 \times 10^{-19}) = 2.07$ eV.

Eqs. (9-1) through (9-3) can be combined into a single equation. The energy associated with an optical radiation of wavelength λ is given by

$$W = \frac{hc}{\lambda} \tag{9-4}$$

which becomes (after substitution of numerical values of c and h):

$$W = \frac{1243}{\lambda \text{ in nanometers}} \text{ eV} \tag{9-5}$$

Drill Problem 9.1: Calculate the energy contained in an optical radiation of wavelength: (a) 15 nm; (b) 225 nm; (c) 4×10^4 nm

Drill Problem 9.2: Calculate the wavelength of the optical radiation of energy: (a) 0.7 eV; (b) 125 eV; (c) 1.9×10^{-21} J.

Flux and Illumination

The flow of optical energy (that is, the energy per second) is called the *flux* and is measured in *watts*. In the *visible* region, however, the commonly used unit of flux is the *lumen*. (One lumen equals 1.48×10^{-3} W for a light wavelength of 555 nm.) The *flux per unit area on a receiving surface* is called *illumination* or *irradiance*. The unit of illumination is W/m², but the commonly used unit for visible light is *lumens*/m². The unit of *footcandle*, which is *lumens*/ft², is also used in practice. As an example, the level of illumination that is due to the sun on the surface of the earth in the summer months is 1.3×10^5 lm/m².

PHOTODETECTORS

A *photodetector* (also called a *photoconductor* or a *photoresistor*) has the property that its *resistivity changes as a function of the incident illumination*. It is used as part of a *circuit activated by an external voltage source and serves to control the total resistance in the circuit*. A pure (intrinsic or undoped) semiconductor in which electrons and holes can be generated by the incident illumination serves as a photoconductor. It was mentioned in Chapter 3 that the valence electrons in a semiconductor crystal form covalent bonds between neighboring atoms. The valence electrons can be released from their bonds by imparting to them a sufficiently high energy to overcome the covalent bond. That incident optical radiation should have enough energy to release a large number of electrons from the covalent bonds (with a corresponding generation of holes since the release of an electron from a covalent bond is equivalent to the generation of a hole) in order to make the material act as a photoconductor. The *minimum energy* that the incident radiation should have is called the *energy gap* of the material and denoted by E_g (usually expressed in electronvolts).

Since the energy of optical radiation is *inversely* proportional to wavelength [Refer to Eq. (9-5)], it follows that the wavelength of the incident radiation must be *less than* that corresponding to E_g. That is, the wavelength obtained by using the energy gap value E_g in Eq. (9-5) represents the *maximum wavelength* λ_{max} that will activate the photoconductor. From Eq. (9-6), we get

$$E_g \text{ in eV} = \frac{1243}{\lambda_{max} \text{ in nm}}$$

or

- λ_{max} in nm = $(1243/E_g \text{ in eV})$ (9-6)

As an example, consider two of the most commonly used photodetector materials: cadmium sulfide (CdS) and cadmium selenide (CdSe). The energy gaps of the two materials are E_g = 2.42 eV (CdS) and E_g = 1.74 eV (CdSe). Eq. (9-6) then gives:

CdS: λ_{max} = (1243/2.42) = 513 nm (green region)

CdSe: λ_{max} = (1243/1.74) = 714 nm (red region)

Theoretically, a photodetector should be equally sensitive to optical radiation of all wavelengths below λ_{max} for that material. For example, CdS may be expected to be equally sensitive to all wavelengths below 513 nm. It is found, however, that, in practice, a photodetector material is sensitive only to a narrow range of wavelengths and does not respond to any wavelength outside that range. Within the range of wavelengths over which the material acts effectively as a photodetector, it is found to have a peak response at a particular wavelength. The peak spectral response of CdS is at 550 nm (green) and CdSe at 690 nm (red). It is necessary to match the wavelength of the incident radiation to the peak spectral response of a given photodetector in practical applications.

Drill Problem 9.3: Lead sulfide (PbS) has an energy gap of 0.37 eV, and lead selenide (PbSe) has an energy gap of 0.26 eV. They can be used as photodetectors. Calculate the maximum wavelength of the optical radiation for the two materials.

Variation of Resistance With Illumination

Consider a sample of a photodetector material illuminated by an optical radiation matching its peak spectral response. The resistance of the sample, when *no illumination* is incident on it, is called its *dark resistance*. The dark resistance of a sample of photodetector material is very high, with values typically in the megohm range. When the sample is illuminated, its resistance decreases: the brighter the incident illumination, the lower the resistance. If the resistance of the sample is plotted as a function of the incident illumination, using a *logarithmic* scale for *both* variables, the resulting graph is found to be close to a straight line, as indicated in Fig. 9-1. This implies that the *ratio* of the *values of the resistance* at two different illuminations is governed

Electronics: Circuits and Systems

by the *ratio of the two illumination levels*. More specifically, if R_A is the resistance of the sample at an illumination level L_A and R_B the resistance at an illumination level L_B, then we have

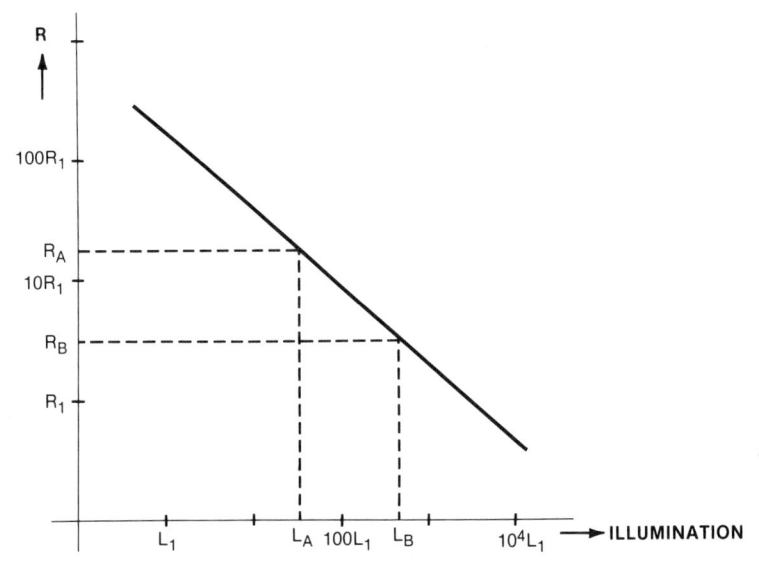

Fig. 9-1. Variation of the resistance of a photodetector with illumination. Note that a *logarithmic* scale is used on *both* axes.

$$\frac{R_A}{R_B} = \left(\frac{L_B}{L_A} \right)^m$$

where,
m is the slope of the straight line in Fig. 9-1.
For CdS, the value of m is quite close to 1 so that we can write

$$\frac{R_A}{R_B} = \frac{L_B}{L_A} \quad \text{(CdS)} \tag{9-7}$$

It is seen that, for CdS, the ratio of the resistance at two illumination levels equals the inverse ratio of the two illumination levels themselves. [For CdSe, the value of m is approximately 0.8. We will use m = 1 and Eq. (9-7) in our calculations for the sake of simplicity.]

The variation of the resistance of a photodetector as a function of the incident illumination forms the basis of a number of applications, some of which are discussed in the following examples.

Example 9.1

A circuit combining a photodetector and an op amp (Fig. 9-2) can be used to generate an output voltage signal that is inversely proportional to the incident illumination.

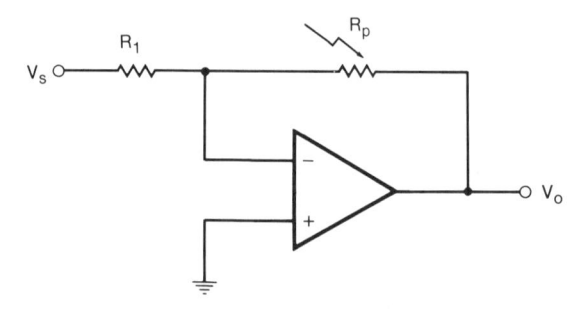

Fig. 9-2. Circuit using a photodetector.

Using the (inverting) op amp formula (Chapter 8), the gain of the given circuit is

$$\frac{V_o}{V_s} = -\frac{R_p}{R_1} \tag{9-8}$$

Suppose the value of R_P is R_A at an illumination level L_A and R_B at an illumination level L_B. Then, from Eq. (9-7),

$$\frac{R_A}{R_B} = \frac{L_B}{L_A} \tag{9-9}$$

If the output voltage of the circuit is V_{oA} at illumination L_A and V_{oB} at illumination L_B, then Eq. (9-8) leads to

$$\frac{V_{oA}}{V_s} = -\frac{R_A}{R_1}$$

and

$$\frac{V_{oB}}{V_s} = -\frac{R_B}{R_1}$$

which gives, in conjunction with Eq. (9-11),

$$\frac{V_{oA}}{V_{oB}} = \frac{R_A}{R_B}$$

$$= \frac{L_B}{L_A}$$

Electronics: Circuits and Systems   549

Therefore, the output voltage varies in inverse proportion to the incident illumination for a fixed value of V_s, which may be due to a dc battery.

Example 9.2

The combination of a photodetector and a capacitor can be used to provide an automatic exposure control circuit for a camera. (Such a circuit was already discussed in Chapter 2. Here we relate its operation to the level of incident illumination.) The arrangement is shown in Fig. 9-3. At t = 0, a button is pressed, which opens the shutter of the camera and simultaneously closes the switch S in the circuit of Fig. 9-3. The capacitor charges toward the battery voltage V_B. When the capacitor voltage reaches a prescribed level, say $0.8V_B$, it activates the mechanism that closes the shutter of the camera. The capacitor voltage decays to zero during this last step.

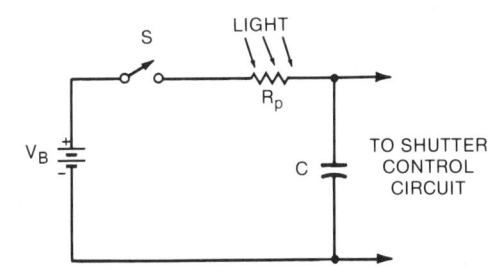

Fig. 9-3. Basic automatic exposure control circuit.

Since the time taken for the capacitor voltage to reach $0.8V_B$ depends upon the time constant R_pC, and since R_p varies with the incident illumination, the interval of time for which the shutter remains open depends upon the incident illumination also. From the charging curve of an RC circuit in Chapter 2, the time taken for the capacitor to charge from zero to $0.8V_B$ is 1.6 time constants or $1.6R_pC$.

Suppose the values of R_p are R_A and R_B at two levels of illumination L_A and L_B, respectively. Then,

$$\frac{\text{exposure interval at illumination } L_A}{\text{exposure interval at illumination } L_B} = \frac{1.6R_AC}{1.6R_BC}$$

$$= \frac{R_A}{R_B}$$

$$= \frac{L_B}{L_A}$$

by using Eq. (9-7).

The exposure interval is seen to be *inversely* proportional to the incident illumination.

Drill Problem 9.4: Consider the circuit of Fig. 9-2 with $R_1 = 10$ MΩ and $V_s = -5$ V. Assume that the photodetector in the circuit has a resistance of 18 MΩ at an illumination of 0.1 fc. Calculate the output voltage of the circuit at each of the following levels of illumination: (a) 0.1 fc; (b) 1 fc; (c) 10 fc.

Drill Problem 9.5: Suppose the photodetector of the previous drill problem is used in the exposure control circuit of Fig. 9-3 with C = 0.005 μF, calculate the exposure intervals for the three levels of illumination specified in the previous drill problem.

Speed of Response of a Photodetector

The resistance of a photodetector does not respond instantaneously to changes in the illumination incident on it. There is a delay involved between the change in the illumination and the attainment of the new value of resistance by the photodetector. The behavior of a photodetector is quite similar to that of a capacitor charging or discharging through a resistance. Fig. 9-4 shows the response of a photodetector as a function of time when the illumination level has been increased or decreased, and these curves have the same shape as the charging and discharging curves of an RC circuit.

A time constant is associated with a photodetector, and it takes approximately five time constants for the photodetector to adjust to the new level of illumination. The values of the time constant for practical photodetectors are 100 ms for CdS, 10 ms for CdSe, 400 μs for PbS, and 10 μs for PbSe. The speed of response of a photodetector, as determined by its time constant, affects the frequency at which the incident illumination can change. If the incident illumination is in the form of a pulse, then the variation in the resistance of the photodetector will be as shown in Fig. 9-5.

In order to permit the photodetector to reach the new value of resistance when the level of illumination is changed, it is necessary for the duration of the incident pulse to be approximately five time constants. If the incident illumination is in the form of a pulse train (which will be the case where the illumination is due to a light source that is turned on and off repeatedly), then the frequency of the pulse train is limited by the constraint that the pulse duration be a minimum of approximately five time constants. In the case of CdS, the time constant is 0.1 s, which means that the pulse duration should be at least 0.5 s. The incident illumination cannot vary at a frequency higher

Electronics: Circuits and Systems

(A) Illumination level decreased.

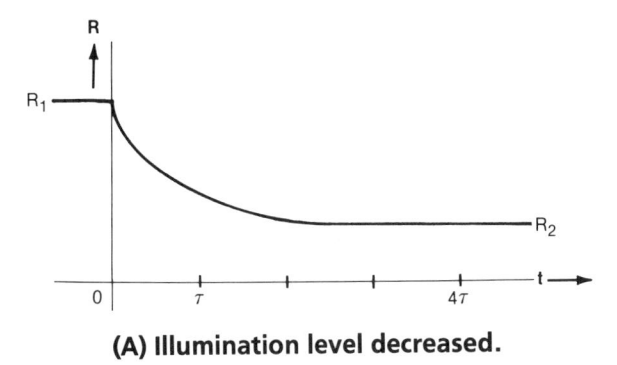

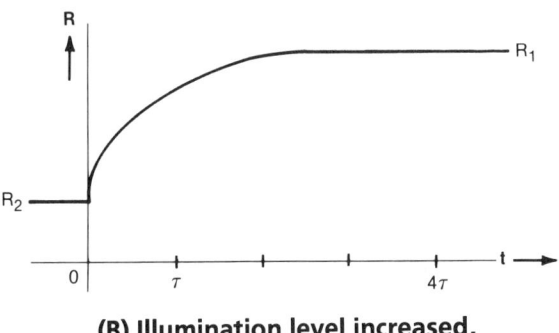

(B) Illumination level increased.

Fig. 9-4. Response of a photodetector to changes in the level of illumination.

than one cycle per second. When an application requires the turning on and off of the illumination at a much higher rate than one cycle per second, it is necessary to use photodetector materials such as PbS and PbSe.

Example 9.3

A chopper wheel is interposed between the light source and the photodetector in the circuit of Fig. 9-6. The chopper rotates at the rate of ten revolutions per second. The time constant of the photodetector is 10 ms. The dark resistance of the photodetector is 100 kΩ, and its resistance under steady illumination from the light source is 0.5 kΩ. Determine and sketch the output voltage v_o.

Solution

The incident illumination on the photodetector is a pulse train of a frequency ten cycles per second, as shown in Fig. 9-7.

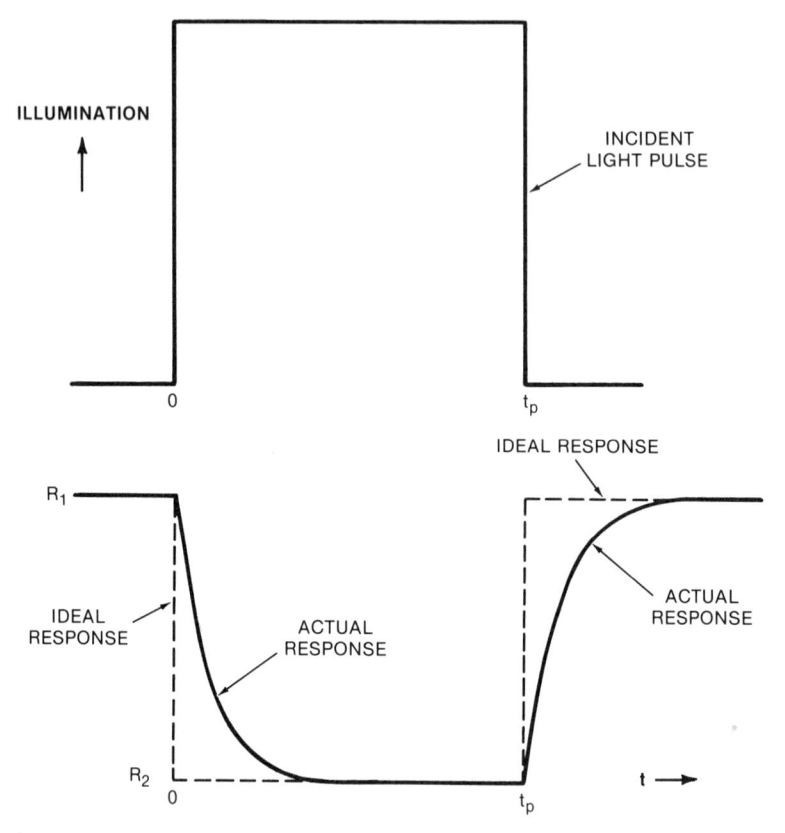

Fig. 9-5. Response of a photodetector to an input pulse of illumination.

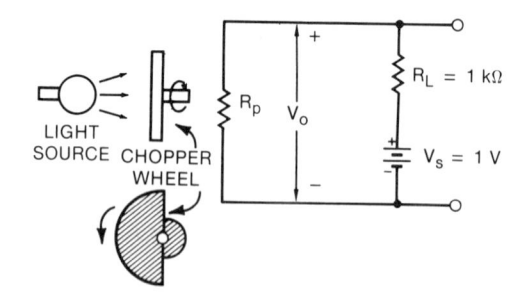

Fig. 9-6. Circuit for Example 9.3.

Let us first calculate the output voltage levels when the chopper is blocking the illumination and when the chopper is transmitting the illumination to the

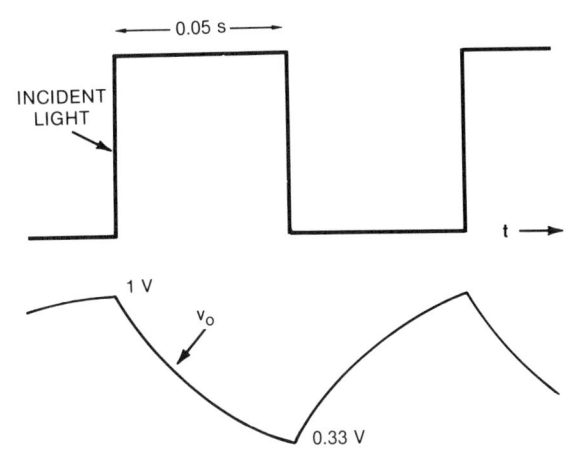

Fig. 9-7. Input and output of the circuit of Example 9.3.

photodetector. The output voltage v_o is obtained by using the voltage-divider formula:

$$v_o = \frac{R_p}{R_p + R_L} V_s$$

$$= \frac{R_p}{R_p + 1}$$

where,
R_p is in kΩ.
The two levels of output are, therefore,

$$v_o = \frac{100}{101}$$

$$= 1 \text{ V (dark)}$$

and

$$v_o = \frac{0.5}{1.5}$$

$$= 0.33 \text{ V (bright)}$$

The output will, therefore, switch back and forth between 1 V and 0.333 V except that the changes will not be instantaneous but governed by the time constant of the photodetector. The output waveform can be drawn by using the charging and discharging curves of RC circuits in Chapter 2 and is as shown in Fig. 9-7.

Drill Problem 9.6: Calculate the maximum frequency at which the incident illumination can turn on and off for each of the following photodetector materials: (a) CdSe; (b) PbS; (c) PbSe.

Structure of a Photodetector

There are two requirements to be met in the practical fabrication of a photodetector. First, the surface area of the device should be large enough to take full advantage of the incident illumination. The second requirement is that the resistance of the device should be high enough to be comparable to the other resistances present in the circuit. To have a high resistance, the photodetector should be in the form of a long, narrow strip since resistance is directly proportional to length and inversely proportional to the area of the cross-section. To have a large surface area without making the device too large in its physical dimensions, the strip is folded in the form shown in Fig. 9-8.

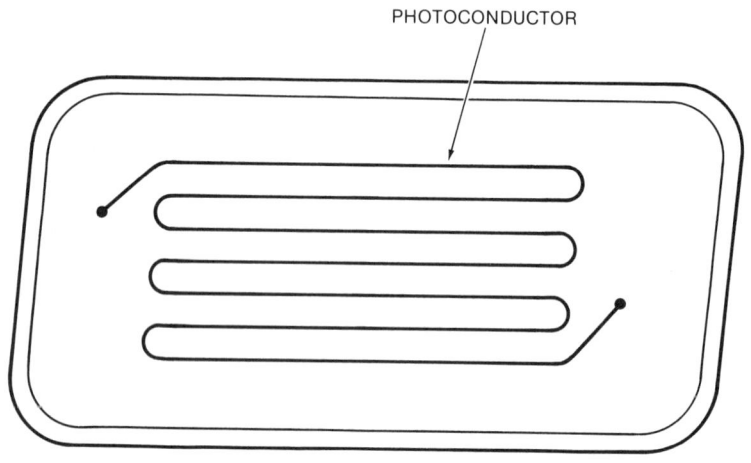

Fig. 9-8. Structure of a photodetector.

CdS is the preferred photodetector in most applications where the speed of response is not important. It shows a larger variation of resistance for a given change in the level of illumination than CdSe and has a much higher dark resistance of CdSe.

PHOTOVOLTAIC CELLS

A *photovoltaic cell* generates a voltage under the influence of incident illumination. It is an *active* device, unlike the photodetector, which is a passive device.

Electronics: Circuits and Systems

The photovoltaic cell is *pn junction* designed so that the depletion region at the junction is modified by the incident illumination. As was discussed in Chapter 3, the migration of electrons from the n-side to the p-side and holes from the p-side to the n-side results in a layer of positive and negative ions that forms the space charge or depletion region. The potential difference due to the space charge is ϕ, but this is not available for external use since it is cancelled by the contact potentials at the metal-semiconductor contacts. [Refer to Eq. (3-1).] Now, suppose illumination is incident on the junction. Electrons are released from covalent bonds of the semiconductor atoms. Since each released electron leaves a hole in its place, there is a generation of electron-hole pairs at the junction. Under the constant level of illumination, there will be (on the average) a fixed number of electron-hole pairs generated per second per unit volume in the area of the junction due to the incident illumination. These electrons and holes migrate to the n-side and p-side, respectively, due to the influence of the electric field in the junction. Their migration increases the number of ions in the space charge region with a consequent widening of the depletion region and a correspondingly larger potential difference. The *increase* in the potential difference caused by the incident illumination, V_{ph}, is available for driving an external load. (Fig. 9-9) Solar cells are based on the previous principle. The voltage V_{ph} developed by a solar cell under bright sunlight is around 0.4 V, and the current that would flow in a short circuit placed across the cell is around 5 mA. To generate a significant amount of power by means of solar cells, it is necessary to connect a sufficient number of cells in series. It is also necessary to choose the load on the cell network to maximize the power transferred to the load since the amount of power involved is small.

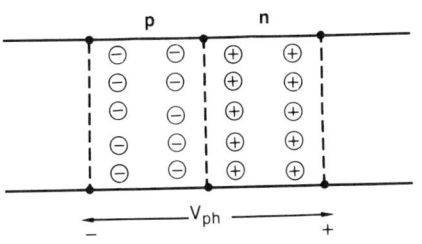

Fig. 9-9. Voltage generated at a pn junction by an incident illumination.

The two most commonly used materials for photovoltaic cells are silicon and selenium. Silicon cells are responsive to the wavelength range 700 nm to 1000 nm, and selenium cells respond to the range 450 nm to 700 nm. The voltage generated by a photovoltaic cell is found to vary linearly with the load resistance when the load resistance is very low. For high values of load resistance, the voltage generated is proportional to the logarithm of the incident illumination.

PHOTODIODES

The principle of operation of a *photodiode* is the same as that of the photovoltaic cell; in fact, a photodiode can be used as a photovoltaic cell. But the photodiode is usually used in a *photoconductive* mode in which it acts as a variable *current source*. When operating in a photoconductive mode, the photodiode is kept under *reverse bias* by means of an external battery as shown in Fig. 9-10. As was discussed in connection with the photovoltaic cell, there is a generation of electron-hole pairs in the pn junction due to the incident illumination, and the electrons so generated move from p to n while the holes move from n to p. This migration of electrons and holes (generated by the incident illumination) sets up a current I_p in the direction n to p in the diode. For a given reverse bias, the current increases as the illumination increases. But, for a given illumination, the current is essentially a constant independent of the reverse bias across the photodiode. The photodiode can, therefore, be used as *a source of constant current whose value depends upon the incident illumination*. The i-v characteristics of a photodiode are of the form shown in Fig. 9-11. Note that the horizontal axis is labeled in terms of the *magnitude* of the reverse bias.

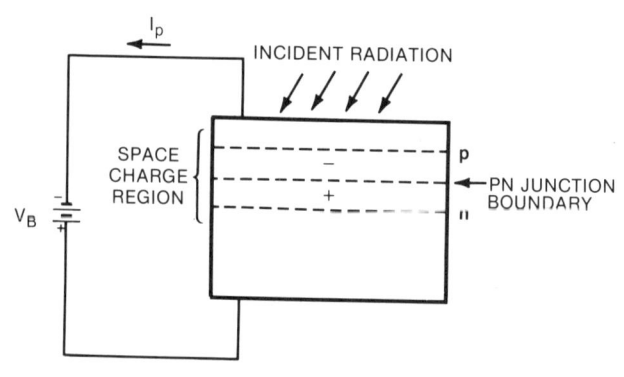

Fig. 9-10. Reverse-biased pn junction is used as a photodiode.

Increasing the Efficiency of a Photodiode

In order to have the photodiode respond efficiently to an incident radiation, it is necessary that the electrons and holes generated in the device flow through the device and the external circuit as was discussed previously. But, when electrons and holes are generated by the incident radiation, there is the

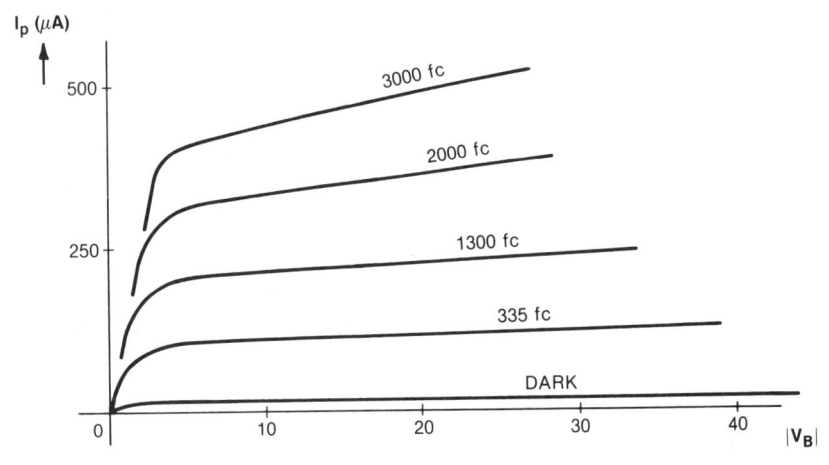

Fig. 9-11. Output characteristics of a photodiode for different illumination levels.

possibility that they *recombine* with other holes and electrons. Such recombinations do not contribute to current flow in a device, and should, therefore, be reduced as much as possible.

It is found that the probability of recombination is higher in the region *outside* the depletion (or space-charge) region of the diode than within the space-charge region itself. Therefore, it is necessary to make the depletion region as wide as possible and the regions outside the depletion region as narrow as possible.

In the structure of a practical photodiode, the p region (refer to Fig. 9-10) is made extremely thin to minimize the probability of an incident photon being absorbed in that region before it can enter the depletion region.

The width of the depletion region of a photodiode is made as large as possible by means of one or both the following approaches. The *width* of a depletion region in a reverse-biased junction *increases* as the *reverse bias increases*. But the reverse bias cannot be increased without limit, since it will eventually cause the device to break down, which sets an upper limit to the reverse bias that can be applied to the diode and to the use of the reverse bias to increase the efficiency of the photodiode.

The other approach to increasing the width of the depletion region is to *increase the resistivity* of the semiconductor material. The width of a depletion region is proportional to the square root of the resistivity of the material. A *low* doping level (that is, the concentration of the impurity material added to a semiconductor material) leads to a high resistivity. Therefore, the doping level of the n-region of the photodiode is made quite small to extend the depletion region farther into it. Unfortunately, such a cure is not without its disadvantages. A metallic lead has to be attached to the n-side of the diode (the bottom terminal in Fig. 9-10), and the contact there should be an *ohmic* contact and not a *rectifying* contact. A rectifying contact will make the junction of the metallic lead and the n-side of the diode behave like a diode also, which is not a desirable situation. To have an ohmic contact at the n-side of the device, the doping level has to be high. (It may be recalled that in the fabrication of npn transistors in ICs, an n+ doping was necessary before aluminum contacts were attached to the n-type collectors.) Therefore, the doping level of the n-region of a photodiode cannot be reduced to a point where the contact between that region and the metallic leads becomes rectifying. The solution to the problem lies in modifying the structure of the photodiode as discussed in the following section.

PIN Photodiode

A layer of *pure* or *intrinsic* semiconductor is interposed between the p- and the n-layers of the photodiode, leading to a *pin* diode, whose basic structure is indicated in Fig. 9-12. An intrinsic semiconductor has a very high resistivity since there is only a sparse supply of electrons and holes in the absence of acceptor or donor-type impurities. Therefore, the depletion region extends far into the intrinsic layer of the pin diode. Even when there is no reverse bias, the depletion region can extend halfway into the intrinsic layer.

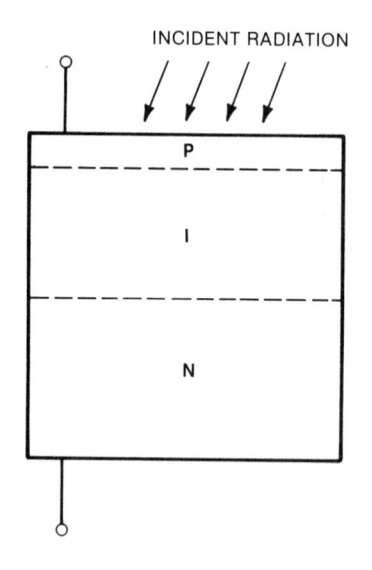

Fig. 9-12. PIN photodiode.

When a reverse bias is applied, the depletion region extends deeper into the intrinsic layer. At some value of the reverse bias, the depletion region touches the edge of the n-region. The voltage needed to lead to that condition is called the *punch-through voltage* and is only a few volts (e.g., 5 V). (The punch-through voltage does not have the same connotation as the breakdown voltage, which may be as high as 200 V for a pin diode.) A pin diode is usually operated at a reverse bias much higher than the punch-through voltage in order to make the intrinsic layer fully depleted. The diode then operates as a good linear device with a high speed of response.

Silicon photodiodes respond to radiations of wavelength between 200 nm to 1100 nm with the peak response occurring at or below 990 nm. The typical operating voltages are between 10 V and 110 V. They are fabricated with surface areas in the range 1 mm^2 to 10 mm^2.

Speed of Response of Photodiodes

The speed with which the current in a photodiode rises to its final (steady-state) value when an input pulse of illumination is applied depends upon a number of factors. The response is much faster when the device is under reverse bias than under zero bias. For example, the typical rise time of a pin diode is around 300 ns under zero bias but drops to less than 1 ns with a reverse bias of 20 V. The manner in which the illumination hits the diode also affects the speed. When the illumination hits the very center of the sensitive area, the effective resistance is larger than when the illumination is applied to portions close to the aluminum contact ring. The speed is higher in the latter case due to the smaller resistance encountered. The load resistance affects the speed also since the time constant is affected by it in conjunction with the junction and packaging capacitances of the device. Rise times of 1 ns are typical with pin photodiodes with a load resistance of 50 Ω and a 20-V reverse bias.

Photodiode Amplifier Circuits

The symbol of a photodiode is shown in Fig. 9-13A. The same symbol is used for both pin and pn photodiodes. Note the direction of the current: it is from *n* to *p* (which is opposite to the direction used in the conventional diodes).

A basic linear amplifier circuit using a photodiode and an op amp is shown in Fig. 9-13B. (The resistor R_2 serves to reduce the offset in the output voltage). The point P is at virtual ground and the current through R_F is I_p since the input current to the op amp is zero. Therefore,

$$V_o = R_F I_p$$

and, for a given level of illumination, I_p is a constant and hence V_o is also constant. The current I_p (and hence the voltage V_o) increases linearly with the level of illumination.

560   Optoelectronic Devices

(A) Photodiode symbol.   (B) Photodiode in a circuit.

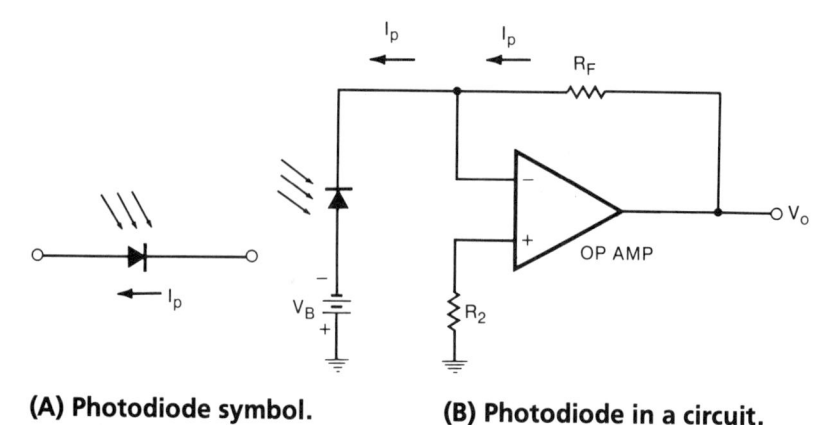

Fig. 9-13. Symbol of a photodiode and a circuit using a photodiode.

Drill Problem 9.7: Suppose, in the circuit of Fig. 9-13B, the photodiode and battery are both reversed in polarity. How will the output voltage vary as a function of the level of illumination?

Photodiodes respond with measurable currents when the illumination is very bright but their sensitivity is too low to be useful at low levels of illumination. Even at light levels to which a human eye can respond without much difficulty, the photodiode current may be in the order of 10^{-16} A, which can be drowned out by leakage currents and noise. A much higher sensitivity is available from a *phototransistor*.

PHOTOTRANSISTORS

The reverse biased collector-to-base junction of a transistor can be operated like a photodiode as indicated in Fig. 9-14. The current I_p is the base current of the transistor and is controlled by the incident illumination. The collector current is β times I_p and the phototransistor is much more sensitive to incident illumination than a photodiode. But, the current amplification also applies to the dark current (when no light is incident on the device), and the leakage current (that is, the collector current present even when there is no light incident) of a phototransistor is much higher than that of a photodiode. The region of linearity of operation of a phototransistor is limited by the variation of β with current and temperature. The application of the phototransistor is, therefore, limited to situations where linearity is not critical.

Fig. 9-14. Basis of a phototransistor.

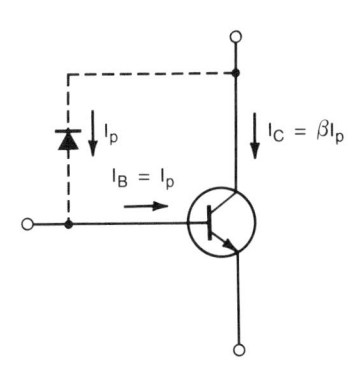

The symbol of a phototransistor and its current-voltage characteristics are shown in Fig. 9-15. Instead of using the base current as the third parameter as we did with ordinary transistors, the incident illumination level (in mW/cm^2) is used in the phototransistor characteristics. To get a "feel" for the illumination levels, a subminiature incandescent lamp produces approximately 5 mW/cm^2, a neon glow lamp slightly less than 1 mW/cm^2, and an LED approximately 0.5 mW/cm^2 (when the distance from the light source to the detector is 5 cm in all the previous cases), whereas, the illumination on the earth's surface due to the sun is about 100 mW/cm^2 on a clear day. It is seen that a phototransistor can produce appreciable currents even at low levels of illumination. In fact by combining two transistors in a *Darlington connection* as indicated in Fig. 9-15C, the output current can be increased to much higher levels than a single phototransistor. The leakage current (with no incident illumination) is also increased by the same high gain, of course.

Speed of Response of Phototransistors

Phototransistors are much slower in responding to pulses than photodiodes. Whereas, the response time of a photodiode is around 1 ns, that of a phototransistor is in the order of several *microseconds*. The reason for the slow response is the necessity to make the collector-base junction large to capture the photons from the incident radiation. The collector-base junction capacitance may be in the range of 20 pF, which is increased even further due to the Miller effect (discussed in Chapter 7), thus leading to large time constants.

Applications of Phototransistors

The phototransistor can be used as a switch that turns on or off depending upon whether the incident illumination is above or below a specified threshold level. A combination of its switching action with other electronic components leads to a wide variety of applications of the phototransistor.

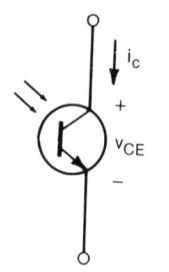

(A) Symbol of a phototransistor.

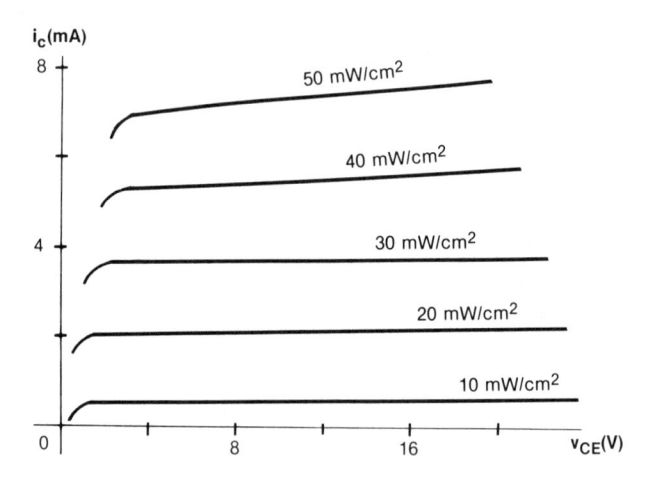

(B) Output characteristics.

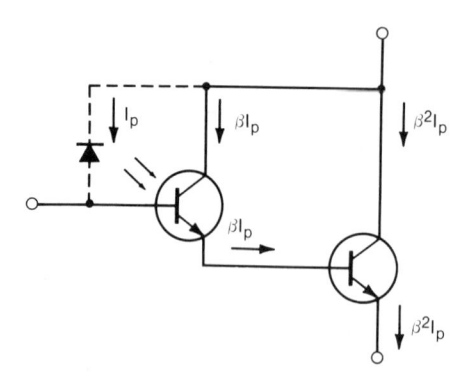

(C) Photo-Darlington connection.

Fig. 9-15. Symbol of a phototransistor and its current-voltage characteristics.

Electronics: Circuits and Systems

The connection of a phototransistor is similar to that of an ordinary npn transistor. The main difference is that the input signal is the light incident on the collector-base junction of the phototransistor. The base terminal is, therefore, not usually connected to an external circuit component.

Light-Beam Detector

A basic circuit to detect the presence (or absence) of light above a prescribed threshold is shown in Fig. 9-16. The transistor T_2 operates in one of two modes: it is either off; that is, no collector current flows; or it is conducting heavily (saturated). When it is off, the output voltage is equal to V_{CC}. When it is conducting heavily, almost all of V_{CC} is across the resistor R_{C2}, and the output voltage is zero. The current in the phototransistor (I_{E1}) depends upon the level of the incident illumination. The resistor R_1 can be adjusted to set the level of the current I_{EI} at which transistor T_2 goes into saturation. R_1 also can be a variable resistor so that the level can be changed easily. In the circuit shown, the output is *low* when the input illumination is *above* a certain value, and the output is high when the illumination is below the threshold. By adding an inverter stage (which inverts high to low and low to high) to the given circuit, we can have a circuit that gives a high output for bright light and a low output for dim light.

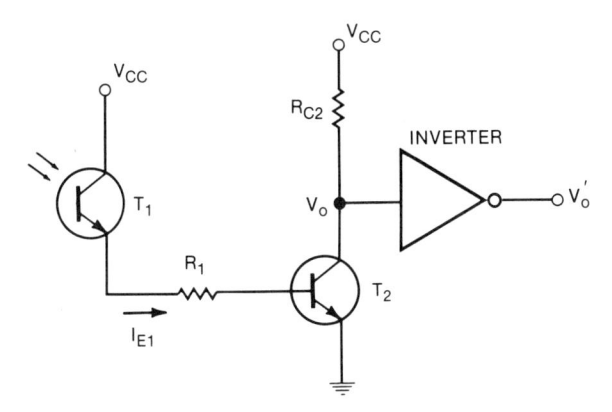

Fig. 9-16. Light-beam detector circuit.

Electronic Slave Flash Control Circuit

When the light from one electronic flashgun (the master flash) is required to trigger a second flashgun (the slave flash), a control circuit of the form shown in Fig. 9-17 can be used.

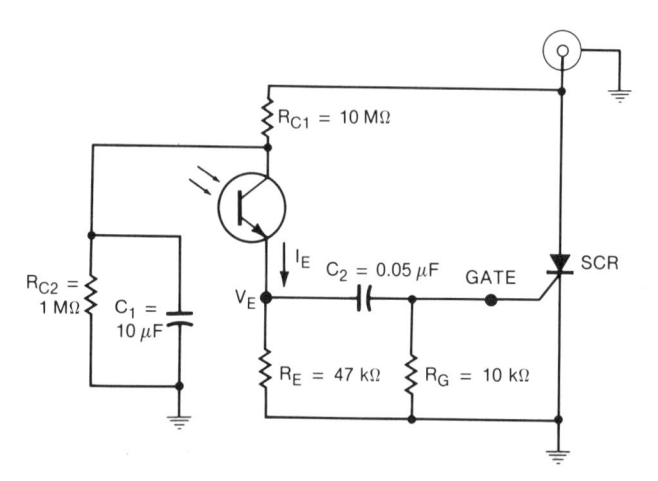

Fig. 9-17. Master-slave flash control circuit.

The circuit utilizes an SCR, which was discussed in Chapter 4. A brief review of the SCR is presented here. The SCR is a pnpn device as shown in Fig. 9-18. If the anode is at a positive potential with respect to the cathode while the gate is at zero potential, there is only a very small current flow in the device. But if the gate is kept at a positive voltage (with respect to the cathode), the p_2n_2 junction is forward biased, and the resulting large current in that junction pulls current through the entire structure. The SCR conducts heavily (it has been "fired"). Once the conduction starts, the SCR cannot be turned off except by applying a zero or reverse bias between the anode and the cathode, or by reducing the current to zero by some means or other. The gate voltage has no control over the SCR's response once it has triggered the SCR into action.

Consider now the circuit of Fig. 9-17. The dc power supply for the circuit is usually taken from the socket of the flash itself. The flash socket usually has 200 V available, and it is reduced to approximately 18 V through the voltage-divider arrangement R_{C1} and R_{C2}. When there is no light incident on the phototransistor, it does not conduct.

$$I_E = 0$$

The voltage at the gate terminal is zero. The voltage on the capacitor C_1 is roughly 18 V.

When the master flash is fired, the light from it is incident on the phototransistor and makes it conduct. The voltage of 18 V on C_1 is now discharged through the collector and emitter of the phototransistor and through the circuit connected to the emitter. The surge of current through the resistor R_E makes the emitter voltage V_E rise suddenly, and since the voltage on the

Electronics: Circuits and Systems

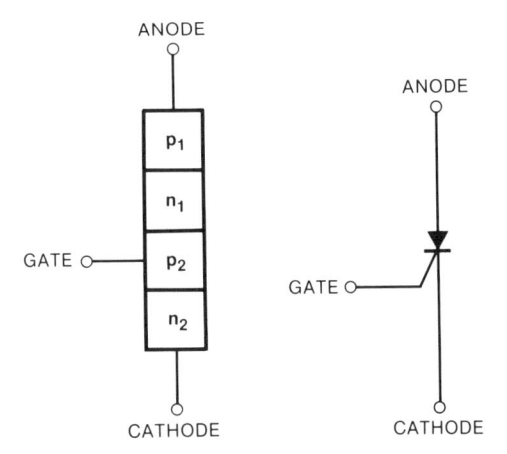

Fig. 9-18. Structure and symbol of an SCR.

capacitor C_2 cannot change instantaneously, the voltage at the gate terminal rises by the same amount as V_E. The gate voltage is positive enough to fire the SCR, which in turn triggers the slave flash.

Window Detector

The window detector circuit (which was discussed in Chapter 8 in connection with comparators) can be used to detect if the incident illumination is between two prescribed levels. Such a circuit can be used, for example, to sort colored and clear bottles. The circuit shown in Fig. 9-19 represents a window detector for optical signals.

In the circuit shown, the upper comparator has a reference voltage V_1 due to the voltage-divider network

$$V_1 = V_{CC} \times \frac{R_2 + R_3}{R_1 + R_2 + R_3} \tag{9-10}$$

which is connected to the noninverting input terminal. The reference voltage V_2 of the lower comparator is given by

$$V_2 = V_{CC} \times \frac{R_3}{R_1 + R_2 + R_3} \tag{9-11}$$

which is connected to the inverting input terminal.

The outputs of the comparators will, therefore, behave as follows:

$V_A > V_1$: V_{o1} is low; V_{o2} is high; V_o is low.

$V_2 < V_A < V_1$: V_{o1} is high; V_{o2} is high; V_o is high.

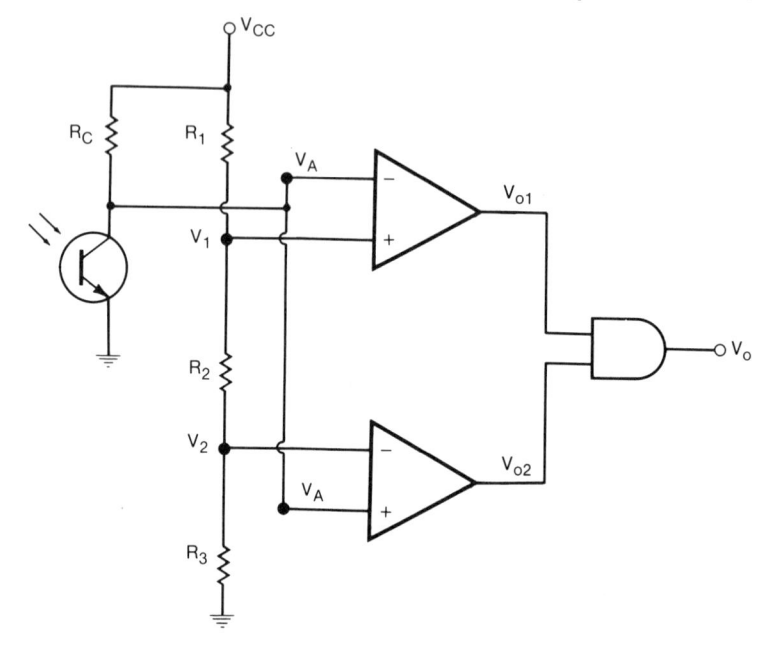

Fig. 9-19. Window detector circuit with an optical signal.

$V_A < V_2$: V_{o1} is high; V_{o2} is low; V_o is low.

Considering the phototransistor, the voltage V_A is given by

$$V_A = V_{CC} - R_C I_C$$

where,

the collector current I_C is controlled by the radiation incident on the collector-base junction of the transistor.

Under a bright illumination, the current is large and V_A is low (due to the voltage drop in R_C). Under low illumination, the voltage V_A is high. Given the range of values of illumination over which the circuit output should be high, the window detector can be designed as illustrated in the following example.

Example 9.4

Design a window detector that has a high output when the incident illumination is between 2 mW/cm² and 8 mW/cm². A phototransistor is available for which the collector current is 10 mA at 2 mW/cm² illumination, and 20 mA at 8 mW/cm² illumination. Use a 40-V battery and $R_C = 1.5$ kΩ.

Solution

First we calculate the values of V_A, the collector voltage, at the boundary values of illumination. When $I_C = 10$ mA,

Electronics: Circuits and Systems

$$V_A = V_{CC} - R_c I_c = 40 - (1.5 \times 10)$$
$$= 25 \text{ V}$$

When $I_c = 20$ mA,

$$V_A = V_{CC} - R_c I_c = 40 - (1.5 \times 20)$$
$$= 10 \text{ V}$$

The circuit output should be high when V_A is between 10 V and 25 V. From the discussion of the comparator portion of the circuit, the circuit output is high (that is, both V_{01} and V_{02} are high) when $V_2 < V_A < V_1$. Therefore, for the present situation,

$$V_1 = 25 \text{ V}$$

and

$$V_2 = 10 \text{ V}$$

The reference voltages to be connected to the two comparators are $V_1 = 25$ V to be connected to the noninverting input terminal of the upper comparator and $V_2 = 10$ V to be connected to the inverting input terminal of the lower comparator. Using Eqs. (9-10) and (9-11) with $V_{CC} = 40$ V, we get

$$\frac{R_2 + R_3}{R_1 + R_2 + R_3} = 0.625$$

and

$$\frac{R_3}{R_1 + R_2 + R_3} = 0.25$$

which leads to

$$\frac{R_2 + R_3}{R_3} = (0.625/0.25) = 2.5$$

If we choose R_3 arbitrarily as 10 kΩ, then we get $R_2 = 15$ kΩ and $R_1 = 15$ kΩ. The complete circuit is shown in Fig. 9-20.

Drill Problem 9.8: Suppose the phototransistor of the last example is replaced by one with $I_C = 5$ mA at 2 mW/cm^2 and 20 mA at 12 mW/cm^2. Keeping R_C and V_{CC} the same as before, calculate the new values of the resistors R_1, R_2, and R_3 so as to have a high output between 2 and 12 mW/cm^2.

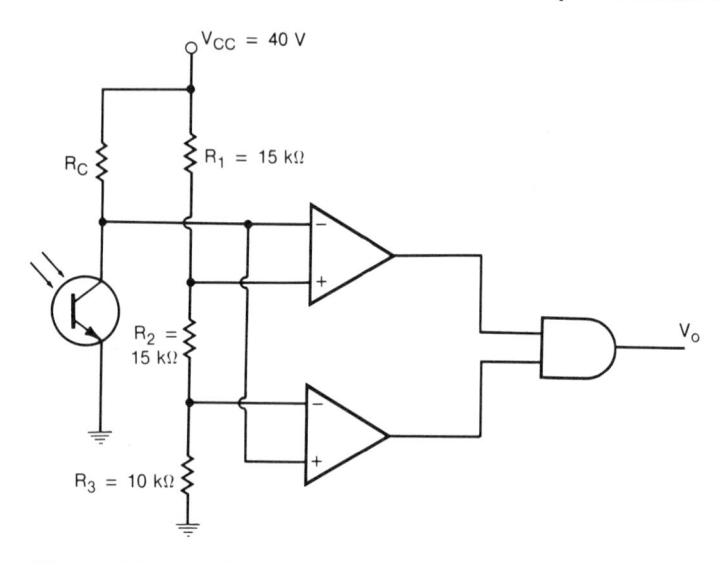

Fig. 9-20. Window detector circuit of Example 9.4.

Photo-FETs

A junction field-effect transistor (JFET) can be used as a phototransistor (called photo-FET) by using the reverse-biased drain-to-gate junction (which is analogous to the collector to base junction of a bipolar transistor) as a photodiode. A typical circuit using a photo-FET is shown in Fig. 9-21. The incident illumination causes a current to flow in the *gate terminal* as indicated. Note that this current was ignored (as being negligible) in our discussion of FETs in Chapter 6. But here it forms the significant current.

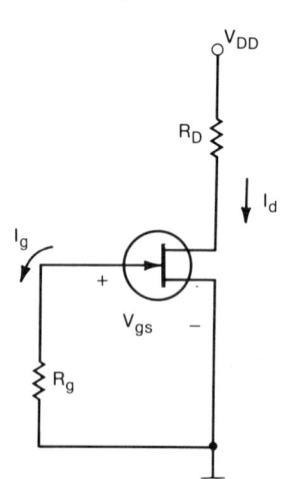

Fig. 9-21. Photo-FET.

The gate-to-source voltage in the circuit is

$$V_{gs} = I_g R_g$$

and the drain current is related to the gate-to-source voltage through the transconductance g_m. Therefore,

$$I_d = g_m V_{gs} = g_m R_g I_g$$

The resistance R_g (which can be quite large) is seen to control the output current of the photo-FET. It is possible to obtain active current gains ($g_m R_g$) of very high values in a photo-FET.

The applications of a photo-FET are similar to those already discussed in connection with the BJT phototransistors.

LIGHT-EMITTING DIODES (LEDs)

Light-emitting diodes (LEDs) are pn junctions that emit optical radiation generated by the recombination of electrons and holes when the junction is forward biased. In a forward-biased junction, electrons diffuse from the n-region to the p-region, and holes diffuse from the p-region to the n-region. When the electrons enter the p-region, they become minority carriers in that region, and a number of such electrons will recombine with the holes (which are the majority carriers) available there. Similarly, a number of holes diffusing into the n-region will recombine with the electrons available in that region. The recombination of a minority carrier and a majority carrier causes a release of energy from the minority carrier—the energy that the minority carrier has acquired from the forward bias at the junction.

The released energy can appear in the form of heat (nonradiative release of energy) or in the form of a photon (radiative release of energy). Both types of release of energy occur in any semiconductor material but certain special materials have a high degree of probability of radiative release of energy. Junctions made of such semiconductor materials act as light-emitting diodes. Gallium arsenide (GaAs) is one such material and the optical radiation emitted has a wavelength of 885 nm (infrared). Gallium phosphide (GaP) and gallium arsenide phosphide (GaAsP) produce optical radiation in the *visible* spectrum. In the case of GaAsP, the actual wavelength emitted can be controlled by varying the proportion of phosphorus to arsenic, and visible radiation of different colors can be obtained. Red, yellow, and green LEDs are commercially available in a variety of sizes and shapes. Besides the familiar type with a circular light-emitting surface, there are those with a rectangular light-emitting surface and light-bar modules.

The radiation efficiency of LEDs varies from 1% to 12%. The efficiency is higher at longer wavelengths (that is, it is higher for the red end of the spectrum than for the green end of the spectrum), but the sensitivity of the human eye is higher for the shorter wavelengths so that the two compensate for each other. The low values of efficiency are of no concern in terms of the visibility of the light emitted by an LED. A GaAsP LED (with the ratio of arsenic to phosphorus of 0.56 to 0.44) emits a wavelength of 650 nm (red) and has an efficiency of only 0.1%. But the luminance achieved is quite adequate, and it can be seen easily even in brightly lit rooms.

LEDs are operated with a forward bias around 2 V with a forward current of 10 to 20 mA. Their response time (to pulse inputs) is in the range 90 to 200 ns. LEDs have several advantages over incandescent lamps: resistance to vibrations, low power consumption, long life, low heat generation, and fast response. They are used in a wide variety of display systems as well as in communication systems. Some of the applications are discussed in the following section.

Bar Graph Display

A circuit that converts the level of an input (electrical) signal into a visible bar graph where the height of the graph is a measure of the amplitude of the input signal is shown in Fig. 9-22.

Each LED is connected to the output of a comparator in such a way that when the output of the comparator becomes *high*, the diode becomes forward biased and lights up. The reference voltages to the comparators are obtained by means of a voltage divider, and they are all connected to the noninverting inputs. The input signal is connected to the noninverting input terminals of all the comparators.

For the output of a particular comparator to be high (to activate the LED), the input voltage V_{in} must be *greater than* the reference signal fed to that comparator. As the input signal increases from zero volts to V_{CC}, it can be seen that it will first exceed V_1 (lighting up D_1), then V_2 (lighting up both D_1 and D_2), and so on until it exceeds V_4 when all the LEDs will be lit. By choosing the number of LEDs and the voltage taps on the voltage divider, the display can be made as smooth a representation of the input signal as needed for a particular application.

In the particular circuit shown, if we choose $V_{CC} = 25$ V and $R_1 = R_2 = R_3 = R_4 = R_5$, then the reference voltages V_1, V_2, V_3, and V_4 are 5, 10, 15, and 20 V, respectively. As the input signal varies, the height of the display will vary in steps corresponding to 5-V intervals of the input signal.

Drivers for LEDs

When more than one LED has to be activated by the same signal (which is the case in LED arrays), it is necessary to provide a sufficient amount of current to drive the requisite number of LEDs. Drivers use either bipolar or

Electronics: Circuits and Systems

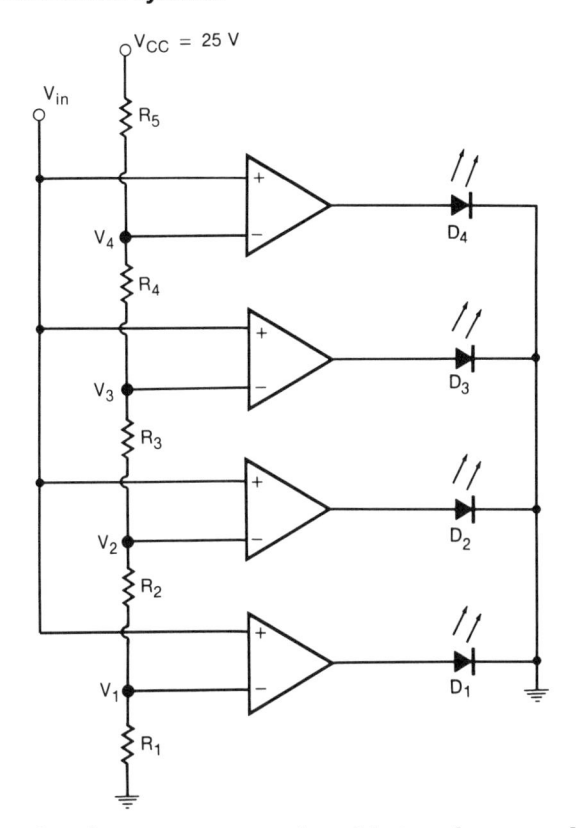

Fig. 9-22. Circuit to convert a signal into a bar graph display arrangement.

field-effect transistors, and Darlington configurations when the number of LEDs to be driven becomes quite large. Four driver circuits are shown in Fig. 9-23.

In the anode drivers, Figs. 9-23A and B, the anodes of the diodes are made high to activate them, whereas, the cathode drivers pull the cathodes of the diodes to a low value in order to activate them.

In the case of the *npn drivers*, Figs. 9-23A and C, current flows through the diodes when the input signal is *high*. That is, the *npn drivers have active high inputs*.

In the case of the *pnp drivers*, Figs. 9-23B and D, current flows through the diodes when the input signal is *low*. Note that in a pnp transistor, the base has to be at a lower potential than the emitter in order to forward bias that junction. The *pnp drivers have*, therefore, *active low inputs*. They are useful when the display is to be lit for low input signal levels rather than high input signals.

572 Optoelectronic Devices

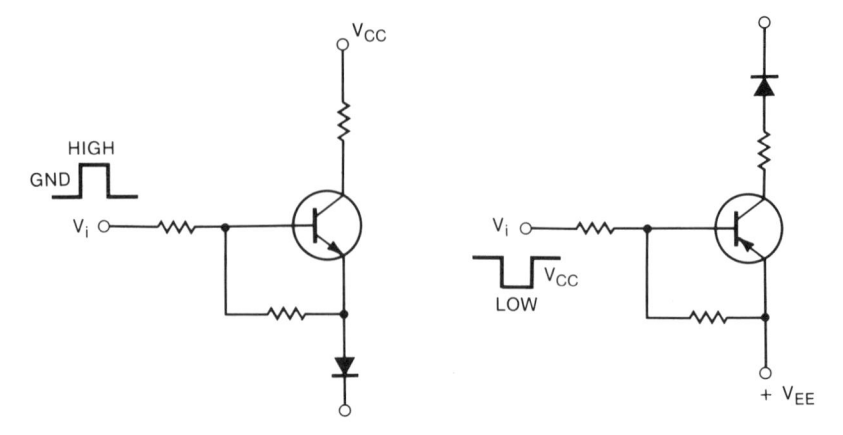

(A) Npn driver with anode high.

(B) Pnp driver with anode high.

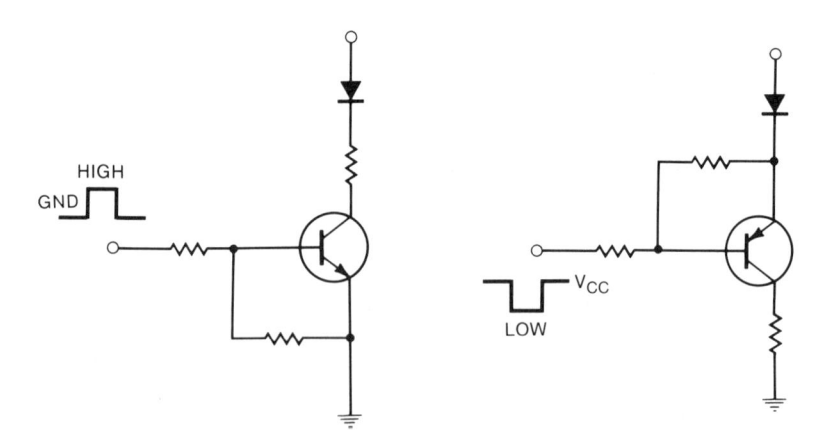

(C) Npn driver with cathode low.

(D) Pnp driver with cathode low.

Fig. 9-23. Driving schemes for LEDs.

XY Addressable LED Arrays

A rectangular array of LEDs is used in digital display systems, in which a particular LED in the array can be selected (or *addressed*) to be lit. Such an addressable array is shown in Fig. 9-24. If a current pulse is sent through the Y_2 line and through the X_3 line (by means of suitable drivers not shown in the present diagram), then the LED in the second row and third column will be lit.

Electronics: Circuits and Systems 573

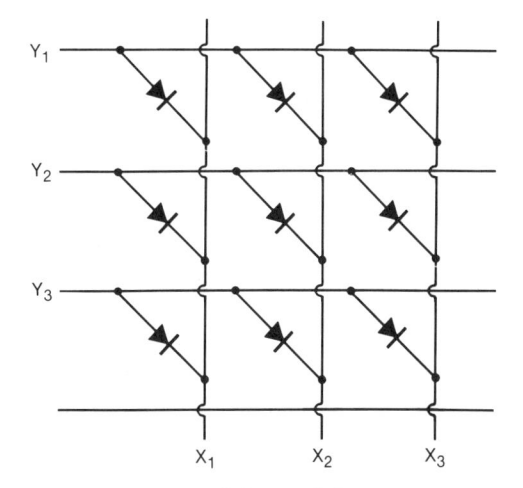

Fig. 9-24. XY addressable LED array.

It is also possible to activate two or more LEDs in a given row by sending the current pulse through the desired columns and through that particular row.

It is a common scheme to use *strobing* in such displays. That is, current pulses are sent through the different rows in succession at a regular rate in order to view successively the LEDs lit in the different rows, which is called row strobing. Column strobing can be used instead of row strobing also.

OPTICALLY COUPLED ISOLATORS

It is possible to combine an LED and a phototransistor such that the LED is activated by an input signal and the light emitted by the LED activates the phototransistor. Such an arrangement is known as an *optically coupled isolator* or *optoisolator* and is shown in Fig. 9-25. The current in the

Fig. 9-25. Optoisolator.

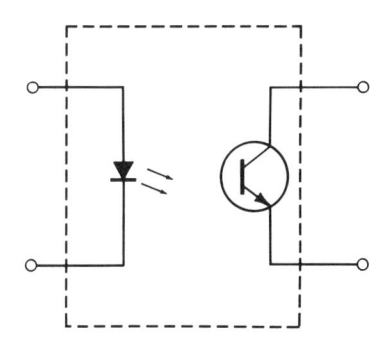

phototransistor is controlled by the input signal current. The optoisolator couples two circuits but provides excellent isolation between them.

Optoisolators are used as an interface between sensors (e.g., microswitches) that are situated in electrically noisy environments and digital control equipment. The electrical noise is not transmitted by the optoisolator and the function of the digital control equipment is unaffected by the noise. Other applications include coupling of signals between digital logic families, transfer of voltage and current from high-voltage circuits to low-voltage circuits, and transmission of signals from computers to peripheral equipment.

Optoisolators provide a far superior electrical isolation than transformers or other buffer stages; they are resistant to mechanical vibrations such as contact bounce, and since the transfer of signal is unilateral (from the LED to the phototransistor), they do not permit any changes in load conditions to affect the input circuit.

SUMMARY SHEETS

BASIC PRINCIPLES AND RELATIONSHIPS OF LIGHT

frequency $f = (c/\lambda)$

where,
c is the velocity of light or 2.998×10^8 m/s,
λ is the wavelength of light specified either in angstroms (Å) or nanometers (nm),
$1 \text{ nm} = 10^{-9} \text{ m} = 10$ Å.

Optical Radiation

Wavelength range: 10 nm to 10^5 nm
Visible range: 380 nm to 750 nm
Ultraviolet: below 380 nm
Infrared: above 750 nm

Photons

Packets of radiant energy.
Energy of a photon $W = hf$ J
h = Planck's constant = 6.63×10^{-34} J·s

Electronvolt

An *electronvolt* (eV) is the energy acquired by an electron when accelerated through 1 volt.

$$1 \text{ eV} = 1.6 \times 10^{-19} \text{ J}$$

Energy of a photon can also be expressed in the form

$$W = \frac{1243}{\lambda \text{ in nm}} \text{ eV}$$

Flux

Flux is a measure of the flow of optical energy.
Unit of flux = lumen (in the visible region). *Illumination* (or *irradiance*) is the flux per unit area of a receiving surface.
Unit of illumination = footcandle (or lumens/ft²).

PHOTODETECTORS

(Also called photoconductors or photoresistors)

A photodetector's resistivity changes as a function of the illumination incident on it. The incident light energy releases a number of electron-hole pairs in a semiconductor material.

Energy Gap

E_g is the minimum energy needed to release an electron from the covalent bond of the semiconductor material.
λ_{max} = wavelength of the optical radiation with energy E_g
 = $(1243/E_g$ in eV$)$ nm
λ_{max} = 513 nm for CdS (green) and 714 nm for CdSe (red)

Peak response of CdS is at 550 nm (green) and that of CdSe is at 690 nm (red).

Variation of Resistance With Illumination

Dark resistance = resistance of a material with no illumination.

Dark resistance is usually in the hundreds of kilohm to megohm range. The resistance decreases as the illumination level increases.

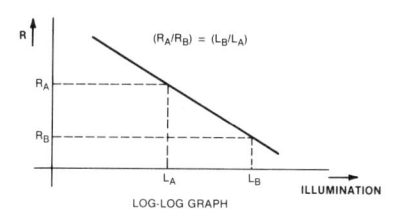

LOG-LOG GRAPH

$(R_A/R_B) = (L_B/L_A)$

Speed of Response of a Photodetector

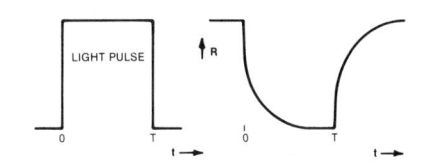

Time constant values: 100 ms (CdS); 10 ms (CdSe); 400 μs (PbS); 10 μs (PbSe).

The rate at which an incident illumination can be turned on and off is determined by the time constant associated with the photodetector. The pulse duration should be at least five time constants.

CdS is the preferred photodectector when the speed of response is *not* important, since it is more sensitive to changes in illumination and also has a higher dark resistance than CdSe.

PHOTOVOLTAIC CELLS

The *photovoltaic cell* is an *active* device and *generates* a voltage when illuminated. It is a pn junction whose depletion region is modified by the incident illumination. The incident illumination leads to the creation of electron-hole pairs at the junction whose migration results in a widening of the depletion region. There is a corresponding increase in the potential difference associated with the depletion region. This *increase*, V_{ph}, is available for driving an external load.

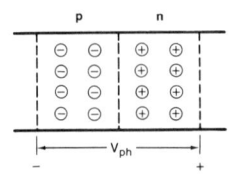

V_{ph} = 0.4 V for a solar cell under bright sunlight and the short circuit current is 5 mA.

It is necessary to connect a number of cells in series to generate a significant amount of power and to match the load to the cell to maximize the transfer of power from the cell to the load.

Commonly Used Photovoltaic Cells

Silicon: responsive to the range of wavelengths 700 nm to 1000 nm.

Selenium: responsive to the range of wavelengths 450 to 700 nm.

PHOTODIODES

Photodiodes operate in the same manner as photovoltaic cells except that they are kept under reverse bias (by means of an external battery). The current I_p is due to the flow of electrons (from p to n) and holes (from n to p) generated by the incident illumination. Photodiodes are used in the *photoconductive mode*.

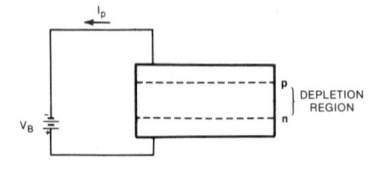

PIN Photodiode

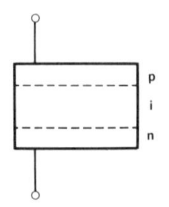

An intrinsic layer is inserted between the p and n regions. The high resistivity of the intrinsic layer makes the depletion region extend far into it. This minimizes the probability of a photon being absorbed before entering the depletion region where it can be effective.

Punch-through voltage is the voltage needed to make the depletion region touch the edge of the n region. Pin photodiodes are usually operated at a reverse bias much higher than the punch-through voltage.

The diode operates as a linear device with a high speed of response.

Silicon photodiodes: response to wavelengths 200 nm to 1100 nm, with peak response at 990 nm. Typical operating voltages: 10 to 110 V. Surface areas: 1 to 10 mm².

Speed of Response of Photodiodes

Typical rise time is 1 ns for pin photodiodes with a load resistance of 50 at a 20-V reverse bias.

Speed of response is affected by such factors as reverse bias (faster under more reverse bias); resistance encountered by the illumination (faster when the resistance is smaller); and the load resistance (faster for smaller load resistance).

Photodiode Amplifier Circuit

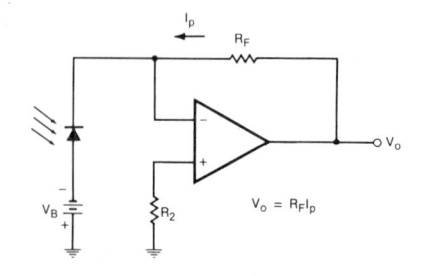

$V_o = R_F I_p$

Electronics: Circuits and Systems

$$V_o = R_F I_P$$

V_o and I_P increase linearly with the level of illumination.

Photodiodes have too low a sensitivity to be useful at low levels of illumination.

PHOTOTRANSISTORS

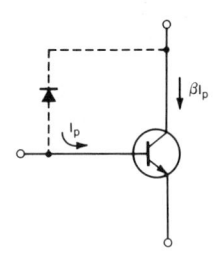

The reverse-biased collector-base junction serves as a photodiode. The transistor provides current gain but the current gain also applies to the dark current and leakage current. The device has only a small region of linearity. The use of phototransistors is limited to areas where linearity is not critical.

Photo-Darlington

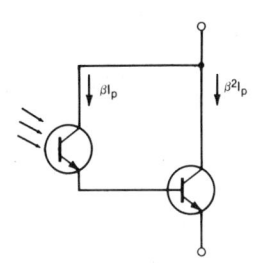

Speed of Response of Phototransistors

Rise time is in the order of several *microseconds* (much slower than photodiodes). The high capacitance between the collector and base and Miller effect are responsible for the slow response.

Photo-FETs

The reverse-biased drain-gate junction of a JFET serves as a photodiode. A gate current flows due to the photodiode action. Active current gains of high values can be attained.

LIGHT-EMITTING DIODES (LEDs)

LEDs are pn junctions that emit optical radiation when forward biased. The radiation is due to the energy released by the recombination of electrons and holes in the pn junction.

GaAs (gallium arsenide): 885 nm

GaP (gallium phosphide) and GaAsP (gallium arsenide phosphide) produce visible radiation.

The wavelength of the radiation produced in GaAsP can be made to vary by changing its chemical composition. Red, green, and yellow LEDs are commercially available.

Radiation efficiency: 1% to 2%. But the luminance available is quite adequate even in brightly lit rooms.

Forward bias: 2 V, with current of 10 to 20 mA.

Response time: 90 to 200 ns.

Drivers for LEDs

npn drivers: active high inputs.
pnp drivers: active low inputs.

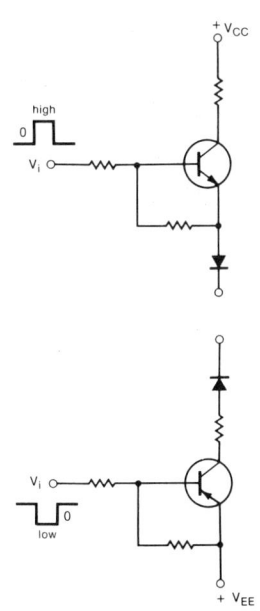

XY Addressable LED Arrays

LEDs can be arranged in the form of a matrix such that any particular LED (or a row or a column) can be selected by sending appropriate signals.

OPTICALLY COUPLED ISOLATORS

An LED is activated by an input signal and the light emitted by the LED activates a phototransistor in an optoisolator. The optoisolator is used as an interface between sensors and control equipment.

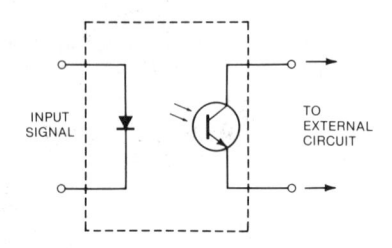

Electronics: Circuits and Systems

ANSWERS TO DRILL PROBLEMS

9.1 (a) 82.3 eV; (b) 5.52 eV; (c) 3.11×10^{-2} eV.
9.2 (a) 1776 nm; (b) 9.94 nm; (c) 1.05×10^5 nm.
9.3 (a) 3360 nm; (b) 4781 nm.
9.4 $V_o = 0.5R_p$ (with R_p in MΩ); (a) 9 V; (b) 0.9 V; (c) 0.09 V.
9.5 Time of exposure $= 1.6R_pC = 0.008R_p$ (with R_p in MΩ). (a) 144 ms; (b) 14.4 ms; (c) 1.44 ms.
9.6 Using five time constants as the minimum pulse duration (a) $t_p = 50$ ms, T = 100 ms, f = 10 Hz; (b) $t_P = 2$ ms, T = 4 ms, f = 250 Hz; (c) $t_p = 50$ μs, T = 0.1 ms, f = 10 kHz.
9.7 $V_o = R_F I_p$. The output is negative but the *magnitude* of V_o increases linearly with I_p.
9.8 High output needed when V_A is between 10 V and 32.5 V. Therefore, V_1 = 32.5 V and V_2 = 10 V. Choosing R_3 = 10 kΩ as before, we get R_1 = 7.5 kΩ, R_2 = 22.5 kΩ, R_3 = 10 kΩ.

PROBLEMS

Basic Principles and Relationships

9.1 Calculate the energy associated with light of the following wavelengths: (a) 900 nm; (b) 550 nm; (c) 2×10^{-6} m.
9.2 Calculate the wavelength in nm corresponding to the following energy levels: (a) 0.7 eV. (b) 1.1 eV. (c) 5×10^{-14} J.
9.3 Given the energy gaps of materials X, Y, and Z to be 0.85 eV, 2.7 eV, and 0.01 eV, respectively. Calculate the maximum wavelength of optical radiation to which they can respond.
9.4 Change the level at which the capacitor in the circuit of Fig. 9-3 activates the exposure control mechanism (so that the shutter closes) to $0.5V_B$. Repeat the calculations of Drill Problem 9.5 for this new situation.
9.5 The circuit of a simple light meter is indicated in Fig. 9-P5, in which the change in the resistance of the photodetector causes a change in the current and the deflection of the meter movement. Use a CdS photodetector with a resistance of 100 MΩ at 0.01 fc. The meter is to be designed so that it can work up to a maximum illumination of 10 fc. The available meter movement has a full scale deflection at 200 μA. Design the light meter.
9.6 Discuss the operation of the circuit shown in Fig. 9-P6. Consider the two cases (a) when there is no light incident on the photodetector; and (b) when there is a sufficient amount of illumination incident on it so as to make it a very low resistance. Show the appropriate waveforms.

580   Optoelectronic Devices

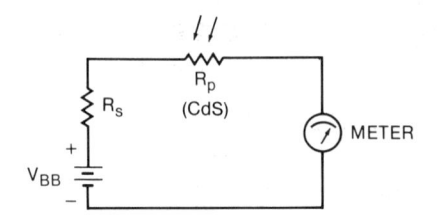

Fig. 9-P5.

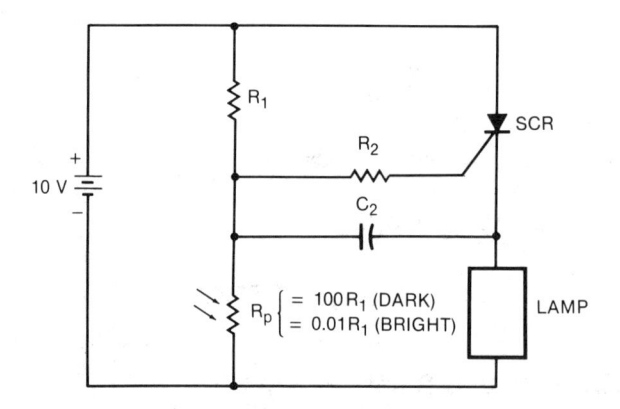

Fig. 9-P6.

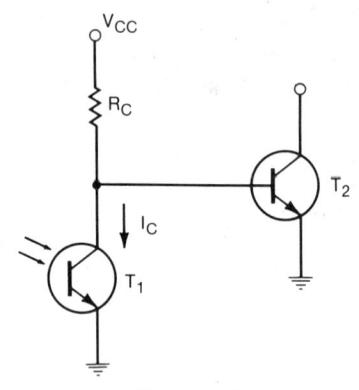

Fig. 9-P7.

9.7   The phototransistor in the circuit of Fig. 9-P7 has the specifications: $I_C = 10$ mA at 2mW/cm^2 and 20 mA at 8 mW/cm^2. Design the circuit so that the transistor T_2 is cut off (that is, its base voltage is significantly less than 0.7 V) when the illumination level exceeds 8 mW/cm^2, and the base voltage of T_2 is $0.5V_{CC}$ when the illumination level reaches 2 mW/cm^2.

Electronics: Circuits and Systems

LEDs

9.8 A phototransistor has the specifications: $I_C = 5$ mA at 2 mW/cm^2 illumination and 20 mA at 12 mW/cm^2. Assume a linear relationship between the phototransistor current and the illumination level (in mW/cm^2). It is required to design a bar graph display that will show the level of illumination incident on the transistor quantized in five equal levels from 2 to 12 mW/cm^2. Assume a power supply of 20 V is available, and the smallest resistance to be used is 1 kΩ. Design the circuit.

CHAPTER 10

LOGIC CIRCUITS

The circuits discussed so far have the property that their output voltages could assume any value between a maximum and a minimum. The output of a BJT amplifier, for instance, may be adjusted to be anywhere between 0 and V_{CC} by applying the proper input voltage. Such circuits are called *analog circuits*, since their output values are analogous to the input levels. There are, on the other hand, circuits designed to operate in *one of two* states: a switch is either on or off, a pair of relay contacts is either open or closed, and an amplifier can be designed so that its output has only two values, high or low. The two digits 0 and 1 are commonly used to denote the two states of such circuits and they are, therefore, called *digital circuits*.

The principles of digital circuits are derived from the theory of mathematical logic, since logical statements also can be only in one of two states: *true* or *false*. Consequently, digital circuits are also called *logic circuits*. The first systematic treatment of logic was done by the British mathematician George Boole in two papers published in the mid-nineteenth century. Claude Shannon applied the concepts of Boolean algebra to the analysis and design of relay switching circuits in 1939 while he was a graduate student at Massachusetts Institute of Technology. His later work and that of others since then have led to the growth of the subject of switching algebra, which is a special class of Boolean algebra.

The application of digital circuits was confined in the early days to relay networks and digital computers. But the advent of IC technology, especially MOSFETs, has made it possible to fabricate extremely complex and powerful digital subsystems in tiny packages. The small size and high versatility of digital ICs have led to the widespread use of digital circuits and systems even in areas that used to be the strongholds of analog circuits. Digital ICs can be classified into four categories: *SSI (small-scale*

Electronics: Circuits and Systems

integration), *MSI (medium-scale integration)*, *LSI (large-scale integration)*, and *VLSI (very large-scale integration)*. SSI circuits contain several gates on 1 chip, MSI circuits contain 10 to 100 gates on a chip and perform complete logic functions, LSI circuits contain more than 100 gates on a chip and perform complex logic functions, and VLSI circuits contain thousands of gates on a single chip and are complete digital subsystems. The trend in modern logic design of digital systems is to use the largest level of integrated circuits to conserve space and simplify the approach to the design. SSI packages are, therefore, usually relegated to the design of logic circuits needed to interconnect the more complex systems. Nevertheless, an understanding of the principles of logic design at the SSI level is important since it forms the basis of the intelligent use of the more complex logic packages.

PRINCIPLES OF LOGIC CIRCUITS

The inputs and outputs of digital circuits are restricted to be at one of two voltage levels, such as for example, 0 volts and 5 volts. The two levels are simply referred to as *low* (abbreviated L) and *high* (abbreviated H) regardless of their actual numerical values. The input-output characteristics of an actual logic circuit are defined in terms of the voltage levels L and H by the designer and the manufacturer.

In logic, the symbols 0 (referred to as logic 0) and 1 (logic 1) are used to denote the two states of a circuit. It is, therefore, necessary to assign the logic levels 0 and 1 to the voltage levels L and H in order to apply the principles of logic to actual digital circuits. The assignment of logic 1 and logic 0 to the voltage levels can follow any one of the three conventions: *positive* logic convention, *negative* logic convention, and *mixed* logic convention.

Positive Logic Convention

The assignment of logic 1 to the high (or H) voltage level and logic 0 to the low (or L) voltage level in a circuit is known as the positive logic convention.

$$H = 1 \text{ and } L = 0 \text{ (positive logic)}$$

Negative Logic Convention

The assignment of logic 0 to the high (H) voltage level and logic 1 to the low (L) level in a circuit is known as the negative logic convention.

$$H = 0 \text{ and } L = 1 \text{ (negative logic)}$$

Mixed Logic Convention

The use of positive logic in some parts of a circuit and negative logic in other parts of the same circuit is known as mixed logic. That is, H is not always logic 1 nor is it always logic 0 in a circuit. Even though this may appear arbitrary and confusing at first, the mixed logic convention leads to a very simple design procedure as will be seen later in this chapter.

It is important to remember that the convention we choose for a particular circuit does not affect the *physical behavior* of that circuit in any way. The response of an actual circuit is completely in terms of the H and L voltage levels. The convention chosen simply alters *our interpretation of the logical function* of the circuit.

Switching Variables

An algebraic symbol (such as A, B, . . ., X, Y, Z) is used for denoting each input and output signal in a digital circuit, and is called a *switching variable*. A switching variable can, at any given instant of time, assume only one of two possible values: 0 or 1.

Basic Operations in Logic

There are three basic operations in switching algebra: *logical sum, logical product*, and *complementation* (or *inversion*). The digital circuits that perform the logical operations are called *gates*. The internal structure of the various gates will not be considered at this point, and our focus will be on their input-output relationships.

Logical Sum (OR Operation)

The operation of addition in logic is governed by the following rules:

$$0 + 0 = 0$$
$$0 + 1 = 1$$
$$1 + 0 = 1$$
$$1 + 1 = 1 \tag{10-1}$$

If X and Y are two switching variables, their *logical sum* is denoted by (X + Y). The operation of the logical sum is also called the *OR operation*, and the OR gate is represented by the symbol in Fig. 10-1. Since each variable has two possible values, 0 or 1, there are four possible combinations of values of X and Y as shown in Table 10-1. The last column of the table shows the values of (X + Y), which were obtained by using the rules in Eq. (10-1).

Electronics: Circuits and Systems

Fig. 10-1. OR gate.

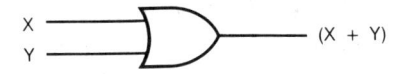

Table 10-1. Truth Table for a Two-Input OR Gate

X	Y	(X + Y)
0	0	0
0	1	1
1	0	1
1	1	1

Table 10-1 and others like it are called *truth tables* since, in logic, a "1" is taken to mean "the statement is true" and a "0" is taken to mean "the statement is false." The truth table of the OR operation (Table 10-1) shows that:

$$(X + Y) = 1$$

if X = 1, or Y = 1, or both X and Y are 1.

$$(X + Y) = 0$$

if, and only if, X and Y are both 0.

The OR operation conforms to the *associative law*:

$$X + (Y + Z) = (X + Y) + Z = X + Y + Z$$

That is, the inputs may be grouped in any manner, and the sum is not affected. The outputs of the three circuits of Fig. 10-2 are identical.

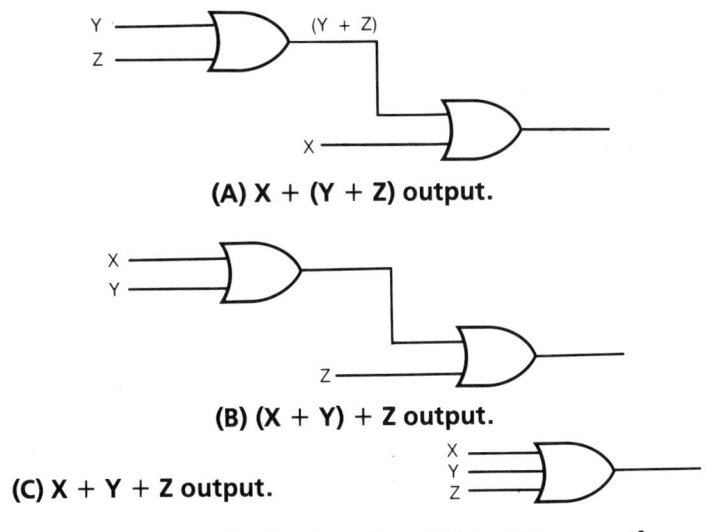

(A) X + (Y + Z) output.

(B) (X + Y) + Z output.

(C) X + Y + Z output.

Fig. 10-2. Associative law is valid in OR operation.

In the case of the logic sum of three or more variables, the output of the OR gates is 0 if, and only if, all the inputs are 0. Otherwise, the output is 1.

If two or more of the inputs of an OR gate are tied to the same variable X, then those inputs act as if they are a single input X, as illustrated in Fig. 10-3.

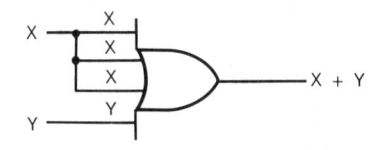

Fig. 10-3. Two or more inputs of an OR gate tied to the same input.

If any of the inputs or an OR gate are tied to a logic 1 level, then the output of the gate is *equal to 1* independent of the values of the other inputs. That is, in logic (1 + anything) = 1.

Drill Problem 10.1: Set up the truth tables for each of the circuits shown in Fig. 10-2 and show that the output T is the same in all three cases.

Logical Product (AND Operation)

The operation of multiplication in logic is governed by the following rules:

$$0 \cdot 0 = 0$$
$$0 \cdot 1 = 0$$
$$1 \cdot 0 = 0$$
$$1 \cdot 1 = 1 \qquad (10\text{-}2)$$

If X and Y are two switching variables, their logical product is denoted by XY. The operation of producing this logical product is also called the *AND operation*, and the AND gate is represented by the symbol in Fig. 10-4. The truth table of the AND operation is shown in Table 10-2. It is seen that (XY) = 1 if, and only if, both variables are 1.

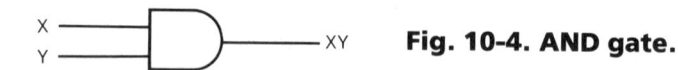

Fig. 10-4. AND gate.

Table 10-2. Truth Table of the AND Operation

X	Y	(X Y)
0	0	0
0	1	0
1	0	0
1	1	1

Electronics: Circuits and Systems

The AND operation conforms to the *associate law*:

$$X(YZ) = (XY)Z = XYZ$$

That is, the inputs may be grouped in any manner, and the product is not affected. The outputs of the three circuits of Fig. 10-5 are identical.

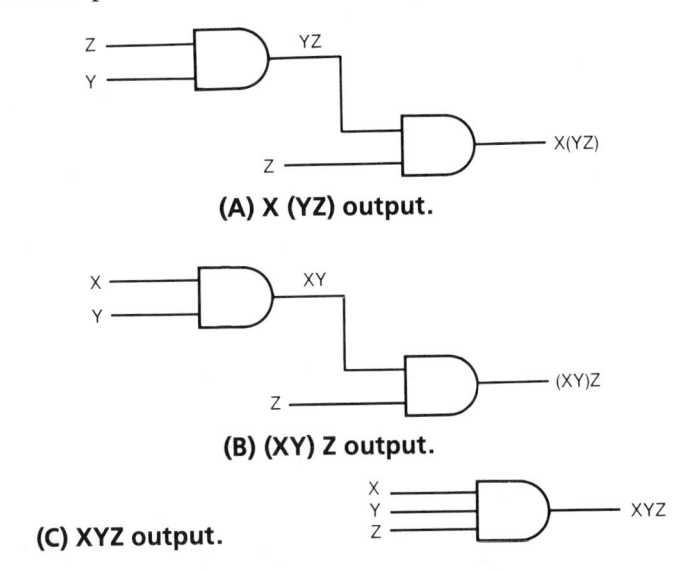

(A) X (YZ) output.

(B) (XY) Z output.

(C) XYZ output.

Fig. 10-5. Associative law is valid in AND operation.

In the case of the logical product of three or more variables, the product is 1, if and only if, *all* the inputs are 1. Otherwise the output is 0.

If two or more of the inputs of an AND gate are tied to the same variable X, then those inputs act as they are a single input X, as illustrated in Fig. 10-6.

Fig. 10-6. Two inputs of an AND gate tied to the same input.

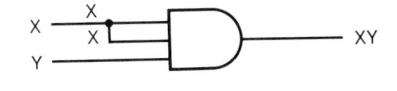

If any of the inputs of an AND gate are tied to a logic 0 level, then the output of the gate is *equal to 0* independent of the values of the other inputs.

Drill Problem 10.2: Set up the truth tables of the circuits of Fig. 10-5 and show that the output T is the same for all three.

Logical Complementation (NOT Operation)

The complement of a switching variable X is denoted by $\overline{X}$ and has the property:

$$\overline{X} = 1 \text{ when } X = 0$$

$$\overline{X} = 0 \text{ when } x = 1$$

That is, $\overline{X}$ assumes exactly the opposite value of X. Logical complementation is also called *negation* or *inversion*. The gate performing complementation is called the NOT gate or *inverter* and has the symbol shown in Fig. 10-7. The small circle (or bubble) denotes complementation in logic diagrams.

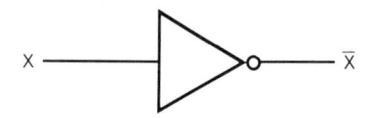

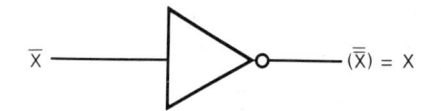

Fig. 10-7. Inverter (NOT gate).

When a variable is subjected to an *even* number of successive complementations, the result is the original variable itself.

Two special relationships involving complementation are of importance:

$$X + \overline{X} = 1$$

$$X\overline{X} = 0 \tag{10-3}$$

Commercially Available OR and AND Gates

Manufacturers of digital circuits *name* the various gates using a *positive* logic convention. Their manuals also give the tables of operations of the various gates in terms of the voltage levels L and H. Packages of OR and AND gates are available with two, three, or four gates per package and are referred to as *dual*, *triple*, and *quad*, respectively. The number of input terminals per gate varies between the packages: each gate in a quad package has two input terminals, each gate in a triple has three input terminals, and each gate in a dual package has four input terminals. When a logic function requires more inputs than available for a gate in a given package, two or more gates can be used as needed.

Electronics: Circuits and Systems 589

Example 10.1
A quad two-input OR package is indicated in Fig. 10-8A. Show how the logic function T = (A + B + C + D) can be realized.

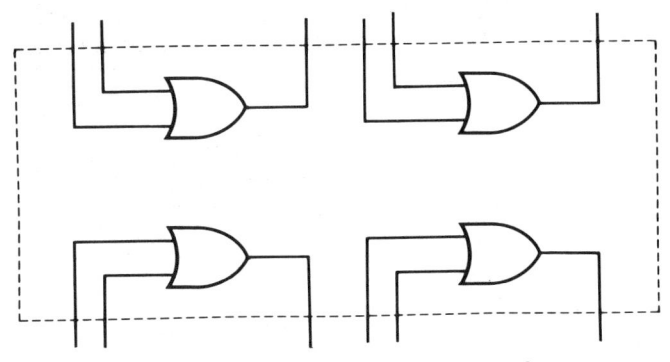

(A) Quad two-input OR gate package.

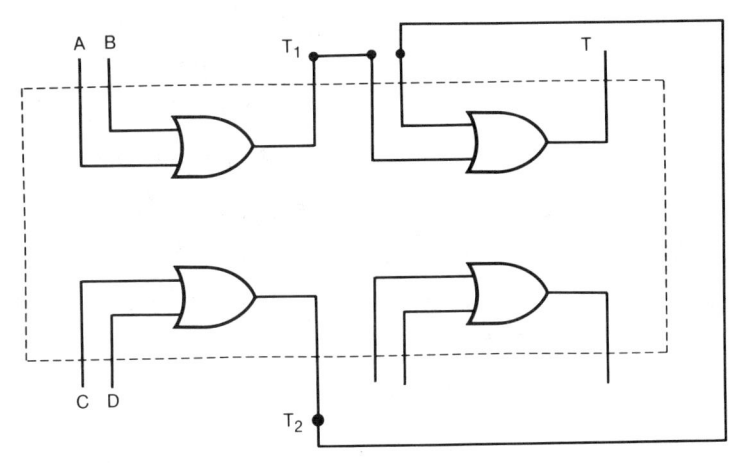

(B) Solution showing how logic function T = (A + B + C + D) can be realized.

Fig. 10-8. OR package for Example 10.1.

Solution
Since each gate can handle only two inputs, we first generate two sums: T_1 = A + B and T_2 = C + D using two of the four gates in the package. Then T_1

and T_2 are added by using one of the remaining gates to obtain T. The arrangement is shown in Fig. 10-8B.

Drill Problem 10.3: A triple three-input AND package is shown in Fig. 10-9. Show the connections needed to produce the logic function T = U V W X Y Z when U, V, W, X, Y, and Z are available as inputs.

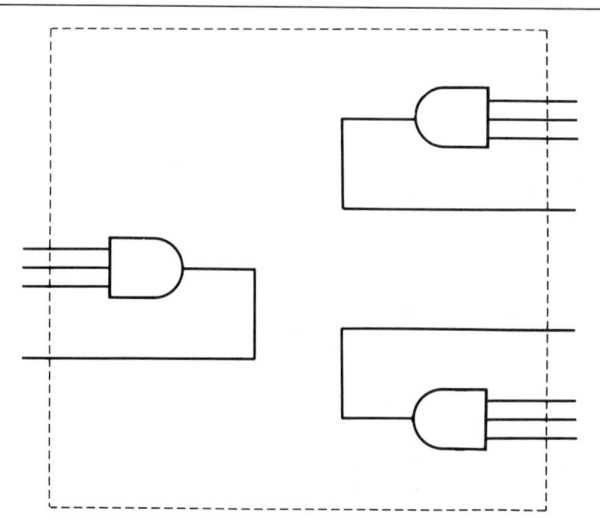

Fig. 10-9. Triple three-input AND gate package.

Example 10.2

Show how the logical sum of five inputs can be obtained by means of a single dual four-input OR package (Fig. 10-10).

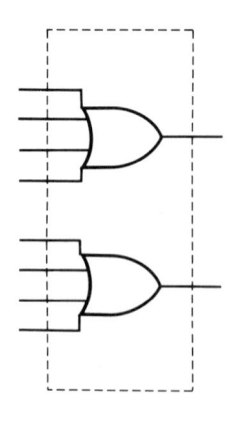

Fig. 10-10. Dual four-input OR gate package.

Electronics: Circuits and Systems

Solution
Either of the two arrangements of Fig. 10-11 can be used. In one case, several inputs of an OR gate are tied together. In the other case, logic 0 is applied to the unused inputs.

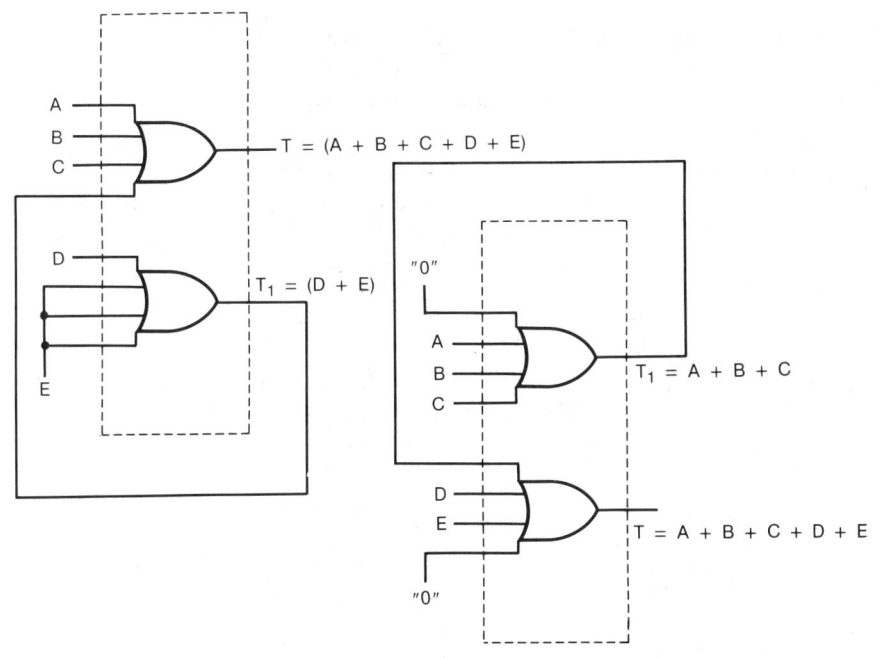

Fig. 10-11. Realization of T = A + B + C + D + E using a dual four-input OR package.

Drill Problem 10.4: Show how a triple three-input AND gate package (Fig. 10-9) can be used to produce the product of four inputs A, B, C, D.

Positive and Negative Logic

The manufacturer's manual gives the table of operations of a gate in terms of the voltage levels L and H. The logical operation of the gate can be viewed through either a positive or a negative logic convention. It will be seen that *changing the convention changes the logical function of a given gate.*

Consider the table of operations of a two-input gate given in Table 10-3A. If the positive logic convention is used to convert the table of operations to a truth table, we obtain Table 10-3B, which is seen to be the truth table of the

OR operation. The given gate is, therefore, a positive OR gate (which is what the manufacturer would call it). If the negative logic convention is used to convert the table of operations to a truth table, we obtain Table 10-3C, which is seen to be the truth table of an AND operation. The given gate is, therefore, a negative AND gate. Therefore, an OR gate can be used as an AND gate by changing the logic convention used. It should be stressed that the *physical operation* of the gate is always as given by Table 10-3A, but the *logical interpretation* of its operation can be changed by choosing one convention or the other.

Drill Problem 10.5: The table of operations of a gate is given in Table 10-4. Convert it to a truth table by using (a) positive logic convention; and (b) negative logic convention. In each case, identify the logic operation being performed.

Table 10-3. Table of Operations of a Two-Input Gate With Positive and Negative Logic Conversion to Truth Tables

X	Y	T
L	L	L
L	H	H
H	L	H
H	H	H

(A)

X	Y	T
0	0	0
0	1	1
1	0	1
1	1	1

(B)

X	Y	T
1	1	1
1	0	0
0	1	0
0	0	0

(C)

Table 10-4. Table of Operations of a Gate for Drill Problem 10.5

X	Y	T
L	L	L
L	H	L
H	L	L
H	H	H

NAND AND NOR OPERATIONS

When a common-emitter BJT or common-source FET amplifier is used in the realization of a logic gate, the output is high when the input is low and the output is low when the input is high. That is, an inversion is automatically introduced in the logic operation performed by such a gate. Therefore, in such circuits, we obtain an AND followed by a NOT or an OR followed by a NOT due to the characteristics of the circuit itself. The *AND followed by*

Electronics: Circuits and Systems

NOT operation is called a *NAND* operation, and the *OR followed by NOT* operation is called a *NOR operation*. NAND and NOR gates are among the most commonly available SSI logic components.

NAND Operations

The NAND function of two variables A and B is denoted by (A ↑ B) and is defined by

$$A \uparrow B = \overline{(AB)} \qquad (10\text{-}4)$$

The truth table of the NAND operation is shown in Table 10-5, and its symbol is shown in Fig. 10-12. It is seen to be the symbol of an AND gate to which a bubble (denoting complementation) is attached.

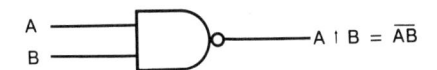

Fig. 10-12. NAND gate symbol.

Table 10-5. Truth Table of the NAND Operation

A	B	$T = \overline{(AB)}$
0	0	1
0	1	1
1	0	1
1	1	0

The NAND function does *not* conform to the associative law. That is, when dealing with more than two variables, the grouping of the variables is important. Consider the three situations shown in Fig. 10-13.

In Fig. 10-13A, the output is given by

$$T_1 = (A \uparrow B) \uparrow C$$
$$= \overline{(AB)} \uparrow C$$

That is, the product AB is inverted first and then NANDed with C. The resulting truth table is shown in Table 10-6A.

In Fig. 10-13B, the output is given by

$$T_2 = A \uparrow (B \uparrow C)$$
$$= A \uparrow \overline{(BC)}$$

That is, the product BC is inverted first and then A is NANDed with $\overline{(BC)}$. The resulting truth table is shown in Table 10-6B.

(A) $T_1 = \overline{(AB)} \uparrow C$.

(B) $T_2 = A \uparrow \overline{(BC)}$.

(C) $T_3 = (A \uparrow B \uparrow C) = \overline{ABC}$.

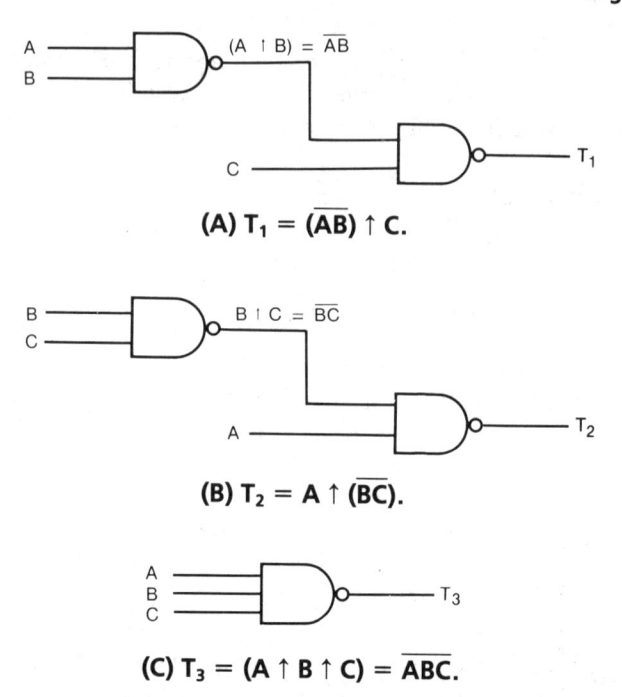

Fig. 10-13. Associative law is *not* valid in NAND operation. The three configurations lead to three different logic functions.

Table 10-6. Truth Tables for the Three Different Logic Functions in Fig. 10-13

A	B	C	$\overline{(AB)}$	$T_1 = \overline{(AB)\uparrow C}$	$\overline{(BC)}$	$T_2 = A\uparrow\overline{(BC)}$	$T_3 = \overline{(ABC)}$
0	0	0	1	1	1	1	1
0	0	1	1	0	1	1	1
0	1	0	1	1	1	1	1
0	1	1	1	0	0	1	1
1	0	0	1	1	1	0	1
1	0	1	1	0	1	0	1
1	1	0	0	1	1	0	1
1	1	1	0	1	0	1	0
				(A)		(B)	(C)

In Fig. 10-13C the output is given by

$$T_3 = A \uparrow B \uparrow C$$
$$= \overline{(ABC)}$$

Electronics: Circuits and Systems

That is, the whole product ABC is inverted. The resulting truth table is shown in Table 10-6C. It can be seen that the three functions T_1, T_2, and T_3 are not identical. Recall that in the case of AND and OR functions, three similar situations (Figs. 10-2 and 10-5) led to the same output.

The property that NAND operations do not conform to the associative law makes it a difficult function to work with algebraically since we cannot manipulate them as freely as AND and OR functions. But, fortunately, the design of a NAND network can be readily accomplished by first using AND and OR gates in a network and using some simple rules, as will be seen later in this chapter.

Use of a NAND Gate as a NOT Gate

If all the inputs of a NAND gate are tied together and are to a single input X, then the output is $\overline{X}$ as can be easily verified from Fig. 10-14.

Fig. 10-14. Use of a NAND gate as an inverter.

Usually, a NAND gate used as an inverter is shown by the simpler symbol of Fig. 10-14B.

Drill Problem 10.6: Obtain the outputs T_c and T_4 of the NAND circuit shown in Fig. 10-15 in the form of a truth table.

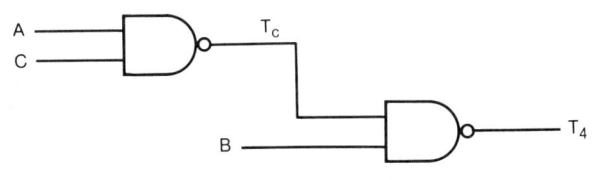

Fig. 10-15. Network for Drill Problem 10.6.

Drill Problem 10.7: Verify that the connection in Fig. 10-14A acts as a NOT gate by setting up the truth table and evaluating the inverted product of the inputs.

NOR Operations

The NOR function of two variables A and B is denoted by (A ↓ B) and defined by

$$A \downarrow B = \overline{(A + B)} \qquad (10\text{-}5)$$

Table 10-7. Truth Table of the NOR Operation

A	B	T = A ↓ B
0	0	1
0	1	0
1	0	0
1	1	0

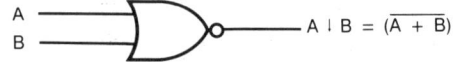

Fig. 10-16. NOR gate symbol.

The truth table of the NOR operation is shown in Table 10-7, and its symbol is shown in Fig. 10-16. It is seen to be the symbol of an OR gate to which a bubble (denoting complementation) is attached.

The NOR function does *not* conform to the associative law. Therefore, the grouping of the variables is important. The situation with NOR is quite similar to that with NAND. The outputs of the three arrangements shown in Fig. 10-17 are not equal to one another as can be verified by setting up the truth tables.

Drill Problem 10.8: Set up the truth table for each of the circuits in Fig. 10-17. In Figs. 10-17A and B, set up a column for the output of gate G_1 and then a column for the output T. Compare the output columns of the three circuits and verify that they are not identical. (In logic functions, equality of two functions requires that they be the same for *all* possible combinations of the inputs.)

Just as the manipulation of NAND functions is made difficult by their nonconformity to associative law, the NOR functions pose the same difficulty. Again, the design of NOR networks is accomplished by an initial design using AND and OR gates and then using some conversion rules.

Use of a NOR Gate as a NOT Gate

If all the inputs of a NOR gate are tied together and are connected to a single input X, then the output is $\overline{X}$ as can be readily verified by setting up the truth table of Fig. 10-18. Usually a NOR gate used as an inverter is shown by the simpler symbol of Fig. 10-18B.

Electronics: Circuits and Systems

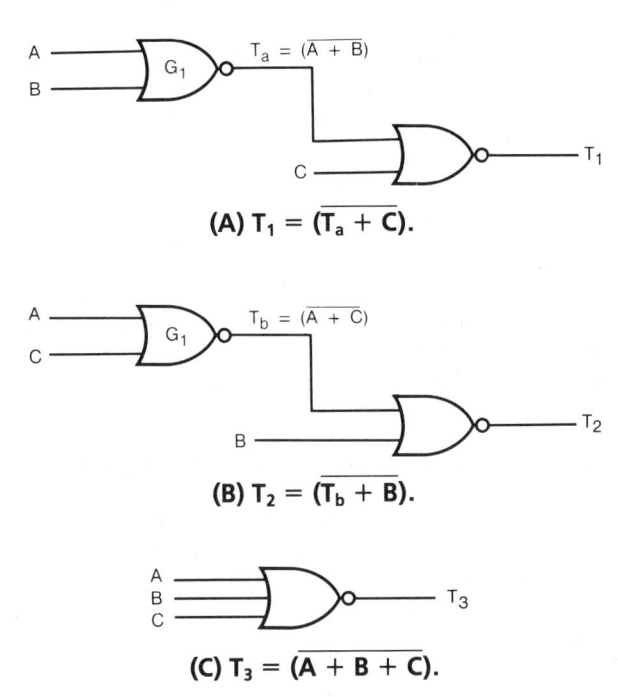

$T_a = \overline{(A + B)}$

(A) $T_1 = \overline{(T_a + C)}$.

$T_b = \overline{(A + C)}$

(B) $T_2 = \overline{(T_b + B)}$.

(C) $T_3 = \overline{(A + B + C)}$.

Fig. 10-17. Associative law is *not* valid in NOR operation. The three configurations lead to three different logic functions.

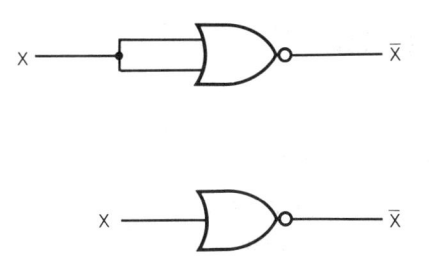

Fig. 10-18. Use of a NOR gate as an inverter.

Drill Problem 10.9: Verify that the connection of Fig. 10-18A acts as a NOT gate by setting up the truth table.

Drill Problem 10.10: The table of operations of a gate is shown in Table 10-8. Set up the truth tables using (a) positive logic convention; and (b) negative logic convention. Identify each logic operation.

Table 10-8. Table of Operations of a Gate for Drill Problem 10.10

A	B	T
L	L	H
L	H	H
H	L	H
H	H	L

Drill Problem 10.11: The table of operations of a gate is shown in Table 10-9. Set up the truth tables using (a) positive logic convention; (b) negative logic convention. Identify each logic operation.

Table 10-9. Table of Operations of a Gate for Drill Problem 10.11

A	B	T
L	L	H
L	H	L
H	L	L
H	H	L

EXCLUSIVE OR AND EXCLUSIVE NOR FUNCTIONS

Exclusive OR Function

The *exclusive OR function* of two variables A and B is denoted by $(A \oplus B)$ and has the property:

$(A \oplus B) = 1$ if $A = 1$ and $B = 0$, or if $A = 0$ and $B = 1$.

That is, the exclusive OR sum of two variables is 1 if, and only if, *one* of the inputs is one (and the other input zero). The truth table of the exclusive OR function is shown in Table 10-10, and its symbol in Fig. 10-19. The symbol is a modified version of the OR gate.

Table 10-10. Truth Table of the Exclusive OR Function

A	B	$A \oplus B$
0	0	0
0	1	1
1	0	1
1	1	0

Fig. 10-19. Exclusive OR gate symbol.

The difference between the exclusive OR and OR functions is for the case when A and B are both 1: $A \oplus B = 0$ but $A + B = 1$.

The exclusive OR function conforms to the associative law. That is, the grouping of the variables is not important. The outputs of the two circuits of Fig. 10-20 are the same as can be seen from the truth tables in Table 10-11.

Table 10-11. Truth Table for the Outputs of Fig. 10-20

A	B	C	$T_a = A \oplus B$	$T_1 = T_a \oplus C$	$T_b = B \oplus C$	$T_2 = A \oplus T_b$
0	0	0	0	0	0	0
0	0	1	0	1	1	1
0	1	0	1	1	1	1
0	1	1	1	0	0	0
1	0	0	1	1	0	1
1	0	1	1	0	1	0
1	1	0	0	0	1	0
1	1	1	0	1	0	1

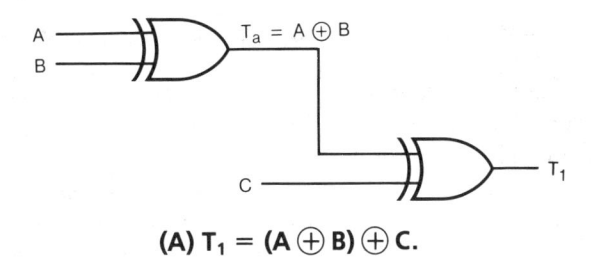

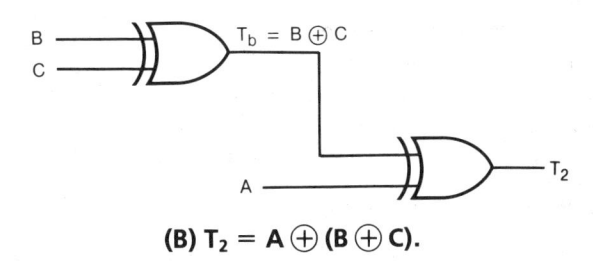

(A) $T_1 = (A \oplus B) \oplus C$.

(B) $T_2 = A \oplus (B \oplus C)$.

Fig. 10-20. Associative law is valid in exclusive OR operation.

Drill Problem 10.12: The circuit of Fig. 10-21 has the output [B $\oplus$ (A $\oplus$ C)]. Set up the truth table with columns for (A $\oplus$ C) and T_3. Verify that the output T_3 is the same as the outputs obtained in Fig. 10-20.

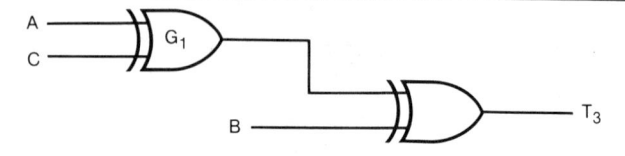

Fig. 10-21. Network for Drill Problem 10.12.

When the exclusive OR sum of three or more variables (Fig. 10-22) is formed, the sum has the following property:

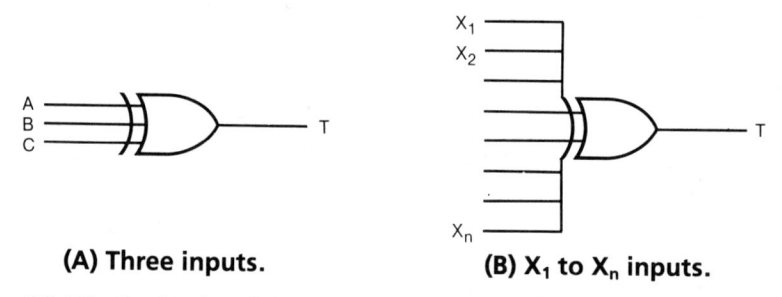

(A) Three inputs.   **(B) X_1 to X_n inputs.**

Fig. 10-22. Exclusive OR operation on three or more variables.

The exclusive OR sum of three or more variables is one if, and only if, an *odd number* of inputs equals one. It is zero if *none* of the inputs is one or an *even number* of inputs is one.

For example, if we consider the exclusive OR sum of seven variables, it will be one when one, or three, or five, or all seven inputs equal one, and will be zero when none, or two, or four, or six inputs equal one. Referring to Table 10-11, it can be seen that the property of the exclusive OR sum is satisfied for the case of three variables.

The property of exclusive OR functions (namely, the sum is one when an odd number of inputs equals one) is useful in performing *odd parity tests* in the transmission of sequences of binary digits. A sequence of digits is said to have an *odd parity* in ones if the number of ones in it is odd, and an *even parity* in ones if the number of ones in it is even. For example,

0 1 0 0 0 1 1 0 1 0 1 1 1 1 0 has an even parity in ones

0 1 1 0 0 1 1 1 1 0 0 1 0 1 1 has an odd parity in ones

By passing the sequence of zeros and ones through an exclusive OR gate (or

Electronics: Circuits and Systems

gates), the parity in ones can be tested. Parity tests are used in the transmission of digital data as a means of checking the occurrence of (a single) error arising during the transmission. If the parity of the sequence at the receiver is different from that at the transmitter, then it is concluded that an error has occurred.

The exclusive OR gate is an important component of adders in arithmetic logic circuits, which will be discussed in a later chapter.

Exclusive NOR Function

The *exclusive NOR function* of two variables A and B is denoted by $A \odot B$ and has the property:

$A \odot B = 1$ if, and only if, $A = B = 0$, or $A = B = 1$

That is, in the case of *two* variables, the function $A \odot B$ is one only when A and B have the same values. The exclusive NOR function of *two* variables is, therefore, called the *coincidence* function. (This definition is *not* valid for three or more variables.)

The symbol of an exclusive NOR gate is shown in Fig. 10-23, which is seen to be the exclusive OR gate to which a bubble has been attached.

Fig. 10-23. Exclusive NOR gate symbol.

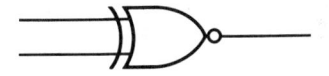

For three or more variables, the exclusive NOR function leads to a value of one if none or an even number of the inputs equals one, and a value of zero if an odd number of the inputs is one.

Drill Problem 10.13: For each of the circuits shown in Fig. 10-24, set up a truth table. In each case, set up a column showing the values of the output of gate G_1 and the output T.

We have discussed a variety of logic operations and gates, and it may at first appear that there is probably too much variety. From a practical point of view, the design of logic networks is usually performed by using the AND, OR, and NOT gates. The networks so obtained can be readily converted to NAND (or NOR) networks by using some simple rules. Therefore, the AND, OR, and NOT gates can be taken as of primary importance. The NAND gates, being very widely used, are also important. The exclusive OR gates may be viewed as special-purpose gates used in certain special situations.

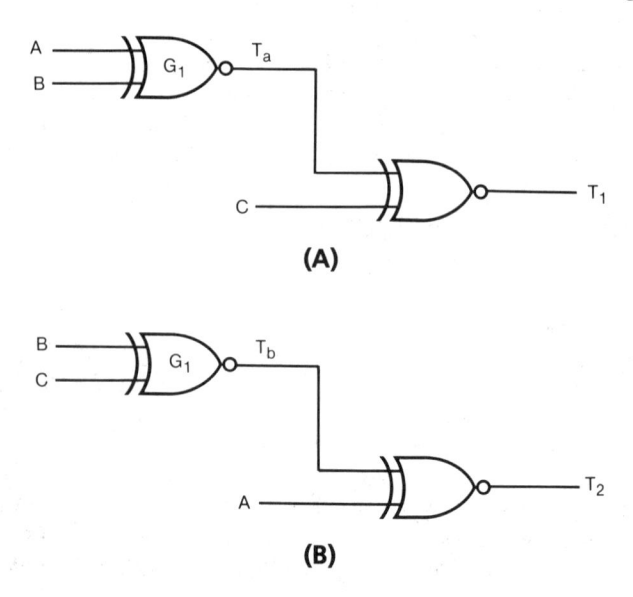

Fig. 10-24. Networks for Drill Problem 10-13.

A summary of the three operations AND, OR, and NOT is given as an aid to the discussion in the following sections.

AND: The output of an AND gate is one if, and only if, all its inputs are one. Otherwise, the output is zero.

OR: The output of an OR gate is one if one or more of the inputs are one. It is zero only when all its inputs are zero.

NOT: The output of a NOT gate is opposite in value to the input. A NOT gate can have *only a single input,* whereas, AND and OR gates have two or more inputs.

TWO-LEVEL POSITIVE LOGIC NETWORKS

The logic network has a number of levels, with the output gate(s) representing level 1 and the other levels numbered as we proceed step by step toward the input. Fig. 10-25A shows the level-numbering scheme for a logic network. The two-level network (Fig. 10-25B) is probably the most commonly used structure since it offers some advantages. The design of a two-level network for a given logic function is straightforward. In a two-level network, each signal passes through the same number of gates thus producing equal delay, which prevents problems that might occur due to unequal delays. Also, the number of levels is only two, and the delay is shorter than for networks with three or more levels.

Electronics: Circuits and Systems

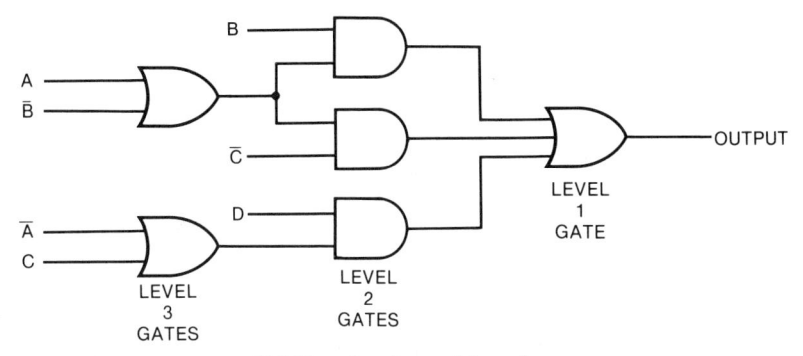

(A) Numbering of levels.

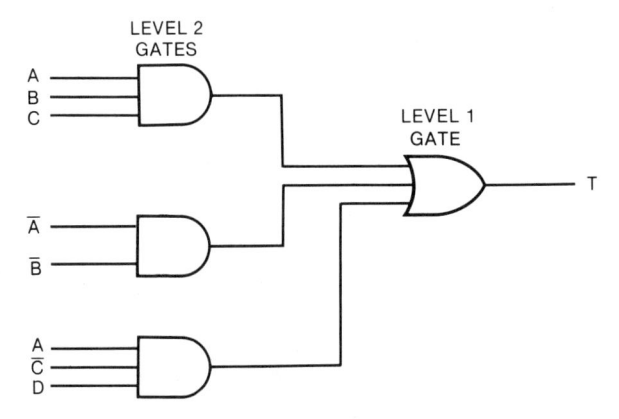

(B) Two-level (sum-of-products) logic network.

Fig. 10-25. Multiple-level logic networks.

Analysis of Two-Level Positive Logic Networks

The *analysis* of a logic network involves the determination of the output function (or functions) of a given network. The output function may be in the form of an algebraic switching expression or a truth table. The procedure of analysis consists of starting from the input level, identifying the output of each gate, and working through the different levels to the output terminal.

Example 10.3

Determine the output function of the circuit in Fig. 10-26 and set up the truth table.

604   Logic Circuits

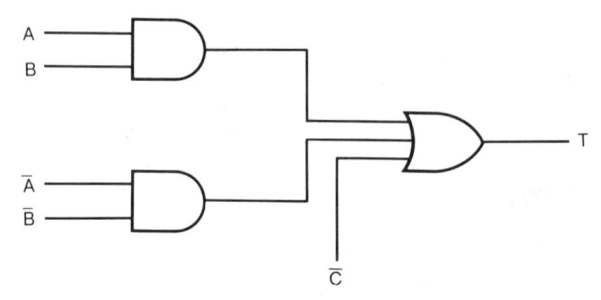

Fig. 10-26. Network for Example 10.3.

Solution

Starting at level 2, the output of each gate is first labeled as indicated in Fig. 10-27 and then the output of the gate in level 1 is determined. The output is seen to be

$$T = AB + \overline{A}\overline{B} + \overline{C}$$

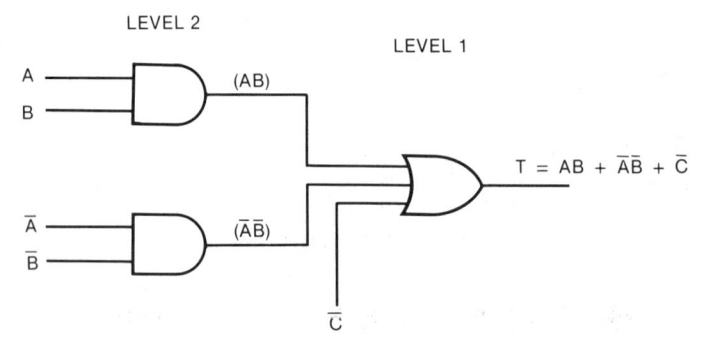

Fig. 10-27. Analysis of the network of Example 10.3.

Table 10-12. Truth Table for Example 10.3 Solution

A	B	C	AB	$\overline{A}\overline{B}$	$\overline{C}$	T
0	0	0	0	1	1	1
0	0	1	0	1	0	1
0	1	0	0	0	1	1
0	1	1	0	0	0	0
1	0	0	0	0	1	1
1	0	1	0	0	0	0
1	1	0	1	0	1	1
1	1	1	1	0	0	1

Electronics: Circuits and Systems

The truth table is shown in Table 10-12, where separate columns have been set up for the three terms in T for the sake of convenience.

Drill Problem 10.14: Determine the output function of each of the circuits in Fig. 10-28 and set up its truth table.

Design of Two-Level Positive Logic Networks

The basic principles of the design of a logic network will be illustrated through an example.

Example 10.4

Three judges of a contest cast votes of "yes" or "no" on each contestant. A candidate who receives a majority of yes votes will be selected as one of the finalists. Design a logic network whose inputs are the judges' votes and whose output will indicate if a contestant has received a majority of yes votes.

Solution

The first step in design is to identify the switching variables and define what logic 0 and logic 1 mean for the variables. In the present problem, there will be three input variables, which will be denoted by A, B, C, corresponding to the three judges. Let logic 1 imply that the vote is "yes" and logic 0 stand for a "no" vote. When the output $T = 1$, it will be taken to mean that the candidate got a majority of "yes" votes.

The next step is to set up the truth table that translates the desired operation to logic values of the output: under what combinations of values of the inputs should T be one? For three inputs, there are 2^3 or eight possible combinations of values, as shown in Table 10-13. According to the statement of the problem, T should have a value of one for any row in which two or more inputs have a value of one.

A logic function, such as the output T of the network being designed, can have only one of two values, zero or one. In fact, the fundamental postulate of switching algebra states that if a switching variable (or function) is not one it must be zero and vice versa. This property can be used in the design of a logic network provided the following rules are observed. Consider *each row of the truth table for which the logic function T equals one.* Set up a product term (using the procedure to be presented later) for each such row. Then, the function so obtained will be one only for those combinations of values of the inputs for which T is one in the truth table; and the function will be zero for all other combinations of values of the inputs in the truth table.

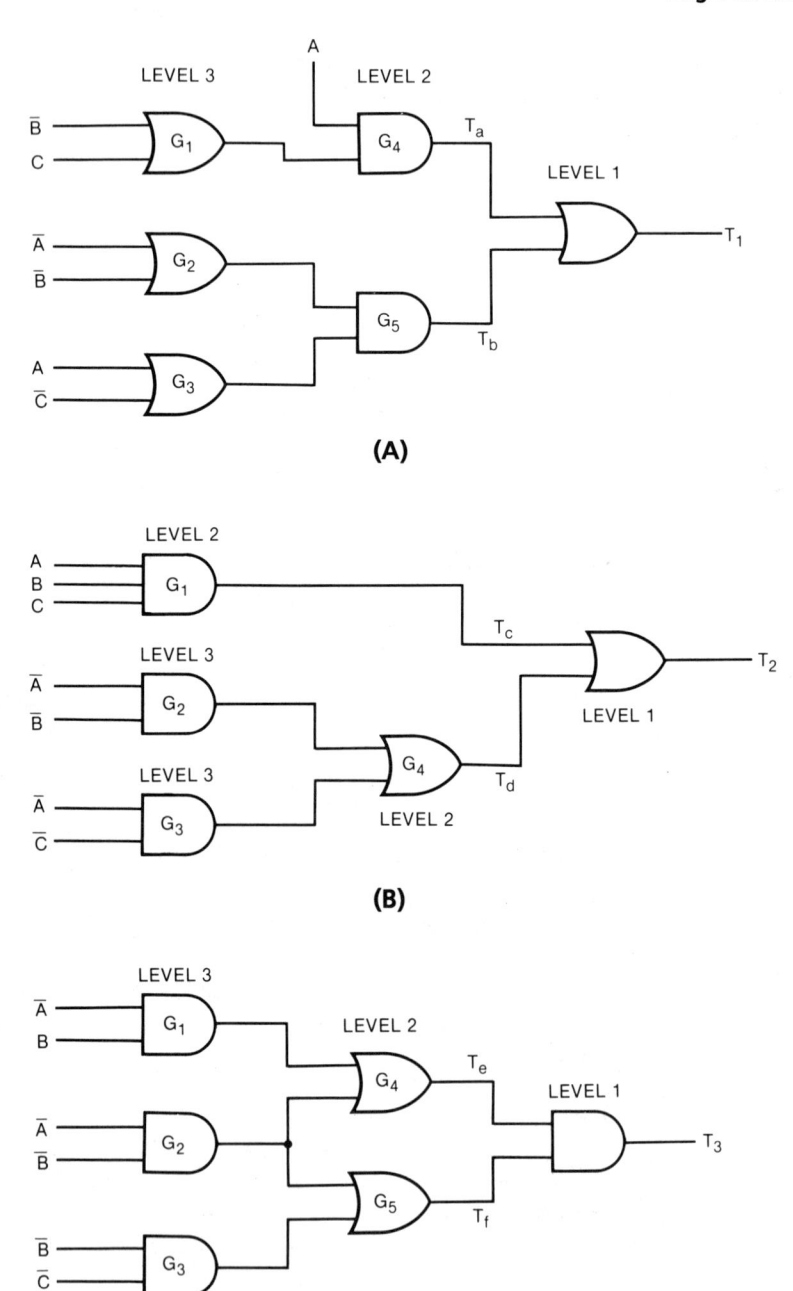

Fig. 10-28. Networks for Drill Problem 10.14.

Table 10-13. Truth Table for Three Inputs for Example 10.4 Solution

A	B	C	T
0	0	0	0
0	0	1	0
0	1	0	0
0	1	1	1
1	0	0	0
1	0	1	1
1	1	0	1
1	1	1	1

For the present problem, we have T = 1 for the following combinations of values of the inputs.

$T = 1$ when $A = 0$, $B = 1$, $C = 1$. This requires a term involving the three variables (or their complements), which has a value of 1 when $A = 0$, $B = 1$, and $C = 1$. The product term $\bar{A}BC$ can be seen to have a value of 1 if $A = 0$ (which corresponds to $\bar{A} = 1$), $B = 1$, and $C = 1$. Therefore, $\bar{A}BC$ is one of the terms in the output function of the present problem. (Fig. 10-29A.)

$T = 1$ when $A = 1$, $B = 0$, and $C = 1$. This requires the product term $A\bar{B}C$, which can be seen to have a value of 1 when $A = 1$, $B = 0$, and $C = 1$. (Fig. 10-29B.)

$T = 1$ when $A = 1$, $B = 1$, and $C = 0$. This requires the product term $AB\bar{C}$. (Fig. 10-29C.)

$T = 1$ when $A = 1$, $B = 1$, and $C = 1$. This requires the product term ABC. (Fig. 10-29D.)

The output function of the network being designed should have a value of one when *any* of the four product terms obtained previously equals one. Therefore, the four product terms should be combined through an OR operation to obtain the function T. The logic network is set up as follows: use an AND gate to produce each of the product terms obtained before, and combine the outputs of the AND gates through an OR gate. (Fig. 10-29E.) The output of the OR gate is the desired function.

Drill Problem 10.15: Consider each of the four product terms obtained in the previous example. For each term, show that it has a value of zero except for the one combination of values of the inputs for which it was chosen. This property of each product term obtained from a truth table guarantees that the output function obtained in the previous procedure will be zero except for the four conditions for which it should be one.

Drill Problem 10.16: *Analyze* the network obtained in the previous example (Fig. 10-29), and verify that it performs the specified function.

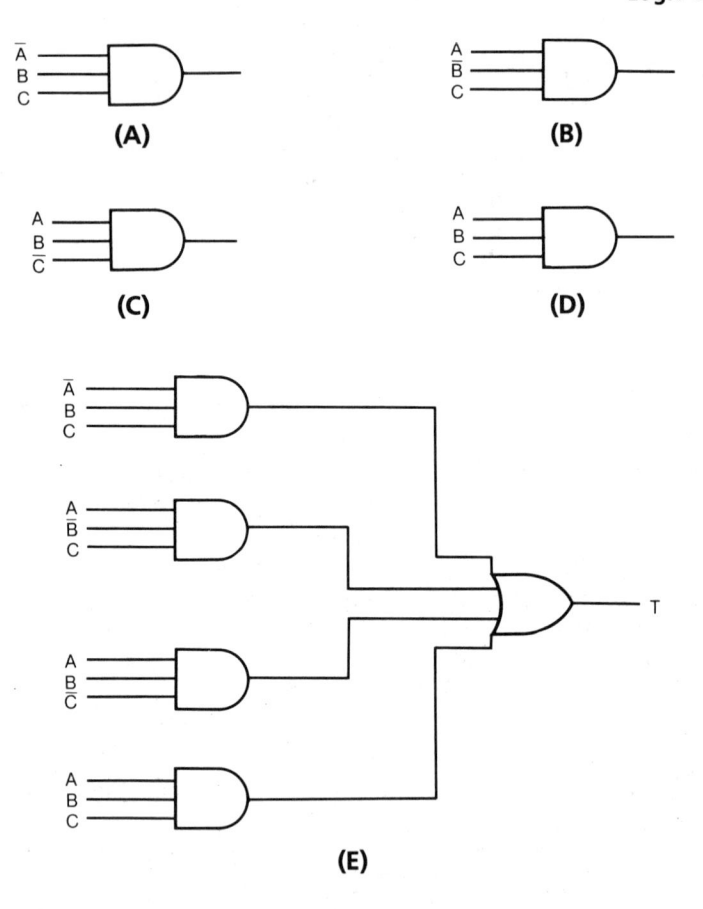

Fig. 10-29. Design of a two-level logic network of Example 10.4.

The design procedure can be summarized as follows:
1. Define the input variables and the logic 1 and logic 0 assignments for each. Define the output variable in a similar manner.
2. Set up the truth table. Translate the problem statement to decide for which row of the truth table the output should be one and for which row it should be zero. (There may occur cases in which the output for a particular row is not defined because the particular set of input conditions may not actually occur. These may be ignored.)
3. For each row in which the output is one, write a product term of the input variables (or their complements) such that the product is one for the particular set of values of the input variables in that row. The rule to be used in forming the product term is

If the input variable is one, then use the *uncomplemented* variable (A, B, C, etc.);

If the input variable is 0, then use the *complemented* variable ($\overline{A}$, $\overline{B}$, $\overline{C}$, . . .).

4. Set up an AND gate for each product term and connect the outputs of the AND gates to an OR gate. The output of the OR gate is the desired output.

Example 10.5

Two switches A and B independently control the operation of a light. A third switch C acts as a master switch in such a way that whenever it is on, the light can never go on. When the master switch is off, the light goes on if either A or B or both are on. Design the logic network needed to control the operation of the light.

Solution

The three input variables are already specified. Let a logic 1 represent the "on" condition and logic 0 the "off" condition for the inputs and the output T.

The output must be 0 whenever C = 1. For the other rows, T = 1 when A = 1 or B = 1 or both are 1. The truth table is as shown in Table 10-14.

Table 10-14. Truth Table for Example 10.5 Solution

A	B	C	T
0	0	0	0
0	0	1	0
0	1	0	1
0	1	1	0
1	0	0	1
1	0	1	0
1	1	0	1
1	1	1	0

The following product terms are obtained for the rows in which T = 1.

$\overline{A}B\overline{C}$ for the row A = 0, B = 1, C = 0

$A\overline{B}\overline{C}$ for the row A = 1, B = 0, C = 0

$AB\overline{C}$ for the row A = 1, B = 1, C = 0

The logic network is shown in Fig. 10-30.

Drill Problem 10.17: Design a logic network with two inputs A and B, whose output is one whenever A equals B and zero whenever A is not equal to B.

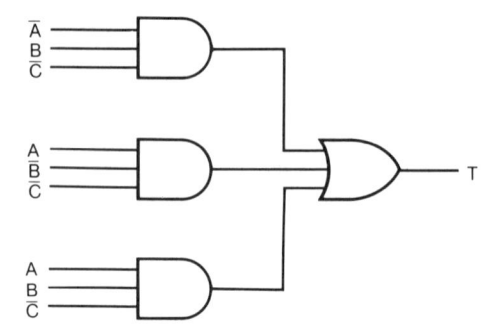

Fig. 10-30. Network of Example 10.5.

Drill Problem 10.18: Three persons are available for a mission: Peter, Paul, and Mary. The following constraints are enforced on the selection of the team that will actually go. Either both Peter and Paul go, or neither goes. Mary must be definitely part of the team. Obtain a logic network that will give an output of one whenever the correct combination of participants is chosen.

Drill Problem 10.19: Obtain a two-level logic network with three inputs A, B, C, and an output T such that T = 1 whenever an odd number of inputs has a value of one and zero otherwise. (This network is of course what a three-input exclusive OR gate may contain.)

Two-Level Sum of Products Networks

The networks designed in the form of a two-level network with a number of AND gates in level 2 and an OR gate in level 1 are also called *two-level sum of products networks* since the output functions of such networks will be in the form of a sum of products. There are other structures available for a logic function but the two-level network can be obtained methodically as outlined previously. Minimization procedures can also be used for two-level networks so as to minimize the number of gates. But the reduction of the number of gates *per se* has lost its importance since the advent of IC logic families. In using ICs, it is much more important to reduce the number of *chips* rather than the number of gates. (A chip represents a single package containing a number of gates.) There is no methodical approach to the minimization of the number of chips for a particular design, and only a trial-and-error procedure will work in the case of simple networks. For complex networks, it is more efficient to use MSI and LSI packages (e.g., multiplexers, programmable logic arrays, and ROMs) rather than use SSI gate packages.

TWO-LEVEL NAND NETWORKS

When a logic network has been designed in a two-level sum of products form, it can be converted to a two-level NAND network easily. *Every* gate in the two-level sum of products network (the AND gates as well as the OR gate) is replaced by a NAND gate. The inputs to the level-2 gates are left unchanged. (If there is an input signal *directly* connected to the level-1 gate, that input is replaced by its complement.)

The NAND networks for Examples 10.4 and 10.5 are shown in Fig. 10-31.

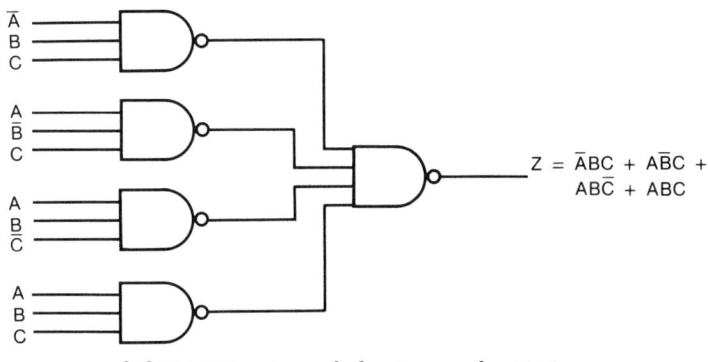

(A) NAND network for Example 10.4.

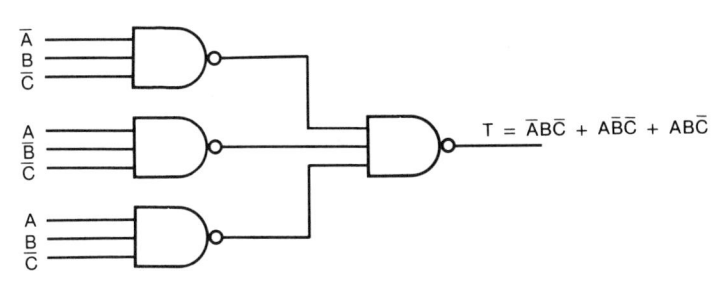

$Z = \bar{A}BC + A\bar{B}C + AB\bar{C} + ABC$

(B) NAND network for Example 10.5.

$T = \bar{A}B\bar{C} + A\bar{B}\bar{C} + AB\bar{C}$

Fig. 10-31. Two-level NAND networks.

The formal proof of the validity of the previous procedure is beyond the scope of the present treatment. But the fact that the outputs of the two networks in Fig. 10-31 are the same as those of the networks obtained earlier can be established by analysis and setting up the truth tables.

Drill Problem 10.20: Obtain the NAND networks of the logic functions of Drill Problems 10.17, 10.18 and 10.19.

MIXED LOGIC NETWORKS

The design procedures of the last two sections assumed a positive logic convention. The use of mixed logic convention where both positive and negative logic conventions are used in the same network offers some advantages especially when using certain types of gates.

All signals in an electronic switching network are at one of two voltage levels: high (H) or low (L). If a particular signal in a network is treated as *logic 1 when it is high*, it is said to have an *active high assignment*. When a particular signal in a network is treated as *logic 1 when it is low*, it is said to have an *active low assignment*.

$$H = 1 \quad \text{active high assignment}$$
$$L = 1 \quad \text{active low assignment}$$

If all the signals in a network are active high, then positive logic convention is being used. If, on the other hand, all signals in a network are active low, then negative logic convention is being used. If some signals in a network are active high and the others are active low, then mixed logic convention is being used. The attractive aspect of mixed logic design is that the same gate can be made to perform different logic operations by varying the active-high and active-low assignments of its inputs and outputs.

The notation to identify which signals are assigned active high and which active low involves the placing of a ".H" or a ".L" after the symbol (or name) of a variable. For example, X.H denotes a signal X, which is active high, while B.L denotes a signal B, which is active low.

The logical operation of a device depends upon whether its inputs are active high or active low, and whether its output is active high or active low. A *triangular polarity mark* will be placed on a terminal when the signal associated with it is *active low*. The absence of a triangular polarity mark will mean that the signal associated with that terminal is active high.

Examples of the notation are shown in Fig. 10-32.

A given gate can be used as an AND gate and as an OR gate by a suitable selection of the active levels of its inputs and outputs, as shown in the following example.

Example 10.6

The table of operations of a gate is given in Table 10-15. Use the following mixed logic assignments and obtain the corresponding truth tables: (a) both inputs active high but output active low; (b) both inputs active low but output active high. Identify the logical function for each assignment.

Electronics: Circuits and Systems

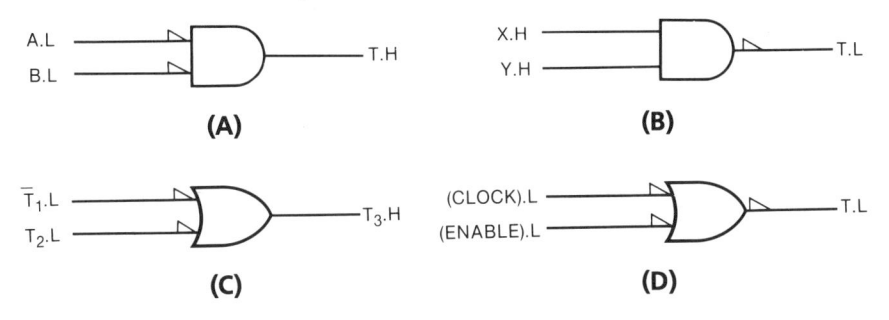

```
A.L ──┐
      ├──D──── T.H
B.L ──┘
      (A)

X.H ──┐
      ├──D▷──── T.L
Y.H ──┘
      (B)

T̄₁.L ──┐
       ├──D)──── T₃.H
T₂.L ──┘
      (C)

(CLOCK).L ──┐
            ├──D)▷──── T.L
(ENABLE).L ─┘
      (D)
```

Fig. 10-32. Notation for mixed logic networks. ".L" denotes an active low assignment and ".H" an active high assignment. The triangular polarity mark attached to any terminal of a gate symbol means that an active low input must be fed to it, or the output from the terminal should be treated as active low.

Table 10-15. Table of Operations of a Gate for Example 10.6

A	B	T
L	L	H
L	H	L
H	L	L
H	H	L

Solution

The truth table for (a) is obtained by making $H = 1$, $L = 0$ for the input columns but $H = 0$, $L = 1$ for the output column. The truth table for (b) is obtained by using the opposite assignments. The resulting truth tables are shown in Table 10-16.

Table 10-16. Truth Tables for Example 10.6

(A) A.H, B.H, T.L:

A	B	T
0	0	0
0	1	1
1	0	1
1	1	1

(B) A.L, B.L, T.H:

A	B	T
1	1	1
1	0	0
0	1	0
0	0	0

It is seen that for assignment (a) the given gate acts as an OR gate and for assignment (b) it acts as an AND gate. The given gate is a positive NOR gate as can be easily verified by using positive logic convention. This example shows that a positive NOR gate can be used either as an AND gate or as an

OR gate in mixed logic. When a gate is used in this (altered) manner, we use the conventional AND and OR symbols for the gate to indicate what *function it is performing*. Triangular polarity marks are placed at the active low terminals in the symbols to indicate the active levels of the signals at the different terminals. The symbols for the present case are shown in Fig. 10-33.

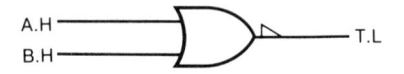

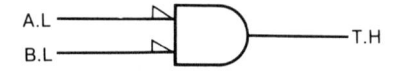

Fig. 10-33. Use of a positive NOR gate to perform OR and AND operations in mixed logic.

Drill Problem 10.21: The table of operations of a gate is shown in Table 10-17. Verify that it is a positive NAND gate by setting up a truth table with positive logic convention. Use the following mixed logic assignments and obtain the truth tables: (a) both inputs active high, output active low; (b) both inputs active low, output active high. Identify the logic function in each case.

Table 10-17. Gate Operations for Drill Problem 10.19

A	B	T
L	L	H
L	H	H
H	L	H
H	H	L

The aim of mixed logic design is to use any given gate as an AND gate and as an OR gate even when the given gate is a positive NAND or a positive NOR gate. Since NAND and NOR gates are used commonly in logic design but are difficult to work with from a design viewpoint, converting them to AND and OR gates through mixed logic assignments eases the design task considerably. A given logic expression can be converted to a network quickly and easily through mixed logic design even when NAND and NOR gates are being used.

To use mixed logic in the previous manner, one important requirement

Electronics: Circuits and Systems

must be met: the input connected to a terminal of a gate must be compatible with the presence or absence of the triangular polarity mark at that terminal. That is, an *active low signal* must be connected to a terminal *with a triangular polarity mark*, and an *active high signal* must be connected to a terminal *without a triangular polarity mark*. Only when this *compatibility requirement* is met can we treat the network as an interconnection of AND and OR gates.

Consider the OR gate shown in Fig. 10-34. Suppose the available input signals are A.H and B.L. Then, the input A should *not* be directly connected to one of the inputs but we need to obtain A.L first. Therefore, the available A.H must be converted to A.L before connecting the A input to the gate's input terminal. Such a change is accomplished through a voltage level changer. The voltage level changer is the same as the logic inverter (NOT gate). An inverter has the table of operations given in Table 10-18.

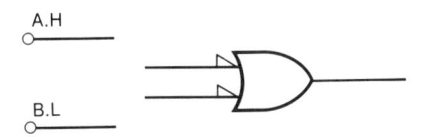

Fig. 10-34. Compatibility question in mixed logic.

Table 10-18. Table of Operations for an Inverter

Input X	Output Y
H	L
L	H

If the mixed logic assignment X.H. and Y.L is used for the inverter, the resulting truth table is as shown in Table 10-19.

Table 10-19. Truth Table for an Inverter With the Mixed Logic Assignment X.H and X.L

X.H	Y.L
1	1
0	0

With the previous assignment, the input and output of the gate are *logically equal* to each other but their active levels are opposite. The inverter can, therefore, be used as a level changer: an active high input is changed to an active low output and an active low input is changed to an active high output but the logical value is not changed in either case. The mixed logic symbol of the voltage-level changer is shown in Fig. 10-35.

X.H ▷— X.L

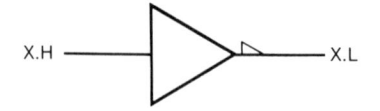

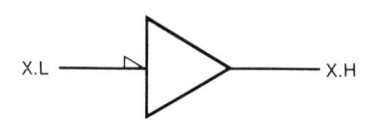

X.L ▷— X.H

Fig. 10-35. Voltage level changer.

Returning to Fig. 10-34, we insert a voltage level changer between A.H and the input of the OR gate to satisfy the compatibility requirement. The resulting connection is shown in Fig. 10-36.

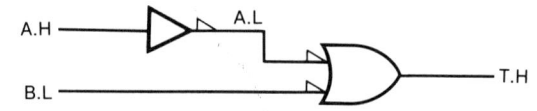

A.H ▷ A.L
B.L ───────── T.H

Fig. 10-36. Proper connection of the inputs of Fig. 10-34.

Design Procedure

The two gates that will be used in the following discussion of mixed logic design are

the 7400 package that contains four 2-input (positive) NAND gates

and

the 7402 package, which contains four 2-input (positive) NOR gates.

A single gate in a package will be labeled as ¼ 7400 or ¼ 7402.

The same gates can be used as voltage level changers also by tying their inputs together. Thus, each voltage level changer will also be a ¼ 7400 or a ¼ 7402 gate.

(Other gates are available and can be used in mixed logic design but we are confining ourselves to the 7400 and 7402 for convenience.)

For convenience of reference, Fig. 10-37 shows the mixed logic assignments for the 7400 and 7402 gates. (Note that these are the results obtained in Example 10.6 and Drill Problem 10.19.)

The design procedure will be illustrated by means of two examples.

Example 10.7

Design a logic network to obtain the function $T = A B + A C$ under the constraint that the available inputs are A.L, B.H, C.H. The desired output is T.L. Use 7400 gates.

Electronics: Circuits and Systems

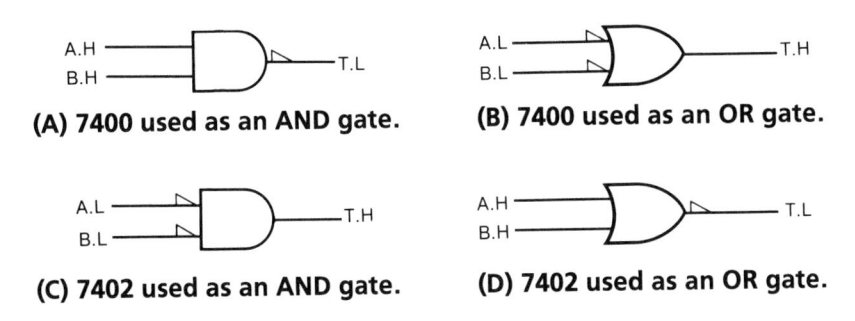

(A) 7400 used as an AND gate. Inputs A.H, B.H; output T.L

(B) 7400 used as an OR gate. Inputs A.L, B.L; output T.H

(C) 7402 used as an AND gate. Inputs A.L, B.L; output T.H

(D) 7402 used as an OR gate. Inputs A.H, B.H; output T.L

Fig. 10-37. Mixed logic assignments of 7400 and 7402 gates.

Solution

Using Fig. 10-37 as a reference, two gates are set up to form the products AB and AC. A voltage-level changer is necessary for the A input. This step is shown in Fig. 10-38A. The outputs of these gates are then fed into a *7400* gate connected as an OR gate. No level changers are necessary since the compatibility requirement is already met. The final network is shown in Fig. 10-38B.

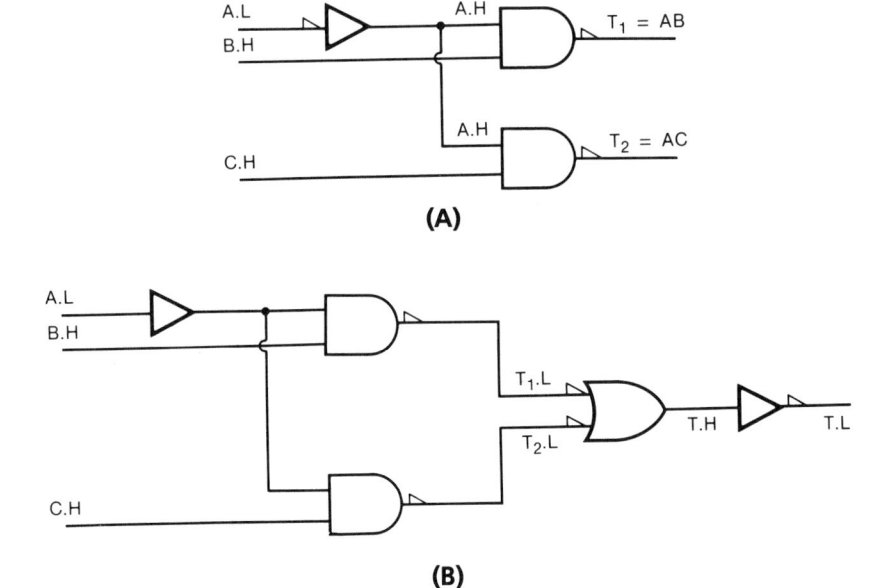

(A) shows A.L/B.H through inverter producing A.H, combined with A.H to produce $T_1 = AB$; and A.H with C.H produces $T_2 = AC$.

(B) Final network with inputs A.L, B.H, C.H; intermediate signals $T_1.L$ and $T_2.L$ feeding OR gate producing T.H, then inverter to T.L.

Fig. 10-38. Design of a mixed logic network of Example 10.7.

Example 10.8

Redesign the network of the previous example using 7402 gates.

Solution

The procedure is exactly the same as before, and the design is shown in Fig. 10-39.

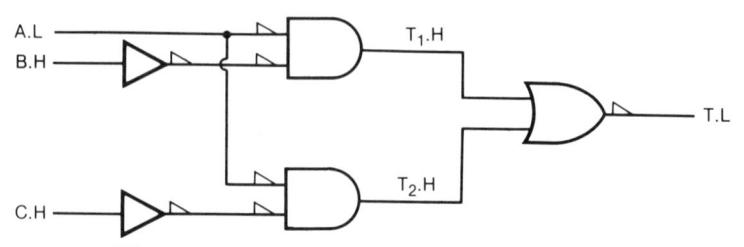

Fig. 10-39. Network of Example 10.8.

Drill Problem 10.22: Design a network to obtain the function $T = A\bar{B} + \bar{A}B$ using 7400 gates. Assume that the inputs A, $\bar{A}$, B, $\bar{B}$ are available: A.H, $\bar{A}$.L, B.H, $\bar{B}$.L. The desired output is T.H.

Drill Problem 10.23: Redesign the network of the previous drill problem with 7402 gates.

Drill Problem 10.24: Design the proper network to obtain the function $T = AB + \bar{A}\bar{B}$ using 7400 gates. Assume the available inputs to be the same as in Drill Problem 10.28. The output may be active high or low.

Drill Problem 10.25: Redesign the network of the previous example using 7402 gates.

Example 10.9

Obtain a realization of the function $T = A + B + C + D$ using (a) 7400 gates; and (b) 7402 gates. All inputs are available active high. The output may be active high or low.

Solution

Since each gate has only two inputs, the given function is first subdivided in the form

$$T = (A + B) + (C + D)$$

The network in each case will have the basic structure: two OR gates in level 2 for producing (A + B) and (C + D), and another OR gate in level 1 to produce T.

Electronics: Circuits and Systems

In each network, voltage level changers are inserted as needed to satisfy the compatibility requirement: the output appearing at a terminal with a triangular polarity mark can be connected only to an input with a triangular polarity mark.

The two solutions are shown in Fig. 10-40.

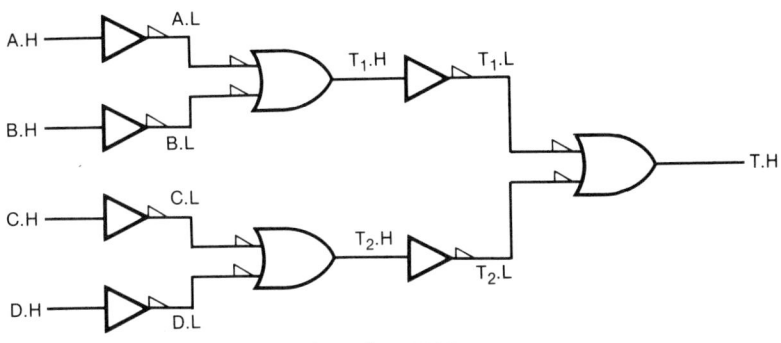

(A) Using the 7400 gates.

(B) Using the 7402 gates.

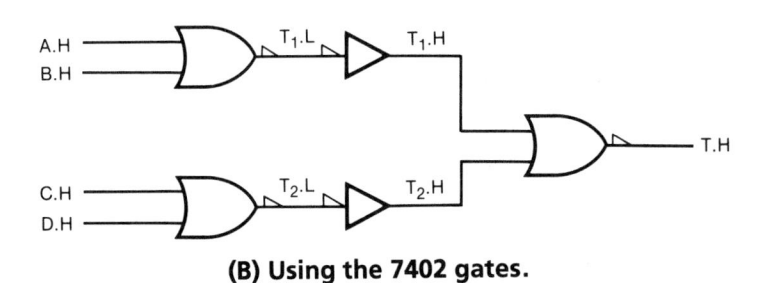

Fig. 10-40. Realization of T = A + B + C + D in mixed logic of Example 10.9.

Drill Problem 10.26: Assuming that all available inputs are active low, design a network using 7400 gates to realize the function T = ABCD.

Drill Problem 10.27: Redesign the network in the previous Drill Problem using 7402 gates.

The design of a multilevel (three or more levels) logic network using mixed logic is illustrated in the following example.

Example 10.10

Design a logic network to obtain the function:

$$T = AB(C + D) + \overline{AB}$$

Inputs available: A.L, $\overline{\text{A}}$.H, B.L, $\overline{\text{B}}$.H, C.H, D.H.
Output desired: T.H
Obtain the network using (a) 7400 gates; and (b) 7402 gates.

Solution

The function is first subdivided into two-input functions (since we are using two-input gates). The following functions are formed first: AB, (C + D), $\overline{\text{AB}}$. Then, the product AB(C + D) is formed by using the outputs AB and (C + D) available from the previous step. Then, the function T is formed.

This *basic structure* is the same for both networks, but the insertion of voltage level changers in order to satisfy the compatibility requirement will be different in the two cases. The two networks are shown in Fig. 10-41.

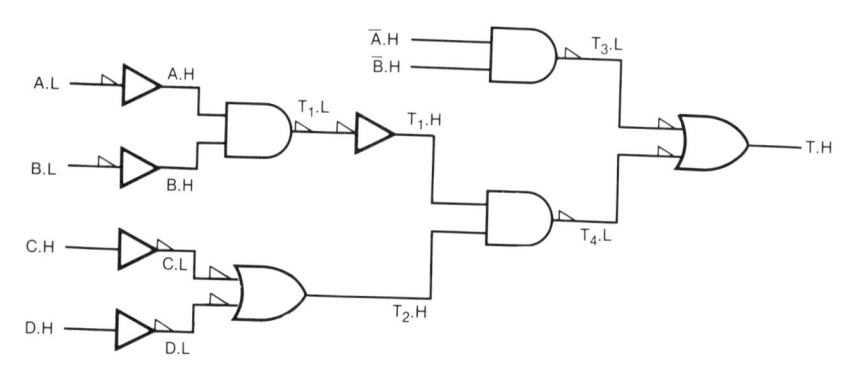

(A) Using the 7400 gates.

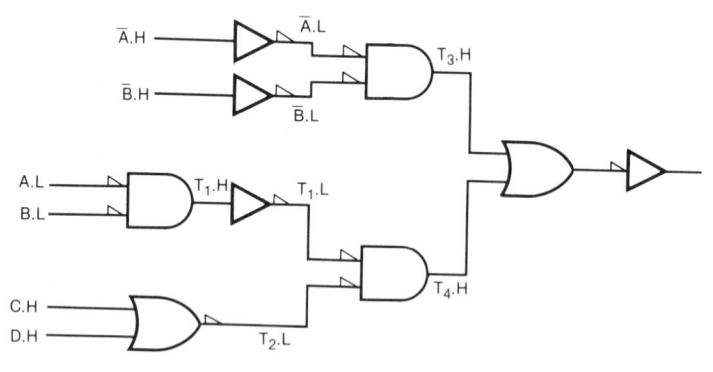

(B) Using the 7402 gates.

Fig. 10-41. Mixed logic networks of Example 10.10.

Drill Problem 10.28: Design a network to obtain the function T = [(A + B + CD) ($\overline{A}$ + $\overline{B}$)] using (a) 7400 gates; (b) 7402 gates. Assume availability of inputs and desired output are the same as in the last example.

The goal in logic design is to reduce the number of IC chips needed to realize the given logic function, which is frequently possible by using the right chip. For example, consider the two networks of Example 10.9. The number of gates needed in the 7400 realization was nine, and the number needed in the 7402 realization was five. Since each chip has four gates, the number of chips needed was three for the 7400 network and two for the 7402 network. (Note that each voltage level changer is obtained from a 7400 or a 7402 gate by tying its two input terminals together to the same input.) For the particular function in Example 10.9, therefore, a 7402 realization is more economical than a 7400 realization. (How do the answers to Drill Problems 10.24 and 10.25 compare?)

Even though the discussion up to this point was based on the use of one type of gate for the whole network, this is not a necessary or even a desirable restriction. Sometimes, a judicious mixture of different types of chips will lead to the most economical solution. Unfortunately, the reduction in the number of chips requires a series of trial-and-error efforts coupled with intelligent judgment; there is no methodical approach to the reduction of the number of chips in mixed logic design.

Mixed Logic Design Versus Positive Logic Design

The emphasis on mixed logic design in the previous discussion should not give the impression that mixed logic design is the only approach or even the most desirable approach to logic design. It is more important to focus on what advantages mixed logic design offers under what circumstances. If we are working with positive NAND and positive NOR packages (such as the 7400, 7402, and others like them), mixed logic is a blessing because it permits us to treat those gates as if they were AND and OR gates, and the design procedure is made quite simple. Conventional positive logic design using NAND and NOR gates is extremely cumbersome especially in networks with three or more levels: both the algebra and the rules of design are complicated. Furthermore, any modification of the network (in order to improve the design) becomes rather difficult in such cases. Mixed logic is unquestionably the best procedure when the packages at our disposal are positive NAND and NOR functions and when the network to be designed has three or more levels. If a network is being built using positive AND and OR gates, mixed logic does not offer any particular advantages and positive logic design is equally attractive.

Summary of Mixed Logic Design Procedure

The function is subdivided into smaller functions as determined by the number of inputs available on each gate. The subdivided functions are logical sums or logical products.

A basic structure is set up or visualized in the form of AND and OR gates to obtain the desired function. The actual structure is then obtained by placing the given gate packages in the positions of the AND and OR gates, and inserting voltage level changers as required by the compatibility condition.

Different combinations of packages should be tried in order to arrive at the most economical solution. The question governing such trials is whether voltage level changers can be eliminated by using a different chip.

Analysis of Mixed Logic Networks

The operations involved in a mixed logic network are AND and OR functions. Therefore, the logic function of a given network can be obtained in the same manner as discussed for positive logic networks. *The voltage level changers and the presence of triangular polarity marks in a mixed logic network do not in any way affect the logic functions* (so long as the compatibility condition is satisfied throughout the network). They are, therefore, ignored.

Example 10.11

Determine the output function of the network in Fig. 10-42.

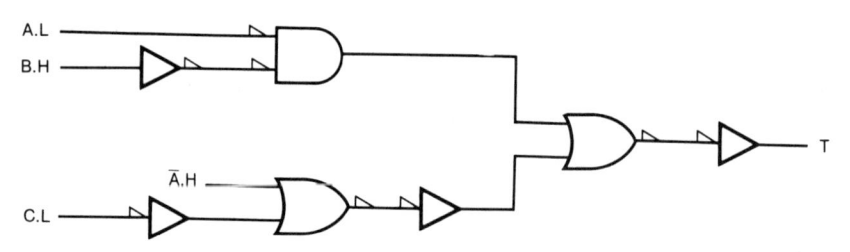

Fig. 10-42. Network for Example 10.11.

Solution

The diagram of Fig. 10-43 shows the signals at various points of the network. The voltage level changers and triangular marks are shown in dashed lines to indicate that they are to be ignored. The active level designations ".H" and ".L" are omitted for the sake of convenience.

The output of the network is seen to be

$$T = AB + (\bar{A} + C)$$

Electronics: Circuits and Systems

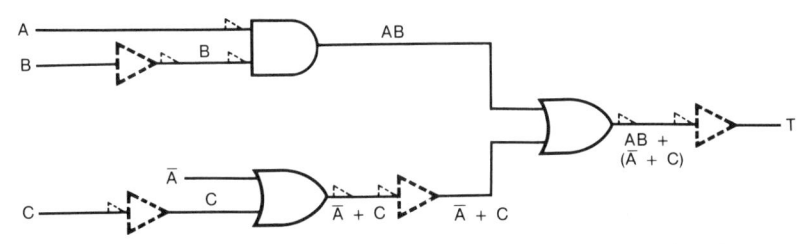

Fig. 10-43. Steps in the analysis of the network of Example 10.11.

Drill Problem 10.29: Write the output function of each of the logic networks in Fig. 10-44.

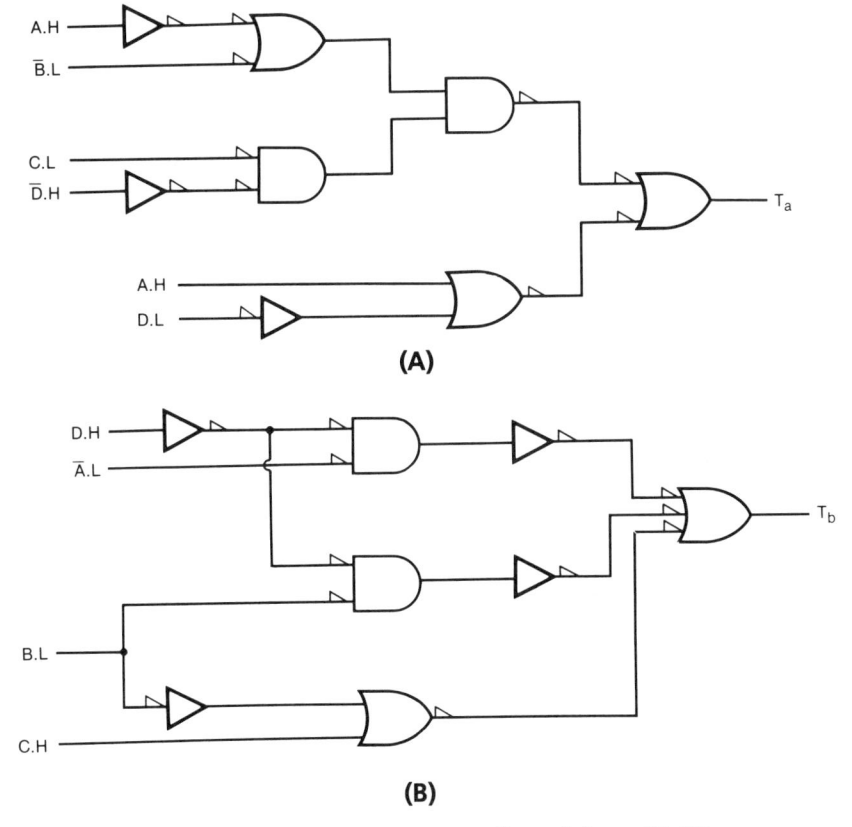

Fig. 10-44. Networks for Drill Problem 10.29.

KARNAUGH MAPS

The *Karnaugh map* (abbreviated K-map) provides a convenient and effective means of visualizing logic functions and is a useful tool in logic design. Even though K-maps are most commonly used for the minimization of two-level logic networks, our attention will be confined to their use as a visual representation of logic functions.

A K-map is made up of cells, the number of cells being 2^n where n is the number of variables in a logic function: the number of cells is four for two variables, eight for three variables, and sixteen for four variables. Each cell corresponds to one particular combination of values of the variables and that combination can be considered as its *address*. A logic function is represented on the K-map by placing a one on each cell for whose address the function is one and a zero in each cell for whose address the function is zero.

Two-Variable Map

A two-variable map has four cells arranged as shown in Fig. 10-45. The two values of A define the rows and the two values of B the columns. The addresses of the cells (in terms of 00, 01, . . .) are shown in the diagram.

Fig. 10-45. Karnaugh maps (two-variable functions).

Consider the function

$$T = AB + \overline{AB}$$

T = 1 when (AB) = 1, which occurs when A = 1, B = 1. A 1 is, therefore, placed in the "11" cell. T is also 1 when $(\overline{AB})$ = 1, which occurs when A = 0, B = 0. Therefore, a 1 is placed in the "00" cell, which exhausts the conditions for which T = 1 for the given function. Therefore, the remaining cells have zeros placed in them. (Fig. 10-45B.)

Given a two-variable map with ones and zeros in the different cells, the corresponding function can be written as illustrated in the following example.

Electronics: Circuits and Systems

Example 10.12

For the logic function shown in Fig. 10-46, we write a product term for each cell occupied by a 1 and then obtain the sum of such products. The procedure is analogous to that used with positive logic networks. Here the first 1 cell has A = 0, B = 1, which gives the product term $\overline{A}B$. The other 1 cell has A = 1, B = 0, which leads to the product term $A\overline{B}$. Therefore, the map represents the function

$$T = \overline{A}B + A\overline{B}$$

Fig. 10-46. K-map of the function T = $\overline{A}B$ + $A\overline{B}$.

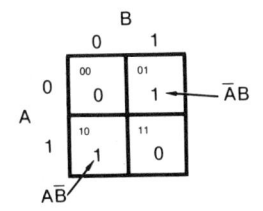

Three-Variable Map

A three-variable map has eight cells arranged in either of the two forms shown in Fig. 10-47. In the map of Fig. 10-47A, the values of A define the rows, and the values of the combination BC define the columns. (The labeling is exactly the opposite for the map in Fig. 10-47B.) The labeling of the columns of map (a) is seen to be in the following order: 00, 01, 11, 10. That is, as we go from one column to the next, only one variable changes in value at a time. This property is useful in the minimization of logic functions.

(A) Values of A define the rows, and BC defines the columns.

(B) Values of A define the columns, and BC defines the rows.

Fig. 10-47. Three-variable Karnaugh maps.

The address of each cell is shown in the diagrams. Suppose we wish to represent the logic function

$$T = \overline{A}B\overline{C} + A\overline{B}\overline{C} + \overline{A}\overline{B}C$$

on the three-variable map. The product term $\overline{A}B\overline{C}$ is 1 for A = 0, B = 1, C = 0, or the address of the corresponding cell is 010. A 1 is placed in the "010" cell. Similarly, the term $A\overline{B}\overline{C}$ leads to the 100 cell and $\overline{A}\overline{B}C$ to the 001 cell. The map of the previous function is shown in Fig. 10-48A. The map of another function

$$T = \overline{A}BC + A\overline{B}C + AB\overline{C} + ABC$$

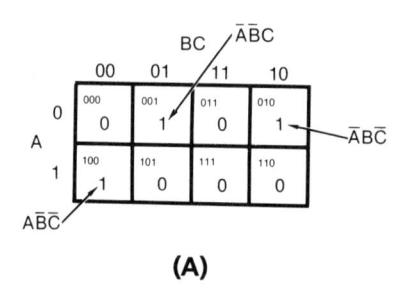

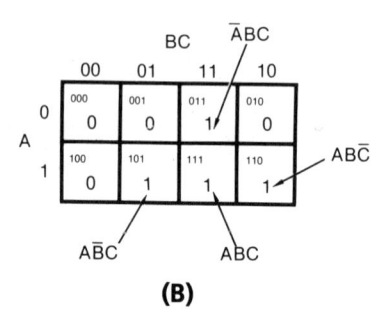

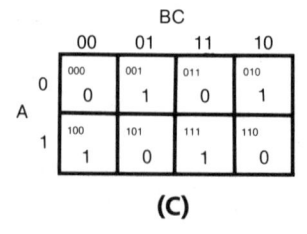

Fig. 10-48. Map representation of several three-variable functions.

is shown in Fig. 10-48B. This is the function that was obtained in Example 10.4.

For a given map, the logic function is obtained by setting up a product term corresponding to the address of each cell occupied by a 1 and then summing all the terms. As an example, the function represented by the map in Fig. 10-48C is

$$T = \overline{AB}\overline{C} + \overline{A}B\overline{C} + A\overline{B}\overline{C} + ABC$$

The map is used in lieu of a truth table in logic design. Each cell of the map is like a row in the truth table, and the cell is filled with a 1 or a 0 in the same manner as the truth table was set up for a given logic design problem.

Example 10.13

A logic network with three inputs A, B, C is to have an output of 1 whenever an odd number of inputs have a value of 0. Otherwise, the output should be 0.

Solution

An *odd* number of inputs having a value of 0 means that any *one* of the inputs or all three inputs must be 0 in order to make the output 1. The input combinations of interest are 011, 101, 110, and 000. The map of the resulting function is shown in Fig. 10-49A and the logic network is obtained by translating 011 to $\overline{A}BC$, 101 to $A\overline{B}C$, 110 to $AB\overline{C}$, and 000 to $\overline{A}\overline{B}\overline{C}$, and adding these product terms. The network is shown in Fig. 10-49B.

Drill Problem 10.30: Draw the K-maps of the functions obtained in Drill Problems 10.16, 10.17, and 10.18.

Drill Problem 10.31: Write the logic functions represented by the maps shown in Fig. 10-50.

Drill Problem 10.32: Design a logic network with three inputs A, B, C and an output T such that T = 1 under either of the following conditions: (a) A = 1 and *one* of the other two variables = 1; (b) A = 0 and *one* of the other variables = 0.

Four-Variable Map

A four-variable map has sixteen cells arranged in four rows by four columns. The combinations of values of AB define the rows and the combinations of values of CD define the columns, as indicated in Fig. 10-51. (The assignment could also be the other way around: AB for the columns and CD for the rows.) The address of each cell is given by the combination of values of

(A) K-map.

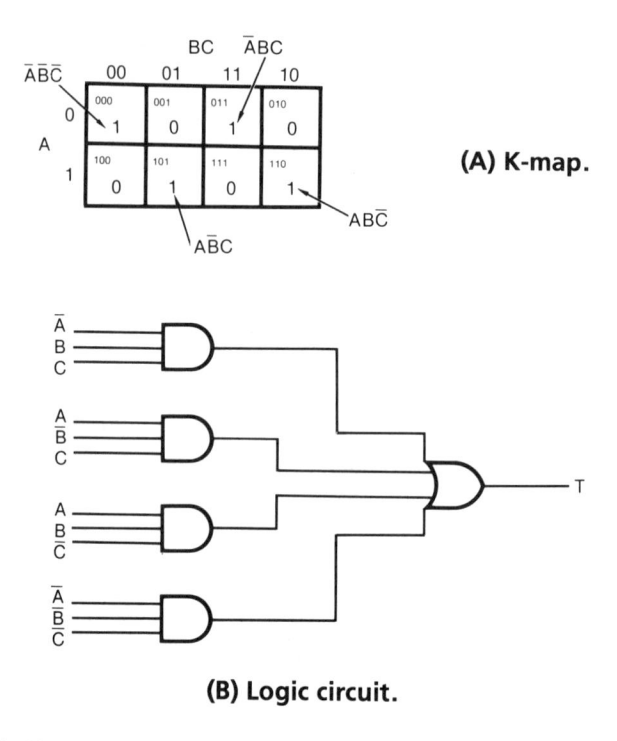

(B) Logic circuit.

Fig. 10-49. Use of a K-map in the design of two-level logic networks of Example 10.13.

(A) **(B)**

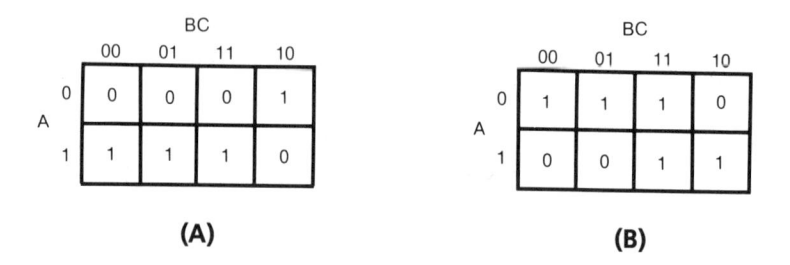

Fig. 10-50. Maps for Drill Problem 10.31.

ABCD for that cell. The addresses of selected cells are shown in the diagram.

Given a logic function

$$T = \overline{A}B\overline{C}D + A\overline{B}\,\overline{C}\,\overline{D} + ABC\overline{D} + \overline{A}BC\overline{D} + A\overline{B}C\overline{D}$$

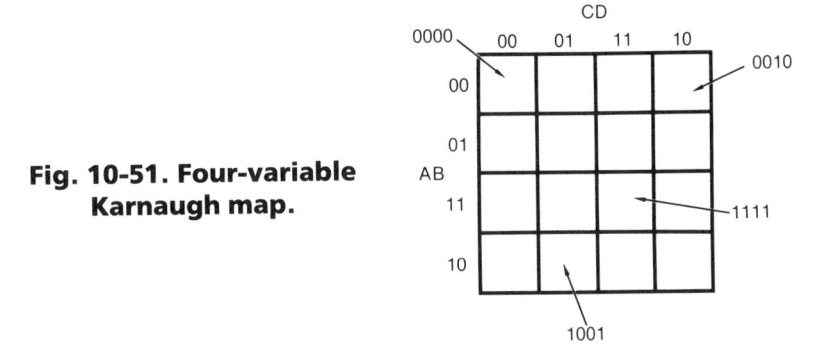

Electronics: Circuits and Systems

Fig. 10-51. Four-variable Karnaugh map.

the cell for each product term is located and a 1 is placed in it. The term $\overline{A}B\overline{C}D$ leads to the 0101 cell, for example. The cells not occupied by 1 are filled with 0 (or may be left blank) as in Fig. 10-52A. The function for a given map is obtained by taking the address of each cell occupied by a 1 and writing the product term, and then summing all such product terms. The cells occupied by a 0 are ignored. (Once the function has been designed to be 1 for the required conditions using the previous procedure, it will automatically be 0 for the other conditions.) As an example, the function for the map in Fig. 10-52B is

$$T = \overline{A}\overline{B}\overline{C}\overline{D} + \overline{A}\overline{B}C\overline{D} + \overline{A}B\overline{C}D + A\overline{B}\overline{C}\overline{D} + \overline{A}B\overline{C}\overline{D} + ABCD$$

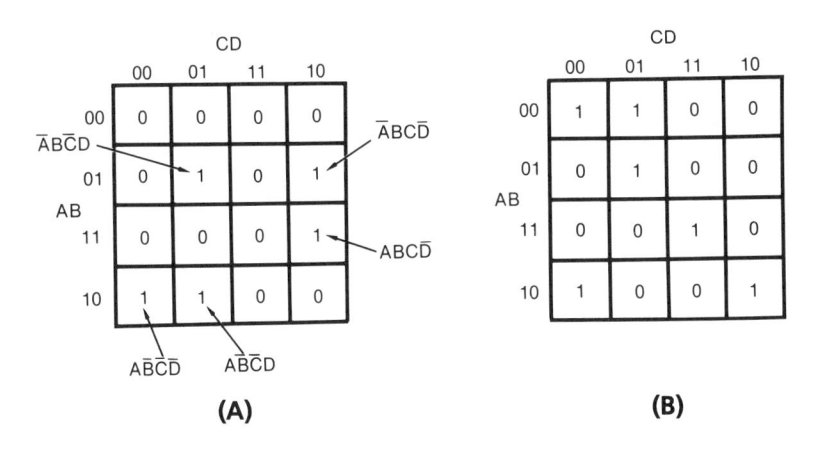

(A)   (B)

Fig. 10-52. Map representations of four-variable logic functions.

SUMMARY SHEETS

PRINCIPLES OF LOGIC CIRCUITS

Two levels of signal: high (H) and low (L).
Positive logic convention: H = 1; L = 0.
Negative logic convention: H = 0; L = 1.
Mixed logic convention: Positive logic convention is used in some parts of a network and negative logic convention in others. The physical behavior of a component or circuit is not affected in any way by the particular convention used; it is only our interpretation of the logical function of the component or circuit that changes with the convention chosen.

Switching Variables: An algebraic symbol used for denoting each input and output signal in a digital circuit is called a switching (or logical) variable.

BASIC OPERATIONS IN LOGIC

Logical Sum (OR Operation)

$0 + 0 = 0; 0 + 1 = 1; 1 + 0 = 1; 1 + 1 = 1$

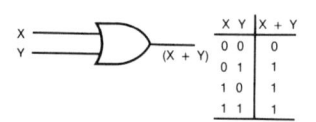

Associative Law

$X + (Y + Z) = (X + Y) + Z$
$= X + Y + Z$

Logical Product (AND Operation)

$0 \cdot 0 = 0; 0 \cdot 1 = 0; 1 \cdot 0 = 0; 1 \cdot 1 = 1$

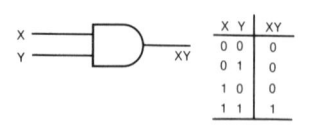

Associative Law

$X(YZ) = (XY)Z = XYZ$

Logical Complementation

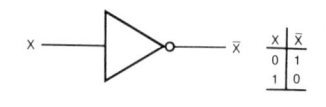

$X + \overline{X} = 1$ and $X\overline{X} = 0$

NAND OPERATION

NAND = AND followed by NOT

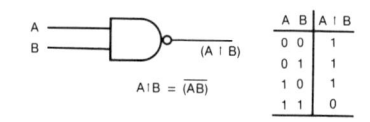

$A \uparrow B = \overline{(AB)}$

NAND operation does not satisfy the associative law. That is

$(A \uparrow B) \uparrow C \neq A \uparrow (B \uparrow C) \neq A \uparrow B \uparrow C$

NAND gate as a NOT gate

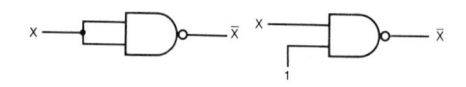

NOR OPERATION

NOR = OR followed by NOT

$A \downarrow B = \overline{(A + B)}$

NOR Operation does not satisfy the associative law. That is

$(A \downarrow B) \downarrow C \neq A \downarrow (B \downarrow C) \neq A \downarrow B \downarrow C$

NOR gate as a NOT gate

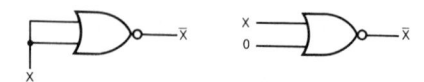

EXCLUSIVE OR FUNCTION

Two Variables

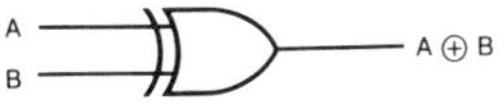

A B	A⊕B
0 0	0
0 1	1
1 0	1
1 1	0

Associative Law

A⊕(B⊕C) = (A⊕B)⊕C = A⊕B⊕C

Three or More Variables

The exclusive OR sum of three or more variables is one if, and only if, an odd number of the variables equals one. It is zero of none or an even number of the inputs equals one.

Odd and Even Parity Checks

If the number of ones in a sequence of zeros and ones is odd, the sequence is said to have an odd parity of ones. If the number is even, then there is an even parity of ones. The exclusive OR function is useful in checking for odd or even parity.

EXCLUSIVE NOR FUNCTION

Two variables

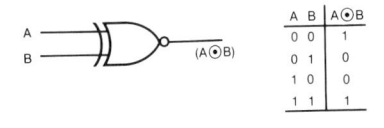

A B	A⊙B
0 0	1
0 1	0
1 0	0
1 1	1

TWO-LEVEL POSITIVE LOGIC NETWORKS

The output gate is numbered as level 1 and the other levels are numbered going toward the input terminals of the network.

Analysis involves the identification of the output of each gate in the network in terms of its input variables or functions and obtaining the output of the network in terms of the input variables to the network.

Design consists of translating a given problem statement to a logic function by using the steps outlined in the text of this chapter. The resulting network will be a two-level sum of products network.

TWO-LEVEL NAND NETWORKS

A two-level sum of products network can be readily converted to a two-level NAND network by replacing each gate in the two-level network by a NAND gate and leaving all the inputs the same (except when a signal is directly connected to the level-one gate, which is replaced by its complement).

MIXED LOGIC NETWORKS

Active high assignment: H = 1, L = 0.
Active low assignment: H = 0, L = 1.
Positive logic convention: All signals in a network are active high.
Negative logic convention: All signals in a network are active low.
Mixed logic convention: Some signals are active high and others are active low.
Triangular polarity marks: A terminal with such a mark indicates that the signal at that terminal has been assigned *active low*. Absence of such a mark indicates an active high assignment.

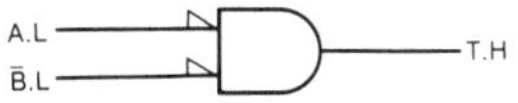

Using 7400 (NAND) Gate as AND and as OR

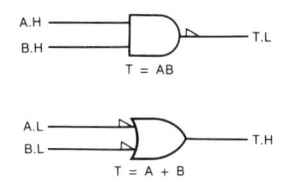

Using 7402 (NOR) Gate as AND and as OR

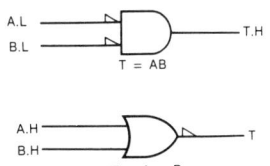

MIXED LOGIC DESIGN

A given gate is used as an AND and as an OR gate by using the proper mixed logic assignments.

Compatibility Requirement: An active low signal must be connected to a terminal with a triangular polarity mark. An active high signal must be connected to a terminal not having a triangular polarity mark. A *voltage level changer* is used as needed in order to fulfill the compatibility requirement.

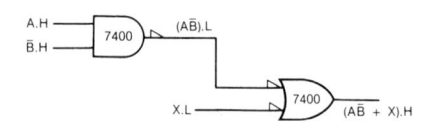

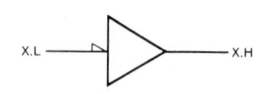

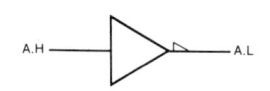

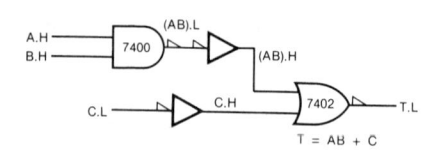

The design procedure is essentially the same as in the two-level sum of products form except that the compatibility requirement must be carefully enforced.

Analysis: So long as the compatibility requirement is met in the network, ignore all the voltage level changers and treat each of the other gates as AND and OR operations. Write the output function in the same manner as in two-level positive logic networks. The mixed logic design approach gives a certain degree of versatility to logic design. It is the best procedure when the gates available are NAND and NOR gates and the network has three or more levels.

KARNAUGH MAPS

Two-Variable Map

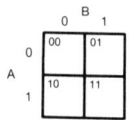

Three-Variable Map

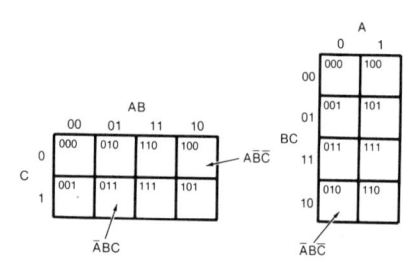

Four-Variable Map

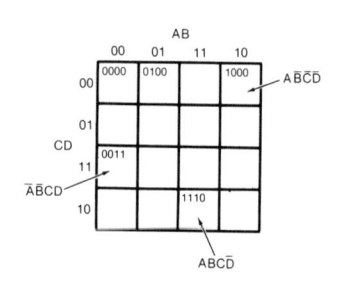

Conversion of the Address of a Cell to Product Term and Vice Versa

1 in the address ⟷ uncomplemented literal symbol

0 in the address ⟷ complemented literal symbol

$T = \bar{A}\bar{B}C + A\bar{B}C + AB\bar{C}$

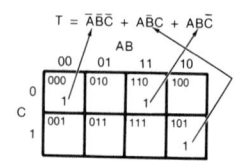

Conversion of a Function to a Map and Vice Versa

Place a one in the cell in the map corresponding to each product term in the given function.

Take each cell in the map occupied by a one and write the corresponding product term. Connect all the product terms so obtained by the sum operation.

ANSWERS TO DRILL PROBLEMS

10.1 See Table 10-DP1.

Table 10-DP1. Answer to Drill Problem 10.1

X	Y	Z	(Y + Z)	$T_1 = X + (Y + Z)$	(X + Y)	$T_2 = (X + Y) + Z$	$T_3 = X + Y + Z$
0	0	0	0	0	0	0	0
0	0	1	1	1	0	1	1
0	1	0	1	1	1	1	1
0	1	1	1	1	1	1	1
1	0	0	0	1	1	1	1
1	0	1	1	1	1	1	1
1	1	0	1	1	1	1	1
1	1	1	1	1	1	1	1

10.2 See Table 10-DP2.

Table 10-DP2. Answer to Drill Problem 10.2

X	Y	Z	(YZ)	X(YZ)	(XY)	(XY)Z	XYZ
0	0	0	0	0	0	0	0
0	0	1	0	0	0	0	0
0	1	0	0	0	0	0	0
0	1	1	1	0	0	0	0
1	0	0	0	0	0	0	0
1	0	1	0	0	0	0	0
1	1	0	0	0	1	0	0
1	1	1	1	1	1	1	1

10.3 See Fig. 10-DP3. (Other connections are also possible.)

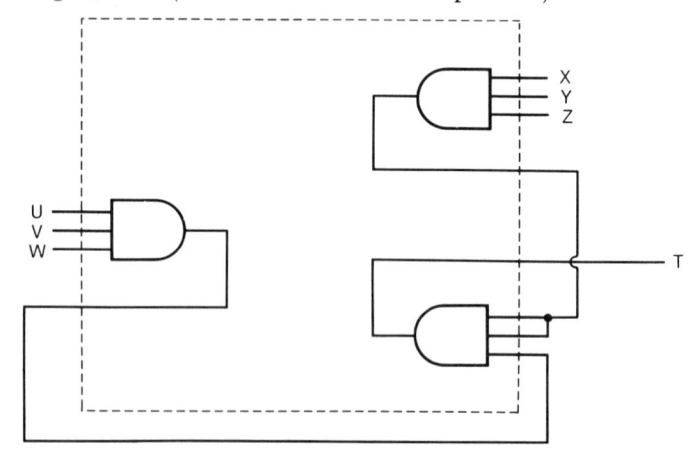

Fig. 10-DP3.

Electronics: Circuits and Systems   635

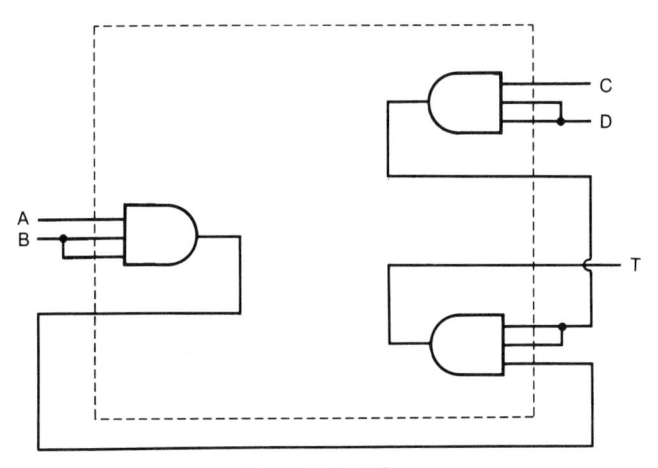

Fig. 10-DP4.

10.4 See Fig. 10-DP4. (Other connections are also possible.)
10.5 See Table 10-DP5. (a) AND gate; (b) OR gate.

Table 10-DP5. Answer to Drill Problem 10.5

(A)

X	Y	T
0	0	0
0	1	0
1	0	0
1	1	1

(B)

X	Y	T
1	1	1
1	0	1
0	1	1
0	0	0

10.6 See Table 10-DP6.

Table 10-DP6. Answer to Drill Problem 10.6

A	B	C	$T_c = \overline{(AC)}$	$T_4 = \overline{(T_c B)}$
0	0	0	1	1
0	0	1	1	1
0	1	0	1	0
0	1	1	1	0
1	0	0	1	1
1	0	1	0	1
1	1	0	1	0
1	1	1	0	1

10.7 See Table 10-DP7.

Table 10-DP7. Answer to Drill Problem 10.7

X	Y = $\bar{X}$	XY	$\overline{XY}$ = $\bar{X}$
0	0	0	1
1	1	1	0

10.8 See Table 10-DP8. The T_1, T_2, and T_3 columns are seen to differ in at least one case. Therefore, they are not equal.

Table 10-DP8. Answer to Drill Problem 10.8

A	B	C	$T_a = \overline{(A + B)}$	$T_1 = \overline{(T_a + C)}$	$T_b = \overline{(A + C)}$	$T_2 = \overline{(T_b + B)}$	$T_3 = \overline{(A + B + C)}$
0	0	0	1	0	1	0	1
0	0	1	1	0	0	1	0
0	1	0	0	1	1	0	0
0	1	1	0	0	0	0	0
1	0	0	0	1	0	1	0
1	0	1	0	0	0	1	0
1	1	0	0	1	0	0	0
1	1	1	0	0	0	0	0

10.9 See Table 10-DP9.

Table 10-DP9. Answer to Drill Problem 10.9

X	Y = $\bar{X}$	(X + Y)	$\overline{(X + Y)}$ = $\bar{X}$
0	0	0	1
1	1	1	0

10.10 See Table 10-DP10. (a) Positive NAND gate; (b) Negative OR gate.

Table 10-DP10. Truth Tables for Drill Problem 10.10

Pos NAND Gate

A	B	T = A ↑ B
0	0	1
0	1	1
1	0	1
1	1	0

(A) Pos Logic

Neg NOR Gate

A	B	T = A ↓ B
1	1	0
1	0	0
0	1	0
0	0	1

(B) Neg Logic

Electronics: Circuits and Systems

10.11 See Table 10-DP11. (a) Positive OR gate; (b) Negative NAND gate.

Table 10-DP11. Truth Tables for Drill Problem 10.11

Pos NOR Gate

A	B	T = A ↓ B
0	0	1
0	1	0
1	0	0
1	1	0

(A) Pos Logic

Neg NAND Gate

A	B	T = A ↑ B
1	1	0
1	0	1
0	1	1
0	0	1

(B) Neg Logic

10.12 See Table 10-DP12. The output column matches that in Table 10-11.

Table 10-DP12. Truth Table for Drill Problem 10.12

A	B	C	(A ⊕ C)	B ⊕ (A ⊕ C)
0	0	0	0	0
0	0	1	1	1
0	1	0	0	1
0	1	1	1	0
1	0	0	1	1
1	0	1	0	0
1	1	0	1	0
1	1	1	0	1

10.13 See Table 10-DP13. The two outputs T_1 and T_2 are equal!

Table 10-DP13. Truth Table for Drill Problem 10.13

A	B	C	$T_a = (A \odot B)$	$T_1 = (T_a \odot C)$	$T_b = (B \odot C)$	$T_2 = (T_b \odot A)$
0	0	0	1	0	1	0
0	0	1	1	1	0	1
0	1	0	0	1	0	1
0	1	1	0	0	1	0
1	0	0	0	1	1	1
1	0	1	0	0	0	0
1	1	0	1	0	0	0
1	1	1	1	1	1	1

10.14 (a) $T_1 = A(\overline{B} + C) + (\overline{A} + B)(A + \overline{C})$. See Table 10-DP14A. (b) $T_2 = ABC + (\overline{AB} + \overline{AC})$. See Table 10-DP14B. (c) $T_3 = (\overline{AB} + \overline{AB})(\overline{AB} + \overline{BC})$. See Table 10-DP14C.

Table 10-DP14. Truth Tables for Drill Problem 10.14

(A)

A	B	C	$\bar{B}+C$	$T_a = A(\bar{B}+C)$	$(\bar{A}+B)$	$(A+\bar{C})$	$T_b = (\bar{A}+B)(A+\bar{C})$	$T_1 = T_a + T_b$
0	0	0	1	0	1	1	1	1
0	0	1	1	0	1	0	0	0
0	1	0	0	0	1	1	1	1
0	1	1	1	0	1	0	0	0
1	0	0	1	1	1	1	1	1
1	0	1	1	1	1	1	1	1
1	1	0	0	0	0	1	0	0
1	1	1	1	1	0	1	0	1

(B)

A	B	C	$T_c = ABC$	$\overline{AB}$	$\overline{AC}$	$T_d = \overline{AB} + \overline{AC}$	T
0	0	0	0	1	1	1	1
0	0	1	0	1	0	1	1
0	1	0	0	0	1	1	1
0	1	1	0	0	0	0	0
1	0	0	0	0	0	0	0
1	0	1	0	0	0	0	0
1	1	0	0	0	0	0	0
1	1	1	1	0	0	0	1

(C)

A	B	C	$A\bar{B}$	$\overline{AB}$	T_e	BC	T_f	T
0	0	0	0	1	1	1	1	1
0	0	1	0	1	1	0	1	1
0	1	0	1	0	1	0	0	0
0	1	1	1	0	1	0	0	0
1	0	0	0	0	0	1	1	0
1	0	1	0	0	0	0	0	0
1	1	0	0	0	0	0	0	0
1	1	1	0	0	0	0	0	0

10.15 The work is self-evident.

10.16 T = $\bar{A}BC + A\bar{B}C + AB\bar{C}$ + ABC. The truth table of this function will be found to be identical to Table 10-13.

10.17 See Table 10-DP17. T = $\overline{AB}$ + AB. See Fig. 10-DP17.

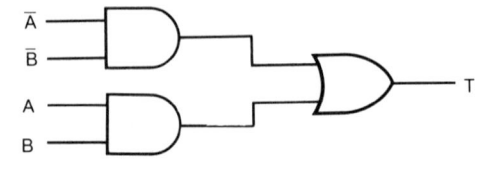

Fig. 10-DP17.

Electronics: Circuits and Systems 639

Table 10-DP17. Truth Table for Drill Problem 10.17

A	B	T
0	0	1
0	1	0
1	0	0
1	1	1

10.18 Let A = Peter goes, B = Paul goes, and C = Mary goes. The output must be 0 whenever C = 0 (which means Mary does not go). When C = 1, the output is 1 if both A and B are 1, or neither A nor B is 1 (which means the same as both A and B are 0). See Table 10-DP18. T = $\overline{A}\overline{B}C$ + ABC. See Fig. 10-DP18.

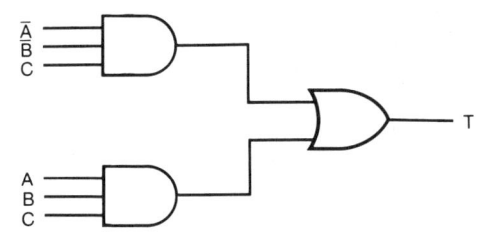

Fig. 10-DP18.

Table 10-DP18. Truth Table for Drill Problem 10.18

A	B	C	T
0	0	0	0
0	0	1	1
0	1	0	0
0	1	1	0
1	0	0	0
1	0	1	0
1	1	0	0
1	1	1	1

10.19 See Table 10-DP19. T = $\overline{A}\overline{B}C$ + $\overline{A}B\overline{C}$ + $A\overline{B}\overline{C}$ + ABC. See Fig. 10-DP19.

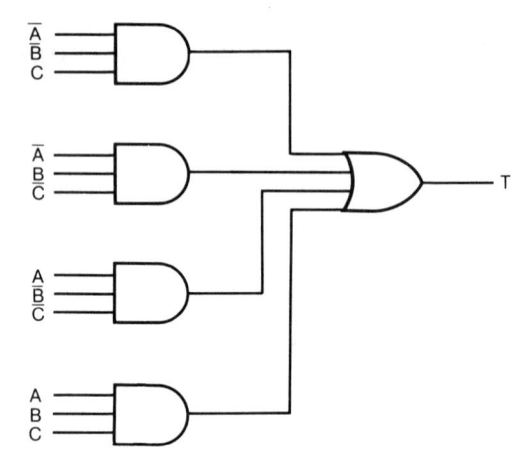

Fig. 10-DP19.

Table 10-DP19. Truth Table for 10.19

A	B	C	T
0	0	0	0
0	0	1	1
0	1	0	1
0	1	1	0
1	0	0	1
1	0	1	0
1	1	0	0
1	1	1	1

10.20 See Fig. 10-DP20.

Electronics: Circuits and Systems

(A)

(B)

(C)

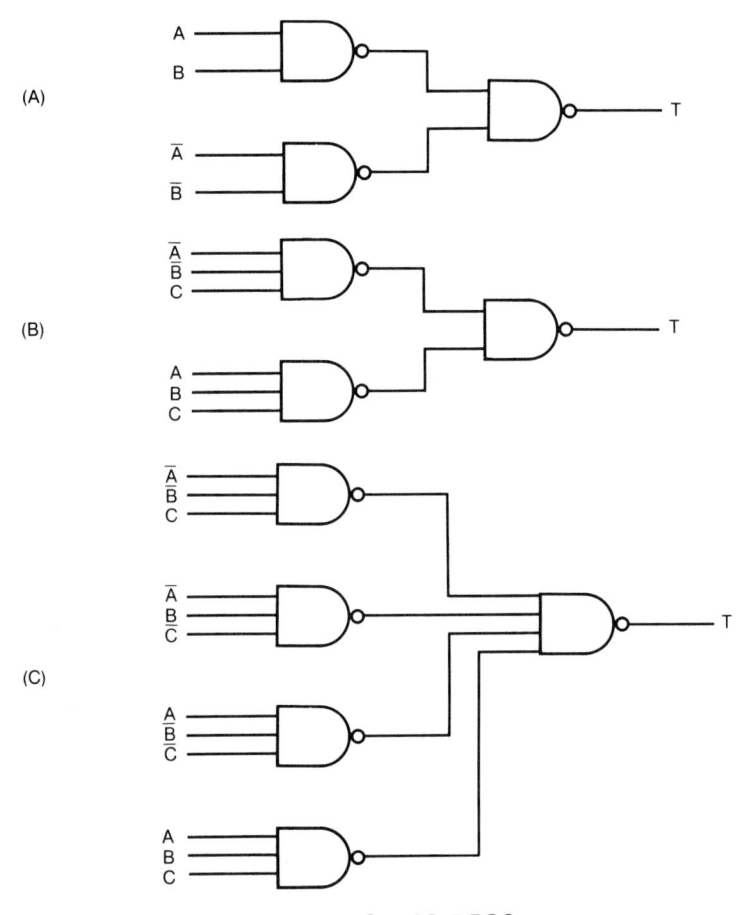

Fig. 10-DP20.

10.21 See Table 10-DP21. (a) AND operation. (b) OR operation.

Table 10-DP21. Truth Tables for Drill Problem 10.21

A	B	T		(A)	A	B	T		(B)	A	B	T
0	0	1			0	0	0			1	1	1
0	1	1			0	1	0			1	0	1
1	0	1			1	0	0			0	1	1
1	1	0			1	1	1			0	0	0

10.22 See Fig. 10-DP22.

642   Logic Circuits

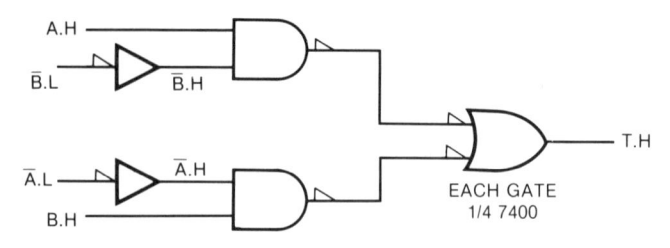

Fig. 10-DP22.

10.23 See Fig. 10-DP23.

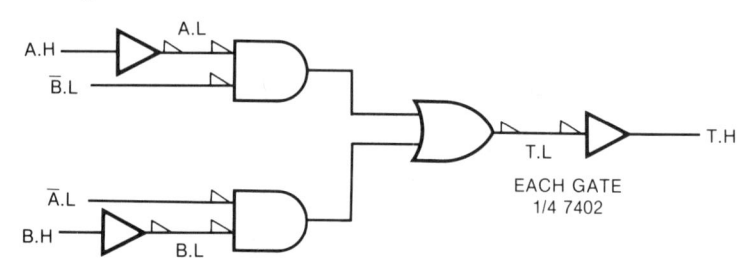

Fig. 10-DP23.

10.24 See Fig. 10-DP24.

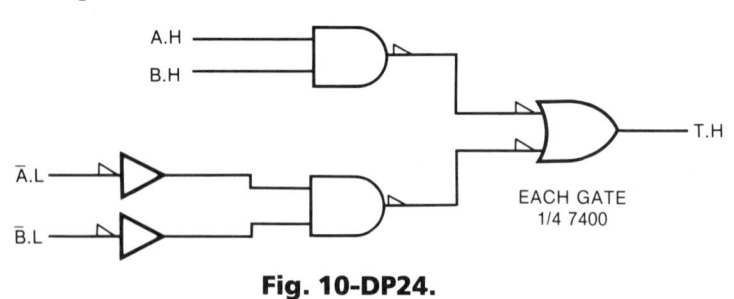

Fig. 10-DP24.

10.25 See Fig. 10-DP25.

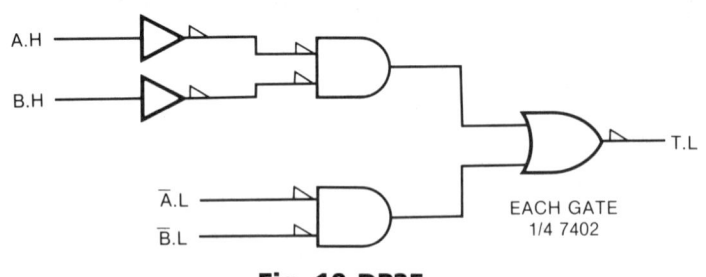

Fig. 10-DP25.

10.26 See Fig. 10-DP26.

Electronics: Circuits and Systems 643

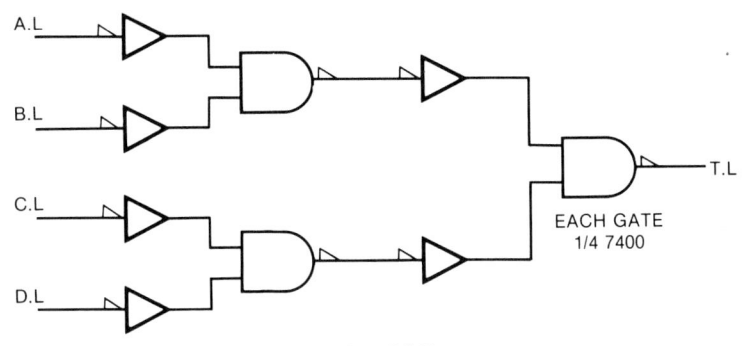

Fig. 10-DP26.

10.27 See Fig. 10-DP27.

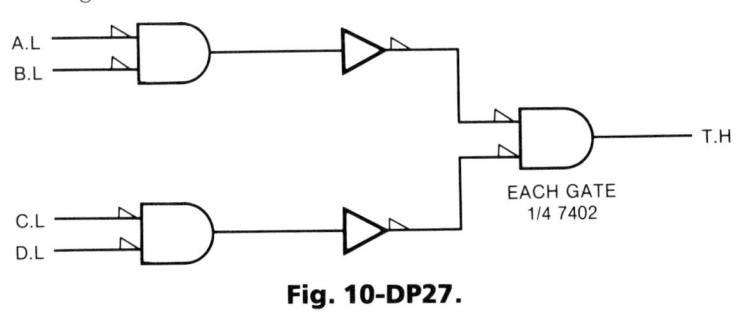

Fig. 10-DP27.

10.28 See Fig. 10-DP28.

Fig. 10-DP28.

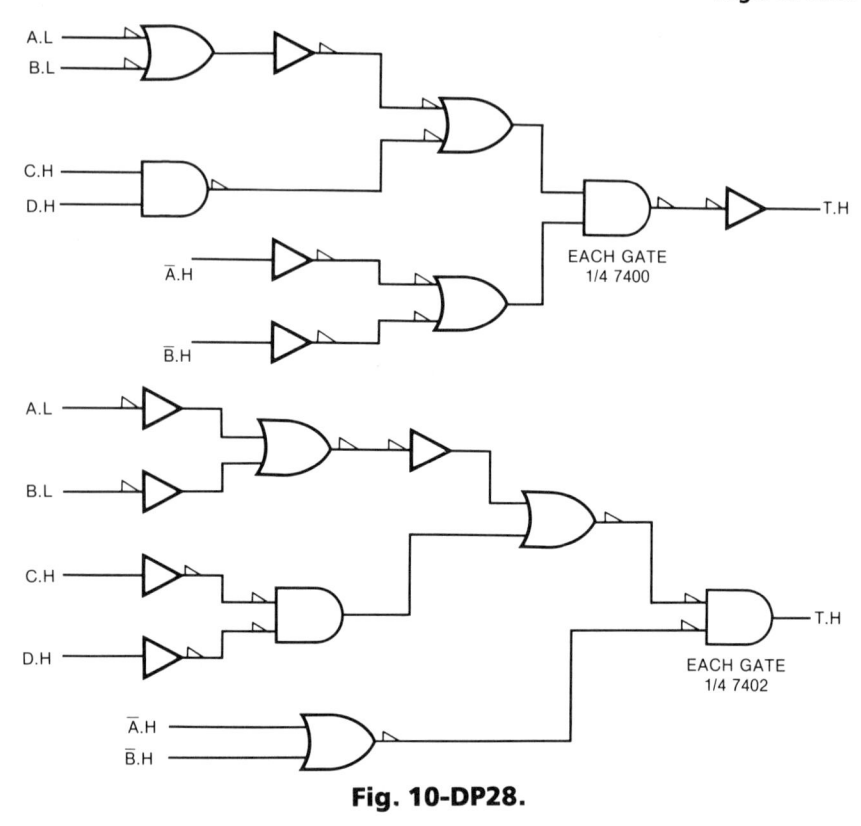

10.29 (a) $T_a = (A + B)C\overline{D} + (A + D)$; (b) $T_b = \overline{A}D + BD + B + C$.

10.30 See Fig. 10-DP30.

Fig. 10-DP30.

Electronics: Circuits and Systems 645

10.31 (a) $T_a = \overline{AB}\overline{C} + A\overline{BC} + A\overline{B}C + ABC$; (b) $T_b = \overline{ABC} + \overline{AB}C + \overline{A}BC + ABC + AB\overline{C}$.

10.32 See Fig. 10-DP32. $T = \overline{AB}\overline{C} + \overline{A}B\overline{C} + A\overline{B}C + AB\overline{C}$.

Fig. 10-DP32.

	BC			
A	00	01	11	10
0	0	1	0	1
1	0	1	0	1

PROBLEMS

Principles of Logic Circuits

10.1 Assuming that the only logic gate packages available are quad two-input AND and OR gates, obtain a network arrangement for each of the logic functions given. Each input is available in both the uncomplemented and complemented form. (a) $T_a = AB(\overline{C} + D) + \overline{A}\overline{B}C$; (b) $T_b = (A + B + \overline{CD})(\overline{A} + \overline{B} + C)$; (c) $T_c = \overline{ABCD} + A\overline{B}\overline{C} + AD$; (d) $T_d = (\overline{A} + B + C + \overline{D})(A + \overline{B} + D)$.

10.2 For each of the logic networks shown in Fig. 10-P2, write the output logic function.

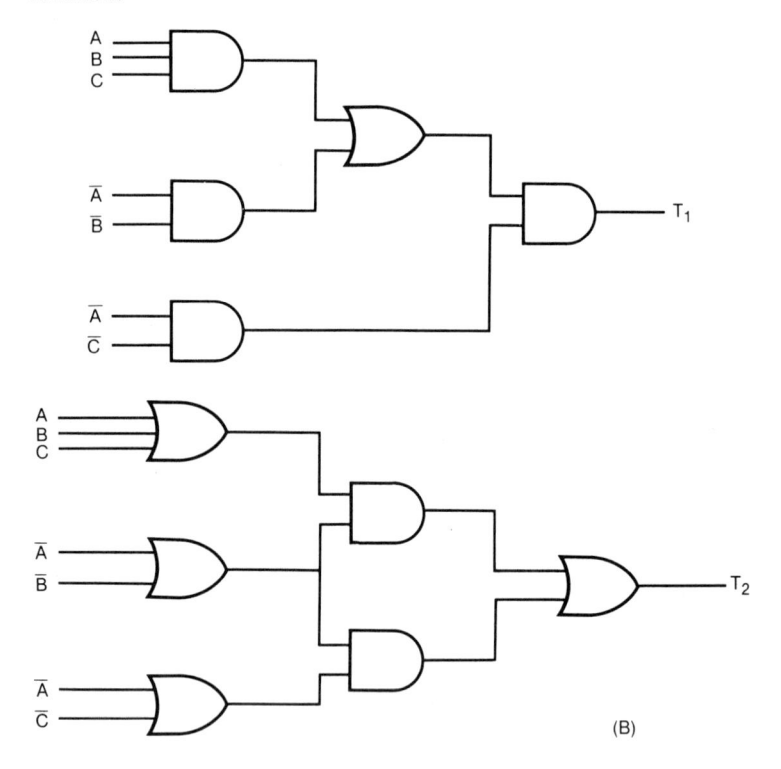

Fig. 10-P2.

NAND and NOR Operations

10.3 Set up the truth table of each of the NAND networks shown in Fig. 10-P3. For the sake of convenience set up separate columns for the outputs of the individual gates before obtaining the column of the output.

Electronics: Circuits and Systems

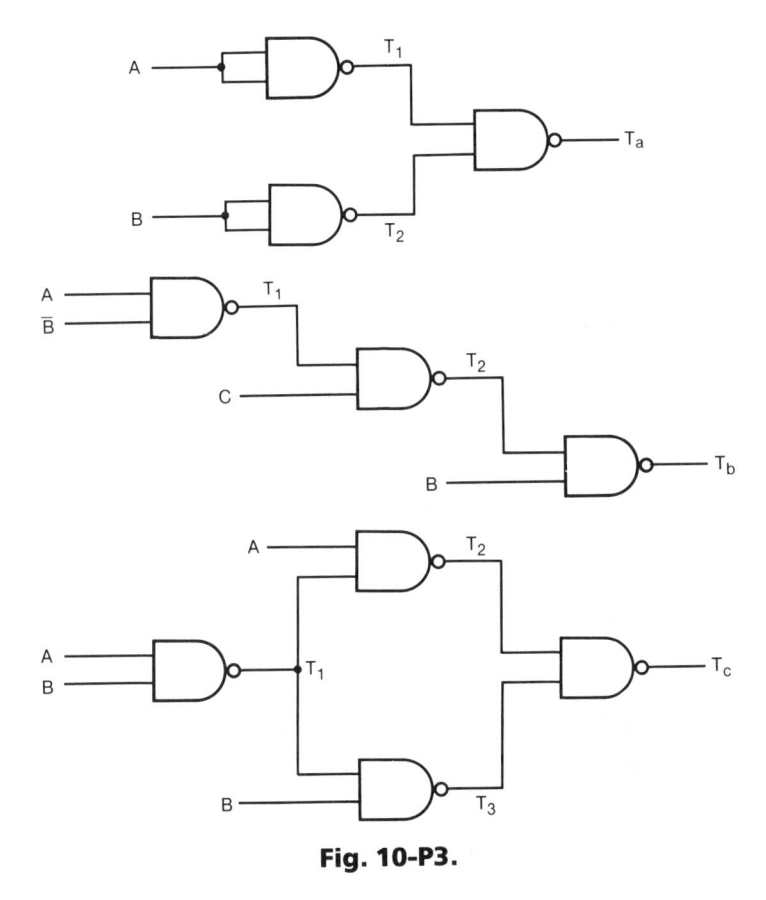

Fig. 10-P3.

10.4 Set up the truth table of each of the NOR networks that would be obtained if each NAND gate of Fig. 10-P3 were replaced by a NOR gate.

10.5 Set up the truth table of each network shown in Fig. 10-P5.

Exclusive OR and Exclusive NOR Functions

10.6 Replace each of the NAND gates of the networks in Fig. 10-P3 by exclusive OR gates and obtain the corresponding truth table.

10.7 Consider a two-input exclusive OR gate with inputs A and B. What will be the output in each of the following cases: (a) A and B input terminals are tied together, and a single input X is applied; (b) A and B are mutual complements (that is, an input X is applied to A and $\overline{X}$ to B; (c) the input B is grounded, and a signal X is applied to A; (d) the input B is tied to logic 1 and a signal X is applied to A.

10.8 Repeat the previous problem for an exclusive NOR gate.

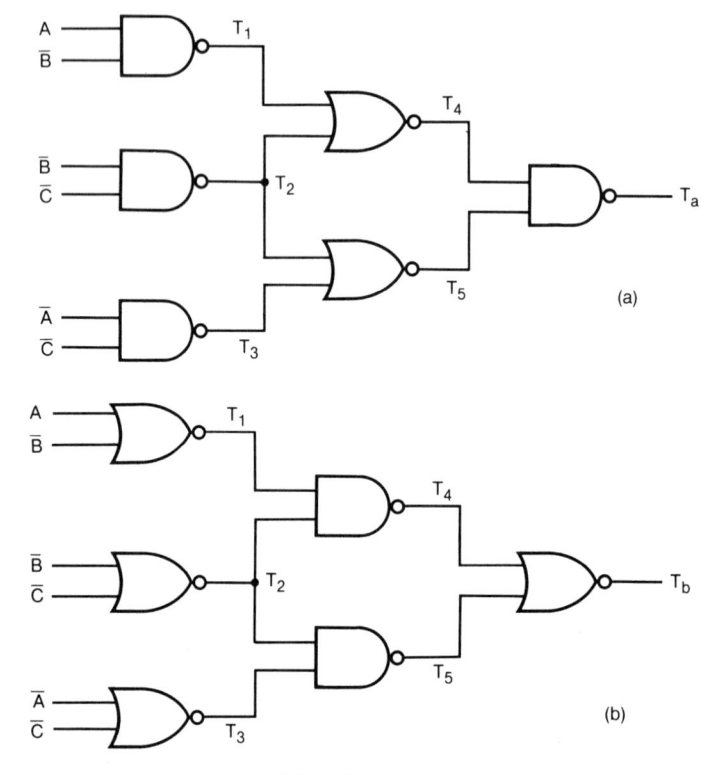

Fig. 10-P5.

Two-Level Positive Logic Networks

10.9 Determine the output function of each of the networks in Fig. 10-P9 and set up its truth table.

10.10 There are two inputs X and Y to a logic network. The output is to be one under the following conditions. If X = 1 then Y must be 1 in order to make the output 1. If X = 0, the output must be 1 regardless of what Y is. (This function is known as *material implication:* X implies Y.) Design the network.

10.11 A logic network has two inputs X and Y. It has *two* outputs T_1 and T_2. T_1 is to be one whenever X is greater than or equal to Y, while T_2 is to be one whenever X is less than or equal to Y. Design the network. Once the two-output network is designed, what would you do if an additional output T_3 is needed to indicate when X = Y?

10.12 Two persons A and B are available to go on a mission. For each of the following conditions, set up a single gate or a logic network (as appropriate). (a) If A goes, B does not go; and if B goes, A does not go. (b) If A goes, B does not go; but if A does not go, it makes no difference

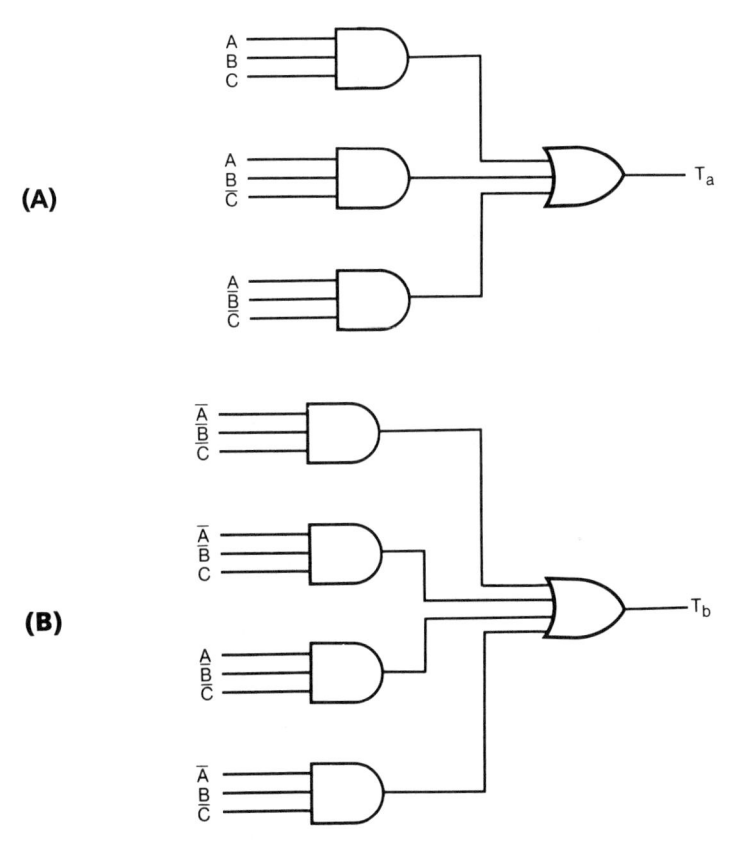

Fig. 10-P9.

whether B goes or not. (c) Either both A and B go, or neither goes. (d) Neither B nor A goes. (Note that a network is desired for *each* of the previous conditions *separately*.)

10.13 Design a logic network with three inputs X, Y, and Z such that the output is 1, if and only if, the product (XY) has a value greater than or equal to the product (YZ).

10.14 In a process control system, it is necessary to turn the system off when the conditions in the fluid storage tank reach specified critical thresholds. The three factors to be considered are the fluid level in the tank, the temperature of the fluid, and the pressure of the fluid. Each of these components has a specified threshold (or critical) value. If neither the temperature nor the pressure exceeds its threshold level, the system is safe and can be allowed to function. If the temperature exceeds its threshold level, the system must shut off. If the pressure exceeds its threshold level, but both the fluid level and the temperature are below their threshold levels, the system can be allowed to function. Design a logic network to control the shutting off of the system.

Two-Level NAND Networks

10.15 A professor uses the following procedure to determine whether a student passes her courses or not. She gives a set of quizzes, a final examination, and requires the submission of a term paper. Her rules for giving a passing grade are as follows. (1) If you flunk the final exam, you fail the course. (2) You must submit a satisfactory term paper or pass the quiz-component of the course in order to receive a passing grade. Design a two-level NAND network whose output will indicate when a student passes the course.

10.16 Obtain a two-level NAND network to realize each of the following logic functions. (a) $T_a = ABC + \overline{A}\overline{B}C + \overline{A}B\overline{C} + A\overline{B}\overline{C}$; (b) $T_b = A\overline{B}C\overline{D} + \overline{A}BC\overline{D} + \overline{A}BCD$.

Mixed Logic Networks

10.17 The table of operations of a gate is given in Table 10-P17. Identify its function based on positive-logic assignment. Suppose both its inputs are active high and its output is active low. Set up the truth table. Can the resulting function be identified as any of the standard logic functions introduced in this chapter? Suppose both its inputs are active low and output active high. Set up the truth table. Can the resulting function be identified?

Table 10-P17. Table of Operations for Problem 10-17

A	B	T
L	L	L
L	H	H
H	L	H
H	H	L

10.18 Consider each of the networks of Fig. 10-P2 (Problem 10.2). Obtain a mixed logic network for each assuming that only 7400 gates are to be used and each input shown is available active high. The output may be active high or low.

10.19 Repeat the previous problem with 7402 gates.

10.20 Redesign the network of Problem 10.14 using (a) 7400 gates; and (b) 7402 gates. Assume that the inputs (level, temperature, and pressure) are active high.

10.21 For each of the logic functions given, obtain a mixed logic realization using (I) 7400 gates and (II) 7402 gates. (a) $T_a = (\overline{AB} + \overline{C})(A\overline{C} + B)$.

Electronics: Circuits and Systems

Inputs available: A.H, B.H, C.H, $\overline{A}$.L, $\overline{B}$.L, $\overline{C}$.L. Output may be active-high or low. (b) T = $(\overline{A} + B)\overline{C} + (A + \overline{C})\overline{B}$. Same input and output specifications as in part (a). (c) T_c = ABCDE. Available inputs: A.L, B.L, C.L, D.L, E.L. Output is to be active low. (d) T_d = (A + B + C + D + E). Same input and output specifications as in part (c).

10.22 Repeat Problem 10.21 if both 7400 and 7402 gates can be mixed together. Compare the chip count with the previous solutions.

10.23 Determine the output functions of the networks shown in Fig. 10-P23.

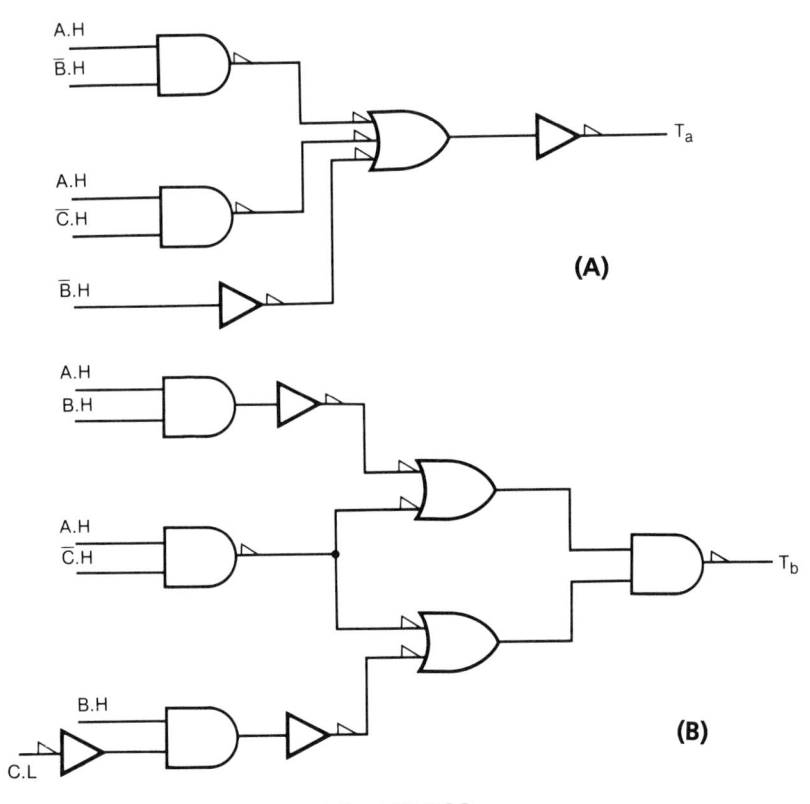

Fig. 10-P23.

Karnaugh Maps

10.24 Set up the Karnaugh maps for the functions of Problems 10.10, 10.11, and 10.12.

10.25 Set up the Karnaugh maps for the functions of Problems 10.14 and 10.15.

10.26 Set up a two-level logic network for each of the maps shown in Fig. 10-P26.

(A)

	BC 00	01	11	10
A 0	0	0	0	1
A 1	1	1	1	0

(B)

	BC 00	01	11	10
A 0	1	1	0	0
A 1	0	0	1	1

Fig. 10-P26.

(C)

	BC 00	01	11	10
A 0	0	0	1	1
A 1	1	1	0	0

10.27 Obtain the K-map of a function of four variables A, B, C, D, which is one whenever three or more variables have a value of one. Draw the two-level network for the function.

CHAPTER 11

DIGITAL LOGIC FAMILIES

Logic gates are available in the form of integrated-circuit (IC) packages and classified as families according to the type of devices and configurations used. The IC logic families are *transistor-transistor logic (TTL), integrated injection logic (I^2L), emitter-coupled logic (ECL),* and *complementary MOS logic (CMOS)*. The basic principle underlying the various logic circuits is that a transistor (BJT or FET) can be used as a device with two distinct states of operation: conducting or nonconducting. The output voltage of a transistor can be made high or low, which is the desired characteristic in a logic gate. Even though the basic principle of operation of the different logic families is the same, the actual structure and other factors such as voltage levels and speed of response are quite different between them. These factors should be taken into consideration when selecting logic gates and interconnecting different logic gates and networks.

TRANSISTOR AS A SWITCH

Consider the basic BJT circuit shown in Fig. 11-1. There are four regions of operation of the circuit: *cutoff, normal active, saturation,* and *reverse active.*

Cutoff Operation

When the input voltage v_i is *less than the minimum* value of V_{BE} (taken as 0.7 V) needed to make the transistor conduct, the transistor is said to be in the *cutoff* region. No currents flow, $I_C = 0$, the voltage drop in $R_C = 0$, and

$$V_o = V_{CC} \qquad \text{(cutoff)} \qquad (11\text{-}1)$$

This represents the *high* output level.

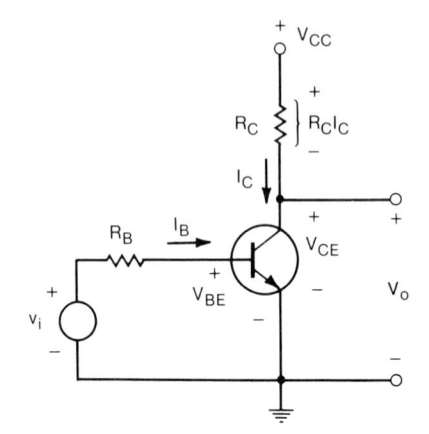

Fig. 11-1. Basic BJT circuit.

Normal Active Operation

When the input voltage v_i exceeds the minimum V_{BE} needed for conduction, the transistor conducts and enters the *normal active* region, which is the region of operation of transistors in amplifier circuits. In this region,

$$I_B = \frac{v_i - V_{BE}}{R_B} \quad (11\text{-}2)$$

$$I_C = \beta I_B \quad (11\text{-}3)$$

and the output voltage is smaller than V_{CC} due to the voltage drop in R_C:

$$V_o = V_{CC} - R_C I_C \quad (11\text{-}4)$$

Saturation Operation

As the input voltage v_i increases, both the base current and collector current increase, while the output voltage decreases as given by the last three equations. But the output voltage cannot decrease without limit. The output voltage is also the collector-to-emitter voltage of the transistor, and in an npn transistor, V_{CE} cannot become negative. In fact, V_{CE} can actually decrease only to a few tenths of a volt and stop at that value. For the sake of convenience, we will take the minimum value reached by V_{CE} as zero. Therefore,

$$V_{o(min)} = 0 \quad (11\text{-}5)$$

Substitution of $V_o = 0$ in Eq. (11-4) gives $I_C = (V_{CC}/R_C)$, which is the *maximum* value the collector current can reach. That is, in Fig. 1-11,

$$I_{C(max)} = (V_{CC}/R_C) \quad (11\text{-}6)$$

Once the collector current reaches its maximum value and the output voltage reaches its minimum value, no further changes can occur in them even though the input voltage may continue to increase. The transistor is now in the *saturation* region, and the output voltage is at its lowest level.

At the *boundary between the normal active operation and saturation*, the base-and-collector currents are still related by Eq. (11-3). But after saturation is established, I_B continues to increase while I_C remains constant at $I_{C(max)}$. Therefore, in the *saturation region, the base current is larger than that needed to maintain the collector current in the transistor*. That is, in the saturation region, Eq. (11-3) changes to an inequality

$$(\beta I_B) > I_{C(max)}$$

and combining this inequality with Eq. (11-6), we have

$$I_B > \frac{V_{CC}}{\beta R_C} \quad \text{(saturation)} \tag{11-7}$$

while I_B is still related to the input voltage v_i by Eq. (11-2).

It is customary to drive the transistor deeply into saturation by making I_B significantly larger than $(V_{CC}/\beta R_C)$. We will see, however, that this tends to decrease the speed of response of the circuit.

Example 11.1

In the circuit of Fig. 11-1, let $R_B = 50$ kΩ, $R_C = 2$ kΩ, and $V_{CC} = 12$ V. $\beta = 40$. Determine the two regions of operation for (a) $v_i = 0$; and (b) $v_i = 10$ V.

Solution

(a) $v_i = 0$: Since v_i is less than 0.7 V, the transistor is in cutoff. $I_C = 0$ and $V_o = V_{CC} = 12$ V.

(b) $v_i = 10$ V: The transistor conducts and from Eq. (11-2),

$$I_B = \frac{v_i - V_{BE}}{R_B}$$

$$= \frac{10 - 0.7}{40}$$

$$= 0.233 \text{ mA}$$

To check whether the transistor is in saturation, we start by *assuming* saturation and then testing if the inequality, (11-7), is satisfied. The value of $(V_{CC}/\beta R_C)$ is 0.150 mA, and since $I_B = 0.233$ mA, the inequality is satisfied, and the transistor is indeed in saturation.

$$I_C = (V_{CC}/R_C)$$
$$= 6 \text{ mA}$$

and

$$V_o = 0$$

Drill problem 11.1: Let $R_C = 5$ kΩ, $R_B = 20$ kΩ, $V_{CC} = 10$ V, and $\beta = 10$ in the circuit of Fig. 11-1. The input voltage is (a) $v_i = 0$; and (b) $v_i = 10$ V. Determine the operating conditions for the two inputs.

Reverse Active Operation

If the transistor is operated so that its emitter-to-base junction is reverse biased while its collector-to-base junction is forward biased, the transistor is in its *reverse active* region. This operation occurs in the TTL circuit and will be discussed later. The reverse action region is not used directly as one of the states of a logic circuit.

The two states of interest in the circuit of Fig. 11-1 are

cutoff: v_i is low, $V_o = V_{CC}$ (high)

saturation: v_i is high, $V_o = 0$ (low)

The circuit, therefore, acts as an *inverter* (or a NOT gate or a voltage level changer). The same general principles as those discussed previously are used in the more complex logic circuits (such as NAND gates). Almost all logic circuits use the cutoff and saturation regions as the two states, but in some circuits (ECL) the transistor does not enter saturation but operates in the normal active region when it is not cut off. Before studying the actual circuits of other logic gates, let us examine the problem of delays in the propagation of signals in logic gates. Again, we will use the inverter as the vehicle of our discussion but the discussion applies to the more complex gates as well.

PROPAGATION DELAYS

Consider an input pulse applied to the inverter as shown in Fig. 11-2A. Then, the collector current and output voltage will vary as a function of time as indicated in Figs. 11-2B and C. There is a delay between the application of the 10-V step input at $t = 0$ and the instant at which the transistor actually enters saturation. This delay (for the transistor to go from H to L) is specified as t_{pHL} by the manufacturer. At the end of the input pulse, there is a delay between the turning off of the input pulse and the instant at which the transistor actually cuts off. This propagation delay (for the transistor to go from L to H) is specified as t_{pLH} by the manufacturer.

Electronics: Circuits and Systems

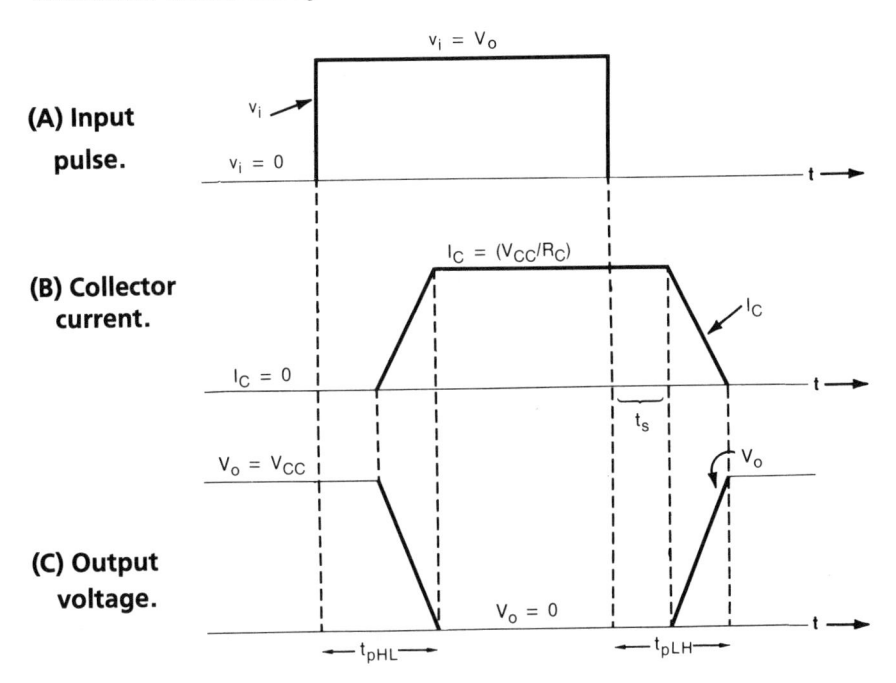

(A) Input pulse.

(B) Collector current.

(C) Output voltage.

Fig. 11-2. Propagation delays in an inverter; t_{pHL} represents the delay in the output switching from low to high, and tpLH represents the delay in switching from low to high.

Let us examine (qualitatively) the causes for the propagation delays. Consider the inverter just before $t = 0$ (or $t = 0-$) as in Fig. 11-3A. The base-emitter junction is reverse biased and acts like a capacitance C_{BE} due to the space charge stored in the reverse-biased junction.

Consider now the instant just after $t = 0$ (or $t = 0+$), as in Fig. 11-3B. The situation is analogous to the charging of an RC circuit: the voltage across C_{BE} cannot change instantaneously but only gradually at a rate controlled by the time constant $R_B C_{BE}$. When the voltage V_{BE} across C_{BE} reaches a value of 0.7 V, the transistor starts conducting and enters the normal active region.

When the transistor is in the normal active region, the collector current increases in direct proportion to the base current [Eq. (11-3)], and there is a corresponding decrease in the output voltage. Eventually, the collector current reaches its maximum value of (V_{CC}/R_C), and the output voltage reaches its minimum value of zero. The transistor is now in saturation.

Even though neither I_C nor V_o changes any further after saturation has been reached, there is some action taking place inside the transistor itself. The

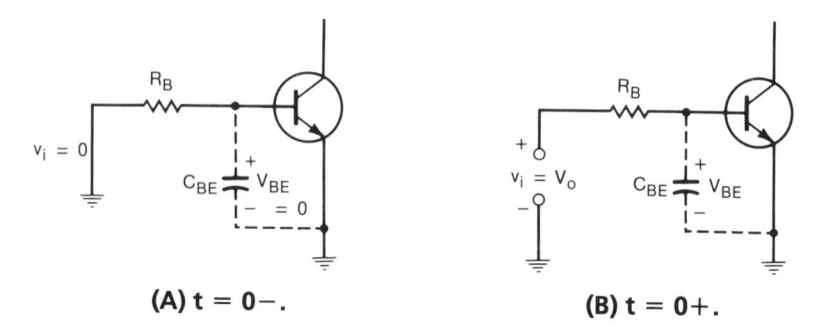

Fig. 11-3. Base-to-emitter capacitance causes the delay in the response of the circuit when the input switches from low to high (and output from high to low).

base current continues to increase as v_i is increasing toward the 10-V level, and this increase requires the injection of electrons from the emitter into the base region. Since the collector current cannot increase above $I_{C(max)}$, these electrons are *stored as excess charge in the base region.* This storage of excess charge is essential for ensuring the saturation of the transistor.

Now, consider the time T at which the input pulse is reduced to 0. Before the collector current can begin to decrease (and head toward cutoff), the excess charge stored in the base must be swept out. The time needed to clear the excess stored charge is called *storage time,* t_s. After the excess charge has been cleared, the transistor enters the normal active region, and the collector current begins to decrease linearly with the base current while the output voltage increases linearly [Eqs. (11-3) and (11-4)]. Eventually, $I_C = 0$ and $V_o = V_{CC}$. The transistor is now cut off.

It is seen that the presence of the excess stored charge in the base during saturation is a key factor in the turning off of the transistor. The more deeply the transistor is in saturation, the longer the storage time and the worse the delay. This slow response is the disadvantage of saturated operation. But saturation offers the advantage of making the low level of the output as low as possible. This increases the difference between the low and high output levels of the logic circuit, and such a difference leads to good noise margins in a logic circuit.

NOISE MARGINS

Even though the low level of a logic gate's output is nominally zero, and the high level is V_{CC} in the previous discussion, the actual voltage levels change when the logic gate is connected to other (similar) gates. The currents drawn by the load tend to produce a significant variation in both the high and low

Electronics: Circuits and Systems

voltage levels. The question arises, how *small* can the output voltage become and still be considered high and how large can it become and still be called low?

Consider the interconnection of two gates (Fig. 11-4). If the output of gate 1 is high, the smallest value V_o will have is specified by the manufacturer as $V_{OH(min)}$ called the *output high voltage*. Gate 2 should respond to this voltage as a high input. The manufacturer specifies the smallest value of the input that will be treated as high by a gate in the form of V_{IH} called the *input high voltage*. Clearly, $V_{OH(min)}$ and V_{IH} must satisfy the condition

$$V_{OH(min)} \geq V_{IH}$$

for proper operation of the interconnected gates. The *difference* [$V_{OH(min)}$ − V_{IH}] is a measure of the margin available for the proper operation of the

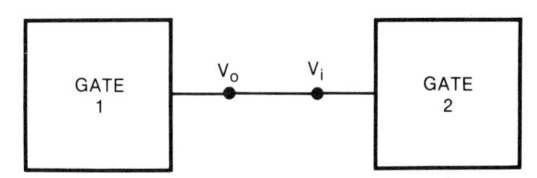

Fig. 11-4. Cascading of two gates.

interconnection. This difference is known as the *high-state noise margin* (or *noise immunity*) V_{NH}:

$$V_{NH} = [V_{OH(min)} - V_{IH}] \qquad (11\text{-}8)$$

If the output of gate 1 were at $V_{OH(min)}$, and a *negative* noise spike were to occur in the line between the two gates, then the input signal of gate 2 could go below the value of V_{IH} and cause malfunctioning of the system. Therefore, V_{NH} represents the immunity from noise spikes available in the high state.

Now consider the output of gate 1 being low. The largest value that V_o will have in this state is specified by the manufacturer as $V_{OL(max)}$ or the *output low voltage*. Gate 2 should respond to this voltage as a low input. The manufacturer specifies the largest value of the input that will be treated as low by a gate as V_{IL} or the *input low voltage*. Clearly, V_{IL} and $V_{OL(max)}$ must satisfy the condition

$$V_{IL} \geq V_{OL(max)}$$

for proper operation of the interconnected gates. The difference [V_{IL} − $V_{OL(max)}$] is a measure of the margin for the proper operation of the interconnection. This difference is known as the *low-state noise margin* (or *noise immunity*) V_{NL}:

$$V_{NL} = [V_{IL} - V_{OL(max)}] \tag{11-9}$$

V_{NL} gives the largest positive noise spike that can be tolerated by the system in the low state.

Example 11.2

The following specifications are given for a TTL gate: $V_{IH} = 2.0$ V, $V_{IL} = 0.8$ V, $V_{OH(min)} = 2.4$ V, $V_{OL(max)} = 0.4$ V. Calculate the noise margins.

Solution

$V_{NH} = (2.4 - 2) = 0.4$ V. $V_{NL} = (0.8 - 0.4) = 0.4$ V

Drill Problem 11.2: Calculate the noise margins of a gate for which $V_{IH} = 2.0$ V, $V_{IL} = 0.8$ V, $V_{OH(min)} = 2.5$ V, $V_{OL(max)} = 0.5$ V.

Drill Problem 11.3: Given that, for a certain gate, $V_{OL(max)} = 0.01$ V, $V_{OH(min)} = 4.99$ V, $V_{NL} = 1.0$ V, and $V_{NH} = 1.5$ V, determine the input low voltage and the input high voltage.

It can be deduced from the previous discussion that the noise margins would be higher in logic circuits where the H and L output levels are well separated from each other, which is the basic advantage of saturated logic circuits.

TRANSISTOR-TRANSISTOR LOGIC

The *transistor-transistor logic (TTL, T^2L)* is the fastest saturating logic gate family available in the form of ICs. It has typical turn-on and turn-off delay times of 5 ns (each), and a worst-case noise margin of 0.4 V at either end. It is one of the most popular digital IC logic components used in logic design.

The basic TTL circuit is shown in Fig. 11-5. The transistor T_1 is fabricated with multiple-emitter regions, and each of these (in conjunction with the base) acts as a diode. Even though only two emitters are shown in the diagram, actual gates can have four or more emitters in T_1.

When any input to the gate is low, that particular diode is forward biased and causes T_1 to conduct. T_1 will conduct when one or more inputs are low and will be cut off when all the inputs are high.

Case 1: At Least One Input Is Low

When any of the inputs are low, the corresponding emitter-base junction of T_1 becomes forward biased and that transistor conducts. From Fig. 11-5B, we have

voltage at the emitter of $T_1 = 0.7$ V

Electronics: Circuits and Systems

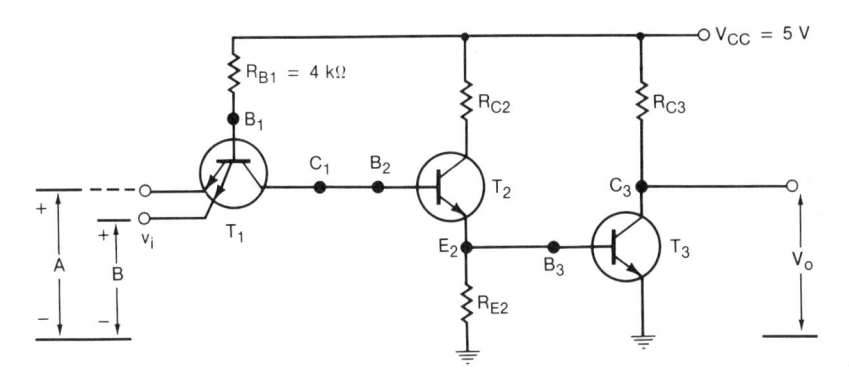

(A) TTL circuit.

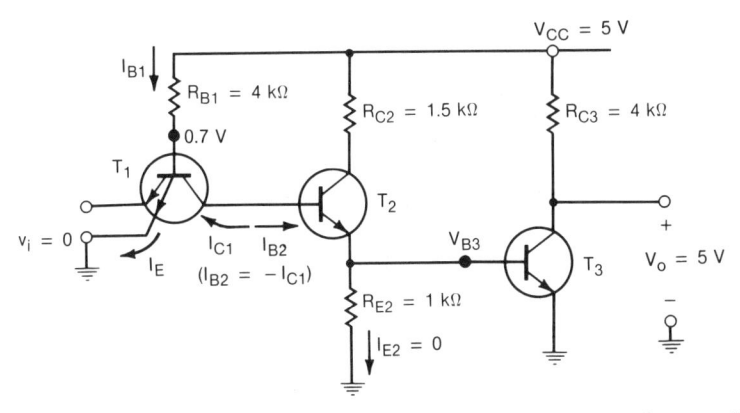

(B) Conditions when one or more inputs are low. T_1 conducts while T_2 and T_3 are cutoff. $V_o = V_{CC}$ (high output).

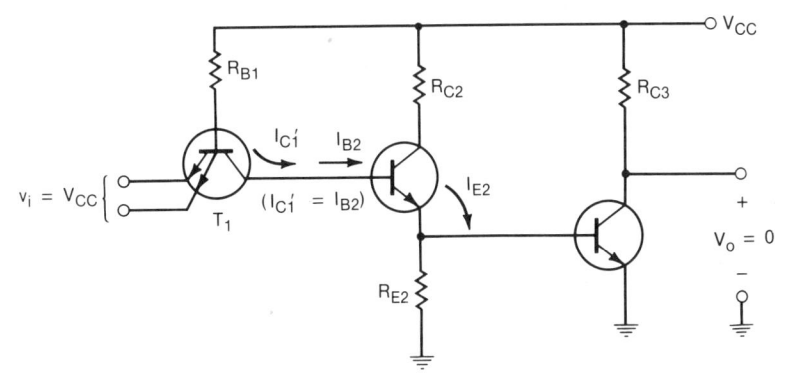

(C) Conditions when all inputs are high. T_2 and T_3 conduct (and are in saturation). T_1 is in the reverse active mode, and its collector current I'_{C1} is in a direction opposite to the normal direction. $V_o = 0$ (low output).

Fig. 11-5. TTL gate.

$$\text{voltage across } R_{B1} = (V_{CC} - 0.7) = 4.3 \text{ V}$$

$$\text{base current } I_{B1} \text{ of transistor } T_1 = (4.3/R_{B1}) = 1.08 \text{ mA}$$

The base current is quite large, and so we can expect T_1 to be saturated. Consider the collector current I_{c1}, which must be in the direction indicated in the diagram since T_1 is forward biased. But, I_{c1} is the negative of I_{B2} (the base current of T_2) in the circuit. Therefore, when I_{c1} flows in the direction indicated, there is a base current *away* from the base of T_2 and so T_2 is cut off. The current I_{c1} is then due to the current in the reverse-biased emitter-base junction of T_2, and is in the order of *nanoamperes*. Thus, with $I_{B1} = 1.08$ mA and I_{c1} in the nanoampere range, we have

$$I_{B1} > (I_{C1}/\beta)$$

which establishes the fact that T_1 is in saturation.

Since T_2 is cut off, we have

$$I_{E2} = 0 \text{ and } V_{B3} = 0$$

Therefore, there is not a sufficient voltage at the base of T_3 to turn it on. T_3 is, therefore, cut off also, which makes its collector current I_{c3} zero. So we have

$$V_o = V_{CC}$$

and the circuit is in the high output state.

Case 2: All Inputs Are High

When all the inputs to the circuit are high, the emitter-base junctions of T_1 are all reverse biased. But the power supply V_{CC} makes the p-type base of T_1 positive with respect to the collector and applies a forward bias across the collector-base junction. The transistor T_1 is thus forced into the *reverse active* mode: its emitter-base junction is reverse biased while its collector-base junction is forward biased. When a transistor is in the reverse active mode, its collector current flows in a direction opposite to the normal direction: that is, it comes *out* of the collector instead of flowing into the collector. The direction of the collector current I'_{c1} in T_1 is in the same direction as I_{B2}, the base current of T_2. Therefore, T_2 starts conducting and goes into saturation. The voltage developed by its emitter current I_{E2} across R_{E2} is large enough to turn T_3 on and drive it into saturation also. Therefore,

$$V_o = 0$$

and the circuit is in the low output state.

The table of operations of the circuit is shown in Table 11-1. It is seen to be a *NAND gate under positive logic convention*.

Table 11-1. Table of Operations of the Circuit

A	B	V_o
L	L	H
L	H	H
H	L	H
H	H	L

The high speed of response of the TTL gate is due to the connection of T_1 and T_2 in the circuit. When T_2 is saturated (which is when all inputs are high), excess charge is stored in its base, which has to be swept out before T_2 can switch to cutoff when one or more inputs become low. When any input becomes low, it turns T_1 on. There is a transient flow of current in the collector of T_1, which is quite large and is in a *direction opposite to* I_{B2}. This current sweeps out the excess stored charge in the base of T_2 very quickly and speeds up the onset of cutoff in T_2 and hence T_3. Thus, there is a considerable reduction of the storage time and the turn-off delay time.

Open Collector TTL Gate

TTL gates are available in which the resistor R_{C3} is omitted so that the circuit is as shown in Fig. 11-6A. Such gates are called *open collector* gates and are useful in some applications. The user has to provide an external resistor R_L (along with V_{CC}) as indicated.

One of the special advantages of the open collector gates is that their output (collector) terminals can be tied together and then to a single resistor R_L and V_{CC} as indicated in Fig. 11-6B. In such a connection, if the output of any single gate goes low, it will make V_o low regardless of the state of the other gates. The output V_o is high only when all the gates have a high output. Therefore, we have an *AND operation between the outputs of the individual gates* by the simple expedient of tying all the outputs together. Such a connection is called *wired-AND* operation, and its symbol is shown in Fig. 11-6C. Note that an AND operation is obtained here without an actual AND gate being present.

The value of R_L in using open collector gates is in the range of a few hundred to a few thousand ohms and is not particularly critical.

Loading Considerations

The normal use of a TTL gate to drive a number of other similar gates is shown in Fig. 11-7. The number of load gates that can be driven by the TTL gate and still maintain proper output voltage levels V_{OH} and V_{OL} is called its *fan-out*. When the output of the driving gate is high, the transistor T_3 is cut off and the emitter-base junctions of the input transistors in the load gates are reverse biased as indicated in Fig. 11-7B. Only a small current flows through

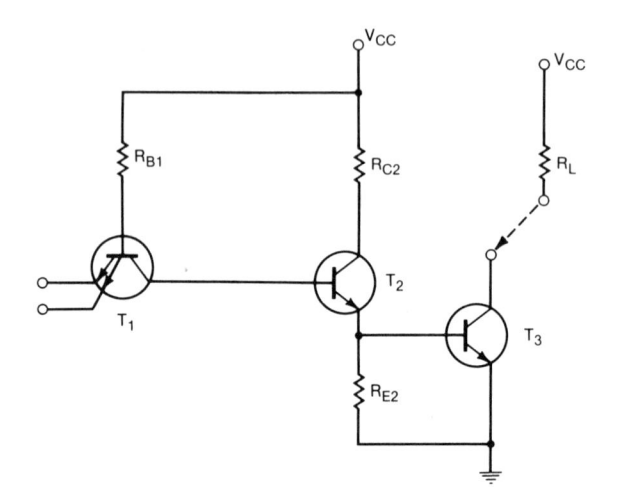

(A) V_{CC} and R_L are to be added to the IC chip by the user.

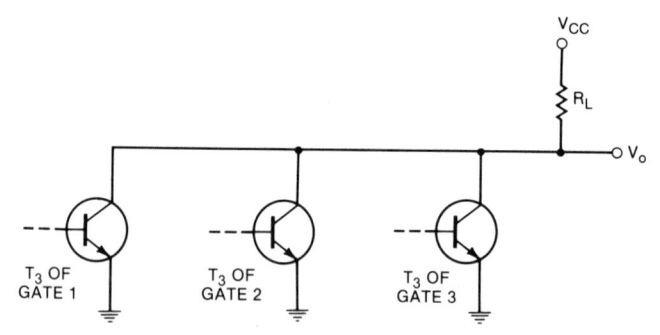

(B) Collectors of the output transistor (T_3) of a number of such gates can be directly connected to a single resistor R_L and a power supply V_{CC}.

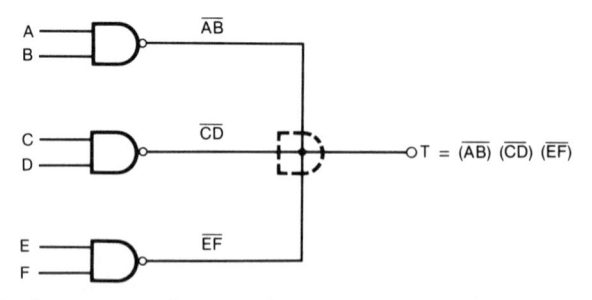

$$T = \overline{(AB)} \, \overline{(CD)} \, \overline{(EF)}$$

(C) Such a connection produces a wired-AND operation.

Fig. 11-6. Open collector TTL gate.

the transistor T_3 of the driving gate. The high output state does not present any problems.

Now consider the case where the output of the driving gate is low. The transistor T_3 is in saturation. The emitter-base junctions of the input transistors of the load gates are forward biased. The current flowing in T_3 is the sum of two components: the current I_{C3} in R_C and the currents sent through the forward biased junctions of the load gates, as indicated in Fig. 11-7C. The current in T_3 is a cause for concern if it gets too large. Even though we have assumed that a saturated transistor is like a short circuit from collector to emitter in our discussion, the fact is that there is a small resistance from its collector to emitter, and this resistance produces a voltage drop from the collector to emitter. As the current in T_3 increases, the voltage drops, and consequently, the output voltage may rise above the permitted value of the input low voltage V_{IL} of the load gates, which causes a malfunction of the load gates. Thus, the fan-out is limited by the current flowing through T_3 when it is in saturation. One of the contributors to the current in T_3 is I_{C3}, which can be unnecessarily large of R_{C3} is small. On the other hand, a large R_{C3} is undesirable also as will be seen shortly.

Totem Pole Output Stage

The presence of R_{C3} in the TTL gate of Fig. 11-5 was seen to present a problem if it is too small. But, it also reduces the speed of response of the gate if it is too large.

When the output of a TTL gate is low, there is an effective load capacitance C_L (typically 15 pF) due to the junction capacitances of the load gates and stray capacitance. If the output is now made to switch from low to high (by making one of the inputs switch to low), the transition requires a charging of the capacitance C_L. The time constant of the charging process is $R_{C3}C_L$ as can be seen from Fig. 11-8. If R_{C3} is made large (in order to limit the current I_{C3} as discussed previously), then the time constant for the low to high transition becomes large and slows down the transition. Thus, whether large or small, R_{C3} creates difficulties and a different output arrangement is necessary. The alternative output stage is called a *totem pole output stage*, and most TTL gates are fabricated with such an output stage. As shown in Fig. 11-9, R_{C3} has been replaced by a transistor T_4, a diode D_4, and a small resistor R_{C4}.

Consider the case when the output of the gate is low. As seen earlier, T_2 and T_3 are in saturation, and the collector voltage of T_2, which is also the base voltage of T_4, is too low to turn T_4 on. In fact, the purpose of the diode D_4 is to raise the minimum level of the base voltage needed at T_4 to turn it on and ensure the cutoff at T_4 when T_2 is saturated. Since T_4 is cut off, very little current is drawn by T_3 in the low output state. Thus, the limitation on the fan-out due to the large current in T_3 in the low output state in the earlier TTL gate is significantly alleviated by the totem pole output stage.

Now consider the case when the output switches from low to high. As discussed earlier, T_2 and T_3 turn off. When T_2 cuts off, its collector voltage increases and turns T_4 on. Since T_3 is off and T_4 is on, a charging path for the capacitance is made available from V_{CC} through R_{C4}, T_4, and D_4. The resistances of T_4 and D_4 are small (since they are in saturation), and the time constant is, therefore, $R_{C4}C_L$, which is quite small since R_{C4} is only about 100 ohms. The totem pole stage speeds up the low to high transition of the gate considerably.

The only time the TTL gate with a totem pole output stage draws any appreciable current is when its output is low and a number of load gates are connected to it. As indicated in Fig. 11-10, the TTL gate acts as a *current sink* for the currents fed to T_3 through the forward biased junctions of the input

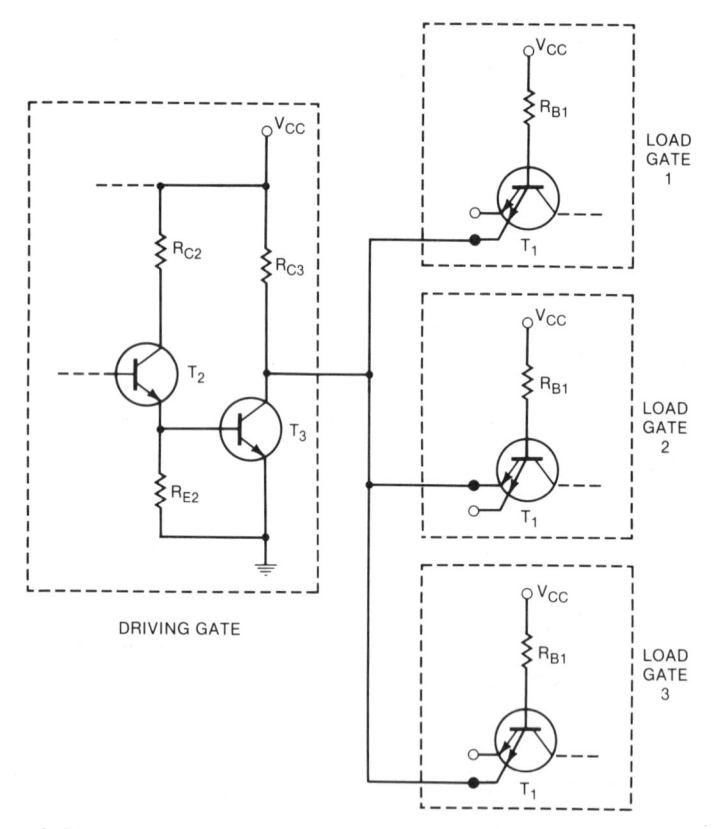

(A) Load on a TTL gate is a number of similar gates.

Fig. 11-7. Loading of a

Electronics: Circuits and Systems

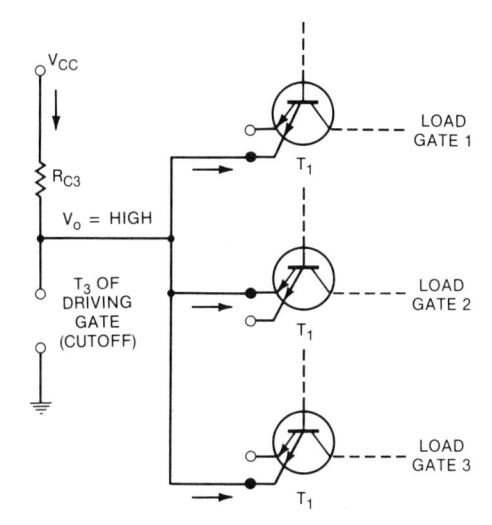

(B) Conditions when the driving gate's output is high. The currents in the load gates are quite small since their emitter-base junctions are reverse biased.

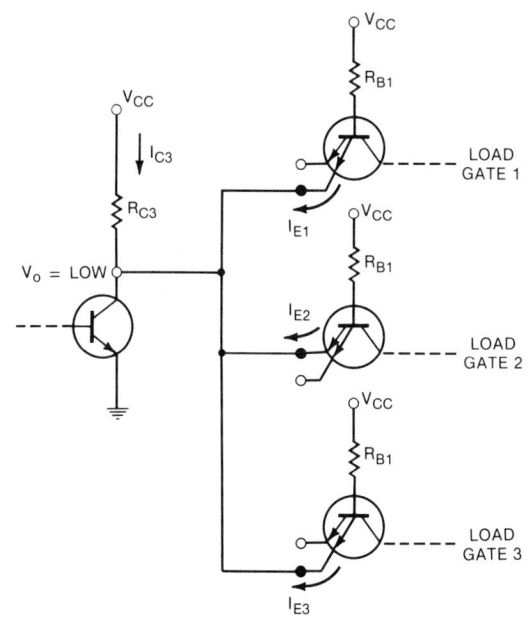

(C) Conditions when the driving gate's output is low. The transistor T_3 of the driving gate acts as a sink for the emitter currents of the load gates that are all forward biased.

TTL gate.

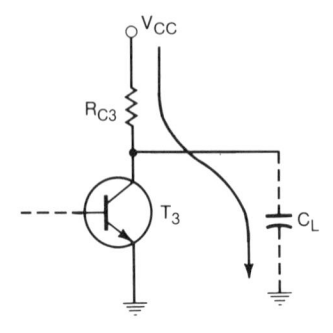

Fig. 11-8. Charging of the junction capacitance when the output of a TTL gate switches from low to high.

transistors of the load gates. The fan-out of a TTL gate is limited by its current sinking capability, as will be discussed shortly.

One important constraint in using TTL gates with totem pole output stages is that *they cannot be tied together to perform a wired-AND operation* like the open collector TTL gates. Consider two TTL gates wired together as indicated in Fig. 11-11.

If gate 1 is in the high output state and gate 2 is in the low output state, T_4 of gate 1 and T_3 of gate 2 are in saturation (while the other two transistors are off). A path for current flow from V_{CC} (of gate 1) is available through T_4 of gate 1 and T_3 of gate 2. This current is quite large since it is limited only by R_{C4} in gate 1. It may be several times larger than the current permitted in T_3 of gate 2 and can easily damage that transistor. Therefore, TTL gates with totem pole outputs cannot function safely as wired-AND gates.

The manufacturer specifies whether a particular TTL gate is open collector. If such a specification is not made, then the gate is taken as having totem pole output stages.

Interpretation and Use of the Data Sheet

An understanding of the data sheet provided for a TTL gate by the manufacturer is important for a proper use of the gate especially in connecting loads to it. The data sheet of the 74LS00 gate is included in the Appendix and will be used to explain the various specifications.

The typical supply voltage (V_{CC}) is 5 V (which is the case for all TTL gates).

The voltages V_{IH}, V_{IL}, V_{OH}, and V_{OL} were discussed in connection with noise margins. V_{IH} represents the smallest input voltage that will be treated as a high input; V_{IL} is the largest input voltage that will be treated as a low input; V_{OH} whose minimum limit is given represents the smallest output voltage that will be available from the gate when it is in the high output state; V_{OL} whose maximum value is given represents the largest output voltage that will be available from the gate when it is in the low output state.

Along with the specifications of V_{OH} and V_{OL}, two currents I_{OH} and I_{OL} are

Electronics: Circuits and Systems 669

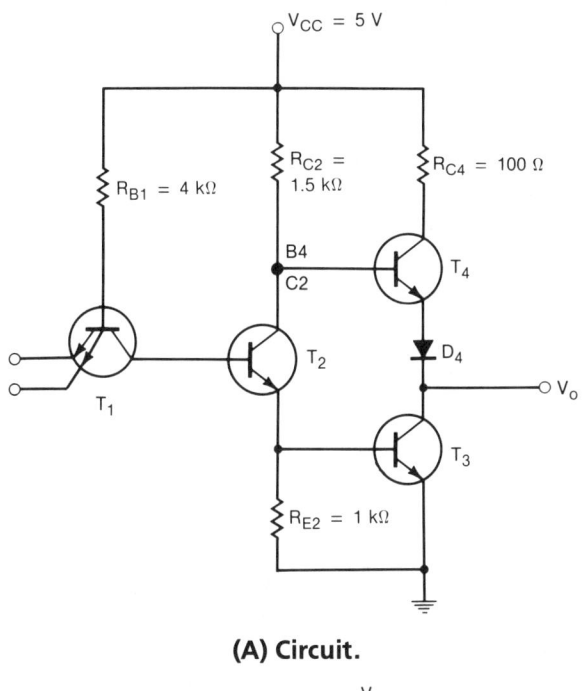

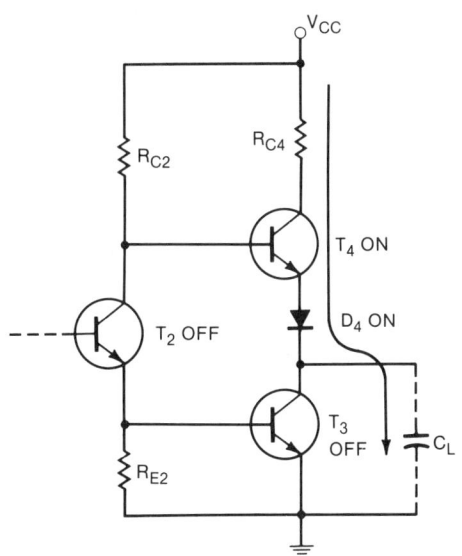

(A) Circuit.

(B) Charging of the junction capacitance when the output switches from low to high. The time constant is small.

Fig. 11-9. Totem pole output stage in a TTL gate.

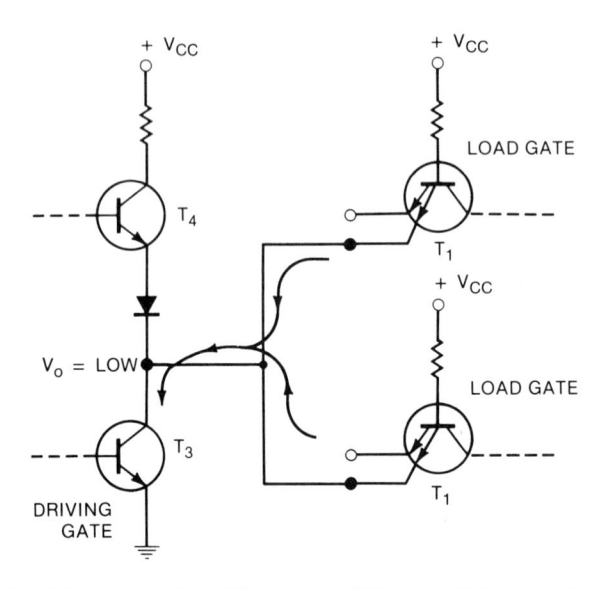

Fig. 11-10. Totem pole TTL gate with conditions when the driving gate's output is low. The transistor T_3 acts as a sink for the currents drawn through the forward biased emitter-to-base junctions of the load gates.

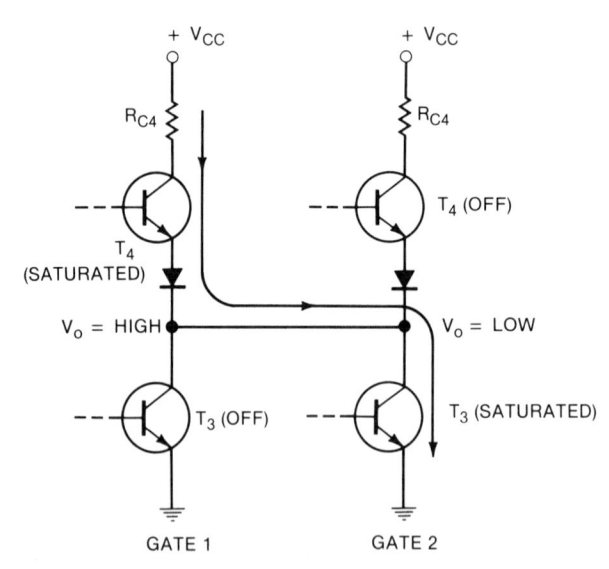

Fig. 11-11. Direct interconnection of TTL gates with totem pole output stages leads to the flow of a large current and is not permitted.

given under test conditions. The *output high current* I_{OH} is the largest output current when the gate is in the high output state without the output voltage going below $V_{OH(min)}$. The *output low current* I_{OL} is the largest current that can be sunk by the gate when it is in the low output state without the output voltage exceeding $V_{OL(max)}$. The fan-out of the gate is limited by the value of I_{OL}.

The *input high current* I_{IH} is the input current when the input is high, and the *input low current* I_{IL} is the input current when the input is low. The negative sign in the value of I_{IL} simply indicates that the current is coming out of the input side and can be ignored.

Let us make some calculations using the data of the 74LS00 gate.

$V_{IH} = 2.0$ V and $V_{OH} = 2.7$ V (for the XC series). Therefore, the high-state noise margin is 0.7 V.

$V_{IL} = 0.8$ V and $V_{OL(max)} = 0.5$ V. Therefore, the low-state noise margin is 0.3 V.

When a driving gate is in the low output state, $I_{OL} = 4.0$ mA. That is, the total current fed to the driving gate by the (forward biased) input transistors of the load gates must not exceed 4.0 mA. The current fed by each load gate when its input is low is $I_{IL} = 0.36$ mA. Therefore, the maximum number of 74LS00 gates that can be connected as load is (4.0/0.36), which is approximately 10.

When a driving gate is in the high output state, $I_{OH} = 400$ μA. That is, the total current drawn by the load gates must not exceed 400 μA. The current drawn by each load gate when its input is high is $I_{IH} = 20$ μA. Therefore, the maximum number of 74LS00 gates that can be connected as load is (400/20) = 20.

The fan-out is limited to ten gates (using the worst case, which is the low output state in this case).

Drill Problem 11.4: For a particular TTL gate, the following data are given: $V_{IH} = 2.0$ V; $V_{IL} = 0.8$ V; $V_{OH} = 2.5$ V; $V_{OL} = 0.4$ V; $I_{OH} = 0.25$ mA; $I_{IH} = 40$ μA; $I_{OL} = 16$ mA; $I_{IL} = 1.6$ mA. Calculate the noise margins and the fan-outs for high and low states.

When two or more *inputs* of a load gate are tied together as indicated in Fig. 11-12, the loading on the driving gate is affected in the following manner: for a low output state, the tied inputs act as a single load; but in the high output state, *each* input (whether tied to others or not) acts as a separate load. In the arrangement shown, for example, the driving gate has a fan-out of six in the high but three gates in the low state.

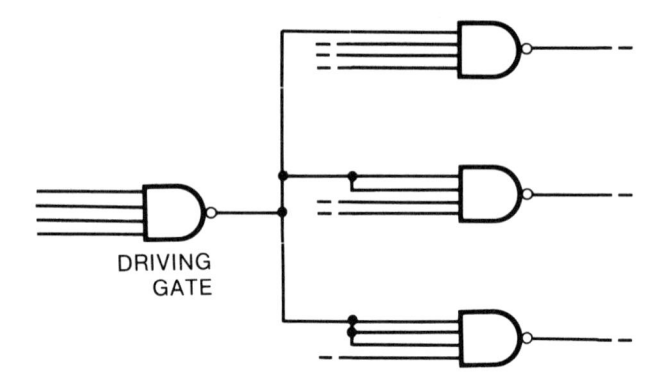

Fig. 11-12. Fan-out considerations when two or more inputs of some load gates are tied together.

Any unconnected input of a TTL gate (that is, left open) behaves as if it were connected to a high input. It is, however, undesirable to leave an input terminal open. When using a gate with more input terminals than needed for a given function, there are two alternatives: connect two or more inputs together (as discussed in the previous chapter) or connect the unused terminals to V_{CC} (usually through a 1-kΩ resistor).

The data sheet sometimes includes a specification of an *input clamp diode voltage*. The inputs of a TTL gate are often connected through clamp diodes

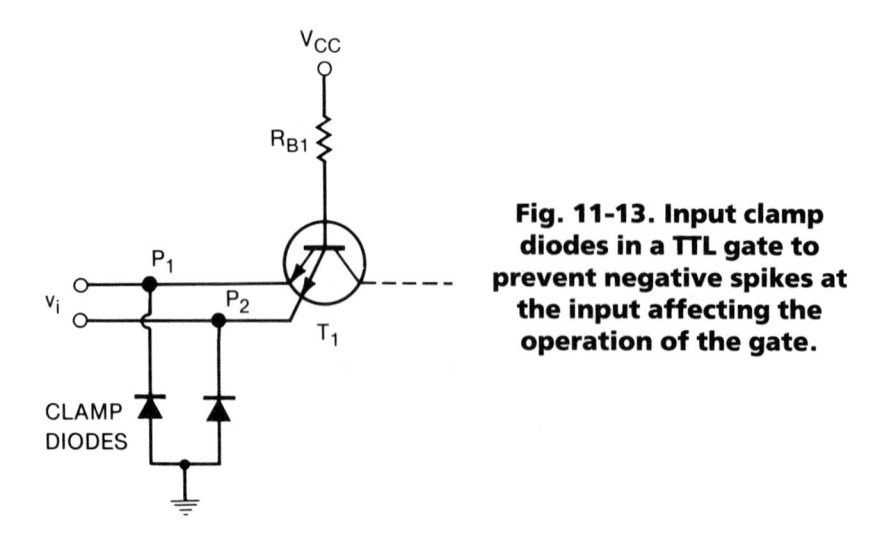

Fig. 11-13. Input clamp diodes in a TTL gate to prevent negative spikes at the input affecting the operation of the gate.

Electronics: Circuits and Systems   673

to ground as indicated in Fig. 11-13, and their function is to prevent the input voltage from becoming too negative if negative noise spikes should occur. In the 74LS00 gate, the input voltage cannot go below -0.65 V because of the clamp diodes conducting if there is a large negative spike.

THREE-STATE LOGIC

The TTL gate with the totem pole output stage is often modified by the addition of some transistors and a diode to form a *tri-state logic gate* or *three-state* logic gate. This does not indicate that there are three logic levels of operation as the name "tri-state" might imply, but a third condition is introduced in which the circuit appears to be a high impedance and is essentially isolated from the other gates in the network.

A tri-state logic gate is shown symbolically by adding an additional *control terminal* as in Fig. 11-14. When the control input is low, the gate behaves like a normal TTL gate: its output is high (which means that the lower transistor T_3 in the totem pole is off and the upper transistor T_4 is on), or low (that is, the lower transistor T_3 is on and the upper transistor is off), depending upon the input conditions. But when the control input is high, *both the transistors T_3 and T_4 are off,* and the gate has a high output impedance. Its output is neither high nor low regardless of the values of its inputs.

Fig. 11-14. Three-state (tri-state) logic gate.

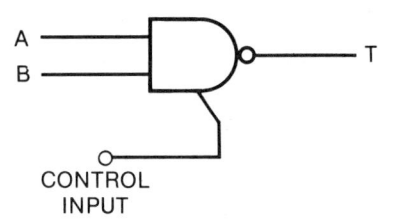

The tri-state gate is useful in the interconnection of two or more gates to a common bus as shown in Fig. 11-15. Only one control input is made low at any given instant. The other gates will then have no influence on the signal on the bus, and the output of the gate with the low control input will appear on the bus. By cycling the control input among the several gates, the bus can be made to carry the outputs of the different gates in turn. Such a bus system is used for the transmission of data on a data bus in computer systems.

The control input is also referred to as the *enable input*. Tri-state buffer gates are available that transmit the input signal to the output of the gate when the control (or enable) input is low. When the control (or enable) input is high, the gate acts as a high-output impedance circuit. The symbol of a tri-state buffer is shown in Fig. 11-16.

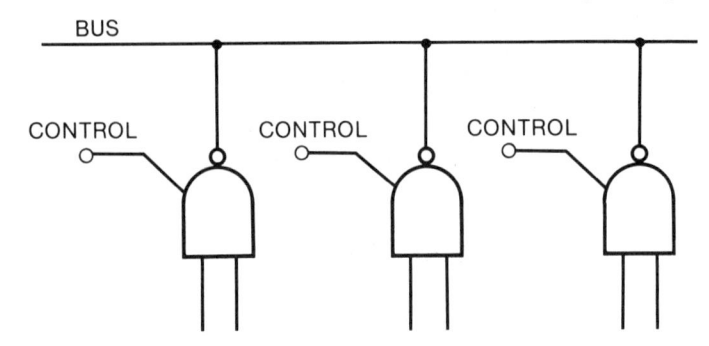

Fig. 11-15. Interconnection of tri-state gates to a common bus.

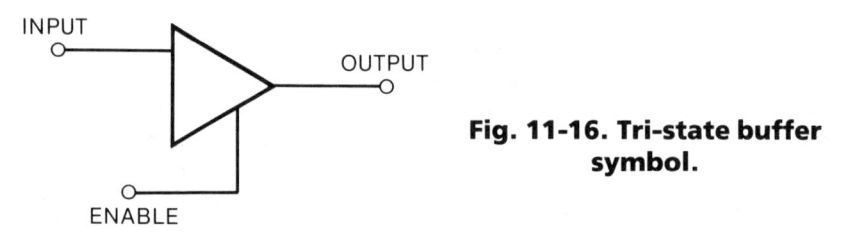

Fig. 11-16. Tri-state buffer symbol.

INTEGRATED INJECTION LOGIC

Integrated injection logic (or I²L) gates can be fabricated at much higher densities in ICs than TTL, and they also permit a trade-off between power consumption and propagation delay. They can be operated at a low power consumption with a large delay (or slow speed), or they can be operated at a high power consumption with a small delay (or high speed).

Regions of the different transistors in a logic circuit can be merged in IC fabrication so as to reduce the number of fabrication steps and also the area needed by the different regions. This merging is the reason for the adjective "integrated" in the name of I²L gates. Also, a current is injected into the bases of the transistors in the circuit through a current source; hence the word "injected."

Consider the circuit shown in Fig. 11-17. It can be seen to be two logic inverters whose bases are tied to each other and whose emitters are connected to ground. The input v_i is obtained from the output of a circuit similar to the one shown here. When the input is high, it presents an *open circuit* to ground. The two transistors now conduct and the current I_O is large enough to saturate the transistors. When the input is low, it presents a short circuit to ground that causes the transistors to cut off. The current I_O now flows through R_{C1} to ground.

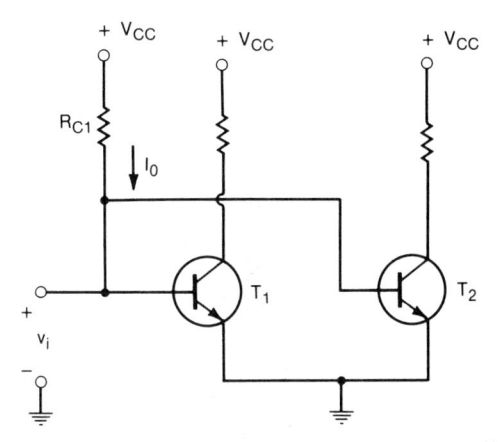

Fig. 11-17. Circuit to introduce the principle of I²L gates.

The current I_O depends upon the value of the input signal in the circuit shown but it is desirable to have a current I_O that is a constant and independent of v_i. A constant current source using a pnp transistor is shown in Fig. 11-18. The current I_O is given by

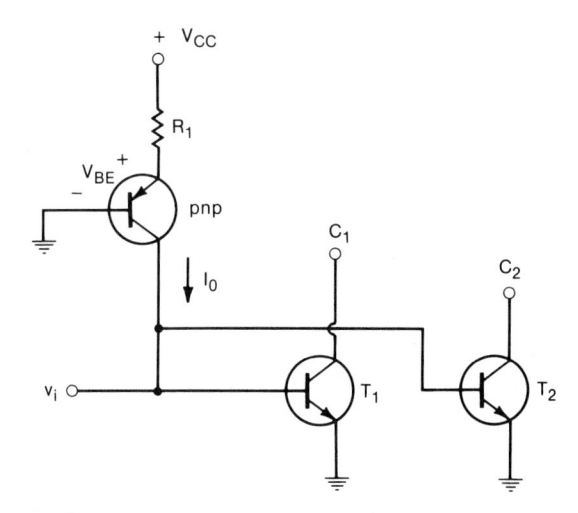

$$I_O = \alpha \frac{(V_{CC} - V_{BE})}{R_1}$$

Fig. 11-18. I²L gate. The pnp transistor acts as a constant-current source.

which is independent of the input signal v_i. The use of a pnp transistor for the current source facilitates the merging of its regions with those of the npn transistors T_1 and T_2 as shown in the following discussion.

Consider the possible mergers of the semiconductor regions in the circuit of Fig. 11-18. The base of the pnp transistor and the emitters of the npn transistors are merged since they are all n-type and all connected to ground. The collector of the pnp transistor and the bases of the two npn transistors are also merged since they are all p-type and connected together. The IC scheme is shown in Fig. 11-19. The n epitaxial layer serves as the base of the pnp transistor and the emitters of the npn transistors. The p-region forming the collector of the pnp transistor and the bases of the npn transistors is usually a long p-type diffusion rail in the fabrication process and called the *injector rail*. The collectors of the npn transistors are n+ type wells in the injector rail.

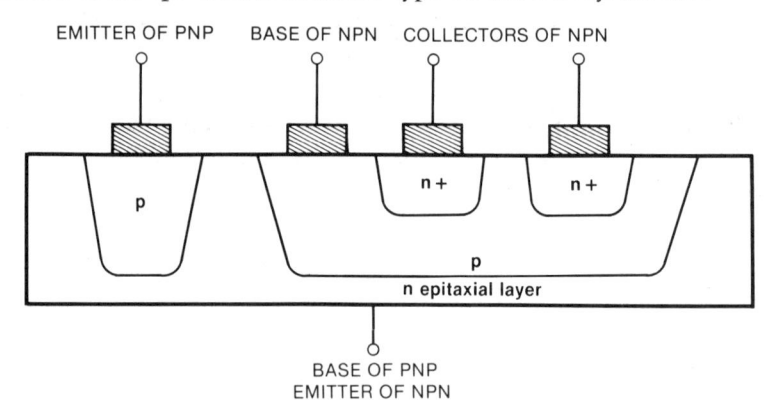

Fig. 11-19. IC version of the I²L gate. Several regions are merged (or integrated) so as to reduce the real estate needed on the chip.

The resistor R_1 and the power supply V_{CC} are connected externally to the chip. The circuit model of the I²L is usually shown using a *multiple-collector transistor* as indicated in Fig. 11-20. (More than two collectors can be easily fabricated.)

The extension of the previous scheme to the fabrication of a logic circuit to perform positive OR and positive AND operations is shown in Fig. 11-21.

The logic levels of the I²L gate are short circuit to ground when the output transistor is saturated and open circuit when the output transistor is not conducting. Usually, the output voltage when the transistor is not conducting is limited to about 0.7 V since such an output often feeds the bases of other transistors. The logic swing of the I²L gate is, therefore, only about 0.7 V.

When a transistor in the I²L gate switches from on to off or from off to on,

Electronics: Circuits and Systems

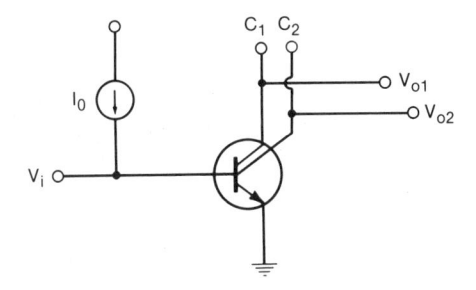

Fig. 11-20. Circuit model of an I²L gate. The gate has two inverters due to the two collectors.

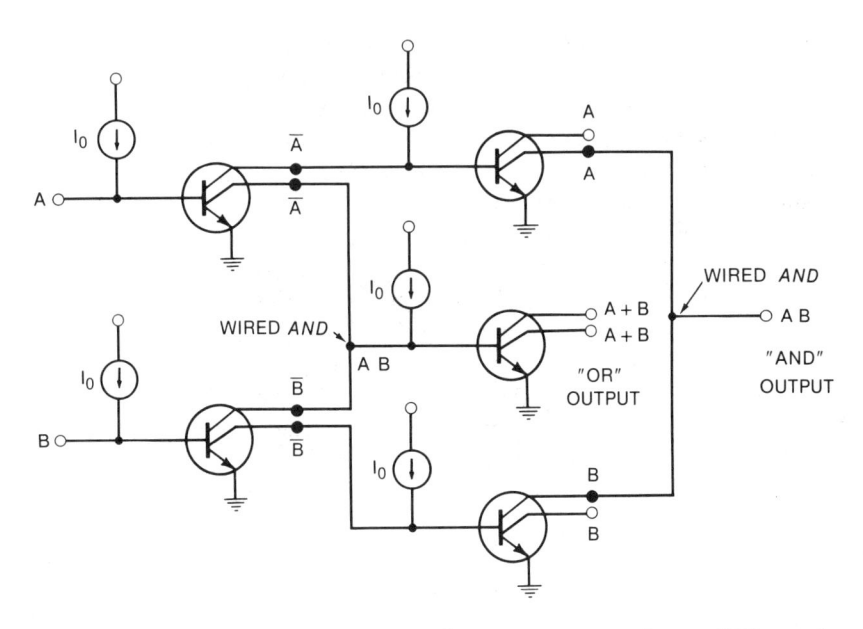

Fig. 11-21. Interconnection of I²L gates to produce different logic functions. Note that both (A + B) and (AB) are available in this configuration.

there is propagation delay due to the capacitances present in the circuit (which is the same sort of situation as in TTL gates) but in I²L the charging and discharging do not take place through resistors. The injected currents perform the necessary transition. If a fast response is needed, then large currents need to be injected with a consequent increase in power consumption. On the other hand, if low power consumption is desired, then the response of the circuit is slower. A figure of merit, which is the product of delay and power, is useful in considering the trade-off of power and propagation delay. A delay-power product of 1 picojoule can mean a power consumption of 50 microwatts/gate at a delay of 20 ns. Other combinations of power consumption and delay can be used. The power consumption is controlled by the injector current, which can be altered by simply changing the resistor R_1 connected externally to the chip. Thus, the speed of operation can be varied after the chip has been fabricated.

I^2L packages are fabricated in LSI (large-scale integration) form and contain more than 100 gates/chip.

EMITTER-COUPLED LOGIC

In the *emitter-coupled logic (ECL)*, transistors operate in the normal active region when conducting instead of in the saturation region. Since there is no excess stored charge in the base region (which would be required in saturation), the speed of response of the ECL gates is higher than that of the TTL. The basic circuit of an ECL gate is the differential amplifier (discussed in Chapter 8) where the emitters are connected to each other, hence the name *emitter coupled*. Consider the differential amplifier in Fig. 11-22. V_{ref} is a reference voltage connected to the base of T_2, and v_i is the input signal at the base of T_1. If v_i is greater than V_{ref} by about 0.1 V, then T_1 conducts and T_2 is cut off, and the voltage V_{o2} becomes equal to V_{CC}. The circuit components are chosen so that T_1 remains in the normal active region under this condition. V_{o1} is less than V_{CC} by the drop in R_{C1}. On the other hand, if v_i is less than V_{ref} by about 0.1 V, T_1 is cut off and T_2 conducts. The output V_{o1} equals V_{CC}, and the circuit components are chosen such that T_2 is in the normal active region. V_{o2} is less than V_{CC} due to the drop in R_{C2}. Therefore, we have

$v_i > V_{ref}$: V_{o1} is low; V_{o2} is high

$v_i < V_{ref}$: V_{o1} is high; V_{o2} is low

In the circuit of Fig. 11-22, the high and low voltages at the output are different from those at the input, which are usually fairly close to the V_{ref} value. The circuit is modified by the addition of two level shifting stages that are emitter-follower circuits. Note that $V_{CC} = 0$, and the power supply is from a negative voltage source $-V_{EE}$. The circuit of Fig. 11-23 acts as a positive OR and a positive NOR gate of the inputs. The availability of two outputs that are mutual logic complements from a single circuit is an attractive feature of the ECL gate. The output levels of the gate shown are -1.7 V (low) and -0.8 V (high), which is not a very large swing. Consequently, the noise margin is small also, being 0.2 V. The delay is 2 ns, which is less than any TTL gate, and the power consumption is 25 mW, which is higher than for a TTL.

Electronics: Circuits and Systems

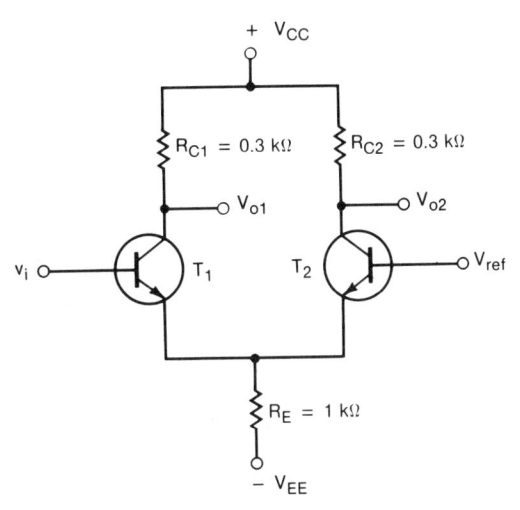

Fig. 11-22. Differential amplifier.

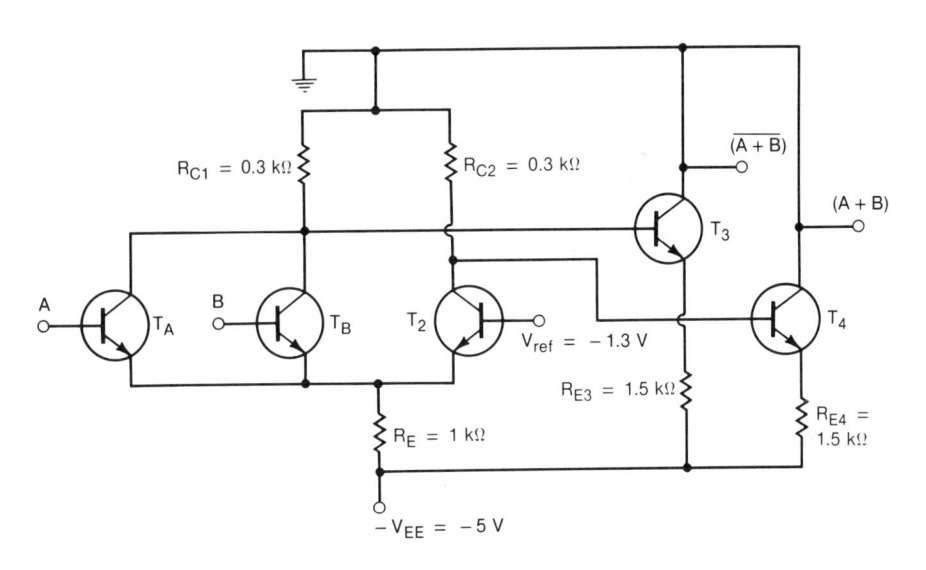

Fig. 11-23. Emitter-coupled logic (ECL) circuit. Note that both (A + B) and $\overline{(A + B)}$ are available in this gate.

It should be noted that the levels of an ECL gate are not compatible with those of a TTL gate, and the two cannot be interconnected directly.

COMPLEMENTARY MOSFET (CMOS) LOGIC

We will start with a brief review of the basic operating characteristics of the *enhancement-mode MOSFET*.

The NMOS, or n-channel enhancement-mode MOSFET, is analogous to the npn transistor. The NMOS symbol is shown in Fig. 11-24A. The NMOS has a threshold voltage V_T, which is positive. When the gate-to-source voltage, V_{GS}, is greater than the threshold voltage, the transistor conducts. The current I_D increases as V_{GS} increases and eventually a saturation condition is reached just as in the BJT. Thus, for an NMOS,

$V_{GS} > V_T$: NMOS conducts   (V_T positive)

$V_{GS} < V_T$: NMOS is cut off

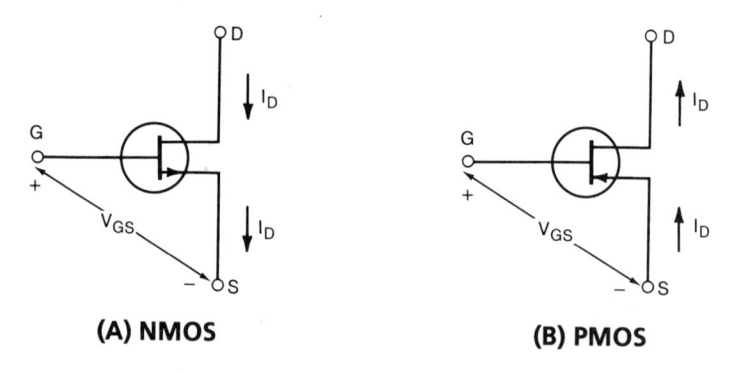

(A) NMOS　　　　　**(B) PMOS**

Fig. 11-24. NMOS and PMOS transistors.

The PMOS, or p-channel enhancement-mode MOSFET, is analogous to the pnp transistor. The PMOS symbol is shown in Fig. 11-24B. The PMOS has a threshold voltage V_T', which is negative. The PMOS conducts when V_{GS} is less than its threshold voltage, becoming saturated at a sufficiently negative value of V_{GS}. For the PMOS, we have

$V_{GS} < V_T'$: PMOS conducts   (V_T' negative)

$V_{GS} > V_T'$: PMOS is cut off

It is possible to fabricate an enhancement-mode NMOS and an enhancement-mode PMOS on the same IC chip by connecting their drain regions together as indicated in Fig. 11-25. The composite device is called a complementary MOSFET (CMOS) since one MOS is conducting when the other is cut off when the same signal is applied to both their gate terminals. The basic CMOS inverter circuit is also shown in Fig. 11-25. The two gates

Electronics: Circuits and Systems

are connected together and receive the same input signal. Note that the *source* terminal of the PMOS is connected to the power supply V_{DD}.

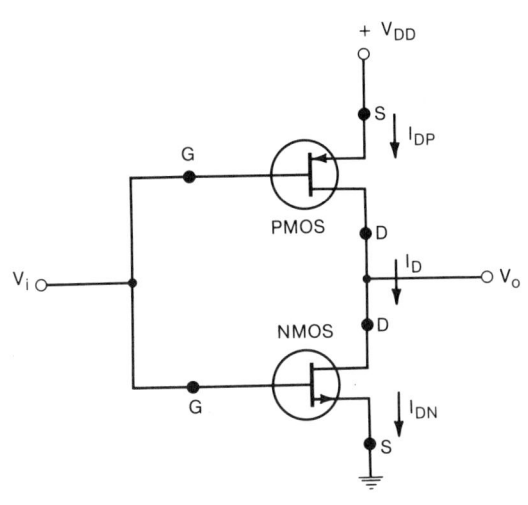

Fig. 11-25. CMOS gate. Note that the *source* terminal of the PMOS is connected to the power supply and that the drains of the two MOS devices are connected together.

The input signal has a high level, which is at V_{DD}, and a low level, which is at ground.

Case 1: Input Is Low

The gate-to-source voltage of the PMOS is $-V_{DD}$ since the gate is at ground and the source terminal is at V_{DD}. Therefore, the PMOS is in the conducting mode. The gate-to-source voltage of the NMOS is at zero since both the gate and the source are at ground. Therefore, the NMOS is cut off (Fig. 11-26A).

Since the drain current of the PMOS is constrained to be the same as that of the NMOS by the interconnection of the two drains, the *drain current of the PMOS is forced to be zero* by the nonconducting NMOS. The PMOS is, therefore, in the conducting mode but its drain current is zero. These conditions result in making the drain-to-source voltage of the PMOS equal zero. That is, we have for the *low input* case:

$$V_{GSP} = -V_{DD}. \quad I_{DP} = I_{DN} = 0. \quad V_{DSP} = 0 \text{ (PMOS)}.$$

The output voltage is V_{DD} since there is no drop in the PMOS. The output is high when the input is low.

Case 2: Input Is High

The situation is now exactly the opposite for case 1. The PMOS is cut off since its gate-to-source voltage is zero, and this forces the drain current to be zero. The NMOS is in the conducing mode since its gate-to-source voltage is V_{DD} (Fig. 11-26B) but its drain current is forced to be zero by the noncon-

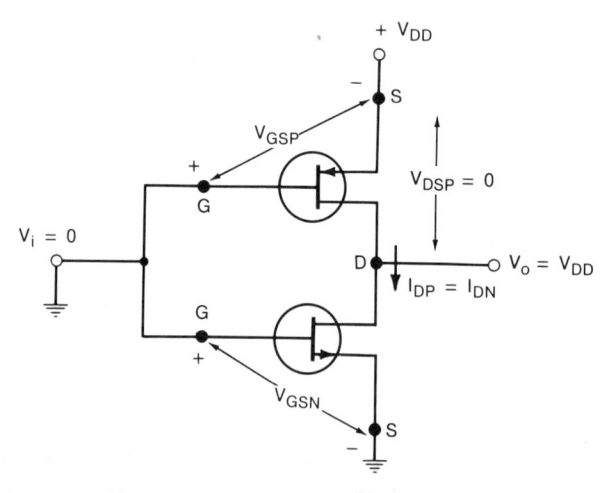

(A) Conditions in a CMOS inverter with a low input. The NMOS is cut off, and the PMOS is conducting but its current is zero. The output is high.

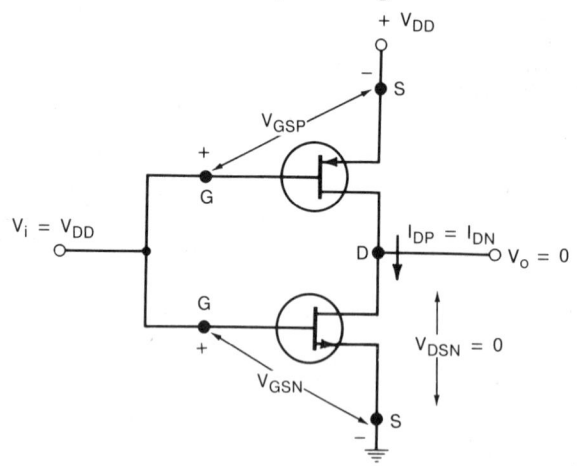

(B) Conditions in a CMOS inverter with a high input. The PMOS is cut off, and the NMOS is conducting but its current is zero. The output is low.

Fig. 11-26. CMOS gate used as an inverter.

Electronics: Circuits and Systems

ducting PMOS. Therefore, the drain-to-source voltage of the NMOS is zero. That is, for the *high input case*,

$$V_{GSN} = V_{DD}. \quad I_{DN} = I_{DP} = 0. \quad V_{DSN} = 0 \quad (NMOS).$$

and the output voltage, being equal to V_{DSN}, is zero. The output is low when the input is high.

The circuit is Fig. 11-25 acts as an inverter: a high output occurs for a low input and a low output for a high input as in Fig. 11-26.

The interesting feature of the CMOS circuit is that it *draws virtually no current* when it is in either of the two states: high or low. Therefore, the steady-state power consumption is zero for the circuit. Current flows in the circuit only when the output switches from low to high or from high to low. The switching currents are due to the presence of capacitances due to the load gates connected to the given gate. As indicated in Fig. 11-27, current is *drawn from the power supply only when the output switches from low to high* in order to charge the load capacitance form 0 to V_{DD}. The current flow from the power supply is in the form of spikes as shown in Fig. 11-27. The average power supplied by the battery is V_{DD} times the *average current* due to the spikes. The average current due to the spikes increases as the frequency of the input signal increases. As the switching of the input becomes more frequent, the spike's area increases, and the average value of the current increases as indicated in Fig. 11-28. The average power consumed by a CMOS circuit is, therefore, a function of the switching rate of the signal.

The propagation delay in a CMOS gate is due to the charging and discharging of the load capacitance, which is typically 5 pF per load gate. The time constant involved is the product of the load capacitance and the resistance of the MOS device. The resistance of the MOS device is found to depend upon the value of V_{DD}: a higher value of V_{DD} leads to a smaller value of the resistance of the MOS device. The delay in a gate is, therefore, smaller for a larger power supply. The typical delay times are 50 ns with $V_{DD} = 5$ V and 25 ns with $V_{DD} = 10$ V.

The voltage levels of a CMOS gate are 0 and V_{DD}, and V_{DD} can be chosen anywhere from 3 to 15 V. The highest input that will be accepted as low, $V_{IL(max)}$, by a CMOS gate is 30% of V_{DD}, and the lowest input that will be accepted as high, $V_{IH(mim)}$, is 70% of V_{DD}.

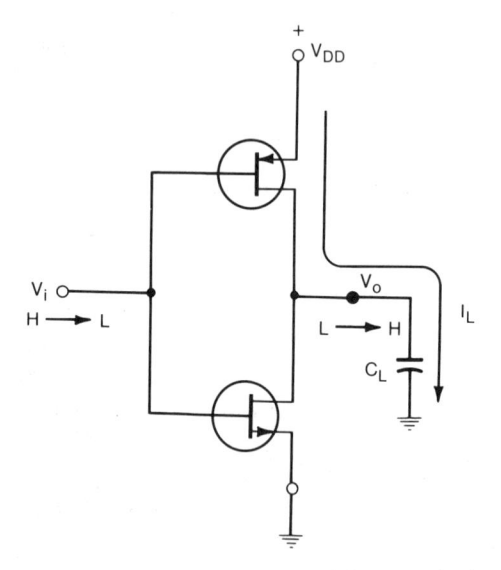

(A) The only time there is current flow is when the output switches from low to high.

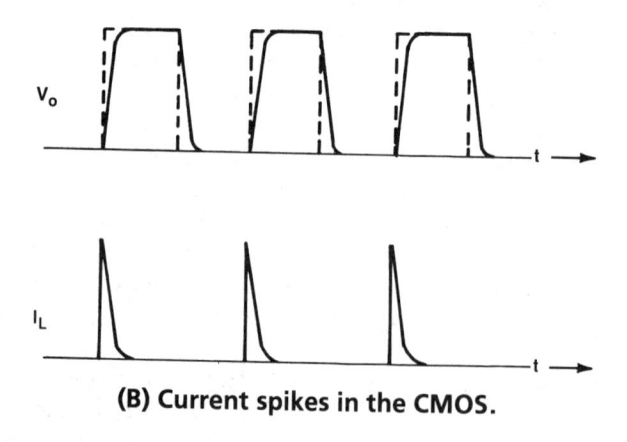

(B) Current spikes in the CMOS.

Fig. 11-27. Current flow in a CMOS inverter.

Electronics: Circuits and Systems

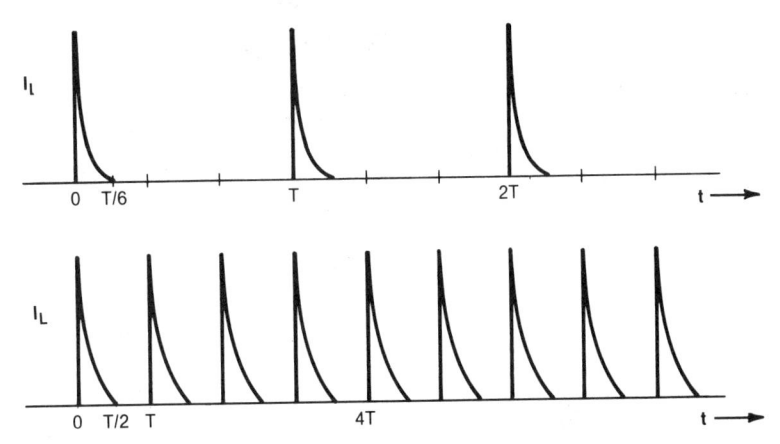

Fig. 11-28. Current spikes in a CMOS gate at different frequencies. The average value of I_L increases as the frequency increases.

SUMMARY SHEETS

TRANSISTOR AS A SWITCH

Four Regions of Operations of a Transistor

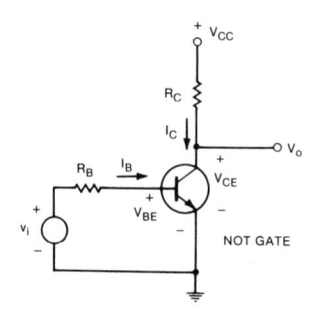

Cutoff: V_{BE} below the minimum needed for conduction. $I_C = 0$ and $V_o = V_{CC}$.

Normal Active Region: $V_{BE} = 0.7$ V. $I_C = I_B$. V_o lies between 0 and V_{CC}.

Saturation: I_C is at its maximum value of (V_{CC}/R_C). $I_C < (\beta I_B)$. $V_{CE} \approx 0$ and $V_o = 0$.

Reverse Active Region: The emitter-base junction is reverse biased and the collector-base junction is forward biased. The collector becomes the emitter and vice versa. This mode occurs in the TTL gate.

NOT Gate (Inverter)

Two states:
cutoff—output is high ($= V_{CC}$)
saturation—output is low (≈ 0)
Output state is the opposite of the input level.

PROPAGATION DELAYS

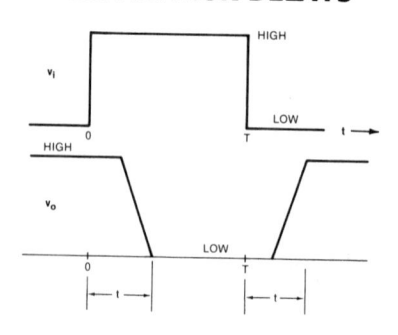

t_{pHL} is the delay between the input and the output pulses when the output goes from *high* to *low*.

t_{pLH} is the delay between the input and the output pulses when the output goes from *low* to *high*.

Causes of Propagation Delay

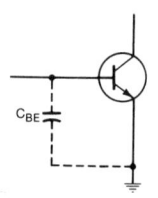

High Output to Low Output Transition

The capacitance C_{BE} (base-emitter junction) has to charge from 0 (or a negative value) to 0.7 V so as to make the transistor conduct. After the transistor starts conducting, I_B and I_C increase. When saturation is reached, I_C has reached its maximum value and stays constant at their value, but I_B continues to increase as determined by the input signal. This causes excess charge to be stored in the base region.

The delay in the high to low transition is due to the charging of C_{BE} to 0.7 V and the building up of I_C to its maximum value.

Low Output to High Output Transition

The excess charge stored in the base region must first be swept out before the transistor enters the normal active region. Then the collector current starts decreasing, and eventually reaches a value of 0. The transistor is now cut off.

The delay in the low to high transition is due to the sweeping out of the excess stored charge from the base region and the reduction of I_C to zero.

NOISE MARGINS

$V_{OH(min)}$ is the smallest *output* of a gate that can be treated as high.

V_{IH} is the smallest *input* to a gate that can be treated as high.

Electronics: Circuits and Systems

High State Noise Margin

$$V_{NH} = V_{OH(min)} - V_{IH}$$

$V_{OL(max)}$ is the largest *output* of a gate that can be treated as low.

V_{IL} is the largest *input* to a gate that can be treated as low.

Low State Noise Margin

$$V_{NL} = V_{IL} - V_{OL(max)}$$

Noise margins are higher in saturated logic circuits since the high and low levels are well separated from each other.

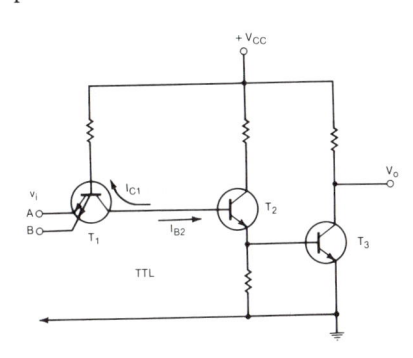

Open Collector TTL Gate

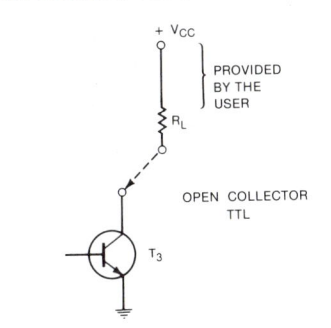

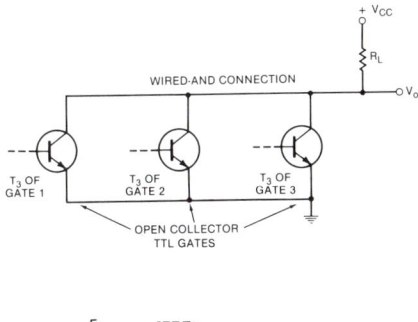

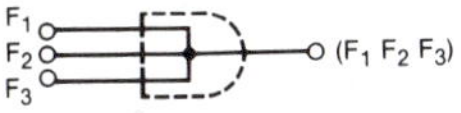

TRANSISTOR-TRANSISTOR LOGIC (TTL)

One or More Inputs Low

T_1 is saturated. T_2 is cut off. T_3 is cut off. $V_o = V_{CC}$ (High output).

All Inputs High

T_1 is in the reverse active mode. The reverse collector current in T_1 drives a base current into T_2. T_2 is saturated. T_3 is saturated. $V_o = 0$. The TTL circuit acts as a positive NAND gate.

The speed of response of the TTL is high due to the reverse flow of current in the collector of T_1 that helps sweep out the excess stored charge in the base of T_2 and turn it off.

Loading Considerations

The load on a TTL gate is a number of other similar gates.

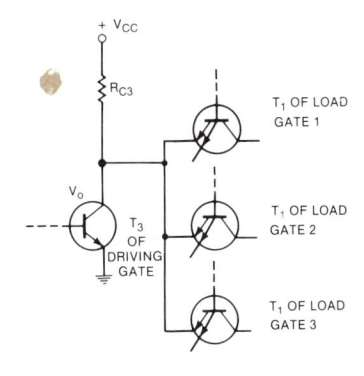

High output state presents no problems,

Digital Logic Families

since only a small current flows through the output transistor T_3 of the driving gate.

Low output state leads to a limit on the number of load gates (or fan-out). The output transistor T_3 must *sink* the currents in the input transistors T_1 of all the load gates. The current through T_3 can cause the low output voltage to rise above the maximum value $V_{OL(max)}$.

Totem Pole Output Stage

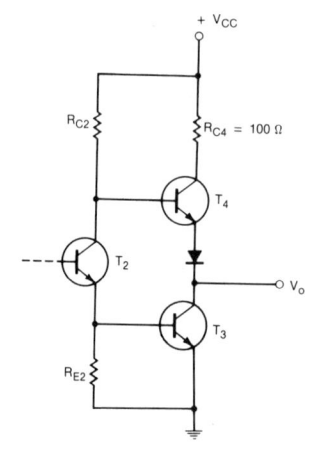

Low output: T_4 is the cutoff and the current drawn by T_3 is small, which increases the fan-out capability of the TTL gate.

Low to High Transition

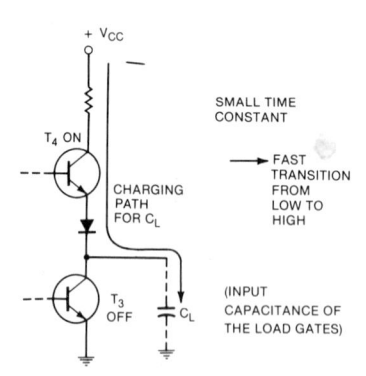

TTL gates with a totem pole output stage cannot be used in the wired-AND connection, since it would result in the flow of a very large current in the transistor T_3 of a load gate with a low output state.

Fan-Out When Two or More Inputs of a Load Gate Are Tied Together

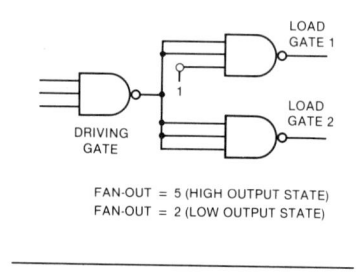

FAN-OUT = 5 (HIGH OUTPUT STATE)
FAN-OUT = 2 (LOW OUTPUT STATE)

Unconnected Inputs of a TTL Gate

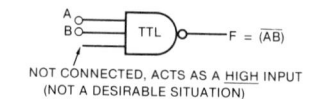

$F = \overline{(AB)}$

NOT CONNECTED, ACTS AS A <u>HIGH</u> INPUT
(NOT A DESIRABLE SITUATION)

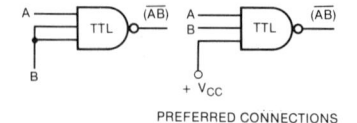

PREFERRED CONNECTIONS

Input Clamp Diodes

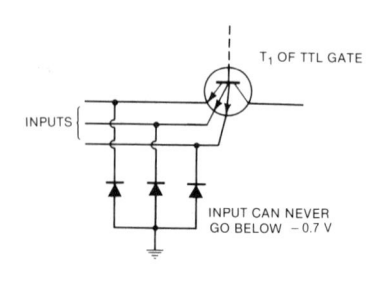

T_1 OF TTL GATE

INPUT CAN NEVER GO BELOW -0.7 V

Electronics: Circuits and Systems

THREE-STATE LOGIC

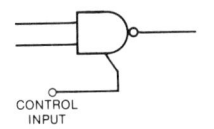

When the control input is low, the gate acts as a normal TTL. When the control input is high, both the transistors in the totem pole stage are cut off, and the TTL gate has a high output resistance.

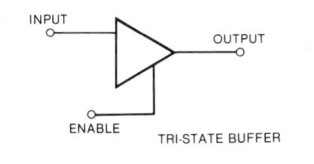

TRI-STATE BUFFER

INTEGRATED INJECTION LOGIC (I^2L)

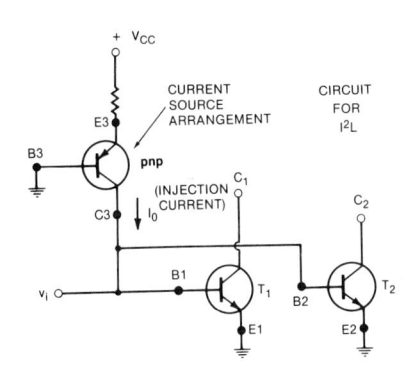

Semiconductor regions of different components are *integrated* into a single region, and an *injection* of current into the bases of the transistors is used.

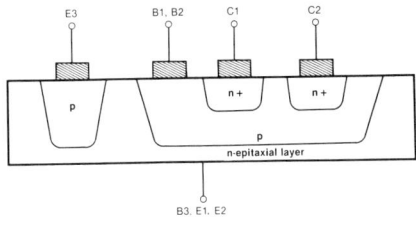

IC Fabrication of I^2

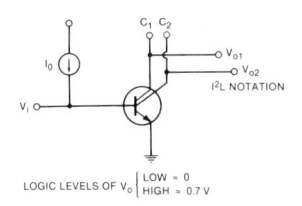

LOGIC LEVELS OF V_o $\begin{cases} \text{LOW} \approx 0 \\ \text{HIGH} \approx 0.7 \text{ V} \end{cases}$

Logic Levels of V_O

Propagation Delay

Charging and discharging of the capacitances take place by injection of currents.
LARGE CURRENT
(LARGE POWER → FAST RESPONSE DISSIPATION)
SMALL CURRENT
(LOW POWER → SLOW RESPONSE DISSIPATION)

EMITTER-COUPLED LOGIC (ECL)

Transistors operate in the normal active region when conducting, instead of being saturated. The speed of response is higher than in TTL. An emitter-coupled diff amp circuit forms the basis of the ECL gates.

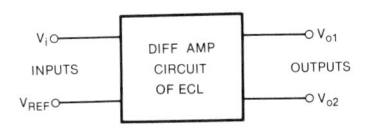

$V_i > V_{ref}$: V_{o1} low, V_{o2} high

$V_i < V_{ref}$: V_{o1} high, V_{o2} low

Small noise margin: about 0.2 V.
Power consumption larger than TTL.

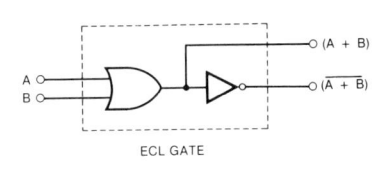

ECL GATE

COMPLEMENTARY MOS (CMOS) GATES

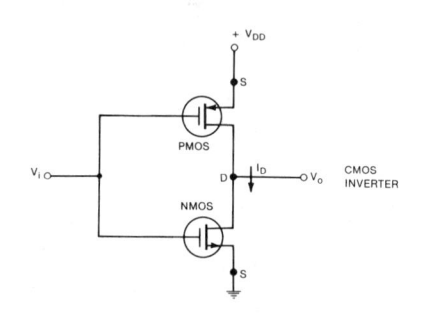

Input low: PMOS is *on*, NMOS is *off*. $I_D = 0$. $V_o = V_{DD}$. High output.

Input high: PMOS is *off*, NMOS is *on*. $I_D = 0$. $V_o = 0$. Low output.

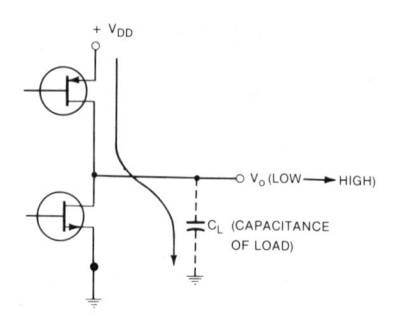

Current is drawn by the circuit only when the output switches from low to high. No current is drawn during the normal steady states. Average power consumed by the CMOS depends upon the frequency at which the input signal switches.

The propagation delay of the CMOS gate is smaller when a larger power supply is used. Typical values: 50 ns with $V_{DD} = 5$ V and 25 ns with $V_{DD} = 10$ V.

Electronics: Circuits and Systems

ANSWERS TO DRILL PROBLEMS

11.1 (a) $V_i = 0$: $I_C = 0$, $V_o = 10$ V. (b) $V_i = V$: $I_B = 0.465$ mA. $I_C = 2$ mA. Saturation condition is satisfied. $V_o = 0$.
11.2 $V_{NH} = 0.5$ V. $V_{NL} = 0.3$ V.
11.3 $V_{IH} = 3.49$ V. $V_{IL} = 1.01$ V.
11.4 $V_{NH} = 0.5$ V. $V_{NL} = 0.4$ V. Low state fan-out $= (16/1.6) = 10$. High state fan-out $= (0.25/40 \times 10^{-3}) = 6.25$.

PROBLEMS

Transistor as a Switch

11.1 In the circuit of Fig. 11-1, let $V_{CC} = 5$ V and $R_B = 50$ kΩ. If the transistor is to go into saturation when $v_i = 2.5$ V and its $\beta = 50$, determine the minimum value of R_C.

11.2 In the circuit of Fig. 11-1, let $V_{CC} = 15$ V, $R_B = 80$ kΩ, $R_C = 4$ kΩ and $\beta = 250$. Calculate the range of values of v_i for which the transistor will be (a) cut off; (b) in normal active mode; and (c) saturated.

Propagation Delays and Noise Margins

11.3 The manufacturer's data sheet for a particular AND gate has the following specifications: high-level input voltage $V_{IH} = 2$ V (min); low-level input voltage $V_{IL} = 0.8$ V (max); high-level output voltage $V_{OH} = 2.4$ V (min) and 3.4 V (typical); low-level output voltage $= 0.2$ V (typical) and 0.4 V (max); $t_{pLH} = 17.5$ ns (typical) and 27 ns (max); $t_{pHL} = 12$ ns (typical) and 19 ns (max). Calculate the noise margins (including worst cases). Suppose an input pulse (of sufficient amplitude) is applied at $t = 0$ and turned off at $t = 100$ ns to the gate. Draw a sketch of the output and input pulses on the same graph.

11.4 A particular gate has the following specifications. The minimum permitted high-level input is -1.25 V; the maximum permitted low-level input is -1.75 V. The guaranteed low-level output is less than -1.5 V, and the guaranteed high-level output is at least -1 V. Calculate the noise margins.

TTL

11.5 Consider the circuit shown in Fig. 11-P5. Discuss its operation by considering both inputs at 0 V, both inputs at 5 V, and one input at 0 V while the other is at 5 V.

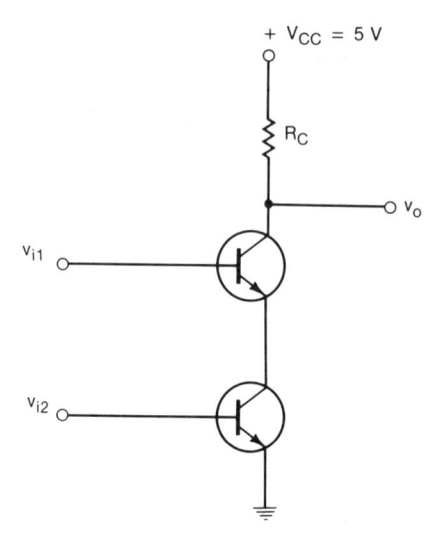

Fig. 11-P5.

11.6 Repeat the work of the previous problem for the circuit shown in Fig. 11-P6.

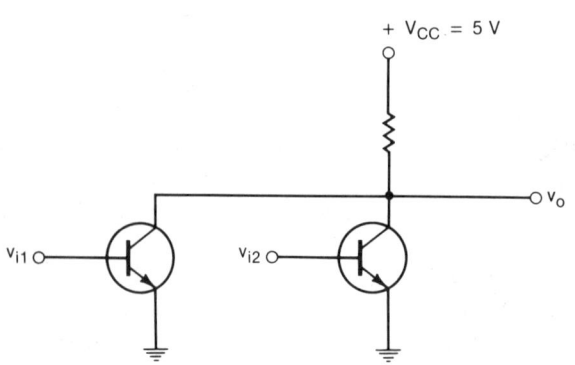

Fig. 11-P6.

11.7 Determine the logic operation of the circuit shown in Fig. 11-P7 (where each NAND gate is an open collector TTL).

11.8 Consider the circuit of Fig. 11-P8 using a tri-state NAND gate. Determine its output when (a) A = high, B = high, enable = low; (b) A = high, B = low, enable = high; (c) A = B = high, enable = high; (d) A = B = low, enable = low.

11.9 For the gate mentioned in Problem 11.3, the following additional specifications are available: Input current at maximum input voltage = 1 mA; high-level input current = 40 μA; low-level input current = 1.6 mA (magnitude). High-level output current = 800 μA and low-level output current = 16 mA. Calculate the fan-outs for high and low states.

Electronics: Circuits and Systems 693

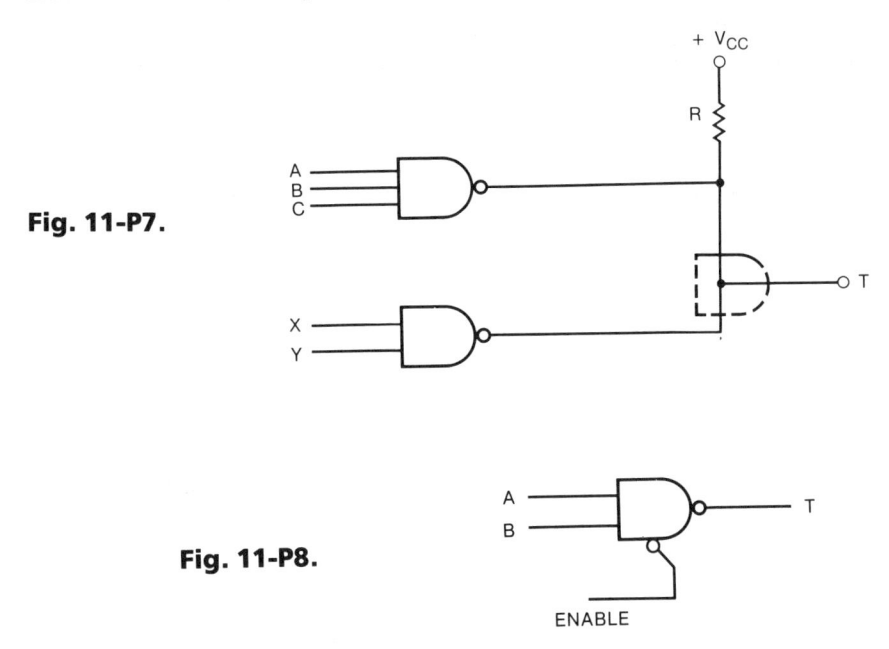

Fig. 11-P7.

Fig. 11-P8.

ENABLE

I²L and CMOS

11.10 Determine the logic expression for the output T in the I²L circuit of Fig. 11-P10.

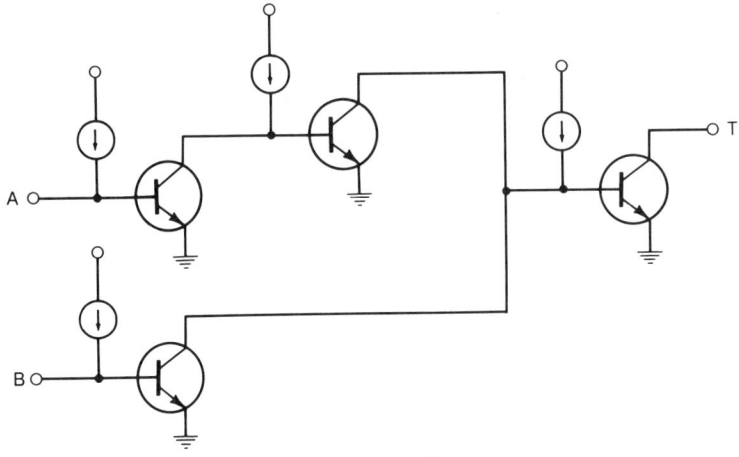

Fig. 11-P10.

11.11 Set up an I²L configuration to realize the logic function $T = AB + \overline{C}$. Also set up the truth table of the function.

11.12 Determine the logic operation performed by the CMOS circuit shown in Fig. 11-P12.

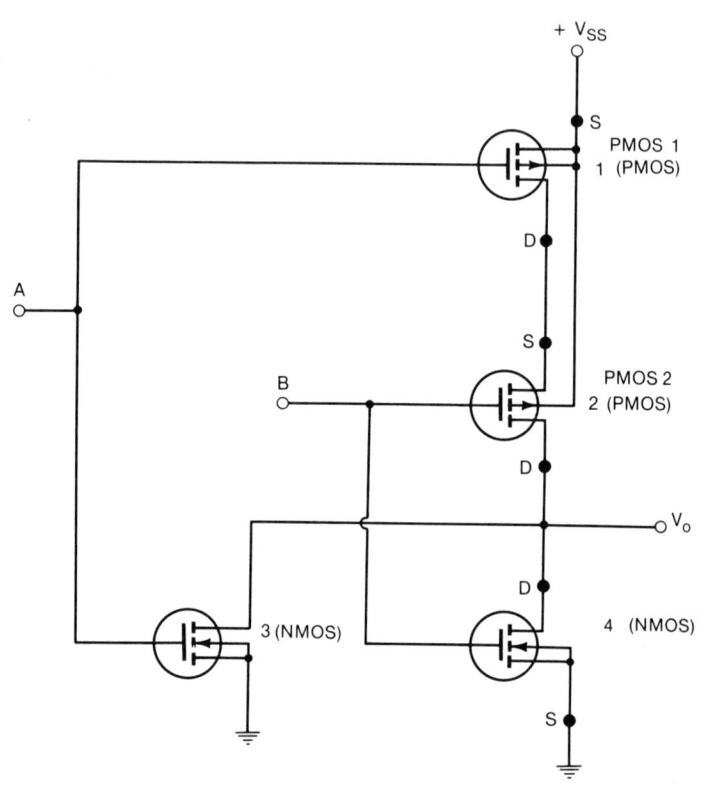

Fig. 11-P12.

CHAPTER 12

LOGIC PACKAGES AND MEMORIES

The modern approach to the design of logic networks and digital systems is to take advantage of the logic packages available in the form of medium- and large-scale integrated circuits (MSI and LSI) whenever possible instead of using SSI gate packages. The use of logic packages speeds up the design procedure and also reduces the physical size of the network. The troubleshooting and repairing of such networks are also greatly simplified. Some of the logic packages that will be considered in this chapter are *multiplexers, encoders* and *decoders, ROMs, RAMs,* and *programmable logic arrays.* A few other packages will be considered in the next chapter.

A digital system requires the storage of information that is used during the operation of the system. Information to be stored could be, for example, data used in the computations and sets of instructions to be carried out during the operation of the system. *Memory* is, therefore, needed in a digital system and memory units of different types are used. Some memory units can only "read"; that is, information already stored in them can be read as needed, but the stored information cannot be routinely changed during the operation of a system. Other memory units permit both reading and writing of information; that is, the stored information can be readily altered during the operation of the system. *Read-only memories (ROMs)* and *random-access memories (RAMs)* are the two most commonly used memory units in a digital system.

MULTIPLEXERS (MUX)

A *digital multiplexer* (abbreviated *MUX*) is a logic package primarily intended to transmit a multiplicity of (digital) signals to a single line. Suppose there are several lines in a system, each of which carries a digital signal. These signals are to be placed on a single line, one at a time. In the scheme of Fig. 12-1, the signals on the *data input* lines are to be transmitted to the single output line, and which signal is so transmitted at any given instant is controlled by means of the signals applied to the *data select lines* (also called *address lines*). The network that performs this operation is the multiplexer.

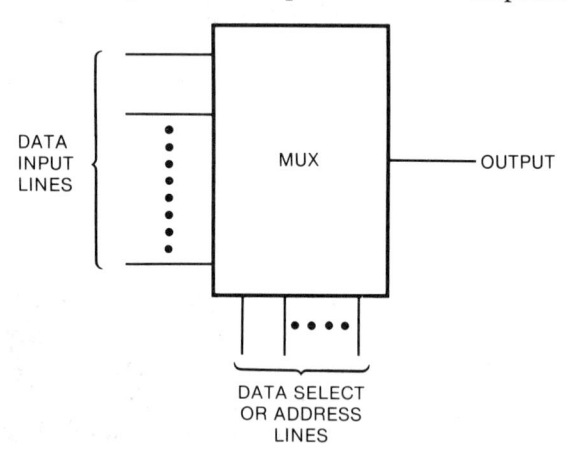

Fig. 12-1. Digital multiplexer (MUX).

Consider the four-input MUX in Fig. 12-2, where D_3, D_2, D_1, and D_0 are the data input lines and A_1 and A_0 are the address lines. The MUX has two outputs T and $\overline{T}$, which are complements of each other. The values of the signals on the address lines determine which input is transmitted to the output line T, as indicated in Table 12-1. For example, consider $A_1A_0 = 10$. Then any signal (0 or 1) present on the input line D_2 will appear on the output line T (and the complement of that signal on the output line $\overline{T}$). By varying the signals on the address lines, any of the data input signals can be placed on the output line. The MUX forms an important part of any digital system where a number of different lines have to interact with a single line.

Table 12-1. Output Table for Four-Input Multiplexer

Address Inputs A_1	A_0	MUX Output T
0	0	Data on line D_0
0	1	Data on line D_1
1	0	Data on line D_2
1	1	Data on line D_3

Multiplexers are available as two-input, four-input, eight-input, and sixteen-input packages, where the number used refers *only to the data input*

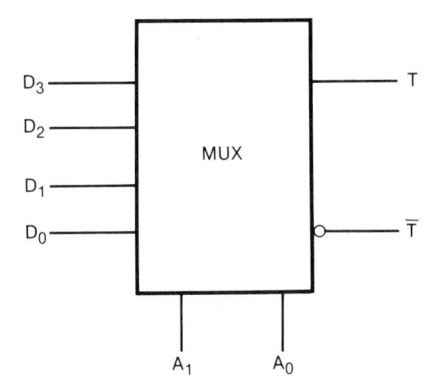

Fig. 12-2. Four-input MUX.

lines. The number of address lines is related to the number of data input lines by the relationship:

$$2^{(\text{no. of address lines})} = \text{no. of data input lines}$$

Thus, a sixteen-input MUX will have four address lines, since $2^4 = 16$.

To determine the output of a MUX at any given instant, arrange the values of the signals on the address lines in the order $A_3A_2A_1A_0$. Convert the resulting binary number to a decimal digit. If the decimal digit so obtained is k, then the data on line D_k will be transmitted to the output at that instant. As an example, consider the eight-input MUX shown in Fig. 12-3. Suppose the signals on the address lines at a given instant are $A_2A_1A_0 = 1\ 1\ 0$. The binary number 110 is equal to the decimal digit 6. Therefore, D_6 will be transmitted to the output line T at that instant (and $\overline{D_6}$ to $\overline{T}$). On the other hand, if we wish to transmit the signal on D_3 to the output, then we obtain the binary equivalent of 3, which is 011. Therefore, the address signals must be $A_2A_1A_0 = 0\ 1\ 1$.

Fig. 12-3. Eight-input MUX.

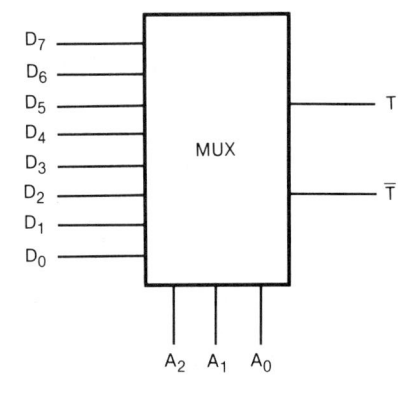

Drill Problem 12.1: For each of the following address signals in an eight-input MUX, find the values of T and $\overline{T}$ (in terms of the appropriate data input): $A_2A_1A_0 =$ (a) 0 0 0; (b) 1 0 1; (c) 1 1 1.

Drill Problem 12.2: In an eight-input MUX, what values should the address signals have to make T equal to (a) the signal on D_2; and (b) the signal on D_4?

Drill Problem 12.3: Set up a complete table for the eight-input MUX similar to Table 12-1.

Expansion Capacity of a MUX

Multiplexers are available with an additional control input, called *enable*, as indicated in Fig. 12-4. The enable is *active low:* when the enable signal is low, the unit works as a normal MUX, but when the enable signal is high, the unit is inoperative and goes into the high impedance state of a three-state logic network as discussed in the previous chapter.

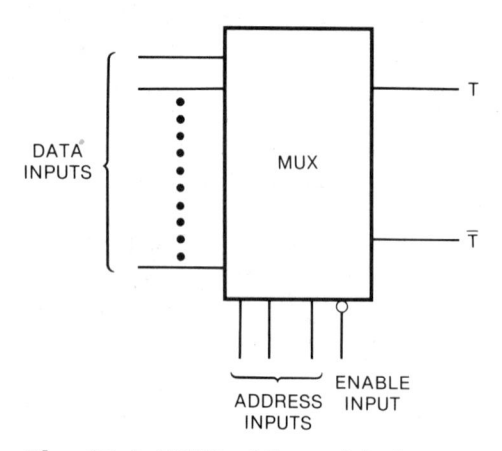

Fig. 12-4. MUX with enable input.

The presence of the enable input permits the interconnection of two or more MUX packages with a given capacity so as to obtain a MUX of larger capacity. The interconnection of two eight-input MUXs to form a sixteen-input MUX is shown in Fig. 12-5. The address signals are X_3, X_2, X_1, X_0. Of these, X_2, X_1, X_0 are connected to the address lines A_2, A_1, A_0 of the two MUX units. The signal X_3 is connected directly to the enable line of MUX 1, but passed through an inverter before being connected to the enable line of MUX 2. The data signals are $B_0, B_1, \ldots, B_{15}$. The first eight of these signals are

Electronics: Circuits and Systems

connected to the data input lines of MUX 1 and the remaining eight are connected to the data input lines of MUX 2. When X_3 is low ($X_3 = 0$), MUX 1 is enabled (since its enable is low) but MUX 2 is disabled (since its enable is high). The output of the network is one of the eight inputs B_0 through B_7 as determined by the address signals $X_2X_1X_0$. When X_3 is high, MUX 1 is disabled but MUX 2 is enabled, and the output of the network is one of the inputs B_8 through B_{15} as determined by the address signals $X_2X_1X_0$. For example, consider the following two situations. Let the address signals of the system be $X_3X_2X_1X_0 = 1\ 1\ 0\ 1$. Since $X_3 = 1$, MUX 2 is enabled. The address seen by that unit is $A_2A_1A_0 = X_2X_1X_0 = 1\ 0\ 1$, which equals the decimal digit 5. Therefore, D_5 of MUX 2, which is the data signal B_{13}, appears on the output. On the other hand, let $X_3X_2X_1X_0 = 0\ 0\ 0\ 1$. Since $X_3 = 0$, MUX 1 is enabled. The address seen by it is $A_2A_1A_0 = X_2X_1X_0 = 0\ 0\ 1$, which equals the decimal digit 1. Therefore, the output of the network is D_1 of MUX 1, or the data signal B_1.

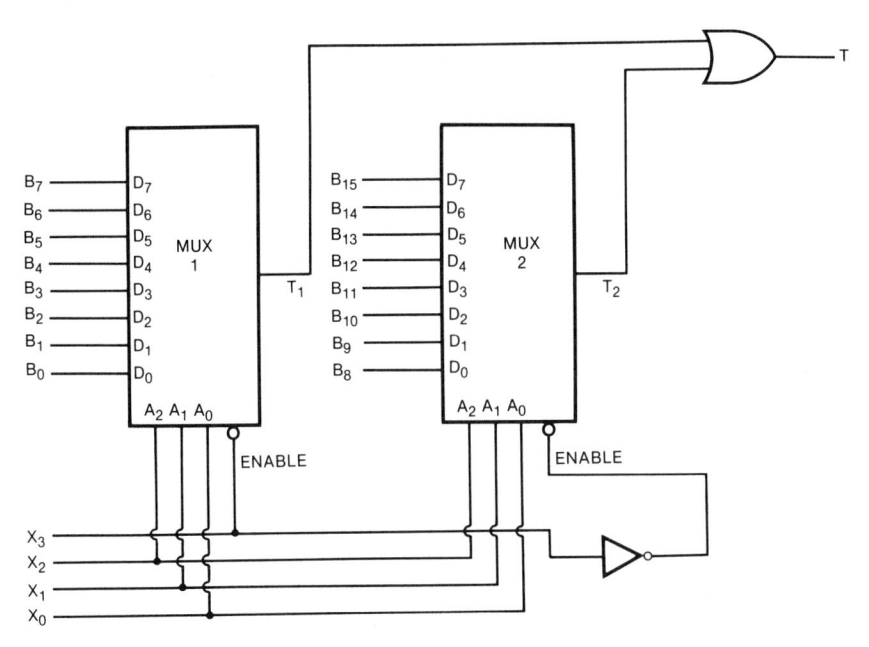

Fig. 12-5. Interconnection of MUXs to expand the available capacity.

Drill Problem 12.4: For each of the following sets of values of $X_3X_2X_1X_0$ in Fig. 12-5, state which of the data signals (B_0 through B_{15}) will appear as the output of the network: (a) 0 1 0 1; (b) 1 0 1 1; (c) 1 1 1 1; (d) 0 0 1 1.

Use of a MUX in Logic Design

A single MUX with *n-address lines* can be used to realize any logic function of (n + 1) variables without any additional gates. For example, a four-input MUX (with two address lines) can be used to generate any logic function of three variables. Similarly, an eight-input MUX with three address lines can be used to generate any logic function of four variables.

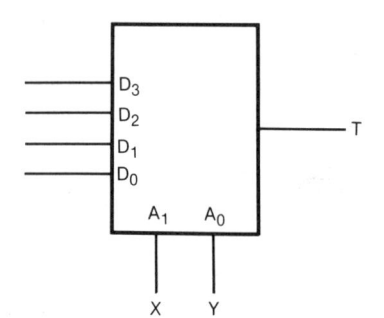

Fig. 12-6. Four-input MUX (for logic design).

Consider a three-variable function to be realized with a four-input MUX. Choose two of the three variables, say X and Y, as the inputs to the address lines of the MUX, as indicated in Fig. 12-6. Then, the inputs to be connected to the data input lines will be the third variable Z, or $\bar{Z}$ (0, or 1) as determined from the truth table of the function to be realized. The rows of the truth table are first partitioned into groups of two according to the values of X Y: 0 0, 0 1, 1 0, 1 1, as indicated in Table 12-2.

Table 12-2. Truth Table for Three Variables for a Four-Input MUX.

X	Y	Z	T	
0	0	0	0	} Group with XY = 0 0
0	0	1	0	
0	1	0	1	} Group with XY = 0 1
0	1	1	1	
1	0	0	0	} Group with XY = 1 0
1	0	1	1	
1	1	0	1	} Group with XY = 1 1
1	1	1	0	

Consider the group with X Y = 0 0. For this address, the MUX output will be equal to D_0. The input to line D_0 must, therefore, be chosen so as to match the MUX output to the outputs in the truth table in that group. In the present example, it is seen that T = 0 for both the cases Z = 0 and Z = 1 in the group. Therefore, if we make D_0 = 0, then T = 0 when X Y = 0 0 (whether Z = 0 or 1), which conforms to the given truth table. Therefore, D_0 = 0.

Consider the group with X Y = 01. The MUX output will be D_1 for this group. It is seen that T = 1 for both Z = 0 and Z = 1 in this group. Therefore, making D_1 = 1 will cause T to be 1 when X Y = 0 1, which conforms to the truth table. Therefore, D_1 = 1.

Consider the group X Y = 1 0. The MUX output will be D_2. T is seen to be 0 when Z = 0 and 1 when Z = 1 in this group. That is, T has the same value as Z in the group. Therefore, we should connect Z to the line D_2. That is, D_2 = Z.

Finally, consider the group X Y = 1 1. The MUX output will be D_3. T is seen to be 1 when Z = 0 and 0 when Z = 1. That is, T has exactly the opposite value to Z, which means that T has the same value as $\bar{Z}$ for this group. Therefore, we should connect $\bar{Z}$ to the line D_3. That is, D_3 = $\bar{Z}$.

The final connections to the MUX are shown in Fig. 12-7.

Fig. 12-7. MUX realization of the logic function of Table 12-2.

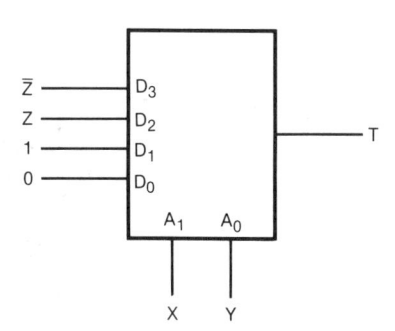

Summary of Design Procedure (Four-Input MUX)

Given a logic function of three variables X, Y, Z, connect X and Y to the address lines A_1 and A_0, respectively. Separate the given truth table into groups of two rows according to the values of X Y. Each group then designates one data input line: X Y = 0 0 corresponds to the data input line D_0, and so on.

Compare the values of the output T in the truth table with those of Z in each group separately. If T = 0 for both values of Z in the group, then make the input to the corresponding data input line equal to 0. If T = 1 for both values of Z in the group, then make the input to the corresponding data input line equal to 1. If T has the same value as Z in the group, then make the input to the corresponding data input line equal to Z. If T has a value opposite to Z in the group, then make the input to the corresponding data input line equal to $\overline{Z}$. Note that all four possible situations have been covered.

Example 12.1

Obtain a MUX realization of the function in Example 10.4, repeated in the following section for convenience.

Table 12-3. Logic Function for Example 12.1. to Obtain MUX Realization

X	Y	Z	T
0	0	0	0
0	0	1	0
0	1	0	0
0	1	1	1
1	0	0	0
1	0	1	1
1	1	0	1
1	1	1	1

Solution

The groups are already shown partitioned.

Group X Y = 0 0: T = 0 in both cases. Therefore, $D_0 = 0$.

Group X Y = 0 1: T has the same value as Z in this group. Therefore, $D_1 = Z$.

Group X Y = 1 0: T has the same value as Z in this group also. Therefore, $D_2 = Z$.

Group X Y = 1 1: T = 1 in both cases. Therefore, $D_3 = 1$.

The final connections are shown in Fig. 12-8.

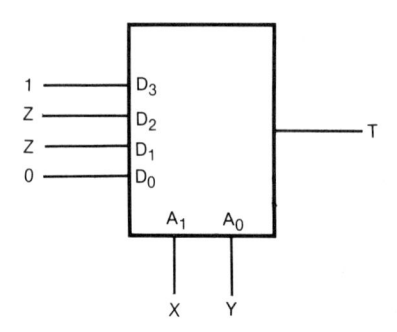

Fig. 12-8. MUX realization of the logic function of Example 12.1.

Drill Problem 12.5: For the truth tables of Example 10.5, and Drill Problems 10.18 and 10.19, obtain single MUX realizations.

The design procedure outlined can be readily extended to the use of an eight-input MUX for generating any four-variable logic function. Three of the variables, W, X, Y, are connected to the address lines of the MUX. The truth table is partitioned into *eight* groups corresponding to W X Y = 0 0 0, 0 0 1, . . ., 1 1 1. These groups correspond to the data input lines D_0, D_1, . . ., D_7, respectively. The connections to the data input lines follow exactly the same rules as those outlined earlier.

Example 12.2

Obtain a single eight-input MUX realization of the logic function given in Table 12-4.

Table 12-4. Table for Example 12.2

W	X	Y	Z	T
0	0	0	0	1
0	0	0	1	0
0	0	1	0	0
0	0	1	1	0
0	1	0	0	1
0	1	0	1	1
0	1	1	0	1
0	1	1	1	1
1	0	0	0	0
1	0	0	1	1
1	0	1	0	1
1	0	1	1	0
1	1	0	0	0
1	1	0	1	0
1	1	1	0	1
1	1	1	1	1

Solution

The groups are already shown partitioned.

Group W X Y = *0 0 0:* T is seen to have a value opposite to Z in this group. Therefore, $D_0 = \bar{Z}$. *Group* W X Y = *0 0 1:* T = 0 for both values of Z in this group. Therefore, $D_1 = 0$. *Group* W X Y = *0 1 0:* T = 1 for both values of Z in this group. Therefore $D_2 = 1$. The other five inputs are determined in a similar manner. The final connections are shown in Fig. 12-9.

```
  1  ──── D₇
  0  ──── D₆
  Z̄  ──── D₅
  Z  ──── D₄
  1  ──── D₃        ──── T
  1  ──── D₂
  0  ──── D₁
  Z̄  ──── D₀
          A₂ A₁ A₀
           │  │  │
           W  X  Y
```

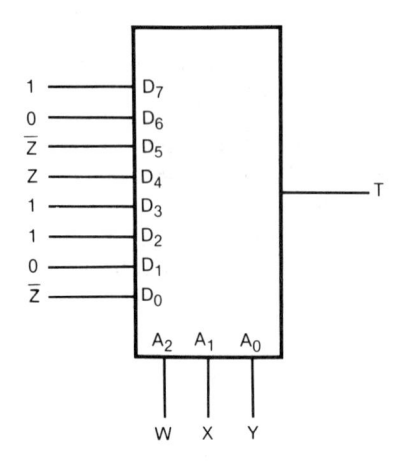

Fig. 12-9. MUX realization of the logic function of Example 12.2.

Drill Problem 12.6: Obtain a single eight-input MUX realization of a logic network with four inputs, whose output is 1 whenever an odd number of inputs has a value of 1. (This is the odd parity checker mentioned in Chapter 10.)

Note that once the odd parity checker is designed using a single MUX, an even parity checker is also made available immediately because of the other output in the MUX. If the output T checks the odd partiy, then $\overline{T}$ checks the even parity. The availability of two complementary outputs in a MUX can often be exploited in the previously described manner.

A *demultiplexer* performs the reverse function of the multiplexer: it has a single data input line, and the data can be routed to *one* of a number of output lines by changing the signals on the address lines. The number of output lines is two raised to the power of the number of address lines. The demultiplexer serves as a data distribution network. For example, it can be used to route a clock signal to different parts of a digital system.

DECODERS AND ENCODERS

The transmission of data in a digital system requires the uses of coded representations of numbers in the form of a sequence of ones and zeros, which will correspond to the absence or presence of a pulse, respectively. One common coding scheme is the *BCD code (binary-coded decimal)* in which a sequence of 4 binary digits (or 4 *bits*) is used to represent each of the decimal digits 0 through 9, as shown in Table 12-5.

Table 12-5. BCD Code

A_3	A_2	A_1	A_0	Decimal Digit
0	0	0	0	0
0	0	0	1	1
0	0	1	0	2
0	0	1	1	3
0	1	0	0	4
0	1	0	1	5
0	1	1	0	6
0	1	1	1	7
1	0	0	0	8
1	0	0	1	9

A *decoder* receives the coded representation of a number and decodes it into the number represented by the input bits. An *encoder*, on the other hand, receives a number and generates the coded representation of that number.

The number of inputs to a decoder equals the number of bits in the code. The number of outputs is determined by how many different numbers are covered by the coding scheme. In the *BCD-to-decimal* decoder (Fig. 12-10), there are four inputs A_3, A_2, A_1, A_0, and ten outputs 0 through 9. In terms of Table 12-5, the bit A_3 denotes the left-most column and A_0 the right-most column. When a 4-bit input is received, the output representing the corresponding number is made active while all the other outputs are inactive. Only one input will become active at any given time. There are two versions available: the *active high output* decoder and the *active low output* decoder. The latter can be identified by the small circles (or bubbles) or triangular polarity marks on the output terminals in the manufacturer's data sheets.

Fig. 12-10. BCD decoder block diagram.

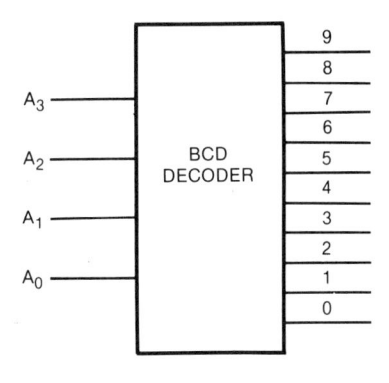

As an example, if the input is $A_3 A_2 A_1 A_0 = 0\ 1\ 0\ 0$, then the output line 4 (from Table 12-5) of the decoder will become active.

Another commonly used decoder is the *BCD-to-seven-segment decoder*, which converts an input BCD code into the signals needed to light up the segments of a seven-segment visual display.

In the case of an *encoder*, the desired coding scheme must first be specified. Then, the desired logic network can be realized in the form of a *diode matrix*. Consider the coding scheme (know as a *Gray code*) for representing the numbers 0 through 7 by means of a 3-bit code, as in Table 12-6.

Table 12-6. Gray Code

Decimal Digit	Gray Code		
0	0	0	0
1	0	0	1
2	0	1	1
3	0	1	0
4	1	1	0
5	1	1	1
6	1	0	1
7	1	0	0

The diode matrix of the encoder of Table 12-6 is shown in Fig. 12-11. The number to be encoded is activated by pressing the appropriate switch (akin to pressing keys on a keyboard of a calculator). The switch connects the 5-V battery to one or more diodes in the matrix. The current flowing in the diode causes a voltage of 4.3 V (which is one diode drop below the battery voltage) at the corresponding output line. As an example, suppose switch S_4 is pressed. Then, the two diodes on line 4 are forward biased, and the resulting currents I_2 and I_1 make the outputs G_2 and G_1 high (4.3 V). The current $I_0 = 0$ since there is no diode linking line 6 with the output line G_0, and G_0 is, therefore, low (0 V). Therefore, the output code is 110 when switch S_4 is pressed, which conforms to Table 12-6. It can be verified in a similar manner that the outputs obtained by pressing any one of the switches conform to the coding scheme in Table 12-6.

Even though diodes are used in the previous encoder example, transistors (BJT and FET) are often used instead. The advantage of the matrix-type realization of an encoder is that it can be made *programmable* by the user. The manufacturer makes a diode matrix by placing a diode linking *every input line with every output line*. The user decides on a coding scheme and disconnects the diodes not needed for that scheme. The disconnection is performed by blowing the fuses provided with the diodes selectively. The form of a *programmable encoder* is shown in Fig. 12-12. Such programmable encoders are an integral part of programmable ROMs to be discussed in the next section.

Electronics: Circuits and Systems   **707**

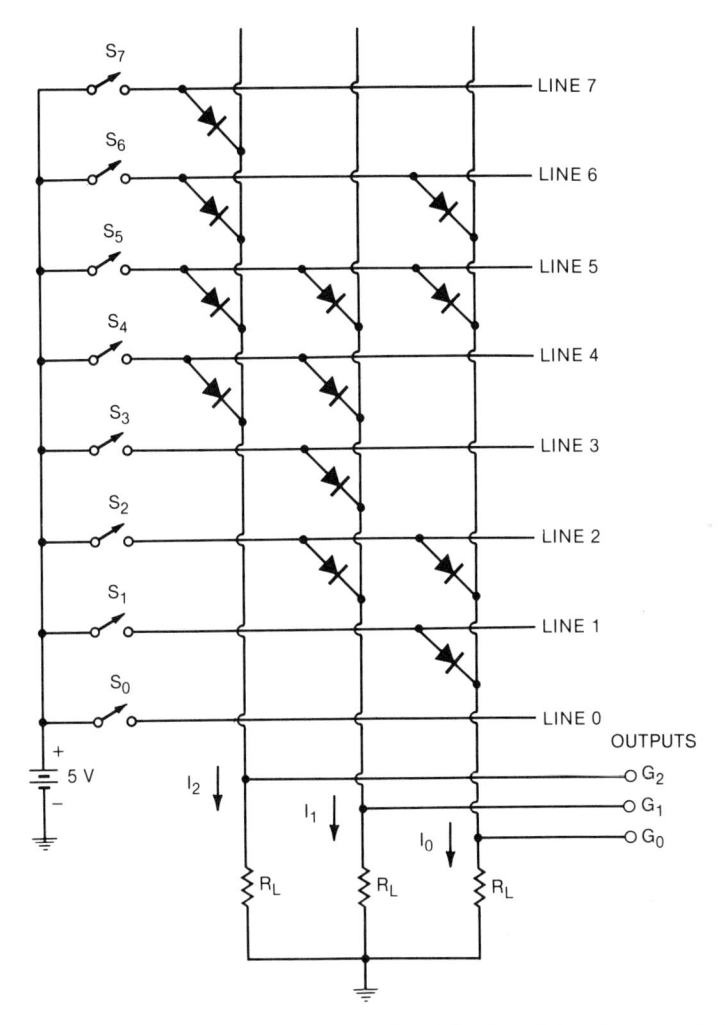

Fig. 12-11. Diode matrix encoder (decimal to Gray code).

Drill Problem 12.7: Set up an encoder matrix that will convert a number (0 through 9) into its BCD code.

READ-ONLY MEMORIES

LSI techniques have led to the development of semiconductor memory packages with extremely high capacity compressed into a very small physical

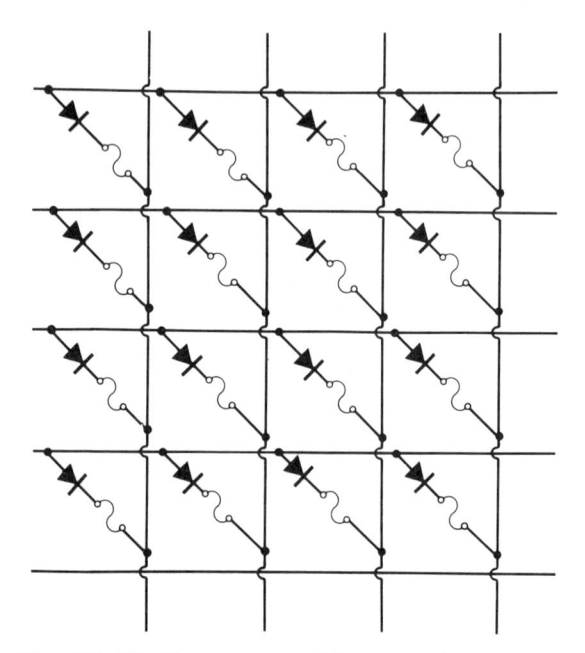

Fig. 12-12. Programmable encoder matrix.

space. Such memory packages have revolutionized the design of digital systems by making them more versatile and powerful than they used to be before the advent of LSI memory packages. VLSI techniques are pushing the capacities to higher and higher values. The availability of memory packages has also led to the use of more hardware in digital systems in places that were traditionally the domain of software programming.

Memory refers to the ability to store information in the form of binary digits 0 and 1 in a digital system. The information stored may be data to be processed or programs for computations to be performed on the data. In some memories, it is possible *only to read* information that is already stored in them. Such units are called *read-only memories*, or ROMs. In others information can be written into them routinely as well as retrieved. They are called *read-write memories*, a common example of which is the *random-access memory*, or RAM.

A ROM is a logic network with n inputs and m outputs, and the relationship between the inputs and outputs can be just about anything we want. For example, the inputs may be coded representations of angles, and the outputs the coded representations of a trigonometric function of the input angles. The ROM is often used to serve as a *lookup table* and in other similar manners. A ROM is used in modern sewing machines to store sewing patterns that can be

Electronics: Circuits and Systems

brought into action by the pressing of an appropriate button. ROMs can also be designed as part of a foreign-language phrase finder.

The internal structure of a ROM consists of a decoder network followed by an encoder network, as indicated in Fig. 12-13. The inputs $x_0, x_1, \ldots, x_{n-1}$ are treated as if they were coded representations of the numbers 0 through r. The decoder portion has a standard structure in all ROMs. The outputs 0 through r (only one of which is active at any time) are then encoded into m-bit words by the encoder in such a way as to implement the proper relationship between the inputs and outputs. The encoder portion of a ROM is a matrix (of diodes or transistors) of the form discussed in the last section. The user of the ROM or the logic designer provides the manufacturer with a truth table listing the m-bit code words for the numbers 0 through r. The manufacturer fabricates an IC mask to transfer the truth table onto the encoder portion of the ROM. Such a ROM is called a *mask-programmed ROM* and can perform only the function for which it was originally designed.

Example 12.3

Design a ROM to convert BCD code to excess-three code shown in Table 12-7.

Table 12-7. Table for Example 12.3

BCD Code				Excess-Three Code			
0	0	0	0	0	0	1	1
0	0	0	1	0	1	0	0
0	0	1	0	0	1	0	1
0	0	1	1	0	1	1	0
0	1	0	0	0	1	1	1
0	1	0	1	1	0	0	0
0	1	1	0	1	0	0	1
0	1	1	1	1	0	1	0
1	0	0	0	1	0	1	1
1	0	0	1	1	1	0	0

Solution

The decoder portion of the ROM converts the BCD coded inputs into the numbers 0 through 9. Note that only one output of the decoder is active for any given input pattern. Each of the numbers 0 through 9 is then converted into the corresponding excess-three code by the encoder. The operation is indicated in Table 12-8.

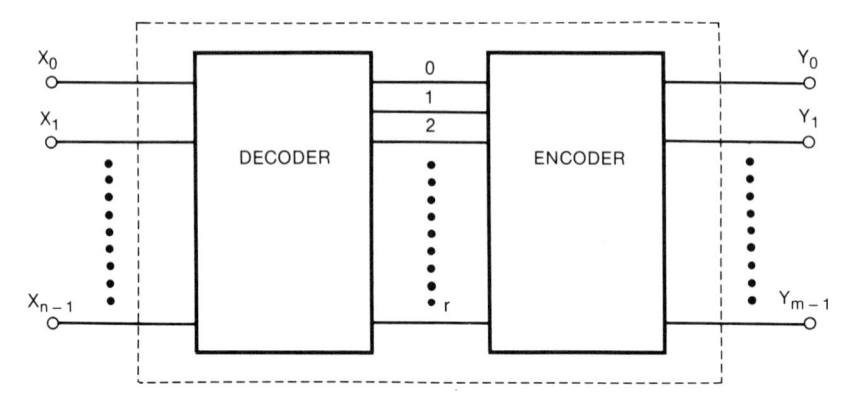

Fig. 12-13. Internal networks of a ROM.

Table 12-8. Table for Solution to Example 12.3

ROM Inputs (BCD Code)	Decoder Output or Encoder Input	ROM Outputs (Excess-Three Code)
0 0 0 0	0	0 0 1 1
0 0 0 1	1	0 1 0 0
0 0 1 0	2	0 1 0 1
0 0 1 1	3	0 1 1 0
0 1 0 0	4	0 1 1 1
0 1 0 1	5	1 0 0 0
0 1 1 0	6	1 0 0 1
0 1 1 1	7	1 0 1 0
1 0 0 0	8	1 0 1 1
1 0 0 1	9	1 1 0 0

Decoder Section   Encoder Section

Capacity of a ROM

If n is the number of inputs, and m the number of outputs of a ROM, then its capacity is defined as

$$(2^n \times m) \text{ bits}$$

The unit *kilobit*, or Kbit, is usually used, where *kilo* means *1024* bits (rather than 1000 as is normally associated with *kilo*). As examples, a ROM with eight inputs and four outputs has a capacity of $2^8 \times 4 = 1024$ or 1 Kbit, while a ROM with thirteen inputs and eight outputs has a capacity of $(2^{13} \times 8)$, or 64K bit. (Sometimes the unit *byte* is used, with 1 byte = 8 bits.)

Programmable ROMs

When the manufacturer uses the specific truth table provided by the customer and makes the ROMs, they are called *mask-programmed* ROMS. They have a high setting up cost due to the need for custom-made masks in their manufacture. They are not economical except in large quantities (several thousands or more).

Programmable ROMs (PROMs) or *field programmable ROMs* are available, in which the encoder portion of the ROM has diodes (or transistors) between every input line and every output line as discussed in the last section. In the *fusible-link* variety of PROMs, the user selectively melts the fuses not needed for the particular program to be incorporated in the ROM. Such PROMs can be programmed *only once* by the user, after which its function remains unalterable. Such PROMs are inexpensive.

Erasable PROMs (EPROMs) are more expensive than the previous variety of PROMS, but they permit the user to erase the program already written, write a new one, and repeat this process as often as needed. The erasure is done by means of shining ultraviolet light on the encoder matrix and such EPROMs have a quartz window for this purpose. Electrically alterable ROMs (EAROMs) are available in which the erasure is accomplished electrically. EPROMs and EAROMs are ideal for building small systems and especially in the construction of prototype systems.

Capacities of all varieties of ROMs are being constantly increased, and the memory market is highly competitive.

The ROM finds numerous applications not only in a wide variety of special input to output transformations, such as lookup tables in calculators, controllers in systems such as automobiles, but also in microcomputers. The ROM is used for storing fixed data in the form of 8-bit or 16-bit numbers in microprocessors. Such data are called *words*. A ROM that stores 8-bit words is indicated in Fig. 12-14. The address of a word is the row number (0, 1, . . ., r) in which it is stored. The set of values y_7, y_6, . . ., y_0 in a row represents the word stored at that address. The 0 and 1 bits in a word actually correspond to the absence or presence, respectively, of a diode link in the encoder matrix. EPROMs are commonly used in microprocessor systems so that the stored words can be changed periodically in order to modify the operation of the system. The change in the stored words is accomplished by means of an *EPROM programmer*.

Expansion of ROM Capacity

PROMs are provided with a *chip-select* or *chip-enable control input* that facilitates the interconnection of two or more units to obtain a larger memory capacity. The chip-enable control is an *active low input*. When the enable signal is low, the ROM is enabled and words stored in it can be read on the

712   **Logic Packages and Memories**

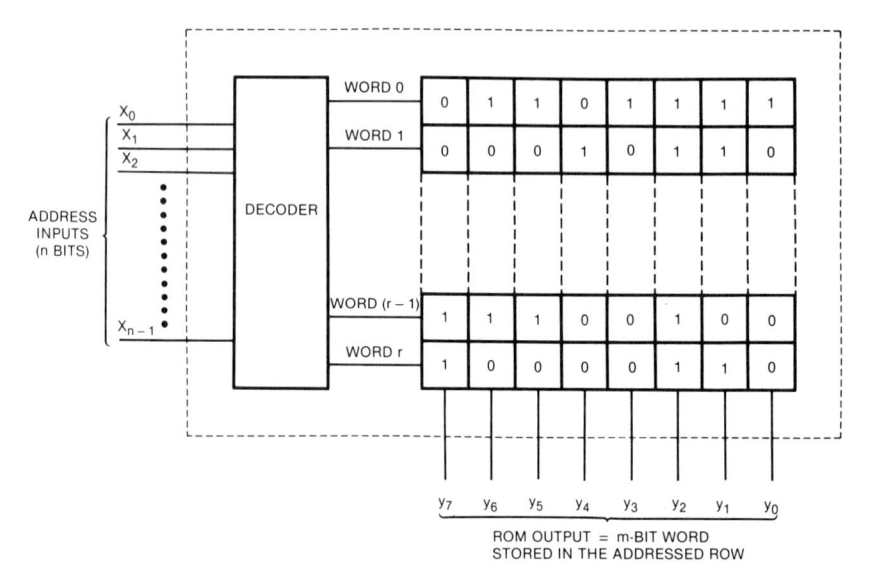

Fig. 12-14. Storage of words in a ROM.

output bus, but when the enable signal is high, the ROM is disabled and appears as a high impedance unit to the output bus (recall the three-state logic gates discussed in the last chapter).

Suppose we wish to store 512 words, each of which has 8 bits, when the available ROMs have a capacity of 256 × 8 bits. Two ROMs are interconnected as indicated in Fig. 12-15. If the words to be stored are labeled W_0, $W_1, \ldots, W_{255}, W_{256}, \ldots, W_{511}$, the first 256 words W_0 through W_{255} are stored in PROM A, while the remaining 256 words W_{256} through W_{511} are stored in PROM B.

Note that with this arrangement, the *address* of each word requires 9 bits, since $2^9 = 512$, and there are 512 stored words. The input address bits to the system are $A_8, A_7, \ldots, A_0$. Of these, the A_8 bit is fed to the *chip-enable control input* of the two PROMs (through a voltage level changer or inverter in the case of PROM B). The remaining address bits $A_7, A_6, \ldots, A_0$, are directly connected to both ROMs.

Consider the address input: 1 0 0 1 0 1 1 0 1; $A_8 = 1$ in this case, and PROM A is disabled (since its enable signal is high) while PROM B is enabled (since its enable signal is made low by the inverter). Looking at the remaining 8 bits of the address, we have 0 0 1 0 1 1 0 1, which is the binary representation of 45. Therefore, the word stored in *location 45* of *PROM B* is read on the output bus. This word is W_{301} (since the word stored in location k in PROM B is W_{256+k}). On the other hand, if we wish to transfer the word W_{163} to the output bus, then converting 163 to binary, we get 1 0 1 0 0 0 1 1.

Electronics: Circuits and Systems 713

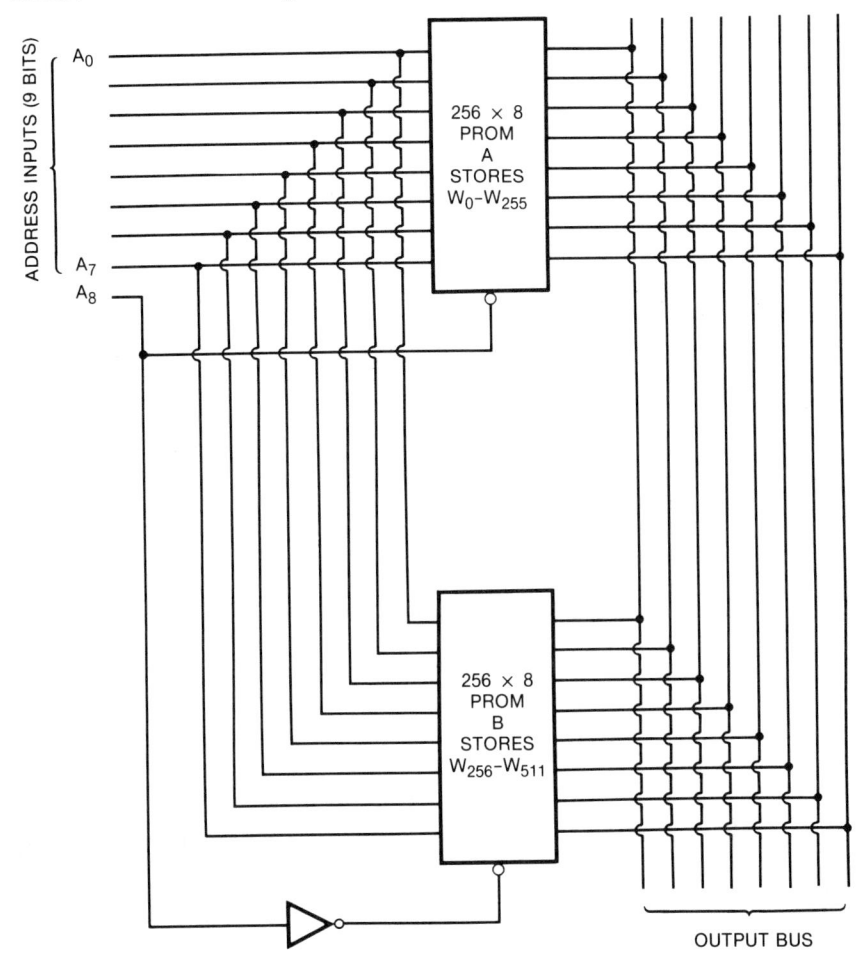

Fig. 12-15. Interconnection of ROMs to increase available storage.

Since W_{163} is to be found in PROM A, A_8 must be made 0. Therefore, the address bits for word W_{163} will be 0 1 0 1 0 0 0 1 1.

Drill Problem 12.8: Determine the words that will be read on the output bus of the system in Fig. 12-15 in each of the following sets of address bits: (a) 0 0 0 0 1 1 1 0 1; (b) 1 0 0 0 1 1 1 0 1; (c) 0 1 1 1 1 1 1 1 1; and (d) 1 1 1 1 1 1 1 1 1.

The same procedure can be extended to the interconnection of more than two ROMs in order to increase the storage capacity. For example, if 1024 8-

bit words are to be stored by using four 256-×-8-bit ROMs, the number of address bits needed is 10 (since $2^{10} = 1024$). The last 8 bits of the address ($A_7, \ldots, A_0$) are directly connected to each PROM. The highest 2 bits (A_9 and A_8) have to be transformed into an appropriate enable signal to activate the relevant PROM. One scheme to generate the appropriate enable signal is to use a one-line to four-line demultiplexer as indicated in Fig. 12-16. The 1 input to the demultiplexer is routed to one of its output terminals as determined by the values of A_9A_8. It is seen that the appropriate word is addressed through this arrangement.

Drill Problem 12.9: For each of the sets of address bits given, determine which output of the demultiplexer in Fig. 12-16 is made low, which PROM is enabled, and which word is read on the output bus of the system. (a) 0 1 1 1 0 0 1 1 1 1; (b) 1 0 1 1 1 1 0 0 0 1; and (c) 1 1 1 1 0 0 0 0 1 0.

RANDOM-ACCESS MEMORY (RAM)

The information stored in a ROM cannot be altered casually but only by means of special-purpose programmers, which first reinstate all the diodes in the ROM's encoder matrix and then selectively disconnect (or "burn") the unwanted diodes. The ROM's use is, therefore, restricted to storing specific programs in a digital system and does not extend to cases where we wish to change the stored words routinely by pressing buttons on a keyboard, for example. In the latter case, it is necessary to use *read-write memories* of which magnetic core memories and semiconductor memories are the most commonly used.

Magnetic core memories are *nonvolatile:* that is, they retain the stored information even when the power is turned off. On the other hand, semiconductor memories are *volatile* since the stored information is lost when the power is turned off. But, in many digital systems, a standby power supply (which is quite small) is usually provided so as to maintain the stored information in semiconductor memories even when the main power is turned off.

A memory unit in which any of the stored words can be addressed directly at random without having to perform a sequential search is called a *random-access memory* (RAM). The term RAM has become associated exclusively with semiconductor memories through common usage. RAMs can be classified as *static RAMs* and *dynamic RAMs*. A static RAM makes use of flip-flops that can maintain a given output level for an indefinite period of time so long as the power is on. A dynamic RAM, on the other hand, depends upon charges stored in capacitors. The charge on the capacitors tends to leak, so it

Electronics: Circuits and Systems

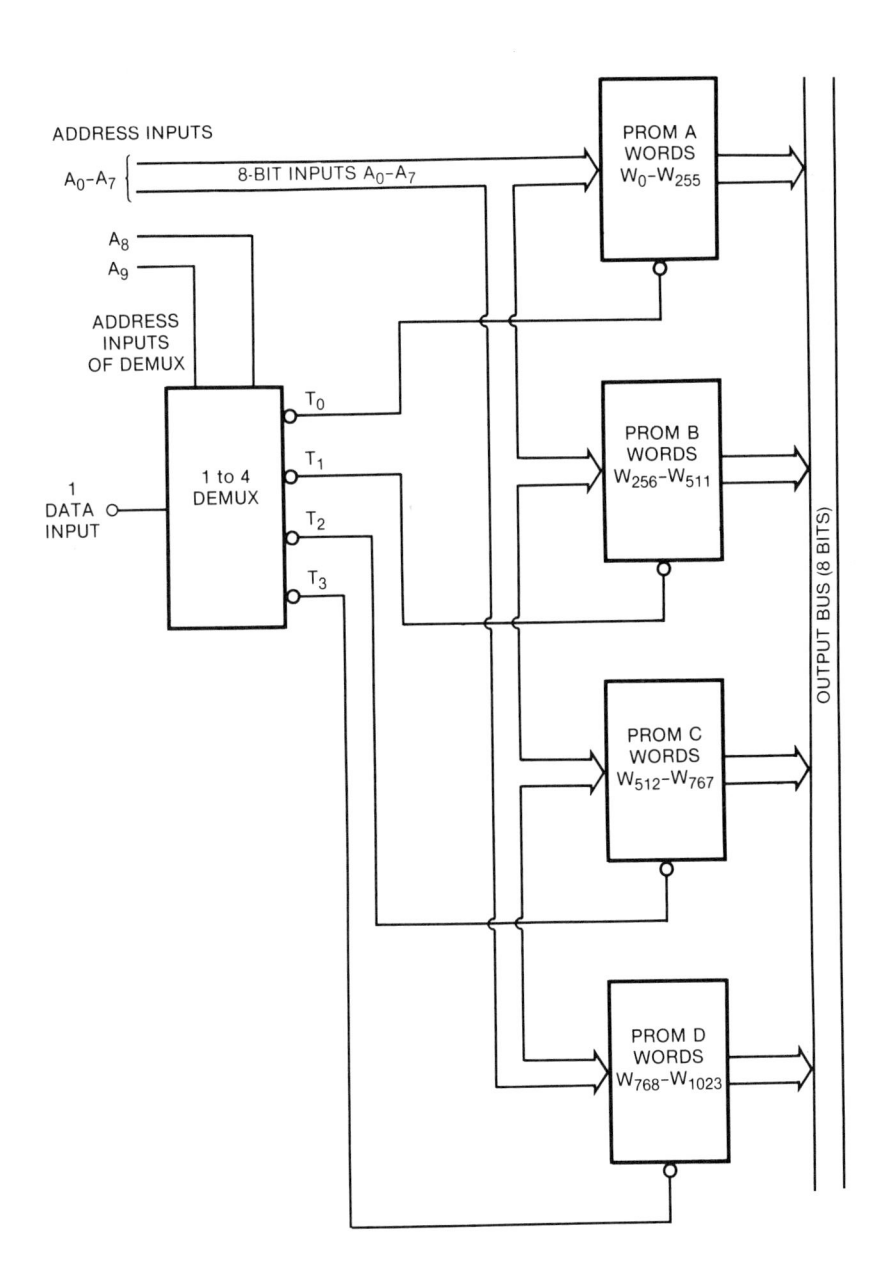

Fig. 12-16. Interconnection of four ROMS to increase available storage capacity.

is necessary to continually refresh the stored charge. A dynamic RAM is, therefore, characterized by the need for a *refresh cycle*.

Static RAMs

Static RAMs make use of *type D flip-flops** (or D FF). As indicated in Fig. 12-17, a D FF has a data input terminal D and two output terminals Q and $\overline{Q}$. The two outputs are complements of each other: $\overline{Q}$ always assumes a value opposite to Q. We will concentrate on the output Q in our discussion. The writing operation in a D FF involves applying the desired data (0 or 1) on the D input. The bit on the D input line gets transferred to the output Q. A periodic pulse train, called the clock pulse train, is applied to the CLK input of the D FF. When CLK = 0, the output does not change but remains stable at the value it already has. When CLK = 1, the bit present on the D input just before the occurrence of a clock pulse gets transferred to the Q output terminal at the end of the clock pulse. (In some D FF, the CLK is active low and the previous situations are reversed. An active low CLK is identifiable through the small circle in the block diagram of a FF in the manufacturer's data sheet.)

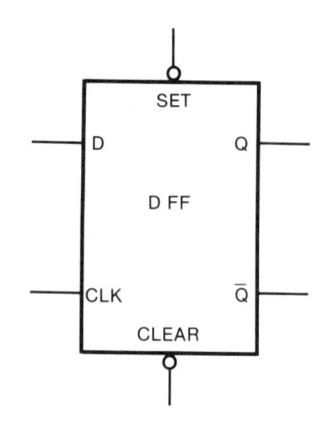

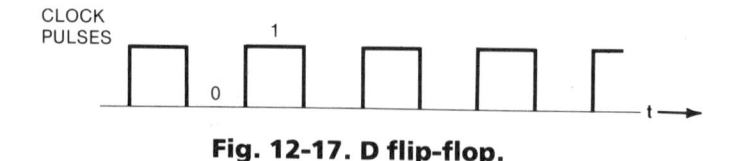

Fig. 12-17. D flip-flop.

---

* A general discussion of flip-flops is presented in the next chapter.

Electronics: Circuits and Systems

Most flip-flops also have preset (set) and pre-clear (clear) input terminals, which permit us to override the data on the D input terminal and to set the output directly at 1 or at 0. These two controls are usually active low.

An arrangement of several D FFs (usually 4, 8, or 16) in the form shown in Fig. 12-18A is called a *register* and is used to store a word (of 4, 8, or 16 bits). For the sake of simplicity, a register is shown in the form of a single block as in Fig. 12-18B.

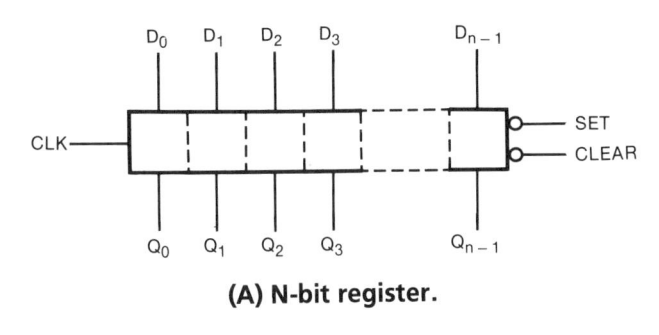

(A) N-bit register.

(B) Four-bit register.

Fig. 12-18. Register.

A static RAM consists of an array of registers along with several control inputs. The typical size of a static RAM can be, for example, 1K × 4, that is, 1024 words each of which is 4 bits. For the sake of discussion, consider a 16 × 4 static RAM, whose internal organization is shown in Fig. 12-19. There are 16 registers each of which stores a 4-bit word. Since there are sixteen registers, we need four address lines. The address inputs A_3, A_2, A_1, A_0 are decoded by a four-input decoder that sends a 1 signal to the appropriate register. Once a register is selected, the word already stored in it can be read off on the output lines or a new word can be written in the register. If the R-W control input is made 1, a *read* operation occurs and the stored word appears on the output lines. If the R-W control input is made 0, a write operation occurs and the bits on the data input lines get stored as the new word in the register. RAM chips are usually provided with *chip enable* or *chip select* control inputs, which permit the interconnection of two or more RAM chips to obtain a memory of larger capacity. An example of the interconnection of two 16 × 4 RAMs to obtain a 32 × 4 memory is shown in Fig. 12-20.

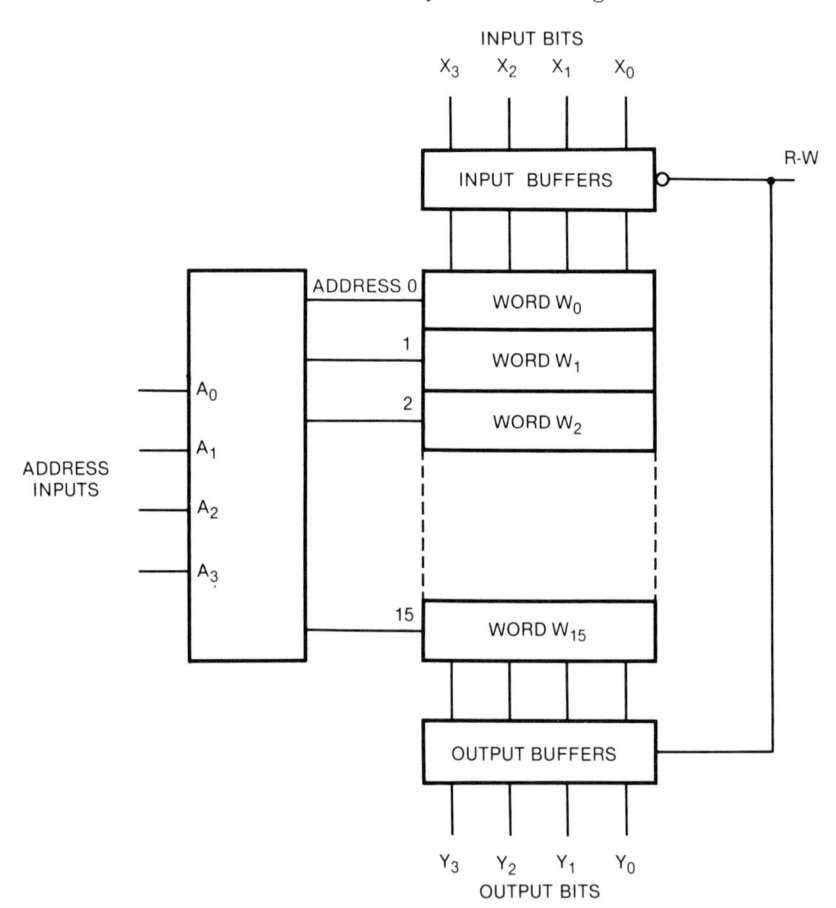

Fig. 12-19. Internal organization of a 16 × 4 static RAM.

Static RAMs are made using BJTs or MOSFETs. The BJT RAMs are faster but more expensive than the MOS RAMs. The speed with which stored words can be read in a static RAM is expressed in terms of the *access time*. The address inputs have to be maintained stable for a duration equal to the access time in order to ensure that the word stored in the address register appears on the output lines. For BJT RAMs, the access time is about 40 ns, while it can be as high as 500 ns for MOS RAMs.

Electronics: Circuits and Systems 719

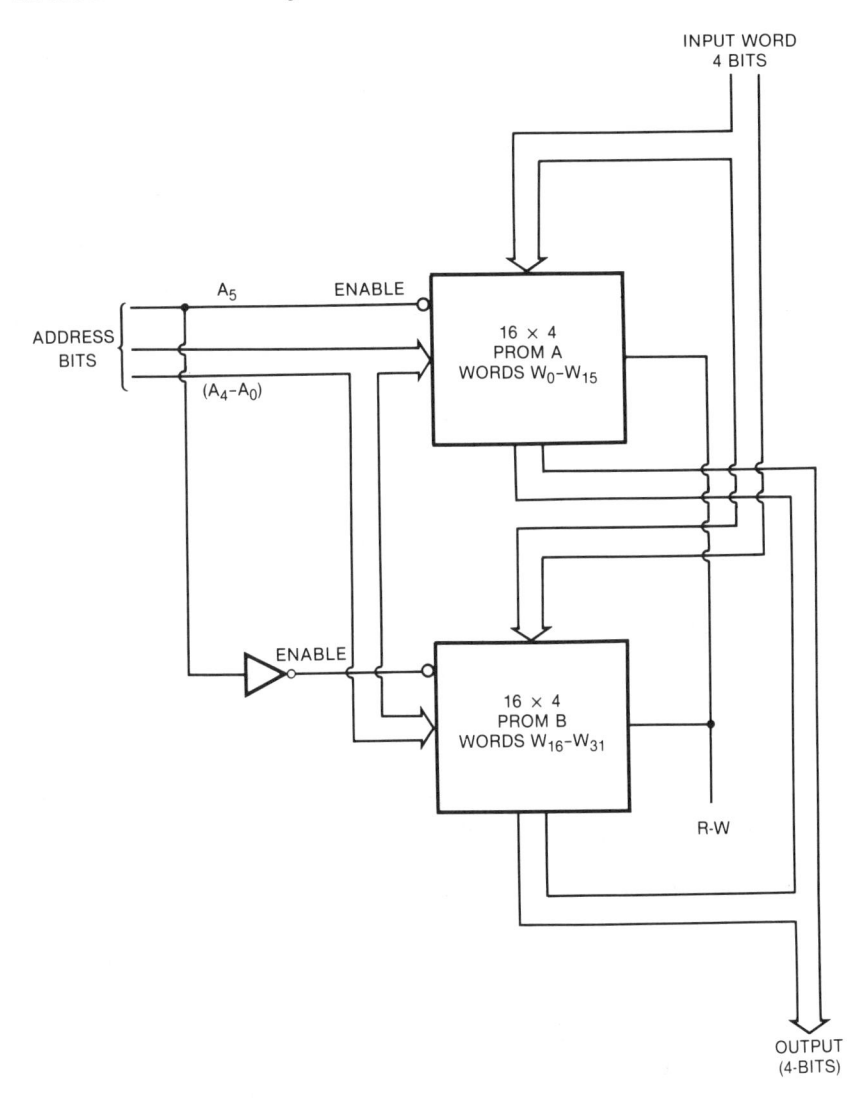

Fig. 12-20. Interconnection of two RAMS to increase available capacity.

Dynamic RAMs

A dynamic RAM is a matrix-like interconnection of memory cells arranged in M rows and N columns. Each memory cell is a circuit made up of one or more MOS transistors and a capacitor (which may be just the substrate capacitance in a MOS device). Data are stored in the form of a charge (corresponding to bit 1) or absence of a charge (corresponding to bit 0) on the capacitor. An individual cell in the memory can be addressed by addressing the row and the column in which it is situated. For example, consider a dynamic RAM with 64 rows and 64 columns so that the total memory capacity is 4K bits. Then, the number of address lines needed for the 64 rows is 6 (since $2^6 = 64$) and the number of address lines needed for the columns is also 6. The total number of address lines on the chip will be 12, with the first 6 (A_0 through A_5) reserved for row addressing and the other 6 (A_6 through A_{11}) for column addressing. The block diagram form of a dynamic RAM chip of 4K-bit capacity is shown in Fig. 12-21. Besides the 12 address lines, there are several control inputs, one data input D_{in} and one data output D_{out}. The data input line is used to write a data bit into a cell, and the data output line is used for reading the data bit stored in a cell. It should be noted that the input data must be available in a serial format (or converted to that format) for writing into the RAM. Similarly, the output data will be read serially from the RAM also. The CS is the *chip-select* control input bit stored in a cell. The CS is the chip-select control input (which is active low) and WE is the *write enable* control input that acts exactly like the R-W control input of the static RAM. In some RAMs, the number of pins in the package is reduced by using the same inputs for both row addressing the column addressing but using a multiplexing of their operations so that they address the row half the time and column the other half the time.

As mentioned earlier, a refresh cycle is necessary in order to maintain the charge on each capacitor at the correct value. In most dynamic RAMs, the refresh operation occurs automatically during the normal operation of reading or writing. In some cases, special refresh controllers are used.

Comparing the dynamic RAM with the static RAM, each has certain advantages and certain disadvantages. Since dynamic RAMs do not use flip-flops, they have a higher memory capacity for a given sized chip so that the cost per bit is lower. The savings in space by using dynamic RAMs is considerable. But, the disadvantage of the dynamic RAM is that the words stored are of 1 bit width (since each cell stores 1 bit only). This means that if we wish to use a memory of 8-bit words, we need to use 8 dynamic RAMs, with each storing one of the bits in each word. In such a situation, the static RAM has a decided advantage. Also the circuitry involved in using static RAMs is much simpler than for dynamic RAMs with their need for refresh cycles.

OTHER TYPES OF MEMORY UNITS

Since RAMs have the disadvantage of being volatile (except where a standby power supply is provided to prevent the data from disappearing), magnetic core memories are normally used in computer systems for perma-

Electronics: Circuits and Systems

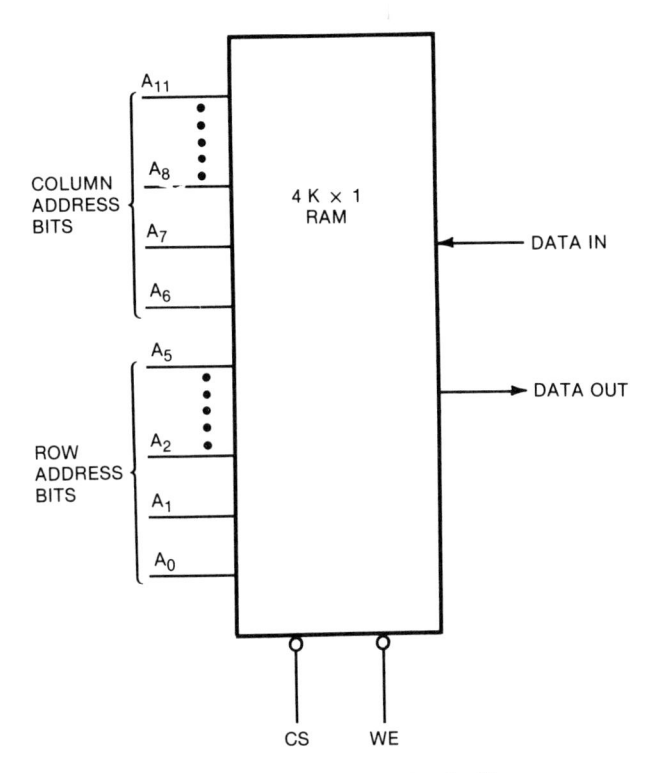

Fig. 12-21. Dynamic RAM block diagram.

nent storage of data. These are a matrix-like array of small toroidal-shaped rings made of ferrite. Each ring is 50 mils in diameter or smaller. The ferrite can be magnetized in one direction or the other corresponding to the storage of bit 1 or 0 by sending currents through conductors passing through the center of the toroid. A matrix has typically 64 rows and 64 columns (or 4096 ferrite rings), and a large number of these matrices can be stacked so as to have a very large memory capacity.

In microcomputer systems where the physical size of the computer does not permit the inclusion of large internal memory units, auxiliary memory units are used externally to the computer, and the information can be transferred to and fro between the computer memory and the external memory. Tape recorders, floppy disks, and magnetic disks (hard disks) are the commonly used types of external memory systems. The floppy disk is a flexible disk that is enclosed in a protective cardboard cover with an access slot for the read-record head. The data are stored in sectors on a number of concentric tracks on the disk. The floppy disk is usually driven at 360 rpm. The average access time is about 0.3 s and is much faster than that in any tape drive system.

SUMMARY SHEETS

MULTIPLEXERS (MUX)

Four-Input MUX

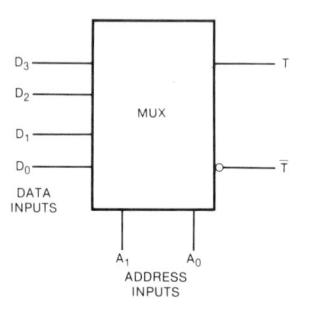

```
D₃ ───┐         ┌─── T
D₂ ───┤         │
      │   MUX   │
D₁ ───┤         │
D₀ ───┤         ├─── T̄
DATA             
INPUTS
       A₁   A₀
       ADDRESS
       INPUTS
```

A₁ A₀	T
00	D₀
01	D₁
10	D₂
11	D₃

Enable Input

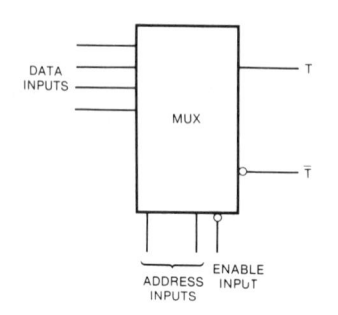

```
DATA ─────┐         ┌─── T
INPUTS ───┤         │
          │   MUX   │
          │         ├─── T̄
          └─────────┘
         ADDRESS  ENABLE
         INPUTS   INPUT
```

Enable = low: MUX is activated.

Enable = high: MUX is inoperative and presents a high resistance.

Expansion of MUX Capacity

The interconnection of a number of MUX packages can be used to simulate a MUX of larger capacity. One or more of the data inputs can act as the enable inputs to the different packages.

MUX in Logic Design

A logic function of (n + 1) variables can be designed with a single MUX with n-address (or 2^n data) lines.

For three variables X, Y, Z, choose X and Y as the address lines. Partition the truth table into groups of XY = 00, 01, 10, and 11. Compare the values of Z with those of the output function in the truth table. This determines the connections to be made to the four data lines, which will be Z, or $\bar{Z}$, or 0, or 1.

Demultiplexers

A demultiplexer has a single data input line and the data can be routed to any one of the output lines by selecting the address signals.

DECODERS AND ENCODERS

Decoder

A decoder translates the coded representation of a number into the number itself.

Number of inputs is the number of bits in the code.

Number of outputs is the number of different numbers covered by the coding scheme.

BCD-to-Decimal Decoder

Treat the 4-bit input as a binary representation of a decimal digit. The output line corresponding to that digit will be made active.

Encoder

An encoder produces the coded representation of the input number.

Diode Matrix Encoders

An array of diodes (with the number of rows = number of words to be encoded and number of columns = number of bits in the code) is used to form the encoder. A 1 bit is obtained when a diode is present and a 0 when the diode is absent.

Programmable Encoders

The matrix array is fabricated with all diodes present. Fusible links are removed for the diodes not needed in a particular application.

Electronics: Circuits and Systems

READ-ONLY MEMORIES (ROM)

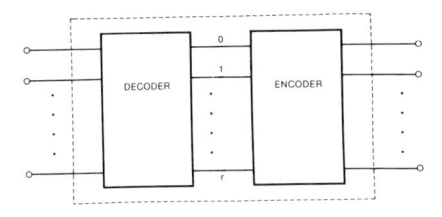

Inputs are decoded by the decoder into numbers 0 through r (only one of which will be active at any given time). The encoder generates a different coded representation of the numbers 0 through r.

ROM Capacity

Number of inputs = n; number of outputs = m. Capacity = $2^n \times m$ bits, 1 kilobit = 1024 bits, 1 byte = 8 bits.

Programmable ROMs

Mask programmed ROM: The manufacturer has produced a ROM to perform a specific function as desired by the customer. The function cannot be changed by the customer.

Programmable ROM: The customer can alter the function of the ROM by selectively melting the fusible links in the encoder.

Erasable PROM: The function of the ROM can be changed more than once. EPROMs use ultraviolet radiation to erase a stored program, while EAPROMs use electrical signals to erase the program.

Expansion of ROM Capacity

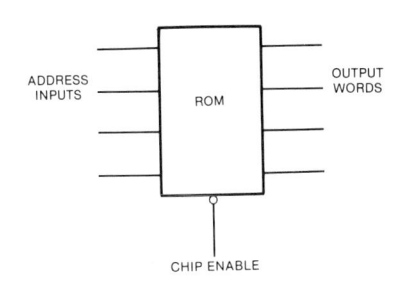

The chip enable or chip select control of a ROM permits the interconnection of two or more ROMs so as to simulate a single ROM of a larger capacity.

RANDOM-ACCESS MEMORY (RAM)

Random-Access Memory (RAM) can be used for both read-and-write operations. Information stored in it can be routinely altered in the course of normal operation.

Nonvolatile memories: Magnetic memories in which the stored information is not lost when the power is turned off.

Volatile memories: Semiconductor memories in which the stored information is lost when the power is turned off. (Standby power supply is used in modern digital systems so as to retain the stored information when the main power is turned off.)

Random-Access Memory: The information stored in any location can be accessed directly (at random) without having to proceed sequentially.

Static RAM: Data are stored as outputs of type D flip-flops. BJT RAMs are faster but more expensive than MOS RAMs. Access times for BJT RAMs are about 40 ns while they are around 500 ns for MOS RAMs.

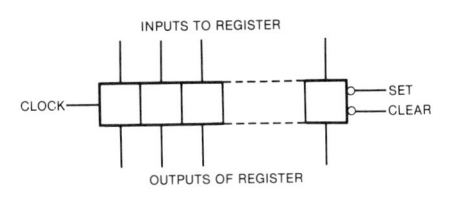

Dynamic RAM: Data are stored in the form of charges on capacitors. It is necessary to recharge the capacitors in order to compensate for the leakage of charge, which is done by a refresh cycle.

A *dynamic RAM* is in the form of a matrix of memory cells. Dynamic RAMs have a larger memory capacity than static RAMs have a given chip size—hence a lower cost per bit. The circuitry of static RAMs is, however, much simpler since a refresh cycle is not needed.

ANSWERS TO DRILL PROBLEMS

12.1 (a) $T = D_0$, $\overline{T} = \overline{D_0}$; (b) $T = D_5$; (c) $T = D_7$.
12.2 (a) 010; (b) 100.
12.3 See Table 12-DP3.

Table 12-DP3. Answer to Drill Problem 12.3

Address Inputs			MUX Output
A_2	A_1	A_0	T
0	0	0	Data on line D_0
0	0	1	Data on line D_1
0	1	0	Data on line D_2
0	1	1	Data on line D_3
1	0	0	Data on line D_4
1	0	1	Data on line D_5
1	1	0	Data on line D_6
1	1	1	Data on line D_7

12.4 (a) B_5; (b) B_{11}; (c) B_{15}; (d) B_3.
12.5 See Fig. 12-DP5.
12.6 The T column of the table (like that in Table 12-4) will have the following entries: 0, 1, 1, 0, 1, 0, 0, 1, 1, 0, 0, 1, 0, 1, 1, 0. The MUX inputs will be: $D_0 = Z$, $D_1 = \overline{Z}$, $D_2 = \overline{Z}$, $D_3 = Z$, $D_4 = \overline{Z}$, $D_5 = Z$, $D_6 = Z$, $D_7 = \overline{Z}$.
12.7 See Fig. 12-DP7.
12.8 (a) W_{29}; (b) W_{285}; (c) W_{255}; (d) W_{511}.
12.9 (a) T_1 low, PROM B enabled. W_{463} is read. (b) T_2 low, PROM C enabled. W_{753} is read. (c) T_3 low, PROM D enabled. W_{962} is read.

(A)

```
C̄ ──── D₃
C̄ ────
                  ──── T
C̄ ────
0 ──── D₀
       A₁  A₀
       │   │
       A   B
```

(B)

```
C ──── D₃
0 ────
                  ──── T
0 ────
C ──── D₀
       A₁  A₀
       │   │
       A   B
```

(C)

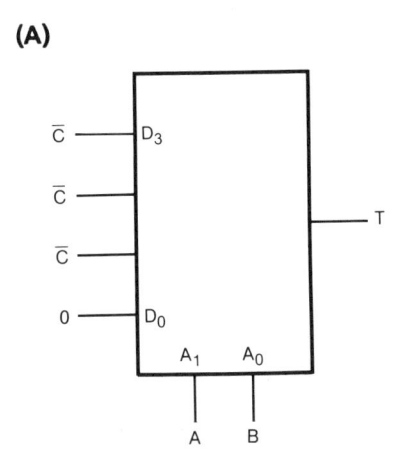

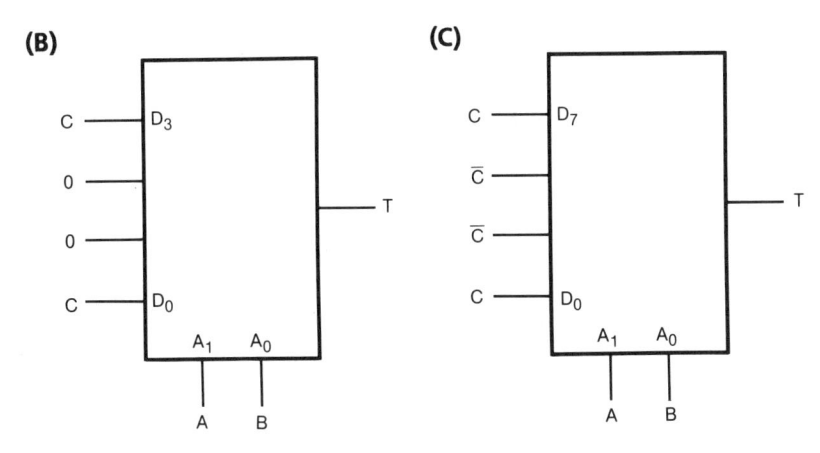

```
C  ──── D₇
C̄ ────
C̄ ────            ──── T
C  ──── D₀
       A₁  A₀
       │   │
       A   B
```

Fig. 12-DP5.

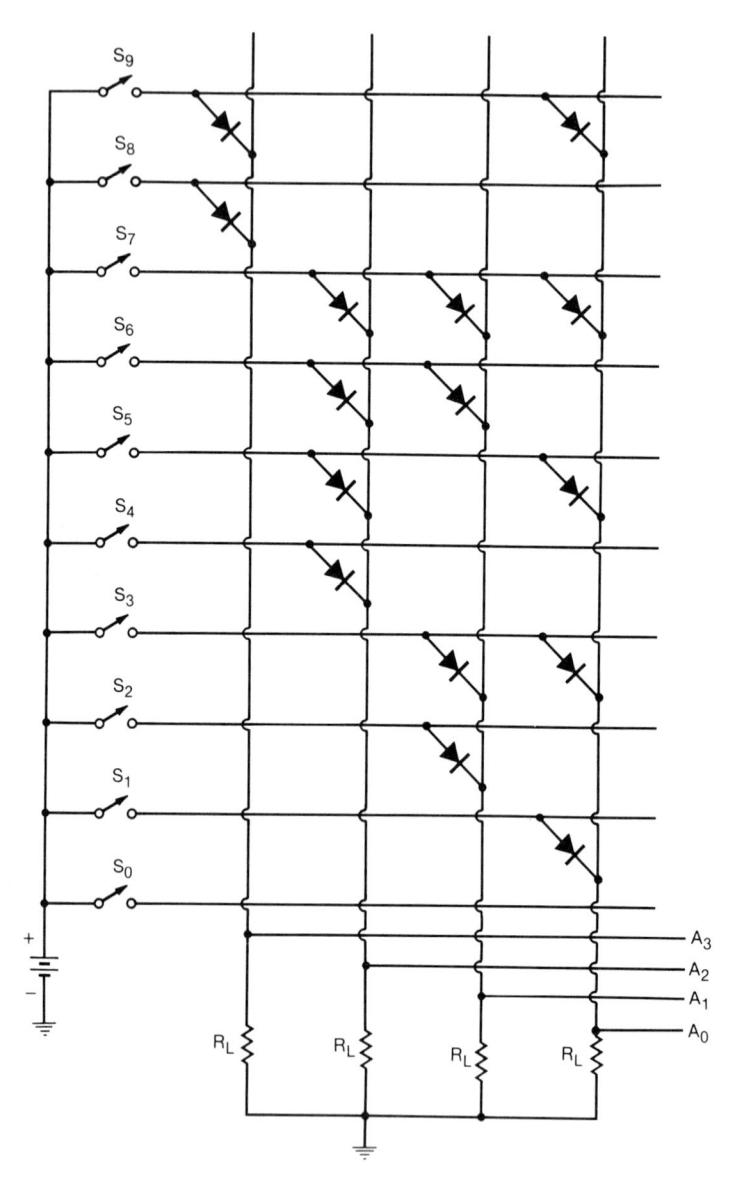

Fig. 12-DP7.

PROBLEMS

Multiplexer

12.1 Set up a two-level NAND network (using available inputs $D_0, \ldots, D_3$, A_1, A_0 and complements) that will function as a four-input MUX.

12.2 Four 3-bit numbers $W_2W_1W_0$, $X_2X_1X_0$, $Y_2Y_1Y_0$, and $Z_2Z_1Z_0$ are fed into a system. The function of the system is to transmit one of the numbers to the output as determined by the values of two control signals A_1A_0. When $A_1A_0 = 11$, the number $W_2\,W_1\,W_0$ is to be transmitted, for example. Set up (in block diagram form) a system of MUXs that will perform the desired function.

12.3 Obtain a single MUX realization of the network in Problem 10.14 (Chapter 10).

12.4 Set up a single MUX network to realize the logic function of Problem 10.15.

12.5 It is possible to realize a logic function of more variables than can be designed with a single MUX by using some external logic gates. Consider the networks shown in Fig. 12-P5. Obtain the truth table of each of the networks shown.

12.6 A four-output *demultiplexer* has a single input X and two control (or address) inputs A_1 and A_0. There are four output lines Z_3, Z_2, Z_1, Z_0. Depending upon the values of A_1A_0, the input signal is transmitted to the appropriate output line. Set up network that will perform this function.

(A)

(B)

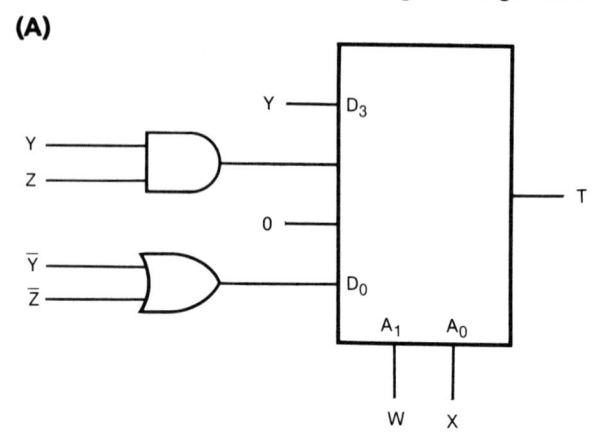

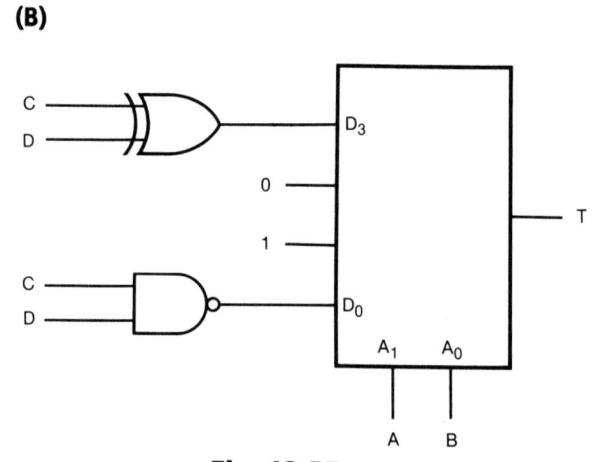

Fig. 12-P5.

Decoders and Encoders

12.7 A decoder can be used for realizing a given logic function since it has outputs corresponding to all the rows of a truth table. Use a 3-bit binary decoder and an OR gate to realize the logic function of Drill Problem 10.16.

12.8 Table 12-P8 shows a particular coding scheme (called a *cyclic* code) of the decimal digits 0 through 9. Obtain an encoder matrix for this coding scheme.

Table 12-P8. Cyclic Code for Problem 12.8

Decimal Digit	C_3	C_2	C_1	C_0
0	0	0	0	0
1	0	0	0	1
2	0	0	1	1
3	0	0	1	0
4	0	1	1	0
5	1	1	1	0
6	1	0	1	0
7	1	0	0	0
8	1	1	0	0
9	0	1	0	0

ROMs

12.9 Design a ROM to convert the cyclic code of the previous problem to BCD code.

12.10 Design a ROM to convert the BCD code to the cyclic code given in Problem 12.8.

12.11 Set up a PROM network to store 1024 words each 8-bit long if each PROM has a capacity of 512 × 4 bits. As part of your solution indicate the PROM that will be enabled and the word addressed for the following address signals: (a) 1 0 0 0 1 1 0 0 0 0; (b) 0 1 0 0 0 0 1 1 1 0.

CHAPTER 13

ARITHMETIC LOGIC, COUNTERS, AND SHIFT REGISTERS

Some of the most common and important applications of digital logic components and networks are in the area of *arithmetical operations*. Arithmetic networks are used not only in numerical computations but in a large variety of data processing operations also. The arithmetical operations in a digital system use *binary numbers*, which are strings of ones and zeros, since such digits can be represented, respectively, by the absence or presence of a pulse. Even though the actual operations involve binary numbers, it is often convenient for us to use alternative representations as well since binary strings tend to be quite long and errors easily crop up when we are reading or transferring binary numbers with a large number of digits. Two systems that are commonly used are the *octal* and *hexadecimal number systems*. It should be remembered that the logic networks work only with the binary representations, which means that entries made in other systems must be translated by some interface logic networks into binary numbers.

BINARY NUMBER SYSTEM

A number system in which k distinct digits 0, 1, . . . , (k − 1) are used in the representation of numbers is called a *mod-k system* since k is the *modulus* of the system. The decimal system that we normally use is a *mod-10 system* since there are 10 digits 0 through 9 in it. The binary number system uses only the two digits 0 and 1 and is, therefore, a *mod-2 system*. The *octal system* has a modulus of 8, and the *hexadecimal system* a modulus of 16.

When an n-digit number is written in a system, the position of each digit has a *weight*. Confining our attention to whole numbers (or integers) for the time being, the weight of each position is a power of the modulus k. In the decimal system, for example, the right-most digit (which is the least significant digit) has a weight of 10^0 or 1, the next position has a weight of 10^1, and so on. The nth digit position has a weight of 10^{n-1}. For example, the number 9758 actually stands for

$$9 \times 10^3 + 7 \times 10^2 + 5 \times 10^1 + 8 \times 10^0$$

In the binary system, the right-most bit in a whole number is the least significant bit, or LSB, (*bit* stands for *binary digit*) and has a weight of 2^0. Then working towards the left, the weights of the successive positions are 2^1, 2^2, 2^3, ..., 2^{n-1}, which correspond to 2, 4, 8, 16, ... The bit in the left-most position is called the most significant bit (MSB).

Given any integer in the binary form, its decimal equivalent can be determined by multiplying each bit by the appropriate power of 2 and adding the products.

Example 13.1

Convert each of the following binary numbers to its decimal equivalent: (a) 1 1 0; (b) 1 0 1 1; (c) 1 1 0 1 0 1 1 0.

Solution

(a) $(1\ 1\ 0)_2 = 1 \times 2^2 + 1 \times 2 + 0 \times 2^0 = (6)_{10}$ where the subscripts 2 and 10 are used to denote the particular modulus being used. (b) $(1\ 0\ 1\ 1)_2 = 2^3 + 2 + 1 = (11)_{10}$; (c) $(1\ 1\ 0\ 1\ 0\ 1\ 1\ 0)_2 = 2^7 + 2^6 + 2^4 + 2^2 + 2 = (214)_{10}$.

Drill Problem 13.1: Convert each of the following binary numbers to its decimal equivalent: (a) 1 1; (b) 1 1 0 1; (c) 1 0 1 1 0 1; (d) 1 0 1 0 1 0 1 0.

The conversion of any decimal integer to its equivalent binary representation can be done in different ways. One procedure involves a successive division by two (which is equivalent to successive division by powers of two), and this procedure is illustrated in the following example.

Example 13.2

Convert the decimal integer 349 to an equivalent binary number.

Solution

	Quotient	Remainder	
Divide 349 by 2	174	1	(This is the LSB.)
Divide 174 by 2	87	0	
Divide 87 by 2	43	1	
Divide 43 by 2	21	1	
Divide 21 by 2	10	1	

Divide 10 by 2	5	0	
Divide 5 by 2	2	1	
Divide 2 by 2	1	0	
Divide 1 by 2	0	1	(This is the MSB.)

This is, the given number and each successive quotient are divided by two, and the digits obtained as the remainders in the steps form the bits in the binary number. The division stops when the quotient becomes 0. In the present example, we get

$$(349)_{10} = (1\ 0\ 1\ 0\ 1\ 1\ 1\ 0\ 1)_2$$

The LSB is the first remainder bit and the final remainder bit is the MSB in the equivalent binary number.

Drill Problem 13.2: Convert each of the following decimal numbers to its equivalent binary form: (a) 45; (b) 117; (c) 1986.

Now consider the binary representation of fractions. A fraction in binary form is written in a similar manner to that used in decimal systems: 0.11010 for example. Starting with the point, the right-most digit after the point has a weight of $(1/2)$, the next bit has a weight of $(1/2)^2$, and the successive bits have a weight of $(1/2)^3$, $(1/2)^4$, and so on. The equivalent decimal number can be determined by multiplying each digit in the fraction by its weight and adding the results.

Example 13.3

Convert each of the following binary fractions to its decimal equivalent: (a) 0.01; (b) 0.101101.

Solution

(a) $(0.01)_2 = 0 \times (1/2) + 1 \times (1/2)^2 = (0.25)_{10}$; (b) $(0.1\ 0\ 1\ 1\ 0\ 1)_2 = (1/2) + (1/2)^3 + (1/2)^4 + (1/2)^6 = 0.703125$.

Drill Problem 13.3: Convert each of the binary fractions given below to its decimal equivalent: (a) 0.1; (b) 0.001; (c) 0.1 1 0 1 0 1.

Given a decimal fraction, the equivalent binary form is found by a successive *multiplication* by two. After each product is formed, the integer portion (1 or 0) provides the bit needed in the binary representation. The fraction part of the product is multiplied again by two and the process is repeated until the fraction part becomes zero.

Example 13.4

Convert each of the following decimal fractions to its binary equivalent: (a) 0.75; (b) 0.333; (c) 0.142.

Solution

(a) The given number is 0.75.

	Integer part	Fraction part	
Multiply 0.75 by 2	1	0.5	(1 is the MSB)
Multiply 0.5 by 2	1	0.0	

Since the fraction part is 0, we stop. We have

$$(0.75)_{10} = (0.11)_2$$

(b) The given number is 0.333.

	Integer part	Fraction part	
Multiply 0.333 by 2	0.	0.666	(MSB = 0)
Multiply 0.666 by 2	1.	0.332	
Multiply 0.332 by 2	0.	0.664	
Multiply 0.664 by 2	1.	0.328	
Multiply 0.328 by 2	0.	0.656	
Multiply 0.656 by 2	1.	0.312	

At this point we begin to wonder if this will ever stop. Except in cases where the given decimal fraction is the sum of a series of powers of (1/2), the process is interminable! We have to decide on a permissible error and stop after a certain number of digits that will lead to that or a smaller error. If we stopped the process as it was, we have

$$(0.333)_{10} = 0.0\ 1\ 0\ 1\ 0\ 1$$

In order to determine the error, we first convert the binary number 0.010101 to decimal: 0.328125 and the error is $[(0.333 - 0.328125)/0.333] \times 100 = 1.46\%$

(c) The given number is 0.142. Using the previous procedure, we obtain:

$$(0.142)_{10} = 0.0\ 0\ 1\ 0\ 0\ 1\ 0\ 0\ 0\ 1$$

Drill Problem 13.4: Convert each of the following decimal fractions to its equivalent binary form (to the number of bits specified when the process does not terminate).
(a) 0.15625; (b) 0.689 (4 bits); (c) 0.2593 (8 bits).

When a binary number has an integer part and a fraction part, then the conversion to its decimal counterpart is done by using the principles employed in Examples 13.1 and 13.3 to the integer portion and the fraction portion, respectively. Similarly, when a decimal number has an integer part and a fraction part, the methods explained in Examples 13.2 and 13.4 are used, respectively, on the integer part and the fraction part separately.

Example 13.5
Convert each of the following binary numbers to the decimal form: (a) 101.0111; (b) 10.10111; (c) 0.1010111.

Solution
(a) $(101.0111)_2 = 2^2 + 2^0 + (1/2)^2 + (1/2)^3 + (1/2)^4 = 5.4375$

(b) $(10.10111)_2 = 2^1 + (1/2) + (1/2)^3 + (1/2)^4 + (1/2)^5 = 2.71875$

(c) $(0.1010111)_2 = (1/2) + (1/2)^3 + (1/2)^5 + (1/2)^6 + (1/2)^7 = 0.6796875$

Drill Problem 13.5: Convert each of the following binary numbers to its equivalent decimal form: (a) 101.0101; (b) 11.111; (c) 1.01011.

Example 13.6
Convert each of the following decimal numbers to an equivalent binary number: (a) 12.34; (b) 1986.0625.

Solution
In each case, we convert the integer part and the fraction part separately into binary form and then write the composite result. Since the conversion of the integer part to binary is quite different from the procedure for converting the fraction part, care must be exercised to keep them separate.

(a) The given number is $(12.34)_{10}$.
Using the procedure of Example 13.2 on the integer 12, we get: 1100.
Using the procedure of Example 13.4 on the fraction part 0.34, we get 0.0 1 0 1 0 1 1 1. (The process does not terminate.) Therefore,

$$(12.34)_{10} = 1 1 0 0.0 1 0 1 0 1 1 1$$

(b) The given number is $(1986.0625)_{10}$.

$(1986)_{10} = 11111000010$

$(0.0625) = .0001$ (terminates)

$(1986.0625) = 1 1 1 1 1 0 0 0 0 1 0.0 0 0 1$

Drill Problem 13.6: Convert each of the following decimal numbers to its binary equivalent. (a) 43.792; (b) 876.54; (c) 1024.03125. (When the process does not terminate, use 6 bits after the decimal point.)

OCTAL AND HEXADECIMAL NUMBER SYSTEMS

It is not difficult to see that a long string of ones and zeros (such as the binary equivalent of 1986.0625 in the last example) is somewhat cumbersome for the human user even though a digital network does not have any problems dealing with such strings. The octal and hexadecimal number representations allow us to shorten the binary number representations to more manageable sizes without changing the inherent binary character of the numbers. Hexadecimal keyboards are frequently used in microprocessor systems.

The modulus of an octal number system is 8, and it uses the digits 0 through 7. In the case of whole numbers, the weights of the positions are (starting with the least significant digit) 8^0, 8^1, 8^2, . . . , 8^{n-1}. In the case of fractions, the weights of the positions are (starting with the most significant digit) $(1/8)$, $(1/8)^2$, $(1/8)^3$, ... The conversion of an octal number to its decimal equivalent is done by multiplying the digit in each position by the appropriate power of 8 or (1/8) and adding the results.

Example 13.7
Convert each of the following octal numbers to its decimal equivalent. (a) 13.7; (b) 174.432.

Solution
(a) The given number is $(13.7)_8$.

> Integer part: $(13)_8 = 1 \times 8 + 3 = (11)_{10}$.
>
> Fraction part: $(0.7)_8 = 7 \times (1/8) = (0.875)_{10}$.
>
> Therefore, $(13.7)_8 = (11.875)_{10}$.

(b) The given number is $(174.432)_8$.

> Integer part: $(174)_8 = 1 \times 8^2 + 7 \times 8 + 4 = (124)_{10}$.
>
> Fraction part: $(0.432)_8 = 4 \times (1/8) + 3 \times (1/8)^2 + 2 \times (1/8)^3$
>
> $= 0.55078125$.
>
> Therefore, $(174.432)_8 = (124.55078125)_{10}$.

Drill Problem 13.7: Convert each of the following octal numbers to its decimal equivalent: (a) 617; (b) 0.5625; (c) 25.643.

The conversion of an octal number to its binary equivalent is extremely simple: each of the digits in the octal number is expanded into a 3-bit binary equivalent and the resulting binary strings give the binary equivalent.

Example 13.8
Convert each of the following octal numbers to its binary equivalent: (a) 174; (b) 0.5625; (c) 17.456.

Solution
(a) The given number is $(174)_8$ = 001 111 100.
                                                   1 7 4

Therefore, $(174)_8$ = $(0\ 0\ 1\ 1\ 1\ 1\ 1\ 0\ 0)_2$.

(b) The given number is $(0.5625)_8$ = .101 110 010 101.
                                                       5   6 2   5

Therefore, $(0.5625)_8$ = $(.1\ 0\ 1\ 1\ 1\ 0\ 0\ 1\ 0\ 1\ 0\ 1)_2$.

(c) The given number is $(17.456)_8$ = $001\ 111.100\ 101\ 110_2$.
                                               1 7   4 5 6

Therefore, $(17.456)_8$ = $(001111.100101110)_2$.

Drill Problem 13.8: Convert each of the following octal numbers to its binary equivalent: (a) 246.43; (b) 1476.325.

The conversion of a decimal integer to its octal equivalent follows a procedure similar to the one used in the conversion of a decimal to binary. The given number and each successive quotient is divided by eight, and the remainder obtained in each step forms the digits of the equivalent octal number. The digit obtained as the remainder after the first division is the last significant digit, and the remainder after the last division is the most significant digit.

Example 13.9
Convert each of the following decimal numbers to its octal equivalent: (a) 1984; (b) 4943.

Solution
(a) The given number is 1984.

	Quotient	Remainder	
Divide 1984 by 8	248	0	(This is LSB)
Divide 248 by 8	31	0	
Divide 31 by 8	3	7	
Divide 3 by 8	0	3	(This is MSB)

Therefore, $(1984)_{10}$ = $(3700)_8$.

(b) The given number is 4943.

Electronics: Circuits and Systems

	Quotient	Remainder	
Divide 4943 by 8	617	7	(This is LSB)
Divide 617 by 8	77	1	
Divide 77 by 8	9	5	
Divide 9 by 8	1	1	
Divide 1 by 8	0	1	

Therefore, $(4943)_{10} = (11517)_8$.

Drill Problem 13.9: Convert each of the following decimal numbers to its octal equivalent: (a) 1776; (b) 1492; (c) 24576.

The conversion of a decimal fraction to its octal equivalent is done by using a procedure similar to the one for converting decimal fractions to their binary equivalents. The given decimal fraction is multiplied by eight. The integer part of each product forms a digit in the octal equivalent arranged in the same order as obtained in the process. The fraction part of each product is multiplied by eight and the previous process is repeated. Again, the process may not terminate unless the given number is the sum of a series of powers of (1/8).

Example 13.10

Convert each of the following fractions to its octal equivalent: (a) 0.6893; (b) 0.1246.

Solution

(a) The given number is 0.6893.

	Integer part	Fraction part
Multiply 0.6893 by 8	5.	0.5144
Multiply 0.5144 by 8	4.	0.1152
Multiply 0.1152 by 8	0.	0.9216
Multiply 0.9216 by 8	7.	0.3728
Multiply 0.3728 by 8	2.	0.9824
Multiply 0.9824 by 8	7.	0.8592

and it goes on.

We have $(0.6893)_{10} = (0.540727)_8$ (error less than 0.001%).

(b) The given number is 0.1246.

	Integer part	Fraction part
Multiply 0.1246 by 8	0.	0.9968
Multiply 0.9968 by 8	7.	0.9744
Multiply 0.9744 by 8	7.	0.7952
Multiply 0.7952 by 8	6.	0.3616
Multiply 0.3616 by 9	2.	0.8928
Multiply 0.8928 by 8	7.	0.1424

Therefore, $(0.1246)_{10} = (0.077627)_8$ (error 0.02%).

Drill Problem 13.10: Convert each of the following decimal fractions to its octal equivalent: (a) 0.3986; (b) 0.1125.

When a decimal number has both an integer part and a fraction component, each part is treated separately for conversion into the octal form.

Example 13.11

Convert each of the following decimal numbers to octal form. (a) 123.4875; (b) 1956.129.

Solution

(a) The given number is $(123.4875)_{10}$; integer part: $(123)_{10} = (173)_8$; fraction part: $(0.4875)_{10} = (0.37146)_8$. Therefore, $(123.4875)_{10} = (173.37146)_8$.
(b) The given number is 1956.129; integer part: $(1956)_{10} = (3644)_8$; fraction part: $(0.129)_{10} = (0.1020304)_8$. Therefore, $(1956.129)_{10} = (3644.1020304)_8$.

Drill Problem 13.11: Convert each of the following decimal numbers to its octal equivalent: (a) 1023.04; (b) 2759.642.

The conversion of a binary number into octal is extremely simple. Starting with the decimal point, we go to the left and group the bits three at a time. Each such group is converted into an octal number (0 through 7). A similar procedure is used by proceeding to the right from the decimal point. The octal digits obtained in this manner give the equivalent octal number.

Example 13.12

Convert each of the binary numbers given below to its equivalent octal number: (a) 101101; (b) 0.100111; (c) 1011.01101; (d) 10111011011.1101010111.

Solution

(a) $(1\ 0\ 1\ 1\ 0\ 1)_2 =\ \ 5\quad 5\ \ = (55)_8.$
101 101

(b) $(0.1\ 0\ 0\ 1\ 1\ 1)_2 = 0\ .\ \ 4\quad 7\ = (0.47)_8.$
100 111

(c) $(1\ 0\ 1\ 1.0\ 1\ 1\ 0\ 1)_2 =\ \ 1\quad 3\ \ .\ \ 3\quad 2\ = (13.32)_8.$
001 011$\quad$011 010

Note that extra zeros are supplied (without, of course, altering the value of the given number) in order to complete the groups of three.

(d) $(1\ 0\ 1\ 1\ 1\ 0\ 1\ 1\ 0\ 1\ 1\ .\ 1\ 1\ 0\ 1\ 0\ 1\ 0\ 1\ 1\ 1)_2$
 = 2 7 3 3 . 6 5 3 4
 010 111 011 011   110 101 011 100
 = $(2733.6534)_8$

Drill problem 13.12: Convert each of the following binary numbers to its equivalent octal form: (a) 1 0 0 0 0 1; (b) . 1 0 0 0 0 1; (c) 1 0 1 1 1 1 . 0 0 1 1 0 1 0 1; (d) 1 1 1 1 1 0 0 0 0 1 0 . 0 0 0 0 0 1.

The modulus of the *hexadecimal* number system is 16. It needs 15 distinct digits! Since we have only 10 numerical digits (0 through 9), the remaining 6 are formed with letters from the alphabet: A through F. The letters stand for the numbers 10 through 15 in the decimal system:

$$A = (10)_{10} \quad B = (11)_{10} \quad C = (12)_{10}$$
$$D = (13)_{10} \quad E = (14)_{10} \quad F = (15)_{10}$$

Thus, a number in the hexadecimal system will be in alphanumerical form: 89AB4F, for example.

In the case of whole numbers, the weights of the positions in a hexadecimal system are (starting with the least significant digit) 16^0, 16, 16^2, 16^3, ..., 16^{n-1}. In the case of fractions, the weights of the positions are (starting with the most significant digit) $(1/16)$, $(1/16)^2$, $(1/16)^3$, ...

The conversion of a hexadecimal (whole) number to its decimal equivalent is done by multiplying each digit by the weight of its position and adding the results.

Example 13.13

Convert each of the following hexadecimal numbers to its decimal equivalent: (a) 8A9F; (b) 0.AB6; (c) FAB.CDE.

Solution

(a) The given number is 8A9F.
$$(8A9F)_{16} = 8 \times 16^3 + \underset{(A)}{10} \times 16^2 + 9 \times 16 + \underset{(F)}{15}$$
$$= (35487)_{10}$$

(b) The given number is 0.AB6.
$$(0.AB6)_{16} = 10 \times (1/16) + 11 \times (1/16)^2 + 6 \times (1/16)^3$$
$$= 0.669433594$$

(c) The given number is FAB.CDE and $(FAB.CDE)_{16} = 15 \times 16^2 + 10 \times 16 + 11 + 12 \times (1/16) + 13 \times (1/16)^2 + 14 \times (1/16)^3 = 4011.8042$.

Drill Problem 13.13: Convert each of the following hexadecimal numbers to its decimal equivalent: (a) 8A6; (b) D92; (c) 0.6A; (d) F9.E6.

The conversion of a hexadecimal number to its binary equivalent is extremely simple: each digit in the hexadecimal number is expanded into a 4-bit binary equivalent and the resulting binary strings give the binary number.

Example 13.14
Convert each of the following hexadecimal numbers to its binary equivalent: (a) 8A9F; (b) FAB.CDE.

Solution
(a) The given number is 8A9F and $(8A9F)_{16}$ = 1000 1010 1001 1111
                                                 8   A   9   F
$= (1 0 0 0 1 0 1 0 1 0 0 1 1 1 1 1)_2$.

(b) The given number is FAB.CDE and $(FAB.CDE)_{16}$
= 1111 1010 1011.1100 1101 1110
   F   A   B   C   D   E
$= (1 1 1 1 1 0 1 0 1 0 1 1.1 1 0 0 1 1 0 1 1 1 1 0)_2$.

Drill Problem 13.14: Convert each of the following hexadecimal numbers to its binary equivalent: (a) 895.23; (b) 9A.2F.

The conversion of decimal numbers (whole numbers and fractions) to hexadecimal forms is done in a manner similar to the case of conversion to octal forms already discussed. The following two examples illustrate the procedure.

Example 13.15
Convert each of the following decimal numbers to its hexadecimal number: (a) 1984; (b) 4943.

Solution
(a) The given number is 1984.

	Quotient	Remainder	
Divide 1984 by 16	124	0	(This is LSB)
Divide 124 by 16	7	12	(or C)
Divide 7 by 16	0	7	(This is MSB)

Therefore, $(1984)_{10} = (7C0)_{16}$.

(b) The given number is 4943.

Electronics: Circuits and Systems

	Quotient	Remainder	
Divide 4943 by 16	308	15 or F	(This is LSD)
Divide 308 by 16	19	4	
Divide 19 by 16	1	3	
Divide 1 by 16	0	1	(This is MSD)

Therefore, $(4943)_{10} = (134F)_{16}$.

Drill problem 13.15: Convert each of the following decimal numbers to hexadecimal form: (a) 1776; (b) 1472; (c) 47521.

Example 13.16

Convert each of the following fractions to its hexadecimal equivalent: (a) 0.6893; (b) 0.0139.

Solution

(a) The given number is 0.6893.

	Integer part	Fraction part
Multiply 0.6893 by 16	11. (B)	0.0288
Multiply 0.0288 by 16	0.	0.4608
Multiply 0.4608 by 16	7.	0.3728
Multiply 0.3728 by 16	5.	0.9648
Multiply 0.9648 by 16	15. (F)	0.4368

Therefore, $(0.6893)_{10} = (0.B075F)_{16}$.

(b) The given number is 0.0139.

	Integer part	Fraction part
Multiply 0.0139 by 16	0.	0.2224
Multiply 0.2224 by 16	3.	0.5584
Multiply 0.5584 by 16	8.	0.9344
Multiply 0.9344 by 16	14. (E)	0.9504
Multiply 0.9504 by 16	15. (F)	0.2064

Therefore, $(0.0139)_{10} = (0.038EF)_{16}$.

Drill Problem 13.16: Convert each of the following decimal numbers to its hexadecimal equivalent: (a) 0.2759; (b) 0.4752.

The conversion of a binary number to hexadecimal is quite simple. Starting with the decimal point, we go to the left and group the bits four at a time. Each such group is converted into a hexadecimal number (0 through F). A similar procedure is used by proceeding to the right from the decimal point.

The hexadecimal digits obtained in this manner give the equivalent hexadecimal number.

Example 13.17
Convert the following binary numbers to the hexadecimal forms.
(a) 1 1 0 0 1 0 0 1; (b) 1 0 0 0 0 1 0 1 0 1 0 1; (c) 0.1 0 0 1 1 1 0 0 1 1 0; (d) 1 0 0 0 1 0 1 1 0 0 1 0 1.1 0 1 0 0 1 1 0 1.

Solution
(a) $(1 1 0 0 1 0 0 1)_2$ = 1100 1001 = $(C 9)_{16}$
                                                                     C   9

(b) $(1 0 0 0 0 1 0 1 0 1 0 1)_2$ = 0010 0001 0101 0101 = $(2155)_{16}$
                                                                    2   1   5   5

where two zeros are added at the beginning in order to get the 4-bit group.

(c) $(0.1 0 0 1 1 1 0 0 1 1 0)_2$ = 0.1001 1100 1100 = $(0.9CC)_{16}$
                                                       9   C   C

where a 0 is added at the end to complete the 4-bit group.

(d) 1 0 0 0 1 0 1 1 0 0 1 0 1.1 0 1 0 0 1 1 0 1 = 0001 0001 0110 0101 . 1010 0110 1000 = $(1165.A68)_{16}$

Drill Problem 13.17: Convert each of the following binary numbers to the hexadecimal equivalent: (a) 1 0 1 1 1 1 0 1 0 1; (b) 1 0 1 0 1 1 1 0 0 0.0 0 1 0 1 0 1 0 1 1 0 1.

ADDERS

The *addition of two numbers* is the most important *arithmetical* operation in a digital system since it forms the basis of other arithmetical operations such as multiplication and subtraction. When more than two numbers are to be manipulated, the operation is broken down into a series of steps each of which manipulates only two numbers at a time. The addition of numbers in the binary system involves the same sort of steps used in the conventional decimal system. When two digits are added, there is a *sum* digit as well as a *carry* digit. For example, in the sum (8 + 9), we get 17 in which the 1 represents the carry digit (needed for the addition of the next higher digits of the numbers), and the 7 is the sum digit.

Electronics: Circuits and Systems

Half-Adders

Table 13-1. Addition of Two Single-Digit Binary Numbers and the Truth Table for a Half-Adder

(A)	A	B	Sum	(B)	A	B	Carry	Sum
	0	0	0		0	0	0	0
	0	1	1		0	1	0	1
	1	0	1		1	0	0	1
	1	1	10		1	1	1	0

Consider first the addition of two single-digit binary numbers, the rules of which are shown in Table 13-1A. (Note that it is *not* the same as the truth table of the *logical* sum discussed in Chapter 11.) It is seen from Table 13-1A that there is a carry of one in the sum $(1 + 1)$, and otherwise, the carry is zero. Let us rewrite the table showing two output columns: one for the carry digit and the other for the sum digit, as shown in Table 13-1B. A logic circuit with two inputs A and B (the digits to be added) and two outputs C (for carry) and S (for sum) can be designed by viewing the latter table as *two truth tables:* one for carry and the other for sum. Such a circuit is known as a *half-adder*. It is seen from the truth tables that:

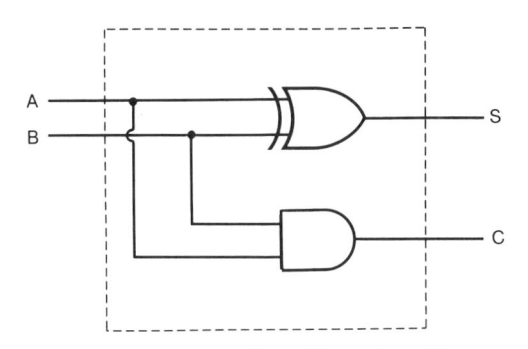

Fig. 13-1. Half-adder circuit.

$$C = 1$$

when A and B are both 1.
   Therefore,

$$C = AB$$

and an AND gate produces the carry digit.

$$S = 1$$

when $A = 0$ and $B = 1$, or $A = 1$ and $B = 0$.

This is seen to be the *exclusive OR* operation on A and B. Therefore, an exclusive OR gate will produce the sum digit. The logic circuit of the half-adder is shown in Fig. 13-1.

The half-adder cannot, of course, handle numbers with more than one digit since it has only two inputs, and there is no provision for a carry input.

Consider now the addition of two multidigit binary numbers. The addition of bits in any position will involve the sum of a carry bit brought in from the addition of the previous bits as well as the bits of the two numbers being added. For example, consider the addition of two 4-bit binary numbers: 1 1 1 0 and 0 1 1 1. The procedure is shown in Table 13-2. A top row has been added for the carry digits generated in each position. The bottom row has the sum digits. Starting with the LSB position, there is no carry into that position, which is taken as a carry of 0. In the LSB position, then we have the sum (0 + 1 + 0) which gives a sum of one and generates a carry of zero. This carry digit is entered in the second bit position of the carry digits' row in Table 13-2. The sum to be done in the second bit position is (0 + 1 + 1), which leads to a sum digit of zero and generates a carry of one, which is entered in the third bit position of the carry digits row. The sum in the third bit position is (1 + 1 + 1), which leads to a sum digit of one and generates a carry of one, which is entered in the fourth bit position of the carry digits' row. The sum in the MSB position is (1 + 0 + 1), which leads to a sum digit of 0 and generates a carry of 1. This carry bit really belongs to the sum of the two numbers since there are no more digits in them. In the digital adder, this last carry will appear as an *overflow* bit.

Table 13-2. Addition of Two 4-Bit Binary Numbers

Carry generated from the addition of the LSB position

Carry from the addition in the second bit position

Carry digits		1	1	0	0
		0	1	1	1
Numbers being added		1	1	1	0
Sum digits	1	0	1	0	1

The truth table for the addition required in *each bit position* in forming the sum of any two numbers is shown in Table 13.3.

Drill Problem 13.18: Obtain the sum of each of the following pairs of binary numbers. (Do the sum of *binary!*) (a) (1 0 1 0) + (0 1 0 0); (b) (1 0 0 1) + (1 0 1 1).

Table 13-3. Truth Table for a Full-Adder

C_{in}	A	B	C_{out}	S
0	0	0	0	0
0	0	1	0	1
0	1	0	0	1
0	1	1	1	0
1	0	0	0	1
1	0	1	1	0
1	1	0	1	0
1	1	1	1	1

Full-Adders

The circuit that realizes the truth table in Table 13-3 is called a *full-adder*. The inputs to a full-adder are the carry from the addition of the previous bit position, C_{in}, and the bits A and B of the two numbers being added. The outputs are the sum digits S, and the carry digit C_{out}. The output functions of the full-adder are obtained from Table 13-3 and are given by

Carry output: $C_{out} = \overline{A}BC_{in} + A\overline{B}C_{in} + AB\overline{C}_{in} + A B C_{in}$
Sum output: $S = A \oplus B \oplus C_{in}$

The expression for the sum was obtained by using the property of an *exclusive OR sum:* it is one whenever an odd number of inputs are one.

The circuit of a full-adder is shown in Fig. 13-2, and the connection of four such units to form a *parallel adder* for the addition of two 4-bit numbers is shown in Fig. 13-3. Several 4-bit parallel adders can be cascaded so as to obtain the sum of numbers with more than 4 bits. Fig. 13-4 shows the interconnection of two 4-bit parallel adders to add two 8-bit numbers.

In a parallel adder, the sum and carry of any output stage cannot be determined until it receives the carry input from the previous stage. That is, the carry information has to *ripple* through the different stages of the adder. It is, therefore, called a *carry-ripple adder*. The time taken for the carry to ripple through the m stages of a parallel adder sets the upper limit to its speed. The inputs due to the two numbers being added by the parallel adder must be kept constant until the propagation of the carry information is completed. When the number of bits to be added becomes large, the parallel adder tends to be too slow. *High-speed adders* use the so-called *look-ahead carry circuits* where the carry digits are calculated and *directly* fed to the individual stages. Such circuits achieve a high speed at the expense of more complex circuitry than the parallel (carry-ripple) adder.

REPRESENTATION OF NEGATIVE NUMBERS

The subtraction of two binary numbers can be performed by circuits similar to the half-adder and full-adder. But it is desirable to use a single circuit

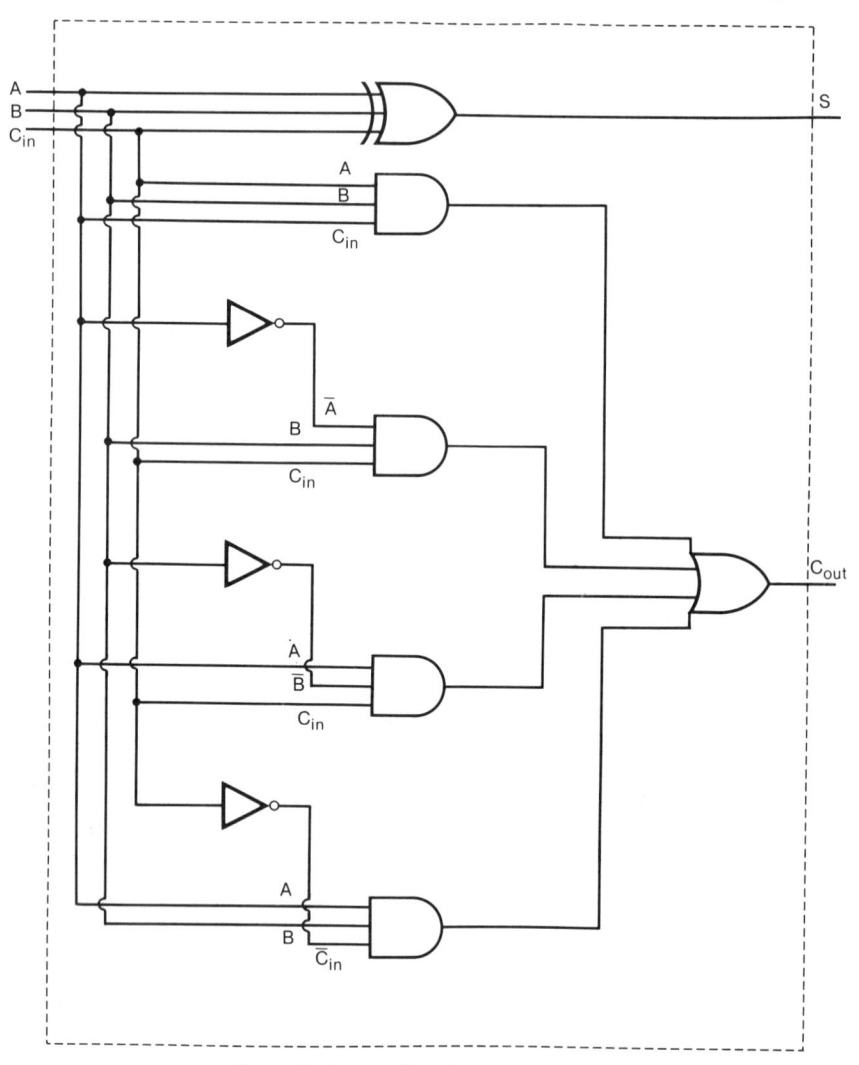

Fig. 13-2. Full-adder circuit.

for performing both the addition and subtraction operations. In order to use the adder circuit to perform subtraction as well, we can think of the subtraction of a number B from a number A as the addition of the number (−B) to the number A:

$$A - B = A + (-B)$$

It is, therefore, necessary to have some means of representing negative numbers in such a manner that [A + (−B)] can be performed by means of an

Electronics: Circuits and Systems

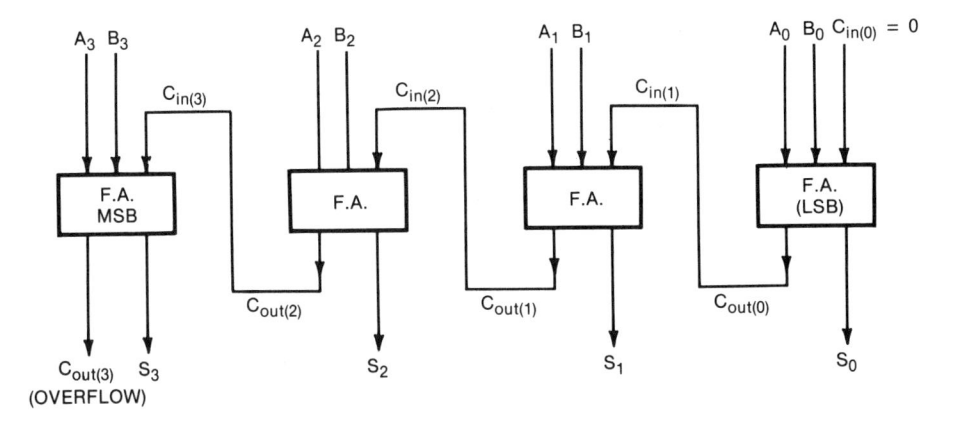

Fig. 13-3. Four-bit parallel adder.

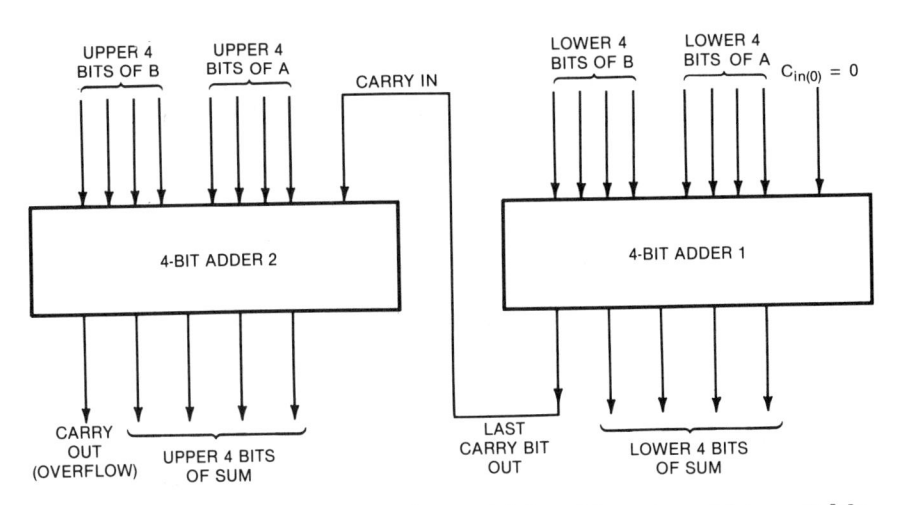

Fig. 13-4. Interconnection of two 4-bit adders to add two 8-bit numbers.

adder. There are two systems, called the *ones complement* and the *twos complement* systems, for representing negative numbers such that subtraction can be done by using adders. The twos complement system is the one used in most modern computers.

First consider the notation for positive numbers. Given a number, say 14, which is 1110 in binary, we attach an additional bit 0, called the *sign bit*, to the left of the binary equivalent. Thus, 14 is represented by 01110, 9 by 01001, and so on. This notation is called the *sign-and-magnitude notation* and consists of a *sign bit of 0 followed by the true binary representation of the given number*.

A *sign bit of 1* will indicate that the number being represented is *negative*. Even though a negative number can be (and frequently is) represented in a sign-and-magnitude notation, it is not useful in performing subtraction by means of adders.

Ones Complement System

In the *ones complement system*, the representation of a negative number is done by first taking the positive number, writing the positive number in the sign-and-magnitude convention, and then replacing each one by a zero and each zero by a one. For example, consider the number (−14). The positive number 14 has the sign-and-magnitude form: 01110

Replacing each 0 by a 1 and each 1 by a 0, we obtain: 1 0 0 0 1 as the ones complement representation of (−14).

Example 13.18

Represent each of the following negative numbers in the ones complement notation. (a) −12; (b) −15; (c) −68.

Solution

(a) +12 → 0 1 1 0 0. Therefore (−12) → 1 0 0 1 1.
(b) +15 → 0 1 1 1 1. Therefore, (−15) → 1 0 0 0 0.
(c) +68 → 0 1 0 0 0 1 0 0. Therefore, (−68) → 1 0 1 1 1 0 1 1.

> **Drill Problem 13.19:** Find the ones complement representation of the following numbers: (a) −4 (use 5 bits including the sign bit); (b) −11; (c) −57 (use 8 bits including the sign bit).

If a negative number is given in its ones complement form, then the number it represents is found by first replacing the zeros by ones and the ones by zeros, finding the positive number so obtained, and then writing the corresponding negative number.

Example 13.19

Find the negative number represented by each of the ones complement strings given: (a) 1 0 1 0 0; (b) 1 1 1 0 0 1; and (c) 1 0 0 0 0 0.

Solution

(a) Given 1 0 1 0 0, replace 0 by 1 and 1 by 0 to get 0 1 0 1 1, which represents + 11. The given number = −11.
(b) Given 1 1 1 0 0 1, replace 0 by 1 and 1 by 0 to get 0 0 0 1 1 0, which denotes + 6. The given number = −6.

Electronics: Circuits and Systems

(c) Given 1 0 0 0 0 0. Replace 0 by 1 and 1 by 0 to get 0 1 1 1 1 1, which denotes + 31. The given number = −31.

Drill Problem 13.20: Find the numbers represented by the following ones complement numbers: (a) 1 0 1 1 1; (b) 1 0 1 0 1 1 0; (c) 1 1 0 0 0 1.

Twos Complement System

The twos *complement representation* of a negative number is obtained by first writing the ones complement notation of that number and then *adding one* to it. For example, consider −14. The ones complement notation of −14 is 1 0 0 0 1. Adding one to this will give 1 0 0 1 0. Therefore, 1 0 0 1 0 represents −14 in the twos complement system.

Example 13.20

Find the twos complement representations of the following numbers: (a) −12; (b) −15; and (c) −68.

Solution

These are the same numbers as used in Example 13.18, where their ones complement representations were found.

(a) −12 has the ones complement: 1 0 0 1 1
                        Add 1:   1
                        Sum: 1 0 1 0 0

Therefore, the twos complement notation for (−12) is 1 0 1 0 0.

(b) −15 in the ones complement system is 1 0 0 0 0. Adding 1, we get 1 0 0 0 1 as the twos complement representation of (−15).

(c) −68 is in the ones complement: 1 0 1 1 1 0 1 1
                        Add 1:   1
                        Sum: 1 0 1 1 1 1 0 0

Therefore, (1 0 1 1 1 1 0 0) is the twos complement representation of (−68).

Drill Problem 13.21: Find the twos complement representations of the negative numbers in Drill Problem 13.19.

Given a twos complement notation of a number, the determination of that number consists of replacing the ones by zeros and the zeros by ones, finding the positive number obtained, *adding one* to the positive number, and then writing the resulting number as a negative number.

Example 13.21

Find the numbers represented by the following twos complement numbers: (a) 1 0 0 0 1; (b) 1 0 1 0 0; (c) 1 0 1 1 0 1 0 1 1 1.

Solution

(a) The given number is 1 0 0 0 1. Replace 0 by 1 and 1 by 0 to get 0 1 1 1 0, which denotes +14. Add 1 to get +15. Therefore, the given number = −15.
(b) The given number is 1 0 1 0 0. Replace 0 by 1 and 1 by 0 to get 0 1 0 1 1 which is +11. Add 1 to get +12. Therefore, the given number = −12.
(c) The given number is 1 0 1 1 0 1 0 1 1 1. Replace 0 by 1 and 1 by 0 to get 0 1 0 0 1 0 1 0 0 0, which is +296. Add 1 to get 297.
Therefore, the given number = −297.

Drill Problem 13.22: Find the number represented by each of the following twos complement representations: (a) 1 1 1 0 1; (b) 1 1 0 0 0 0 0; and (c) 1 0 0 0 1 0 0 0 1 0.

The *positive* number is always represented by a sign bit of 0 followed by the true binary representation of its magnitude regardless of the system being used. That is, the number +14, for example, is 0 1 1 1 0 in the sign and magnitude convention, the ones complement convention, and the twos complement convention.

ADDITION AND SUBTRACTION IN THE TWOS COMPLEMENT SYSTEM

As mentioned earlier, subtraction is performed by using an adder circuit in a digital system. That is, if A and B are two numbers and it is required to find (A − B), then A and (−B) are *both* put in the twos complement form and added. We will see whether this scheme works by considering different cases.

In order to perform addition or subtraction of numbers in the twos complement system, it is necessary that *all numbers have the same number of bits*. In our discussion we will use a total of 5 bits of which the first bit is the sign bit. Any number is then written in the form of a 5-bit number including the sign bit. For example, +1 will be 0 0 0 0 1. The largest number we can handle with a sign bit plus 4 bits is 15. The discussion can, however, be readily extended to numbers needing more bits.

Case 1: Addition of Two Positive Numbers

A = +9 and B = +5. (A + B) is to be determined by means of an adder.

```
+ 9 in twos complement form:  0 1 0 0 1
+ 5 in twos complement form:  0 0 1 0 1
                    Add:      0 1 1 1 0   which is +14 and
                                          the answer is correct.
```

Case 2: Subtraction of a Smaller Number From a Larger Number

A = +9 and B = +5. (A − B) is to be determined by means of an adder. We express +9 and (−5) in twos complement form and add.

$$+9 \text{ in twos complement form:} \quad 0\ 1\ 0\ 0\ 1$$
$$-5 \text{ in twos complement form:} \quad \underline{1\ 1\ 0\ 1\ 1}$$
$$\text{Add:} \quad 1\ 0\ 0\ 1\ 0\ 0$$

There is an additional bit generated, called an *overflow*. If the overflow is disregarded, then we have 0 0 1 0 0, which is +4 and the answer is correct.

Case 3: Subtraction of a Larger Number From a Smaller Number

A = +5 and B = +9. (A − B) is to be determined by means of an adder. Express +5 and (−9) in twos complement form and add.

$$+5 \text{ in twos complement form:} \quad 0\ 0\ 1\ 0\ 1$$
$$-9 \text{ in twos complement form:} \quad \underline{1\ 0\ 1\ 1\ 1}$$
$$\text{Add:} \quad 1\ 1\ 1\ 0\ 0$$

The sign bit 1 indicates that the number is negative. We see that 1 1 1 0 0 represents the number (−4) by using the conversion procedure discussed in Example 13.21.

Again the answer is correct.

Case 4: Addition of Two Negative Numbers

A = −9 and B = −5. (A + B) is to be determined by means of an adder.

$$-9 \text{ in twos complement form:} \quad 1\ 0\ 1\ 1\ 1$$
$$-5 \text{ in twos complement form:} \quad \underline{1\ 1\ 0\ 1\ 1}$$
$$\text{Add:} \quad 1\ 1\ 0\ 0\ 1\ 0$$

There is an overflow. If we ignore the overflow bit, we have 1 0 0 1 0, which is a negative number and is found to be equal to −14, which is the correct answer.

It appears from these examples, then, that if we express both numbers in twos complement form, the (A + B) and [A + (−B)] can be determined by an adder circuit. The discarding of the overflow can be automatic since the register in which the answer is stored is limited to 5 bits in situations like the ones discussed previously; the overflow simply gets lost.

Drill Problem 13.23: Perform the following operations using addition in the twos complement system: (a) $(8 + 6)$; (b) $(8 - 6)$; (c) $(-8 + 6)$; (d) $(-8 - 6)$; and (e) $(8 - 8)$.

This procedure does give rise to some difficulty when the two numbers being added are such that their magnitudes add up to 16 or more, as for example $(10 + 9)$ and $(-11 - 13)$. A logic circuit that provides corrective action in such cases can be designed, and this circuit is connected to the adder circuit.

BCD CODE AND BCD ADDERS

The representation of numbers in pure binary form is not always most convenient because of the human mind's prejudice in favor of the decimal number system. An alternative scheme is the *binary coded decimal*, or *BCD*, system. The decimal digits 0 through 9 are coded in binary form using 4 bits: 0000 through 1001. In the case of numbers greater than nine, *each digit is separately coded into a four-digit binary*. Some examples are

```
      45:   0100  0101
             4     5

    1984:   0001  1001  1000  0100
             1     9     8     4
```

The BCD code offers the advantage that we can separate the given number into blocks of 4 bits and convert each block to a decimal digit. It is a popular code because of this advantage. BCD adders are available so that numbers in BCD format can be added directly without first having to convert them to straight binary form. When a multidigit BCD number is to be added to another, it is necessary to use a *separate adder* for each digit in the BCD numbers being added. The arrangement for adding two 4-digit decimal numbers in BCD code is shown in Fig. 13-5.

ARITHMETIC LOGIC UNIT

The *ALU* (or *arithmetic logic unit*) is a logic package that can perform a wide variety of arithmetical as well as logical operations on two-input words. It is a vital component in any computer system, large or small. The 74181 ALU is a high-speed unit containing 75 gates on an IC chip and can perform 16 arithmetical and 16 logical operations on two 4-bit input words. The block diagram representation of the 74181 ALU is shown in Fig. 13-6. It has 14

Electronics: Circuits and Systems

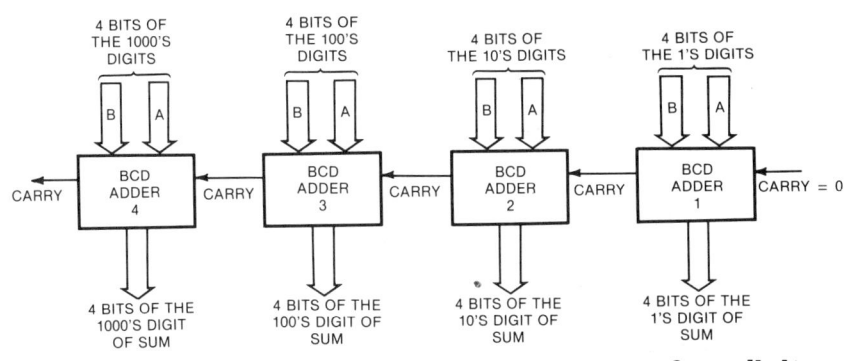

Fig. 13-5. Network of BCD adders to add two four-digit decimal numbers.

input terminals: 4 for the input word A, 4 for the input word B, 4 for a function selection, one for carry input, and one for mode control. It has 8 output terminals: 4 for function outputs, one for comparator output, 2 for carry outputs.

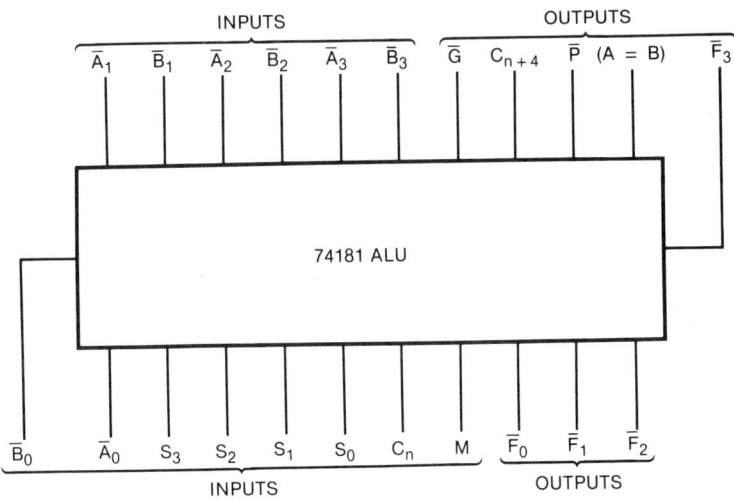

Fig. 13-6. Arithmetic logic unit (ALU) in block diagram form.

When the mode control M is low, the ALU performs 16 different arithmetical operations and when the mode control M is high, it performs 16 different logical functions. We will discuss some of these functions. (For a complete table, refer to manufacturers' data sheets such as the National Semiconductor Corp. *Digital Integrated Circuits* catalog.)

The two input words will be treated as two 4-bit numbers in arithmetical operations. Addition, subtraction, formation of twos complements, incrementing and decrementing of A, and transfer of a number to the output terminals are some of the 16 arithmetical operations that can be performed by the ALU. (The increment and decrement operations are useful in increasing and decreasing a number stored in a register so as to keep track of the program steps in a computer.) In the case of logical operations, when the mode control is high, the inputs are treated as pairs of logic variables (A_0 and B_0, A_1 and B_1, etc.) and functions such as AND, OR, exclusive OR, and exclusive NOR can be performed on the input pairs. The ALU can also be used as a comparator: the A = B output is high when the two input words are identical. The carry input and output are operative only during arithmetical operations.

The combination of an ALU with a control circuit that sends the signals to its function select lines in the correct order forms the *central processing unit* (or *CPU*) of a small computer. In a microprocessor, the CPU is contained in a single IC chip.

COUNTERS

Counters are digital circuits that count the pulses received at their inputs. The pulses may occur at a regular rate or at a random rate. Counters perform a variety of useful functions. The operation of a digital clock, for example, is based upon the counting of the pulses produced by a generator and the dividing of the count in such a way as to indicate seconds, minutes, hours, and days. Counters are also used as frequency dividers and to control various timing and sequencing operations in digital systems. Counters can be designed so as to have any desirable number of counting steps. A *mod-16* counter, for example, has 16 counting steps: 0, 1, 2, . . ., 15. A *mod-10* counter counts from 0 through 9. At the end of the counting cycle, the sequence is repeated. In order to keep track of the count, memory is necessary in a counter, and this memory is provided by means of *flip-flops*. The number of flip-flops is determined by the number of steps in the counting cycle. A mod-16 counter needs 4 bits to count from 0000 to 1111 and, consequently, needs four flip-flops. A mod-10 counter also needs four flip-flops. When the input pulses are directly counted to all the flip-flops so as to activate them simultaneously, the counter is called a *synchronous counter*. In some counters, however, the pulses activate only one flip-flop directly. Each of the other flip-flops is activated by the preceding flip-flop in such counters that are, therefore, called *asynchronous counters*. Synchronous counters are faster in their response than asynchronous counters, but their circuitry is also more complex. We will first discuss the flip-flops themselves before studying the circuitry and applications of counters.

SET-RESET FLIP-FLOPS (SR FLIP-FLOPS)

A *flip-flop* is an electronic circuit that has two stable output states: one state in which the output is high and the other in which the output is low. It is more formally called a *bistable multivibrator*. The name flip-flop comes from the fact that it can jump from one output state to the other (on applying the proper control signals). A flip-flop has one or more input terminals that serve as the *controls* and has two output terminals that are mutual complements (Fig. 13.7). Q and $\overline{Q}$ are always in opposite states. A variety of flip-flops is available, distinguished by the manner in which the inputs control the output states.

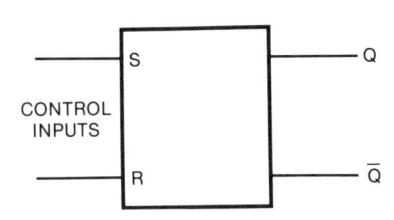

Fig. 13-7. Flip-flop.

The basic flip-flop is the *set-reset flip-flop*, or SR flip-flop, also called *set clear*, or SC, flip-flop. The term *set* means setting the output Q to a value of one (or high), and the term *reset* or *clear* means resetting the output Q to a value of 0 (or low). (In all cases, $\overline{Q}$ assumes the value opposite to Q, and we will concentrate on the output Q in our discussion.)

The SR flip-flop has two control inputs S and R, and its operation is as given in the following table:

S	R	Output Q
0	0	Q stays at the value it already had.
0	1	Q is *reset* to 0.
1	0	Q is *set* to 1.
1	1	This input condition is not permitted.

It should be noted that the output state of the flip-flop can be made to stay the same (by making S = R = 0) or change (by making either S or R, but not both, equal 1). The fact that a flip-flop can either maintain its present output state or change it is the basis for its capacity to remember or act as a memory device.

An SR flip-flop can be obtained by interconnecting NAND gates as shown in Fig. 13-8.

Consider the case S = R = 0. Then, the inputs to gate 1 are $\overline{S}$ = 1 and whatever the present value of $\overline{Q}$ may be. The output of gate 1 is, therefore, the complement of the present value of $\overline{Q}$, that is, Q. Similarly, the output of gate 2 is the complement of the present value of Q, that is, $\overline{Q}$. For example, if

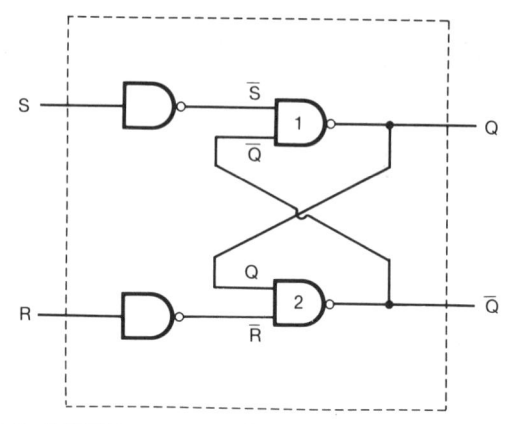

Fig. 13-8. NAND gate realization of the SR flip-flop.

Q is at a value of 1 and S = R = 0, the Q stays at 1, and $\overline{Q}$ stays at 0. *No change occurs in the output state of the flip-flop when S = R = 0.*

When S = 0 and R = 1, the input $\overline{R}$ to gate 2 is forced to 0 and this forces the output of that gate to the value 1. That is, $\overline{Q}$ becomes 1. The inputs to gate 2 are both forced to 1 (since $\overline{S} = 1$ from the input condition and $\overline{Q} = 1$ from the action of gate 2), and the output of gate 1 is, therefore, forced to a value of 0; that is Q = 0. *Therefore, when S = 0 and R = 1, the output Q is reset to 0.*

When S = 1 and R = 0, an argument similar to the previous one will show that Q is forced to a value of 1, which in turn forces $\overline{Q}$ to a value of 0. Therefore, *when S = 1 and R = 0, the output gets set to 1.*

Now we come to the case of the "forbidden" input condition: S = R = 1. Suppose the output Q is at 0 (and $\overline{Q} = 1$) and we make S = R = 1. Then the input $\overline{S}$ of gate 1 is made 0, which causes the output of gate 1 to become 1. That is, Q becomes 1. But look at the other gate: one of its inputs $\overline{R}$ is also made 0 by the input condition, and therefore its output is also made 1. That is, $\overline{Q} = 1$ also. So, we have a situation where both Q and $\overline{Q}$ have the same value, which is an ambiguous logical condition to work with. Consequently, the input condition S = R = 1 is not allowed in the SR flip-flop. The design procedure used in connection with SR flip-flops automatically prevents the occurrence of S = R = 1.

Example 13.22

The S and R inputs of a flip-flop are as shown in Fig. 13-9. Sketch the output waveform Q. Assume Q = 0 to start with.

Electronics: Circuits and Systems

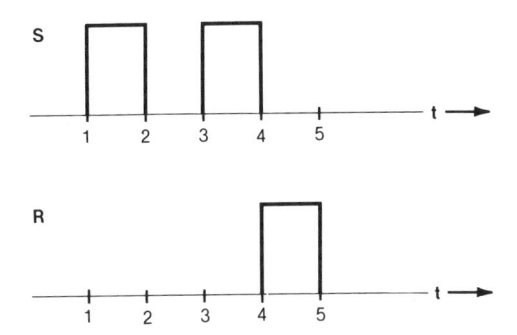

Fig. 13-9. Control inputs for Example 13.22.

Solution

From t = 1 to t = 2, S = 1 and R = 0, which is a *set* signal and Q switches to 1. From t = 2 to t = 3, S and R are both 0. This is a "stay put" signal and Q stays at 1. From t = 3 to t = 4, S = 1 and R = 0. This is a *set* signal, and Q (which is already at 1) is set to 1. That is, Q continues at 1. From t = 4 to t = 5, S = 0 and R = 1. This is a reset signal, and Q is reset to 0. The resulting output waveform is shown in Fig. 13-10. (The output $\overline{Q}$ is the complement of Q and can be readily obtained from the Q waveform.)

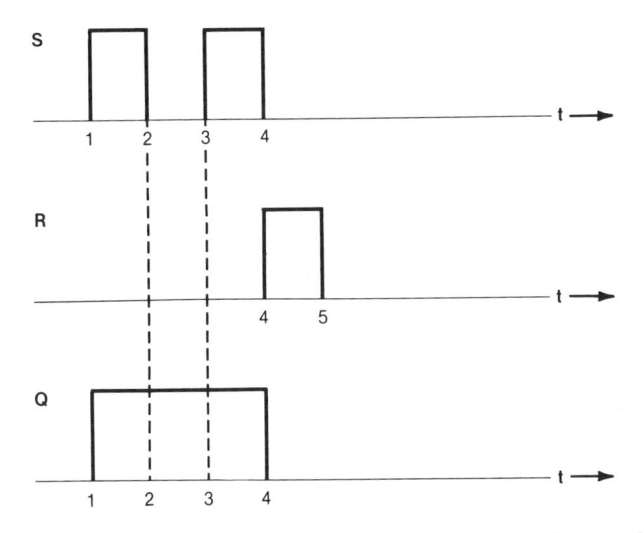

Fig. 13-10. Control inputs and output waveforms in Example 13.22.

Drill Problem 13.24: Sketch the output waveform of an SR flip-flop when the control waveforms are as shown in Fig. 13-11.

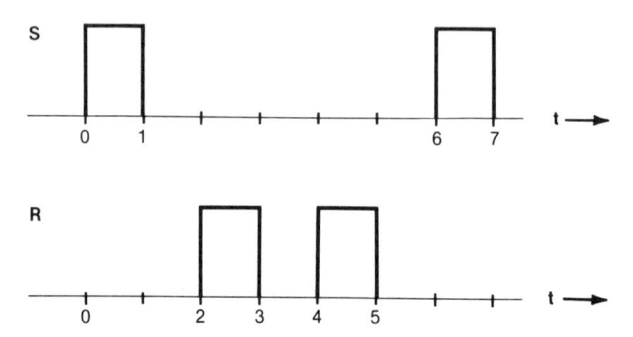

Fig. 13-11. Control inputs for Drill Problem 13.24.

Clocked SR Flip-Flops

It is desirable to make the operation of a flip-flop coincide with the occurrence of a control pulse called the *clock pulse*. That is, in addition to the set and reset controls, a third control, called a *clock*, is used. The reason for the desirability of the clock pulse controlling the operation of a flip-flop will become clear when we discuss counter circuits. Even though the term *"clock pulse"* implies a periodic pulse train, such a regular occurrence of the clock pulses is not a necessary condition; they may occur at random.

The modification of the circuit of the SR flip-flop using NAND gates (Fig. 13-8) in order to have a clocked SR flip-flop is shown in Fig. 13-12. It can be seen that when CLK = 0, the inputs $\bar{S}$ and $\bar{R}$ to the NAND gates 1 and 2 are both 1, and the outputs Q and $\bar{Q}$ maintain the values that they already had, regardless of the values of S and R. On the other hand, when CLK = 1, then the gates 1 and 2 receive the appropriate values of $\bar{S}$ and $\bar{R}$ (as determined by the input control signals), and the flip-flop acts in the normal fashion. (S = R = 1 is still not permitted, however.) The table of operation of the SR flip-flop given earlier is still valid except that when CLK = 0, no action takes place.

Example 13.23

The control and CLK signals of an SR flip-flop are shown in Fig. 13-13. Sketch the output waveform.

Electronics: Circuits and Systems

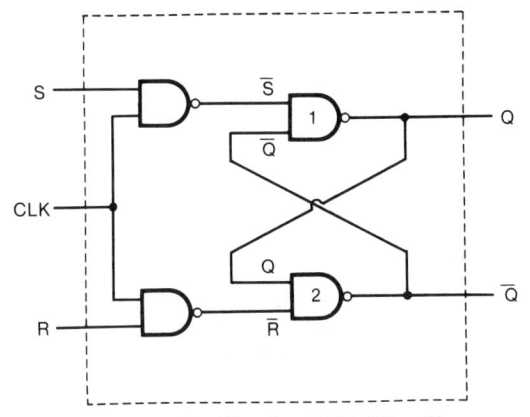

Fig. 13-12. Clocked SR flip-flop.

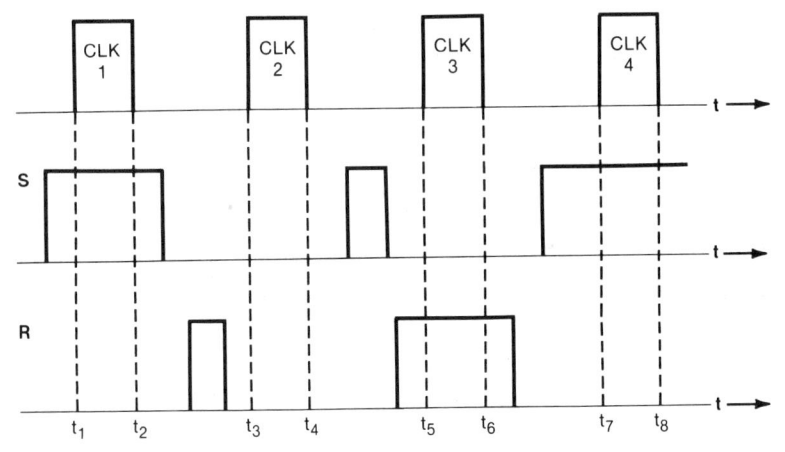

Fig. 13-13. Clock and control inputs for Example 13.23.

Solution

The values of S and R in *between* the CLK pulses should be ignored. The flip-flop cannot respond to them.

When CLK 1 occurs, S = 1, R = 0, and Q becomes 1. Q stays at 1 until CLK 2 occurs. At CLK 2, S = R = 0, and no change is effected. Therefore, Q stays at 1. At CLK 3, S = 0, R = 1, and Q becomes 0. At CLK 4, S = 1, R = 0, and Q switches to 1. The resulting output waveform is shown in Fig. 13-14.

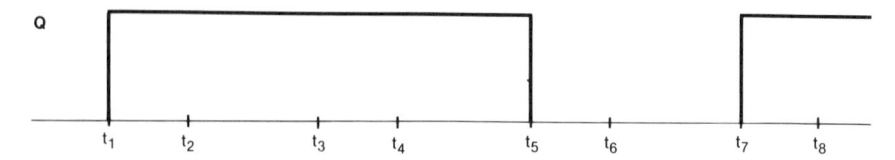

760   Arithmetic Logic, Counters, and Shift Registers

Fig. 13-14. Output waveform of Example 13.23.

Drill Problem 13.25: Sketch the output of a clocked SR flip-flop when the control and CLK signals are as shown in Fig. 13-15.

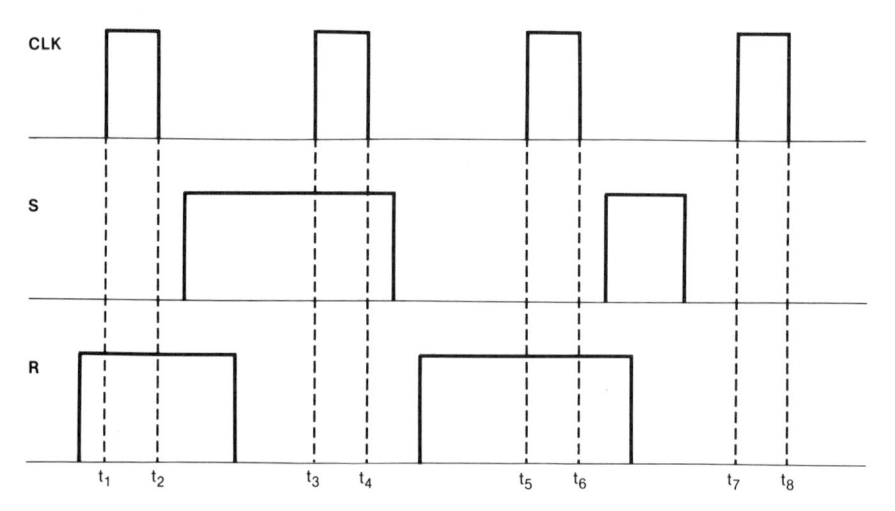

Fig. 13-15. Clock and control inputs for Drill Problem 13.25.

The SR flip-flop (whether clocked or not) has two disadvantages. The first is the constraint that the inputs cannot both be equal to one simultaneously, which reduces the flexibility available to the designer. The second is that if the values of S and R were to *change in the middle of a clock pulse*, then the outputs would also change. Therefore, the values of S and R must maintain their values from slightly before the beginning of a clock pulse, through the duration of the clock pulse and slightly after the end of the clock pulse in order to ensure the stable operation of the flip-flop. These two disadvantages are overcome by using modified versions of the flip-flop: the *edge-triggered* and *master-slave* JK and D flip-flops. (There is also a master-slave SR flip-flop, but in the *JK flip-flop* the constraint on the inputs not being allowed to be one simultaneously is lifted.)

Edge-Triggered JK Flip-Flops

A JK flip-flop is a modification of the SR flip-flop, in which *J acts like the set input*, and *K acts like the reset input* with one exception: when *both J and K are one*, the *output always switches to the opposite value*. The table of operation of a JK flip-flop is, therefore, as shown in the following:

J	K	Output Q
0	0	Q does not change.
0	1	Q is reset to 0.
1	0	Q is set to 1.
1	1	Q flips to the opposite value.

A *clocked JK flip-flop* acts in the same manner as a clocked SR flip-flop (except that J = K = 1 is permitted): when CLK is low, no action takes place, and when CLK is high, the flip-flop responds to the JK input according to the previous table.

In order to avoid the problem of the JK inputs changing while the CLK is high, JK flip-flops are available in the *edge-triggered* version. The *edge* here refers to the leading edge (when the pulse goes from zero to one) or the trailing edge (when the pulse goes from one to zero) of the clock pulse as indicated in Fig. 13-16. A *positive edge-triggered flip-flop* responds to the leading edge (zero-to-one transition) of the clock pulse, that is, the values of J and K at the instant when the clock pulse goes from zero to determine the response of the flip-flop. The values of J and K at any other points of the clock pulse have no effect on the operation. Fig. 13-17 shows the response of a positive edge-triggered flip-flop to a set of given JK signals and clock pulses. Note that the values of J and K are ignored except at the zero-to-one transition instants of the clock pulses.

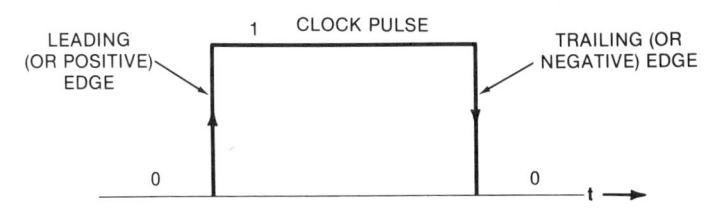

Fig. 13-16. Leading and trailing edges of a clock pulse.

The symbol of a positive edge-triggered flip-flop is shown in Fig. 13-18, where the triangle (or wedge) indicates that it is a positive edge-triggered flip-flop.

Drill Problem 13.26: The control signals and clock pulses of a positive edge-triggered flip-flop are shown in Fig. 13-19. Sketch the output waveform.

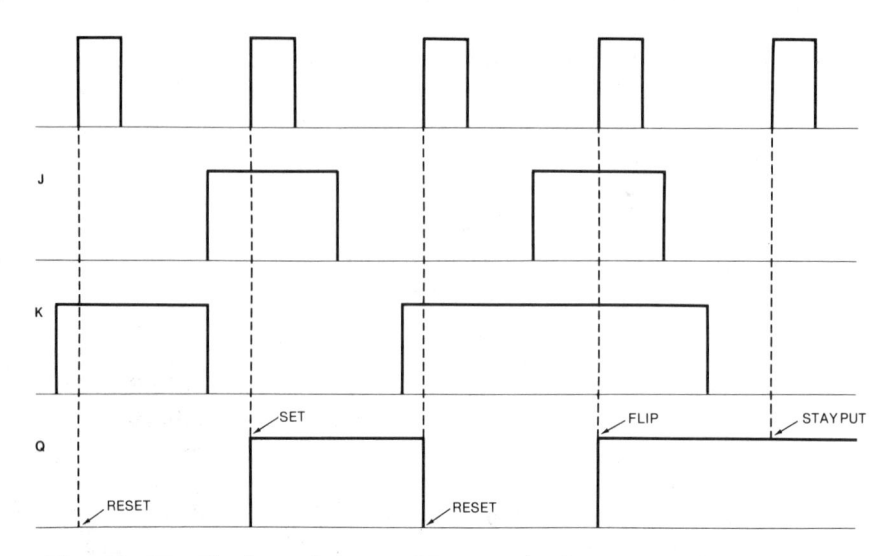

Fig. 13-17. Clock and control inputs and output of a positive edge-triggered flip-flop.

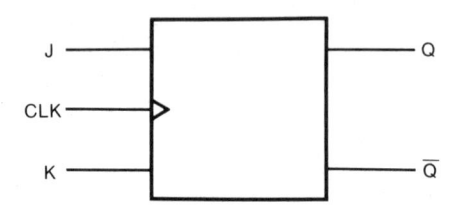

Fig. 13-18. Symbol of a positive edge-triggered flip-flop.

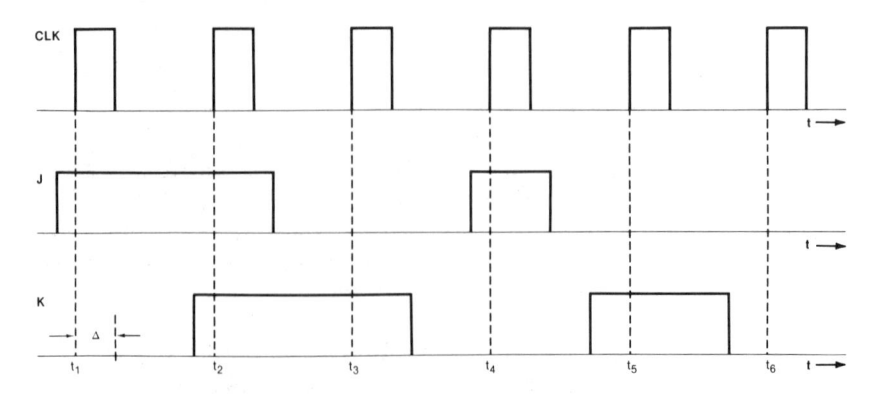

Fig. 13-19. Control and clock inputs for Drill Problem 13.26.

Electronics: Circuits and Systems 763

Negative edge-triggered flip-flops are also available, in which the flip-flop responds to the JK signals present at the *trailing* edge *(one-to-zero transition)* of the clock pulse. The response of such a flip-flop to the control signals and clock pulses of Fig. 13-17 will be as shown in Fig. 13-20. A comparison of this response with that of a positive edge-triggered flip-flop will show how the two flip-flops differ in their response characteristics. The symbol of a negative edge-triggered flip-flop is shown in Fig. 13-21, where the presence of the little circle at the CLK input terminal implies negative edge triggering.

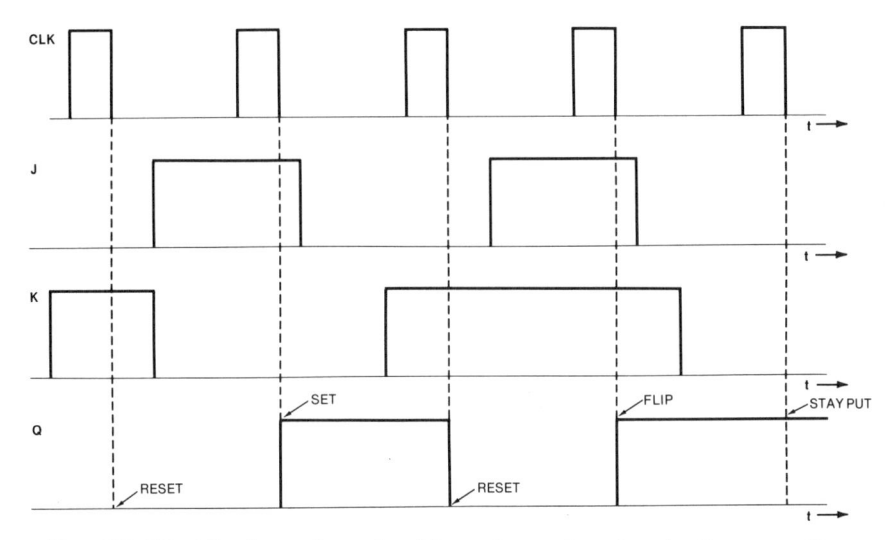

Fig. 13-20. Clock and control inputs and output of a negative edge-triggered flip-flop.

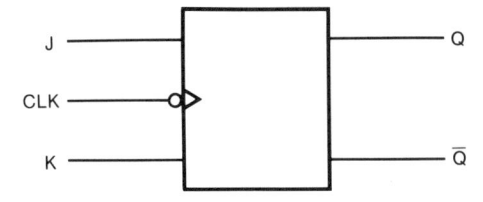

Fig. 13-21. Symbol of a negative edge-triggered flip-flop.

Drill Problem 13.27: If the control signals and the clock pulses of Fig. 13-19 are applied to a negative edge-triggered flip-flop, sketch the output waveform.

Master-Slave Flip-Flops

A *master-slave flip-flop (M/S flip-flop)* derives its name from the fact that it is actually made up of two cascaded flip-flops, one of which acts as a "master" and the other as a "slave."

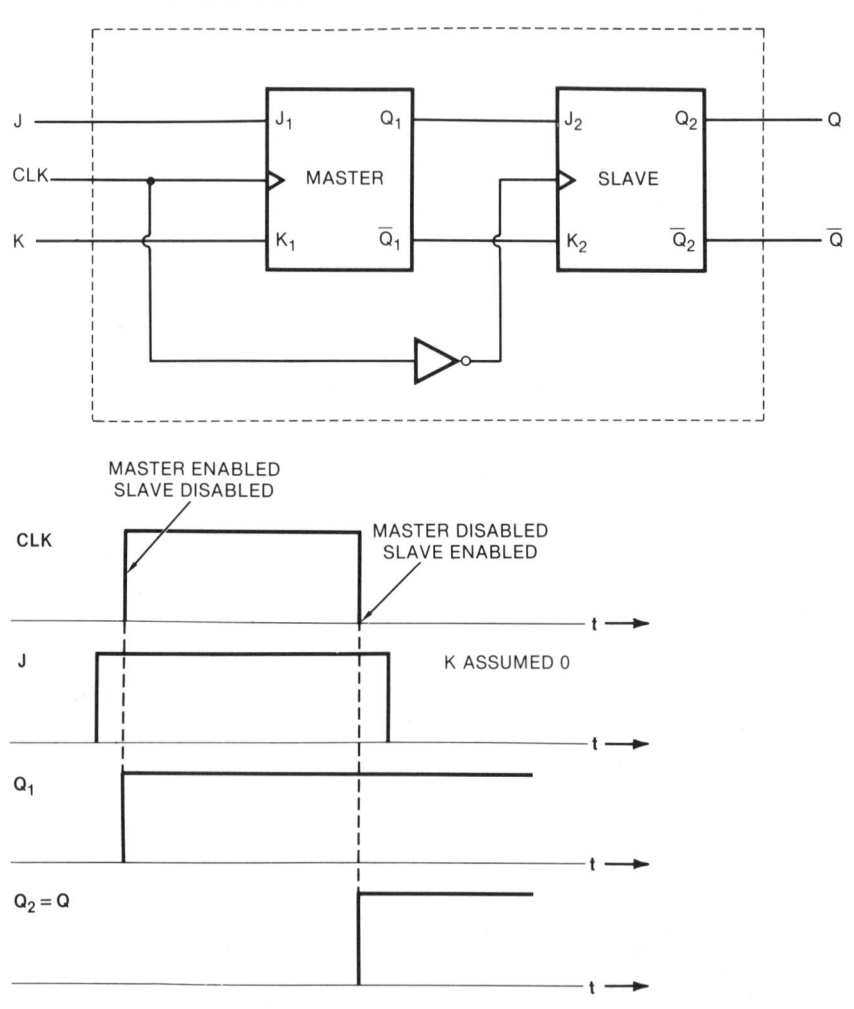

Fig. 13-22. Master-slave (M/S) flip-flop and associated waveforms.

Consider the arrangement shown in Fig. 13-22, where the CLK input is directly connected to the master but is inverted and connected to the slave. When CLK = 1, the master is enabled, and the slave is disabled. The JK sig-

Electronics: Circuits and Systems **765**

nals present when CLK = 1 determine the output of the master. When CLK becomes 0, the master is disabled but the slave is enabled. The output of the master is transferred to the output of the slave. Thus, the response of the M/S unit will be *a change in the output at the negative-going transition of the clock pulse but determined by the values of JK while the clock pulse is high.* Note that from the user's point of view, the internal operation (of transfer from master to slave) is of no concern; only the overall response is important. Therefore, an M/S flip-flop is shown only as a single block (as if it were a single flip-flop), which is the same as that used for a negative edge-triggered flip-flop. The overall response of an M/S flip-flop to the control signals and the clock pulses of Fig. 13-17 are shown in Fig. 13-23.

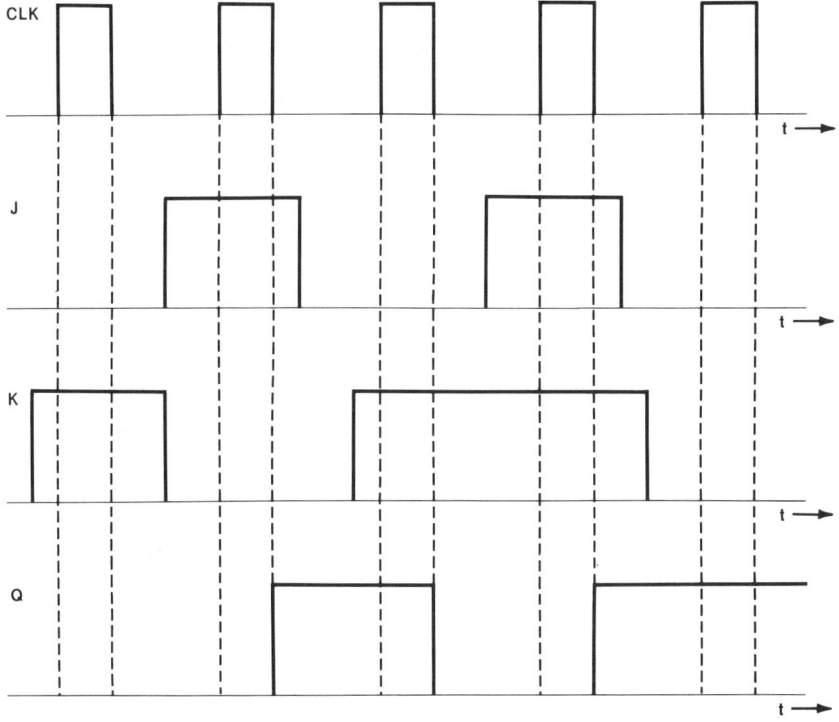

Fig. 13-23. Clock and control inputs and the output waveforms of an M/S flip-flop.

Drill Problem 13.28: Sketch the output waveform of an M/S flip-flop if the control signals and clock pulses of Fig. 13-19 are applied to it.

The only difference between the overall response of an M/S flip-flop and a negative edge-triggered flip-flop is that in the M/S flip-flop, the output is determined by the JK values present *before the one-to-zero transition* of the clock pulse occurs, but in the negative edge-triggered flip-flop, the output is determined by the JK values present *at the instant of the one-to-zero transition* itself. *If the JK values do not change when CLK = 1 and they are allowed to stay at their values a little past the one-to-zero transition of the clock pulse, then we cannot tell the difference between an M/S flip-flop and a negative edge-triggered flip-flop.*

We will generally assume that edge-triggered flip-flops (JK or its variations) are used in the circuits in our discussion. Commercially available flip-flops have additional input terminals marked *preset* and *clear*. These permit us to make the output of the flip-flop one (preset) or zero (clear) by making the appropriate input active. These two inputs override any JK signals and act independently of the clock pulses. The diagrams of Fig. 13-24 show the two edge-triggered flip-flops with preset and clear inputs.

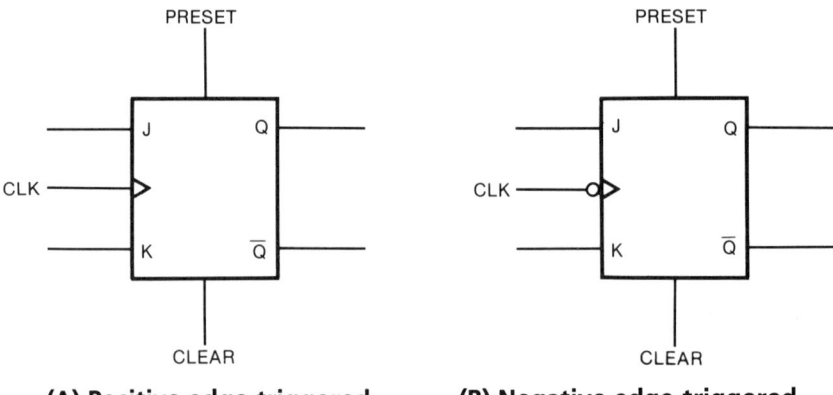

(A) Positive edge-triggered flip-flop.

(B) Negative edge-triggered flip-flop.

Fig. 13-24. Edge-triggered flip-flops with preset and clear inputs.

Special Cases of JK Flip-Flops

Two cases are of interest in which J and K are made to depend upon each other.

Case 1: J = K. When the two control inputs are tied together to the same input, the response of the flip-flop is to stay put when the input is zero and to flip to the other state when the input is one. Such a connection is used frequently in counter circuits.

Electronics: Circuits and Systems

Drill Problem 13.29: Assuming J = K, sketch the output of a negative edge-triggered flip-flop for the control signal and clock pulse shown in Fig. 13-25. Assume Q = 0 to start with.

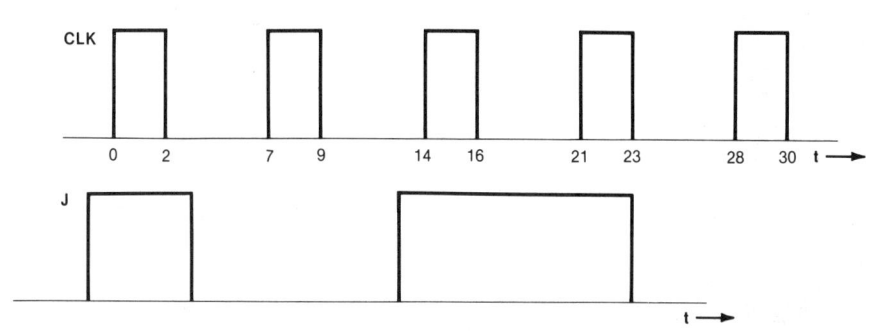

Fig. 13-25. Clock and control inputs for Drill Problem 13.29 and 13.30.

Case 2: J and K are mutual complements. The connection shown in Fig. 13-26 makes J and K complements of each other. Such a flip-flop is known as a *D flip-flop*. (Recall that these were used in the last chapter for registers.) The D flip-flop is shown symbolically in Fig. 13-26B.

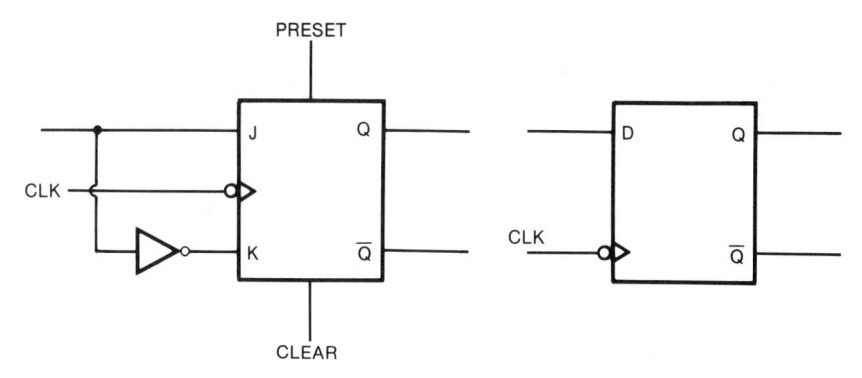

(A) JK flip-flop connected as a D flip-flop.

(B) Symbol of a D flip-flop.

Fig. 13-26. D flip-flop.

When CLK is 1, the *output of the flip-flop will become 1 if D = 1 and will become 0 if D = 0*.

The D-type *latch* is slightly different from the D flip-flop (which is assumed to be edge triggered or M/S). The output of a D latch changes its output with the value of D when the clock pulse is present. If the value of D should change when the clock pulse is present, the D *latch* output will change correspondingly, but the output of a D *flip-flop* will respond only to the value of D at the edge of the clock pulse.

> **Drill Problem 13.30:** Suppose the J control signal and clock pulse of Fig. 13-25 are applied to a D flip-flop. Sketch the output.

Summary of the Response Characteristics of Flip-Flops

For the sake of convenience of reference, the important response characteristics of flip-flops are summarized in this section. We will generally assume that *negative edge-triggered flip-flops* are used. (The discussion that follows will be equally valid for positive edge-triggered flip-flops and M/S flip-flops also.) Note that no change in the output can occur when CLK = 0. Also any change in the output will be noticed when the clock pulse has returned from one to zero.

JK Flip-Flop: The output response is according to the following table.

J	K	Output Q
0	0	No change.
0	1	Reset to 0.
1	0	Set to 1.
1	1	Flip to opposite value.

D Flip-Flop: The output assumes the *same value as* the control *input D*.

J = K = 1: Even though this is not really a special type of flip-flop, it is a frequently used connection in counters. The output *flips at every clock pulse.*

In all cases, look at the values of J and K at the trailing edge of the clock pulse to determine the next output.

APPLICATION OF FLIP-FLOPS

Frequency Dividers

Given a clock pulse train of frequency f (that is, f pulses per second), pulse trains at frequencies (f/2), (f/4), (f/8), . . . can be produced by using JK flip-flops. The circuit shown in Fig. 13-27 produces pulse trains at (f/2) and (f/4).

Since J = K = 1 for both flip-flops, the output of each will switch to the opposite value whenever its CLK input goes from one to zero. The input pulse train feeds the CLK input of the first flip-flop. Its output will change whenever the input clock pulses have a trailing edge. Therefore, the output of the

Electronics: Circuits and Systems   **769**

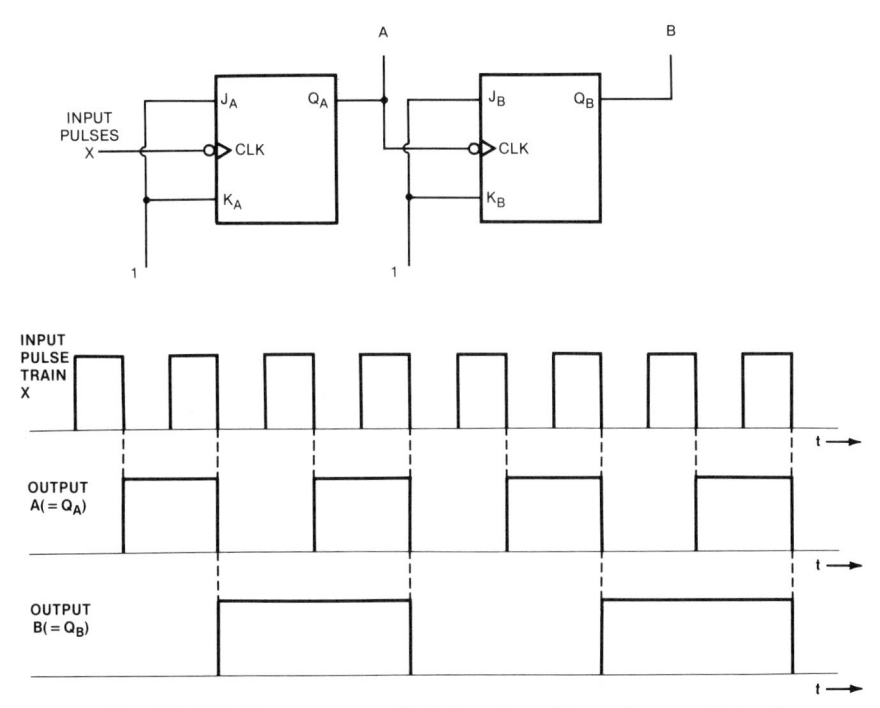

Fig. 13-27. Frequency divider circuit and associated waveforms.

first flip-flop will be a pulse train at one half the frequency of the given pulse train. The output of the first flip-flop feeds the CLK input of the second flip-flop. The output of the second flip-flop will switch to the opposite value whenever its CLK input has a trailing edge that corresponds to the *one-to-zero transitions of the output of the first flip-flop*. Therefore, the output of the second flip-flop will be a pulse train at one half the frequency of the output of the first flip-flop, which means one fourth the frequency of the input pulse train. The relevant waveforms of the circuit are shown in Fig. 13-27.

Ripple Counters

The circuit used for frequency division can also be used as a counter. Such a counter is known as an *asynchronous* or *ripple* counter. A counter that counts from zero to seven repeatedly is shown in Fig. 13-28. If the waveforms are set up as was done in the frequency divider circuit (Fig. 13-28), and the values of the output C, B, and A are tabulated at the end of each clock pulse, it will be found that the outputs CBA follow the sequence: 000, 001, 010, . . . , 111, 000.

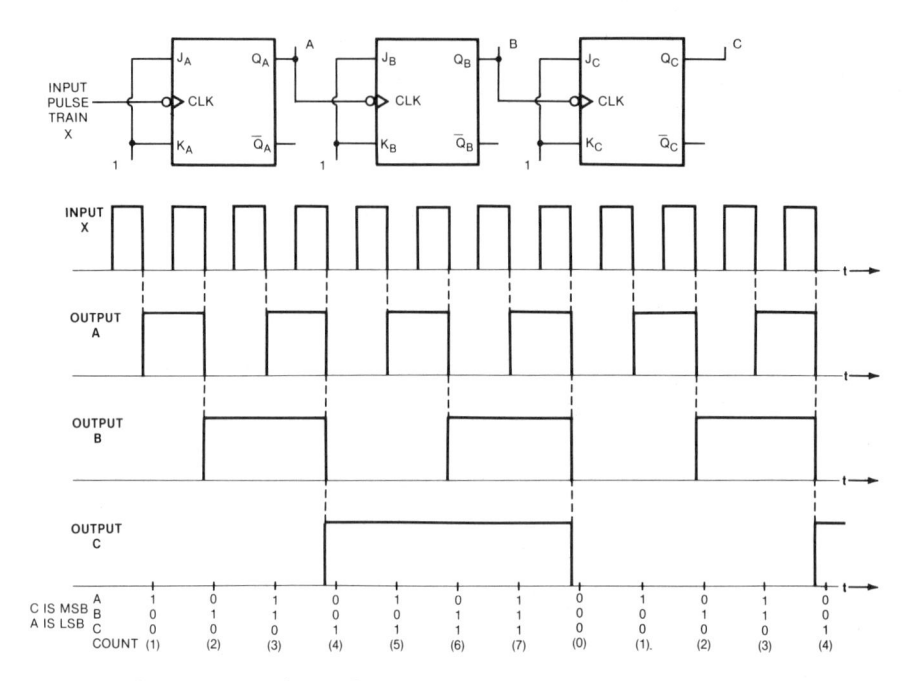

770   Arithmetic Logic, Counters, and Shift Registers

Fig. 13-28. Three-bit ripple counter and associated waveforms.

Drill Problem 13.31: In the circuit of Fig. 13-28, make the following changes: Feed the $\overline{Q}_A$ output of flip-flop A to the CLK input of flip-flop B instead of the Q_A output. Similarly, feed the $\overline{Q}_B$ output of flip-flop B to the CLK input of flip-flop C instead of the Q_B output. Draw the relevant waveforms. (Note that a one-to-zero transition of $\overline{Q}$ corresponds to a zero-to-one transition of Q.) By examining the values of the outputs CBA, determined the counting sequence.

The ripple counter is obviously an extremely simple circuit and has the further advantage in that extending it to larger counting cycles is simply the addition of more flip-flops without any additional logic circuitry. It has one disadvantage that makes it impractical for increasing the counting cycle beyond a certain point. The response of each flip-flop has to wait until the previous flip-flop has changed its output. That is, the change in the input of the first flip-flop in the counter must ripple through all the intermediate flip-flops before the last flip-flop can respond to it. Since there is a definite delay in each flip-flop, the total delay between the change in the input to the first flip-flop and the response of the last flip-flop increases as the number of flip-flops in the counter increases. It is evident that the input cannot change faster than the total delay needed for the counter, as otherwise there will be incorrect responses in the circuit.

Synchronous Counters

In order to overcome the previously mentioned disadvantage of the ripple counters, the clock pulse train in a counter can be *directly* connected to the CLK inputs of all the flip-flops in a counter. Such a counter is said to be *synchronous*. The synchronous counters can handle pulse trains of higher frequencies than the ripple counters since the delay is now due to only one flip-flop regardless of the total number of flip-flops in the circuit. But the synchronous counter requires additional logic circuitry to make it work unlike the ripple counter, which was formed by a direct interconnection of the flip-flops.

The circuit of a mod-8 synchronous counter is shown in Fig. 13-29; it counts from 000 to 111, jumps to 000, and repeats the cycle.

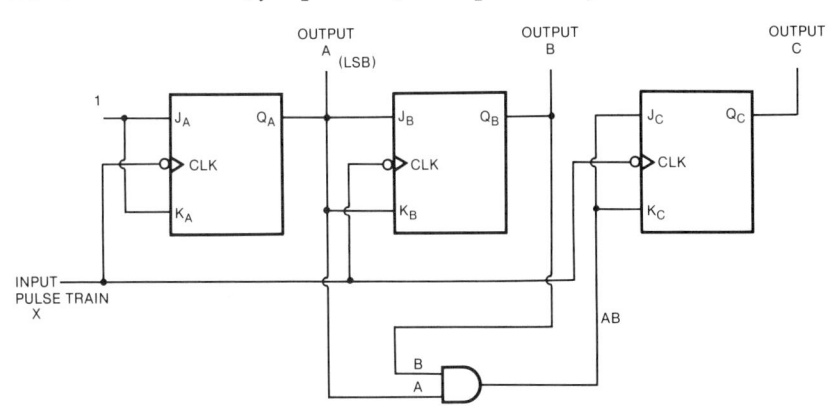

Fig. 13-29. Mod-8 synchronous counter.

Counters of various sizes and descriptions are available commercially. Among them are *up-down counters* that can count up (from 0 to some number N) or down (from N to 0), *BCD counters* that count from 0 to 9, and so on. Since the BCD counter is commonly used, let us look at the capabilities of a commercially available BCD counter.

The 74190 is an up-down decade counter that can count up (from 0 to 9) or down (from 9 to 0). It can be preset to any count from 0 to 9. The BCD counter block diagram with the input and output terminals is shown in Fig. 13-30. The enable is an active low input: when enable is low, the counter operates, and when enable is high, the counter is inhibited. When enable = low and down-up input is high, the counter counts down. When enable = low and down-up input is low, the counter counts up. The outputs Q_D, Q_C,

Q_B, Q_A are the 4-bit counts (with Q_D as the MSB and Q_A as the LSB). The data inputs A, B, C, D are provided for entering information asynchronously for presetting the count to any desired value.

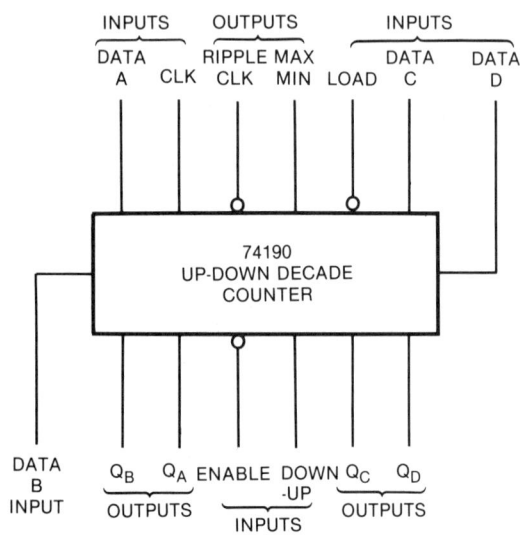

Fig. 13-30. Up-down decade counter.

Counters of this type can be cascaded to expand the count to larger values. For example, we can count from 0 to 99 by cascading two such counters, to 999 with three such counters, and so on. The ripple output in the counter is provided for facilitating such cascading. The ripple output is connected to the enable input of the next counter when parallel clocking is used (that is, the clock pulse train being counted is directly connected to all the counters). In cascaded connections, the lowest counter will produce an output for every input clock pulse. When the count of the lowest counter reaches nine, it produces a ripple output that will enable the next higher counter. Thus the second counter will count every tenth clock pulse and produce an output for every hundredth clock pulse. Thus, the BCD counter can also be used as a "divide-by-10" circuit.

Example 13.24

A municipal parking lot has space for 250 cars. It is necessary to design a digital system that will keep track of and display the number of cars in the lot at any given time. When the lot is full, a sign is to be displayed to that effect to prevent more cars from coming in. The system should be such that the parking lot attendant can manually alter the number on the display in case of some malfunctioning of the system. We will look at the design of the system.

Solution

Since this is a design problem, there is no single solution. We will proceed by making some initial assumptions, coming up with a partial solution, then making necessary refinements to fit the constraints, and finishing off with a system that can meet with all the requirements.

We will assume the following arrangement in the parking lot. There is a single entry and a single exit. At the entry and at the exit are metal loops buried in the ground that send a signal to the digital system being designed. When an entry signal is received by the system, it should count up, and when an exit signal is received, it should count down. Our first pass at the design is to get three up-down decade counters (like the 74190 for example). Three counters are necessary since the count goes up to 250. Every signal received by the system (up or down) will be counted by the system, but provision must be made (through the down-up input of the counter) to count in the correct direction. The units counter (which counts from 0 to 9) can be permanently enabled, the ripple output of the unit's counter is made to enable the tens counter, and the ripple output of the tens counter is made to enable the hundreds counter. The down-up input is required to be high for counting down (from the data sheet of the counter). The initial design state is shown in Fig. 13-31. This system will count in the proper direction but it will not do much else.

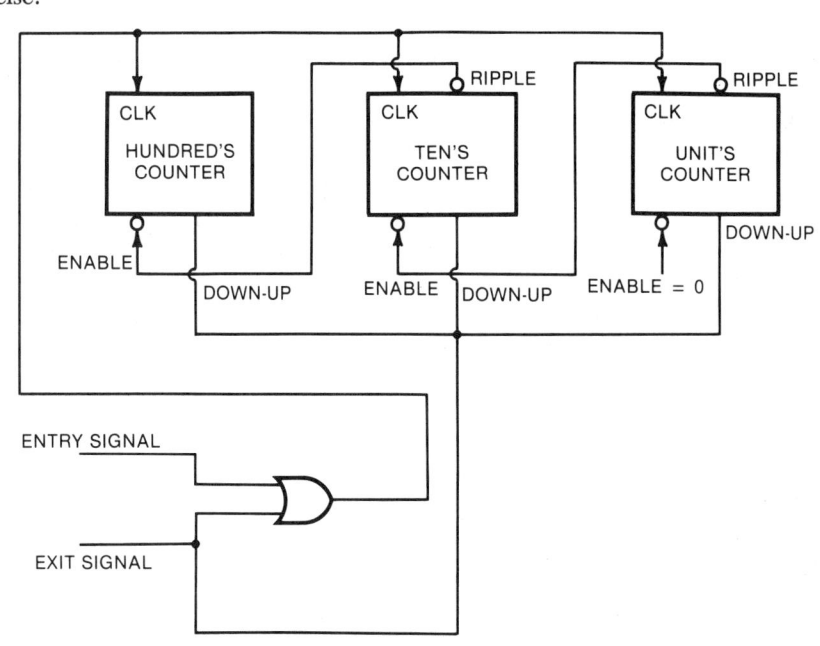

Fig. 13-31. Initial pass at the design in Example 13.24.

The next step is to take care of the "lot-full" display and to prevent more cars from coming in when the lot is full. We have to be aware that people will try to drive in even if the sign is on and prevent them from doing so. A gate can be set up *ahead* of the entry sensor and disable the gate (prevent it from opening) when the lot is full. A logic circuit is designed that will produce an output of $Z = 1$ when the count reaches 250 (the capacity of the lot) and also disable the gate. This scheme is indicated in Fig. 13-32A. The logic circuit can be designed using all the 12 inputs and an appropriate set of AND gates. But a simple circuit is enough by using the physical constraints of the system. The bits H_B (in the hundreds counter) and T_C and T_A (in the tens counter) reach a value of one when the count reaches 250. It is true that the count may go over 250, and these bits can still have a value of one, but if we can prevent any more cars from coming in, then the condition $H_B = T_C = T_A = 1$ is alone enough to ensure that the count is 250. So a single three-input AND gate is sufficient to activate the lot-full sign and disable the gate.

The count of the number of cars in the lot is to be displayed, which is done by means of seven-segment LED displays that are connected to the counters through BCD to seven-segment display drives (which are commercially available), as shown in Fig. 13-33.

The attendant should be able to manually change the display. Suppose three pairs of buttons are provided on the control panel, which when pressed send a series of pulses to the system The hundreds portion of the display is controlled by "hundreds-up" and "hundreds-down" buttons, and similar buttons control the tens and units displays. When the attendant presses the appropriate button, that particular part of the display is to be affected. The pulses produced through these buttons should have the same effect as the signals from the entry and exit sensors. A modification is, therefore, necessary of our initial system, and this is shown in Fig. 13-34A, where the signals from the up buttons (producing a manual-up signal) have the same effect as the entry signal, and the signals from the down buttons (producing a manual-down signal) have the same effect as an exit signal. Finally, the manual signals should enable the correct part of the display (only), which is done by providing three logic circuits, one of which is shown in Fig. 13-34B. The hundreds-up and hundreds-down signals are made to enable the hundreds counter along with the ripple output of the tens counter (which was in the initial design). Two other circuits similar to this one are needed for the tens and the units enable signals. For the units enable, instead of a ripple input, we will use a zero signal; that is, the zero shown as the input to the enable of the units counter in Fig. 13-31 will be combined through the OR gate with the units-up and units-down signals (which are inverted first). The inverters are necessary since enable has to be low to make the counters count.

The final design will be a combination of the diagrams of Figs. 13-32, 13-

Electronics: Circuits and Systems 775

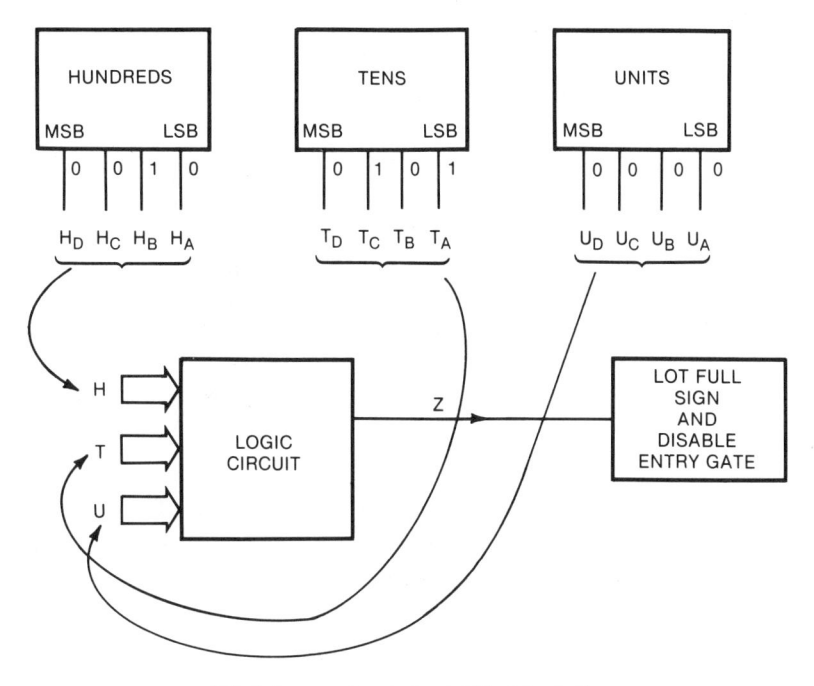

(A) Scheme to detect if lot is full.

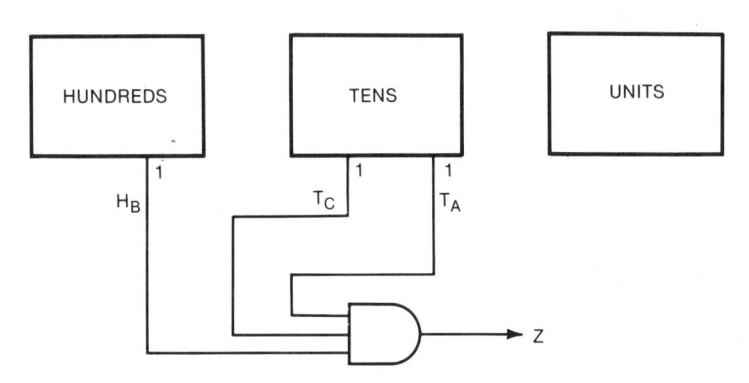

(B) AND gate use to see if condition has been met.

Fig. 13-32. Logic circuit for producing the "lot-full" signal.

```
┌──────────┐  ┌──────────┐  ┌──────────┐
│ HUNDRED'S│  │  TEN'S   │  │  UNIT'S  │
│ COUNTER  │  │ COUNTER  │  │ COUNTER  │
└──────────┘  └──────────┘  └──────────┘
```

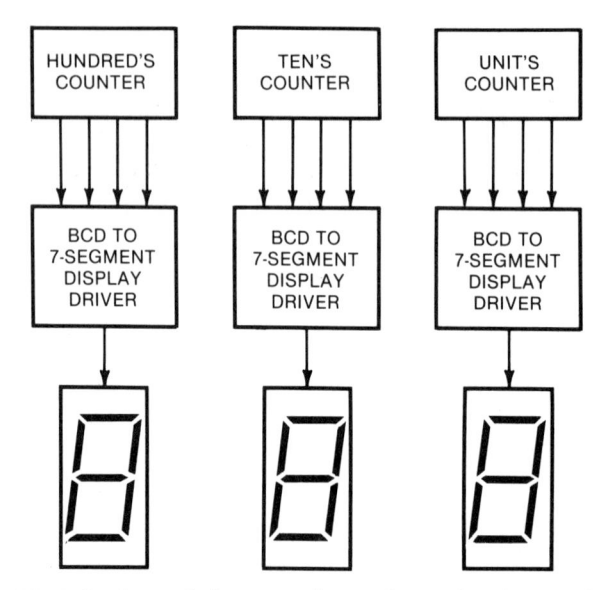

Fig. 13-33. Display of the number of cars in the parking lot.

33, and 13-34. The chances are, however, that there are some questions that are still unanswered (the "what-if?" questions) but the design has at least met the basic requirements of the problem.

Example 13.25

The design of a basic digital watch requires the use of appropriate counters also. The block diagram of the basic system is shown in Fig. 13-35.

It is assumed that a pulse generator is available that produces pulses at a rate of 1000 pulses per second (pps). This frequency is divided successively by 10 (using BCD or decade counters). The decade counter will produce a ripple output at every tenth input pulse and thus provide a "divide-by-10" response. When the pulse frequency is down to 1 pps, it is then counted first by a BCD counter (to keep track of seconds) and then by a mod-6 counter (to keep track of tens of seconds). At every 60 second interval, the mod-6 counter will produce an enable signal to activate the minutes section. The minutes section is exactly the same form as the seconds section. The tens of hours section is only mod-2 since the hours go only up to 12 (in the US). The mod-2 counter counts 0 and 1 repeatedly. The outputs of the seconds, minutes, and hours section are fed to seven-segment displays through appropriate display drivers. A colon display is also activated by the 1-pps signal to indicate that the watch is working. This design is a rather primitive one. Many improvements (and

Electronics: Circuits and Systems

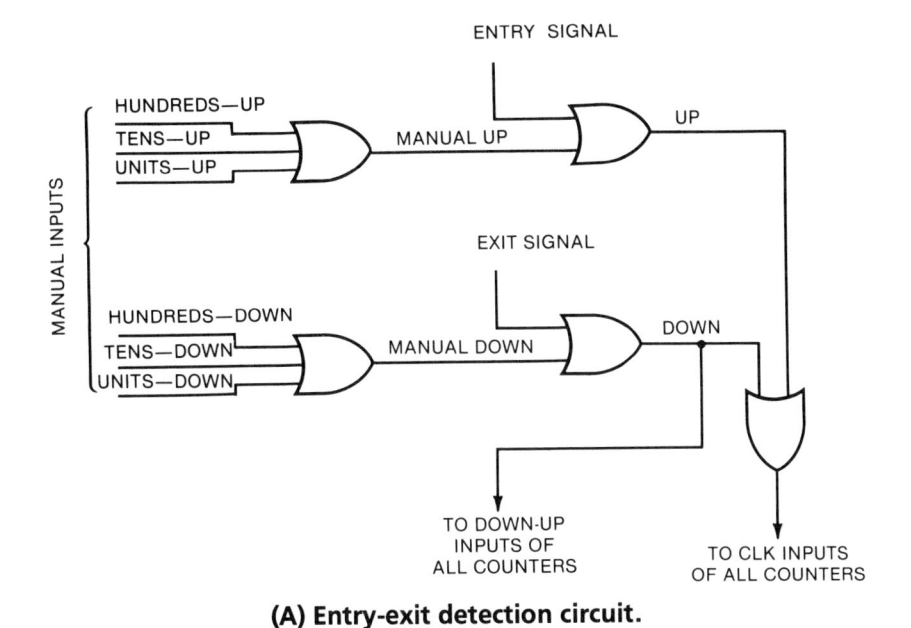

(A) Entry-exit detection circuit.

(B) Circuit to enable correct part of display.

Fig. 13-34. Manual control for the parking lot circuit.

additions) are possible. For example, the signals from the counters can be multiplexed and fed to the display driver that will permit us to use a single driver through a demultiplexer. The persistence of vision of the user will make the display look steady even though the digits are not displayed continuously.

It should be mentioned that complete LCD watch units are available on a *single* chip with much more capability than the basic system discussed.

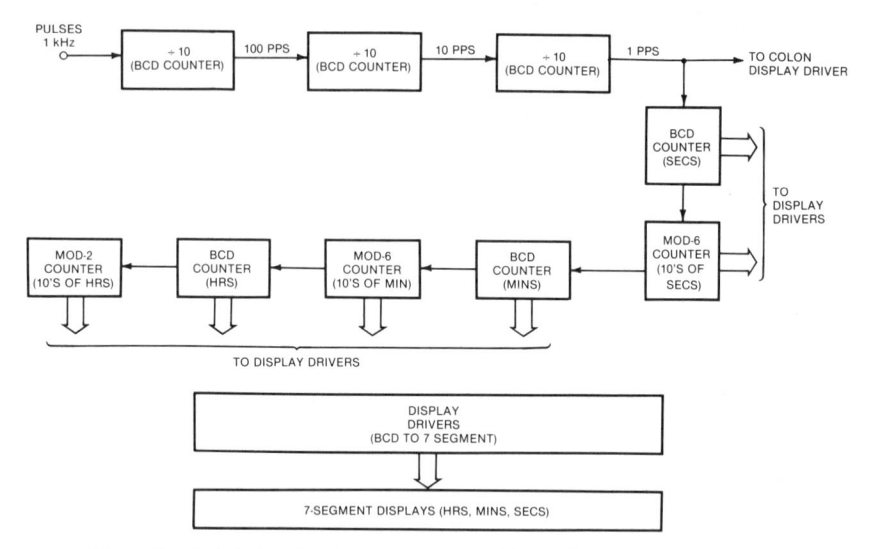

Fig. 13-35. Block diagram of a digital watch circuit.

Switch Debouncing

The application of a high or low input signal to a digital circuit frequently involves the use of a mechanical switch as indicated in Fig. 13-36. As the wiper of the switch is moved from one contact to the other, the input to the circuit switches from high to low and vice versa. Ideally, when the wiper is moved from one contact to the other, it should establish an electrical connection until the operator moves it back to the first contact. But practical switches are apt to bounce; that is, the wiper makes and breaks connection with the new contact position. The bouncing can lead to making and breaking connections a large number of times and can last for several milliseconds. The input to the circuit keeps changing from high to low and low to high due to the bouncing and the circuit responds to all these transitions as if they were legitimate input changes. A counter, for example, will count all the changes produced by bouncing. In order to avoid such spurious actions of a digital circuit, it is necessary to have mechanical switches that are bounce-free.

A *switch debouncing circuit* combines a *single-pole double-throw (SPDT) switch* and an SR flip-flop as shown in Fig. 13-37. The output Q of the debouncing circuit is used as the input to any digital circuit. We will assume that when the switch is moved to a new contact (A or B), its bouncing will be confined to making and breaking connections with that new contact only and that it will not bounce back and forth between the old and the new contacts.

Suppose the switch is initially at contact A. Then $\overline{S}$ = low or $\overline{R}$ = high. This makes the output Q = high and $\overline{Q}$ = low.

Electronics: Circuits and Systems

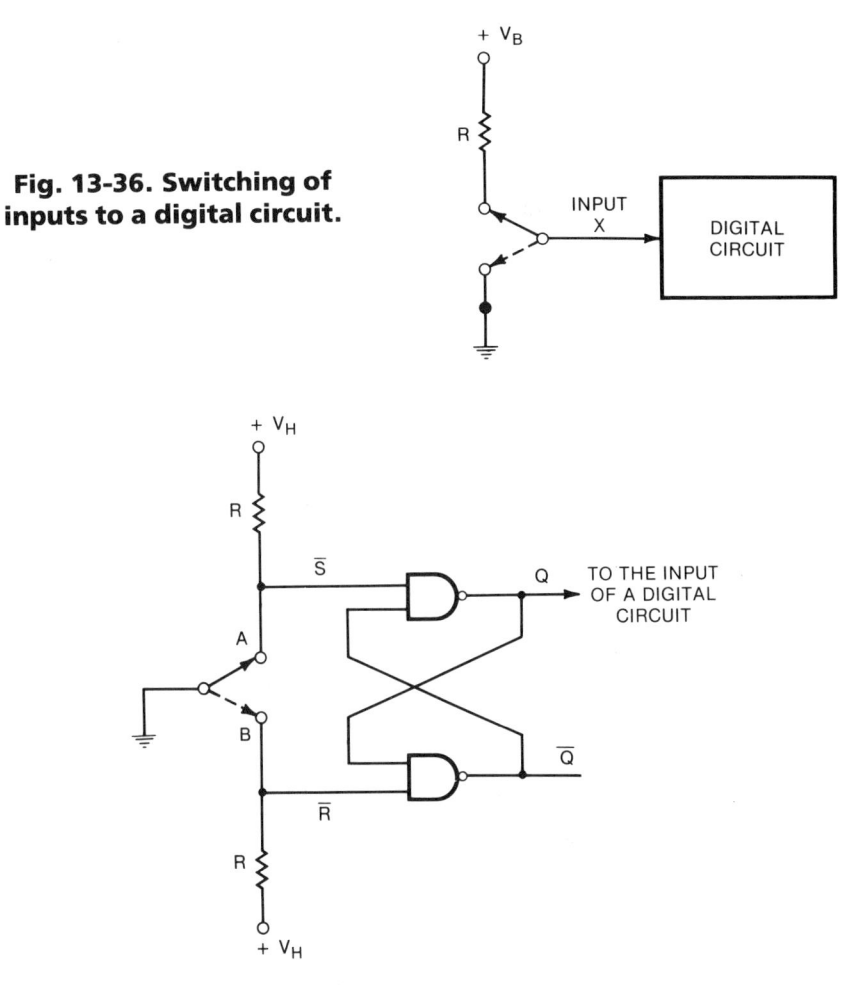

Fig. 13-36. Switching of inputs to a digital circuit.

Fig. 13-37. Switch debouncing circuit.

Now let the switch be moved to contact B. When it is at B, we have $\overline{S}$ = high and $\overline{R}$ = low, which makes Q = low and $\overline{Q}$ = high. Now, if the switch bounces off the contact B and breaks connection with it, we have $\overline{S}$ = high and $\overline{R}$ = high, which *keeps* Q at low and $\overline{Q}$ at high. Thus, Q is stable at its new value (low) regardless of how many times the switch bounces off and on at contact B (so long as it does not bounce all the way back to contact A). The value of Q (which acts as the input to a digital circuit) becomes low when the switch makes its initial connection at contact B and stays at low in spite of the switch's bouncing off and on at that contact.

It can be shown that a similar stable situation occurs when the switch is moved from contact B to contact A. The value of Q switches from low to high as soon as the initial connection is made at contact A and remains stable in spite of the bouncing at contact A.

Bounce-free switches are commercially available incorporating a flip-flop and the pull-up resistors R. It is generally assumed that any switches used in digital circuits are of the bounce-free variety.

SHIFT REGISTER

The shifting of the data stored in a register (either to the right or to the left) is necessary in some manipulations of the data such as, for example, the multiplication and division of numbers. It is also necessary to take data bits coming into a digital system in *serial* form (that is, one bit at a time in succession) and make the data bits available in *parallel* form (that is, all bits available at the same time). This transformation of serial-to-parallel format is necessary if a parallel adder is being used and the numbers are being received serially. The opposite operation, parallel-to-serial conversion, is also necessary in some cases. Such operations are performed by a cascaded connection of D flip-flops, called the *shift register*.

The basic 4-bit shift register is shown in Fig. 13-38. Four D flip-flops are connected so that the output Q of each serves as the D input of the next. The data bits are fed at the D input of the first flip-flop. The shift operation occurs at the trailing edge (one-to-zero transition) of the clock pulses (which are sometimes called *shift pulses*). It can be seen that since the output of a flip-flop becomes equal to the value of its D input at the negative edge of a clock pulse, the output of each flip-flop in the shift register becomes equal to the output of the preceding flip-flop after each clock pulse. This is a *shift-right* operation. The timing diagrams are shown in Fig. 13-38.

Drill Problem 13.32: Sketch the A, B, C, D waveforms of a 4-bit shift register when the input X is as shown in Fig. 13-39.

Shift registers are available in different sizes, the 4 bit and 8 bit being the most common. The larger shift registers have only a serial-in/serial-out capability; that is, they can receive data only serially (one bit at a time) and their outputs can also be read only serially from the last flip-flop. Smaller shift registers have the ability to mix serial and parallel formats: the inputs can be read in serial or in parallel (that is, all 4 bits are directly fed to the 4 registers), and the outputs can also be read in serial or in parallel.

Some shift registers can shift data either to the right or to the left, the direction being controlled by a mode signal. The diagram of a right-shift/left-shift register is shown in Fig. 13-40.

Electronics: Circuits and Systems 781

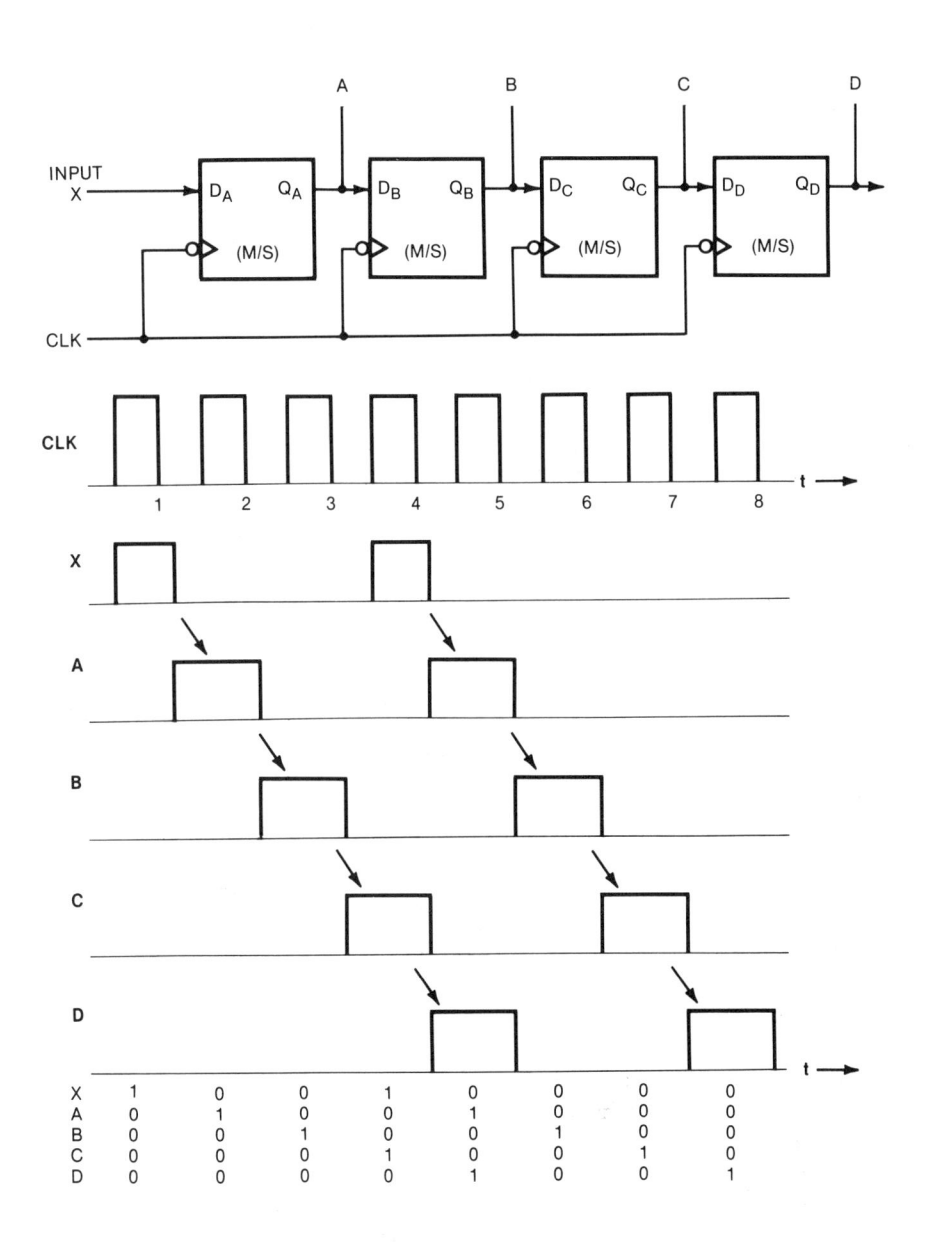

Fig. 13-38. Four-bit shift register and associated timing diagrams.

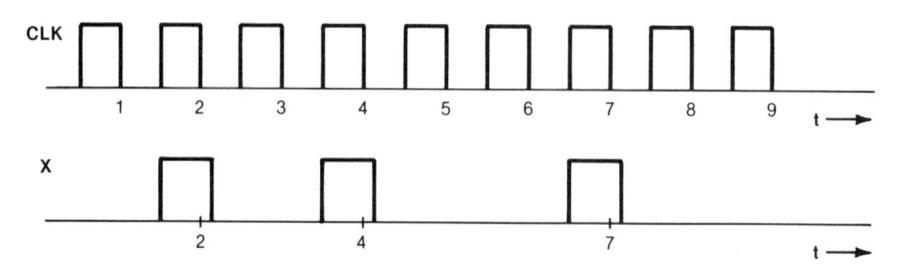

Fig. 13-39. Clock and X input of a shift register for Drill Problem 13.32.

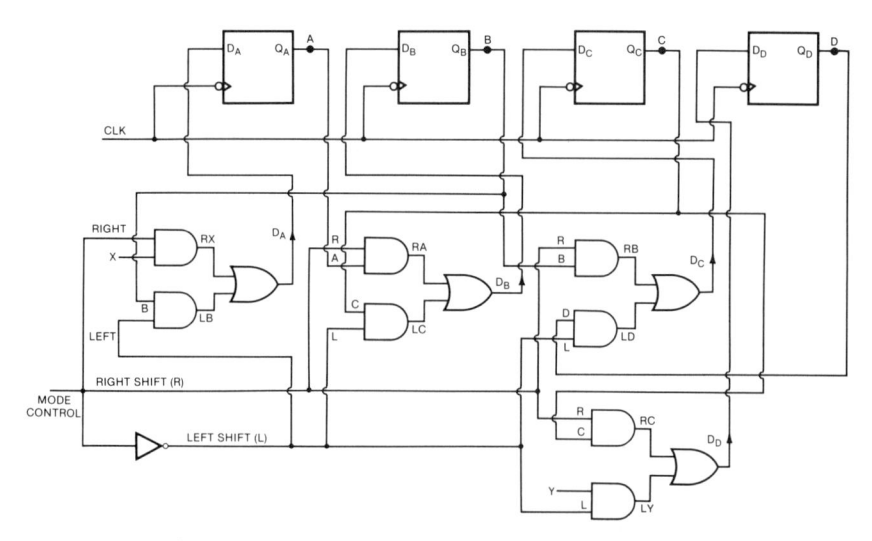

Fig. 13-40. Right-shift/left-shift register.

The normal use of a shift register is to transfer data from one register to another or to change the data format from serial to parallel and vice versa. The use of shift registers to add two 4-bit numbers using a single full-adder is shown in Fig. 13-41. The two 4-bit numbers X and Y to be added are stored in shift registers. The bits are fed into the full-adder serially. As each pair of bits is added, the sum digit is fed to the sum register and the carry bit to the carry register. As each clock pulse occurs, the sum of the X and Y bits in the last register of the two shift registers and the carry bit from the carry register are added in the full-adder, and the bits in the registers X and Y shift one cell to the right so as to be next in line. Frequently, we wish to record the numbers X and Y as well as the sum S. In such cases, X and Y can be stored by feeding the X_4 and Y_4 bits back to the X_1 and Y_1 positions at each clock pulse, which will make X and Y available at the end of the addition cycle.

Electronics: Circuits and Systems   783

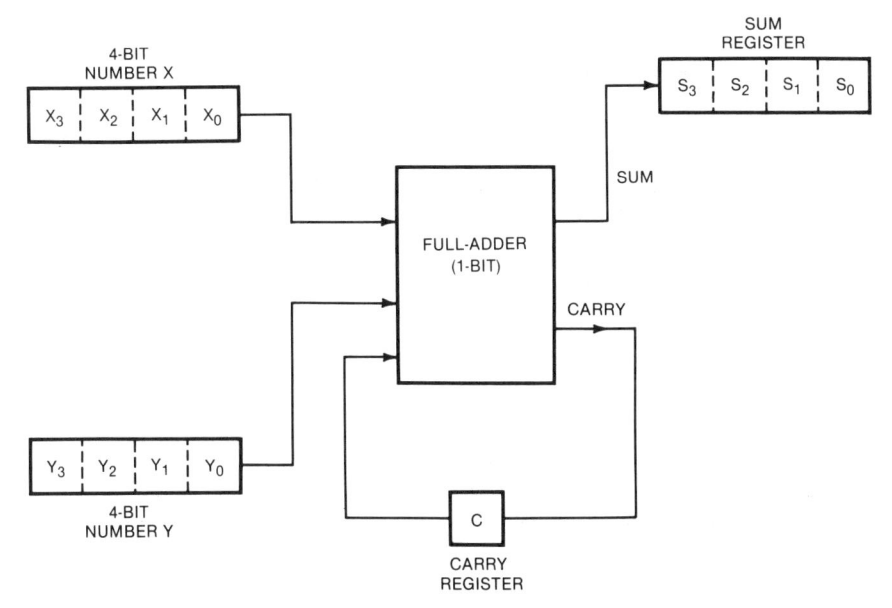

Fig. 13-41. Addition of two 4-bit numbers using a single full-adder.

SHIFT REGISTER WITH FEEDBACK

The applicable area of shift registers can be expanded by using a feedback link from the outputs to the input through some logic gates. Such a connection makes the shift register go through a *counting cycle*. Even though the counting cycle is not in the normal order (0, 1, 2, . . .), the arrangement has the advantage that the *number of steps* in the counting sequence can be altered by simply altering the feedback logic circuit. Such circuits are used as counters, random sequence generators, and generators of linear recurring sequences.

The principle of operation of a shift register with feedback is that the input X (Fig. 13-42) can be generated by means of a logic circuit that is activated by one or more of the outputs of the flip-flops in the register. As the bits in the register shift to the right, the value of X (zero or one) shifts into flip-flop 1. We can control whether X should be zero or one by means of the feedback circuit.

The use of a 3-bit shift register to generate a sequence with eight steps is shown in Fig. 13-43. Suppose A B C = 0 0 0 to start with. The output X of the logic circuit can be shown to be 1 in this case. Therefore, when the next clock pulse occurs, X = 1 will move into register A, while the 0 that was in A

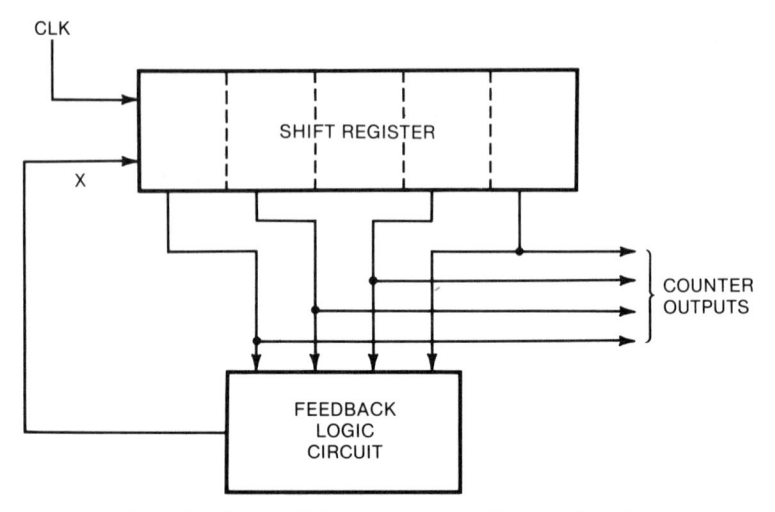

784   Arithmetic Logic, Counters, and Shift Registers

Fig. 13-42. Shift register with feedback.

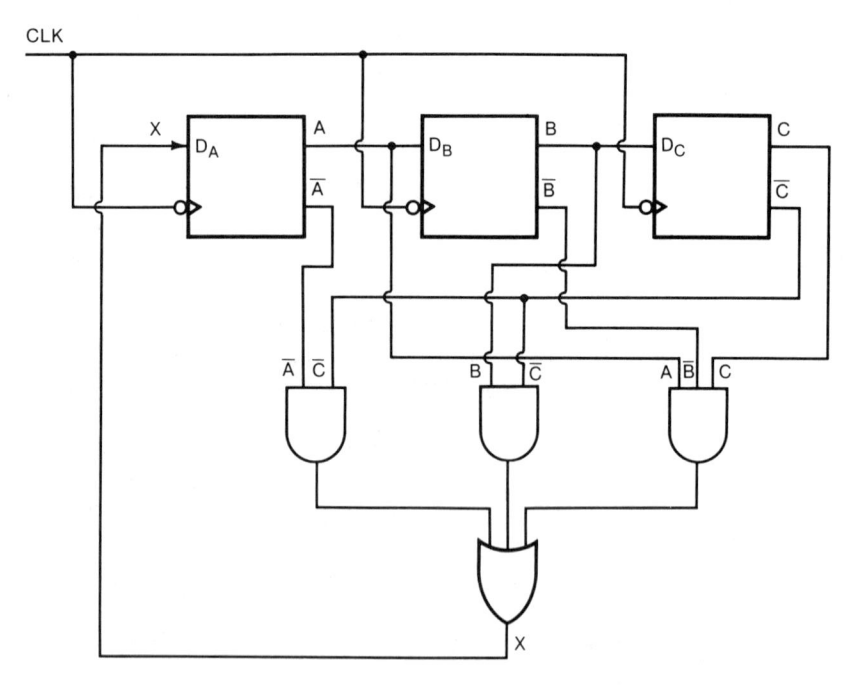

Fig. 13-43. Three-bit counter using a shift register and a feedback network.

Electronics: Circuits and Systems

will move to B, and the 0 that was in B will move to C. The register will now have A B C = 1 0 0, which will make X = 0 through the logic circuit. At the next clock pulse, then, A will become 0, B will become 1, and C will become 0. The sequencing of the numbers in the circuit is shown in the following tabular form:

A	B	C	X
0	0	0	1
1	0	0	0
0	1	0	1
1	0	1	1
1	1	0	1
1	1	1	0
0	1	1	0
0	0	1	0
0	0	0	1	(cycle repeats)

In each row of this table, X is determined by the values of A, B, C applied to the logic circuit, A is the value of X from the previous row, the value of B is the value of A from the previous row, and the value of C is the value of B from the previous row.

It may at first appear that since the counting sequence is not in the order 0 to 7 (as one would usually like it) but is $0 \rightarrow 4 \rightarrow 2 \rightarrow 5 \rightarrow 6 \rightarrow 7 \rightarrow 3 \rightarrow 1 \rightarrow 0$, this is not a particularly worthwhile arrangement. But, it has the advantage of simplicity, and as mentioned earlier, the number of steps in the counting sequence can be easily changed by changing only the logic circuit in the feedback loop. In the 3-bit shift register, for example, counting sequences with any number of steps from one to eight, can be obtained. Such sequences are often useful in digital systems in cycling a process through a specific period of time with the period of time at the control of the user.

SUMMARY SHEETS

BINARY NUMBER SYSTEM

Weights of the Digit Positions
Starting with the least significant bit (LSB) and working toward the higher bits: 2^0, 2^1, 2^2, ..., 2^n.

Binary-to-Decimal Conversion
Given the number $b_n b_{n-1} \ldots b_2 b_1 b_0$ in binary. Decimal equivalent $= b_n 2^n + b_{n-1} 2^{n-1} + \ldots + b_2 \times 2^2 + b_1 \times 2 + b_0$.

Decimal-to-Binary Conversion
Continuous Division by 2: Divide the given number by 2. Save the remainder. Divide the quotient by 2, and repeat the process until the quotient becomes 0. The remainders form the binary digits of the equivalent with the first remainder being the LSB and the last the MSB.

Binary Fractions
Starting with the "decimal" point, the weights are (moving toward the right): $(½)$, $(½)^2$, $(½)^3$, ...

Binary-to-Decimal Conversion (Fractions)
Given the binary number $0.a_1 a_2 a_3 \ldots$, Decimal equivalent $= a_1 \times 0.5 + a_2 \times 0.5^2 + a_3 \times 0.5^3 + \ldots$.

Decimal-to-Binary Conversion (Fractions)
Multiply the given number by 2. Save the integer part of the product (0 or 1). Multiply the fraction part by 2. Continue the process until the fraction part becomes 0 (or when this is not possible, stop after a specified number of digits). The binary equivalent is then the various integer parts obtained: the first integer is the MSB and the last integer is the LSB. (The MSB is the digit immediately after the decimal point.)

When a binary number has an integer part and a fraction part, each part is treated separately to obtain the decimal equivalent. The same is true when going from decimal binary also.

OCTAL AND HEXADECIMAL SYSTEMS

Octal Number System
Weights of the Digit Positions
Integers: 8^0, 8^1, 8^2, ...
Fractions: $(⅛)$, $(⅛)^2$, $(⅛)^3$, ...
Conversion From Octal to Decimal: Treat the integer and fraction parts separately. Multiply each position by the corresponding weight and add.
Conversion From Octal to Binary: Replace each digit by its 3-bit binary equivalent.
Conversion From Decimal to Octal: The procedure is similar to that used to convert decimal to binary, except that a division by 8 (or multiplication by 8 for fractions) is used.
Conversion From Binary to Octal: Starting from the decimal point, work in either direction. Group the binary digits in groups of three and replace each group by its octal equivalent. (Use zeros to fill out groups if needed.)

Hexadecimal System
Symbols A, B, C, D, E, F are used to denote the decimal numbers 10, 11, 12, 13, 14, and 15, respectively.

Weights of the Digit Positions
Integers: 1, 16, 16^2, 16^3, ...
Fractions: $(1/16)$, $(1/16)^2$, $(1/16)^3$, ...
Conversion From Hexadecimal to Decimal: Multiply each position by the corresponding weight and add.
Conversion From Hexadecimal to Binary: Replace each hexadecimal digit by its 4-bit binary equivalent.
Conversion From Decimal to Hexadecimal: The procedure is similar to those used for binary and octal systems, except that division by 16 (or multiplication by 16 for fractions) is used.
Conversion From Binary to Hexadecimal: Starting with the decimal point, work in either direction and group the binary digits into groups of four (using zeros to fill out groups as needed). Replace each 4-bit group by its hexadecimal equivalent.

ADDERS

Half-Adder
There is no carry input. There are two outputs: the sum of the two input bits and a carry.

Electronics: Circuits and Systems

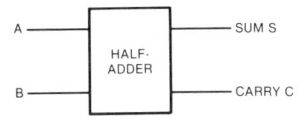

Full-Adder
There are three inputs: the two bits being added and the carry input from the previous bit position. There are two outputs: the sum of the three inputs and a carry.

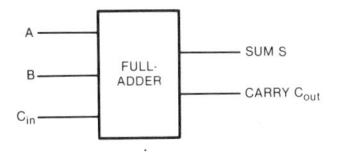

NEGATIVE NUMBERS

Sign Bit
The binary representation of a number uses n bits: $(n-1)$ magnitude bits and the n-th bit (in the left-most position) representing the sign.

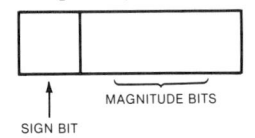

SIGN BIT = 0: POSITIVE NUMBER
SIGN BIT = 1: NEGATIVE NUMBER

A positive number is always represented as a sign bit (0) followed by the magnitude bits.

Ones Complement Representation of Negative Numbers
First write the positive counterpart in (sign + magnitude) bits form. Replace each one by zero and each zero by one to get the ones complement representation.

+10: 01010 → −10: 10101

To find the negative number represented by a given ones complement representation, replace ones by zeros and zeros by ones. The resulting binary number is the positive counterpart. Attach a minus sign.

11011 → 00100 = +4 → −4

Twos Complement Representation of Negative Numbers
First obtain the ones complement representation of the given negative number, and then add one to it.

$$+12: 01100 \rightarrow \begin{array}{r} 10011 \\ +\quad 1 \\ \hline 10100 \end{array} = (-12)$$

To find the negative number corresponding to a given twos complement representation, replace ones by zeros and zeros by ones and *add* 1 to obtain the positive counterpart. Attach a minus sign.

$$11010 \rightarrow 00101 = 7 \rightarrow (-7) \\ +\ 1 \\ \rightarrow -6$$

ADDITION AND SUBTRACTION IN THE TWOS COMPLEMENT SYSTEM

$(A - B) = A + (-B)$ Negative numbers are
$-A + B = (-A) + B$ put in their twos
$-A - B = (-A) + (-B)$ complement forms.

Except in those cases where the magnitudes of the two numbers add up to 16 or more, this procedure can lead automatically to the correct results. Corrective logic circuitry is necessary to handle the exceptional cases.

BCD CODE AND BCD ADDERS
Representation of a (Decimal) Number
Code each decimal digit separately as a 4-bit binary number.

532 : 0101 0011 0010
        5   3   2

Conversion of BCD to Decimal
Group the BCD into blocks of four bits and convert each block to a decimal digit.

1001 0110 0001 → 961
 9   6   1

When numbers are to be added in the BCD code, a separate adder is used for each block of 4 bits, and a suitable interconnection is made between the different adders to handle the carry information.

Arithmetic Logic, Counters, and Shift Registers

ARITHMETIC LOGIC UNIT

The ALU is a high-speed logic package designed to perform arithmetical operations (such as addition, subtraction, etc.) as well as logical operations (such as AND, OR, etc.).

The combination of an ALU with a control circuit forms the central processing unit (CPU) of a small computer. The CPU is contained in a single chip in a microprocessor.

COUNTERS

Mod-N Counter

A mod-n counter counts from 0 to (n−1) and repeats the sequence. It has n counting steps.

Synchronous Counter

The clock pulses being counted are directly connected to all the flip-flops in the synchronous counter. Each flip-flop is, therefore, directly activated by the pulses being counted.

Asynchronous (or Ripple) Counter

The clock pulses being counted are directly connected to only the first flip-flop in the counter. Each of the other flip-flops is activated by a pulse produced by the preceding flip-flop.

SR FLIP-FLOP

S	R	Q
0	0	NO CHANGE
0	1	RESETS TO 0
1	0	SETS TO 1
1	1	NOT ALLOWED

SR Flip-Flop Using NAND Gates

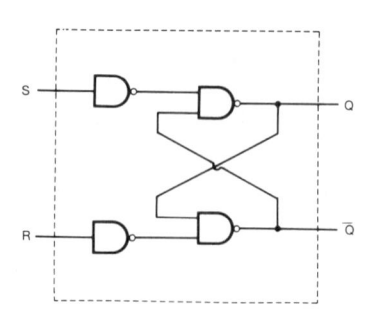

Clocked SR Flip-Flops

No action takes place in the absence of a clock pulse. When a clock pulse occurs, the output changes in accordance with the S and R values.

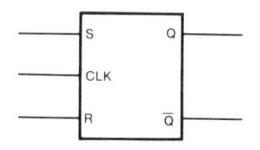

JK FLIP-FLOP

J	K	Q
0	0	NO CHANGE
0	1	RESET TO 0
1	0	SETS TO 1
1	1	FLIPS TO OPPOSITE VALUE

Clocked JK Flip-Flop

No action takes place in the absence of a clock pulse. When a clock pulse occurs, the output changes as dictated by the values of J and K.

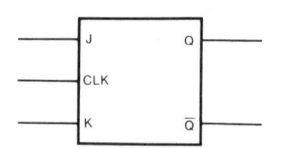

Positive Edge-Triggered Flip-Flops

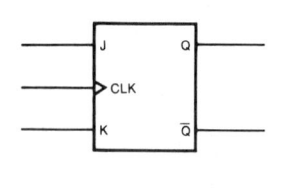

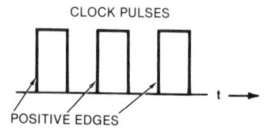

CLOCK PULSES

POSITIVE EDGES

J, K VALUES AT THE INSTANTS OF THE POSITIVE EDGES DETERMINE THE RESPONSE.

Electronics: Circuits and Systems

Negative Edge-Triggered Flip-Flops

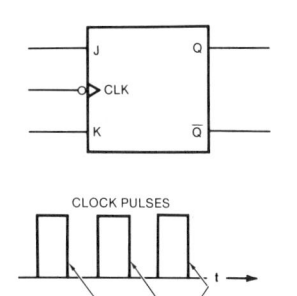

CLOCK PULSES

NEGATIVE EDGES

J, K VALUES AT THE INSTANTS OF THE NEGATIVE EDGES DETERMINE THE RESPONSE.

Master-Slave Flip-Flops

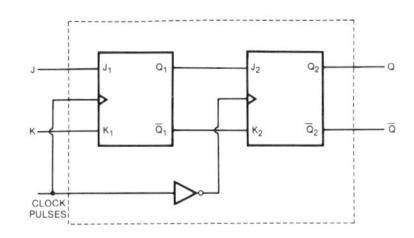

When CLK = 1, the master is enabled and the slave disabled. When CLK = 0, the master is disabled and the slave enabled. The information from the master is transferred to the slave.

From an *overall response* point of view (which is all we are interested in), the M/S flip-flop behaves very much like a negative edge-triggered flip-flop.

Preset and Clear Inputs

These are provided so as to permit the user to make the output of a flip-flop 0 or 1 regardless of the actual values of J and K.

Special Cases of JK Flip-Flops

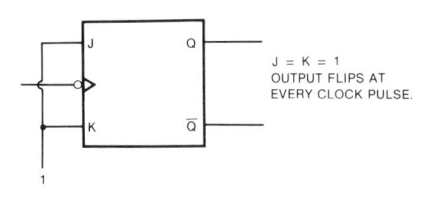

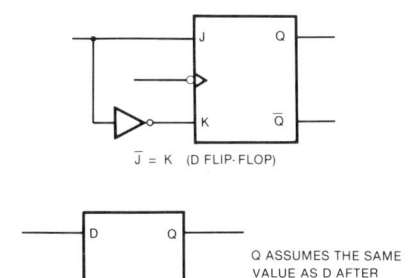

J = K = 1
OUTPUT FLIPS AT EVERY CLOCK PULSE.

$\bar{J} = K$ (D FLIP-FLOP)

Q ASSUMES THE SAME VALUE AS D AFTER A CERTAIN DELAY.

Q assumes the same value as D after a certain delay.

D-Type Latch

In the D-type latch, the output will change if the D input changes while the clock pulse is present. In the D flip-flop, on the other hand, the output will depend only upon the value of D at the edge of the clock pulse.

APPLICATIONS OF FLIP-FLOPS

Frequency Division

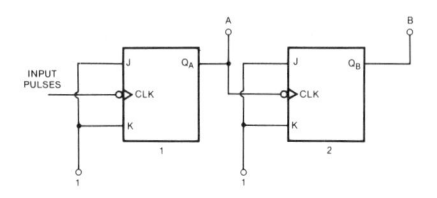

Output of flip-flop 1 is at one half the frequency of the incident clock pulses, and the output of flip-flop 2 is at one fourth the frequency of the incident clock pulses.

Ripple Counters

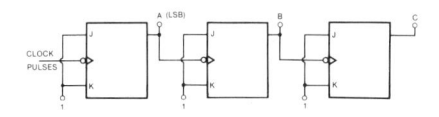

The outputs CBA follow the counting sequence: 0, 1, . . . , 7, 0, 1 . . .

Arithmetic Logic, Counters, and Shift Registers

Mod-8 Synchronous Counter

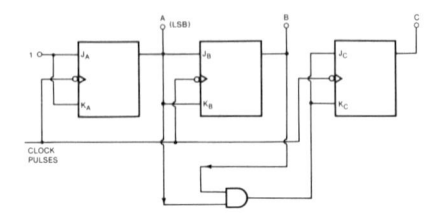

SWITCH DEBOUNCING

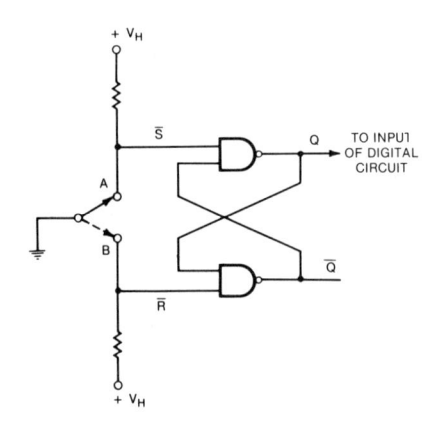

Q switches to low when the switch makes its initial contact at B and stays at low even if there is a bouncing on and off that contact.

Q switches to high when the switch makes its initial contact at A and stays high even if there is a bouncing on and off that contact.

SHIFT REGISTERS

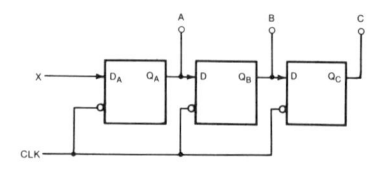

At the occurrence of each clock pulse, the output of each flip-flop shirts to the next one.

Shift Registers With Feedback

The number of steps in the counting sequence can be varied by varying the feedback logic circuit.

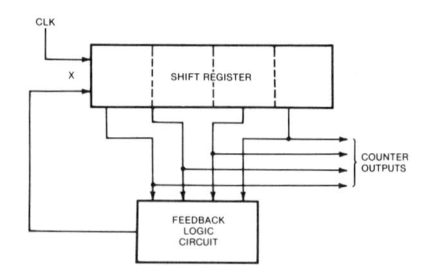

Electronics: Circuits and Systems

ANSWERS TO DRILL PROBLEMS

13.1 (a) 3; (b) 13; (c) 45; (d) 170.
13.2 (a) 101101; (b) 1110101; (c) 11111000010.
13.3 (a) 0.5; (b) 0.125; (c) 0.828125.
13.4 (a) 0.00101; (b) 0.1011; (c) 0.01000010.
13.5 (a) 5.3125; (b) 3.875; (c) 1.34375.
13.6 (a) 101011.110010; (b) 1101101100.1000101; (c) 10000000000.00001.
13.7 (a) 399; (b) 0.72387695; (c) 21.81835938.
13.8 (a) 10100110.100011; (b) 1100111110.011010101.
13.9 (a) 3360; (b) 2724; (c) 60000.
13.10 (a) 0.314061; (b) 0.0714631.
13.11 (a) 1777.0243656; (b) 5307.510550.
13.12 (a) 41; (b) 0.41; (c) 57.152; (d) 3702.02.
13.13 (a) 2214; (b) 3474; (c) 0.4140625; (d) 249.8984375.
13.14 (a) 100010010101.00100011; (b) 10011010.00101111.
13.15 (a) 6F0; (b) 5C0; (c) B9A1.
13.16 (a) 0.46A16E; (b) 0.79A6B50B.
13.17 (a) 2F5; (b) 2B8.2AD.
13.18 (a) 1110; (b) 10100.
13.19 (a) 11011; (b) 10100; (c) 11000110.
13.20 (a) -8; (b) -41; (c) -14.
13.21 (a) 11100; (b) 10101; (c) 11000111.
13.22 (a) -3; (b) -32; (c) -478.
13.23 (a) 01110; (b) 00010 (The overflow is discarded.); (c) 11110; (d) 10010 (The overflow is discarded.); (e) 00000 (The overflow is discarded.).
13.24 See Fig. 13-DP24.

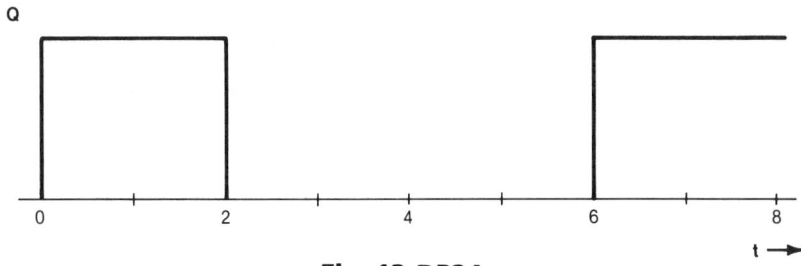

Fig. 13-DP24.

13.25 See Fig. 13-DP25.

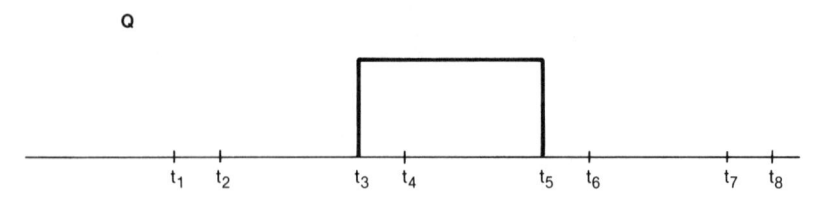

792 Arithmetic Logic, Counters, and Shift Registers

Fig. 13-DP25.

13.26 See Fig. 13-DP26.

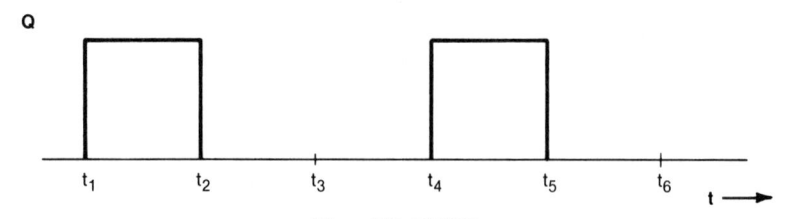

Fig. 13-DP26.

13.27 See Fig. 13-DP27.

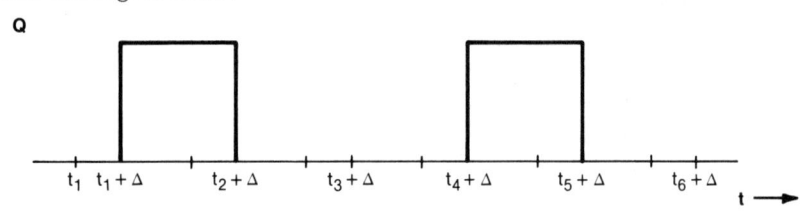

Fig. 13-DP27.

13.28 Same as in Fig. 13-DP27.
13.29 See Fig. 13-DP29.

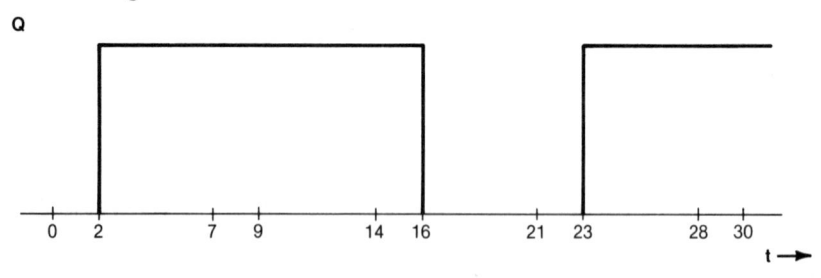

Fig. 13-DP29.

13.30 See Fig. 13-DP30.

Electronics: Circuits and Systems

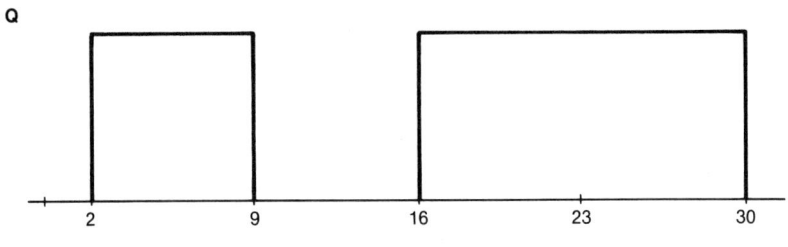

Fig. 13-DP30.

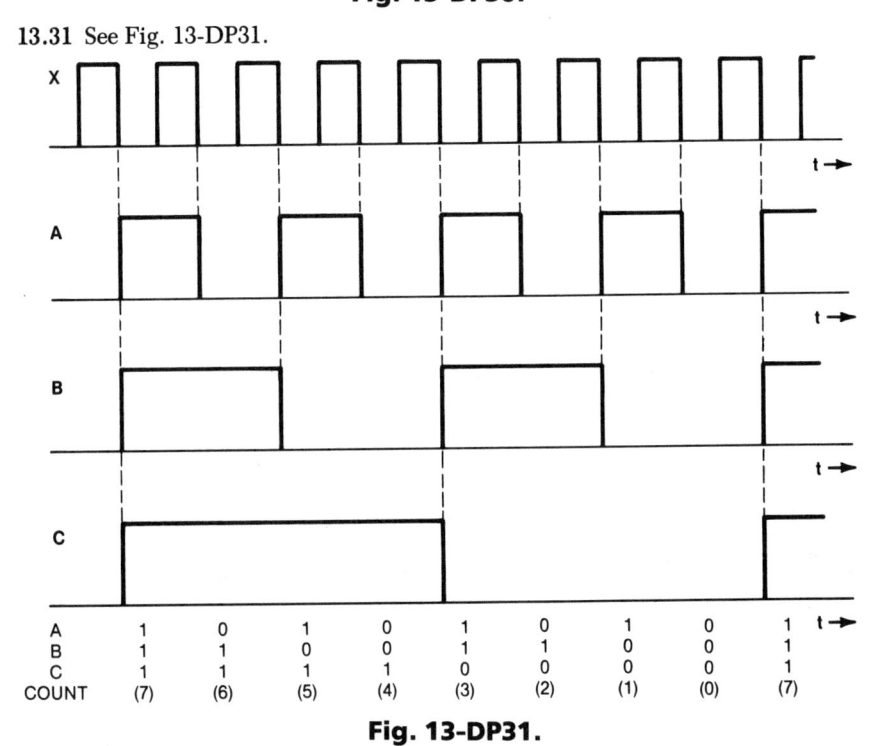

13.31 See Fig. 13-DP31.

A	1	0	1	0	1	0	1	0	1
B	1	1	0	0	1	1	0	0	1
C	1	1	1	1	0	0	0	0	1
COUNT	(7)	(6)	(5)	(4)	(3)	(2)	(1)	(0)	(7)

Fig. 13-DP31.

13.32 See Fig. 13-DP32. (Some delay in the response of each flip-flop has been introduced in the diagrams so as to avoid ambiguity.)

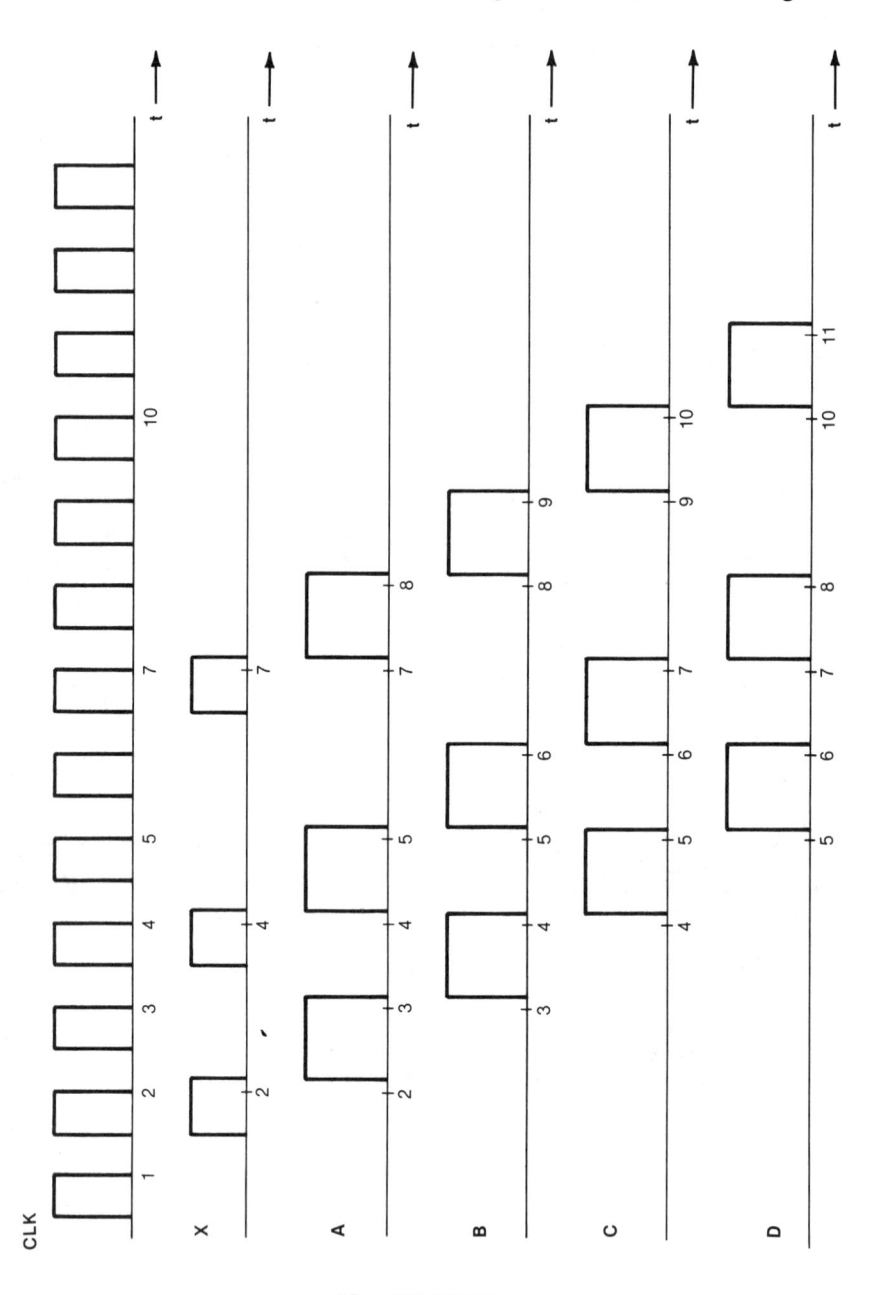

Fig. 13-DP32.

PROBLEMS

Binary Number System

13.1 Convert each of the following binary numbers to its decimal equivalent: (a) 10110; (b) 111010; (c) 1111101; (d) 110011010101.

13.2 Convert each of the following decimal numbers to its binary equivalent: (a) 1029; (b) 1956; (c) 4096; (d) 127.

13.3 Convert each of the following binary fractions to its decimal equivalent: (a) 0.111; (b) 0.0101; (c) 0.1100101; (d) 0.000111.

13.4 Convert each of the following decimal fractions to its equivalent binary form: (a) 0.35; (b) 0.00625; (c) 0.128. (When the process does not terminate, use 8 bits. Calculate the percent error in such cases.)

13.5 Convert each of the following decimal numbers to its equivalent binary form: (a) 19.56; (b) 1.956; (c) 195.6; (d) 127.9375.

13.6 Convert each of the following binary numbers to its equivalent decimal form; (a) 10111.01001; (b) 10.101; (c) 1.0101; (d) 11111.1111.

Octal and Hexadecimal Systems

13.7 Convert each of the following octal numbers to its decimal equivalent: (a) 4006; (b) 347; (c) 127.

13.8 Convert each of the following octal numbers to its decimal equivalent: (a) 40.06; (b) 3.47; (c) 34.7; (d) 1.27.

13.9 Convert each of the following decimal numbers to its octal equivalent: (a) 4096; (b) 347; (c) 1812.

13.10 Convert each of the following decimal fractions to its octal equivalent: (a) 0.987; (b) 0.1212; (c) 0.1365.

13.11 Convert each of the following decimal numbers to its octal equivalent: (a) 9.87; (b) 98.7; (c) 555.1212.

13.12 Convert each of the following binary numbers to its octal equivalent: (a) 1110.1101; (b) 10101011.111000110; (c) 10.01.

13.13 Convert each of the following hexadecimal numbers to its decimal equivalent: (a) 4F3; (b) ABC; (c) DEF; (d) 8921.

13.14 Convert each of the following hexadecimal numbers to its binary equivalent: (a) ADF; (b) 9821; (c) A9F2.

13.15 Convert each of the following hexadecimal numbers to its binary equivalent: (a) A.DF; (b) AD.F; (c) 98.21; (d) A.9F2.

13.16 Convert each of the following decimal numbers to its equivalent hexadecimal form: (a) 9821; (b) 2001; (c) 9642.

13.17 Convert each of the following decimal fractions to its equivalent hexadecimal form: (a) 0.9821; (b) 0.2001; (c) 0.9642.

13.18 Convert each of the following decimal numbers to its hexadecimal equivalent: (a) 275.9642; (b) 475.2164; (c) 65535.0665.

13.19 Convert each of the following binary numbers to its hexadecimal equivalent: (a) 10101110.10101011; (b) 111011.110101.

13.20 Convert each of the following hexadecimal numbers to its *octal* equivalent: (a) A4F; (b) A.4F; (c) 0.A4F.

13.21 Convert each of the following *octal* numbers to its hexadecimal equivalent: (a) 1776; (b) 1212; (c) 17.76; (d) 1.212.

Adders

13.22 Perform the following additions of binary numbers: (a) (10010 + 11101); (b) (101010 + 111110).

13.23 Two half-adders can be combined to form a full-adder and such a circuit is shown in Fig. 13-P23. Set up the truth tables and verify that the sum output S and carry output C_{out} conform to the truth table of a full-adder.

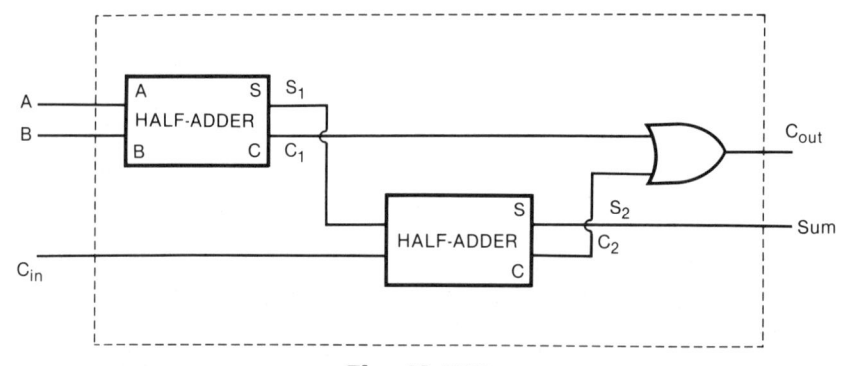

Fig. 13-P23.

13.24 Consider the 4-bit parallel adder. Let the two numbers to be added be 1010 and 1101. Show the inputs and outputs of each block in the full-adder while performing this addition.

Negative Number Representations

13.25 Find the ones complement notation of each of the following numbers. In all cases, use 8 bits including the sign bit. (a) −85; (b) +119; (c) −32; (d) −126; (d) +91.

13.26 Find the (decimal) number represented by each of the following ones complement numbers: (a) 10010111; (b) 01101101; (c) 11110000; (d) 01101010.

13.27 Find the twos complement representation of each of the following numbers. In all cases, use 8 bits including the sign bit. (a) −97; (b) +49; (c) −124; (d) +99; (e) −101.

Electronics: Circuits and Systems

13.28 Find the decimal number represented by each of the following twos complement numbers: (a) 10101011; (b) 01011111; (c) 11100100; (d) 10000010.

13.29 Perform the following operations using the twos complement notation: (a) $(9 - 7)$; (b) $(-9 + 3)$; (c) $(-8 - 2)$; (d) $(2 - 9)$.

13.30 Perform the following operations using the twos complement notation. (You will find that these create some difficulties since the magnitude of the result exceeds 15. Consider possible remedies): (a) $(-10 - 9)$; (b) $(-15 - 9)$; (c) $(13 + 10)$.

13.31 Consider the operation of subtraction using the *ones complement* system (instead of the twos complement). Use the same example problems as in Fig. 13-P29. Are there any general rules that can be developed (similar to those in the twos complement system)?

13.32 Suppose a 4-bit parallel *binary* adder is available and we wish to use it to add two 4-bit *BCD-coded* decimal digits. Discuss how this can be done.

Counters and Applications of Flip-Flops

13.33 Two NOR gates can be interconnected to obtain an SR flip-flop. Using the NAND interconnection of Fig. 13-8 as a guide for getting started, obtain an SR flip-flop made up by interconnecting NOR gates.

13.34 The S and R inputs of a clocked SR flip-flop (along with the clock pulses) are shown in Fig. 13-P34. Sketch the output Q.

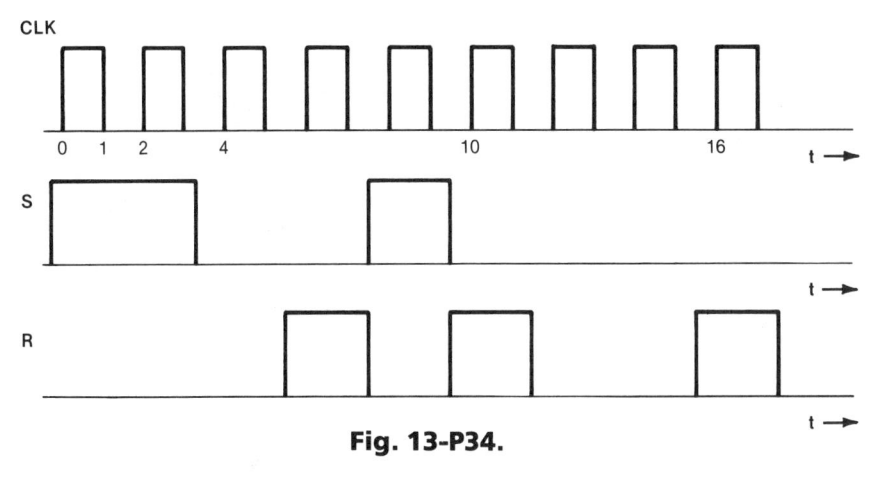

Fig. 13-P34.

13.35 Suppose the S and R inputs of the previous problem are treated, respectively, as the J and K inputs of a positive edge-triggered JK flip-flop. Sketch the output Q.

13.36 If the inputs of the previous problem are applied to a negative edge-triggered flip-flop, sketch the output Q.

798   Arithmetic Logic, Counters, and Shift Registers

13.37 The clock pulses and the control inputs of a JK flip-flop are shown in Fig. 13-P37. Sketch the output in each of the following cases: (a) the flip-flop is positive edge triggered; (b) the flip-flop is negative edge triggered; (c) the flip-flop is M/S.

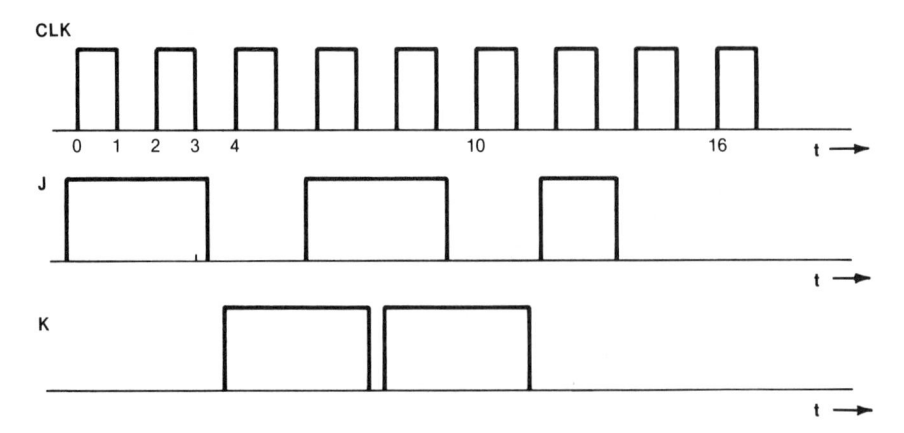

Fig. 13-P37.

13.38 Suppose the J input shown in Fig. 13-P37 (previous problem) is applied to both the J and K inputs of a flip-flop. Repeat the work asked for in the previous problem for a negative edge-triggered FF.

13.39 Suppose the J input to a flip-flop is as shown in Fig. 13-P37 and the K input is made equal to $\bar{J}$. Repeat the work asked for in the previous problem.

13.40 Suppose the J input of a flip-flop is as shown in Fig. 13-P37 and the K input is tied to high. Repeat the work asked for in the previous problem.

13.41 Consider the synchronous mod-8 counter shown in Fig. 13-29. Assume a series of clock pulses is fed to the counter. Determine and sketch the waveforms of the outputs of the flip-flops and verify the count sequence.

Switch Debouncing

13.42 If the NAND gates of the switch debouncing circuit of Fig. 13-37 are replaced by NOR gates, how will the circuit perform? Can you think of any modifications if the circuit does not work as intended?

Shift Registers

13.43 Sketch the A, B, C, D output waveforms of a 4-bit shift register if the input X is as shown in Fig. 13-P43.

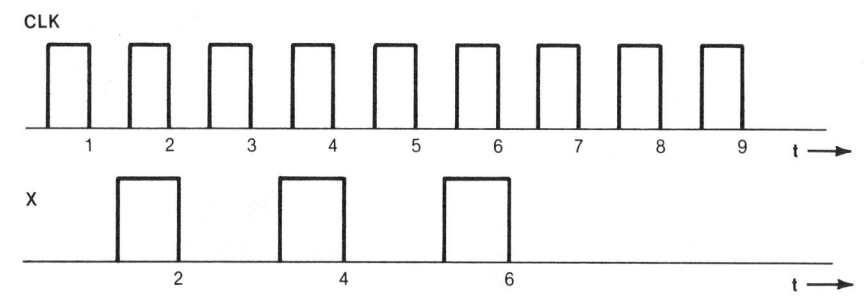

Fig. 13-P42.

13.44 A *Johnson counter* uses a shift register in which the *inverted* output of the last flip-flop is fed back as the input to the first flip-flop. Draw the block diagram of a 3-bit Johnson counter and discuss its counting operation.

13.45 Consider the shift-left/shift-right shift register shown in Fig. 13-40. Discuss the operation of the circuit by first making R = 1 and then R = 0. In each case, evaluate the inputs D of the different flip-flops. Use the X input of the block diagram you have drawn for Problem 13.44 to show that shifting does occur in the intended direction.

13.46 If the *uncomplemented* output of the last flip-flop in a shift register is fed back as input to the first flip-flop draw the block diagram of the circuit and discuss its counting operation. (This is known as a ring counter). Assume that the counter starts with a one output in the last flip-flop.

CHAPTER 14

ANALOG COMMUNICATION TECHNIQUES

Electronic communication involves the transmission of signals from one point to another. In order to transmit the signals efficiently and effectively, certain modifications of the signal have to be made. Such modifications might be shifting the signal to a different range of frequencies from its original range, conversion of signals to pulses, and other similar transformations. The type of modification is dictated by the constraints on the available bandwidth for the signal in the given communication system and considerations such as the need to send a number of different signals simultaneously over the same communication channel. When signals are sent in the form of *continuous waveforms*, the term *analog* is used since the value of the signal being transmitted at any instant is analogous to the original signal. When *samples* of signals are transmitted, the resulting communication is called *pulse communication*. The special case of communication in which each pulse is first encoded in the form of a binary string (zeros and ones) is referred to as *digital data transmission*. Digital data transmission has become extremely important over the last decade because of the proliferation of digital systems and the numerous advantages of digital signal processing over analog signal processing. We will examine the analog communication systems in this chapter and consider pulse and digital transmission in the next chapter.

The signals involved in communication systems may arise from a number of different sources. Even though signals may vary widely depending on their origins, they can be classified into a few categories. A signal may be *periodic* or *nonperiodic*. A signal may be *deterministic* or *random*. There is also a classification based on the finiteness of the power or energy contained in the signal. Many of these points were discussed in Chapter 1 in some detail, and we will only review them briefly here.

Electronics: Circuits and Systems

POWER AND ENERGY IN A SIGNAL

We normally speak of a signal as being a voltage that varies with time. It is standard practice to use a resistor of 1 Ω as the reference load in calculation of power and energy, which is called *normalized power* or *normalized energy*. The power content of a signal v(t) at any given time is, therefore, simply [v(t)]². The energy content of the signal at any instant is the area under the power curve in the interval from t = 0 to the given instant of time. These two concepts are shown illustrated in Fig. 14-1.

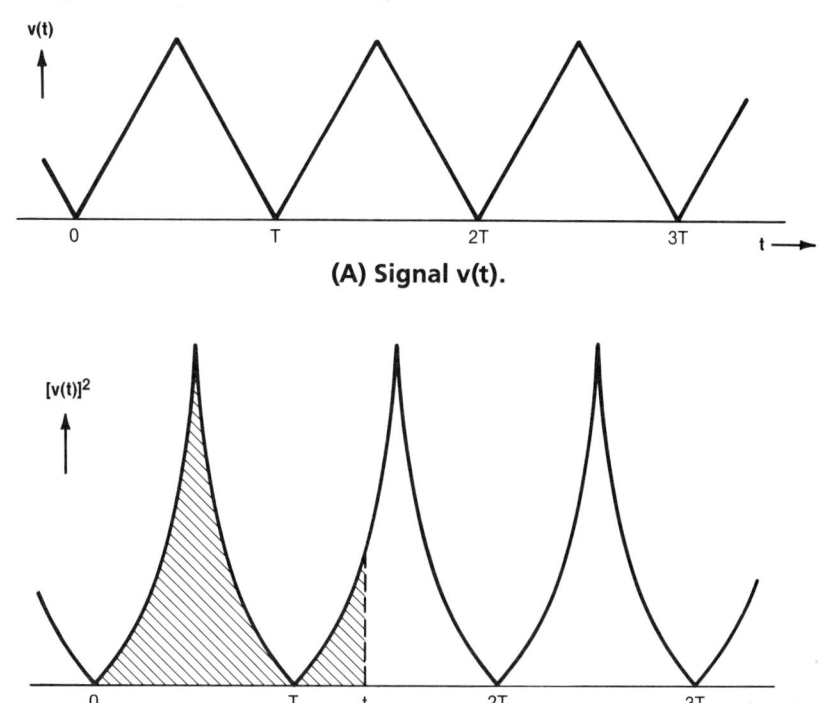

(A) Signal v(t).

(B) Curve gives the instantaneous value of the power in the signal v(t), and the area (shown shaded) gives the energy delivered in the interval 0 to t.

Fig. 14-1. Power and energy in a signal.

For a *sinusoidal* signal, that is, a signal that can be expressed by a sine or a cosine function of time, the *average power* is given by

$$P_{avg} = \frac{1}{2} \times (\text{peak value})^2 \tag{14-1}$$

A sinusoidal signal of peak value 10 V, for example, has an average power of 50 W.

When a signal is *periodic but not sinusoidal*, it can be decomposed into a number of sinusoidal components called the *harmonics*. If the period of the signal is T seconds, and if we define

$$f_0 = (1/T) \text{ Hz} \tag{14-2}$$

then, the nonsinusoidal signal can be written as the sum of a dc component, a component at f_0 (or the fundamental frequency), and compoments at multiples of f_0 (which are the second, third, . . . , harmonics). The amplitudes (that is, peak values) of the various components can be evaluated mathematically or graphically or through measuring instruments. The plot showing the amplitudes of the components in the nonsinusoidal signal is called the *frequency spectrum* of the signal. An example of a nonsinusoidal signal and its frequency spectrum is shown in Fig. 14-2. The total average power in the nonsinusoidal signal is given by the sum of the average power contained by each component. The *power spectrum* of a nonsinusoidal signal is obtained by squaring the amplitude of each line in the frequency spectrum and dividing by two. The power spectrum of the signal in Fig. 14-2A is shown in Fig. 14-2C. The power spectrum is important in the determination of the bandwidth required for the transmission of the signal. Usually some percentage (90 or 95%) of the total power in the signal is prescribed to be transmitted by a system, and the bandwidth needed can be estimated by examining the power spectrum.

When a signal is *nonperiodic but finite in duration*, its frequency spectrum becomes a continuous curve instead of being a line spectrum as in periodic signals. The frequency spectrum of a single rectangular pulse is shown in Fig. 14-3. It can be seen that the spread of the frequency spectrum is inversely proportional to the duration of the pulse, which means that a narrower pulse will contain a wider band of frequencies of appreciable amplitude than a wider pulse. The bandwidth needed for the transmission of a pulse (or series of pulses) will vary inversely in proportion to the duration of the pulse.

The area under the curve of the *square* of the frequency spectrum of a nonperiodic signal of finite duration is a measure of the energy content of the signal. The curve obtained by taking the area of the square of the frequency spectrum is called the *energy spectrum*, an example of which is shown in Fig. 14-3C. *The energy contained in a band of frequencies between f_1 and f_2 is given by the area of the energy spectrum between those two frequencies.*

When a signal continues indefinitely and its value at any instant cannot be precisely determined from measurements already made, it is a *random signal*. A *power spectrum* can usually be obtained for such signals by means of measurements or computation on extensive samples of the signal. The *power* spec-

Electronics: Circuits and Systems 803

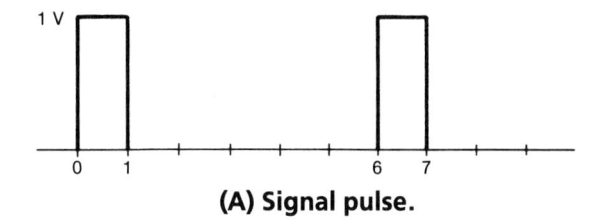

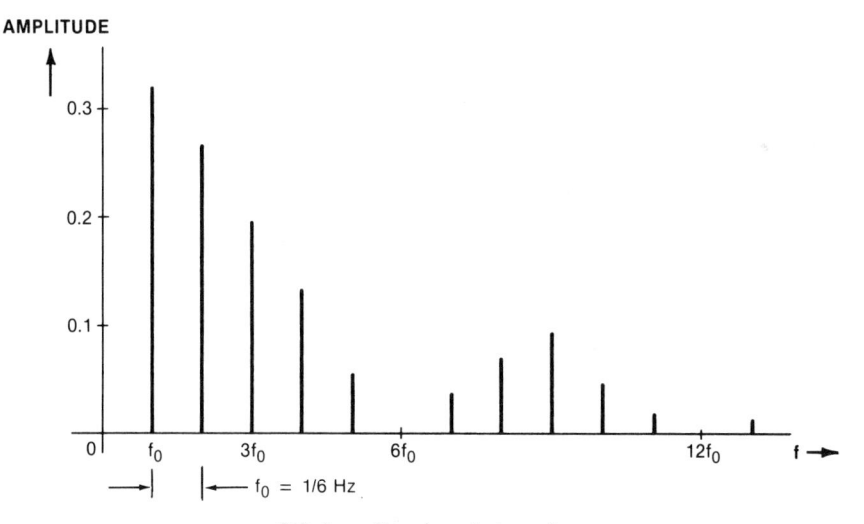

(A) Signal pulse.

(B) Amplitude of signal. $f_0 = 1/6$ Hz

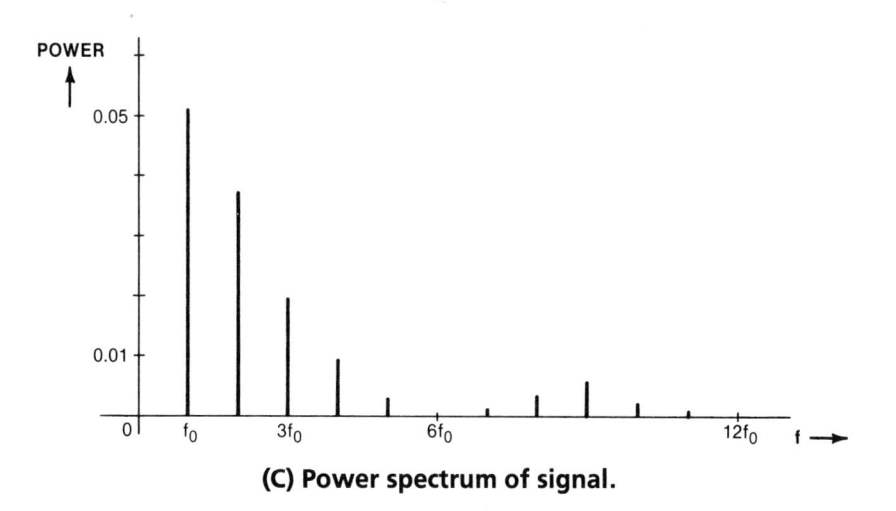

(C) Power spectrum of signal.

Fig. 14-2. Amplitude and power spectra of a pulse train.

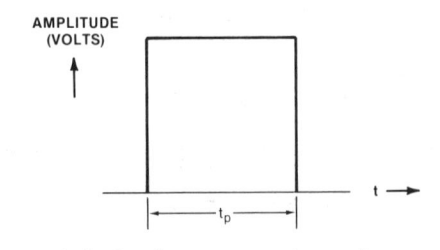

(A) Single rectangular pulse.

(B) Amplitude.

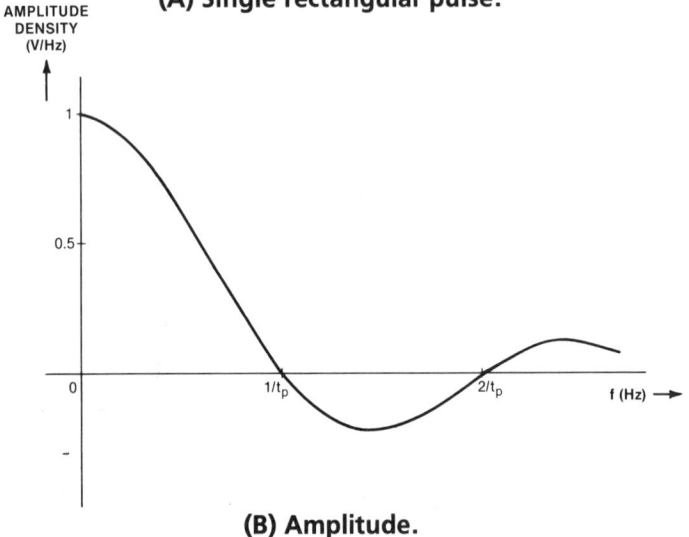

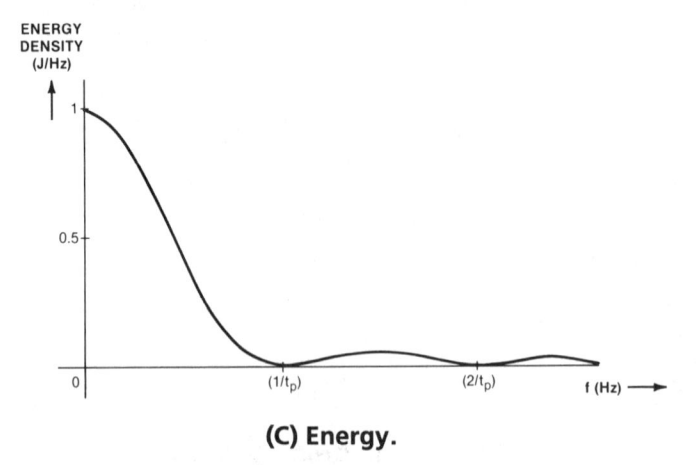

(C) Energy.

Fig. 14-3. Amplitude density and energy density spectra of a single pulse.

Electronics: Circuits and Systems   805

trum has the property that the *area contained under it between two given frequencies f_1 and f_2 is the average power contained by the signal in that band of frequencies*.

In some signals, the average power is a finite quantity although the total energy may be infinitely large, which is true in any signal that seems to last forever (or at least continue for an unpredictable length of time). In such cases, the average power, which is the area of the square of the signal over some period of time T averaged over T, is a finite quantity. But since the signal lasts for a very long time, the energy, which is the average power multiplied by the duration of the signal, is possibly infinite. Such signals are called *power signals*. In the case of other signals, such as pulses of short duration, the average power is zero since the averaging is done over a long period of time T, which is much larger than the duration of the signal. But the energy contained in such signals is finite because of their finite duration. Such signals are called *energy signals*. From our earlier discussion, we can see that pulses of short duration belong to the class of energy signals, and periodic and random signals belong to the class of power signals.

MODEL OF A COMMUNICATION SYSTEM

The function of a communication system is essentially to transmit a signal from the sender to the receiver. The model of a communication system, which will be useful in our discussion, is of the form shown in Fig. 14-4. The signal source generates the signal, which may be in a variety of forms. The signal will be assumed to be in the form of electrical signals in most communication systems even though optical signals are frequently used. The signal is subjected to some form of modulation or encoding depending upon the particular communication technique being used. Then, it is transmitted through the channel, which may be a cable or telephone lines or simply free space. At the receiving end, the signal is demodulated or detected, processed through some amplifier circuitry, and then is reproduced in a form suitable to the particular receiver. During transmission, random disturbances are introduced in the signal due to a number of causes. Such disturbances are called *noise*, and one of the important problems in communication is the recovery of the signal in the presence of noise.

We will be concentrating on the modulation and demodulation of signals in this chapter.

MODULATION OF A SIGNAL

The process of modulation requires the modification of a *carrier* waveform by means of the signal. For example, the carrier amplitude may be made to

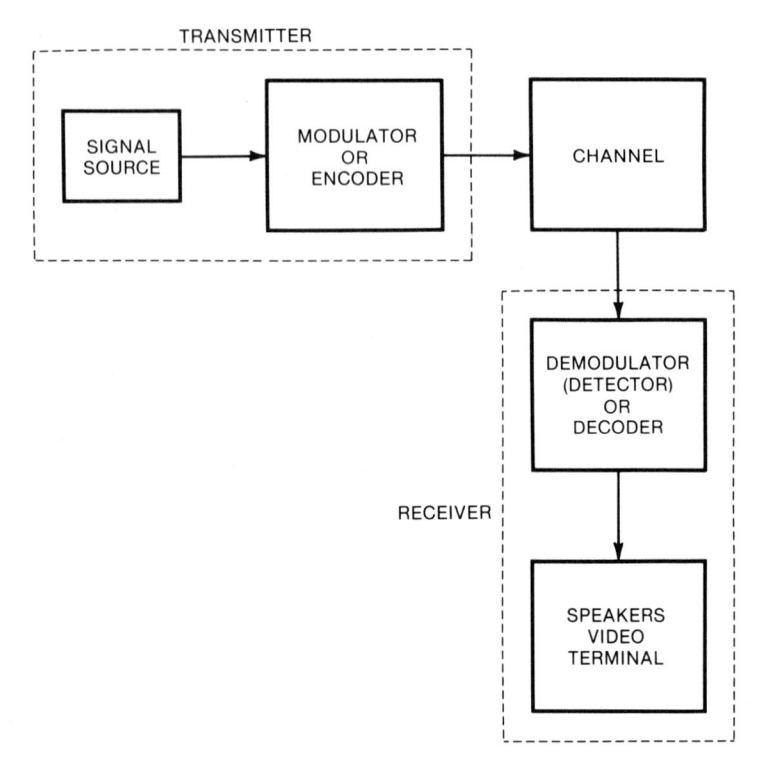

Fig. 14-4. Model of a communication system.

vary in proportion to the amplitude of the signal or the frequency of the carrier may be made to vary in proportion to the amplitude of the signal. There are several important reasons why modulation is necessary in communication. The antennas of a communication system should have dimensions that are of the same order of magnitude as the wavelength of the signals. If signals in the audio frequency range were used directly, then the sizes of the antennas will be prohibitively large. If f is the frequency of the signal, its wavelength is given by (c/f), where c is the velocity of electromagnetic waves (or light). For a signal with a frequency of 10 kHz, the wavelength is 30 kilometers and antennas of such dimensions are impractical. But if the audio signal were to modulate a carrier with a frequency of 1 MHz, then the wavelength will be 300 meters, which is much less than 30 km. It should be mentioned that there are communication systems (especially in the case of underwater communication) in which a signal is transmitted directly without a carrier of higher frequency. But in such cases, the antennas (occupying vast areas) are situated in isolated regions.

Another advantage of modulation is the ability to transmit a number of different signals simultaneously over the same channel, which is called frequency division multiplexing and is accomplished by using carriers of different frequencies for the different signals, which eliminates the overlapping of the frequency bands occupied by the different signals. The use of high-frequency carriers permits the multiplexing of a large number of signals over a specified bandwidth. As an example, the AM broadcast band in the US is from 550 kHz to 1600 kHz, which accommodates 100 stations spaced 10 kHz apart from the adjacent stations. The bandwidth needed for television is about 6 MHz and a much higher carrier frequency that those used in AM is needed for TV signals. The carrier frequencies used for TV are 80 MHz and higher. Increasing the carrier frequency to the ultrahigh-frequency range (470 MHz and higher), many more channels of TV stations can be multiplexed.

Different Types of Modulation

There are basically two types of analog (or continuous wave) modulation: amplitude modulation (AM) and angle modulation. These are further subdivided: in AM, we have single sideband (SSB), double sideband suppressed carrier (DSB-SC), and double sideband large carrier, which is simply called AM; in angle modulation, we have frequency modulation (FM) and phase modulation (PM). Each modulation scheme has some advantages and some disadvantages as will be seen later.

Any modulation scheme starts with a carrier, which is a sinusoidal waveform of much higher frequency that the signal. The carrier has a certain amplitude A, a frequency f_c and a phase angle θ. In amplitude modulation, the amplitude of the carrier waveform is made to vary in direct proportion to the amplitude of the signal. The modulated carrier will, therefore, show a variation in its amplitude but its frequency and phase angle will not be affected. In frequency modulation, the frequency of the carrier is made to vary in direct proportion to the amplitude of the signal but its amplitude is not affected. In phase modulation, the phase angle of the carrier is made to vary in direct proportion to the amplitude of the signal, but again the amplitude of the carrier is not affected. Fig. 14-5 shows a modulating signal, the waveform of a carrier amplitude modulated by the given signal, and the waveform of the carrier frequency modulated by the signal. Note that in the case of the FM signal, the zero crossings (that is, the points at which it crosses the horizontal axis) become more frequent as the modulating signal becomes stronger.

Time Domain and Frequency Domain

The waveforms of Fig. 14-5 are what we will see on the screen of an oscilloscope set up to observe the modulating signal or the AM or FM signal in the

(A) Signal.

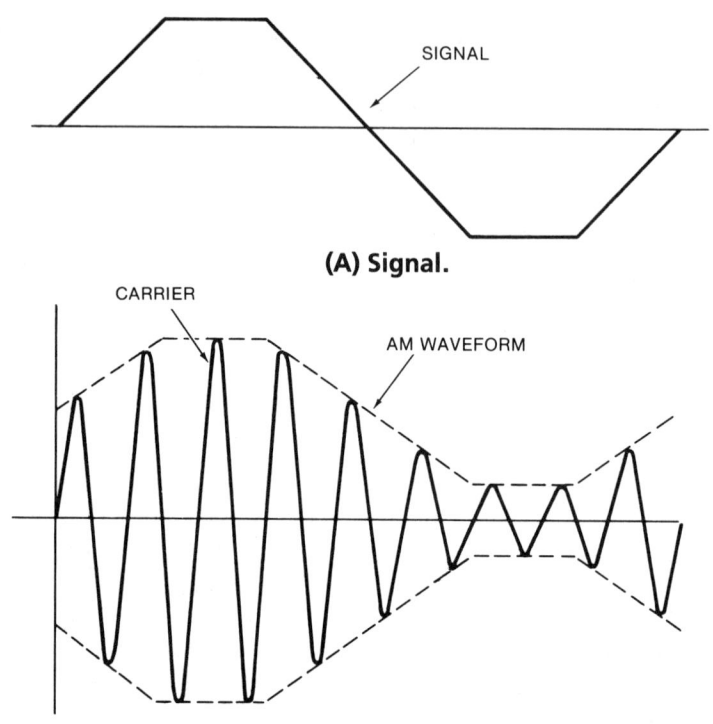

(B) In AM, the amplitude varies as a function of the signal, but the carrier frequency does not change.

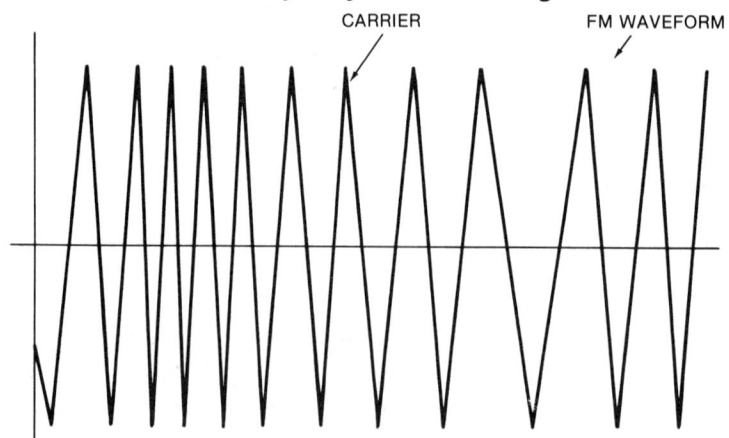

(C) In FM, the amplitude remains constant, but the frequency changes as a function of the strength of the signal.

Fig. 14-5. Waveforms of amplitude modulated (AM) and frequency modulated (FM) carriers.

Electronics: Circuits and Systems

time domain. As the modulating signal becomes complex and random (which is the way almost all interesting signals are in practice), the time domain waveforms appear like a meaningless jumble and are not directly useful to us.

A signal can be studied from the frequency domain viewpoint also. That is, we can examine what frequencies are present in a given signal by looking at its frequency spectrum. The frequency spectrum is more useful in the analysis and design of communication systems since it leads to the determination of the bandwidth needed for the transmission of a signal. As mentioned earlier, the frequency spectrum (or the power spectrum) of even highly complex and random signals can be estimated fairly accurately through measurements or computations.

What we have then is this: the *operations* such as modulation occur in the *time domain* but our interest lies in the *frequency domain*.

FREQUENCY TRANSLATION

One common operation used in modulation systems is the *multiplication* of two signals in the *time domain*. The result of *multiplication in the time domain* is found to be a *shift in frequencies* in the *frequency domain*.

Consider two sinusoidal signals of frequencies f_1 and f_2. Suppose the two signals are multiplied together as indicated in Fig. 14-6. Then the output of the multiplier will be (v_1v_2) in the time domain. The output signal can be shown (mathematically) to consist of *two frequencies* in the frequency domain: the *sum* of the two original frequencies ($f_1 + f_2$) and the *difference* between the two original frequencies ($f_1 - f_2$).* (In order to avoid negative frequencies, *subtract the smaller* of the two frequencies *from the larger* to obtain the difference frequency.)

The frequency spectra of the two input signals and the product signal are shown in Fig. 14-6.

As an example, consider two signals of frequencies 10 kHz and 25 kHz. When these two are multiplied, the product signal will consist of the two frequencies: $(10 + 25) = 35$ kHz and $(25 - 10) = 15$ kHz.

---

* The proof is based upon the (well-known) trigonometrical identity: $(A \cos X)(B \cos Y) = (AB/2)[\cos(X + Y) + \cos(X - Y)]$. The two signals v_1 and v_2 can be written as

$$v_1 = V_{m1} \cos(2\pi f_1 t)$$

$$v_2 = V_{m2} \cos(2\pi f_2 t)$$

where,
V_{m1} and V_{m2} are the peak values. Then, the previous trigonometrical identity leads to

$$v_1 v_2 = (V_{m1} V_{m2}/2)[\cos 2\pi(f_1 + f_2)t + \cos 2\pi(f_1 - f_2)t]$$

which shows the presence of *two* frequency components in the product signal: $(f_1 + f_2)$ and $(f_1 - f_2)$. Since $\cos(f_1 - f_2)t = \cos(f_2 - f_1)t$, it does not matter if we take the difference frequency as $(f_1 - f_2)$ or $(f_2 - f_1)$, and we choose whichever gives a positive frequency.

Analog Communication Techniques

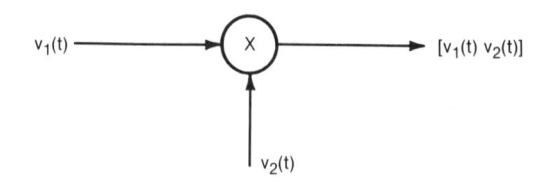

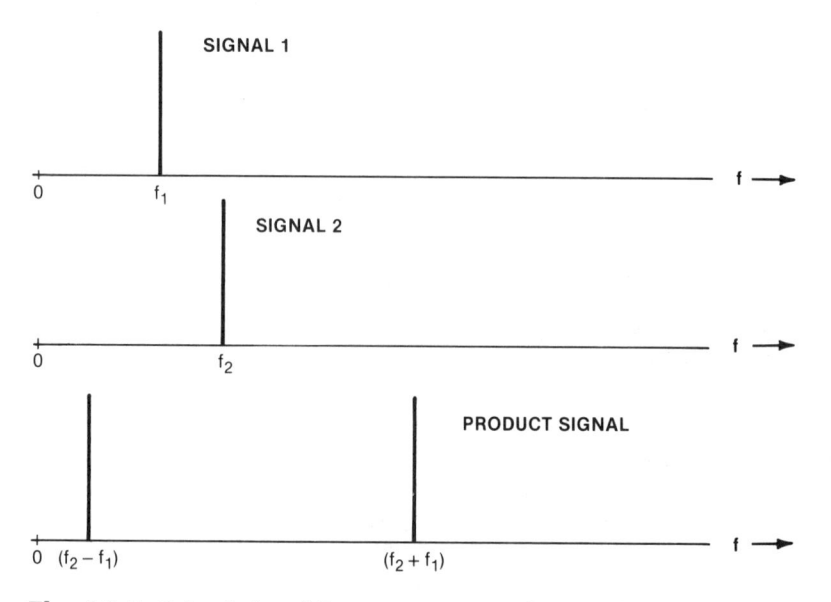

Fig. 14-6. Principle of frequency translation. The product signal contains the sum and difference frequencies.

The property that the product of two signals leads to a relocation of frequencies in the frequency domain forms the basis of *frequency translation:* a signal of a given frequency can be shifted up or down in the frequency axis by multiplying it by another signal of appropriate frequency.

Example 14.1

Given a signal v_1 of frequency 14 kHz, calculate the frequency of a signal v_2 to multiply v_1 in order to obtain a frequency component at (a) 15 kHz; (b) 50 kHz; and (c) 5 MHz.

Solution

Let the frequency of the signal v_2 be f_2 kHz. Then, the product signal $v_1 v_2$ will consist of the frequencies:

$(f_2 + 14)$ kHz and $(f_2 - 14)$ kHz

In each given case, we can equate the desired frequency to either the sum frequency $(f_2 + 14)$ or the difference frequency $(f_2 - 14)$. Therefore, there are two possible answers for each case.

(a) $(f_2 + 14) = 15$ gives $f_2 = 1$ kHz. In this case, the product signal will consist of 13 kHz and 15 kHz. $(f_2 - 14) = 15$ gives $f_2 = 29$ kHz and in this case, the product signal will consist of 14 kHz and 43 kHz.

(b) The two values of f_2 are 36 kHz and 64 kHz. In the former case, the product signal will consist of 22 kHz and 50 kHz; in the latter case, the product signal will consist of 50 kHz and 78 kHz.

(c) The two values of f_2 are 4986 kHz and 5014 kHz. In the former case, the product signal will consist of 4972 kHz and 5000 kHz; in the latter case, the product signal will consist of 5000 kHz and 5028 kHz.

Drill Problem 14.1: Given a signal v_1 of frequency 8 kHz to be multiplied by a signal v_2 of frequency f_2. Calculate the frequencies present in the product signal for each of the following values of f_2: (a) 4 kHz; (b) 40 kHz; and (c) 4 MHz.

Drill Problem 14.2: Given a signal v_1 of frequency 2.5 kHz. Calculate the frequency f_2 of a signal v_2 to multiply v_1 so as to produce the following frequency pairs: (a) 25 kHz and 30 kHz; (b) dc and 5 kHz; (c) 2 kHz and 7 kHz; and (d) 650 kHz and 655 kHz.

Drill Problem 14.3: Given a signal v_1 of frequency 3.3 kHz. It is multiplied by a signal v_2 of frequency f_2 so that the product signal contains a frequency component at (a) 30 kHz; (b) 550 kHz; and (c) 91.5 MHz. In each case, determine the two possible values of f_2, and for each choice of f_2 list the frequencies present in the product signal.

When one or both the signals being multiplied contain more than one frequency, each frequency component of *one* signal mixes with each frequency component of the *other* signal to produce the sum and difference frequency pair. As an example, let a signal v_1 consist of two frequencies f_{a1} and f_{b1} and a signal v_2 consist of two frequencies f_{a2} and f_{b2}. When these two signals are multiplied, the frequency spectrum of the product signal is obtained as follows:

Mix f_{a1} with f_{a2} to get $(f_{a1} + f_{a2})$ and $(f_{a1} - f_{a2})$

Mix f_{a1} with f_{b2} to get $(f_{a1} + f_{b2})$ and $(f_{a1} - f_{b2})$

Mix f_{b1} with f_{a2} to get $(f_{b1} + f_{a2})$ and $(f_{b1} - f_{a2})$

Mix f_{b1} with f_{b2} to get $(f_{b1} + f_{b2})$ and $(f_{b1} - f_{b2})$

(In all difference frequencies, always subtract the smaller from the larger.)

Example 14.2

A signal v_1 is made up of the two frequencies 50 Hz and 3 kHz. It is multiplied by a signal v_2 consisting of two frequencies 1 kHz and 9 kHz. Determine all the frequencies present in the product signal.

Solution

Mix 50 Hz (or 0.05 kHz) with 1 kHz. We get 1.05 kHz and 0.95 kHz.

Mix 0.05 kHz with 9 kHz and get 9.05 kHz and 8.95 kHz.

Mix 3 kHz with 1 kHz to get 2 kHz and 4 kHz.

Mix 3 kHz with 9 kHz to get 6 kHz and 12 kHz.

The product signal will consist of eight frequency components: 0.95, 1.05, 2, 4, 6, 8.95, 9.05 and 12 kHz (listed in ascending order).

Drill Problem 14.4: Determine all the frequencies present in the product of two signals v_1 and v_2 if v_1 consists of two frequencies 750 kHz and 6250 kHz and v_2 has the single frequency 3500 Hz.

Drill Problem 14.5: Determine all the frequencies present in the product of two signals v_1 and v_2 if v_1 consists of two frequencies 15 kHz and 25 kHz while v_2 consists of two frequencies 1.65 kHz and 455 kHz.

Interesting signals in practice are made up of a continuous band of frequencies rather than discrete frequencies like those considered previously. When a signal with a continuous frequency spectrum over a given bandwidth is multiplied with a signal of a single frequency (called the carrier frequency), the product signal consists of *two bands of frequencies* that can be obtained by an extension of the principles discussed earlier.

Example 14.3

The frequency spectrum of a particular signal is as shown in Fig. 14-7, and the signal is multiplied by a single frequency carrier signal of $f_c = 100$ kHz. Obtain the frequency spectrum of the product signal.

Solution

First consider the lower edge of the frequency spectrum of the given signal: $f_1 = 1$ kHz. Mix this with f_c to obtain the sum and difference frequencies: ($f_c + f_1$) and ($f_c - f_1$) or 101 kHz and 99 kHz.

Next, consider the upper edge of the frequency spectrum of the given sig-

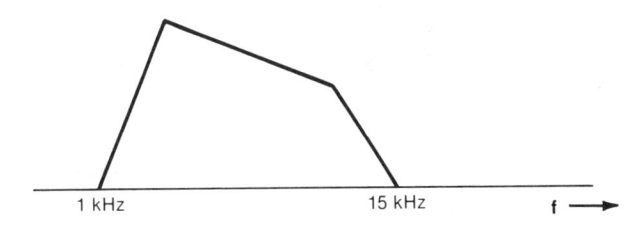

Fig. 14-7. Spectrum of the signal of Example 14.3.

nal: $f_2 = 15$ kHz. Mix it with f_c to obtain the sum and difference frequency pair: $(f_c + f_2)$ and $(f_c - f_2)$ or 115 kHz and 85 kHz.

Now, if we consider the frequency spectrum of the signal as a whole, it can be seen that the *band due to the sum frequencies must lie between the two extreme sum frequencies* calculated, and the *band due to the difference frequencies must lie between the two extreme difference frequencies* calculated.

Therefore, the frequency spectrum of the product signal will be made up of two bands: one from 101 kHz to 115 kHz (due to the sum frequencies) and the other from 85 kHz to 99 kHz (due to the difference frequencies). Note the correspondence in Fig. 14-8 between the edges of the frequency spectrum of the given signal and those of the two bands of the product signal. *The two bands of the product signal are mirror images of each other about the frequency f_c of the carrier.*

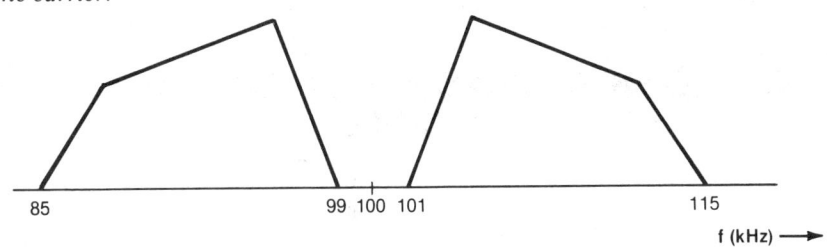

Fig. 14-8. Spectrum of the product signal of Example 14.3.

Drill Problem 14.6: Obtain the frequency spectrum of the product of a carrier of frequency 95 kHz and each of the signals whose frequency spectra are shown in Fig. 14-9.

The two bands of frequencies in the product signal are called *sidebands:* the one above f_c is the *upper sideband* and the one below f_c is the *lower sideband*. Note that each sideband completely replicates the signal by itself. Also, note that the frequency spectrum of the product signal does not contain the carrier frequency (the carrier has been "suppressed"). The process of multi-

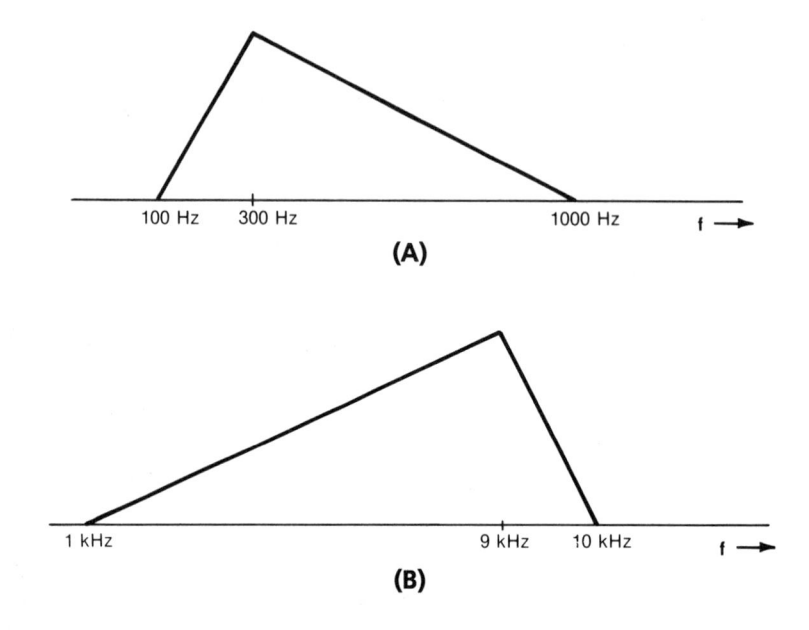

Fig. 14-9. Spectra of signals for Drill Problem 14.6.

plying a carrier frequency by a signal leads to the form of amplitude modulation called *double-sideband suppressed carrier* (DSB-SC).

DOUBLE-SIDEBAND SUPPRESSED-CARRIER AM (DSB-SC)

Consider the simple case of a signal of frequency f_m, called the *modulating signal* (Fig. 14-10A), and a carrier of frequency f_c (Fig. 14-10B). The carrier frequency is usually chosen to be much higher than the signal frequency in order to translate the signal's frequency spectrum to a high frequency.

If we were to observe the product signal in the *time domain* through an oscilloscope, the waveform will be as shown in Fig. 14-10C. The frequency spectrum of the product signal will be as shown in Fig. 14-10F as already discussed. The simple situation of a single-frequency signal modulating a carrier is known as *tone modulation*.

When the modulating signal has a frequency spectra as in Fig. 14-10G, the DSB-SC signal will have the frequency spectra as shown in Fig. 14-10H. There are two sidebands: the *lower sideband* extending from $(f_c - f_2)$ to $(f_c - f_1)$ and the *upper sideband* extending from $(f_c + f_1)$ to $(f_c + f_2)$. The two sidebands are mirror images of each other.

A DSB-SC system should have a bandwidth wide enough to transmit both

Electronics: Circuits and Systems

the sidebands fully. From Fig. 14-10H, we see that the bandwidth needed must extend from ($f_c - f_2$) to ($f_c + f_2$). Therefore, the bandwidth needed for DSB-SC is

$$(f_c + f_2) - (f_c - f_2) = 2f_2$$

That is, the *bandwidth of a DSB-SC system is twice the highest frequency contained in the modulating signal.*

The *demodulation*, that is, the recovery of the signal from a DSB-SC waveform, is accomplished by another multiplication step. Consider a DSB-SC signal with frequencies $f_c + f_m$ and $f_c - f_m$ multiplied by the carrier frequency f_c as shown in Fig. 14-11. The output of the multiplier will have the sum and difference frequencies of the two input signals; that is, the output signal will consist of two sum and difference pairs:

$$[f_c + (f_c - f_m)], [f_c - (f_c - f_m)]$$
$$[f_c + (f_c + f_m)], [(f_c + f_m) - f_c]$$

which gives the frequencies: $2f_c - f_m$, f_m, $2f_c + f_m$, and f_m. The frequency spectrum of the output of the multiplier is as shown in Fig. 14-11B.

Of the components in the multiplier output, we are interested in the f_m component since it represents the original modulating signal. It is seen from the frequency spectrum that the f_m component is at the low end of the frequency axis while the others are at the high end. Therefore, a low-pass filter (like the RC filters discussed in Chapter 2) can be used to recover the original signal. The demodulator for a DSCB-SC waveform is, therefore, a multiplier followed by an LPF as indicated in Fig. 14-11C. The cutoff frequency of the LPF must be at least the signal frequency f_m.

When the input DSB-SC waveform contains two sidebands rather than two discrete frequencies, the demodulation process is as shown in Fig. 14-12.

The demodulation of a DSB-SC waveform requires an oscillator at the receiving end that has exactly the same frequency (and phase) as the carrier at the modulating end of the system. Such a demodulation is called *synchronous*. When synchronism is not perfect, distortion occurs at the receiving end. In order to guarantee synchronism, a small pilot carrier signal is often transmitted along with the two sidebands in some communication systems.

Drill Problem 14.7: A DSB-SC input is made up of two frequencies 850 kHz and 950 kHz. The carrier frequency at the demodulator is 1000 kHz. Sketch the frequency spectra at the output of the multiplier in the demodulator. If the LPF has a cutoff frequency of 200 kHz, sketch the output of the LPF.

Drill Problem 14.8: The input to a DSB-SC demodulator has the frequency spectrum shown in Fig. 14-13 and the local carrier is at 2 MHz. The LPF has a cutoff frequency of 1 MHz. Sketch the frequency spectra at the output of the multiplier and the LPF.

(A) Modulating signal.

(B) Unmodulated carrier.

(C) DSB-SC waveform.

(D) Spectrum for modulating signal.

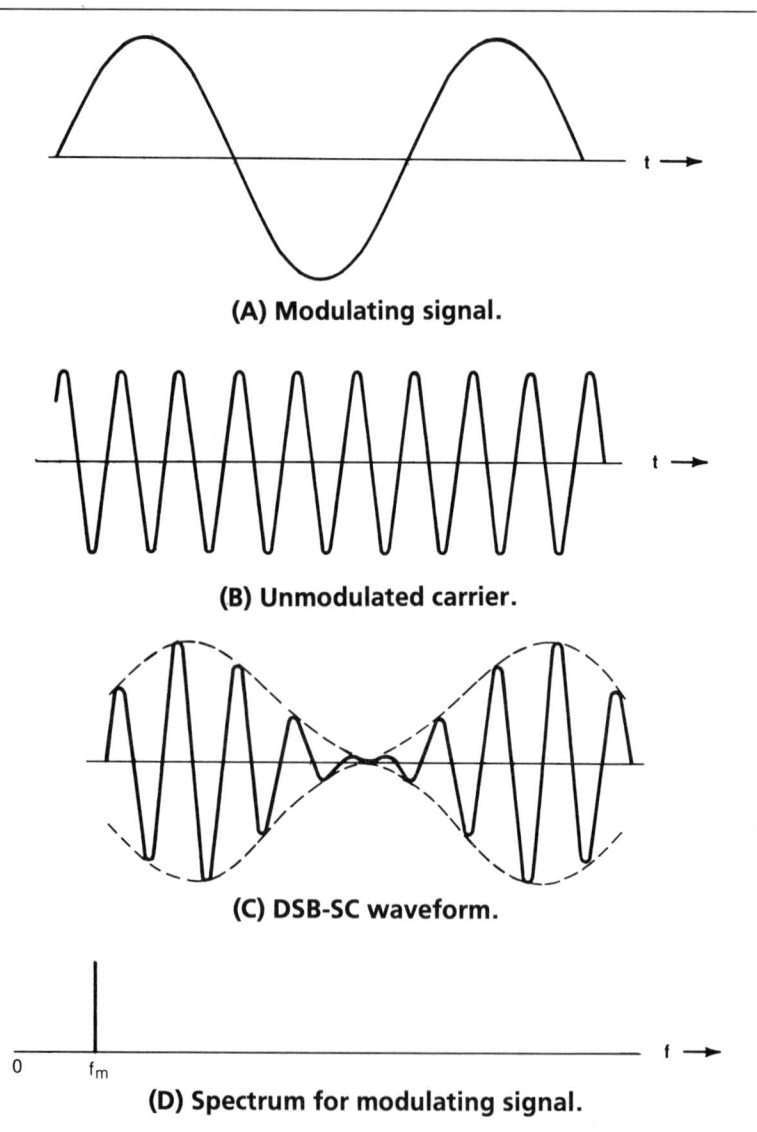

Fig. 14-10.

GENERATION OF PRODUCT SIGNALS

The multiplication of two signals in the time domain required in the modulation and demodulation of DSB-SC signals is usually done by a special circuit called the *balanced modulator*. It uses two identical nonlinear devices (such as transistors) in a push-pull configuration, and its output equals the product

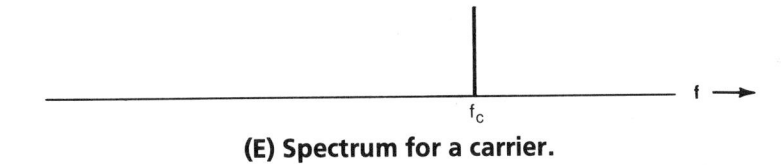

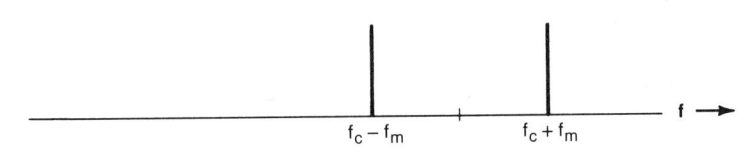

(E) Spectrum for a carrier.

(F) Spectra in the DSB system for a sinusoidal signal.

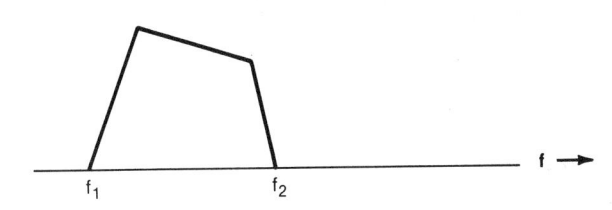

(G) Modulating signal spectra.

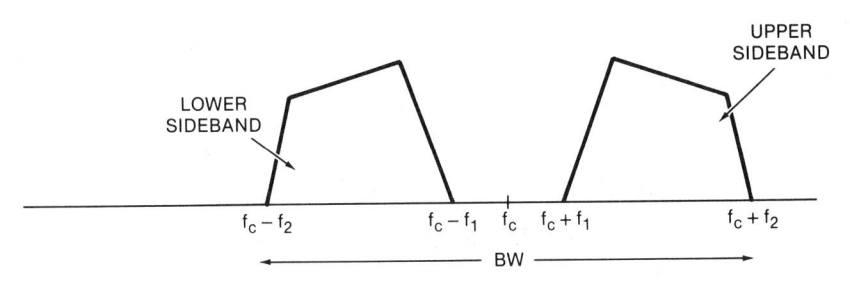

(H) Spectra in the DSB system for a complex signal.

DSB-SC AM.

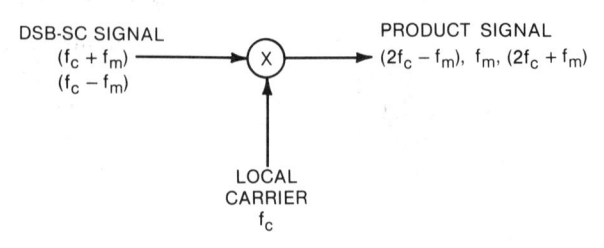

(A) Multiplier.

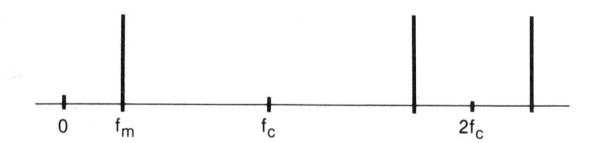

DSB-SC SIGNAL $(f_c + f_m)$, $(f_c - f_m)$ → X → PRODUCT SIGNAL $(2f_c - f_m)$, f_m, $(2f_c + f_m)$

LOCAL CARRIER f_c

(B) Frequency spectrum of the output.

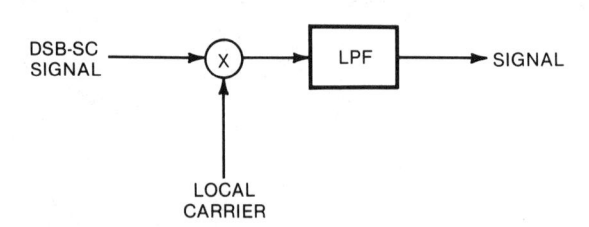

0, f_m, f_c, $2f_c$

(C) Multiplier followed by a low-pass filter.

DSB-SC SIGNAL → X → LPF → SIGNAL

LOCAL CARRIER

Fig. 14-11. Demodulation of a DSB-SC signal.

of its two input signals. The block diagram representation of a balanced modulator is shown in Fig. 14-14. Note that the output signal of the balanced modulator will contain the sum and difference frequencies of the two inputs.

SINGLE-SIDEBAND AM (SSB)

The bandwidth of a double-sideband system was seen to be twice the highest frequency contained in the modulating signal. If we consider the frequency spectrum of a DSB-SC signal, there are two sidebands each containing the same information as the other. It is clear that one of the sidebands should be sufficient to provide all the information that was contained in the

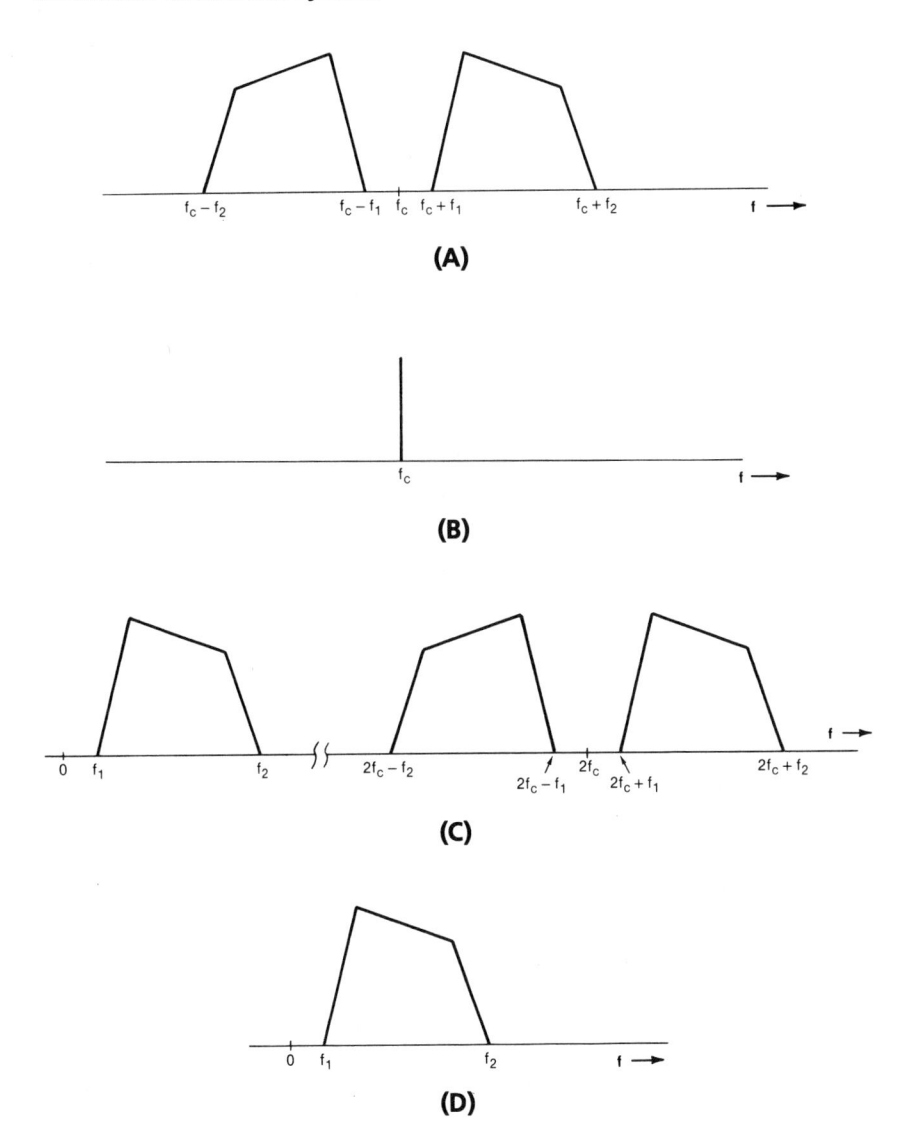

Fig. 14-12. Spectra in the demodulation system of a DSB-SC receiver.

original modulating signal, and the other sideband is not really necessary. If we transmit only one sideband, rather than both, then the bandwidth needed for the transmission can be reduced by a factor of two, which is the basis of single-sideband (SSB) amplitude modulation.

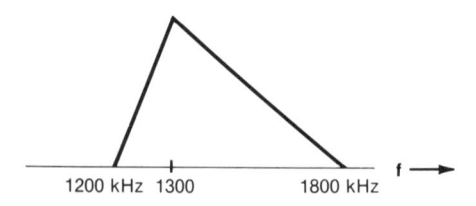

Fig. 14-13. Spectrum of the signal for Drill Problem 14.8.

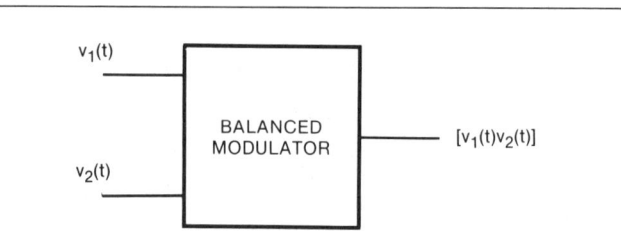

Fig. 14-14. Balanced modulator (block representation).

Generation of SSB-AM

There are two methods of generating SSB-AM. In one method, a filtering scheme is used in conjunction with a DSB-SC system, and the other involves introducing 90° phase shifts so as to suppress one of the sidebands.

Filtering of DSB-SC to Produce SSB

The generation of SSB-AM can be viewed merely as a matter of filtering a DSB-SC waveform so as to retain one of the two sidebands. A bandpass filter can be used, as indicated in Fig. 14-15A, which passes one of the sidebands and suppresses the other. If we wish to transmit the upper sideband, then the bandpass filter must have a *constant gain* in the range of frequencies ($f_c + f_1$) to ($f_c + f_2$), and its gain must decrease substantially when we get to the frequency ($f_c - f_1$), the upper edge of the sideband to be eliminated. For example, let $f_c = 10$ MHz, $f_1 = 100$ Hz, and $f_2 = 3$ kHz. Then, the bandpass filter must have a constant gain from 10000.100 kHz to 10003 kHz and must be down to a small value at 9999.9 kHz. These kinds of specifications are too stringent on a bandpass filter. It is usually necessary to use a series of bandpass filters and mixers in order to obtain the kind of filter response needed for an SSB system, which is the disadvantage of the system of generating SSB: the equipment needed becomes highly complex and increases the cost of the system.

Electronics: Circuits and Systems

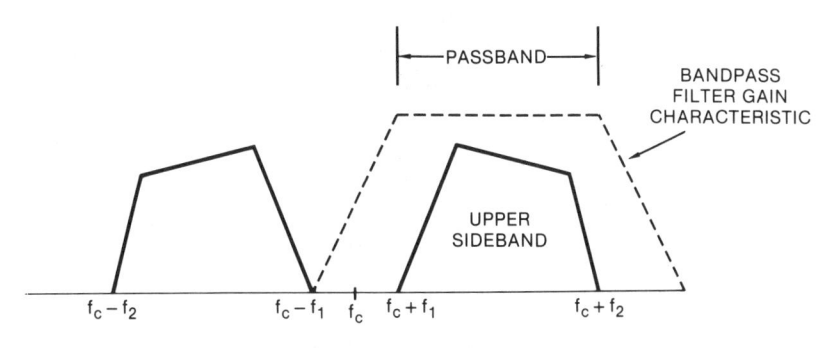

(A) Filtering method.

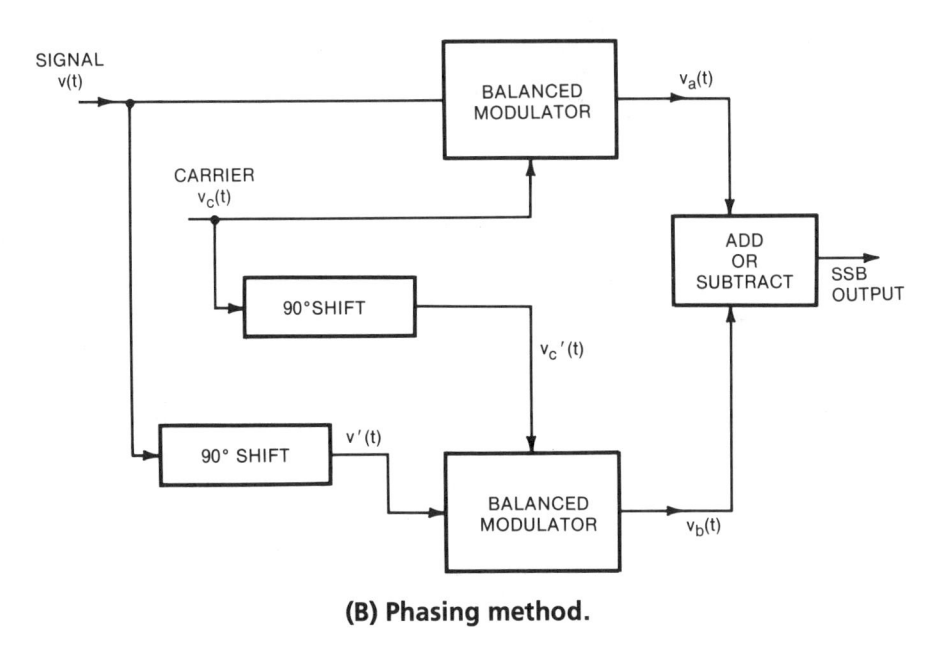

(B) Phasing method.

Fig. 14-15. Generation of SSB signals.

Phasing Method for Generation of SSB

The *phasing method* for generating the SSB signal utilizes two product signals: one is the product of the carrier and the modulating signal, and the other is the product of the carrier shifted in phase by 90° and the modulating signal shifted in phase by 90°. The scheme is shown in block diagram form in

Fig. 14-15B. When the two product signals are added (or subtracted) the resulting signal contains only one of the sidebands.

Consider a modulating signal

$$v(t) = V_m \cos(2\pi f_1 t)$$

and a carrier

$$v_c(t) = V_c \cos(2\pi f_c t)$$

Then, in Fig. 14-15B,

$$v'(t) = V_m \sin(2\pi f_1 t)$$

and

$$v'_c(t) = V_c \sin(2\pi f_c t)$$

since a 90° phase shift on a cosine function produces a sine function. The outputs of the two balanced modulators will, therefore, be

$$v_a(t) = V_m V_c \cos(2\pi f_1 t) \cos(2\pi f_c t)$$

and

$$v_b(t) = V_m V_c \sin(2\pi f_1 t) \sin(2\pi f_{c_1} t)$$

By using appropriate trigonometrical identities, it can be verified that*

$$[v_a(t) + v_b(t)] = V_m V_c \cos 2\pi (f_c - f_1)t$$

and

$$[v_a(t) - v_b(t)] = V_m V_c \cos 2\pi (f_c + f_1)t$$

Thus, the output of the system contains either the upper or the lower sideband. The phasing method has the advantage that it does not have the stringent filtering requirement of the previous method. But it too suffers from several disadvantages. One of the difficulties, for example, is that all frequency components in the modulating signal must be shifted in phase by precisely 90° without, however, altering their amplitudes unequally. Because of this and other practical difficulties, the phasing method is not as popular as the filtering method for generating SSB. In general, SSB communication is used when conservation of bandwidth is such a prime consideration that cost and complexity of equipment and even loss of quality are not of critical concern.

Demodulation of SSB

The demodulation of SSB follows exactly the same procedure as DSB-SC, as indicated in Fig. 14-16. The local carrier has the same frequency as the carrier used at the modulating end of the system.

---
* $\cos(A + B) = [\cos A \cos B - \sin A \sin B] \cos(A - B) = [\cos A \cos B + \sin A \sin B]$

Electronics: Circuits and Systems

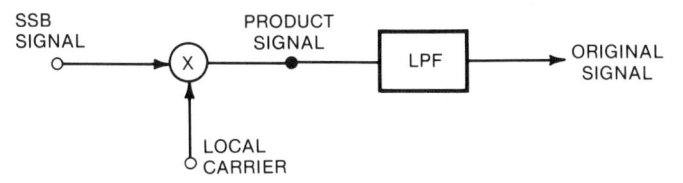

Fig. 14-16. Demodulation of SSB signals.

Example 14.4

The frequency spectrum of an SSB signal is shown in Fig. 14-17A and the local carrier at the demodulator has a frequency of $f_c = 6.8$ MHz. The frequency spectra at the multiplier output and the LPF output are shown in Figs. 14-17C and D. The LPF output is the original modulating signal.

The demodulation of an SSB signal requires a local carrier that must be identical in frequency and phase to the carrier at the transmitting station. If there is a drift in frequency or phase, a distortion is introduced at the receiver, and the distortion may be sufficiently serious to actually cancel the signal. A pilot carrier (which is a highly attenuated sample of the original carrier) is, therefore, transmitted along with the single sideband, and the pilot carrier is used to synchronize the local carrier (at the demodulator) with the carrier at the transmitter.

The following examples show the use of SSB in various applications of communication.

Example 14.5

The system of Fig. 14-18 permits the transmission of two different messages on a single carrier. The basic idea is to use the lower sideband SSB on one and the upper sideband SSB on the other. The diagrams explain the operation of the system.

Example 14.6

The system of Fig. 14-19 produces a garbled version of a message so as to introduce secrecy in transmission. It is seen that the spectrum of the output signal is the *reverse* of the spectrum of the input signal and is consequently "garbled."

The restoration of the original signal in the previous system is accomplished through the use of the system in Fig. 14-20.

Drill Problem 14.9: Sketch the frequency spectra at points P, Q, R, and S of the system in Fig. 14-20 and verify the fact that the final output is the same as the original signal used in the system of Fig. 14-19.

(A) Spectrum of the incoming SSB signal.

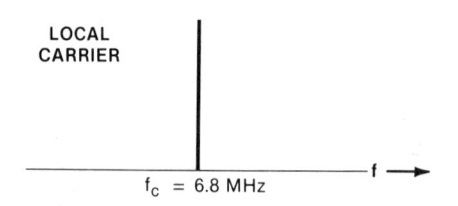

6.8001 MHz — 6.8050 MHz

(B) Local carrier.

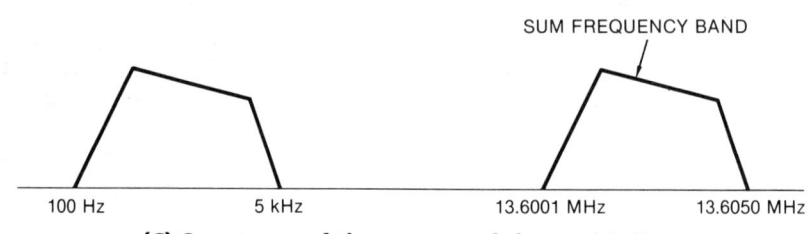

$f_c = 6.8$ MHz

(C) Spectrum of the output of the multiplier.

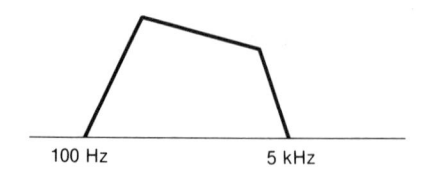

100 Hz — 5 kHz, SUM FREQUENCY BAND 13.6001 MHz — 13.6050 MHz

(D) Spectrum of the LPF output (original signal).

100 Hz — 5 kHz

Fig. 14-17. Spectra in an SSB demodulator.

VESTIGIAL-SIDEBAND AM (VSB-AM)

It was seen that the bandpass filter needed in SSB systems is subject to rather stringent specifications and tends to become complex and expensive. On the other hand, transmission of both sidebands increases the bandwidth needed in a communication system. As a compromise, one complete sideband

Electronics: Circuits and Systems

and a portion (or vestiges) of the other sideband can be transmitted in such a way that the bandwidth needed is only slightly larger than SSB, and the specifications of the bandpass filter are not as stringent as in SSB systems. The vestigial-sideband AM system or VSB-AM is used when the bandwidth requirements of double sidebands become too large as, for example, in TV systems.

In VSB systems, the filtering has to satisfy an important condition. Even though we stated that one sideband is fully transmitted along with a portion of the other, this is not actually the case. If one sideband is transmitted completely and a portion of the other sideband is also sent, then at the receiving end a portion of the signal will get reinforced (due to the extra contribution from the portion of the other sideband) while the rest of the signal will be at the proper level, which leads to an undesirable distortion. This distortion can be seen from the frequency spectra shown in Fig. 14-21. The two sidebands are as shown in Fig. 14-21A and if we transmit the upper sideband fully and a portion of the lower sideband as indicated in Fig. 14-21B, then at the receiver, the portion shown hatched in Fig. 14-21B will make an additional contribution to the low-frequency end of the signal (over and above that due to the upper sideband). This will boost the low-frequency end of the signal and cause distortion. In order to prevent such a distortion, VSB filters have to be designed to have specific characteristics, as indicated in Fig. 14-22. Even with such filtering, phase distortion cannot be avoided, and VSB is not used in systems where phase distortion is undesirable.

CONVENTIONAL AM (DOUBLE-SIDEBAND—LARGE CARRIER)

The major drawback in the type of AM systems discussed up to now (DSB-SC, SSB, and VSB) is that a local carrier is needed at the demodulator, which should be in perfect synchronism with the carrier at the transmitting end. Conventional AM on the other hand does not need a local carrier for demodulation, and in fact, the demodulating circuit is extremely simple. In the conventional AM systems (which are simply called AM without any qualifying adjective) a large carrier is also transmitted with the two sidebands.

Consider a modulating signal v_m as in Fig. 14-23A and a carrier of much higher frequency than the signal frequency as in Fig. 14-23B. In conventional AM, the amplitude of the carrier is made to vary in direct proportion with the amplitude of the modulating signal so that the modulated carrier (or the AM signal) has the waveform shown in Fig. 14-23C. It is seen that the envelope of the AM waveform is an exact replica of the modulating signal. A modulation index m is defined by

$$m = \frac{V_{max} - V_{min}}{V_{max} + V_{min}}$$

where,

V_{max} and V_{min} are the maximum and minimum values of the *envelope*.

The modulation index can be greater than, equal to, or less than one. The AM waveforms due to a sinusoidal signal for the three cases $m < 1$, $m = 1$, and $m > 1$ are shown in Fig. 14-24. It is seen that whereas the envelope is an exact replica of the signal when $m < 1$ and $m = 1$, it is not so when $m > 1$. Overmodulation must be avoided if the recovery of the signal depends upon

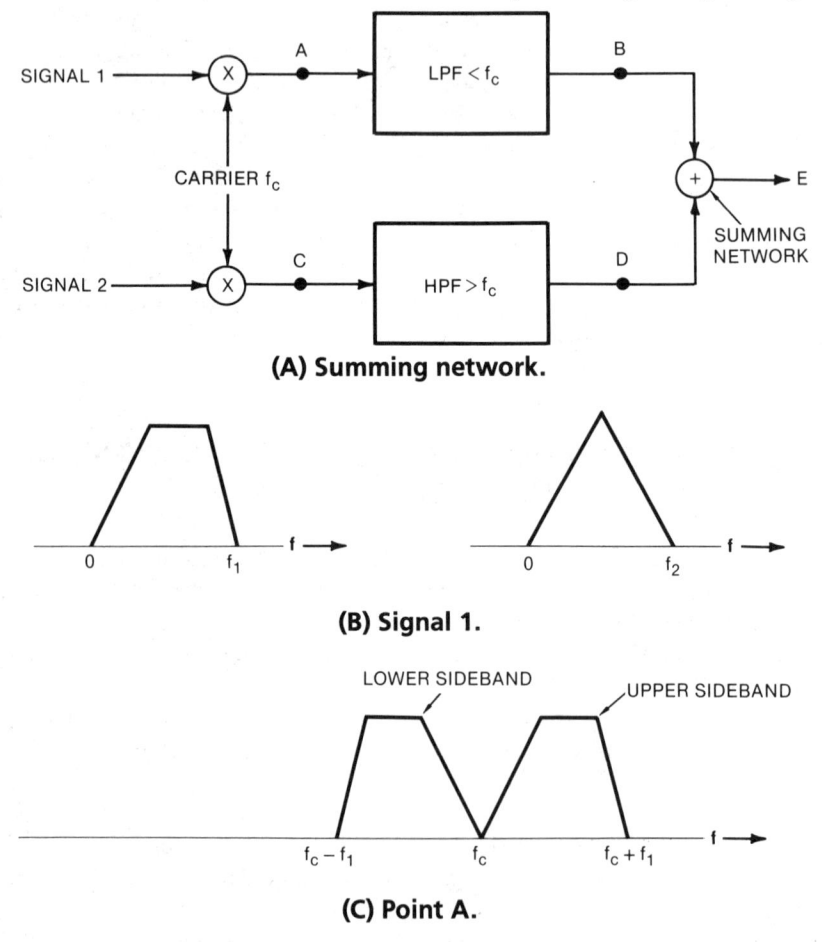

(A) Summing network.

(B) Signal 1.

(C) Point A.

Fig. 14-18. Transmission of two

using the envelope of the AM waveform, which is usually the case. Usually m is made somewhat smaller than one so as to ensure the prevention of overmodulation.

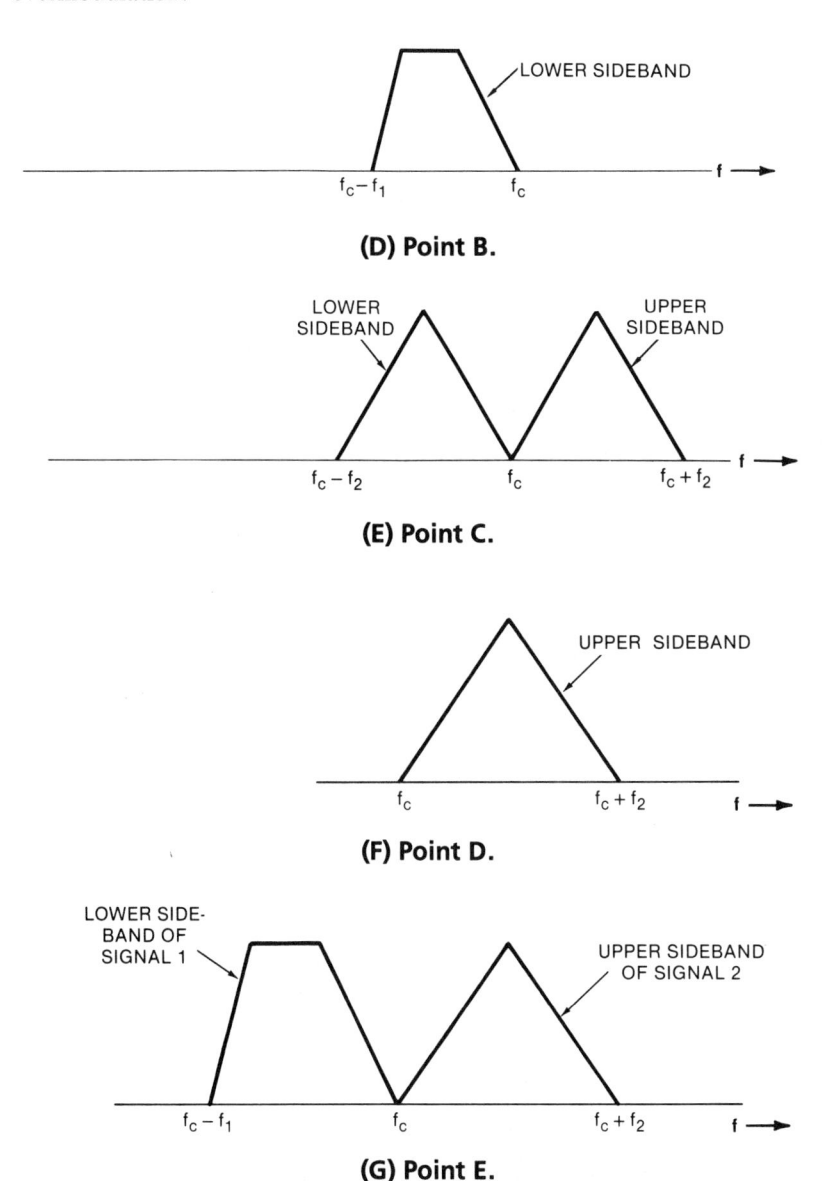

(D) Point B.

(E) Point C.

(F) Point D.

(G) Point E.

signals on the same carrier.

828   Analog Communication Techniques

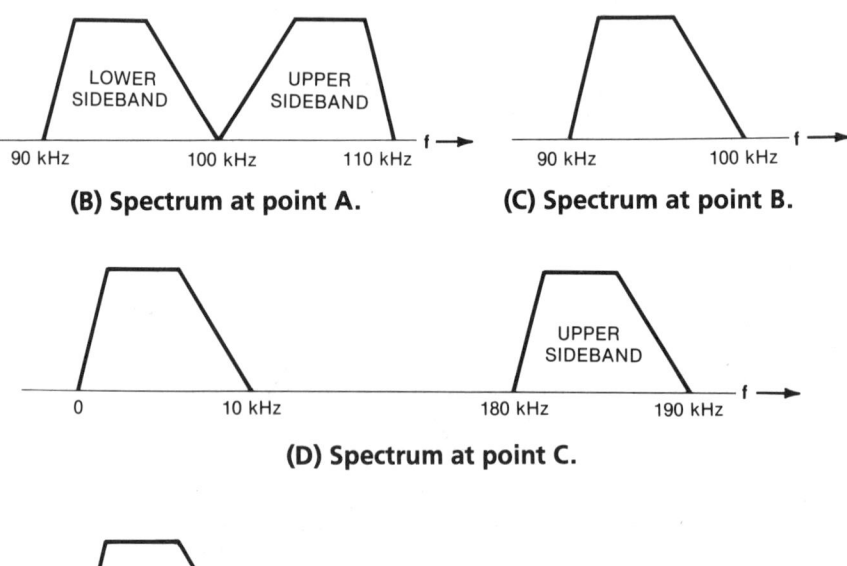

Fig. 14-19. Message garbling system.

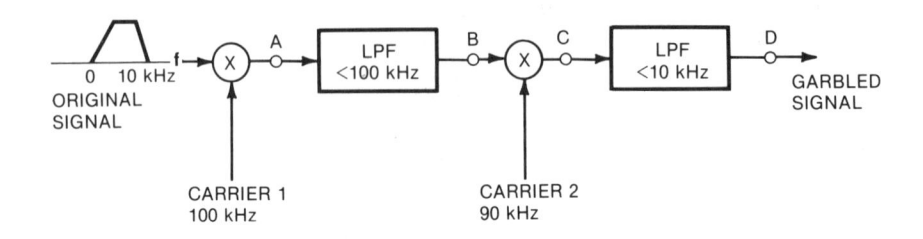

(A) Original signal to the system.
(B) Spectrum at point A.
(C) Spectrum at point B.
(D) Spectrum at point C.
(E) Spectrum at point D is the reverse of input.

Spectrum of an AM Signal

The only difference between AM and DSB-SC is the presence of the carrier frequency component in AM. The frequency spectrum of an AM signal will,

Electronics: Circuits and Systems

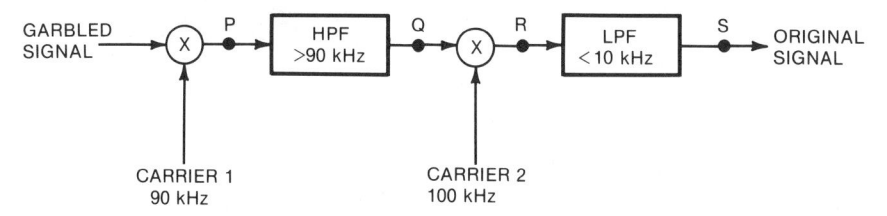

Fig. 14-20. Recovery of the original signal from the garbled signal of the system in Fig. 14-19.

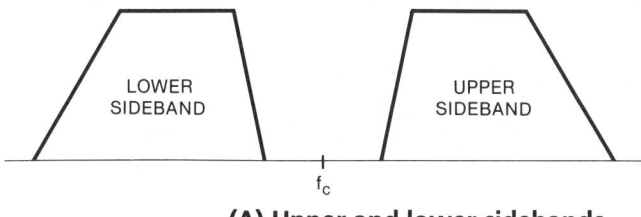

(A) Upper and lower sidebands.

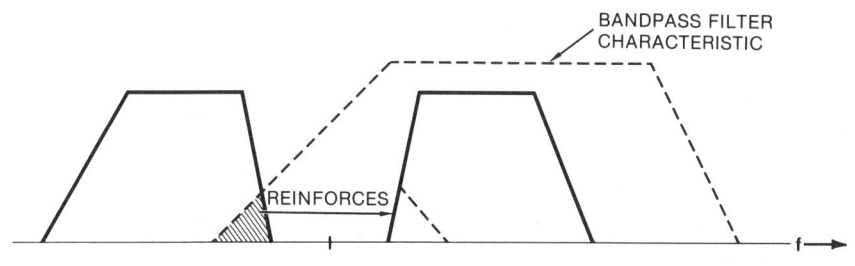

(B) Upper sideband transmitted fully with a portion of the lower sideband.

Fig. 14-21. VSB-AM concept. A reinforcement of the low-frequency end occurs due to the filtering shown.

therefore, consist of two side frequencies ($f_c + f_m$) and ($f_c - f_m$) as well as the carrier frequency itself. In the case of a signal with a continuous spectrum, the frequency spectrum of the AM signal will consist of the two sidebands and the carrier frequency. These two situations are shown in Fig. 14-25.

Power Considerations in AM

Up to this point, we did not pay attention to the relative amplitudes of the side-frequency components in an AM signal, but they are important when we consider the power distribution in an AM signal. Consider the case of a signal of a single frequency f_1 modulating a carrier of amplitude A_c. If the modula-

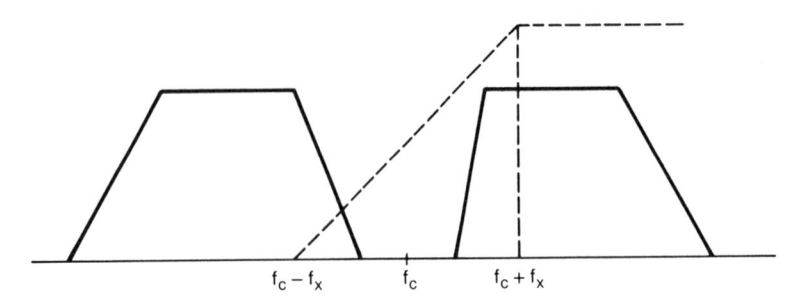

Fig. 14-22. Special filter characteristic needed in VSB-AM.

tion index is m, then the amplitudes of the three spectral components can be shown to be: carrier amplitude = A_c and the amplitude of each side-frequency component = $(mA_c/2)$, as indicated in Fig. 14-26.[*] Since *normalized power* = $[0.5 \times (\text{peak value})^2]$ from Eq. (14-1), we have

power in the carrier component = $0.5A_c^2$

power in *each* side-frequency component = $0.5[mA_c/2]^2$

$$= 0.125m^2A_c^2$$

total power in the side-frequency components = $0.25m^2A_c^2$ (14-3)

and

total power in the AM signal = $(0.5A_c^2)(1 + 0.5m^2)$ (14-4)

For example, if $A_c = 10$ V and $m = 0.8$,
power in the carrier = 50 W
power in each side-frequency component = 8 W
power in the AM signal = 66 W

Comparing the power in the side-frequency components (which represents the "useful" power) with the total power in the AM signal, we have from Eqs. (14-3) and (14-4),

---

[*] An AM signal can be written in the form

$$A_c(1 + m \cos 2\pi f_1 t) \cos 2\pi f_c t$$

where,
A_c is the amplitude of the unmodulated carrier,
m is the modulation index,
f_c is the carrier frequency,
f_1 is the frequency of the modulating signal.
The above expression leads to

$$A_c \cos 2\pi f_c t + (mA_c/2)[\cos 2\pi(f_c + f_1)t + \cos 2\pi(f_c - f_1)t]$$

so that the relative amplitudes are A_c for the carrier and $(mA_c/2)$ for each of the side-frequency components.

Electronics: Circuits and Systems 831

$$\left(\frac{\text{power in side-frequency components}}{\text{total power}}\right) = \frac{0.25m^2 A_c^2}{0.5 A_c^2 (1 + 0.5m^2)} \quad (14\text{-}5)$$

$$= \frac{m^2}{2 + m^2}$$

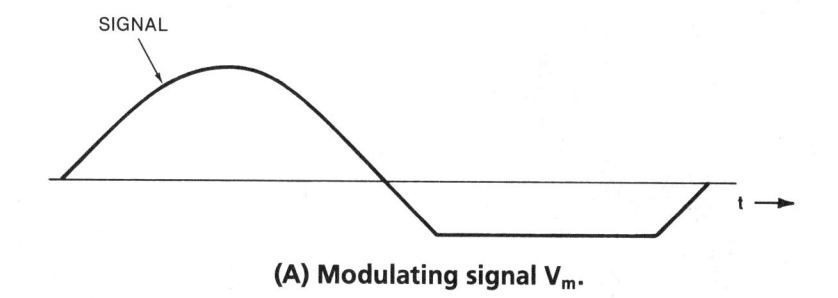

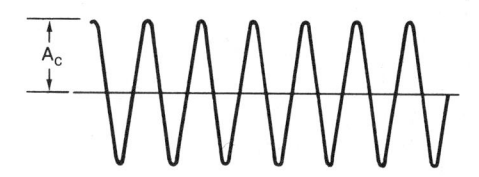

(A) Modulating signal V_m.

(B) Carrier at much higher frequency.

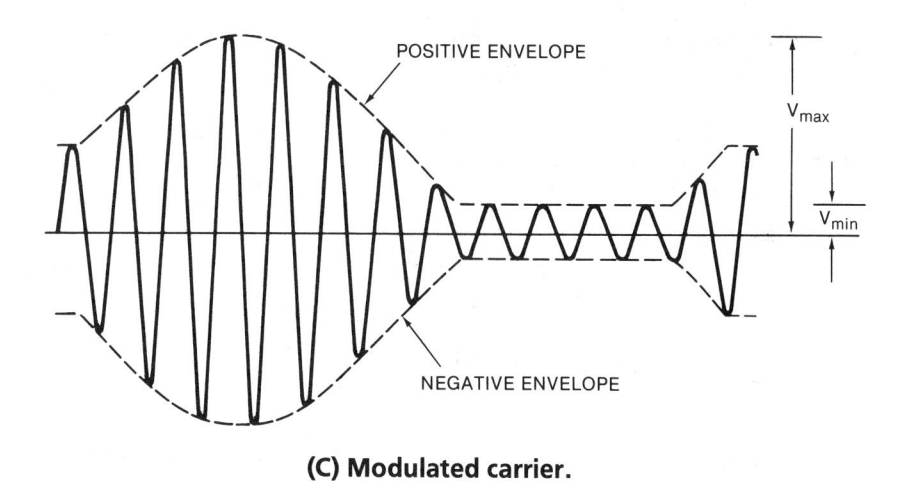

(C) Modulated carrier.

Fig. 14-23. Conventional AM (DSB-large carrier).

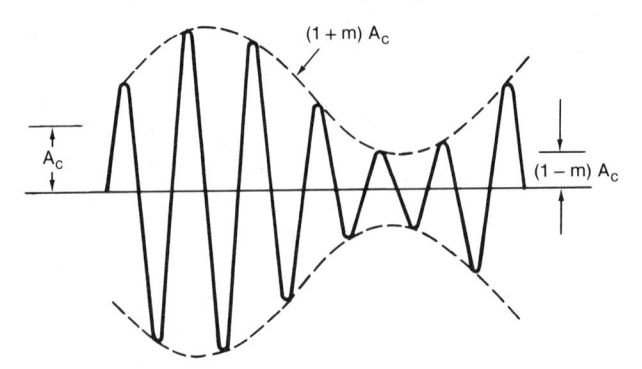

(A) When m is less than one.

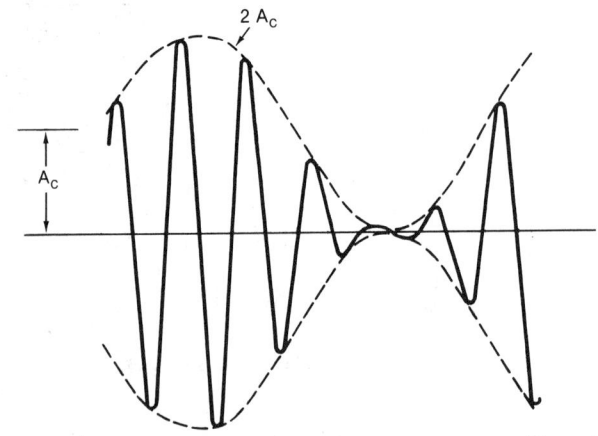

(B) Or equal to one the envelope is an exact replica of the modulating signal.

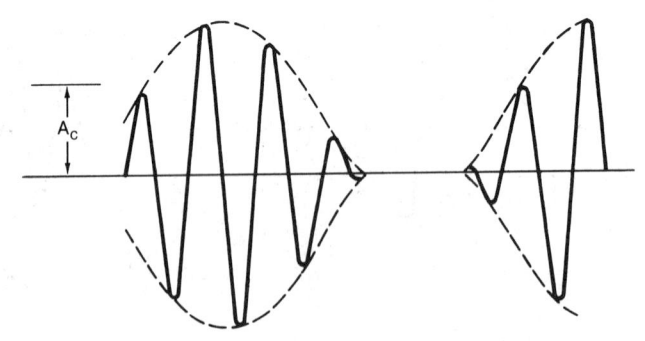

(C) But when m is greater than one, overmodulation occurs, and the envelope is not a replica of the modulating signal.

Fig. 14-24. Effect of the modulation index on the AM waveform.

Electronics: Circuits and Systems

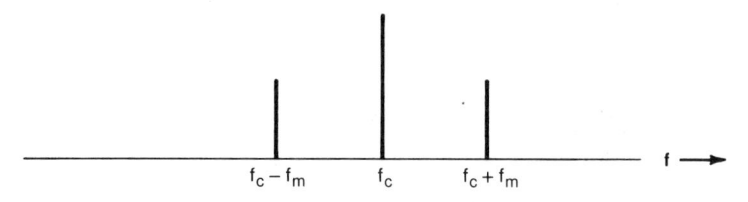

(A) Spectrum of an AM signal.

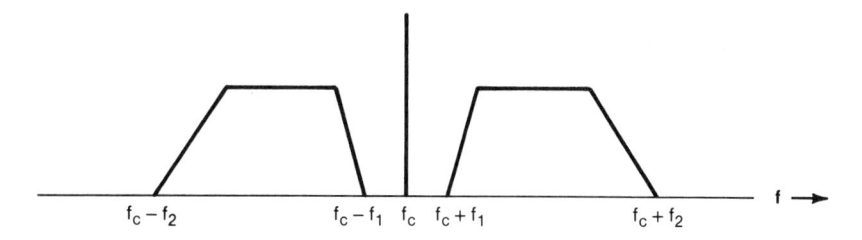

$f_c - f_m$ f_c $f_c + f_m$

(B) Frequency spectrum of a signal with a continuous spectrum.

$f_c - f_2$ $f_c - f_1$ f_c $f_c + f_1$ $f_c + f_2$

Fig. 14-25. Spectra in an AM system.

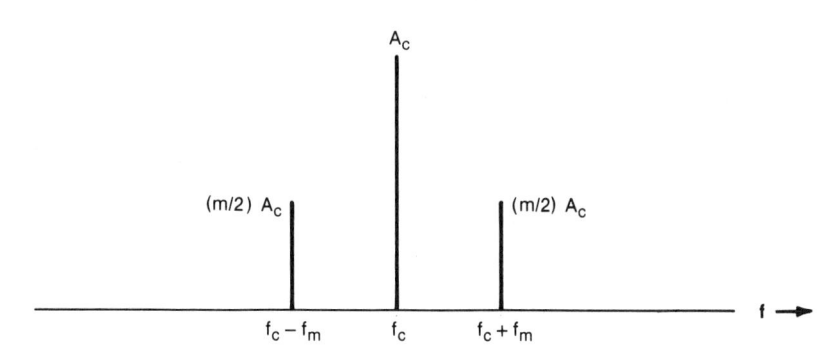

A_c

$(m/2) A_c$ $(m/2) A_c$

$f_c - f_m$ f_c $f_c + f_m$

Fig. 14-26. Relative amplitudes in an AM signal.

The maximum value of this ratio is one third when m = 1. That is, at best, only one third of the total transmitted power is useful in carrying the signal itself. The remainder is used to transmit the large carrier.

Example 14.7

An AM radio station has a transmission power of 50 kW. (This is the power allowed for so-called clear-channel stations that can be heard over a large distance. Most AM radio stations do not operate at such a high power.) The modulation index is 95%. Calculate the power components and the ratio of the power in the side frequencies to the total power transmitted.

Solution

From Eq. (14-5), we have

$$\frac{\text{power in the side-frequency components}}{\text{total power in AM signal}} = \frac{0.95^2}{2 + 0.95^2}$$

$$= 0.31$$

Therefore,

$$\text{power in side-frequency components} = 0.31 \times 50 \text{ kW}$$

$$= 15.5 \text{ kW}$$

$$\text{power in the carrier} = (50 - 15.5) \text{ kW}$$

$$= 34.5 \text{ kW}$$

It can be seen that the efficiency of an AM system (in terms of power) is rather poor. In the DSB-SC, SSB, and VSB systems, all the transmitted power is used to carry the signal itself. But, as mentioned earlier, the advantage of the AM system is in the ease of demodulation that will be discussed shortly.

The generation of an AM signal usually involves placing the modulating signal in series with the power supply battery of a transistor (or vacuum tube), and applying the carrier frequency signal at the input to the device. If the carrier frequency is much greater than the signal frequency, then the variation in the signal has the effect of shifting the bias slowly about the Q point so that the output current and voltage are modulated by the modulating signal. Another approach to the generation of AM is to first obtain DSB-SC using a square-law device and then adding a sufficiently large carrier component so as to have an AM signal.

Demodulation of an AM Signal

The fact that the envelope of an AM signal is a replica of the signal to be recovered leads to a simple method of demodulating it. Consider the simple diode circuit shown in Fig. 14-27A. If an AM signal is applied as its input, then the output will consist of the positive half cycles of the AM signal as shown in Fig. 14-27B. The output is a series of half-sine pulses at the carrier frequency, with the envelope of the pulse train representing the signal to be recovered. The addition of a capacitor to the circuit, as shown in Fig. 14-27C, has the effect of smoothing out the output so that the oscillations of the carrier component are essentially eliminated. The output of the circuit will now be of the form shown in Fig. 14-27D. The time constant $R_L C_L$ has to be chosen carefully. It should be large enough so that the circuit does not respond to the high frequency variations of the carrier, but it must be small enough so that it does respond to the variations of the envelope itself. Therefore, the condition to be satisfied is

$$(1/f_c) \ll (R_L C_L) \ll (1/f_m)$$

This is one of the reasons for making the carrier frequency much larger than the signal frequency.

SUPERHETERODYNE AM RECEIVERS

The commercial AM broadcast band in the U.S. ranges from 550 kHz to 1600 kHz. Each radio station is permitted a bandwidth of 10 kHz, which means that each sideband is limited to 5 kHz. Therefore, the modulating signal is also limited to 5 kHz, which accounts for the poor quality of music in an AM radio. The block diagram of an AM receiver, the so-called superheterodyne receiver, is shown in Fig. 14-28.

One of the important design constraints in the reception of AM (or any other radio) signals is that a bandpass filter is needed with the following specifications: its center frequency equals the carrier frequency of the station being tuned to; its bandwidth must be equal to the bandwidth allotted to each station; over the passband, the gain must be as flat as possible; and outside the passband, the gain must quickly reduce to zero. Clearly, it will be difficult to design a single bandpass filter that can have a variable center frequency and still satisfy all the previous constraints. In order to overcome this difficulty, an *intermediate frequency* (or *IF*) is introduced in the receiver. The signal from any station is first shifted to the IF range where it is properly amplified and then passed on to the demodulator (or detector) circuit.

For AM radio, the IF range is from 450 to 460 kHz. The signal from a desired radio station is obtained by tuning the RF amplifier stage to the carrier frequency f_c of the station. The tuning action also sets the frequency of a local oscillator to $(f_c + 455 \text{ kHz})$. The circuit in the mixer generates the product of the signal from the RF amplifier, and the new carrier at $f_c + 455$ kHz and the band from 450 to 460 kHz is filtered and transmitted to the IF amplifier. The diode detector demodulates the signal, which is then amplified and passed on to the speakers.

As an example, suppose the AM radio station broadcasts at 1180 kHz. The RF amplifier is tuned to 1180 kHz and passes the band 1175 to 1185 kHz to the mixer. At the same time, the local oscillator generates a local carrier of frequency 1635 kHz. The band 1175 to 1185 kHz is mixed with the carrier at 1635 kHz, thus generating two bands: the sum frequency band (1175 + 1635) to (1185 + 1635) kHz, and the difference frequency band (1635 − 1175) to (1635 − 1185). Filters inside the mixer suppress the sum frequency band and transmit the difference frequency band, 450 to 460 kHz to the IF amplifier.

(A) Simple diode circuit.

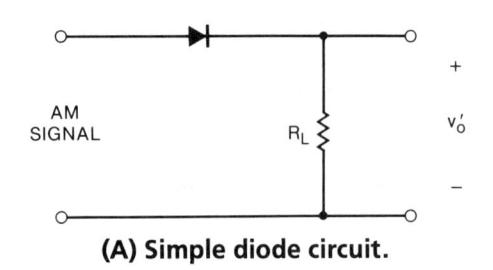

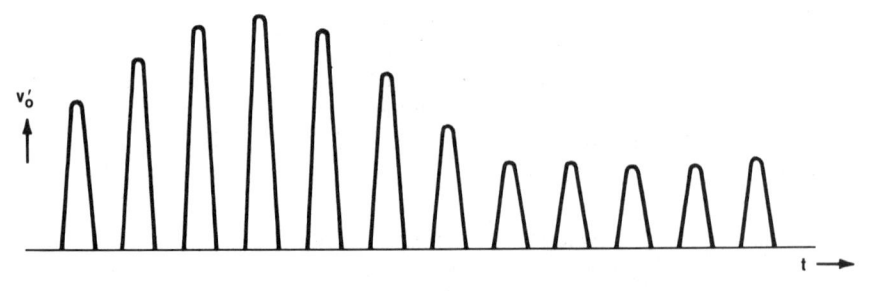

(B) Output of the diode circuit.

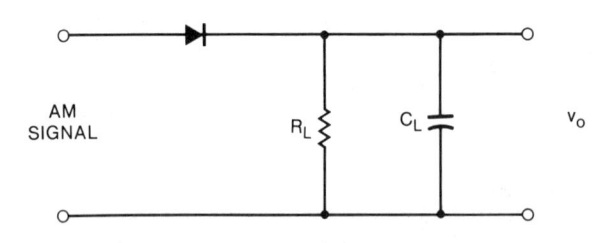

(C) Diode circuit with a capacitor.

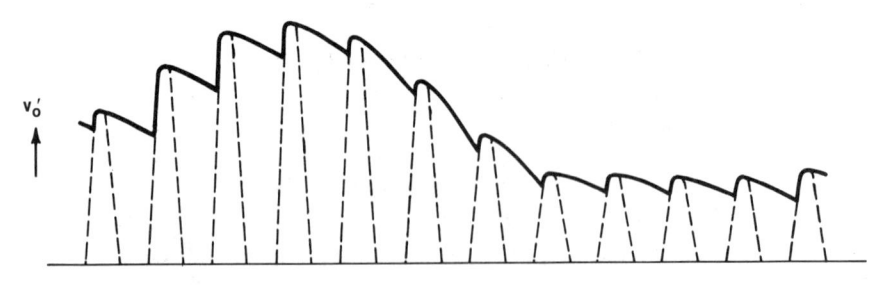

(D) Output of the diode circuit with a capacitor.

Fig. 14-27. Envelope detector for AM signals.

Electronics: Circuits and Systems 837

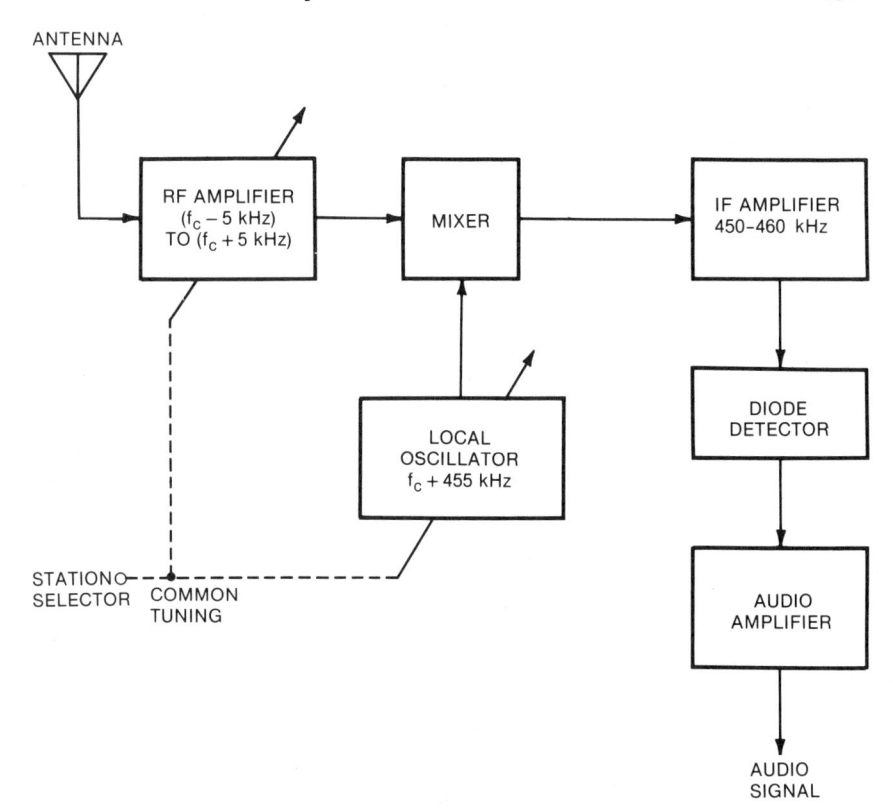

Fig. 14-28. Superheterodyne AM receiver.

The name *superheterodyne* arises from the fact that there is a mixing operation with a different carrier frequency than that of the radio station's carrier frequency (such a mixing is known as *heterodyning*), and since the local carrier frequency is *above* that of the radio station, the adjective "super" is used. The superheterodyne principle is used for FM receivers also.

Drill Problem 14.10: An AM radio station transmits at a total power of 1 kW. If its modulation index is 0.8, determine the power contained in the carrier and in the side frequency components. Draw a frequency spectrum when $f_m = 1$ kHz and $f_c = 100$ kHz.

Drill Problem 14.11: Repeat the previous drill problem when the modulation index is 0.5.

Drill Problem 14.12: An AM radio station operates at 950 kHz. Draw the frequency spectrum of the output of the RF amplifier of the AM receiver. Calculate the local carrier frequency. Draw the frequency spectrum of the two sidebands generated by the mixer and the frequency spectrum of the output of the IF amplifier.

Drill Problem 14.13: To what extent will the results of the previous drill problem change if you considered a radio station at 1450 kHz instead of 950 kHz?

FREQUENCY MODULATION

The carrier in a communication system can be written in the form [$A_c \cos(2\pi f_c t)$]. The quantity ($2\pi f_c t$) varies from 0° to 360° as t varies from 0 to ($1/f_c$) and can be considered as the total *phase* of the carrier, θ. Therefore, the carrier can be written as $A_c \cos \theta$. In the unmodulated carrier, $\theta = 2\pi f_c t$; that is, θ is proportional to the time t. A plot of θ against t will be a straight line with a slope of ($2\pi f_c$) for an unmodulated carrier as shown in Fig. 14-29A.

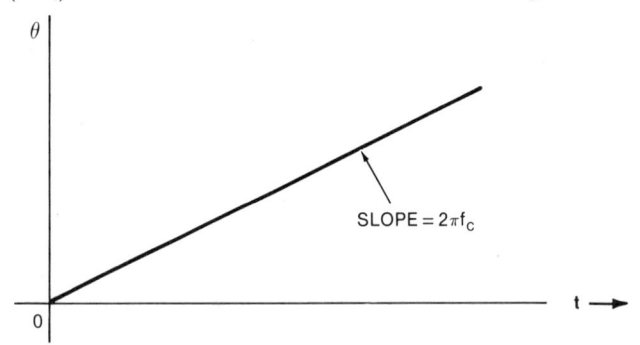

(A) In the unmodulated carrier, the plot of θ against t will be a straight line.

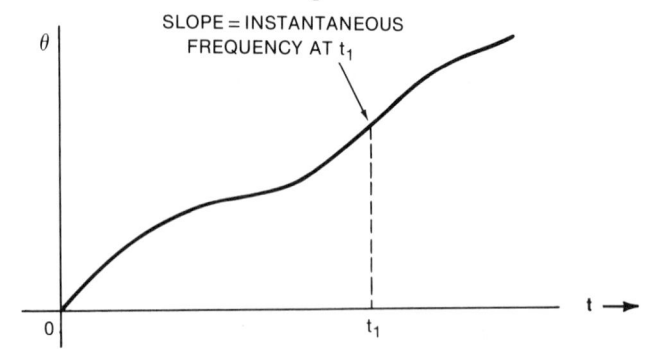

(B) Slope of the phase-angle curve gives the instantaneous frequency of the signal.

Fig. 14-29. Concept of instantaneous frequency.

Now consider the possibility that θ does not vary linearly with time but is of the form shown in Fig. 14-29B. Then, the *slope of the curve at any given instant of time* can be considered as an *instantaneous frequency*, which is a variable quantity since the slope of the curve in Fig. 14-29B is not a constant. In *frequency modulation* systems, the *amplitude of the signal* is made to vary the *instantaneous frequency* (or slope of the curve of Fig. 14-29B) of the carrier. When the signal is positive the instantaneous frequency is made higher than f_c, and when the signal is negative, the instantaneous frequency is made lower than f_c. Thus, the variation in the instantaneous frequency provides the information about the modulating signal.

When the modulating signal is positive, the instantaneous frequency is greater than f_c and the *increase* in f_c is made proportional to the amplitude of the signal. The *largest increase* in the instantaneous frequency permitted in a system is called *frequency deviation*. Suppose the carrier frequency is 10 MHz when there is no modulating signal and the instantaneous frequency is permitted to vary by 1 kHz per volt of the modulating signal. This means that if the signal were 2.5 V, then the instantaneous frequency would be (10 MHz + 2.5 kHz). If the largest signal that will occur in the system is 10 V in this example, then the highest instantaneous frequency will be (10 MHz + 10 kHz). The frequency deviation is, therefore, 10 kHz. If, on the other hand, we were to vary the instantaneous frequency by 10 kHz per volt of the modulating signal, then the highest instantaneous frequency (when the largest signal is 10 V) would be (10 MHz + 100 kHz). We can deduce that the latter system will need a larger bandwidth (since the frequency deviation is 100 kHz) than the former system (with a frequency deviation of 10 kHz), which is indeed the case. It can be shown that the bandwidth needed for FM transmission is approximately *twice the frequency deviation*.

The two important concepts introduced can be briefly summarized as follows:

1. The instantaneous frequency is the rate at which the phase θ of the carrier is made to vary with time. We may think of it as the number of times a waveform crosses the time axis per second at any given instant.

2. The frequency deviation is the maximum increase in the instantaneous frequency permitted above the carrier frequency when the largest modulating signal occurs. The choice of frequency deviation affects the bandwidth needed for FM transmission.

Modulation Index in FM

The modulation index in frequency modulation is defined as

$$\beta_{FM} = \frac{\text{frequency deviation}}{\text{frequency of the modulating signal}} \qquad (14\text{-}6)$$

For example, if a 5-kHz signal modulates a 100-kHz carrier so that the carrier frequency varies between 70 kHz and 130 kHz (as the signal amplitude goes from a negative minimum to a positive maximum), then the frequency deviation is (130 − 100) = 30 kHz, and the modulation index is

$$\beta_{FM} = \frac{30}{5}$$

$$= 6$$

Conversely, if a modulating signal has a frequency of 15 kHz and we wish to make $\beta_{FM} = 5$, then the resulting frequency deviation is (5 × 15) = 75 kHz.

Spectrum of an FM Signal

In the case of AM, we saw that if a single-frequency signal is used to modulate a carrier, then there were *two* side-frequency components about the carrier frequency. But, in FM, even when the modulating signal has only a single frequency, it is found to generate a large number of side-frequency components. In fact, theory predicts that the number of such side-frequency components is infinite even with a single-frequency signal. But from a practical viewpoint, the number of significant side-frequency components is finite. The number of side-frequency components of significance is dependent upon the modulation index β_{FM}.

The sketches in Fig. 14-30 show the frequency spectra of an FM signal for a signal of given frequency f_m and different values of β_{FM}. The following points should be noted:

1. The spacing between the successive lines in the spectra is f_m in all cases. The spectral lines are situated at f_c, $f_c + f_m$, $f_c - f_m$, $f_c + 2f_m$, $f_c - 2f_m$, ...

2. The amplitude of the component at f_c in the spectrum is *not* always larger than the others. In fact, for certain values of β_{FM}, the carrier component may actually be absent from the FM spectrum. The presence or absence of the carrier component in an FM spectrum is of no significance.

3. The number of side-frequency components of significant amplitude increases as β_{FM} is increased. For $\beta_{FM} = 0.2$, there are only two significant side-frequency components, whereas for $\beta_{FM} = 10$, the number goes up to 12. The *bandwidth* becomes larger as β_{FM} increases: it is only 2 f_m for $\beta_{FM} = 0.2$, but $24f_m$ for $\beta_{FM} = 10$.

Bandwidth of an FM Signal

It is impossible to develop a theoretical measure of the bandwidth of an FM signal since the heights of the spectral lines are not found to change in an

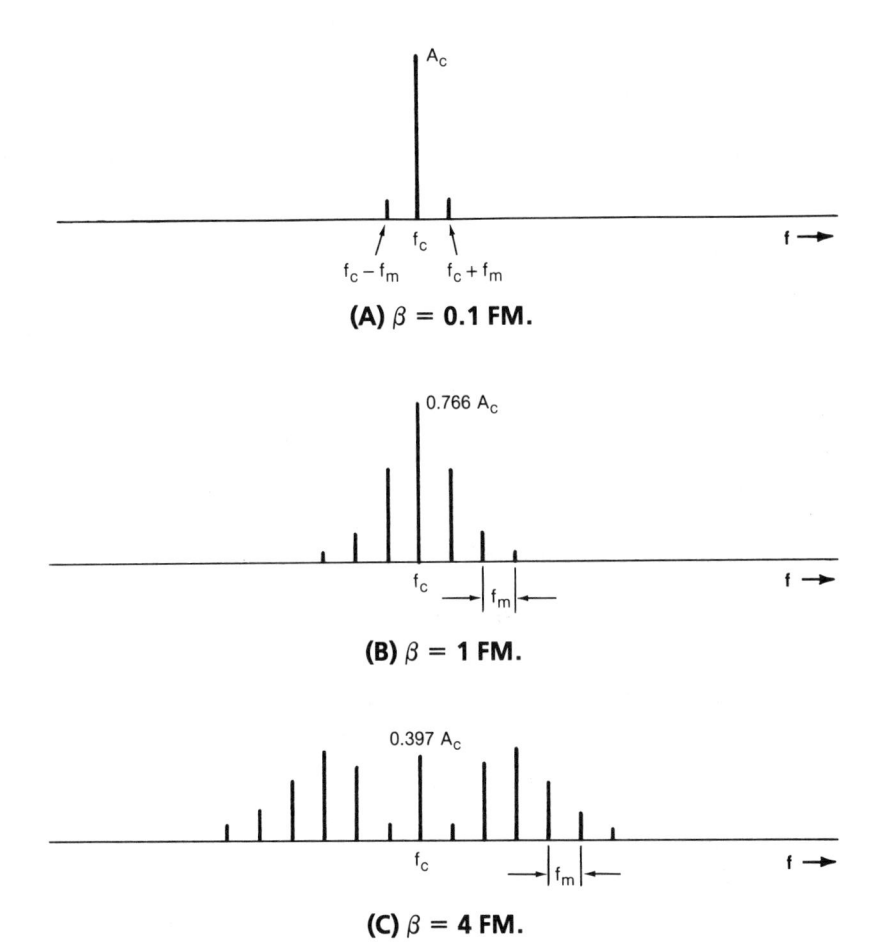

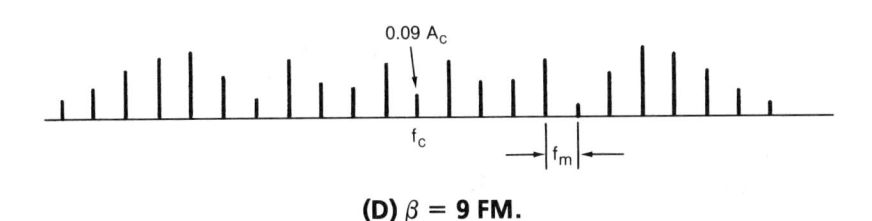

Fig. 14-30. Frequency spectra of FM signals with fixed f_m for different values of the modulation index. Note the proliferation of significant side-frequency components as β_{FM} increases.

(A) $\beta = 0.1$ FM.

(B) $\beta = 1$ FM.

(C) $\beta = 4$ FM.

(D) $\beta = 9$ FM.

orderly manner as we go from small to large values of β_{FM}. Only a *rule of thumb* is used. It is found that the number of significant side-frequency components is approximately $(\beta_{FM} + 1)$ on either side of f_c. That is, if $\beta_{FM} = 6$, for example, then there are 7 side-frequency components on either side of f_c, which are significant. (The significance is determined by the contribution to the power in the FM signal.) The rule for determining the bandwidth of an FM signal is called *Carson's rule*, which states that

$$\text{bandwidth of an FM signal} = 2(\beta_{FM} + 1)f_m \qquad (14\text{-}7)$$

where,
f_m is the frequency of the modulating signal.

For example, if $\beta_{FM} = 6$ and $f_m = 10$ kHz, then the bandwidth needed is 140 kHz. Carson's rule is reasonably accurate for values of $\beta_{FM} > 5$.

Drill Problem 14.14: The instantaneous frequency of a carrier is made to vary (from its normal value of 91.5 MHz) by 10 kHz per volt of the modulating signal. If the largest amplitude of the signal is given as 9 V, calculate the frequency deviation.

Drill Problem 14.15: If in the previous drill problem, the frequency deviation is to be kept at 40 kHz, what is the largest signal permissible?

Drill Problem 14.16: A modulating signal of frequency $f_m = 15$ kHz frequency modulates a carrier of 10 MHz. Calculate the modulation index for each of the values of frequency deviations given: (a) 10 kHz; (b) 100 kHz; (c) 1 MHz.

Drill Problem 14.17: How will the results of the previous drill problem be affected if the *carrier* frequency were changed to (a) 5 MHz; (b) 125.6 MHz?

Drill Problem 14.18: How will the results of Drill Problem 14.16 be changed if the *signal* frequency were changed to 5 kHz?

Drill Problem 14.19: The frequency of a modulating signal is 15 kHz. For each of the following values of β_{FM}, calculate the bandwidth needed for transmission of the FM signal (using Carson's rule): (a) 0.8; (b) 4; (c) 40.

Drill Problem 14.20: The frequency of a modulating signal is 20 kHz. For each of the following values of frequency deviation, calculate (from Carson's rule) the bandwidth needed for FM transmission: (a) 25 kHz; (b) 250 kHz; (c) 750 kHz.

Electronics: Circuits and Systems

For large values of the modulation index, that is when $\beta_{FM} \gg 1$, Carson's rule can be approximated by

$$\text{bandwidth needed} = 2\,\beta_{FM} f_m$$
$$= 2 \times \text{frequency deviation} \qquad (14\text{-}8)$$

The frequency deviation permitted in commercial FM is 75 kHz, and the bandwidth allotted to each station is 150 kHz. The modulating signal extends from roughly 50 Hz to 15 kHz (in the case of music). At the low end, with $f_m = 50$ Hz, we have

$$\beta_{FM} = \frac{\text{frequency deviation}}{50}$$

$$= 1500$$

Therefore, the number of significant side frequencies for the 50-Hz modulating signal is about 3000 (twice the value of β_{FM}) and the allotted bandwidth of 150 kHz is adequate for the low end of the signal being broadcast.

At the high end of the signal, $f_m = 15$ kHz, and we have

$$\beta_{FM} = \frac{\text{frequency deviation}}{15 \times 10^3}$$

$$= 5$$

The frequency spectrum for $\beta_{FM} = 5$ shows the presence of 7 significant side-frequency components on either side of the carrier. That is, the bandwidth needed for transmitting the high end of the signal frequencies is $(2 \times 7 \times f_m)$ = 210 kHz. But, the permitted bandwidth is only 150 kHz. Therefore, a certain amount of high-frequency information is lost in FM broadcasting if the frequency deviation is kept the same throughout the signal's bandwidth.

Narrowband FM

When the modulation index is small, less than 0.5, it is found that there are only two significant side-frequency components: one at $(f_c + f_m)$ and the other at $(f_c - f_m)$ and the frequency spectrum resembles that of AM. This type of FM is called *narrowband FM*. Narrowband FM was, in fact, the basis of the first practical FM system (in a laboratory setting) demonstrated by Armstrong. Narrowband FM can be obtained by starting with AM and introducing a 90° phase shift between the carrier and the side-frequency components. Once a narrowband FM has been generated, it can be converted to wideband FM by stepping up the frequency deviation through a series of frequency multipliers.

Direct Generation of FM

The basic principle involved in the direct generation of FM is to make the frequency of oscillation of an oscillator vary in proportion to the strength of a modulating signal. A basic arrangement is shown in Fig. 14-31. The element in the box marked X is a *variable* reactance (L or C) whose value varies with the voltage across it. One device whose reactance can be varied as a function of the voltage applied to it is the *varactor diode* discussed in Chapter 4. A circuit using a varactor diode is shown in Fig. 14-32.

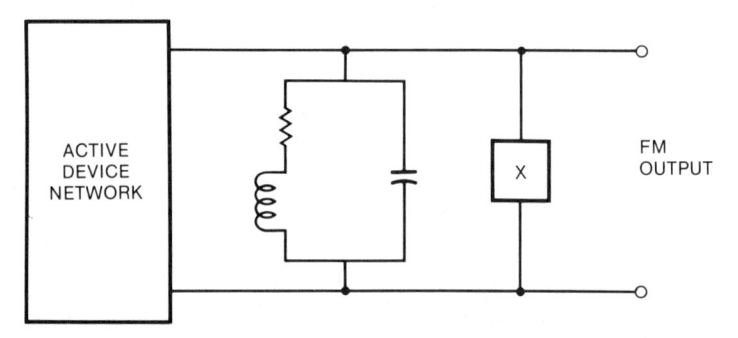

Fig. 14-31. Principle of generation of FM signals.

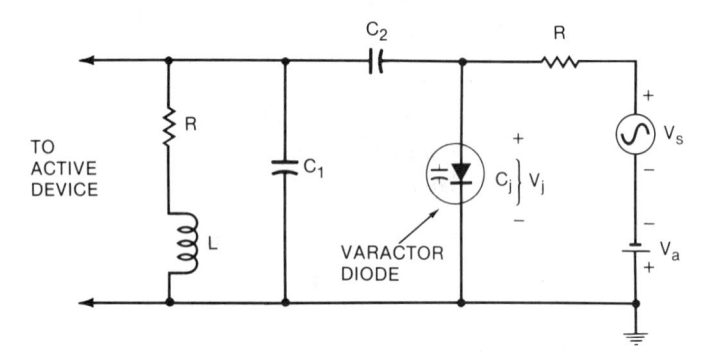

Fig. 14-32. Generation of FM signals using a varactor diode.

The varactor is a reverse-biased pn junction diode. In the circuit of Fig. 14-32, the effective bias V_j on the varactor is

$$(V_a + V_s)$$

where,

V_a is a fixed reverse bias,

v_s is the modulating signal.

The varactor acts as a capacitor whose capacitance C_j is *inversely proportional to the square root of the voltage V_j across it*. Therefore, the frequency

Electronics: Circuits and Systems

of oscillation of the circuit varies with the signal voltage. If the amplitude of the modulating signal is kept small, then the change in the frequency of oscillation is found to be linear with respect to the amplitude of the modulating signal.

The disadvantage of this circuit is that it does not permit large frequency deviations. But, it is possible to obtain a reasonably high value of frequency deviation by initially using a very high carrier frequency and then heterodyning down to a lower carrier frequency.

Demodulation of an FM Signal

The demodulation of an FM signal requires a circuit that produces an *output voltage proportional to the difference between the carrier frequency and the instantaneous frequency of the FM signal*. A circuit with such a property is called a *discriminator*. A simple form of discriminator is the *resonant circuit* as shown in Fig. 14-33. It is seen that the output voltage of the resonant circuit has a range over which it is proportional to the frequency. The incoming current i(t) in the circuit is an FM signal. The constants are arranged so that f_c is in the center of the linear portion as indicated in Fig. 14-33B. The output voltage varies linearly with the difference between the instantaneous frequency and the carrier frequency. The frequency deviation must be small compared with f_c in order to use the above circuit, and also ($f_0 - f_c$) must be much greater than the frequency deviation. For example, if $f_c = 10$ MHz and frequency deviation is 75 kHz, then a choice of f_0 at 10.2 MHz will be a suitable choice.

A more complicated circuit than the simple resonant circuit, and one that is widely used in the reception of FM signals, is the *Foster Seely discriminator*, whose description is outside the scope of this book.

FM Receivers

The commercial FM broadcast band in the US extends from 88 to 108 MHz, with each station permitted 150 kHz plus a guard band of 50 kHz between stations. Heterodyning is used in FM receivers just as in AM, and the block diagram of an FM receiver is shown in Fig. 14-34. The operation of the various components in this system is similar to those in the AM receiver. The one new component is the *limiter*, which is used to clip off the amplitude variations in an FM signal introduced during its propagation from the transmitter to the receiver. Since amplitude variations in an FM signal do not have any significance, limiting is necessary.

NOISE IN COMMUNICATION SYSTEMS

Electrical noise is any unwanted voltage appearing at the receiving end of a communication system. If the signal power at the receiver is not sufficiently

(A) Resonant circuit.

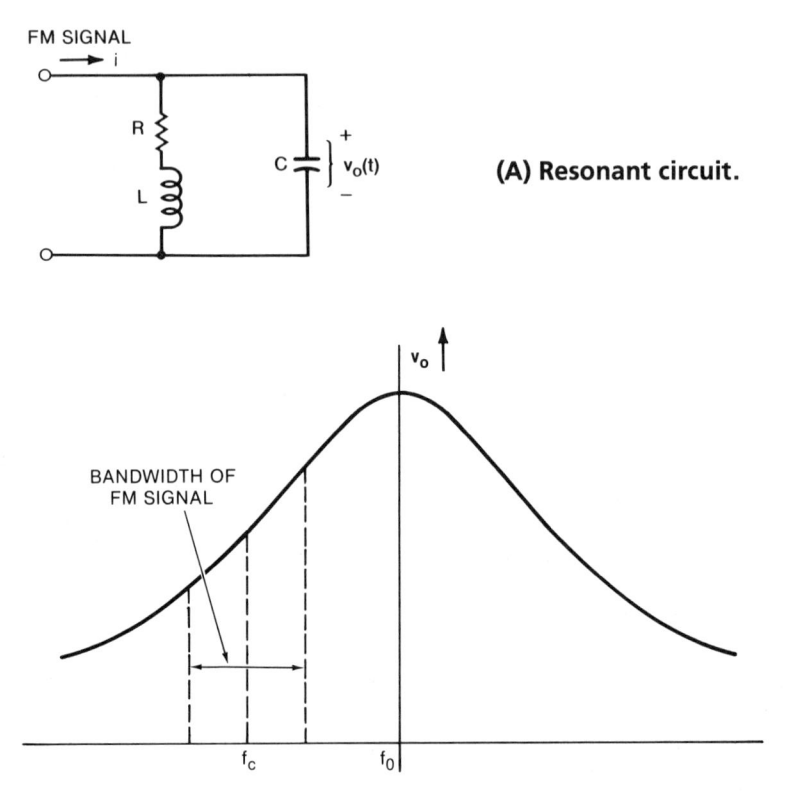

(B) Constants arranged so that f$_c$ is in the center of the linear portion.

Fig. 14-33. Resonant circuit can be used to produce an output voltage proportional to frequency deviation from the carrier frequency.

high compared with the noise power, then a partial or even total loss of information is bound to occur. In order to study the effect of noise in a system, it is necessary to define the *power spectral density* of noise.

Noise Power Spectral Density

It is possible to experimentally measure the amount of power generated by a source of noise over a very narrow band of frequencies centered at any given frequency. That is, we can choose any frequency f and measure the noise power generated by a noise source over a very narrow range of frequencies about f. Then, we define the *noise power spectral density* at the frequency f as (noise power over the narrow band divided by the range of the narrow band). For example, if the noise power generated over a band of 300 Hz cen-

Electronics: Circuits and Systems

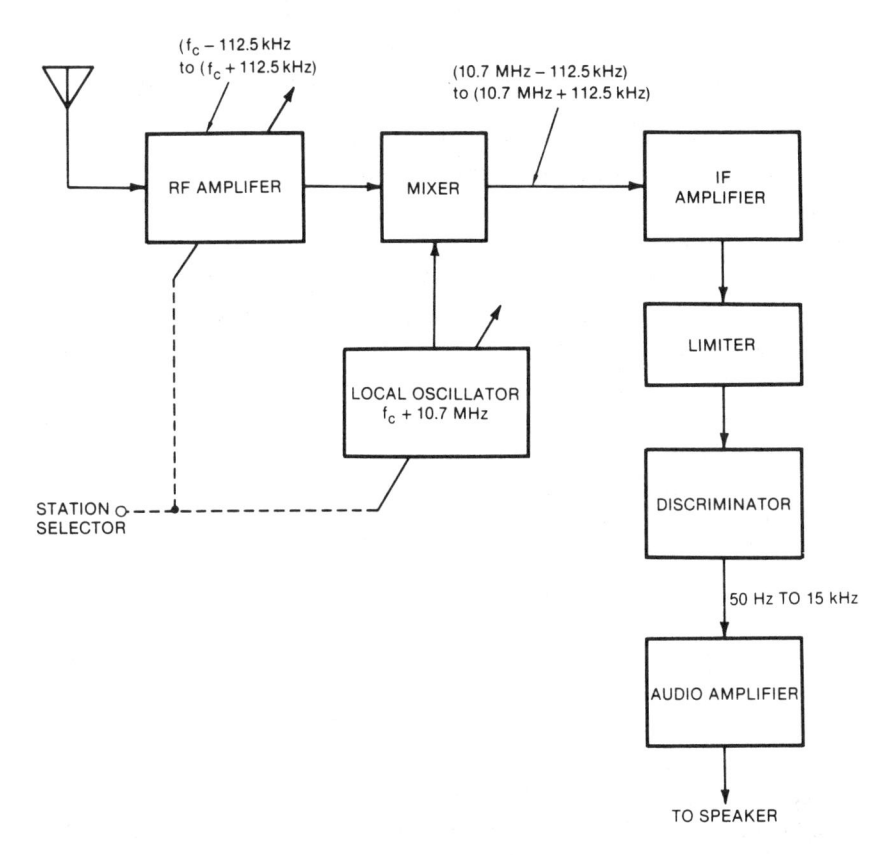

Fig. 14-34. Superheterodyne FM receiver.

tered at 5 kHz is 10^{-6} W, then the noise power spectral density at 5 kHz is ($10^{-6}/300$) W/Hz. A typical noise power spectral density curve is shown in Fig. 14-35.

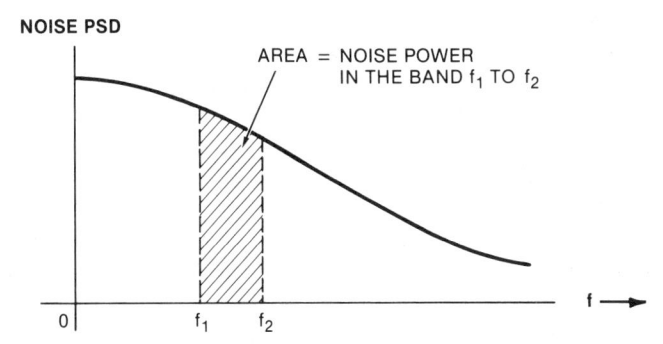

Fig. 14-35. Noise power spectral density.

Once the noise power spectral density curve is available (either through theoretical considerations or experimental measurements), then the noise power over any range of frequencies f_1 to f_2 can be calculated by determining the area of the power spectral density curve between f_1 and f_2, which is why the noise power spectral density curve is important.

White Noise

A type of noise that has a constant value of power spectral density at all frequencies is called *white noise*. If N_0 is the power spectral density due to white noise, then it can be seen from Fig. 14-36 that the noise power contained in a band of frequencies f_1 to f_2 is simply $[N_0 (f_2 - f_1)]$ W.

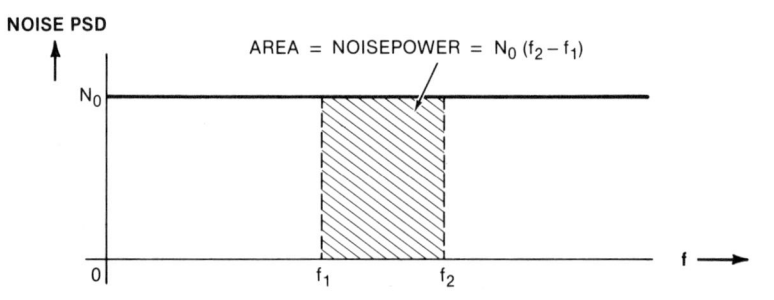

Fig. 14-36. White-noise power spectral density.

The noise power due to white noise increases in direct proportion to the bandwidth being considered, which can be a limiting constraint on the extent of the bandwidth in some communication systems.

White noise is commonly assumed since it is computationally most convenient. Many noise sources are found to satisfy the constant power spectral density property of white noise.

Thermal Noise

The flow of current in a resistor (under constant voltage) is constant only as a statistical average. The free electrons in a conductor have a random motion, and this randomness gives rise to a random current or voltage, which is called *thermal noise* or *Johnson noise*. The thermal noise power spectral density is found to be a constant at all frequencies and, consequently, thermal noise can be treated as white noise.

The thermal noise power density can be shown to be (4kTR) W/Hz, where R is the resistance, T is the absolute temperature, and k is Boltzmann's constant (1.38×10^{-23} J/°K). For example, a 10 kΩ resistor has a noise power spectral density of 1.66×10^{-16} W/Hz at 300°K.

Equivalent Noise Temperature

The concept of *equivalent noise temperature* is useful in comparing different sources of noise even when the sources are not actual resistors.

The equivalent noise temperature T_n is defined by

$$T_n = \frac{\text{noise power available in a frequency range}}{k \times \text{(frequency range used)}}$$

The equivalent noise temperature has nothing to do with the ambient temperature. Examples of noise temperature are: atmospheric noise, which has a temperature of 10^8 degrees K at night and 10^7 degrees K in the daytime, and galactic noise, which has a noise temperature of 10^5 to 10^6 degrees K. These figures are valid at a frequency of 10 MHz.

Equivalent noise temperature can also be used to describe the noisiness of an amplifier. Some very low-noise amplifiers have an equivalent noise temperature of $10°$K to $30°$K while standard broadcast receivers may have noise temperatures of $1000°$K.

Shot Noise

The type of noise due to the random motion of current carriers in a semiconductor junction or the emission of electrons from a heated cathode in vacuum tubes is known as *shot noise*. Shot noise power spectral density is constant at all frequencies, and it can be treated as white noise also.

SIGNAL-TO-NOISE RATIO AND NOISE FIGURE

The ratio of the power due to the signal and the power due to the noise in a system is called the *signal-to-noise ratio* (or SNR). It is usually expressed in decibels.

SNR in decibels = 10 log [signal power/noise power]   (14-9)

In any two port, there is an SNR for the input and an SNR for the output port. If the two port itself is noise free, then the two SNR values will be equal. Otherwise, the SNR at the output will be less than the SNR at the input. A figure of merit for a two-port system is the *noise figure* NF:

NF in decibels = (SNR at input in dB) − (SNR at output in dB)

For a noise-free system, NF = 0 dB. But otherwise, NF will be greater than 0 dB since SNR at input will be greater than that at the output.

In multistage amplifiers, the worst culprit in terms of NF is the first stage. By keeping the noise figure of the first stage as low as possible and making the gain of the first stage as high as possible, the NF of a multistage amplifier can be kept low.

Effect of Noise in Different Modulation Systems

The receiver circuit in a communication system receives not only the modulated carrier but also noise introduced during the transmission. In amplitude modulation systems such as SSB and DSB-SC, the SNR in the DSB-SC system is better by a factor of two than the SSB system. In envelope detection of conventional AM, it is found that when the SNR falls below a certain level (called the *threshold level*), the signal is completely lost in the detection process. Therefore, for reception of AM signal (using envelope detection) the SNR has to be above a threshold level, which can be calculated. Threshold effect is not present in synchronous detection of amplitude modulated signals.

An FM system provides an improvement in SNR over an AM system by a factor of $3\beta_{FM}^2$, where β_{FM} is the modulation index. By choosing a large modulation index, the SNR of an FM system can be made far superior to an AM system, but the price we pay is the larger bandwidth, which increases as β_{FM} increases. For example, if $\beta_{FM} = 5$, then the SNR of the FM system will be 75 times as high as that of the AM system. But the bandwidth of the FM system is, by Carson's rule, 12 f_m, whereas the bandwidth of the AM system is only 2 f_m, where f_m is the frequency of the modulating signal.

A continued increase of bandwidth in an FM system will not, however, cause an indefinite increase in its SNR. The reason is that as the bandwidth increases, so does the total noise power (which is the area under the noise power spectral density over the bandwidth of transmission). If the signal power is fixed and the bandwidth is increased, the noise power eventually becomes comparable to the signal power and the improved SNR of an FM system is no longer a reality. When the input SNR of the FM receiver falls below a certain level, the noise takes over and completely wipes out the signal at the receiver. The receiver cannot demodulate the signal below this minimum threshold. The threshold SNR level is 13 dB for FM.

Deemphasis and Preemphasis in FM

The noise power spectral density at the input to an FM receiver is found to be small at low frequencies and large at high frequencies. The shape of the noise power spectral density at the input of an FM receiver is as shown in Fig. 14-37. The noise power spectral density varies as the square of the frequency. This characteristic of the noise power can be used to improve the performance of an FM system.

The signal in FM systems usually extends up to 15 kHz. But the frequency spectrum of an actual signal is not uniform over the entire 15-kHz range but tends to be smaller at high frequencies. Voice signals, for example, have only negligible components above 3 kHz. Even in the case of music, the energy content of the signals is considerably smaller at the high-frequency end of the 15-kHz band. Since frequency deviation in an FM system depends upon the

NOISE PSD

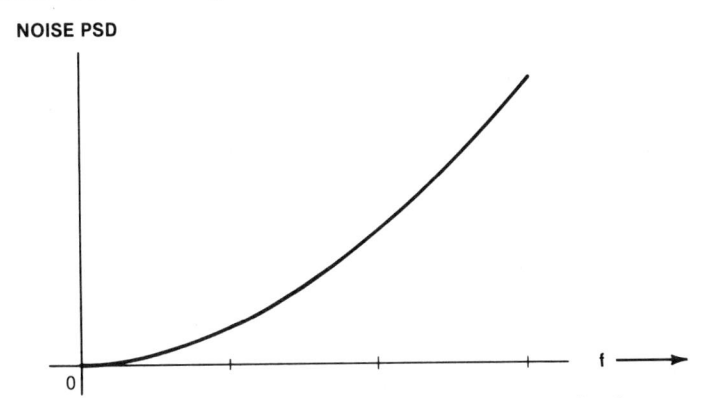

Fig. 14-37. Noise power spectral density at the input of an FM receiver.

strength of the signal, the frequency deviation produced by the high-frequency signals is not as large as can be accommodated by the allotted bandwidth.

Thus, the signal and noise in an FM system have opposing characteristics: the signal is weak at high frequencies but the noise is strong at high frequencies. Therefore, the SNR of an FM system can be improved by using preemphasis and deemphasis. At the transmitting station, the *high-frequency signals* are artificially *boosted* so as to make them use up the full frequency deviation allowed by the 150-kHz bandwidth. At the receiver, after the discriminator stage, the high-frequency signals are attenuated to what they were in the original audio signal source, so as to restore the fidelity. This attenuation at the high-frequency end affects the noise power also. Thus, the noise power is reduced in the frequency range where it would normally be most damaging. The boosting of the high-frequency portion of the signal at the transmitter is called preemphasis, and the attenuation at the receiver to reverse the preemphasis is called deemphasis. It is possible to obtain a 13- to 16-dB improvement in the SNR by using preemphasis and deemphasis in an FM system.

SUMMARY SHEETS

REVIEW OF SIGNALS

Normalized power = Power consumed by a standard load of 1 ohm

AC signals
$P = 0.5(\text{peak value})^2$

Periodic Nonsinusoidal Signals

The signal contains dc components, and components at nf_0 (n = 1, 2, 3, ...), where f_0 is the fundamental frequency. $f_0 = (1/T)$. Refer to Chapter 1 (summary sheet) for the amplitude and power spectra of periodic signals.

Total power in a periodic signal

> P = power due to dc component + powers due to the different harmonic components each taken independently of the others.

Nonperiodic Signals

Refer to Chapter 1 (summary sheet) for the amplitude and energy spectra of pulses.

A pulse of *shorter duration* (a narrower pulse) requires a wider bandwidth for transmission than a pulse of longer duration.

The bandwidth is inversely proportional to the duration of a pulse.

MODEL OF A COMMUNICATION SYSTEM

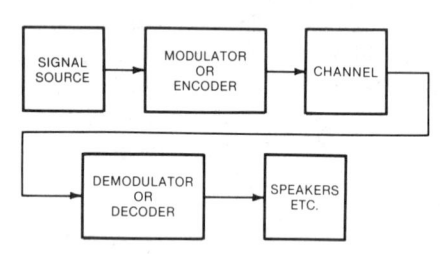

SIGNAL SOURCE → MODULATOR OR ENCODER → CHANNEL

DEMODULATOR OR DECODER → SPEAKERS ETC.

The signal is made to modulate a carrier at the sending end and is recovered by a demodulation process at the receiving end. In the case of digital signals, encoding is used instead of modulation and decoding instead of demodulation.

Modulation of a carrier of high frequency by a signal is necessary in order to use antennas of practical dimensions (to be of the same order of magnitude as the wavelength of the carrier) in a communication system. Also, it permits the *frequency division multiplexing* of a number of signals over a specified bandwidth.

Different Types of Modulation

Amplitude Modulation (AM)—The modulating signal causes a variation of the amplitude of the carrier wave in proportion to the strength of the signal.

Frequency Modulation (FM)—The modulating signal causes a variation of the instantaneous frequency of the carrier wave in proportion to the strength of the signal. The amplitude of the carrier is not affected, but the number of zero crossings per second is.

Phase Modulation—This is a variation of the FM case. The phase angle of the carrier wave is made to vary in proportion to the strength of the signal. The amplitude of the carrier is not affected.

These three types of modulation are called *continuous wave* (or *cw*) modulation.

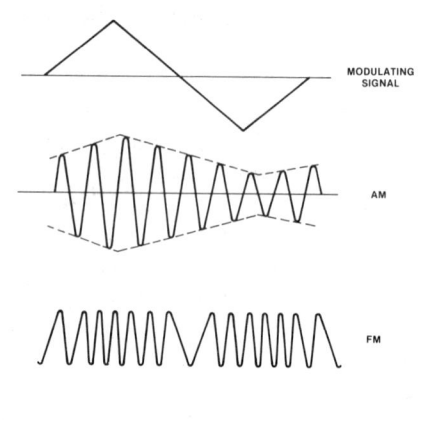

MODULATING SIGNAL

AM

FM

Time Domain and Frequency Domain

Actual signal processing operations occur in the time domain (that is, as a function of time). What we see on an oscilloscope, for example, is the time domain representation.

Frequency domain viewpoint concentrates on the frequency components or the band of frequencies present in signals. The frequency spectrum is an extremely useful tool in the frequency domain. The frequency spectra of even highly complex and random signals can be esti-

Electronics: Circuits and Systems

mated fairly accurately through measurements or computations.

Actual signal processing takes place in the time domain but analysis (and design) takes place in the frequency domain.

FREQUENCY TRANSLATION

When two signals are multiplied in the time domain, there is a shifting (or translation) of the frequencies of the two signals.

The frequency of a given signal can be shifted up or down by multiplying the signal by another signal of an appropriate frequency.

When two signals (each containing a single frequency) are multiplied, the product signal contains two frequency components: the sum of the two original frequencies and the difference between the two original frequencies.

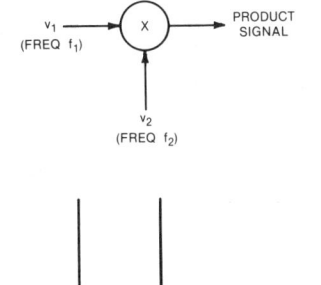

When signals (each of which contains more than one frequency) are multiplied, the product signal contains all the sum frequencies and difference frequencies obtained by taking each frequency present in one of the signals and pairing it with each frequency present in the other signal.

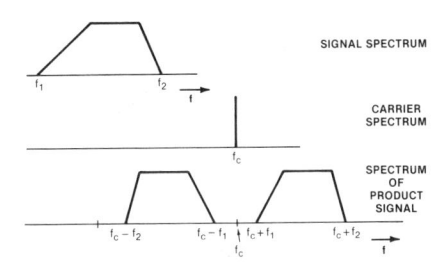

When a signal containing a continuous band of frequencies is multiplied by a signal of a single frequency (the carrier waveform), the product contains two bands of frequencies obtained by shifting the band of the signal above and below the carrier frequency. The lower and upper sidebands are mirror images of each other about the carrier frequency.

AMPLITUDE MODULATION

The carrier frequency is assumed to be much higher than that of the modulating signal.

DSB-SC AM (Double Sideband Suppressed Carrier AM)

When the signal has a single frequency f_m, the DSB-SC signal has two frequency components ($f_c + f_m$) and $f_c - f_m$) but does not contain f_c.

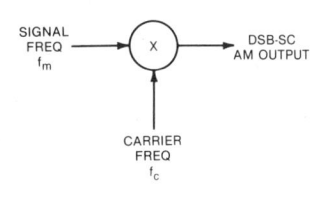

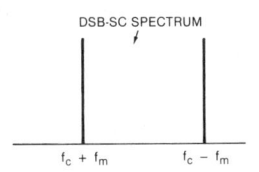

When the signal has a band of frequencies from 0 to f_m, the DSB-SC signal has two sidebands: the lower sideband from ($f_c - f_m$) to f_c and the upper sideband from f_c to ($f_c + f_m$) but does not contain the carrier component.

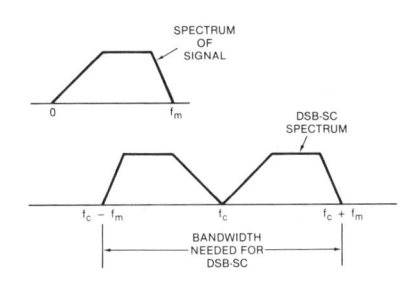

bandwidth of DSB-SC AM = twice the bandwidth of the signal

Demodulation of DSB-SC AM needs synchronous detection. The DSB-SC AM waveform is multiplied by a local carrier of the same frequency and in phase with the carrier at the sending end. A low-pass filter transmits the recovered signal to the receiver.

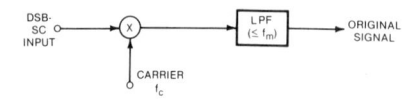

The generation of product signals is by means of a *balanced modulator* that produces the sum and difference frequencies of the two inputs.

SSB-AM (SINGLE-SIDEBAND AM)

Since each sideband in the DSB-SC AM waveform contains all the information present in the signal, it is not necessary to transmit both sidebands. SSB-AM suppresses the carrier and one of the sidebands and transmits the other.

bandwidth of SSB = bandwidth of the signal

Generation of SSB-AM

1. *Filtering Method:* A DSB-SC waveform is first generated (by using a balanced modulator), and one sideband is filtered out for transmission. The constraints on the filter are extremely stringent, and it is usually necessary to have several stages of filtering. The coat and complexity of the system become quite high.

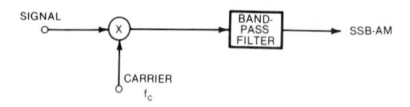

2. *Phasing Method:* Two multipliers are used. One multiplies the given signal by the carrier and the other multiplies the given signal shifted by 90° by the carrier shifted by 90°. These two product signals are added (or subtracted), and the sum (or difference) contains only one sideband.

The phasing method does not need complex equipment. But the shifting of all the frequency components by 90° without however affecting their amplitudes is difficult and even impractical.

SSB-AM is used when the conservation of the bandwidth is the overriding concern.

Demodulation of SSB-AM

Demodulation of SSB-AM uses the same procedure as that of DSB-SC AM. A pilot carrier is usually transmitted in order to synchronize the carrier at the receiving end with that at the sending end.

VSB AM (VESTIGIAL-SIDEBAND AM)

Since the filtering of one sideband for transmission while suppressing the other one completely (as needed in SSB) is extremely difficult, the filtering constraints can be relaxed if a portion of the other sideband is also transmitted. This is the principle underlying VSB. The filter used in a VSB system must have a gain characteristic as shown.

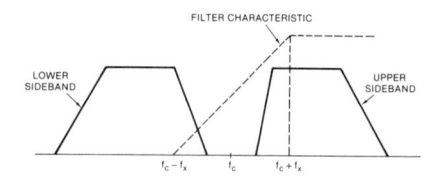

The bandwidth needed in VSB is somewhat larger than in SSB much less than in DSB-SC.

AM (DSB-LC OR DOUBLE-SIDEBAND LARGE CARRIER AM)

A large sample of the carrier is transmitted along with the two sidebands. In the time domain, the amplitude of the carrier is made to vary in proportion to the amplitude of the modulating signal.

$$\text{Modulation index } m = \frac{V_{max} - V_{min}}{V_{max} + V_{min}}$$

When $m > 1$, the envelope of the modulated waveform no longer resembles the modulating signal. Even though the signal can still be recovered, it will require synchronous detection (like DSB-SC or SSB). If envelope detection is to be used (which is desirable because of its simplicity), m must be kept ≤ 1 (usually < 1).

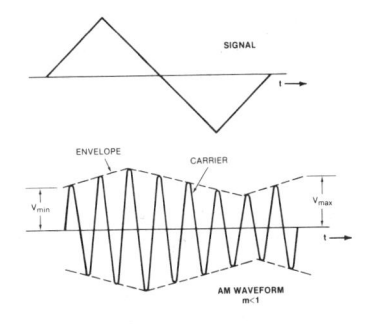

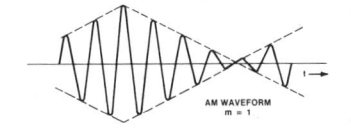

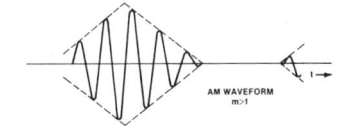

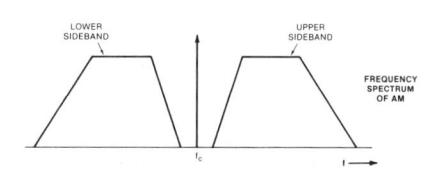

Electronics: Circuits and Systems

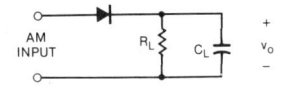

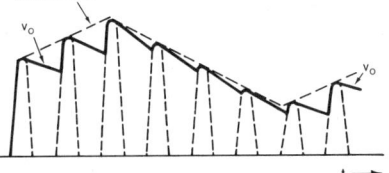

The time constant of the detector should be small enough to permit the output to follow the variations of the envelope.

$$(1/f_c) \ll R_L C_L \ll (1/f_m)$$

SUPERHETERODYNE RECEIVER

Any incoming AM band is first shifted to an intermediate frequency (IF) band occupying the range 450 kHz to 460 kHz, by mixing the incoming band with the carrier produced by a local oscillator.

frequency of the local oscillator = $(f_c + 455)$ kHz

where,

f_c is the carrier frequency of the station to which the receiver is tuned.

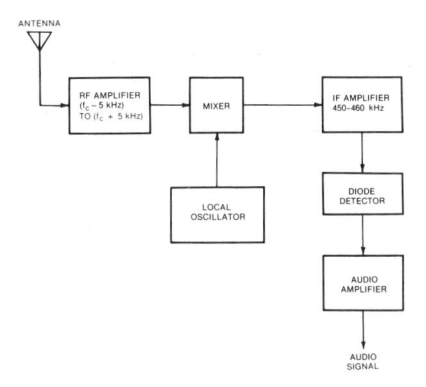

POWER CONSIDERATIONS IN AM

Power in carrier = $0.5 A_c^2$

Power in each side frequency component = $0.125 m^2 A_c^2$

Total power in the side frequency components = $0.25 m^2 A_c^2$

Total power in the AM waveform = $0.5 A_c^2 (1 + 0.5 m^2)$

$$\frac{\text{Power in side frequencies}}{\text{Total power transmitted}} = \frac{m^2}{2 + m^2}$$

DEMODULATION OF AM—ENVELOPE DETECTION

FREQUENCY MODULATION

The instantaneous frequency of a carrier waveform is made to vary in proportion to the amplitude of the modulating signal in FM

The *instantaneous frequency* is the rate at which the phase θ of the carrier waveform varies at any given instant.

The largest increase in the instantaneous frequency (caused by the strongest signal) over the

Analog Communication Techniques

normal carrier frequency value is called *frequency deviation*. The bandwidth needed for the transmission of FM signals depends directly on the frequency deviation.

Modulation Index β_{FM}:

$$\beta_{FM} = \frac{\text{Frequency deviation}}{\text{(Frequency of the modulating signal)}}$$

FM Spectra

The spectrum of an FM signal (when the modulating signal has a single frequency) is a line spectrum with the lines occurring at intervals of f_m (= the signal frequency) on either side of the carrier frequency f_c.

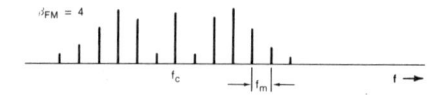

The amplitude of the component at f_c is not always larger than the others (as was the case in AM) and in fact may even be zero for some values of the modulation index. The number of significant side-frequency components increases as β_{FM} increases, and the bandwidth needed becomes larger as β_{FM} increases.

Bandwidth of an FM Signal Carson's Rule

$$\text{Bandwidth} = 2(\beta_{FM} + 1) f_m$$

where,
f_m is the frequency of the modulating signal.
When β_{FM} is $>> 1$,

$$\text{Bandwidth} \approx 2\beta_{FM} f_m$$

$$= 2 \times \text{frequency deviation}.$$

The bandwidth used in FM broadcasting systems is 150 kHz. The modulating index varies from 1500 for the low end of the signal bandwidth (at 50 Hz) to 5 for the high end of the signal (at 15 kHz).

Narrowband FM

When $\beta_{FM} < 0.5$, there are only two significant side-frequency components: $(f_c + f_m)$ and $(f_c - f_m)$, and the bandwidth needed is only 2 f_m (same as in AM). Narrowband FM (which can be generated in a manner similar to that of AM signals) can be converted to wideband FM by stepping up the frequency deviation through a series of frequency multipliers. This method was the one used by Armstrong.

Direct Generation of FM

For signals of small amplitude, a varactor diode can be used to control the frequency of oscillation of a resonant circuit. The varactor diode acts as a capacitance whose value is controlled by the voltage applied to it. A discriminator circuit is used for the demodulation of an FM signal. The output voltage of the discriminator is proportional to the difference between the carrier frequency and the instantaneous frequency of the FM signal.

FM Receivers

The principle of operation of an FM receiver is somewhat similar to the AM receiver except for the use of a discriminator to perform the demodulation.

NOISE IN COMMUNICATION SYSTEMS

Noise Power Spectral Density

If the noise power in a narrow band of frequencies is divided by the bandwidth of the narrow band, the ratio gives a measure of the *noise power spectral density*. By centering the narrow band at various frequencies, it is possible to experimentally obtain the noise power spectral density curve of a noise source.

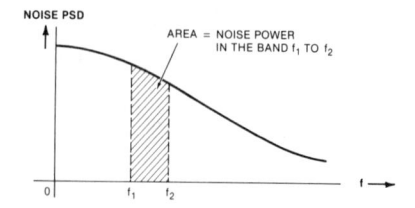

White Noise

When the power spectral density is a constant for all frequencies, the noise is called *white noise*.

Electronics: Circuits and Systems

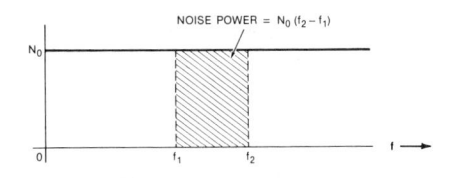

The total noise power in white noise increases in direct proportion to the bandwidth. White noise is usually assumed in computations in communication systems.

Thermal Noise

The random flow of electrons in a conductor due to thermal agitation causes a noise component, which is called *thermal noise* or *Johnson noise*. Thermal noise has a noise power spectral density given by (4 k T R) W/Hz and is a constant at all frequencies (like white noise).

Equivalent Noise Temperature

Equivalent noise temperature is used for comparing different sources of noise that may or may not be resistors.

$$T_n = \frac{\text{noise power in a frequency range}}{k \times (\text{frequency range})}$$

k = Boltzmann's constant

= 1.38×10^{-23} joules/°K

The equivalent noise temperature has no relationship to the ambient temperature. Atmospheric noise has a temperature of 10^8 degrees K at night. Low-noise amplifiers have equivalent noise temperatures in the range 10 to 30°K, while standard broadcast receivers have noise temperatures in the range of 1000°K.

Shot Noise

The random motion of current carriers in a semiconductor junction or the flow of electrons from a heated cathode in vacuum tubes causes *shot noise*. Shot noise can be treated as white noise also.

SIGNAL-TO-NOISE RATIO (SNR)

SNR in dB = 10 × log (signal power/noise power)

The SNR at the output of a two port will be less than that at the input unless the two port itself is noise free. In the latter case, the two values of SNR will be equal.

noise figure of a two port = (input SNR in dB) − (output SNR in dB)

If the two port is noise free, the noise figure is 0 dB; otherwise it is greater than 0 dB.

In multistage amplifiers, the noise figure of the first stage has the most serious effect. The noise figure can be kept low by making the noise figure of the first stage as close to zero as possible and increasing the gain of the first stage as much as possible.

Effect of Noise in Modulation Systems

SNR of DSB-SC AM is better than the SNR of SSB AM by a factor of two.

When the SNR of an AM waveform falls below a threshold level in envelope detection systems, the signal cannot be recovered. The threshold effect is, however, not present when synchronous detection is used. SNR of FM systems = $3\beta\,^2_{FM} \times$ SNR of AM. An increase in SNR by using large values of β_{FM} is attained at the cost of a wide bandwidth. An unlimited increase in the bandwidth of an FM system does not result in an indefinite increase in the SNR since the noise power continues to increase while the signal power remains constant. FM systems exhibit a threshold effect also. The minimum SNR needed is 13 dB.

Deemphasis and Preemphasis in FM

Signals in FM systems show a frequency spectrum that is stronger at low frequencies than at high frequencies. The noise power spectral density at the input to an FM receiver is found to vary as the square of the frequency: it gets stronger as the frequency increases.

In order to take advantage of the opposing characteristics of the signal spectrum and the noise power spectrum, the high-frequency signals are artificially boosted at the transmitting station. At the receiver, the high-frequency components are attenuated so as to return them to their normal level. This attenuation also reduces the noise power from the high frequencies. The boosting of the high frequencies at the transmitter is called preemphasis, and the attenuation at the receiver is called deemphasis. An improvement of 13 to 16 dB in SNR can be attained by using preemphasis and deemphasis in FM systems.

ANSWERS TO DRILL PROBLEMS

14.1 (a) 4 kHz, 12 kHz; (b) 32 kHz, 48 kHz; (c) 3.992 MHz, 4.008 MHz.
14.2 (a) 27.5 kHz; (b) 2.5 kHz; (c) 4.5 kHz; (d) 652.5 kHz.
14.3 (a) $f_2 = 26.7$ kHz: 30 kHz and 23.4 kHz; or $f_2 = 33.3$ kHz: 30 kHz and 36.6 kHz; (b) $f_2 = 546.7$ kHz: 550 kHz and 543.4 kHz, or $f_2 = 553.3$ kHz: 550 kHz and 556.6 kHz. (c) $f_2 = 91.4967$ MHz: 91.5 MHz and 91.4934 MHz, or $f_2 = 91.5033$ MHz: 91.5 MHz and 91.5066 MHz.
14.4 2750, 4250, 9750 (all in kilohertz).
14.5 13.35, 16.65, 23.35, 26.65, 430, 440, 470, 480 (all values in kilohertz).
14.6 See Fig. 14-DP6.

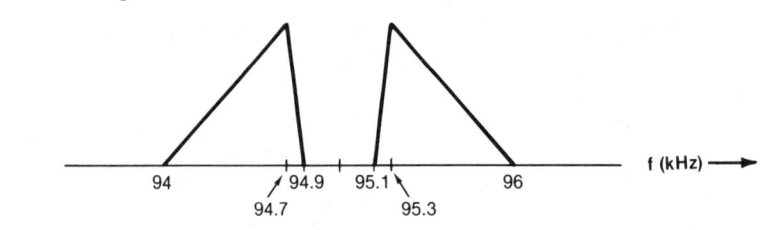

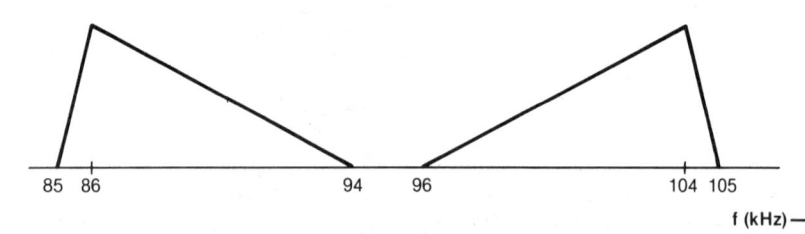

Fig. 14-DP6.

14.7 See Fig. 14-DP7.

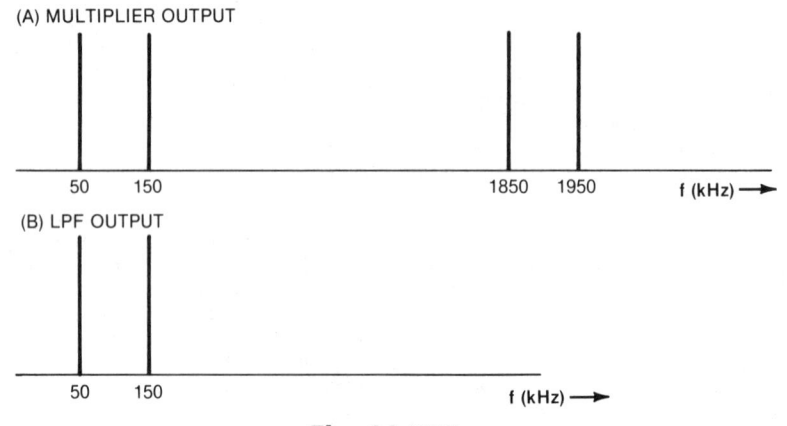

Fig. 14-DP7.

Electronics: Circuits and Systems

14.8 See Fig. 14-DP8.

(A) MULITPLIER OUTPUT

200 800   3200 3800   f (kHz) →

(B) LPF OUTPUT

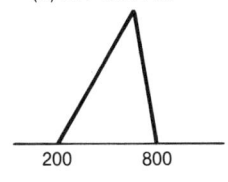

200 800

Fig. 14-DP8.

14.9 See Fig. 14-DP9.

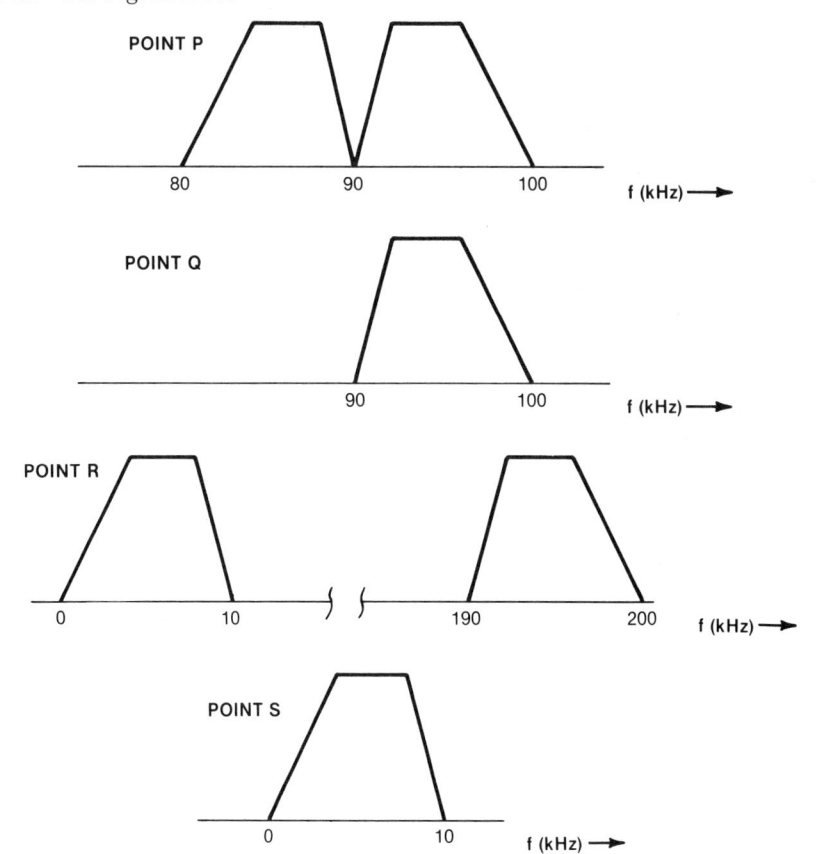

POINT P

80   90   100   f (kHz) →

POINT Q

90   100   f (kHz) →

POINT R

0   10   190   200   f (kHz) →

POINT S

0   10   f (kHz) →

Fig. 14-DP9.

14.10 $A_c^2 = 1515$. Power in carrier component = 758 W; power in the side-frequency components = 242 W.
14.11 889 W (carrier) and 111 W (side-frequency components)
14.12 See Fig. 14-DP12.

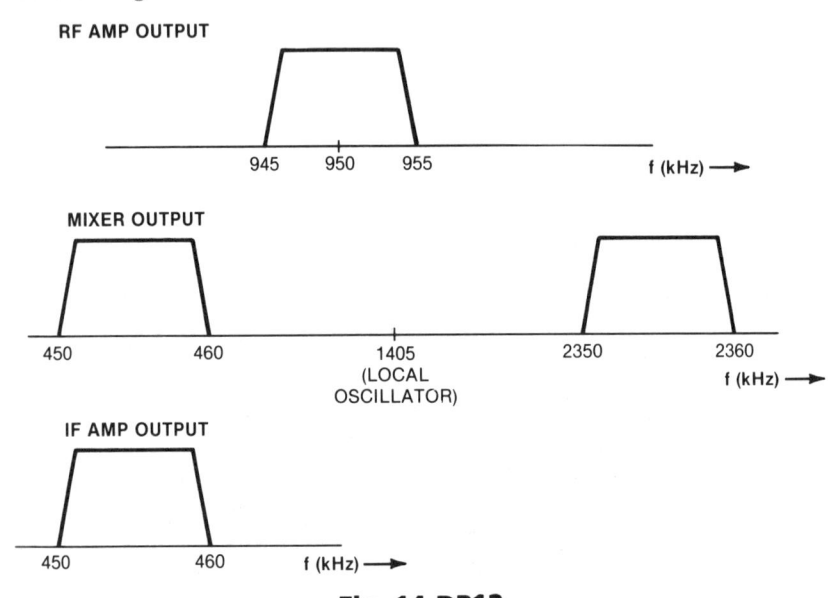

RF AMP OUTPUT

945 950 955   f (kHz) →

MIXER OUTPUT

450   460   1405   2350   2360
            (LOCAL   f (kHz) →
         OSCILLATOR)

IF AMP OUTPUT

450   460   f (kHz) →

Fig. 14-DP12.

14.13 Local-oscillator frequency = 1905 kHz. Mixer produces the bands 450 to 460 kHz and 3350 to 3360 kHz. Output of the IF amp same as before.
14.14 90 kHz.
14.15 4 V.
14.16 (a) 0.667; (b) 6.67; (c) 66.7.
14.17 No change from the values already obtained.
14.18 (a) 2; (b) 20; (c) 200.
14.19 (a) 54 kHz; (b) 150 kHz; (c) 1200 kHz (or 1230 kHz).
14.20 (a) 90 kHz; (b) 540 kHz; (c) 1540 kHz.

PROBLEMS

Power and Energy in Signals

14.1 Calculate the power (normalized 1-ohm load) for each of the following sinusoidal signals: (a) peak value = 20 V; (b) *peak-to-peak value* = 120 V; (c) RMS value = 110 V. How would these values change (that is, by what factor) if the load resistance was changed to 300 ohms? to 0.1 ohm?

14.2 A periodic nonsinusoidal signal is known to have the following components: no dc component, 10 V at f_0, 6.3 V at 2 f_0, 3.1 V at 3 f_0, 1.15 V at 5 f_0. All the other harmonics are either absent or negligibly small. Draw the power spectrum of the signal and calculate the total power in the signal (normalized load).

Frequency Translation

14.3 A signal consisting of two frequencies f_1 and 2 f_1 is multiplied by another signal consisting of two frequencies f_2 and 2 f_2. Plot the frequency spectrum (do not worry about the amplitude values) for each of the following two cases: (a) $f_2 = 500$ Hz, $f_1 = 200$ Hz; (b) 2 $f_2 = 2 f_1 = 500$ Hz.

14.4 A signal of the frequency 35 kHz is to be multiplied by another signal of frequency f_1 so that one of the resulting frequencies is at (a) 10 kHz; (b) 35 kHz; (c) 70 kHz. In each case, find both possible values of f_1 and also list the frequency of the other side-frequency component.

14.5 A signal extends from 50 Hz to 3000 Hz. Draw the spectrum when the signal is multiplied by a sinusoidal signal whose frequency is (a) 3000 Hz; (b) 3 MHz.

14.6 A signal extends from 1 kHz to 11 kHz. It is required to multiply it by a sinusoidal signal of frequency f_1 so that one of the sidebands is situated in the range of (a) 0 to 10 kHz; (b) 100 to 110 kHz; (c) 9.9987 MHz to 10.0087 MHz. In each case, find both possible values of f_1 and list the range of the other sideband.

DSB-SC AM

14.7 A DSB-SC signal consists of two frequencies: 1.008 MHz and 493.5 kHz. It is multiplied by a local carrier of frequency 1.2 MHz. List the frequencies present in the multiplier output. What bandwidth should the LPF have to pass one set of side-frequency components?

14.8 A DSB-SC signal has its sidebands extending from 940 kHz to 960 kHz and 970 kHz to 995 kHz. It is known that the original signal extended from 5 kHz to 25 kHz. Calculate the frequency of the local carrier at the demodulator and the cutoff frequency of the LPF.

14.9 Discuss the effect of the frequency of the local carrier at the demodulator of a DSB-SC system being different from that of the original carrier by 1%. Use a signal spectrum from 5 kHz to 20 kHz and $f_c = 100$ kHz in your discussion.

SSB-AM

14.10 An SSB-AM signal extends from 950 kHz to 954 kHz. It is known that the original signal extended from 0.1 kHz to 4.1 kHz. Calculate the frequency of the local carrier and the cutoff frequency of the LPF at the demodulator. If there is more than one answer, find both.

14.11 Repeat Problem 14.9 for the SSB case.

14.12 Consider the system of Example 14.5 in which two different signals are transmitted on a single carrier. Devise a scheme for demodulation and recovery of the two signals.

DSB-Large Carrier AM

14.13 If the carrier amplitude in an AM signal is 10 V and the modulation index is 0.75, calculate the power in the carrier, the power in each side-frequency component, and the total power in the AM signal.

14.14 If the total power in an AM signal is 10 kW and the power in the carrier component is 8 kW, calculate the modulation index. If the modulation index in the system is increased to 0.85 but the total power in the AM signal is kept the same as before, calculate the power in the carrier and the power in the side-frequency components. If the modulation index is kept at 0.85 but the carrier power restored to 8 kW, calculate the total power in the AM signal.

14.15 An AM signal is said to be overmodulated when m is made greater than 1. Such a signal has the same distribution of carrier frequency and side-frequency components as the AM signal with m less than 1. But, an overmodulated AM signal cannot be demodulated with an envelope detector (as can be seen from Fig. 14-24C). In order to recover the original signal, we have to use the same sort of system that was used for DSB-SC system. Discuss the recovery of the original signal in an overmodulated AM waveform by using appropriate frequency spectra.

Frequency Modulation

14.16 A signal of frequency 2 kHz frequency modulates a carrier of frequency 90 MHz. If the peak value of the signal is 5 V and the modulation index is 7, calculate the rate of variation of the instantaneous frequency. Also calculate the frequency deviation. If the signal frequency were changed to 10 kHz, how would this affect your results? If the carrier frequency were doubled, with the signal frequency kept at 2 kHz, how would this affect your results?

14.17 Calculate the modulation index in the following cases: (a) f_m = 10 kHz, frequency deviation = 156 kHz; (b) frequency deviation = 75 kHz and bandwith needed for transmission is 160 kHz; (c) f_m = 12.5 kHz and bandwidth needed for transmission = 100 kHz.

14.18 Calculate the bandwidth needed to transmit a signal that extends from 100 Hz to 10 kHz in each of the following systems: (a) DSB-SC; (b) SSB; (c) AM (DSB-LC); (d) FM with a modulation index of 0.5; (e) FM with a modulation index of 60; (f) FM with a frequency deviation of 5 kHz; (g) FM with a frequency deviation of 150 kHz.

14.19 The broadcasting of stereophonic FM uses a composite waveform that frequency-modulates a carrier at the transmitter. The composite signal is produced as follows. Two microphones are used to produce a left signal L(t) and a right signal R(t). An adder is used to produce the sum of the two signals, and a subtractor to produce the difference between the two signals in the time domain. The difference signal is multiplied by a 38-kHz carrier through a balanced modulator. The output of the balanced modulator, the sum signal, and a pilot carrier of a 19-kHz frequency are all added in a network to produce the composite signal. Draw a block diagram of this system. Assume an arbitrary spectrum for the sum signal and another for the difference signal (assume the highest frequency to be 15 kHz) and show the spectra at the different points of the system.

14.20 The composite signal in a stereophonic FM system (mentioned in the previous problem) consists of a band extending from 0 to 15 kHz, a pilot carrier at 19 kHz, and two sidebands of the difference signal one extending from 23 kHz to 38 kHz and the other from 38 to 53 kHz. At the FM receiver, the FM demodulator produces and recovers this composite signal. The composite signal is fed simultaneously to three filters: an LPF of cutoff frequency 15 kHz, a bandpass filter that passes 23 kHz to 53 kHz, and a narrowband filter centered at 19 kHz. Devise a scheme to recover the left and right signals. Assume a frequency doubler is available. Draw a block diagram of the system and show appropriate frequency spectra.

CHAPTER 15

PULSE MODULATION AND DIGITAL COMMUNICATION

The modulation schemes discussed in the previous chapter are classified as continuous wave (or cw) modulation since the transmitted waveform contains information about every instantaneous value of the modulating signal. Another system of modulation, called *pulse modulation*, uses only *samples* of the modulating signal taken at regular intervals, which are then used to modulate a carrier that is in the form of a pulse train. There are different forms of pulse modulation, and in one of them, the samples of the signal are encoded into a sequence of zeros and ones before being transmitted. Such systems are called *digital communication systems* since communication occurs in the form of the digits 0 and 1. Digital communication has become widespread since the 1970s to the point where it is displacing cw communication systems. The reason for the popularity of digital communication is the possibility of fabricating extremely complex digital systems on small chips using LSI and VLSI techniques. The processing of digital signals also has some advantages over cw communication as will be seen in this chapter.

SAMPLING THEOREM

The possibility of transmitting a signal by means of its samples, which can then be used to reconstruct the original signal at the receiving end, is based upon the *sampling theorem*. The sampling theorem is applicable only to band-limited signals; that is, signals whose frequency spectra have a *finite* upper limit. The upper limit may be as high as necessary to accommodate a given signal source, but it must be finite. Such an assumption is not too restrictive, since almost all signals are confined to a finite bandwidth because of their being processed through some sort of filtering (through amplifiers or RC filters, for example). If a signal is not band-limited, then transmission in the form of samples will lead to a distortion at the receiver.

Suppose a signal has a spectrum extending up to a frequency f_m Hz. Let the period corresponding to the maximum frequency be denoted by T_m; that is,

$$T_m = (1/f_m)$$

Then, in order to uniquely recover a signal from its samples, we must take *at least two samples for every interval of T_m seconds*. This means that the number of samples per second, called the *sampling rate*, must be at least ($2/T_m$), which equals $2f_m$ *samples per second*. A signal with a maximum frequency of 5 kHz needs a sampling rate of 10^4 samples per second, and a signal with a maximum frequency of 6 MHz requires a sampling rate of 12×10^6 samples per second.

A formal statement of the sampling theorem is: a band-limited signal having no spectral components higher than a frequency f_m is determined uniquely from its values taken at uniform intervals spaced less than $(1/2f_m)$ seconds apart.

Sampling at regularly spaced intervals is not necessary for the validity and applicability of the sampling theorem, but we will assume a uniform rate for the sake of convenience. The sampling rate of $2f_m$ samples per second is called the *Nyquist rate* and represents the *minimum* rate of sampling. If a signal is sampled less frequently than the Nyquist rate, then the reconstructed signal will exhibit distortion as we will see shortly. Sampling at exactly the Nyquist rate is sufficient, but it is usually desirable to use a somewhat higher rate so as to make the specifications of the receiver more reasonable than when the exact Nyquist rate is used. For example, in time division multiplex telephony, the voice signal is first filtered so as to limit its bandwidth to 3.2 kHz. The Nyquist rate is then 6.4×10^3 samples/sec or 6.4 kHz. But the actual rate used is 8 kHz.

In the previous discussion, it has been implicitly assumed that the spectrum of the signal extends from dc to f_m. In the case of signals that extend from some low frequency f_L (not dc) to some high frequency f_H, it is *possible to use a Nyquist rate* of $[2(f_H - f_L)]$ or *twice the bandwidth of the signal*. The use of twice the bandwidth as the Nyquist rate is valid provided certain relationships between the range of values of f_L to f_H and the sampling frequency are satisfied.

INFORMAL JUSTIFICATION OF THE SAMPLING THEOREM

A nonmathematical justification of the validity of the sampling theorem will be presented here. Rigorous mathematical proofs of the theroem can be found in advanced text books.

866 Pulse Modulation and Digital Communication

Consider a signal s(t) being multiplied by a pulse train p(t) as shown in Fig 15-1. The period of the pulse train is T_p, the output of the multiplier will consist of a series of pulses whose heights are proportional to the amplitudes of the signal at t = 0, T_p, $2T_p$, ... The output shown in Fig. 15-1D is the series of samples of the original signal being transmitted. Now we wish to figure out how the original signal can be reconstructed from these samples.

Let us look at the frequency spectra involved in the system of Fig. 15-1. Suppose the spectrum of the signal s(t) extends from dc to some frequency f_m as shown in Fig. 15-2A. The pulse train p(t) is a *periodic* signal of period T_p. From our discussion of signals in Chapter 1, the spectrum of a periodic signal consists of spectral *lines* situated at dc, the fundamental frequency $f_p = (1/T_p)$, the second harmonic $2f_p$, and so on. For the sake of convenience, let us assume the lines in the spectrum of p(t) to be of the same height as shown in Fig. 15-2B. The uniform heights are correct when the pulses in p(t) have an extremely short duration, but the discussion can be shown to be valid even when the spectral lines have different heights.

In the previous chapter, we saw that when a signal of a given frequency spectrum is multiplied by a carrier, the spectrum of the product consists of a lower sideband and an upper sideband centered at the carrier frequency. In the present system, the signal s(t) is multiplied not by a carrier of single frequency but by a pulse train. Since the pulse train contains a whole lot of frequencies, *each* of those frequencies will mix with the spectrum of the input signal and produce a pair of sidebands. The spectrum of the output of the multiplier will, therefore, consist of a series of pairs of sidebands situated at dc, f_p, $2f_p$, ..., as shown in Fig. 15-2C, where we have assumed f_p to be greater than $2 f_m$.

We observe that the spectrum of the output of the multiplier is many repetitions of the signal spectrum. In particular, we note that the original spectrum of the signal itself is available from dc to f_m (due to the sideband at dc). So all we have to do to recover the original signal is to pass the sampled signal through a low-pass filter with a cutoff frequency of f_m.

This discussion provides an informal justification of the idea that a signal can be sampled by means of a pulse train (using a multiplier and a pulse train), and the original signal can be recovered at the receiver by using an LPF.

The frequency f_p of the pulse train represents the *sampling rate* since a sample is taken at each pulse. Consider the three cases shown in Fig. 15-3: $f_c = 2f_m$, $f_c > 2f_m$, and $f_c < 2f_m$.

It is seen that when $f_p = 2f_m$, a replica of the spectrum of the signal is present in the output of the multiplier, and the signal can, therefore, be recovered by using an LPF with a cutoff frequency f_m. This is, however, only a *theoretical* possibility since the LPF will have to be ideal and *completely* suppress any

Electronics: Circuits and Systems

(A) S(t) multiplied by p(t).

(B) Period is T_p.

(C) Output of multiplier.

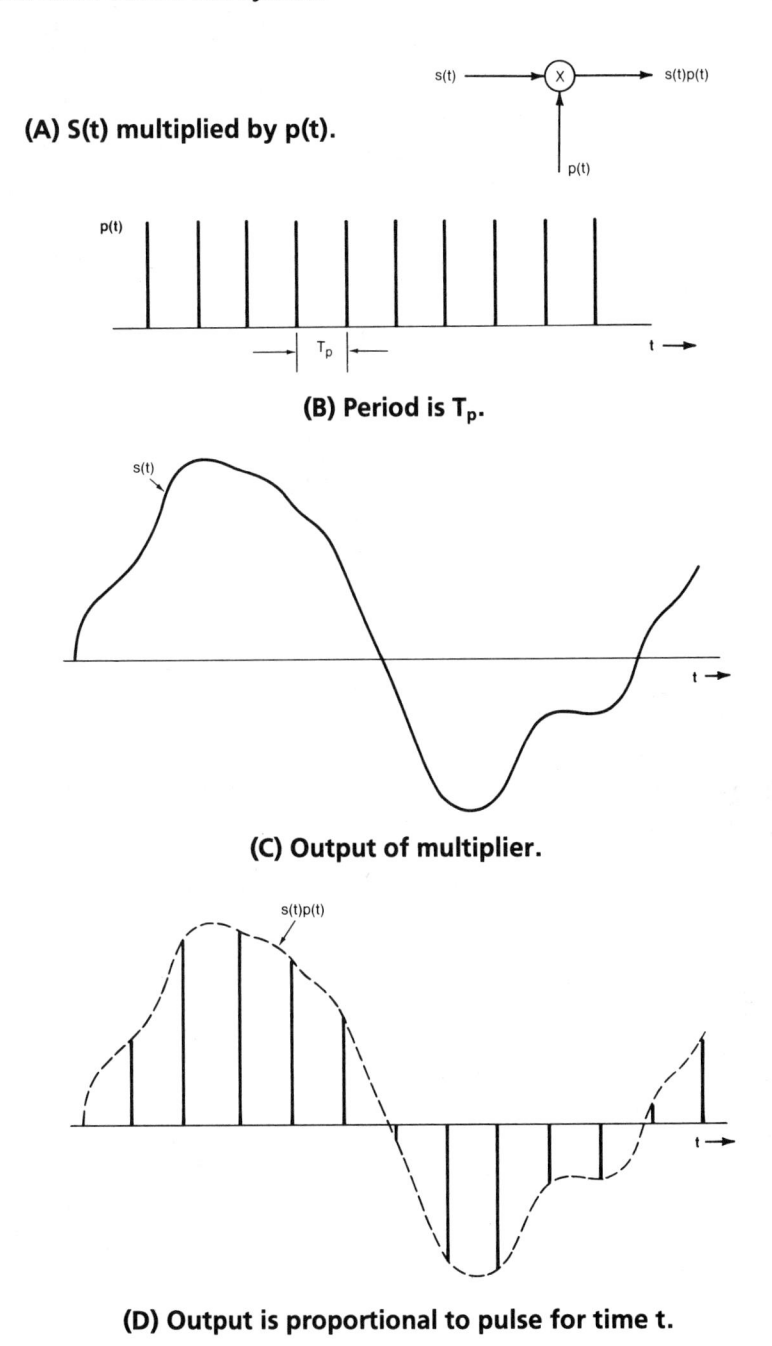

(D) Output is proportional to pulse for time t.

Fig. 15-1. Sampling of a signal by a periodic pulse train.

(A) Spectrum of the signal s(t).

(B) Spectrum of the pulse train p(t).

(C) Spectrum of the product s(t)p(t).

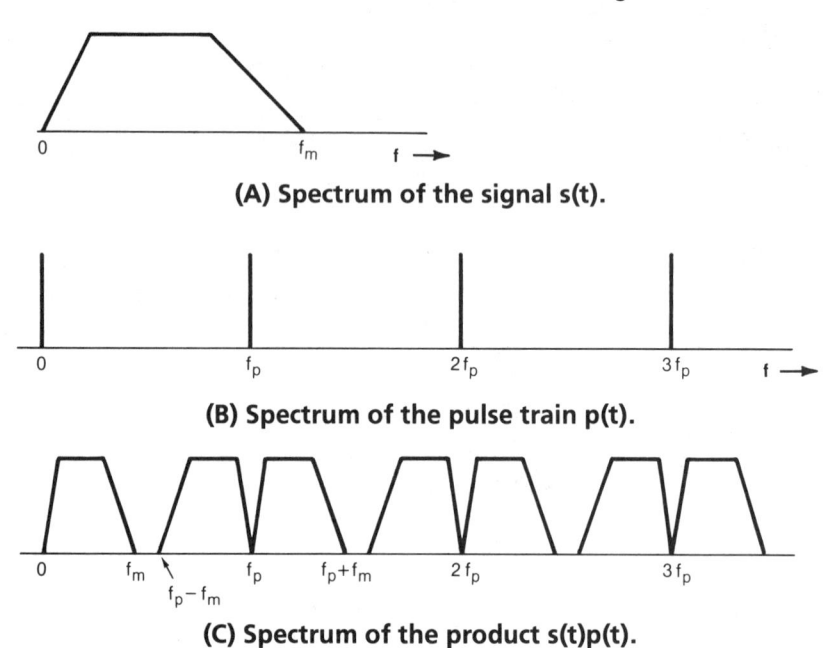

Fig. 15-2. Frequency spectra in the sampling process.

frequency above f_m. A more practical situation is seen to occur when the sampling rate f_p is made greater than $2f_m$. The LPF has the same cutoff frequency f_m as before but it need not be ideal.

Considering the case when the sampling rate f_p is less than $2f_m$, suppose an LPF with a cutoff frequency f_m is used. The spectrum passed by such a filter will be as shown in Fig. 15-3D. It is seen that part of the high-frequency end of the spectrum is folded back into a lower-frequency part of the spectrum. This means that the output of the spectrum will have distortion (relative to the original signal) at the high-frequency end. The folding back of the high-frequency portion present in this case is known as *aliasing*. Note that the distortion by aliasing cannot be eliminated by any filtering process.

It is, therefore, evident that the (theoretical) minimum sampling rate is $2f_m$, or twice the highest frequency present in the signal.

We have shown (albeit informally) that a signal sampled at a minimum rate of twice its highest frequency can be recovered completely (by means of an LPF), which is the crux of the sampling theorem.

Drill Problem 15.1: A signal has a spectrum extending from dc to 4.5 MHz. Calculate the minimum sampling rate needed to transmit the signal using pulse modulation.

Electronics: Circuits and Systems

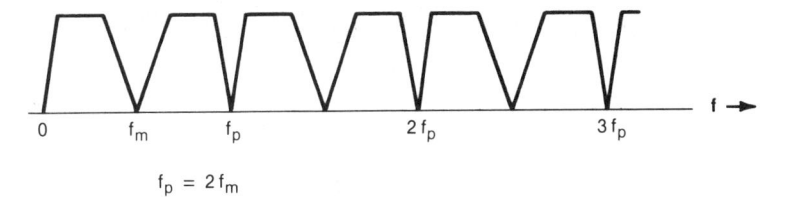

(A) Replica of the spectrum signal. ($f_p = 2f_m$)

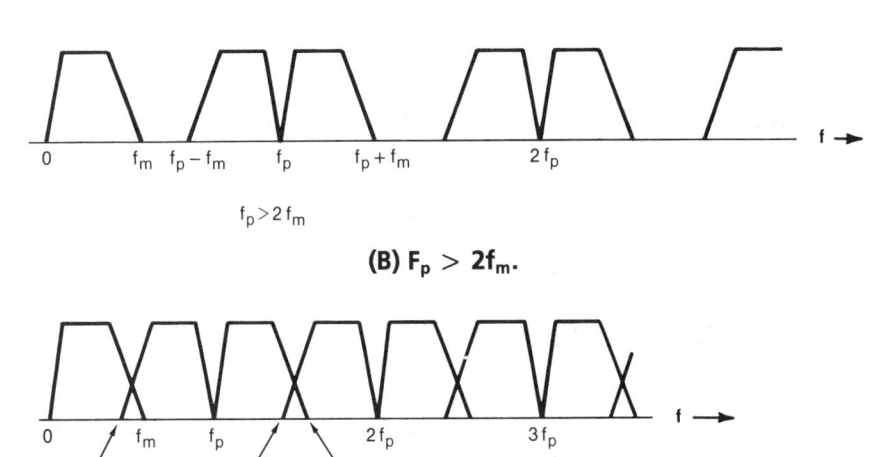

(B) $F_p > 2f_m$.

(C) When the sampling rate is less than $2f_m$.

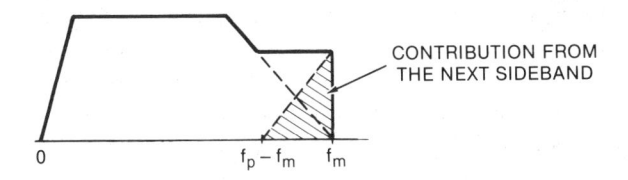

(D) Distortion (aliasing) occurs.

Fig. 15-3. Effect of the sampling rate on the frequency spectrum of the sampled signal.

Drill Problem 15.2: What is the limit on the bandwidth of a signal that can be transmitted by using a sampling rate of 5 kHz?

Drill Problem 15.3: The spectrum of a signal is shown in Fig. 15-4. The signal is sampled at a rate of 10 kHz. Sketch the spectrum of the product s(t)p(t). Also sketch the spectrum of the output of the LPF if the cutoff frequency of the LPF is 4 kHz.

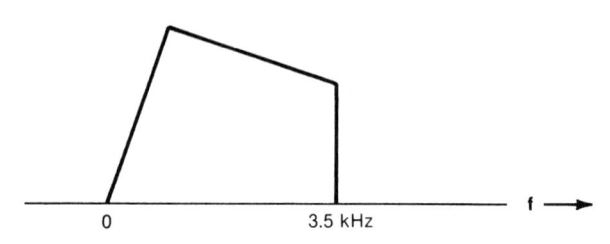

Fig. 15-4. Spectrum of the signal in Drill Problem 15.3.

Drill Problem 15.4: Change the sampling rate in the previous drill problem to 6 kHz and repeat the sketches.

PULSE-AMPLITUDE MODULATION

The series of pulses whose heights are proportional to the instantaneous values of the signal s(t) in the system of Fig. 15-1 represents a *pulse-amplitude modulation* (or PAM) waveform. In practice, pulses have a finite duration rather than the near zero width indicated in Fig. 15-1C. When nonzero width pulses are use to obtain a PAM waveform, there are two types of sampling that are possible: *natural sampling* and *flat-top sampling* as shown in Fig. 15-5. In natural sampling, the sampled waveform consists of pulses each of which is an exact replica of the signal over the entire sampling interval. But, in flat-top sampling, the samples have the same height through each sampling interval, the height being equal to the amplitude of the sample at the beginning of the interval. Flat-top sampling has the advantage that it requires relatively simple circuitry to generate it. For example, a capacitor can be charged to the amplitude of the signal at the beginning of the sampling interval and made to hold it for a prescribed length of time after which it discharges and awaits the next pulse. But, flat-top sampling has the disadvantage that the signal recovered by means of a low-pass filter has a certain amount of distortion. The distortion is due to the frequency transmission properties of a low-pass filter and can be minimized if the *duration τ of* each pulse in a p(t) is made much less than $(1/f_m)$. But, the reduction of τ to a value too low creates the problem of each sample containing an insufficient amount of energy. The more practical solution to the distortion in flat-top sampling is to use a large enough τ to provide sufficient energy in each sample and use a type of compensating network, called the *equalizer*, to follow the LPF. The equalizer

Electronics: Circuits and Systems

has a frequency transmission characteristic that is the inverse of that of the LPF.

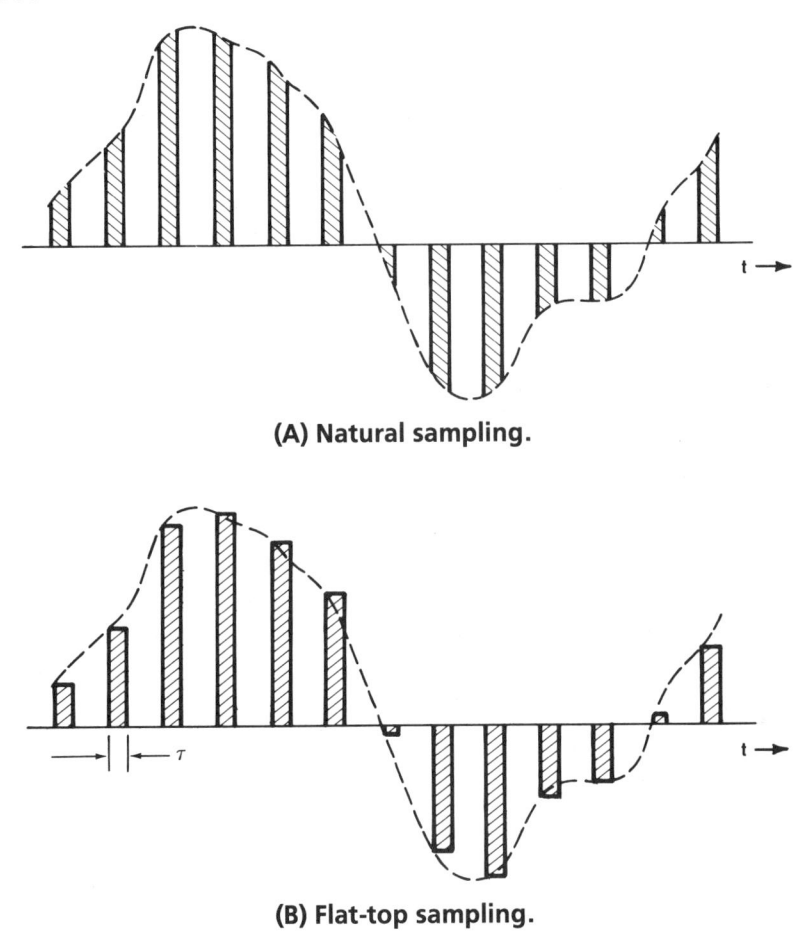

(A) Natural sampling.

(B) Flat-top sampling.

Fig. 15-5. Sampling used to obtain PAM waveform.

TIME-DIVISION MULTIPLEXING

A number of different signals can be transmitted through a single transmission channel by means of *time-division multiplexing*. Samples from the different message sources are interspersed as shown in Fig. 15-6. The number of message sources that can be multiplexed can be seen to depend upon the period T_p of the pulse train and the duration of each pulse. The larger the period T_p, the larger the number of message sources that can be multiplexed. But the value of T_p (which is the reciprocal of the sampling rate) has a minimum value of $(1/2\ f_m)$.

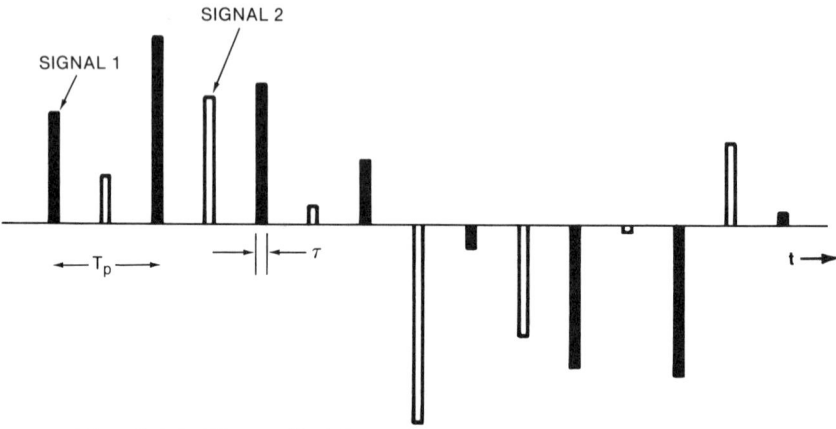

Fig. 15-6. Time-division multiplexing of two signals.

In time-division multiplexing, the total bandwidth needed for transmitting N message sources is N times the bandwidth needed for each message source.

At the receiving end, the samples of each message source have to be first separated out and then demodulated (by passing through an LPF). There is a need for synchronization between the transmitter and receiver in order to separate the samples of the different message sources correctly. A sync pulse is, therefore, transmitted along with the message samples, and it is used to synchronize a very stable local clock at the receiver.

PULSE-TIME MODULATION

The modulation of the pulse train p(t) by the signal s(t) in a pulse-modulation system can be in the form of varying the duration (or the width) of each pulse in proportion to the amplitude of the signal. Such a system is called *pulse-duration modulation* (PDM) or *pulse-width modulation* (PWM). An alternative scheme is to make the starting instant of each pulse shift by varying intervals in proportion to the amplitude of the signal. Such a system is called *pulse-position modulation* (PPM). These two schemes lead to pulse trains of the form shown in Fig. 15-7.

IC timing modules are available for the generation of PDM and PPM.

The demodulation of PDM and PPM signals is usually accomplished by first converting those signals to generate a PAM waveform (that is, pulses of different heights) that are then passed through an appropriate LPF to recover the original signal.

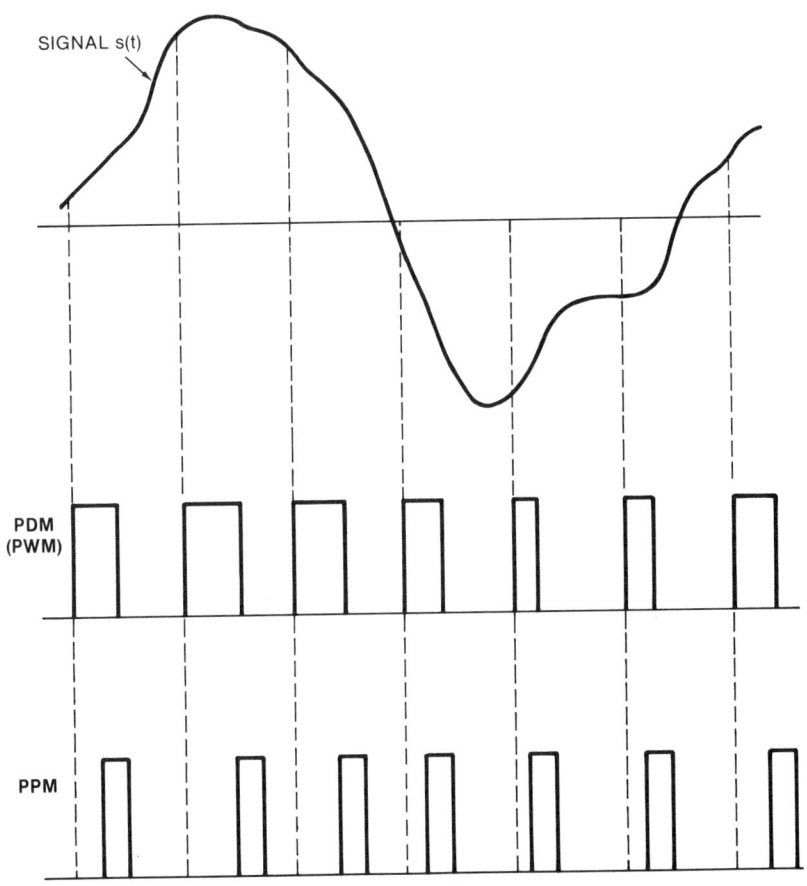

Fig. 15-7. Pulse-duration modulation (PDM) and pulse-position modulation (PPM).

PULSE-CODE MODULATION

Pulse-code modulation (PCM) involves the conversion of the samples to be transmitted into a coded form (made up of a sequence of zeros and ones) and then transmitting the code words. It forms the basis of digital (data) communication, and it offers several attractive features. The signals are either the presence of a pulse (corresponding to the digit 1) or the absence of a pulse (corresponding to the digit 0). Since only the presence of a pulse is to be detected and not the height or width, the pulses can be reshaped at various points in a long-distance communication system so as to compensate for any deterioration caused by the transmission. The processing of the transmitted

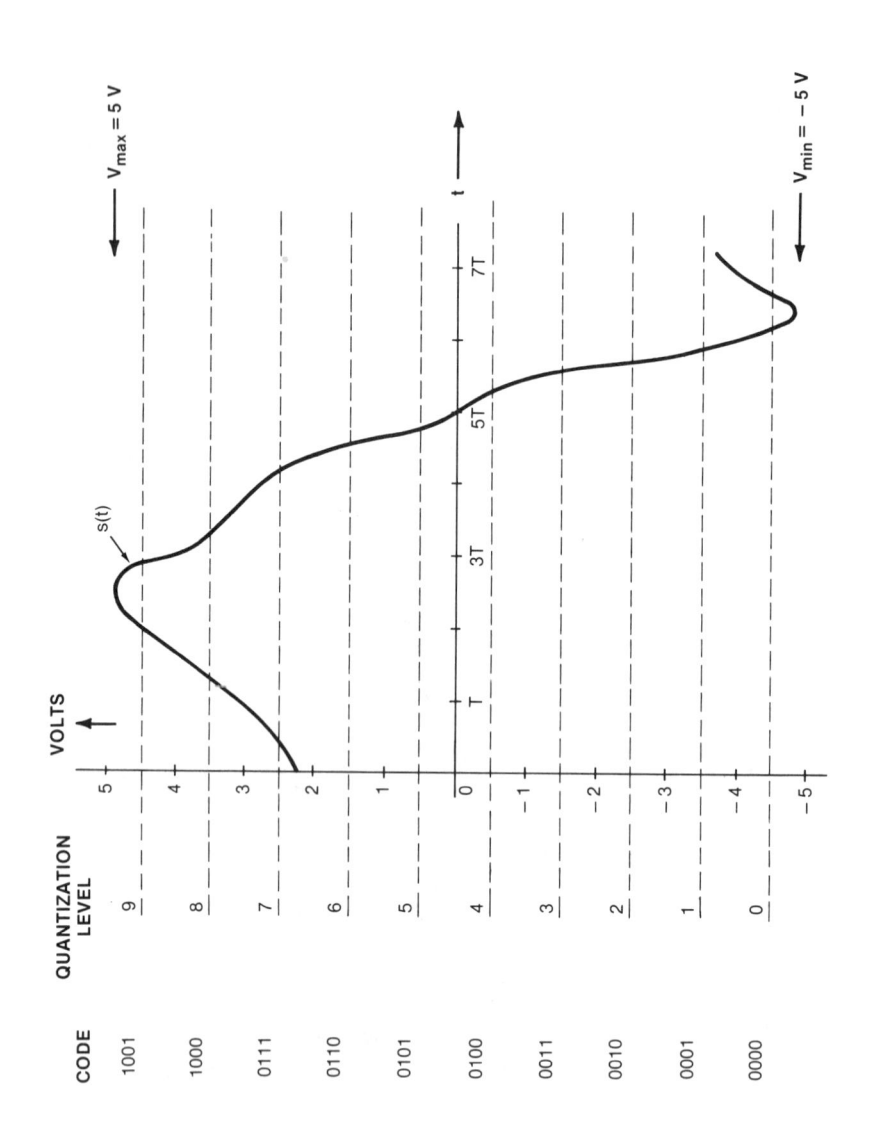

(A) Quantization levels and code words.

Fig. 15-8. Pulse-code

Electronics: Circuits and Systems 875

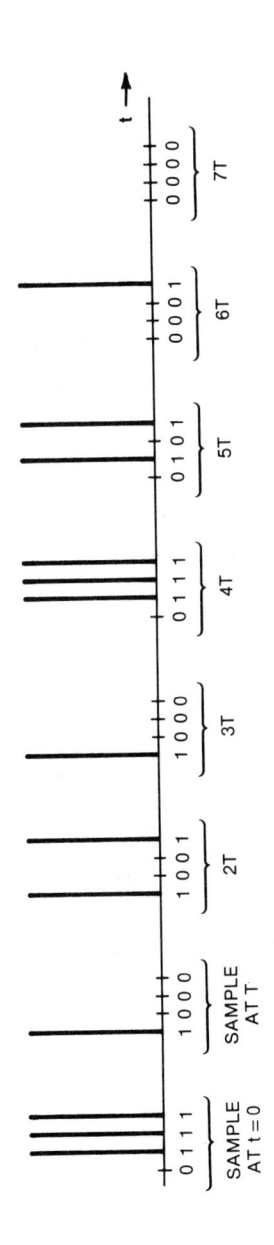

(B) Code words transmitted for the sample signal.

modulation (PCM).

signals can be done through high-speed computers that can handle more information than other types of communication systems. The noise and interference in PCM systems can be minimized by using appropriate coding schemes.

Consider a signal v(t) with a minimum and a maximum value of, respectively, V_{min} and V_{max}. The range [$V_{max} - V_{min}$] is first divided into k voltage intervals. The middle of each level is called a *quantization level,* and the quantization levels are numbered 0, 1, . . . , (k-1) starting with the lowest level. These numbers 0, 1, . . . , (k-1) are then coded into m-bit sequences of zeros and ones (called m-bit code words) using some form of encoding scheme. The number of bits m and the number of quantization levels k are related by the equation:

$$2^m = k$$

When 2^m is not exactly equal to k, the next higher value of m is chosen.

The instantaneous value of a signal at each sampling instant is then transmitted as an m-bit code word corresponding to the quantization level closest to it.

As an example, suppose a signal has a minimum value of −5 and a maximum value of +5, and we choose to divide the 10 V range into 10 intervals, each of width 1 V, as shown in Fig. 15-8. The quantization levels are −4.5, −3.5, . . . , 3.5, and 4.5 V as indicated, and they are numbered, respectively, 0, 1, 2, . . . , 9. Since there are 10 quantization levels, the number of bits needed is 4 (2^3 is only 8 and so we need to go to 4, since $2^4 = 16$, which is higher than 10). The coding scheme is shown in Table 15-1.

Table 15.1 Pulse-Code Modulation Coding Scheme

Quantization Level	Voltage	Code
0	−4.5	0000
1	−3.5	0001
2	−2.5	0010
3	−1.5	0011
4	−0.5	0100
5	0.5	0101
6	1.5	0110
7	2.5	0111
8	3.5	1000
9	4.5	1001

For the particular signal shown in Fig. 15-8 suppose the sampling instants are 0, T, 2T, . . . Then, the instantaneous amplitudes at the sampling instants are coded by going to the nearest quantization level. These code words are transmitted, and the transmitted waveform will be as shown in

Electronics: Circuits and Systems

Fig. 15-8B. As an example, at t = 3T, the signal is 3.8 V. The closest quantization level is 3.5 V, and the code transmitted is 1000 as given by Table 15-1.

Drill Problem 15.5: The signal shown in Fig. 15-9 is quantized by using the levels and codes of Table 15-1. Sketch the transmitted PCM waveform.

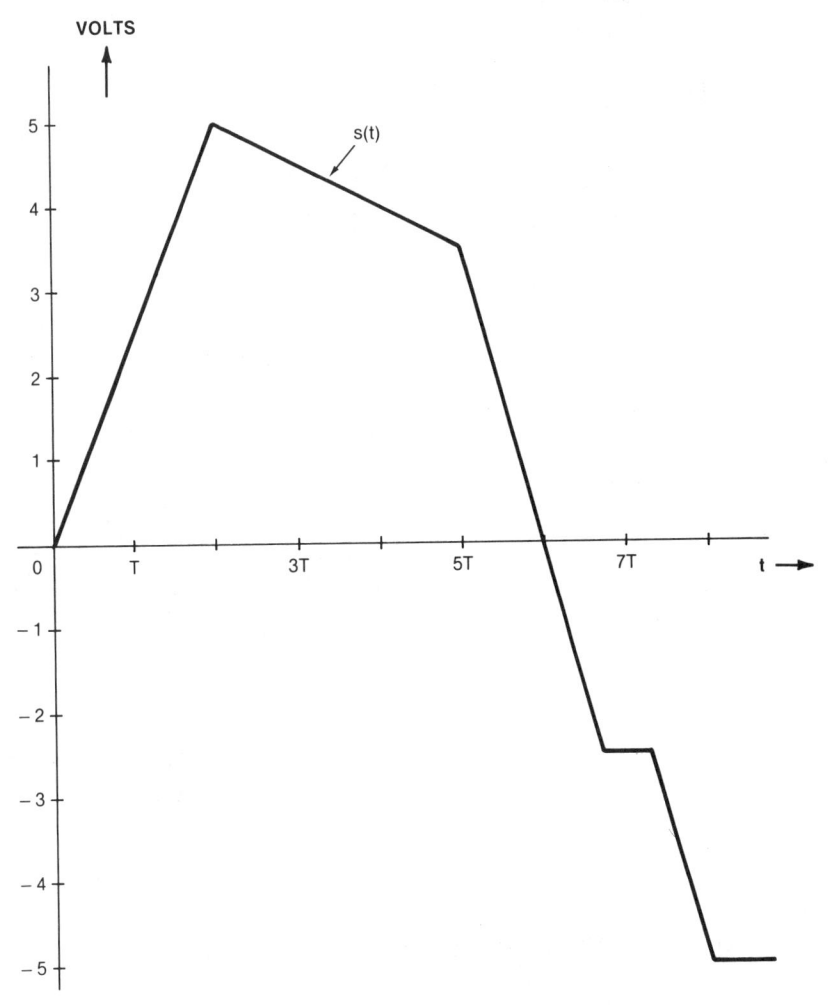

Fig. 15-9. Signal for Drill Problem 15.5.

878   **Pulse Modulation and Digital Communication**

At the receiver in the PCM system, the presence or absence of pulses in each code word is detected, and the digital word is converted into an analog voltage level by using digital-to-analog converters. The output of the D/A converter is a PAM signal, which is passed through an LPF to recover the original signal. For the input PCM waveform of Fig. 15-8B the output of the D/A converter will be a staircase function as shown in Fig. 15-10. We see that some of the finer details in the original signal are lost due to quantization, which leads to a quantization error.

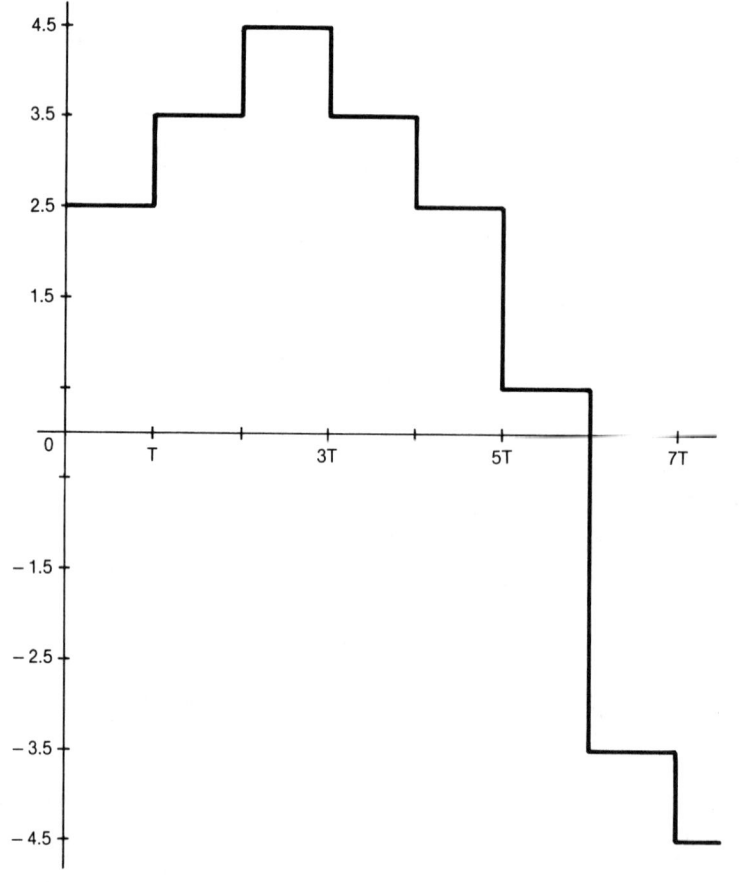

Fig. 15-10. D/A converter output in the PCM receiver.

Drill Problem 15.6: Sketch the output of the D/A converter for the PCM waveform of Drill Problem 15.5.

Electronics: Circuits and Systems

The complete block diagram of a PCM system is shown in Fig. 15-11.

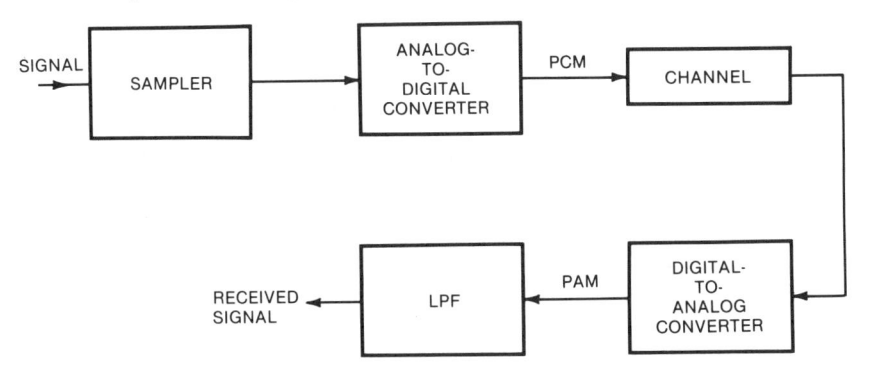

Fig. 15-11. Block diagram of a PCM receiver.

Quantization Error in PCM

We can see from the waveform in Fig. 15-10 that we cannot distinguish between voltage levels separated by small values in the original signal in a PCM system because of the quantization. For example, the voltage values of 1.2 and 1.9 will both be taken as being at the quantization level 6 in Table 15-1 and transmitted as 0110. At the receiving end, there is no means of distinguishing between the two, and both will be converted to 1.5 V by the D/A converter. An error caused by the quantization process is called the *quantization error*.

Quantization error is defined as the difference between the actual signal voltage at the sampling instant and the quantization level closest to it.

The quantization error can be related to a quantization noise power and a signal-to-noise ratio (SNR) due to quantization can be determined. The quantization error and SNR can be reduced by increasing the number of quantization levels for the voltage range in a given signal. For example, with 16 levels, the SNR is found to be 34.9 dB, while 128 levels lead to an SNR of 52.9 dB. The price we pay for the higher SNR is the larger number of bits per code word to accommodate the larger number of quantization levels. The length of the code word affects the bandwidth needed for the transmission of PCM signals as will be seen in the following discussion.

Bandwidth of a PCM System

First consider a PAM system in terms of bandwidth requirements. If a signal has a bandwidth of f_m Hz (that is, its spectrum extends from dc to f_m), then there are $2f_m$ samples per second to be transmitted. In a PAM system, the bandwidth needed is f_m since the received signal is immediately passed through an LPF of cutoff frequency f_m at the receiver.

Pulse Modulation and Digital Communication

Therefore, in a PAM system, the number of pulses transmitted per second is $2f_m$, and the bandwidth needed is one half of that number, or f_m.

Now consider the PCM system. Again the number of *samples* per second is $2f_m$. Suppose the number of quantization levels is k, and the number of bits per code word is m, where m and k are related by

$$2^m = k$$

or

$$m = \log_2 k$$

Then, each transmitted sample requires m pulses to denote the m bits. Therefore,

$$\text{number of pulses transmitted} = 2mf_m \text{ per sec}$$
$$= 2f_m [\log_2 k]$$

By using the fact that the transmission of a number of pulses per second requires a bandwidth of one half that number, we get

$$\text{bandwidth of the PCM system} = mf_m \text{ Hz} \qquad (15\text{-}1)$$
$$= [\log_2 k]f_m \text{ Hz} \qquad (15\text{-}2)$$

Either of the two equations can be used in the calculation of the bandwidth. Remember that when $\log_2 k$ is not a whole number, the next higher value of m is used.

As an example consider a signal band limited to 10 kHz and let the number of quantization levels k = 16, then m = 4. The bandwidth needed is 40 kHz. If the number of quantization levels is increased to k = 100, then m must be 7 (since 2^6 is 64 and 2^7 is 128). The bandwidth needed becomes 70 kHz.

Drill Problem 15.7: Calculate the bandwidth needed in a PCM system using 5-bit code words and signals band limited to 15 kHz.

Drill Problem 15.8: Repeat the calculation of the previous drill problem, if the number of quantization levels is to be 200.

Drill Problem 15.9: A PCM system has a bandwidth of 100 kHz. If a signal of bandwidth 3.2 kHz is to be transmitted, calculate the number of bits per word that can be accommodated.

Consider the effect of the bandwidth on the number of quantization levels. If B denotes the bandwidth, then from Eqs. (15-1) and (15-2)

$$B = mf_m$$
$$= [\log_2 k]f_m$$

Electronics: Circuits and Systems

which leads to

$$k = 2^{(B/f_m)} \qquad (15\text{-}3)$$

Therefore, the available increase in the number of quantization levels for a given increase in bandwidth is exponential. A change in the bandwidth by a factor of three permits us to change the number of quantization levels by a factor of eight. Since a large number of quantization levels leads to a large SNR, it follows that the trade-off of bandwidth to SNR is quite favorable in a PCM system compared with other modulation schemes. A comparison of FM systems (in cw modulation) and PCM shows that at low signal-to-noise ratio levels of the input to a receiver, the PCM is a better system than the FM system. But, as the signal-to-noise ratio level increases, the FM systems are found to be superior to PCM.

The most important advantage of PCM is that repeater stations can be used in a long-distance communication system to reshape and retransmit the pulses so as to compensate for their degradation introduced by propagation. Therefore, assuming that there has been no error in the detection of the pulses at any repeater station, there is no accumulation of noise from one section of the system to the next. By increasing the number of repeater stations, it is possible to keep the effect of noise at a desired low level.

The detection of a pulse in the received signal is based on whether or not the received signal exceeds a specified *threshold level* in each slot where a bit of the code word is occurring. The choice of the threshold level is arrived at through statistical considerations of the noise in the system and the probability of error.

Companding in PCM Systems

The quantization error (when the quantization intervals are equal in size) is much more serious at low voltage levels than at high voltage levels. For example, in the scheme of Table 15.1, a voltage value of 0.1 V in the signal will be treated as a 0.5-V quantization level and will lead to a 0.5-V output at the receiver, which represents a 500% error, whereas a 4.1 V signal will be treated as 4.5 V, which is only a 10% error.

Therefore, it is desirable to use *unequal* voltage intervals in the process of quantization: smaller intervals for the low voltage range and large intervals for the high voltage range. The process of tapering the interval sizes from small at low signal levels to large at high signal levels is called *compressing*. At the receiver, the intervals have to be restored to the normal equal levels by reversing the compressing operation. This reversal of compression is called expanding. The combined operation of compressing at the transmitter and expanding at the receiver is called *companding* and is used in some communication systems, especially in telephone systems. It has been found empirically that speech signals are at levels below one fourth of the rms value for 50% of the time and companding is a desirable operation.

The compression of the quantization levels is performed by means of a circuit with nonlinear transfer characteristics as shown in Fig. 15-12. As can be seen from the diagram, the large amplitude inputs undergo a smaller gain than the small amplitude signals. The expanding is obtained by using a circuit with exactly the inverse characteristics of the compressing circuit, as shown in Fig. 15-12B.

(A) Compression of the quantization levels through a nonlinear circuit.

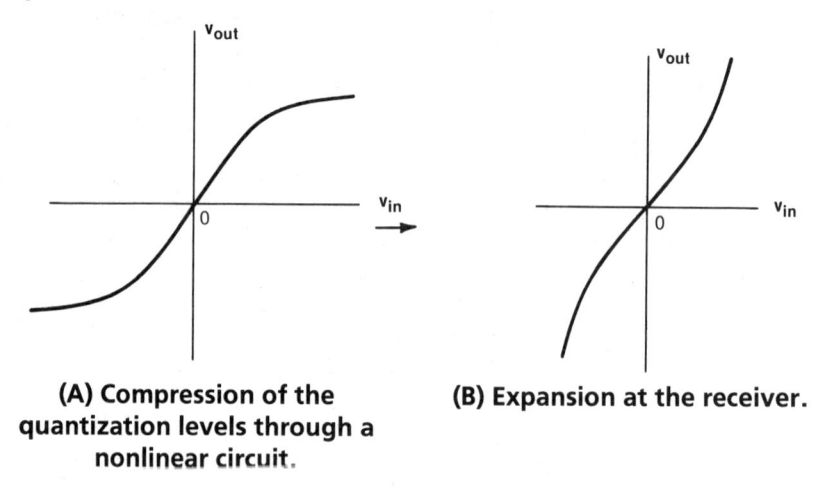

(B) Expansion at the receiver.

Fig. 15-12. Companding in a PCM system.

Baseband PCM Transmission

Baseband transmission refers to the sending of the digits of the code words directly from a transmitter to a receiver. The digits 1 and 0 can be sent, respectively, as high and low voltage levels as in the conventional binary system. Baseband transmission requires links between the transmitter and receiver in the form of telephone lines or optical fibers.

A variety of schemes is used for sending the digits 0 and 1 besides the conventional binary format where a 0 is transmitted as a *low* level and a 1 as a *high* level. Several such schemes are shown in Fig. 15-13. In each case, the digital signal being transmitted is the sequence: 0 1 0 0 1 1 1 0 1 0. The situation in Fig. 15-13A is called *NRZ-L* or *nonreturn-to-zero level* and is seen to be the conventional binary form. Each 1 is represented by a high pulse and each 0 by a low pulse, and the pulse width is the same for a 0 and a 1.

The situation in Fig. 15-13B is called *RZ*, or *return to zero*. A 0 is transmitted by a low pulse for the duration of the 0 signal. When a 1 is to be transmitted, a high pulse of half the width of a normal pulse is inserted in the interval.

(A) NRZ-L.

(B) RZ.

DATA-BIT INTERVAL

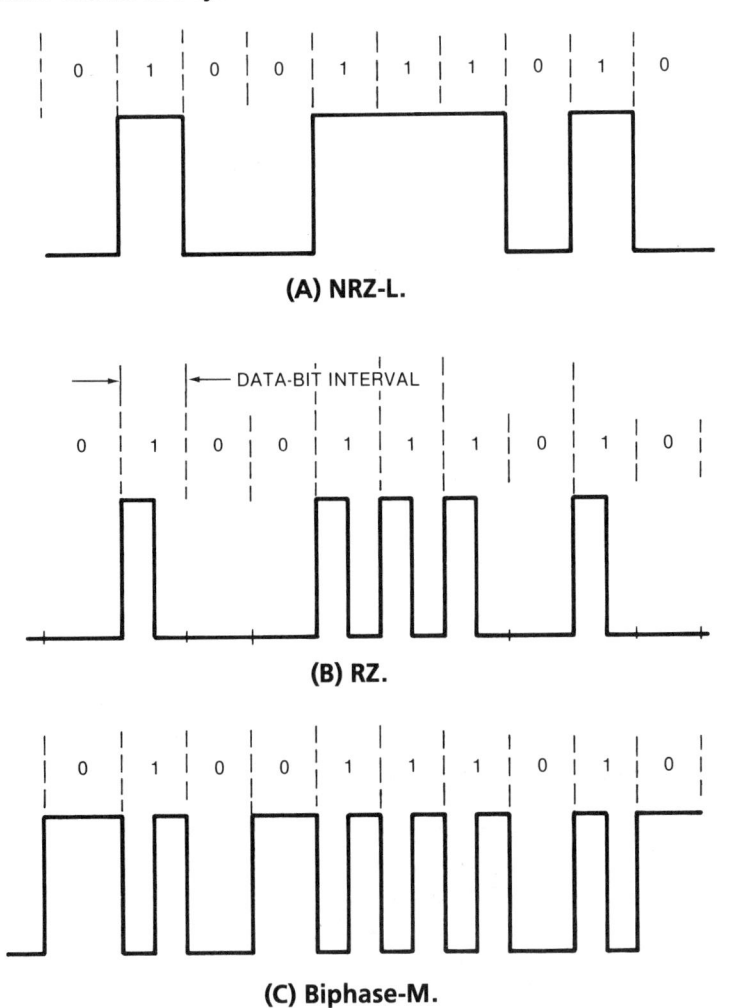

(C) Biphase-M.

Fig. 15-13. Different formats for transmitting digital data.

The scheme in Fig. 15-13C is called *biphase-M* or *biphase mark*. (The term *mark* is used to denote a binary 1, and the term *space* is used to denote a binary 0 in digital communication systems.) In the biphase-M system, there is *always* a change in the pulse level at the beginning of the pulse interval. If a 0 is to be transmitted, the pulse stays constant through the width of the pulse. If a 1 is to be transmitted, there is a *second* transition at the middle point of the interval.

Each of the previous formats has its advantages and disadvantages. For example, the *shortest* pulses in the NRZ format have a duration equal to the width of the data bit interval, but the shortest pulses in the RZ and biphase formats have a width of half that interval. Since the bandwidth of a system is inversely proportional to the pulse duration, we can see that the NRZ format will need only half the bandwidth of the other two formats. On the other hand, the RZ and NRZ formats require the system to be able to respond to dc since the pulse level may not change between data bit intervals in these two formats. But in the biphase format, there is always at least one transition in each interval, and the transmission of the biphase format does not require dc coupling.

Drill Problem 15.10: Sketch the waveforms of the transmitted PCM signals for the sequence: 1 0 1 1 0 0 1 1 1 0 1 using each of the three formats NRZ-L, RZ, and biphase-M.

Radio-Frequency Transmission of PCM Signals

Baseband PCM signals have spectra extending from dc up to some frequency determined by the pulse duration. As such they cannot be transmitted except through telephone wires or some such links. For transmission through free space in the form of electromagnetic waves, it is necessary to use some form of radio-frequency (RF) modulation schemes in conjunction with PCM signals. The three methods used widely in digital communications are *amplitude shift keying (ASK)*, *frequency-shift keying (FSK)* and *phase-shift keying (PSK)*. In each of the three schemes, a carrier at a high frequency (the RF carrier) is modulated in some form by the data bits 0 and 1 in the PCM signal to be transmitted.

In ASK, the data bit 1 is transmitted as a burst of the RF carrier and the bit 0 as the absence of an RF carrier. In FSK, the carrier frequency is increased by f_0 when the data bit 1 is transmitted and decreased by f_0 when the data bit 0 is transmitted. In PSK, the frequency of the carrier is kept constant but a 180° phase shift is introduced whenever there is a change in the incoming data bits from 0 to 1 or from 1 to 0. The diagrams of Fig. 15-14 illustrate the three keying schemes.

The generation of amplitude shift keying (ASK) signals uses exactly the same method as that of the DSB-SC signal studied in the last chapter. The PCM baseband signal (made up of the bits 0 and 1) is multiplied by an RF carrier, and the product is the desired ASK signal.

Frequency-Shift Keying (FSK)

In FSK, two frequencies are transmitted: a burst of frequency $(f_c + f_0)$ when a 1 is to be transmitted and a burst of frequency $(f_c + f_0)$ when a 0 is to be transmitted, where f_0 is a constant offset from the carrier frequency. A

(A) Signal to be transmitted.

(B) Amplitude shift keying.

(C) Frequency shift keying.

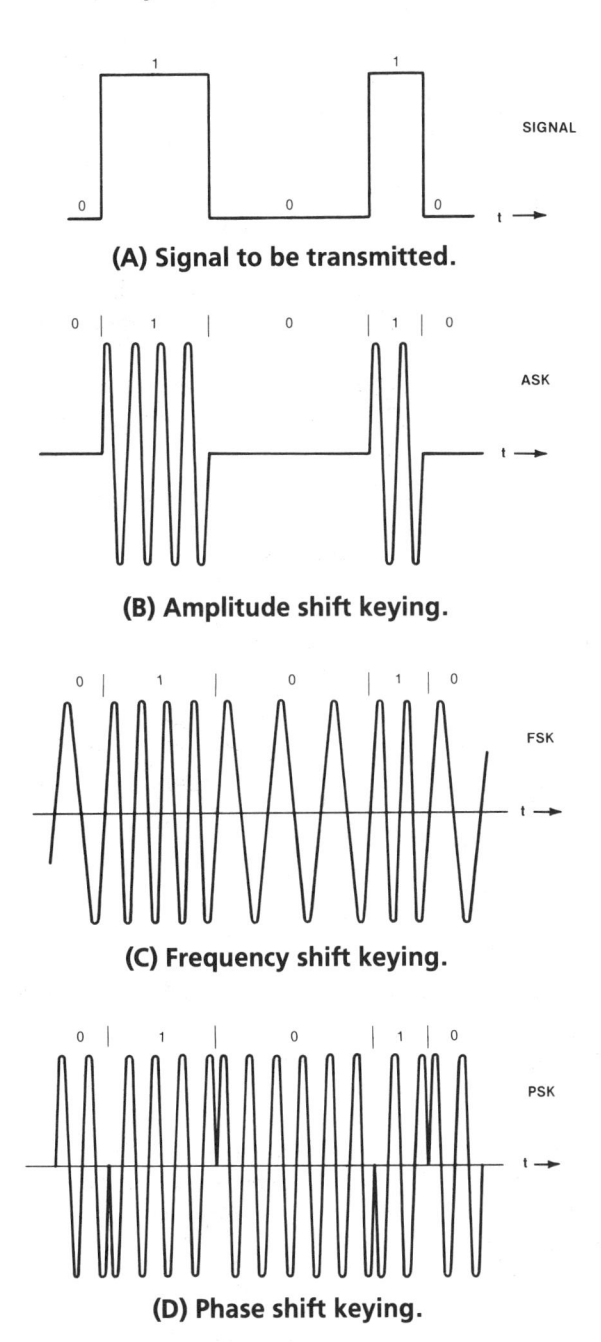

(D) Phase shift keying.

Fig. 15-14. Radio-frequency transmission of PCM signals.

voltage-controlled oscillator (VCO) is used to produce oscillations at two distinct frequencies as indicated in Fig. 15-15.

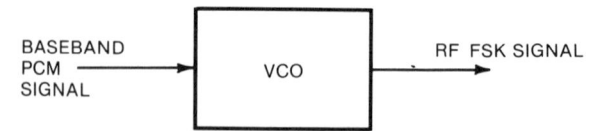

Fig. 15-15. Generation of FSK signal using a voltage-controlled oscillator (VCO).

The demodulation of the FSK signal requires the translation of the two frequencies $(f_c + f_0)$ and $(f_c - f_0)$ into two voltage levels. A demodulator for FSK is shown in Fig. 15-16.

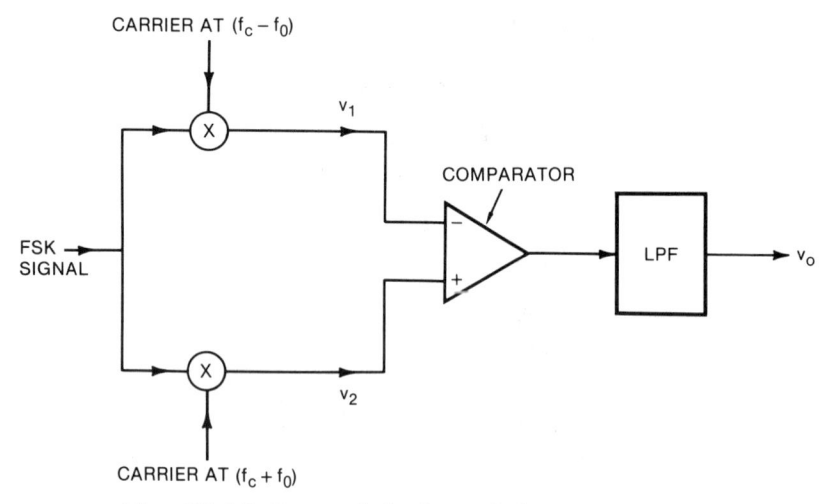

Fig. 15-16. Demodulation of the FSK signal.

The two voltages v_1 and v_2 are product signals, and as such, each will contain the sum and difference frequencies of their inputs. If the incident signal has a frequency $(f_c + f_0)$, then v_1 will consist of the two frequencies $2f_c$ and $2f_0$, while v_2 will consist of the two frequencies: dc and $2(f_c + f_0)$. The comparator output will contain a *positive* dc component (and other components as well) since the inverting input has no dc component while the noninverting input has a dc component. The LPF will pass only the dc component, and the output will be positive. If the incident signal has a frequency $(f_c - f_0)$, then the previous situation is reversed and the output of the comparator will have a negative dc component that will be transmitted to the output v_0. The two levels of v_0 can be interpreted as the two levels of PCM signals. The PCM signals are then converted to PAM signals by means of a D/A converter and passed through an LPF to recover the original (analog) signal.

Phase-Shift Keying

In PSK systems, the frequency of the RF carrier is not changed but a 180° phase angle is introduced whenever there is a change in the incoming data bits. The system is quite simple since all that is needed is a multiplier as shown in Fig. 15-17. The output is $+ V_o \cos(2\pi f_c t)$ when the incoming data bit is 1 and $- V_o \cos(2\pi f_c t)$ when the incoming data bit is 0.

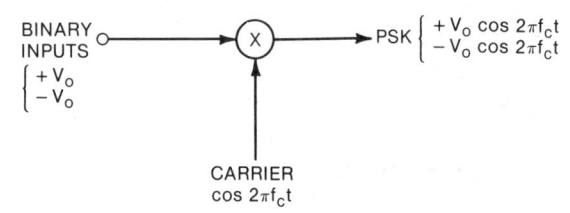

Fig. 15-17. Generation of PSK signals.

The demodulator for PSK is shown in Fig. 15-18. The output of the square-law device contains a dc component and a component at $2f_c$ (since the input signal of frequency mixes with itself in a square-law device). The $2f_c$ component is passed by the bandpass filter, and a carrier of frequency f_c is generated by the frequency divider. This carrier is multiplied by the PSK signal itself, and the product will contain the sum and difference frequencies: dc and $2f_c$. The dc component will be positive when the input is $+V_o \cos(2\pi f_c t)$, and negative when the input is $-V_o \cos(2\pi f_c t)$. The positive and negative dc components correspond to the levels of PCM signals. This PCM signal is converted to PAM signals by a D/A converter and then passed through an LPF to recover the original signal.

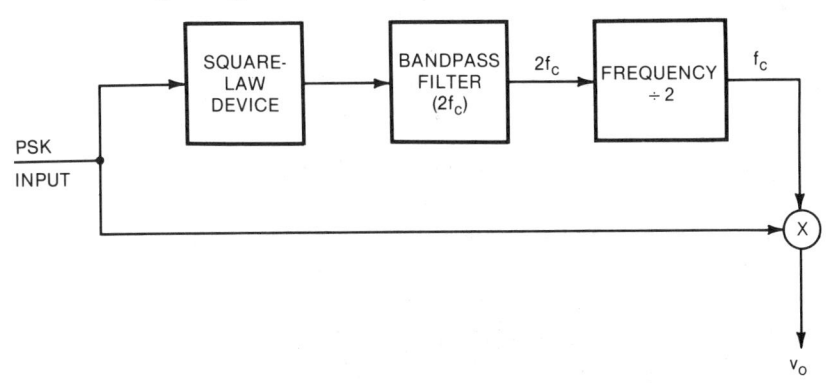

Fig. 15-18. Demodulation of PSK signals.

DELTA MODULATION

Delta modulation (DM) is similar to PCM in the sense that information about the signal is sent in the form of pulses but the pulses do not directly represent the amplitude (or a quantization level) of the signal. On the other hand, the transmitted pulses indicate whether a signal is *increasing or decreasing* in amplitude from one interval to the next.

The DM system generates a staircase function that is compared with the input signal. If the staircase function is *higher* than the input signal in a sampling interval, then the next step of the staircase is made *downward* and at the same time a pulse (of a very short duration) of *negative* polarity is transmitted. If the staircase function is *lower* than the signal in a sampling interval, then the next step of the staircase is made *upward* and a *positive* pulse is transmitted. As can be expected, a feedback between the output and the input is necessary to generate the staircase function.

Staircase Function

The block diagram of a DM generating system is shown in Fig. 15-19. The two inputs to the differential amplifier (or comparator) are the actual signal $s(t)$ to be transmitted, and a staircase function $\bar{s}(t)$ that is generated by the system itself. Assume for a moment that a step of the staircase function is already present. If the signal is higher than the step of the staircase function, then the output $\Delta(t)$ will be positive. This positive output is used to multiply the input pulse from the pulse generator by +1. The output of the sign multiplier $p_0(t)$ is a positive pulse, as shown in Fig. 15-19B. The positive pulse output of the sign multiplier is integrated and forms the next upward step. On the other hand, if the signal is lower than the step of the staircase function, then the output $\Delta(t)$ of the comparator is negative. This causes the input pulse from the generator to be multiplied by a negative sign, and a negative pulse is produced as shown in Fig. 15-19C. The negative pulse is integrated and forms the next downward step of the staircase function.

For a general signal $s(t)$, the staircase function and the output pulse train are as shown in Fig. 15-19D. It is seen that whenever the step in the staircase is higher than the signal, a negative output pulse and a downward step are generated; and whenever the step in the staircase is lower than the signal, a positive output pulse and an upward step are generated. The DM output is, therefore, a series of positive and negative pulses of the same height and duration. Therefore, what is being transmitted is not the coded representation of the signal amplitude itself as in PCM, but information about the *difference* between the signal and its staircase approximation $\Delta(t)$. This is the reason for the name *delta* modulation.

At the receiving end of the system, Fig. 15-20, the quantizer reshapes each

Electronics: Circuits and Systems

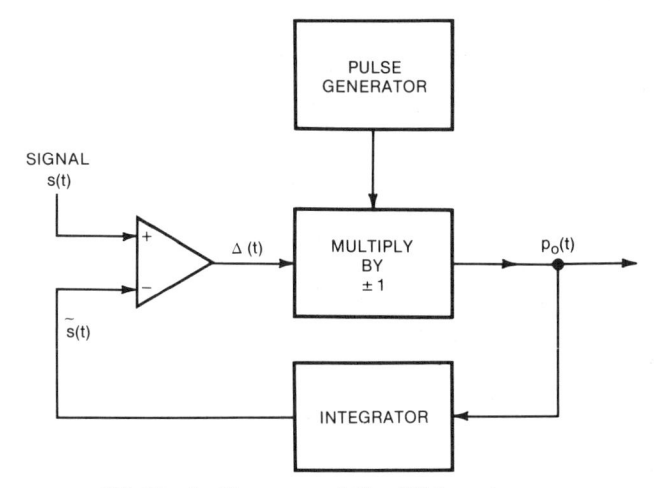

(A) Block diagram of the DM system.

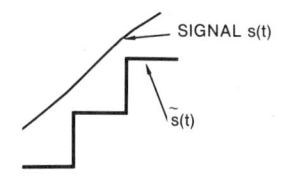

(B) When the signal is higher than the step in the staircase function, the comparator generates a positive step.

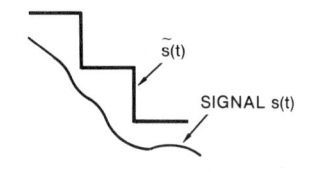

(C) When the signal is lower than the step in the staircase function, the comparator generates a negative step.

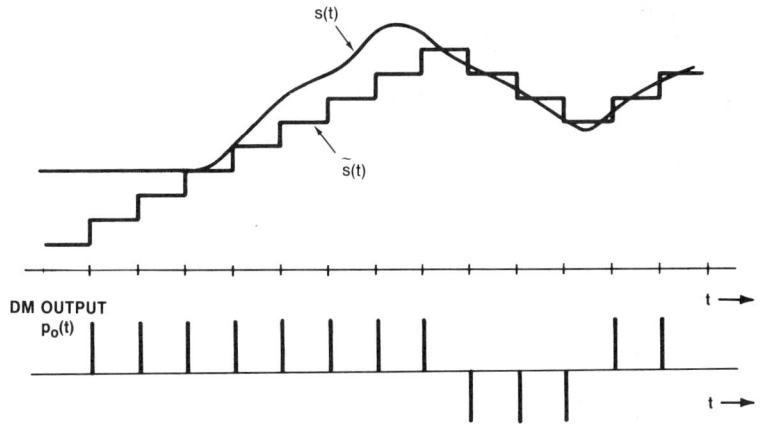

(D) A sample signal and corresponding DM output.

Fig. 15-19. Generation of delta modulated signals.

(A) Demodulator.

(B) Output.

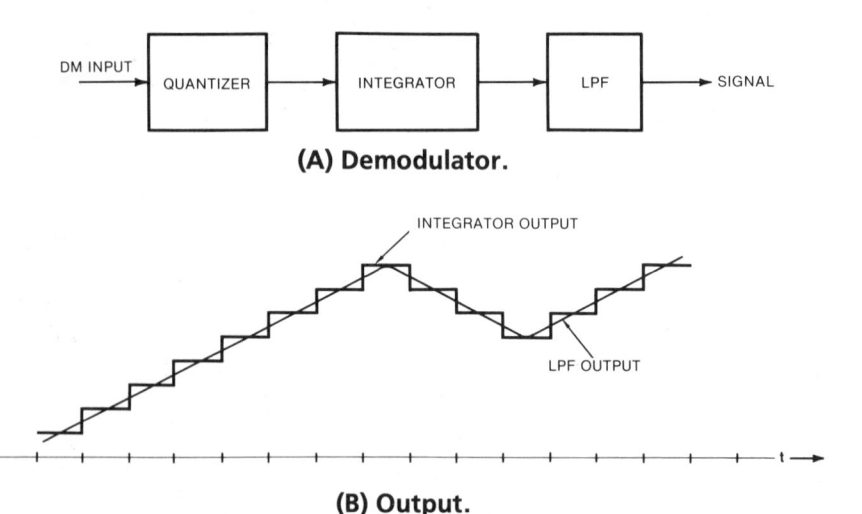

Fig. 15-20. Demodulation of a DM signal.

pulse which is then integrated to produce a staircase function similar to the one at the transmitting end. The staircase function is passed through an LPF to produce the original signal (or a close approximation thereof). The output signal differs from the original signal because of the effects of the stepwise approximation introduced by the staircase function.

It is possible to run into difficulties in the DM system when steps of fixed height are used in the staircase function, which is shown in Fig. 15-21, where the *slope* of the signal is greater than the rate at which the steps of the staircase function can build up. (A similar problem exists with signals of negative slope also.) The system is said to be "slope overloaded" in such cases. This difficulty can be remedied by using *adaptive* delta modulation.

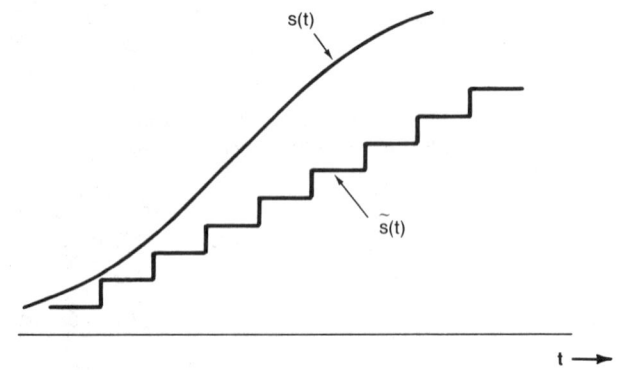

Fig. 15-21. Slope overloading in a DM system.

Adaptive Delta Modulation

In an adaptive DM system, the *height of each step* in the staircase function is made to *vary in proportion to the difference between the actual signal and the staircase signal*. When the difference between the staircase step and the signal is small, the next step is made short; but when the difference between the staircase step and the signal is large, then the next step is made high. The variation in the height of the steps is attained by using a *variable-gain amplifier* in the feedback link between the output of the DM system and the staircase input terminal of the comparator. The *gain* of the amplifier is small when the input to it is small, and the gain is high when its input is high. Therefore, if the input signal to the DM system is increasing at a fast rate, the steps are made high enough to make the staircase function catch up with the signal rapidly, thus reducing or even eliminating the slope-overload effect. It is, of course, necessary to have an adaptive scheme at the demodulator also.

The bandwidth requirements of a DM system are higher than for a PCM system if the same *quality* of signal transmission is to be achieved. For example, for good-quality speech transmission, DM requires 100 kHz but PCM requires 64 kHz. But, if some compromise can be made in the quality of signal transmission, then the bandwidth of DM systems can be reduced below that for PCM. The primary advantage of DM over PCM is that it requires much less complicated hardware to implement.

INFORMATION THEORY

In the discussion up to this point, the digital signals being transmitted were representations of analog signals. But there are many systems in which the digital signals represent actual data being manipulated. For example, in computer networks, alphabetical characters and numerical data are being transmitted from one computer to another. Transmission of this type where the digital signals represent *data* is usually referred to as *digital data transmission*, which is essentially the same sort of situation as digital communication with the only difference being that there is no need to reconstruct a signal at the receiving end. The received bits are handled directly by a computer at the receiving end. In the design of digital data transmission systems, the concepts of information theory became important since that theory provides the basis for calculations of channel capacity and the design of efficient coding schemes.

Measure of Information

When an event is totally predictable, there is no information associated with it. On the other hand, when a certain degree of unpredictability is involved in the outcome of an event, then significant information is received when the outcome is known. In fact, the information is a maximum when there are a number of possible outcomes and all of them are equally probable: the throw of a pair of dice, for example. Probability must, therefore, play an important part in the definition of information. Another property that should be satisfied by a measure of information is that if there are two *independent* events, then the total information received from them must equal the sum of the information due to each event taken separately.

Based on arguments similar to the previous one, a measure of information is defined as follows. Suppose an event has n possible outcomes: $x_1, x_2, \ldots, x_n$ and let the probability of these outcomes be, respectively, $p_1, p_2, \ldots, p_n$. Note that the sum of all the p's will be 1. Then the information associated with the event is denoted by H and defined by

$$H = p_1 \log_2 \frac{1}{p_1} + p_2 \log_2 \frac{1}{p_2} + \ldots + p_n \log_2 \frac{1}{p_n} \text{ bits} \quad (15\text{-}4)$$

where,

the logarithm is taken to the base 2.

For example, consider the tossing of a fair coin. This event has two outcomes, heads or tails, each of which has a probability of (1/2). Therefore, the average information associated with the tossing of a coin is

$$H = \frac{1}{2}\log_2(2) + \frac{1}{2}\log_2(2)$$

$$= 1 \text{ bit}$$

As another example, suppose a message source produces six possible messages with the probability values: 0.25, 0.25, 0.125, 0.125, 0.125, and 0.125. Then the average information given by the message source is

$$H = 0.25 \log_2(4) + 0.25 \log_2(4) + 0.125 \log_2(8) + 0.125 \log_2(8) + 0.125 \log_2(8) + 0.125 \log_2(8)$$

$$= 2.5 \text{ bits/message}$$

(Logarithms to the base 2 are not usually found in a calculator. To find the logarithm to the base 2, take the logarithm to the base 10 and multiply the result by 3.322.)

Drill Problem 15.11: The throwing of a die has six possible outcomes, each of which has a probability of (1/6). Calculate the average information associated with the throwing of a die.

Drill Problem 15.12: The probability of a newborn baby being a male is 0.48 and that of its being a female is 0.52. Calculate the average information associated with the birth of a child.

Electronics: Circuits and Systems

Information Rate

If a message source generates r messages per second and the average information associated with a message is H bits/message, then the *information rate* of the source is defined by

$$R = rH \text{ bits/sec} \tag{15-5}$$

For example, if a message source produces 500 messages per second and the average information is 2.5 bits/message, then the information rate of that source is 1250 bits/sec.

Example 15.1

A signal is band-limited to 10 kHz and sampled at the Nyquist rate. Each sample is quantized into eight levels for PCM transmission. A statistical analysis of the occurrence of the various levels shows that their probabilities of occurrence are: level 0 has a probability of 0.25; levels 1, 3, 5, 6, and 7, each has a probability of 0.125; levels 2 and 4, each has a probability of 0.0625.

Calculate the information rate associated with the signal.

Solution

Since the bandwidth of the signal is 10 kHz, the Nyquist rate is 2×10 kHz $= 20$ kHz. Therefore, there are 20×10^3 samples per second.

The average information associated with a sample is, from Eq. (15-4):

$H = 0.25 \log_2(4) + 5 \times 0.125 \log_2(8) + 2 \times 0.0625 \log_2(16)$

$= 2.875 \text{ bits/sample}$

Therefore, the information rate of the signal is

$R = (20 \times 10^3)(2.875)$

$= 57.5 \times 10^3 \text{ bits/sec}$

Drill Problem 15.13: A message source generates three different messages at the rate of 400 messages/sec. The probabilities associated with the messages are 0.5, 0.35, and 0.15. Calculate the information rate associated with the source.

Drill Problem 15.14: Recalculate the information rate of the above source if the probabilities of the three messages were made equal to each other.

Drill Problem 15.15: The number of quantization levels used in a PCM system is 16, and each level is known to have a probability of (1/16). If the sampling rate of the signal is 8 kHz, calculate the information rate of the signal.

CHANNEL CAPACITY AND SHANNON'S THEOREM

Information is transmitted from one point to another through some sort of a channel, a familiar example of which is the telephone line used to transmit data from one computer to another. Channels can be noisy or noiseless depending upon whether they introduce random disturbances that make a transmitted symbol appear to be different at the receiving end, or not. Channels also have a bandwidth restriction. A cable for example has a distributed resistance and capacitance, which set an upper limit to the frequency that can be transmitted through it. As information is transmitted in the form of the bits zero and one through a channel, it is naturally important that they be detected correctly at the receiving end. But there is always a possibility of error, and it is necessary to make the probability of error as small as possible. A number of factors can affect the probability of error. For example the kind of coding used for the words being transmitted may affect the probability of error. Also, if the source connected to the channel is producing information at a faster rate than can be handled by the channel, then the probability of error can be expected to increase. A measure of the maximum rate of information that can be handled by a channel is expressed in terms of the *channel capacity*.

Noiseless Channel

If the channel is noiseless, for example, then there is no error introduced by the channel itself. In this case, the maximum information rate that the channel can handle is the maximum rate at which the source can produce the information. For example, suppose a source generates N different messages and each message occupies an interval of T seconds, then there are (1/T) messages per second. The average information per message is a maximum when all the messages have the same probability of occurrence (as can be shown mathematically). Therefore, if each message occurred with a probability of (I/N), we have

$$\text{maximum average information per message} = N \times [\frac{1}{N} \log_2 (N)]$$

$$= \log_2 (N)$$

from Eq. (15-4). The maximum rate of information that the source can produce is, therefore,

Electronics: Circuits and Systems 895

$$C = [\log_2 (N)] \times (1/T) \text{ bits/sec} \qquad (15\text{-}6)$$

since there are (1/T) messages per second and each message has an average information of $\log_2 (N)$ bits.

Eq. (15-6) defines the channel capacity of a *noiseless channel*. For example, if a source produces 100 different messages, each message occupying 10 ms, the channel capacity of a noiseless channel connected to the source is

$$C = [\log_2 (100)] \times \frac{1}{10 \times 10^{-3}}$$

$$= 664.4 \text{ bits per sec}$$

If the actual source connected to the channel is producing N messages with unequal probabilities, then the channel capacity given by Eq. (15-6) can be approached by using special coding schemes for the messages.

Drill Problem 15.16: Calculate the channel capacity of a noiseless channel connected to a source that generates 26 different symbols and the code for each symbol occupies 5 ns.

Noisy Channel

The channel capacity of a noisy channel is limited to two factors: bandwidth and the signal-to-noise ratio. As the rate of information to be transmitted increases, the average time for each bit decreases, which means that the pulse duration decreases also. A shorter pulse duration requires a larger bandwidth for transmitting the signal. Therefore, we can see that if the rate of information increases, the bandwidth needed for the transmission of the signal can exceed the bandwidth of a given channel. Thus, the bandwidth sets a limit on the maximum rate of information that can be handled without a high probability of error. As the signal transmitted through the channel is contaminated by the noise (due to the random disturbances in the channel), a probability of error in the detection of the signal is introduced: a transmitted one may appear as a zero at the receiving end or a transmitted zero may appear as a one at the receiving end due to the noise. The probability of error must, therefore, depend upon the signal-to-noise ratio (S/N) where S represents the signal power and N the noise power. Since we are interested in the maximum rate of information that can be handled by a channel with as small a probability of error as possible, we can expect that the signal-to-noise ratio (S/N) will affect the channel capacity also.

The channel capacity C is related to the bandwidth B and the signal-to-noise ratio (S/N) through the equation:

$$C = B \log_2 [1 + (S/N)] \text{ bits/sec} \qquad (15\text{-}7)$$

where,
the bandwidth B is in hertz.

Eq. 15-7 is valid when the noise introduced by the channel is additive (that is, it adds directly to the noise rather than being multiplicative) and white.

It is seen that the channel capacity increases the proportion to the bandwidth. But, increasing the bandwidth indefinitely is not beneficial since the noise power introduced by the channel is directly proportional to the bandwidth also (as can be seen from the discussion of white noise in Chapter 14). Thus, as the bandwidth B increases, the signal-to-noise ratio (S/N) decreases and cuts down some of the gain achieved in the channel capacity.

Example 15.2

Calculate the channel capacity of a channel of bandwidth 500 kHz for the following values of the signal-to-noise ratio: (a) 20 dB; (b) 60 dB.

Solution

In each case, the signal-to-noise ratio has to be first converted to an actual ratio from the given decibel value.

(a) SNR = 20 dB leads to (S/N) = antilog_{10} (20/10) = 100.
The channel capacity is, therefore, from Eq. (15-7):

$$C = 500 \times 10^3 \log_2 [1 + 100]$$

$$= 3.33 \times 10^6 \text{ bits/sec}$$

(b) SNR of 60 dB leads to (S/N) = antilog_{10} (60/10) = 10^6.

$$C = 500 \times 10^3 \log_2 (10^6)$$

$$= 9.97 \times 10^6 \text{ bits/sec}$$

Example 15.3

The noise power present in a channel is 4×10^{-21} W/Hz. The signal power being transmitted is 10^{-6} W. For each of the following bandwidth values, calculate the channel capacity: (a) 1 MHz; (b) 50 kHz.

Solution

(a) Bandwidth = 1 MHz. Noise power = $4 \times 10^{-21} \times 10^6 = 4 \times 10^{-15}$ W.

$$(S/N) = [10^{-6}/4 \times 10^{-15}] = 2.5 \times 10^8$$

$$C = 10^6 \log_2 [1 + 2.5 \times 10^8] = 2.79 \times 10^7 \text{ bits/sec}$$

(b) Bandwidth = 50 kHz. Noise power = $4 \times 10^{-21} \times 50 \times 10^3 = 2 \times 10^{-16}$ W.

$$(S/N) = [10^{-6} / 2 \times 10^{-16}] = 5 \times 10^9$$

$$C = 50 \times 10^3 \log_2 [1 + 5 \times 10^9] = 1.61 \times 10^6 \text{ bits/sec}$$

Comparing the two values, the channel capacity changes by a factor of 17.3 while the bandwidth changes by a factor of 20.

Drill Problem 15.17: Calculate the channel capacity of a channel with a bandwidth of 1 MHz for the following values of the signal-to-noise ratio: (a) 5.7 dB; (b) 57 dB; (c) 570 dB.

Drill Problem 15.18: The noise present in a channel is 4×10^{-21} W/Hz. For each of the following bandwidths of the channel, calculate the channel capacity if the signal power is 10^{-12} W: (a) 50 kHz; (b) 5 MHz; (c) 500 MHz.

Information Rate and Shannon's Theorem

The information rate that can be handled by a channel is given by Shannon's theorem: (the same Shannon who was mentioned in connection with logic circuits):

Given a channel of capacity C bits/sec, it can handle an information rate of R bits/sec with an arbitrarily small probability of error provided R is less than or equal to C.

That is, so long as the information rate of a source is less than or equal to the capacity of a channel, and then the error probability can be reduced to as small a value as we please by means of proper coding techniques.

On the other hand, if the information rate of the source is greater than the channel capacity, then the probability of error becomes close to 1 (or 100%) regardless of how sophisticated the coding scheme is.

Shannon's theorem in conjunction with Eq. (15-7) for the capacity of a noisy channel can be used to determine the bandwidth of a channel for a given information rate.

Example 15.4

A TV picture may be considered as being composed of approximately 3×10^5 picture elements. Suppose each of these picture elements can assume 10 possible (distinct) brightness levels for proper contrast. Assume that the ten brightness levels have equal probabilities of occurrence. The number of picture frames transmitted per second is 30. The desired signal-to-noise ratio is 30 dB for a satisfactory reproduction of the picture.

Calculate the bandwidth of the video channel.

Solution

A lot of information is given in the problem statement. There are two groups of values: one needed to calculate the information rate and the other needed for the bandwidth.

First let us calculate the information rate, which is obtained by calculating the average information per picture frame and multiplying that by the rate of transmission.

number of levels in each picture element = 10

probability of occurrence of each level = (1/10)

since the levels are stated as occurring with equal probabilities.
Therefore, from Eq. (15-4),
Average information/picture element

$$= 10 \times [\frac{1}{10} \times \log_2 (10)]$$

$$= 3.32 \text{ bits}$$

Since each picture frame contains 3×10^5 picture elements,

average information/picture frame $= 3 \times 10^5 \times 3.32$

$$= 9.96 \times 10^5 \text{ bits/frame}$$

The rate of transmission is 30 frames/sec. Therefore,

information rate $= 30 \times 9.96 \times 10^5 = 2.99 \times 10^7$ bits/sec

The channel capacity must, therefore, be at least 2.99×10^7 bits/sec.

Now we calculate the bandwidth needed to give the previous channel capacity for the specified SNR of 30 dB.

SNR of 30 dB leads to $(S/N) = 10^3$. From Eq. (15-7) then

$$C = 2.99 \times 10^7$$

$$= B \log_2 [1 + 10^3]$$

or

$$B = 3 \text{ MHz}$$

The bandwidth needed is, therefore, 3 MHz.

Example 15.5

The facsimile of a picture is to be transmitted over a standard telephone line. The following data are available: number of elements/picture = 2.25 ×

Electronics: Circuits and Systems

10^6, number of brightness levels/picture element = 12, with all levels having the same probability of occurrence. Bandwidth of the telephone circuit = 3 kHz, signal-to-noise ratio desired = 40 dB.

Calculate the time taken to transmit a picture.

Solution

This is essentially the reverse of the previous example. We can first calculate the channel capacity from the data as well as the average information per picture and hence the time for transmitting a picture.

Calculation of Channel Capacity

SNR of 40 dB gives (S/N) = 10^4. Using Eq. (15-7) we get

$$C = 3 \times 10^3 \log_2 (1 + 10^4) = 3.99 \times 10^4 \text{ bits/sec}$$

The information rate must not exceed 3.99×10^4.

Calculation of Average Information/Picture

Since the number of brightness levels per picture element is 12 and each has a probability of occurrence of (1/12), the average information per picture element is, from Eq. (15-4):

$$\text{average information/picture element} = 12 \times [\frac{1}{12} \log_2 (12)]$$

$$= 3.58 \text{ bits}$$

The number of elements per picture is 2.25×10^6. Therefore,

$$\text{average information/picture} = 2.25 \times 10^6 \times 3.58$$

$$= 8.07 \times 10^6 \text{ bits}$$

Calculation of the Time of Transmission

The (maximum) rate of information is 3.99×10^4 bits/sec, which is the channel capacity. The average information/picture is 8.07×10^6 bits. Therefore,

$$\text{time of transmission/picture} = \frac{8.07 \times 10^6}{3.99 \times 10^4}$$

$$= 202 \text{ seconds or } 3.37 \text{ minutes}$$

Drill Problem 15.19: In the television picture (Example 15.4), suppose the number of brightness levels is increased to 16 again with equal probabilities of occurrence, calculate the new bandwidth of the video channel.

Drill Problem 15.20: Suppose the video channel has a bandwidth of 6 MHz and the data given about the picture are the same as in Example 15.4. What will be the SNR in the channel?

Drill Problem 15.21: In the facsimile transmission (Example 15.5) change the number of brightness levels to 16 and recalculate the time of transmission.

Drill Problem 15.22: What bandwidth of the channel will be needed in the facsimile transmission problem if the time of transmitting the picture is to be 2 minutes? Assume all the other data are as given in Example 15.5.

DATA TRANSMISSION

Data in the form of numbers and alphabetical characters are transmitted between computers both in large systems (such as, for example, those used in a space mission) and small systems (such as, for example, those used in an automated-teller machine system in a bank). Data transmission can be *synchronous* or *asynchronous*. In a *synchronous* data transmission system, the bits making up the code word for each character are sent at a periodic rate in synchronism with a master clock pulse train in the system. In an *asynchronous* data transmission system, the code word may be initiated at any arbitrary instant of time. The asynchronous system is also called the *start/stop* system since each code word is preceded by a start pulse to indicate the starting of a character and a stop pulse to indicate the end of the character.

One of the basic considerations in a digital data transmission system is the encoding of each character (numerals, the letters of the alphabet, and punctuation marks). Several different standard coding schemes are available, the most widely used one being the ASCII (pronounced "askee") code, where ASCII is the acronym for *American Standard Code for Information Interchange*. In the ASCII code, each character is coded in the form of a 7-bit code. This means that 128 different characters can be transmitted in this coding system, since $2^7 = 128$. The beginning of each character in an asynchronous system is indicated by a *start space*, which is always a 0 pulse. Then the 7 bits of the code of the character are transmitted with the least significant bit first and most significant bit last. The seventh bit is followed by an odd parity check bit that is followed by a *stop mark* that is always a 1 pulse of duration 1.42 to 2 data bit intervals. An example of a character transmitted in the previous format is shown in Fig. 15-22. The character being transmitted is an M whose ASCII 7-bit code is 1 0 0 1 1 0 1.

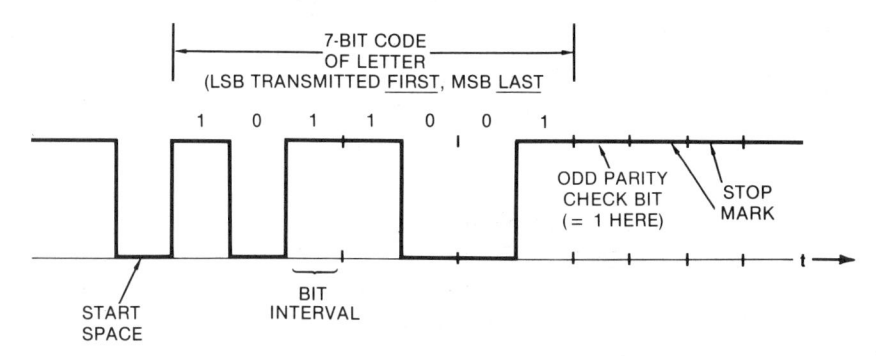

Fig. 15-22. ASCII code for the letter M.

If you were sitting at a computer terminal and typed the key M, the signal being sent from the terminal to the computer will be of the form shown in Fig. 15-22.

The information rate in data transmission can be calculated if we know the number of characters being transmitted per second and the number of information bits per character. For example, in the ASCII code, the number of information bits is ten: seven data bits, one start bit, one stop bit, and one parity bit. In one ASCII system, the number of *characters* transmitted per second is specified as ten. The information rate of the system is, therefore, 100 bits/sec. If, on the other hand, the number of characters transmitted per second is 30, then the information rate is 300 bits/sec.

Example 15.6

A particular coding scheme (called the Bell System Baudot code) uses 1 start bit, 5 data bits, and a stop bit that occupies 1.42 data bit intervals. Therefore, the number of data bit intervals per character is 7.42. If each data bit interval is 22 ms, then the transmission of each character takes (7.42 × 22 ms) = 163.2 ms. When a series of characters is being transmitted, suppose an additional 1 data bit interval is allowed between characters, the time per character becomes (8.42 × 22 ms) = 185.2 ms. In order to calculate the information rate, the number of *information* bits per character is: 1 start bit + 5 data bits forming the code of the character being transmitted + 1 stop bit = 7 bits. Therefore, the information rate is 37.8 bits/second.

DIGITAL-TO-ANALOG AND ANALOG-TO-DIGITAL CONVERSION

Since most signals occur in the form of a continuous waveform in practice (or as *analog* signals), they have to be converted to a digital form in a digital communication or a digital processing system. As was seen in the section on

PCM, the analog signals are quantized, and each quantization level is converted to a code that has a digital format. The conversion of each quantization level to a digital code word is done by means of analog-to-digital converters (A/D converters or ADC). In most communication systems, the final signal at the output of a receiver has to be in an analog form: visual image or sound for example. The conversion of digital signals to analog signals is accomplished by means of digital-to-analog converters (D/A converters or DACs). The principles of D/A and A/D conversion as well as some of the converter circuits will be discussed in this section.

Digital-to-Analog Conversion

The basic principle involved in the conversion of a digital code word (such as 1 1 0 1 0 1 1 0) is to treat it as the binary representation of a number in the decimal system. Then if the equivalent decimal number is determined and a signal at a level proportional to the decimal number is produced, we have converted the digital code word to an analog signal. This principle underlies the *binary weighted D/A conversion*.

The decimal equivalent of a binary number involves the multiplication of each binary by an appropriate power of 2 and then adding the products. For example, the 8-bit word 1 1 0 1 0 1 1 0 is converted to its decimal equivalent by multiplying the most significant digit by 2^7, the next lower bit by 2^6, and so on until the least significant bit is multiplied by 2^0.

$$(1 \times 2^7) + (1 \times 2^6) + (0 \times 2^5) + (1 \times 2^4) + (0 \times 2^3) + (1 \times 2^2) + (1 \times 2^1) + (0 \times 2^0) = 214$$

The formation of a weighted sum can be done by means of an op amp connected as a summing amplifier. As a specific example, consider a 4-bit code word: $b_3 b_2 b_1 b_0$. If these 4 bits are used as the inputs to a summing op amp (Fig. 15-23), the output of the op amp will be

$$V_o = -[\frac{R_F}{R_3} b_3 + \frac{R_F}{R_2} b_2 + \frac{R_F}{R_1} b_1 + \frac{R_F}{R_0} b_0]$$

If the resistances R_3, R_2, R_1, R_0 are chosen so as to satisfy the following equations:

$$R_3 = (1/2^3)R_0$$
$$R_2 = (1/2^2)R_0$$
$$R_1 = (1/2R_0)$$

the output voltage is given by

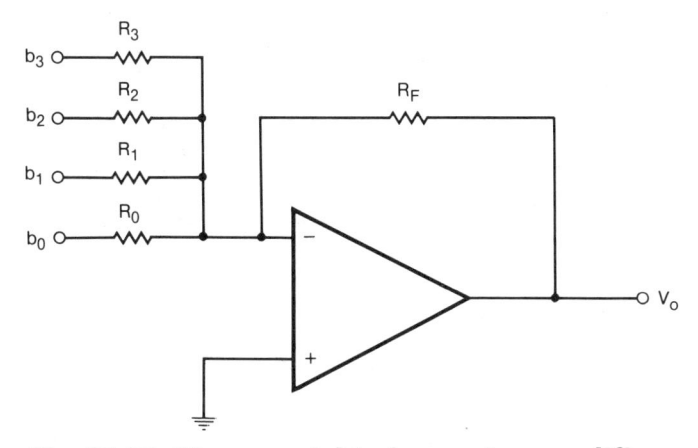

Fig. 15-23. Binary weighted summing amplifier.

$$V_o = -\frac{R_F}{R_0} [2^3 b_3 + 2^2 b_2 + 2 b_1 + b_0]$$

which is proportional to the binary weighted sum of the inputs. The output voltage is, therefore, proportional to the decimal equivalent of the input code word viewed as a binary number.

It is necessary to make the inputs to the op amp high or low as determined by the input code word. This can be accomplished by means of a set of *single-pole double-throw* (SPDT) switches as in Fig. 15-24. When an input bit is 1, that particular switch is moved to the V_B contact, and when the input bit is 0, the switch is moved to the ground contact. An electronic version of the SPDT switch uses a CMOS inverter in conjunction with a voltage-follower circuit (Fig. 15-25). The input to the CMOS inverter is $\bar{b}$, the complement of the input bit b; $\bar{b}$ is low when the input bit b is high, and $\bar{b}$ is high when the input bit b is low. When $\bar{b}$ is low, the upper FET conducts and the lower FET is cut off. The output voltage V_o is then at 5 V. Thus, when the input bit is high, V_o is high also. It can be seen by a similar argument that when the input bit is low, V_o is low also. Therefore, the output V_o is at the same level as the input bit. Each input to the summing op amp will be derived from the output of an electronic SPDT switch of the above type. The *binary weighted D/A converter* for a 4-bit input code is shown in Fig. 15-26, where each box contains an electronic SPDT switch.

The previous approach can be extended to any size digital code word. It can also be used for code words that are not weighted in a straight binary scale by choosing the appropriate factors between the resistors. One disadvantage of the previous circuit is the wide range of values of resistances needed.

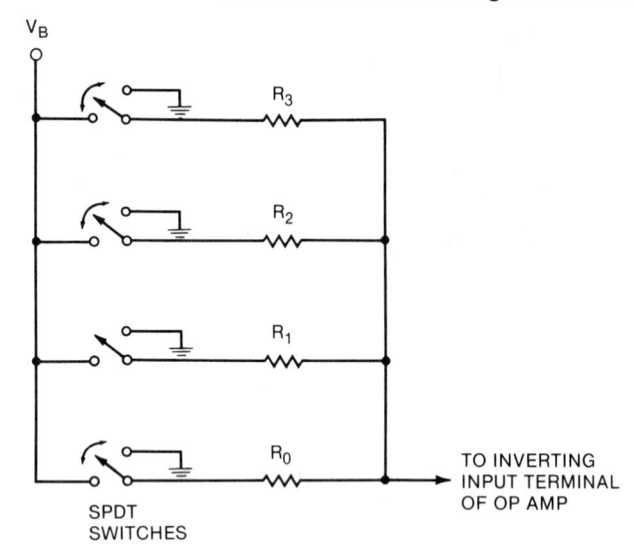

Fig. 15-24. Single-pole double-throw (SPDT) switches needed for the binary weighted D/A converter.

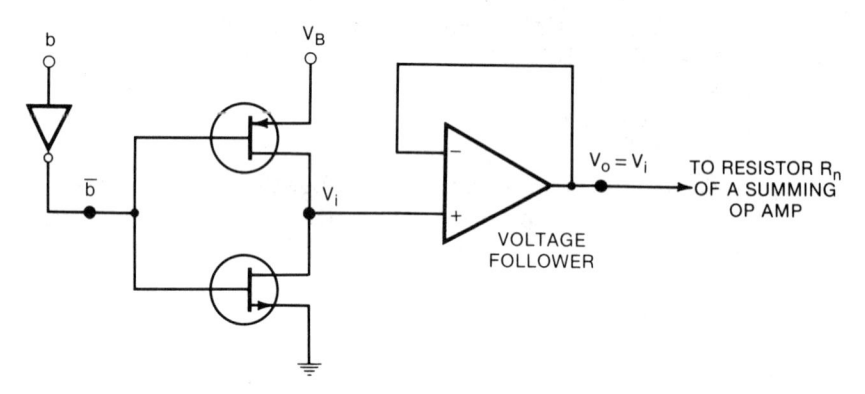

Fig. 15-25. Electronic SPDT switch.

For example, in an 8-bit converter, the resistances may range from 10 kΩ to 1280 kΩ in a circuit, which may be too wide a range to be practically feasible. The *R-2R Ladder D/A Converter* circuit of Fig. 15-27 overcomes the previous difficulty by requiring only two different resistances R and 2R, but the number of resistors used is twice the number of input bits. An analysis of the R-2R ladder shows that its output is given by

$$V_o = \frac{b_3}{2} + \frac{b_2}{2^2} + \frac{b_1}{2^3} + \frac{b_0}{2^4}$$

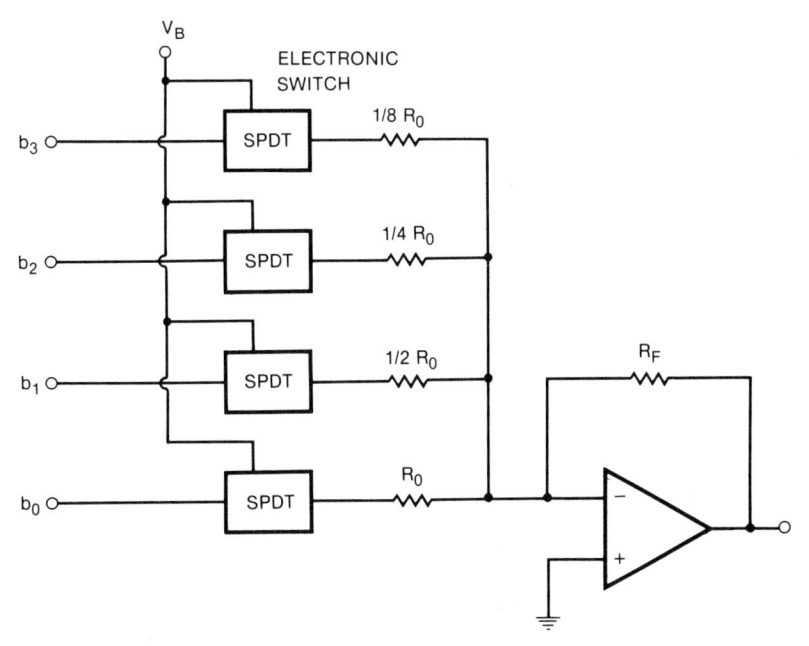

Fig. 15-26. Binary weighted D/A converter.

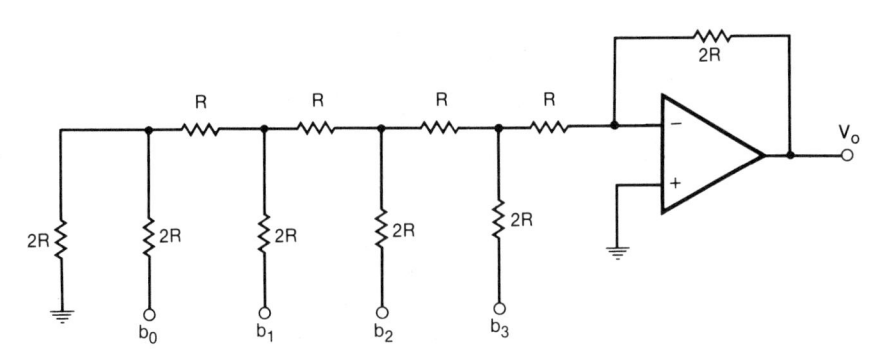

Fig. 15-27. R-2R ladder D/A converter.

which can be put in the form

$$V_o = \frac{2^3 b_3 + 2^2 b_2 + 2 b_1 + b_0}{2^4}$$

Thus, the output voltage is proportional to the decimal equivalent of the input binary code. As in the previous circuit, the input bits are fed to the ladder circuit through electronic SPDT switches.

The important specifications of a D/A converter are resolution, linearity, accuracy, and settling time. *Resolution* (expressed as an *m-bit resolution*) is a measure of the smallest possible change in the output voltage, which is $(1/2^m)$ times the full range of the output voltage. A 10-bit converter, for example, has a resolution of $(1/2^{10}) = 9.77 \times 10^{-4}$ times the full range of the output voltage. The output of a D/A converter should vary linearly with the increase in the value of the input code word. *Linearity* depends upon the precision of the resistors used in the circuit. The converter has 100% *accuracy* when its actual analog output equals the ideal output for each binary code word input. Factors affecting the accuracy are linearity (or lack thereof), uncertainty in the reference voltages used, amplifier gain, and amplifier offsets. When the switches in the D/A converter open and close due to changes in the input bits, abrupt voltage changes are caused and the resulting transients persist for a certain length of time due to the capacitances in the circuit. The settling time is the time taken for the output to reach the new steady-state value. Typical values of setting time vary from 25 ns to 100 s.

Analog-to-Digital Conversion

In analog-to-digital conversion, the amplitude of a signal (which varies between a maximum and a minimum) into the digital code corresponding to the quantization level. The first consideration is, therefore, the number of levels into which a signal of a given voltage range is to be subdivided. The choice of the number of quantization levels is based on factors such as quantization error and the cost and complexity of the system. The larger the number of quantization levels, the smaller the quantization error but the higher the cost and complexity of the system.

Consider a signal that varies from 0 to a maximum V_m and let the quantization levels be 0, $(V_m/8)$, $(2V_m/8)$, ..., $(7V_m/8)$ (Fig. 15-28). Suppose a straight binary code is assigned to the eight levels. Then, whenever the instantaneous value of the signal lies between two quantization levels, the three-code bit to be generated corresponds to the level just below the value of the signal. For example, if the value of the signal at some instant is $0.6\ V_m$, then it is between $(4V_m/8)$ and $(5V_m/8)$, and the code to be generated is 1 0 0 (corresponding to $4V_m/8$).

A number of analog-to-digital converters is available, three of which will be discussed here.

Parallel Encoders

The determination of the two quantization levels between which a given sample of a signal lies can be done by means of comparators as in the parallel encoder of Fig. 15-29. The reference voltages are the voltages of the quantization levels obtained by a voltage-divider arrangement. Whenever the input

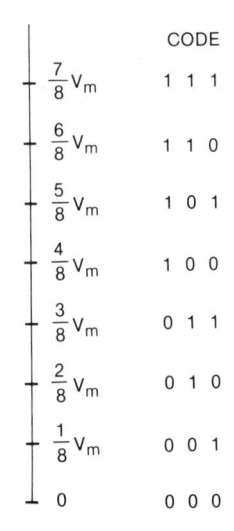

Level	CODE
$\frac{7}{8}V_m$	1 1 1
$\frac{6}{8}V_m$	1 1 0
$\frac{5}{8}V_m$	1 0 1
$\frac{4}{8}V_m$	1 0 0
$\frac{3}{8}V_m$	0 1 1
$\frac{2}{8}V_m$	0 1 0
$\frac{1}{8}V_m$	0 0 1
0	0 0 0

Fig. 15-28. Quantization levels and codes.

voltage exceeds a particular quantization level, then a high output is produced by all the comparators at and below that level. The outputs of the bank of comparators are encoded into binary code through the encoder (which can be a commercially available unit).

The disadvantage of the parallel encoder is that the number of comparators equals the number of quantization levels. As the number of bits in the coding scheme increases, the number of comparators increases by a *factor of two* for *each additional bit*. This makes converters of this type with more than 8-bit outputs quite expensive.

The *counting A/D converter* shown in Fig. 15-30 is far less complex than the parallel encoder but it is much slower. A single comparator is used, in which the noninverting input receives the input signal and the inverting input receives a reference voltage through a feedback loop. The reference voltage is not a constant as will be seen from the following discussion.

Just before each sample of a signal is received, the value of the reference voltage V_R is set to 0. Since the signal V_a is positive, the comparator produces a high output. This enables the AND gate to send the pulses from the clock, which activates the three-bit counter. The counter counts the clock pulses, and the outputs of the counter are converted to an analog signal through a D/A converter in the feedback loop. As more and more pulses are counted, the value of the reference voltage V_R increases in the form of a staircase function. The different waveforms in the circuit are shown in Fig. 15-31. As V_R builds up, it will at some point exceed the input signal V_a, and this will make the output of the comparator low. This will make the output of the AND gate

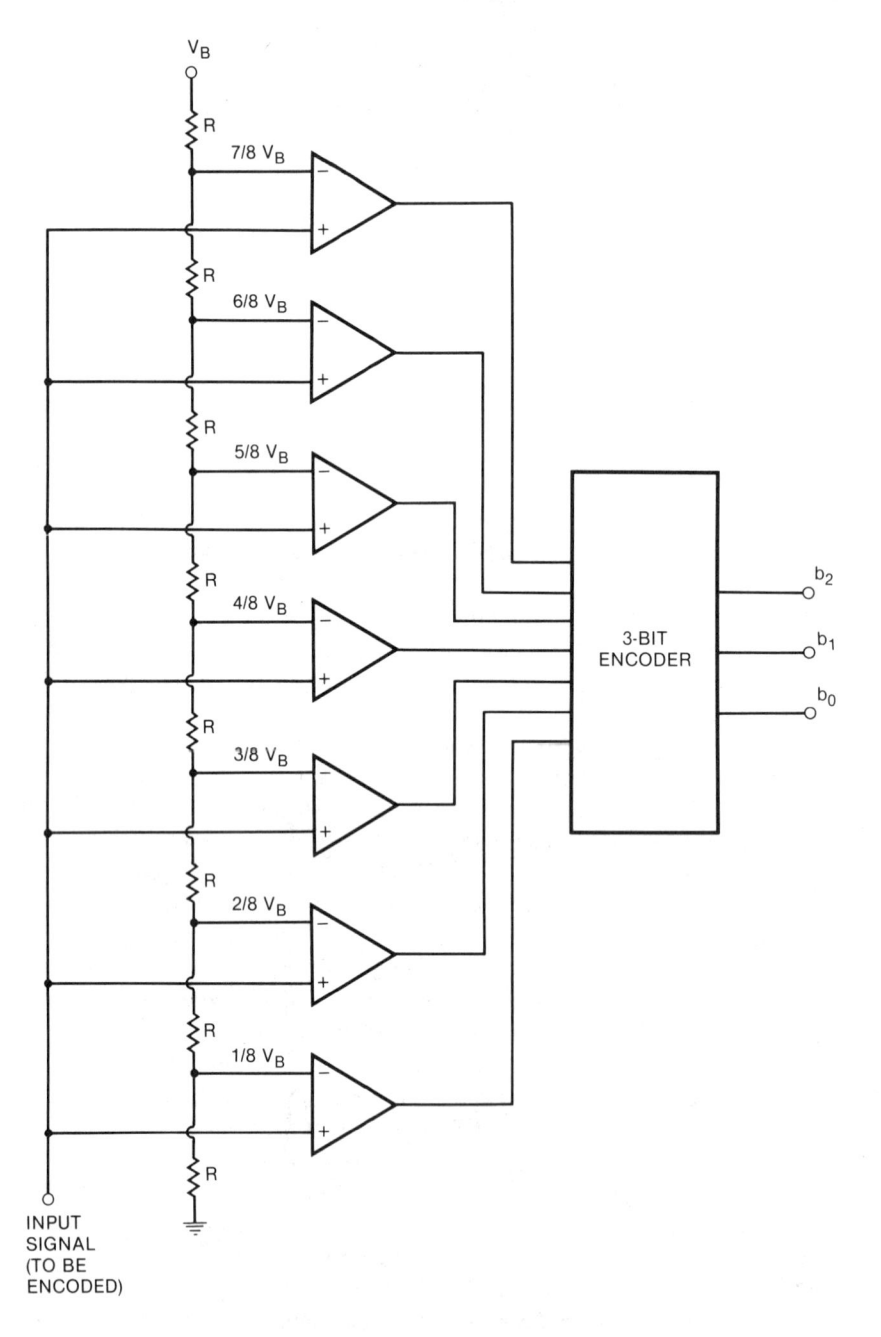

Fig. 15-29. Parallel encoder A/D converter.

Electronics: Circuits and Systems

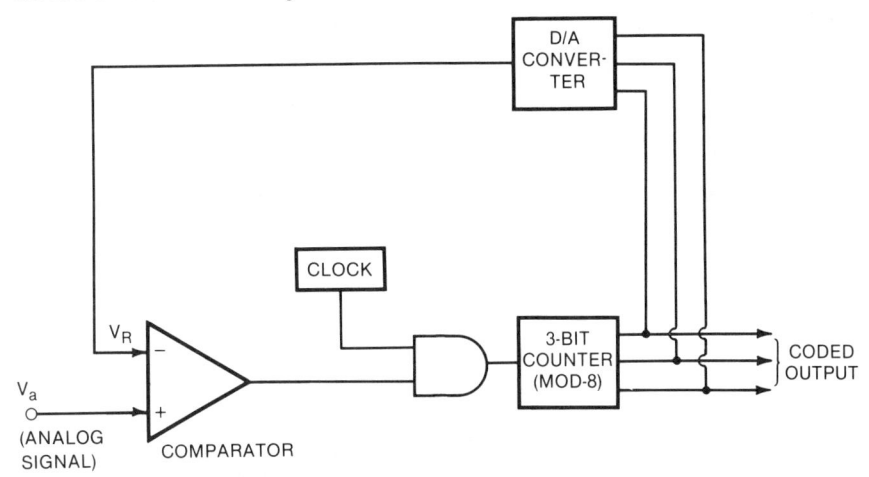

Fig. 15-30. Counting A/D converter.

low also, and the counter will stop counting. Its output will now be fixed at the last count it made. The output of the circuit will be the three-bit code word for the particular signal sample being encoded. In the example of Fig. 15-31, the count stops after the sixth clock pulse, and the code word will, therefore, be 1 1 0.

An A/D converter that does not use feedback is the *dual-slope integrator*, which is extensively used whenever speed is not important. It is used, for example, in digital multimeters.

The principle of operation of the dual-slope integrator is based on the property of a capacitor: if the *current* in a capacitor is kept *constant*, then the voltage across it increases (or decreases depending upon the direction of the current) linearly in the form of a *ramp function* as shown in Fig. 15-32.

Consider the sample V_a of an analog signal to be encoded. Let the voltage V_a be made to generate a constant current I_0 such that

$$I_0 = K_1 V_a$$

where,

K_1 is a constant.

This current I_0 is made to flow through a capacitor for a *fixed length of time* T_1 (Fig. 15-33). The voltage V_c will increase linearly from 0 to a value V_m given by

$$V_m = \frac{I_0 T_1}{C} = \frac{K_1 T_1}{C} V_a$$

At time T_1, the charging of the capacitor is stopped, and a discharge operation is initiated. The discharge is also through a constant current, and the

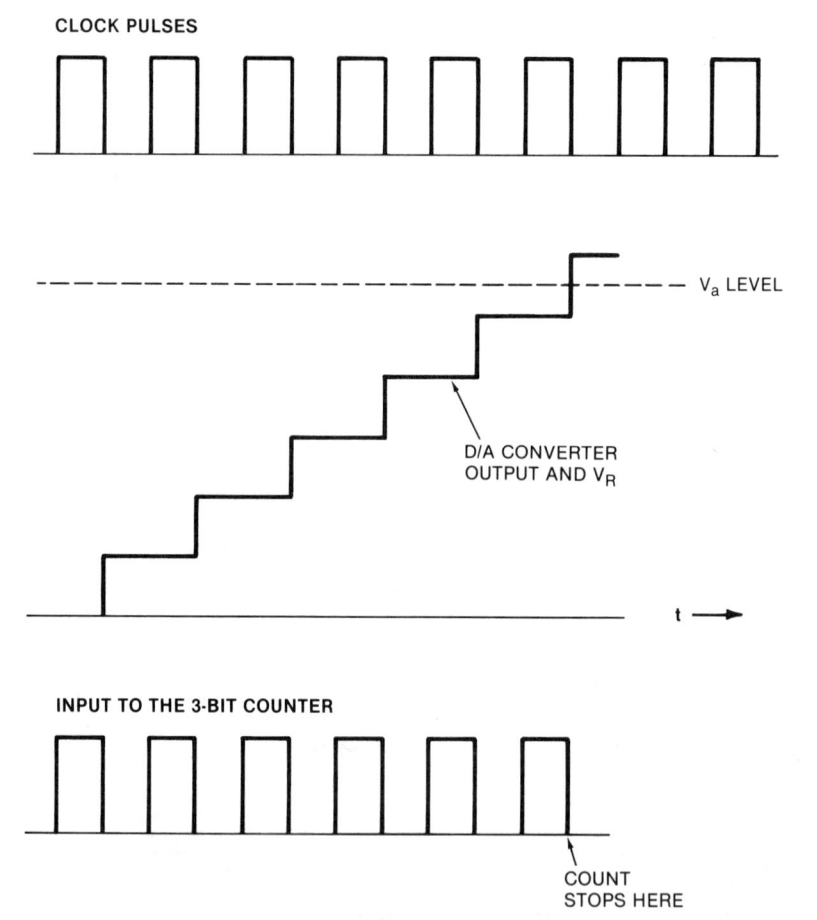

Fig. 15-31. Generation of digital output in the counting A/D converter.

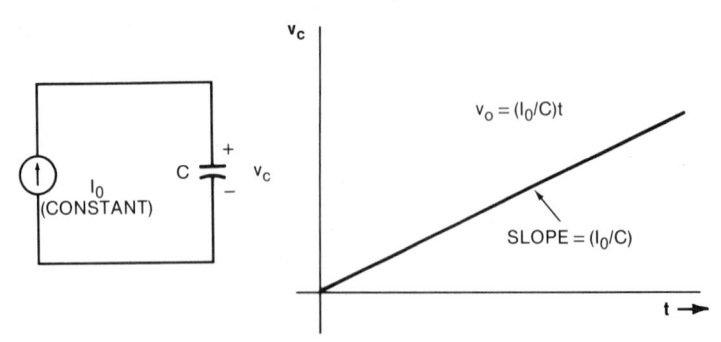

Fig. 15-32. Charging of a capacitor by a constant current. The capacitor voltage is a ramp function.

Electronics: Circuits and Systems

capacitor voltage decreases linearly from V_m to 0. The time T_2 needed to completely discharge the capacitor will be proportional to the voltage V_m as can be seen from the two graphs of Fig. 15-33. (Note that the slope of the discharge curve is always the same whereas the charging portion has different slopes depending upon V_a.)

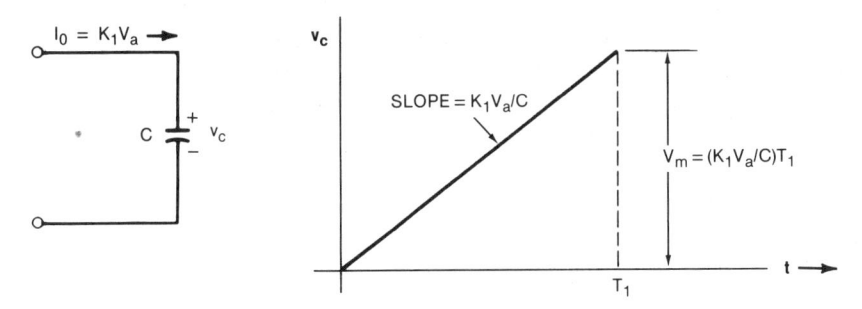

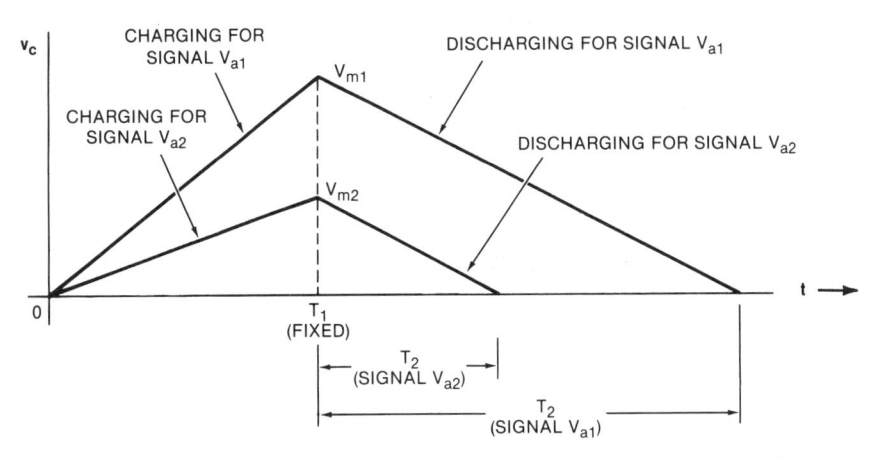

Fig 15-33. Principle underlying the dual-slope integration technique of A/D conversion. The capacitor is always charged for the *same length* of time T_1. It is always discharged at the *same rate*.

The time T_2 is seen to be proportional to the voltage V_a, the voltage level of the signal to be encoded. If the time T_2 is converted to a digital code (by means of a counter), then the analog signal is converted to a digital code word.

Summarizing the previous discussion, we make the sample of an analog signal charge a capacitor at a constant rate for a fixed length of time. At the end of the charging interval, we make the capacitor discharge to 0 at a constant rate and convert the time of discharge to a digital code word.

The charging of the capacitor with a constant current proportional to the voltage V_a is done by means of an op amp circuit as shown in Fig. 15-34. When the input switch is at V_a, a constant current (V_a/R) flows through the capacitor, and the output voltage V_o increases linearly. At $t = T_1$ (which can be controlled by some sort of a counter and an electronic single-pole double-throw switch) the input terminal is switched to a voltage $-V_R$. This reverses the current flow in the capacitor, and the capacitor starts discharging. The discharge current is $(-V_R/R)$ and is a constant. The time for discharge to 0 is measured by means of a counter circuit whose output will represent the code word for the analog input level.

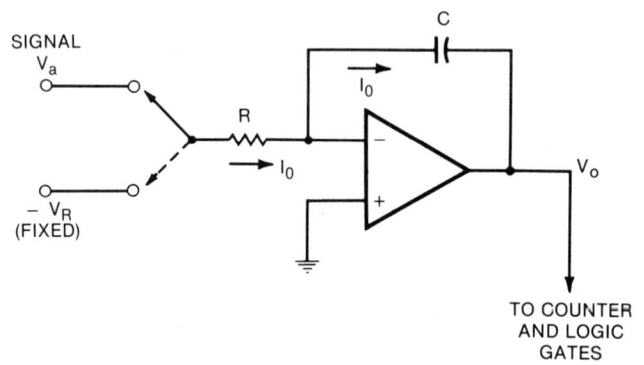

Fig. 15-34. Dual-slope integrator.

SUMMARY SHEETS

PULSE MODULATION

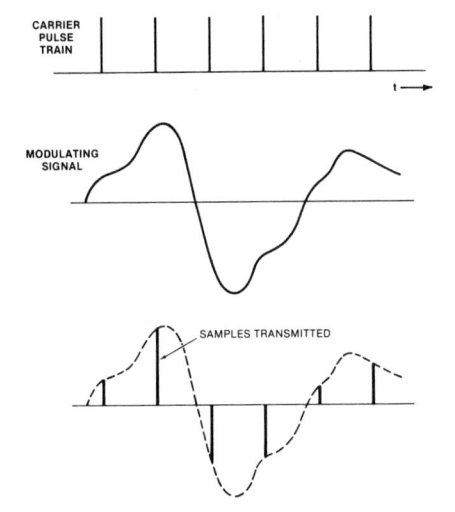

Samples of the modulating signal are made to modulate a carrier, which is a pulse train. In digital communication systems, the samples are encoded into a sequence of zeros and ones before being transmitted.

SAMPLING THEOREM

When a signal is band-limited, that is, it has a finite upper limit to its frequency content, it can be uniquely determined from samples of its values taken at uniform intervals spaced less than $(1/2f_m)$ seconds apart. The sampling rate of $2f_m$ samples per second is known as the *Nyquist rate*.

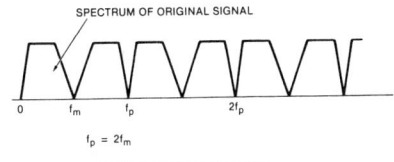

If the sampling rate is less than the Nyquist rate, then aliasing occurs and the original signal is not accurately reproduced at the receiver.

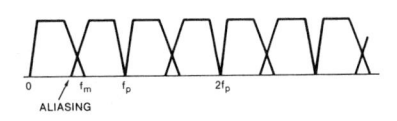

PULSE-AMPLITUDE MODULATION

The height of the pulses in the carrier are modulated by the samples of the signal.

Natural Sampling

The samples are exact replicas of the signal over the sampling interval.

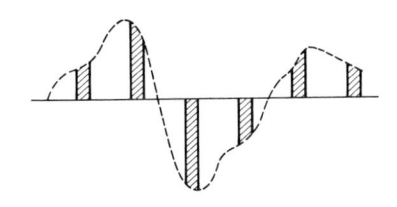

Flat-Top Sampling

Each sample maintains a constant amplitude over the sampling interval.

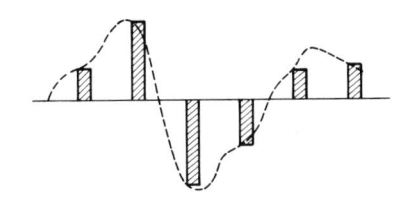

Flat-top sampling requires less complex circuitry (such as sample and hold circuits) but there is some distortion at the receiving end.

Time Division Multiplexing

The samples from a number of different signals are interspersed so that they can all be transmitted over a single channel.

Bandwidth needed to transmit N message sources = N × bandwidth needed for each source.

A synchronization pulse (sync pulse) is transmitted along with the message samples so as to synchronize a local clock at the receiver with that at the transmitter. After the samples of each message source are separated at the receiver, then an LPF can be used to recover the original message.

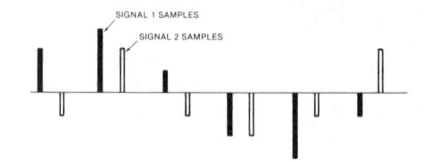

PULSE-TIME MODULATION

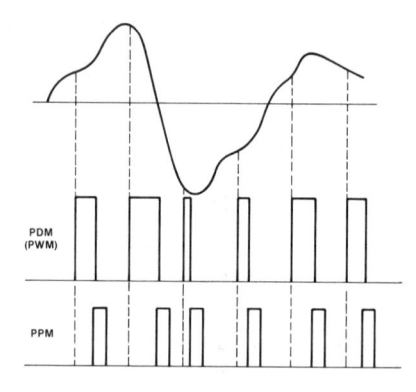

Pulse-Duration Modulation (PDM)
The width of each pulse in the carrier is increased or decreased in proportion to the polarity and amplitude of the sample of the signal.

Pulse Position Modulation (PPM)
The starting instant of each pulse is shifted in time with the shift depending upon the polarity and amplitude of the sample of the signal.

The demodulation of PDM and PPM signals usually requires the conversion of the signals to PAM (that is, pulses of different heights) and then passed through an LPF.

PULSE-CODE MODULATION

The samples are first encoded into sequences of zeros and ones. A zero corresponds to the absence of a pulse, and a one to the presence of a pulse. The processing of the signals is done through high-speed computers with a larger information handling capacity than other types of communication systems. The noise and interference in PCM systems can be minimized by using sophisticated coding schemes.

Quantization Levels
The range of the signal voltage is divided into k voltage intervals of equal amplitude. The middle point of each interval is called a quantization level. The quantization levels are numbered 0 through (k-1), and these numbers are coded into m-bit words.

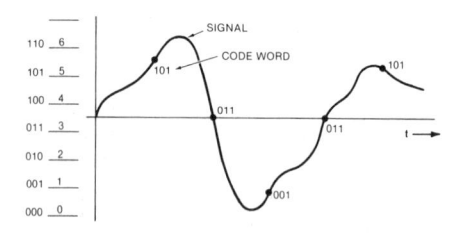

number of bits per word = m

number of quantization levels = k

$2^m = k$ (If 2^m is not exactly equal to k, then choose the next higher value of m.)

Block Diagram of a PCM System

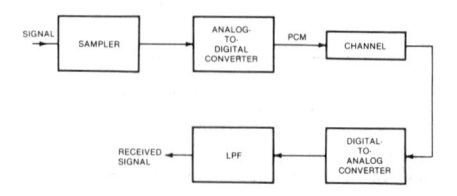

Quantization Error in PCM
Quantization error is the difference between the actual signal voltage at the sampling instant and the quantization level close to it.

An SNR can be associated with the quantization error. An increase in the number k of quantization level for a given signal range will increase the SNR but at the expense of more bits per word and a corresponding need for a wider bandwidth of the transmission channel.

Bandwidth of a PCM System
Bandwidth of a PAM system = f_m (or the highest frequency in the signal to be transmitted)

Electronics: Circuits and Systems

bandwidth of a PCM system $= [\log_2 k] f_m$ Hz

$\qquad\qquad\qquad\qquad\quad = m f_m$ Hz

The trade-off between bandwidth and SNR is quite favorable in a PCM system since a change in bandwidth by a factor of n leads to a change in the number of available quantization levels by a factor of 2^n.

At low SNR levels of the input signal to a receiver, the PCM is a better system than the FM system; but at high SNR levels of the input to the receiver, the FM system is found to be superior to PCM.

Since PCM systems require only the *detection* of the presence or absence of a pulse at the receiver without having to reproduce an exact replica of the sample, it is possible to reshape and retransmit the pulses by providing repeater stations at various intervals.

Companding in a PCM System

The quantization levels are made unequal: smaller intervals for the low voltage range and larger intervals for the high voltage range. This is known as *compressing*. Since the quantization error is more significant at low voltage levels, compressing leads to a reduction of the quantization error.

At the receiver, the intervals are restored to normal levels by using an *expanding* operation.

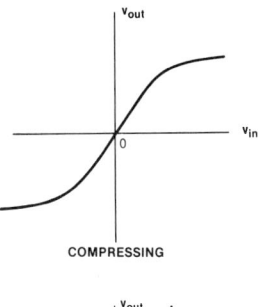

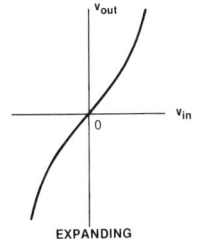

COMPRESSING

EXPANDING

The combination of compressing and expanding is termed *companding*. Companding is used particularly in telephone communication.

Baseband PCM Transmission

The digital code words of the samples are directly transmitted to the receiver in a baseband PCM system. The system requires a link in the form of telephone lines or optical fibers between the transmitter and receiver.

Different Coding Schemes in Baseband System

NRZ-L (Nonreturn-to-zero level): Each one is a high pulse and each zero is a low pulse. The pulse width is the same for both.

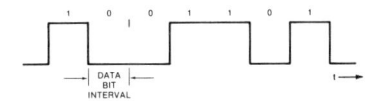

RZ (Return-to-zero level): A zero is transmitted as a low pulse. When a one is to be sent, a high pulse of half the width of normal pulse is inserted in the interval.

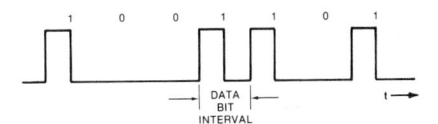

Biphase-M (Biphase Mark): There is always a change in the pulse level at the start of a pulse interval. If a zero is to be transmitted, the pulse stays at a constant level. If a one is to be transmitted, there is a second transition at the midpoint of the interval.

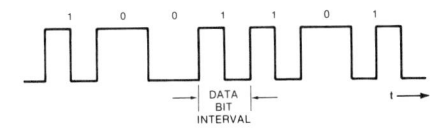

RF TRANSMISSION OF PCM SIGNALS

When PCM signals are to be transmitted through free space, RF modulation is necessary.

Amplitude Shift Keying (ASK)

A one is transmitted as a burst of an RF carrier and zero as the absence of a carrier.

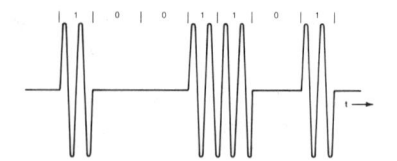

The same method as in DSB-SC AM is used for the generation of ASK signals: the PCM baseband signal is multiplied by an RF carrier.

Frequency-Shift Keying (FSK)

Two different frequencies are transmitted: a burst of frequency $(f_c + f_0)$ to represent a one and a burst of frequency $(f_c - f_0)$ to represent a zero; f_0 is a constant offset from the carrier frequency f_c. The two frequencies are produced by means of a voltage-controlled oscillator (VCO).

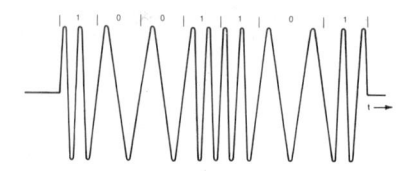

Demodulation of FSK

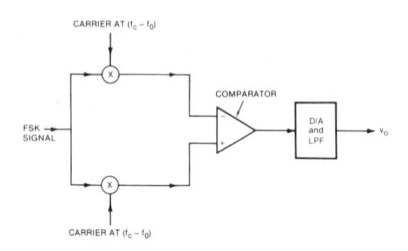

An input at frequency $(f_c + f_0)$ will produce a positive dc component in the comparator output, while an input at $(f_c - f_0)$ will produce a negative dc component in the comparator output. The PCM signals from the output of the comparator are converted to PAM by means of a digital-to-analog converter, and passed through an LPF to recover the original signal.

Phase-Shift Keying (PSK)

A one is transmitted as the normal carrier while a zero as the carrier inverted (that is, with a 180° phase shift) in the PSK system.

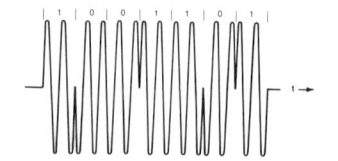

PSK Demodulator

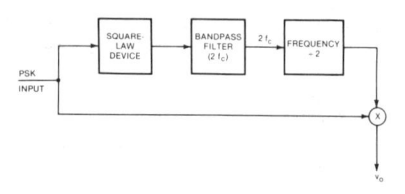

The dc component in the PSK demodulator output will be positive when the input has no phase shift and negative when the input has a 180° phase shift. Again, a conversion to PAM and an LPF are used to recover the original signal.

DELTA MODULATION (DM)

A series of pulses is transmitted in the DM system: a positive pulse if the signal amplitude is higher in the present sampling interval than in the previous one and a negative pulse for the opposite situation.

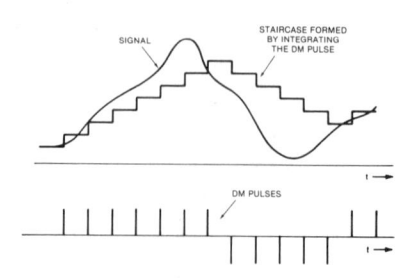

At the demodulator, the incoming pulses are integrated to produce a staircase approximation to the original signal.

Slope Overloading

If the slope of the signal is greater than the rate at which the steps of the staircase function build up in a DM system, or if the slope of the signal is negative and steeper than the rate at which the steps of the staircase function decrease, then slope overloading occurs.

Electronics: Circuits and Systems

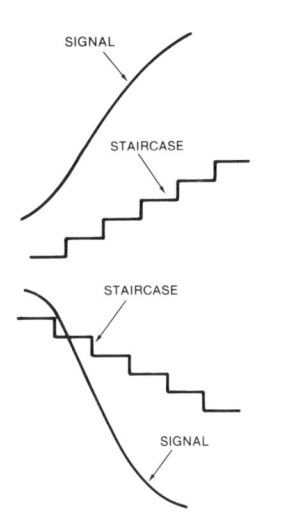

Adaptive DM is used to overcome such difficulties.

Adaptive DM
The height of each step in the staircase function varies in proportion to the actual signal and the staircase function. If the input signal is increasing at a fast rate, the steps are made higher. The demodulation system requires an adaptive process also.

INFORMATION THEORY

A Measure of Information
If an event has n possible outcomes, whose probabilities of occurrence are $p_1, p_2, \ldots, p_n$, then the information associated with the event is

$$H = p_1 \log(1/p_1) + p_2 \log(1/p_2) + \ldots + p_n \log(1/p_n)$$

where,
the logarithm is to the *base of 2*.

INFORMATION RATE
number of messages per second from a source = r
information associated with each message = H
information rate R = r H bits/second

CHANNEL CAPACITY
A channel can be noise free; that is, no error is introduced by the channel; or it can be noisy and there is a probability of error associated with it. A channel has only a finite bandwidth.

Channel capacity is a measure of the maximum rate of information that can be handled by the channel.

Noiseless Channel
Channel capacity = maximum rate at which a source produces information.

$$C = (\log_2 N)/T \text{ bits/second}$$

N = number of different messages from the source
T = interval occupied by the message

Noisy Channel
The channel capacity of a noisy channel is limited by two factors: bandwidth and SNR.

$$C = B \log_2 [1 + (S/N)] \text{ bits/second}$$

B = bandwidth

(S/N) = signal-to-noise ratio

Shannon's Theorem
If C is the capacity of a channel, then it can handle an information rate of R bits/second with an arbitrarily small probability of error, provided $R \leq C$.

That is, if $R \leq C$, then the probability of error can be made as small as we please by means of proper coding techniques. If, on the other hand, $R > C$, then the probability of error approaches 100% regardless of how elaborate a coding scheme is used.

Shannon's theorem and the formula for the channel capacity of a noisy channel can be used to calculate the bandwidth of a channel to handle a given information rate.

DATA TRANSMISSION
Data in the form of alphanumerical characters are sent between computers. Synchronous data transmission utilizes clock pulses to synchronize the bits of each word between the transmitter and receiver. In an asynchronous or start/stop system, the code word may be initiated at any time (without reference to any clock pulse) and a start pulse indicates the starting of a character, and a stop pulse the end of the character.

ASCII Code

Each character is represented by a 7-bit word. A start space (a zero pulse) is used to indicate the beginning of a character. This is followed by the 7 bits of the code word that are followed by an odd parity bit. A stop mark (a one pulse) indicates the end of the character.

Digital-to-Analog Conversion

Binary Weighted D/A Converter: Each digital code word is treated as a binary number and the output is proportional to that number.

$$V_o = \frac{R_F}{R_O}[2^3 b_3 + 2^2 b_2 + 2 b_1 + b_0]$$

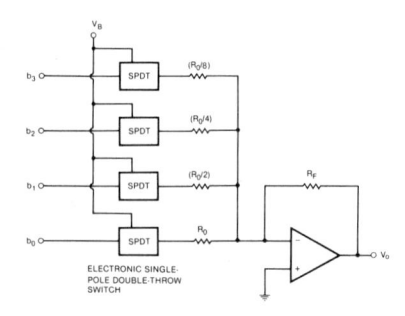

R-2R Ladder D/A Converter:

$$V_O = (2^3 b_3 + 2^2 b_2 + 2 b_1 + b_0)/2^4$$

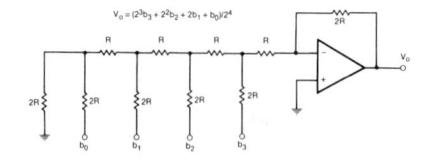

Analog-to-Digital Conversion

Parallel Encoders: Whenever the input voltage exceeds a particular quantization level, then a high output is produced by all the comparators at and below that level.

number of comparators = number of quantization levels

The number of comparators increases by a factor of two for each additional bit.

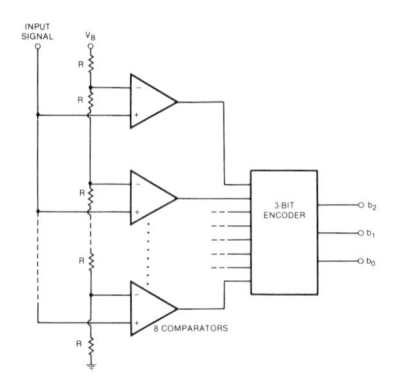

Counting A/D Converters

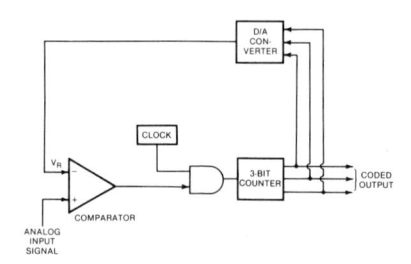

$V_R = 0$ to start with

As the pulses from the clock are counted, V_R increases. When V_R exceeds the signal value, the counter stops counting. The output of the circuit is the three-bit code word for the signal being sampled.

Dual-Shape Integrator

The voltage on the capacitor increases linearly from zero to

$$V_m = (K_1 T_1/C) V_a$$

where,

T_1 is a *fixed* length of time.

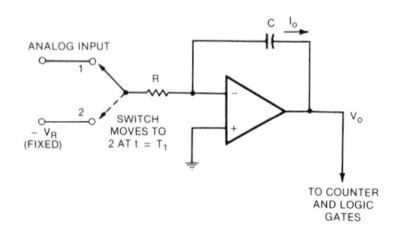

At $t = T_1$, the capacitor starts discharging. The time T_2 needed for completely discharging the capacitor is proportional to V_a. The slope of the discharge portion is the same in all cases.

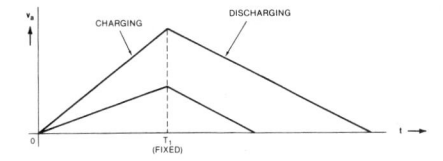

ANSWERS TO DRILL PROBLEMS

15.1 9 MHz
15.2 2.5 kHz
15.3 See Fig. 15-DP3.

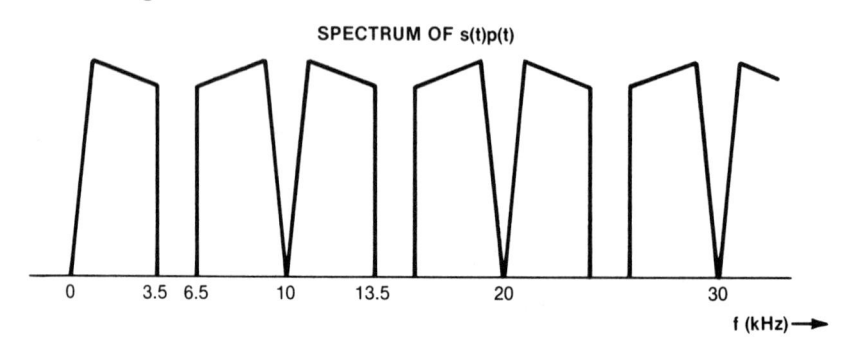

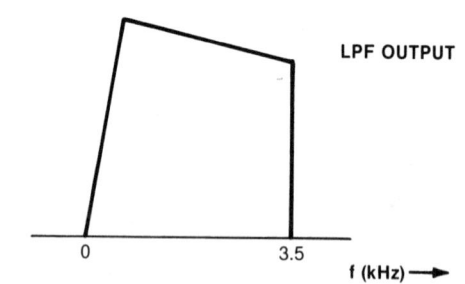

Fig. 15-DP3.

15.4 See Fig. 15-DP4.
15.5 Code words are 0101, 0111, 1001, 1001, 1000, 1000, 0101, 0010, 0000.
15.6 See Fig. 15-DP6.
15.7 75 kHz.
15.8 $k = 200$. $m = 8$. Bandwidth = 120 kHz.
15.9 $m = (100/3.2)$. $m = 31$ bits.
15.10 See Fig. 15-DP10.
15.11 2.6 bits.
15.12 0.998 bits.
15.13 1.443 bits/message. Rate = 577 bits/second.
15.14 632 bits/s.
15.15 32×10^3 bits/s.
15.16 4.7 bits/symbol. $C = 940 \times 10^6$ bits/s.
15.17 (a) 2.24×10^6 bits/s. (b) 18.95×10^6 bits/s.
(c) 189×10^6 bits/s.

SPECTRUM OF s(t)p(t)

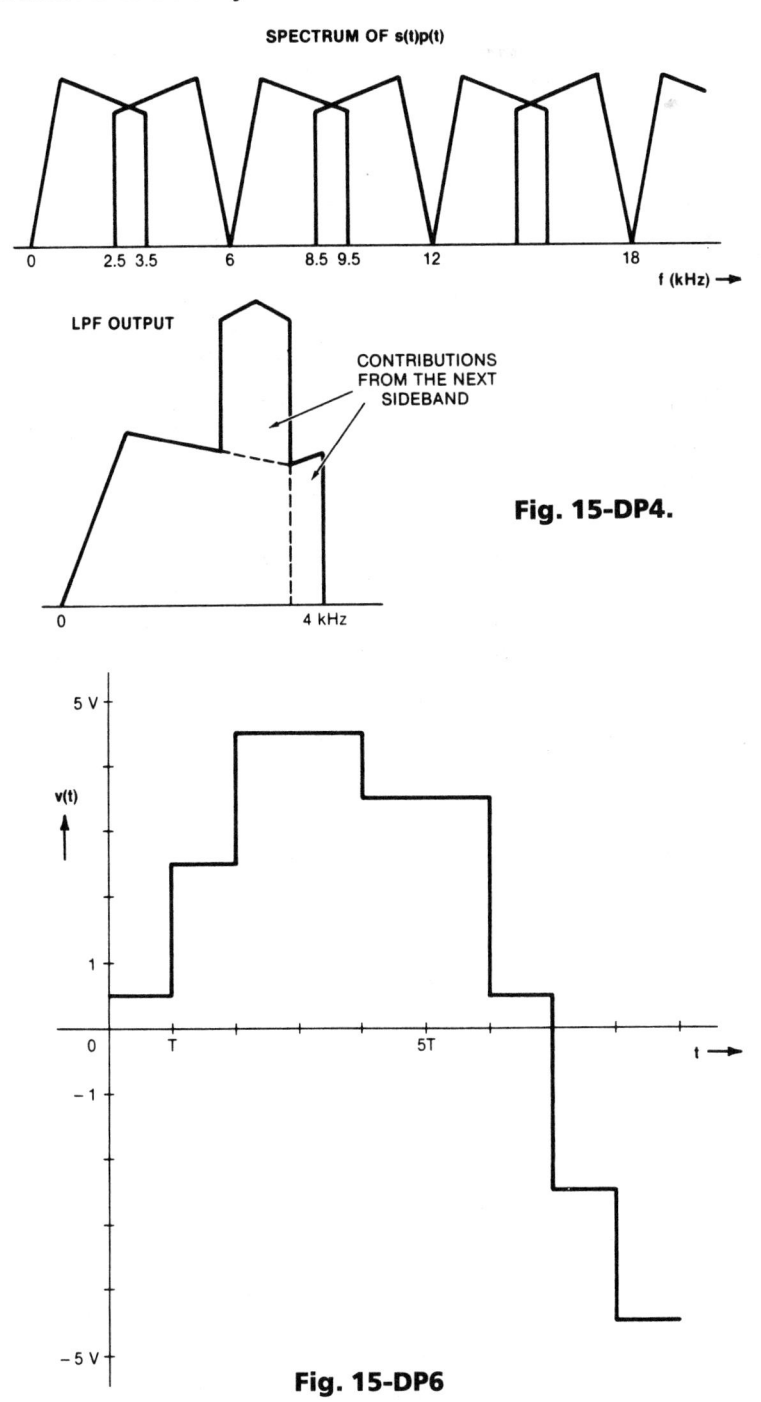

LPF OUTPUT

CONTRIBUTIONS FROM THE NEXT SIDEBAND

Fig. 15-DP4.

Fig. 15-DP6

Fig. 15-DP10

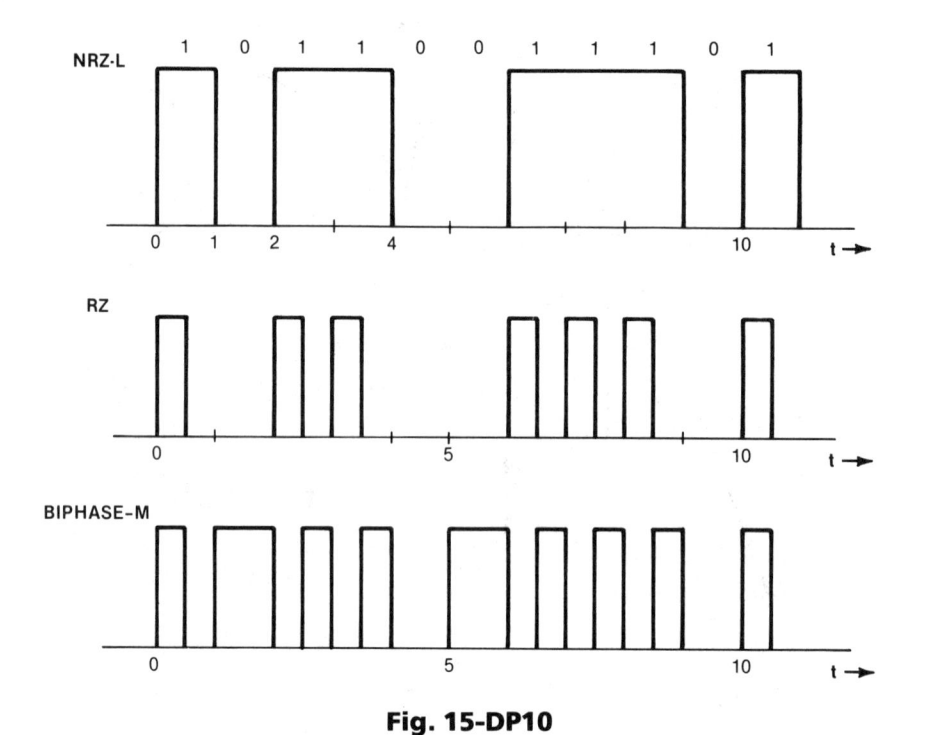

15.18 (a) 614.5×10^3 bits/s. (b) 28.4×10^6 bits/s.
(c) 292×10^6 bits/s.
15.19 Information rate = 3.6×10^7 bits/s. B = 3.6 MHz.
15.20 (S/N) = 30.6 or 14.8 dB.
15.21 Average information per picture = 9×10^6 bits. Time = 225 s.
15.22 C = 6.725×10^4 bits/s. B = 5.06 kHz.

PROBLEMS

Sampling Theorem

15.1 An analog signal has a spectrum extending from dc to 25 kHz. What is the minimum rate of sampling required in order to be able to reconstruct the signal from the samples? In practical systems, it is necessary to provide a "guard band", which is the gap between the sidebands between f_m and ($f_p - f_m$) in Fig. 15-3B in the transmission system. Suppose for the signal under consideration, we desire a guard band of 5 kHz, what should be the new sampling rate?

15.2 A sampled data system operates at a sampling rate of 50×10^4 samples/second. What is the bandwidth of the signal that can be transmitted through the system? What will be the bandwidth if a guard band of 10 kHz is required?

15.3 An analog signal has a spectrum as shown in Fig. 15-P3. It is sampled at the following sampling rates: (a) $2f_m$; (b) $1.5f_m$; (c) $4f_m$. In each case, draw the spectrum of the multiplier output (Fig. 15-1A).

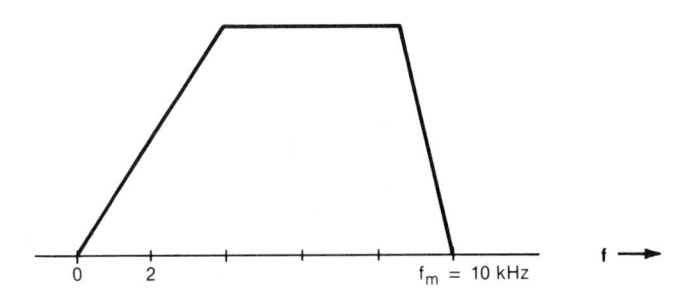

Fig. 15-P3

PCM

15.4 An analog signal is known to range from -7.5 to $+7.5$ V. It is to be quantized into 15 levels (corresponding to $-7, -6, \ldots, 0, 1, 2, \ldots, 7$ V). Assign the code 0000 to -7, 0001 to $-6, \ldots$, 1110 to $+7$ V. Set up a table showing the coding scheme. Prepare a list of the transmitted code words for the signal shown in Fig. 15-P4. Sketch the output of the D/A converter in the receiver for the previous PCM signal.

15.5 When an analog signal is restricted to *positive* values, the system is said to be *unipolar* (in contrast with the *bipolar* case where a signal can be either positive or negative). Consider an analog signal that ranges from zero to slightly under 16 V. It is to be quantized into 16 quantization levels (corresponding to 0.5, 1.5, 2.5, . . ., 15.5 V). Assign the code 0000 to 0.5, 0001 to 1.5, . . ., 1111 to 15.5. Set up a table showing the coding

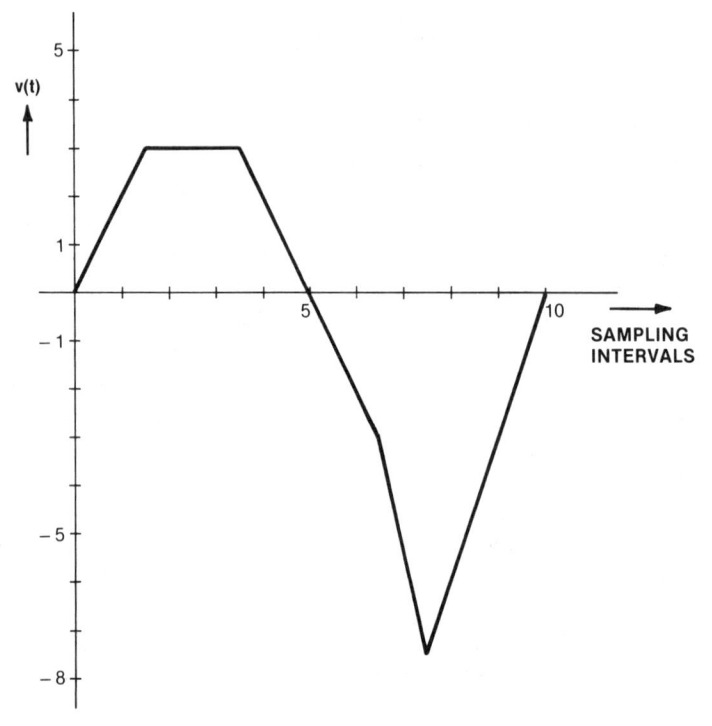

Fig. 15-P4

scheme. Prepare a list of the transmitted code words for the signal shown in Fig. 15-P5. Sketch the output of the D/A converter in the receiver for the previous PCM signal.

15.6 Calculate the bandwidth of each of the following PCM systems: (a) signal band limited to 25 kHz; number of quantization levels = 10; (b) signal band limited to 25 kHz; number of quantization levels = 20 (c) signal band limited to 5 kHz; number of quantization levels = 10; (d) signal band limited to 5 kHz; number of quantization levels = 20. In each case compare the answers with the bandwidth needed for transmitting the same signal through AM (double sideband—large carrier) and FM (with a modulation index of 10).

15.7 Sketch the waveforms of the transmitted PCM signals in NRZL, RZ, and Biphase M for each of the following sequences. (Assume a zero level just before the first bit.) (a) 1 1 1 1 0 0 1 1; (b) 0 0 0 0 1 1 0 0.

Information Theory

15.8 A message source produces four distinct messages with their probabilities being: 0.6, 0.3, 0.08, and 0.02. Calculate the average information associated with the source. Compare this value with the value that would

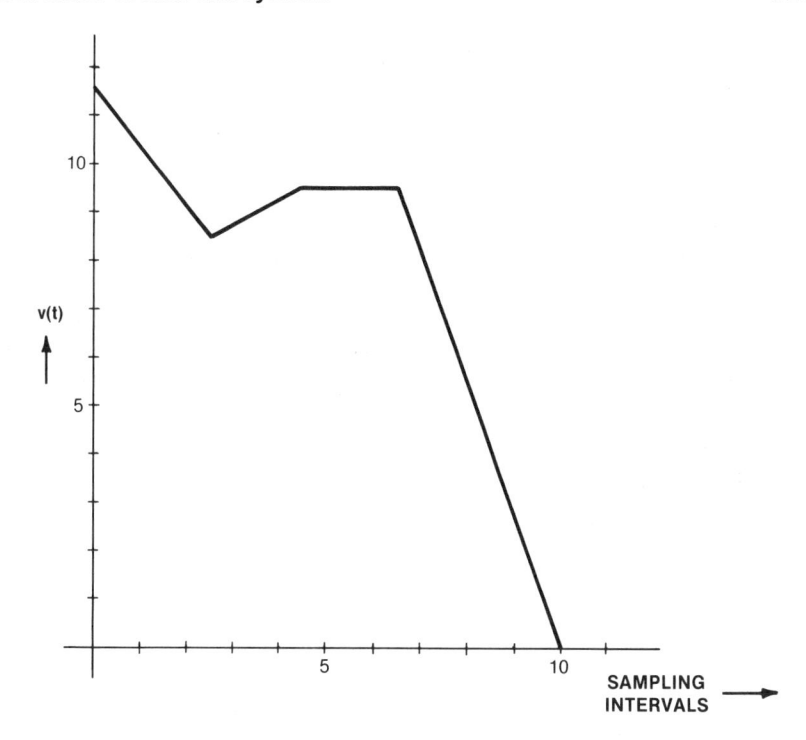

Fig. 15-P5

be obtained if the messages were equiprobable.

15.9 Suppose the four messages of the previous problem are transmitted in the form of binary pulses (00 for A, 01 for B, 10 for C, and 11 for D), with each pulse occupying 5 ms and a gap of 5 ms being provided between successive messages. Calculate the information rate of the system. Compare this with the rate if the messages are equiprobable.

15.10 An analog signal is quantized into eight levels, with all levels occurring with equal probability. In the transmitted signals, each binary digit requires 100 ns, and there is an interval of 20 ns between successive binary digits and 50 ns between successive samples. Calculate the information rate of the system.

15.11 The transmission of a picture takes 2 ms. There are 5×10^4 elements to be transmitted for each picture, and the number of brightness levels (for each element) is 16. Calculate the information rate.

15.12 Calculate the channel capacity of a noiseless channel in each of the last three problems.

15.13 Suppose the messages of Problem 15.9 are to be transmitted through a noisy channel and the desired SNR is 40 dB. Calculate the bandwidth of the channel.

15.14 Repeat the work of the previous problem for the system in Problem 15.10 with an SNR of 30 dB.

15.15 Consider the system of Problem 15.11. Calculate the SNR that will result when the information is transmitted over a channel whose bandwidth is (a) 100 kHz; (b) 1 MHz; (c) 10 MHz.

15.16 The noise present in a channel is 10^{-22} W/Hz. Calculate the channel capacity if the signal power is 10^{-10} W and the channel bandwidth is (a) 100 kHz; (b) 1 MHz; (c) 10 MHz.

15.17 The picture information described in Problem 15.11 is to be transmitted over a telephone circuit of a 4 kHz, bandwidth, and the desired SNR is 60 dB. Calculate the time needed to transmit one picture. Repeat if the SNR is reduced to 20 dB.

15.18 Repeat Problem 15.17 if the number of brightness levels in the picture is changed to 32.

Appendix A

ANSWERS TO PROBLEMS

CHAPTER 1

1.1 -4 A; 16 A; 12 A; -11.25 A.
1.2 -50 V; -30 V; 14 V; -48 V.
1.3 80 W, 640 W, 432 W, 1012.5 W, 250 W, 180 W, 98 W, 384 W.
1.4 800 W. 17.32 V; 8.66 A.
1.5 (a) 20 Hz; (b) 100 MHz; (c) 5×10^{10} Hz.
1.6 (a) 80 ms; (b) 0.8 μs; (c) 5 μs.
1.7 $18°$.
1.8 0.267 μs.
1.9 $V_m = 50$ V. $V_{rms} = 35.4$ V
1.10 (a) 1.36 A, 1.93 A; (b) 10.9 A, 15.4 A; (c) 45.4 mA, 64.3mA.
1.11 $P_{R1} = 576$ W. $P_{R2} = 2$ W. $P_{R3} = 66.7$ W. $P_{R4} = 13.5$ W. P_{R5} cannot be calculated. $P_{R6} = 32$ W.
1.12 20 kHz; 100 kHz; 0.05 ms.
1.14 7.683 W; 20.492 W.
1.15 (a) f_0; (b) $2f_0$; (c) $4f_0$. The choice of the bandwidth does not depend upon the resistance.
1.16 Bandwidth = f_0.
1.18 (a) 14 joules; (b) 14 joules; (c) 48 joules.
1.19 (a) 29.2%; (b) 83.3%.
1.20 (a) 3.333 A; (b) 16 A; (c) 17.5; (d) 6720 W.
1.21 8.94 V; 1.12 A.
1.22 15 mV; 30 μW; 0.6 V; 18 mW; 15; 600.
1.23 (a) 2 V; (b) 5 mA; (c) 120 V; (d) 1800 W.
1.24 (a) 200 V; (b) 12.5 A; (c) 2500 W; (d) 2×10^5; (e) 5000.

1.25 (a) 10 V, 1 mA; (b) 10 V, 1 mA; (c) 5 V, 0.5 mA; (d) 0.2 V, 0.02 mA; (e) 0.1 mV, 10 μA.
1.26 (a) $R_s > 100$ MΩ; (b) $R_s < 0.1$ MΩ.
1.27 (a) 11.2 dB; (b) 1.21 dB; (c) −8.79 dB; (d) −48.8 dB.
1.28 (a) 8.32×10^7; (b) 6.19; (c) 0.161; (d) 1.20×10^{-8}.
1.29 8.54 mW.
1.30 3.
1.31 (a) 26.0 dB; (b) 46.1 dB; (c) 26.1 dB; (d) −2.04 dB; (e) −4.08 dB.
1.32 (a) 0.319; (b) 0.565; (c) 3.133; (d) 1.77.
1.33 1.26 nW; 1.26 mW; 5 W; 50 W.
1.34 377 mV; 15 V; 7.52 V.
1.35 (a) 400, 52 dB; (b) 66.7, 36.5 dB; (c) 40, 32 dB; (d) 2.67×10^4, 44.3 dB; (e) 1.6×10^4, 42 dB.
1.36 (a) 4.99 mA; (b) 19.9 μW; (c) 25 A; (d) 5000 W.

CHAPTER 2

2.1 30 A; 7 A.
2.2 215 V; 140 V.
2.3 (a) 20 V, −20 V, 60 V, −70 V; (b) −60 V, −20 V, 140 V, 290 V.
2.4 (a) −10 V; (b) 4.05 V; (c) −45 V.
2.5 (a) 2 A; (b) 1 mA.
2.6 (a) 22.2 V, 77.8 V; (b) 20 V, 60 V; (c) 100 V, 100 V; (d) 50 V, 0.5 V.
2.7 (a) 150 V, 60 V, 3 A; (b) 30 V, 60 V, 3 A.
2.8 (a) 40 V, 140 V, 0.667 A; (b) 200 V, 300 V, 12.5 A; (c) 100 V, 200 V, 6.67 A; (d) 11.1 V, 111 V, 0.111 A.
2.9 $R_1 = 21.5$ Ω (minimum); $R_2 = 53.5$ Ω (minimum)
2.10 (a) 15.6 Ω; (b) 37.5 Ω; (c) 10 Ω; (d) 10 Ω.
2.11 (a) 7.78 A, 2.22 A; (b) 15 A, 5 A; (c) 25 mA, 25 mA; (d) 100 A, 1 A.
2.12 (a) 25 A, 35 A, 50 V; (b) 80 A, 90 A, 80 V.
2.13 $R_1 = 35.7$ Ω (max); $R_2 = 22.7$ Ω (max).
2.14 (a) 100 Ω, 0.8 A, 4 V; (b) 100 Ω, 0.5 A, 0.5 V.
2.15 (a) 5 Ω, 400 V, 4 A; (b) 1 Ω, 50 V, 0.5 A.
2.16 0.3 V, 2.25 μJ
2.17 39.7 μF, 0.0218 coulomb.
2.18 (a) 0.6 s; (b) 160 V; (c) 3 s; (d) 0.512 J.
2.19 12 Ω, 250 μF.
2.20 (a) 0.9 ms; (b) 2 ms; (c) 6.4 ms.
2.21 (a) 19.7 V; (b) 32 V; (c) 50 V.
2.22 (a) 10.4 s; (b) 6.1 s; (c) 2 s.

Electronics: Circuits and Systems　　　　　　　　　　　　　　　　　　　　　**929**

2.23　0.1 μF
2.24　(a) 9.2 ms; (b) 2.2 ms.
2.25　(a) 8944 V; (b) 6000 V; (c) 1210 V.
2.26　(a) 0.433 ms; (b) 1.44 ms.
2.27　40 ms.
2.29　2.5 J; 31.6 A.
2.30　(a) 40 ms; (b) 20 A; (c) 200 ms.
2.31　240 Ω; 1.44 H.
2.32　(a) 1.1 ms; (b) 2.6 ms; (c) 8 ms.
2.33　(a) 2.5 mJ; (b) 6.7×10^{-8} s.
2.34　(a) 0.635 mJ; (b) 3.12×10^{-5} J.
2.35　(a) 120 $\underline{/0°}$ V; (b) 16 $\underline{/-43°}$ A; (c) 32 $\underline{/37°}$ A.
2.36　V_2 leads V_1 by 75°. V_1 leads I_a by 9°. I_b leads V_1 by 119°. V_2 leads I_a by 66°. I_b leads V_2 by 44°. I_b leads I_a by 110°.
2.37　(a) 10.7 $\underline{/75°}$ Ω; (b) 19.2 $\underline{/0°}$ Ω; (c) 10 $\underline{/60°}$ A; (d) 84 $\underline{/115°}$ V; (e) 84 $\underline{/-25°}$ V.
2.38　(a) 31.8 MΩ; (b) 0.318 MΩ; (c) 3.18 kΩ; (d) 3.18×10^{-3} Ω.
2.39　(a) 25.1 Ω; (b) 251 Ω; (c) 2.51×10^7 Ω.
2.40　(a) 61.3 $\underline{/-11.7°}$ Ω, 1.63 $\underline{/11.7°}$ A, 97.8 $\underline{/11.7°}$ V, 20.2 $\underline{/-78.3°}$ V.
　　　(b) 87.5 $\underline{/-46.7°}$ Ω, 1.14 $\underline{/46.7°}$ A, 68.4 $\underline{/46.7°}$ V, 72.6 $\underline{/-43.3°}$ V;
　　　(c) 637 $\underline{/-85°}$ Ω, 0.157 $\underline{/85°}$ A, 9.42 $\underline{/85°}$ V, 100 $\underline{/-5°}$ V.
2.41　(a) 100 $\underline{/5.7°}$ Ω, 1 $\underline{/-5.7°}$ A, 100 $\underline{/-5.7°}$ V, 10 $\underline{/84.3°}$ V; (b) 141.4 $\underline{/45°}$ Ω, 0.707 $\underline{/-45°}$ A, 70.7 $\underline{/-45°}$ V, 70.7 $\underline{/45°}$ V; (c) 1010 $\underline{/84.3°}$ Ω, 0.099 $\underline{/-84.3°}$ mA, 9.9 $\underline{/-84.3°}$ V, 99.5 $\underline{/5.7°}$ V.
2.42　(a) 15.4 MHz, 54.1 Ω; (b) 6.37 MHz, 70.7 Ω; (c) 2.64 MHz, 130 Ω.
2.43　(a) 329 Hz, 54.1 Ω; (b) 796 Hz, 70.7 Ω; (c) 1.92 kHz, 131 Ω.
2.44　(a) 0.140; (b) 0.815; (c) 1.
2.45　707 Hz.
2.46　(a) 0.988; (b) 0.537; (c) 0.0635.
2.47　31.8 Hz.
2.48　(a) 15.9 MHz; (b) 3.51 MHz.
2.50　(a) 20 V, 22 Ω; (b) 90 V, 9 Ω.
2.51　(a) 16 V, 10 Ω; (b) 5.88 V, 43.5 Ω.
2.52　10 V, 22.5 V, 30 V, 35 V, 38.6 V, 41.2 V.
2.53　(a) 16 Ω; (b) 1.67 kΩ.
2.54　(a) 250 kΩ; (b) 25 Ω; (c) 0.444 Ω.
2.55　(a) $(N_1/N_2) = 7.91$; (b) $(N_2/N_1) = 15.8$; (c) $(N_1/N_2) = 0.0316$.
2.56　0.316, 86.6

CHAPTER 4

- **4.1** (a) 3.33 V; (b) 5.33 V.
- **4.2** (a) 3.33 V; (b) 5.33 V.
- **4.3** (a) 6.66 V; (b) 10.7 V.
- **4.4** 10 V, 16 V.
- **4.5** $v_o = +5$ when v_{in} exceeds $+5$ V; otherwise $v_0 = v_{in}$.
- **4.7** $v_o = v_{in}$ in the range $-5 < v_{in} < 7$.
- **4.8** *Diode D1:* $v_{D1} = 0$ when $v_{in} > 7$ V; $v_{D1} = (v_{in} - 7)$ when $-5 < v_{in} < 7$; $v_{D1} = -12$ V when $v_{in} < -5$ V.
- **4.10** Output = 10 except in the intervals (15, 20), (25, 30), (45, 50), (55, 60).
- **4.12** Output = 10 in the intervals (0, 5), (10, 15), (30, 35), (40, 45).
- **4.14** $v_o = 10$ V in the intervals (0, 15), (20, 25), (30, 35), (40, 45), (50, 55), (60, 75).
- **4.15** v_o is in the form of 10-V pulses, each of duration 5 seconds, starting at 0, 10, 20, . . .
- **4.16** 0.107 kΩ. $V_{in(min)} = 7.01$ V.
- **4.17** $V_{in(min)} = 3.32$ V.
- **4.18** (a) Z_1 acts as a forward biased diode, while Z_2 is in the zener mode. (b) Neither diode is in the zener mode. (c) Z_1 is in the zener mode and Z_2 acts as a forward biased diode.
- **4.20** 51 ms.

CHAPTER 5

- **5.1** (a) 50.7 dB; (b) 25.3 dB; (c) −3.44 dB; (d) −0.158 dB; (e) 56.0 dB.
- **5.2** (a) 631; (b) 3.98×10^5; (c) 1; (d) 6.02×10^3; (e) 0.451.
- **5.3** 16.8 dB.
- **5.4** 4.45 dB, 46 dB, 25.2 dB.
- **5.5** 61.9 dB, 40 dB, 51 dB.
- **5.6** (a) 0.897 mA, 9.44 V; (b) 2 mA, 4.5 V.
- **5.7** 40 kΩ, 0.778 kΩ.
- **5.8** (a) cutoff; (b) normal active, but saturation possible; (c) cutoff; (d) saturation.
- **5.9** 0.588 mA, 8.82 V.
- **5.10** 27.2 dB, 40.9 dB, 34 dB.
- **5.11** 53.3 dB, 16.4 dB, 34.9 dB.
- **5.12** 45.4 dB.
- **5.13** 48.5 dB, 22.3 dB, 35.4 dB.
- **5.14** 39.4 dB, 33.1 dB, 36.3 dB.
- **5.15** 29.9 dB, 31.6 dB, 30.7 dB.
- **5.16** 22.2 dB, 20.1 dB, 21.1 dB.

5.17 267 kΩ, 29.4 Ω.
5.18 625, 1.79 kΩ

CHAPTER 6

6.1 (a) ≥ 3.5 V; (b) ≥ 3.5 V; (c) ≥ 4.5 V; (d) ≥ 4 V.
6.2 (a) ≤ 0; (b) ≤ 5.5 V; (c) ≤ 4 V.
6.3 4.1 V, .227 mA, 33.7 mA.
6.4 −5.73 V, 72 mA.
6.5 −4.15 V, 17.8 mA.
6.6 $I_D = 0.75 (V_{GS} - 2)^2$ mA.
6.7 $I_D = 0.052 (V_{GS} + 16.4)^2$ mA.
6.10 1.87 mA, 4.41 V.
6.11 0.0866 mA, 16.9 V.
6.12 0.416 kΩ, 0.417 kΩ.
6.16 4 mA, −16 V.
6.18 10.8 mA, −9.16 V.
6.23 11 dB, 40 dB, 25.6 dB.
6.24 5.33 dB, 39.6 dB, 22.4 dB.
6.25 6.19 dB, 51.3 dB, 28.7 dB.
6.26 Voltage gains: 3.17 dB, 2.73 dB; current gains: 47.7 dB, 47.3 dB; power gains: 25.4 dB, 25 dB.
6.27 Voltage gains: 2.28 dB, −1.21 dB; current gains: 28.2 dB, 24.8 dB; power gains: 15.2 dB, 11.8 dB.

CHAPTER 7

7.1 (a) 3.4 Hz; (b) 13.6 Hz; (c) 3.25 Hz.
7.2 (a) 1.95 Hz; (b) 1.95 Hz; (c) 1.75 Hz.
7.3 1.47 μF, 0.398 μF, 68.9 μF, 122 Hz.
7.4 0.0637 μF, 0.318 μF, 19.7 μF, 122 Hz.
7.5 (a) 258 pF; (b) 43.8 pF; (c) 4782 pF; (d) 496 pF.
7.6 (a) 285 kHz; (b) 498 kHz.
7.7 (a) 7.47 MHz; (b) 50.2 kHz.
7.8 Gain = 102; cutoff frequencies: 21 Hz, 253 kHz.
7.9 Gain = 10.3; cutoff frequencies: 8.47 Hz, 8.67 MHz.
7.10 Gain = 89; C_S = 15.3 μF, C_1 = 0.532 μF (min), C_2 = 0.245 μF (min); 436 kHz.

7.12 628 Hz.
7.13 5 MHz.
7.14 1.89 MHz.
7.15 7.04 MHz; 9.09 MHz; 4.07 MHz.

CHAPTER 8

8.1 (a) -50; (b) -0.02; (c) -0.25; (d) -4.
8.2 (a) 51; (b) 1.02; (c) 1.25; (d) 5.
8.3 125 MΩ.
8.4 124.5 MΩ.
8.5 100 kΩ; 0.75 V; -200 kΩ (not possible with a passive component).
8.7 $-0.05V_s$ mA.
8.8 $R_1R_3 = R_2R_4$.
8.10 $(V_{EE} - V_1)/R_1$.
8.13 (a) -10 V; (b) 10 V; (c) 0; (d) 0.
8.15 (a) -10 V; (b) 10 V; (c) 0.5 V; (d) -0.5 V.
8.18 (a) 0.01; (b) 0.2.
8.19 (a) 3.16×10^7; (b) 6×10^5
8.20 $v_o = -5$ V when $v_i > 10$ V; $v_o = 5$ V when $5 < v_i < 10$; $v_o = v_i$ when $v_i < 5$.
8.22 V_o is high at low light levels. V_o is low at bright levels.
8.24 8.75, 1.25 V. Output switches from high to low at 0.072T and at 1.072T; it switches from low to high at 0.49T.
8.27 Output switches from 0 to 20 V at 16.7 ms and from 20 to 0 at 56 ms.
8.28 (a) 96.8; (b) 9.97; (c) 1.11.
8.29 0.245.
8.30 29.5 dB, 49.5 dB, 68.6 dB.

CHAPTER 9

9.1 (a) 1.38 eV, 2.21×10^{-19} J; (b) 2.26 eV, 3.62×10^{-19} J; (c) 0.622 eV, 9.95×10^{-20} J.
9.2 (a) 1776 nm; (b) 1130 nm; (c) 0.00398 nm.
9.3 1462 nm; (b) 460 nm; (c) 1.243×10^5 nm.
9.4 62.4 ms, 6.24 ms, 0.624 ms.
9.5 $V_{BB(min)} = 20$ V. $R_s = 0.005 (V_{BB} - 20)$ MΩ.
9.7 R_C (in kΩ) $= 0.05V_{CC}$.

CHAPTER 10

10.2 (a) $(ABC + \overline{AB})\overline{AC}$; (b) $(A + B + C)(\overline{A} + \overline{B}) + (\overline{A} + B)(A + \overline{C})$.
10.3 (a) The T_a column should read: 0 1 1 1 (from top to bottom); (b) T_b column: 1 1 0 1 1 1 0 1; (c) T_c column: 0 1 1 0.
10.4 (a) T_a column: 0 0 0 1; (b) T_b column: 0 1 0 0 0 1 0 0; (c) T_c column: 1 0 0 1.
10.5 (a) $T_a = 1$ for all eight input combinations; (b) $T_b = 0$ for all eight input combinations.
10.6 (a) $T_a = 0$; (b) T_b column 1 0 1 0 0 1 0 1; (c) T_c column: 0 1 0.
10.7 (a) 0; (b) 1; (c) X; (d) $\overline{X}$.
10.8 (a) 1; (b) 0; (c) $\overline{X}$; (d) X.
10.9 (a) T_a column: 0 0 0 0 1 0 1 1; (b) T_b column: 1 1 1 0 1 0 0 0.
10.10 $T = \overline{XY} + \overline{X}Y + XY$.
10.11 $T_1 = \overline{XY} + X\overline{Y} + XY$; $T_2 = \overline{XY} + \overline{X}Y + XY$. T_3 can be obtained by using an exclusive NOR gate with inputs T_1 and T_2.
10.12 (a) $A \oplus B$; (b) $\overline{AB} + \overline{A}B + A\overline{B}$; (c) $\overline{(A \oplus B)}$.
10.14 $Z = \overline{LTP} + \overline{L}TP + L\overline{T}P$.
10.17 (a) $A \oplus B$; (b) $A \odot B$; (c) $A \oplus B$.
10.23 (a) $(A\overline{B} + A\overline{C} + B)$; (b) $(AB + A\overline{C})(A\overline{C} + BC)$.

CHAPTER 11

11.1 2.78 kΩ.
11.2 (a) 0.7 V; (b) $0.7 < v_i < 1.9$; (c) 1.9 V.
11.3 $V_{HL} = 0.4$ V (worst case), 1.4 V (typical); $V_{NL} = 0.4$ V (worst case), 0.6 V (typical).
11.4 $V_{NL} = 0.25$ V; $V_{HL} = 0.25$ V.
11.5 A positive NAND gate.
11.6 A positive NOR gate.
11.7 $\overline{(A\ B\ C)}\ \overline{(X\ Y)}$.
11.8 (a) T = low. (b) High impedance output state − T is not affected by A or B; (c) Same as (b); (d) T = high.
11.9 Fan-out = 20 (high output), 10 (low output).
11.12 A positive NOR gate.

CHAPTER 12

12.1 $T = \bar{A}_1\bar{A}_0 D_0 + \bar{A}_1 A_0 D_1 + A_1\bar{A}_0 D_2 + A_1 A_0 D_3$.
12.3 $A_1 = L$, $A_0 = T$, $D_0 = 1$, $D_1 = 0$, $D_2 = \bar{P}$, $D_3 = 0$.
12.4 $A_1 = Q$, $A_0 = F$, $D_0 = 0$, $D_1 = P$, $D_2 = 0$, $D_3 = 1$.
12.5 (a) Output column of the truth table: 1 1 1 0 0 0 0 0 0 0 1 0 0 1 1;
(b) output column: 1 1 1 0 1 1 1 1 0 0 0 0 0 1 1 0.
12.7 Feed the output lines 1, 7 to an OR gate.

CHAPTER 13

13.1 (a) 22; (b) 58; (c) 125; (d) 3285.
13.2 (a) 1 0 0 0 0 0 0 0 1 0 1; (b) 1 1 1 1 0 1 0 0 1 0 0; (c) 1 0 0 0 0 0 0 0 0 0 0 0; (d) 1 1 1 1 1 1 1.
13.3 (a) 0.875; (b) 0.3125; (c) 0.7890625; (d) 0.109375.
13.4 (a) 0.0 1 0 1 1 0 0 1; (b) 0.0 0 0 0 0 0 0 1; (c) 0.0 0 1 0 0 0 0 0.
13.5 (a) 1 0 0 1 1.1 0 0 0 1 1 1 1; (b) 1.1 1 1 1 0 1 0 0;
(c) 1 1 0 0 0 0 1 1.1 0 0 1 1 0 0 1; (d) 1 1 1 1 1 1 1.1 1 1 1.
13.6 (a) 23.28125; (b) 2.625; (c) 1.3125; (d) 31.9375.
13.7 (a) 2054; (b) 231; (c) 87.
13.8 (a) 32.09375; (b) 3.609375; (c) 28.875; (d) 1.359375.
13.9 (a) 10000; (b) 533; (c) 3424.
13.10 (a) 0.77126; (b) 0.07603366; (c) 0.1057065.
13.11 (a) 11.675341; (b) 142.546314; (c) 1053.07603366.
13.12 (a) 16.61; (b) 253.706; (c) 2.2.
13.13 (a) 1267; (b) 2748; (c) 3567; (d) 35105.
13.14 (a) 1 0 1 0 1 1 0 1 1 1 1; (b) 1 0 0 1 1 0 0 0 0 0 1 0 0 0 0 1;
(c) 1 0 1 0 1 0 0 1 1 1 1 1 0 0 1 0.
13.15 (a) 1 0 1 0.1 1 0 1 1 1 1 1; (b) 1 0 1 0 1 1 0 1.1 1 1 1;
(c) 1 0 0 1 1 0 0 0.0 0 1 0 0 0 0 1; (d) 1 0 1 0.1 0 0 1 1 1 1 1 0 0 1 0.
13.16 (a) 265D; (b) 7D1; (c) 25AA.
13.17 (a) 0.FB6AE7; (b) 0.3339C0; (c) 0.F6D5CF.
13.18 (a) 113.F6D5CF; (b) 1DB.3765FD; (c) FFFF.11.
13.19 (a) AE.AB; (b) 3B.D4.
13.20 (a) 5117; (b) 12.236; (c) 0.5117.
13.21 (a) 3FE; (b) 28A; (c) F.F8; (d) 1.45.
13.25 (a) 1 0 1 0 1 0 1 0; (b) 0 1 1 1 0 1 1 1; (c) 1 1 0 1 1 1 1 1;
(d) 1 0 0 0 0 0 0 1; (e) 0 1 0 1 1 0 1 1.
13.26 (a) −104; (b) 109; (c) −15; (d) 106.

Electronics: Circuits and Systems

13.27 (a) 1 0 0 1 1 1 1 1; (b) 0 0 1 1 0 0 0 1; (c) 1 0 0 0 0 1 0 0;
(d) 0 1 1 0 0 0 1 1; (e) 1 0 0 1 1 0 1 1.
13.28 (a) −85; (b) 95; (c) −28; (d) −126.
13.31 When there is an overflow bit of 1, add it to the result already obtained to get the correct answer.
13.33 Omit the two inverters (to change S to $\overline{S}$ and R to $\overline{R}$) and feed S and R directly to the level-1 gates. Q is the output of the NOR gate to which R is applied.

CHAPTER 14

14.1 (a) 200 W; (b) 1800 W; (c) 1.21×10^4 W. For a 300-Ω load, each value is smaller by a factor of 300. For a 0.1-Ω load, each value is larger by a factor of 10.
14.2 75.3 W.
14.3 (a) 100, 300, 600, 700, 800, 900, 1200, 1400 Hz; (b) 0, 250, 500, 750, 1000 Hz.
14.4 (a) f_1 = 25 kHz, other side frequency = 60 kHz; or f_1 = 45 kHz, other side frequency = 80 kHz. (b) f_1 = 0, other side frequency is not present; or f_1 = 70 kHz, other side frequency = 105 kHz. (c) f_1 = 35 kHz, other side frequency = dc; or f_1 = 105 kHz, other side frequency = 140 kHz.
14.6 (a) f_1 = 11 kHz; (b) f_1 = 111 kHz or 99 kHz; (c) f_1 = 10009.7 kHz or 9997.7 kHz.
14.7 0.192, 0.7065, 1.6935, 2.208 MHz. LPF should cut off at a little above 0.7065 MHz.
14.8 f_c = 965 kHz; LPF cutoff at 25 kHz.
14.10 949.9 kHz, 954.1 kHz. LPF cuts off at 4.1 kHz.
14.12 Use an LPF and an HPF to separate the two sidebands and multiply by a local carrier f_c. Two LPFs will then recover the original signals.
14.13 50 W in carrier, 7.03 W in each side frequency, 64.06 W in the total AM signal.
14.14 m = 0.707. 2.65 kW in the side frequency components and 7.35 kW in the carrier at m = 0.85. Total power = 10.89 kW.
14.16 Frequency deviation = 14 kHz. Instantaneous frequency varies at the rate of (14/5) kHz/volt. For a 10 kHz signal, the instantaneous frequency varies at the rate of (70/5) kHz/volt. The value of the carrier frequency does not affect the previous results.
14.17 (a) 15.6; (b) 15; (c) 3.
14.18 (a) 20 kHz; (b) 10 kHz; (c) 20 kHz; (d) 30 kHz; (e) 1220 kHz; (f) 30 kHz; (g) 320 kHz.

CHAPTER 15

15.1 50 kHz. 55 kHz with the guard band.
15.2 250 kHz. 245 kHz with the guard band.
15.6 (a) 100 kHz, 50 kHz (AM), 550 kHz (FM); (b) 125 kHz, 50 kHz (AM), 550 kHz (FM); (c) 20 kHz, 10 kHz (AM), 110 kHz (FM); (d) 25 kHz, 10 kHz (AM), 110 kHz (FM).
15.8 1.368 bits. 2 bits.
15.9 91.2 bits/second. 133.3 bits/s.
15.10 7.32×10^6 bits/s.
15.11 10^8 bits/s.
15.12 (a) 133.3 bits/s; (b) 7.32×10^6 bits/s; (c) 10^8 bits/s.
15.13 10 Hz.
15.14 734 kHz.
15.15 (a) 3010 dB; (b) 300 dB; (c) 30 dB.
15.16 (a) 2.326×10^6 bits/s; (b) 19.9×10^6 bits/s; (c) 1.66×10^8 bits/s.
15.17 (a) 2.5 seconds; (b) 7.52 seconds.
15.18 (a) 3.14 seconds; (b) 9.4 seconds.

APPENDIX B

MANUFACTURERS' DATA SHEETS

A sampling of data sheets on various semiconductor devices and circuits is provided in the following pages. The intention is to make the reader aware of the general range of specifications of the devices and circuits rather than presenting a comprehensive list. The latest information from the different manufacturers should be consulted before actually proceeding with the design of a circuit or a system.

The data sheets have been reprinted with the kind permission of the following manufacturers.

Motorola Semiconductor Products, Inc.
Box 20912, Phoenix, Arizona, 85036

Fairchild Camera and Instrument Corp.,
313 Fairchild Drive
Mountain View, California, 94042

Texas Instruments, Inc.
Dallas, Texas

2N1842A thru 2N1850A

MOTOROLA

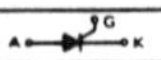

REVERSE BLOCKING TRIODE THYRISTOR

... designed primarily for half-wave ac control applications, such as motor controls, heating controls and power supplies, or wherever half-wave silicon gate-controlled, solid-state devices are needed.

- Glass Passivated Junctions with Center Gate Geometry for Greater Parameter Uniformity and Stability
- Blocking Voltage to 500 Volts
- Junction Temperature Rated @ 125°C

SILICON CONTROLLED RECTIFIERS

16 AMPERE RMS
25 – 500 VOLTS

MAXIMUM RATINGS (T_C = 125°C unless otherwise noted)

Rating	Symbol	Value	Unit
*Peak Repetitive Forward or Reverse Blocking Voltage (1)	V_{DRM} or V_{RRM}		Volts
2N1842A		25	
2N1843A		50	
2N1844A		100	
2N1845A		150	
2N1846A		200	
2N1847A		250	
2N1848A		300	
2N1849A		400	
2N1850A		500	
*Non-Repetitive Peak Reverse Voltage	V_{RSM}		Volts
2N1842A		35	
2N1843A		75	
2N1844A		150	
2N1845A		225	
2N1846A		300	
2N1847A		350	
2N1848A		400	
2N1849A		500	
2N1850A		600	
*Average On-State Current (T_C = 80°C)	$I_{T(AV)}$	10	Amp
*Peak Non-Repetitive Surge Current (One cycle, 60 Hz, preceded and followed by rated current and voltage)	I_{TSM}	125	Amp
Circuit Fusing (T_J = -65 to +125°C, t = 1.0 to 8.3 ms)	$i^2 t$	80	$A^2 s$
*Peak Gate Power	P_{GM}	5.0	Watts
*Average Gate Power	$P_{G(AV)}$	0.5	Watt
*Peak Forward Gate Current	I_{GM}	2.0	Amp
*Peak Gate Voltage – Forward	V_{FGM}	10	Volts
Reverse	V_{RGM}	5.0	
*Operating Junction Temperature Range	T_J	-65 to +125	°C
*Storage Temperature Range	T_{stg}	-65 to +125	°C

THERMAL CHARACTERISTIC

Characteristic	Symbol	Max	Unit
Thermal Resistance, Junction to Case	$R_{\theta JC}$	2.0	°C/W

(1) V_{DRM} and V_{RRM} for all types can be applied on a continuous dc basis without incurring damage. Ratings apply for zero or negative gate voltage. Devices should not be tested for blocking capability in a manner such that the voltage supplied exceeds the rated blocking voltage.

*Indicates JEDEC Registered Data.

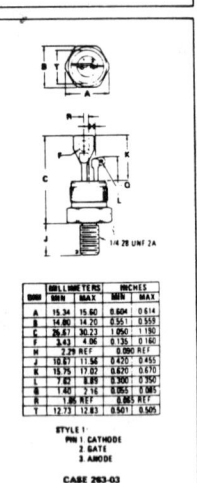

MILLIMETERS / INCHES

DIM	MIN	MAX	MIN	MAX
A	15.34	15.60	0.604	0.614
B	14.00	14.20	0.551	0.559
C	26.67	30.23	1.050	1.190
F	3.43	4.06	0.135	0.160
H	2.29 REF		0.090 REF	
J	10.67	11.56	0.420	0.455
K	15.75	17.02	0.620	0.670
L	7.62	8.89	0.300	0.350
Q	1.40	2.16	0.055	0.085
R	1.85 REF		0.065 REF	
T	12.73	12.83	0.501	0.505

STYLE 1
PIN 1 CATHODE
2 GATE
3 ANODE

CASE 263-03

Electronics: Circuits and Systems 939

2N1842A thru 2N1850A

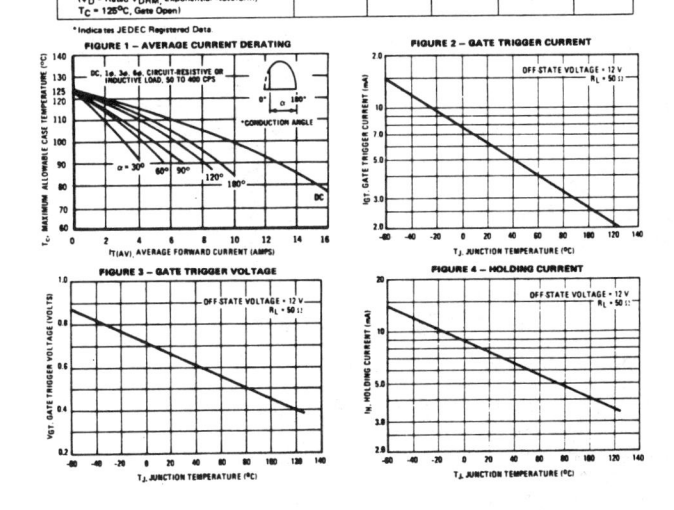

ELECTRICAL CHARACTERISTICS ($T_C = 125°C$ unless otherwise noted.)

Characteristic	Symbol	Min	Typ	Max	Unit
*Average Forward or Reverse Blocking Current (V_D = Rated V_{DRM} or V_R = Rated V_{RRM}, gate open, $T_C = 125°C$)	$I_{D(AV)}, I_{R(AV)}$				mA
2N1842A		–	–	22.5	
2N1843A		–	–	19	
2N1844A		–	–	12.5	
2N1845A		–	–	8.5	
2N1846A		–	–	6.0	
2N1847A		–	–	5.5	
2N1848A		–	–	5.0	
2N1849A		–	–	4.0	
2N1850A		–	–	3.0	
Peak Forward or Reverse Blocking Current (V_D = Rated V_{DRM} or V_R = Rated V_{RRM}, gate open, $T_C = 125°C$)	I_{DRM}, I_{RRM}	–	–	6.0	mA
*Peak On-State Voltage (I_{TM} = 31.4 A peak, Pulse Width ≤ 1.0 ms, Duty Cycle ≤ 2.0%)	V_{TM}	–	–	2.5	Volts
Gate Trigger Current, Continuous dc (V_D = 12 Vdc, R_L = 50 Ω) *(V_D = 12 Vdc, R_L = 50 Ω, T_C = –65°C)	I_{GT}	–	6.0	80 150	mA
Gate Trigger Voltage, Continuous dc (V_D = 12 Vdc, R_L = 50 Ω) *(V_D = 12 Vdc, R_L = 50 Ω, T_C = –40°C) *(V_D = 12 Vdc, R_L = 50 Ω, T_C = –65°C) *(V_D = Rated V_{DRM}, R_L = 50 Ω, T_C = 125°C)	V_{GT}	0.25	0.65 – – –	– 3.5 3.7 –	Volts
Holding Current (V_D = 12 Vdc, Gate Open)	I_H	–	7.0	–	mA
Critical Rate of Rise of Off-State Voltage (V_D = Rated V_{DRM}, Exponential Waveform, T_C = 125°C, Gate Open)	dv/dt	–	30	–	V/µs

*Indicates JEDEC Registered Data

FIGURE 1 – AVERAGE CURRENT DERATING

FIGURE 2 – GATE TRIGGER CURRENT

FIGURE 3 – GATE TRIGGER VOLTAGE

FIGURE 4 – HOLDING CURRENT

Manufacturers' Data Sheets

MOTOROLA

2N3903
2N3904

CASE 29-02, STYLE 1
TO-92 (TO-226AA)

GENERAL PURPOSE TRANSISTOR

NPN SILICON

MAXIMUM RATINGS

Rating	Symbol	Value	Unit
Collector-Emitter Voltage	V_{CEO}	40	Vdc
Collector-Base Voltage	V_{CBO}	60	Vdc
Emitter-Base Voltage	V_{EBO}	6.0	Vdc
Collector Current — Continuous	I_C	200	mAdc
Total Device Dissipation @ T_A = 25°C Derate above 25°C	P_D	625 2.8	mW mW/°C
*Total Device Dissipation @ T_C = 25°C Derate above 25°C	P_D	1.5 12	Watts mW/°C
Operating and Storage Junction Temperature Range	T_J, T_{stg}	−55 to +150	°C

*THERMAL CHARACTERISTICS

Characteristic	Symbol	Max	Unit
Thermal Resistance, Junction to Case	$R_{\theta JC}$	83.3	°C/W
Thermal Resistance, Junction to Ambient	$R_{\theta JA}$	200	°C/W

*Indicates Data in addition to JEDEC Requirements.

ELECTRICAL CHARACTERISTICS (T_A = 25°C unless otherwise noted.)

Characteristic	Symbol	Min	Max	Unit
OFF CHARACTERISTICS				
Collector-Emitter Breakdown Voltage(1) (I_C = 1.0 mAdc, I_B = 0)	$V_{(BR)CEO}$	40	—	Vdc
Collector-Base Breakdown Voltage (I_C = 10 μAdc, I_E = 0)	$V_{(BR)CBO}$	60	—	Vdc
Emitter-Base Breakdown Voltage (I_E = 10 μAdc, I_C = 0)	$V_{(BR)EBO}$	6.0	—	Vdc
Base Cutoff Current (V_{CE} = 30 Vdc, V_{EB} = 3.0 Vdc)	I_{BL}	—	50	nAdc
Collector Cutoff Current (V_{CE} = 30 Vdc, V_{EB} = 3.0 Vdc)	I_{CEX}	—	50	nAdc
ON CHARACTERISTICS				
DC Current Gain(1) (I_C = 0.1 mAdc, V_{CE} = 1.0 Vdc) 2N3903 2N3904	h_{FE}	20 40	— —	—
(I_C = 1.0 mAdc, V_{CE} = 1.0 Vdc) 2N3903 2N3904		35 70	— —	
(I_C = 10 mAdc, V_{CE} = 1.0 Vdc) 2N3903 2N3904		50 100	150 300	
(I_C = 50 mAdc, V_{CE} = 1.0 Vdc) 2N3903 2N3904		30 60	— —	
(I_C = 100 mAdc, V_{CE} = 1.0 Vdc) 2N3903 2N3904		15 30	— —	
Collector-Emitter Saturation Voltage(1) (I_C = 10 mAdc, I_B = 1.0 mAdc) (I_C = 50 mAdc, I_B = 5.0 mAdc)	$V_{CE(sat)}$	— —	0.2 0.3	Vdc
Base-Emitter Saturation Voltage(1) (I_C = 10 mAdc, I_B = 1.0 mAdc) (I_C = 50 mAdc, I_B = 5.0 mAdc)	$V_{BE(sat)}$	0.65 —	0.85 0.95	Vdc
SMALL-SIGNAL CHARACTERISTICS				
Current-Gain — Bandwidth Product (I_C = 10 mAdc, V_{CE} = 20 Vdc, f = 100 MHz) 2N3903 2N3904	f_T	250 300	— —	MHz

MOTOROLA SEMICONDUCTORS   SMALL-SIGNAL DEVICES

2N3903, 2N3904

ELECTRICAL CHARACTERISTICS (continued) (T_A = 25°C unless otherwise noted.)

Characteristic		Symbol	Min	Max	Unit
Output Capacitance (V_{CB} = 5.0 Vdc, I_E = 0, f = 1.0 MHz)		C_{obo}	—	4.0	pF
Input Capacitance (V_{BE} = 0.5 Vdc, I_C = 0, f = 1.0 MHz)		C_{ibo}	—	8.0	pF
Input Impedance (I_C = 1.0 mAdc, V_{CE} = 10 Vdc, f = 1.0 kHz)	2N3903 2N3904	h_{ie}	1.0 1.0	8.0 10	k ohms
Voltage Feedback Ratio (I_C = 1.0 mAdc, V_{CE} = 10 Vdc, f = 1.0 kHz)	2N3903 2N3904	h_{re}	0.1 0.5	5.0 8.0	$\times 10^{-4}$
Small-Signal Current Gain (I_C = 1.0 mAdc, V_{CE} = 10 Vdc, f = 1.0 kHz)	2N3903 2N3904	h_{fe}	50 100	200 400	—
Output Admittance (I_C = 1.0 mAdc, V_{CE} = 10 Vdc, f = 1.0 kHz)		h_{oe}	1.0	40	μmhos
Noise Figure (I_C = 100 μAdc, V_{CE} = 5.0 Vdc, R_S = 1.0 k ohms, f = 10 Hz to 15.7 kHz)	2N3903 2N3904	NF	— —	6.0 5.0	dB
SWITCHING CHARACTERISTICS					
Delay Time	(V_{CC} = 3.0 Vdc, V_{BE} = 0.5 Vdc, I_C = 10 mAdc, I_{B1} = 1.0 mAdc)	t_d	—	35	ns
Rise Time		t_r	—	35	ns
Storage Time	(V_{CC} = 3.0 Vdc, I_C = 10 mAdc, I_{B1} = I_{B2} = 1.0 mAdc)	t_s	—	175 200	ns
Fall Time	2N3903 2N3904	t_f	—	50	ns

(1) Pulse Test: Pulse Width ≤ 300 μs, Duty Cycle ≤ 2.0%.

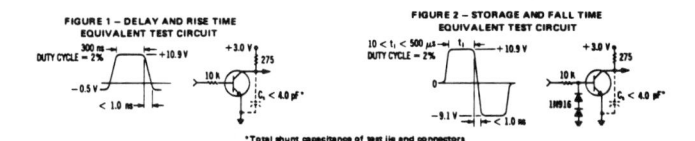

FIGURE 1 – DELAY AND RISE TIME EQUIVALENT TEST CIRCUIT

FIGURE 2 – STORAGE AND FALL TIME EQUIVALENT TEST CIRCUIT

*Total shunt capacitance of test jig and connectors

TYPICAL TRANSIENT CHARACTERISTICS
— T_J = 25°C --- T_J = 125°C

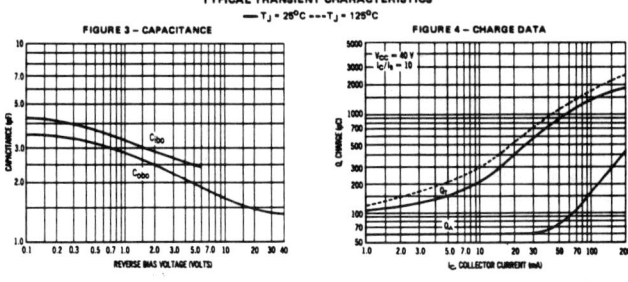

FIGURE 3 – CAPACITANCE

FIGURE 4 – CHARGE DATA

SMALL-SIGNAL DEVICES   MOTOROLA SEMICONDUCTORS

2N3905
2N3906

PNP SILICON ANNULAR♦ TRANSISTORS

PNP SILICON SWITCHING & AMPLIFIER TRANSISTORS

.... designed for general purpose switching and amplifier applications and for complementary circuitry with types 2N3903 and 2N3904.

- High Voltage Ratings — BV_{CEO} = 40 Volts (Min)
- Current Gain Specified from 100 μA to 100 mA
- Complete Switching and Amplifier Specifications
- Low Capacitance — C_{ob} = 4.5 pF (Max)

*MAXIMUM RATINGS

Rating	Symbol	Value	Unit
Collector-Base Voltage	V_{CB}	40	Vdc
Collector-Emitter Voltage	V_{CEO}	40	Vdc
Emitter-Base Voltage	V_{EB}	5.0	Vdc
Collector Current	I_C	200	mAdc
Total Power Dissipation @ T_A = 60°C	P_D	250	mW
Total Power Dissipation @ T_A = 25°C Derate above 25°C	P_D	350 2.8	mW mW/°C
Total Power Dissipation @ T_C = 25°C Derate above 25°C	P_D	1.0 8.0	Watt mW/°C
Junction Operating Temperature	T_J	+150	°C
Storage Temperature Range	T_{stg}	-55 to +150	°C

THERMAL CHARACTERISTICS

Characteristic	Symbol	Max	Unit
Thermal Resistance, Junction to Ambient	$R_{\theta JA}$	357	°C/W
Thermal Resistance, Junction to Case	$R_{\theta JC}$	125	°C/W

*Indicates JEDEC Registered Data.
♦Annular semiconductors patented by Motorola Inc.

STYLE 1
PIN 1. EMITTER
   2. BASE
   3. COLLECTOR

DIM	MILLIMETERS		INCHES	
	MIN	MAX	MIN	MAX
A	4.450	5.200	0.175	0.205
B	3.180	4.190	0.125	0.165
C	4.320	5.330	0.170	0.210
D	0.407	0.533	0.016	0.021
F	0.407	0.482	0.016	0.019
K	12.700	-	0.500	-
L	1.150	1.390	0.045	0.055
N	-	1.270	-	0.050
P	6.350	-	0.250	-
Q	3.430	-	0.135	-
R	2.410	2.670	0.095	0.105
S	2.030	2.670	0.080	0.105

CASE 29-02
(TO-92)

© MOTOROLA INC., 1973   DS 5128 R2

Electronics: Circuits and Systems

*ELECTRICAL CHARACTERISTICS ($T_A = 25°C$ unless otherwise noted.)

Characteristic	Fig. No.	Symbol	Min	Max	Unit
OFF CHARACTERISTICS					
Collector-Base Breakdown Voltage ($I_C = 10\ \mu Adc, I_E = 0$)		BV_{CBO}	40	–	Vdc
Collector-Emitter Breakdown Voltage (1) ($I_C = 1.0\ mAdc, I_B = 0$)		BV_{CEO}	40	–	Vdc
Emitter-Base Breakdown Voltage ($I_E = 10\ \mu Adc, I_C = 0$)		BV_{EBO}	5.0	–	Vdc
Collector Cutoff Current ($V_{CE} = 30\ Vdc, V_{BE(off)} = 3.0\ Vdc$)		I_{CEX}	–	50	nAdc
Base Cutoff Current ($V_{CE} = 30\ Vdc, V_{BE(off)} = 3.0\ Vdc$)		I_{BL}	–	50	nAdc
ON CHARACTERISTICS (1)					
DC Current Gain	15	h_{FE}			
($I_C = 0.1\ mAdc, V_{CE} = 1.0\ Vdc$) 2N3905 / 2N3906			30 / 60	– / –	
($I_C = 1.0\ mAdc, V_{CE} = 1.0\ Vdc$) 2N3905 / 2N3906			40 / 80	– / –	
($I_C = 10\ mAdc, V_{CE} = 1.0\ Vdc$) 2N3905 / 2N3906			50 / 100	150 / 300	
($I_C = 50\ mAdc, V_{CE} = 1.0\ Vdc$) 2N3905 / 2N3906			30 / 60	– / –	
($I_C = 100\ mAdc, V_{CE} = 1.0\ Vdc$) 2N3905 / 2N3906			15 / 30	– / –	
Collector-Emitter Saturation Voltage	16, 17	$V_{CE(sat)}$			Vdc
($I_C = 10\ mAdc, I_B = 1.0\ mAdc$)			–	0.25	
($I_C = 50\ mAdc, I_B = 5.0\ mAdc$)			–	0.4	
Base-Emitter Saturation Voltage	17	$V_{BE(sat)}$			Vdc
($I_C = 10\ mAdc, I_B = 1.0\ mAdc$)			0.65	0.85	
($I_C = 50\ mAdc, I_B = 5.0\ mAdc$)			–	0.95	
SMALL-SIGNAL CHARACTERISTICS					
Current-Gain — Bandwidth Product		f_T			MHz
($I_C = 10\ mAdc, V_{CE} = 20\ Vdc, f = 100\ MHz$) 2N3905 / 2N3906			200 / 250	– / –	
Output Capacitance ($V_{CB} = 5.0\ Vdc, I_E = 0, f = 100\ kHz$)	3	C_{ob}	–	4.5	pF
Input Capacitance ($V_{BE} = 0.5\ Vdc, I_C = 0, f = 100\ kHz$)	3	C_{ib}	–	1.0	pF
Input Impedance	13	h_{ie}			k ohms
($I_C = 1.0\ mAdc, V_{CE} = 10\ Vdc, f = 1.0\ kHz$) 2N3905 / 2N3906			0.5 / 2.0	8.0 / 12	
Voltage Feedback Ratio	14	h_{re}			$\times 10^{-4}$
($I_C = 1.0\ mAdc, V_{CE} = 10\ Vdc, f = 1.0\ kHz$) 2N3905 / 2N3906			0.1 / 1.0	5.0 / 10	
Small-Signal Current Gain	11	h_{fe}			–
($I_C = 1.0\ mAdc, V_{CE} = 10\ Vdc, f = 1.0\ kHz$) 2N3905 / 2N3906			50 / 100	200 / 400	
Output Admittance	12	h_{oe}			μmhos
($I_C = 1.0\ mAdc, V_{CE} = 10\ Vdc, f = 1.0\ kHz$) 2N3905 / 2N3906			1.0 / 3.0	40 / 60	
Noise Figure	9, 10	NF			dB
($I_C = 100\ \mu Adc, V_{CE} = 5.0\ Vdc, R_S = 1.0\ k\ ohm$, $f = 10\ Hz$ to $15.7\ kHz$) 2N3905 / 2N3906			– / –	5.0 / 4.0	
SWITCHING CHARACTERISTICS					
Delay Time	1, 5	t_d	–	35	ns
Rise Time ($V_{CC} = 3.0\ Vdc, V_{BE(off)} = 0.5\ Vdc$, $I_C = 10\ mAdc, I_{B1} = 1.0\ mAdc$)	1, 5, 6	t_r	–	35	ns
Storage Time 2N3905 / 2N3906	2, 7	t_s	– / –	200 / 225	ns
Fall Time ($V_{CC} = 3.0\ Vdc, I_C = 10\ mAdc$, $I_{B1} = I_{B2} = 1.0\ mAdc$) 2N3905 / 2N3906	2, 8	t_f	– / –	60 / 75	ns

*Indicates JEDEC Registered Data. (1) Pulse Width = 300 μs, Duty Cycle = 2.0 %.

FIGURE 1 – DELAY AND RISE TIME EQUIVALENT TEST CIRCUIT

FIGURE 2 – STORAGE AND FALL TIME EQUIVALENT TEST CIRCUIT

*Total shunt capacitance of test jig and connectors.

MOTOROLA

2N5245, 2N5246, 2N5247

CASE 29-02, STYLE 23
TO-92 (TO-226AA)

JFET HIGH-FREQUENCY AMPLIFIER

N-CHANNEL — DEPLETION

Rating	Symbol	Value	Unit
Drain-Gate Voltage	V_{DG}	30	Vdc
Gate-Source Voltage	V_{GS}	−30	Vdc
Gate Current	I_G	50	mA
Total Device Dissipation @ $T_A = 25°C$ (Free Air) Derate above 25°C	P_D	360 2.88	mW mW/°C
Total Device Dissipation @ $T_C = 25°C$ Derate above 25°C	P_D	500 4.0	mW mW/°C
Lead Temperature (1/16" from Case for 10 Seconds)	T_L	260	°C
Storage Temperature Range	T_{stg}	−65 to −150	°C

Refer to 2N4416 for graphs.

ELECTRICAL CHARACTERISTICS ($T_A = 25°C$ unless otherwise noted.)

Characteristic		Symbol	Min	Max	Unit
OFF CHARACTERISTICS					
Gate-Source Breakdown Voltage ($I_G = -1.0$ μA, $V_{DS} = 0$)		$V_{(BR)GSS}$	−30	—	Vdc
Gate Reverse Current ($V_{GS} = -20$ V, $V_{DS} = 0$)		I_{GSS}	—	−1.0	nA
Gate 1 Leakage Current ($V_{G1S} = -20$ V, $V_{DS} = 0$, $T_A = 100°C$)		I_{G1SS}	—	−0.5	μA
Gate Source Cutoff Voltage ($V_{DS} = 15$ V, $I_D = 10$ mA) 2N5245 2N5246 2N5247		$V_{GS(off)}$	−1.0 −0.5 −1.5	−6.0 −4.0 −8.0	Vdc
ON CHARACTERISTICS					
Zero-Gate-Voltage Drain Current ($V_{DS} = 15$ V, $V_{GS} = 0$, Pulsed. See Note 1) 2N5245 2N5246 2N5247		I_{DSS}	5.0 1.5 8.0	15 7.0 24	mA
SMALL-SIGNAL CHARACTERISTICS					
Forward Transfer Admittance ($V_{DS} = 15$ V, $V_{GS} = 0$, $f = 1.0$ kHz) 2N5245 2N5246 2N5247		$|y_{fs}|$	4500 3000 4500	7500 6000 8000	μmhos
Input Admittance ($V_{DS} = 15$ V, $V_{GS} = 0$) (100 MHz) (400 MHz)		$Re(y_{is})$	— —	100 1000	μmhos
Output Admittance ($V_{DS} = 15$ V, $V_{GS} = 0$, $f = 1.0$ kHz) 2N5245 2N5246 2N5247		$|y_{os}|$	— — —	50 50 70	μmhos
Output Conductance ($V_{DS} = 15$ V, $V_{GS} = 0$) 2N5245 (100 MHz) 2N5246 2N5247 2N5245 (400 MHz) 2N5246 2N5247		$Re(y_{os})$	— — — — — —	75 75 100 100 100 150	μmhos

SMALL-SIGNAL DEVICES   MOTOROLA SEMICONDUCTORS

2N5245, 2N5246, 2N5247

ELECTRICAL CHARACTERISTICS (continued) (T_A = 25°C unless otherwise noted.)

Characteristic		Symbol	Min	Max	Unit
Forward Transconductance (V_{DS} = 15 V, V_{GS} = 0, f = 400 MHz) 2N5245 2N5246 2N5247		Re(y_{fs})	4000 2500 4000	— — —	µmhos
Input Capacitance (V_{DS} = 15 V, V_{GS} = 0, f = 1.0 Mhz)		C_{iss}	—	4.5	pF
Reverse Transfer Capacitance (V_{DS} = 15 V, V_{GS} = 0, f = 1.0 MHz)		C_{rss}	—	1.0	pF
Input Susceptance (V_{DS} = 15 V, V_{GS} = 0)	(100 MHz) (400 MHz)	$I_M(Yis)$	— —	3.0 12.0	mmho
FUNCTIONAL CHARACTERISTICS					
Noise Figure (V_{DS} = 15 V, I_D = 5.0 mA, R'_G = 1.0 kΩ)		NF	— —	2.0 4.0	dB
Common Source Power Gain (V_{DS} = 15 V, I_D = 5.0 mA, R'_G = 1.0 kΩ)	2N5245 (100 MHz) 2N5245 (400 MHz)	G_{ps}	18 10	— —	dB
Output Susceptance (V_{DS} = 15 V, V_{GS} = 0)	(100 MHz) (400 MHz)	$I_M(Yos)$	— —	1000 4000	µmho

Note 1: tp = 100 ms, Duty Cycle = 10%.

MOTOROLA SEMICONDUCTORS   SMALL-SIGNAL DEVICES

MOTOROLA

2N5265 thru 2N5270

CASE 20-05, STYLE 5
TO-72 (TO-206AF)

**JFET
GENERAL PURPOSE**

P-CHANNEL — DEPLETION

MAXIMUM RATINGS

Rating	Symbol	Value	Unit
Drain-Source Voltage	V_{DS}	60	Vdc
Drain-Gate Voltage	V_{DG}	60	Vdc
Reverse Gate-Source Voltage	V_{GSR}	60	Vdc
Drain Current	I_D	20	mAdc
Forward Gate Current	I_{GF}	10	mAdc
Total Device Dissipation @ T_A = 25°C Derate above 25°C	P_D	300 2.0	mW mW/°C
Junction Temperature Range	T_J	−65 to +175	°C
Storage Temperature Range	T_{stg}	−65 to +200	°C

ELECTRICAL CHARACTERISTICS (T_A = 25°C unless otherwise noted.)

Characteristic	Symbol	Min	Max	Unit
OFF CHARACTERISTICS				
Gate-Source Breakdown Voltage (I_G = 10 µAdc, V_{DS} = 0)	$V_{(BR)GSS}$	60	—	Vdc
Gate Reverse Current (V_{GS} = 30 Vdc, V_{DS} = 0) (V_{GS} = 30 Vdc, V_{DS} = 0, T_A = 150°C)	I_{GSS}	— —	2.0 2.0	nAdc µAdc
Gate Source Cutoff Voltage (V_{DS} = 15 Vdc, I_D = 1.0 µAdc) 2N5265, 2N5266 2N5267, 2N5268 2N5269, 2N5270	$V_{GS(off)}$	— — —	3.0 6.0 8.0	Vdc
Gate Source Voltage (V_{DS} = 15 Vdc, I_D = 0.05 mAdc) 2N5265 (V_{DS} = 15 Vdc, I_D = 0.08 mAdc) 2N5266 (V_{DS} = 15 Vdc, I_D = 0.15 mAdc) 2N5267 (V_{DS} = 15 Vdc, I_D = 0.25 mAdc) 2N5268 (V_{DS} = 15 Vdc, I_D = 0.4 mAdc) 2N5269 (V_{DS} = 15 Vdc, I_D = 0.7 mAdc) 2N5270	V_{GS}	0.3 0.4 1.0 1.0 2.0 2.0	1.5 2.0 4.0 4.0 6.0 6.0	Vdc
ON CHARACTERISTICS				
Zero-Gate-Voltage Drain Current (V_{DS} = 15 Vdc, V_{GS} = 0) 2N5265 2N5266 2N5267 2N5268 2N5269 2N5270	I_{DSS}	0.5 0.8 1.5 2.5 4.0 7.0	1.0 1.6 3.0 5.0 8.0 14	mAdc
SMALL-SIGNAL CHARACTERISTICS				
Forward Transfer Admittance (V_{DS} = 15 Vdc, V_{GS} = 0, f = 1.0 kHz) 2N5265 2N5266 2N5267 2N5268 2N5269 2N5270	$|y_{fs}|$	900 1000 1500 2000 2200 2500	2700 3000 3500 4000 4500 5000	µmhos
Output Admittance Common Source (V_{DS} = 15 Vdc, V_{GS} = 0, f = 1.0 kHz)	$|y_{os}|$	—	75	µmhos

SMALL-SIGNAL DEVICES　　　　　　　　　　　MOTOROLA SEMICONDUCTORS

MOTOROLA

3N169
3N170
3N171

CASE 20-03, STYLE 2
TO-72 (TO-206AF)

MOSFET SWITCHING

N-CHANNEL — ENHANCEMENT

Refer to 2N4351 for graphs.

MAXIMUM RATINGS

Rating	Symbol	Value	Unit
Drain-Source Voltage	V_{DS}	25	Vdc
Drain-Gate Voltage	V_{DG}	±35	Vdc
Gate-Source Voltage	V_{GS}	±35	Vdc
Drain Current	I_D	30	mAdc
Total Device Dissipation @ T_A = 25°C Derate above 25°C	P_D	300 1.7	mW mW/°C
Total Device Dissipation @ T_C = 25°C Derate above 25°C	P_D	800 4.56	mW mW/°C
Junction Temperature Range	T_J	175	°C
Storage Temperature Range	T_{stg}	−65 to +175	°C

ELECTRICAL CHARACTERISTICS (T_A = 25°C unless otherwise noted.)

Characteristic	Symbol	Min	Max	Unit
OFF CHARACTERISTICS				
Drain-Source Breakdown Voltage (I_D = 10 μAdc, V_{GS} = 0)	$V_{(BR)DSX}$	25	—	Vdc
Zero-Gate-Voltage Drain Current (V_{DS} = 10 Vdc, V_{GS} = 0) (V_{DS} = 10 Vdc, V_{GS} = 0, T_A = 125°C)	I_{DSS}	— —	10 1.0	nAdc μAdc
Gate Reverse Current (V_{GS} = −35 Vdc, V_{DS} = 0) (V_{GS} = −35 Vdc, V_{DS} = 0, T_A = 125°C)	I_{GSS}	— —	10 100	pAdc
ON CHARACTERISTICS				
Gate Threshold Voltage (V_{DS} = 10 Vdc, I_D = 10 μAdc) 3N169 3N170 3N171	$V_{GS(Th)}$	0.5 1.0 1.5	1.5 2.0 3.0	Vdc
Drain-Source On-Voltage (I_D = 10 mAdc, V_{GS} = 10 Vdc)	$V_{DS(on)}$	—	2.0	Vdc
On-State Drain Current (V_{GS} = 10 Vdc, V_{DS} = 10 Vdc)	$I_{D(on)}$	10	—	mAdc
SMALL-SIGNAL CHARACTERISTICS				
Drain-Source Resistance (V_{GS} = 10 Vdc, I_D = 0, f = 1.0 kHz)	$r_{ds(on)}$	—	200	Ohms
Forward Transfer Admittance (V_{DS} = 10 Vdc, I_D = 2.0 mAdc, f = 1.0 kHz)	$|y_{fs}|$	1000	—	μmhos
Input Capacitance (V_{DS} = 10 Vdc, V_{GS} = 0, f = 1.0 MHz)	C_{iss}	—	5.0	pF
Reverse Transfer Capacitance (V_{DS} = 0, V_{GS} = 0, f = 1.0 MHz)	C_{rss}	—	1.3	pF
Drain-Substrate Capacitance ($V_{D(SUB)}$ = 10 Vdc, f = 1.0 MHz)	$C_{d(sub)}$	—	5.0	pF
SWITCHING CHARACTERISTICS				
Turn-On Delay Time	$t_{d(on)}$	—	3.0	ns
Rise Time	t_r	—	10	ns
Turn-Off Delay Time	$t_{d(off)}$	—	3.0	ns
Fall Time	t_f	—	15	ns

(V_{DD} = 10 Vdc, $I_{D(on)}$ = 10 mAdc, $V_{GS(on)}$ = 10 Vdc, $V_{GS(off)}$ = 0, R_G = 50 Ohms) See Figure 1

MOTOROLA SEMICONDUCTORS   SMALL-SIGNAL DEVICES

MOTOROLA

3N157
3N158

CASE 20-03, STYLE 2
TO-72 (TO-206AF)

**MOSFET
AMPLIFIER AND SWITCHING**

P-CHANNEL — ENHANCEMENT

MAXIMUM RATINGS

Rating	Symbol	Value	Unit
Drain-Source Voltage*	V_{DS}	±35	Vdc
Drain-Gate Voltage*	V_{DG}	±50	Vdc
Gate-Source Voltage*	V_{GS}	±50	Vdc
Drain Current*	I_D	30	mAdc
Total Device Dissipation @ T_A = 25°C Derate above 25°C*	P_D	300 1.7	mW mW/°C
Junction Temperature Range*	T_J	−65 to +175	°C
Storage Channel Temperature Range*	T_{stg}	−65 to +175	°C

*JEDEC Registered Limits

ELECTRICAL CHARACTERISTICS (T_A = 25°C unless otherwise noted.)

Characteristic	Symbol	Min	Typ	Max	Unit
OFF CHARACTERISTICS					
Drain-Source Breakdown Voltage (I_D = −10 μAdc, $V_G = V_S = 0$)	$V_{(BR)DSX}$	−35	—	—	Vdc
Zero-Gate-Voltage Drain Current (V_{DS} = −15 Vdc, V_{GS} = 0) (V_{DS} = −35 Vdc, V_{GS} = 0)	I_{DSS}	— —	— —	−1.0 −10	nAdc μAdc
Gate Reverse Current* (V_{GS} = +25 Vdc, V_{DS} = 0) (V_{GS} = +50 Vdc, V_{DS} = 0)	I_{GSS}	— —	— —	+10 +10	pAdc nAdc
Input Resistance (V_{GS} = −25 Vdc)	R_{GS}	—	1×10^{12}	—	Ohms
Gate Source Voltage* (V_{DS} = −15 Vdc, I_D = −0.5 mAdc) 3N157 3N158	V_{GS}	−1.5 −3.0	— —	−5.5 −7.0	Vdc
Gate Forward Current* (V_{GS} = −25 Vdc, V_{DS} = 0) (V_{GS} = −50 Vdc, V_{DS} = 0) (V_{GS} = −25 Vdc, V_{DS} = 0, T_A = +55°C) (V_{GS} = −50 Vdc, V_{DS} = 0, T_A = +55°C)	$I_{G(f)}$	— — — —	— — — —	−10 −1.0 −10 −1.0	pAdc nAdc nAdc μAdc
ON CHARACTERISTICS					
Gate Threshold Voltage* (V_{DS} = −15 Vdc, I_D = −10 μAdc) 3N157 3N158	$V_{GS(Th)}$	−1.5 −3.0	— —	−3.2 −5.0	Vdc
On-State Drain Current* (V_{DS} = −15 Vdc, V_{GS} = −10 Vdc)	$I_{D(on)}$	−5.0	—	—	mAdc
SMALL-SIGNAL CHARACTERISTICS					
Forward Transfer Admittance* (V_{DS} = −15 Vdc, I_D = −2.0 mAdc, f = 1.0 kHz)	$|y_{fs}|$	1000	—	4000	μmhos
Output Admittance* (V_{DS} = −15 Vdc, I_D = −2.0 mAdc, f = 1.0 kHz)	$|y_{os}|$	—	—	60	μmhos
Input Capacitance* (V_{DS} = −15 Vdc, V_{GS} = 0, f = 140 kHz)	C_{iss}	—	—	5.0	pF
Reverse Transfer Capacitance* (V_{DS} = −15 Vdc, V_{GS} = 0, f = 140 kHz)	C_{rss}	—	—	1.3	pF
Drain-Substrate Capacitance ($V_{D(SUB)}$ = −10 Vdc, f = 140 kHz)	$C_{d(sub)}$	—	—	4.0	pF
Noise Voltage (R_S = 0, BW = 1.0 Hz, V_{DS} = −15 Vdc, I_D = −2.0 mAdc, f = 100 Hz) (R_S = 0, BW = 1.0 Hz, V_{DS} = −15 Vdc, I_D = −2.0 mAdc, f = 1.0 kHz)	e_n	— —	300 120	— 500	$NV/\sqrt{Hz}$

*JEDEC Registered Limits

SMALL-SIGNAL DEVICES   MOTOROLA SEMICONDUCTORS

MOTOROLA

3N128

CASE 20-03, STYLE 7
TO-72 (TO-206AF)

MOSFET AMPLIFIER

N-CHANNEL — DEPLETION

MAXIMUM RATINGS

Rating	Symbol	Value	Unit
Drain-Source Voltage	V_{DS}	-20	Vdc
Drain-Gate Voltage	V_{DG}	+20	Vdc
Gate-Source Voltage	V_{GS}	±10	Vdc
Drain Current	I_D	50	mAdc
Total Device Dissipation @ T_A = 25°C Derate above 25°C	P_D	330 2.2	mW mW/°C
Operating and Storage Junction Temperature Range	T_J, T_{stg}	-65 to -175	°C

ELECTRICAL CHARACTERISTICS (T_A = 25°C unless otherwise noted.)

Characteristic	Symbol	Min	Max	Unit
OFF CHARACTERISTICS				
Gate-Source Breakdown Voltage(1) (I_G = -10 μAdc, V_{DS} = 0)	$V_{(BR)GSS}$	-50	—	Vdc
Gate Reverse Current (V_{GS} = -8.0 Vdc, V_{DS} = 0) (V_{GS} = -8.0 Vdc, V_{DS} = 0, T_A = 125°C)	I_{GSS}	— —	0.05 5.0	nAdc
Gate Source Cutoff Voltage (V_{DS} = 15 Vdc, I_D = 50 μAdc)	$V_{GS(off)}$	-0.5	-8.0	Vdc
ON CHARACTERISTICS				
Zero-Gate-Voltage Drain Current(2) (V_{DS} = 15 Vdc, V_{GS} = 0)	I_{DSS}	5.0	25	mAdc
SMALL-SIGNAL CHARACTERISTICS				
Forward Transfer Admittance (V_{DS} = 15 Vdc, I_D = 5.0 mAdc, f = 1.0 kHz)	$|y_{fs}|$	5000	12,000	μmhos
Input Admittance (V_{DS} = 15 Vdc, I_D = 5.0 mAdc, f = 200 MHz)	$Re(y_{is})$	—	800	μmhos
Output Conductance (V_{DS} = 15 Vdc, I_D = 5.0 mAdc, f = 200 MHz)	$Re(y_{os})$	—	500	μmhos
Forward Transconductance (V_{DS} = 15 Vdc, I_D = 5.0 mAdc, f = 200 MHz)	$Re(y_{fs})$	5000	—	μmhos
Input Capacitance (V_{DS} = 15 Vdc, I_D = 5.0 mAdc, f = 1.0 MHz)	C_{iss}	—	7.0	pF
Reverse Transfer Capacitance (V_{DS} = 15 Vdc, I_D = 5.0 mAdc, f = 1.0 MHz)	C_{rss}	0.05	0.35	pF
FUNCTIONAL CHARACTERISTICS				
Noise Figure (V_{DS} = 15 Vdc, I_D = 5.0 mAdc, f = 200 MHz)	NF	—	5.0	dB
Power Gain (V_{DS} = 15 Vdc, I_D = 5.0 mAdc, f = 200 MHz)	P_G	13.5	23	dB

(1) Caution: Destructive Test, can damage gate oxide beyond operation.
(2) Pulse Test: Pulse Width = 300 μs, Duty Cycle = 2.0%.

SMALL-SIGNAL DEVICES   MOTOROLA SEMICONDUCTORS

FAIRCHILD
A Schlumberger Company

μA741
Operational Amplifier

Linear Products

Description
The μA741 is a high performance Monolithic Operational Amplifier constructed using the Fairchild Planar epitaxial process. It is intended for a wide range of analog applications. High common mode voltage range and absence of latch-up tendencies make the μA741 ideal for use as a voltage follower. The high gain and wide range of operating voltage provides superior performance in integrator, summing amplifier, and general feedback applications.

- NO FREQUENCY COMPENSATION REQUIRED
- SHORT-CIRCUIT PROTECTION
- OFFSET VOLTAGE NULL CAPABILITY
- LARGE COMMON MODE AND DIFFERENTIAL VOLTAGE RANGES
- LOW POWER CONSUMPTION
- NO LATCH-UP

Connection Diagram
10-Pin Flatpak

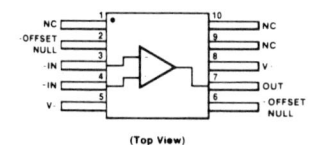

(Top View)

Order Information

Type	Package	Code	Part No.
μA741	Flatpak	3F	μA741FM
μA741A	Flatpak	3F	μA741AFM

Connection Diagram
8-Pin Metal Package

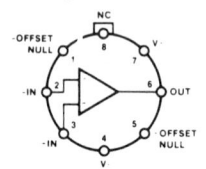

(Top View)

Pin 4 connected to case

Order Information

Type	Package	Code	Part No.
μA741	Metal	5W	μA741HM
μA741A	Metal	5W	μA741AHM
μA741C	Metal	5W	μA741HC
μA741E	Metal	5W	μA741EHC

Connection Diagram
8-Pin DIP

(Top View)

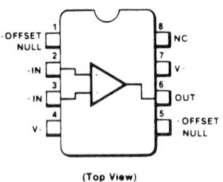

Order Information

Type	Package	Code	Part No.
μA741C	Molded DIP	9T	μA741TC
μA741C	Ceramic DIP	6T	μA741RC

Equivalent Circuit

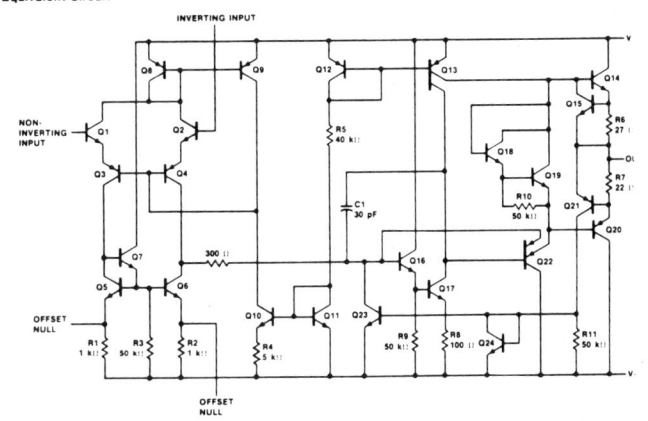

Notes
1. Rating applies to ambient temperatures up to 70°C. Above 70°C ambient derate linearly at 6.3 mW/°C for the metal package, 7.1 mW/°C for the flatpak, and 5.6 mW/°C for the DIP.
2. For supply voltages less than ±15 V, the absolute ma input voltage is equal to the supply voltage.
3. Short circuit may be to ground or either supply. Rating to +125°C case temperature or 75°C ambient tempera

μA741

μA741 and μA741C
Electrical Characteristics $V_S = \pm 15$ V, $T_A = 25°C$ unless otherwise specified

Characteristic	Condition	μA741 Min	μA741 Typ	μA741 Max	μA741C Min	μA741C Typ	μA741C Max	Unit
Input Offset Voltage	$R_S \leq 10$ kΩ		1.0	5.0		2.0	6.0	mV
Input Offset Current			20	200		20	200	nA
Input Bias Current			80	500		80	500	nA
Power Supply Rejection Ratio	$V_S = +10, -20$ $V_S = +20, -10$ V, $R_S = 50$ Ω		30	150		30	150	μV/V
Input Resistance		.3	2.0		.3	2.0		MΩ
Input Capacitance			1.4			1.4		pF
Offset Voltage Adjustment Range			±15			±15		mV
Input Voltage Range					±12	±13		V
Common Mode Rejection Ratio	$R_S \leq 10$ kΩ				70	90		dB
Output Short Circuit Current			25			25		mA
Large Signal Voltage Gain	$R_L \geq 2$ kΩ, $V_{OUT} = \pm 10$ V	50k	200k		20k	200k		
Output Resistance			75			75		Ω
Output Voltage Swing	$R_L \geq 10$ kΩ				±12	±14		V
	$R_L \geq 2$ kΩ				±10	±13		V
Supply Current			1.7	2.8		1.7	2.8	mA
Power Consumption			50	85		50	85	mW
Transient Response (Unity Gain)	Rise Time	$V_{IN} = 20$ mV, $R_L = 2$ kΩ, $C_L \leq 100$ pF		.3			.3		μs
	Overshoot			5.0			5.0		%
Bandwidth (Note 4)			1.0			1.0		MHz
Slew Rate	$R_L \geq 2$ kΩ		.5			.5		V/μs

Notes
4. Calculated value from $BW(MHz) = \frac{0.35}{\text{Rise Time }(\mu s)}$
5. All $V_{CC} = 15$-V for μA741 and μA741C.
6. Maximum supply current for all devices
   25°C = 2.8 mA
   125°C = 2.5 mA
   −55°C = 3.3 mA

Electronics: Circuits and Systems

MOTOROLA

MRD500 MRD510

PIN SILICON PHOTO DIODE

100 VOLT
PHOTO DIODE
PIN SILICON

100 MILLIWATTS

... designed for application in laser detection, light demodulation, detection of visible and near infrared light-emitting diodes, shaft or position encoders, switching and logic circuits, or any design requiring radiation sensitivity, ultra high-speed, and stable characteristics.

- Ultra Fast Response — (<1.0 ns Typ)
- High Sensitivity — MRD500 (1.2 μA/mW/cm² Min)
  MRD510 (0.3 μA/mW/cm² Min)
- Available With Convex Lens (MRD500) or Flat Glass (MRD510) for Design Flexibility
- Popular TO-18 Type Package for Easy Handling and Mounting
- Sensitive Throughout Visible and Near Infrared Spectral Range for Wide Application
- Annular◆ Passivated Structure for Stability and Reliability

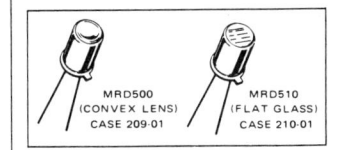

MRD500 (CONVEX LENS) CASE 209-01

MRD510 (FLAT GLASS) CASE 210-01

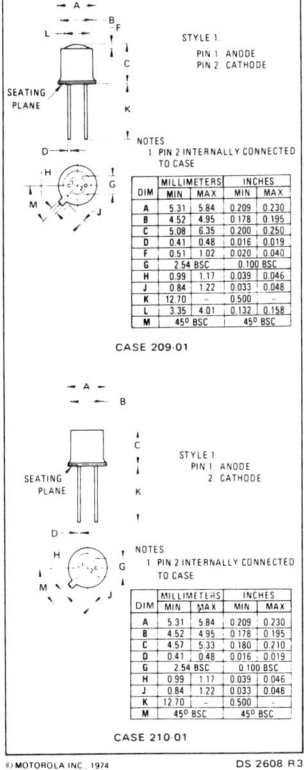

STYLE 1
PIN 1. ANODE
PIN 2. CATHODE

NOTES
1. PIN 2 INTERNALLY CONNECTED TO CASE

CASE 209-01

DIM	MILLIMETERS		INCHES	
	MIN	MAX	MIN	MAX
A	5.31	5.84	0.209	0.230
B	4.52	4.95	0.178	0.195
C	5.08	6.35	0.200	0.250
D	0.41	0.48	0.016	0.019
F	0.51	1.02	0.020	0.040
G	2.54 BSC		0.100 BSC	
H	0.99	1.17	0.039	0.046
J	0.84	1.22	0.033	0.048
K	12.70	–	0.500	–
L	3.35	4.01	0.132	0.158
M	45° BSC		45° BSC	

MAXIMUM RATINGS ($T_A = 25°C$ unless otherwise noted)

Rating	Symbol	Value	Unit
Reverse Voltage	V_R	100	Volts
Total Power Dissipation @ $T_A = 25°C$ Derate above 25°C	P_D	100 0.57	mW mW/°C
Operating and Storage Junction Temperature Range	T_J, T_{stg}	–65 to +200	°C

FIGURE 1 — TYPICAL OPERATING CIRCUIT

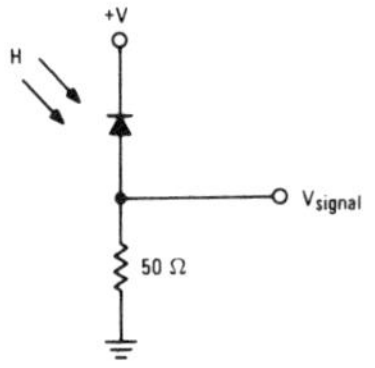

STYLE 1
PIN 1. ANODE
2. CATHODE

NOTES
1. PIN 2 INTERNALLY CONNECTED TO CASE

DIM	MILLIMETERS		INCHES	
	MIN	MAX	MIN	MAX
A	5.31	5.84	0.209	0.230
B	4.52	4.95	0.178	0.195
C	4.57	5.33	0.180	0.210
D	0.41	0.48	0.016	0.019
G	2.54 BSC		0.100 BSC	
H	0.99	1.17	0.039	0.046
J	0.84	1.22	0.033	0.048
K	12.70	–	0.500	–
M	45° BSC		45° BSC	

CASE 210-01

◆Trademark of Motorola Inc.

© MOTOROLA INC. 1974   DS 2608 R3

MRD500 • MRD510

STATIC ELECTRICAL CHARACTERISTICS ($T_A = 25°C$ unless otherwise noted)

Characteristic	Fig. No.	Symbol	Min	Typ	Max	Unit

Dark Current
($V_R = 20$ V, $R_L = 1.0$ megohm; Note 2)
   $T_A = 25°C$
   $T_A = 100°C$
	4 and 5	I_D	–	–
				14
					2.0
					–
						nA

Reverse Breakdown Voltage ($I_R = 10 \mu A$)	–	BV_R	100	200	–	Volts
Forward Voltage ($I_F = 50$ mA)	–	V_F	–	–	1.1	Volts
Series Resistance ($I_F = 50$ mA)	–	R_s	–	–	10	ohms
Total Capacitance ($V_R = 20$ V; f = 1.0 MHz)	6	C_T	–	–	4	pF

OPTICAL CHARACTERISTICS ($T_A = 25°C$)

Characteristic	Fig. No.	Symbol	Min	Typ	Max	Unit
Light Current ($V_R = 20$ V, Note 1) MRD500 MRD510	2 and 3	I_L	6.0 1.5	9.0 2.1	– –	μA
Sensitivity at 0.8 μm ($V_R = 20$ V, Note 3) MRD500 MRD510	–	$S(\lambda = 0.8 \mu m)$	–	6.6 1.5	–	$\mu A/mW/cm^2$
Response Time ($V_R = 20$ V, $R_L = 50$ ohms)	–	$t_{(resp)}$	–	1.0	–	ns
Wavelength of Peak Spectral Response	7	λ_s	–	0.8	–	μm

NOTES:

1. Radiation Flux Density (H) equal to 5.0 mW/cm² emitted from a tungsten source at a color temperature of 2870 K.
2. Measured under dark conditions. (H ≈ 0).
3. Radiation Flux Density (H) equal to 0.5 mW/cm² at 0.8 μm

MOTOROLA Semiconductor Products Inc.

Electronics: Circuits and Systems

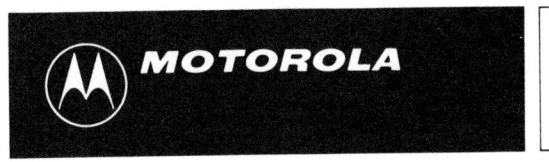

MOTOROLA

MRD300 MRD310

NPN SILICON HIGH SENSITIVITY PHOTO TRANSISTOR

... designed for application in industrial inspection, processing and control, counters, sorters, switching and logic circuits or any design requiring radiation sensitivity, and stable characteristics.

- Popular TO-18 Type Package for Easy Handling and Mounting
- Sensitive Throughout Visible and Near Infra-Red Spectral Range for Wider Application
- Minimum Light Current 4 mA at H = 5 mW/cm^2 (MRD 300)
- External Base for Added Control
- Annular ♦ Passivated Structure for Stability and Reliability

50 VOLT PHOTO TRANSISTOR NPN SILICON

250 MILLIWATTS

MAXIMUM RATINGS (T_A = 25°C unless otherwise noted)

Rating (Note 1)	Symbol	Value	Unit
Collector-Emitter Voltage	V_{CEO}	50	Volts
Emitter-Collector Voltage	V_{ECO}	7.0	Volts
Collector-Base Voltage	V_{CBO}	80	Volts
Total Device Dissipation @ T_A = 25°C Derate above 25°C	P_D	250 1.43	mW mW/°C
Operating Junction and Storage Temperature Range	T_J, T_{stg}	−65 to +200	°C

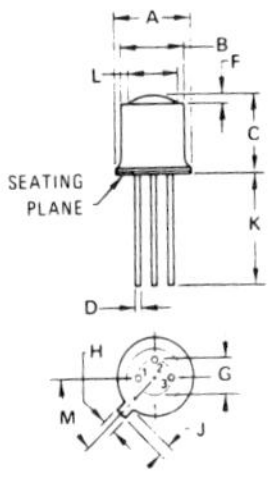

FIGURE 1 — LIGHT CURRENT versus IRRADIANCE

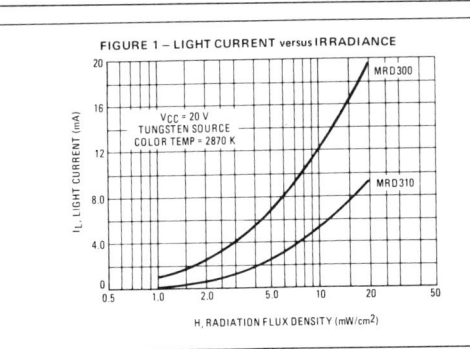

V_{CC} = 20 V
TUNGSTEN SOURCE
COLOR TEMP = 2870 K

I_L, LIGHT CURRENT (mA)

H, RADIATION FLUX DENSITY (mW/cm^2)

STYLE 1:
PIN 1. EMITTER
2. BASE
3. COLLECTOR

NOTES:
1. LEADS WITHIN .13 mm (.005) RADIUS OF TRUE POSITION AT SEATING PLANE, AT MAXIMUM MATERIAL CONDITION.
2. PIN 3 INTERNALLY CONNECTED TO CASE.

DIM	MILLIMETERS MIN	MAX	INCHES MIN	MAX
A	5.31	5.84	0.209	0.230
B	4.52	4.95	0.178	0.195
C	5.08	6.35	0.200	0.250
D	0.41	0.48	0.016	0.019
F	0.51	1.02	0.020	0.040
G	2.54 BSC		0.100 BSC	
H	0.99	1.17	0.039	0.046
J	0.84	1.22	0.033	0.048
K	12.70	−	0.500	−
L	3.35	4.01	0.132	0.158
M	45° BSC		45° BSC	

CASE 82-01

♦Annular Semiconductors Patented by Motorola Inc.

© MOTOROLA INC. 1974

DS 2601 R2

MRD300 • MRD310

STATIC ELECTRICAL CHARACTERISTICS ($T_A = 25°C$ unless otherwise noted)

Characteristic	Symbol	Min	Typ	Max	Unit
Collector Dark Current ($V_{CC} = 20$ V, $H \approx 0$) $\quad T_A = 25°C$ $\qquad T_A = 100°C$	I_{CEO}	– –	– 4.0	25 –	nA μA
Collector-Base Breakdown Voltage ($I_C = 100 \mu A$)	BV_{CBO}	80	–	–	Volts
Collector-Emitter Breakdown Voltage ($I_C = 100 \mu A$)	BV_{CEO}	50	–	–	Volts
Emitter-Collector Breakdown Voltage ($I_E = 100 \mu A$)	BV_{ECO}	7.0	–	–	Volts

OPTICAL CHARACTERISTICS ($T_A = 25°C$ unless otherwise noted)

Characteristic	Device Type	Symbol	Min	Typ	Max	Unit
Light Current ($V_{CC} = 20$ V, $R_L = 100$ ohms) Note 1	MRD300 MRD310	I_L	4.0 1.0	7.5 2.5	– –	mA
Light Current ($V_{CC} = 20$ V, $R_L = 100$ ohms) Note 2	MRD300 MRD310	I_L	– –	2.5 0.8	– –	mA
Photo Current Rise Time (Note 3) ($R_L = 100$ ohms, $I_L = 1.0$ mA peak)		t_r	–	–	2.5	μs
Photo Current Fall Time (Note 3) ($R_L = 100$ ohms, $I_L = 1.0$ mA peak)		t_f	–	–	4.0	μs

NOTES:
1. Radiation flux density (H) equal to 5.0 mW/cm^2 emitted from a tungsten source at a color temperature of 2870 K.
2. Radiation flux density (H) equal to 0.5 mW/cm^2 (pulsed) from a GaAs (gallium-arsenide) source at $\lambda \approx 0.9$ μm.
3. For unsaturated response time measurements, radiation is provided by pulsed GaAs (gallium-arsenide) light-emitting diode ($\lambda \approx 0.9$ μm) with a pulse width equal to or greater than 10 microseconds (see Figure 6) $I_L = 1.0$ mA peak.

MOTOROLA Semiconductor Products Inc.

Electronics: Circuits and Systems

FAIRCHILD
A Schlumberger Company

93422
256 x 4-Bit Static
Random Access Memory

Bipolar Division

TTL Bipolar Memory

Description
The 93422 is a 1024-bit read/write Random Access Memory (RAM), organized 256 words by four bits. It is designed for high speed cache, control and buffer storage applications. The 93422 is available in two speeds, "standard" speed and an "A" grade. The device includes full on-chip decoding, separate Data inputs and non-inverting Data outputs, as well as two Chip Select lines.

- Commercial Address Access Time
  93422 — 45 ns Max
  93422A — 35 ns Max
- Military Address Access Time
  93422 — 60 ns Max
  93422A — 45 ns Max
- Fully TTL Compatible
- Features Three State Outputs
- Power Dissipation — 0.46 mW/Bit Typ
- Power Dissipation Decreases with Increasing Temperature

Pin Names
A_0–A_7	Address Inputs
D_0–D_3	Data Inputs
$\overline{CS}_1$	Chip Select Input (Active LOW)
CS_2	Chip Select Input (Active HIGH)
$\overline{WE}$	Write Enable Input (Active LOW)
$\overline{OE}$	Output Enable Input (Active LOW)
O_0–O_3	Data Outputs

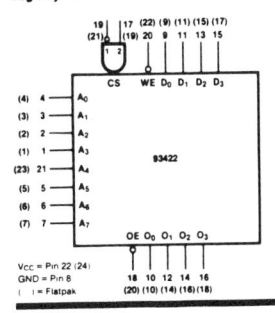

Connection Diagrams
22-Pin DIP (Top View)

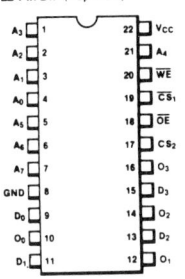

```
A3  [ 1      22 ] Vcc
A2  [ 2      21 ] A4
A1  [ 3      20 ] WE
A0  [ 4      19 ] CS1
A5  [ 5      18 ] OE
A4  [ 6      17 ] CS2
A7  [ 7      16 ] O3
GND [ 8      15 ] D3
D0  [ 9      14 ] O2
O0  [ 10     13 ] D2
D1  [ 11     12 ] O1
```

24-Pin Flatpak (Top View)

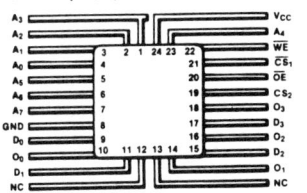

Logic Symbol

Vcc = Pin 22 (24)
GND = Pin 8
( ) = Flatpak

93422

Logic Diagram

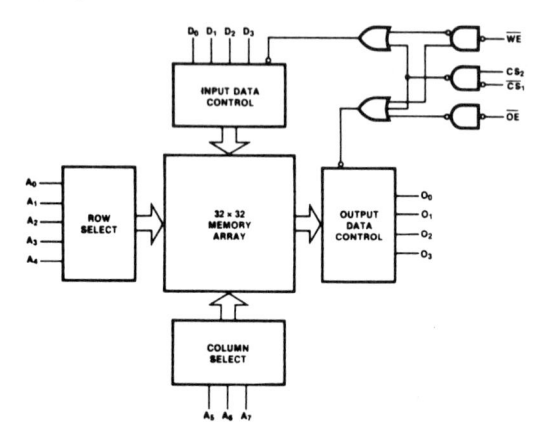

Functional Description

The 93422 is a fully decoded 1024-bit Random Access Memory organized 256 words by four bits. Word selection is achieved by means of an 8-bit address, A_0 through A_7.

Two Chip Select inputs, inverting and non-inverting, are provided for logic flexibility. For larger memories, the fast chip select access time permits the decoding of the chip selects from the address without increasing address access time.

The read and write operations are controlled by the state of the active LOW Write Enable ($\overline{WE}$) input. When $\overline{WE}$ is held LOW and the chip is selected, the data at D_0-D_3 is written into the addressed location. Since the write function is level-triggered, data must be held stable for at least $t_{WSD(min)}$ plus $t_{W(min)}$ plus $t_{WHD(min)}$ to insure a valid write. To read, $\overline{WE}$ is held HIGH and the chip selected. Non-inverted data is then presented at the outputs (O_0-O_3).

The 93422 has 3-state outputs which provide active pull-ups when enabled and high output impedance when disabled. This allows optimization of word expansion in bus organized systems.

Electronics: Circuits and Systems

93422

Truth Table

Inputs					Outputs	
OE	CS₁	CS₂	WE	D₀–D₃	3-State	Mode
X	H	X	X	X	HIGH Z	Not Selected
X	X	L	X	X	HIGH Z	Not Selected
L	L	H	H	X	O₀–O₃	Read Stored Data
L	L	H	L	L	HIGH Z	Write "0"
L	L	H	L	H	HIGH Z	Write "1"
H	L	H	H	X	HIGH Z	Output Disabled
H	L	H	L	L	HIGH Z	Write "0" (Output Disabled)
H	L	H	L	H	HIGH Z	Write "1" (Output Disabled)

H = HIGH Voltage Level (2.4 V)
L = LOW Voltage Level (.5 V)
X = Don't Care (HIGH or LOW)
High Z = High-Impedance

DC Characteristics: Over operating temperature ranges (Notes 1, 2)

Symbol	Characteristic	Min	Typ	Max	Unit	Condition
V_{OL}	Output LOW Voltage		0.3	0.45	V	V_{CC} = Min, I_{OL} = 8 mA
V_{IH}	Input HIGH Voltage	2.1	1.6		V	Guaranteed Input HIGH Voltage for All Inputs[6]
V_{IL}	Input LOW Voltage		1.5	0.8	V	Guaranteed Input LOW Voltage for All Inputs[6]
V_{OH}	Output HIGH Voltage	2.4			V	V_{CC} = Min, I_{OH} = −5.2 mA
I_{IL}	Input LOW Current		−150	−300	μA	V_{CC} = Max, V_{IN} = 0.4 V
I_{IH}	Input HIGH Current		1.0	40	μA	V_{CC} = Max, V_{IN} = 4.5 V
				1.0	mA	V_{CC} = Max, V_{IN} = 5.25 V
V_{IC}	Input Diode Clamp Voltage		−1.0	−1.5	V	V_{CC} = Max, I_{IN} = −10 mA
I_{OFF}	Output Current (HIGH Z)			50	μA	V_{CC} = Max, V_{OUT} = 2.4 V
				−50		V_{CC} = Max, V_{OUT} = 0.5 V
I_{OS}	Output Current Short Circuit to Ground			−70	mA	V_{CC} = Max, Note 4
I_{CC}	Power Supply Current		95	150	mA	Commercial	V_{CC} = Max
			95	170		Military	All Inputs GND

Notes

1. Operating specification with adequate time for temperature stabilization and transverse airflow exceeding 400 linear feet per minute. Conformance testing performed instantaneously where $T_A = T_J = T_C$. Correlated temperatures, typically 25°C and 100°C, and limits may be used to guarantee performance.
2. Typical values are at V_{CC} = 5.0 V, T_A = +25°C and maximum loading.
3. The maximum address access time is guaranteed to be the worst case bit in the memory using a pseudorandom testing pattern.
4. Short circuit to ground not to exceed one second.
5. T_W measured at t_{WSA} = Min, t_{WSA} measured at t_W = Min.
6. Static condition only.

3342
Quad 64-Bit
Static Shift Register

MOS Memory Products

Description

The 3342 is a static shift register in quad 64-bit organization. An on-chip clock generator provides appropriate internal clock phases from a single external TTL-level clock input. Passive on-chip input pull-up resistors allow direct TTL compatibility on all inputs. The outputs are capable of driving a single TTL load directly without the need for external components. The 3342 is manufactured with p-channel silicon gate technology. It is available in ceramic or plastic 16-pin dual in-line packages in the commercial temperature range, 0°C to +70°C.

- SINGLE TTL-COMPATIBLE EXTERNAL CLOCK
- DIRECT TTL COMPATIBILITY
- 1.5 MHz OPERATION GUARANTEED
- LOW CLOCK CAPACITANCE
- INPUT OVERVOLTAGE PROTECTION
- EXTERNAL RECIRCULATE CONTROL
- 16-PIN CERAMIC OR PLASTIC DUAL IN-LINE PACKAGE

Pin Names

D_1-D_4	Data Inputs
REC_1-REC_4	Recirculate Inputs
CP	Clock Pulse Input
Q_1-Q_4	Data Outputs
V_{SS}	+5 V Power Supply
V_{DD}	0 V Power Supply
V_{GG}	−12 V Power Supply

Absolute Maximum Ratings

All Inputs Including Clock (Note 1)	−20 V to +0.3 V
V_{GG} (Note 1)	−20 V to +0.3 V
V_{DD} and Outputs (Note 1)	−7.0 V to +0.3 V
Output Current when Output is LOW (Note 2)	10 mA
Storage Temperature	−55°C to +150°C
Operating Temperature	0°C to +70°C

Notes

1. All voltages with respect to V_{SS}.
2. LOW logic level is the most negative level and HIGH logic level is the most positive.

Stresses greater than those listed under "Absolute Maximum Ratings" may cause permanent damage to the device. This is a stress rating only, and functional operation of the device at these or any other conditions above those indicated in the operational sections of this specification is not implied. Exposure to absolute maximum rating conditions for extended periods may affect device reliability.

Logic Symbol

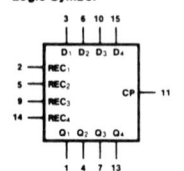

V_{SS} = Pin 16
V_{DD} = Pin 8
V_{GG} = Pin 12

Connection Diagram
16-Pin DIP

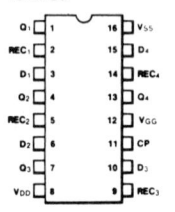

(Top View)

Package	Outline	Order Code
Ceramic DIP	6Z	D
Plastic DIP	9B	P

TYPES SN54ALS00A, SN74ALS00A
QUADRUPLE 2-INPUT POSITIVE-NAND GATES

D2661, APRIL 1982

- Package Options Include Both Plastic and Ceramic Chip Carriers in Addition to Plastic and Ceramic DIPs
- Dependable Texas Instruments Quality and Reliability

description

These devices contain four independent 2-input NAND gates. They perform the boolean functions $Y = \overline{A \cdot B}$ or $Y = \overline{A} + \overline{B}$ in positive logic.

The SN54ALS00A is characterized for operation over the full military temperature range of $-55\,°C$ to $125\,°C$. The SN74ALS00A is characterized for operation from $0\,°C$ to $70\,°C$.

FUNCTION TABLE (each gate)

INPUTS		OUTPUT
A	B	Y
H	H	L
L	X	H
X	L	H

logic symbol

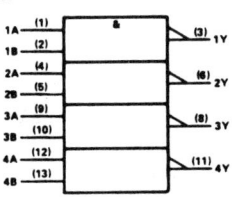

Pin numbers shown are for J and N packages.

SN54ALS00A . . . J PACKAGE
SN74ALS00A . . . N PACKAGE
(TOP VIEW)

```
1A  [1  U 14] VCC
1B  [2    13] 4B
1Y  [3    12] 4A
2A  [4    11] 4Y
2B  [5    10] 3B
2Y  [6     9] 3A
GND [7     8] 3Y
```

SN54ALS00A . . . FH PACKAGE
SN74ALS00A . . . FN PACKAGE
(TOP VIEW)

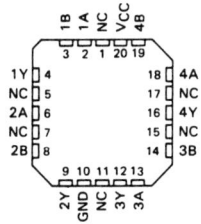

NC — No internal connection

Copyright © 1982 by Texas Instruments Incorporated

TEXAS INSTRUMENTS
INCORPORATED
POST OFFICE BOX 225012 • DALLAS, TEXAS 75265

TYPES SN54ALS00A, SN74ALS00A
QUADRUPLE 2-INPUT POSITIVE-NAND GATES

absolute maximum ratings over operating free-air temperature range (unless otherwise noted)

Supply voltage, V_{CC} .... 7 V
Input voltage .... 7 V
Operating free-air temperature range: SN54ALS00A .... $-55°C$ to $125°C$
                                 SN74ALS00A .... $0°C$ to $70°C$
Storage temperature range .... $-65°C$ to $150°C$

recommended operating conditions

		SN54ALS00A MIN	NOM	MAX	SN74ALS00A MIN	NOM	MAX	UNIT
V_{CC}	Supply voltage	4.5	5	5.5	4.5	5	5.5	V
V_{IH}	High-level input voltage	2			2			V
V_{IL}	Low-level input voltage			0.8			0.8	V
I_{OH}	High-level output current			-0.4			-0.4	mA
I_{OL}	Low-level output current			4			8	mA
T_A	Operating free-air temperature	-55		125	0		70	°C

electrical characteristics over recommended operating free-air temperature range (unless otherwise noted)

PARAMETER	TEST CONDITIONS		SN54ALS00A MIN	TYP‡	MAX	SN74ALS00A MIN	TYP‡	MAX	UNIT
V_{IK}	$V_{CC} = 4.5$ V,	$I_I = -18$ mA			-1.5			-1.5	V
V_{OH}	$V_{CC} = 4.5$ V,	$I_{OH} = -0.4$ mA	2.5	3.4					V
	$V_{CC} = 4.5$ V	$I_{OH} = -0.4$ mA				2.7	3.4		
V_{OL}	$V_{CC} = 4.5$ V,	$I_{OL} = 4$ mA		0.25	0.4		0.25	0.4	V
	$V_{CC} = 4.5$ V	$I_{OL} = 8$ mA					0.35	0.5	
I_I	$V_{CC} = 5.5$ V,	$V_I = 7$ V			0.1			0.1	mA
I_{IH}	$V_{CC} = 5.5$ V,	$V_I = 2.7$ V			20			20	µA
I_{IL}	$V_{CC} = 5.5$ V,	$V_I = 0.4$ V			-0.1			-0.1	mA
I_O §	$V_{CC} = 5.5$ V,	$V_O = 2.25$ V	-15		-70	-15		-70	mA
I_{CCH}	$V_{CC} = 5.5$ V,	$V_I = 0$ V		0.5	0.85		0.5	0.85	mA
I_{CCL}	$V_{CC} = 5.5$ V,	$V_I = 4.5$ V		1.5	3		1.5	3	mA

‡ All typical values are at $V_{CC} = 5$ V, $T_A = 25°C$.
§ The output conditions have been chosen to produce a current that closely approximates one half of the true short-circuit output current, I_{OS}.

switching characteristics (see note 1)

PARAMETER	FROM (INPUT)	TO (OUTPUT)	$V_{CC} = 5$ V, $C_L = 15$ pF, $R_L = 500$ Ω, $T_A = 25°C$ 'ALS00A TYP	$V_{CC} = 4.5$ V to 5.5 V, $C_L = 50$ pF, $R_L = 500$ Ω, $T_A =$ MIN to MAX SN54ALS00A MIN	MAX	SN74ALS00A MIN	MAX	UNIT
t_{PLH}	A or B	Y	4	3	14	3	11	ns
t_{PHL}	A or B	Y	3	2	10	2	8	ns

NOTE 1: For load circuit and voltage waveforms, see page 1-12.

TEXAS INSTRUMENTS
INCORPORATED
POST OFFICE BOX 225012 • DALLAS, TEXAS 75265

Electronics: Circuits and Systems

TYPES SN54ALS02, SN74ALS02
QUADRUPLE 2-INPUT POSITIVE-NOR GATES

D2661, APRIL 1982

- Package Options Include Both Plastic and Ceramic Chip Carriers in Addition to Plastic and Ceramic DIPs
- Dependable Texas Instruments Quality and Reliability

description

These devices contain four independent 2-input NOR gates. They perform the boolean functions $Y = \overline{A+B}$ or $Y = \overline{A} \cdot \overline{B}$ in positive logic.

The SN54ALS02 is characterized for operation over the full military temperature range of $-55\,°C$ to $125\,°C$. The SN74ALS02 is characterized for operation from $0\,°C$ to $70\,°C$.

SN54ALS02 . . . J PACKAGE
SN74ALS02 . . . N PACKAGE
(TOP VIEW)

1Y	1	14	V_CC
1A	2	13	4Y
1B	3	12	4B
2Y	4	11	4A
2A	5	10	3Y
2B	6	9	3B
GND	7	8	3A

FUNCTION TABLE (each gate)

INPUTS		OUTPUT
A	B	Y
H	X	L
X	H	L
L	L	H

SN54ALS02 . . . FH PACKAGE
SN74ALS02 . . . FN PACKAGE
(TOP VIEW)

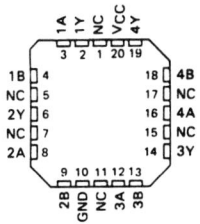

logic symbol

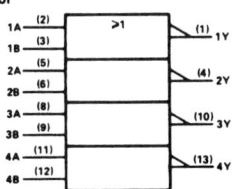

Pin numbers shown are for J and N packages.

NC — No internal connection

Copyright © 1982 by Texas Instruments Incorporated

TEXAS INSTRUMENTS
INCORPORATED
POST OFFICE BOX 225012 • DALLAS, TEXAS 75265

TYPES SN54ALS02, SN74ALS02
QUADRUPLE 2-INPUT POSITIVE-NOR GATES

absolute maximum ratings over operating free-air temperature range (unless otherwise noted)

Supply voltage, V_{CC} .... 7 V
Input voltage .... 7 V
Operating free-air temperature range: SN54ALS02 .... −55 °C to 125 °C
                                      SN74ALS02 .... 0 °C to 70 °C
Storage temperature range .... −65 °C to 150 °C

recommended operating conditions

		SN54ALS02			SN74ALS02			UNIT
		MIN	NOM	MAX	MIN	NOM	MAX	
V_{CC}	Supply voltage	4.5	5	5.5	4.5	5	5.5	V
V_{IH}	High-level input voltage	2			2			V
V_{IL}	Low-level input voltage			0.8			0.8	V
I_{OH}	High-level output current			−0.4			−0.4	mA
I_{OL}	Low-level output current			4			8	mA
T_A	Operating free-air temperature	−55		125	0		70	°C

electrical characteristics over recommended operating free-air temperature range (unless otherwise noted)

PARAMETER	TEST CONDITIONS		SN54ALS02			SN74ALS02			UNIT
			MIN	TYP‡	MAX	MIN	TYP‡	MAX	
V_{IK}	V_{CC} = 4.5 V,	I_I = −18 mA			−1.5			−1.5	V
V_{OH}	V_{CC} = 4.5 V,	I_{OH} = −0.4 mA	2.5	3.4					V
	V_{CC} = 4.5 V	I_{OH} = −0.4 mA				2.7	3.4		
V_{OL}	V_{CC} = 4.5 V,	I_{OL} = 4 mA		0.25	0.4		0.25	0.4	V
	V_{CC} = 4.5 V	I_{OL} = 8 mA					0.35	0.5	
I_I	V_{CC} = 5.5 V,	V_I = 7 V			0.1			0.1	mA
I_{IH}	V_{CC} = 5.5 V,	V_I = 2.7 V			20			20	µA
I_{IL}	V_{CC} = 5.5 V,	V_I = 0.4 V			−0.1			−0.1	mA
I_O §	V_{CC} = 5.5 V,	V_O = 2.25 V	−30		−112	−30		−112	mA
I_{CCH}	V_{CC} = 5.5 V,	V_I = 0 V		0.86	2.2		0.86	2.2	mA
I_{CCL}	V_{CC} = 5.5 V,	V_I = 4.5 V		2.16	4		2.16	4	mA

‡ All typical values are at V_{CC} = 5 V, T_A = 25 °C.
§ The output conditions have been chosen to produce a current that closely approximates one half of the true short-circuit output current, I_{OS}.

switching characteristics (see note 1)

PARAMETER	FROM (INPUT)	TO (OUTPUT)	V_{CC} = 5 V, C_L = 15 pF, R_L = 500 Ω, T_A = 25 °C	V_{CC} = 4.5 V to 5.5 V, C_L = 50 pF, R_L = 500 Ω, T_A = MIN to MAX				UNIT
			'ALS02	SN54ALS02		SN74ALS02		
			TYP	MIN	MAX	MIN	MAX	
t_{PLH}	A or B	Y	6	3	14	3	12	ns
t_{PHL}	A or B	Y	5	3	11	3	10	ns

NOTE 1: For load circuit and voltage waveforms, see page 1-12.

TEXAS INSTRUMENTS
INCORPORATED
POST OFFICE BOX 225012 • DALLAS, TEXAS 75265

INDEX

A

Ac
  circuits, 118-126,150
    capacitors, 122-123
    impedance, 120-122
    inductors, 123-124
    phasor representation, 118-120
    resistors, 122
    series RC, 124-126
  power rectification, 225-235
  signals, 27-35, 72
    peak value, 28-29
    phase angle, 29-32
    power relationships, 32-35
    rms, 28-29
    switch, SCR, 251
  operation
    normal, 654
    reverse, 656, 686
Adaptive, delta, modulation, 891, 917
Adders, 742-745, 786-787
  BCD, 752, 753, 787
  full-, 745, 746, 787
  half-, 743-744, 786-787
Addition, twos complement system, 750-752, 787
Algebraic determination, Q point, 343-345, 381
AM
  double-sideband
    large-carrier conventional, 825-835, 853-854
    suppressed-carrier, 814-817, 853-854
  power, considerations, 829-834, 855
  receivers, superheterodyne, 835, 837-838, 855
  sideband
    single-, 818-824, 854
    vestigial-, 824-825, 854
  signal
    demodulation, 834-835, 836, 855
    spectrum, 828-829, 833

Amplifiers
  bandwidth, increase, 446-447, 451
  BJT, 274-329
    analysis, 275-276, 418-434
    biasing, 283-293, 320
    common-base, 294, 317-319, 321, 322-323
    common-collector, 294-295, 312-317, 321, 322
    common-emitter, 293-294, 295-312, 321-322
    comparison, 319, 323
    communication techniques, 800-863
    configurations, 293-295, 321
    two-port model, 276-283, 320
  circuits, photodiode, 559-560, 576-577
  coupling capacitors, 396
  cutoff frequency, upper determination, 413-415, 449
  differential, 486-501, 529-530
    circuits, 493-495, 529-530
  feedback, 518-527, 531
    connections, 521-523, 531
  FET, 330-392
    biasing, 342-356, 381
    circuit design, 356-361, 381
    common-drain, 375-376, 382
    common-source, 361-374, 382
    constant K, 340-341, 381
    properties, 331-335, 379
    Q point, 343-353, 381
    transfer characteristics, 335-340, 380
  frequency response, 393-455
    bypass capacitors, 397
    curve, 394-395
    factors that affect, 395-396, 448
    interelectrode capacitances, 397
    lower cutoff frequency, 397-409, 448-449
    principles, 331-335
  high-frequency model, 409-415
    capacitance C_F, 411

965

Index

Amplifiers—cont
  high frequency model
    capacitance C_1, 411
    Miller effect, 411-413, 449
  logarithmic, 480-482, 529
  negative feedback, 523, 531
  nonsinusoidal, periodic signals, 72
  operational, 456-518, 528-530
    comparators, 501-506, 530
    ideal, 457-486, 528
    stages, 457
  parameters, 458
  pulse-modulation, 870-871, 913-914
  RC-coupled pulse response, 434-446
  response, pulse, 393-455
  spectrum, 37-40
  stability, 525-526, 532
  variation gain, 394-397
  to-digital conversion, 901, 906, 918
Analysis
  BJT amplifier, 275-276, 418-434
    biasing circuit, 284-290
  Fourier, 35-37
  mixed logic networks, 622-623, 632
  two-level positive logic networks, 603-605, 631
AND
  gates, commercially available, 588-591
  operation, logical product, 586-587
Angle, phase, 29-32
Answers, problems, 927-937
Applications
  flip-flops, 768-780, 790
  phototransistors, 561
Arithmetic logic, 730-754
  unit, 752-754, 788
ASCII code, 900-901, 98

B

Balance, pn junction, 173
Bandwidth
  FM signal, 840, 842-843, 856
  increase, amplifier, 446-447, 451
  PCM system, 879-881, 914-915
Bar graph display, 570
Barrier, potential, 173
Baseband, PCM transmission, 882-887, 915
Base diffusion, IC fabrication, 206, 208
BCD
  adders, 752, 753, 787
  code, 752, 753, 787
Bias
  forward, pn junction, 175, 176
  pn junction, reverse, 175-178
Biasing
  BJT amplifier, 283-293, 320
  FET amplifier, 353-361, 381
    design, 356-361, 381
BJT
  amplifiers, 274-329
    analysis, 275-276, 418-434
    biasing, 283-293, 320

BJT—cont
  amplifiers
    common-base, 294, 317-319, 321, 322-323
    common-collector, 294-295, 312-317
    common-emitter, 293-294, 295, 312, 321-322
    comparison, 319, 323
    configurations, 293-295, 321
    two-port model, 276-283, 320
  ICs, 204
  parameters, 416-417, 449
Binary number system, 730-734, 786
Bipolar junction transistor, 178-186, 215-216
Bridge, Wheatstone, 97-99
Bypass capacitor, 397
  lower cutoff frequency, 401-409, 448-449

C

Capacitance
  C_F, high-frequency model, amplifier, 411
  C_1, high-frequency model, amplifier, 411
  interelectrode, 397
Capacitors, 99-110, 149
  ac circuits, 122-123
  bypass, 397
  charging, 100-103
  connection
    parallel, 110-111
    series, 110
  discharging, 105-106
  energy stored, 100
  filter, 233-235
  ICs, 202
  universal charging curve, 104-105
  voltage continuity, 100
Capacity
  channel, 894-900, 917
  expansion ROM, 711-714, 723
  ROM, 710
Cells, photovoltaic, 554-556, 576
Channel
  capacity, 894-900, 917
  noiseless, 894-895, 917
  noisy, 895-897, 917
Charging
  capacitor, 100-103
  curve, universal, 104-105
Circuits
  ac, 118-126
    capacitors, 122-123
    impedance, 120-122
    inductors, 123-124
    resistors, 122
  amplifier
    differential, 493-495, 529-530
    photodiode, 559-560, 576-577
  diodes, 260
  electric, 83-165
  integrated, 199-213
  logic, 582-652
    principles, 583-598, 630

Electronics: Circuits and Systems

Circuits—cont
  op amp, 461, 528-529
    inverting, 461-467, 528
    noninverting, 467-473, 528
    summing, 473-476, 528
  parallel
    connection capacitor, 110-111
    encoders, 906-912
    resistors, 93-94, 96, 97
  RC, 106-110
  RL, 117-118
  sample and hold, 479-480, 529
  series
    connection capacitors, 110
    RC, ac, 124-126
    resistors, 89-91, 96, 97
  voltage follower, 476-478, 528
  window detector, 504-506
Clipping circuits, diodes, 236-241
Clocked SR flip-flop, 758-760, 788
CMOS logic, 680-685, 690
CMRR; *see* common-mode rejection ratio.
Code
  ASCII, 900-901, 918
  BCD, 752, 753, 787
  pulse-, modulation, 873-887, 914-915
Common-
  base BJT amplifier, 293-294, 322-323
  collector BJT amplifier, 294-295, 312-317
    emitter follower, matching, 316-317
    input resistance, emitter follower, 313-315
    output resistance emitter follower, 315
  drain FET amplifier, 375-376, 382
  emitter BJT amplifier, 293-294, 295, 312, 321-322
  small-signal analysis, 297-312, 321-322
  mode, 489-491
    operation, 495-497
    rejection ratio, 491-493, 529
  source FET amplifier, 361-374, 382
Communication
  analog techniques, 800-863
  digital, 891-926
  system
    model, 805, 852
    noise, 845-849, 856-857
Companding PCM system, 881-882, 915
Comparators, 501-506, 530
Complement system
  ones, 748-749, 787
  twos, 749-750, 787
    addition, 750-752, 787
    subtraction, 750-752, 787
Complementary MOSFET (CMOS), 207, 210
Computational advantages, decibel, 64-67
Configurations, BJT, amplifier, 293-295, 321
Connections, amplifier, feedback, 521-523, 531
Considerations
  AM power, 829-834, 855

Considerations—cont
  loading, 663-665, 687
Constant-
  current source, 482-483, 529
    realization, 498-501
  K determination, 340-341, 381
Contacts, semiconductor, 173-175
Continuity, voltage, 100
Convention
  logic
    mixed, 584
    negative, 583
    positive, 583
Conventional AM, double-sideband, large-carrier, 825-835, 854
Conversion
  analog-to-digital, 901, 906, 918
  digital-to-analog, 901-906, 918
Converter, voltage-to-current, 478-479, 528
Counters, 754, 788
  ripple, 769-771, 790
  synchronous, 771-778, 788
Coupling capacitors, amplifiers, 396
Current
  dividers, 94-96
  gain, 67-71
    BJT amplifiers, 319
  inductors
    buildup, 112-113
    continuity condition, 112
    decay, 115-116
  mirror, ICs, 212-213
  sinusoidal, 27-28
  transistor, 183-185
Curve, frequency response, amplifiers, 394-395
Cutoff frequency
  high-pass filters, 128-130
  lower, 397-409, 448-449
    bypass capacitor, 401-409, 448-449
    input coupling capacitor, 400, 448-449
    output coupling capacitor, 400-401, 448-449
  upper
    amplifiers, determination, 413-415, 449
Cutoff operation, 653, 686

D

Data
  sheet
    manufacturers', 938-961
    use of, 668, 671-672
  transmission, 900-901, 917-918
dB; *see* Decibels.
Dc
  signals, 25-27
    power relationships, 26-27
    resistors, 26
  static switching, SCRs, 253, 254
Debouncing, switch, 778-780, 790
Decibel, 61-71, 73
  computational advantages, 64-67

968 Index

Decibel—cont
  gain
    current, 67-71
    voltage, 67-71
Decoders, 704-706, 722
Deemphasis, 850-851, 857
Delays, propagation, 656-658, 686
Delta
  modulation, 888-891, 916-917
    adaptive, 891, 917
Demodulation
  signal
    AM, 834-835, 836, 855
    FM, 845, 846, 855
    SSB, 822-824, 854
Depletion region, pn junction, 171
Design
  logic MUX, 700-701, 722
  procedure
    four-input MUX, 701-704
    networks, mixed logic, 616-621, 632
    two-level positive logic networks, 605-610, 631
Detector
  light-beam, 563
  window, 565-568
Determination, constant K, 340-341, 381
Devices, optoelectronic, 543-581
Differential
  amplifiers, 486-501, 529-530
    circuits, 493-495, 529-530
  mode, 489-491
  operation, 496, 497-498
Diffusion, ICs, 201
Digital
  communication, 891-926
  logic families, 653-694
  -to-analog conversion, 901-906, 918
Diodes, 223-273
  ac power rectification, 225-235
  circuits, 260
  clipping circuits, 236-241
  ICs, 204
  ideal, 259
    approximation, 224-225
  light-emitting, 569-573, 577
    bar graph display, 570
    drivers, 570-572, 577
    OR gates, 241-245
  peak inverse voltage, 232-233
  pn junction, 178, 223-224
  Schottky, 211
  silicon-controlled rectifiers, 248-258
  varactor, 248, 261
  zener, 245-247, 260-261
Direct generation FM, 844-845, 856
Discharging capacitor, 105-106
Dividers
  current, 94-96
  frequency, 768-769, 790
  voltage, 91-92
Division, time-, multiplexing, 871-872, 913-914

Domain
  frequency, 807-809, 852-853
  time, 807-809, 852-853
Doped semiconductors, 167-170
  N-type, 167-168, 169
  P-type impurity, 168-170
Double-sideband
  large-carrier, conventional AM, 825-835
  suppressed-carrier AM, 814-817, 853-854
Drivers, LEDs, 570-572, 577
DSB-SC, 814-816, 853-854
Dynamic RAMs, 719-720, 723

E

Edge-triggered JK flip-flops, 761-763, 789
Effect, noise, different modulation systems, 850, 857
Electric circuits, 83-165
Electronic slave flash control circuit, 563-565
Emitter-
  coupled logic, 678-679, 689-690
  follower BJT amplifier
    input resistance, 313-315
    matching, 316-317
    output resistance, 315
Encoders, 704-707, 722
  parallel, 906-912
Energy signals, 50-52, 801-805
  spectrum pulse, 46-50
  stored, capacitor, 100
Enhancement NMOS, 194-195
Epitaxial growth, IC fabrication, 205, 208
Equivalent
  noise temperature, 849, 856-857
  Thevenin, 134-147
    circuit, 151
Etching, ICs, 201
Exclusive
  NOR function, 601-602, 631
  OR function, 598-601, 631
Expansion capacity
  MUX, 698-700, 722
  ROM, 711-714, 723

F

Fabrication, ICs, 200-201, 204-205, 208-209
Feedback
  amplifiers, 518-527, 531
  connections, 521-523, 531
  positive oscillators, 524, 525, 531
  shift register, 783-785, 790
FET
  amplifiers, 330-392
    biasing, 353-356, 381-382
    biasing circuit design, 356-361, 381
    common-drain, 375-376, 382
    common-source, 361-374, 382
    constant K, 340-341, 381
    principles, 331-335
    properties, 331-335, 379
    Q point, 343-353, 381
    transfer characteristics, 335-340, 380

FET—cont
  parameters, 417, 449
  photo-, 568-569, 577
  resistor, voltage-controlled, 376-377, 378, 382-383
Field-effect transistor, see FET.
Figure, noise, 849-851, 857
Filters, 126-134, 150-151
  capacitors, 233-235
  high-pass, 126-130
    cutoff frequency, 128-130
  low-pass, 130-131
  power supplies, 233
  RC, 133-134
    high-pass, 435-440, 450
    low-pass, 440-446, 450-451
  response
    frequency, 131-133
    pulse, 131-133
Filtering, SSB-AM, generation, 820-821, 854
Flash control circuit, electronic slave, 563-565
Flip-flops
  applications, 768-780, 790
  edge-triggered JK, 761-763, 789
  JK, special cases, 766-768, 789
  master-slave, 764-766, 789
  response characteristics, 768
  set-reset, 755-780, 788
  SR, clocked, 758-760, 788
Flux, 545, 575
FM
  generation, direct, 844-845, 856
  modulation index, 839-840, 855
  narrowband, 843, 856
  receivers, 845, 847, 856
  signal
    band width, 840, 842-843, 856
    demodulation, 845, 846, 855
    spectrum, 840, 841, 855-856
Follower, voltage, circuit, 476-478, 528
Fourier analysis, 35-37
Four-
  input MUX, design procedure, 701-704
  variable map, 627-629, 632
Forward bias pn junction, 175, 176
Frequency
  cutoff, lower, 397-409, 448-449
  dividers, 768-769, 790
  domain, 807-809, 852-853
  modulation, 838-845, 855-856
  response
    amplifiers, 393-455
    amplifiers, bypass capacitor, 397
    amplifiers, coupling capacitors, 396
    amplifiers, curve, 394-395
    amplifiers, factors that affect, 395-396, 448
    amplifiers, interelectrode capacitances, 397
    filters, 131-133
  -shift keying, 884, 886-887, 916

Frequency—cont
  translation, 809-814, 853
FSK, 884, 886-887, 916
Full-adders, 745, 746, 787
Function
  exclusive
    NOR, 601-602, 631
    OR, 598-601, 631
  staircase, 888-890

G

Gain
  current, 67-71
  margin, 526, 527, 532
  quantities, two-port system, 52-55
  voltage, 67-71
Gates
  AND, OR, commercially available, 588-591
  open collector, 663, 687
Generation
  direct, FM, 844-845, 856
  product signals, 817-818
  SSB-AM, 820-822, 854
    filtering, 820-821, 854
    phasing, 821-822, 854
Generator, sine-wave, square-wave, 506, 507
Graphical determination, Q point, 345-353, 381

H

Half-adders, 743-744, 786-787
Handling precautions, MOSFETs, 377, 383
Hexadecimal number systems, 735-739, 786
High-
  frequency model amplifier, 409-415
    capacitance C_1, 411
    cutoff frequency, upper, determination, 413-415, 449
    Miller effect, 411-413, 449
  pass filters, 126-130
    cutoff frequency, 128-130
    RC, 435-440, 450
HPF; see high-pass filters.

I

ICs, 199-213, 218
  advantages, 213
  BJTs, 204
  capacitors, 202
  contacts
    nonohmic, 202, 204
    ohmic, 202, 204
  current mirror, 212-213
  diodes, 204
  disadvantages, 213
  fabrication, 200-201, 204-206, 208-209
    base diffusion, 206, 208
    diffusion, 201
    emitter diffusion, 206-207, 209
    epitaxial growth, 205, 208
    etching, 201

Index

ICs—cont
  fabrication
    isolation diffusion, 205, 208
    mask generation, 201
    metallization, 207, 209
    oxidation, 201
    photolithography, 201
    pre-ohmic etching, 207, 209
    wafer, 205
    resistors, 202
Ideal
  diode, 259
    approximation, 224-225
  op amps, 457-486, 528
  transformers, matching, Thevenin equivalent, 141-147
Illumination, 545, 575
  variation, resistance, 546-550, 576
Impedance, ac circuits, 120-122
Increase, bandwidth, amplifier, 446-447, 451
Increasing efficiency, photodiode, 556-560
Index, FM, modulation, 839-840, 855
Inductors, 111-118, 149-150
  circuits
    ac, 123-124
    RL, 117-118
  current
    buildup, 112-113
    buildup, universal, 113-115
    continuity condition, 112
    decay, 115-116
    decay, universal curve, 115-116
  stored energy, 111-112
Informal justification, sampling theorem, 854-870
Information
  measure, 891-893, 917
  rate, 893-894, 897-900, 917
  theory, 891-894, 917
Injection, integrated, logic, 674, 678, 689
Input
  coupling capacitor, lower cutoff frequency, 400, 448-449
  resistance
    BJT amplifiers, 319
    emitter follower, common-collector BJT amplifier, 313-315
    two-port system, 55-61
Integrated
  circuits; *see* ICs.
  injection logic, 674-678, 689
Integrators, op amp, 483-486, 529
Interelectrode capacitances, 397
Interpretation of data sheet, 668, 671-672
Inverting op amp circuits, 461-467, 528
Isolation diffusion, IC fabrication, 205, 208
Isolators, optically coupled, 573-574, 578

J

JFET; *see* junction field-effect transistor.
JK flip-flops
  edge-triggered, 761-763, 789

JK flip-flops
  special cases, 766-768, 788
Junction
  field-effect transistor, 186-192, 216-217
    currents, 180
    n-channel, 187
    n-channel operation, 187-189
    output characteristics, 191, 192
    p-channel, 190-191
    pinch-off condition, 190
    symbols, 191
  pn, 170-178

K

Karnaugh maps, 624-629, 632
  four-variable, 627-629, 632
  three-variable, 625-627, 632
  two-variable, 624-625, 632
KCL; *see* Kirchhoff's current law.
KCV; *see* Kirchhoff's voltage law.
Kirchhoff's laws, 83-88, 148
  current, 83-84, 148
  voltage, 84-88, 148
Keying, frequency-shift, 884, 886-887, 916

L

Loading considerations, TTL gates, 663-665, 687
LEDs, 569-573, 577
  arrays, XY addressable, 572-573, 578
  bargraph display, 570
  drivers, 570-572, 577
  optoisolator, 573-574, 578
Light-
  beam detector, 563
  emitting diodes, 569-573, 577
    bar graph display, 570
    drivers, 570-572, 577
    XY addressable arrays, 572-573, 578
  principles, 543-545, 575
  relationships, 543-545, 575
Logarithmic amplifiers, 480-482, 529
Logic
  arithmetic, 730-754
  circuits, 582-652
    principles, 583-598, 630
  CMOS, 680-685, 690
  convention
    mixed, 584
    negative, 583
    positive, 583
  design
    mixed, procedure, 622, 632
    mixed vs. positive, 621
    MUX, 700-701, 722
  digital, families, 653-694
  emitter-coupled, 678-679, 689-690
  integrated injection, 674-678, 689
  memories, 707-721
    random-access, 714-720, 723
    read-only, 707-714, 723
  negative, 591-592
  networks, mixed, 612-624, 631-632

Electronics: Circuits and Systems

Logic—cont
  operations, 584-586, 630
  packages, 695-707
    multiplexers, 695-704, 722
  positive, 591-592
  three-state, 673-674, 689
  transistor-transistor, 660-672, 687-688
  unit, arithmetic, 752-754, 788
Logical
  complementation, NOT operation, 588
  product, AND operation, 586-587
  sum, OR operation, 584-586
Lower cutoff frequency, 397-409, 448-449
  bypass capacitor, 401-409, 448-449
  input coupling capacitor, 400, 448-449
  output coupling capacitor, 400-401, 448-449
Low-pass filter, 130-131
  RC, 440-446, 450-451
LPF; *see* low-pass filter.

M

Manufacturers' data sheets, 938-961
Maps
  Karnaugh, 624-629, 632
    four-variable, 627-629, 632
    three-variable, 625-627, 632
    two-variable, 624-625, 632
Margins
  gain, 526, 527, 532
  noise, 658-660, 686-687
  phase, 526, 527, 532
Mask generation, ICs, 201
Master-slave flip-flops, 764-766, 789
Matching, emitter follower, BJT amplifier, 316-317
Measure information, 891-893, 917
Memories
  logic, 707-721
  random-access, 714-720, 723
  read-only, 707-714, 723
Memory units, 720-721
Metallization, IC fabrication, 207, 209
Metal-oxide semiconductor, field-effect transistor, 192-199, 217-218
  advantages, 199
  complementary, 207, 210
  enhancement NMOS, 194-195
  handling precautions, 377, 383
  n-channel
    depletion-mode, 196-197
    enhancement-mode, 193-194
  p-channel enhancement mode, 195-196
Miller effect high-frequency model amplifier, 411-413, 449
Mixed
  logic
    convention, 584
    design procedure, 622, 632
    networks, 612-624, 631-632
    networks, analysis, 622-623, 631
    networks, design procedure, 616-621, 631-632

Mixed—cont
  vs. positive logic design, 621
Mode
  common, 489-491
  differential, 489-491
Model communication system, 805, 852
Modulation
  delta, 888-891, 916-917
    adaptive, 891, 917
  different types, 807, 852
  frequency, 838-845, 855-856
  index, FM, 839-840, 855
  pulse, 864-891, 913
    amplitude, 870-871, 913-914
    code, 873-887, 914-915
    time, 872-873, 914
  signal, 805-809
  systems, noise effect, 850, 857
MOSFET, 192-199, 217-218
  advantages, 199
  complementary, 207, 210
  enhancement NMOS, 194-195
  handling precautions, 377, 383
  n-channel depletion-mode, 196-197
  p-channel enhancement-mode, 195-196
Multiplexers, 695-704, 722
Multiplexing, time-division, 871-872, 913-914
MUX, 695-704, 722
  expansion capacity, 698-700, 722
  four-input design procedure, 701-704
  logic design, 700-701, 722

N

NAND
  gate as NOT gate, 595, 630
  networks, two-level, 611
  operations, 592-598, 630
Narrowband, FM, 843, 856
N-channel
  JFET, 187
    currents, 180
    operation, 187-189
  MOSFET
    depletion-mode, 196-197
    enhancement-mode, 193-194
Negative
  feedback effects, 523, 531
  logic, 591-592
    convention, 583
  numbers representation, 745-750, 787
Networks, mixed logic, 612-624, 631-632
NMOS, enhancement, 194-195
Noise
  communication systems, 845-849, 856-857
  effect, different modulation systems, 850, 857
  figure, 849-851, 857
  margins, 658-660, 686-687
  power, spectral density, 846-848, 856
  shot, 849, 857
  temperature, equivalent, 849, 856-857
  thermal, 848, 856

972 Index

Noise—cont
  white, 848, 856
Noiseless channel, 894-895, 917
Noisy channel, 895-897, 917
Noninverting op amp circuits, 467-473, 528
Nonohmic contacts, ICs, 202, 204
Nonperiodic signals, 45-52
  energy, 50-52
  spectrum, 46-50
Nonsinusoidal periodic signals, 35-44, 72-73
NOR
  exclusive function, 601-602, 631
  gate, as NOT gate, 596-598
  operations, 592-598, 630
Normal active operation, 654, 686
NOT
  gate
    NAND gate as, 595
    NOR gate as, 596-598
  operation, logical complementation, 588
Npn transistor, 180-183
  output characteristics, 185-186
N-type impurity, doped semiconductors, 167-168, 169
Number system
  binary, 730-734, 786
  hexadecimal, 735-739, 786
  octal, 735-739, 786

O

Octal number systems, 735-739, 786
Ohmic contacts, ICs, 202, 204
Ohm's law, 72
Ones complement system, 748-749, 787
Op Amp
  circuits, 461, 528-529
    inverting, 461-467, 528
    noninverting, 467-473, 528
    summing, 473-476, 528
  ideal, 457-486, 528
  integrators, 483-486, 529
  stages, 457
Open collector gate, 663, 687
Operation
  common-mode, 495-497
  cutoff, 653, 686
  differential-mode, 496, 497-498
  NOR, 592-598, 630
  normal active, 654
  logic, 584-586
  saturation, 654-656, 686
Operational amplifier, 456-518, 528-530
  comparators, 501-506, 530
  ideal, 457-486, 528
  stages, 457
Optical radiation, 543-545, 575
Optically coupled isolators, 573-574, 578
Optoelectronic devices, 543-581
Optoisolator, 573-574, 578
OR
  exclusive function, 598-601, 631
  gates
    commercially available, 588-591

OR—cont
  gates
    diodes, 241-245
    operation logical sum, 584-586
Oscillators, positive feedback, 524, 525, 531
Output
  characteristics, JFETS, 191, 192
  coupling capacitor, lower cutoff frequency, 400-401, 448-449
  resistance, 319
  resistance emitter follower, common-collector BJT amplifier, 315
  stage, totem pole, 665-668, 669, 670, 688
Oxidation, ICs, 201

P

Packages
  logic, 695-707
  multiplexers, 695-704, 722
Parallel
  connection capacitors, 110-111
  encoders, 906-912
  resistors, 93-94, 96, 97
Parameters
  amplifier, 458
  transistor, 415-418, 449-450
    BJT, 416-417, 449
    FET, 417, 450
P-channel
  enhancement-mode MOSFET, 195-196
  JFET, 190-191
PCM
  baseband transmission, 882-887, 915
  quantization error, 879, 914
  signals, radio-frequency transmission, 884, 885, 915-916
  system
    bandwidth, 879-881, 914-915
    companding, 881-882, 915
Peak
  inverse voltages, diodes, 232-233
  value, 28-29
Periodic signals
  amplitude spectrum, 37-40
  nonsinusoidal, 35-44, 72-73
    spectrum amplitude, 72
    spectrum power, 72-73
  power, 40-43
  spectrum, 43-44
Phase
  angle, 29-32
  margin, 526, 527, 532
Phasing, generation, SSB-AM, 821-822, 854
Phasor representation, ac circuit, 118-120
Photodetectors, 545-554, 575-576
  response speed, 550-554, 575-576
  structure, 554
Photodiodes, 556-560, 576-577
  amplifier circuits, 559-560, 576-577
  increasing efficiency, 556-560, 576-577
  PIN, 558-559, 576
  response speed, 559, 576

Photo-FETs, 568-569, 577
Photolithography, ICs, 201
Photons, 543-545, 575
Phototransistors, 560-569, 577
  applications, 561
    flash control circuits, 563-565
    light-beam detector, 563
    photo-FETs, 568-569, 577
    window detector, 565-568
  optoisolator, 573-574, 578
  response speed, 561, 577
Photovoltaic cells, 554-556, 576
PIN photodiode, 558-559, 576
Pinch-off condition, JFETs, 190
Pn junction, 170-178, 214-215, 223-224
  balance, 173
  bias
    forward, 175, 176
    reverse, 175-178
  depletion region, 171
  diode, 178
  metal contacts, 173-175
  potential barrier, 173
Pnp transistor, 183
Positive
  feedback, oscillators, 524, 525, 531
  logic, 591-592
    convention, 583
    two-level network, 602-611
      analysis of, 603-605, 631
      design of, 605-610, 631
Potential barrier, 173
Potentiometers, 93
Power
  ac, rectification, 225-235
  AM considerations, 829-834, 855
  equation, 72
  gain, BJT amplifiers, 319
  noise, spectral density, 846-848, 856
  periodic signals, 40-43
  relationships
    ac, 32-35
    dc, 26-27
  signal, 801-805
  spectrum
    periodic signals, 43-44
    periodic signals, nonsinusoidal, 72-73
Power supplies, filters, 233
Preemphasis, 850-851, 857
Pre-ohmic etching, IC fabrication, 207, 209
Principles
  light, 543-545, 575
  logic circuits, 583-598, 630
Problems, answers, 927-937
Product signals, generation, 817-818
Programmable ROMs, 711, 723
Propagation delays, 656-658, 686
P-type impurity doped semiconductors, 168-170
Pulse-
  amplitude modulation, 870-871, 913-914
  code modulation, 873-887, 914-915
    bandwidth, 879-881, 914-915

Pulse- —cont
  code modulation
    baseband transmission, 882-887, 915
    companding, 881-882, 915
    quantization error, 879, 914
  modulation, 864-891, 913
  response
    amplifiers, 393-455
    filters, 131-133
    RC-coupled, 434-436
    signals, 73
  time modulation, 872-873, 914

Q

Q point determination
  algebraic, 343-345, 381
  graphical, 345-353, 381
Quantization error, PCM, 879, 914

R

Radiation, optical, 543-545, 575
Radio-frequency transmission, PCM signals, 884, 885, 915
RAM, 714-720, 723
  dynamic, 719-720, 723
  static, 716-719, 723
Random-access memory, 714-720, 723
Rate information, 893-894, 897-900, 917
Ratio
  rejection, common-mode, 491-493, 529
  signal-to-noise, 849-851, 857
RC
  circuit, 106-110
  filters, 133-134
    high-pass, 435-440, 450
    low-pass, 440-446, 450-451
  coupled amplifier, pulse response, 434-436
Read-only memories, 707-714, 723
Receivers
  AM, superheterodyne, 835, 837-838, 855
  FM, 845, 847, 856
Recombination, 170
Rectification, ac power, 225-235, 259
Register, shift, 780-786, 790
  with feedback, 783-785, 790
Rejection ratio, common-mode, 491-493, 529
Relationships
  light, 543-545, 575
  power, ac, 32-35
Representation, negative numbers, 745-750, 787
Response
  characteristics, flip-flops, 768
  frequency, amplifiers, 393-455
    curve, 394-395
  speed
    photodector, 550-554, 575-576
    photodiodes, 559, 576
    phototransistor, 561, 577
Resistance, variation, 546-550, 576

Index

Resistors, 26, 72, 89-97, 148
  ac circuits, 122
  current dividers, 94-96
  FET, voltage-controlled, 376-377, 378, 382-383
  ICs, 202
  parallel, 93-94, 96, 97
  potentiometers, 93
  series, 89-91, 96, 97
  voltage dividers, 91-92
Reverse
  active operation, 656, 686
  bias pn junction, 175-178
Ripple counters, 769-771, 790
RL circuit, 117-118
Rms value, 28-29
ROM, 707-714
  capacity, 710
    expansion, 711-714, 723
  programmable, 711, 723
Root-mean-square; *see* rms.

S

Sample and hold circuit, 479-480, 529
Sampling theorem, 864-870, 913
  informal justification, 854-870
Saturation operation, 654-656, 686
Schmitt trigger, 506-518, 530
Schottky
  diodes, 211
  transistors, 211
SCRs; *see also* silicon-controlled rectifiers.
SCRs, 261
  triacs, 255-258
Semiconductors, 166-198, 214
  contacts, 173-175
  doped, 167-170
  metal-oxide, field-effect transistor, 192-199
  pn junction, 170-178
  recombination, 170
  thermal generation, 170
Series
  connection capacitors, 110
  RC ac circuits, 124-126
  resistors, 89-91, 96, 97
Set-reset flip-flops, 755-780, 788
Shannon's theorem, 894, 897-900, 917
Shift register, 780-786, 790
  with feedback, 783-785, 790
Shot, noise, 849, 857
Sideband
  double-
    large-carrier, conventional, 825-835, 853-854
    suppressed-carrier, 814-817, 853-854
  single-, 818-824, 854
  vestigial-, 824-825, 854
Signals, 25-83
  ac, 27-35, 72
    peak value, 28-29
    phase angle, 29-32
    power relationships, 32-35

Signals—cont
  ac
    rms, 28-29
  AM
    demodulation, 834-835, 836, 855
    spectrum, 828-829, 833
  dc, 25-27
    power relationships, 26-27
  energy, 801-805
  FM
    bandwidth, 840, 842-843, 856
    demodulation, 845, 846, 855
    spectrum, 840, 841, 855-856
  Fourier analysis, 35-37
  modulation, 805-809
  nonperiodic, 45-52
    energy, 50-52
    energy spectrum, 46-50
  periodic
    amplitude spectrum, 37-40
    nonsinusoidal, 35-44, 72-73
    power, 40-43
    power spectrum, 43-44
  power, 801-805
  product generation, 817-818
  pulse, 73
  -to-noise ratio, 849-851, 857
Silicon-controlled rectifiers, 248-258
  ac switch, 251
  dc static switching, 253, 254
Sine-wave, square-ware generator, 506, 507
Single-sideband AM, 818-824, 854
sinusoidal current, 27-28
Small-signal analysis, common-emitter BJT amplifier, 297-312
Source, constant-current, 482-483, 529
  realization, 498-501
Spectral density noise power, 846-848, 856
Spectrum
  amplitude, 37-40
  energy pulse, 46-50
  power, periodic signal, 43-44
  signal
    AM, 828-829, 833
    FM, 840, 841, 855-856
Speed, response
  photodector, 550-554, 575-576
  photodiodes, 559, 576
  phototransistors, 561, 577
SR flip-flop, clocked, 758-760, 788
SSB-AM, 818-824, 854
  demodulation, 822-824, 854
  generation, 820-822, 854
  filtering, 820-821, 854
  phasing, 821-822, 854
Stability, amplifier, 525-526, 532
Stages, op amp, 457
Staircase function, 888-890
State, three-, logic, 673-674, 689
Static RAMs, 716-719, 723
Stored energy, inductors, 111-112
Structure, photodetector, 554

Electronics: Circuits and Systems

Subtraction, twos complement system, 750-752, 787
Summing, op amp circuits, 473-476, 528
Sum of products, two-level networks, 610, 631
Superheterodyne, AM receivers, 835, 837-838, 855
Suppressed-carrier, double-sideband, AM, 814-817, 853-854
Switch
  as transistor, 653-656, 686
  debouncing, 778-780, 790
Switching variables, 584
Symbols
  JFETs, 191
  transistors, 185
Synchronous counters, 771-778, 788
System
  communication model, 805, 852
  communication, noise, 845-849, 856-857
  complement
    ones, 748-749, 787
    twos, 749-750, 787
  PCM companding, 881-882, 915
  two-port, 52-61, 73
    gain quantities, 52-55
    input resistance, 55-61

T

Techniques, analog, communication, 800-863
Temperature
  efects, transistor, 290-293, 321
  equivalent noise, 849, 856-857
Theorem
  sampling, 864-870, 913
    informal justification, 854-870
  Shannon's, 894, 897-900, 917
Theory, information, 891-894, 917
Thermal
  generation, 170
  noise, 848, 856
Thevenin equivalent, 134-147
  circuit, 151
  ideal transformers, matching, 141-147
  maximum power transfer, 140-141
Three
  -state logic, 673-674, 689
  -variable map, 625-627, 632
Thyristor, 261
Time
  -division multiplexing, 871-872, 913-914
  domain, 807-809, 852-853
  pulse-, modulation, 872-873, 914
Totem pole output stage, 665-668, 669, 670, 688
Transistor
  as swich, 653-656, 686
  bipolar junction, 178-186
  currents, 183-185
  JFET
    n-channel operation, 187-189
    output characteristics, 191, 192

Transistor—cont
  JFET
    p-channel, 190-191
    symbols, 191
  junction field-effect, 186-192
  n-channel JFET, 187-189
  npn, 180-183
  output characteristics, 185-186
  parameters, 415-418, 449-450
    BJT, 415-417, 449
    FET, 417, 450
  photo-, 560-569, 577
  pnp, 183
  Schottky, 211
  symbols, 185
  temperature effects, 290-293, 321
  -transistor logic, 660-672, 687
Translation, frequency, 809-814, 853
Transmission
  data, 900-901, 917-918
  PCM
    baseband, 882-887, 915
    radio-frequency, 884, 885, 915-916
Triacs, 255-258, 261
Two-
  level
    NAND networks, 611
    positive logic networks, 602-611, 631
    positive logic networks, analysis of, 603-605, 631
    positive logic networks, design of, 605-610, 631
    sum of products networks, 610, 631
  port model
    BJT amplifier, 276-283, 320
  port systems, 52-61, 73
    gain quantities, 52-55
    input resistance, 55-61
  variable map, 624-625, 632
Twos complement system, 749-750, 787
  addition, 750-752, 787
  subtraction, 750-752, 787
Types, different, modulation, 807, 852

U

Unit
  logic, arithmetic, 752-754, 788
  memory, 720-721
Universal
  charging curve, 104-105
  current
    buildup curve, 113-115
    decay curve, 115-116
Upper cutoff frequency, determination, 413-415, 449
Use, data sheet, 668, 671-672

V

Varactors, 261
  diode, 248
Value
  peak, 28-29
  rms, 28-29

Variable
 four-, map, 627-629, 632
 three-, map, 625-627, 632
 two-, map, 624-625, 632
Variables, switching, 584
Variation, resistance, 546-550, 575
Vestigial-sideband AM, 824-825, 854
Voltage
 continuity, 100
 -controlled resister FET, 376-377, 378, 382-383
 dividers, 91-92
 follower circuits, 476-478, 528
 gain, 67-71
  BJT amplifiers, 319
 -to-current converter, 478-479, 528

VSB-AM, 824-825, 854

W

Wafer, IC fabrication, 205
Wheatstone bridge, 97-99
White noise, 848, 856
Window detector, 565-568
 circuit, 504-506

X

XY addressable LED arrays, 572-573, 578

Z

Zener diode, 245-247, 260-261